Twenty-first Coastal Engineering Conference

Proceedings of the International Conference

VOLUME 2

June 20-25, 1988
Costa del Sol-Malaga, Spain

Conference held under the auspices of the
Coastal Engineering Research Council of the
American Society of Civil Engineers

Organized by the Colegio de Ingenieros de
Caminos, Canales y Puertos de Espana

and cosponsored by the
Directorate General for Ports and Coasts of the
Spanish Ministry of Public Works
Polytechnical Universities of
Madrid, Barcelona, Santander,
and Valencia

Edited by Billy L. Edge

Published by the
American Society of Civil Engineers
345 East 47th Street
New York, New York 10017-2398

ABSTRACT

This book contains over 220 papers presented at the 21st International Conference on Coastal Engineering. The book is divided into six parts: theoretical and observed wave characteristics; coastal processes and sediment transport; coastal structures and related problems; coastal, estaurine and environmental problems; case studies; and ship motions. The individual papers include such topics as the effects of wind, waves, storms and currents, erosion, sedimentation, and beach nourishment. Special emphasis is given to case studies of completed engineering projects. With the inclusion of both the theoretical and the practical, these papers provide the civil engineer with a broad range of information on coastal engineering.

ISBN 0-87262-687-3
ISSN 0893-8717
Manufactured in the United States of America.

CONTENTS

PART II
COASTAL PROCESSES AND SEDIMENT TRANSPORT

PART III
COASTAL STRUCTURES AND RELATED PROBLEMS

PART IV
COASTAL, ESTUARINE AND ENVIRONMENTAL PROBLEMS

PART V
CASE STUDIES OF SOLUTIONS TO COASTAL ENGINEERING PROBLEMS

PART VI
SHIP MOTIONS

CHAPTER 74

Dynamic Wave Setup

Jen-Men Lo, M. ASCE[1]

Abstract

The wave setup for a given wave spectrum was re-evaluated with the radiation stress term including the low-frequency terms. This setup which is referred to here as "dynamic wave setup", was compared with the steady setup, which is generated by including only the non-periodic radiation stress terms.

The results of the study showed that the dynamic wave setup is greater than the steady wave setup, sometimes almost double its value. Therefore, the dynamic wave setup is important for engineering applications, particularly in the study of storm surges and coastal engineering problems.

Introduction

Longuet-Higgins and Stewart (1963) showed theoretically from their concept of radiation stress, and Bowen et al. (1968) verified it in the laboratory, that there is a water level setdown outside the surf zone followed by a water level setup within the surf zone because of the presence of the wave motion.

Longuet-Higgins and Stewart (1963) considered only the monochromatic waves, but Collins (1972), Battjes (1974), and Wu et al. (1978) re-evaluated wave setup by considering the radiation stress for a given spectrum.

The radiation stress derived for a wave spectrum should include the interaction between any wave component and itself as well as the interaction between any wave component and other wave components. However, the studies carried out by Collins (1972), Battjes (1974), and Wu et al. (1978) considered only the self-interaction terms.

Lo (1981) derived the radiation stress term for any given wave spectrum. In this derivation both non-periodic and long-period oscillation (low frequency) terms were included. The low-frequency terms were generated from the non-linear transfer of energy of the phase difference between different wave components.

The purpose of this study is to re-evaluate wave setup for a given wave spectrum with the radiation stress term including the low-frequency terms. This setup which is referred to here as "dynamic wave setup" was compared with the

[1] Research Scientist, Kuwait Institute for Scientific Research, Hydraulics and Coastal Engineering Group, P. O. Box 24885, 13109, Safat, Kuwait.

steady wave setup, which was generated by including only the non-periodic radiation stress terms.

Governing Equations for the Surf Beat Wave on a Sloping Bottom

The motion is considered irrotational and the fluid incompressible. Integrating the continuity and momentum equations from the bottom, $Z = -h$, to the free surface, $Z = \eta$ (water surface displacement), imposing the Leibnitz rule, and substituting the no-flow bottom boundary condition (BBC), the kinematic free surface boundary condition (KFSBC), and applying the time average over one incoming wave period one can obtain the governing equations for the sloping bottom surf beat wave. For convenience, only a uniform bottom slope will be considered.

Leibnitz Rule. The Leibnitz rule of differentiation will be used to obtain a derivative from within an integral. It is stated as follows

$$\int_{\alpha(x)}^{\beta(x)} \frac{\partial f(x,z)}{\partial x}\, dz = \frac{\partial}{\partial x}\int_{\alpha(x)}^{\beta(x)} f(x,z)dz - f(\beta,z)\,\frac{\partial \beta(x)}{\partial x} + f(\alpha,z)\frac{\partial \alpha(x)}{\partial x} \tag{1}$$

where f (x, z) is any function; x is a variable of integration; z is a dummy variable of integration; α and β are limits of integration.

Boundary Conditions. At the bottom, $Z = -h$, a no-flow boundary condition exists. In vector form

$$\vec{V} \,.\, \vec{n} = 0 \tag{2}$$

where

$$\vec{V} = u\vec{i} + v\vec{j} + w\vec{k} \tag{3}$$

$$\vec{n} = \frac{\left(\frac{\partial h}{\partial x}\vec{i} + \frac{\partial h}{\partial y}\vec{j} + \vec{k}\right)}{\sqrt{\left(\frac{\partial h}{\partial x}\right)^2 + \left(\frac{\partial h}{\partial y}\right)^2 + 1}} \tag{4}$$

(Unit vector normal to the bottom)

where $\vec{i}, \vec{j}, \vec{k}$ are unit vectors in the x, y, z directions; $\vec{V}$ is the velocity vector and u, v, w are the x, y and z components of the velocity vector.
BBC can be rewritten as

$$u\frac{\partial h}{\partial x} + v\frac{\partial h}{\partial y} + w = 0, \quad \text{at } z = -h(x,y)$$

At the free surface, the vertical velocity (w) must account for the changes in the instantaneous water surface (η)

$$w = \frac{D\eta}{Dt} = \frac{\partial\eta}{\partial t} + u\frac{\partial\eta}{\partial x} + v\frac{\partial\eta}{\partial y}\,, \quad \text{at } z = \eta \tag{5}$$

This is the KFSBC.

Continuity Equation. The general form of the continuity equation in three dimensions is

$$\frac{\partial\rho}{\partial t} + \frac{\partial\rho u}{\partial x} + \frac{\partial\rho v}{\partial y} + \frac{\partial\rho w}{\partial z} = 0 \tag{6}$$

Integrating each term with respect to depth between Z = -h and Z = η, and imposing the Leibnitz Rule

$$\frac{\partial}{\partial t}\int_{-h}^{\eta}\rho dz - \rho_{\eta}\frac{\partial\eta}{\partial t} - \rho_{-h}\frac{\partial h}{\partial t} + \frac{\partial}{\partial x}\int_{-h}^{\eta}\rho u dz - (\rho u)_{\eta}\frac{\partial\eta}{\partial x} - (\rho u)_{-h}\frac{\partial h}{\partial x} +$$

$$\frac{\partial}{\partial y}\int_{-h}^{\eta}\rho v dz - (\rho v)_{\eta}\frac{\partial\eta}{\partial y} - (\rho v)_{-h}\frac{\partial h}{\partial y} + (\rho w)_{\eta} - (\rho w)_{-h} = 0 \tag{7}$$

Invoking the boundary conditions BBC and KFSBC, the above equation simplifies to

$$\frac{\partial}{\partial t}\int_{-h}^{\eta}\rho dz - \rho_{-h}\frac{\partial h}{\partial t} + \frac{\partial}{\partial x}\int_{-h}^{\eta}\rho u dz + \frac{\partial}{\partial y}\int_{-h}^{\eta}\rho v dz = 0 \tag{8}$$

Considering the depth (relative to still water level) as constant in time, and the problem as two-dimensional (no dependency on the y - direction), the above equation becomes:

$$\frac{\partial}{\partial t}\int_{-h}^{\eta}\rho dz + \frac{\partial}{\partial x}\int_{-h}^{\eta}\rho u dz = 0 \tag{9}$$

Consider the horizontal water particle velocity as the superposition of two components,

$$u = u_I + u_s \tag{10}$$

where

u_I = the horizontal water particle velocity of the incoming wave

u_s = the horizontal water particle velocity of the surf beat wave which is uniform over depth since the water depth is relatively shallow in comparison to the surf beat wave length.

Thus, for a fluid of constant density ρ, the continuity equation becomes

$$\frac{\partial}{\partial t}(h + \eta) + \frac{\partial}{\partial x}u_s(h + \eta) + \frac{\partial}{\partial x}\int_{-h}^{\eta} u_I \; dz = 0 \tag{11}$$

The time averages are defined

$$\overline{F} = \frac{1}{T}\int_{o}^{T} F dt \tag{12}$$

then

$$F_s = \overline{F}_s = \frac{1}{T}\int_{0}^{T} F_s dt$$

where T is the incoming wave period, and F_s represents a general surf beat variable. Time averaging the continuity equation, we obtain

$$\frac{\partial\overline{\eta}}{\partial t} + \frac{\partial}{\partial x}u_s(h + \overline{\eta}) + \overline{(\frac{\partial}{\partial x}\int_{-h}^{\eta} u_I)} \; dz = 0 \tag{13}$$

where $\overline{\eta} = \eta_s$, is the water surface displacement of the surf beat wave. The time-averaged water surface displacement of the incoming wave, $\overline{\eta}_I$, is zero.

Following the definitions of Phillips (1966) for the total flux

$$M_x = \overline{\int_{-h}^{\eta} \rho u_I dz} \tag{14}$$

and for the mean transport velocity

$$u_I = \frac{M_x}{\rho(h + \overline{\eta})} \tag{15}$$

The continuity equation can be written as

$$\frac{\partial\eta_s}{\partial t} + \frac{\partial}{\partial x}(u_s + u_I) \quad (h + \eta_s) + \frac{\partial}{\partial x}\overline{(u_I\eta_I)}_{z=0} = 0 \tag{16}$$

where

$$\frac{\partial}{\partial x}(u_s + u_I)\eta_s \quad << \frac{\partial\eta_s}{\partial t}$$

The final result is then

$$\frac{\partial \eta_s}{\partial t} + \frac{\partial}{\partial x} \quad (u_s + u_I)h + \frac{\partial}{\partial x}\overline{(u_I\eta_I)}_{z=0} = 0 \tag{17}$$

Equation of Motion. Without considering the shear stresses, the general form of the equations of motion are:

x Direction.

$$\frac{\partial u}{\partial t} + \frac{\partial u^2}{\partial x} + \frac{\partial uv}{\partial y} + \frac{\partial uw}{\partial z} = \frac{1}{\rho}\frac{\partial P}{\partial x} \tag{18}$$

z Direction.

$$\frac{\partial w}{\partial t} + \frac{\partial uw}{\partial x} + \frac{\partial vw}{\partial y} + \frac{\partial w^2}{\partial z} = -\frac{1}{\rho}\frac{\partial P}{\partial z} - g \tag{19}$$

where p = pressure, and g = gravitational constant.

Following the same procedure as outlined before the final equation of motion for the case of two dimensional sloping bottom surf beat wave becomes

$$\frac{\partial(u_s + \bar{u}_I)h}{\partial t} = -gh\frac{\partial \eta_s}{\partial x} - \frac{1}{\rho}\frac{\partial S_{xx}}{\partial x} - \frac{\partial}{\partial t}\overline{(u_I\eta_I)}_{z=0} \tag{20}$$

where

$$S_{xx} = \rho\int_{-h}^{0}\overline{(u_I^2 - w_I^2dz)} + \frac{1}{2}\ \rho g\bar{\eta}^2 \tag{21}$$

It should be noted that the sloping bottom surf beat wave problem is governed by equation of motion (Eq. 20), and the continuity equation (Eq. 17).

Irregular Wave Conditions

The irregular sequence of linear waves can be represented as an infinite sum of simple harmonic waves

$$\eta(x,\ t) = \sum_n a_n \cos\ (k_n x - \sigma_n t + \epsilon_n) \tag{22}$$

where

$$a_n = \sqrt{2\rho_\eta(\sigma)\Delta\sigma}$$

$P_\eta(\sigma)$ = the energy density spectrum of the irregular sea, varying with wave angular frequency σ, and

ϵ_n = the phase angle (radians) for the nth wave component.

When small amplitude wave theory is used to estimate the flow regime in a wave system from the surface profile, the linear relationship between the surface profile and the velocity potential results in

$$\phi(x, z, t) = -\sum_n \frac{a_n g}{\sigma_n} \frac{\cosh k_n (h+z)}{\cosh k_n h} sin\ (k_n x - \sigma_n t + \epsilon_n); \tag{23}$$

the dispersion relationship is given by

$$\sigma_n^2 = gk_n\ tanh\ k_n h \tag{24}$$

also the horizontal and vertical water particle velocity components are given by

$$u\ (x, z, t) = \sum_n \frac{a_n k_n g}{\sigma_n} \frac{\cosh k_n(h+z)}{\cosh k_n h} \cos\ (k_n x - \sigma_n t + \epsilon_n) \tag{25}$$

and

$$w\ (x, z, t) = \sum_n \frac{a_n k_n g}{\sigma_n} \frac{\sinh k_n(h+z)}{\cosh k_n h} \sin\ (k_n x - \sigma_n t + \epsilon_n) \tag{26}$$

After some algebraic manipulation the following equations result:

$$\eta^2 = \sum_{n=1}^{N} a_n^2 \cos^2 \theta_n + \sum_{m=1}^{N-1} \sum_{n=m+1}^{N} a_m a_n \left[\cos\ (\theta_m - \theta_n) + \cos\ (\theta_m + \theta_n)\right], \tag{27}$$

$$u^2 = \sum_{n=1}^{N} A_n^2 \cos^2 \theta_n + \sum_{m=1}^{N-1} \sum_{n=m+1}^{N} A_m A_n \left[\cos\ (\theta_m - \theta_n) + \cos\ (\theta_m + \theta_n)\right], \tag{28}$$

$$w^2 = \sum_{n=1}^{N} B_n^2 \sin^2 \theta_n + \sum_{m=1}^{N-1} \sum_{n=m+1}^{N} B_m B_n \left[\cos\ (\theta_m - \theta_n) + \cos\ (\theta_m + \theta_n)\right] \tag{29}$$

where N is the total number of the wave components, and

$$\theta_n = k_n x - \sigma_n t + \epsilon_n$$

$$A_n = - a_n \sigma_n \frac{\cosh k_n(h + z)}{\sinh k_n h}$$

$$B_n = - a_n \sigma_n \frac{\sinh k_n(h + z)}{\sinh k_n h}$$

The sum of the phases are not considered for the surf beat wave problem, because they are in the high frequency region. Then

$$S_{xx} = \frac{\rho}{z} \sum_{n=1}^{N} \frac{a_n^2 \sigma_n^2}{\sinh^2 k_n h} h + \frac{1}{4} \rho g \sum_{n=1}^{N} a_n^2$$

$$+ \rho \sum_{m=1}^{N-1} \sum_{n=m+1}^{N} \frac{a_m a_n \sigma_m \sigma_n \sinh (k_m - k_n) h}{(k_m - k_n) \sinh k_n h \sinh k_m h} \cos (\theta_m - \theta_n)$$

$$+ \frac{\rho g}{z} \sum_{m=1}^{N-1} \sum_{n=m+1}^{N} a_m a_n \cos (\theta_m - \theta_n) \qquad (30)$$

and

$$\overline{(u_I \eta_I)}_{z=o} = \frac{1}{2} \sum_{m=1}^{N} \sum_{n=1}^{N} a_m a_n \sigma_n \frac{\cosh k_n h}{\sinh k_n h} \cos (\theta_m - \theta_n) \qquad (31)$$

Method of Solutions

The governing equations are

1. Equation of Motion

$$\frac{\partial q}{\partial t} = - gh \frac{\partial \eta_s}{\partial x} - \frac{1}{\rho} \frac{\partial S_{xx}}{\partial x} - \frac{\partial}{\partial t} \overline{(u_I \eta_I)}_{z=0} \qquad (32)$$

2. Continuity Equation

$$\frac{\partial \eta_s}{\partial t} + \frac{\partial q}{\partial x} = - \frac{\partial}{\partial x} \overline{(u_I \eta_I)}_{z=0} \qquad (33)$$

where

$$q = (u_s + u_I)h$$

The complete mathematical solution of these equations is complicated. However, these equations can be solved numerically by applying finite difference techniques to the method of characteristics. This procedure accounts for bottom slope (wave reflection) and the transient nature of the problem.

Due to the depth variations and associated changes in the slopes of the characteristic lines, the characteristic lines in the x-t plane form a curved network and the results require interpolation to give values on a regular rectangular grid system. This problem can be resolved through the introduction of the following dimensionless variables, which will transform the curved characteristics to straight lines:

1. $Y = \frac{\sqrt{x}}{x_2}$

2. $t' = \frac{t}{T}$

3. $Z = \frac{\eta_s C_2 \sqrt{Y}}{U^2 T}$

4. $Q = \frac{q}{U^2 T \sqrt{Y}}$ (34)

5. $r = 1 - \sqrt{\frac{h_1}{h_2}}$

6. $F_1 = \frac{1}{u^2}\left[\frac{1}{\rho}\frac{\partial S_{xx}}{\partial x} + \frac{\partial}{\partial t}(u_I \eta_I)_{z=0}\right]$

7. $F_2 = -\frac{C_2}{u^2}\left[\frac{\partial}{\partial x}(u_I \eta_I)_{z=0}\right]$

where

$$T = (x_2 - x_1) / \overline{C}$$

$$\overline{C} = (C_1 + C_2) / z$$

$$C_1 = \sqrt{gh_1} = \sqrt{gmx_1}$$

$$C_2 = \sqrt{gh_2} = \sqrt{gmx_2}$$

$$U = 0.22815 \ \ gm \ \ h_{b_{(max)}}$$

x_1 is the distance from shore (x = 0) to the wall boundary; x_2 is the distance from shore to the offshore boundary, and m is the beach slope.

By using the above transformations, the equation of motion and continuity equation are rewritten in the form

$$\frac{\partial Q}{\partial t'} + \gamma\left[\frac{\partial z}{\partial Y} - \frac{z}{2Y}\right] = -\frac{F_1}{\sqrt{Y}} \tag{35}$$

$$\frac{\partial z}{\partial t'} + \gamma\left[\frac{\partial Q}{\partial Y} - \frac{Q}{2Y}\right] = -\frac{F_2}{\sqrt{Y}} \tag{36}$$

The sum of these equations gives

$$\frac{d(Q+z)}{dt'} = \frac{\gamma}{2Y}[z-Q] - \frac{F_1}{\sqrt{Y}} - F_2\sqrt{Y} \quad \text{for} \quad \gamma = \frac{dY}{dt'} \tag{37}$$

and the difference of these equations gives

$$\frac{d(Q-z)}{dt'} = \frac{\gamma}{2Y}[z+Q] - \frac{F_1}{\sqrt{Y}} + F_2\sqrt{Y} \quad \text{for} \quad \gamma = -\frac{dY}{dt'} \tag{38}$$

A path in the Y, t' plane having a constant slope $\pm\gamma$ is the characteristic line and the values of $Q \pm Z$ along such lines are in accordance with equations 37 and 38. These equations are solved by finite different methods.

1. *Boundary Condition.* At the shore, where the water depth equals zero, this is a singular point. For convenience, a wall will be assumed for the calculations, i.e., the no-flow boundary condition at the wall is q = 0.
In deep water, the forcing function of the surf beat wave is very small. Therefore, the boundary condition at offshore is a radiative condition; the surf beat wave will propagate offshore and no surf beat wave will enter the system under consideration from offshore.

2. *Initial Condition.* The sloping bottom surf beat problem is a transient problem. The transient nature is due to the increase in the surf beat wave height during propagation toward shore under the action of the forcing function, S_{xx}. At time zero, no wave occurs on the water surface, then later an incoming wave group propagates from infinity to shore with a surf beat wave force at the speed of the incoming wave velocity due to the forcing function induced by the incoming wave.

3. *Changing Breaking Zone.* A spilling wave breaking was assumed in this model, the incoming wave breaking height and local depth are related by

$$a_b = kh$$

The value of k is constant. In this study, the constant k will be taken as 0.39 and the wave height in the surf zone will depend only on the water depth. For an irregular wave field, the location of wave breaking will vary with the water depth. Therefore, the wave breaking process occurs in an area instead of at only one location as the regular wave breaking process.

Comparison of the Steady Wave Setup with the Surf Beat Wave Simulation Model

A given deep water wave spectra and the surf beat wave simulation model is applied to compare the difference between the surf beat wave and the steady wave setup. Figure 1 shows the assumed deep water wave spectra. The width of the energy frequency band ϵ defined by Rice (1945) is equal to 0.16. The incoming wave periods range from 8 to 12 seconds, and ten wave components with equal frequency increment are selected from the given spectrum. The significant wave height is 4.76 ft. In the simulation model, the bottom slope is 0.02, the offshore boundary water depth is 80 ft and the shore boundary water depth is 1.2 ft. The time step is 1.383 s.

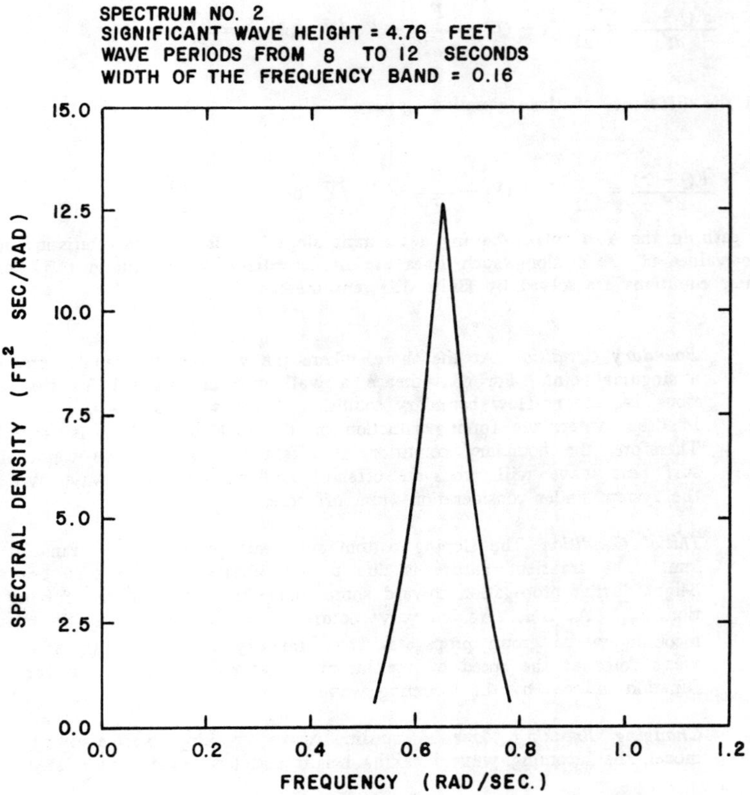

Figure 1. Deep water incoming wave spectrum for the study of the correlation between the surf beat and the wave envelope.

Figure 2 presents the comparisons of the steady wave setup with the surf beat wave on the shore boundary, where the steady wave setup is calculated from

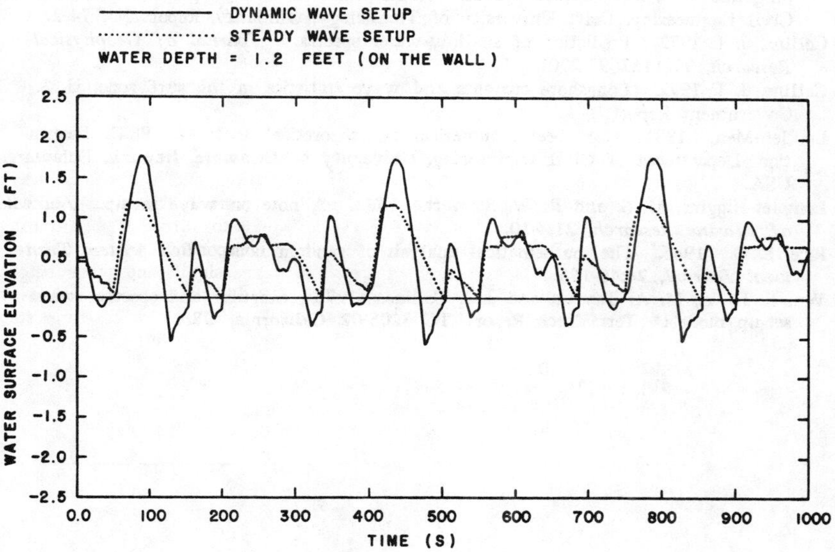

Figure 2. Comparison of the steady wave setup with the dynamic wave setup (simulated by using the wave spectrum in Figure 1).

$$\overline{\eta} = -\frac{K^2 h_b}{16} - \frac{\frac{3}{8}K^2}{1 + \frac{3}{8}K^2}(h_1 - h_b) \tag{39}$$

and h_1 is the water depth at the shore boundary, h_b is the wave breaking depth, K is a constant (here it is 0.78). It is clear that the steady setup and the surf beat wave are almost in phase; however, the crest of the surf beat wave is higher than the steady wave setup height. This setup due to the existence of the surf beat wave is referred to as the "dynamic wave setup", because it includes the contribution from both the non-periodic and low frequency radiation stress terms, and the steady wave setup is generated by including only the non-periodic radiation stress terms.

Conclusions

The results of the study showed that the dynamic wave setup is greater than the steady wave setup, sometimes almost double its value. This is due to the existence of the surf beat wave. Therefore, the dynamic wave setup is important for engineering applications, particularly in the study of storm surges and coastal engineering problems.

References

Battjes, J. A. 1974. Computation of set-up, longshore currents run-up and overtopping due to wind-generated waves. Communications of Hydraulics Department of Civil Engineering, Delft University of Technology, (Holland), Report No. 74-2.

Collins, J. I. 1972. Prediction of shallow-water spectra. *Journal of Geophysical Research*, **77**(11):2693-2707.

Collins, J. I. 1972. Longshore currents and wave statistics in the surf zone, U. S. Government Report.

Lo, Jen-Men. 1981. Surf beat : numerical and theoretical analyses. Ph.D. dissertation, Department of Civil Engineering, University of Delaware, Newark, Delaware, USA.

Longuet-Higgins, M. S. and R. W. Stewart. 1963. A note on wave set-up. *Journal of Marine Research*, **21**:4-10.

Rice, S. O. 1945. The mathematical analysis of random noise. *Bell System Technical Journal*, **24**:46-156.

Wu, F. H. Y., M. A. Tayfun and J. I. Collins. 1978. Singular and spectral wave set-up models. Tetra Tech Report, TC 3205-02, California, USA.

CHAPTER 75

Pore pressure induced by wind waves in sand bed

S.R. Massel * and L. Kaczmarek **

Abstract

The two projects (LUBIATOWO 79 and LUBIATOWO 86) were aimed at study pore pressure behaviour in natural sand bed in the coastal zone of the Baltic sea under the action of storm waves. During both projects, the wave induced pore pressures at the various levels in the sea bed were measured. The collected data were used to verify the applicability of the various theoretical approaches. In the conclusion, the range of the application of the particular analytical method is given.

1. Introduction

The problem of wave induced pressures and stresses in natural sea beds is important with regard to the design of foundations for various ocean and nearshore structures, such as gravity type breakwaters, offshore oil storage tanks and drilling rigs. The subject is also important when one considers the floatation of buried pipelines and the burial of rubble mounds by waves.

The wave-induced flow in a porous bed experiences a flow resistance and transmits forces, seepage forces, to the soil skelton. From a hydrodynamic point of view the flow induced by waves in a porous bed is of interest since energy dissipation, i.e., wave attenuation, is associated with this flow. Early papers (Putnam, 1949; Reid and Kajiura, 1957; Liu, 1973; Moshagen and TØrum, 1975) have focused on a permeable sea bed with a fixed rigid grain skeleton. Moshagen and TØrum (1975) considered the wave-induced flow in a porous medium under the assumption of a compressible pore fluid but an incompressible soil. They found that the inclusion of pore fluid compressibility in the analysis of wave-induced pore pressures in a porous soil significantly altered the vertical seepage forces acting on the soil. Yamamoto et al. (1978) have developed a quasi-static model of a poro-elastic sea bed using the equations developed by Biot (1941). They obtained solutions which, when best fit to laboratory data of pore pressure, showed reasonable

* Professor, Director of Institute of Hydroengineering, 80-288 Gdańsk, Poland.

** Research Assistant, Institute of Hydroengineering, 80-288 Gdańsk, Poland

for both magnitude and phase. Mei and Foda (1980) have shown that the boundary layer approximation well known in fluid dynamics, greatly facilitates the solution of Biot's equations for a variety of boundary conditions.

However, besides the numerous theoretical solutions, the lublished laboratory and field data are very scarce. Dunlap et al. (1978), Bennett and Faris (1979), and Cross et al. (1979) have presented the results of the experiments on the pore pressure generation in the sediments of Mississippi and coastal zone of Pacific. However, they are mainly related to the long-term changes. The relation of the observed pressure to the surface oscillations was not considered. The major objective of the present study is to describe the field measurement projects, i.e. LUBIATOWO 79 and LUBIATOWO 86 which were aimed at studying pore pressure hebaviour in natural sand bed in the coastal zone of the Baltic Sea. The pore pressure measuring system is described and experimental results are compared with results of various analytical approaches.

2. Field experiments

2.1. Scope of the work

The field experiments were carried out at the Coastal Research Staion in Lubiatowo (Poland) in the Southern Baltic (water depth $h \approx 6 \div 7$ m).

A set of over one hundred records were taken for various wave conditions at one profile. Simultaneous measurements of the sea surface oscillations, water pressures at two points in the water layer and pore pressures at four points in the soil layer were carried out (Fig. 2.1).

2.2. Porous layer characteristics

According to the geological information in the vicinity of the point of measurements, the thickness of the rather uniform sand is equal to ~ 6.5 m. The soil is of fine-grained type ($D_{10} = 0.135 \cdot 10^{-3}$ m, $D_{20} = 0.135 \cdot 10^{-3}$ m, $D_{60} = 0.150 \cdot 10^{-3}$ m, $D_{90} = 0.180 \cdot 10^{-3}$ m). Because the uniformity coefficient is less than 6 and the contents of the fraction $\leqslant 0.074 \cdot 10^{-3}$ m is less than 5%, the fine-grained sand can be classified as a poorly graded and fairly clean.

The density of soil or the porosity was not measured "in situ". Thus, only some estimation can be made. For very dense packing the porosity is about 26%, while for very loose it is about 47%. The large forces which were needed to push the pore pressure gauge indicated a rather dense packing of the sand at the measuring profile. Therefore, the porosity of sand is likely to be in the range $30 \div 40\%$.

Moreover, the experimental data on the permeability of the bed material were lacking. Thus, we adopted an empirical formulae to estimate the permeability from the grain-size distribution characteristics. The conclusion from these estimates is that the permeability coefficient is in the range of $10^{-5} \div 10^{-4}$ m/s.

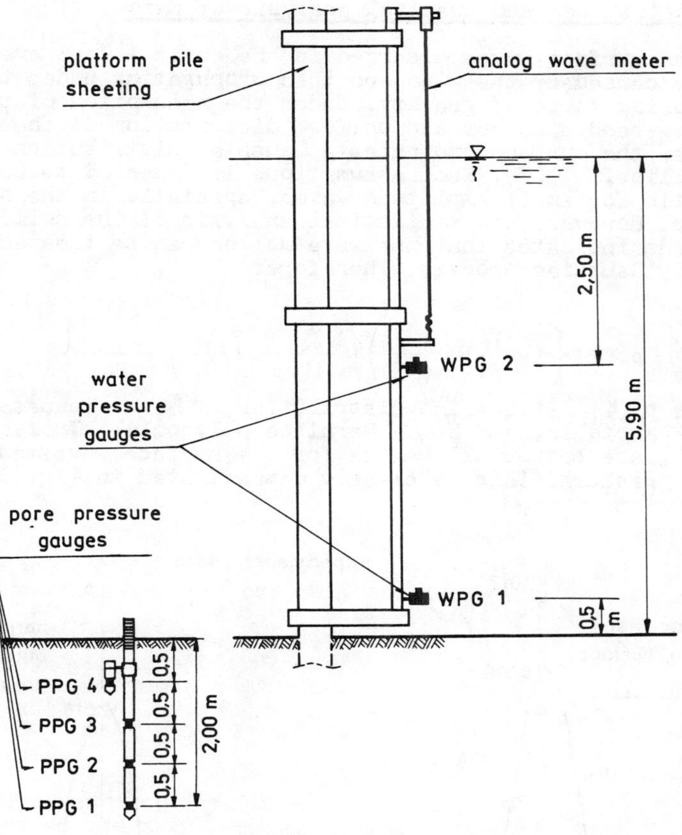

Fig. 2.1. Measurement gauges lay-out

2.3 Instrumentation and measurements

All sensors were arranged in one profile and the measurement signals were transferred by cables to the Station building on shore. Due to access convenience, the gauges were attached to a framework fixed to the sheet-pile section of the platform.

The data were recorded in the real time in digital form. The system of transmission and acquisition of data consists of the modules of PCM transmitter and receiver, Camac system, microcomputer BIZON 580, PC computer and analog and digital recorder. The technical description of the instrumentation and transmission modules can be found elsewhere (Massel and Moshagen, 1980, Kaczmarek, 1987) and it will not be repeated here.

3. Statistical and spectral analysis of data

The surface waves observed in the coastal zone are the waves caused by the wind and then propagating under the restoring force of gravity. Under the assumption of the narrow-band spectrum and uniform distribution of the random phase, the surface ordinates a Gaussian distribution (Phillips, 1977). Such assumptions is expected to be realistic for small amplitude waves, specially in the deep water. However, the statistical analysis of the collected records indicates that the wave motion can be treated as nearly Gaussian process. Therefore:

$$p(\zeta) = p_n(\zeta)\left\{1 + \frac{\gamma_1}{3!} H_3\left(\frac{\zeta}{\sigma_\zeta}\right) + \frac{\gamma_2}{4!} H_4\left(\frac{\zeta}{\sigma_\zeta}\right) + \cdots\cdots\right\} \qquad (3.1)$$

where $p_n(\zeta)$ - Gaussian distribution, γ_1 - skewness, γ_2 - kurtosis, and H( ) - Hermitte polynomial. In fact γ_1 and γ_2 are both small values for the surface waves and the pore pressure. This is clearly demonstrated in Fig. 3.1. in

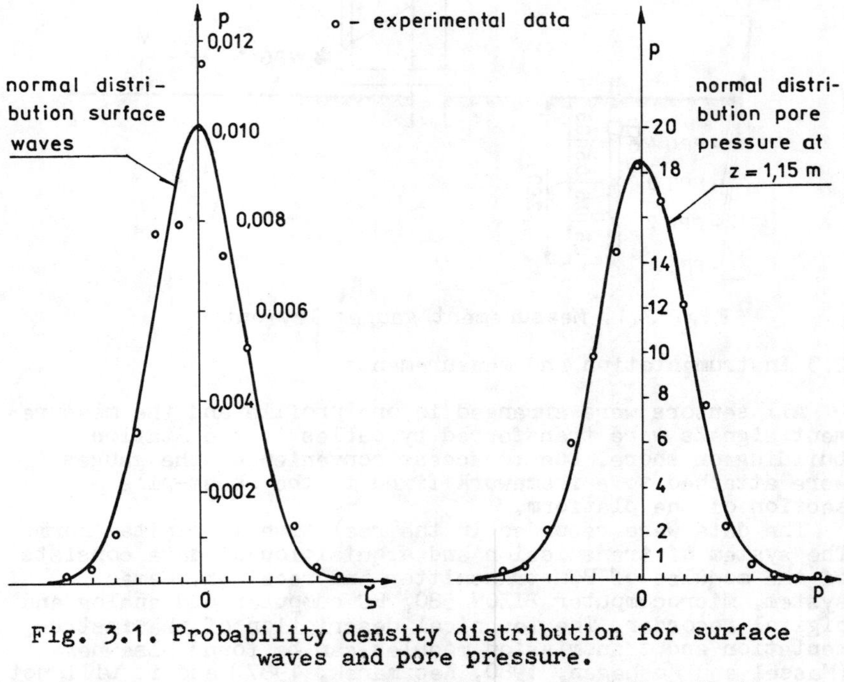

Fig. 3.1. Probability density distribution for surface waves and pore pressure.

which the corresponding probability density distributions are compared.

For the farther analysis we assume additionally that the system: surface waves - pore pressure is the linear system. Hence:

$$S_p(\omega) = \left|H(\omega)\right|^2 \cdot S_\zeta(\omega) \qquad (3.2)$$

in which:

$S_\zeta(\omega)$ - spectral density for the surface elevation,

$S_p(\omega)$ - spectral density for the pore pressure, and

$H(\omega)$ - transfer function.

As the empirical spectral density functions are known, the transfer function can be easy obtained. In the Figs. 3.2 and 3.3, the function $H_{p\zeta}(\omega)$ for the pore-pressure gauges at two points: z = -1.0 m (one meter below the bottom surface) and z = -1.5 m (one and half meter below the bottom surface). As should be expected, the function $H_{p\zeta}(\omega)$ is the decreasing function of the frequency. It should be noted, that during the measurements, the spectral peak frequency for the surface waves was in the interval of $0.80 \leq \omega_p \leq 1.50$ [rd/s], while the mean wave heights ($\bar{H}$) were of order ($0.50 \div 1.5$ m).

The calculations of the coherence function $\gamma^2_{p\zeta}$ (not shown here) indicate that the high values of the $\gamma^2_{p\zeta}$ function are related to the frequency band:

$$(0.6 \div 0.7) \leq \frac{\omega}{\omega_p} \leq (1.8 \div 2.) \qquad (3.3)$$

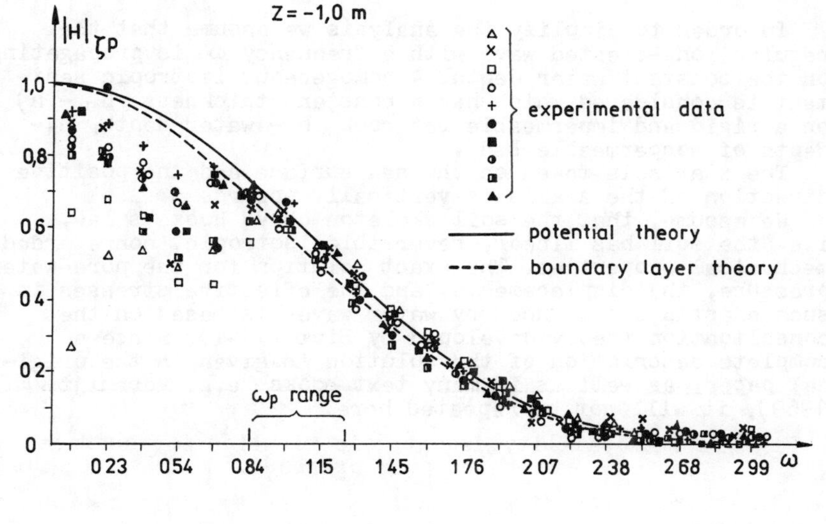

Fig. 3.2. Transfer function $H_{p\zeta}(\omega)$ for pressure gauge at z = -1.0 m

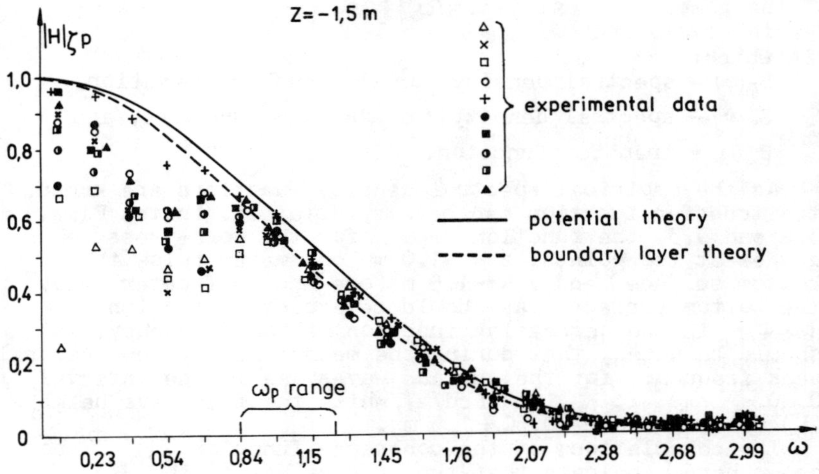

Fig. 3.3. Transfer function $H_{p\zeta}(\omega)$ for pressure gauge at z = − 1.5 m

Outside that band the value of the coherence function is very small. Thus, the usefulness of the theoretical linear model for the pore-pressure decaying is restricted to the frequency band given above.

4. Validity of the potential theory approach

In order to simplify the analysis we assume that the regular long-crested wave with a frequency ω is propagating on the constant water depth. A homogeneous isotropic sediment is considered which has a constant thickness (h_n − h) on a rigid and impermeable bed rock h − water depth, h_n− depth of nonpermeable bed .

The x axis is taken on the sea surface and the positive direction of the z axis is vertically upward.

We assume, that the soil skeleton obeys Hooke's law, i.e. the soil has linear, reversible isotropic, nonretarded mechanical properties. The exact solution for the pore-water pressure, the displacements, and the effective stresses in such elastic bed induced by water waves is based on the consolidation theory developed by Biot (1941). Since a complete description of the solution is given in the original paper, as well as in many text books (e.g. Verruijt, 1969), it will not be repeated here.

The general solution simplyfies considerably in the special case, when $G/E'_w \to 0$, in which G - shear modulus of soil, i.e.

$$G = \frac{E_s}{2(1+\nu)} \qquad (4.1)$$

where E - Young's modulus and ν - Poisson ratio.

The apparent bulk modulus of pore-water containing a some amount of air (E'_w) is related to the true bulk modulus of elasticity of water ($E_w = 2.3\cdot10^9$ N/m) by the relation:

$$\frac{1}{E'_w} = \frac{1}{E_w} - \frac{1-S_r}{P_o} \qquad (4.2)$$

where S_r - the degree of saturation and P_o is the absolute pore-water pressure.
The condition $G/E'_w \to 0$ corresponds to the soil completely saturated with pore-water which does not contain gas. Since the value of G for the fine sand as in LUBIATOWO experiments is about $10^7 \div 10^8$ N/m², the stiffness ratio, G/E'_w becomes :

$$G/E'_w \approx 10^{-2} \div 10^{-1} \qquad (4.3)$$

The general solution of the boundary value problem under the condition $G/E'_w \to 0$ yields (Yamamoto et al., 1978; Massel, 1981):

$$\nabla^2 p = 0 \qquad (4.4)$$

and

$$p \approx \cosh k(z+h_n) , \qquad (4.5)$$

in which k - wave number related to the frequency ω by the classical dispersion relation.
The result (4.4) is the same as that obtained by Massel (1976)(see also Putnam, 1949, Reid and Kajiura, 1957) who assumed that the soil is rigid and water is incompressible (the potential solution). The pressure attenuation for this case is small and independent of the permeability of the soil. From the theory given in Massel' (1976) paper, the following transfer function can be deduced:

$$H_{p\zeta}(\omega) = \frac{\cosh k(z+h_n)}{\cosh k(h_n-h)\cdot\cosh kh} \qquad (4.6)$$

when the pore-pressure is expressed in the water column. The solution (4.6) is shown in the Figs. 3.3 and 3.4 by the solid lines. They confirm the general agreement between theory and experiment; especially in the frequency band, in which the coherence function $\gamma^2 \geqslant 0.75$. Out of this range, for the low frequencies, the experimental transfer function is smaller than the theoretical one. Reason for such behaviour is not quite clear.

5. Others theoretical solutions

The solution of Biot's equations for the various boundary conditions is greatly facilitated by the boundary layer approximation introduced by Mei and Foda (1980).

In this approximation, two regions with different physical mechanisms are distinguished. In the interior of the soil the water and solid skeleton velocities are assumed to be of the same order. On the other hand, near the bottom surface, the drainage is much easier and relative fluid velocity is higher. The solution of the problem consists of the solution of the typical elasto-static problem and the boundary layer correction being appreciable only within this layer.

The results of the calculation, according Mei and Foda (1980) approximation, are shown in Figs. 3.3. and 3.4. by the dotted lines. The agreement with the experimental data is quite good.

It should be noted that the calculations using the full Yamamoto et al. (1978) solution for true G/E'_w value were carried out. The results of calculations are in complete agreement with those of the boundary layer approximation; therefore they are not shown in Figs. 3.3 and 3.4.

6. Conclusions

In recent years the understanding of the interaction of the wave motion and the porous sea and the ability to predict the poro-pressure has increased considerably. A lot of the theoretical solutions were obtained. They include the effects of the stiffness of porous medium and the bulk modulus of water elasticity on the pore pressure attenuation, displacements and stresses in the porous media.
However, the field data are not numerous. In the paper, the two field experiments were reported. The pore pressure attenuation was measured at a few levels in the porous media under the wind waves action. The collected data were processed and the statistical and spectral characteristics were found. The comparison of the experimental data with the various theoretical solutions demonstrates that the boundary layer approximation and the full Yamamoto et al. (1978) solutions properly reflects the observed physical mechanisms. The potential theory only slightly overpredicts the experimental values. However, it should be noted that this conclusion is true for the soil conditions similar to that in LUBIATOWO experiments (fine-grained sand).

7. References

Bennett, R.H., and Faris, J.R. (1979). "Ambient and dynamic pore pressure in fine-grained submarine sediments: Mississippi Delta". Appl. Ocean Res., 1(3).

Biot, M.A. (1941). "General theory of three-dimensional consolidation". Jour. Appl. Phys., 12, 155-164.

Cross, R.H. et al. (1979). "Attenuation of wave-induced pore pressure in sand". Proc. Speciality Conf. Civil Eng. in Oceans. San Francisco.

Dunlap, W.A. et al. (1978). "Pore pressure measurements in underconsolidated sediments. Offshore Techn. Conf., No 3168.

Kaczmarek, L. (1987) "Wind wave induced pore pressure in a natural sea-bed". Tech. Rep. CPBR9.5/51, 28 pp.

Liu, P.L.F., (1973) "Damping of water waves over porous bed". Proc. ASCE, Jour. Hydr. Div., 99, 2263-2271.

Massel, S.R. (1976) "Gravity waves propagated over permeable bottom. Proc. ASCE, Jour. Waterways, Herb. and Coastal Eng., 101/WW1/, 49-57.

Massel, S.R. (1981) "Hydrodynamics in ocean engineering". PWN Warszawa, 150 pp. in Polish .

Massel, S.R., and Moshagen, H. (1980). Wind wave induced pore pressures in a natural sea bed". Measurements at Lubiatowo Coastal Research Station". Techn. Rep., River and Harbour Lab., Trondheim - Institute of Hydroengineering, Gdańsk, 34 pp.

Mei, C.C., and Foda, M.A. (1980) "Boundary layer theory of waves in a poro-elastic sea bed". Proc. Inter. Symp. on Soil under Cyclic and Transient Loading, Swansea, 609-618.

Moshagen, H., and Tørum, A. (1975) "Wave induced pressures in permeable seabeds". Proc. ASCE, Jour. Waterways, Harb. and Coastal Eng., Div., 101, 49-58.

Phillips, O.M. (1977). "The dynamics of the upper ocean". Camb. Univ. Press, 336 pp.

Putnam, J.A. (1949) "Loss of wave energy due to percolation in a permeable sea bottom". Trans. Am. Geoph. Un., 30, 349-356.

Reid, R.O., and Kajiura, K. (1957). "On the damping of gravity waves over a permeable sea bed". Trans. Am. Geoph. Un., 38, 662-666.

Yamamoto, T., Koning, H.L. Sellmeijer, H., and Van Jijum, E., (1978). "On the response of a poro-elastic bed to water waves . Jour. Fluid Mech., 193-206.

Verruit, A. (1969). "Elastic storage of aquifers". In "Flow through porous media" ed. R.J.M. De Wiest . Academic Press.

CHAPTER 76

PROBABILITY OF WAVE BREAKING IN THREE-DIMENSIONAL SEAS.

D. Mather, M.W.G. Snook, C.A. Greated, *
W.J. Easson,** and I.G. Bryden. ***

Introduction

Current guidance notes for the design of offshore or coastal structures to withstand breaking waves are based on the behaviour of long-crested, monochromatic waves. It has been acknowledged for some time that although useful for first approximations and in some cases conservative load estimation, the adoption of long-crested wave theories and design principles for a short-crested sea environment may be inappropriate.

Halliwell & Machen (1981) have shown that in shallow water, isolated breaking waves under short-crested conditions can occur with greater elevations and hence potentially greater destructive power than existing design codes predict. Similarly work by Easson & Greated (1984) has highlighted deficiencies in wave theories to predict particle velocities and accelerations within waves by direct non-contact measurement of these properties, the implications being that such theories may significantly underestimate the total wave loading encountered under breaking waves. These observations call into question our ability to predict accurately and so design to withstand wave breaking effects in short-crested environments.

This paper reviews some of the present approaches to the prediction of wave breaking or whitecapping occurrences and compares data

* Department of Physics, University of Edinburgh, The King's Buildings, Mayfield Road. EDINBURGH. EH9 3JZ. Scotland.
** Department of Mechanical Engineering, University of Edinburgh, Mayfield Road, EDINBURGH. Scotland.
*** Department of Offshore Engineering, Heriot-Watt University, Riccarton, EDINBURGH. Scotland.

collected offshore with simulations conducted in wave basin facilities in Edinburgh, which illustrate that wave breaking can occur at wave steepnesses significantly below those which are commonly assumed to represent the onset of breaking implying that design procedures must recognise and allow for this.

Survey of Reported Studies

Several authors have investigated the statistical properties of wave breaking by the visual observation of sea or model sea conditions to characterise properties of breaking or whitecapping wave crests.

Holthuijsen & Herbers (1986) observed the occurrence of whitecaps at a wave recording buoy in open sea conditions off the Dutch coast, the situation being such as to effectively represent a deep sea environment. Concurrent with the wave data, a visual identification time history technique for identifying the occurrence of whitecaps at the buoy permitted their correlation with wave elevation. By this method, each wave crest between zero-downcrossings could be individually distinguished as a whitecap or an undisturbed, plain wave. Their work compared the marginal and joint probability distributions of a number of wave characteristics (height, period, steepness, etc.) for both wave types together with a brief look into breaking occurrances within wave groups and as a function of wind speed. Unfortunately, although directional measurements were made at the site by the buoy, the nature of the resulting wave spectra were not presented.

Certain of the present authors recently pursued a similar series of investigations under model conditions using a video time logging technique (Mather et.al. 1988) to identify the occurrance of whitecaps on the resulting time history and other wave crest instabilities which were judged to be breaking waves. Under these idealised conditions - free of the effects of wind input, non-stationary wave conditions and currents - it was shown that no single geometric property of directionally spread waves recorded by a conventional omni-directional probe, could be used to distinguish breaking from plain waves.

These appear to be the only experiments reported which visually identify the breaking of waves and correlate them to elevation time histories in a three-dimensional environment, that is to say other than in a wave flume.

Papers by Kjeldsen & Myrhaug (1980) and Kjeldsen et.al. (1981) have established a number of wave geometry properties determined from single point measurements which are useful first descriptions of individual waves. In addition to the conventional definition of Wave Steepness (see Appendix), they also define Crest Front and Rear Steepnesses, Horizontal and Vertical Crest Assymetries. They established the advantages to be gained from using an analysis based on the Zero-Downcrossings rather than Crest or Zero-Upcrossing Period, from the standpoint of physically representative parameters and visual interpretation. They justify their opinion that the joint probability density function of Crest Front Steepness and Wave Height (evaluated between zero-downcrossings i.e. Crest to Preceding Trough) provides the most relevant characteristic for the useful forecasting of breaking waves (at high values of each property) but no guidance is given as to the quantitative regression line useful for this purpose.

Snyder and Kennedy (1983) investigated the probability of occurrance of whitecapping over an area at a fixed instant of time as distinct from other authors who considered temporal characteristics at discrete points. They argue the virtues of a variety of plausible wave and sea-surface characteristics which may potentially influence the occurrance of wave breaking. They assume that a single property may be adopted to form the basis of their theory and suggest on heuristic grounds that this must be a scalar quantity and exhibit relative maximum values only at wave crests. This leads to the identification of wave elevation, vertical surface acceleration and surface curvature as candidate properties. The use of elevation as an identifier is readily dismissed since breaking can be observed over a range of wave heights, just as some of the largest waves do not necessarily break. The argument used to dismiss the adoption of surface curvature is that of the practical difficulties to be encountered in quantitative description, analytical treatment and especially its measurement. Such problems of description and analysis stages are not insurmountable as work by Longuet-Higgins (1962) and Glazman (1985) demonstrates, whilst those of measurement do not seem to have been addressed in recent years in the context of water waves, despite the advances in topographic and remote-sensing methods.

X-Band microwave backscatter experiments by Kwoh & Lake (1984) operating with wavelengths of approximately 3cm, reported that backscattering events occurred in discrete bursts which were highly correlated with wave breaking events, contributions to these signals coming from the "background waveform" (the dominant source) and parasitic capilliaries. Experimenting with a scanning laser-slope gauge they point out that scattering begins slightly before breaking under relatively smooth surface conditions, whereas conventional microwave backscatter detection relies on Bragg-Wave size surface roughness and that at peak response time the only high frequency feature in their tests are small radii wave crests implying that this small scale curvature may be a causal property of wave breaking. Srokosz (1988) however has pointed out that turbulence behind a breaking wave also leads to locally high curvatures and thus cannot reliably be applied as a characteristic, prerequisite to breaking.

The remaining identifier, that of vertical component surface acceleration, has so far seemed to yield the most promising identification results both in experiments (Snyder et.al. 1986) and theoretically [Longuet-Higgins 1969, Ochi & Tsai 1983, Srokosz 1984, Tung & Huang 1987 , Huang et.al. 1986).

Huang et.al. (1986) developed an analytical description of whitecap coverage using the acceleration threshold criterion applied to probabilistic expressions of wave crest amplitude. As with other theories, they apply the restriction that wave breaking only occurs at wave crests in their model and so do not generalise the total coverage to include the effect of entrained foaming water which can be distinguished visually. The model is based on the Wallops spectrum (Huang et.al. 1981), containing the influence of the wind frictional velocity u_*, but the method could easily be applied to the more widely adopted Pierson-Moskowitz or JONSWAP system. In the work of Snyder et.al. the probabilistic description was obtained without adopting the (empirical) u_* and so is to be preferred as an analytic predictive tool although the empirical relationship in the Huang & Long model is probably no less valid and indeed may be preferred for its relative simplicity.

The work of Ochi & Tsai (1983) was one of the earliest to develop a probabilistic function for wave breaking occurrances, basing their efforts on the need to establish a more reliable statistical relationship between wave height and wave period, of those waves which would break. Their Figure 3 illustrates the difficulty in applying a single numerical relationship between these properties either as an average or an upper limiting criteria.

Model Simulation of Sea States

The Fluid Dynamics Unit at Edinburgh University has obtained wave data in the form of elevation plus two subsurface particle velocity components from a Southern North Sea platform site.

To extract the requisite spectral information from these records, the "Extended" Maximum Likelihood Method, due to Isobe et.al. (1984) was implemented. Sample resulting omni-directional point spectra of wave elevation and particle velocities together with the directionally spread variation of elevation spectra are shown in Figure 1. The growth in the spectral peak and its spread to lower frequencies as the storm developed was clearly visible when comparing successive sample spectra.

To investigate the probabilistic properties of wave breaking in the sea states computed from the site data, wave basin reproduction of these results was undertaken in the Edinburgh University Wide Wave Basin. This facility is described elsewhere [Salter, 1981] but briefly occupies a working area of 23m x 5m with 1.2m depth and with eighty, 0.3m width wavemakers occupying one long side. The wave generation system currently implemented is the "mixed frequency snake" summation of wavefronts permitting specification of the reproduced seas' RMS amplitude and discrete spectral form together with a discrete comb type directional spreading function.

The task of evaluating the directional wave spectrum in the wave basin poses the same problems of the same task at full size. In our experiments, the same analysis technique (the EMLM) was used as for determining the site directional spectrum, except that rather than elevation / velocity component combinations of data being analysed, it was decided to construct and use an array of conventional fixed wave probes. Seven probes were used and arranged as two regular, common-axis Mercedes-Benz style stars each of differing arm radius. This arrangement is more valuable in situations where the principal wave direction is unknown rather than for our application in a wave basin of known direction but the experience gained will, we feel, be of value when more fully assessing the analysis procedure at a later date.

Based upon visual inspection of the site spectra, we chose to attempt our simulations with the peak enhanced JONSWAP spectral form combined with the Cosine-power directional spreading function over a restricted range of angular values considered appropriate for modelling purposes ($\pm 45^{\circ}$) and beyond which the site data fell off to less than three percent of the peak value. The success of the simulations hinges on finding suitable parameters of the existing formulations, to match in a least squares way, the model form to the site results (Figures 2,3).

For the purposes of simulation this approach is not ideal, and the

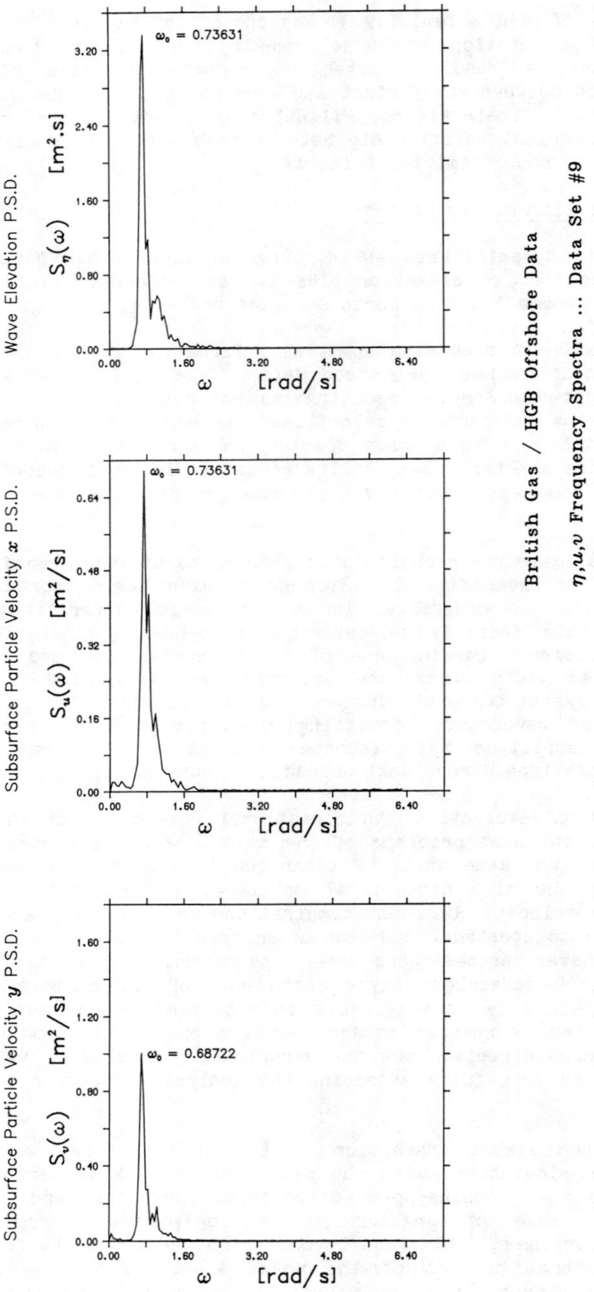
British Gas / HGB Offshore Data
η,u,v Frequency Spectra ... Data Set #9
Wave Elevation P.S.D.
$S_\eta(\omega)$ [m².s]
ω_0 = 0.73631
3.20
2.40
1.60
0.80
0.00
0.00
1.60
3.20
4.80
6.40
ω [rad/s]
Subsurface Particle Velocity x P.S.D.
$S_u(\omega)$ [m²/s]
ω_0 = 0.73631
0.64
0.48
0.32
0.16
0.00
0.00
1.60
3.20
4.80
6.40
ω [rad/s]
Subsurface Particle Velocity y P.S.D.
$S_v(\omega)$ [m²/s]
ω_0 = 0.68722
1.60
1.20
0.80
0.40
0.00
0.00
1.60
3.20
4.80
6.40
ω [rad/s]

FIGURE 1

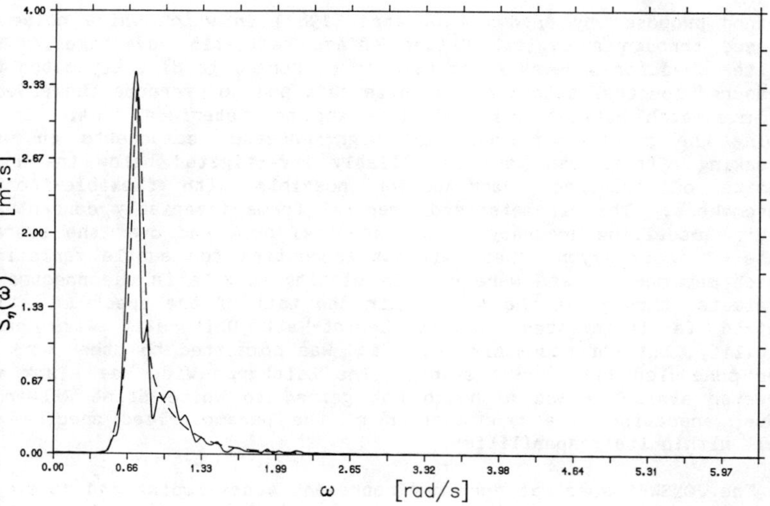

FIGURE 2

Wave Elevation Parameterised Spectra

British Gas / HGB Data Set #9

Site Data .... Full Line

Parameterised JONSWAP ($\gamma = 3.75$) ...... Dashed

Peak Frequency .... $\omega_0 = 0.73631$ rad/s

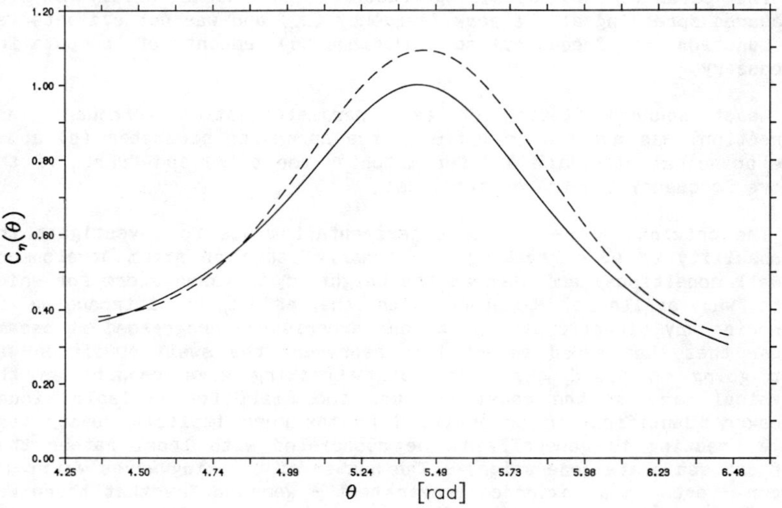

FIGURE 3

Wave Elevation Parameterised Spectral Spreading

method proposed by Bryden & Greated (1984) in which white noise is passed through a digital filter offers realistic advantages. The filter used for wavemaker control can be formed to directly match the measured spectral output of the site data and so overcome the effects of mis-match between the real data and parameterised form. In so doing the greater influence of high-frequency components on wave breaking effects can be more reliably investigated below the upper limits of frequency reproduction possible with flexible-fronted wavemakers. The parameterised spectral forms invariably concentrate their modelling accuracy at the spectral peak and over the average rate of decay beyond that, without accounting for subtle variations which may occur - and were readily distinguishable in the spectra we evaluated throughout the storm - in the tail of the spectrum. This method is implemented in the Heriot-Watt University wave basin facility, but unfortunately this task was committed to other work at the scheduled time for testing. The Edinburgh Wide Wave Tank was however available and although not geared to White Noise Filtering wave generation, the reproduction of the parameterised spectra was well within its' capabilities.

The JONSWAP spectral form may represent a developing and developed sea-state by virtue of the peak enhancement factor (γ). In order to match the parameterisation to the measured spectra, the following justifications were adopted.....

- Peak frequencies will be the same $\omega_{oJ} = \omega_o$
- Range of frequencies must match $\omega_{maxJ} = \omega_{max}$
- Total power in the spectrum must match $\int S_J(\omega)d\omega = \int S(\omega)d\omega$
- Peak spectral energies must be the same $S_J(\omega_o) = S(\omega_o)$

The parameterised spreading function was based solely on the measured spreading at the peak frequency (ω_o) and was not evaluated as a function of frequency so reducing the amount of computation necessary,

Least squares fitting of each parameterisation (frequency and direction) was adopted to optimise the spreading parameter (s) about the principal direction and for matching the third and fourth of the above frequency spectra constraints.

The original scope of the experimentation was to investigate the probability of wave breaking at an early stage of storm development (swell conditions) and then at the height of the same storm for which data was available, together with the effect of introducing or ignoring any directionality. As our experiments progressed it became clear that the record selected to represent the swell condition was not going to yield any significant breaking wave results as the physical size of the model sea was too small for reliable visual breaker identification in addition to the more implicit reason that wave breaking is generally to be associated with local rather than distant sea state generation. The number of breaking waves which did occur - even in an extended length test - were so few that there was too low a level of confidence in the results to yield an adequate statistic. As a consequence of the tests were re-defined to examine the effect on the breaking wave statistic of an increase in the energy content of the evaluated spectral form at the peak of the storm.

Four situations were reproduced from the parameterisations of available offshore wave spectral data. The first reproduced the omni-directional spectrum in a uni-directional (monochromatic) form, as is common practise, whilst the second introduced spreading according to the Cosine-power function suitably scaled to maintain the constant total power of the spectrum. The third and fourth tests used the same spectral form, peak frequency and frequency range but with an increase in the area under the spectral curve corresponding to a rise in the RMS waveheight of 40% based on a Rayleigh distribution. The technique adopted for identifying the geometric properties of the individual breaking waves is described elsewhere [Mather et.al. (1988)] and briefly, relates the time locations of visually identified breaking waves (from a video replay), to the wave record obtained from the reference wave probe. In this way each breaking crest is "flagged" to allow further investigation of whatever geometric or temporal property may be of interest (e.g. steepness, crest assymetry, etc.)

Test Results

Probability density function histograms of geometry properties for each of the lower enrgy sea-states are shown in Figures 4,5, overlaying in each case the probability density histograms of breaking waves and all waves (plain and breaking). The properties illustrated were chosen from consideration of the results of Kjeldsen et.al. (1981), which were those most promising readily computed geometric characteristics likely to distinguish breaking from non-breaking waves. Review of our results led us to conclude that for each of the properties evaluated, breaking waves could occur over their whole range of realisable values.

In many instances the low number of samples of breaking waves lead to seemingly spurious results detached from the main body of the probability distribution, for example in all instances of elevation. Also clear is that the highest waveheight, longest periods, highest elevations or greatest steepnesses that do occur do not always break. This characteristic is particularly unsettling in each case of wave steepness, where the steepest identified breaking waves only exist up to about 75% of the values attained by plain waves. This may be a consequence of the fact that a visually identified breaker has to posess some whitecap which may have formed some relatively long time (at model scales) before passing the wave elevation probe. As a result the elevation will have diminished and the corresponding steepness reduced. Plain waves may on the other hand retain their form all the way up to breaking and inspection of the scales for each wave steepness shows that no waves occurred above values of between 0.055, 0.060 in each seastate of differing average wave height and director spreading. The values obtained for wave steepness on an individual wave basin were all significantly below the maximum value limited by breaking proposed by Hallam et.al. (1978) of 0.142. Small numbers of plain (i.e. non-breaking) waves did approach this value but with so many waves, this number became sufficiently insignificant as not to be apparent on the histograms.

The same problem is encountered when plotting the joint probability distribution (contour plots, Figures 6,7) in that as a wave breaks, its elevation and thus waveheight is reduced so that identified individual wave breaking occurrances on these figures are shifted away from their values at the point of breaking to that noted in each case,

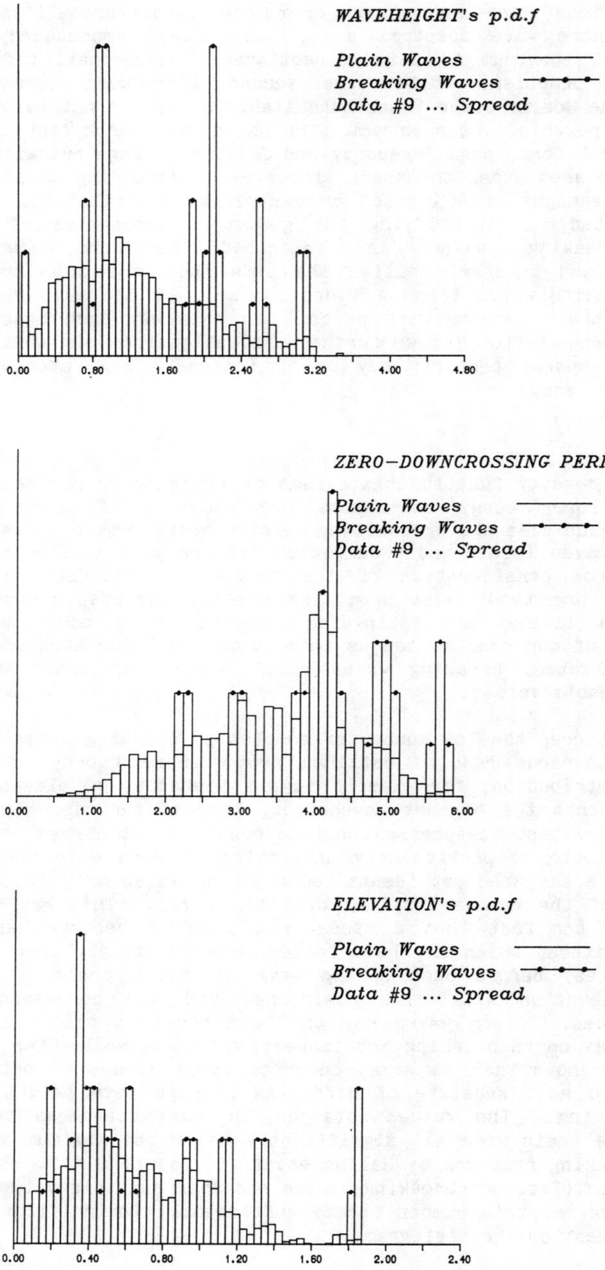

WAVEHEIGHT's p.d.f
Plain Waves
Breaking Waves
Data #9 ... Spread
0.00
0.80
1.60
2.40
3.20
4.00
4.80
ZERO-DOWNCROSSING PERIOD's p.d.f
Plain Waves
Breaking Waves
Data #9 ... Spread
0.00
1.00
2.00
3.00
4.00
5.00
6.00
ELEVATION's p.d.f
Plain Waves
Breaking Waves
Data #9 ... Spread
0.00
0.40
0.80
1.20
1.60
2.00
2.40

FIGURE 4 abc

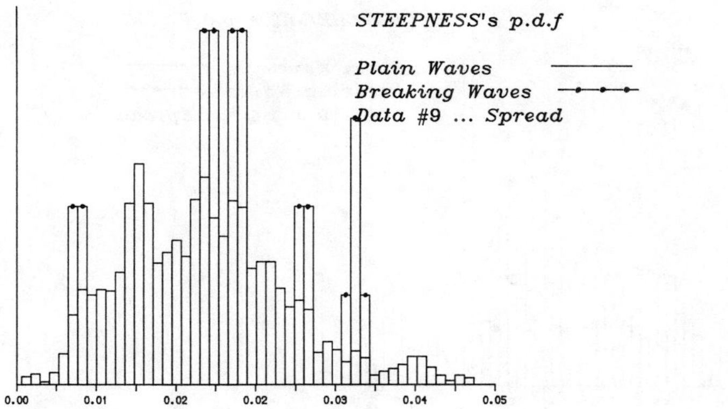
STEEPNESS's p.d.f
Plain Waves
Breaking Waves
Data #9 ... Spread
0.00
0.01
0.02
0.02
0.03
0.04
0.05

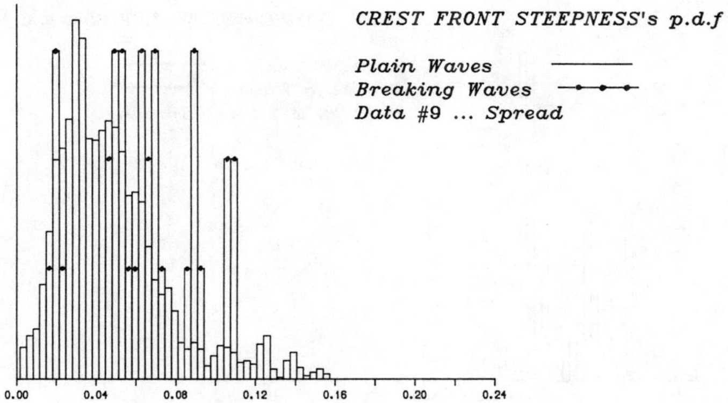
CREST FRONT STEEPNESS's p.d.f
Plain Waves
Breaking Waves
Data #9 ... Spread
0.00
0.04
0.08
0.12
0.16
0.20
0.24

FIGURE 4 de

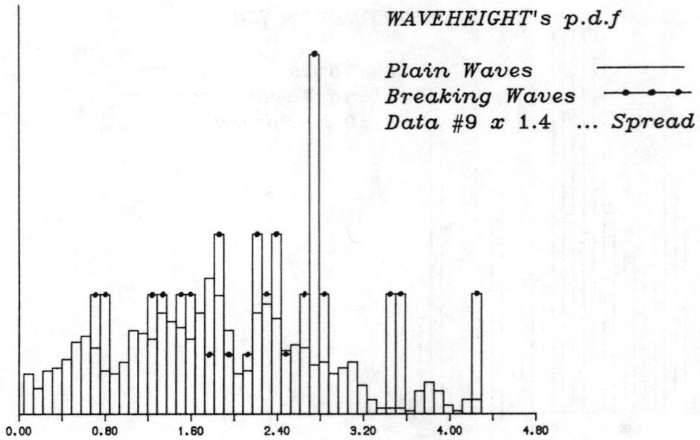
WAVEHEIGHT's p.d.f
Plain Waves
Breaking Waves
Data #9 x 1.4 ... Spread
0.00
0.80
1.60
2.40
3.20
4.00
4.80

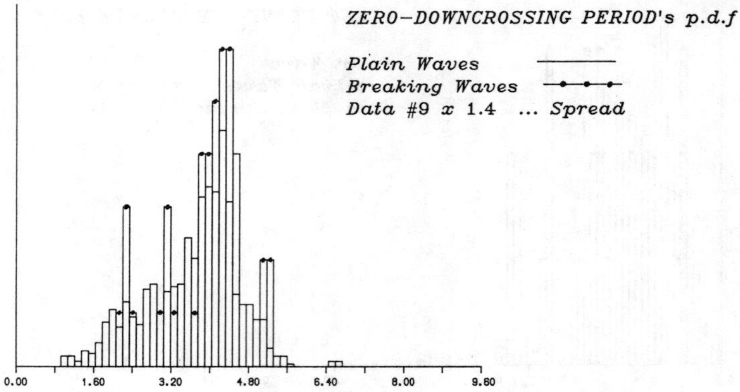
ZERO-DOWNCROSSING PERIOD's p.d.f
Plain Waves
Breaking Waves
Data #9 x 1.4 ... Spread
0.00
1.60
3.20
4.80
6.40
8.00
9.60

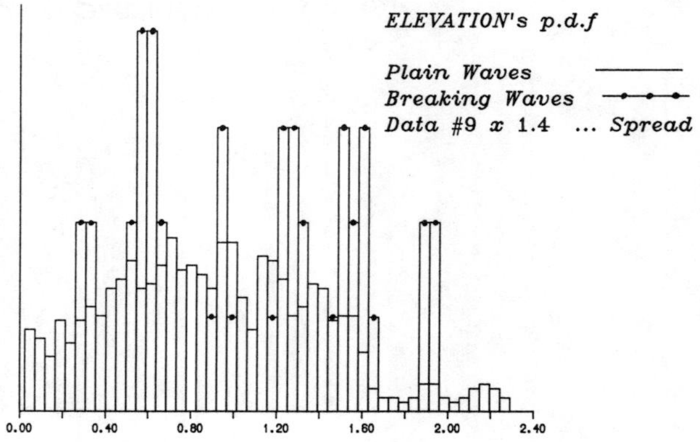
ELEVATION's p.d.f
Plain Waves
Breaking Waves
Data #9 x 1.4 ... Spread
0.00
0.40
0.80
1.20
1.60
2.00
2.40

FIGURE 5 abc

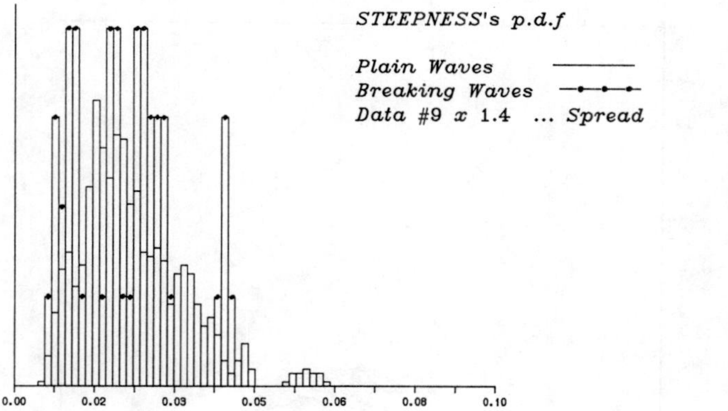
STEEPNESS's p.d.f
Plain Waves
Breaking Waves
Data #9 x 1.4 ... Spread
0.00
0.02
0.03
0.05
0.06
0.08
0.10

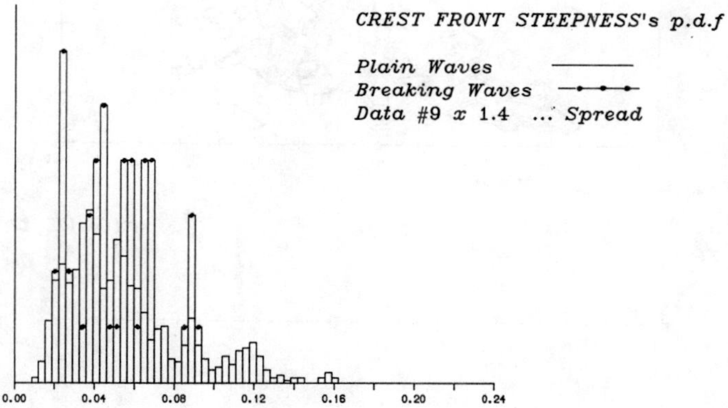
CREST FRONT STEEPNESS's p.d.f
Plain Waves
Breaking Waves
Data #9 x 1.4 ... Spread
0.00
0.04
0.08
0.12
0.16
0.20
0.24

FIGURE 5 de

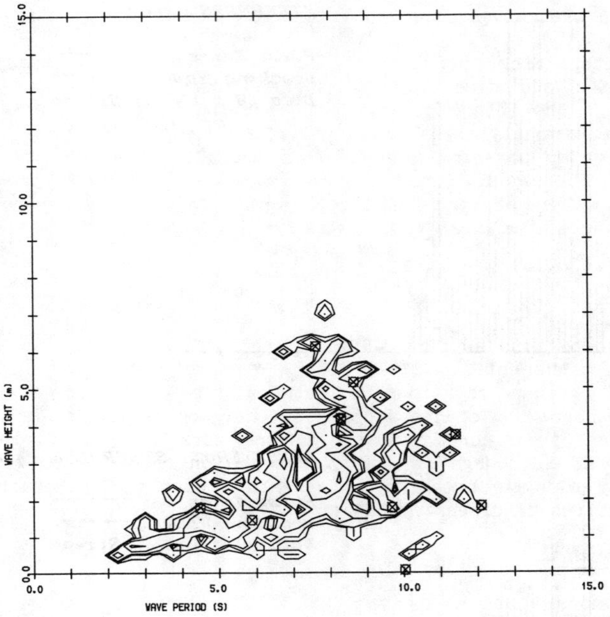

FIGURE 6

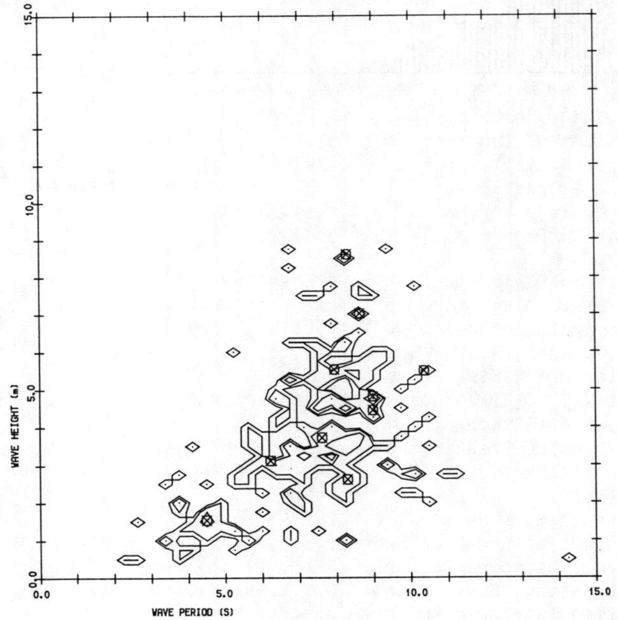

FIGURE 7

these being the values observed at the elevation probe.

Two interesting points of note arise when comparing the monochromatic and directionally spread plots in each case of RMS waveheight. The introduction of a directional spread eliminates the occurrances of outlying, extreme value combinations of waveheight and period over the observed test periods though this observation cannot necessarily be assumed to be true when considering extreme value occurences. This appears to be accompanied in each case by a drift of values towards the regression line equivalent to the maximum wave steepness.

Conclusions

These tests lend support to earlier work by Holthuijsen & Herbers (1986) and Mather et.al. (1988) that in directional seas there seems to be no obvious relationship between the occurence of visually identifable breaking events and any of the geometrically defined wave characteristics in current use. Future work in this field at the University of Edinburgh is to concentrate on attempts to relate the probability of wave breaking to sea-state spectral characteristics and simulations of directionally spread sea surfaces.

Acknowledgement

The authors wish to acknowledge British Maritime Technology Ltd. for financial support under the CASE scheme for Mr. Mather and are grateful to British Gas plc for access to the offshore site data. Many thanks also to Professor Stephen Salter and Mr. David Skyner at the University of Edinburgh and Dr. Meric Srokosz at the Institute of Oceanographic Science, Wormley for assistance during different stages of the preparation of this paper.

References

Bryden, I.G. & Greated, C.A.
Generation of Three-Dimensional Random Waves,
J. Phys. D. Appl. Phys., Vol.17 (1984) Pp.2351-2366.

Easson W.J. & Greated, C.A.
Breaking Wave Forces and Velocity Fields
Coastal Engineering, Vol.8 (1984) Pp.233-241.

Glazman, R.E.
Statistical Characterisation of Sea Surface Geometry for a Wave Slope Field Discontinuous in the Mean Square.
J. Geophysical Res., Vol.91 (1986) Pp.6629-6641

Hallam, M.G. Heaf, N.J. & Wootton, L.R.
Dynamics of Marine Structures.
Report UR8, CIRIA Underwater Engineering Group (1978).

Halliwell, A.R. & Machen, P.C.
Short-Crested Breaking Waves in Water of Limited Depth
Proc. Instn. of Civil Engrs. Vol.71 Part 2, Sept. (1981) Pp.663-674.

Holthuijsen, L.H. & Herbers, T.H.C.
Statistics of Breaking Waves Observed as Whitecaps in the Open Sea
J. Phys. Oceanography, Vol16, Feb. (1986) Pp.290-297.

Huang, N.E., Long, S.R., Tung, C.C., Yuen, Y. & Bliven, L.F.
A Unified Two-Paramter Wave Spectral Model for a General Sea State
J. Fluid Mechanics, Vol.112 (1981) Pp.203-224.

Huang, N.E., Bliven, L.F., Long, S.R., & Tung, C.C.
An Analytic Model for Oceanic Whitecap Coverage
J. Phys. Oceanography, Vol.16 (1986) Pp.1597-1604
Isobe, M., Kondo, K. & Horikawa, K.
Extension of MLM for Estimating Directional Wave Spectrum
Proc. Symp or Description and Modelling of Directional Seas
June (1984), Denmark.
Kjeldsen S.P. & Myrhaug, D.
Wave-Wave Interaction, Current-Wave Interactions and Resulting Extreme
Waves and Breaking Waves.
Proc. 17th Conf. Coastal Engineering, Sydney (1980).
Kjeldsen, S.P., Lystad, M. & Myrhaug D.
Forecast of Breaking Waves on the Norwegian Continental Shelf
Project Report - Ships in Rough Seas (1981) N.H.L.
Longuet-Higgins, M.S.
The Statistical Geometry of Random Surfaces
Hydrodyn. Instab. Proc. Symp. Appl. Math. (1962) Pp.105-143.
Longuet-Higgins, M.S.
On Wave Breaking and the Equilibrium Spectra of Wind-Generated Waves
Proc. Roy. Soc. A., Vol.310, (1969) Pp.151-159.
Mather, D., Bryden, I.G. & Greated, C.A.
Analysis of Visual and Time Series Data from Breaking Wave Fields
Proc. Conf. Offshore Mechanics and Arctic Engineering, Feb. (1988), Houston.
Ochi, M.K. & Tsai, C.H.
Prediction of Occurrence of Breaking Waves in Deep WAter
J. Phys. Oceanography, Vol. 13 Nov. (1983) Pp.2008-2019.
Salter, S.H.
Absorbing Wave-Makers and Wide Tanks
Proc. Conf. on Directional Wave Spectra Applications, (1981) New York.
Snyder, R.L. & Kennedy, R.M.
On the Formation of Whitecaps by a Threshold Mechanism: Part I
J. Phys. Oceanography, Vol.13, Aug. (1983) Pp.1482-1492
Srokosz, M.A.
On the Probability of Wave Breaking in Deep Water
J. Phys. Oceanography, Vol.16, Feb. (1986) Pp.382-385
Srokosz, M.A.
Private Communication (1988).
Tung, C.C. & Huang, N.E.
The Effect of Wave Breaking on the Wave Energy Spectrum
J. Phys. Oceanography, Vol.17, Aug. (1987) Pp.1156-1162.

CHAPTER 77

A NEW PARAMETER FOR WAVE BREAKING WITH OPPOSING CURRENT ON SLOPING SEA BED

Shigeki Sakai*
Ken-ichi Hirayama**
Hiroshi Saeki***

INTRODUCTION

Opposing currents affect the wave breaking processes. The condition of wave breaking caused by opposing currents in deep water is described by the ratio of the wave celerity to the velocity of the opposing current(Yu(1952)). For the wave breaking caused by the opposing currents in shallow water on a flat bed, the equation given by Miche(1951) for the breaking criteria without current remains available (Iwagaki et al.(1980)). However, it is the breaking of shoaling wave in the presence of opposing current on a uniform slope, which is of concern in this paper.

Sakai et al.(1981) showed that the wave breaking affected by opposing currents on a sloping sea bed is characterized by a normalized unit width discharge q*, as well as an incident wave steepness Ho/Lo and a slope of sea bed S, where q* is defined as $q^* = q/g^2T^3$; q: a unit width discharge, g; the gravitational acceleration, T: a wave period. They proposed diagrams for the breaker height and the breaking depth as a function of these three parameters. The breaker index curves in their diagrams show the relationships between the breaker height (or the breaking depth) and q* for waves with particular values of Ho/Lo and S, and it is much more convenient to give a general expression for the breaker indices for arbitrary conditions of q*, Ho/Lo and S.

The purpose of this study is to formulate the effects of these parameters on wave breaking, based on the results of systematic experiments performed by the authors, and to derive an empirical and simplified expression for the breaker height and depth in the presence of opposing currents.

---

* Dept. of Civil Eng., Univ. of Delaware, Newark, DE 19716, USA (presently, Dept. of Civil Eng., Iwate Univ., 4-3-5 Ueda, Morioka, 020 Japan)
** Dept. of Civil Eng., Iwate Univ., 4-3-5 Ueda, Morioka, 020 Japan
*** Dept. of Civil Eng., Hokkaido Univ., Kita-13, Nishi-8, Sapporo, 060 Japan

The original idea is to choose the ratios of the breaker height (or depth) with opposing currents to the breaker height (or depth) without currents, and to search for an individual functional relationship between the ratios and each parameter (q*,Ho/Lo and S). Then, with the help of statistical approaches, these relationships are reorganized into a more simple relationship between the ratios and a new parameter.

EXPERIMENTS

Experiments were conducted by using two wave channels, one of which is 26m long, 1m deep and 0.36m wide; the other is 24m long, 0.8m deep and 0.36m wide. These have the same systems of wave generation and of water circulation. The slopes of the bottom were 1/50, 1/30 and 1/15.

Data acquisition of wave profiles was started when the first wave was generated, and the wave height was measured using five wave profiles from the sixth to the tenth, in order to avoid the reflecting waves. An incident wave height and length were calculated with the linear theory for wave with uniform current. The breakpoint was defined as a point which has the highest wave height. Experimental conditions of the unit width discharge, the wave period and the incident wave height are shown in table 1, and the ranges of the normalized unit width discharge and the incident wave steepness on each sea bed slope are given in Appendix A.

Table 1 Experimental conditions

unit width discharge	q : 0.0 ~ 790.0 ($cm^3/s/cm$)
wave period	T : 0.83 ~ 2.40 (s)
incident wave height	H_0: 1.1 ~ 24.0 (cm)

NEW PARAMETER

In the present study, the breaker height (or depth) in the presence of the opposing current is examined, as compare to the breaker height (or depth) in the absence of the current. In this approach, the breaker height (or depth) in the absence of the current is expected to be described as a function of the incident wave steepness and the sea bed slope. The results of the present experiments without currents give the relationships between the breaker height (or depth) and the incident wave steepness for each sea bed slope, as shown in Appendix B-1 and B-2. In stead of scattering of data, the average lines were obtained for each cases. In the following discussion these average lines will be used to estimate the breaker height (or depth) without current. In these figures, the breaker index curves proposed by Goda(1970) are also illustrated. The present results are consistent with Goda's curves.

The influence of opposing currents appears in the change of breaking depth more significantly than in that of breaker height, and the normalized unit width discharge q* explains the effects of opposing currents on wave transformation(Sakai and Saeki(1984)). The first step is to examine the relationship between Rh and q*, where Rh is the ratio of breaking depth with and without opposing currents. Figure 1 shows the relationships between Rh and q* for given Ho/Lo on the slope

1/30. Rh depends on q* linearly, and then the relationship is expressed by the regression line for each condition of Ho/Lo. Since the slopes of regression lines (SRL)h depend on Ho/Lo, the relationship between (SRL)h and Ho/Lo is examined. Figure 2 shows that (SRL)h is inversely proportional to Ho/Lo on a logarithmic scale in cases where Ho/Lo<0.05. (SRL)h also depends on the slope of sea bed. As shown in Figure 3, (SRL)h is proportional to the slope of sea bed S on a logarithmic scale. Then, the slope of regression line (SRL)h is described as follows:

$$(SRL)_h \propto (H_0/L_0)^{-a} \cdot S^b = ((H_0/L_0) \cdot S^{-c})^{-a} \quad (1)$$

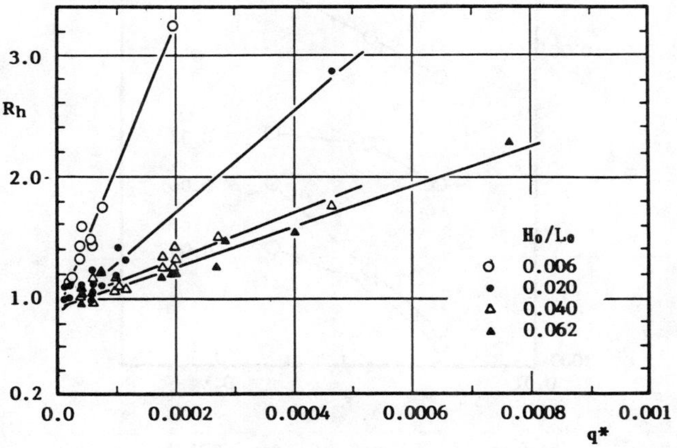

Figure 1 Relationship between Rh and q* (S=1/30)

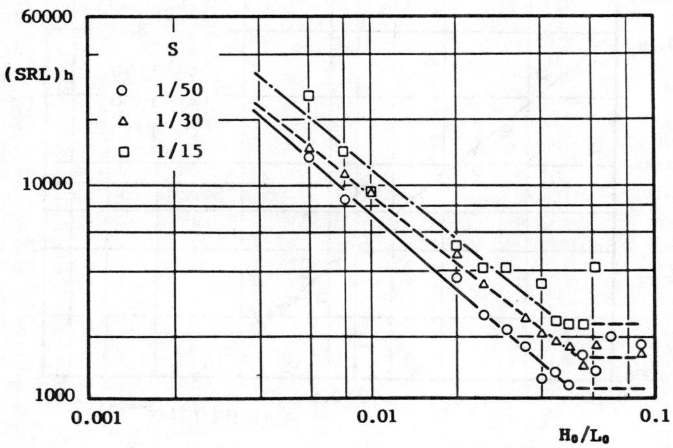

Figure 2 Effect of Ho/Lo on (SRL)h

To explain the combined effects of Ho/Lo and S on (SRL)h, the optimum value of c was found to be between 0.11 and 0.76. Figure 4 shows the relationship between (SRL)h and the combined parameter of Ho/Lo and S, with c=0.41, which gives the highest coefficient of correlation. Using the slope of regression line in this figure, (SRL)h can be formulated as a function of Ho/Lo and S.

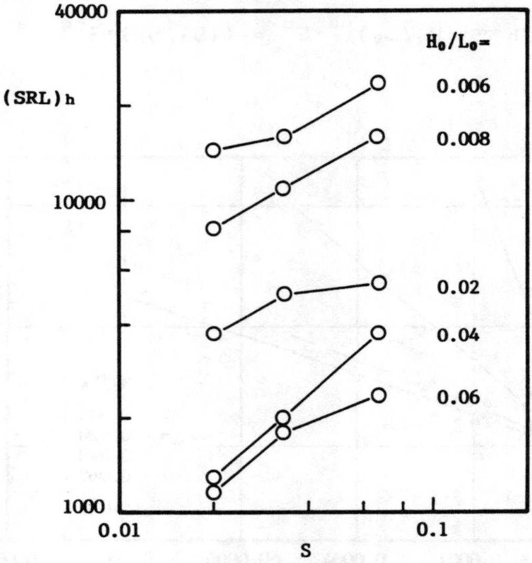

Figure 3 Effect of S on (SRL)h

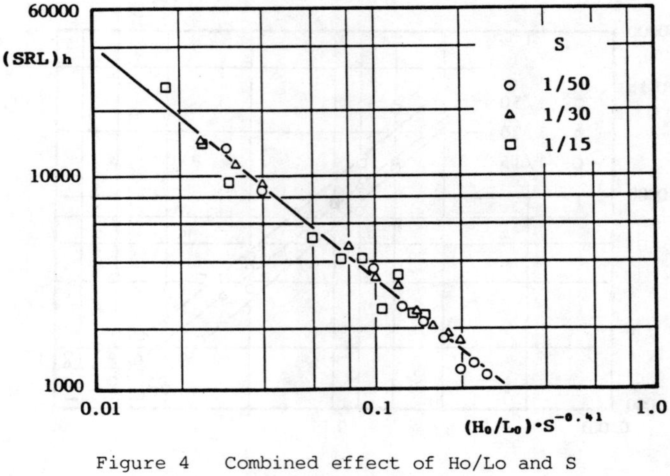

Figure 4 Combined effect of Ho/Lo and S on (SRL)h, with c=0.41

The relationships between RH and q* are illustrated in Figure 5, where RH means the ratio of breaker height with and without opposing currents. The slope of regression line (SRL)H was examined as to be related to Ho/Lo and S. A similar relationship was also obtained as follows:

$$(SRL)_H \propto (H_0/L_0)^{-a'} \cdot S^{b'} = ((H_0/L_0) \cdot S^{-c'})^{-a'} \quad (2)$$

The optimum value of c' is between 0.14 and 0.65. Figure 6 gives (SRL)H expressed by the combined parameter of Ho/Lo and S, with c'=0.37.

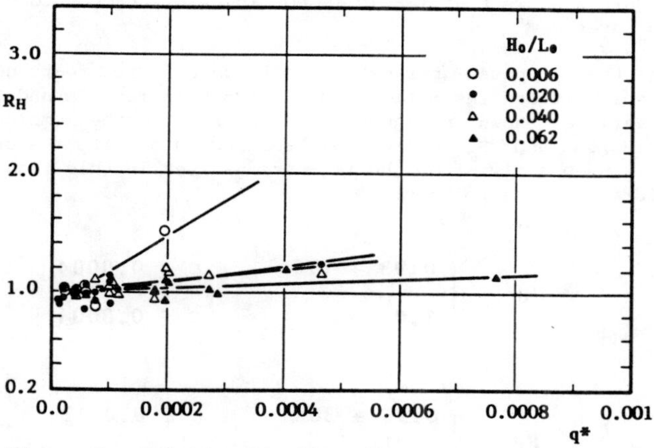

Figure 5 Relationship between RH and q* (S=1/30)

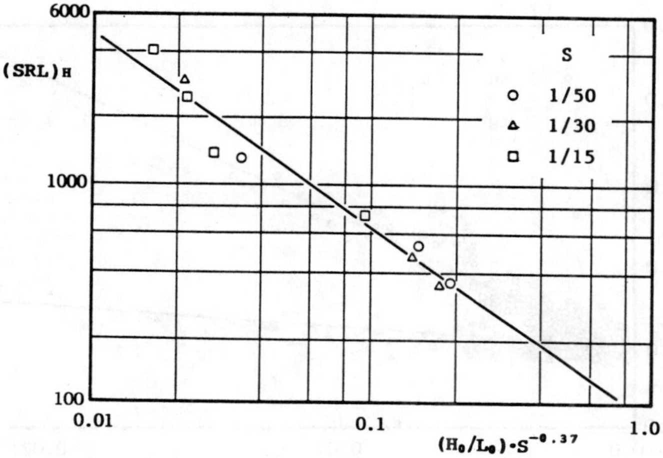

Figure 6 Combined effect of Ho/Lo and S on (SRL)H, with c'=0.37

For a given Ho/Lo and S, both Rh and RH are proportional to q* on a normal scale(see Figures 1 and 5). If the product of q* and (SRL)h (or (SRL)H) is chosen as a parameter, the relationship between Rh(or RH) and this parameter will be described by a single line. Since the range of optimum value of c overlaps with that of c', and it will be more convenient to define a single parameter, a new parameter was defined for both Rh and RH, as follows:

$$\varepsilon = \begin{cases} \dfrac{q^*}{H_0/L_0}\sqrt[4]{S} & H_0/L_0 \leqq 0.05 \\ \dfrac{q^*}{0.05}\sqrt[4]{S} & H_0/L_0 > 0.05 \end{cases} \quad (3)$$

For waves with Ho/Lo larger than 0.05, Ho/Lo is assumed to be 0.05, since Figure 2 shows that (SRL)h is not inversely proportional to Ho/Lo for such cases.

Using the proposed parameter, all data were rearranged into a relationship between the ratios and ε. Both ratios (Rh and RH) depend on ε linearly as shown in Figure 7, and ε explains the combined effects of q*, Ho/Lo and S on the wave breaking on a uniform slope, fairly well. Consequently, from the regression lines in this figure, Rh and RH are formulated as follows:

$$R_h = \begin{cases} 0.93 + 170\varepsilon & \varepsilon \geqq 0.0004 \\ 1.0 & \varepsilon < 0.0004 \end{cases} \quad (4)$$

$$R_H = \begin{cases} 0.96 + 30\varepsilon & \varepsilon \geqq 0.0013 \\ 1.0 & \varepsilon < 0.0013 \end{cases} \quad (5)$$

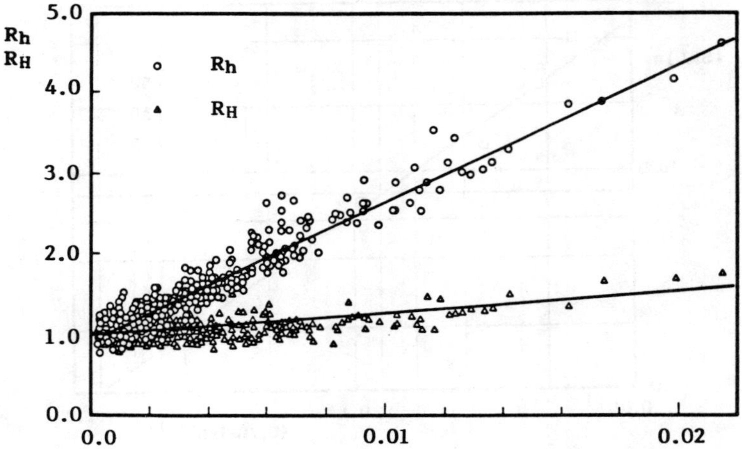

Figure 7 Expressions for Rh and RH by ε

CONCLUSIONS

In the present study, the influences of opposing current on wave breaking were formulated as functions of a new parameter ε, which accounts for the combined effects of the unit width discharge, the incident wave steepness and the sea bed slope. These formulations give the ratios of the breaker height(or depth) with and without opposing currents. Several results have been reported for wave breaking without currents. For example, Goda(1970) proposed diagrams for the relationships between the ratio of the breaker height to the incident wave height and the incident wave steepness. For the breaking depth or the maximum breaker height in a given depth, Weggel(1972) described the ratio of water depth to wave height at breakpoint as a function of breaker height, incident wave length and sea bed slope. Goda(1975) also gave an another expression derived from his diagrams. These formulations and diagrams have been widely used to predict the breaker height and depth without currents. The present formulations gives the breaker height(or depth) with opposing currents, by using the breaker height(or depth) without current predicted by the previous studies.

REFERENCES

Goda,Y.(1970), "A Synthesis of Breaker Indices", Transactions of the Japanese Society of Civil Engineers, JSCE, Vol.2, Pt.2.

Goda,Y.(1975), "Irregular wave deformation in the surf zone", Coastal Engineering in Japan, JSCE, Vol.18, 13-26

Iwagaki,Y. et al.(1980), "Fundamental Study on Breaking of Waves due to Currents"(in Japanese), Proc. of 27th Japanese Conference on Coastal Engineering, JSCE, 30-34

Miche,M.(1951), "Le Pouvoir Reflechissant des ouvrages Maritimes, Exposes a l'action de la Houle", Annales Ponts et Chaussees, 12 Annee, 285-319

Sakai,S. et al.(1981), "Effects of Opposing Current on Wave Breaking on Slope"(in Japanese), Proc. 29th Japanese Conference on Coastal Engineering. JSCE, 138-143.

Sakai,S. and H. Saeki(1984), "Effects of Opposing Currents on Wave Transformation on Sloping Sea Bed", Proc. of 19th Coastal Engineering Conference, ASCE, 1132-1148

Weggel,J.R.(1972), "Maximum Breaker Height", Journal of the Waterways, Harbors and Coastal Engineering Division, ASCE, Vol.98, WW4, 529-548

Yu, Y.Y.(1952), "Breaking of Waves by Opposing Current", Transaction of AGU, AGU, Vol.33, No.1, 39-41

Appendix A Ranges of normalized unit width discharge and incident wave steepness

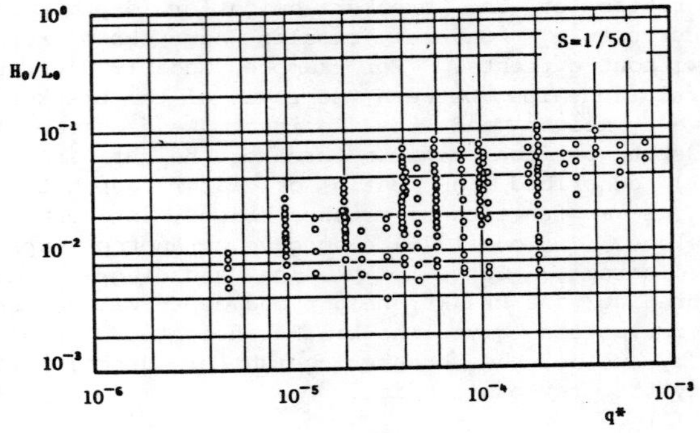

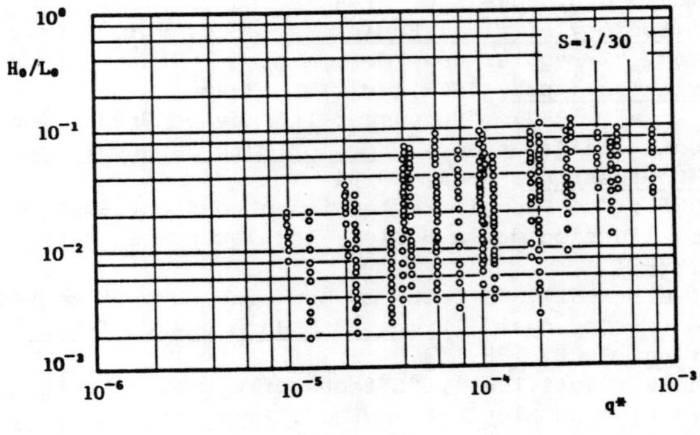

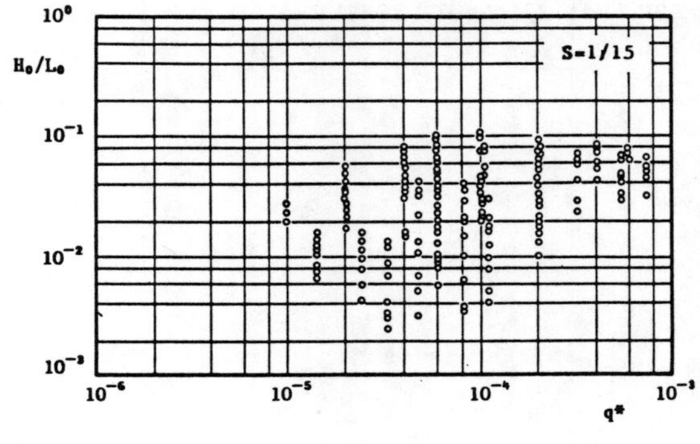

Appendix B-1 Breaker height without current

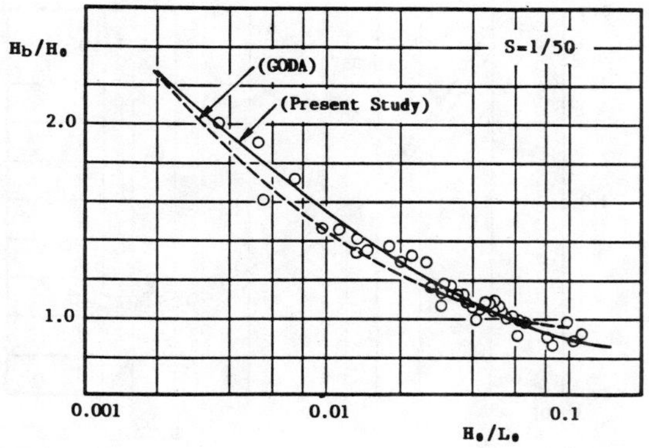

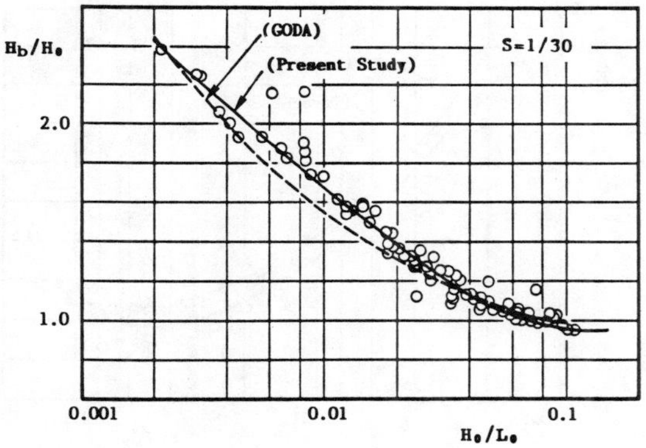

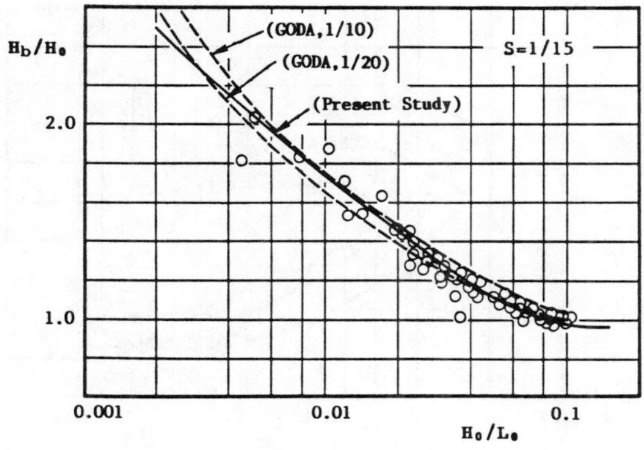

Appendix B-2 Breaking depth without current

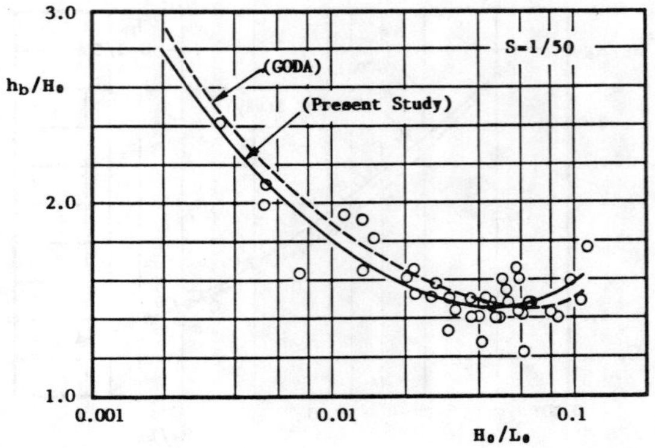

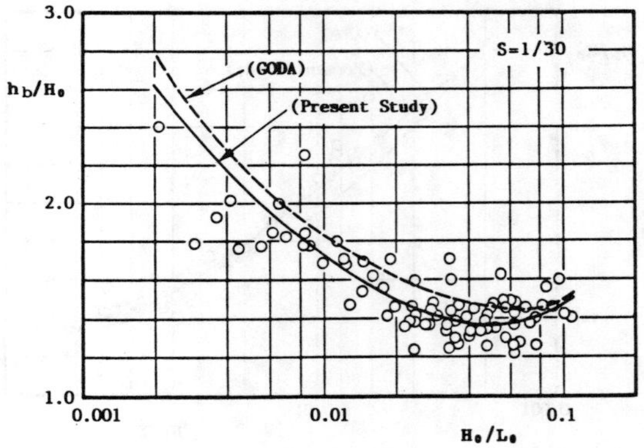

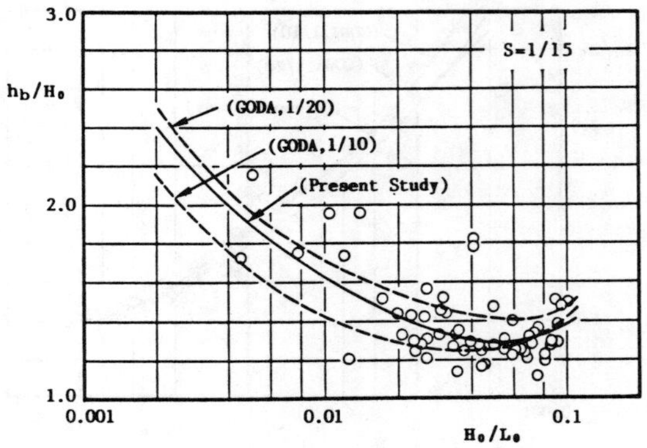

CHAPTER 78

WHAT IS THE SLOPE OF EQUILIBRIUM RANGE IN THE FREQUENCY SPECTRUM OF WIND WAVES?[1]

Paul C. Liu[2]

Abstract An effort to empirically assess the slope of the equilibrium range in a wind-wave frequency spectrum with a large number of data recorded in the Great Lakes did not serve to clarify the uncertainty between a -4 or a -5 frequency exponent representation. The uncertainty is further compounded by indications that the slope is not necessarily unique, it tends to vary with wave momentum. For sufficiently well-developed wind waves the exponent appears to cluster between -3 and -4. For practical applications the f^{-4} equilibrium range is perhaps an effective approximation. What the correct slope is for the equilibrium range, or even whether or not a unique slope exists, remains elusive and has yet to be satisfactorily substantiated.

Introduction

The slope of the equilibrium range on the high frequency side of a wind-wave frequency spectrum has been subjected to continued uncertainy over the last 35 years. In the decade of 1950s, there were three separate spectral representations with slopes of -6, -7, and -5 for the high frequency side of the spectrum developed respectively by Neumann (1953), Darbyshire (1955), and Roll and Fischer (1956). Phillips (1958) resolved these differences by inferring that a range of frequencies on the high frequency side of the spectrum will be dynamically saturated between energy input and dissipation processes, independent of wind stress, and proportional to the -5th power of frequency based on dimensional considerations. Oceanographers and coastal engineers have adhered to this f^{-5} equilibrium range concept since then, as many observations appeared to substantiate it. Recent observations, however, show evidence that the frequency spectrum is actually proportional to f^{-4} on the high frequency side rather than the widely used f^{-5}. Phillips (1985) re-examined the processes of energy input from wind stress, wave breaking and wave-wave interactions and revised his earlier concept by concluding

1 GLERL Contribution NO. 629.

2 Oceanographer, NOAA/Great Lakes Environmental Research Laboratory, 2205 Commonwealth Blvd., Ann Arbor, Michigan 48105

theoretically that the frequency spectrum **S(f)** is in fact proportional to gu_*f^{-4}, where u_* is the wind friction velocity, as Toba (1973) and others have found empirically. While the ocean-wave community has gradually accepted this new concept (e.g., Battjes *et al.*, 1987), there are still measurements showing f^{-5} to be a good representation. So what is the slope of equilibrium range in the frequency spectrum of wind waves? In this paper this questioni is explored, using a large number of measured wind-wave spectra in the Great Lakes to make a detailed examination of the equilibrium range slope of a wave spectrum.

Data

Eight NOAA Data Buoy Center (NDBC) buoys are deployed in the Great Lakes (Figure 1) during the ice-free period of the year, providing measurements of air and surface water temperature, wind speed and direction at 5 m above the water surface, and surface wave frequency spectra. The data used here were recorded by the NDBC buoys during 1981-1986. The wave spectra were calculated with 24 degree of freedom and a 0.4 Hz high frequency cutoff, from 20 minutes of hourly accelerometer measurements. A detailed description of the NDBC buoy measurement system is given in Hamilton (1986). A comparison of wave measurements obtained from a NDBC buoy and a Waverider buoy is presented in Schwab and Liu (1985).

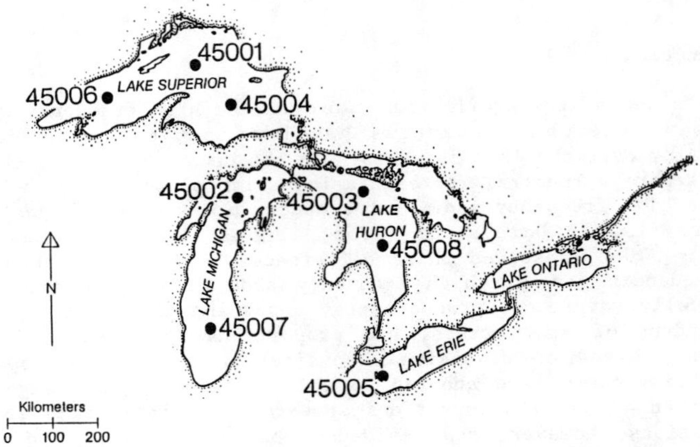

Figure 1. Location map of the eight NDBC buoys in the Great Lakes.

With the wind speed and the air and water temperatures given from the measurements, the wind friction velocity u_* used in this study was readily obtained using the formula for overwater roughness length by Charnock (1955) with the formulation of stability length by Businger *et al.* (1971), assuming a constant neutral drag coefficient of 1.6 x

10^{-3}. This method is basically similar to various other methods of estimating u_* as shown in Liu and Schwab (1987).

Results and Discussions

An episode of well-defined wave growth was examined first. The episode was recorded from the Northern Lake Michigan Buoy 45002 on November 23, 1984. As shown on the left hand side of Figure 2, the wind was blowing steadily from the southwest direction with wind speeds increasing continuously from 7 ms^{-1} to over 15 ms^{-1}. Because of the relatively calm conditions prior to the episode, the wave field developed steadily with significant wave height growing from less than 1 m to 4.5 m over a 19-hour interval. The wave spectra recorded during the episode are plotted on the right hand side of Figure 2. The pure wave growth is exhibited by the continuous shifting of the front faces at the low frequency side of the spectra toward lower frequencies. The high frequency side of the spectra bands together to signify the equilibrium range. To see the slope of the equilibrium range we draw a f^{-5} line and a f^{-4} line along the banded part of the spectra. It is clear that either of the two straight lines could be considered good representation of the equilibrium range. Thus we are not able to visually delineate precisely which one provides a better representation.

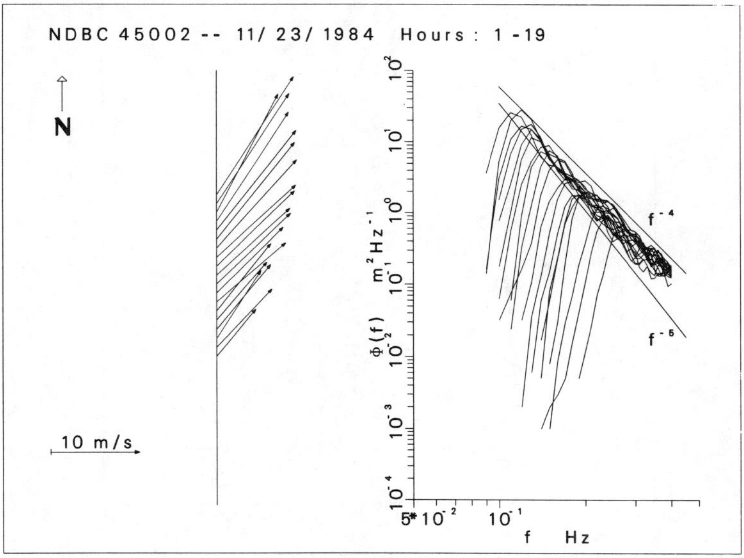

Figure 2. Wind vectors and wave spectra plots of a growing wave episode in Northern Lake Michigan.

Normalizations

Several normalization schemes for studying spectral representation are available in the literature. One, proposed by Toba (1973) simply normalizes the spectral density $\Phi(f)$ by the peak energy density $\Phi(f_p)$ and the frequency f by the peak energy frequency f_p. A scheme adopted in Liu (1971) and others uses the same frequency normalization, but nondimensionalizes the spectral density by a factor of f_pE^{-1} where E is the total energy integrated from the spectrum. Another scheme normalizes the spectral parameters with wind friction velocity u_* and acceleration of gravity g to yield normalized density and frequency as $\Phi(f)g^2u_*^{-5}$ and fu_*g^{-1} respectively. Phillips (1985) and several other authors have used this u_* normalization. An application of these three normalizations to the November 23, 1984 episode is shown in Figure 3. In the graphs the -4 as well as -5 slope lines are also drawn for comparison. Again, as in the case of the dimensional spectral plots, it is difficult to distinguish the best representation for the equilibrium range.

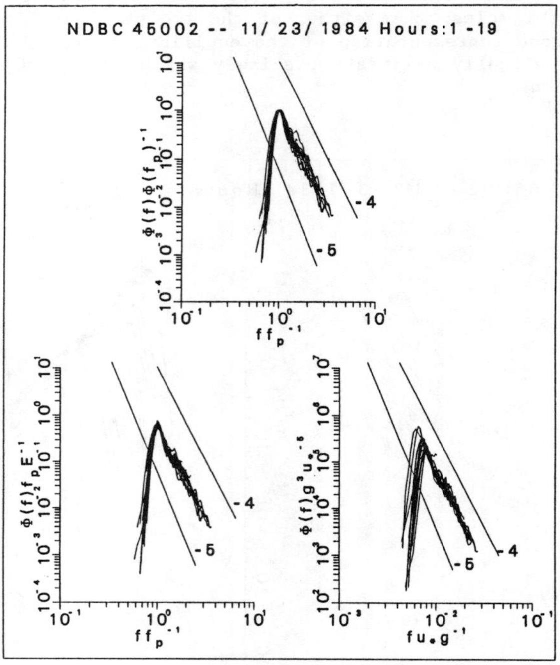

Figure 3. An application of three different normalizations to the growing wave episode spectra; -4 and -5 slope lines are included for comparison.

A frequency spectrum with f^{-5} equilibrium range is generally represented by

$$\Phi(f) = \beta(2\pi)^{-4}g^2f^{-5} \quad (1)$$

on the high frequency side. If (1) is valid then the normalized spectral density $\Phi(f)(2\pi)^4f^5g^{-2}$ should be independent of frequency and approaching the value β at the high frequency side. Similarly a frequency spectrum with f^{-4} equilibrium range can be represented by

$$\Phi(f) = \alpha(2\pi)^{-3}gu_*f^{-4} \quad (2)$$

In order for (2) to be the correct representation, the parameter $\Phi(f)(2\pi)^3f^4(gu_*)^{-1}$ should also be independent of frequency and approaching the value α at the high frequency side.

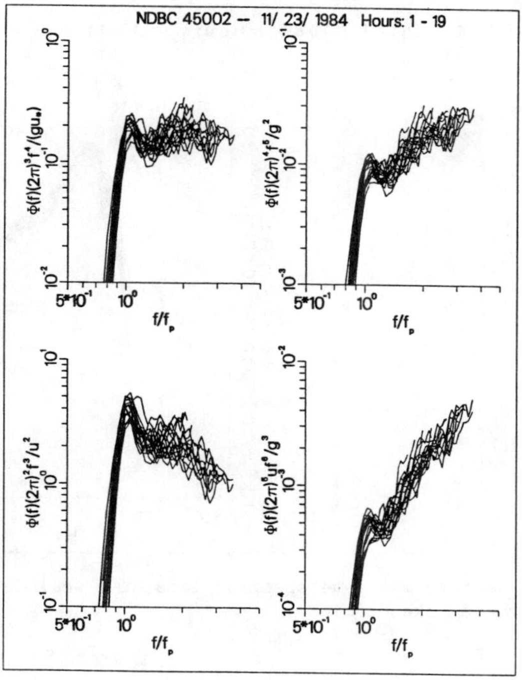

Figure 4. Normalizing the growing wave episode spectra with frequency exponents of 3, 4, 5, and 6 respectively shown clockwise from the lower left.

Rather than confining the exponent to merely -4 or -5, the formulations (1) and (2) can be generalized to allow any exponent i by

$$\Phi(f) = \alpha_i(2\pi)^{-i+1}g^{i-3}u_*^{5-i}f^{-i} \quad (3)$$

and then expect the normalized parameter $\Phi(f)(2\pi)^{i-1}g^{-i+3}u_*^{-5+i}f^i$ to be constant and independent of frequency at the high frequency side if a given i is the valid exponent. Figure 4 shows the normalizations deduced from (3) for the same November 23, 1984 episode for i = 3, 4, 5, and 6. The results are now clearly distinctive. Only the bundle of f^{-4} normalizations shows a reasonably constant trend at the high frequency side. Thus this episode of wave growth spectra emerged to support the f^{-4} formulation.

Yet another normalization scheme is given in Donelan *et al.* (1985), which averages the high frequency side of the spectral density between $1.5f_p$ and $3f_p$ and uses it to normalize the whole spectrum. This scheme avoids the use of u_*, but the results are essentially the same as those in Figure 4.

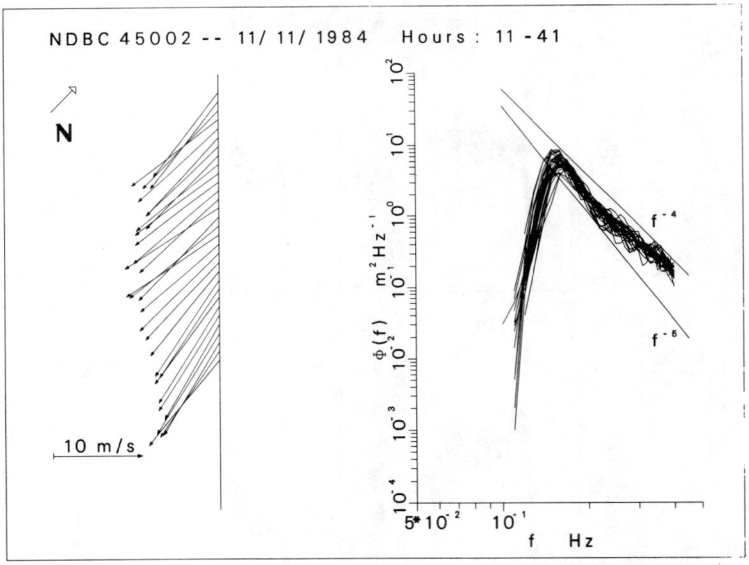

Figure 5. Wind vectors and wave spectra plots of a well-developed wave episode in Northern Lake Michigan.

An Episode of Well-Developed Waves

If all the field measurement could provide results that are similar to Figure 4, then the results would lead to evident conclusions. Expecting to show the universality of the f^{-4} equilibrium range sustained in Figure 4, another episode measured from the same Northern Lake Michigan buoy only 12 days earlier was selected. This was an episode of well-developed waves with winds blowing from the north at virtually constant wind speeds of 12 ms^{-1} over 30 hours. This wind field generated significant wave heights of 2 m to 2.8 m during the episode. As shown in Figure 5 all the spectra were banded together over all frequency components as expected for a fully-developed wave field. The same normalization deduced from (3) that led to Figure 4 is applied to this episode and presented in Figure 6. These results are also distinctive. The constant trend of the normalization at the high frequency side is displayed, however, by the f^{-3} normalization rather than the f^{-4}.

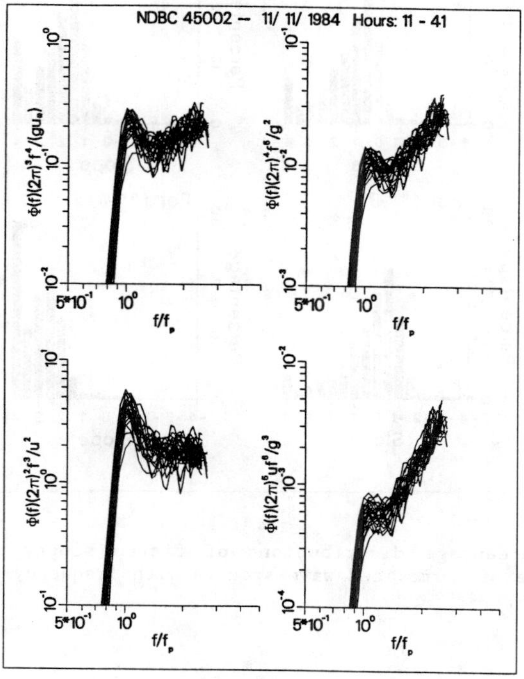

Figure 6. Normalizing the well-developed wave episode spectra with frequency exponents of 3, 4, 5, and 6 respectively shown cloclwise from the lower left.

These two episodes as presented in Figures 4 and 6 are not typical. Rather these were ideal cases selected from a large amount of available data to provide definitive and consistent results. The results presented in Figures 4 and 6 are, however, clearly inconsistent.

An immediate question arrises: Is this inconsistency due to instrumentation? This seems not to be the case, since both episodes were recorded by the same NDBC buoy 45002 during a time period the gauge was known to be working properly - and 45002 is one of the earliest and most persistently operating gauge in the Great Lakes. It is conceivable that the slope of equilibrium range in the wave spectra is not unique, and that the growing and well-developed waves may render different slopes.

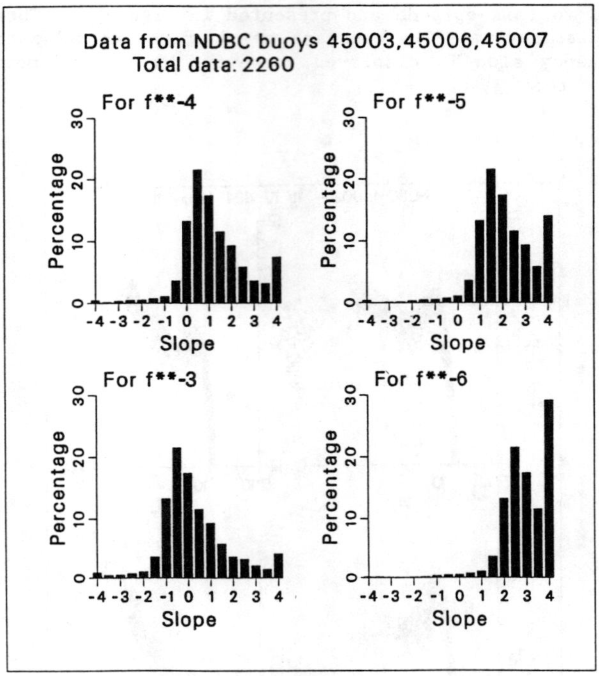

Figure 7. Percentage distributions of fitted slopes at the high frequency side of normalized wave spectra with frequency exponents of 3, 4, 5, and 6.

Examining Large Data Set

To further explore the "non-unique" nature of the equilibrium range slope in the wind wave spectra, many more data of selected episodes had to be examined. To do this systematically we arbitrarily combined 40 days of data that had been measured during October and November of 1984 from each of the three buoys 45003, 45006, and 45007 yielding a sample data set covering a variety of field measurement conditions. Spectra with peak-energy frequency higher than 0.25 hz were eliminated, to ensure the data were mainly well-developed or sustained growth waves. It is not practical to present graphs similar to Figures 4 and 6 for over 2200 items of data, but the same normalizations given in Figures 4 and 6 were applied to the data set and subsequently a linear regression line was fitted to each normalized spectrum between the frequency range of $1.5f_p$ and $3f_p$. A valid f^{-i} representation would render mostly "near zero" slopes to signify a constant trend over the normalized equilibrium range. Figure 7 presents the percentage distributions of these slopes, analogous to the four graphs given in Figures 4 and 6. Note that even though the four figures in Figure 7 were obtained from four separate normalizations, they were in fact the same distribution sliding over different slope abscissa. In general, both f^{-3} and f^{-4} normalizations provide the most near zero slopes. This is consistent with the different results obtained in Figures 4 and 6, and it appears that the equilibrium range slope may very well not be unique.

To this point, in the normalization with f^{-i}, the equilibrium range frequency exponent i has been set to be 3, 4, 5, or 6. Alternatively, this requirement can be relaxed by simply fitting a linear regression line between the $1.5f_p$ and $3f_p$ range to the original wave spectra. The slope of the fitted line which may or may not be an integer, denoted by the exponent n as in f^{-n}, thus represents the actual measured equilibrium range slope of the spectra. The resulting percentage distribution of these exponents obtained from applying this approach to the combined data set is shown in Figure 8. Again, there is no single prevalent exponent, but the larger percentages lie between exponents 3.5 and 5.5.

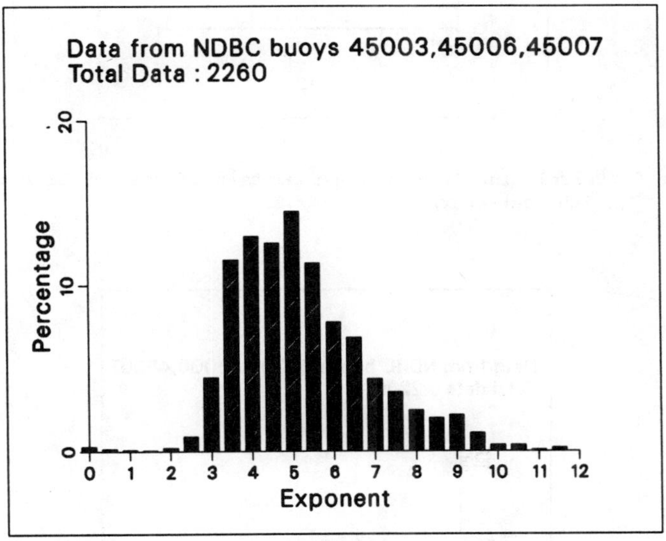

Figure 8. Percentage distribution of the exponents n obtained by fitting the high frequency side of the original wave spectra with a logarithmic line of the form f^{-n}.

All of these results are consistent in that they lack clear indications of the existence of a unique equilibrium range slope. The inconsistent results shown in Figures 4 and 6 can now be readily justified as being characteristic of the non-uniqueness of the equilibrium range slope.

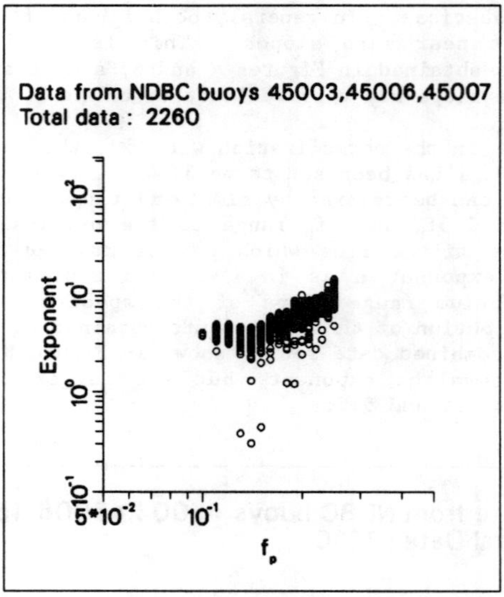

Figure 9. Correlating frequency exponents n of the equilibrium range, as in f^{-n}, with peak-energy frequency f_p.

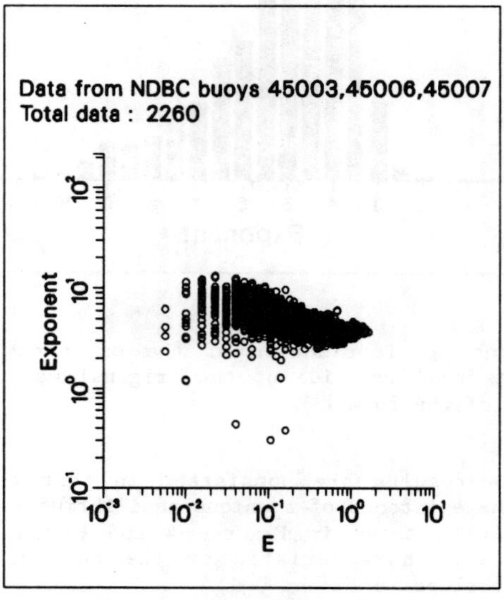

Figure 10. Correlating frequency exponents n of the equilibrium range, as in f^{-n}, with total energy E.

Correlating the Equilibrium Range Frequency Exponent

Having been generally established that the frequency exponent of equilibrium range is not unique in this extensive data set, it is of interest to examine briefly the variability of the exponent. Meaningful correlations are lacking between the exponent and commonly used nondimensional parameters; however, the exponent tends to correlate well with dimensional parameters. Figures 9, 10, and 11 show respectively the exponent's correlation with peak-energy frequency, total spectral energy, and wave momentum. Basically the figures show that the exponent, at early wave growth stage, exibits a decreasing trend versus decreasing peak-energy frequency, increasing wave energy, and consequently, increasing wave momentum. At the stage of well-developed or fully-developed waves the exponent appears to be unchanging and eventually settled between the absolute values of 3 and 4.

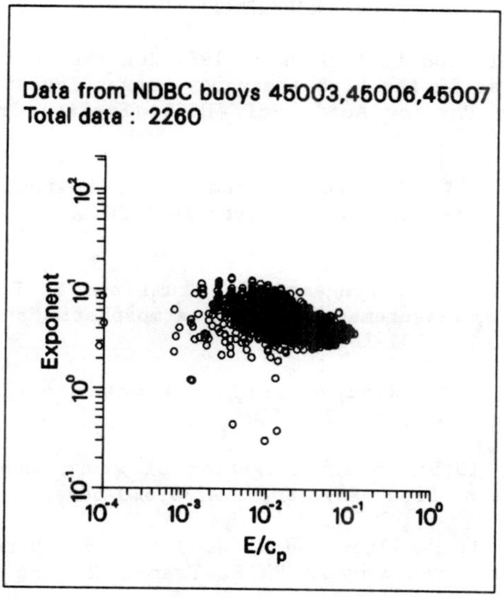

Figure 11. Correlating frequency exponents n of the equilibrium range, as in f^{-n}, with wave momentum E/c_p.

The non-uniqueness of the equilibrium range slope have been recognized by other investigators previously. Spectral representations with non-predetermined frequency exponent have beed proposed for the high frequency side of the spectrum. Huang *et al.* (1981) deduced that the exponent was a function of the significant wave slope. Barenblatt and Leykin (1981) showed that the exponent relates to $g/(f_p u_*)$. The present analyses, however, do not seem to support either of these earlier results.

Concluding Remarks

This study was originally expected to simply delineate the applicability of f^{-4} or f^{-5} representation for the equilibrium range of the wind wave spectra. Examining over 2200 well-developed storm wave spectra recorded in the Great Lakes has not resolved the uncertainty. There are well documented cases for f^{-4}, f^{-5}, f^{-3}, and others, but the data seem not to converge and lead to a unique frequency exponent. As most of the exponents lie in the -3.5 and -5.5 range and between -3 and -4 for the fully developed cases, a representation of f^{-4} is perhaps a useful approximation for practical applications. What the correct slope is for the equilibrium range, or even whether or not a unique slope emerges, remains elusive and has yet to be satisfactorily substantiated.

REFERENCES

Barenblatt, G. I. and I. A. Leykin, 1981: On the self-similar spectra of wind waves in the high frequency range. *Izvestiya, Atmospheric and Oceanic Physics* Acad. Sci. USSR (English Translation), 17, 35-41.

Battjes, J. A., T. J. Zitman, and L. H. Holthuijsen, 1987: A reanalysis of the spectra observed in JONSWAP. *J. Phys. Oceanogr.*, 17, 1288-1295.

Businger, J. A., J. C. Wyngaard, Y. Izumi and E. F. Bradley, 1971: Flux-profile measurements in the atmospheric surface layer. *J. Atmos. Sci.*, 28, 181-189.

Charnock, H., 1955: Wind stress on a water surface. *Quart. J. Roy.Meteorol. Soc.*, 81, 639-640.

Darbyshire, J., 1955: An investigation of storm waves in the North Atlantic Ocean. *Proc. Roy. Soc.*, A230,560-569.

Donelan, M. A., J. Hamilton and W. H. Hui, 1985: Directional spectra of wind-generated waves. *Phil. Trans. R. Soc. Lond.*, A 315, 509-562.

Hamilton, G. D., 1986: National Data Buoy Center programs. *Bull. Am. Meteorol. Soc.*, 67, 411-415.

Huang, N. E., S. R. Long, C. C. Tung, Y. Yuan, and L. F. Bliven, 1981: A unified two-parameter wave spectral model for a general sea state. *J. Fluid Mech.*, 112, 203-224.

Liu, P. C., 1971: Normalized and equilibrium spectra of wind waves in Lake Michigan. *J. Phys. Oceanogr.*, 1,249-257.

Liu, P. C. and D. J. Schwab, 1987: A comparison of methods for estimating u_* from given u_z and air-sea temperature differences. *J. Geophys. Res.*, 92C, 6488-6494.

Neumann, G., 1953: On ocean wave spectra and a new method for forecasting wind generated sea. *Tech. Memo. 43*, U. S. Beach Erosion Board, 42p.

Phillips, O. M., 1958: The equilibrium range in the spectrum of wind-generated ocean waves. *J. Fluid Mech.*, 4, 426-434.

Phillips, O. M., 1985: Spectral and statistical properties of the equilibrium range in wind-generated waves. *J. Fluid Mech.*, 156, 505-531.

Roll, H. U. and G. Fischer, 1956: Eine kritische Bemerkung zum Neumann Spektrum des Seeganges. *Deuts. Hydrogr. Z.*, 9, 9-14.

Schwab, D. J. and P. C. Liu, 1985: Intercomparison of wave measurements obtained from a NOMAD buoy and from a Waverider buoy in Lake Erie. *Proceedings Oceans 85*, MTS and IEEE Ocean Engineering Society, 1131-1137.

Toba, Y., 1973: Local balance in the air-sea boundary processes. III. On the spectrum of wind waves. *J. Oceanogr. Soc. Japan*, 29, 209-220.

CHAPTER 79

SURF BEAT GENERATION ON A MILD-SLOPE BEACH

Hemming A. Schäffer[1] and Ib A. Svendsen[2]

ABSTRACT: Two dimensional generation of surf beats by incident wave groups is examined theoretically. An inhomogeneous wave equation describes the amplitude of the surf beat wave. The forcing function is the modulation of the radiation stress. The short waves are amplitude modulated both outside and inside the surf zone causing the long wave generation to continue right to the shore line. Resonant generation as shallow water is approached is included. The analytical solution is evaluated numerically and shows a highly complicated amplitude variation of the surf beat depending on the parameters of the problem.

1. INTRODUCTION

Field observations show that on mild slope beaches a significant amount of wave energy occurs at frequencies far below the peak frequency of the incoming sea waves. The existence of such long waves were first reported by Munk (1949) who also speculated that the components with period around 2 minutes were caused by variation in height of the surf and he coined the name "surf beats." Also Tucker (1950) found long waves of 1-5 minute period with a height that increase linearly with the height of the short period sea, and a time lag corresponding approximately to the time it would take for a wave group to reach the breaker zone or beach and for a long wave generated there to be reflected back to the observation point. Longuet-Higgins and Stewart (1962,64) suggested that while the short waves are destroyed by breaking, the set-down wave generated by and following the wave groups outside the surf zone as a forced wave is reflected at the beach and propagates seaward as a free wave.

Since the first recording, numerous observations have shown that the energy of surf beat can actually be very substantial and in some cases even exceed that of the high frequency wind waves (Wright et al., 1982) and the amplitudes at the shoreline can be comparable to that of wind waves (Guza & Thornton, 1982,85).

Although no final proof has been established, it seems widely accepted today that the surf beats are generated by mechanisms in the nearshore region and that they are associated with modulations of the amplitude of the short-period incoming waves. However, several possible ways in which energy can be transferred from the high

[1]Visiting Scholar, Dept. of Civil Engineering, Univ. of Delaware, Newark, DE 19716 (present address Inst. Hydrodyn. & Hydr. Eng., DtH, Denmark).
[2]Professor, Dept. of Civil Engineering, Univ. of Delaware, Newark, DE 19716 (mailing address for this paper).

frequency wind waves to the surf beats have been considered with more or less decisive results. It is likely that several possible mechanisms can be active either separately or at the same time. Closely related to the question of generation is the nature of the long wave motion in the nearshore region. In particular, it has been discussed extensively whether the nearshore long wave motion is dominated by forced or free waves; by progressive or standing waves, and whether it consists of crosshore directed (essentially two-dimensional) waves or trapped edge waves, which are three dimensional reflection-refraction wave patterns.

The present work assumes that the surf beat can exist as a two-dimensional motion. It is inspired by the work of Symonds et al. (1982) who investigated a mechanism for generating two dimensional surf beat that had not been considered previously. That mechanism is based on the fact that waves initially of different height will break at different distances from the shore line and (more important) have a different height at breaking. The surf beats are generated by the variation which this breaking pattern causes in the total radiation stress in the region between the extreme seaward and shoreward positions of the breakpoint. The varying breakpoint generates a shoreward moving long wave with the frequency of the wave groups, and since the forcing takes place only in the breaker region the long wave is a free wave through the rest of the surf zone. At the shoreline this wave is fully reflected so that a standing long wave is formed in the surf. Outside the surf zone the reflected wave continues seaward as a progressive wave.

In the model of Symonds et al., the groupiness of the waves is totally destroyed at the breaking point which is why there is no surf beat generation in the actual surf zone. A saturation model with a constant wave height to water depth ratio is used for the waves in the surf zone.

In the work presented in the following, we allow the wave groups to be maintained all the way to the shoreline. Measurements as well as experiments with the surf zone model developed by Svendsen (1984) indicates that if there is a variation in wave height at the breaker point the waves will remain different through the surf zone. For simplicity, we model this by introducing a breaker height variation at a fixed breaking point. This will represent the other extreme relative to the situation studied by Symonds et al. Observations show that the true picture probably represents a combination of the two: a varying breakpoint with some groupiness left in the surf zone. A somewhat similar problem was studied by Foda & Mei (1981) using a multiple scale expansion and different assumptions about the relative magnitude of the wave components involved. Our result contains a resonant interaction which was also included in the analysis by Freilich & Guza (1984) although under different assumptions about the magnitude of the bottom slope.

2. <u>DESCRIPTION OF MODEL CONDITIONS</u>

Since the phenomena associated with wave groups are quite complicated it may be useful to give a more detailed description of the situation which is created mathematically in the following.

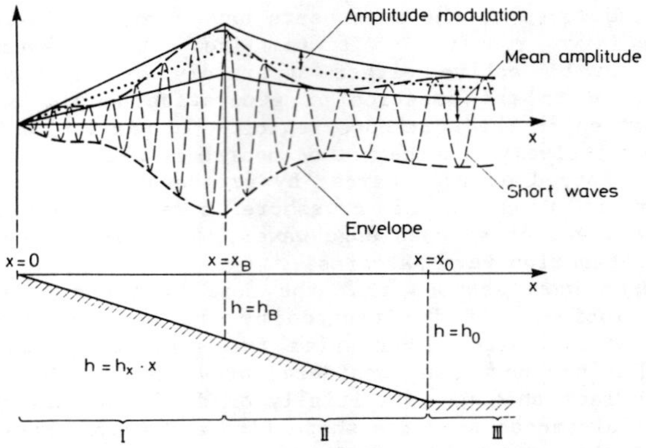

Fig. 1: Sketch of bottom topography and short wave variation considered.

Region III

The situation is depicted in Fig. 1. We consider a region of constant depth which could be a continental shelf, say, (region III in the figure). In the present study we simplify by letting the groupiness of the waves be caused by two regular incoming wave trains with slightly different wave number $k_0(1+\epsilon)$ and $k_0(1-\epsilon)$ the second with a much smaller amplitude than the first. This implies that the surf beat wave becomes a simple wave of constant period which allows us to seek analytical solutions to the problem. The basic equations used, however, can also be applied to the general case of a time varying wave height as in a train of irregular Stokes waves, but this would require a numerical solution.

In region III we assume a quasi-uniform state. The modulation of the waves causes a variation in radiation stress which generates a set down wave, and we assume that the length of the wave height modulation is much larger than the depth so that the set down wave is a long wave. This set down wave and its propagation into the shore region was neglected by Symonds et al. (1982).

Present in region III is also an outgoing long wave which represents the results of the transformations and reflections taking place in the nearshore region. The amplitude and phase of that wave is one of the unknowns of the problem.

Region III is primarily introduced to be able to establish well defined and reasonably simple seaward boundary conditions for the problem.

Region II

In region II the depth is decreasing sufficiently gently that we can assume local solutions for short and long wave components. Thus the short wave will be shoaling towards the breaking point x_B. During this process, however, the balance between the variation in radiation stress and set down wave is constantly changing which causes the forced set down wave to grow. As the short waves approach breaking their group velocity approaches that of the forced wave creating a state of almost resonance. We describe this dynamical process (in contrast to the equilibrium situation in region III) by a WKB approximation.

For most values of the governing parameters we find that in spite of the resonant interaction the energy transferred from the short wave to the forced wave is relatively small, and we therefore omit in the present paper to take the effect of this energy flux into account in the development of the short waves. There is no principle difficulty though in extending the formulation to include this effect as well.

At the transition point x_0 between region III and II the solutions for the two regions are matched by requiring continuity in mass and momentum flux for the long wave solution. These conditions cause a partial reflection at x_0 of the set down waves, whereas it is assumed that no such reflection occurs for the short waves (either because of sufficiently deep water or because the transition between horizontal and sloping bottom actually takes place sufficiently smoothly to allow the short waves to adjust).

Region I

At x_B the short waves are assumed to break. Since there is a temporal variation in the height of the waves reaching this point, we have a time-varying breaker height at x_B. The dissipation of energy in the surf zone is proportional to H^3 and therefore, although the wave height variation will decrease shorewards, those waves that are highest at the (fixed) breakpoint will always remain the highest, etc. Thus this mechanism implies that the groupiness of waves remains present all the way to the shoreline. It also implies that the generation process responsible for the modification of the forced long wave continues through the surf zone.

Again we have chosen to use the simplest possible description of the processes involved. Thus a saturation model based on $a=\gamma h$, where h is the water depth, is used to describe the wave height variation in the surf zone of the short waves. Essentially the specification of wave height as a fraction of depth replaces solution of the energy equation for the short waves. Since the short wave height varies as the groups propagate shorewards, this means that the parameter γ is $\gamma(x,t)$.

The matching at x_B again requires continuity in the mass and momentum flux for the long wave motion. The abrupt shift in the rate of change of the radiation stress represents a source of difference at the two sides of the matching point.

At the shoreline the long wave is fully reflected and radiated seaward as a free wave. The background for this assumption is discussed below and involves some aspects that, although known from other areas of wave dynamics, do not seem to have been applied before to analyze the behavior of surf beats near the shoreline.

3. THE GOVERNING EQUATIONS

The governing equations for the surf beat are derived from the depth integrated and time averaged equations for waves and currents (see e.g. Phillips, 1977; Mei, 1983). In those equations the set-up/set-down will then correspond to the long wave surf beat and the time-varying current will represent the particle motion in that long wave. Since the equations are derived under the assumption that the current velocity is uniform over depth, their use implies that we assume the surf beat is everywhere a long wave (as already indicated). On the other hand, the equations contain all relevant non-linear terms in the current discharge and set-up (including the momentum flux due to the mass flux Q_S in the waves). Thus the description of the long wave component of the wave motion is actually equivalent to that given by the nonlinear shallow water equations.

The total water particle velocity u is split into

$$u = u_W + U \tag{3.1}$$

where U is the velocity in the current, u_w that of the oscillatory motion so that $\overline{u_w} = 0$ below wave trough level ($\overline{}$ meaning average over a wave period).

The relevant equations are then the continuity equation

$$\frac{\partial b}{\partial t} + \frac{\partial \overline{Q}}{\partial x} = 0 \tag{3.2}$$

where b is the elevation of the mean water surface and

$$\overline{Q} = Q_c + Q_s \quad ; \quad Q_s = \overline{\int_{-h_o}^{\eta} u_w dx} \quad ; \quad Q_c = \int_{-h_o}^{c} U dz \tag{3.3}$$

η is the total surface elevation measured from a horizontal reference level. The equation of shore-normal (x) momentum is

$$\frac{\partial \overline{Q}}{\partial t} + \frac{\partial}{\partial x}\left(\frac{\overline{Q}^2}{h} + \frac{1}{\rho} S_{xx}\right) + gh\frac{\partial b}{\partial x} + \frac{\overline{\tau_b}}{\rho} = 0 \tag{3.4}$$

in which h is the local depth (including the long wave surface elevation), S_{xx} is the shore wave radiation stress and $\overline{\tau}_b$ the mean bottom shear stress (which we will neglect here).

Since we are particularly interested in the surface elevation b of the long waves we first eliminate $\overline{Q}$ from the linear terms of (3.2) and (3.4) by cross differentiation. This yields

$$\frac{\partial^2 b}{\partial t^2} - \frac{\partial}{\partial x}\left(gh\frac{\partial b}{\partial x}\right) - \frac{\partial}{\partial x}\left(\frac{\overline{Q^2}}{h} + \frac{1}{\rho} S_{xx}\right) = 0 \tag{3.5}$$

It is assumed that the total motion of any point consists of a quasisteady regular wave on which we superimpose a small perturbation (the amplitude modulation) that varies in time. Thus all wave averaged quantities have the form

$$f(x,t) = f_0(x) + f_1(x,t) \tag{3.6}$$

In particular the depth h is

$$h(x,t) = h_0(x) + b_0(x) + b_1(x,t) \tag{3.7}$$

By assuming that there is a steady state basic solution of wave height variation it can be inferred that for this solution there is no net mass flux. Hence by continuity $Q_0 = 0$ and

$$\overline{Q}(x,t) = Q_1(x,t) \tag{3.8}$$

When these assumptions are substituted into (3.5) we get the following equation for b_1

$$\frac{\partial^2 b_1}{\partial t^2} - \frac{\partial}{\partial x}\left[g(h_0+b_0)\frac{\partial b_1}{\partial x} + gb_1\frac{\partial b_0}{\partial x}\right] - \frac{\partial^2 S_{xx,1}}{\partial x^2} = 0 \tag{3.9}$$

where $S_{xx,1}$ represents the variation in S_{xx} due to the wave height modulation for the short waves (vide (3.6)).

Finally, realizing that at most points $b_0 \ll h_0$ we neglect terms proportional to $b_0 b_1$ so that the governing equation for the surf beat amplitude b_1 becomes

$$\frac{\partial^2 b_1}{\partial t^2} - \frac{\partial}{\partial x}\left(gh_0 \frac{\partial b_1}{\partial x}\right) = \frac{1}{\rho}\frac{\partial^2 S_{xx,1}}{\partial x^2} \tag{3.10}$$

This is a wave equation with a forcing term which represents the effect the wave height modulation has on the slowly varying surf beat. It is the same equation that was used by Symonds et al. (1982) but only in the region of the breaker point variation. Chu & Mei (1970) and Mei & Benmoussa (1984) derive the same equation for 3D by a multiple scale expansion of slowly varying Stokes waves.

Following Mei & Benmoussa (1984) we let the short wave motion be composed of two waves with almost the same wave numbers

$$\begin{aligned} \eta_1 &= \frac{1}{2}\, a_1 \exp\left(i\int k_1 dx + \omega_1 t\right) + * \\ \eta_2 &= \frac{1}{2}\, \delta a_1 \exp\left(i\int k_2 dx + \omega_2 t\right) + * \end{aligned} \tag{3.11}$$

where * means complex conjugate.

In region III we have k_1 and k_2 constant and we define k_0 and ϵ so that

$$k_1 = k_0(1+\epsilon) \quad ; \quad k_2 = k_0(1-\epsilon) \qquad \text{for } x > x_0 \tag{3.12}$$

The equivalent change in wave frequencies are given by the dispersion relation. We have

$$\omega_g = \frac{1}{2}(\omega_1 - \omega_2) \simeq \frac{\partial \omega}{\partial k} \cdot \epsilon k_0 = c_{g0}\, \epsilon k_0 = \frac{c_{g0}}{c_0}\, \epsilon\omega_0 \qquad \text{for } x > x_0 \tag{3.13}$$

where c_{g0} is the group velocity in region III for the wave with wave number k_0 and frequency ω_0 and $c_0 = \omega_0/k_0$. Equation (3.13) also defines ω_0 as approximately $(\omega_1+\omega_2)/2$.

The total short wave motion $\eta = \eta_1 + \eta_2$ can then be written

$$\eta = \frac{1}{2} A \exp\left\{i\left(\left(\int k_0 dx + \omega_0 t\right)\right)\right\} + * \tag{3.14}$$

where A is a complex amplitude. With

$$\Omega_0 = \frac{c_{g0}}{c_0}\,\omega_0 \quad , \quad K_0 = k_0 \cdot \frac{c_{g0}}{c_g} \tag{3.15}$$

we get

$$A = a_1\left[\exp\left(i\epsilon\left(\int K_0 dx + \Omega_0 t\right)\right) + \delta \exp\left(-i\epsilon\left(\int K_0 dx + \Omega_0 t\right)\right)\right] \tag{3.16}$$

or

$$\left.\begin{aligned} A &= a_1\left(e^{i\theta} + \delta e^{-i\theta}\right) \\ \text{where} \quad \theta &= \epsilon\left(\int K_0 dx + \Omega_0 t\right) \end{aligned}\right\} \tag{3.17}$$

The radiation stress S_{xx} for the short waves is then given by

$$\frac{1}{\rho} S_{xx} = \frac{g|A|^2}{2}\left[\frac{2c_g}{c} - \frac{1}{2}\right] \tag{3.18}$$

If we therefore write S_{xx} according to (3.6) we see that

$$\frac{1}{\rho} S_{xx,0} = \frac{g}{2} a_1^2 (1+\delta^2)\left[\frac{2c_g}{c} - \frac{1}{2}\right] \tag{3.19}$$

$$\frac{1}{\rho} S_{xx,1} = g a_1^2 \delta \left[\frac{2c_g}{c_0} - \frac{1}{2}\right] \exp(2i\theta) \tag{3.20}$$

Inside the surf zone the more general expression applies:

$$\frac{1}{\rho} S_{xx} = g|A|^2 P \tag{3.21}$$

In the present context, however, we will, for simplicity, allow P to be approximated by $(2c_g/c_0 - 1/2)$ since this only changes the results quantitatively not qualitatively.

The variation of b_1 and $S_{xx,1}$ is described by

$$b_1 = b_a \exp(2i\epsilon\Omega_0 t) \tag{3.22}$$

$$S_{xx,1} = S_a \exp(2i\epsilon\Omega_0 t) \tag{3.23}$$

so that

$$S_a = a_1^2 \delta \; \rho g\left(\frac{2c_g}{c_0} - \frac{1}{2}\right) \exp\left(2i\epsilon\int K_0 dx\right) \tag{3.24}$$

which substituted into (3.10) yields

$$\frac{\partial}{\partial x}\left(h_0 \frac{\partial b_a}{\partial x}\right) + 4\epsilon^2 \frac{\Omega_0^2}{g} b_a = -\frac{1}{\rho g}\frac{\partial^2 S_a}{\partial x^2} \tag{3.25}$$

This is the equation we solve.

Variation of Carrier Wave Amplitude

The two wave components a_1 and $a_1\delta$ which together form the carrier wave will show a variation that in region II corresponds to a simple shoaling under conservation of energy flux. Thus in region II we have

$$a_1 = \left(\frac{c_{g\infty}}{c_g}\right)^{1/2} a_\infty \quad ; \quad a_\infty = a_1(h{=}\infty) \tag{3.26}$$

In the surf zone (region I) the wave heights actually ought to be determined by one of the surf zone models developed in recent years as e.g. Svendsen (1984). This implies solving the energy and momentum equations using realistic descriptions for the relevant wave properties such as radiation stress, energy flux, and energy dissipation. If we assume that the wave height modulation in the wave groups is moderate, the breaker type will be virtually the same for all waves. Except perhaps for violently plunging waves (for which we know very little) this means that if two waves initially have different heights at the breaker point, the highest will remain so throughout the surf zone. Essentially this further implies that some groupiness is conserved also beyond the breaker point. How much of the original wave height modulation that actually is maintained will depend on how the variation in wave height influences the position of the breaker point, an effect we, as mentioned earlier, have neglected here by using a fixed breaking point.

The above mentioned groupiness would result also from the wave model by Svendsen (or other wave models based on solving the energy equation). For simplicity, however, we choose a simpler representation of this physical feature by using a modified saturation model. In the surf zone we let

$$|A| = \gamma h \tag{3.27}$$

where $\gamma^2 = \gamma_0^2 + \gamma_1^2 e^{2i\theta}$. Particularly for $|A|^2$ we get

$$\begin{aligned} |A|^2 &= |A_0|^2 + |A_1|^2 \\ &= \left(\gamma_0^2 + |\gamma_1|^2 \left(e^{2i\theta} + *\right)/2\right) h^2 \end{aligned} \tag{3.28}$$

The values of γ_0 and γ_1 are then according to the saturation hypothesis determined by the breaker heights of the waves. Since before breaking we have

$$|A|^2 = (a_\infty^2(1+\delta^2) + a_\infty^2\delta(e^{2i\theta} + *)) \frac{c_{g\infty}}{c_g} \tag{3.29}$$

we get

$$\gamma_0^2 = \frac{a_\infty^2(1+\delta^2)}{c_{g_B} h_B^2} c_{g\infty} \tag{3.30}$$

and

$$\gamma_1^2 = \frac{2a_\infty^2\delta}{c_{g_B} h_B^2} c_{g\infty} \tag{3.31}$$

where index ∞ refers to deep water, index $_B$ to the values at the breaking point.

4. MATCHING AND BOUNDARY CONDITIONS

To obtain a solution over the three regions described in Section 2, we need a boundary condition at the outer end of region III, matching conditions between II and III and between I and II, and a boundary condition at the shore line. Since (3.25) is a second order equation a total 2x3 conditions are required to establish the solutions in the three regions.

Seaward Radiation Condition

The seaward boundary condition is a radiation condition stating that there are no free waves propagating towards the shore, only the bounded (and known) set down wave.

Matching Conditions

At x_0 and x_B the propagation conditions change, which causes changes in the constants of the general solution. We ensure continuity at those points by requiring that

$$\left[b_a\right]_{x_0^-}^{x_0^+} = 0 \quad ; \quad \left[b_a\right]_{x_B^-}^{x_B^+} = 0 \tag{4.1a,b}$$

This corresponds to assuming continuity in mass flux across x_0 and x_B. Similarly, continuity in momentum flux can be obtained by applying (3.4) on the two sides of each matching point in combination with (4.1a,b). This yields

$$\left[\frac{\partial b_a}{\partial x}\right]_{x_0^-}^{x_0^+} = -\frac{1}{\rho g h(x_0)}\left[\frac{\partial S_a}{\partial x}\right]_{x_0^-}^{x_0^+} \tag{4.2a}$$

$$\left[\frac{\partial b_a}{\partial x}\right]_{x_B^-}^{x_B^+} = -\frac{1}{\rho g h_B}\left[\frac{\partial S_a}{\partial x}\right]_{x_B^-}^{x_B^+} \tag{4.2b}$$

These relations indicate that db_a/dx will show a discontinuity at the two matching points, of which the one at h_B is by far the most significant.

Boundary Condition at the Shore Line

The model assumes that the short waves are entirely destroyed by breaking. No similar mechanism, however, is available for the long forced wave which will approach the shore line with a finite amount of energy. The wave must therefore be fully reflected there.

It is worth noticing that near the shoreline we actually stress the assumptions underlying the solution, which are that the bottom slope is gentle (i.e., $\Lambda = h_x L/h$ remains small) and the amplitude to depth ratio γ_s is small. As $h \to 0$, however, wave length of the long wave will go to zero as $h^{\frac{1}{2}}$. Hence the slope parameter Λ will grow as $h^{-\frac{1}{2}}$ for a constant h_x. Similarly, the amplitude to depth ratio of the long wave does not remain small.

The growth of Λ and γ_s near the shoreline indicates that the motion is more appropriately described by the nonlinear shallow water equations. However, the basic equations (3.2) & (3.4), and therefore also (3.5), remain valid even under those conditions since no assumptions as to the magnitude of Λ or γ_s have been invoked at the derivation of those equations. As mentioned previously, (3.2) and (3.4) correspond to this approximation, so that, close to the shore those two equations represent the appropriate description. Further away from the shore the NLSW-equations can then be matched to the linear solution as e.g. the wave equation (see Carrier, 1966).

In the present case (3.10) is a linearized form of (3.5) since some approximations have been introduced to get from (3.5) to (3.10). Therefore (3.10) does not represent a proper approximation as $h \to 0$.

This also applies to the WKB approximation used to transform (3.10) to (3.25), since that approximation also requires $\Lambda << O(1)$ corresponding to small changes in depth over a wave length. It does of course also require $\gamma_s << 1$.

For the time being we have chosen to disregard the problem and simply assume that the linearized equation (3.22) applies to the shoreline. The requirement of full reflection then corresponds to requiring that close to the shoreline the wave motion is a purely standing wave with zero net energy flux. Since this approach is in agreement with the actual physical situation it only means that the description is not accurate near the shoreline. At some distance this solution should be the same as if a more correct matching with the NLSW approximation had actually been made.

5. SOLUTION FOR THE SURF BEAT AMPLITUDE

The complete solution to the homogeneous version of (3.25) can be expressed in terms of two linearly independent Hankel functions representing two waves propagating shoreward and seaward respectively. The complete solution to the inhomogeneous equation (3.25) can then be found by the method of variation of parameters. In region I and II we get (i = 1,2) the complex solution

$$b_{a,i} = \left[H_0^{(1)}(a\sqrt{x})\left[-C_i^{(1)}+\int^x H_0^{(2)}(a\sqrt{x})\frac{d^2Sa}{dx^2}dx\right]\right.$$

$$\left.-H_0^{(2)}(a\sqrt{x})\left[C_i^{(2)}+\int^x H_0^{(1)}(a\sqrt{x})\frac{d^2Sa}{dx^2}dx\right]\right]\frac{i\pi}{2\rho gh_x} \qquad (5.1)$$

where $a = 4\omega_g/\sqrt{gh_x}$. In region III the solution is

$$b_{a,3} = C_3\, e^{i2\omega_g x\sqrt{gh_o}} + b_{a,o} \qquad (5.2)$$

where $b_{a,o}$ is the incoming, forced set-down wave, for which we have

$$b_{a,o} = -\frac{1}{\rho}\frac{S_a}{gh_o - c_{go}^2} \qquad (5.3)$$

The five constants $C_{1,2}^{(1)}$, $C_{1,2}^{(2)}$ and C_3 are then determined by the four matching conditions and by the boundary condition at the shoreline.

The latter requires that the amplitude of the two wave components are the same and hence yields directly

$$C_1^{(1)} = C_1^{(2)} \qquad (5.4)$$

The four other constants are found numerically by solution of the flow complex equations resulting from applying the matching equations. The details are left out here.

6. NUMERICAL RESULTS

After determining the equations for the integration constants in (5.1) (5.2) and (5.3) numerical results have been calculated by evaluating the $\partial^2 S_a/\partial x^2$ term by numerical differentiation and similarly the integrals by numerical quadrature.

The problem under study has a substantial number of independent parameters. Clearly the bottom slope h_x is a parameter, but since the bottom steepness that the waves actually "feel" depends on the waterdepth to wavelength ratio the relevant measure of the bottom slope can be shown from the solution to be $\Lambda = h_x L/h$ where L is the local wave length defined as cT. Λ is assumed to be small in order to allow the waves to adjust to the local depth as assumed in the basic equations. In the solution Λ occurs in connection with the matching process at h_o and h_B which leads to the two parameters

$$\Lambda_o = \frac{h_x L_o}{h_o} \qquad \& \qquad \Lambda_B = \frac{h_x L_B}{h_B} \qquad (6.1)$$

The value of Λ_B is actually determined by the assumed breaker index γ.

Other (small) parameters are the ratio of group wave number to carrier wave number

$$\epsilon = (k_1 - k_2)/2k_o \qquad (6.2)$$

and the steepness of the carrier wave system

$$\epsilon_s = a_\infty\ \omega_0^2/g \tag{6.3}$$

The small parameter

$$\delta = a_2/a_1 \tag{6.4}$$

describes the weakness of the amplitude modulation causing the wave groups. Hence the problem is characterized by a total of five (small) parameters. In addition the dimensionless carrier wave frequency $\omega_0^2 h/g$ is of course important for the carrier wave description and particularly the value of $\omega_0^2 h_0/g$ connects the scale of the bottom geometry (h_0) to the length scale of the waves (g/ω_0^2).

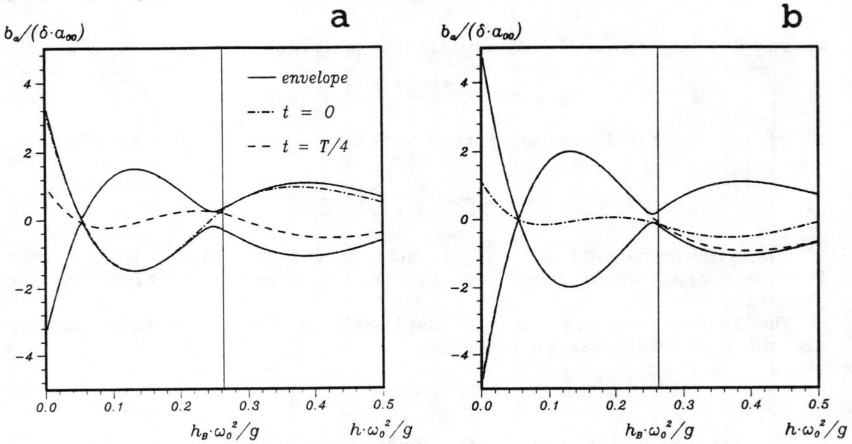

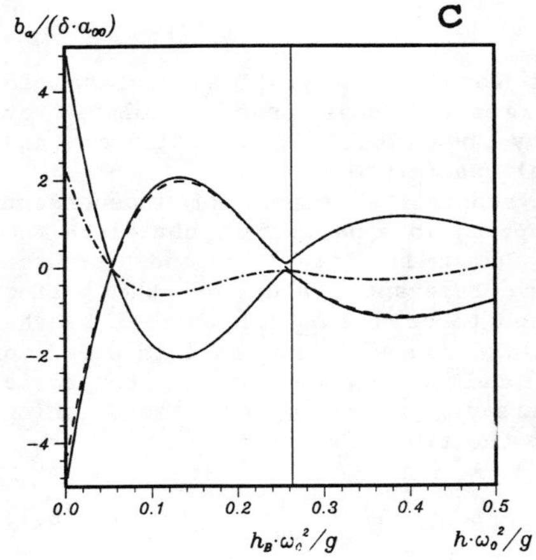

Fig. 2:
Envelope of surf beat motion for three different values of $H_0 = \omega_0^2 h_0/g$.
a: H_0=0.5, R=0.90
b: H_0=1.0, R=4.74
c: H_0=5.0, R=37.5.
In all cases ϵ_s=0.10 and $2h_x\omega_0/\omega_g$ = 0.20. The vertical line corresponds to $h_B\omega_0^2/g$. Instantaneous profiles at t=0, T/4 are also shown.

Fig. 2 shows the variation of the solution for the envelope b_a for a selection of parameter values. The three parts a, b & c corresponds to three different values of the parameter $H_0 = \omega_0^2 h_0/g$ (0.5, 1.0 and 5.0). In all three cases all other parameters have been kept unchanged so that the three figures indicate the effect of increasing the depth h_0 in front of the slope (see Fig. 1) relative to the wave length of the carrier wave system.

Since it is difficult to compare the wave patterns as a whole we have chosen here to focus on the ratio between the amplitude of the wave propagating seawards in the constant region depth over the amplitude of the incoming set-down wave in the same region. This "reflection coefficient" R is a measure of how much the original set-down wave is amplified by energy transferred from the short wave system.

The variation of R we see in Fig. 2 is mainly due to the fact that the amplitude of the free (reflected) wave varies as $h_0^{-1/4}$ whereas the set-down wave decreases as h_0^{-1} as h_0 increases. Notice that for H_0 = 1.0 and 5.0 the figure does not show the whole slope region.

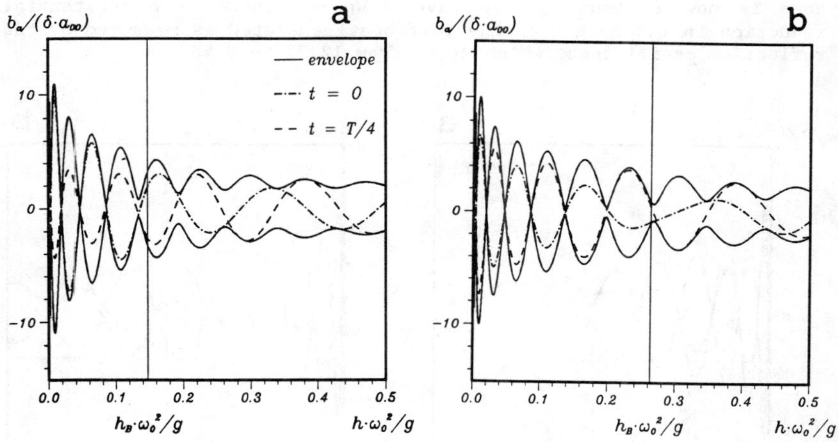

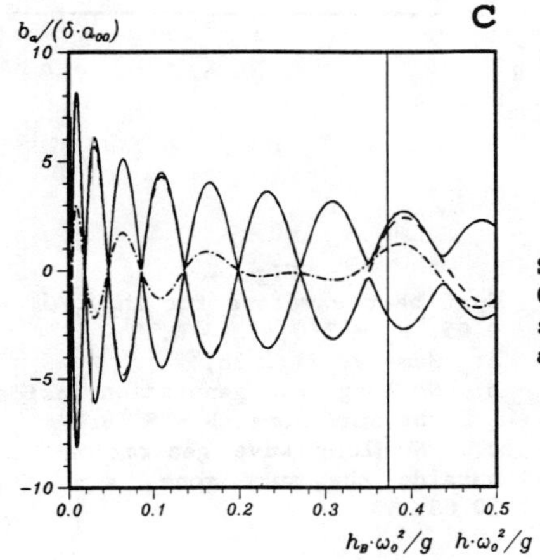

Fig. 3
Surf beat envelope for $2h_x\omega_0/\omega_g$ = 0.05, H_0 = 1.0 and three different short wave steepnesses. The data are a: ϵ_s = 0.05, R = 27.79
b: ϵ_s = 0.10, R = 12.77
c: ϵ_s = 0.15, R = 6.89.

In Fig. 3 the total slope length has been kept constant and the position of the breaker point varied. The reflection coefficients of up to 28 show a significant amplification of the set down wave. On the other hand, the value of R is biggest for the most narrow surf zone.

From those observations one might arrive at the conclusion that the surf beat generation primarily takes place on the slope seaward of the breaker point. This idea might be further supported by the fact that in the surf zone the minimum envelope amplitudes are very small as if the amplitude of the outgoing, reflected wave were nearly the same as the shoreward moving surf beat. (Clearly this becomes a better and better approximation the closer we get to the beach, as should be expected.)

To check this conjecture Fig. 4 shows numerical experiments with the same general data as in Fig. 3b (which is repeated in Fig. 4a for comparison). In Fig. 4b, however, we have artificially suppressed the surf beat generation inside the surf zone by letting the right hand side of (3.25) be zero in region I. Thus the long wave in the surf zone is now a standing free wave. We see there is a substantial reduction in the height of the surf beat generated as measured by the reflection coefficient, which drops from 12.77 to 8.58.

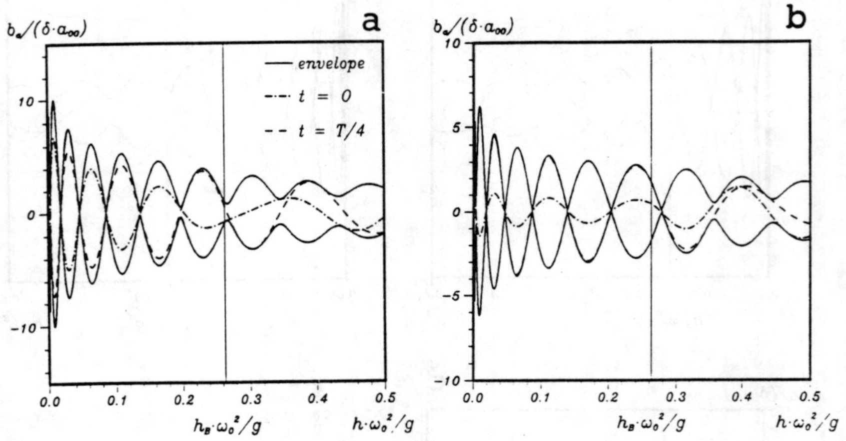

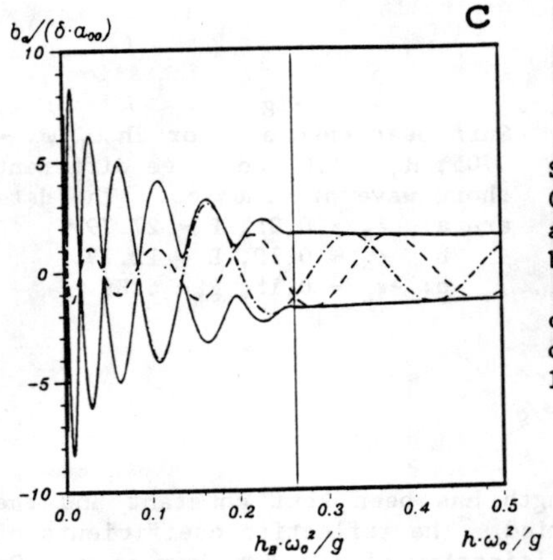

Fig. 4
Surf beat envelope for $2h_x\omega_0/\omega_g$ = 0.05, H_0 = 1.0, ϵ_S = 0.10.
a: Same as Fig. 3b,
b: No long wave generation inside the surf zone, R = 8.58,
c: No long wave generation outside the surf zone, R = 10.08.

Similarly Fig. 4c shows a numerical experiment in which there is no long wave generation at all until in the surf zone. There the long wave motion outside the breaker point is a purely progressive wave moving seaward after having been reflected from the shoreline. This seaward oriented wave is now 10.8 times the set-down wave we would have had in the constant-depth region III had the generation been normal. In Fig. 5a-c the same type of experiment has been repeated for the same wave period and slope width but with the breaker point moved seaward by a change in carrier wave steepness. The picture is seen generally to be the same.

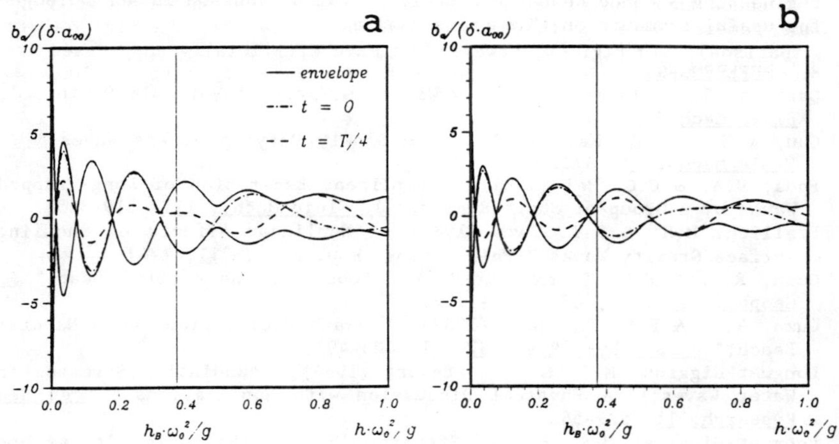

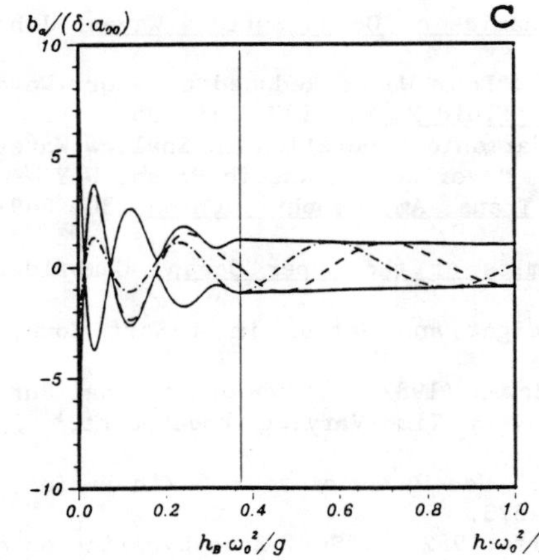

Fig. 5
Same as Fig. 4 except $2h_x\omega_0/\omega_g$ = 0.10, ϵ_s = 0.15, H_0 = 1.0.
a: R = 5.38,
b: no long wave generation inside the surf zone, R = 3.61,
c: no long wave generation outside the surf zone, R = 5.10.

As mentioned in section 2, as the waves move to shallower water the wave number of the wave groups approaches the wave number of the free wave solution to (3.25) and hence the generation process assumes the character of a resonant transfer of energy from the short waves to the shoreward moving long wave. This aspect of the process is equivalent to the resonant interaction first pointed out for water waves by Mei & Unluata (1972). Only here we meet a somewhat more complicated form where the wave number of the driving force instead of being constant is slowly drifting towards the resonant wave number as the waves move shoreward. Although we have not yet studied the details of this situation it is undoubtedly the resonant nature of the process which is responsible for the large amplification indicated by the enormous reflection coefficients. On the other hand, since the nearly resonant interaction only takes place over a finite distance the amplitude of the surf beat of course remains finite.

7. ACKNOWLEDGMENTS

This research was carried out on a contract with the Danish Engineering Science Research Council. Ivar G. Jonsson is acknowledged for useful comments on this presentation.

8. REFERENCES

Carrier, G.F. (1966). "Gravity Waves on Water of Variable Depth," J. Fluid Mech., 24, 4, 641-659.

Chu, V.H. & C.C. Mei (1970). "On Slowly Varying Stokes Waves," J. Fluid Mech., 41, 873.

Foda, M.A. & C.C. Mei (1981). "Nonlinear Excitation of Long-Trapped Waves by a Group of Short Swells," J. Fluid Mech., 111, 319-345.

Freilich, M.H. & R.T. Guza (1984). "Nonlinear Effects on Shoaling Surface Gravity Waves," Phil. Trans. Roy. Soc., 311, 1-41.

Guza, R.T. & E.B. Thornton (1985). "Observations of Surf Beat," J. Geophy. Res., 90, C2, 3161-3172.

Guza, R.T. & E.B. Thornton (1982). "Swash Oscillations on a Natural Beach," J. Geophys. Res., 87, C1, 483-491.

Longuet-Higgins, M.S. & R.W. Stewart (1964). "Radiation Stresses in Water Waves: A Physical Discussion with Applications," Deep Sea Research, 11, 529-562.

Longuet-Higgins, M.S. & R.W. Stewart (1962). "Radiation Stress and Mass Transport in Gravity Waves, with Application to Surf Beats," J. Fluid Mech., 13, 481-504.

Mei, C.C. (1983). The Applied Dynamics of Ocean Surface Waves, John Wiley & Sons, N.Y.

Mei, C.C. & C. Benmoussa (1984). "Long Waves Reduced by Short-Wave Groups Over an Uneven Bottom," J. Fluid Mech., 139, 219-235.

Mei, C.C. & U. Unluata (1972). "Harmonic Generation in Shallow Water Waves," in Waves on Beaches, R.E. Meyer ed., Academic Press, N.Y.

Munk, W.H. (1949). "Surf Beats," Trans. Am. Geophys. Union, 30, 849-854.

Phillips, O.M. (1977). The Dynamics of the Upper Ocean, Cambridge Univ. Press.

Svendsen, I.A. (1984). "Wave Height and Set-up in a Surf Zone," Coastal Eng., 8, 303-329.

Symonds, G., D.A. Huntley & A.J. Bowen (1982). "Two Dimensional Surf Beat: Long Wave Generation by a Time-Varying Breakpoint," J. Geophys. Res., 87, C1, 492-498.

Tucker, M.J. (1950). "Surf Beats: Sea Waves of 1 to 5 Min Period," Proc. Roy. Soc. Lond., A202, 565-573.

Wright, L.D., R.T. Guza & A.D. Short (1982). "Surf Zone Dynamics on a High Energy Dissipative Beach," Mar. Geol., 45, 41-62.

CHAPTER 80

Solitary Wave Transmission through Porous Breakwaters

by

C. Vidal*, M. A. Losada*, R. Medina* and J. Rubio*

Abstract

A semi-empirical theory is formulated to predict wave reflection and transmission at a porous breakwater of rectangular cross section for normally incident solitary waves. The solution is based on the linearized form of the governing equations and on equivalent linearization of the friction loss in the porous structure.

Experimental results of transmission coefficients are presented for a large range of incident wave amplitudes, with several gravel sizes, water depths and breakwater geometries.

Experimental and theoretical results are compared and evaluated; the comparison shows satisfactory agreement for the transmission coefficient.

---

* Universidad de Cantabria. Spain.

Introduction

When a wave or surge reaches a rubble mound breakwater, it will be partly or totally reflected, depending on the permeability of the structure. Part of the flow may penetrate the porous barrier, and consequently the reflection is only partial. The incident wave energy is, in this way, split into reflected and transmitted. Reflection features determine the accessibility of the harbor trough its influence on the wave energy at the entrance, while transmission characteristics assess the effectiveness of the breakwater in protecting the harbor from the energy of incident waves.

An extensive literature has been reported to model the reflection-transmission features of a porous structure. The main approximations made in the literature are: the incident wave is harmonic and normal to the structure, the breakwater is a homogeneous isotropic medium governed by Darcy or Forchheimer type flow, the structure is of rectangular cross section. Sollit and Cross 1972, presented a summary of analytical approaches up to 1972. Madsen 1974 simplified the analysis presenting a linear long-wave model by assuming a Forchheimer type of flow and a linearized coefficient of friction. An extension of his model including trapezoidal cross section was given by Madsen and White, 1976. A statistical approach for random waves passing a perforated and porous breakwater was given by Massel and Mei, 1977. A generalization of Sollit and Cross work for multilayered structure with arbitrary cross section was presented by Sulisz, 1985. Damping of the incident and reflected waves due to bottom friction was considered in the long wave transmission through porous breakwater approach given by Scarlatos and Singh, 1987.

The long wave assumption made by several authors accounts for severe wave condition for most breakwaters. The governing equations for long waves in shallow water are the Boussinesq equations with the solitary wave as a typical solution. Theoretically, solitary waves have the advantage that, although non-linear, they can be described with two parameters: the wave height and the depth: experimentally they propagate with constant shape in constant depth and generally they can be separated from reflected waves. The objective of this study is to develop a semi-empirical model to the leading order for the motion of an incident solitary wave, the reflected wave and transmitted wave resulting from the presence of a normal rectangular cross section porous breakwater. This model will be checked with experimental tests. We assume first, that following ratios are small:

$$ak \ll 1 \quad kh \ll 1 \quad kb \ll 1 \quad O(b) \leq O(h)$$

in which a=wave amplitude, k=wave number, h=depth and b=breakwater width.

Following the procedure used by Mei, Liu and Ippen, 1974 for periodic long waves impinging on slotted or perforated breakwater and also used by Mei, 1983 for solitary waves climbing onto a shelf, three separate regions can be distinguished, within each of which different physical processes are dominant. In the evolution region, which is many wave lengths away from the breakwater, dispersion and nonlinearity are important and Boussinesq equations must be used. In the intermediate region roughly one wave length away from the breakwater, nonlinearity and dispersion are of second order in wave slope. In the neighborhood of the breakwater, where the typical distance is b, the loss friction effects are important but the apparent inertia due to local acceleration is not.

Theoretical formulation

Within the intermediate region, one-dimensional linearized equations are sufficient to the leading order:

$$\eta_t + h\, u_x = 0 \qquad (1)$$

$$u_t + g\, \eta_x = 0 \qquad (2)$$

the governing equations for the motion inside the porous rubble-mound are

$$n\, \eta_t + h\, u_t = 0 \qquad (3)$$

$$(1/n)\, u_t + g\, \eta_x + (C_1\, \mu/\rho d^2 n)\, u + (C_2/d\, n^2)\, u^2 = 0 \qquad (4)$$

where C_1 and C_2 account for the laminar and turbulent friction loss respectively and n is the porosity of the structure. The effect of friction loss is localized within O(h) or O(b) and can be represented as a boundary condition:

$$\eta_a - \eta_b = \frac{C_1\, \mu}{d^2\, n\, g\rho}\, b\, u_b + \frac{C_2}{d\, g\, n^2}\, b\, u_b^2 \qquad (5)$$

where subscripts $()_a$, $()_b$ indicate upstream and downstream respectively, and the apparent inertia (local acceleration) has been ignored in accordance with $kb \ll 1$.

As in the theory for perforated breakwater and for the scattering of a solitary wave at a sudden depth change and equivalent linear condition

$$\eta_a - \eta_b = c_a\, u_b \qquad (6)$$

such that the total square error

$$e = \frac{C_1\ \mu}{d^2\ n\ g\rho}\ b\ u_b + \frac{C_2}{d\ g\ n^2}\ b\ u_b^2 - c_\bullet\ u_b \qquad (7)$$

is minimized over the wave period

$$\overline{e^2} = \int_{-\infty}^{\infty} e^2\ dt$$

the minimum occurs when $\delta\overline{e^2}/\delta c_\bullet = 0$. The optimum value of $c_\bullet$ is:

$$c_\bullet = \frac{C_1\ \mu}{d^2\ n\ g\rho}\ b + \frac{C_2\ \ b}{d\ g\ n^2}\ \frac{\overline{u_b^3}}{u_b^2} \qquad (9)$$

we assume the incident wave is given by the solitary wave

$$n_i = a_i\ \ \text{sech}^2\ \ \lambda_i\ (x - ct)$$

in which

$$\lambda_i = (3\ a_i\ /\ 4\ h^3)^{1/2}$$

$$c\ \ = (g\ h)^{1/2}\ (1 + a_i/2\ h\ )\ \approx\ \ (g\ h)^{1/2}$$

Experimentally it has been observed that the transmitted and the reflected waves are similar in shape (sech^2 profiles) but not solitary waves, that is:

$$n_a = a_i\ \text{sech}^2\ \lambda_i\ (x - c\ t) + a_i\ R\ \text{sech}^2\ \lambda_i\ (x+c\ t) \qquad (13)$$

$$n_b = a_i\ T\ \text{sech}^2\ \lambda_i\ (x - c\ t + \theta) \qquad (14)$$

where R, T represent the reflection and transmission coefficients and θ is a phase shift. To the leading order, this phase shift is negligible when kb << 1. The corresponding horizontal velocities are:

$$u_a = (g\ a_i/c)\left[\text{sech}^2\ \lambda_i\ (x - c\ t) - R\ \text{sech}^2\ \lambda_i\ (x+ct)\right] \qquad (15)$$

$$u_b = (ga_iT\ /\ c)\ \text{sech}^2\ \lambda_i\ \ (x - c\ t) \qquad (16)$$

We remark that the experimental assessment of constant λ_i, can also be deduced from no phase shift and Eq. 1 in order to continuity to be satisfied for all t.

With these velocities, the coefficient c_m is:

$$c_m = \frac{C_1 \mu}{d^2 n g \rho} b + \frac{C_2 b}{d n^2} \frac{a_1 T}{c} \frac{4}{5} \qquad (17)$$

Substituting Eq (13), (14), (16) into the boundary condition Eq (6), we obtain

$$1 + R = T \left(1 + \frac{c_m g}{c}\right) \qquad (18)$$

Condition Eq 1 implies for the continuity of mass flux

$$1 - R = T \qquad (19)$$

Combining Eq (18) with Eq (19), we obtain

$$T = \left[1 + \frac{c_m g}{2 c} \right]^{-1} \qquad (20)$$

$$R = \left[\frac{c_m g}{2 c} \right] \left[1 + \frac{c_m g}{2 c} \right]^{-1} \qquad (21)$$

the structure of Eq 20 also appears in the scattering of sinusoidal waves by perforated breakwater, in the scattering of solitary waves at a sudden depth change or in the scattering of sinusoidal waves by a porous breakwater.

Experimental Tests.

The tests were carried out in the 68 m long, 2 m wide, and 2 m deep wave channel of the water sciences Department of the University of Cantabria. In a length of 20 m, the flume has glass walls to allow visual observations. The incident solitary waves were generated with the procedure sketched by Goring, 1978. With this procedure, roughly one wave length away from the paddle the volume of water pushed by the paddle resembles a solitary wave.

As the solitary wave propagates on the horizontal smooth concrete bottom the frequency dispersion balances the non-linear effects and the waves remain stable without appreciable change. Only side wall and bottom friction cause a decrease of the wave amplitude. The measured profiles of the damped waves at the model test (47 m away from the paddle) are closely represented by the theoretical profile of Boussinesq, 1871. The agreement of both profiles is better on the upper 3/4 (Fig. 1).

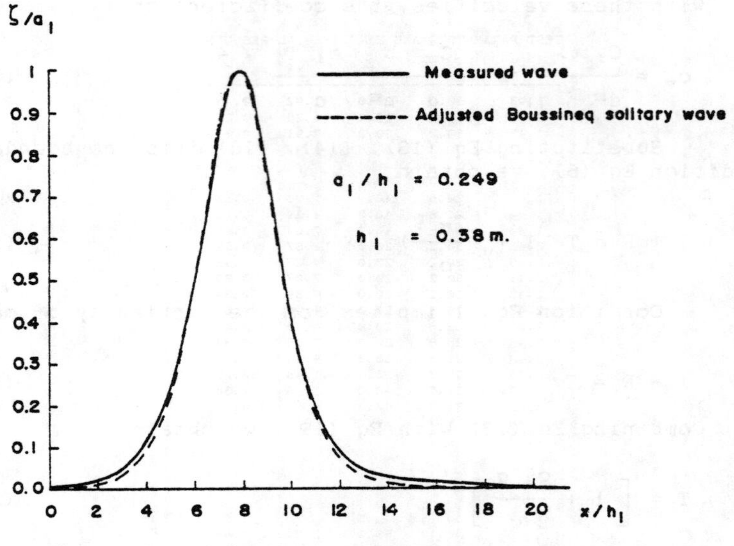

Fig. 1.- Recorded solitary wave and adjusted Boussinesq

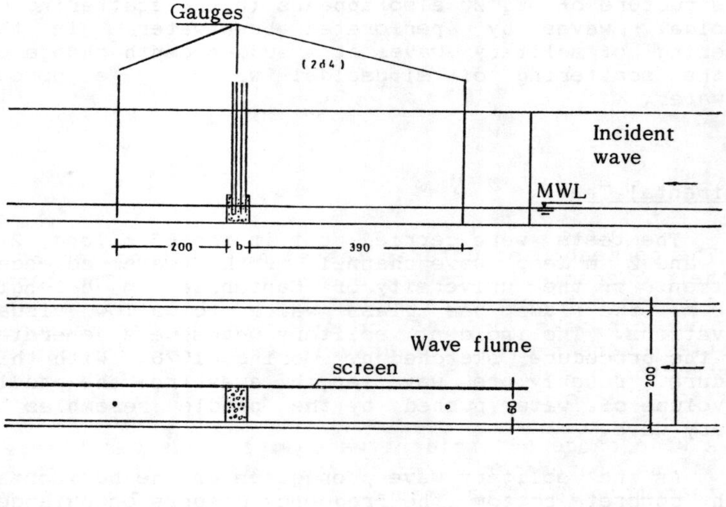

Fig. 2.- Experimental arrangement

TABLE I

TEST PARAMETERS AND EXPERIMENTAL RESULTS

$^*D_{50}$	$^*h$	$^*b$	$^*a_i$	T
1.43	30.0	20.0	1.92	0.45
1.43	30.0	20.0	4.28	0.38
1.43	30.0	20.0	7.91	0.31
1.43	30.0	40.0	1.97	0.34
1.43	30.0	40.0	4.41	0.29
1.43	30.0	40.0	7.97	0.21
1.43	24.6	40.0	1.91	0.30
1.43	24.6	40.0	4.29	0.24
1.43	24.6	40.0	7.64	0.19
2.43	30.2	20.0	1.87	0.57
2.43	30.2	20.0	4.27	0.45
2.43	30.2	20.0	7.07	0.38
2.43	30.2	20.0	9.39	0.34
2.43	25.8	20.0	1.90	0.55
2.43	25.8	20.0	4.40	0.43
2.43	25.8	20.0	7.05	0.34
2.43	25.8	20.0	9.49	0.32
2.43	25.8	20.0	11.46	0.29
2.43	31.7	40.0	1.95	0.50
2.43	31.7	40.0	4.63	0.34
2.43	31.7	40.0	7.64	0.28
2.43	31.7	40.0	10.21	0.25
2.43	24.9	40.0	2.19	0.33
2.43	24.9	40.0	4.74	0.29
2.43	24.9	40.0	7.65	0.23
2.43	24.9	40.0	10.31	0.21
2.43	24.9	40.0	12.37	0.19
3.15	30.1	40.0	2.01	0.46
3.15	30.1	40.0	4.38	0.34
3.15	30.1	40.0	7.22	0.28
3.15	30.1	40.0	9.65	0.25
3.15	25.0	40.0	1.84	0.43
3.15	25.0	40.0	4.23	0.33
3.15	25.0	40.0	6.87	0.26
3.15	25.0	40.0	9.02	0.25
3.15	25.0	40.0	10.78	0.21
3.15	30.1	20.0	1.99	0.57
3.15	30.1	20.0	4.53	0.46
3.15	30.1	20.0	7.38	0.39
3.15	30.1	20.0	9.96	0.37
3.15	25.1	20.0	1.86	0.56
3.15	25.1	20.0	4.36	0.46
3.15	25.1	20.0	6.83	0.38
3.15	25.1	20.0	9.30	0.35
3.15	25.1	20.0	11.18	0.34

* centimetres

The permeable structures tested had rectangular form and their characteristics are shown in table I. The tests were carried out with gravel of D_{50}=1.43, 2.43 cm and small cubic blocks of 3.15 cm side length. The measured porosities were n=0.44 for gravel and n=0.42 for cubes. The structures were 20 to 40 cm wide and water depth varied from 25 to 30 cm.

A wide range of incident wave amplitudes were tested so that the parameter a_i/h_o range was $0.05<a_i/h_o<0.5$. Note that this last parameter (a_i/h_o=0.5) violates the small wave amplitude hypothesis.

The experimental arrangement is shown in figure 2. As the flume width was excessive for the test requirements it was divided longitudinally by a screen so that the model covered a width of only 60 cm. Within this new narrow channel all the measuring equipment was placed as shown in figure 2. Upstream and downstream the model, two gauges were placed in order to measure the incident and the transmitted wave. Inside the permeable structure the pressure and water level were measured at several points depending on the structure width. A complete documentation of the experimental program is presented by Medina, 1988.

The experimental results of transmission coefficients for the different models are presented in table I. The transmission coefficient decreases when the structure width increases, the incident wave amplitude increases and the gravel size decreases. The influence of the water depth on the transmission coefficient is less important, and the transmitted wave amplitude decreases when the water depth decreases. Though the cubic block size is much larger then the gravel size, the transmission coefficient is slightly larger due to the lower porosity.

Comparison of theory and experiment.

For the experimental coefficients C_1, C_2,(Eq.17) we take the expressions given by Muskat,1946,also used by Engelund, 1953 and Madsen, 1974.

$$C_1 = \alpha\ (1-n)^3\ /\ n$$

$$C_2 = \beta\ (1-n)\ /\ n$$

where $\alpha = 1092$, $\beta = 0.81$

The comparison of experimental transmission coefficient, T, and predicted transmission coefficient, T_t, for all the data given in table I is presented in fig. 3. The line T_t=T is also drawn. As is evident from the comparison in fig. 3, the difference is small and the predicted transmission gives a reasonable representation of experimental data. Note that the experiments correspond to values of the parameter $0.05<a_i/h<0.5$ thus the agreement is good even for not too small wave amplitude ratio.

The influence on the transmission coefficient due to the grain size, incident wave amplitude, water depth and breakwater width is represented in figures 4-5 in which theoretical (lines) and measured (symbols) coefficients are drawn.

Figure 4 shows the influence of the breakwater width on the transmission; the transmission decreases when the width increases. Figure 5 shows the influence of the incident wave amplitude; the transmission decreases when the a_i/h ratio increases, we can also observe the agreement between predicted and measured coefficients for a wide range of the ratio a_i/h ($0.05<a_i/h<0.5$).

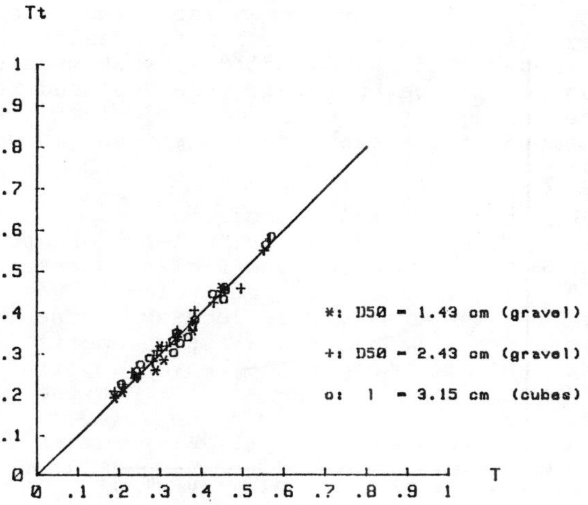

Fig. 3
Experimental versus Theoretical transmission

Summary and Conclusions

Based on a semi-empirical formulation, a simple solution for the reflection and transmission coefficients for a given incident solitary wave has been derived. This solution was based on the leading order of the governing equations and a linearized form of the friction loss in the porous structure. The linearization procedure was a slight modification of the Lorentz condition for sinusoidal waves. The coefficients depend on the geometry of the structure the incident wave amplitude, the water depth and on the properties of the porous medium (porosity and grain diameter).

Experimental results of transmission coefficients have been reported for a wide range of the involved parameters. These results have been compared with the predicted values showing a satisfactory agreement for the transmission coefficient.

Acknowledgments

This study was sponsored partially by the Dirección General de Puertos y Costas and the Comisión Asesora Científica y Técnica.

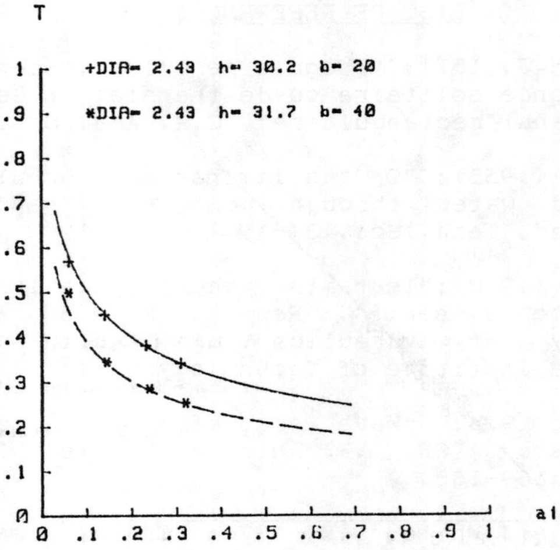

Fig. 4.- Influence of breakwater width on the transmission

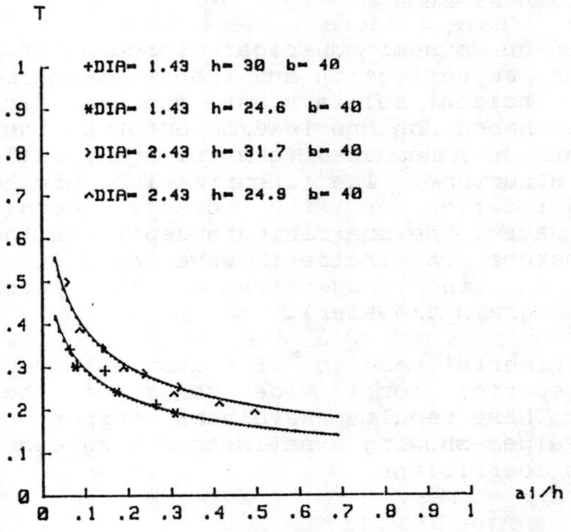

Fig. 5.- Influence of incident wave amplitude on the transmission

LIST OF REFERENCES.

Boussinesq, M.J. 1871; "Thèorie de l'intumescence liquide appelèe onde solitaire ou de translation se propagant dans un canal rectangulaire", C.R. Aca. Sci. 72:755.

Engelund, F. (1953); "On the laminar and turbullent flows of ground water through homogeneus sand", Trans. Danish Acad. Tech. Sci. 3(3).

Goring, D.G. (1978);"Tsunamis- the propagation of long waves onto a sehlf", Rep N. KH-R-38, W.M. Keck Laboratory of hydraulics and water Resources, California Institute of Technology.

Madsen, O.S. 1974; "Wave transmission through porous structures", ASCE, J.W! Harbors Coastal Eng. Div., 100 (WW3):169-188.

Madsen, O.S. and White, S.M. 1976; "Energy dissipation on a rough slope", ASCE, J.W. Harbors Coastal Eng. Div., 102 (WW1):31-48.

Massel, S.R. and Mei, C.C. 1977; "Transmission of random wind waves through perforated or porous breakwater", Coastal Eng., 1:63-78.

Medina, R. 1988; "Sobre la transformaciÒn de ondas solitarias. PropagaciÒn en un escalÒn y en medio poroso", Tesis doctoral, Departamento de Ciencias y Tècnicas del Agua y del Medio Ambiente, Univ. de Cantabria.

Mei, C.C. Liu, P.L.F. Ippen, A.T. (1974); "Quadratic head loss and scattering of long waves", J. Waterway Harbors Coastal Eng. Div. Proc. ASCE 100, 217-239.

Mei C.C. (1983) "The applied dydamics of Ocean surface waves", J. Wiley and sons Inc. N.Y.

Muskat, M. (1946); "The flow of homogeneous fluids through porous media", J.W. Edwards, Inc. Ann Arbor, Michigan.

Scarlatos, P.D. and Singh, V.P. 1987; "Long wave through porous breakwaters", Coastal Eng. 11:141-157.

Sollit, C.K; Cross, R.H. (1972);"Wave transmission through permeable breakwater", Proc. XIII Coastal eng. Cong. ASCE, Vancouver, B.C. Canadà.

Sulisz, W.(1985); "Wave reflection and transmission at permeable breakwaters of arbitrary cross-section", Coastal Engineering, 9:pp 371-386.

CHAPTER 81

TIME-DEPENDENT WAVE SHEAR STRESS

By Suphat Vongvisessomjai[1], M.ASCE

ABSTRACT

A knowledge of bed shear stress induced by waves is required to understand dynamic processes of nearshore morphologies as a results of sediment transport. However, the information on the stress is still incomplete due to lack of measured data. The study analyzes the unsteady horizontally averaged shear stresses measured over mobile beds in a water tunnel. It is found from the analysis that the presence of the third and fifth harmonics in the shear stress is in good agreement with the measured concentration of suspended sediments.

INTRODUCTION

Magnitude and phase of bed shear stress induced by waves over mobile bed depend on characteristics of waves and wave induced ripples on the bed. The magnitude of bed shear stress can be computed from known wave friction factor over mobile bed, Grant and Madsen (1982) and Vongvisessomjai (1987). Lofquist (1986) provided experimental results of the time-dependent drags on naturally rippled sand beds tested in a water tunnel which were obtained from the measured difference between pressure gradients in the test section split longitudinally into two channels, one containing the sand bed and the other having a smooth metallic bottom. Two sands were used, with diameters 0.55 and 0.18 mm. The bottom stress, including skin friction and profile drag components was found to be a fairly complicated function of the phase and not proportional to the square of the simultaneous sinusoidal velocity far above the bed. Average values of the bottom stress and coefficient of energy dissipation were both typically an order to magnitude greater than for a flat surface roughened with fixed grains. The wave friction factors on mobile beds f_{ws}, computed from the coefficient of energy dissipation $\bar{f}_1$ of Lofquist (1986) with the conversion $f_{ws} = (3\pi/4)\,\bar{f}_1$, were found to have the same order of magnitudes as those obtained by Vongvisessomjai (1987).

The time-dependent wave shear stresses obtained by Lofquist (1986) are analyzed and quantified for the phases, and the obtained friction factors together with the earlier results are used to develop a new expression.

---

[1] Professor, Division of Water Resources Engrg., Asian Inst. of Tech., P.O. Box 2754, Bangkok 10501, Thailand.

METHODOLOGY

The velocity field over the self-formed 2 dimensional ripple of $h = \eta \cos(2\pi x/\lambda)$ can be expressed as a sum of the fundamental component (n = 0) and its higher harmonics (n = 1, 2, 3...) caused by the presences of eddies or turbulences generated from the interaction of the fundamental component and the ripples as

$$u(x,z,t) = \sum_n \varepsilon^n u_n(x,z,t) \quad \text{.......(1)}$$

where ε = a small parameter.

The corresponding shear stresses along the ripple bed can be expressed as

$$\tau_o(x,t) = \sum_n \varepsilon^n \tau_{on}(x,t) \quad \text{.......(2)}$$

where subscript o denotes the bed, z = 0.

These velocity field, shear stresses and eddies will cause sediment suspension in the form

$$c(x,z,t) = \underset{\sim}{c}(x,z) + \sum_n \varepsilon^n c_n(x,z,t) \quad \text{.......(3)}$$

These unsteady two dimensional velocity, the unsteady shear stress along the ripple and the suspended sediment concentration are too complicated to be measured and quantified. Only some measurements are presently available for horizontally averaged values of the above parameters which can be expressed respectively as

$$\bar{u}(z,t) = \frac{1}{\lambda}\int_o^\lambda u(x,z,t)dx = \sum_n \varepsilon^n \bar{u}_n(z,t) \quad \text{.......(4)}$$

$$\bar{\tau}_o(t) = \frac{1}{\lambda}\int_o^\lambda \tau_o(x,t)dx = \sum_n \varepsilon^n \bar{\tau}_{on}(t) \quad \text{.......(5)}$$

$$\bar{c}(z,t) = \bar{c}(z) + \sum_n \varepsilon^n \bar{c}_n(z,t) \quad \text{.......(6)}$$

The following analysis is based on the approach of Lofquist (1986) in analyzing his experimental results. The freestream velocity of flow above the bed and beyond its effects is

$$u_\alpha = \left(\frac{2\pi a}{T}\right) \sin \left(\frac{2\pi t}{T}\right) = U \sin \theta \quad \text{.......(7)}$$

where t is the time T is the period and 2a is the orbital diameter of the water motion. Equation (7) defines the maximum velocity, U, and the phase, θ. The bed surface profile, assumed two dimensional and periodic with length λ, has elevation $h(x,\theta)$, where x is the horizontal coordinate. The average bottom stress is then defined by

$$\bar{\tau}_o(\theta) = \frac{1}{\lambda}\int_o^\lambda [\tau_h + (p_h - p_\alpha)\frac{\partial h}{\partial x}]\, dx \quad \text{.......(8)}$$

where τ_h is the local tangential stress on the bed, and $(p_h - p_\alpha)$ is the difference between pressures at the bed surface and at any fixed elevation beyond the influence of the bed. Separate integrals for the two terms in the integrand would resolve $\tau_o(\theta)$ into components expressing the effects of tangential and normal stresses, the latter being a form drag. However, such a separation is generally impossible. A stress coefficient has been made dimensionless for bottom stress by dividing by ρU^2, rather than by $\rho u_\alpha^2(\theta)$, in order to keep it proportional to the bottom stress throughout the cycle where ρ is the density of water,

$$f(\theta) = \frac{2\,\bar{\tau}_o(\theta)}{\rho\, U^2} = f_s(\theta) + f_n(\theta) \qquad (9)$$

The dimensionless local tangential stress over the smooth boundary for laminar flow is

$$f_s(\theta) = \beta_4 \sin(\theta + \pi/4) \quad \text{where} \quad \beta_4 = \sqrt{\frac{\nu}{a^2\omega}} \qquad (10)$$

and the dimensionless normal stress is

$$f_n(\theta) = \beta_1\, \bar{D}(\theta) \qquad \text{where} \quad \beta_1 = \frac{2H}{\rho L U^2} \qquad (11)$$

in which $\omega = 2\pi/T$; H = height of pressure taps from bed; and L = distance between pressure taps.

The measured differential pressure $\bar{D}(\theta)$ is found to be the summation of the fundamental harmonic $D_1(\theta)$ and its higher odd harmonics as

$$\bar{D}(\theta) = D_1(\theta) + D_3(\theta) + D_5(\theta) + \ldots \qquad (12)$$

It was found from the experiments that there existed a shift of the mean level of ripple bed from the initial bed of $\bar{h}$, a correction of $2\bar{h}/a = \beta_3$ is made to Eq. 9 as

$$f(\theta) = \beta_1\bar{D}(\theta) - \beta_3 \cos\theta + \beta_4 \sin(\theta + \pi/4) \qquad (13)$$

ANALYSIS AND RESULTS

Expressions of friction factor on fixed bed are first summarized for comparison with the obtained friction factor on mobile beds from this analysis. The friction factor can then be used to compute the magnitude of the maximum bed shear stress. The analysis of the phase of the shear stress is finally made.

Friction Factor of Fixed Bed.- Jonsson (1966) defined the wave friction factor $f_w = 2\,\bar{\tau}_{om}/(\rho U^2)$ where subscript m denotes the maximum value and expressed it in terms of the relative smoothness of the bed $k_* = a/k_s$ as

$$\frac{1}{4\sqrt{f_w}} + \log\frac{1}{4\sqrt{f_w}} = -0.08 + \log k_* \qquad (14)$$

The above expression was later confirmed by two additional sets of tests in a water tunnel by Jonsson and Carlsen (1976).

Kamphuis (1975) reanalyzed the measured values of the friction factors obtained by Riedel, Kamphuis and Brebner (1972) and proposed an expression for the friction factor also in terms of k_* as

$$\frac{1}{4\sqrt{f_w}} + \log\frac{1}{4\sqrt{f_w}} = -0.35 + \frac{4}{3}\log k_* \quad \ldots\ldots\ldots\ldots(15)$$

Grant and Madsen (1979) derived an expression of friction factor for combined waves and currents in fully rough turbulent flow. A relationship between f_w and k_* could be obtained for the limiting condition of a pure wave motion as

$$f_w = 0.08/[\mathrm{Ker}^2 2(\zeta_c)^{1/2} + \mathrm{Kei}^2 2(\zeta_o)^{1/2}] \quad \text{for } k_* > 1 \quad \ldots\ldots\ldots(16)$$

and $f_w = 0.23$ for $k_* < 1$; in which the dimensionless roughness length $\zeta_o = k_s/(30\ell)$; the characteristic length $\ell = 0.4\ u_*/\omega$; the shear velocity $u_* = \sqrt{f_w/2}U$; Ker and Kei = Kelvin functions of zero order.

Vongvisessomjai's (1984) expression of wave friction factor

$$f_w = 0.287\ k_*^{-2/3} \quad \ldots\ldots\ldots\ldots(17)$$

Friction Factor of Mobile Bed.- Grant and Madsen (1982) presented a model to predict the roughness in unsteady oscillatory flows over mobile beds using data of Carstens et al. (1969). The total roughness was the sum of the form drag component and the sediment transport component:

$$\frac{1}{k_*} = 28\,\frac{\eta}{a}\frac{\eta}{\lambda} + 160(s+K)\,\frac{D}{a}\left[\left(\frac{\psi}{\psi_c}\right)^{1/2} - 0.7\right]^2 \quad \ldots\ldots\ldots\ldots(18)$$

in which $K = 1/2$ for spherical sand; ψ = Shields parameters; and ψ_c = critical value of Shields parameter for initiation motion. Based on the law-of-the-wall, the expression of friction factor of the fixed bed, Eq. 16, was adopted for that of the mobile bed using k_* of Eq. 18.

Profiles of mean and unsteady concentration were derived by Vongvisessomjai (1986) from the mass conservation equation using the diffusion coefficient profile adapted from the eddy viscosity profile of Vongvisessomjai (1984) taking into account changes of the friction factor and boundary layer thickness in the presence of suspension. The theoretical profiles were expressed explicitly as functions of the friction factor, velocity and shear stress profile parameters, and the settling velocity of the sediment relative to the fluid velocity. The tabulated data available and the new experimental data of the mean concentration were used to fit the theoretical profile. The ratio of the friction factor f_{ws}/f_w obtained was then correlated to the densimetric sediment Froude number $Fd_* = U^2/[(s-1)gD]$ and the relative bed smoothness $a_* = a/D$ as

$$\frac{f_{ws}}{f_w} = 0.866\ Fd_*^{-0.50}\ a_*^{0.17} \qquad \text{for fine sand}$$

or

$$f_{ws} = 0.185\ Fd_*^{-0.50}\ a_*^{0.17} \qquad \text{for } f_w \doteq 0.214 \quad \ldots\ldots\ldots\ldots\ldots (19)$$

Vongvisessomjai (1987) based on his friction factors obtained from suspended sediment profile data and those of Carstens et al. (1969) expressed f_{ws} as function of Fd_*, a_* and $D_* = (s-1)gD^3/\nu^2$.

$$f_{ws} = 0.049\ Fd_*^{-0.44}\ a_*^{0.21}\ D_*^{0.22} \quad \ldots\ldots\ldots\ldots\ldots\ldots\ldots\ldots\ldots (20)$$

The data used for the present analysis are from Carstens et al. (1969), Lofquist (1986) and Vongvisessomjai (1986). Table 1 summarizes the sources and characteristics of the data. Note that Vongvisessomjai (1986) used various concentration profile data, D = 0.14 mm from Nakato et al. (1977) and Kennedy and Locher (1972), D = 0.18 mm from Hom-ma et al. (1965) and D = 0.21 mm for his own data.

TABLE 1.–Sources and Characteristics of Data of Wave Friction Factor

Source (1)	T (s) (2)	a (cm) (3)	D (mm) (4)	$k_* = a/(4\eta)$ (5)	f_{ws} (6)	Symbol (7)
Carstens	3.55	8–31	0.19	1.2–15	0.077–0.20	⦶
	3.55	8–45	0.30	1.0–23	0.070–0.38	⊖
	3.55	10–42	0.59	1.0– 8	0.193–0.54	○
Lofquist	4–15	14–55	0.18	1.3–3.2	0.099–0.22	◫
	2–11	17–52	0.55	0.8–1.6	0.128–0.42	□
Author						
a) Author	2.00	9.55	0.21	1.77	0.083	▲
	1.00	3.18		1.14	0.132	
	1.00	6.37		1.59	0.059	
	1.00	9.55		1.59	0.053	
b) Nakato	2.40	10.20	0.14	1.41	0.137	▽
	1.80	7.50		1.56	0.107	
	1.20	5.10		1.42	0.079	
c) Kennedy	1.00	4.82	0.14	2.19	0.085	▼
d) Hom-ma	1.30	5.29	0.18	1.51	0.087	▲
	1.60	5.27		1.41	0.125	
	1.60	7.10		1.73	0.080	
	1.60	9.12		2.05	0.071	
	1.72	6.75		1.63	0.104	
	1.72	9.25		2.02	0.086	

The friction factors of the mobile beds will first be compared with those expression for fixed beds, Eqs. 14-17. This depends on how to define the roughness of the bed, k_s, or the relative smoothness of the bed, $k_* = a/k_s$:

i) Madsen and Grant (1976) used $k_s = D$ and $k_* = a/D$;
ii) Kamphuis (1975) used $k_s = 2D$ and $k_* = a/(2D)$;
iii) Horikawa and Watanabe (1967) used $k_s = 4\eta$ and $k_* = a/(4\eta)$;
iv) Swart (1976) used $k_s = 25\eta(\eta/\lambda)$; and
v) Grant and Madsen (1982) used $k_s = 28\eta(\eta/\lambda)$.

Note that $k_s = 25\eta(\eta/\lambda) \doteq 4\eta$ when $\eta/\lambda = 1/6$ for growing ripple or equilibrium ripple.

Figs. 1a) and 1b) show the comparison plots of friction factors of the mobile beds with those expression for fixed beds, Eqs. 14-17, using the relative smoothness of the bed a) $k_* = a/[25\eta(\eta/\lambda)]$ and b) $k_* = a/D$ respectively. It can be seen in **Fig.** 1a) that the four expressions of friction factors for fixed beds have about the same magnitudes but mostly overestimate the measured values when assuming b) $k_* = a_* = a/D$.

The new expression of f_{ws} when data of Lofquist (1986) with the conversion $f_{ws} = (3\pi/4)\ \bar{f}_1$ are added to data of Carstens et al. (1969) and Vongvisessomjai (1986) is

$$f_{ws} = 0.043\ Fd_*^{-0.31}\ a_*^{0.16}\ D_*^{0.21} \quad \text{.............................(21)}$$

From the above f_{ws}, the Shields parameter which is important to describe the sediment transport can be determined as

$$\psi = \frac{1}{2} f_{ws}\ Fd_* = 0.022\ Fd_*^{0.69}\ a_*^{0.16}\ D_*^{0.21} \quad \text{....................(22)}$$

Since the above friction factor of the mobile bed f_{ws} depends on three dimensionless parameters (D_* Fd_* and a_*) its dependence on these three parameters as well as $\psi_r = Fd_*/a_*^{1/2}$ and D_* or the diameter of the sediment D are shown in **Fig.** 2. The 3 diameters of fine, medium and coarse sands of Carstens et al. (1969) are used as the representatives in **Fig.** 2a) while in **Fig.** 2b) the 6 and 3 densimetric Froude numbers (Fd_*) respectively of the coarse sand and the fine sand of Lofquist (1986) are used as the representatives. Also included in Fig. 2b) are the **9** curves obtained by Lofquist (1986) with the conversion $f_{ws} = (3\pi/4)\ \bar{f}_1$ which show how his curves are smoothed to the present expression, Eq. 21. The additional plot of f_{ws} versus a_* of Carstens et al. (1969) and Vongvisessomjai (1986) has been presented in Fig. 1b).

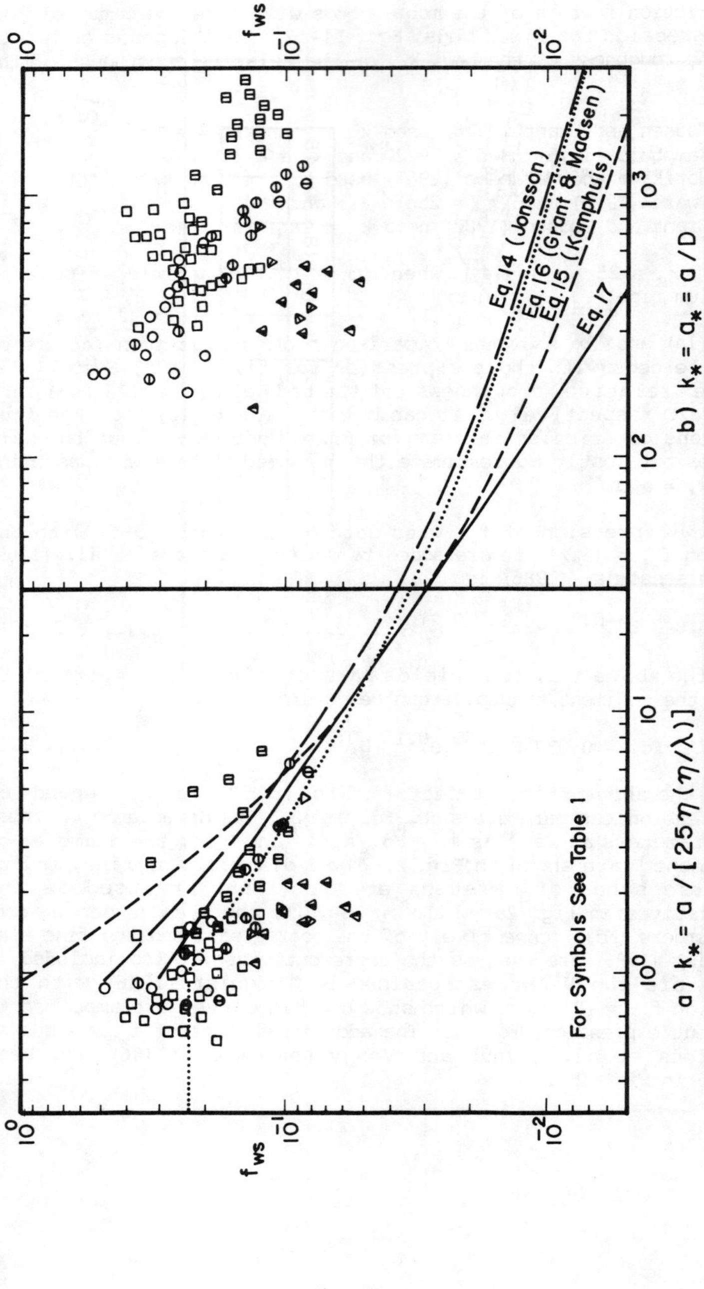

Fig. 1. – Comparison of Measured Friction Factors on Mobile Beds with Expressions of Friction Factors on Fixed Beds

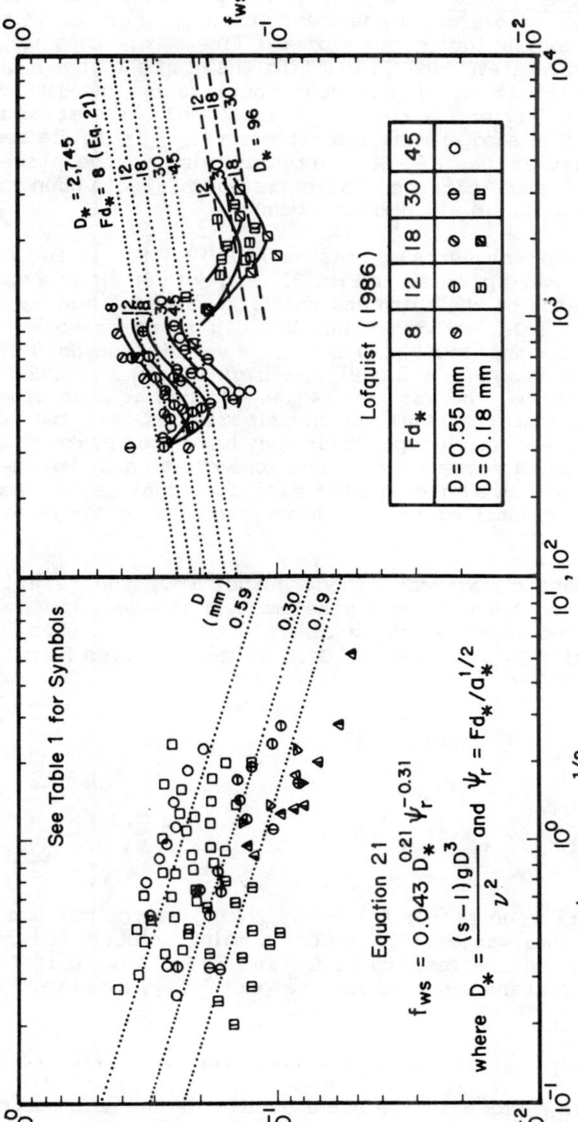

FIG. 2 – Comparison of Measured Friction Factors on Mobile Beds with Equation 21

Phase of Bed Shear Stress.- Lofquist (1986) presented graphically the phase variations of dimensionless shear stress $f(\theta)$ of Eq. 13 for all experimental results for coarse sand and fine sand. Lofquist kindly provided all parameters used in plotting the graphs to the author for further quantification of the contributions by the fundamental harmonic and its higher harmonics. Tables 2 and 3 summarize results of coarse and fine sand experiments respectively. It can be seen from the tables that values of $-\beta_3$ (correction due to the presence of ripples on bed) and β_4 (tangential stress over smooth boundary) are small compared with $\beta_1\bar{D}(\theta)$ (normal stress).

The dimensionless normal stresses $\beta_1\bar{D}(\theta)$ of Eq. 13 for Run Nos. C28, C13 and C01 are plotted in **Fig.** 3. Run No. C28 is an example of high contributions of the third and the fifth harmonics where D_{3m}/D_{1m} = 1.1733 and D_{5m}/D_{1m} = 0.5352; Run No. C13 is an examole of high D_{3m}/D_{1m} = 0.5421 and low D_{5m}/D_{1m} = 0.0746 while Run No. C01 is an example of low D_{3m}/D_{1m} = 0.0581 and high D_{5m}/D_{1m} = 0.2503 where subscript m denotes the maximum value. It can be seen from these three examples that the resultant normal stresses have two peaks in half a cycle $\theta = \pi$ which imply that they have four peaks in a cycle $\theta = 2\pi$. The measured suspended sediment concentration by Hom-ma et al. (1965) and Nakato et al (1977) also exhibited four peaks in a cycle which could be quantified by the theory proposed by Vongvisessomjai (1986).

The densimetric sediment Froude number Fd_*, the relative bed smoothness a_* and the densimetric sediment Froude-Reynolds number D_* which have been used in correlating with f_{ws} are then used in correlating with D_{3m}/D_{1m} and D_{5m}/D_{1m} listed in Tables 2 and 3. The results obtained are

$$\frac{D_{3m}}{D_{1m}} = 2,670\ Fd_*^{1.50}\ a_*^{-1.68}\ D_*^{-0.44} \qquad (23)$$

$$\frac{D_{5m}}{D_{1m}} = 1.49\ Fd_*^{0.60}\ a_*^{-0.64}\ D_*^{0.003} \qquad (24)$$

The wave friction factor $f_{ws} = 2\,\tau_{om}/(\rho U^2)$ is the maximum of the dimensionless shear stress $f(\theta)$ which is mainly contributed by $D_1(\theta)$, $D_3(\theta)$ and $D_5(\theta)$. Using results of f_{ws} and $\beta_1\ D_{1m}$ listed in Tables 2 and 3 for coarse sand and fine sand respectively, a relationship is found

$$f_{ws} = 1.13\ (\beta_1\ D_{1m}) \qquad (25)$$

The fundamental harmonic of the normal stress is the main contribution to the resultant stress, only 13 percent is contributed by higher harmonics and other factors.

The obtained results from this study could be used to further advance on the analytical descriptions of time-dependent sediment transport and suspension.

TABLE 2.-Results of Coarse Sand Experiments (D_*= 2,745)

Run (1)	f_{ws} (2)	$\beta_1 D_{1m}$ (3)	D_{3m}/D_{1m} (4)	D_{5m}/D_{1m} (5)	β_1x100 (6)	$-\beta_3$x100 (7)	β_4x100 (8)	Fd_* (9)	a_* (10)
C01	0.2853	0.2354	0.0581	0.2503	1.472	5.079	0.768	8.21	435
C02	0.2177	0.1794	0.3750	0.1058	1.245	5.079	0.642	18.50	435
C03	0.2483	0.2006	0.1271	0.2545	1.261	5.079	0.710	12.31	435
C04	0.2467	0.2080	0.1069	0.1399	1.354	4.761	0.686	12.35	464
C05	0.2580	0.1934	0.1680	0.1773	1.222	5.540	0.741	12.34	399
C06	0.2106	0.1940	0.2233	0.1713	1.233	5.917	0.766	12.14	374
C07	0.2641	0.2230	0.0514	0.1473	1.223	4.385	0.659	12.36	504
C10	0.2182	0.2292	0.2037	0.1638	1.121	1.475	0.924	8.19	316
C11	0.2691	0.2342	0.3945	0.1482	1.029	2.224	0.835	12.30	316
C12	0.2832	0.2421	0.5674	0.0730	1.367	2.961	0.755	18.35	316
C13	0.2773	0.2598	0.5421	0.0746	1.430	3.698	0.755	18.33	316
C15	0.2552	0.2144	0.0504	0.0950	1.242	0	0.615	12.31	580
C16	0.2891	0.2445	0.0245	0.1412	1.262	0.100	0.616	12.29	580
C17	0.3150	0.2660	0.0580	0.1371	1.323	0.194	0.680	8.25	580
C18	0.2825	0.2330	0.0627	0.1527	1.053	0.294	0.556	18.47	580
C19	0.2396	0.2068	0.5845	0.3035	1.728	0.388	0.491	30.46	580
C22	0.3477	0.2930	0.0861	0.1382	1.388	0	0.621	8.24	697
C23	0.3513	0.2962	0.0708	0.0903	1.449	0	0.733	12.37	697
C24	0.3141	0.2721	0.0224	0.1390	1.428	0	0.507	18.51	697
C25	0.1871	0.1766	0.2487	0.1985	1.580	0	0.507	30.79	697
C26	0.1845	0.1687	0.2672	0.1956	1.244	0	0.643	18.39	465
C27	0.1812	0.1672	0.5655	0.2200	1.053	0.779	0.544	30.70	472
C28	0.1628	0.1586	1.1733	0.5352	1.223	1.403	0.466	46.07	526
C33	0.1779	0.1555	0.2338	0.1995	0.839	0	0.639	18.22	443
C34	0.1977	0.1708	0.1251	0.1341	0.691	0	0.706	12.22	443
C35	0.2356	0.2038	0.0512	0.0921	0.800	0	0.779	8.25	443
C36	0.1496	0.1286	0.5233	0.3463	0.669	0	0.536	30.43	486
C38	0.2113	0.1806	0.2613	0.2012	0.873	0	0.643	17.73	443
C39	0.1911	0.1676	0.3520	0.2608	0.872	0	0.643	17.74	443
C40	0.2085	0.1802	0.3206	0.2077	0.874	0	0.643	17.75	443
C41	0.2160	0.1833	0.2173	0.1020	1.393	0	0.574	18.00	551
C42	0.2361	0.2035	0.4150	0.2355	1.231	0	0.496	30.68	566
C43	0.1986	0.1749	0.8459	0.8166	1.137	0	0.430	45.92	617
C44	0.2196	0.1948	0.1923	0.0167	0.902	0	0.591	18.42	515
C45	0.2771	0.2420	0.0823	0.1599	1.251	0	0.659	12.20	508
C46	0.2773	0.2364	0.0448	0.0267	1.357	0	0.734	8.16	501
C47	0.2146	0.1828	0.3230	0.1022	0.995	0	0.476	30.58	617
C48	0.2467	0.2122	0.3066	0.1553	0.995	0	0.467	30.72	639
C49	0.1835	0.1678	0.4950	0.5842	1.950	0	1.402	45.87	704
C50	0.2526	0.2152	0.0570	0.1200	0.961	0	0.553	18.42	588
C51	0.3169	0.2701	0.0100	0.1379	1.136	0	0.616	12.23	580
C52	0.3454	0.2908	0.0397	0.1237	1.543	0	0.682	8.17	580
C53	0.3091	0.2596	0.0518	0.1134	1.215	0	0.562	12.33	697
C54	0.3951	0.3331	0.0807	0.1250	1.602	0	0.622	8.19	697
C55	0.2828	0.2358	0.0264	0.1538	0.977	0	0.505	18.39	704
C56	0.2290	0.1958	0.2367	0.0769	1.411	0	0.428	30.68	762
C57	0.2375	0.2005	0.2489	0.2170	1.194	0.086	0.368	45.88	842
C58	0.3296	0.2759	0.1000	0.1287	1.239	0	0.503	12.30	871
C59	0.4154	0.3497	0.0592	0.0998	1.918	0	0.557	8.17	871
C60	0.3294	0.2763	0.0641	0.1150	1.628	0	0.453	18.40	878

TABLE 2 (Continued)

Run (1)	f_{ws} (2)	$\beta_1 D_{1m}$ (3)	D_{3m}/D_{1m} (4)	D_{5m}/D_{1m} (5)	β_1x100 (6)	$-\beta_3$x100 (7)	β_4x100 (8)	Fd_* (9)	a_* (10)
C61	0.2476	0.2096	0.0667	0.1745	1.572	0	0.384	30.77	943
C62	0.2120	0.1846	0.1307	0.1911	1.131	0.086	0.719	12.33	424
C63	0.2707	0.2375	0.1637	0.0720	1.456	0.137	0.797	8.19	424
C64	0.1831	0.1732	0.3277	0.0641	0.932	0.253	0.646	18.37	432
C65	0.1864	0.1726	0.5544	0.2273	1.002	0.423	0.548	30.58	464
C66	0.1421	0.1281	0.8019	0.2827	1.192	0.615	0.466	45.71	526
C67	0.2740	0.2562	0.2360	0.0392	1.374	0.210	0.840	12.28	312
C68	0.2241	0.3261	0.1760	0.1457	1.458	0.361	0.930	8.19	312
C69	0.2573	0.2396	0.6632	0.2377	1.260	0.661	0.757	18.47	319
C70	0.2745	0.2571	1.1181	0.4229	1.571	1.101	0.637	30.71	343
C71	0.2585	0.2409	1.5154	0.3462	2.004	1.636	0.542	45.76	385
C72	0.2429	0.2235	0.5336	0.1846	1.644	0.353	0.748	18.63	319
C73	0.2179	0.1859	0.1235	0.1336	0.679	-0.661	0.504	18.56	704
C74	0.1819	0.1655	0.3227	0.1392	0.890	0.149	0.640	18.42	439
C75	0.1751	0.1579	0.2335	0.2849	0.888	0.282	0.622	18.46	464
C76	0.1642	0.1693	0.3863	0.1459	0.849	0.492	0.672	18.35	399
C77	0.1845	0.1876	0.4784	0.1768	0.889	0.694	0.690	18.45	377
C78	0.1767	0.1557	0.2041	0.2302	0.889	0.646	0.599	18.45	501

TABLE 3.-Results of Fine Sand Experiments (D_*=96)

Run (1)	f_{ws} (2)	$\beta_1 D_{1m}$ (3)	D_{3m}/D_{1m} (4)	D_{5m}/D_{1m} (5)	β_1x100 (6)	$-\beta_3$x100 (7)	β_4x100 (8)	Fd_* (9)	a_* (10)
F01	0.1256	0.1086	0.2098	0.0283	1.043	0.386	0.801	18.40	1496
F02	0.1298	0.1094	0.3092	0.1130	1.025	0.423	0.750	18.42	1707
F03	0.1301	0.1138	0.0591	0.0794	1.251	0.715	0.841	12.26	1663
F04	0.1503	0.1286	0.0630	0.0513	1.396	0.927	0.940	8.17	1629
F05	0.1237	0.1042	0.1334	0.2653	1.037	1.302	0.750	18.41	1707
F06	0.1593	0.1308	0.1570	0.1538	1.326	1.838	0.623	30.61	1862
F07	0.1098	0.0933	0.1581	0.1351	1.420	0.087	0.686	18.38	2040
F08	0.1279	0.1075	0.0852	0.0653	0.864	0.150	0.782	12.19	1929
F10	0.1263	0.1056	0.2013	0.0440	1.588	0.270	0.583	30.71	2184
F11	0.1397	0.1194	0.0201	0.0655	1.059	0.041	0.626	18.28	2461
F12	0.1708	0.1441	0.0460	0.0535	0.885	0.070	0.705	12.22	2372
F13	0.1498	0.1294	0.1016	0.0498	1.613	0.124	0.528	30.39	2682
F14	0.1275	0.1184	0.0674	0.0814	1.051	0.171	0.801	18.41	1496
F15	0.1673	0.1434	0.0795	0.0568	1.271	0.301	0.903	12.27	1441
F16	0.1682	0.1389	0.0954	0.1768	1.900	0.522	0.674	30.60	1641
F17	0.1887	0.1674	0.1729	0.1813	0.903	0.123	0.941	18.34	1086
F18	0.1826	0.1672	0.1445	0.1317	0.013	0.211	1.057	12.26	1053
F20	0.2106	0.1989	0.2421	0.4439	1.084	0.084	1.098	18.35	799
F21	0.2233	0.2036	0.0346	0.1108	1.260	0.143	1.231	12.27	776
F23	0.1390	0.1149	0.1626	0.0322	0.903	0.025	0.559	18.42	3070
F28A	0.1199	0.1119	0.1579	0.0023	2.084	0	0.748	18.59	1707
B	0.1378	0.1179	0.1158	0.0196	2.084	0	0.748	18.59	1707
C	0.1053	0.0916	0.1795	0.0636	2.084	0	0.748	18.59	1707
D	0.0092	0.0792	0.2755	0.0656	1.041	0	0.748	18.59	1707

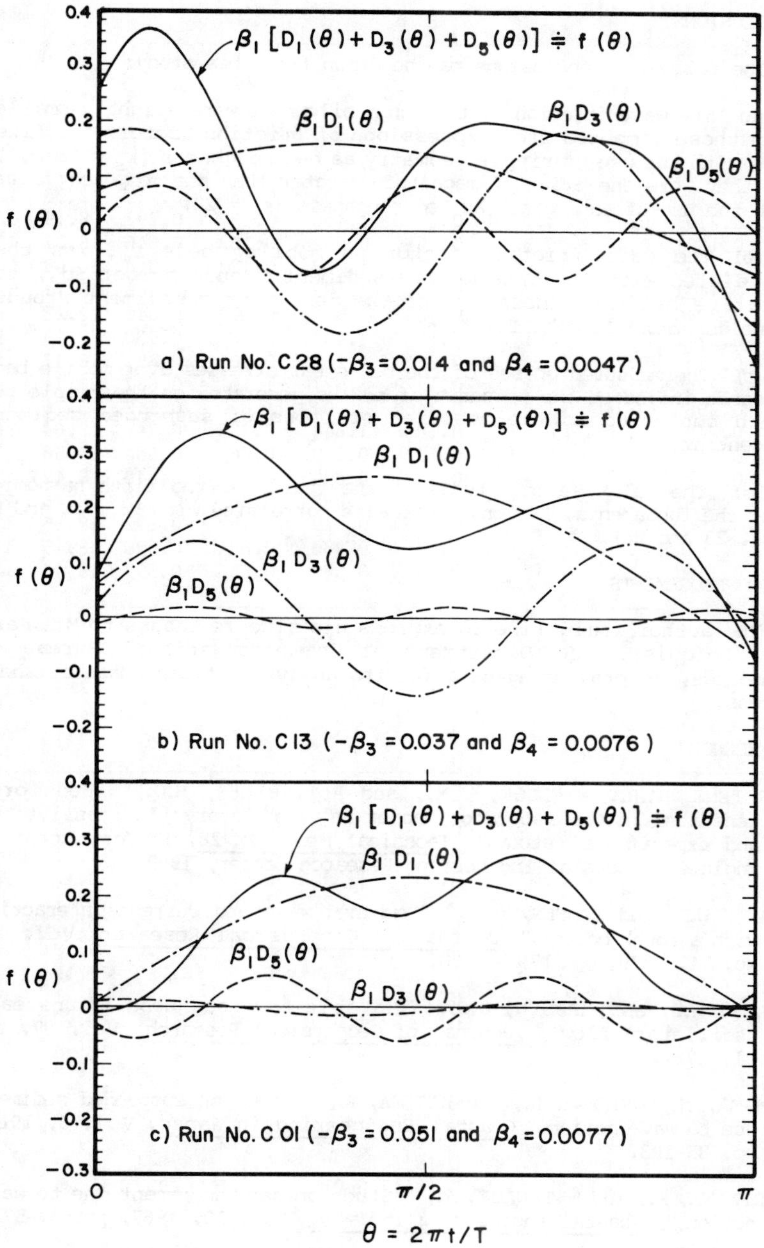

FIG. 3. – Phase of Bed Shear Stress, Eq. 13

CONCLUSIONS

The following conclusions can be drawn from this study:

(a) The wave friction factors on mobile beds were slightly smaller than those computed from expressions of friction factors for fixed beds with the use of ripple geometry as bed roughness [$k_s = 25\eta(\eta/\lambda)$] but they were one order of magnitude greater than the computed values with the use of sand grain as bed roughness ($k_s = 2D$).

(b) The wave friction factor on mobile beds f_{ws} was then correlated with the densimetric sediment Froude number Fd_*, the relative bed smoothness a_* and the densimetric sediment Froude-Reynolds number D_* (Eq. 21).

(c) The measured phases of the bed shear stresses over mobile beds of sands indicated the presence of eddies generated on the ripple bed which conformed to the measured profiles of suspended sediment concentration.

(d) The relative magnitudes of the third and the fifth harmonics over the fundamental harmonic were also correlated with Fd_*, a_* and D_* (Eqs. 23 and 24).

ACKNOWLEDGEMENTS

The author would like to express his sincere thanks to Mr. Karl E.B. Lofquist, U.S. Department of Commerce-National Bureau of Standards, in provinding data for the analysis of the phase of shear stress.

REFERENCES

CARSTENS, M.R., NIELSON, R.M., and ALTINBILEK, H.D., "Bed forms generated in the laboratory under an oscillatory flow: analytical and experimental study," Technical Memo. No. 28, US Army Corps of Engineers, Coastal Engineering Research Center, 1969.

GRANT, W.D. and MADSEN, O.S., "Combined wave and current interaction with a rough bottom," Journal of Geophysical Research, Vol. 84, No. C4, 1979, pp. 1797-1808.

GRANT, W.D. and MADSEN, O.S., "Movable bed roughness in unsteady oscillatory flow", Journal of Geophysical Research, Vol. 87, No. C1, 1982.

HOM-MA, M. HORIKAWA. K., and KAJIMA, R., "Study on suspended sediment due to wave action, Coastal Engineering in Japan, Vol. 8, 1965, pp. 85-103.

HORIKAWA, K., and WATANABE, A., "Study on sand movement due to wave action," Coastal Engineering in Japan, Vol. 10, 1967, pp. 39-57.

JONSSON, I.G., "Wave boundary layers and friction factors," Proceedings of the Tenth International Conference on Coastal Engineering, Vol. 1, 1966, pp. 127-148.

JONSSON, I.G., and CARLSEN, N.A., "Experimental and theoretical investigations in an oscillatory turbulent boundary layer," Journal of Hydraulic Research, Vol. 14, No. 1, 1976, pp. 45-60.

KAMPHUIS, J.W., "Friction factor under oscillatory waves," Journal of the Waterways, Harbors and Coastal Engineering Division, ASCE, Vol. 101, No. WW2, 1975, pp. 135-144.

KENNEDY, J.F. and LOCHER, F.A., "Sediment suspension by water waves," Waves on Beaches and Resulting Sediment Transport, R.E. Meyer, Ed., Academic Press, New York, NY, 1972, pp. 249-295.

LOFQUIST, K.E.B., "Drage on naturally rippled beds under oscillatory flows," Miscellaneous Paper CERC-86-13, US Army Corps of Engineers, Coastal Engineering Research Center, 1986.

MADSEN, O.S. and GRANT, W.D., "Sediment transport in the coastal environment", Report No. 209, Ralph M. Parsons Lab. for Water Resources and Hydrodynamics, Massachusetts Inst. of Tech., 1976.

NAKATO, T., LOCHER, F.A., GLOVER, J.R. and KENNEDY, J.F., "Wave entrainment of sediment from rippled beds," Journal of the Waterways, Harbours and Coastal Engineering Division, ASCE, No. WW1, Vol. 103, 1977, pp. 83-100.

RIEDEL, H.P., KAMPHUIS, J.W., and BREBNER, A., "Measurement of bed shear stresses under waves," Proceddings of the Thirteenth International Conference on Coastal Engineering, Vol. 1, 1972, pp. 587-603.

SWART, D.H., "Predictive equations regarding coastal transports," Proceedings of the Fifteenth International Conference on Coastal Engineering, Vol. 2, 1976, pp. 1113-1132.

VONGVISESSOMJAI, S., "Oscillatory boundary layer and eddy viscosity," Journal of Hydraulic Engrg., ASCE, Vol. 110, No,. 4, 1984.

VONGVISESSOMJAI, S., "Profile of suspended sediment due to wave action," Journal of Waterways, Port Coastal and Ocean Engrg., ASCE, Vol. 112, No. 1, 1986.

VONGVISESSOMJAI, S. "Wave friction factor on sand ripples". Proc. Specialty Conf. on Advances in Understanding of Coastal Sediment Processes, Vol. 1, 1987, pp. 393-408.

CHAPTER 82

A PARAMETRIC HURRICANE WAVE PREDICTION MODEL

Ian R. Young*

ABSTRACT

A spectral wave model based on a numerical solution of the Radiative Transfer Equation is used to create a synthetic data base on wave conditions within hurricanes. The results indicate that both the velocity of forward movement and maximum wind velocity within the storm play an important role in determining both the magnitude of the waves generated and also the spatial distribution of these waves. An equivalent fetch for hurricane wave generation which is a function of these two parameters is proposed. This concept, together with the standard JONSWAP fetch limited growth relationships, provide a simple means for estimating wave conditions within hurricanes.

INTRODUCTION

Hurricane generated waves play a significant role in the design of almost all coastal and offshore structures in tropical and semi-tropical regions. Despite their obvious importance, the complex processes active in their generation are only beginning to be understood. The physics of the generation process within hurricanes is complicated by the rapidly turning winds which generate cross-seas. The nonlinear interactions within such seas have only recently been investigated (Young et al.,1987). Progress has also been hindered by the lack of reliable field data. Buoy measurements represent only point data in the two-dimensional spatial wave field and only recently have provided directional information. More promising are remote sensing techniques such as Synthetic Aperture Radar (SAR) and Radar Altimeters (ALT) which have provided data from a limited number of hurricanes (King and Shemdin, 1978; Gonzalez et al., 1982; McLeish and Ross, 1983; Beal et al., 1986; Holt and Gonzalez, 1986). Although such instruments represent a significant advance, the small number of hurricanes for which data exists cover only a very limited range of storm parameters.

Recent advances in the physics of numerical wave prediction models (Komen et al., 1984; Hasselmann,S. et al., 1985) have enabled the development of operational hurricane prediction models which yield results consistent with the increasing field data base (Young, 1987c). Due to the complexity of these models they are, however, computationally expensive.

---

* Lecturer, Dept. Civil Engineering, University College, Australian Defence Force Academy, Northcott Drive, Canberra, ACT, 2600, Australia.

In order to develop an extensive synthetic data base, the model of Young (1987c) has been run for a wide range of hurricane parameters. The data from these numerical experiments will be used to clarify the wave generation process within hurricanes and to develop a simple parametric model suitable for wave prediction in deep water.

MODEL DESCRIPTION

The model used for the numerical experiments is called ADFA1 (Young, 1987c). It is a Second Generation (SWAMP, 1985) spectral wave model based on a numerical solution of the Radiative Transfer Equation (Hasselmann, 1960). Details of the model have been presented elsewhere (Young, 1987a,c). In particular, the model has a simple but very flexible treatment of the nonlinear source term which is of particular importance under the rapidly turning winds of hurricanes (SWAMP, 1985). An extensive comparison with field data under tropical cyclone conditions off the north-west coast of Australia has been presented by Young (1987b,c). These results clearly validate model performance under such conditions.

The model was used to generate a synthetic data base covering a wide range of hurricane parameters. The three wind field parameters varied were: velocity of forward movement, V_{fm}; maximum wind velocity in storm, V_{max} and radius to maximum winds, R. To cover the three-dimensional parameter space completely would involve a prohibitively large number of runs. It is reasonable to assume that R is simply a scaling parameter and hence for the majority of the experiments was held constant. A total of 43 experiments were conducted, the parameters being shown in Table 1.

Successful numerical modelling of hurricanes represents a conflict of scales. A grid of large geophysical extent is required to ensure all significant atmospheric forcing occurs within the grid and to predict the arrival of remotely generated swell. In order to resolve the eye structure and the associated strong winds, however, a relatively fine grid is required. To accommodate these constraints at acceptable computational expense, ADFA1 uses a system of nested grids.

A course grid with a spatial extent of 750km $\times$ 1050km with a resolution of 30km and a computational time step of 30mins was used to provide boundary conditions to a nested finer grid of extent 360km $\times$ 330km with a 15km resolution and time step of 15mins. Runs were commenced with the hurricane near the southern boundary of the course grid where it was held stationary for a period of two days to enable sea conditions to build up from an initially calm state. The storm was then moved off in a northerly direction at its velocity of forward movement. All results presented are from the finer grid whose south-west corner was located 300km east of the course grid western boundary and 650km north of its southern boundary. In all cases sea conditions had reached steady state by the time the hurricane entered the finer grid.

The hurricane wind field model used for the atmospheric forcing was a slightly modified form of that proposed by Graham and Hudson (1960). The original model had the region of maximum winds in the right rear quadrant whereas the modified form adopted here positions this region in the right forward quadrant, 70° from the direction of forward movement. This modification is consistent with observations from the NOAA hurricane program and has also been adopted in other modelling applications (SWAMP, 1985). In reality, the wind field is almost certainly more complex than the simple model used, with considerable variability between storms. In view of the accuracy with which hurricane parameters can be estimated in practice, however, there is little point in attempting a more sophisticated approach.

HURRICANE WAVE GENERATION

The advent of remote sensing techniques and particularly the Synthetic Aperture Radar (SAR) has provided considerable insight into the directional properties of hurricane generated waves. SAR data from a number of hurricanes (King and Shemdin, 1987; Gonzalez et al., 1982; McLeish and Ross, 1983; Beal et al., 1986; Holt and Gonzalez, 1986) consistently shows swell ahead of hurricanes radiating out in a fan shaped pattern from the centre of the storm. The hurricanes investigated by these experiments have all had slow to moderate velocities of forward movement ($V_{fm} = 2.5 - 5ms^{-1}$). King and Shemdin (1978) have, however, speculated that V_{fm} would have a critical role in determining these trends. Figs. 1 show contour plots of directional wave spectra at various points relative to the centre of the hurricane. Plots are presented for a storm with $V_{max} = 40ms^{-1}$ and values of $V_{fm} = 2.5$ and $12.5ms^{-1}$. Each of the contour plots has been normalized such that the spectral peak has a value of one; contours being drawn at 0.9, 0.5, 0.1 and 0.01. The maximum frequency shown on the polar grid is 0.25 Hz. In each of the spectra a relatively high frequency peak with a broad directional distribution centred about the local wind direction can be seen (ie. locally generated wind-sea). In addition, many of the spectra also exhibit a lower frequency swell peak. For the slower moving storm (Fig. 1a), examination of the spectra ahead of the storm centre clearly show swell radiating out from a region to the right of the storm centre in a similar fashion to that reported in the SAR data. Tracing this swell back along wave rays at the appropriate group velocity reveals that it was generated in the intense wind region to the right of the storm. This swell ($f = 0.08Hz$) has a group velocity of $9.8ms^{-1}$ and as such propagates ahead of the storm ($C_g > V_{fm}$). The more rapidly moving storm (Fig. 1b) has a velocity of forward movement, $V_{fm} = 12.5ms^{-1} > C_g$ and consequently no swell is evident ahead of the storm.

This result confirms the speculation of King and Shemdin (1978) as to the importance of V_{fm} in determining the wave field. V_{max} also plays an important role since it can be expected that the peak frequency of the dominant waves will decrease with increasing V_{max} (ie. C_g increases with increasing V_{max}). For a given V_{max}, if V_{fm} is relatively slow, the dominant waves will "outrun" the storm and appear as swell ahead of the storm. Conversely, if V_{fm} is relatively fast, the waves will be left behind the storm and no swell will be present ahead of the storm.

The maximum significant wave height within the hurricane, $H_s(max)$ as a function of both V_{max} and V_{fm} is shown in Fig. 2. For a given value of V_{max}, $H_s(max)$ gradually increases as a function of V_{fm} until a peak is reached, after which the wave height rapidly decreases. The V_{fm} at which this peak occurs increases with V_{max}. The reason for this behaviour is evident in Fig. 3 which shows the group velocity of the spectral peak frequency of the maximum waves in the storm as a function of both V_{fm} and V_{max}. Maximum wave conditions occur when the waves have a group velocity slightly greater than V_{fm}. Bretschneider (1957) proposed that maximum wave conditions would occur when $V_{fm} = C_g(max)$. Under such

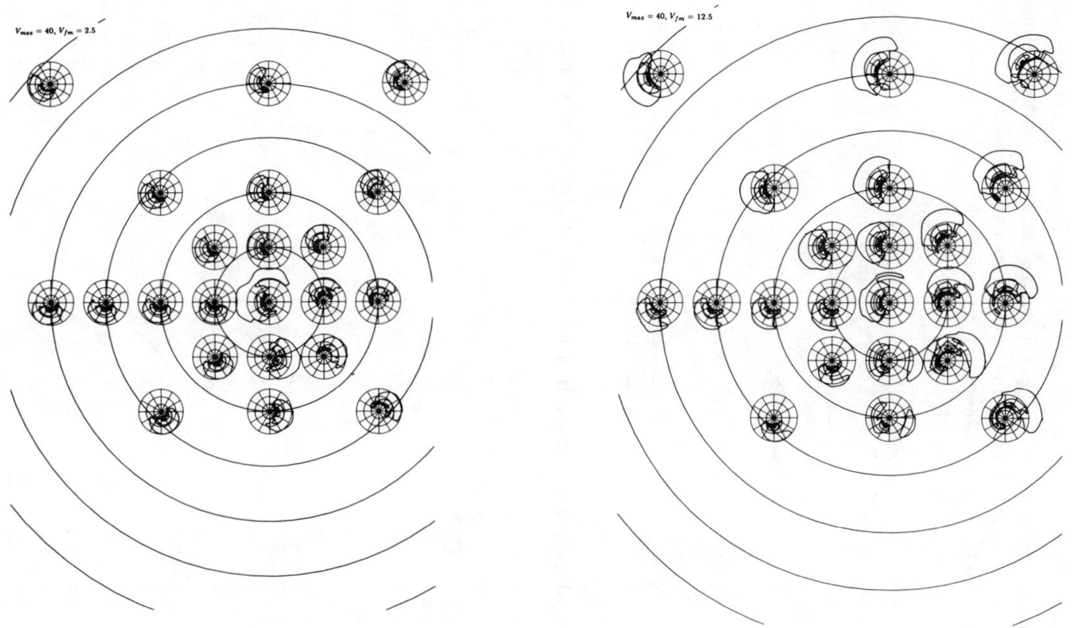

Fig. 1. Contour plots of directional wave spectra at various points within hurricanes with $V_{max} = 40ms^{-1}$ and $V_{fm} = 2.5ms^{-1}$ and $12.5ms^{-1}$. Circular lines are drawn at distance R, 2R, 3R etc. from the storm centre. Note, a southern hemisphere storm is shown.

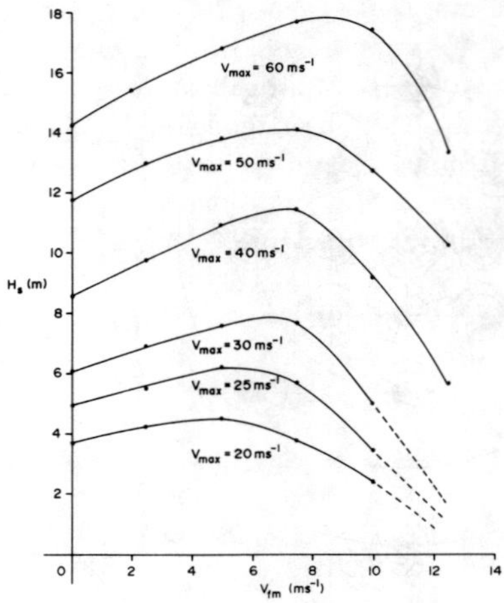

Fig. 2. Maximum significant wave height in a hurricane, $H_s(max)$ as a function of V_{fm} and V_{max}.

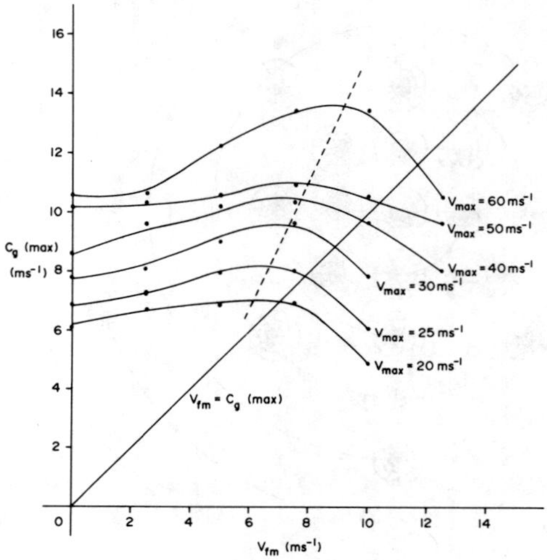

Fig. 3. Group velocity of the maximum waves in a hurricane, $C_g(max)$ as a function of V_{fm} and V_{max}.

a condition waves would move forward with the hurricane and experience an extended fetch. Due to the effect of nonlinear wave-wave interactions, however, there is a continual migration of the spectral peak to lower frequencies (Hasselmann et al., 1973). Consequently, there will be a tendency for the dominant waves to continually move to lower frequencies and "outrun" the storm (King and Shemdin, 1978). Only under fully arisen sea conditions when the peak frequency reaches the Pierson-Moskowitz value (Pierson and Moskowitz, 1958) does the spectral peak migration stop (Komen et al., 1984). As such conditions seldom, if ever, occur in hurricanes, the waves which will dominate are those which remain in the high wind regions for the maximum time.

EQUIVALENT FETCH

Bretschneider (1957) has used the concept of an equivalent fetch within a hurricane to apply fetch limited wave growth relationships to these situations. As already shown, the equivalent fetch must be a function of both V_{max} and V_{fm}.

The JONSWAP (Hasselmann et al., 1973) fetch limited growth relationship is

$$\frac{gH_s}{U_{10}^2} = 0.0016 \left(\frac{gF}{U_{10}^2}\right)^{0.5} \tag{1}$$

where U_{10} is the 10m wind velocity and F the fetch length. Replacing U_{10} by V_{max}, which is an appropriate wind scaling parameter for hurricane conditions, Eq. 1 can be written as

$$\frac{gH_s(max)}{V_{max}^2} = 0.0016 \left(\frac{gF}{V_{max}^2}\right)^{0.5} \tag{2}$$

Applying Eq. 2 to the present data and solving for F, yields the equivalent fetch values shown in Table 1. Initially, only the results for the storms with $R = 30km$ will be considered. The fetch dependence on R, however, will be investigated later. A polynomial approximation to the results yields:

$$F_{30} = aV_{max}^2 + bV_{max}V_{fm} + cV_{fm}^2 + dV_{max} + eV_{fm} + f \tag{3}$$

where $a = -6.525 \times 10^1$, $b = 4.518 \times 10^2$, $c = -3.669 \times 10^3$, $d = 6.570 \times 10^3$, $e = 2.021 \times 10^4$, $f = 2.394 \times 10^4$; V_{max}, V_{fm} both have units of $[ms^{-1}]$ and F_{30} is the equivalent fetch for $R = 30km$ (units of [m]). Eq. 3 is presented in graphical form in Fig. 4. It is clear from this figure that for a given V_{max}, there is a value of V_{fm} which will give the maximum equivalent fetch and hence maximum wave conditions. As V_{max} increases, the value of V_{fm} which produces the maximum equivalent fetch also increases. This occurs since higher values of V_{max} will generate waves with high group velocities and hence a more rapidly moving storm is required to maximize the equivalent fetch.

This rather simple model can also be applied to determine the spectral peak frequency of the maximum waves in the storm, $f_m(max)$. Reformulating the JONSWAP result gives

$$\frac{g}{2\pi f_m(max) V_{max}} = 0.045 \left(\frac{gF}{V_{max}^2}\right)^{0.33} \tag{4}$$

Fig. 5 shows a comparison between the values of $f_m(max)$ generated by the wave model and shown in Table 1 with those calculated from Eq. 4 using the values of

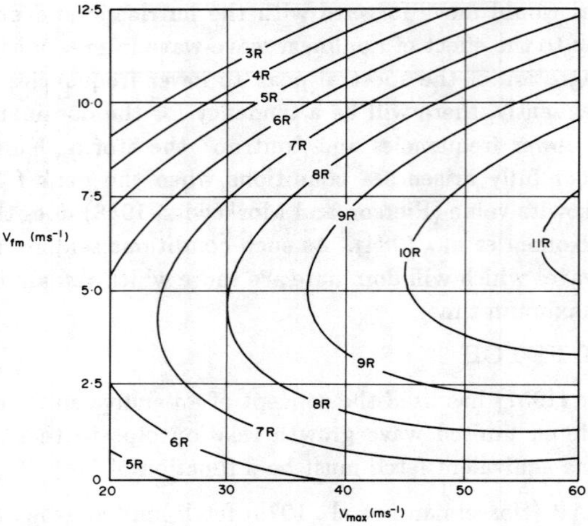

Fig. 4. Contour plot of the equivalent fetch (expressed in terms of R, which equals 30km) as a function of V_{fm} and V_{max}.

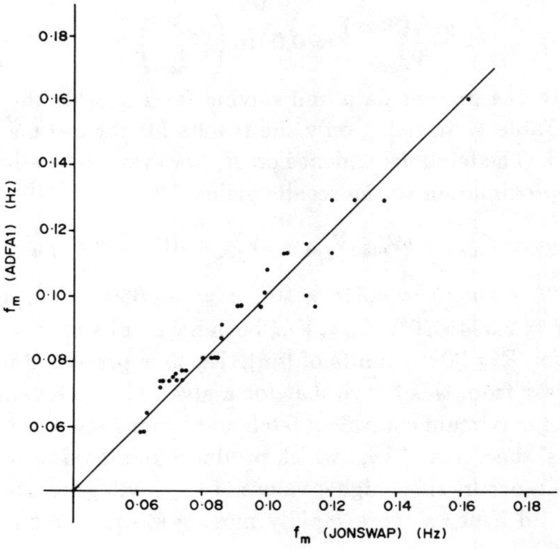

Fig. 5. Peak spectral frequency within the hurricane predicted by the spectral wave model, $f_m(max)$ (ADFA1) verses that predicted by the equivalent fetch assumption, $f_m(max)$ (JONSWAP).

Run No.	V_{fm} (ms^{-1})	V_{max} (ms^{-1})	R (km)	$H_s(max)$ (m)	F	$f_m(max)$ (Hz)
1	0.0	20	30	3.7	4.4R	0.129
2	0.0	25	30	5.1	5.3R	0.113
3	0.0	30	30	6.1	5.3R	0.101
4	0.0	40	30	8.6	5.9R	0.091
5	0.0	50	30	11.8	7.1R	0.077
6	0.0	60	30	14.3	7.3R	0.074
7	2.5	20	30	4.2	5.6R	0.116
8	2.5	25	30	5.5	6.2R	0.108
9	2.5	30	30	6.9	6.8R	0.097
10	2.5	40	30	9.7	7.5R	0.081
11	2.5	50	30	13.0	8.6R	0.075
12	2.5	60	30	15.4	8.4R	0.074
13	5.0	15	30	3.0	5.1R	0.129
14	5.0	20	30	4.5	6.5R	0.113
15	5.0	25	30	6.2	7.9R	0.097
16	5.0	30	30	7.5	8.0R	0.087
17	5.0	40	30	10.9	9.5R	0.077
18	5.0	50	30	13.8	9.7R	0.074
19	5.0	60	30	16.8	10.0R	0.064
20	7.5	20	30	3.7	4.4R	0.113
21	7.5	25	30	5.7	6.6R	0.097
22	7.5	30	30	7.7	8.4R	0.081
23	7.5	40	30	11.4	10.4R	0.076
24	7.5	50	30	14.1	10.2R	0.072
25	7.5	60	30	17.7	11.1R	0.058
26	10.0	20	30	2.4	1.8R	0.160
27	10.0	25	30	3.4	2.4R	0.129
28	10.0	30	30	5.0	3.6R	0.100
29	10.0	40	30	9.1	6.6R	0.081
30	10.0	50	30	12.7	8.2R	0.074
31	10.0	60	30	17.4	10.8R	0.058
32	12.5	40	30	5.6	2.5R	0.097
33	12.5	50	30	10.2	5.3R	0.081
34	12.5	60	30	13.3	6.3R	0.074
35	5.0	40	15	9.3	13.8R	0.081
36	5.0	40	45	11.8	7.4R	0.074
37	5.0	40	60	12.2	5.9R	0.074
38	5.0	25	15	5.2	11.1R	0.110
39	5.0	25	45	6.4	5.6R	0.097
40	5.0	25	60	6.6	4.5R	0.097
41	5.0	60	15	14.5	14.9R	0.074
42	5.0	60	45	17.1	6.9R	0.064
43	5.0	60	60	18.2	5.9R	0.058

Table 1. The synthetic hurricane data base

F from Table 1. The agreement between the two is very good. This is particularly so when it is considered that estimates of f_m from spectral models such as ADFA1 are not particularly robust due to the discrete frequency resolution of such models.

RADIAL DEPENDENCE

As indicated earlier, it was assumed that the radius to maximum wind, R, was a scaling factor. Runs 35 to 43 were included to investigate this influence. Comparing the equivalent fetch values calculated from these runs, and presented in Table 1, with the values for the equivalent runs with $R = 30km$ shows that the scaling relationship is not linear. (ie. Doubling the radius to maximum winds does not double the equivalent fetch.) Fig. 6 shows the equivalent fetch as a function of radius to maximum winds. In order to collapse the data onto one curve, values have been normalized in terms of the equivalent results for a radius to maximum winds of 30km. (ie. the $R = 30km$ run with the same values of V_{fm} and V_{max}). These results can be approximated quite well by the relationship

$$F/F_{30} = 0.75 log_{10} R/(30 \times 10^3) + 1 \tag{5}$$

where F is the equivalent fetch for a hurricane with a radius to maximum winds, R (units of [m]) and F_{30} is the equivalent fetch for the hurricane with the same values of V_{fm} and V_{max} but with a radius to maximum winds of 30km. In practice, F_{30} can be calculated from Eq. 3.

Eq. 5 provides a means of determining F and hence $H_s(max)$ for any value R. Since F is directly related to H_s for given V_{fm} and V_{max}, it would appear to be a far more appropriate spatial scale parameter than R.

PARAMETRIC MODEL

Applying the results in the previous sections, it is possible to develop a relatively simple yet flexible parametric model for hurricane wave prediction in deep water. Given V_{fm}, $V_{max}$and R, an effective radius R' can be defined from Eq.5

$$R' = 22.5 \times 10^3 log_{10} R - 70.8 \times 10^3 \tag{6}$$

where both R and R' have units of metres. Using R', V_{fm} and V_{max} the equivalent fetch, F is determined from Eq.3

$$F/R' = aV_{max}^2 + bV_{max}V_{fm} + cV_{fm}^2 + dV_{max} + eV_{fm} + f \tag{7}$$

where $a = -2.175 \times 10^{-3}$, $b = 1.506 \times 10^{-2}$, $c = -1.223 \times 10^{-1}$, $d = 2.190 \times 10^{-1}$, $e = 6.737 \times 10^{-1}$ and $f = 7.980 \times 10^{-1}$. Again, all values are in standard S.I. units. The maximum significant wave height and associated spectral peak frequency can be determined from the modified JONSWAP relationships, Eqs. 2 and 4.

SPATIAL DISTRIBUTION

The model presented in the previous section provides a means for determining the maximum significant wave height and its associated spectral peak frequency but no insight into the spatial distribution of these quantities within the storm. As with $H_s(max)$, the spatial distribution of significant wave heights is also a function of both V_{fm} and V_{max}. Hence it is not possible to present simply one spatial distribution diagram in a manner similar to that of Bretschneider (1957). Due

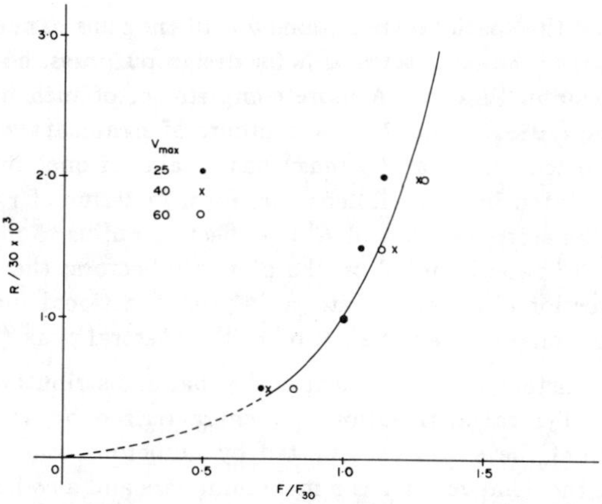

Fig. 6. Dependence of the equivalent fetch, F, on the radius to maximum winds, R. Values are presented in terms of the equivalent fetch for a hurricane with R=30km. The solid line is Eq. 5.

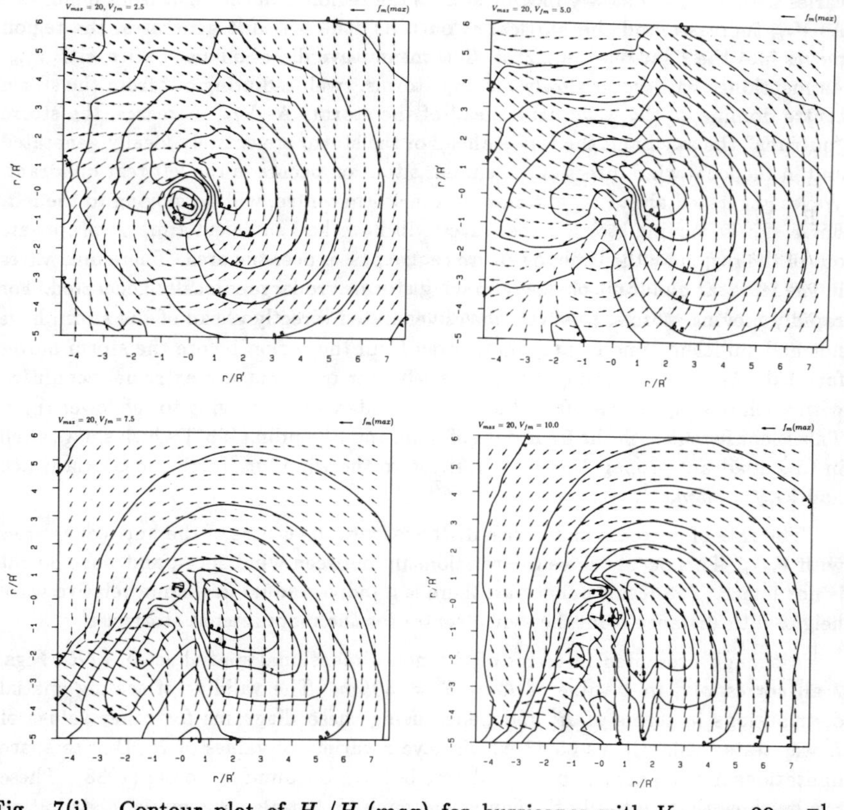

Fig. 7(i). Contour plot of $H_s/H_s(max)$ for hurricanes with $V_{max} = 20ms^{-1}$. Vectors indicate mean direction of wave propagation. Vector lengths are directly proportional to $T_m = 1/f_m$. Scale is shown at top right.

to space limitations, the spatial distributions for all the runs cannot be presented here. Due to the importance of such results for design purposes, however, as many as is practical appear in Figs. 7. A more complete set of such figures has been presented by Young (1988). Figs. 7 show contours of significant wave height; the values being normalized such that $H_s(max)$ has a value of one. Spatial distances have also been presented in a nondimensional form in terms of r/R', where r is the distance from the storm centre and R' the effective radius to maximum winds, as defined by Eq. 6. Superimposed on the plots are vectors; the direction being the mean wave direction obtained by integrating the directional spectrum and the length being proportional to the wave period of the spectral peak ($T_m = 1/f_m$).

A number of consistent trends are clear in the spatial distributions as a function of V_{fm} and V_{max}. Typical distributions are characterized by a crescent shaped region to the right of the storm as reported by numerous other authors. This region occurs since the wind velocity is a maximum here and also because the wind direction is approximately parallel with the storm track. Hence, waves generated in this region move forward with the hurricane and maximize the time for which they experience strong winds. The actual position of the crescent shaped region varies with V_{fm}. For slowly moving storms the region is in the right front quadrant. As V_{fm} increases and the hurricane "outruns" the waves it generates, the region moves into the right rear quadrant. The mean wave direction also varies with V_{fm}. As mentioned earlier, for slowly moving storms, swell radiating out from the storm centre dominates the wave field ahead of the storm. As V_{fm} increases, the storm "outruns" the swell and the waves ahead of the storm are entirely locally generated and in the local wind direction. A similar situation occurs in the left rear quadrant. Waves generated ahead of, but close to the storm centre, will be moving to the left. For a slowly moving storm these waves will have had sufficient time to propagate out of the region by the time the storm centre passes over the area. Therefore, waves in the left rear quadrant of slowly moving storms are largely locally generated. For rapidly moving storms, however, waves generated directly ahead of the storm have not had sufficient time to propagate away from the region before the storm moves forward. As a result, conditions in the left rear quadrant are extremely confused with swell moving to the left and locally generated waves moving to the lower right. This is consistent with the SAR data of King and Shemdin (1987) which shows swell in this quadrant propagating in opposition to the wind for Hurricane Gloria which had $V_{fm} = 8.3ms^{-1}$.

The rate of decrease in H_s with distance from the centre of the hurricane varies with V_{max}. This occurs since the relationship between wind speed and wave height is not linear. The result being that there is a more gradual decline in relative wave height with distance from the storm centre for the more intense storms.

Although presented in terms of the non-dimensional spatial scale, r/R', Figs. 7 all correspond to a value of $R = R' = 30km$. The validity of scaling spatial distributions in terms of R' and hence using these diagrams for other values of R was tested using Runs 35-43 which have a variety of values of R. Due to space limitations, results are not presented here but can be found in Young (1988). These results show that, not only the equivalent fetch, F but also the spatial distribution can be scaled in terms of R'.

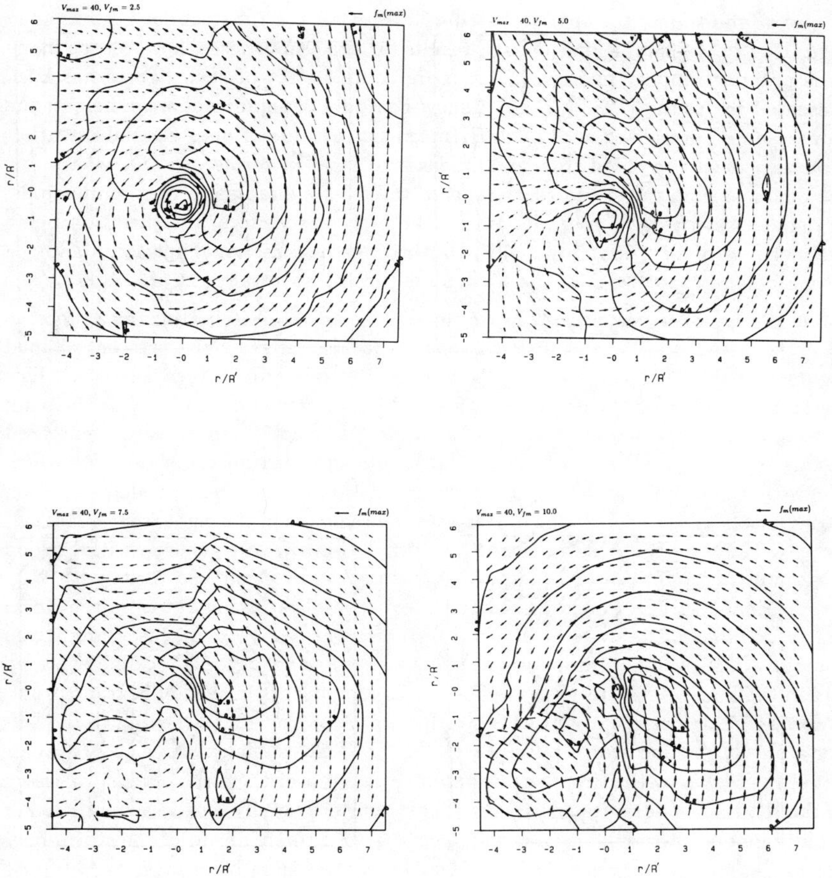

Fig. 7(ii). Contour plot of $H_s/H_s(max)$ for hurricanes with $V_{max} = 40ms^{-1}$. Vectors indicate mean direction of wave propagation. Vector lengths are directly proportional to $T_m = 1/f_m$. Scale is shown at top right.

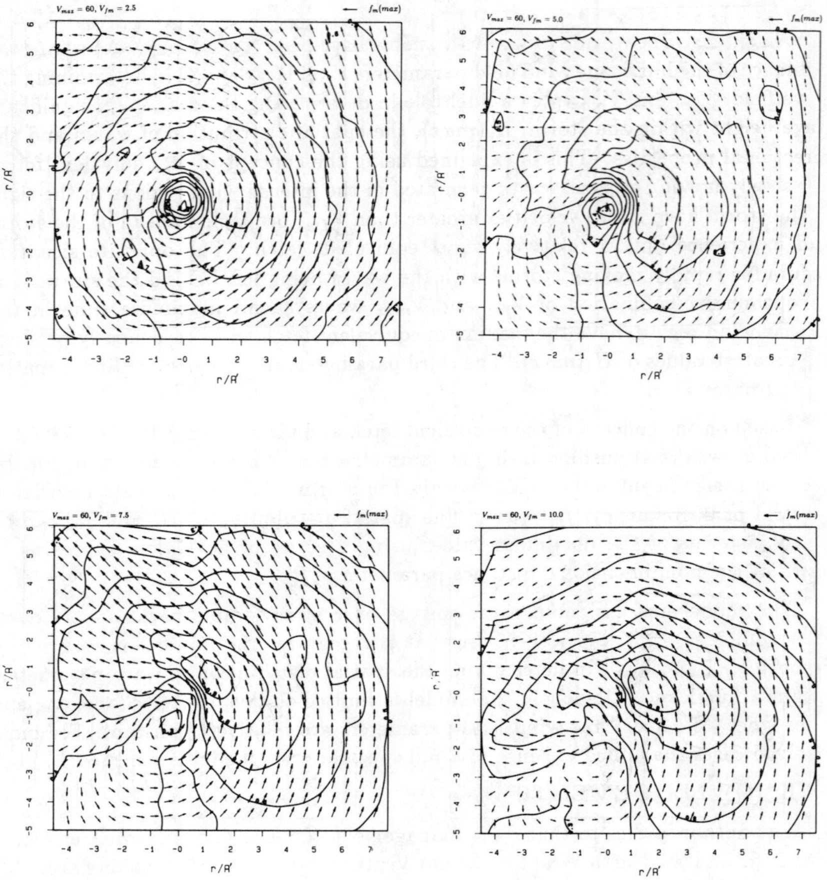

Fig. 7(iii). Contour plot of $H_s/H_s(max)$ for hurricanes with $V_{max} = 60ms^{-1}$. Vectors indicate mean direction of wave propagation. Vector lengths are directly proportional to $T_m = 1/f_m$. Scale is shown at top right.

CONCLUSIONS

An extensive set of numerical experiments have been conducted to determine the influence of the hurricane wind field parameters V_{fm}, V_{max} and R in determining the wave field. V_{fm} and V_{max} play a duel role in determining the maximum significant wave height within the storm, $H_s(max)$, the spatial distribution of waves and the directional properties. This is explained using the concept of an equivalent fetch. For slowly moving storms, waves generated in the intense wind regions to the right of the storm have group velocities greater than V_{fm}, propagate ahead of the storm and experience only a relatively short equivalent fetch. The opposite situation occurs for rapidly moving storms with the waves being left behind the storm. For an optimum combination of V_{fm} and V_{max} waves spend maximum time in the intense wind region, have the maximum equivalent fetch and consequently produce the greatest values of $H_s(max)$. The third parameter, R, acts as a nonlinear spatial scaling parameter.

Based on the concept of the equivalent fetch and the standard JONSWAP fetch limited growth relationships, a simple parametric model has been developed for the maximum significant wave height within the storm, $H_s(max)$ and its associated spectral peak frequency, $f_m(max)$. The spatial distribution of H_s and f_m can be found in terms of the maximum values using a series of field plots expressed in terms of the nondimensional distance parameter r/R'.

The model has a significant advantage over other simple parametric models (Bretschneider, 1957; Ross, 1976) in that it recognizes the important role played by both V_{fm} and V_{max} in determining the spatial distribution of wave parameters within a hurricane. Obviously the model is limited to deep water conditions and cases where the hurricane wind field parameters are relatively constant. For more involved cases the added expense of a full spectral wave model is required.

ACKNOWLEDGEMENTS

The author wishes to thank the management of Woodside Offshore Petroleum Pty. Ltd. and the North West Shelf Joint Venture Participants for making available the tropical cyclone data presented in the paper. The participants in the LNG phase of the project are Woodside Petroleum Pty. Ltd., BHP Petroleum Pty. Ltd., BP Developments Australia Ltd., California Asiatic Oil Company, Japan Australia LNG (MIMI) and Shell Development (Aust.) Pty. Ltd.

REFERENCES

Beal,R.,C., Gerling,T.W., Irvine,D.E., Monaldo,F.,M. and Tilley,D.,G., 1986, "Spatial Variations of Ocean Wave Directional Spectra from the SEASAT Synthetic Aperture Radar", Jnl. Geoph. Res., 91, C2, pp2433-2449.

Bretschneider,C.,L., 1957, "Hurricane Design Wave Practices", ASCE, Jnl. Waterways and Harbors, 83, WW2, No. 1238.

Gonzalez,F.,I., Thompson,T.,W., Brown,W.,E. and Weissman,D.,E., 1982, "SEASAT Wind and Wave Observations of Northeast Pacific Hurricane Iva, August 13, 1978", Jnl. Geoph. Res., 87, C5, pp3431-3438.

Graham,H.E. and Hudson,G.N., 1960, "Surface Winds Near the Centre of Hurricanes (and Other Cyclones)", National Hurricane Research Project, Rep. No. 39,

U.S. Dept. of Commerce, Washington, D.C.

Hasselmann,K., 1960, "Grundgleichungen der Seegangsvorhersage", Schiffstechnik, 7, pp191-195.

Hasselmann,K., et al., 1973, "Measurements of Wind Wave Growth and Swell Decay during the Joint North Sea Wave Project", Herausgegeben vom Deutsch. Hydrograph. Institut, Reihe A, No. 12, 95pp.

Hasselmann,S., Hasselmann,K., Allender,J.H. and Barnett,T.P., 1985, "Computations and Parameterizations of the Nonlinear Energy Transfer in a Gravity-Wave Spectrum. Part II: Parameterizations of the Nonlinear Energy Transfer for Applications in Wave Models", Jnl. Physical Oceanography, 15, pp1378-1391.

King,D.B. and Shemdin,O.,H., 1978, "Radar Observations of Hurricane Wave Directions", 16th Int. Conf. on Coastal Eng., Hamburg, pp209-226.

Komen,G.J., Hasselmann,S. and Hasselmann,K., 1984,"On the Existence of a Fully Developed Wind-Sea Spectrum", Jnl. Physical Oceanography, 14, pp1271-1285.

McLeish,W. and Ross,D.B., 1983, "Imaging Radar Observations of Directional Properties of Ocean Waves", Jnl. Geoph. Res., 88, C7, pp4407-4419.

Pierson,W.J. and Moskowitz,L., 1964, "A Proposed Spectral Form for Fully Developed Wind Seas Based on the Similarity Theory of S.A. Kitaigorodskii", Jnl. Geogh. Res., 69, pp5181-5190.

Ross,D., 1976, "A Simplified Model for Forecasting Hurricane Generated Waves", Bull. Amer. Met. Soc., January. Presented at Conference on Atmospheric and Oceanic Waves, Seattle, Washington, March 29-April 2.

SWAMP Group, 1985, "The Sea Wave Modeling Project (SWAMP): Principal Results and Conclusions", Plenum Press, 256pp.

Young,I.R., 1987a, "A General Purpose Spectral Wave Prediction Model", Dept. Civil Eng., Aust. Defence Force Academy, Res. Rep. No. 16.

Young,I.R., 1987b, "Validation of the Spectral Wave Model ADFA1", Dept. Civil Eng., Aust. Defence Force Academy, Res. Rep. No. 17.

Young, I.R., 1987c, "A Shallow Water Spectral Wave Model",To appear Jnl. Geoph. Res.

Young, I.R., 1988, "A Parametric Model for Tropical Cyclone Waves", Dept. Civil Eng., Aust. Defence Force Academy, Res. Rep. No. 28.

Young,I.R., Hasselmann,S. and Hasselmann,K., 1987,"Computations of the Response of a Wave Spectrum to a Sudden Change in Wind Direction", Jnl. Physical Oceanography, 17, 9, pp1317-1338.

CHAPTER 83

A NUMERICAL MODEL OF NEARSHORE CURRENTS DUE TO IRREGULAR WAVES

Masataka Yamaguchi*

ABSTRACT

This paper presents a numerical model of nearshore currents due to irregular waves. The radiation stress is estimated by a current-depth refraction model for irregular waves, in which the energy dissipation due to wave breaking is modeled through the use of a saturated frequency spectrum in shallow water. The model is in reasonable agreement with measured wave height, mean water level variation and observed nearshore current patterns. Next, the model is applied to the computation of wave transformation and nearshore currents on a uniformly sloping beach and on model topographies with complicated contour lines. Comparison with the results based on a regular wave model shows that wave irregularity has a smoothing effect on cross-shore distributions of wave height, mean water level variation and longshore currents, but that it does not have much effect on nearshore current patterns.

1. INTRODUCTION

In most nearshore current models, a regular wave theory including finite amplitude wave theory (Yamaguchi, 1986) has been used for the calculation of wave transformation in shoaling water and in the surf zone, and the resulting radiation stress which is a driving force of nearshore currents. Since ocean waves are irregular waves approximated by the superposition of many component waves with different frequency and direction, a nearshore current model taking into account the effect of wave irregularity is needed for better understanding and description of coastal phenomena. Nevertheless, it seems that there are no numerical models of nearshore currents due to irregular waves applicable to an arbitrary bottom topography, although analytical models of longshore currents by Collins(1972), Battjes(1972) and Thornton & Guza (1986) and a numerical model of longshore

* Prof., Dept. of Ocean Eng., Ehime Univ., Bunkyocho 3, Matsuyama 790, Ehime Pref., Japan

currents by Hubertz (1984), are available. Therefore, the aim of this study is to present a numerical model of nearshore currents due to irregular waves, in which wave transformation is computed by a current-depth refraction model, and to establish its applicability by comparison with the results of experiments and observations of wave transformation and nearshore currents. Then, based on the numerical computations, the effect of wave irregularity on them and the characteristics of nearshore currents on two-dimensional model topographies are discussed.

2. MODEL DESCRIPTION

(1) Wave transformation model

The wave transformation model is to solve the conservation equation of wave action spectral density ($w = E(k_x,k_y)/\sigma_m$) and the equations for wave number components (k_x,k_y) simultaneously, where $E(k_x,k_y)$ is the wave number spectrum and σ_m the relative angular frequency. The former is written as

$$\frac{\partial w}{\partial t}+\frac{dx}{dt}\frac{\partial w}{\partial x}+\frac{dy}{dt}\frac{\partial w}{\partial y}+\frac{dk_x}{dt}\frac{\partial w}{\partial k_x}+\frac{dk_y}{dt}\frac{\partial w}{\partial k_y}=0 \quad (1)$$

and the latter is

$$\begin{aligned}&\frac{\partial k_x}{\partial t}+(C_g\cos\theta+U)\frac{\partial k_x}{\partial x}+(C_g\sin\theta+V)\frac{\partial k_x}{\partial y}\\&\quad=-\frac{gk^2\,\mathrm{sech}^2kD}{2\sigma_m}\frac{\partial D}{\partial x}-k_x\frac{\partial U}{\partial x}-k_y\frac{\partial V}{\partial x}\\&\frac{\partial k_y}{\partial t}+(C_g\cos\theta+U)\frac{\partial k_y}{\partial x}+(C_g\sin\theta+V)\frac{\partial k_y}{\partial y}\\&\quad=-\frac{gk^2\,\mathrm{sech}^2kD}{2\sigma_m}\frac{\partial D}{\partial y}-k_x\frac{\partial U}{\partial y}-k_y\frac{\partial V}{\partial y}\end{aligned} \quad (2)$$

where C_g is the group velocity of the component wave, θ the wave direction, k the wave number, $D(=h+\eta)$ the total water depth including the mean water level variation η, h the still water depth and (U,V) are the wave-induced nearshore current components.

These equations mean that the wave action spectral density in a nondissipative system is conserved along the characteristics defined by

$$\frac{dx}{dt}=C_g\cos\theta+U,\ \frac{dy}{dt}=C_g\sin\theta+V \quad (3)$$

where the wave number components are computed from

$$\begin{aligned}\frac{dk_x}{dt}&=-\frac{gk^2\,\mathrm{sech}^2kD}{2\sigma_m}\frac{\partial D}{\partial x}-k_x\frac{\partial U}{\partial x}-k_y\frac{\partial V}{\partial x}\\\frac{dk_y}{dt}&=-\frac{gk^2\,\mathrm{sech}^2kD}{2\sigma_m}\frac{\partial D}{\partial y}-k_x\frac{\partial U}{\partial y}-k_y\frac{\partial V}{\partial y}\end{aligned} \quad (4)$$

Thus, the conservation of the wave action spectral density expressed in frequency-direction space is written as

$$w(f,\theta)=(c_g+U\cos\theta+V\sin\theta)\,E(f,\theta)/(2\pi k\sigma_m)=\mathrm{const} \quad (5)$$

and is rewritten in the following equation

$$E_2(f_2,\theta_2)=\left\{\frac{(C_{g1}+U_1\cos\theta_1+V_1\sin\theta_1)/(k_1\sigma_{m1})}{(C_{g2}+U_2\cos\theta_2+V_2\sin\theta_2)/(k_2\sigma_{m2})}\right\}E_1(f_1,\theta_1) \quad (6)$$

where f is the absolute frequency, $E(f,\theta)$ is the directional spectrum and subscripts '1' and '2' denote the variables at different positions. This is the basic relationship used to compute the transformation of directional spectra. The model can estimate the effects of not only depth-refraction and wave shoaling, but also current refraction.

The energy dissipation due to wave breaking is modeled through the use of a saturated frequency spectrum proposed by Kitaigorodskii et al.(1975)

$$E_\infty(f)=\alpha g^2(2\pi)^{-4}f^{-5}\phi(\omega_h)$$
$$\phi(\omega_h)=x^{-2}(1+2\omega_h^2x/\sinh 2\omega_h^2x)^{-1},\quad \omega_h^2=\sigma_m^2D/g,\ x=gk/\sigma_m^2 \quad (7)$$

where α is the equilibrium constant at the high frequency tail and g the acceleration of gravity. Assuming the invariability of angular distribution in directional spectra associated with wave breaking, directional spectra including the effect of depth-controlled wave breaking are re-evaluated by

$$E_{new}(f,\theta)=E_{old}(f,\theta)E_\infty(f)/E_{old}(f) \quad (8)$$

where directional spectra with subscripts 'new' and 'old' mean directional spectra before and after the inclusion of the wave breaking effect respectively.

The radiation stress components (S_{xx}, S_{xy}, S_{yy}) and wave statistics are obtained by the numerical integration of directional spectra with respect to wave direction and frequency using

$$S_{xx}=\int_0^\infty\int_0^{2\pi}\rho g\{n\cos^2\theta+(n-1/2)\}E(f,\theta)\,d\theta df$$
$$S_{xy}=S_{yx}=\int_0^\infty\int_0^{2\pi}\rho gn\cos\theta\sin\theta E(f,\theta)\,d\theta df \quad (9)$$
$$S_{yy}=\int_0^\infty\int_0^{2\pi}\rho g\{n\sin^2\theta+(n-1/2)\}E(f,\theta)\,d\theta df$$

and

$$E(f)=\int_0^{2\pi}E(f,\theta)\,d\theta,\ m_0=\int_0^\infty E(f)\,df,\ m_1=\int_0^\infty fE(f)\,df,\ m_2=\int_0^\infty f^2E(f)\,df$$
$$H_{1/3}=4.00\sqrt{m_0},\ T_{1/3}=1.20\sqrt{m_0/m_2},\ H_{rms}=2.83\sqrt{m_0},\ T_{m01}=m_0/m_1 \quad (10)$$
$$\bar\theta=\tan^{-1}\left\{\int_0^\infty\int_0^{2\pi}E(f,\theta)\cos\theta\,d\theta df\ \Big/\ \int_0^\infty\int_0^{2\pi}E(f,\theta)\sin\theta\,d\theta df\right\}$$

where $n=C_g/C$, C is the wave celerity of the component wave, ρ the density of fluid, m_n the spectral moment, E(f) the frequency spectrum, $H_{1/3}$ the significant wave height, $T_{1/3}$ the significant wave period, H_{rms} the root mean square wave height, T_{m01} the mean wave period and $\bar\theta$ the mean wave direction.

The input directional spectrum at the offshore boundary is given by the Bretschneider-Mitsuyasu-type frequency spectral model with $\cos^{2s}(\theta/2)$ type angular distribution function

$$E(f,\theta)=E(f)\cdot D(f,\theta)$$
$$E(f)=0.257\,(H_{1/3})_0^2(T_{1/3})_0\{(T_{1/3})_0 f\}^{-5}\exp[-1.03\{(T_{1/3})_0 f\}^{-4}]\,\phi(\omega_h) \tag{11}$$
$$D(f,\theta)=\frac{2^{2s-1}\Gamma^2(s+1)}{\pi\Gamma(2s+1)}\cos^{2s}\left(\frac{\theta-\theta_0}{2}\right)$$

where S is the energy concentration factor, θ_0 the principal wave direction, Γ the Gamma function, and subscript '0' indicates the wave conditions in deep water. The input frequency spectrum is modified by the introduction of the shallow water effect. Consequently, significant wave height and wave period in deep water ($(H_{1/3})_0$, $(T_{1/3})_0$) are different from those at the offshore boundary of the computation region located in shallow water ($(H_{1/3})_M$, $(T_{1/3})_M$). In most cases, it is not $(H_{1/3})_0$ and $(T_{1/3})_0$ but $(H_{1/3})_M$ and $(T_{1/3})_M$ that are prescribed beforehand in the computation. The values of $(H_{1/3})_0$ and $(T_{1/3})_0$ to be given as input conditions are determined in a trial and error manner so that $(H_{1/3})_M$ and $(T_{1/3})_M$ computed from Eq.(11) agree approximately with the prescribed $(H_{1/3})_M$ and $(T_{1/3})_M$.

Numerical computation is conducted making use of a piecewise ray method as shown in Fig. 1. The method is to trace a wave ray backward only by one time step and to interpolate wave action spectral density at the tip point of the ray. The wave ray is followed by solving Eq. (3) and Eq. (4) with the Runge-Kutta method. In the ray computation, both input wave direction and current direction have to be reversed. The interpolation is executed in two steps. First, the Lagrange interpolation formula with third order accuracy is applied to the wave action spectral densities at the grid points surrounding the ray point in order to estimate the wave action spectral densities for the prescribed input directions, putting the wave direction at the ray point between them. It is given as

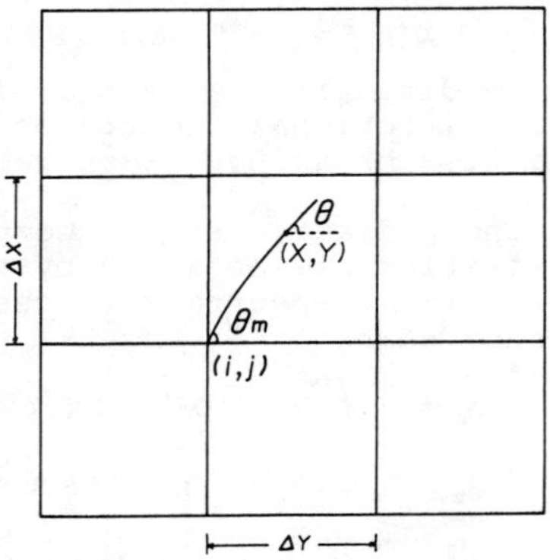

Fig. 1 Computation of direction of propagation.

$$\omega_{i+r,j+s}=\sum_{I=i-1}^{i+2}\sum_{J=j-1}^{j+2}\left(\prod_{k\neq I}\frac{x-x_k}{x_I-x_k}\right)\left(\prod_{l\neq J}\frac{y-y_l}{y_J-y_l}\right)\omega_{IJ}$$
$$r=(x-[x])/\Delta x\ ,\quad s=(y-[y])/\Delta y \tag{12}$$

Second, the wave action spectral density for wave direction at the ray point is obtained by applying a linear interpolation on wave direction to the action densities estimated above.

$$w(f,\theta_m)=w(f,\theta)=w(f,\theta_k)+\{w(f,\theta_{k+1})-w(f,\theta_k)\}\cdot\left(\frac{\theta-\theta_k}{\theta_{k+1}-\theta_k}\right) \tag{13}$$

(2) Nearshore current model

The equations used in the computation of nearshore currents are vertically-integrated continuity and momentum equations and they are expressed as

$$\frac{\partial \rho D}{\partial t}+\frac{\partial M_x}{\partial x}+\frac{\partial M_y}{\partial y}=0$$

$$\frac{\partial M_x}{\partial t}+\frac{\partial}{\partial x}(UM_x)+\frac{\partial}{\partial y}(UM_y)=-\rho gD\frac{\partial \eta}{\partial x}+\frac{\partial}{\partial x}\left(\overline{L}D\frac{\partial U}{\partial x}\right)+\frac{\partial}{\partial y}\left(\overline{L}D\frac{\partial U}{\partial y}\right)-\left(\frac{\partial S_{xx}}{\partial x}+\frac{\partial S_{xy}}{\partial y}+\tau_{bx}\right) \tag{14}$$

$$\frac{\partial M_y}{\partial t}+\frac{\partial}{\partial x}(VM_x)+\frac{\partial}{\partial y}(VM_y)=-\rho gD\frac{\partial \eta}{\partial y}+\frac{\partial}{\partial x}\left(\overline{L}D\frac{\partial V}{\partial x}\right)+\frac{\partial}{\partial y}\left(\overline{L}D\frac{\partial V}{\partial y}\right)-\left(\frac{\partial S_{yx}}{\partial x}+\frac{\partial S_{yy}}{\partial y}+\tau_{by}\right)$$

where $M_x=\rho DU$ and $M_y=\rho DV$.

The Longuet-Higgins expression (1970) is used as the lateral mixing term

$$\overline{L}=N_c\rho l_x\sqrt{gD} \tag{15}$$

where l_x is the distance measured from the real shoreline and N_c(=0.01) the constant. The bottom friction components (τ_{bx},τ_{by}) are obtained with Nishimura's expression (1983) for regular waves, which approximates the usual quadratic formula with high accuracy

$$\tau_{bx}=\rho c_f\{(w+\frac{\overline{u}^2}{w}\cos^2\theta)U+\frac{\overline{u}^2}{w}\sin\theta\cos\theta V\}$$

$$\tau_{by}=\rho c_f\{\frac{\overline{u}^2}{w}\sin\theta\cos\theta U+(w+\frac{\overline{u}^2}{w}\sin^2\theta)V\} \tag{16}$$

$$w=\left(\sqrt{U^2+V^2+\overline{u}^2+2W\overline{u}}+\sqrt{U^2+V^2+\overline{u}^2-2W\overline{u}}\right)/2$$

$$W=U\cos\theta+V\sin\theta,\quad \overline{u}=2u_{max}/\pi,\quad u_{max}=\pi H/T\sinh kD$$

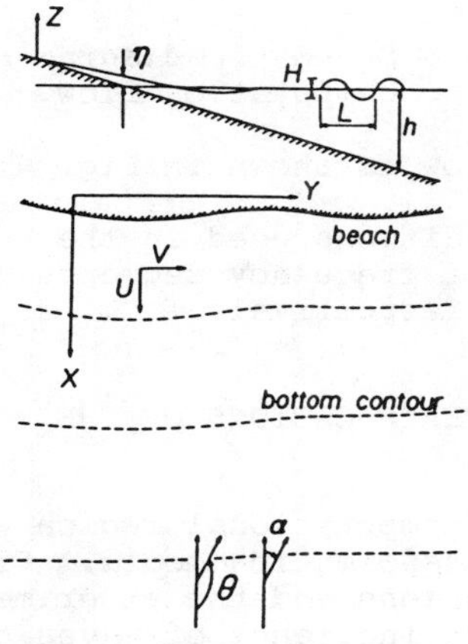

Fig. 2 Coordinate system used in nearshore current model.

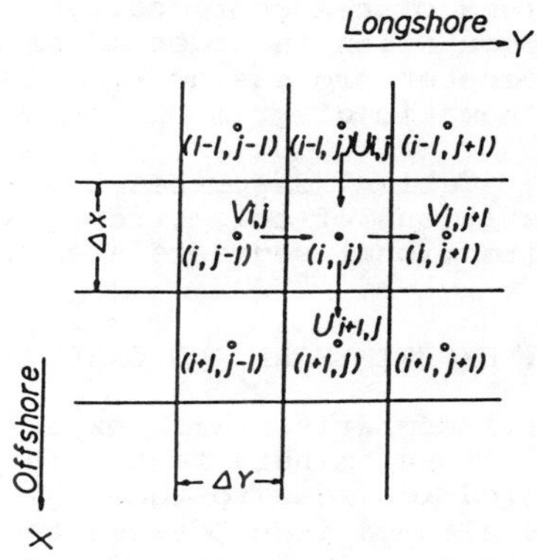

Fig. 3 Configurations of variables in finite difference model.

because their expressions in the case of irregular waves are not known by the author at least, where c_f(=0.01) is the friction factor. The wave-induced velocity is estimated approximately by the small amplitude wave theory using H_{rms}, T_{m01} and $\overline{\theta}$ computed from the wave transformation model.

A finite difference method is used to solve the governing equations of nearshore currents. Fig. 2 is the coordinate system. The x and y axes are taken in the offshore direction, and in the longshore direction respectively and the z axis is taken vertically upward. Finite difference approximation is made by the forward difference in time and by the central difference in space under the configuration of variables shown in Fig. 3. Zero initial condition, fixed offshore and moving onshore boundary conditions, and periodic longshore boundary conditions are imposed respectively.

(3) Flow of computation

The computer program starts by determining the spatial distribution of wave characteristics under the given input conditions of offshore waves and bottom topography, and then nearshore current computation is made using the radiation stress components obtained from the wave transformation model.

A steady state solution is accomplished for 1000 to 2400 iterations of alternating computations of wave transformation and nearshore currents. In this case, computation of wave characteristics is executed every 18 iterations of nearshore current computation in order to save computer processing time. A schematic diagram of computation flow is shown in Fig. 4.

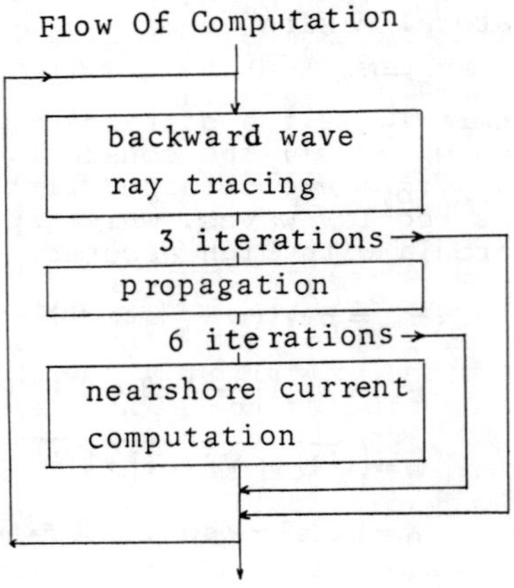

Fig. 4 Schematic diagram of computation flow.

Table 1 describes numerical conditions used in the computations of this study. Numbers of frequency segments and directional segments are 14 and 19 respectively.

3. EXPERIMENTAL AND OBSERVATIONAL VERIFICATIONS OF THE MODEL

(1) Comparison with experiments

A preliminary test in a smaller computational region was tried in order to tune the wave transformation model. Fig.5 is a comparison between the computations and the experiments on wave height variation with normal incidence of waves on a uniformly sloping beach conducted by Battes and Janssen (1978). In order to approximate long-crested incident waves generated in the experiment, the value of 800 is used for

Table 1 Conditions used in the computations.

Model topog-raphy	Num.of grid MxN	Grid dist. (m)	$(H_{1/3})_0$ (m)	$(T_{1/3})_0$ (s)	$(H_{1/3})_M$ (m)	$(T_{1/3})_M$ (s)	α_M (°)	S	Num.of itera-tions
S. B.	24x8	1	0.304	2.15	0.204	1.74	0	800	1000
S. B.	45x8	1	0.304	2.15	0.204	1.74	0	800	2400
S. B.	45x8	1	0.221	2.25	0.147	1.81	0	800	2400
S. B.	23x8	12.5	1.30	4.20	1.00	3.54	30	20	1000
S. B.	23x8	12.5	1.30	4.20	1.00	3.54	30	800	1000
SY. C.	27x20	5	1.41	4.28	1.00	3.51	0	20	1000
ASY. C.	27x20	5	1.41	4.28	1.00	3.51	30	20	1000
SY. V.	27x20	20	2.06	3.95	2.00	3.66	0	20	1200
SY. V.	24x35	20	2.06	3.95	2.00	3.66	30	20	1200
C. B.	24x73	20	2.18	3.95	2.00	3.56	-20	20	1080
Sonu	19x26	5	0.586	5.00	0.300	3.77	0	40	1000

S. B.: straight beach, SY. C.: symmetrical concave topography, ASY. C.: asymmetrical concave topography, SY. V.: symmetrical convex shoreline, C. B. : arbitrarily curved bay, Sonu : bottom topography in Sonu's observation

the energy concentration factor in the directional spectral model and the direction data is taken to range from 162° - 198°. The computation in the surf zone results in slightly greater value than the experiment. Thus, a correction to the saturated frequency spectrum in shallow water being used for the estimation of energy dissipation due to wave breaking is introduced to improve the correspondence with the experiment. The correction factor is determined as

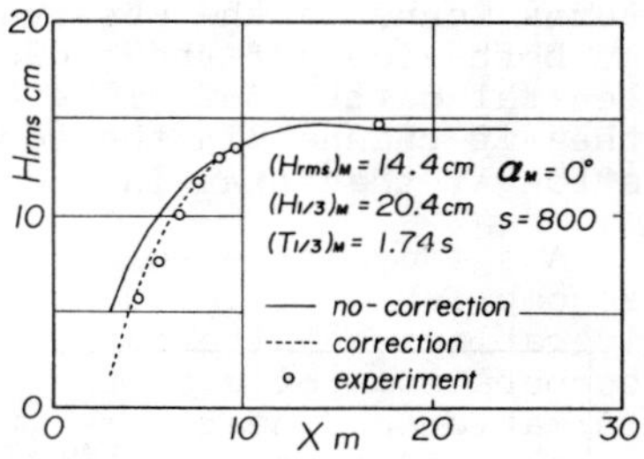

Fig. 5 Tuning of wave transformation model.

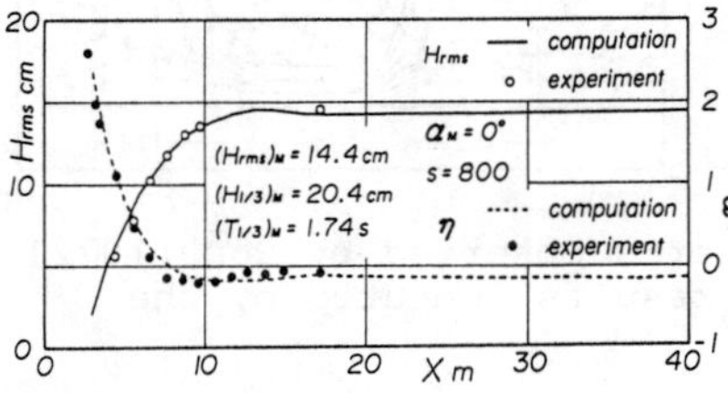

Fig. 6 Comparison of computed and measured wave heights and mean water level variations on a uniformly sloping beach.

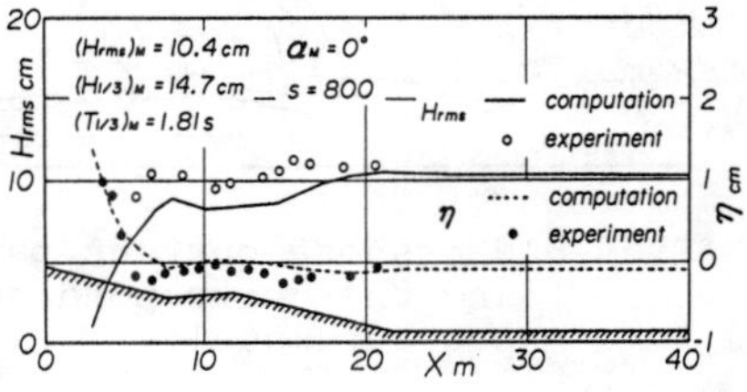

Fig. 7 Comparison of computed and measured wave heights and mean water level variations on a uniformly sloping beach with longshore bar.

$$\beta = \begin{cases} -0.0471 \log^2 \{(H_{rms})_M/D\} - 0.206 \log \{(H_{rms})_M/D\} \\ -0.875 \,;\, (H_{rms})_M/D > 0.484 \\ 1 \qquad ;\, (H_{rms})_M/D \leqq 0.484 \end{cases} \tag{17}$$

Cross-shore distributions of wave height and mean water level variation computed in a larger computational region for the same case are re-compared with the experiment in Fig. 6. As a matter of course, the computation of wave height variation is in good agreement with the experiment. Consequently, the computation of mean water level variation also shows good correspondence with the experiment.

Fig. 7 is another comparison between the computation and the experiment by Battjes and Janssen on a uniformly sloping beach with a longshore bar. The agreement of wave height variation is not always satisfactory compared to that of wave set-up and set-down, because energy dissipation due to wave breaking is not well-formulated in the present model.

(2) Comparison with observation

Fig. 8 illustrates the observed nearshore current patterns by Sonu (1972) and the corresponding results computed by the present model. The model reproduces well the qualitative features of the complicated nearshore current patterns found in the observation, such as the onshore currents on both sides of the region and the rip currents in the central part. But, in a quantitative sense, outflow from the rip channel in the computation seems to be a little stronger than that in the observation.

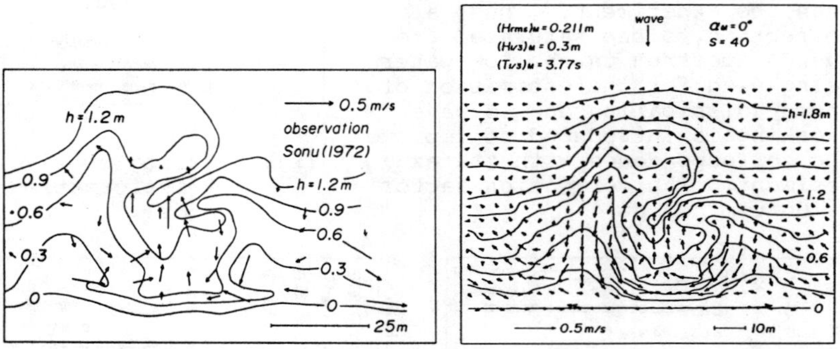

Fig. 8 Nearshore current patterns observed by Sonu(1972) and the corresponding results computed by the present model.

4. COMPUTATIONAL RESULTS AND CONSIDERATIONS

(1) Computations on model topographies

The model is applied to the nearshore current computations on a uniformly sloping beach and on model topographies with two-dimensional variation of bottom contour. The computations are compared with those based on regular wave theory (Yamaguchi et al. 1983 and 1985) and the effect of wave irregularity on wave transformation and nearshore currents is discussed.

(2) Uniformly sloping beach

The cross-shore distributions of wave height, wave direction, mean water level variation and longshore current profile on a plain beach with a slope of 0.016 are drawn in Fig. 9. In the case of irregular waves, the computations using energy concentration factor S= 20 or 800 or neglecting the wave-current interaction are carried out. The effect of wave-current interaction is weak in this case and the effect of the energy concentration factor appears distinctly in longshore current velocity. Compared to the case of regular waves, the variations of wave height and wave direction of irregular waves in the surf zone and in shoaling water is slower. As a result, offshore variations of the resulting radiation stress, wave set-up and set-down and longshore current velocity have a similar tendency, and the magnitudes of mean water level variation and longshore current velocities become smaller.

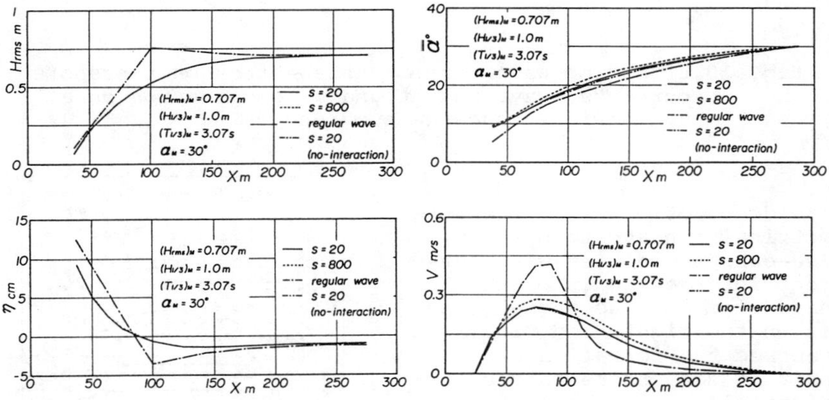

Fig. 9 Cross-shore distributions of wave height, wave direction, mean water level variation and longshore currents on a uniformly sloping beach.

(3) Bottom topography with a concave contour line

The bottom topography model proposed by Noda (1974) is first chosen for the computation of nearshore currents on a two-dimensional topography and it is expressed as

$$h(x,y) = ix[1 + A_0 \exp\{-3(x/20)^{1/3}\} \sin^{10}\{(\pi/\lambda)(y - x\tan\varepsilon)\}] \tag{18}$$

where i is a mean beach slope of 0.025, λ the longshore beach length of 80 m, A_0 the maximum amplitude of the bottom undulation of 20 m and ε the skewness of the bottom undulation of 0° or 30°. The model topographies with ε =0° and with ε =30° are referred to as symmetrical concave topography and asymmetrical concave topography respectively.

Fig. 10 show the nearshore current patterns with normal

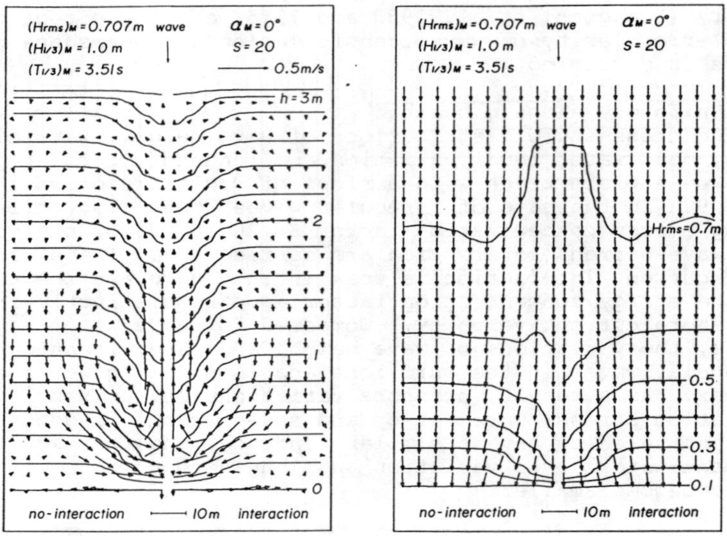

Fig. 10 Effect of wave-current interaction on nearshore currents, wave height and wave direction on a symmetrical concave topography.

incidence of waves on the symmetrical concave topography. The case neglecting wave-current interaction is on the left hand side and the normal case is on the right hand side. We can see the formation of a pair of seaward nearshore current cells and a pair of weak and flat cells near the shoreline, and the neglect of wave-current interaction increases cross-shore current velocity such as in the case of regular waves.

Fig. 11 illustrates the current patterns with oblique incidence of waves on the asymmetrical concave topography. Meandering longshore currents are predominant, but a small and flat circulation cell in the region with concave contour line near the shore is also found.

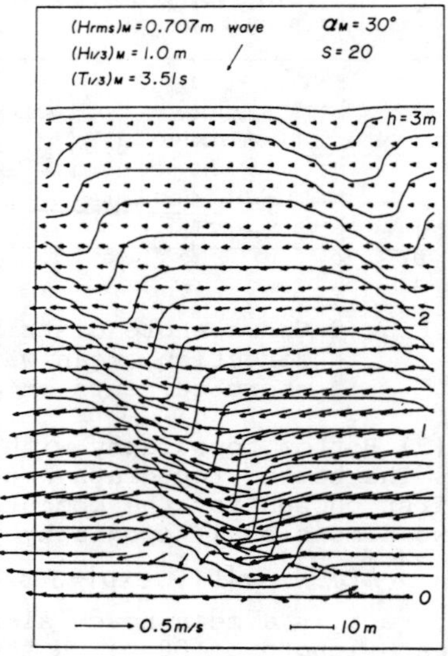

Fig. 11 Nearshore current patterns on asymmetrical concave topography.

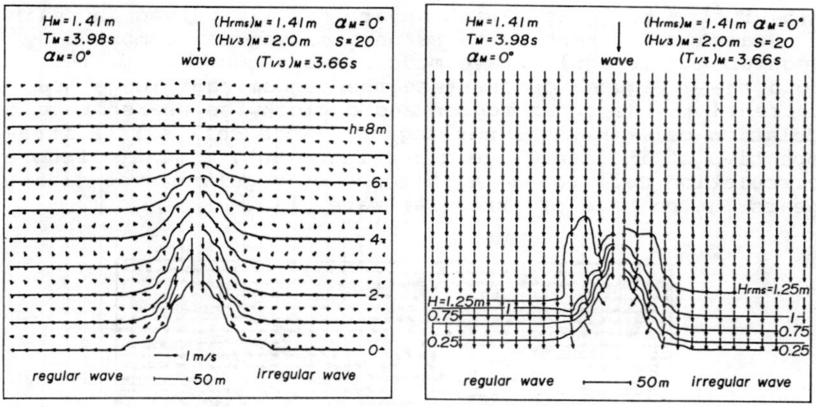

Fig. 12 Effect of wave irregularity on nearshore currents, wave height and wave direction on a model topography with convex shoreline.

(4) Bottom topography with a convex shoreline

Fig. 12 show the spatial distributions of wave height and wave direction, and the corresponding nearshore current patterns on a two-dimensional model topography with convex shoreline. On the left hand side of each figure, the results based on a regular wave theory are given as well. In the computations, root mean wave height and peak period of frequency spectrum and principal wave direction are used as input conditions. In each case, a pair of nearshore current cells with rotational direction contrary to the case of symmetrical concave topography mentioned above are formed and on the center line of the computational region, the onshore current is accelerated toward the shoreline and the current direction turns to be tangential to the contour line. Current patterns in both cases are very similar to each other, but current velocity in the case of irregular waves is smaller by about 30 %.

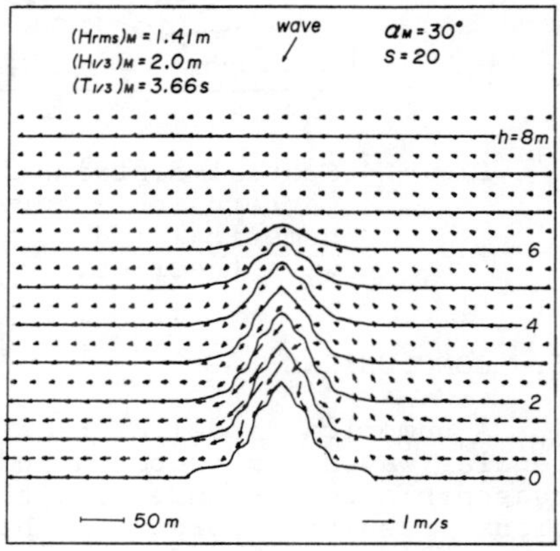

Fig. 13 Nearshore current patterns on a model topography with convex shoreline.

Fig. 13 is the current pattern in the case of obliquely incident waves. Predominant longshore currents develop along the contour line and a weak counterclockwise circulation cell appears near the shoreline in the upwave region.

(5) Bottom topography with an arbitrarily curved shoreline

Nearshore current patterns on a curved bay computed by regular and irregular wave models are shown in Fig. 14. Both these figures demonstrate the predominant longshore currents in the outer region and a clockwise nearshore circulation cell in the inner region. This cell exists irrespective of incident wave direction. But, a weak offshore circulation cell found in the case of regular waves disappears in the case of irregular wave case.

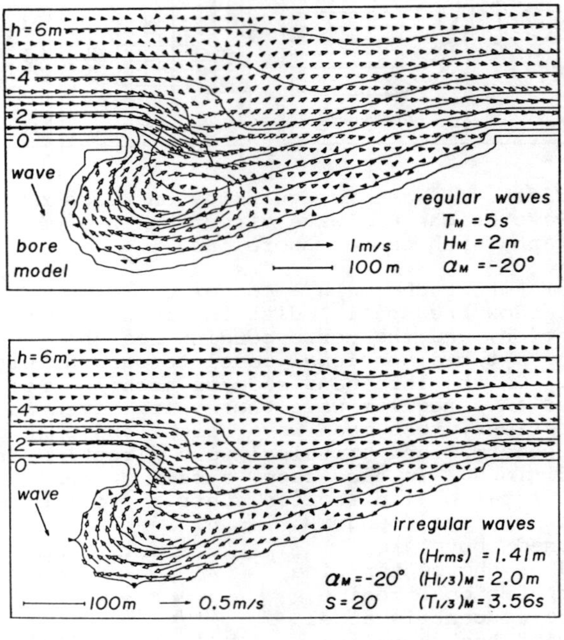

Fig. 14 Spatial distribution of nearshore current patterns on a curved bay computed by the numerical models with regular and irregular wave theories.

5. CONCLUSIONS

A numerical model of nearshore currents induced by irregular waves was proposed and its validity was supported by reasonable correspondence with experimental and observational results. Moreover, the effect of wave irregularity was considered, based on numerical computations. Comparison with the results based on a regular wave model shows that wave irregularity has a smoothing effect on cross-shore distribution of wave height, mean water level variation and longshore currents, but that it does not have much effect on nearshore current patterns.

6. ACKNOWLEDGMENTS

Thanks are due to Messrs. K. Hosono and H. Kawahara, former graduate course students of Ocean Engineering, Ehime University, for their valuable help during the study. A part of this study was accomplished with the support of the Science Research Fund of the Ministry of Education, for which the author expresses his appreciation.

REFERENCES

Battjes, J.A. (1972): Set-up due to irregular waves, Proc. of 13th ICCE, Vol. III, pp. 1993 - 2004.

Battjes, J.A. and J.P.F.M. Janssen (1978): Energy loss and set-up due to breaking of random waves, Proc. of 16th ICCE, Vol. I, pp.569 - 589.

Collins, J.I. (1972): Longshore currents and wave statistics in the surf zone, Rep. No. TC-149-2, Tetra Tech. Inc.

Hubertz, J.M. (1984): Modeling of nearshore driven currents, Proc. of 19th ICCE, Vol. III, pp. 2208 - 2219.

Kitaigorodskii, S.A. et al. (1975): On Phillips' equilibrium range in the spectra of wind generated gravity waves, Jour. Phys. Oceanogr., Vol. 5, pp. 410 - 420.

Longuet-Higgins, M.S. (1970): Longshore currents generated by obliquely incident sea waves Part 2, Jour. Geophys. Res., Vol. 75, No. 33, pp. 6778 - 6801.

Nishimura, H. (1982): Numerical simulation of nearshore circulation, Proc. of 29th Conf. on Coastal Eng. in Japan, pp. 333 - 337 (in Japanese).

Noda, E. (1974): Wave-induced nearshore circulation, Jour. Geophys. Res., Vol. 79, No. 27, pp. 4097 - 4106.

Sonu, C.J. (1972): Field observation of nearshore circulation and meandering currents, Jour. Geophys. Res., Vol. 77, No. 8, pp. 3232 - 3246.

Thornton, E.B. and R.T. Guza (1986): Surf zone longshore currents and random waves: Field data and models, Jour. Geophys. Res., Vol. 16, pp. 1165 - 1179.

Yamaguchi, M. et al. (1983): A numerical solution of nearshore currents taking account of wave-induced mass flux, Proc. of 30th Conf. on Coastal Eng. in Japan, pp. 480 - 484 (in Japanese).

Yamaguchi, M. et al. (1985): A numerical model of nearshore currents applicable to coastal area with an arbitrarily curved shoreline, Proc. of 32nd Conf. on Coastal Eng. in Japan, pp. 248 - 252 (in Japanese).

Yamaguchi, M. (1986): A numerical model of nearshore currents based on a finite amplitude wave theory, Proc. of 20th ICCE, Vol. 1, pp. 849 - 863.

Vicedo Lugo

PART II

COASTAL PROCESSES

Vicedo Lugo

CHAPTER 84

NON-UNIFORM SUSPENDED SEDIMENTS UNDER WAVES

Aronne Armanini (*) Piero Ruol (**)

Abstract

An original mathematical formulation for suspended sediments in a two-dimensional wave boundary layer is presented. The model accounts for non-immediate adaptation of sediments to the hydrodinamic conditions, and allows to include the effect of sorting of the different diameters considered.
The mathematical model is numerically solved through a finite difference scheme.
It is suitable that results compare favourably with experimental data by Staub et alii.

1) Introduction

While in the last decade a number of mathematical formulations regarding the velocity distribution in a wave boundary layer have been proposed (including numerical and analytical solutions [Kesteren and Bakker,1984] [Kajiura,1968]), less effort has been devoted to the problem of predicting the sediment distribution [Justesen and Fredsoe, 1985] [Bakker, 1974]. In particular for aspects concerning bed-boundary conditions and non-uniformity of material.

In the present paper an original formulation of the bed-boundary condition in a two-dimensional wave boundary layer, accounting for non uniform material, is proposed. The problem has been solved through a mathematical model, in which the fluid flow and the sediment concentration equations are solved numerically.

(*) Professor of Hydraulics, Dept.of Eng., University of Trent, Mesiano di Povo, Trento, Italy.

(**) Ph.D.Candidate, Inst. Maritime Constr., University of Padua, Via Ognissanti, 39 Padova, Italy.

2) Equations of motion

The fluid flow equations are usually solved aside from the sediment presence by means of momentum equation.
For a two-dimensional turbulent boundary layer the general equation appears:

$$\frac{\partial u}{\partial t} + u\frac{\partial u}{\partial x} + v\frac{\partial u}{\partial y} = -\frac{1}{\rho}\frac{\partial p}{\partial x} - \frac{\partial}{\partial x}\overline{u'u'} - \frac{\partial}{\partial y}\overline{v'u'} \qquad 1)$$

For a wave boundary layer advection terms, as well as the horizontal diffusion term are neglected; moreover the pressure term is substituted by the inertial term external to the boundary layer; as a consequence, eq.1) reads:

$$\frac{\partial(u-U_0)}{\partial\, t} = \frac{\partial}{\partial y}\left(\varepsilon\frac{\partial u}{\partial y}\right) \qquad 2)$$

where the turbulent diffusion coefficient:

$$-\varepsilon = \frac{\overline{u'v'}}{\partial u/\partial y} \qquad 3)$$

has been introduced.
Upon introduction of the defect velocity $u_d = u - U_0$, eq.2) becomes:

$$\frac{\partial u_d}{\partial t} = \frac{\partial}{\partial y}\left(\varepsilon\frac{\partial u_d}{\partial\, y}\right) \qquad 4)$$

To solve eq. 4), a turbulent closure hypothesis and adequate bed boundary conditions must be assigned.
The following expressions are assumed for the turbulent diffusion coefficient:

$$\varepsilon = l^2\frac{\partial u}{\partial y} \qquad 5)$$

where l is a turbulent mixing length, function of the boundary layer thickness δ, for which the following Escudier formulation [Launder and Spalding,1972] is assumed:

$$l = 0.41\, y \qquad \text{for} \qquad y \leqslant 0.22\,\delta \qquad 6a)$$

$$l = 0.09\,\delta \qquad \text{for} \qquad y > 0.22\,\delta \qquad 6b)$$

The bed boundary conditions are imposed according to the universal law of the wall:

$$\frac{u}{u_*} = \frac{1}{K} \ln(y^+ E) \qquad \text{for} \quad y^+ = \frac{u_* y}{\nu} \geq 11.6 \qquad 7a)$$

$$\frac{u}{u_*} = y^+ \qquad \text{for} \quad y^+ < 11.6 \qquad 7b)$$

the latter being referred to the presence of a viscous sublayer next to the wall, and E being a roughness parameter (E=9 for hydraulically smooth walls).

Eq. 4), 5), and 6) are solved numerically through a Preissman finite difference-scheme [Lyn and Goodwin, 1987]. The first points of the integration mesh were chosen according to either eq. 7a) or 7b). Eq.7b), being a non-linear equation, solving for u* requires an iterative procedure based on Newton-Raphson scheme.

The mathematical model warrants sufficient generality to be applied both under simple waves and under superimposed waves and current conditions; however, only the first case is dealt with in this paper.

The results of the mathematical model have been compared with some experimental data obtained by Jonsson and Carlsen [1976] in an oscillating tunnel.
Comparison between measured and predicted velocities for different phase angles are reported in Fig. 1

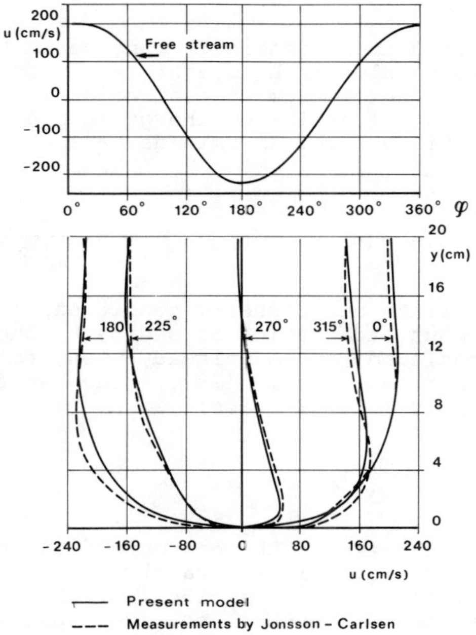

Fig.1. Velocities in the free stream and comparison between measured and predicted ones as a function of height for different phase angles.

3) Equations of sediment transport

3.1) Uniform grain-size hypothesis

The general suspended sediment transport equation in a two-dimensional turbulent boundary layer is written as:

$$\frac{\partial c}{\partial t} + \frac{\partial (u\ c)}{\partial\ x} + \frac{\partial (v-w)\ c}{\partial\ y} = - \frac{\partial\ \overline{u'c'}}{\partial\ x} - \frac{\partial\ \overline{v'c'}}{\partial\ y} \qquad 8)$$

in which w is the fall velocity and c the volume concentration of sediments.
The vertical component of velocity, the horizontal diffusion term and the advection term have been neglected, as usual.
The turbulent diffusion coefficient ε_s for the solid phase was introduced in the form:

$$- \varepsilon_s = \frac{\overline{v'c'}}{\partial c/\partial y} \qquad 9)$$

and was assumed equal to that of the momentum (eq. 3 - 6); therefore eq 8) becomes:

$$\frac{\partial c}{\partial t} = \frac{\partial}{\partial y} (w\ c + \varepsilon \frac{\partial\ c}{\partial\ y}) \qquad 10)$$

The boundary conditions generally considered to define the proper boundary value problem are: zero concentration far from the bed, and bottom concentration (at a reference level y=a), as a function of the grain properties and of the flow conditions (i.e. Θ Shields parameter); The latter appears:

$$c(a) = c_b = f(\theta) \qquad \text{being} \qquad \theta = \frac{u_*^2}{(S-1) g\ D} \qquad 11)$$

A more appropriate bed-boundary condition, accounting for non-instantaneous adaptation of sediment concentration to the flow conditions, has already been proposed by the authors [Armanini and Ruol, 1987]. The net flux, at the reference level y=a is, in fact:

$$D = -(\ w\ c + \varepsilon \frac{\partial c}{\partial y})_a \qquad 12)$$

The first term represents the downward flux and the second one the upward flux (pick-up rate). It is reasonable to assume [Di Silvio and Armanini 1982, 1985] that the first term (deposition) depends on the upstream boundary condition, while the second term (erosion) depends on the flow conditions and is the same as for local equilibrium condition.
Under this assumption the erosion flux (pick-up rate) is

just equal to $(-w\,c_b)$ where c_b is the bottom concentration in a channel flow with the same Shields parameter. In this way the bottom boundary condition becomes:

$$D = -\,w\,[c(a) - c_b] \qquad 13)$$

According to the methodology considered for the solution of the motion equations, eq. 10) is solved numerically through the Preissman finite difference scheme [Lyn and Goodwin, 1987].

In fig.2) the vertical distribution of the concentration under the two bottom boundary conditions is shown. It does appear that the hypothesis of immediate adaptation of the bottom concentration to the equilibrium condition (eq.11), produces a positive gradient of the concentration next to the bed, which is somewhat devoid of physical content, whenever the friction velocity falls under the critical Shields threshold value; this phenomenon does not appear with the suggested boundary condition (eq.13).

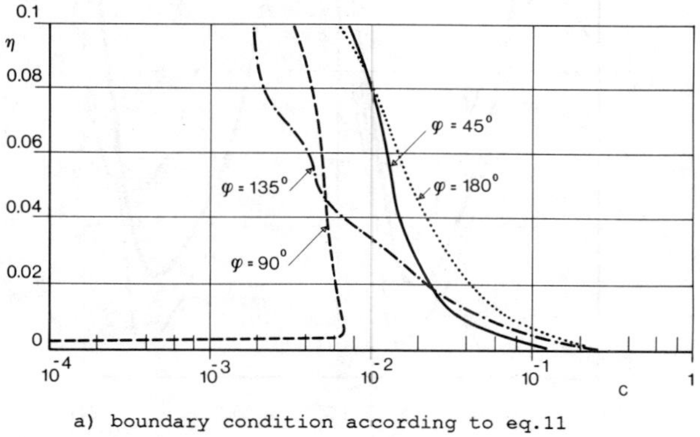

a) boundary condition according to eq.11

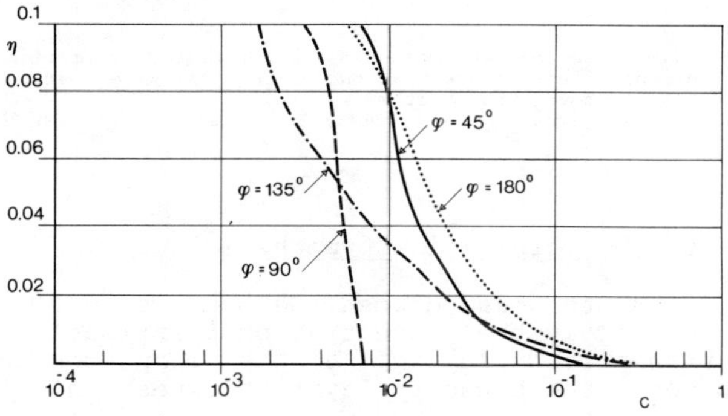

b) boundary condition according to eq.13

Fig.2 Comparison between the two different bed-boundary conditions in determining the concentration c as a function of dimensionless height $\eta = (y-z)/(H-z)$ for mean grain size $D_{50} = 0.19$mm.

In Fig.3 the variation of dimensionless concentration during a wave period is compared with the experimental results of Staub et alii [1984]. In the same figure the simulation carried out by Justesen and Fredsoe [1985] is also reported: the present model seems to yield a better agreement with the experimental data.

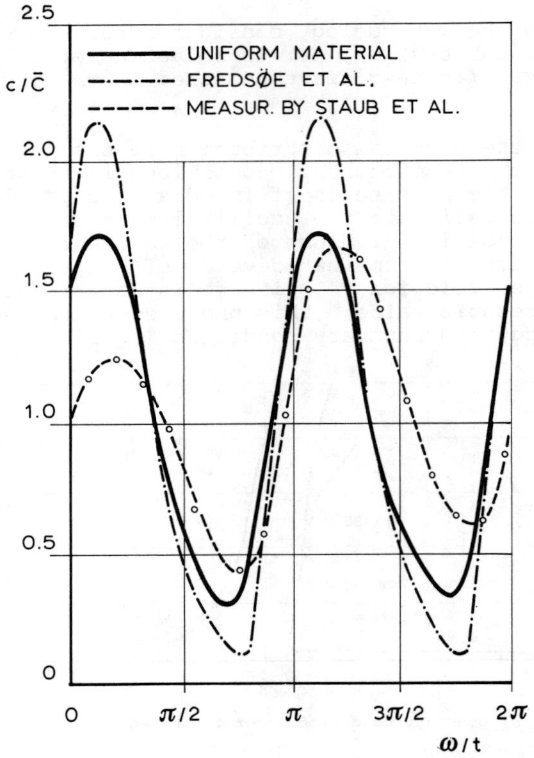

Fig.3 Comparison between measured and predicted variation of dimensionless concentration during one wave period at 1.8cm above the bottom.
(D_{50} = 0.19mm; u_{max} = 1.28m/s; T = 9.1s)

3.2) Non uniform grain size hypothesis

In many cases the mean diameter alone is not able to represent correctly the concentration distribution and the sediment transport. In fig. 4) the mean concentration integrated over the boundary layer thickness and along the period is reported.

From the figure it does appear that, at least in a certain range of diameters (0.12-0.17mm), at a small variation of particle diameter a strong variation of mean concentration is observed.

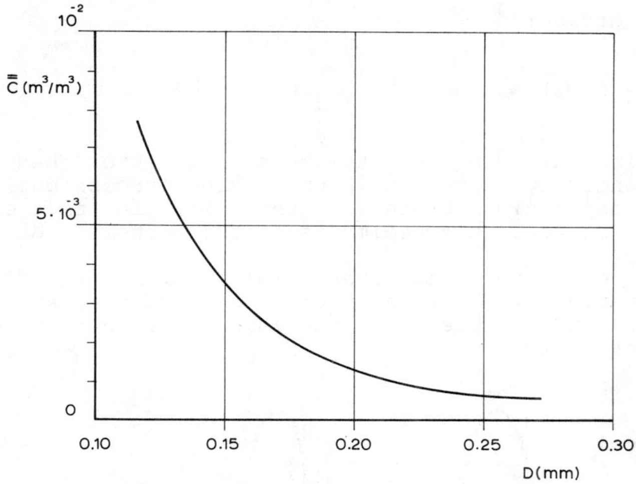

Fig.4 Average concentration (over depth and period) versus grain diameter in the case of uniform material model. (u_{max} = 1.2m/s; T = 9.1s)

In order to overcome these limitations it is then convenient to divide the grain size distribution curve into a discrete number of classes j [Ruol, 1988].
The continuity equation of suspended sediments for each class is then the same as for uniform material (eq.10):

$$\frac{\partial c_j}{\partial t} - w_j \frac{\partial c_j}{\partial y} = \frac{\partial}{\partial y} \left(\varepsilon \frac{\partial c_j}{\partial y} \right) \qquad 14)$$

On the other hand in assigning the bed- boundary conditions it is necessary to account for the mutual interactions amongthe various classes.
The upward flux for each class can be assumed, as for uniform material,the same as in equilibrium condition, but multiplied by the actual concentration for the j-th class on the bed (β_j):

$$\beta_j \, w_j \, c_b(\theta_j) \qquad 15)$$

the downward flux, instead, depends on the effective concentration near the bed. As a consequence, the net flux at the reference level a, becomes:

$$D_j = - w_j \, [\, c(a)_j - \beta_j \, c_b\,(\theta_j)\,] \qquad 16)$$

Eq. 16), which represents the new bed-boundary condition, contains the new variable β_j; one more equation is then necessary to solve the problem. This equation is given by the volume sediment balance in the active layer at the bottom surface:

$$\frac{\partial}{\partial t}(\beta_j \Delta) = -D_j + \beta^*_j \sum_k D_k \qquad 17)$$

in which Δ is the thickness of the bed layer instantaneously involved in the mixing process during the erosion-deposition process, that is, in the exchange process between the sediments of the bottom and of the stream.
It is easy to show that β^*_j equals β_j whenever sedimentation prevails over erosion, and the actual bed concentration above the exchange layer during erosion processes.

Finally it must be:

$$\sum_k \beta_k = 1 \qquad 18)$$

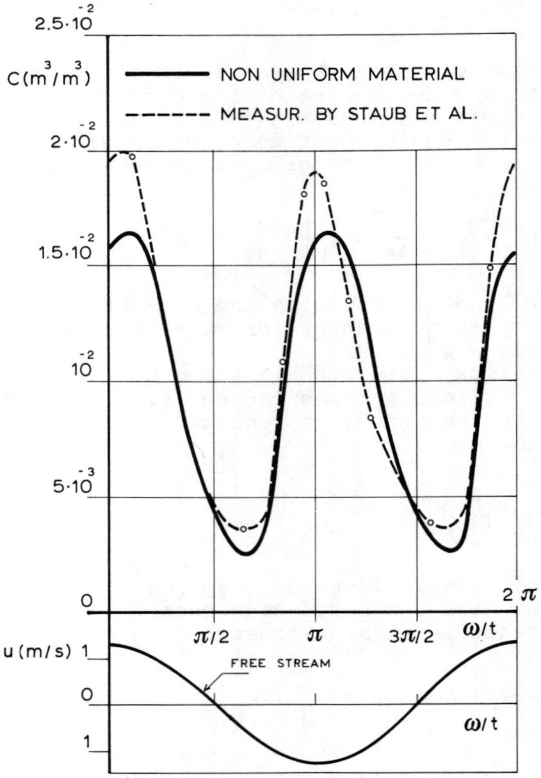

Fig.5 Velocities in the free stream (u_{max} = 1.2m/s; T = 9.1s) and comparison between measured and predicted variation of total concentration during one wave period at y=1.2cm above the bottom. (D_1=0.13mm {20%}; D_2=0.20mm {80%})

In fig.5) the concentration obtained from computations accounting for two grain-size fraction are compared with experimental data by Staub et Al. [1983].

The differences between uniform and non uniform material models are much more pronounced as becomes more evident the non-uniformity of gradation curve. The results shown in the present paper are related to the experimental data by Staub et alii, where the gradation curve are almost uniform. Nevertheless the vertical distribution of concentration is more sensitive to the gradation of the material than the mean concentration.

In fig. 6) a comparison between the results obtained from the two-fraction and from the uniform material models are shown. The vertical distribution of the single diameter concentrations (D_1, D_2), of the total concentration (D_1+D_2), and the vertical distribution of the corresponding uniform material (D_{50}) concentrations are plotted together.

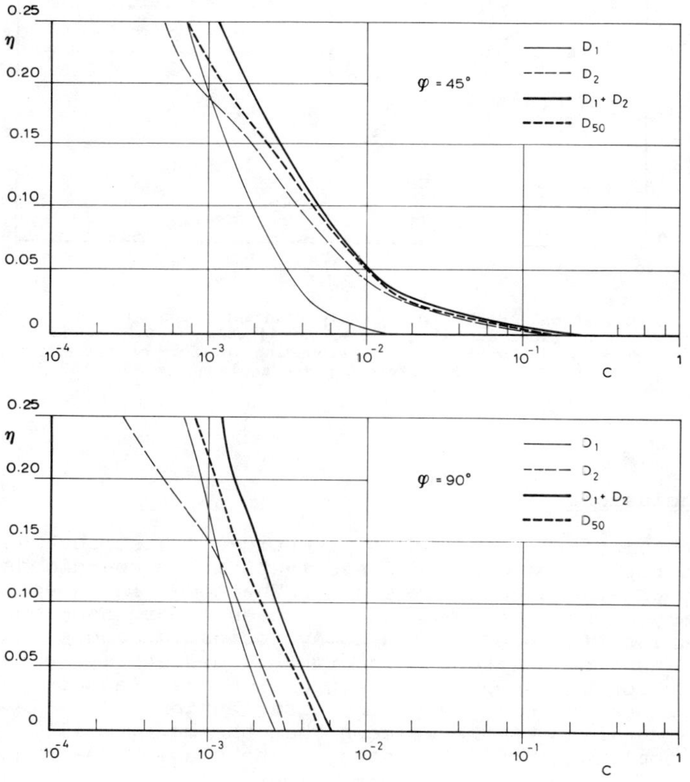

Fig.6a Vertical distribution of the concentration obtained from the two-fraction run (D_1=0.13mm {20%}; D_2=0.20mm {80%}) compared with the corresponding uniform material run (D_{50}=0.19mm) for different phase angles (φ =45°,90°).

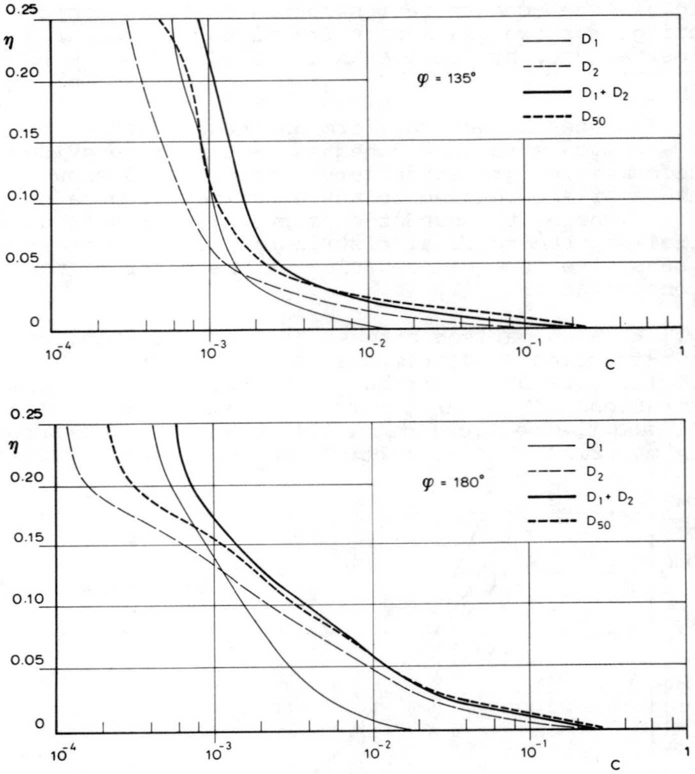

Fig.6b Vertical distribution of the concentration obtained from the two-fraction run (D_1=0.13mm {20%}; D_2=0.20mm {80%}) compared with the corresponding uniform material run (D_{50}=0.19mm) for different phase angles (φ =135°, 180°).

4) Conclusions

A mahtematical model of vertical distribution of concentration of suspended sediments in a two-dimensional wave boundary layer is presented. The model includes a bed boundary condition capable to account for non-immediate adaptation of concentration to hydrodinamic conditions. The model has been extended in order to describe the mutual interaction between different grain size classes in the case of non uniform grain size distribution.
It is opinion of the writers that the model will find a wider applicability if adapted to the sloping beach case by integration over the boundary layer thickness, (in a way already proposed by Armanini and Di Silvio [1982] [1988] for a channel flow). In this case in fact the model should be able to describe the sorting effect induced through the breaking zone.

5) References

[1] Armanini A.,Di Silvio G., 1982:"Sudden morphological modifications along a mountain river simulated by a mathematical model", APD-IAHR, Bandung.

[2] Armanini A.,Di Silvio G., 1985:"Transport of suspended sediments along channels with a trench" XXI Congr. I.A.H.R. vol.3 pp.527-532, Melbourne.

[3] Armanini A.,Di Silvio G., 1988:"A one-dimensional model for the transport of a sediment mixture in non-equilibrium condition" Journ. of Hydraulic Res.Vol.26, n.3 ,pp.275-291.

[4] Armanini A., Ruol P., 1987: "Distribution of suspended sediments in a wave boundary layer" Euromech 215, S.Margherita Ligure pp.80-83.

[5] Bakker W.T., 1974: "Sand concentration in an oscillatory flow". Proc. Coast. Eng. Conf. , ch.66.

[6] Di Silvio G., Armanini A., 1981: "Influence of the upstream boundary conditions on the erosion-deposition process in open channels" XIX Congr. IAHR n.22 New Delhi.

[7] Jonsson G., Carlsen N.A., 1976: "Experimental and theoretical investigation in an oscillatory turbulent boundary layer" Journ. Hydr. Res. 14, n.1 pp.45-59.

[8] Justesen P., Fredsoe J., 1985: " Distribution of turbulence and suspended sediment in the wave boundary layer" Rep. 62, pp.61-67, Inst. Hydrodyn. and Hydr. Eng., Tech. Univ. Denmark.

[9] Kajiura K., 1968: "A model of the bottom boundary layer in water waves". Bull. Earthquake Res. Inst., vol.46.

[10] Kesteren W.G.M., Bakker W.T., 1984: "Near bottom velocities in waves with a current; analytical and numerical computations". Proc. Coast. Eng. Conf. ch.79, pp. 1161-1177.

[11] Launder B., Spalding D.B. 1972: "Mathematical moels of turbulence". Accademic Press. London and New York. 169 p.

[12] Lyn D., Goodwin P., 1987: "Stability of a general Preissman scheme". Journ. of Hydr. Eng., vol.113 n.1.

[13] Ruol P. 1988: "How to represent bed materials in sediment transport models under waves" Proc. Conf. Trasp. sol. nei corsi d'acqua, Trento Italy, 9-10 jun.

[14] Staub C., Svendsen A., Jonsson I., 1983: "Measurements of the instantaneous sediment suspension in oscillatory flow". Prog. Rep. 58, Inst. Hydrodyn. Tech. Univ. Denmark.

[15] Staub C., Jonsson I., Svendsen A., 1984: "Variation of sediment suspension in oscillatory flow" Proc. Coast. Eng. Conf., ch.155 pp. 2310-2321.

CHAPTER 85

INITIAL MOTION IN COMBINED WAVE AND CURRENT FLOWS

J.S. Lee-Young* and J.F.A. Sleath**

Abstract

Measurements are presented of the conditions for the initial motion of sediment under combined steady and oscillatory flow. The measurements were made in a steady flow flume with an oscillating tray set into its bed. The direction of oscillation of the tray was at right angles to the axis of the steady flow flume. Four different grades of sand were tested.

It is found that the critical condition for the initiation of motion is reasonably represented by a critical value of the vector sum of the component shear stresses assuming no nonlinear interaction between the steady and oscillatory flows. The resultant bed shear stress was also calculated with the aid of several combined wave-current models. The results of these various approaches are compared with Shields curve.

Introduction

The conditions under which sediment first begins to move due to a flow of fluid are of great importance in many engineering problems. There have consequently been numerous studies of the initial motion condition in both pure steady flow and in pure oscillatory flow (see, for example, Sleath 1984). On the other hand, there have been surprisingly few investigations of the initial motion condition for combined steady and oscillatory flow. Hammond and Collins (1979) studied the case where the steady and oscillatory flows are collinear. More recently, Katori et al (1984) have made measurements of steady and oscillatory flows at right angles to each other. The aim of the present paper is to study the second of these two cases.

Experimental Layout

The experimental apparatus is shown schematically in Fig. 1. The sediment is contained in a tray set into the bed of the flume. The tray oscillates (see Sleath, 1976,

----

*Research student, Cambridge University Engineering Department, Trumpington Street, Cambridge CB2 1PZ England
**Lecturer, Cambridge University Engineering Department, Trumpington Street, Cambridge CB2 1PZ England

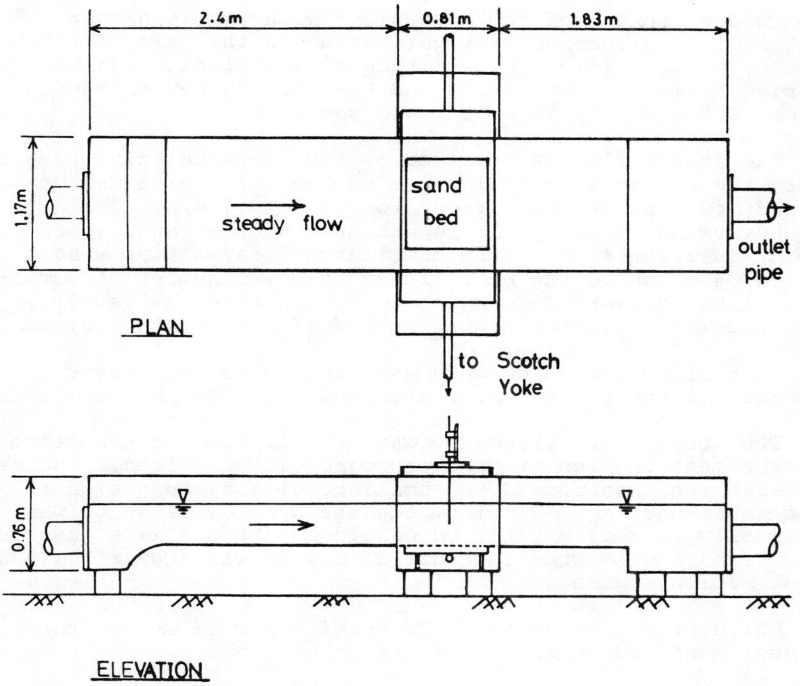

Fig. 1. Schematic drawing of the combined steady and oscillatory flow flume

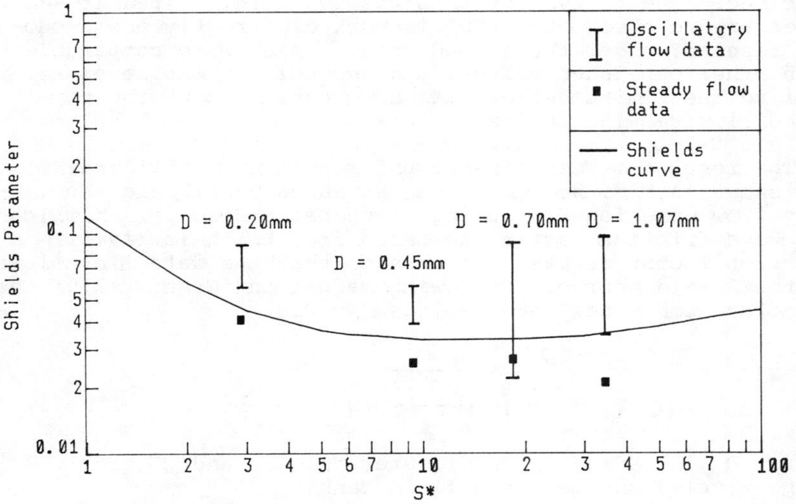

Fig. 2. Results of initial motion experiments under steady flow alone and oscillatory flow alone

for a description of the mechanism) with simple harmonic motion in a direction at right angles to the axis of the steady flow flume. The amplitude of oscillation can be varied from 0 to 0.19 m and the working period ranges from 0.5 secs up to about 6.0 secs.

The steady flow is provided by two re-circulating pumps. With the aid of a control valve on the inlet pipe the mean steady current can be varied from 0 to 0.4 m/s. The steady velocity was monitored throughout the tests at a point just upstream of the oscillating tray and at a height of 0.04 m above the bed. In addition, a number of vertical traverses were made which indicated that the velocity profile was logarithmic up to that height at this position.

Four different sands were used in the tests. Their median diameters were 0.20 mm, 0.45 mm, 0.70 mm and 1.07 mm.

The normal test procedure was as follows. At the start of the test the bed of sand was carefully levelled. The bed of sand was then caused to oscillate at a fixed period T and amplitude A_b. Finally, the steady flow velocity was gradually increased until threshold conditions were observed. The criterion adopted for initial motion was Kramer's (1935) weak motion regime.

The depth of water was held equal to 0.17 m for the finest sand and 0.12 m for the other sands.

Experimental Results

As well as performing initial motion experiments under combined flow conditions, initial motion under, (a), pure steady flow and, (b), pure oscillatory flow was also examined. The reason for this was two-fold. First it was necessary to check that the rig worked correctly and produced results, under these conditions, which were comparable with other published values, and secondly it was necessary to find the limits between which we could expect the combined flow results to lie.

The results of experiments under conditions (a) and (b) are shown in Fig. 2. The shear stresses under pure oscillatory flow were found assuming a quadratic drag law in which the wave friction factor was taken from Kamphuis (1975) friction factor curves. It is seen that the data lies close to the Shields curve. Following Madsen and Grant (1976) the Reynolds number has been replaced in Fig. 2 by

$$S_* = \frac{D}{4\nu}\sqrt{\frac{(\rho_s - \rho)}{\rho}\, g D} \qquad (1)$$

where D is median grain diameter and ρ_s and ρ are, respectively, sediment and fluid density.

The raw results for the critical conditions under which sediment just begins to move in combined steady and oscillatory flow are shown in Figs. 3 - 6. In these figures u_w is the amplitude of the velocity of the oscillating tray and u_c is the depth-averaged steady current just upstream of the test section. In order to guide the eye, curves of the form

$$\left(\frac{u_w}{u_{wo}}\right)^n + \left(\frac{u_c}{u_{co}}\right)^n = 1 \qquad (2)$$

have been drawn through the experimental points. In this equation u_{wo} and u_{co} are the critical values of u_w when the steady current is zero and u_c when the oscillatory velocity is zero, respectively. Values of n between 2 and 4 appear to give the best agreement with the experimental results.

Inertia Effects

One problem with a study of sediment entrainment in an oscillatory tray apparatus is the inertia force acting on the grains of sediment. If we consider spherical particles of diameter D and dry density ρ_s there is an additional force (i.e. in addition to the forces which the particle would experience if it were on a stationary bed in an oscillating flow) equal to

$$(\rho_s - \rho)\frac{\pi D^3}{6}\frac{\partial u}{\partial t}$$

where u is the velocity of the tray.

For the finest sand used in these experiments the additional force due to inertia is negligible compared with the other forces acting on the grains. However, for the coarser sands this is not always the case, particularly at the lowest periods of oscillation. It would be possible to make a correction to the experimental results to allow for this additional force. However, this requires a knowledge of the phase relationships of the forces acting on the particle, the horizontal and vertical directions in which they act and, finally, some estimate of the shape of the grains. When the required correction is large this procedure is not very reliable. Consequently, in the comparison of different models outlined below only those experimental results for which the magnitude of the additional force is less than 30 % of the magnitude of the force which the grains would experience on a stationary bed have been retained. For the purposes of deciding which results to retain the force for a stationary bed has been taken as the vector sum of the forces calculated for oscillatory flow and steady flow separately. The same set of experimental results is used in each of the Figures 8 - 11.

Although the magnitude of the additional force in the retained experimental results may be as much as 30 % of the magnitude of the force for a stationary bed the actual error in the resultant force is generally much less than that.

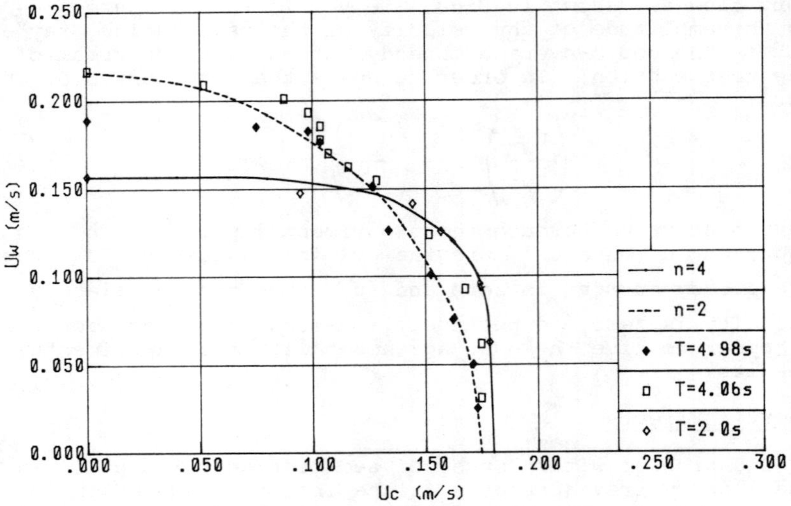

Fig. 3. Critical velocities for D = 0.20 mm

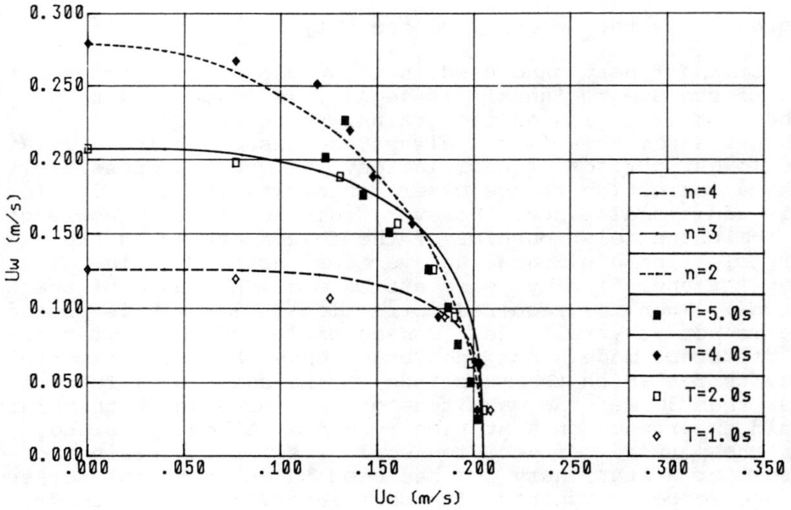

Fig. 4. Critical velocities for D = 0.45 mm

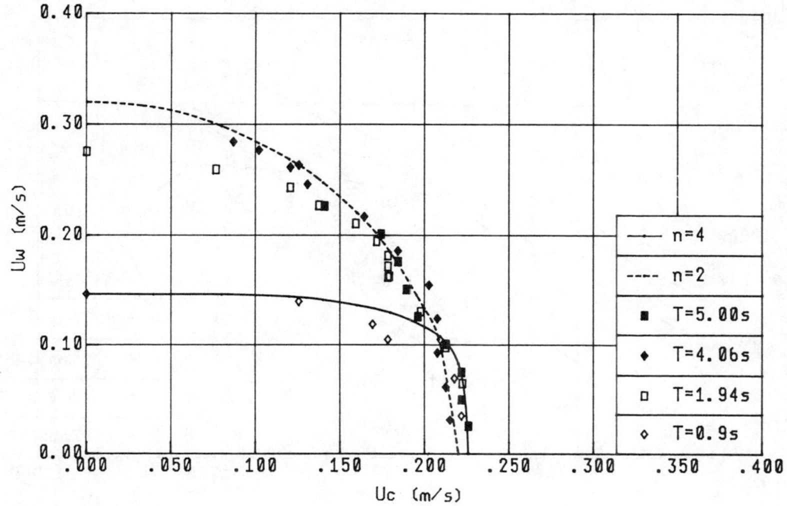

Fig. 5. Critical velocities for D = 0.70 mm

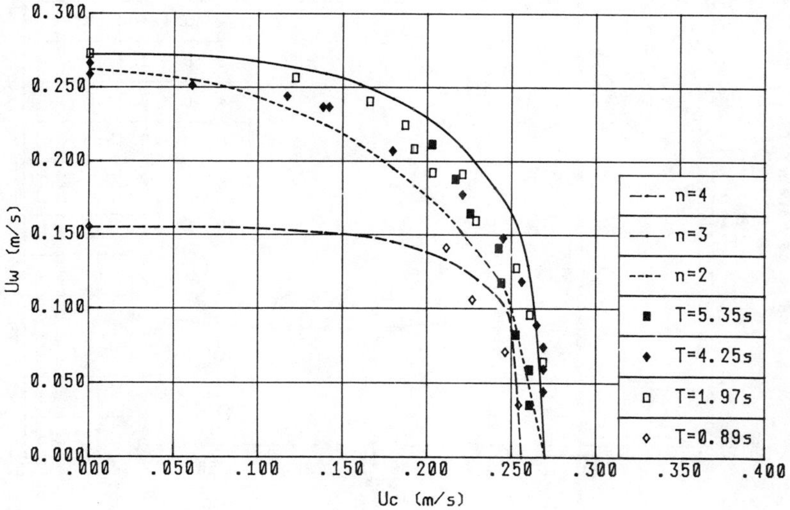

Fig. 6. Critical velocities for D = 1.07 mm

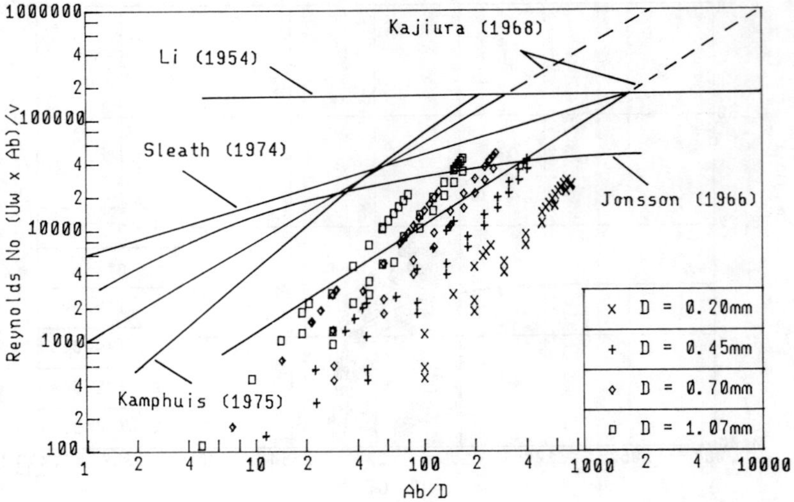

Fig. 7. The oscillatory flow parameters compared with various relationships for transition of an oscillatory flow boundary layer over a flat bed of sand

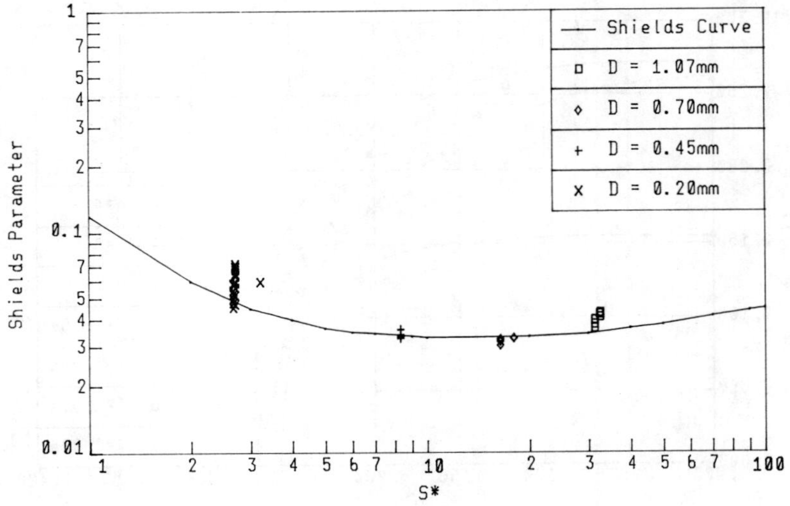

Fig. 8. Initial motion data assuming no non-linear interactions

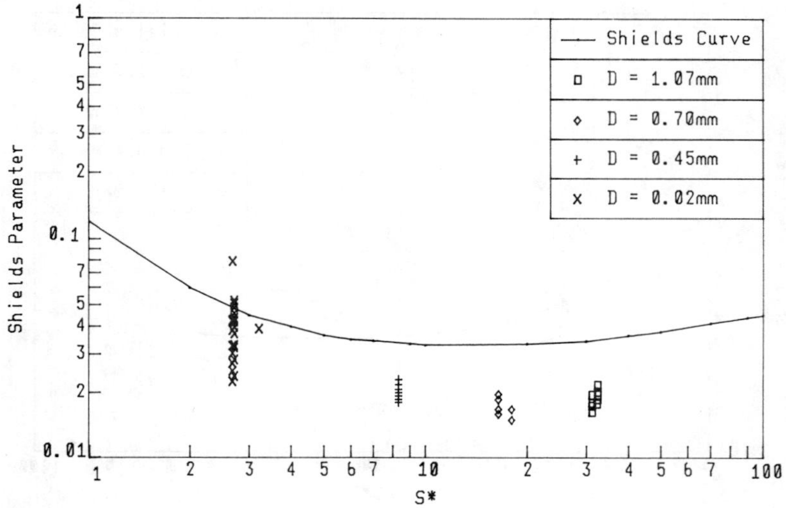

Fig. 9. Initial motion data as analysed by Bijker's (1967) model

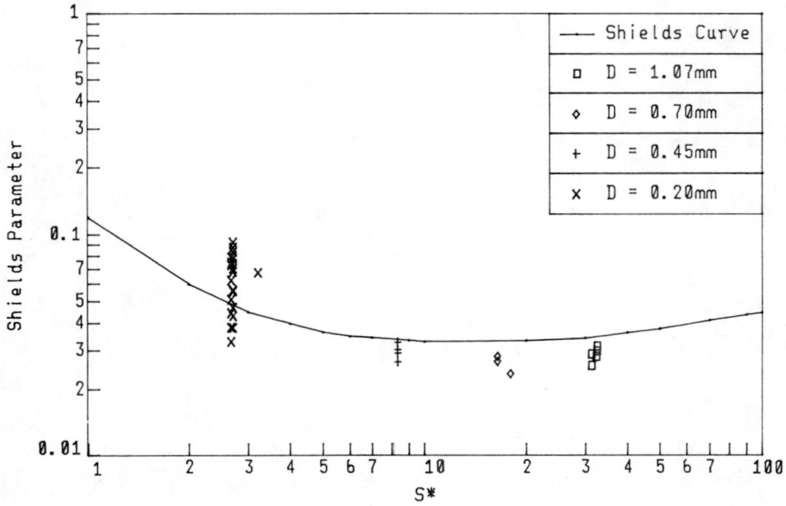

Fig. 10. Initial motion data as analysed by Grant and Madsen's (1979) model

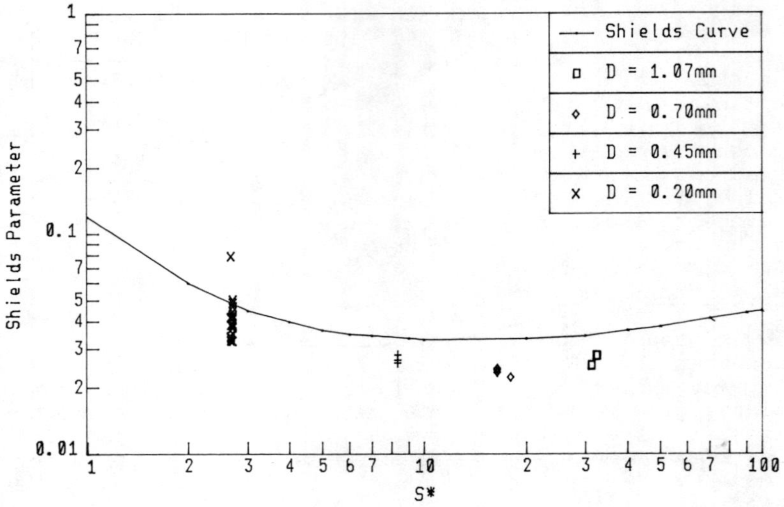

Fig. 11. Initial motion data as analysed by Christoffersen and Jonsson's (1985) model

This is because the additional force is in quadrature with the velocity of the tray and is consequently nearly 90° out of phase with the resultant force on the grains. Also, when the steady current is large this additional force is at a significant angle to the direction of the resultant force which further reduces the overall error. In order to test this conclusion, the comparisons shown in Figs. 8 - 11 were carried out both with a correction for the additional force and also without one. There appeared to be no significant difference between the two sets of results. Consequently, only the uncorrected results are shown in Figs. 8 - 11.

Comparison of Different Models for Combined Shear Stress

It is probably reasonable to assume as a first approximation that the initial motion condition corresponds to a critical value of the bed shear stress. To conform with usual practice we use the term shear stress to mean the horizontal force on the grains per unit area of bed. Lift forces are not entirely negligible but will be ignored for the moment. The problem is how to evaluate the shear stress in a combined flow.

The simplest assumption which we can make is that the resultant shear stress is the vector sum of the shear stress τ_w for the oscillatory flow assuming no effect of the steady current, and that for the steady flow τ_c, assuming no effect of the oscillation, i.e.

$$\tau = (\tau_w^{\,2} + \tau_c^{\,2})^{\frac{1}{2}} \qquad (3)$$

We note, in passing, that shear stress proportional to velocity squared would lead to $n = 4$ in Eq. (2) whereas $n = 2$ would correspond to shear stress directly proportional to velocity.

Eq. (3) relies on the assumption that there is no non-linear interaction between the steady and oscillatory components of the flow. This assumption is more likely to be correct if the flow is laminar rather than turbulent. Fig. 7 shows how the critical values of $u_w A_b/\nu$ and A_b/D for the present experiments compare with the criteria for transition to turbulence proposed for oscillatory flow by various investigators. The fact that so many of the points lie below or only slightly above Kajiura's (1968) curve for the initiation of turbulence is an indication that turbulence generation by the oscillatory flow would have been negligible in most of the tests. On the other hand, the steady flow Reynolds numbers are well above the limit for transition in almost all cases. We conclude that the flow would have been turbulent in almost all of the tests but not strongly so in the immediate vicinity of the bed.

In view of this conclusion about turbulence, and consequently of the improbability of there being no non-linear interaction between the steady and oscillatory components of the flow, it is rather surprising to find in Fig. 8 that the experimental values of Shields parameter calculated with the aid of Eq. (3) all lie close to Shields curve. In Eq. (3), τ_w was calculated from Kamphuis (1975) curves for friction factor and τ_c from the assumption of a logarithmic velocity profile up to the reference point just upstream of the oscillatory tray.

The other methods used to calculate the resultant shear stress are those proposed by Bijker (1967), Grant and Madsen (1979) and Christoffersen and Jonsson (1985). The corresponding Shields diagram plots of the experimental results are shown in Figs. 9, 10, 11. We see that the closest agreement between the experimental results and Shields curve is provided by the methods of Grant and Madsen and of Christoffersen and Jonsson with, perhaps, slightly less scatter using the latter method. Surprisingly, the scatter with all three of these methods is greater than that obtained from Eq. (3).

Conclusions

The results of this experimental study suggest that Shields curve may be used to predict initial motion of sediment in combined steady and oscillatory flow. A simple model, in which the resultant shear stress is calculated on; the assumption that there is no non-linear interaction between the steady and oscillatory flows, gave the closest agreement between Shields curve and the measurements but good agreement was also shown by the models of Grant and Madsen (1979) and Christoffersen and Jonsson (1985).

Appendix

References

Bijker E.W. 1966 The increase of bed shear in a current due to wave motion. Proc. 10th Conf. Coastal Engrg. Tokyo, pp.746-765

Christoffersen J.B. and Jonsson I.G. 1985 Bed friction and dissipation in a combined current and wave motion. Ocean Engrg., 12, No.5, pp.387-423

Grant W.D. and Madsen O.S. 1979 Combined wave and current interaction with a rough bottom. J. Geophys. Res., 84, C4, pp.1797-1808

Hammond T.M. and Collins M.B. 1979 On the threshold of transport of sand-sized sediment under the combined influence of unidirectional and oscillatory flow. Sedimentology (1979), 26, pp.795-812

Jonsson I.G. 1966 Wave boundary layers and driction factors. Proc. 10th Conf. Coastal Engrg. Tokyo, pp.127-148

Kajiura K. 1968 A model of the bottom boundary layer in water waves. Bull. Earthquake Res. Inst. 46, pp.75-123

Kamphuis J.W. 1975 Friction factors under oscillatory waves. Proc. ASCE J. Waterways Harbors Coastal Engrg. Div. 101 (WW2), pp.135-144

Katori S., Sakakiyama T. and Watanabe A. 1984 Measurement of sand transport in a cross unidirectional-oscillatory flow tank. Coastal Engrg. in Japan, 27, pp.193-203

Kramer H. 1935 Sound mixtures and sound movement in fluvial models. Transactions ASCE 100, Paper No. 1909, pp.798-878

Li H. 1954 Stability of oscillatory laminar flow along a wall. BEB Tech. Memo 47

Madsen O.S. and Grant W.D. 1976 Quantitative description of sediment transport by waves. Proc. 15th Conf. Coastal Engrg. II, pp.1093-1112

Sleath J.F.A. 1974 Velocities above rough bed in oscillatory flow. J. Waterways Harbors and Coastal Engrg. Div. ASCE 100 (WW4), pp.287-304.

Sleath J.F.A. 1976 On rolling-grain ripples. J. Hydraulic Research 14 (1), pp.69-81

Sleath J.F.A. 1984 Sea Bed Mechanics. John Wiley & Sons, New York.

CHAPTER 86

Study of Shelf Waves vs Sand drift in NW coast of Taiwan

by

Ho-Shong Hou*, Ph.D., P.E. Professor

* Head, Transportation Engineering Dept., Institute of Transportation, MOC. Taipei, R.O.C.

* Chief, Project office of Transportation Engineering Preliminary Planning, IOT, MOC, R.O.C.

ABSTRACT

For the planning and design of harbor and coastal Engineering, most important affected factors are waves and littoral drift.

This research deals with analysis of the measured waves and the budget of the beach material whether it is deposited or scoured in the North-western coast of Taiwan; and the relationship between wave energy and littoral transport rate. For processing this research the field Survey wave measurement, sand sampling, and echo sounding is necessitated. The objective of this research is to establish the model of waves and littoral transport budget. Therefore, the results of this research are planned to submit to the planning and design of harbor and coastal engineering works of the Taiwan west sandy coast.

The main contains of this research are consisted of

(1) Analysis, observations and investigations of the field oceanographic data, winds, waves, coastal geomorphology and littoral drift of the North-western coast of Taiwan.
(2) Analysis of the shelf waves from the actual measured wave records.
(3) Research of the wave decaying process of the continental shelf waves.
(4) Statistical research of the measured waves and establishment of the practical model of the relationship between the shelf wave energy and the littoral transport rate in the north-west coast of Taiwan.

(5) Plotting time history of significant wave heights and significant wave periods in the Taiwan Strait.

I. Introduction

Based on the wave pattern, the geographical location and the disposition of rivers, the littoral drift moves predominantly from NE to SW direction in section I as shown in Fig. 1. Seven rivers of rapid stream bring tremendous amount of sediments from the high mountain to the nearshore of this section in typhoon season (i.e. from June to September). But for the winter monsoon season, i.e. from October to the next April, the waves induced by NE monsoons migrate littoral drift from north toward south.

Applying the energy approach for unidirectional steady flow derived by Bagnold (1963), the theoretical relationship between the littoral immersed weight transport rate and the alongshore breaking wave energy is found out. It reveals that the relationship is not strictly linear, i.e. the larger part of the alongshore breaking wave energy is supplied for transporting the sediment as the former increases. But for a coast having a steady oceanographical condition, the relationship could be considered as linear relation since the alongshore breaking wave energy is not varying very much.

By using the shelf wave records and the littoral drift quantity obtained from long-term echo-sounding map, the relationship between alongshore breaking wave energy and littoral immersed weight transport rate is found out.

2. Information of Oceanographical Data

For the movable-bed investigation of Lin-Kou Power Plant (Located in the NW coast of Taiwan), the following field investigation is undertaken.

1) Wind & Waves: The monsoon of this area starts from September to April of the next year, the prevailing wind direction is NE-ENE, and the wave direction of the nearshore is NNE, the maximum wave height of strong monsoon is 4M, while the wave period is 9.0 sec. The maximum significant wave height of hurricane attacking $H_{1/3}$= 4.75M, the corresponding period $T_{1/3}$= 12 sec.

2) River sediment discharge: The sediment which affects obviously on the Lin-Kou Power Plant is from the Tansui River, the river sediment transport rate is 6.095 x 10^6 MT/year.

3) Littoral transport: Based on the sounding data of Lin-Kou coastal area, including survey maps of 1971, 1975, 1977, 1978 and 1983; and the sand sampling data, it is recognized that sediment transport toward south.

a) Coastal morphology: The mean beach slope of nearshore none is 1/66, the sea-bed slope of depths from -15M to -20M is 1/300, is flat, the profile near the intake structure is steeper than those of other profiles. Therefore, it is very clear that the effect of wave action is severe, the littoral transport is obvious as Fig. 2.2 shown.

b) Sediment particle: Based on the field measurement, the coastal area of this region spreads over the cobble stone, the distribution of sand particle after Sieve analysis is shown as Fig. 2. From the figure, the sediments of the breaking zone and inner shore zone of the power plant area d_{50}= 0.3-0.4 m/m, while contour of -5M to -10M, d_{50} is 0.2 m/m, therefore the sand particle of the upstream of the plant site is bigger than that of the downstream site, it is recognized that the littoral transport direction is from north to south, as shown in Fig. 2.3.

c) Littoral transport rate: Based on the echo sounding data, using mesh method to calculate the littoral transport rate of the deposit sand around the coastal area of the plant site. It is 37.5 x 10^4 M^3/ year, and could be found out from Fig. 2.4.

d) Tide and tidal current: The tide is belong to semidiurnal tide. The mean high water level of the spring tide is 3.1 M, while the mean low water level of the spring tide is -0.3M. Tidal current is measured about 0.75 m/sec or so.

3. Statistical Analysis of Self Waves

Waves in Taiwan Strait, i.e. self waves, are to be divided into three patterns, i.e., waves in winter monsoon, waves in hurricane and waves generated by convective wind in the summer. Statistical characteristics of significant waves extracted from the meteorological and wave measuring records along the continental self for 4 years are shown in Fig. 3.1 (Tang, 1986).

Wave records obtained from cassette type wave meter are continuously measuring the whole year in the NW coast located in the nearshore area of Tou-Yuan Fig. 3.2, Fig. 3.3, Fig. 3.4, Show that relationship between wave period, wave height vs percent frequency in January (winter), in August (summer) and the whole year's record (all available data). While Fig. 3.5 shows that correlation between wave height and wave period in January and in September. Their range for wave height is from 20 cm to 4m, wave period is from 5 sec to 15 sec or more. The waves in the NW coast are slightly stronger than those of the mid-west coast of Taiwan.

4. The calculation of the alongshore breaking wave energy:

The wave is affected by the effects of shoaling, refraction, bottom friction and percolation. By neglecting the effects of bottom friction and percolation as wave is propagating toward the surf zone, there the breaking wave height is equal to

$$H = H_o K_r K_s \tag{4.1}$$

where K_r and K_s are refraction coefficient and shoaling coefficient respectively. The mathematical representation is as follows

$$K_r = (B_o/B)^{\frac{1}{2}} \tag{4.2}$$

$$K_s = (C_{go}/C_g)^{\frac{1}{2}} \tag{4.3}$$

where B is the separation of the wave rays and prefix "o" represents the characters of deep water. As the bottom slope is smaller than 1/10, then K_r and K_s could be calculated from the four equations derived by Chao, Y.Y. (1970) as follows

$$C^2 = \left(\frac{g}{k}\right) \text{tank } kh \tag{4.4}$$

$$\frac{d}{ds} = \frac{1}{c}\left(\sin\theta\frac{\partial c}{\partial x} - \cos\theta\frac{\partial c}{\partial y}\right) = -\frac{1}{c}\frac{dc}{dB} \tag{4.5}$$

$$H_o^2 (C_g)_o B_o = H^2 \cdot C_g \cdot B = \text{constant} \tag{4.6}$$

$$\frac{d^2 B}{ds^2} - P^{(1)}\frac{dB}{ds} + P^{(2)} B = 0 \tag{4.7}$$

$$P^{(1)} = \frac{1}{c}\left(\cos\theta\frac{\partial c}{\partial x} + \sin\theta\frac{\partial c}{\partial y}\right) \tag{4.8}$$

$$P^{(2)} = \frac{1}{c}\left(\sin^2\theta\frac{\partial^2 c}{\partial x^2} - 2\sin\theta\cos\theta\frac{\partial^2 c}{\partial x\partial y} + \cos^2\theta\frac{\partial^2 c}{\partial y^2}\right) \tag{4.9}$$

where D is water depth, θ is the angle between X axis and wave direction, S is the distance along wave ray and C is the phase velocity. Use numerical method to get

$$D_{n+1} = D_0 + \left(\frac{\partial D}{\partial x}\right)_n dx + \left(\frac{\partial D}{\partial y}\right)_n dy + \frac{1}{2}\left(\frac{\partial^2 D}{\partial x \partial y}\right) dxdx$$

$$+ \left(\frac{\partial^2 D}{\partial x \partial y}\right)_n dxdy + \frac{1}{2}\left(\frac{\partial^2 D}{\partial x \partial y}\right)_n dydy \qquad (4.1)$$

$$P_{n+1} = [(4-2P_n^{(2)} \Delta s^2)/(2-P_n^{(1)} \Delta s^2)] B - [(2+P_n^{(1)} \Delta s)$$

$$/(2-P_n^{(1)} \Delta s)] B_{n-1} \qquad (4.11)$$

The subscript "n" represents the value of the nth calculation, as shown in Fig. 4.1. where X axis is taken parallel to the shoreline. Developing these numerical calculations to get "THE WAVE CHARACTER COMPUTING PROGRAM" (Hou, 1980), the wave characters such as K_r, K_s, H, C_g and θ could be found out at any water depth h=D. Since $\theta = \frac{\pi}{2} - \alpha$ as shown in Fig. 41. then

$$(P_\ell)_b = \frac{1}{8}\rho g H_b^2 (c_g)_b \cos\theta_b \sin\theta_b \qquad (4.12)$$

is the alongshore breaking wave energy. For the self wave energy it is obtained as $(P_\ell)_b = 2.25 \times 10^{10}$ $kgm^2/sec^2/m\text{-}yr$.

5. Relationship between shelf waves and littoral drift in the NW coast of Taiwan.

From the echo sounding data, the littoral transport rate (I_ℓ) of the NW coast is calculated as $37.5 \times 10^4 m^3/yr$.

I_ℓ and $(P_\ell)_b$ are in units of cgs system.

But for the Taiwan Strait where the climate is so steady that the alongshore breaking wave energy fluctuate slightly, the relationship between I_ℓ and $(P_\ell)_b$ could be expressed as

$$I_\ell = K(P_\ell)_b \qquad (5.1)$$

Where K is constant and is equal to 0.14 for the north western coast of Taiwan.

The NW coast has the relation of $I_\ell = 0.14 (P_\ell)_b$. This equation could be applied for the coast of similar oceanographic conditions and beach characteristics to estimate the littoral transport rate. Therefore, the harbor planning and shore protection could be based on.

The relationship between I_ℓ and $(P_\ell)_b$ is not exact linearly proportional to each other, this is proved by Eq(5.1) which is originally prepared by Komar and Inman (1970) there is a upper limit K = 0.77. Adding the data of the author, reanalyzing the total data by the least square method to get a regression line which is expressed as $I_\ell = 0.154 (P_\ell)^{1.0695}$ as shown in Fig. 5.1.

Reference

1) Bagnold, R.A. (1963) "Mechanics of Marine Sedimentation" THE SEA, Interscience Publishers, New York, 3:507-528.

2) Bruun, Per (1966) Tidal Inlets and Littoral Drift, University of Florida.

3) Komar, P.D. and D.L. Inman (1970) "Longshore Sand Transport on Beaches", J. of Geophysical Research, Vol. 76, No.3.

4) Galvin, C.J., Jr. (1972) "Wave Breaking in Shallow Water", Waves on Beaches and Resulting Sediment Transport, edited by Meyer, R.E., 1972, Academic Press.

5) Taiwan Power Co. (1980) "The Research Report of Littoral Transport of Lin-Kou Power Plant".

6) H.S. Hou, C.P. Lee, G.H. Weng, T.J. Liaw (1980), "The Movable-Bed Model Investigation of North-East Sandy Coast of Taiwan". Research Rpt. of NTCMST. Keelung.

7) H.S. Hou, C.-P. Lee and L-H Lin (1980) "Relationship Between Alongshore Wave Energy and Littoral Drift in the Mid-West Coast at Taiwan, R.O.C." 17th ICCE. Sydney, Australia, Sept., 1980.

8) H.S. Hou, T.J. Liaw (1982), "The Movable-Bed Model Investigation of the 2nd Tansui Fishery Harbor Out-Breakwater Construction Stage". Research Rpt. of NTCMST., Keelung.

9) H.S. Hou (1982), "Researches of Harbor Site Investigation, Planning, and Design of Coastal Shipping System in Taiwan, R.O.C.", Special Report No.7 by Institute of Harbor and Marine Technology.

10) H.S. Hou (1983). "Planning and Design of coastal and Harbor Engineering", Special Report No.9 by Institute of Harbor and Marine Technology.

11) H.S. Hou, T.J. Liaw (1983), "Hydraulic Model Experiments Deal with Shoaling Problem of Littoral Transport for Improvement on the Intake and Outlet of Lin-Kou Power Plant", Special Report No. 12 by Institute of Harbor and Marine Technology.

12) H.S. Hou (1985) "Littoral Drift Model Investigation along the Taiwan Coast and Research of the Related Problems of Inlet Planning (II)," NTU-INA-Tech Rept. 216.

13) H.S. Hou (1986), "Case Study - Model Study of Lin Kou Power Plant" Short Course, 20th ICCE, Dynamics of Sand Beaches. Nov. 8-9, 1986, The Olympic Hotel, Taipei, R.O.C.

14) F.L.W. Tang (1986), "Wave Forecasting and Wave Statistics in Taiwan Strait". Short Course, 20th ICCE, Dynamics of Sand Beaches. Nov. 8-9, 1986, The Olympic Hotel, Taipei, R.O.C.

15) H.S. Hou (1987) "Offshore Mechanics and Research of the Related Problems of Port Planning (III)" NTU-INA Tech Rept. 248.

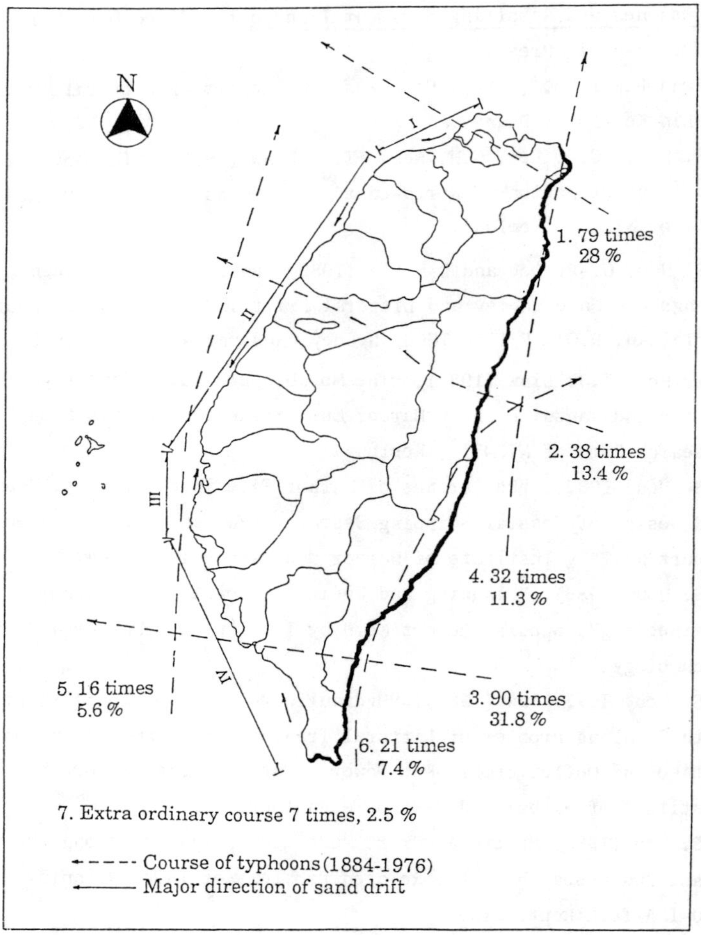

Fig.1 The sketch of the prevailing direction of littoral drift at Taiwan sandy coast

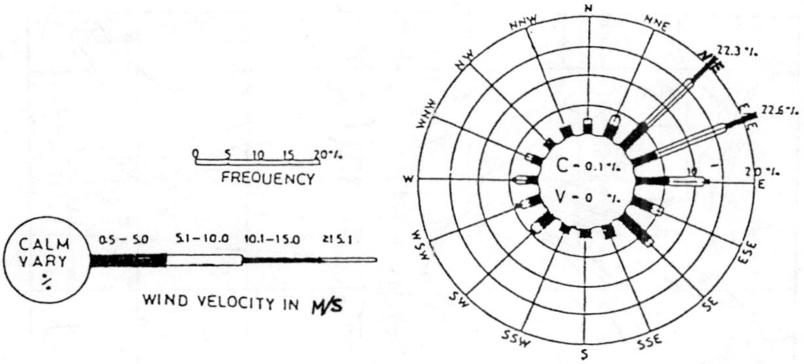

Fig.2.1 Wind Rose of the NW coast

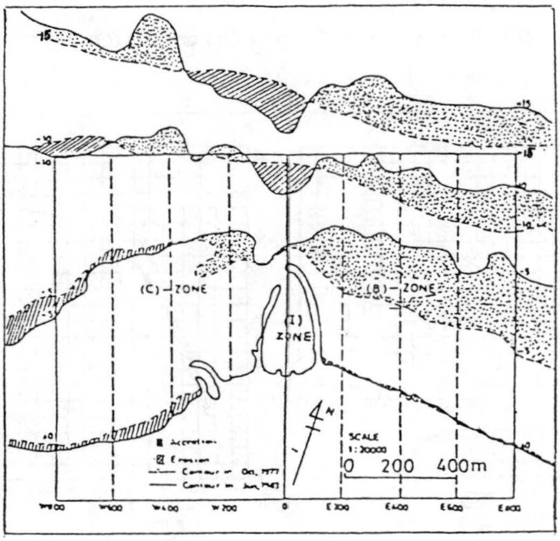

Fig 2.2 Accretion and Erosion of the
Profile near the Linkou

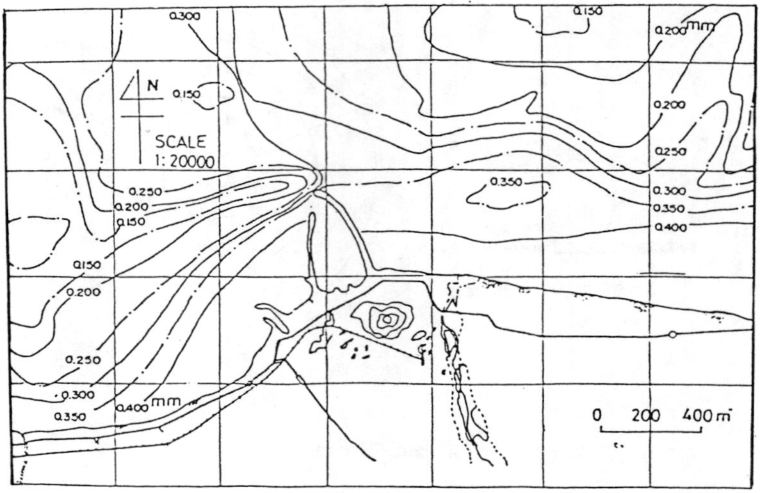

Fig.2.3 Sediment Particle d_{50} Distribution in the Nearshore Area of Linkou (mm)

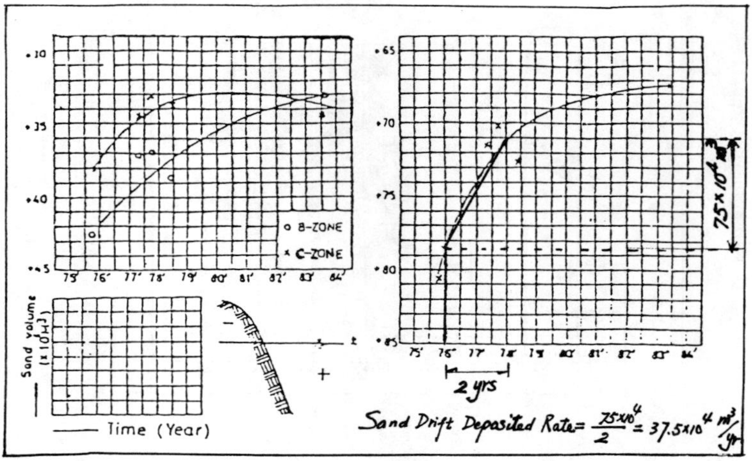

Fig.2.4 Sand Volume Change of Surveyed Zones B, C & B+C

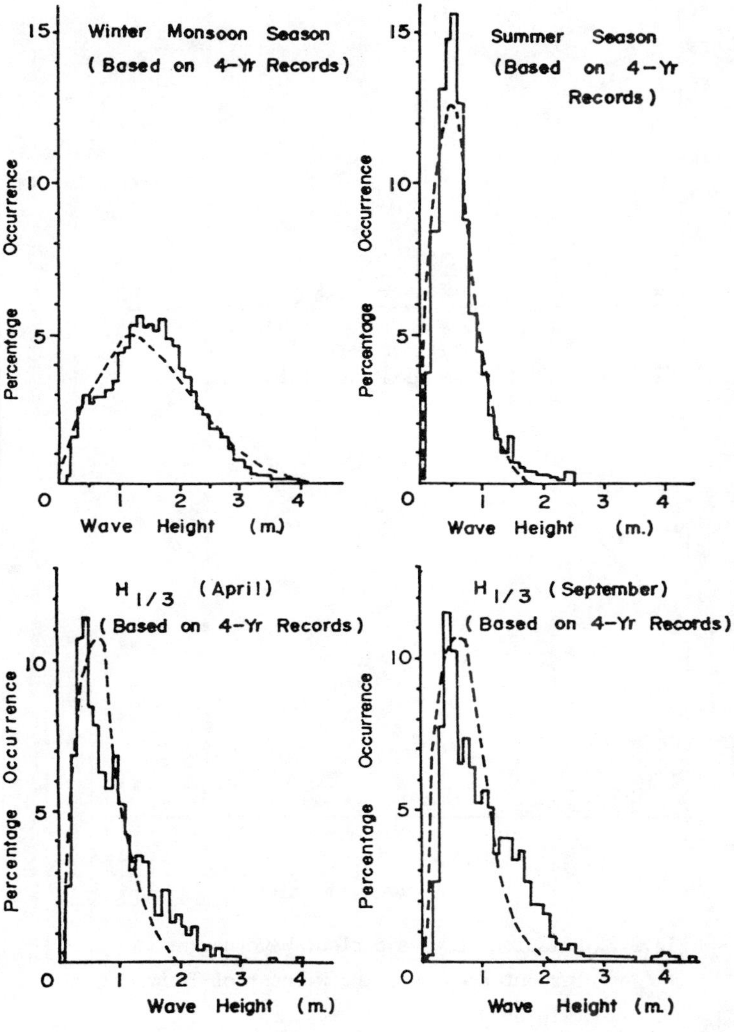

Fig.3.1 Distribution of Significant wave height
(Self Waves)

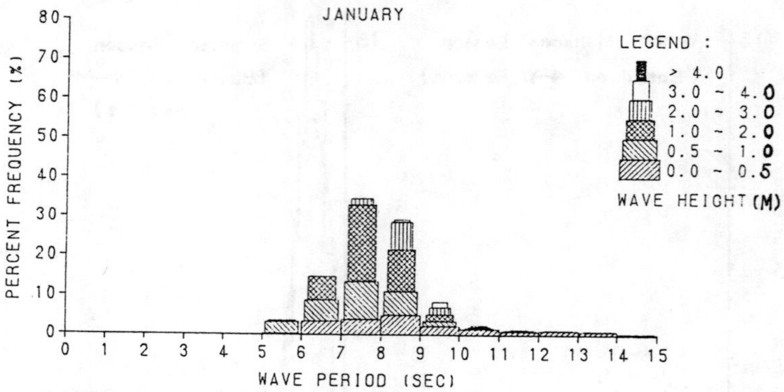

LOCATION : TOU-YUAN (NW coast of Taiwan)

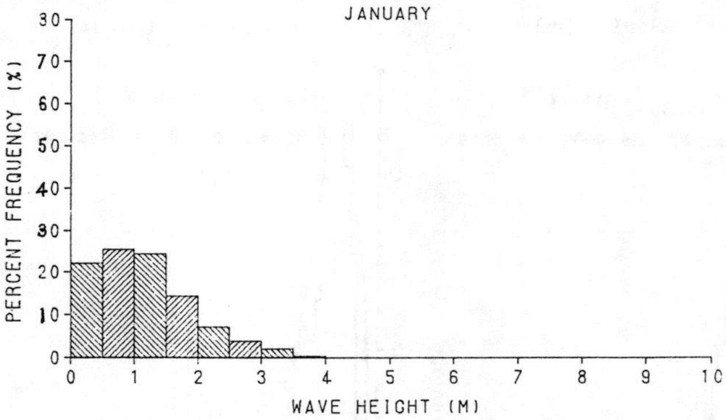

LOCATION : TOU-YUAN (NW coast of Taiwan)

Fig.3.2. Relation of Wave period, Wave height vs Percent Frequency in NW coast of Taiwan (Winter)

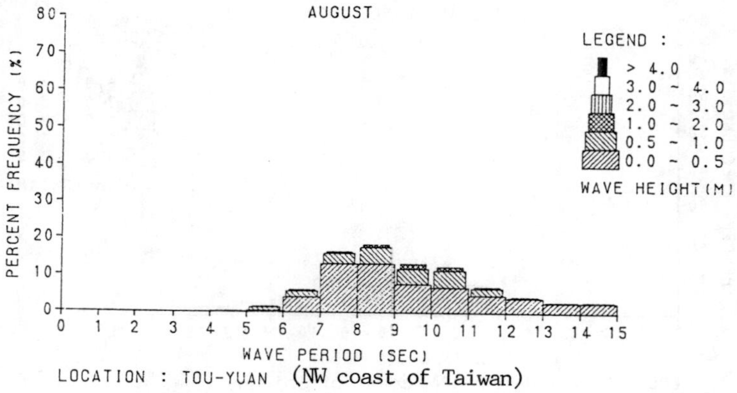

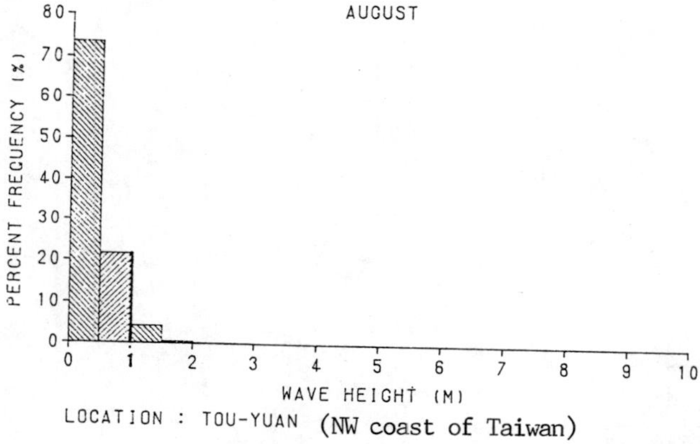

Fig.3.3 Relation of Wave period, wave height vs percent frequency in NW coast of Taiwan (Summer)

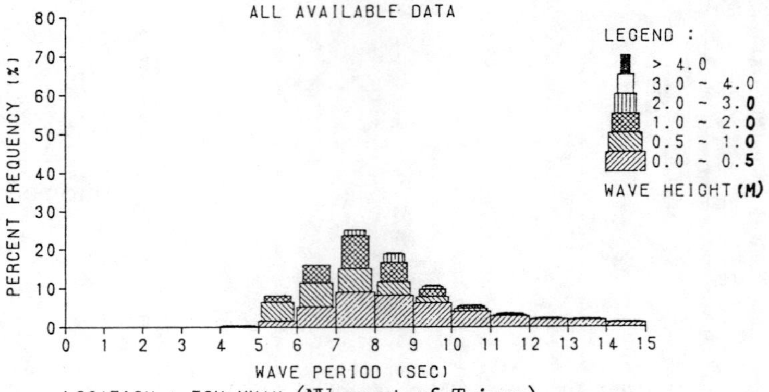

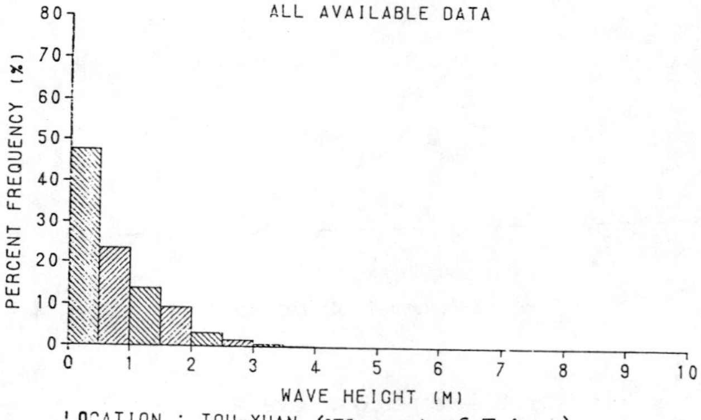

Fig.3.4 Relatiuon of wave period, wave height vs percent frequency in NW coast of Taiwan (the whole year)

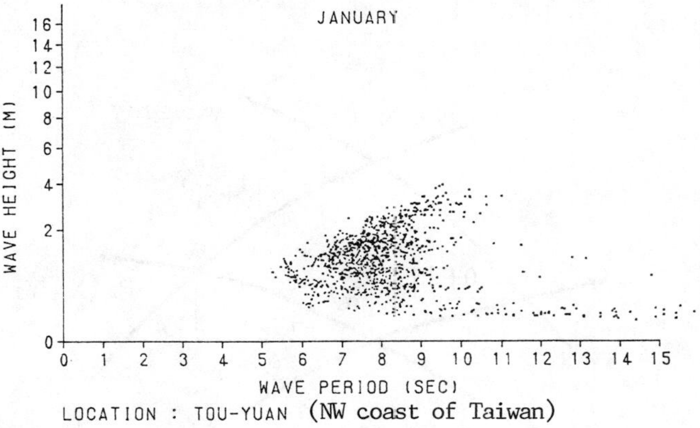

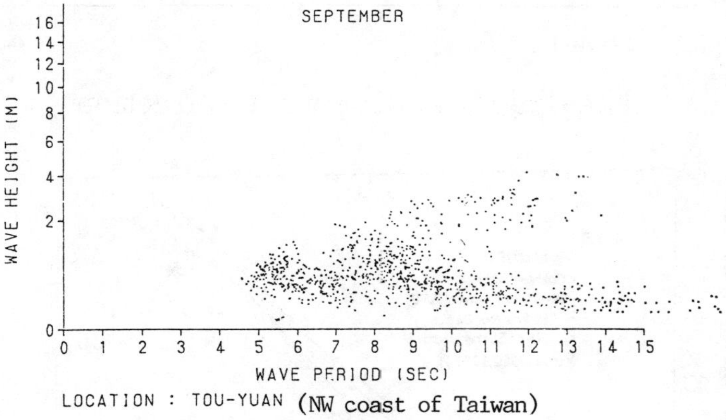

Fig.3.5 Correlation between wave height and wave period in NW coast of Taiwan.

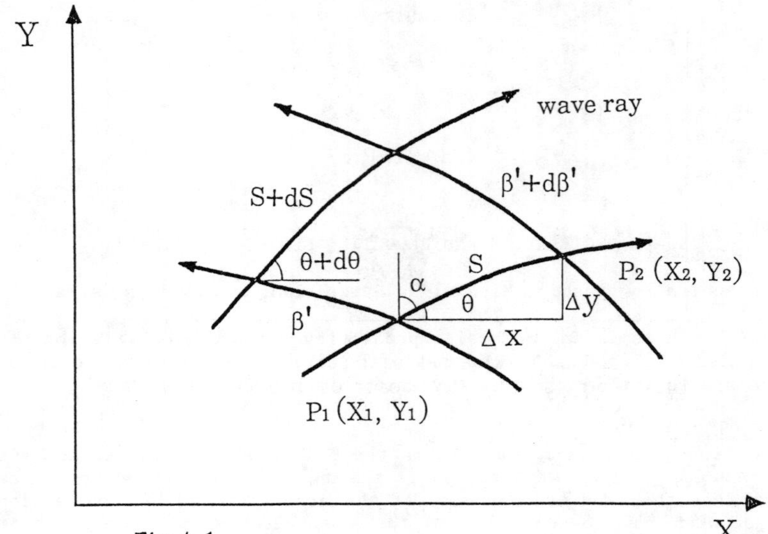

Fig.4.1

The diagram of the wave refraction

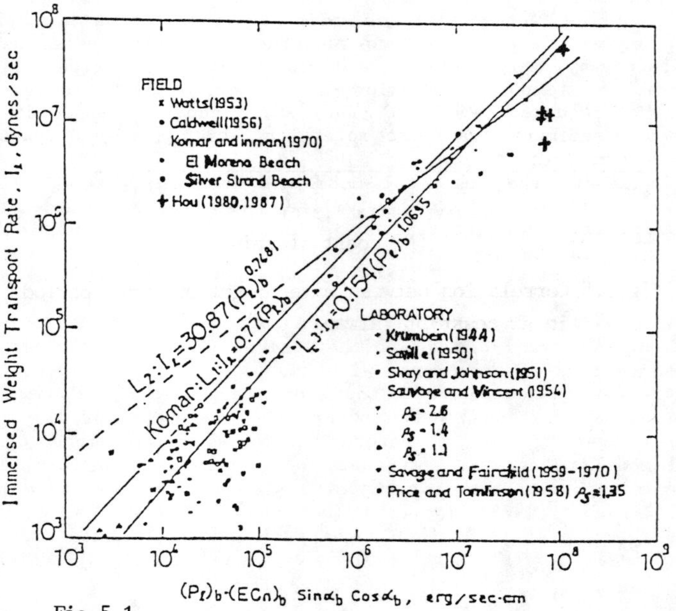

Fig.5.1

The relationship of the alongshore breaking wave energy and the immersed weight transport rate

CHAPTER 87

SWASH OSCILLATION AND RESULTING SEDIMENT MOVEMENT

by Nobuhisa Kobayashi[1], Michael S. Strzelecki[2] and Andojo Wurjanto[3]

ABSTRACT: A numerical model for predicting the swash oscillation on a beach is described and compared with field data on wave setup and swash statistics on a moderately steep beach with a nearshore bar.

INTRODUCTION

The swash zone on a beach forms the boundary zone between the surf zone and backshore. Wave run-up is the upper limit of wave uprush and determines the landward boundary of the area affected by wave action. Few studies have been performed for the hydrodynamics and resulting sediment movement in the swash zone probably because existing hydrodynamic models such as those proposed by Battjes and Stive (1985) and Svendsen et al. (1987) do not account for the variations of hydrodynamic quantities over a wave period which are essential in the swash zone. A quantitative understanding of sediment transport in the swash zone is required for better establishing the landward boundary condition for existing cross-shore sediment transport models such as that proposed by Stive (1986) and De Vriend and Stive (1987). Furthermore, the temporal variation of the horizontal fluid velocity is normally required even in the surf zone to predict the instantaneous sediment transport rate from which the net transport rate can be computed.

In this paper, the numerical model of Kobayashi et al. (1987) developed for predicting the waterline oscillation on the rough impermeable slope of a coastal structure is slightly modified and applied to predict the swash oscillation on a natural beach. Kobayashi and Greenwald (1986,1988) showed that the numerical model could predict the measured temporal variations of hydrodynamic quantities on a 1:3 gravel slope with an impermeable base. Moreover, Kobayashi and Watson (1987) showed that the numerical model could also be applied to coastal structures with smooth slopes by adjusting the friction factor associated with the slope roughness. Kobayashi and Wurjanto (1988) extended the numerical model to predict wave overtopping over coastal structures. These applications of the numerical model to coastal structures were limited to uniform slopes of 1:5 or steeper as well as composite slopes. On the other hand, Kobayashi et al. (1988) modified the numerical model slightly to predict the wave transformation in the surf and swash zones on gentle slopes as well as the wave reflection and swash oscillation on

[1]Assoc. Prof., Dept. of Civil Engrg., Univ. of Delaware, Newark, DE 19716.
[2]Civil Engineer, Planning Div., U.S. Army Corps of Engineers, Baltimore District, P.O. Box 1715, Baltimore, MD 21203.
[3]Grad. Student, Dept. of Civil Engrg., Univ. of Delaware, Newark, DE 19716.

relatively steep beaches. The slight modification was related to the effect of the wave-induced current on the seaward boundary condition used in the numerical model, which influenced the computed mean water level on gentle slopes. The modified numerical model was compared with small-scale test data for monochromatic waves spilling on gentle slopes. The comparison included the comprehensive test results for a 1:40 smooth slope presented by Stive (1980) and Stive and Wind (1982) as well as the undertow measurement for a 1:34.25 smooth slope performed by Hansen and Svendsen (1984). The numerical model was shown to be capable of predicting the development of the wave profile asymmetry about the vertical axis from the symmetric cnoidal wave profile outside the breakpoint to the sawtooth profile in the inner surf zone. The computed shoreline oscillation on the gentle slope showed the dominance of the setup over the swash in accordance with the empirical formulas proposed by Battjes (1974). As a whole, the numerical model was in good agreement with the gentle slope data except that the numerical model based on the finite-amplitude shallow-water equations predicts the depth-averaged velocity only and can not predict shoaling without wave breaking over the horizontal distance which is large relative to the wavelength. In order to take the seaward boundary location far seaward of the breakpoint, the numerical model would need to be matched with a numerical model based on the Boussinesq equations for a sloping bottom (Peregrine, 1967) such as the time domain model of Abbott et al. (1984) and the frequency domain model of Freilich and Guza (1984). In addition, Kobayashi et al. (1988) compared the modified numerical model with the wave reflection and swash excursion measurements for monochromatic waves plunging and surging on a 1:8.14 slope described by Guza and Bowen (1976) and Guza et al. (1984). The agreement between the model and the data was only qualitative probably because the wave reflection data was obtained on the basis of linear standing wave theory and the visual measurements of swash excursion on the relatively steep slope were difficult to define quantitatively. Kobayashi et al. (1988) also conducted small-scale tests for monochromatic and transient grouped waves on a 1:8 slope with and without an idealized nearshore bar at the toe of the 1:8 slope. The numerical model was shown to be capable of predicting the measured shoreline oscillations fairly well.

In the following, the numerical model used by Kobayashi et al. (1988) is described concisely and compared with the field data on swash oscillations on a moderately steep beach with a nearshore bar given by Holman and Sallenger (1985). As a first attempt, incident random waves are approximated by monochromatic waves, although any incident wave train in the time domain could be specified as input at the seaward boundary of the numerical model. This monochromatic approximation may be reasonable for the swash oscillation in the incident frequency band but excludes the swash oscillation in the infragravity frequency band which was not negligible for the field data. The sediment transport mechanics in the swash zone on a natural beach are discussed only briefly at the end of this paper.

ONE-DIMENSIONAL UNSTEADY NUMERICAL FLOW MODEL

Under the assumptions of alongshore uniformity and normally incident waves, the finite-amplitude shallow-water equations for an impermeable beach of arbitrary geometry are expressed as (Kobayashi et al., 1988)

$$\frac{\partial h'}{\partial t'} + \frac{\partial}{\partial x'} (h'u') = 0 \qquad (1)$$

$$\frac{\partial}{\partial t'}(h'u') + \frac{\partial}{\partial x'}(h'u'^2) = -gh'\frac{\partial \eta'}{\partial x'} - \frac{1}{2}f'|u'|u' \qquad (2)$$

where t'=time; x'=horizontal coordinate taken to be positive in the landward direction with $x'=0$ at the seaward boundary of the numerical model; h'=instantaneous water depth; u'=instantaneous depth-averaged horizontal velocity; g=gravitational acceleration; η'=instantaneous free surface elevation above the still water level (SWL); and f'=bottom friction factor which is assumed constant. The prime indicates the dimensional variables which are normalized in the following. The vertical coordinate z' is taken to be positive upward with $z'=0$ at SWL. The arbitrary beach geometry is specified by d_t'=water depth below SWL at $x'=0$ and θ'=local angle of the bed varying with respect to $x'\geq 0$. Denoting the reference wave height and period by H_r' and T_r', respectively, which are assumed to be given, the following dimensionless variables are introduced:

$$t = \frac{t'}{T_r'} \quad ; \quad x = \frac{x'}{T_r'\sqrt{gH_r'}} \quad ; \quad u = \frac{u'}{\sqrt{gH_r'}} \qquad (3)$$

$$z = \frac{z'}{H_r'} \quad ; \quad h = \frac{h'}{H_r'} \quad ; \quad \eta = \frac{\eta'}{H_r'} \quad ; \quad d_t = \frac{d_t'}{H_r'} \qquad (4)$$

$$\sigma = T_r'\sqrt{\frac{g}{H_r'}} \quad ; \quad \theta = \sigma \tan\theta' \quad ; \quad f = \frac{1}{2}\sigma f' \qquad (5)$$

In terms of the normalized coordinate system, the bed is located at

$$z = \int_0^x \theta dx - d_t \quad ; \quad \text{for } x \geq 0 \qquad (6)$$

Substitution of Eqs. 3-5 into Eqs. 1 and 2 yields

$$\frac{\partial h}{\partial t} + \frac{\partial m}{\partial x} = 0 \qquad (7)$$

$$\frac{\partial m}{\partial t} + \frac{\partial}{\partial x}\left(m^2 h^{-1} + \frac{1}{2}h^2\right) = -\theta h - f|u|u \qquad (8)$$

in which $m=uh$ is the normalized volume flux per unit width.

Eqs. 7 and 8 expressed in the conservation-law form of the mass and momentum equations except for the two terms on the right hand side of Eq. 8 are solved in the time domain using the explicit dissipative Lax-Wendroff finite difference method based on a finite-difference grid of constant space size Δx and constant time step Δt as explained by Kobayashi et al. (1987). The initial time $t=0$ for the computation marching forward in time is taken to be the time when the incident wave train specified as input arrives at the seaward boundary, $x=0$, and no wave action is present in the computation domain $x\geq 0$. The landward boundary on the beach is located at the moving shoreline where the water depth is essentially zero. For the computation, the shoreline is defined as the location where h equals an infinitesimal value δ. The shoreline oscillation is computed using the predictor-corrector-smoothing procedure explained by Kobayashi et al. (1987). The seaward boundary at $x=0$ is taken to be located seaward of the breakpoint where the flow at $x=0$ can be assumed to be subcritical and satisfy the condition $u<\sqrt{h}$. Expressing Eqs. 7 and 8 in the characteristic forms, the equation for the characteristics, $\beta=(-u+2\sqrt{h})$, advancing seaward at x=0 is given by Kobayashi et al. (1987)

$$\frac{\partial\beta}{\partial t} + (u-\sqrt{h})\,\frac{\partial\beta}{\partial x} = \theta + \frac{f|u|u}{h} \quad ; \quad \text{along}\ \frac{dx}{dt} = u - \sqrt{h} \tag{9}$$

Eq. 9 is discretized using a simple first-order finite difference to obtain the value of $\beta=(-u+2\sqrt{h})$ (Kobayashi et al., 1987). In addition, the total water depth at the seaward boundary is expressed in the form

$$h = d_t + \eta_i(t) + \eta_r(t) \quad ; \quad \text{at } x = 0 \tag{10}$$

in which η_i and η_r are the free surface variations with respect to t at x=0 normalized by the reference wave height H_r'. The incident wave train seaward of the breakpoint is specified by prescribing the variation of η_i with respect to $t\geq0$. The term $\eta_r(t)$ in Eq. 10 accounts for the difference between the actual value $\eta=(\eta_i+\eta_r)$ at x=0 and the prescribed value η_i. For reflective slopes such as coastal structures, $\eta_r(t)$ may be regarded as the normalized free surface variation associated with the reflected wave train at x=0 (Kobayashi et al., 1987). For dissipative beaches, incident wave reflection may be negligible but $\eta_r(t)$ accounts for the secondary effects excluded from the prescribed variation of $\eta_i(t)$. Kobayashi et al. (1988) used the following approximate expression of $\eta_r(t)$ in terms of the value of β at x=0 computed using Eq. 9:

$$\eta_r(t) \simeq \frac{1}{2}\sqrt{d_t}\,\beta(t) - d_t - C_t \quad ; \quad \text{at } x = 0 \tag{11}$$

with $$C_t = -\frac{1}{2}\sqrt{d_t}\,\bar{u}_t \tag{12}$$

where $\bar{u}_t$=value of the time-averaged horizontal velocity $\bar{u}$ at x=0. Substitution of Eq. 11 into Eq. 10 yields the value of h at x=0 for given $\eta_i(t)$ and estimated C_t. The value of u at x=0 is then obtained from $u=(2\sqrt{h}-\beta)$ at x=0. The nonlinear correction term C_t associated with the time-averaged velocity $\bar{u}_t$ was not included in the numerical model developed for coastal structures by Kobayashi et al. (1987). This term was shown by Kobayashi et al. (1988) to improve the prediction of wave set-down and setup on a gentle slope.

The time-averaged mass and momentum equations corresponding to Eqs. 7 and 8 can be expressed as

$$\bar{m} = \overline{hu} = 0 \tag{13}$$

$$\frac{d}{dx}\left[\overline{hu^2} + \frac{1}{2}\overline{(\eta-\bar{\eta})^2}\right] = -\,\bar{h}\,\frac{d\bar{\eta}}{dx} - f\,\overline{|u|u} \tag{14}$$

where the overbar denotes time averaging and $\bar{\eta}$ is the vertical difference between the mean and still water levels. Use is made of the condition of no flux into the assumed impermeable beach to derive Eq. 13. The left hand side of Eq. 14 is the normalized gradient of the cross-shore radiation stress (e.g., Svendsen et al., 1987) under the assumptions of vertical uniformity and hydrostatic pressure. In this paper, the computed temporal variations of h and u at given location are used to compute the cross-shore variations of $\bar{\eta}$ and $\bar{u}$ without using Eqs. 13 and 14. Rearranging Eq. 13, the time-averaged horizontal velocity $\bar{u}$ can be expressed as

$$\bar{u} = -\,\overline{(\eta-\bar{\eta})(u-\bar{u})}\,(\bar{h})^{-1} \tag{15}$$

For gentle slopes with little wave reflection, $(\eta-\bar{\eta})$ and $(u-\bar{u})$ are expected to be in phase, resulting in $\bar{u}<0$ from Eq. 15. The computed seaward velocity $\bar{u}$ was found to be smaller than the undertow measured below the wave trough since the numerical model does not account for the vertical variation of the time-averaged horizontal velocity (Kobayashi et al., 1988). An approximate value of $\bar{u}_t$ may be found using Eq. 15 with the assumption of incident monochromatic linear long wave at the seaward boundary where the wave height and period are given by H′ and T′, respectively. Under this assumption, $(\eta-\bar{\eta}) \simeq (K_s/2)\cos(2\pi t)$, $(u-\bar{u}) \simeq (\eta-\bar{\eta})/\sqrt{d_t}$ and $\bar{h} \simeq d_t$ at x=0, where $K_s=H'/H_r'$ and $T'=T_r'$ is assumed. For gentle slopes with little wave reflection, $\bar{u}_t$ and C_t may hence be approximated by

$$\bar{u}_t \simeq -K_s^2(8d_t^{3/2})^{-1}; \qquad C_t \simeq K_s^2(16d_t)^{-1} \tag{16}$$

It should be mentioned that Kobayashi et al. (1988) used $K_s=1$ since the reference wave height H_r' was taken to be the wave height H′ at the seaward boundary. For steep slopes with significant wave reflection for which η_r is on the order of unity, it might be more reasonable to assume that $\bar{u}_t \simeq 0$ and $C_t \simeq 0$, but the effect of C_t on η_r in Eq. 11 is generally very small. For the previous comparisons made for coastal structures by Kobayashi et al. (1987), $d_t \gtrsim 3$ and $K_s=1$, so that $C_t \lesssim 0.02$. For gentle slopes, it is necessary to choose a smaller value of d_t so that the seaward boundary is not located too far seaward of the breakpoint (Kobayashi et al., 1988).

COMPARISON WITH SWASH OSCILLATIONS MEASURED ON A BEACH

The numerical model is compared with the wave setup and swash statistics on a moderately steep beach given by Holman and Sallenger (1985). The field data were collected over a 3-week period in October, 1982 at the Army Corps of Engineers Field Research Facility located at Duck, North Carolina. Fig. 1 shows the beach profile on October 26, 1982 which was given as an example profile in their paper.

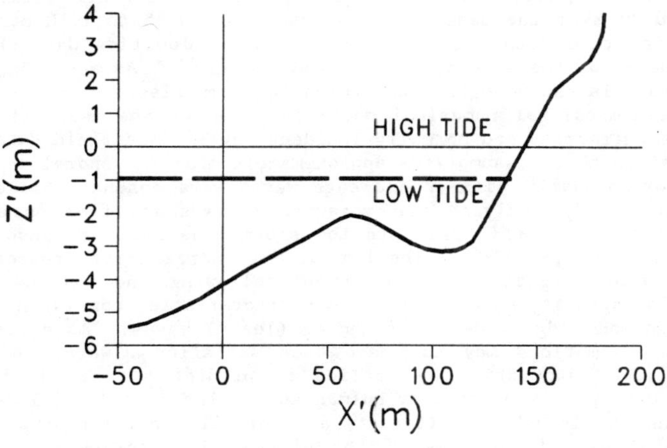

Fig. 1. Beach Profile with High and Low Tides

The slope angle θ_f' of the steep foreshore composed of coarse sand (mean size of 1-2 mm) varied between 5° and 9° (Sallenger et al., 1985). The position and height of the single bar shown in Fig. 1

varied in response to storm events, while the bar morphology varied from linear to crescentric. The incident wave data were collected from a wave-rider buoy in approximately 20-m depth. The significant wave height and peak period based on the measured spectra were given by Holman and Sallenger (1985) and are used herein as the reference wave height H_r' and period T_r' used for the normalization in Eqs. 3-5. During the experiment, $H_r' = 0.4$–4.0m and $T_r' = 6$–16s. No information on wave direction was given in their paper. The tide data was provided by a tide gauge located outside the surf zone for all but the largest storm. The measured tidal range was −0.35m to 1.10m relative to the datum. The swash oscillation data were collected by using longshore-looking time-lapse photography together with large markers. A frame was shot every second for a total run length of 35 min. Sixty-one films were digitized, most at two longshore locations 100 and 150 m from the camera, but some films with apparent longshore variability were digitized more intensively. The digitized time series of the shoreline oscillation for each run were transformed to the vertical component from which the mean and the standard deviation σ_s were calculated. The tidal elevation was subtracted from the calculated mean to find the setup, whereas the significant swash height was taken as $4\sigma_s$. The total runup was defined as the sum of the setup and $2\sigma_s$. The setup and swash statistics were calculated for a total of 154 time series. Holman and Sallenger (1985) normalized the setup, swash height and total runup by the significant wave height H_r' and plotted the normalized setup, swash height and total runup as a function of the surf similarity parameter defined as

$$\xi = \frac{\sigma \tan\theta_f'}{\sqrt{2\pi}} = T_r'\left(\frac{g}{2\pi H_r'}\right)^{1/2} \tan\theta_f' \qquad (17)$$

where σ is defined in Eq. 5 and the foreshore slope $\tan\theta_f'$ ranged from 0.09 to 0.16 for the range of θ_f'=5°-9° indicated by Sallenger et al. (1985). In order to reduce the scatter of the plotted data points, the data were split into three sections corresponding to low, mid and high tides. The cutoff tidal elevations were arbitrarily taken as 0.25 and 0.70 m of the measured tidal range of −0.35 to 1.10 m. The mid and high tide data were similar, while the low tide data showed some influence of the single bar depicted in Fig. 1. As a result, the mid tide data is excluded from the following comparison.

The one-dimensional numerical model is based on the assumption of alongshore uniformity and normally incident waves. The field data may have satisfied these assumptions approximately near the shoreline when the bar morphology was linear and edge waves were absent. Sallenger and Holman (1987) analyzed the measured cross-shore flow during a storm in October, 1982. Prior to the storm, the bar was reasonably linear and shore parallel. The bar became increasingly crescentric when the wave heights were decreasing following the storm. The measured infragravity band spectra had characteristics consistent with either high-mode edge waves or standing (leaky) waves. As a result, the basic assumptions may not have been satisfied always, but the numerical model may still be applied to predict the trend of the scattered data points plotted by Holman and Sallenger (1985) since the assumptions of alongshore uniformity and normally incident waves must have been satisfied for some of the plotted data points. For the following computation, the beach profile shown in Fig. 1 is used as a typical profile neglecting the beach profile changes. The representative tidal elevations relative to the datum for the high and low tide data are simply taken as 0.90m and −0.05m, respectively. The vertical coordinate z' with $z'=0$ at SWL shown in Fig. 1 is that assumed for the high tide. The seaward boundary of the numerical model located at $x'=0$ in Fig. 1 is taken to be sufficiently seaward of the bar so that the normalized incident wave profile $\eta_i(t)$ at $x'=0$ may be specified as input using an appropriate wave theory for an essentially horizontal seabed. The water depth below SWL at $x'=0$ in Fig. 1 is $d_t'=4.15$m for the high tide and $d_t'=3.20$m for the low tide. As a first attempt, the incident random waves measured in approximately 20-m depth are assumed to be represented by the monochromatic wave whose height and period are the significant wave height H_r' and the spectral peak period T_r' used by Holman and Sallenger (1985) to plot their data. Table 1 shows five different cases selected for the subsequent computation to represent the range of the wave conditions associated with the field data except that cases with larger wave heights are excluded to avoid wave breaking seaward of the selected seaward boundary location. The surf similarity parameter ξ

TABLE 1. Five Different Wave Conditions Used for Computation

Case	H_r'(m)	T_r'(s)	σ	ξ
1	2.0	7	15.5	0.73
2	0.9	7	23.1	1.09
3	1.6	12	29.7	1.40
4	0.8	11	38.5	1.82
5	0.8	14	49.0	2.31

is defined in Eq. 17 where the foreshore slope $\tan\theta_f'=0.118$ at the still water shoreline for the beach profile shown in Fig. 1. The range of ξ for the selected cases corresponds to that for the field data. For each case, a shoaling analysis is performed to find the wave height H' at $x'=0$ where the water depth d_t' below SWL is taken as 4.15 m for the high tide and 3.20 m for the low tide. In this paper, the shoaling analysis and the specification of the normalized incident wave profile $\eta_i(t)$ for $t\geq0$ are made using cnoidal wave theory for $U_r\geq26$ and Stokes second-order wave theory for $U_r<26$ in which U_r=Ursell parameter at $x'=0$ defined below (Svendsen and Brink-Kjaer, 1972). Table 2 summarizes the estimated monochromatic wave characteristics at

TABLE 2. Wave Characteristics at Seaward Boundary for High and Low Tides

Case	d_t	K_s	L	U_r	r
H1	2.08	1.03	11.1	61	0.02
H2	4.61	1.07	10.1	24	0.02
H3	2.59	1.58	21.5	283	0.05
H4	5.19	1.35	17.4	79	0.11
H5	5.19	1.58	23.0	161	0.17
L1	1.60	1.16	14.0	141	0.02
L2	3.56	1.10	12.2	46	0.02
L3	2.00	1.93	27.6	734	0.04
L4	4.00	1.56	21.0	171	0.07
L5	4.00	1.85	27.9	359	0.10

the seaward boundary of the numerical model for the five cases listed in Table 1, where the capital letters H and L indicate the high and low tides, respectively. In Table 2, $d_t = d'_t/H'_r$ and K_s, L and U_r are defined as

$$K_s = \frac{H'}{H'_r} \quad ; \quad L = \frac{L'}{d'_t} \quad ; \quad U_r = \frac{H'(L')}{(d'_t)} = \frac{K_s L}{d_t} \qquad (18)$$

where L'=wavelength at x'=0. The assumption of finite-amplitude shallow-water waves in the computation domain $x' \geq 0$ may be appropriate since $L \gg 1$ and $U_r \gg 1$. Table 2 also lists the computed reflection coefficient r for each case. The value of r is estimated as the height of the computed periodic variation of $\eta_r(t)$ divided by K_s, as will be explained later, where the period and height of the periodic variation of $\eta_i(t)$ are equal to unity and K_s, respectively. The computed reflection coefficient r increases with the increase of the surf similarity parameter ξ. This trend is similar to that for coastal structures (Kobayashi and Watson, 1987). However, the value of r in Table 2 with the corresponding value of ξ in Table 1 is smaller than that based on the empirical formula, $r=(0.1\xi) \leq 1$, for smooth plane slopes proposed by Battjes (1974) even if $H'=K_s H'_r$ is used instead of H'_r in Eq. 17 to reduce the value of ξ for each case.

For the subsequent computation, use is made of approximately 400 nodes in the computation domain $x' \geq 0$ in Fig. 1 with the dimensional space size $\Delta x' \simeq 0.46$m between the two adjacent nodes. Correspondingly, the dimensionless space size of the finite difference grid is in the range of $\Delta x = 0.0097$–0.022 for the ten cases listed in Table 2. The number of time steps over one normalized wave period of unity is taken as $(\Delta t)^{-} = 4000$ except that $(\Delta t)^{-} = 5000$ is used for Case L2 due to numerical instability. The computational shoreline is defined by $h=\delta=0.002$ since the increase of this value from $\delta=0.001$ used for smooth steep slopes (Kobayashi and Watson, 1987) tends to improve the numerical stability in the vicinity of the computational shoreline. The measured shoreline oscillation is defined by the physical water depth $h'=\delta'_r$ in which $\delta'_r \simeq 0.5$ cm for the photography technique used by Holman and Sallenger (1985) and $\delta'_r=3.0$ cm for the dual-resistance wire sensor used by Guza and Thornton (1982) whose data are also included in the following comparison. For $H'_r=0.8$–2.0 m as shown in Table 1, $\delta_r=(\delta'_r/H'_r)>\delta=0.002$. The numerical damping coefficients for reducing high frequency numerical oscillations at the rear of breaking wave crests are taken as two for gentle slopes (Kobayashi and DeSilva, 1987; Kobayashi et al., 1988). The bottom friction factor f' is tentatively assumed to be f'=0.05 on the basis of the limited calibration made by Kobayashi and Watson (1987) for small-scale smooth slopes without beach sand, neglecting the scale effects and the effects of moving sediment particles on f' (Kobayashi and Seo, 1985). The previous sensitivity analyses performed by Kobayashi et al. (1987) indicated that the computed results should not be very sensitive to the value of f'.

The computed results for the ten cases listed in Table 2 are given in the thesis of Strzelecki (1988). The computed results for Case H3 are presented as an example in the following. Fig. 2 shows the periodic cnoidal wave profile $\eta_i(t)$ specified at x=0 and the temporal variation of $\eta_r(t)$ computed using Eq. 11 with Eq. 16. The normalized wave period is unity. The detailed variation of $\eta_r(t)$ is shown in Fig. 3. The depression of η_r during the transition period $0 \leq t \lesssim 5$ appears to be related to the depression of the mean water level under large waves (Longuet-Higgins and Stewart, 1962) since the incident wave train initially propagates into the region of no wave action. The temporal variation of $\eta_r(t)$ for $t \gtrsim 10$ consists of steady and oscillatory components. The steady component is the wave set-down

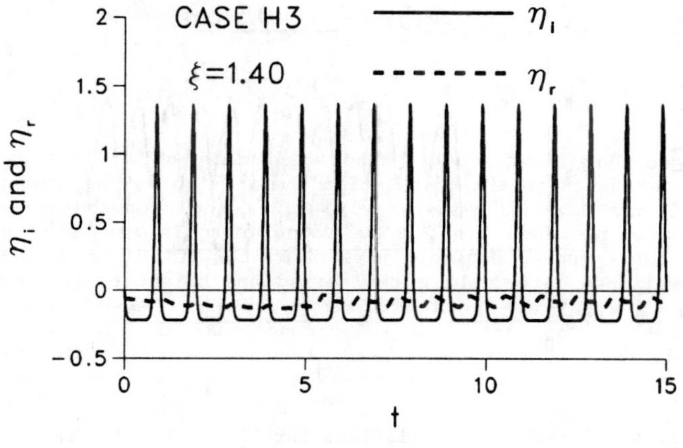

Fig. 2. Specified $\eta_i(t)$ and Computed $\eta_r(t)$ at Seaward Boundary

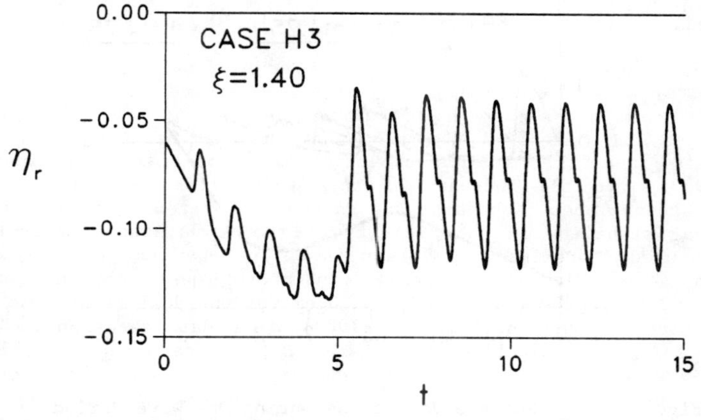

Fig. 3. Detailed Variation of $\eta_r(t)$ at Seaward Boundary

$\overline{\eta}=\overline{\eta_r}$ at x=0 since $\overline{\eta_i}=0$ for the assumed cnoidal wave profile. The oscillatory component is associated with the reflected wave. The reflection coefficient r listed in Table 2 is taken as the height of this oscillatory component divided by the height of $\eta_i(t)$, that is, K_s. Fig. 4 shows the computed shoreline oscillations corresponding to the water depth δ_r'=0.5 and 3cm plotted in the form of the normalized vertical elevation Z_r as a function of t. During wave downrush, the shoreline location is sensitive to the water depth δ_r' used to define its location since a thin layer of water remains on the relatively steep foreshore during wave downrush (Kobayashi et al., 1988). After the initial transient oscillation, the temporal variation of Z_r for t $\gtrsim$ 6 is composed of steady and oscillatory components. The steady component is the normalized setup on the foreshore denoted by $\overline{Z}_r$, while the oscillatory component is the normalized swash about the setup level. The maximum and minimum values of $Z_r(t)$ after the establishment of periodicity are denoted by R and R_d in which R and R_d are the run-up and run-down normalized by the reference wave height H_r', respectively. The value of $(R-R_d)$ is the

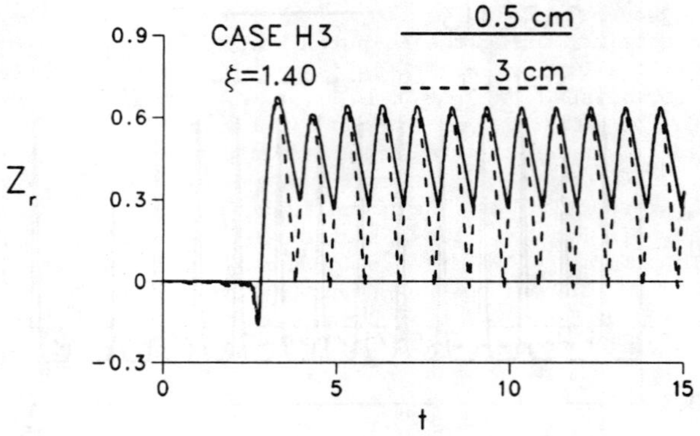

Fig. 4. Shoreline Oscillations for $\delta_r' = 0.5$ and 3 cm

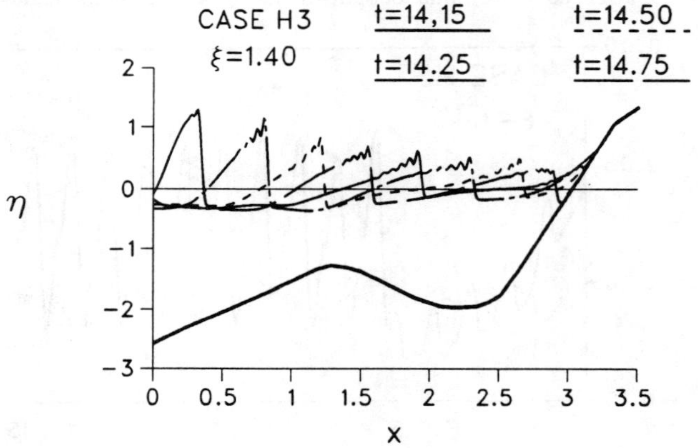

Fig. 5. Free Surface Variations During One Wave Period

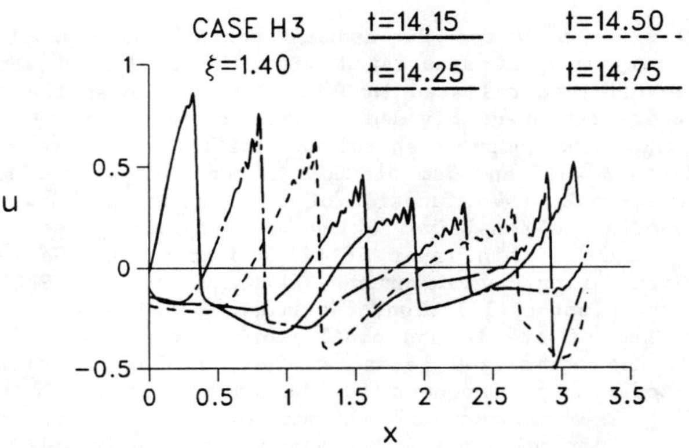

Fig. 6. Horizontal Velocity Variations During One Wave Period

normalized swash height. The computed values of $\overline{Z}_r$, R and R_d for given δ_r' are obtained using the computed variation of $Z_r(t)$ during $t_p \leq t \leq (t_p+1)$. For Case H3, t_p=14 is used since the periodicity is definitely established before t=14. t_p=14 is also found to be adequate for the other cases except for Cases L1 and L2 for which t_p=24 is used to ensure the definite periodicity. Figs. 5 and 6 show the computed cross-shore variations of η and u at t=14, 14.25, 14.5, 14.75 and 15, respectively. The normalized beach profile given by Eq. 6 is also shown in Fig. 5. The computed variations of η and u at t=14 and 15 are identical. The effects of the bar on the variations of η and u landward of the bar are apparent in Figs. 5 and 6. Figs. 7 and 8 show the computed cross-shore variations of the time-averaged free surface elevation $\overline{\eta}$ above SWL and the time-averaged horizontal velocity $\overline{u}$, respectively, where the time averaging is performed for

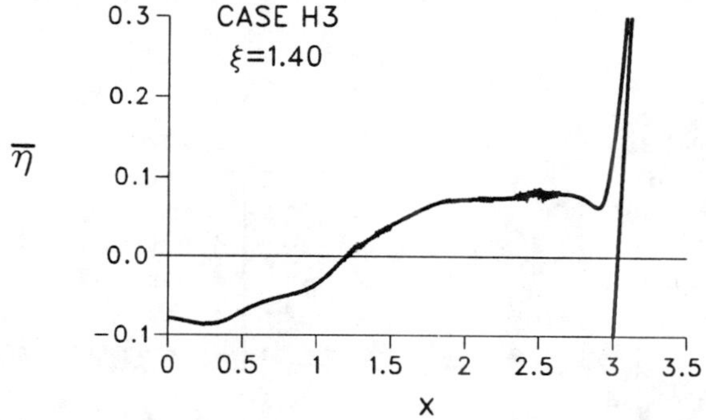

Fig. 7. Cross-Shore Variation of Mean Water Level

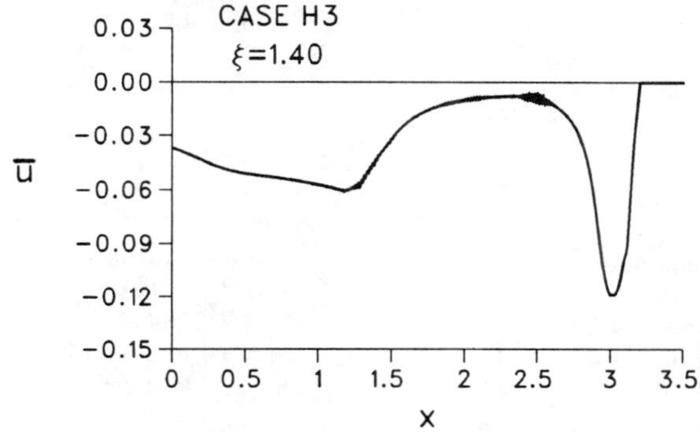

Fig. 8. Cross-Shore Variation of Mean Horizontal Velocity

$t_p \leq t \leq (t_p+1)$. The effects of the bar on the setup and mean velocity landward of the bar are apparent in these figures. Fig. 7 indicates that the mean water level on the steep foreshore will be affected by the large swash oscillation. Fig. 8 suggests that the bar may modify the cross-shore variation of the undertow noticeably, although $\overline{u}$ is not the same as the undertow (Kobayashi et al., 1988).

The values of $\overline{Z}_r$, R and $(R-R_d)$ for δ_r'=0.5 and 3 cm computed for each of the ten cases in Table 2 are plotted in Fig. 9 as a function of the surf similarity parameters ξ for the high and low tides. The data points in Fig. 9 are read from the figures given in Holman and Sallenger (1985) where $\overline{Z}_r$, R and $(R-R_d)$ are assumed to be the same as the nondimensional setup, total runup and significant swash height

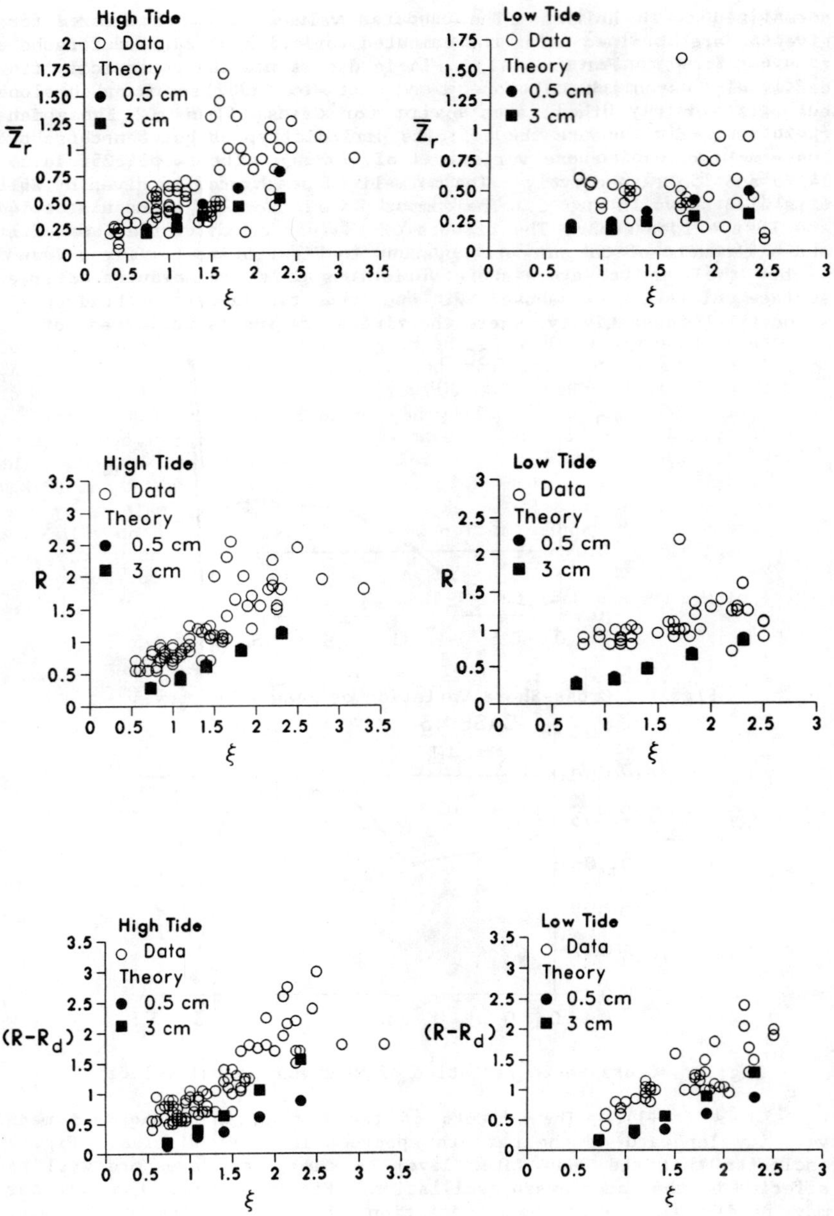

Fig. 9. Comparisons of Measured and Computed Setup, Run-up and Swash Height for High and Low Tides

calculated from the mean and standard deviation of the measured time series of the shoreline location with $\delta_r' \simeq 0.5$ cm on the foreshore slope of approximately 0.1. The data points for the high tide included those measured by Guza and Thornton (1982) on a beach slope of approximately 0.023 using a wire sensor with δ_r'=3 cm. These data points were in the region of $\xi < 1$. The differences between these two data sets were discussed by Guza et al. (1984). The computed results for δ_r'=0.5 and 3 cm suggest that the two measuring techniques will yield large differences in $(R-R_d)$ and $\overline{Z}_r$ for $\xi \geq 2$. Intercalibration of the two techniques on a low-slope beach indicated that the film technique registered a slightly higher mean and a 35% larger standard deviation than the wire sensor. The intercalibration results are consistent with the computed results for $\overline{Z}_r$ but opposite to the computed results for $(R-R_d)$. Considering the subjective interpretation of run-down of the films especially for large values of ξ, it appears to be difficult to specify an appropriate value of δ_r' for the film technique unlike the wire sensor with a specific value of δ_r'. The effects of permeability neglected in the numerical model may not be negligible on the foreshore slope composed of coarse sand (Packwood, 1983). However, the premeability effects should reduce the computed values of R and $(R-R_d)$. The underestimation of R and $(R-R_d)$ by the numerical model is expected to be caused mainly by the monochromatic wave approximation which may be reasonable only for the swash oscillation in the incident frequency band. The computed values of $(R-R_d)$ with δ_r'=3 cm for the high and low tides are plotted on Fig. 10(a) of Holman and Sallenger (1985) which

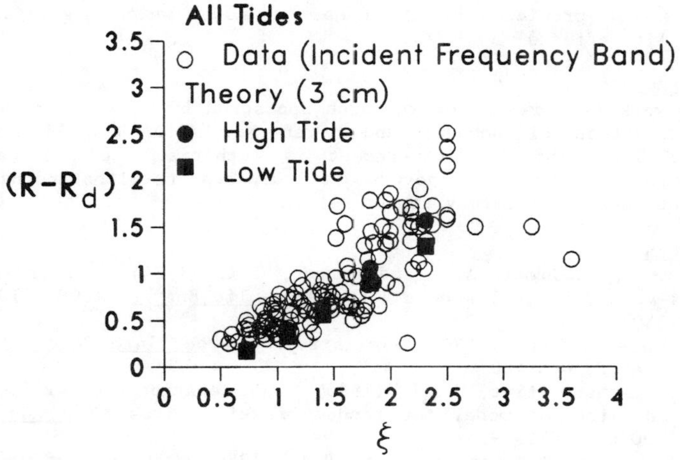

Fig. 10. Comparison with Swash Height in Incident Frequency Band

showed the nondimensional significant swash height in the incident frequency band (frequency ≥ 0.05 Hz) for all tides. The computed points with δ_r'=3 cm follow the trend of the scattered data points well as shown in Fig. 10. However, the computed points with δ_r'=0.5 cm, which are not shown in Fig. 10, are less than those shown in Fig. 10. Figs. 9 and 10 imply that the monochromatic approximation without regard to the swash oscillation in the infragravity frequency band will underestimate the swash height and run-up even on the moderately steep foreshore. Extension of the numerical model to random waves is definitely required, although wave group statistics in shallow water are not well known (Elgar et al., 1984).

SEDIMENT TRANSPORT IN SWASH ZONE

Kobayashi (1988) assembled and synthesized recent publications which might contribute to the improvement of our quantitative capabilities for predicting shoreline changes due to the cross-shore sediment transport in the surf and swash zones on beaches. The comparisons shown above suggest that an accurate prediction of the swash oscillation including the effects of random waves and actual beach profiles is essential for predicting the resulting sediment movement in the swash zone. It should be stated that purely empirical models for beach and dune erosion might work reasonably well for highly erosive storm events if storm surge is more important than swash. Nevertheless, an improved understanding of the hydrodynamics and resulting sediment movement in the swash zone is required for better establishing the landward boundary condition for such an erosion model as well as for predicting accretion and berm building. Even for laboratory data with monochromatic waves, it appears to be difficult to explain the detailed processes of berm building. Kobayashi and DeSilva (1987) applied a Lagrangian sediment transport model to predict the movement of individual bedload particles in the swash zone under monochromatic wave action. The Lagrangian model, which did not include the correction term C_t in Eq. 11, was found to explain observed erosion and bar formation of an initially uniform sand slope but could not predict observed accretion and berm building. The major reason appeared to be the large water velocity during the wave downrush predicted by the numerical model which neglected the permeability effects.

ACKNOWLEDGEMENT

This work is a result of research sponsored by NOAA Office of Sea Grant, Department of Commerce, under Grant No. NA85AA-D-SG033 (Project No. R/OE-3). The U.S. Government is authorized to produce and distribute reprints for government purposes notwithstanding any copyright notation that may appear herein.

REFERENCES

Abbott, M.B., McCowan, A.D. and Warren, I.R. (1984). "Accuracy of short-wave numerical models." J. Hydraulic Engrg., ASCE, 110(10), 1287-1301.

Battjes, J.A. (1974). "Surf similarity." Proc. 14th Coast. Engrg. Conf., ASCE, 466-480.

Battjes, J.A. and Stive, M.J.F. (1985). "Calibration and verification of a dissipation model for random breaking waves." J. Geophys. Res., 90(C5), 9159-9167.

Elgar, S., Guza, R.T. and Seymour, R.J. (1984). "Groups of waves in shallow water." J. Geophys. Res., 87(C3), 3623-3634.

Freilich, M.H. and Guza, R.T. (1984). "Nonlinear effects on shoaling surface gravity waves." Philos. Trans. R. Soc. London, Ser. A, 311, 1-41.

De Vriend, H.J. and Stive, M.J.F. (1987). "Quasi-3D modelling of nearshore currents." Coast. Engrg., 11, 565-601.

Guza, R.T. and Bowen, A.J. (1976). "Resonant interactions for waves breaking on a beach." Proc. 15th Coast. Engrg. Conf., ASCE, 560-579.

Guza, R.T. and Thornton, E.B. (1982). "Swash oscillations on a natural beach." J. Geophys. Res., 87(C1), 483-491.

Guza, R.T., Thornton, E.B. and Holman, R.A. (1984). "Swash on steep and shallow beaches." Proc. 19th Coast. Engrg. Conf., ASCE, 708-723.

Hansen, J.B. and Svendsen, I.A. (1984). "A theoretical and experimental study of undertow." Proc. 19th Coast. Engrg. Conf., ASCE, 2246-2262.

Holman, R.A. and Sallenger, A.H. (1985). "Setup and swash on a natural beach." J. Geophys. Res., 90(C1), 945-953.

Kobayashi, N. and Seo, S.N. (1985). "Fluid and sediment interaction on a plane bed." J. Hydraulic Engrg., ASCE, 111(6), 903-921.

Kobayashi, N. and Greenwald, J.H. (1986). "Prediction of wave runup and riprap stability." Proc. 20th Coast. Engrg. Conf., ASCE, 1958-1971.

Kobayashi, N., Otta, A.K. and Roy, I. (1987). "Wave reflection and runup on rough slopes." J. Wtrway. Port Coast. and Oc. Engrg., ASCE, 113(3), 282-298.

Kobayashi, N. and Watson, K.D. (1987). "Wave reflection and runup on smooth slopes." Proc. Coast. Hydrodynamics, ASCE, 548-563.

Kobayashi, N. and DeSilva, G.S. (1987). "Motion of sediment particles in swash zone." Proc. Coast. Hydrodynamics, ASCE, 715-730.

Kobayashi, N. and Greenwald, J.H. (1988). "Waterline oscillation and riprap movement." J. Wtrway. Port Coast. and Oc. Engrg., ASCE, 114(3), 281-296.

Kobayashi, N. (1988). "Review of wave transformation and cross-shore sediment transport processes in surf zones." J. Coast. Res., 4(3), (in press).

Kobayashi, N. and Wurjanto, A. (1988). "Wave overtopping on coastal structures." J. Wtrway. Port Coast. and Oc. Engrg., ASCE (in press).

Kobayashi, N., DeSilva, G.S. and Watson, K.D. (1988). "Wave transformation and swash oscillation on gentle and steep slopes." J. Geophys. Res., (submitted).

Longuet-Higgins, M.S. and Stewart, R.W. (1982). "Radiation stress and mass transport in gravity waves, with application to 'surf beats'." J. Fluid Mech., 13, 481-504.

Packwood, A.R. (1983). "The influence of beach porosity on wave uprush and backwash." Coast. Engrg., 7, 29-40.

Peregrine, D.H. (1967). "Long waves on a beach." J. Fluid Mech., 27, 815-827.

Sallenger, A.H., Holman, R.A. and Birkemeier, W.A. (1985). "Storm-induced response of a nearshore-bar system." Marine Geology, 64, 237-257.

Sallenger, A.H. and Holman, R.A. (1987). "Infragravity waves over a natural barred profile." J. Geophys. Res., 92(C9), 9531-9540.

Stive, M.J.F. (1980). "Velocity and pressure field of spilling breakers." Proc. 17th Coast. Engrg. Conf., ASCE, 547-566.

Stive, M.J.F. and Wind, H.G. (1982). "A study of radiation stress and set-up in the nearshore region." Coast. Engrg.. 6, 1-25.

Stive, M.J.R. (1986). "A model for cross-shore sediment transport." Proc. 20th Coast. Engrg. Conf., ASCE, 1550-1564.

Strzelecki, M.S. (1988). "Wave setup and swash on a beach and wave transmission over a nearshore bar." Will be presented to the University of Delaware in partial fulfillment of the requirements for the Master's degree in Civil Engineering.

Svendsen, I.A. and Brink-Kjaer, O. (1972). "Shoaling of cnoidal waves." Proc. 13th Coast. Engrg. Conf., ASCE, 365-383.

Svendsen, I.A., Schäffer, H.A. and Hansen, J.B. (1987). "The interaction between the undertow and the boundary layer flow on a beach." J. Geophys. Res., 92(C11), 11845-11856.

CHAPTER 88

TOWARD AN IMPROVED EMPIRICAL FORMULA FOR LONGSHORE SAND TRANSPORT

Nicholas C. Kraus[1], M. ASCE, Kathryn J. Gingerich[2], and Julie Dean Rosati[3], AM. ASCE

ABSTRACT

This paper presents results of two field experiments performed using portable traps to obtain point measurements of the longshore sand transport rate in the surf zone. The magnitude of the transport rate per unit width of surf zone is found to depend on the product of the local wave height and mean longshore current speed, but correlation is much improved by including two correction terms, one accounting for local wave energy dissipation and the other for the fluctuation in the longshore current. The field transport rates are also found to be compatible with laboratory rates obtained under combined unidirectional and oscillatory flow. Total transport rates previously reported for this experiment program are revised with recently determined sand trapping efficiencies.

INTRODUCTION

More than 30 years ago, it was empirically established that the longshore transport of sand on beaches is related to the height and direction of the incident breaking waves (Watts, 1953; Caldwell, 1956). Considerable effort has been made since then to improve empirical predictive capabilities for engineering applications. To a great extent, however, the present field data base rests on the tracer experiments of Komar and Inman (1970) and Kraus et al. (1983), and similar techniques which average over long time intervals and wide spatial extent. Such field programs have been directed mainly toward measuring the total longshore sand transport rate in the surf zone, and variability in the data is high.

Point measurements of local longshore sand transport rates are needed to determine dependencies of the rate on wave type and form, turbulence, current velocity, water depth, grain size, and beach morphology. This paper describes results of two field data collection projects performed by the Coastal Engineering Research Center (CERC), U.S. Army Engineer Waterways Experiment Station, which were aimed at

(1) Senior Research Scientist, (2) Research Physical Scientist, and (3) Hydraulic Engineer, Coastal Engineering Research Center, U.S. Army Engineer Waterways Experiment Station, P.O. Box 631, Vicksburg, MS 39180-0631.

measuring the local longshore sand transport rate by means of portable traps. This method is labor intensive but direct, and the determined rates, as averages over minutes, are compatible with modern engineering methodologies aimed at predicting sand transport and beach evolution. Main emphasis is placed on point measurements of the longshore sand transport rate per unit width of surf zone, with some results summarized for the vertical distribution of transport rates and total surf zone transport rates. Complete descriptions of the experiments and listings of the data are given in CERC technical reports (Kraus, Rosati, and Gingerich, 1989; Kraus, Gingerich, and Rosati, in prep.).

METHODOLOGY

Site

Results pertain to two field data collection projects performed in September 1985 and September 1986 at CERC's Field Research Facility (FRF). The FRF is located along the Outer Banks of North Carolina just north of the village of Duck. The data collection projects were named DUCK85 (Mason, Birkemeier, and Howd, 1987) and SUPERDUCK, respectively. A location map and site description are given by Kraus and Dean (1987). The experiments were conducted on a sandy beach at the north end of the FRF property line, approximately 1000 m from the FRF research pier. Most experiment runs were carried out in the south longshore feeder current of a moderate-sized rip current that frequently appears about 50 m north of the property line. Operation in this area provided unidirectional and quasi-steady longshore currents. In spite of the presence of the rip, the bottom contours where the traps were placed were predominantly plane and parallel. During DUCK85, the steeply sloping foreshore was composed mainly of gravel, whereas a sand substrate characterized the surf zone. Median grain size in the surf zone during both experiments was approximately 0.17 mm.

Equipment and Field Procedure

The longshore sand transport rate was measured with portable traps called streamer traps. The traps used in 1985 are described in Kraus (1987) and Kraus and Dean (1987). Hydraulic efficiency tests performed in a uniform flow flume led to redesign of the streamer nozzle that was then used during SUPERDUCK (Rosati, 1988; Rosati and Kraus, 1988). Figure 1 shows a trap used at SUPERDUCK. The nozzle is made of 1/16-inch stainless steel and is 15 cm long, 2.5 cm high, and 2.5 cm wide. A streamer consisting of 0.107-mm polyester filter cloth approximately 2 m long is attached to the nozzle. The streamer collects sand flowing into it while allowing water to pass through. Streamer nozzles are mounted vertically on stainless steel racks and located forward of the rack by curved steel bars to be upstream of the influence of the rack.

Sand trapping efficiencies of both types of streamer nozzles were determined by Rosati (1988) and used to revise DUCK85 results reported by Kraus and Dean (1987). Table 1 gives values of sand trapping efficiencies determined for both nozzle types. "On-Bed" sand trapping efficiencies were determined for nozzles resting on the bottom in a uniform flow tank with mid-flow speeds in the range of 60 to 66 cm sec^{-1}. This range of flow speeds produced a flat bed condition as found in the surf zone at DUCK85 and SUPERDUCK. "Off-Bed" efficiencies were determined for nozzles located above the bottom in mid-flow speeds ranging from 22 to 76 cm sec^{-1}.

The longshore current velocity was measured with one or two 2-component, 13-cm diameter ball, Marsh-McBirney electromagnetic flow meters mounted on tripods. The meters were connected to a super-minicomputer data logger on shore which gave a graphic display of the record and allowed data analysis upon completion of a run. Wave height and period were measured by the photopole method which has been described by Ebersole (1987) and Ebersole and Hughes (1987). The procedure involved filming the water surface elevation at poles placed at approximately 6-m intervals across the surf zone using as many as eight 16-mm synchronized movie cameras.

Figure 1. Streamer trap used at SUPERDUCK.

Table 1. Sand trapping efficiency factors for streamer nozzles in uniform flow (after Rosati, 1988).

Nozzle Type	On-Bed[a]	Off-Bed[b]
DUCK85 (15×9 cm)	0.13 ± 0.03[c]	0.92 ± 0.05
SUPERDUCK (15×2.5×2.5 cm)	0.68 ± 0.31[d]	1.02 ± 0.03

a) Nozzle resting on bed; collects bedload and suspended load.
b) Nozzle positioned above bed; collects suspended load only.
c) Maximum deviation about the mean value as determined by the 95% confidence limit.
d) Error estimate based on limited number of points.

At DUCK85 a spatial sampling method (SSM) was used, in which traps were simultaneously deployed across the surf zone to measure the cross-shore distribution of the longshore sand transport rate. Trapping intervals were 10 min. Measured waves and currents from the trapping periods were analyzed in corresponding time segments to give estimates of the the local forcing functions. The SSM results are described in Kraus and Dean (1987); however, magnitudes of transport rates reported there are incorrect because of a miscalculation in the original analysis and adjustment of measured quantities by the trapping efficiencies given in Table 1. Revised results of the SSM runs are presented here.

Longshore sand transport rate data collection at SUPERDUCK emphasized a temporal sampling method (TSM) in which traps were interchanged from 3 to 14 times at the same location. Trapping intervals typically were 6 min. Waves and currents were also measured as described above. Data from photopoles located immediately seaward and shoreward of the trap location were analyzed to define the average wave height at the trap and to obtain the cross-shore gradient of wave height at the trap. Typically, traps were centered between photopoles.

Mean water surface elevations were obtained from both the photopole record and a tide gage located on the FRF research pier. The beach profile was surveyed in the immediate vicinity of the experiments at least once a day from a survey station mounted on the pier.

Analysis Technique

The wave and current data were subjected to standard statistical and time series analysis. Sand collected in the streamers was weighed at the beach in a drip-free state that has been shown to be closely correlated with the dry weight of a sample (Kraus and Nakashima, 1986). Samples from selected runs of each experiment day were retained for grain size analysis and drying to determine the empirical coefficient relating drip-free and dry sand weight. Example results of vertical and cross-shore distributions of grain size were described by Kraus and Dean (1987) for the DUCK85 experiments. The median grain size of the trapped sand did not significantly vary with elevation in the water column or across the surf zone.

The streamer trap gives a direct measure of the flux f of sand at each nozzle in units of weight of sand per unit width normal to the transport direction per unit elevation per unit time. Measured fluxes for each streamer were adjusted by dividing by the average trapping efficiencies given in Table 1, and fluxes in areas between nozzles were obtained by interpolation. Integration of the flux through the water column gives the transport rate density i, which can be expressed as the immersed weight transport rate per unit width:

$$i = \int_{-h}^{\eta} f \, dz \,, \qquad [i] = \frac{N}{m * sec} \qquad (1)$$

in which η is the mean water surface elevation, z is the elevation measured from the still-water level, and h is the water depth at the trap. The measured transport rate density, which is obtained as a dry weight of trapped sand, was expressed as a standard immersed weight rate.

For the SSM, transport rate densities were integrated across the surf zone by use of the trapezoid rule to give the total immersed weight longshore sand transport rate I:

$$I = \int_{o}^{X_b} i \, dx \, , \qquad [I] = \frac{N}{sec} \qquad (2)$$

in which x is the cross-shore distance with origin at the mean water line, and X_b is the width of the surf zone. The location of the break point was determined from photopole data as the location of the maximum root mean square (rms) wave height; this quantity is known to within +/- 3 m. If a trap was located outside the surf zone, the seaward limit of integration was extended to that point. Longshore transport immediately outside the breaker zone was found to be very small.

Calculated transport rates were correlated with simple combinations of measured mean wave height H and longshore current speed V, as discussed below, to obtain empirical predictive relations. Most theoretical and empirical expressions for the longshore sand transport rate density i based on either wave energy dissipation or bottom shear stress reduce to a simple dependence on the product HV. The present data set of local sand transport, wave, and current measurements allows detailed examination of the relationship between transport rate and forcing parameters.

RESULTS

Table 2 provides a summary of the range of wave, current, and sand transport conditions comprising the presently analyzed data set. Wave heights and periods given for DUCK85 pertain to breaking waves over a nominal measurement interval of 12 min, and values of the transport rate and current pertain to a 10-min interval. Wave quantities and current speeds for SUPERDUCK represent average values at the traps for a transport collection interval of 6 to 8 min.

Table 2. Range of wave, current, and transport rate conditions.

	H_s[a] (m)	H_{rms} (m)	V (m sec^{-1})	T[b] (sec)	i[c] (kg $m^{-1}min^{-1}$)	i[d] (N $m^{-1}sec^{-1}$)
DUCK85						
Maximum	1.19	0.86	0.33	10.3	8.88	0.89
Minimum	0.83	0.57	0.08	8.9	0.21	0.02
SUPERDUCK						
Maximum	1.05	0.73	0.58	10.1	14.30	1.43
Minimum	0.53	0.42	0.09	6.1	0.44	0.04

a) Significant wave height; b) Average wave period.
c) Dry mass longshore sand transport rate density.
d) Immersed weight longshore sand transport rate density.

Transport Rate Density

Discussion in this section will center on 39 values of the transport rate density obtained in six TSM runs performed during SUPERDUCK. Rms wave height is used in the present analysis, unless otherwise specified, because correlations were always slightly higher with rms wave height than with significant wave height.

Standard formulas for i derived from either a bottom shear stress approach (e.g., Komar, 1971) or a wave energetics approach (e.g., Inman and Bagnold, 1963) reduce to a leading dependence on the product of wave height and longshore current speed if linear shallow water wave theory is employed. Thus, as a first step, measured transport rate densities were plotted with respect to the quantity ρgHV, in which ρ is the density of seawater, and g is the acceleration due to gravity. The result is shown in Figure 2, in which the straight line is a best fit from linear regression analysis. Values of the determined regression equation coefficients and the correlation coefficient squared (r^2) are listed in Table 3. Figure 2 shows that the measured transport rate densities are fairly well described by a purely linear function of HV. However, scatter is relatively great, and the trend in the data suggests a power law dependence on HV.

Qualitative observations made during DUCK85 indicated that the trapped amount of sand depended on the intensity of water agitation occurring at or immediately seaward of a trap. For example, the transport rate appeared to increase in turbulent white water as compared to calmer green water for traps located at approximately the same depth. The white, agitated water was produced by waves breaking at the trap or convected to the trap by waves breaking immediately seaward. The local gradient of the wave height dH/dx was identified

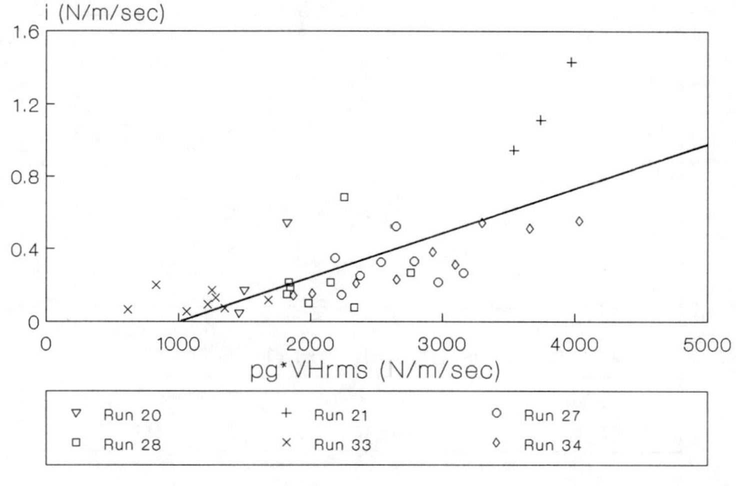

Figure 2. Longshore sand transport rate density versus HV.

Table 3. Summary of regression results for the equation $i = k\ [\rho g\ HV\ (1 + \alpha\ dH/dx + \beta\ S_V/V) + \text{const.}]$

Expresssion	$k \cdot 10^4$	α	β	const. ($N\ m^{-1} sec^{-1}$)	r^2
$H_s V$	1.8	0	0	$-1.2 \cdot 10^3$	.45
$H_{rms} V$	2.5	0	0	$-9.9 \cdot 10^2$	.51
$H_{rms} V(1+\alpha dH/dx)$	2.0	20	0	$-7.7 \cdot 10^2$	.66
$H_{rms} V(1+\alpha dH/dx+\beta S_V/V)$	1.5	20	1.8	$-2.4 \cdot 10^3$	.77

as a readily evaluated measure of water agitation, and the SUPERDUCK TSM experiments were configured to provide this quantity. The gradient of the wave height was calculated from the nearest two poles (i.e., over a 6-m interval). This quantity was usually negative, indicating a decrease in wave height as the waves moved toward shore. However, in some cases the gradient was positive, indicating that broken waves were reforming.

The gradient of wave height was introduced as a correction to the quantity HV in the form of $HV(1 + \alpha\ dH/dx)$ in which the value of the empirical coefficient α was determined by iteration to provide the best linear least squares fit. The resultant plot and regression line are given in Figure 3. Visual agreement and the correlation coefficient are considerably improved over Figure 2, which involved only the product HV.

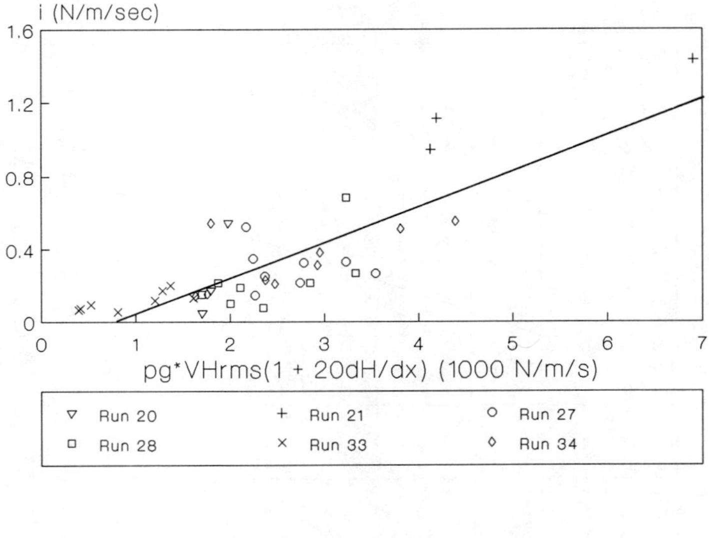

Figure 3. Longshore sand transport rate density versus $HV(1 + \alpha\ dH/dx)$.

The longshore current speed used in the above analysis is the average of a time-varying flow. The sand transport rate should depend on the range of current speed as well as the average. As a measure of the range, the coefficient of variation of the current speed S_V/V was chosen, in which S_V is the standard deviation of the speed during the averaging interval. The coefficient of variation was conceptualized as providing a correction to the leading term HV , and the quantity $HV(1 + \alpha\ dH/dx + \beta\ S_V/V)$ was used for regression. The result is shown in Figure 4, and associated values of determined coefficients are given in Table 3. Grouping of the data points about the regression line is improved over previous plots, and the apparent necessity of using a nonlinear or power law function of HV , as was suggested by Figure 2, is eliminated.

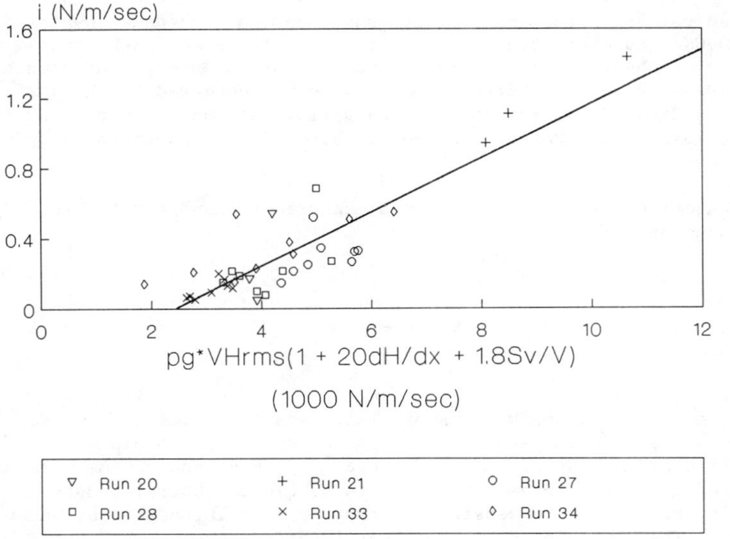

Figure 4. Longshore sand transport rate density versus $HV(1 + \alpha\ dH/dx + \beta\ S_V/V)$.

The correlation lines in Figs. 2, 3, and 4 all intercept the positive x-axis. The value of the intercept is partially an artifact of the use of a straight-line regression analysis. However, the intercept may be interpreted as an effective cutoff for transport of significance in engineering applications, since transport rates lying below this value evidently have a much weaker dependence on the quantity HV than the plotted measured values.

Stepwise correlation analysis indicated that there was no relation between the quantities H, dH/dx, V, and S_V. In a situation where the longshore current is produced by obliquely incident waves, the magnitude of the current speed is proportional to the square root of the wave height. In the present case, V and H were not related because the experiments were performed in or near the feeder current of a rip current. Caution should be taken in general use of the correction term proportional to α , as most TSM measurements were performed on a plateau with a very mild slope. Values of dH/dx ranged from -0.035 to 0.032 and values of S_V/V ranged from 0.02 to 0.37. The present empirical formulation is expected to lead to erroneous results if the value of either of the correction terms exceeds unity.

Comparison with Laboratory Data

Katori, Sakakiyama, and Watanabe (1984) measured the sand transport rate produced in a unique cross-flow tank in which a unidirectional current and an oscillatory current were made to intersect at right angles. Experiments were performed for pure unidirectional flow and for combined unidirectional and uniform flow; here, only the latter measurements are considered. The cross-flow tank replicated surf zone flow conditions, except that turbulence generated from the surface by breaking waves was absent. Uniform flow speeds achieved in the tank reached values in the range of longshore current speeds encountered at SUPERDUCK. However, sand ripples typically appeared in the test section of the tank, evidently because of the absence of turbulence from breaking waves, whereas ripples were not observed in the surf zone during SUPERDUCK. Transport rate densities obtained by Katori et al. were compared with rates obtained at SUPERDUCK to examine magnitudes and trends.

Katori et al. (1984) expressed measured transport rates in nondimensional form as:

$$\Phi = \frac{q}{w_o d} \tag{3}$$

in which q is the bulk transport rate density (units of $m^3 m^{-1} sec^{-1}$), w_o is the sand fall speed, and d is the median grain diameter. Three uniform quartz sands of median diameter 0.2, 0.4, and 0.7 mm were used. Measured transport rates were found to be closely correlated to a quantity called the "dimensionless flow power," introduced by Watanabe (1982). The dimensionless flow power is defined as:

$$\Theta = \frac{(\tau_m - \tau_c)\, V}{\rho (sgd)^{1.5}} \tag{4}$$

in which τ_m is the maximum shear stress at the bottom produced by the combined flow, τ_c is the shear stress for inception of sand movement, V is the magnitude of the steady current, $s = \rho_s/\rho - 1$, and ρ_s is the density of quartz. Katori et al. (1984) obtained the result $\Phi = 1.8\ \Theta^{1.5}$ for transport rates ranging over two orders of magnitude for their complete data set. They argued that the power 1.5 may have been artificially high because of limitations in experiment conditions. One characteristic viewed as a limitation was the unsteadiness of the unidirectional current in the combined flow tests, a condition that actually mimics the longshore current in the field.

In the present study, quantities analogous to Φ and Θ were calculated for the SUPERDUCK data. The shear stress was taken as $\tau_m = 1/2\ \rho f_w\ (U_m^2 + V^2)$, in which f_w is the friction factor introduced by Jonsson (1967), and U_m is the amplitude of the wave orbital velocity at the bottom, which was calculated by shallow water linear wave theory using measured water depth and rms wave height. The critical shear stress in Eq. 4 was assumed to be zero for the field measurements. Katori et al. (1984) had used a combined wave and current friction factor to calculate shear stress.

Nondimensionalized SUPERDUCK transport rates are plotted in Figure 5 together with the data of Katori et al. (1984). The total set of measured transport rates and flow powers spans more than three orders of magnitude. The best-fit power law for the 56 points was

$$\Phi = 0.85\ \Theta^{1.1} \tag{5}$$

with $r^2 = 0.90$. The approximate agreement in trends in the laboratory and field data is surprising because transport in the laboratory was dominated by ripple processes, whereas field transport occurred over a flat bed under highly turbulent flow conditions. This result is encouraging and supports the validity of energetics-based transport concepts. Regression on the 39 SUPERDUCK data points gave $\Phi = 1.14\ \Theta^{0.95}$, with $r^2 = 0.54$. This r^2-value is essentially the same as found for the simple $i \propto HV$ analysis given in Table 3 and is not as good as the agreement produced by introducing corrections for the gradient in wave height and variability in the current (cf. Figure 4). Thus, introduction of the friction factor through Eq. 4 did not provide an improvement over the simple empirical transport relation using only HV. (For completeness, the 15 laboratory points used here were described by $\Phi = 1.10\ \Theta^{1.3}$, with $r^2 = 0.91$.)

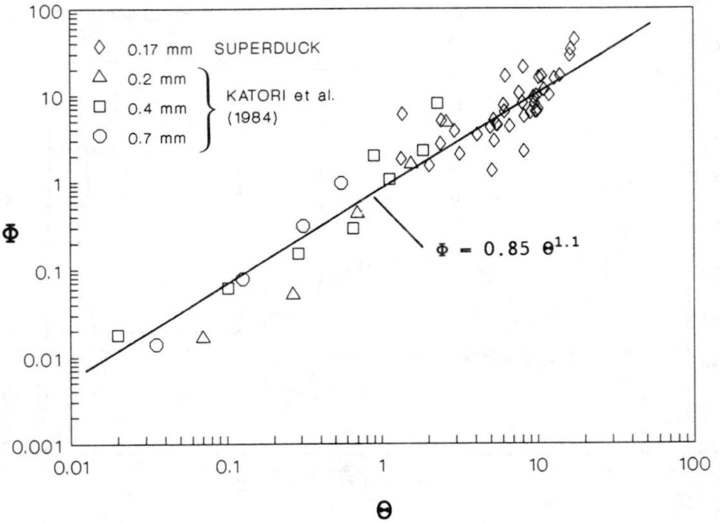

Figure 5. Dimensionless transport rate versus flow power.

Lateral Distribution of the Transport Rate

Figure 6 shows an example of cross-shore transport rate distributions obtained during a 10-min SSM experiment at DUCK85. The lengths of the histograms represent the magnitude of the longshore flux of sand measured at the specified location and elevation. Similar figures have been given by Kraus and Dean (1987); however, the values in Figure 6 have been revised. The flux decreases sharply with increase in elevation, independent of location in the surf zone. This regular decrease in flux was found in all experiments. On the basis of these observations, Kraus and Dean (1987) proposed an expression for the transport rate density of the form:

$$i(x,z) = i_o(x) * p(h,z) \tag{6}$$

in which i_o is a "magnitude function" for the transport rate, assumed to depend on local wave, current, water depth, and beach conditions, z is elevation above the bed, and p is a "shape function," assumed to have a universal form for the surf zone as:

$$p = e^{-\lambda(z/h)} \tag{7}$$

The average value of λ obtained from the DUCK85 data was 3.6 with a standard deviation of 0.86 for 55 trap deployments, whereas for the SUPERDUCK data the average λ was 2.8 with a standard deviation of 0.82 for 44 deployments.

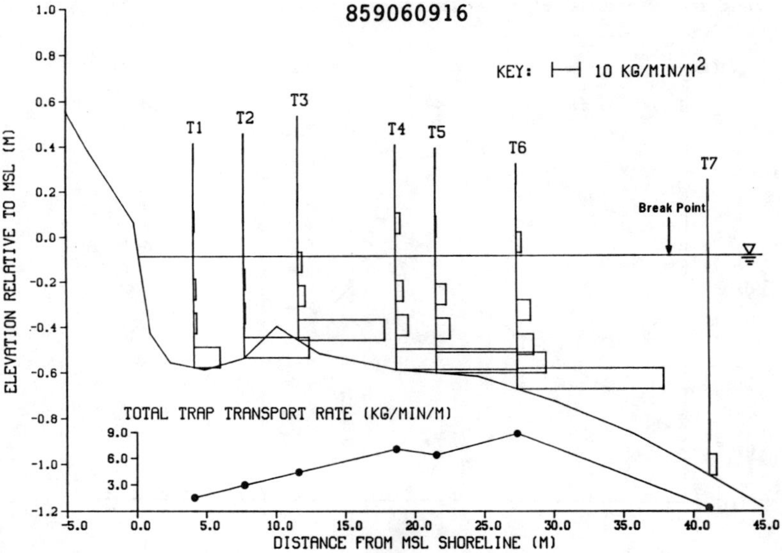

Figure 6. Example of vertical distributions of the longshore sand flux across the surf zone.

These revised values of λ indicate that sand movement at and close to the bed formed the major portion of the transport. This reverses the conclusion of Kraus and Dean (1987) and supports the idea that "bedload" transport predominates over suspended load (Komar, 1978) for the range of wave and current conditions and grain size encountered in these experiments.

The following expression was obtained for the magnitude function in Eq. 6 on the basis of the SUPERDUCK TSM data:

$$i_o = k_o\ [\rho g\ HV\ (1 + \alpha_o\ dH/dx + \beta_o\ S_V/V) + \text{const.}] \qquad (8)$$

where the subscript "o" denotes values determined for $z = 0$ in Eq. 6. This equation had $r^2 = 0.74$, and values of the empirical coefficients were: $k_o = 9.4 \cdot 10^{-4}$; const. $= -3.0 \cdot 10^3$ N $m^{-1} sec^{-1}$; $\alpha_o = 20$; and $\beta_o = 1.8$.

Total Transport Rate

Total longshore sand transport rates measured during DUCK85 were found to be well correlated with a quantity $R = VX_bH_b$ called the discharge parameter (Kraus and Dean, 1987). Since the product X_bH_b is proportional to the cross-sectional water area of the surf zone, R is proportional to the average discharge of water moving alongshore. Figure 7 shows total transport rates plotted as a function of R for eight DUCK85 runs. The surf zone width was typically 30-40 m, and 6 or 7 traps were used. Figure 7 contains one additional data point than an analogous figure given by Kraus and Dean (1987).

Assuming that a linear dependence exists between I and R, the following regression equation plotted as the straight line in Figure 7 is obtained with $r^2 = 0.76$:

$$I = 2.7\ (R - R_c) \qquad (9)$$

in which the intercept $R_c = 3.9\ m^3 sec^{-1}$ is interpreted as a threshold value for significant longshore sand movement. In Eq. 9, I is expressed in N sec^{-1} and R in $m^3 sec^{-1}$.

The longshore current which generated the transport rates plotted in Figure 7 and described by Eq. 9 was not directly related to oblique wave incidence, but was associated with the circulation of a rip current cell. Therefore, predictions from Eq. 9 cannot be directly compared to the CERC formula (Shore Protection Manual, 1984) for the total longshore sediment transport rate. Inman and Bagnold (1963) have given a predictive equation for I based on wave-energetics concepts that was shown by Komar and Inman (1970) to be compatible with the CERC formula. The Inman and Bagnold equation is:

$$I = K'(ECg)_b \frac{V}{U_m} \qquad (10)$$

in which $(ECg)_b$ is the wave energy flux at breaking (evaluated for rms wave height), V is the average longshore current speed and U_m is the maximum wave orbital velocity under breaking waves. The value of the empirical coefficient $K' = 0.28$ determined by Komar and Inman (1970) is consistent with the value of the empirical coefficient commonly used in the CERC formula.

If shallow water linear wave approximations are used, the functional form of Eq. 10 reduces to $H_b{}^2V$, which is essentially equivalent to the discharge parameter R if the surf zone bottom can be approximated by a plane sloping surface. Thus the basic functional forms of Eqs. 9 and 10 are similar, except for the incorporation of an effective threshold in Eq. 9.

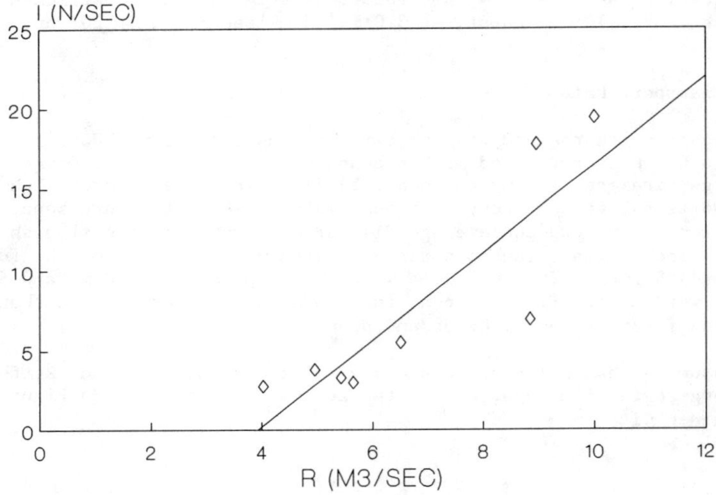

Figure 7. Total immersed weight longshore sand transport rate versus discharge parameter.

CONCLUDING DISCUSSION

These direct measurements of the longshore sand transport rate per unit width of surf zone confirm the functional form of previously derived theoretical models of transport which have a leading dependence on the product of wave height and longshore current speed. The results also indicate that considerably improved correlation between the transport rate and local waves and currents can be obtained by including corrections due to turbulence introduced by breaking waves and the variation in the longshore current speed. [Recently, Roelvink and Stive (personal communication, 1988) have derived a theoretical expression for the transport rate density that includes an explicit contribution from breaking wave-induced turbulence.] The next stage of analysis will include more data from the SUPERDUCK field project and other experiments, and an investigation of higher-order moments of the wave orbital velocity and longshore current.

In experiments involving use of portable traps, the bottom streamer nozzle, which collects both bedload and suspended load near the bottom, contained the major portion of the total transport at the trap. The vertical profile of the transport rate density decreased rapidly with elevation from the bed, and both the estimated transport rate density at the bottom and profile shape could be reasonably approximated by simple functions.

Field sand transport rates showed surprising compatibility with laboratory transport rates generated under intersecting uniform and oscillatory flows, despite the fact that the bed surfaces were quite different (flat in the field and rippled in the laboratory). The nondimensionalized transport rates were reasonably well described by an energetics-based predictive relation which reduces to the functional form of HV. Incorporation of a friction coefficient did not improve correlation above an empirical HV-predictive form; however, the range in wave period, which would change the friction factor, was limited in the field experiments.

The trend in total surf zone sand transport rate as displayed in Figure 7 is supportive of the energetics theory of Inman and Bagnold (1963). It also suggests that a simple empirical transport rate formula for the total transport rate expressed in the basic functional form H_b^2V or H_bX_bV may apply equally as well as more sophistiated formulas based on quantities such as wave energy flux and wave orbital velocity.

ACKNOWLEDGEMENTS

Space does not permit us to properly acknowledge all the individuals who resolutely labored to collect the field data. They are acknowledged in the data reports. We would like to thank Mr. Sadakazu Katori, Coastal Consultants, Inc., Tokyo, Japan, for providing his laboratory data. The experiments described and the resulting data presented herein, unless otherwise noted, were obtained under the work unit "Surf Zone Sediment Transport Processes" of the Shore Protection and Restoration Program of the United States Army Corps of Engineers by the Coastal Engineering Research Center at the Waterways Experiment Station. Permission was granted by the Chief of Engineers to publish this information.

REFERENCES

Caldwell, J. M. 1956. "Wave Action and Sand Movement near Anaheim Bay, California," Tech. Memo. No. 68, Beach Erosion Board, U.S. Army Corps of Engrs., 21 pp.

Ebersole, B. A. 1987. "Measurements and Prediction of Wave Height Decay in the Surf Zone," Proc. Coastal Hydrodynamics, R. A. Dalrymple, ed., Am. Soc. Civil Engrs., pp. 1-16.

Ebersole, B. A. and Hughes, S. A. 1987. "DUCK85 Photopole Experiment," Misc. Pap. CERC 87-18, U.S. Army Engr. Waterways Expt. Station, Coastal Engrg. Res. Center, Vicksburg, Miss., 62 pp.

Inman, D. L. and Bagnold, R. A. 1963. "Littoral Processes," In: The Sea, M. N. Hill, ed., Interscience, New York, pp. 529-533.

Jonsson, I. G. 1967. "Wave Boundary Layer and Friction Factors," Proc. 10th Coastal Engrg. Conf., Am. Soc. Civil Engrs., pp. 127-148.

Katori, S., Sakakiyama, T., and Watanabe, A. 1984. "Measurement of Sand Transport in a Cross Unidirectional-Oscillatory Flow Tank," Coastal Engrg. in Japan, 27, pp. 193-203.

Komar, P. D. 1971. "The Mechanics of Sand Transport on Beaches," J. Geophys. Res., 76(3), pp. 713-721.

Komar, P. D. 1978. "Relative Quantities of Suspension Versus Bed-Load Transport on Beaches," J. Sed. Petrol., 48(3), pp. 921-932.

Komar, P. D., and Inman, D. L. 1970. "Longshore Sand Transport on Beaches," J. Geophys. Res., 75(30), pp. 5914-5927.

Kraus, N. C. 1987. Application of Portable Traps for Obtaining Point Measurements of Sediment Transport Rates in the Surf Zone," J. Coastal Res., 3(2), pp. 139-152.

Kraus, N. C. and Dean J. L. 1987. "Longshore Sediment Transport Rate Distribution Measured by Trap," Proc. Coastal Sediments '87, N. C. Kraus, ed., Am. Soc. Civil Engrs., pp. 881-896.

Kraus, N. C. and Nakashima, L. 1986. "Field Method for Determining Rapidly the Dry Weight of Wet Sand Samples," J. Sed. Petrol., 56(4), pp. 550-551.

Kraus, N. C., Isobe, I., Igarashi, H., Sasaki, T. O., and Horikawa, K. 1983. "Field Experiments on Longshore Sand Transport in the Surf Zone," Proc. 18th Coastal Engrg. Conf., Am Soc. Civil Engrs., pp. 969-988.

Kraus, N. C., Rosati, J. D., and Gingerich, K. J. 1989. "DUCK85 Surf Zone Sand Transport Data Report," Tech. Rep., U.S. Army Engr. Waterways Expt. Station, Coastal Engrg. Res. Center, Vicksburg, Miss.

Kraus, N. C., Gingerich, K. J., and Rosati, J. D. in prep. "SUPERDUCK Surf Zone Sand Transport Data Report," Tech. Rep., U.S. Army Engr. Waterways Expt. Station, Coastal Engrg. Res. Center, Vicksburg, Miss.

Mason, C., Birkemeier, W. A., and Howd, P. A. 1987. "An Overview of DUCK85, A Nearshore Processes Experiment," Proc. Coastal Sediments '87, N. C. Kraus, ed., Am. Soc. Civil Engrs., 818-833.

Rosati, J. D. 1988. "Calibration of a Portable Sediment Trap for Use in the Nearshore," unpub. M.S. Thesis, Dept. of Civil Engrg., Mississippi State Univ., Starkville, Miss.

Rosati, J. D. and Kraus, N. C. 1988. "Hydraulic Calibration of the Streamer Trap," J. Hydraulic Engrg, 114(12), in press.

Shore Protection Manual. 1984. 4th ed., 2 Vols. U.S. Army Engr. Waterways Expt. Station, Coastal Engrg. Res. Center, U.S. Govt. Printing Office, Washington, DC.

Watanabe, A. 1982. "Numerical Models of Nearshore Currents and Beach Deformation, Coastal Engrg. in Japan, 25, pp. 147-161.

Watts, G. M. 1953. "A Study of Sand Movement at South Lake Worth Inlet, Florida," Tech. Memo. No. 42, Beach Erosion Board, U.S. Army Corps of Engrs., 24 pp.

CHAPTER 89

GENERIC TREATMENT OF DUNE EROSION FOR 100-YEAR EVENT

by
Robert J. Hallermeier*
and
Perry E. Rhodes#
Member ASCE

Abstract

Quantitative procedures provide the eroded dune geometries expected during an extreme storm on U.S. Atlantic and Gulf coasts. A criterion for either duneface retreat or dune removal is based on measured erosion cross sections in many cases where the pre-storm dune proved to be a durable barrier to wave effects. After that decision, the eroded profile is constructed using specified planar segments cut into or across the existing dune. This empirical treatment appears appropriately detailed for assessing hazards due to the 100-year event considered within the U.S. National Flood Insurance Program, and an example assessment is outlined. Additional discussion addresses other site-specific factors possibly affecting dune erosion, along with another potential application for this simplified methodology.

INTRODUCTION

The Federal Emergency Management Agency develops a Flood Insurance Study (FIS) for each community participating in the U.S. National Flood Insurance Program (NFIP). The focus of an FIS is expected effects in the flood having a one-percent chance of being equalled or exceeded in any year. This "base flood" is equivalent to the 100-year event for a given site, expected to recur once each 100 years on the average. Open-coast communities are subject to particularly extreme hazards due to storm surge and wave action from large water bodies. Coastal areas of special flood hazard in 100-year events are designated as V zones, having the potential for inundation by water flows with significant velocity. Within a V zone, floodwater depth is sufficient to permit a wave height of at least 3 feet (0.9 meters).

*Senior Coastal Scientist, Dewberry & Davis
#Senior Coastal Engineer and Associate, Dewberry & Davis, 8401 Arlington Boulevard, Fairfax, Virginia 22031 USA

Proper delineation of V zones requires quantitative determination of wave effects associated with the base flood, including coastal erosion.

Coastal sand dunes usually extend above local 100-year flood elevations, but such apparent barriers to flooding may not prove durable in view of the massive erosion associated with extreme storms. Several levels of dune effectiveness as a flooding barrier need to be distinguished, as outlined in a schematic manner by Figure 1. These four cases correspond to the eroded dune providing:

a. a reliable barrier to flood effects from the sea;
b. a nearly complete barrier, with some overtopping;
c. a partial barrier, with some wave transmission; or
d. an ineffective barrier, as relatively deep water and high waves reach past the initial dune site.

Sound analysis of potential flood hazards thus requires an objective estimate of dune geometry during the base flood.

Treatment of expected erosion must be consistent with other NFIP methodologies. First, it should be recognized that the base flood represents a statistical measure rather than the actual effects arising in some particular coastal storm. Second, erosion assessment must conform to the standard FIS transect viewpoint: representative shore-normal profiles are considered. Third, the 100-year still-water flood level (SWFL) has a known elevation on a given coastal transect, evaluated prior to treating additional wave effects expected to occur in the base flood. Fourth, eroded dune geometry must be determined by a uniform but perhaps highly simplified method, appropriately reflecting basic physical principles, as in FIS procedures for treating wave dimensions (National Academy of Sciences, 1977).

This paper outlines objective procedures specifying coastal dune erosion expected to accompany the base flood. These estimates of eroded dune geometry are meant for application at exposed U.S. sites along the Atlantic Ocean and Gulf of Mexico. The present erosion treatment has a broad empirical basis and appears appropriate in the context of other NFIP methodologies for wave effects in extreme events.

DUNEFACE RETREAT AND DUNE REMOVAL

The Federal Emergency Management Agency (1986) assessed usual procedures and developed several recommendations to correct weaknesses in FIS identification of coastal high-hazard areas, or V zones. Recognizing the transient and sensitive nature of coastal sand dunes, it was recommended that the entire primary frontal dune be designated a V zone. Since NFIP regulations prohibit alteration of sand dunes in V zones, this serves to protect natural dunes. Substantial construction standards and adequate insurance rates now automatically apply in this area exposed to high hazards, by means of an NFIP rule effective October 1, 1988 (Federal Emergency Management Agency, 1988).

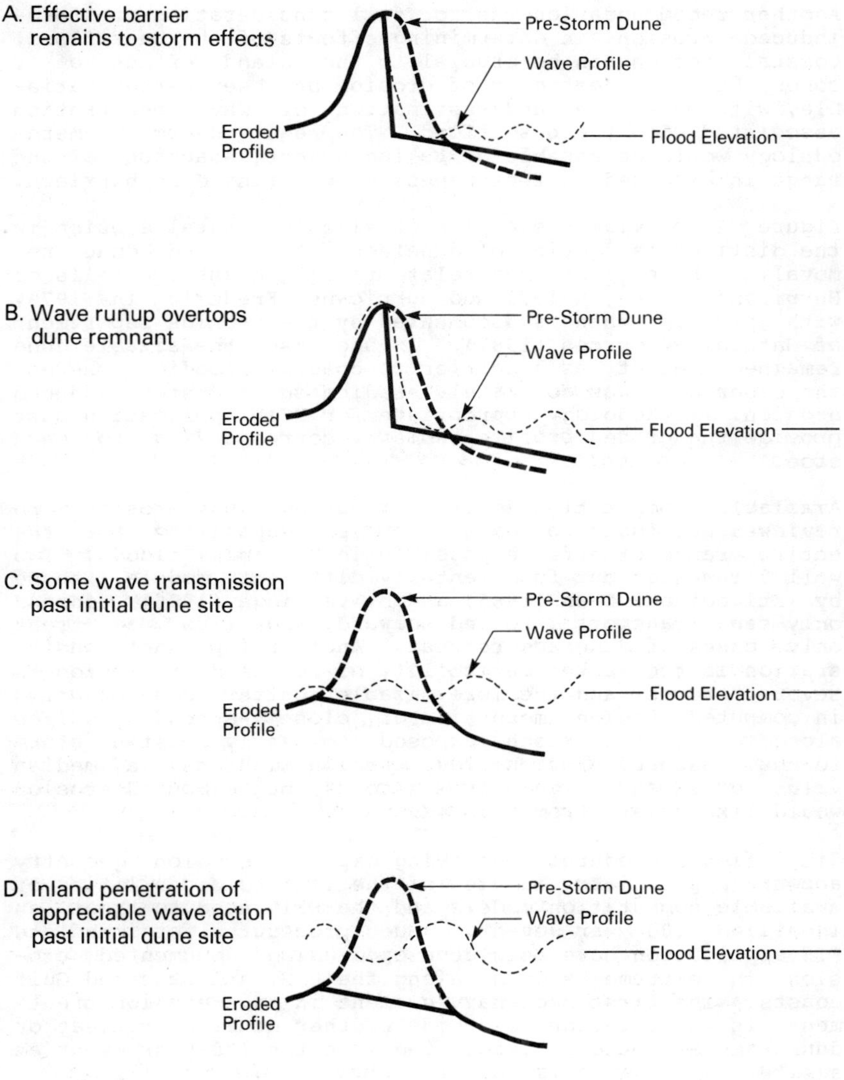

Figure 1. Schematic illustration of possible levels of flood protection provided by eroded frontal dunes

Another recommendation was for full consideration of storm-induced erosion in determining effects of the base flood. Coastal erosion can be crucial to the inland extent of V zones, but FIS treatments of erosion had been rather variable, with occasional underestimation of wave penetration associated with the base flood. The required erosion methodology would be capable of taking objective account of the range in expected effectiveness of existing dune barriers.

Figure 2 provides examples of sizable coastal erosion in the distinct categories of duneface retreat and dune removal. These instances relate to Gulf coast landfalls of Hurricane Eloise in 1975 and Hurricane Frederic in 1979, with profile changes documented by the Florida Department of Natural Resources (1987). In one case, the sizable dune remained intact as a barrier to coastal flooding, whereas the other dune was completely eradicated by storm-induced erosion. Methodology appropriate for NFIP application must provide the eroded profile geometry corresponding to each case.

Available computation models treating dune erosion were reviewed and found to have unproven capability for the entire range of effects possible in U.S. base floods. Two well documented but fundamentally different models, those by Kriebel and Dean (1985) and by Vellinga (1986), address only sand transport directed seaward, and thus can treat only cases of duneface retreat. Another important consideration is the marked variability of actual dune erosion in severe storms and the more sizable scatter to be expected in computed erosion amounts. For closely-spaced profiles along a coastal reach exposed to nearly constant storm forces, measured erosion cross section might have a median value of 2X and a range from X to 3X, but computed erosion would likely vary from 0 to 4X.

Simplified procedures specifying expected erosion geometry appear appropriate in view of the limited capabilities of available computation models and the NFIP need to treat an idealized 100-year event. Such procedures for immediate FIS application have been developed using documented erosion in extreme storms along the U.S. Atlantic and Gulf coasts. The first necessary element in this erosion treatment is a criterion to decide whether duneface retreat or dune removal would be associated with the 100-year event at a site.

EXPECTED DUNEFACE RETREAT IN 100-YEAR EVENT

The central analysis in this methodology development defined the sand reservoir above SWFL required for a dune to remain intact during specific storms. Published data were reviewed to accumulate 38 separate cases of duneface retreat, where pre- and post-storm profiles confirm that all eroded sand was transported seaward. In each case, the existing dune at a site provided a durable and effectively unlimited barrier during the particular storm. The data define a relationship between erosion cross sections and

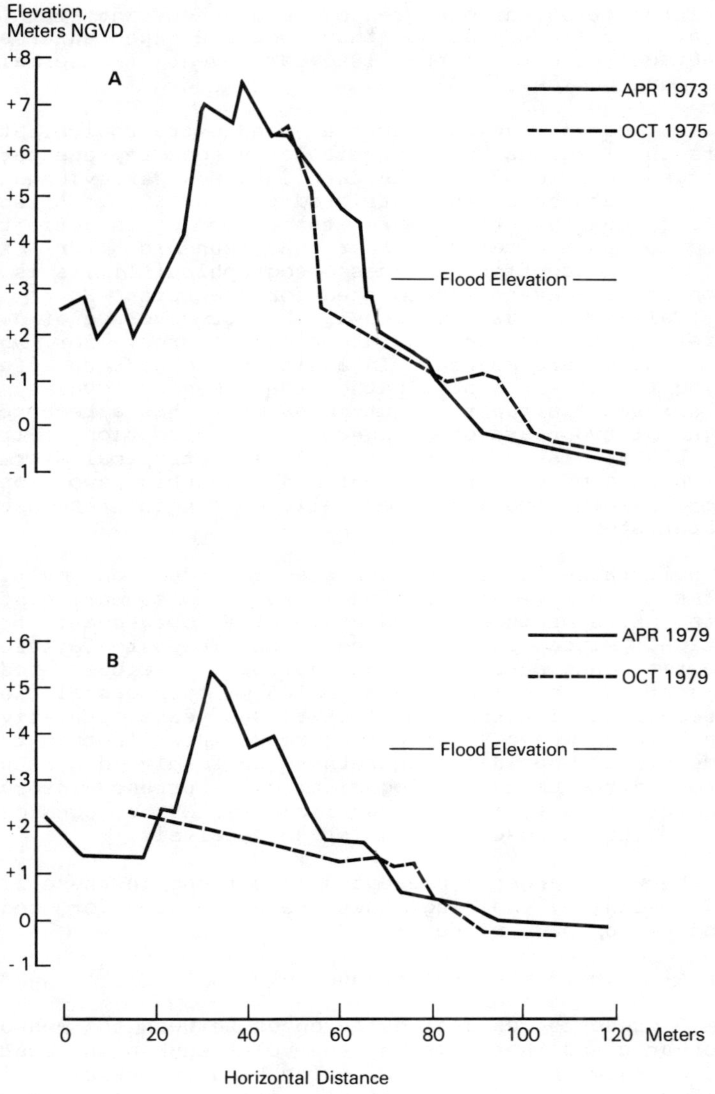

Figure 2. Examples of duneface retreat and dune removal, as documented by Florida Department of Natural Resources (1987). Effects are: A - September 1975 Hurricane Eloise, Profile R-15, Walton County, Florida; B -September 1979 Hurricane Frederic, Profile B-13, Baldwin County, Alabama.

storm intensities, as measured by recurrence interval for SWFL at specific sites, and thus indicate expected erosion in duneface retreat for the 100-year event (Hallermeier, Rhodes, and Buckley, 1988).

These 38 cases include hurricane and extratropical storm impacts in 11 states, with particularly good representation of sites in Florida, North Carolina, New Jersey, and New York. The data base was restricted to the U.S. Atlantic and Gulf coasts, to ensure that flooding is definitely related to storm effects (rather than tsunamis, for example). The exception to these geographical limits is inclusion of five events documented for the Dutch North Sea coast (Vellinga, 1986), to provide better coverage of duneface retreat in extreme extratropical storms, for which U.S. examples are scarce. In addition, to address effects relating to extreme storms rather than seasonal cycles, the data set was limited to events where SWFL has a recurrence interval of two years or longer. The exception (Birkemeier, 1979) is a December 1977 extratropical storm at sites where a more extreme event had occurred two months earlier, likely ensuring a generally meaningful response to the later storm.

About half these cases have changes recorded on multiple profiles, where median erosion is taken to summarize storm effects. Erosion is measured above the open-coast flood elevation, since such elevated change leads to failure of the initial dune barrier. The database includes median cross-sectional erosion from 5 to 105 m^2 for coastal floods with recurrence intervals of 1.25 to 300 years. Relatively common events have much better representation than truly extreme events, but data are rather uniformly distributed when considered using the logarithm of recurrence interval. The number of cases appears sufficient to be considered large for the purposes of statistical analysis.

Over the wide ranges represented, the trend in these field data is definite, yielding this relationship for median erosion in duneface retreat:

$$\text{Erosion } [m^2] = 8\ (\text{Recurrence Interval [yr]})^{0.4} \qquad (1)$$

Figure 3 presents residual differences between the measured erosion area and that given by the relationship in separate cases, along with an empirical error band for results. For the generic 100-year event, results indicate that expected duneface retreat will amount to $(50 \pm 15) m^2$ above local SWFL at U.S. Atlantic or Gulf sites. The uncertainty in this estimate refers to median erosion, not the actual erosion variability along the coast in a specific storm.

Figure 3 reveals little pattern among residual differences, showing this relationship has no distinct empirical defect. Relatively large underestimates of erosion appear somewhat more common than large overestimates, so that the expected duneface erosion has not been summarized as too extreme. However, few U.S. frontal dunes are massive enough to permit the indicated erosion for the 100-year event. This

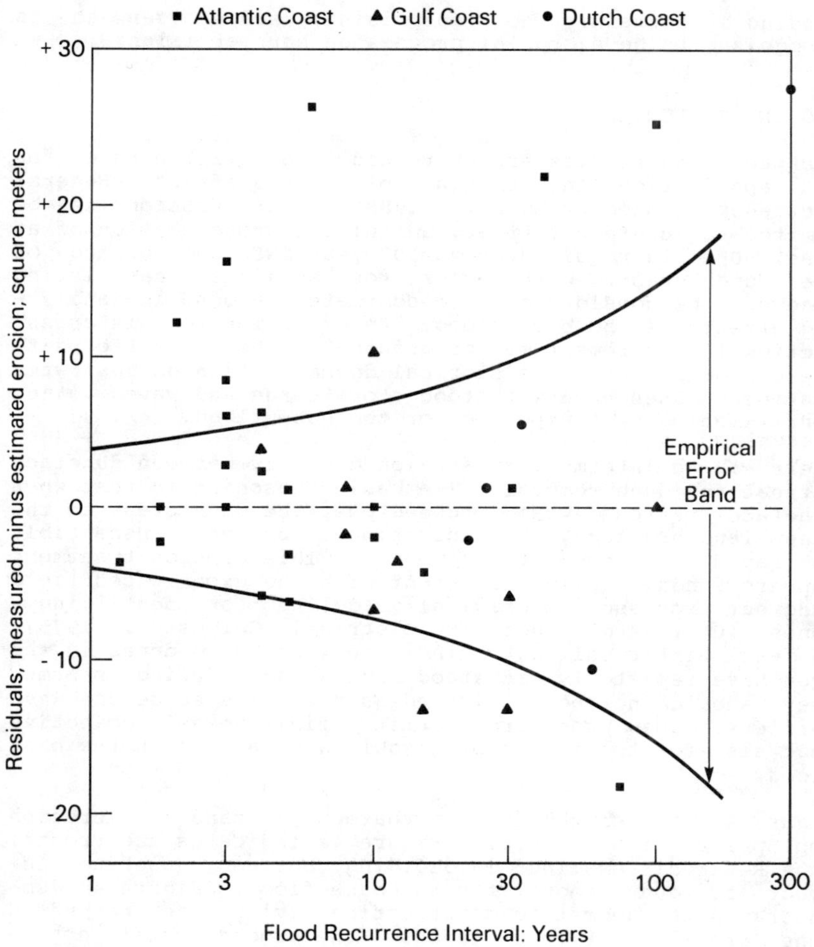

Figure 3. Residuals between measured erosion cross section and that from equation 1, for 38 cases of duneface retreat.

finding is certainly not surprising, since overwash is recognized to be a crucial process on barrier islands.

ERODED GEOMETRIES

The preceding results are expressed as a definite rule for FIS application in treating base flood effects (Federal Emergency Management Agency, 1988). For erosion to be limited to duneface retreat, an initial cross section of at least 50 m^2 is required above 100-year SWFL and seaward of the dune crest. With lesser cross sections, the existing dune will be considered to be completely eroded in analyzing effects of the base flood. An exception to this cross-sectional requirement may be granted at a specific site where authoritative historical documentation demonstrates that sand dunes have withstood storm surge and wave action approximating that expected for the base flood.

There is no intermediate erosion geometry between duneface retreat and dune removal. The basic reasoning is that when duneface retreat would proceed past the dune crest to the steep landward face, the dune remnant becomes susceptible to rapid and complete removal. This erosion treatment appears fundamentally consistent with the experienced field judgment expressed in original guidelines for identifying V zones (U.S. Army Engineer District, Galveston, 1975): "Unless historical data indicate that sand dunes in the area have repeatedly withstood wave attack during storms, they should not be considered as effective surge and wave barriers." New procedures permit definitive and objective analysis for the extent of sizable wave action in the base flood.

Figure 4 outlines the entire treatment of sand dune erosion appropriate in an FIS. Figure 4a indicates the Frontal Dune Reservoir examined in deciding between duneface retreat or dune removal for the base flood. Figure 4b demonstrates the geometric construction giving the retreated dune profile. Figure 4c shows the eroded profile appropriate in cases where the pre-storm dune is completely removed.

In regard to the Frontal Dune Reservoir of Figure 4a, consideration is restricted to sand located above 100-year SWFL and seaward of the steep rear slope. The vertical line as the landward boundary to the frontal dune reservoir seems a proper simplification, given that a steep escarpment occurs in duneface retreat and that sand in the rear dune wedge has some resistance to removal. The indicated reservoir is a straightforward measure for primary dunes with a prominent ridge or with a mound configuration but a marked rear slope. For some complicated dune profiles, judgment may be required to identify the dune segment expected to be fully effective in resisting removal.

The geometry for duneface retreat in Figure 4b represents a simplification of the erosion computation model developed by Delft Hydraulics Laboratory (Vellinga, 1986). Besides a

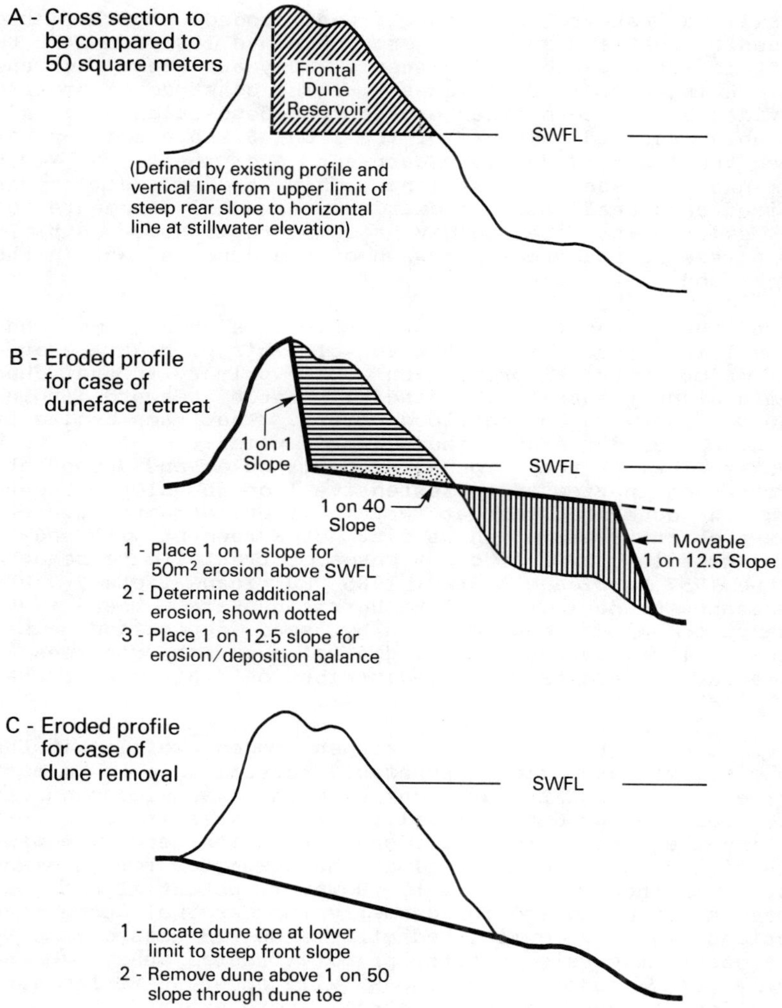

Figure 4. Treatment of sand dune erosion in 100-year event for a coastal Flood Insurance Study.

rectilinear rather than a curvilinear eroded profile, the present treatment specifies erosion above 100-year SWFL to be 50 m^2, as expected in a generic 100-year event on the U.S. Atlantic or Gulf coasts. The constructed profile provides a balance between erosion and deposition, by adjustable seaward extent of the 1 on 40 slope segment between the 1 on 1 duneface escarpment and the 1 on 12.5 terminus to sand deposition. This eroded geometry is spliced onto unaffected landward and seaward segments of the existing profile, to give a complete transect suitable for assessing potential overwash of the dune remnant in the base flood.

Where the frontal dune reservoir is less than 50 m^2, dune removal is effected as in Figure 4c. This procedure erases the major vertical projection of the primary frontal dune from a given transect, yielding a gentle seaward-dipping ramp for storm waves and flood waters. That ramp begins at the dune toe, defined as the seaward point on the steep duneface, with the 1 on 50 slope taken as an appropriate inclination in view of the extensive 1 on 40 slope occurring in duneface retreat with a steep landward barrier. This erosion treatment shows distinct agreement with post-storm profiles for Gulf coast cases of frontal dune removal in the 1957 Hurricane Audrey, 1965 Hurricane Betsy, 1979 Hurricane Frederic, and 1985 Hurricane Kate. There is no attempt to balance dune erosion by sand deposition elsewhere, since current knowledge of overwash processes is inadequate to address that redistribution (Birkemeier, et al., 1987).

There are major differences between eroded profiles in the two distinct categories. Duneface retreat occurs almost entirely above SWFL, but a cut at much lower elevation can take place during dune removal. For a sizable dune not meeting the stated cross-sectional criterion, erosion quantity might amount to three times that expected for duneface retreat in the 100-year event. However, potential alternatives in estimated erosion geometry for marginal dune cross sections may have diminished effects on FIS hazard assessment due to the rule that the primary frontal dune is entirely a V zone. Exceptions will occur where landward elevations in dune removal control V-zone extent.

Figure 4 procedures are intended for immediate FIS usage in treating the base flood at typical sandy sites on the U.S. Atlantic or Gulf coasts. This erosion methodology is efficient and objective, with a broad empirical basis in recorded storm effects on these coasts. Erosion estimates should be finalized with full consideration of documented historical storm effects at the particular study site.

EXAMPLE OF FIS APPLICATION

This example treats dune erosion expected to be associated with the 100-year event for a coastal site in Massachusetts. As shown in Figure 5, the shore topography is divided into two reaches, each represented by a typical

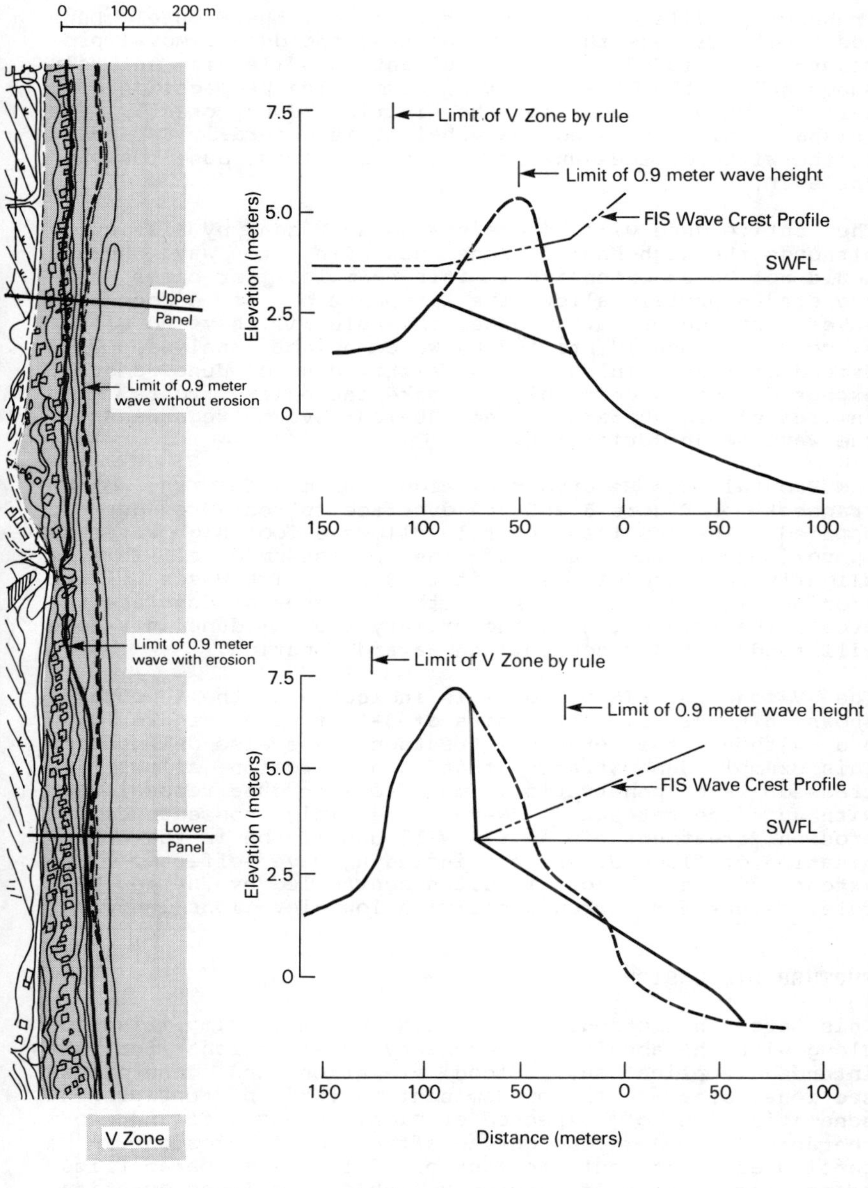

Figure 5 - Coastal topogtraphy (at left) and two reresentative transects in example of dune erosion treatments and FIS V Zone determination.

transect profile. On the upper transect, the Frontal Dune Reservoir is less than 50 m^2 so that the dune removal procedure is employed. The resultant profile is entirely submerged by the 100-year SWFL, permitting propagation of a 3-foot-high wave about to the initial dune crest. The inland limit to 3-foot wave height is generally dependent on the width of the dune, the location of the dune toe, and the SWFL.

The entire dune will be designated as V zone by NFIP rule, although the high-hazard zone according to wave height would not be so extensive in this example. For cases where the eroded profile allows the propagation of 3-foot-high waves past the pre-storm dune, the rule will have no effect since the V zone identified by wave height analysis will extend further inland. In either case of dune removal, expected erosion certainly can make the entire frontal dune an area of high hazard in the 100-year event, regardless of the wave height criterion.

The Frontal Dune Reservoir exceeds 50 m^2 for the other transect of Figure 5 and the duneface retreat procedure is applied. The resultant profile allows 3-foot-high waves to approximately the face of the pre-storm dune. This is slightly landward of the limit to 3-foot-high waves if no erosion is presumed. As with all cases of duneface retreat, the rule defining the primary frontal dune as V zone will fundamentally control the hazard zonation.

The V zone for FIS purposes is indicated on the topography shown in Figure 5. The limits of 3-foot wave heights with and without the erosion treatment are also delineated. This example demonstrates that dune erosion allows increased wave penetration, more so with dune removal than with duneface retreat. However, the main consequence of erosion treatment in an FIS will usually be in the determination of flood elevations including wave effects. The extent of the V zone is often controlled by the new NFIP rule, except for narrow dunes with low elevations landward.

FURTHER DISCUSSION

This erosion methodology offers maximum simplification along with the absolutely necessary distinctions for the intended application. Flood elevation and dune cross-sectional area are the predominant factors in the present schematization of expected erosion. Other site and storm characteristics are thought to affect dune erosion in a particular event, but the lack of definitive parametrizations for effects of ignored variables precludes quantitative considerations as a supplement to the present generic procedures.

One potentially important factor is the storm surge duration, since massive dune erosion must require appreciable time to occur. Also, storm intensity has been characterized by recurrence interval of local flood elevation, but erosion is actually caused by associated waves impacting the dune. A fundamental obstruction to more detailed erosion treatment in the present application is that the base flood is simply a water elevation recurring at a definite frequency; this 100-year event cannot be specified as a certain storm surge and wave sequence. The present erosion methodology has a broad empirical basis, reflecting expected effects in various storms and thus approximating a generic event.

Another notable set of site-specific factors includes dune characteristics such as sand size, consolidation, and vegetation. Those variables must figure in dune resistance to erosion, but general quantitative guidance is not available. The present treatment is based on effects at a variety of sites, and can be expected to be usually applicable to sand dunes established by natural processes. Another presumption is that detailed dune geometry can be summarized by cross-sectional area; any independent effect of dune height can be considered in terms of wave runup and overtopping after erosion assessment is completed.

The Dutch computation model incorporates strong dependences of erosion cross section on both sand size and dune height (van de Graaff, 1986). Effects of those dependences can be compensatory only if higher dunes are naturally associated with coarser sands, which seems unlikely. Thus, the most significant factor ignored in the present treatment appears to be detailed dune characteristics, since storm intensity has been at least partially considered by means of the flood recurrence interval. Still, the relationship in Equation 1 accounts for a majority of the variance in the present data base, demonstrating the usefulness of a simplified erosion treatment. Considering the natural irregularity in erosion amounts, there appears to be little room for improvement by detailed consideration of causative and response variables.

The only site characteristics treated in the Figure 4 procedure are the 100-year SWFL and some aspects of existing shore profile. However, exceptionally wide sand beaches have been documented to limit the extent of dune erosion or wave damages during extreme storms, and such sheltering effects should be taken into account. For duneface retreat, a stable inner surf zone is provided by slope of 1 on 40, and comparable existing shore slope must limit dune erosion by effecting gradual dissipation of storm waves. That shore situation must complicate eroded profile construction with the specified duneface erosion and sand balance, but a simple and consistent adjustment of procedure might limit retreat to the point defined by 1 on 40 slope from usual sea level up to 100-year SWFL. Note that the existing profile considered should be that for the season when extreme storms are expected at the site.

The present erosion procedures are intended for sandy open-coast sites where the basic transect viewpoint is adequate. This excludes application in three-dimensional regions such as an inlet vicinity, or near coastal structures influencing water level, waves and erosion. Even focusing on relatively ideal sites, it appears that only rather generalized erosion treatment might have a direct empirical basis since data are scarce. Present simplifications presume that erosion is mainly independent of site-specific details: existing conditions may reflect some long-term equilibrium, so that a generic response to the 100-year event can be anticipated.

OTHER APPLICATIONS

Without additional investigation, present results cannot be applied with confidence beyond the geographic limits of the data base. This guidance could be useful for quantitative erosion analyses where a simplified viewpoint suffices. Away from the 100-year recurrence interval addressed here, erosion geometries would require empirical modification: bed slopes steeper than 1 on 40 or 50 will be appropriate in dissipating smaller waves of more common storms. However, the erosion climatology of equation 1 has a broad basis and is conducive to other direct applications, for example, in designing artificial sand dunes as storm protection.

Considering a dune project with a 50-year design life, there is 50% risk of a 73-year event as the extreme, according to the binomial distribution usually employed (U.S. Interagency Advisory Committee on Water Data, 1982). That event entails erosion of a frontal dune reservoir of 44.5 m^2 according to equation 1. Likely storm events between dune maintenance operations determine the needed increase of the sand reservoir over that required to withstand the design storm. With a 10-year interval chosen between dune replenishments, notable expected storms (50% risk) have recurrence intervals of 14.9 and 6.1 years, presuming a berm provides protection against storms less severe than a 5-year event. Those two storms could yield additional upper-dune erosion totaling 40 m^2, nearly doubling advisable cross section.

Natural recovery or dune growth by aeolian processes would lessen requirements for placed sand. However, computations refer to erosion expected with established dunes, and should be increased for a safe upper bound to effects with loose sand. It seems clear that such estimates could assist in optimization of interrelated design choices: berm and dune geometries; maintenance interval and project lifetime; etc.

CONCLUDING COMMENTS

Geometries provided for duneface retreat and dune removal are recommendations for appropriate erosion treatments rather than official FIS procedures. Present results are certainly liable to improvement based on advances in detailed erosion modeling and particularly in knowledge of dune removal processes. Fundamental simplifications here currently seem necessary and appropriate in view of the marked variability of storm-induced erosion quantities, limitations apparent in published erosion treatments, and the basic idealization in FIS consideration of base flood effects along representative transects. Given present knowledge of dune erosion processes in extreme storms, only a simple treatment of coastal erosion can be expected to provide sufficiently generic estimates for a 100-year event.

REFERENCES

Birkemeier, W. A. (1979). "Effects of the 19 December 1977 Coastal Storm on Beaches in North Carolina and New Jersey," Shore and Beach (January), pp. 7-15.

Birkemeier, W. A., Kraus, N. C., Sheffner, N. W., and Knowles, S. C. (1987). "Feasibility Study of Quantitative Erosion Models for Use by the Federal Emergency Management Agency in the Prediction of Coastal Flooding," Technical Report CERC-87-8, Waterways Experiment Station, Corps of Engineers, Vicksburg, Mississippi, 102 pp.

Dewberry & Davis (1988). "Basis of Dune Erosion Treatment for Coastal Flood Insurance Studies," report to Federal Emergency Management Agency, Washington, D.C., 95 pp.

van de Graaff, J. (1986). "Probabilistic Design of Dunes; An Example from the Netherlands," Coastal Engineering, Vol. 9, pp. 479-500.

Federal Emergency Management Agency (1986). "Assessment of Current Procedures for the Identification of Coastal High Hazard Areas (V Zones)," Washington, D.C., 133 pp.

Federal Emergency Management Agency (1988). "National Flood Insurance Program; Flood Plain Management Standards; Final Rule," Federal Register, Vol. 53, pp. 16269-16280.

Florida Department of Natural Resources, Division of Beaches and Shores (1987). "Data Bank on Storm Changes in Sequential Coastal Profiles," magnetic tape, Tallahassee, Florida.

Hallermeier, R. J., Rhodes, P. E., and Buckley, M. (1988). "Jeopardy of Frontal Dune Removal in Extreme Coastal Storms," paper submitted for publication in Science.

Kriebel, D. L., and Dean, R. G. (1985). "Numerical Simulation of Time-Dependent Beach and Dune Erosion," Coastal Engineering, Vol. 9, pp. 221-245.

National Academy of Sciences (1977). "Methodology for Calculating Wave Action Effects Associated with Storm Surges," Washington, D.C., 29 pp.

U.S. Army Engineer District, Galveston (1975). "Guidelines for Identifying Coastal High Hazard Zones," Galveston, Texas, 87 pp.

U.S. Interagency Advisory Committee on Water Data (1982). "Guidelines for Determining Flood Flow Frequency," Bulletin #17B of the Hydrology Subcommittee, U.S. Department of the Interior, Geological Survey, Reston, Virginia, 183 pp.

Vellinga, P. (1986). "Beach and Dune Erosion During Storm Surges," Delft Hydraulics Communication No. 372, Delft, The Netherlands, 169 pp.

CHAPTER 90

Calculation of On-offshore Sand Movement and Wave Deformation on Two-dimensional Wave-current Coexistent System

Koichi Kinose* ,Shuji Okushima* and Masahito Tsuru**

Abstract

In this paper, we proposed a method of calculation to predict quantitatively the on-offshore sand movement and the wave deformation on a wave-current coexistent system by assuming a river mouth. And the calculated results were compared with the experimental data obtained for the coexistent system in a two-dimensional wave tank. The distribution of wave height on the breaker zone was analyzed by use of BORE MODEL. It was required for the calculation of the sand transport rate. The model was established on the assumption that the value of energy loss in a breaking wave was equivalent to that of bore. The wave height distribution on the offshore side of breaking point was presumed employing the third order approximate solution of Stokian wave on the coexistent system. The breaking point was obtained by use of Miche's criteria equation. The local sand transport rate could be calculated by use of POWER MODEL. The predominant direction of sand drift was recognized using relations for judgement which were derived from the experimental results. The transformation of sea bottom and river one was estimated on the basis of the calculated distributions of the wave height and the sand transport rate. The results obtained by this analytical method agreed well with the experimental results.

I. Introduction

In order to perform studies about closure of river mouth by sand drift, it is necessary to have a deep understanding of wave deformation and sand movements on a wave-current coexistent system where waves and the steady flow interact. Inman and Bowen[1], Abou-Seida[2] and Tanaka et al.[3] have made experiments on

* National Research Institute of Agricultural Engineering
2-1-2, Kannondai, Tsukuba-shi, Ibaraki, 305 Japan.
** Department of Civil Engineering, Science University of Tokyo

sand transport rates and shapes of ripples under wave-current coexistent system. Watanabe[4)] proposed a bed load formula applicable to coexistent system on the basis of a power model. Jonsson[5)] and Sakai et al.[6)] have calculated wave shoaling on coexistent system. In the present work, we will try as the initial step to construct a model simulating the on-offshore sand movement, the wave deformation due to wave and steady flow, and the two-dimensional bottom transformation. For this analysis, we considered the wave-current coexisting system by assuming a river mouth as shown in Fig.1. In addition, the results will be compared with the experimental data measured in a wave tank.

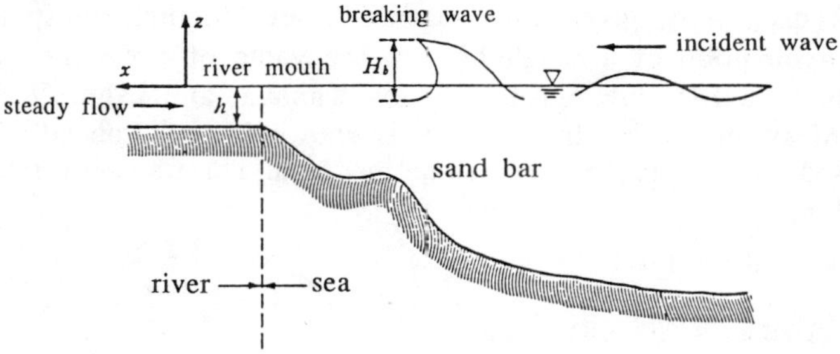

Fig.1 Wave-current coexistent system assumed to be river mouth.

II. Distribution of on-offshore sand transport rate

The sand transport rate on the wave-current coexistent system will be discussed in this chapter for the estimation of sea and river mouth bottom transformation.

2.1 Bed load formula

The local transport rate of bed load at a given point could be estimated by use of the power model proposed by Watanabe[4)]. The transport rate per unit width, q_{net}, can be obtained as the sum of the component q_w caused by wave motion and the component q_f caused by steady flow.

$$q_f = A_f \frac{\tau - \tau_c}{\rho g} U_f$$

$$q_w = A_w \frac{\tau - \tau_c}{\rho g} U_w \qquad (1)$$

where A_w and A_f are non-dimensional constants, respectively, which are to be determined from laboratory or field data, τ is the maximum bed shear stress, τ_c is the critical shear stress, ρ is the fluid density, and g is the acceleration of gravity. U_f is the velocity of the steady flow and U_w is the bottom velocity amplitude of orbital motion induced by wave. The critical shear stress can be obtained from the critical Shields parameter ψ_c[4].

$$\psi_c=0.11 \quad \text{for} \quad d/\delta_L<1/6.5 \quad \text{(fine sand)}$$

$$\psi_c=0.06 \quad \text{for} \quad d/\delta_L>1/4.0 \quad \text{(coarse sand)} \tag{2}$$

$$\psi_c=\frac{\tau_c}{\rho s g d}, \quad \delta_L=\sqrt{\frac{\nu T}{\pi}}$$

where d is the median diameter of the sand, s is the specific gravity of immersed sand, δ_L is the thickness parameter of the boundary layer, ν is the kinematic viscosity of the fluid, and T is the wave period. When the diameter is intermediate between fine sand and coarse sand, the value of ψ_c is determined by interpolation. The value of q_w and q_f given by equation(1) are the net transport rate. On the assumption of a rough bottom, the value of τ can be obtained from the friction law proposed by Tanaka and Shuto[5] for the coexistent system. The friction law is shown in the Appendix.

Based on the above discussion, the local net transport rate, q_{net}, is given by

$$q_{net}=q_f+q_w. \tag{3}$$

2.2 Expression of velocity field

By analogy with the analysis of phenomena in the vicinity of river mouth, the direction of the steady flow is considered to be offshore. Assuming that waves progress up a steady flow, the bottom velocity amplitude of orbital motion induced by the wave can be indicated from a small amplitude wave theory.

$$U_w=\frac{\pi H/L(L/T+U_f)}{\sinh kh} \tag{4}$$

where H is the wave height, h is the water depth, and L, k are the wave length and the wave number, respectively. The values of L and k are given by the following equations.

$$C=\frac{L}{T}+U_f=\left(\frac{gL}{2\pi}\tanh kh\right)^{1/2}$$

$$k=\frac{2\pi}{L} \tag{5}$$

where C is the relative wave celerity to the steady flow. If the velocity of the steady flow exceeds a critical value, the wave does not progress up the steady flow. Its critical velocity U_{fc} can be obtained by

$$U_{fc}=\frac{C_o}{4}\tanh kh. \tag{6}$$

where C_o is the wave celerity at the deep water free from the effect of the steady flow. When $U_f>U_{fc}$, it is assumed that only the steady flow acts as the driving force of sand drift.

On the other hand, the vertically averaged velocity, U_f, is used for the calculation of the sand drift component, q_f. That is,

$$U_f = \frac{q_c}{h} \qquad (7)$$

where q_c is the discharge per unit width.

2.3 Judgment of predominant direction of sand drift

The direction of the sand drift, q_f, is offshore, because the flow direction is considered to be offshore like that in the river mouth. But the predominant direction of q_w does not necessarily coincide with the direction of wave progress. Sunamura[6] has indicated that the direction of sand drift might change under the condition where equation(8) is satisfied.

$$\Phi = \gamma\, U_{ro} \qquad (8)$$

where γ is an empirical constant, Φ is Hallermeier's parameter, and U_{ro} is Ursell parameter. They are expressed as follows by the long wave assumption, respectively.

$$\Phi = \frac{H^2}{s\,d\,h}\ ,\quad U_{ro} = \frac{g\,T^2 H}{h^2} \qquad (9)$$

However, equation(8) is found from an experimental study of an uniformly sloping movable bed in a non-coexistent system where only wave acts. Sunamura has concluded that γ was equal to 0.13 on the surf zone in his experiments.

Taking the effect of the steady flow into consideration, equations(8) and (9) must be modified for the analysis of the wave-current coexistent system. The Ursell parameter on the coexistent system is shown as the following equation by applying the long wave assumption to the wave celerity, C, in the equation(5).

$$U_r = U_{ro}\left(1 - \frac{1}{2}\,\frac{H}{h}\,\frac{U_f}{U_w}\right)^2 \qquad (10)$$

The relative intensity U_f/U_w in the above equation, which indicates a relative intensity ratio of the steady flow to the wave, was introduced for the judgment of the predominant direction of the sand drift, q_w. Using these parameters the judgment can be done as the follows. The zone of sand movement by wave and steady flow is divided into two regions; *pre-breaking* and *post-breaking region.*

In the post-breaking region, the values of Φ and U_r are calculated by use of equations(9) and (10), respectively, and a point, x_h, where the relation of Φ and U_r satisfies equation(11), is searched.

$$\Phi = \gamma\, U_r \qquad (11)$$

Here, the value of U_f/U_w at point x_h is supposed to be β.

The direction of sand drift, q_w, induced by wave motion can be recognized using relations in equation(12).

$$\begin{aligned} &\text{if } \beta>0.15 && \text{then offshore for all over the surf zone} \\ &\text{if } \beta\leq 0.15 \text{ and } U_f/U_w>\beta && \text{then offshore at a given point.} \\ &\text{if } \beta\leq 0.15 \text{ and } U_f/U_w\leq\beta && \text{then onshore at a given point.} \end{aligned} \tag{12}$$

In the pre-breaking region, we employ the relations shown in equation(13) .

$$\begin{aligned} &\text{if } U_f/U_w>0.15 && \text{then offshore at a given point.} \\ &\text{if } U_f/U_w\leq 0.15 && \text{then onshore at a given point.} \end{aligned} \tag{13}$$

The critical value (0.15) of β could be given from the results of our previous experimental study[7].

III. Wave deformation of incoming waves on coexistent system

The wave height variation due to changes of depth and velocity of the steady flow will be considered. The zone of wave deformation is also divided into two regions; *pre-breaking* and *post-breaking region.*

3.1 Wave shoaling and breaking criterion

In the pre-breaking region, we employ the equation given by Sakai, Saeki and Ozaki[8] to calculate the wave deformation. This equation is the third order approximate solution of Stokian wave on the wave-current coexistent system. It is introduced on the basis of a wave energy flux conservation. The solution is shown in the Appendix.

Miche's criterion can be valid for a breaking criterion of regular waves on the coexistent system, as being pointed out by Iwagaki[9] and Sakai[10].

$$\left(\frac{H}{L}\right)_b = \alpha \tanh(kh)_b \tag{14}$$

where subscript b indicates the quantities of the breaking point. The value of α is equal to 0.142 in the case of only wave action. But in the wave-current coexistent system, its value is inclined to decrease according to the increase of the velocity. The breaking point can be calculated by using the ratio of water depth to wave length obtained from the wave height calculation mentioned above.

3.2 Wave height distribution in post-breaking region

In order to obtain the wave height distribution in the post-breaking region, it was assumed that breaking wave has the same process as that of the energy dissipation in bores. The energy loss per unit width of bore is given by

$$\Delta E' = \frac{1}{4}\rho g(h_2 - h_1)^3\sqrt{\frac{g(h_1+h_2)}{2h_1h_2}} . \tag{15}$$

where h_1 and h_2 are the water depths before and after bore, respectively. The energy loss[11] per unit length can be represented by the following equation(16) on the basis of the long wave assumption.

$$h^2 = h_1 \cdot h_2$$

$$\Delta E = \frac{\Delta E'}{L} = \frac{\Delta E'}{CT} = \frac{B}{4} \frac{\rho g H^2}{T} \left(\frac{H}{h}\right) \tag{16}$$

where B is

$$B = \frac{(h_2 - h_1)^3}{H^3} \tag{17}$$

where H is the wave height at a given point in the surf zone.

The wave energy flux taking the mean energy level as a reference on the wave-current coexistent system was given by Jonsson[12]. That is,

$$E = \frac{1}{8} \rho g H^2 (C_g - U_f)\left(1 - \frac{U_f}{C}\right) \tag{18}$$

where C and C_g are the relative wave celerity and the relative wave group celerity to the steady flow, respectively. They can be obtained by use of equation(5). In a shallow water, this equation becomes,

$$E = \frac{1}{8} \rho g H^2 \left(\sqrt{gh} - U_f\right)\left(1 - \frac{U_f}{\sqrt{gh}}\right) \tag{19}$$

Based on the assumptions discussed above, the wave height after breaking is able to be given by the energy flux conservation equation.

$$\frac{dE}{dx} + \Delta E = 0 \tag{20}$$

The x-axis is taken as pointing onshore direction. In order to calculate the wave height distribution in the surf zone for an arbitrary bottom configuration, we performed a numerical solution using equations(16) and (19). In the breaking waves, a turbulent kinetic energy is produced in the surface roller. In addition, turbulence is also induced by the systematic eddy near the breaking point. Consequently, the value of B in equation(16) depends on the location of the surf zone. However, the value on the wave-current coexistent system could not be known in detail. Therefore, the following equations, which were given for a non-coexistent system by Mase[13], had to be used.

i) In cases that $m > 1/20$

$0.9 < h/h_b \leq 1.0$: $B = 1$

$$0.6 < h/h_b \leq 0.9 \ : \ B = 13 - \frac{40}{3}\frac{h}{h_b} \tag{21}$$

ii) In cases that $m<1/20$

$$0.6<h/h_b<1.0 \quad : B=11-10\frac{h}{h_b}$$
$$h/h_b \leq 0.6 \quad : B=5 \tag{22}$$

where h_b is the breaking water depth and m is the beach slope.

IV. Prediction of bottom profile transformation

The continuity equation for the bottom material is indicated as follows.

$$\frac{\partial z_b}{\partial t} = -\frac{\partial h}{\partial t} = -\frac{\partial q_{net}}{\partial x} \tag{23}$$

where z_b is the bottom level taking the still water level as a reference, and t is the elapsed time from the beginning of the wave-current action. In the practical numerical analysis, the following equation[4)] was employed to restraint the appearances of radical irregular bottom profiles and steep gradients of the bottom more than the angle of repose of the immersed sand. These phenomena may not occur in real beaches.

$$\frac{\partial z_b}{\partial t} = -\frac{\partial}{\partial x}\left(q_{net} - \varepsilon_s |q_{net}| \frac{\partial z_b}{\partial x}\right) \tag{24}$$

where ε_s is the positive constant. In addition, we used a finite difference method in which staggered mesh scheme is applied to equation(24). The stability of the calculus is indicated by

$$\Delta t \leq \frac{1}{2} \frac{\Delta x^2}{\varepsilon_s |q_{max}|} \quad . \tag{25}$$

where q_{max} is the maximum value of q_{net}.

V. Comparison between experimental and calculated results

The comparison between the experimental and the calculated results will be discussed in this chapter. Assuming a river mouth as shown in Fig.2, the experiments on the wave-current coexistent system were carried out under the conditions as listed in Table 1. The experimental apparatus consisted of a piston-type wave maker and a pump for the generation of the steady flow in a wave tank(30m long, 1.0m high, and 0.6m wide).

5.1 Method of experiment

A model was formed out of the sand listed in Table 1. It was composed of two parts; the river mouth(4m long, flat) and the sloping beach(12m long, 1/20). The wave height and the change of bottom profile induced by wave and steady flow were measured with spaces of 1cm by use of wave gauge and sand-level-meter which were loaded on the observation truck.

Table 1 Condition of experiments.

Case	(1) q_c(cm²/s)	(2) H_o(cm)	(3) T(s)	(4) h(cm)	(5) (cm/s)	(6) Fr	(7) d_{50}(mm)	(8) S	C_S
A	197.2	6.49	2.2	4.38	45.02	0.69	0.37	1.72	2.96
B	138.2	6.49	2.2	4.86	28.43	0.41	0.37	1.72	2.96
C	35.8	4.68	2.2	6.96	5.13	0.062	0.28	1.44	2.57
D	133.6	17.51	1.6	8.25	16.20	0.18	0.28	1.44	11.86

(1): discharge per unit width
(2): wave height
(3): wave period
(4): depth of river mouth
(5): velocity at river mouth
(6): Froude number
(7): median diameter of sand
(8): specific gravity of immersed sand

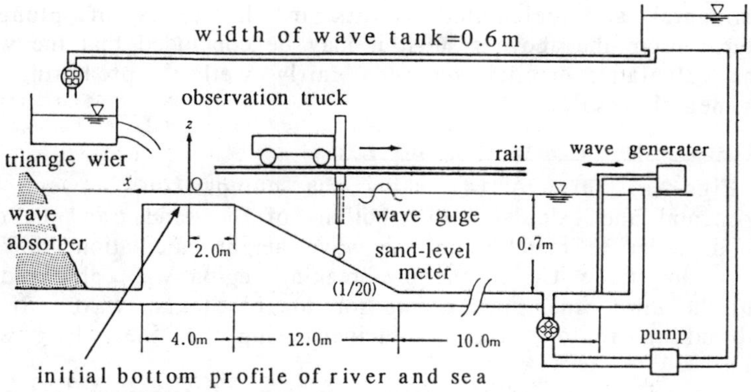

Fig.2 Schematic drawing of experimental apparatus for wave-current coexistent system.

The net transport rate of sand drift was calculated from the profile change at 20-minute intervals by use of the difference equation of equation(23). In Table 1, the waves which caused an accretion of the beach were worked for Cases of A, B and C, and in the Case of D the wave produced an erosion of the beach. In addition, the classification of the wave-type for two-dimensional laboratory beach transformation in terms of a parameter, C_S, was used here. It was proposed by Sunamura and Horikawa[14].

$$\frac{H_o}{L_o} = C_S \cdot m^{-0.27} \cdot \left(\frac{d}{L_o}\right)^{0.67} \tag{26}$$

where H_0 is the deep water wave height and L_0 is the deep water wave length. If Cs< 4, the beach is accreted, while if Cs> 8, the beach is eroded.

5.2 Wave height distribution

Figs.3(a) through (d) show the comparison between the experimental and calculated results of the wave height distribution. In determining the breaking height on the wave-current coexistent system for the numerical calculation, the value of α in equation(14) was taken as 0.132. The bottom profiles at the time of wave height measurement are also shown in the figures. These bottom profiles are used for the calculation of the wave height. Figs.3(a) through (c) and (d) correspond to cases of plunging breaker and spilling breaker, respectively. The calculated results for the wave height near the breaking point in the case of spilling breaker have a tendency to give smaller values than the experimental results as shown in Fig.3(d). But good agreements are found between the experimental and calculated results in the cases of plunging breaker. From the above results, it may be concluded that the wave height calculation model succeeds fairly well in predicting the experimental results.

5.3 Distribution of sand transport rate

Figs.4(a) through (d) show the comparison between the experimental and calculated distributions of the sand transport rate. Here, in order to match with the wave height calculation, the bed velocity amplitude U_w in the pre-breaking region was calculated by use of the third order approximate solution of Stokian wave. In the post-breaking region, U_w is obtained using the linear long wave theory.

The procedures of the calculation for the transport rate, q_{net}, will be illustrated in the following;

1) The local transport rate at a given point is calculated by use of equations(1) through (7) under the assumption of $A_f = |A_w| = 1$.

2) The predominant direction of transport rate, q_w, at a given point is judged with equations(9) through (13). If the direction is offshore, A_w is equal to 1, while if the direction is onshore, A_w is equal to -1.

3) Comparing the peak value of the calculated distribution of transport rate with that of observed distribution and finding the both peak values to be equal each other, then the $A_f (= |A_w|)$ is required one.

4) The final value of the transport rate at the each point is obtained multiplying the value of the transport rate given in procedure *1)* by $A_f (= |A_w|)$ gotten above.

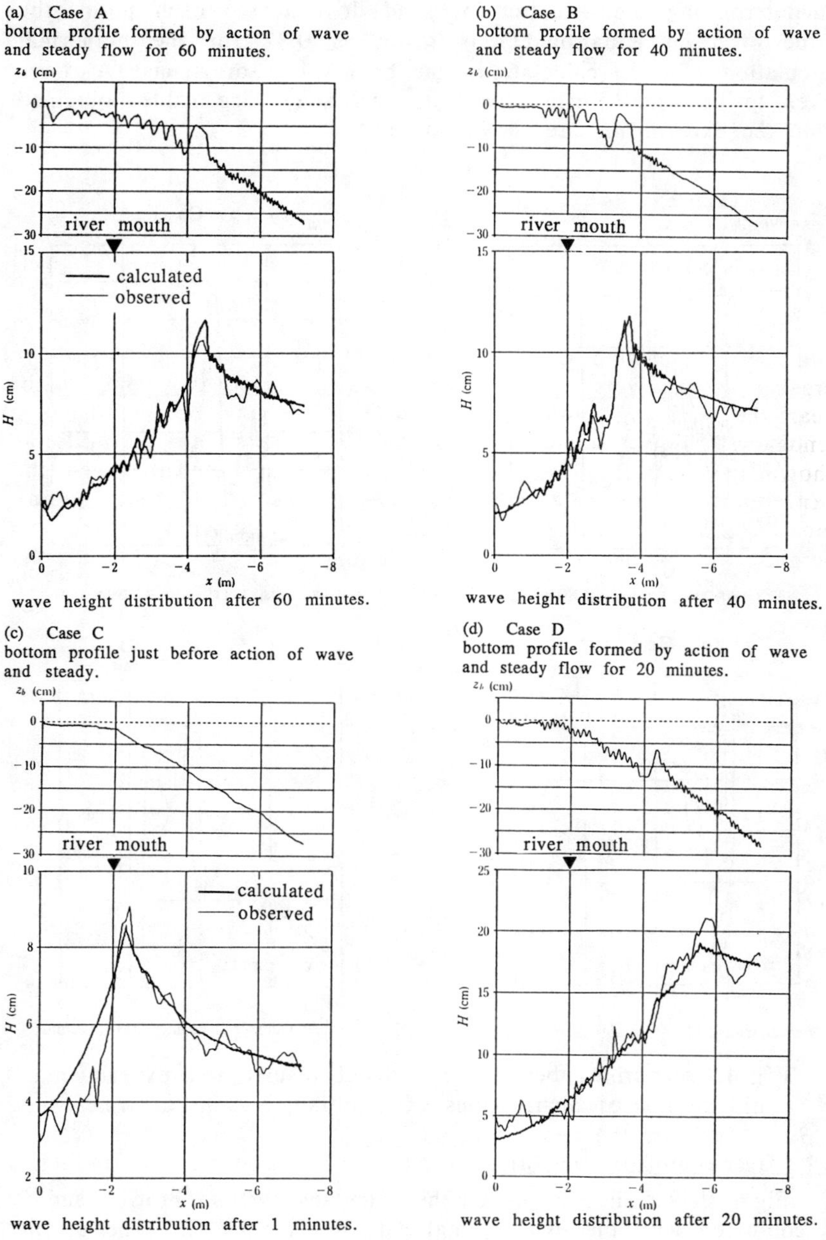

Fig.3 Comparison between calculated and experimental results of wave height.

In determining the direction of sand drift at the each point, the value of γ in equation(11) is taken as 0.32 for the numerical calculation. In the calculations, the both values of A_f and $|A_w|$ are taken to be equal for all over the surf zone. The values obtained from the experiments are shown in Fig.5.

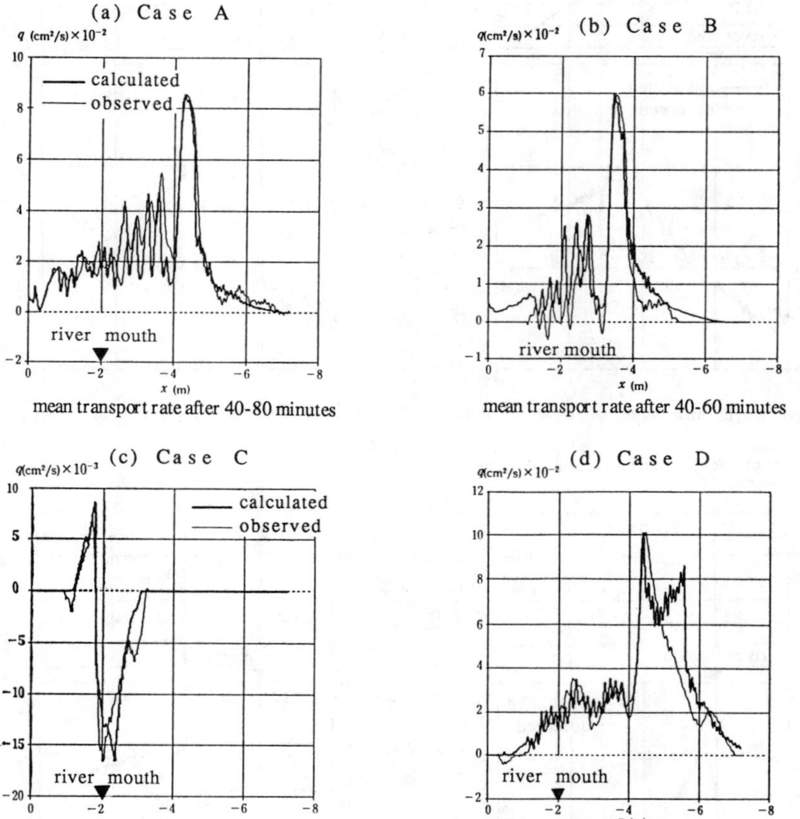

Fig.4 Comparison between calculated results and experimental results of distributions of on-offshore transport rate.

5.4 Beach profile transformation

Fig.6 shows an example of the calculated bottom profile, and it is compared with the experimental data in Case of C. The profile was formed under the action of wave and current for 20 minutes, and the initial profile which was observed just before beginning of wave-current action is shown in Fig.3(c). Using the calculated distributions of the wave height and the sand transport rate, the prediction of bottom transformation was done at intervals of 1cm from the difference equation of equation(24) under the conditions of ε_s=2.0 and Δt=10sec. The value of A_f or A_w for the sand transport rate calculation was obtained from the experimental results as shown in Fig.5. In average, the profile calculated by use of this model agrees well with that obtained from the experiment.

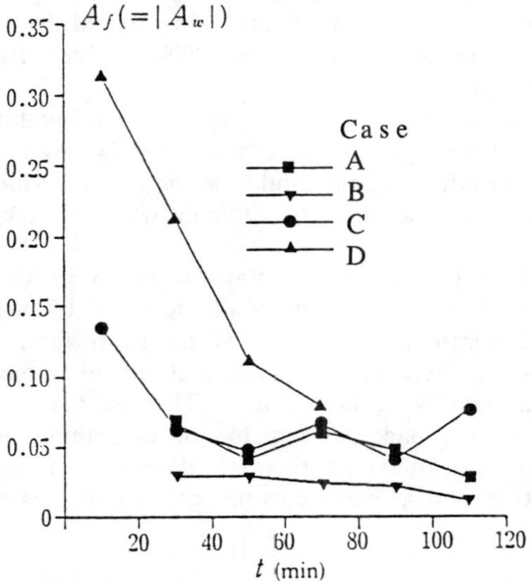

Fig.5 Change of value A_f(=| A_w |) with elapsed time.

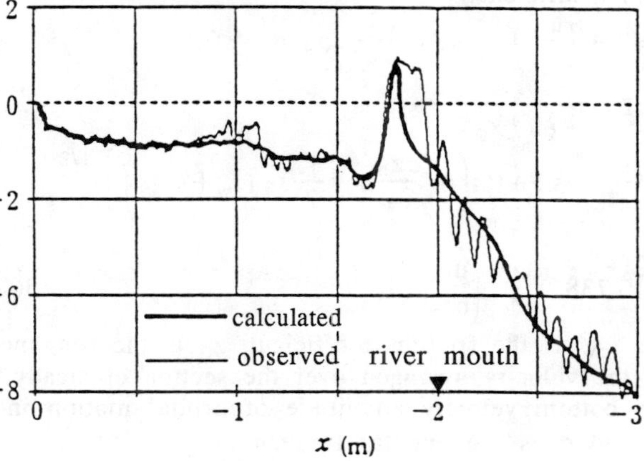

initial profile of bottom is shown in Fig.3(c)

Fig.6 An exsample of calculation for beach transformation due to wave and current.

VI. Conclusion

The conclusions in this study are indicated as follows;

1) The local transport rate of sand drift in the both regions of pre-breaking and post-breaking could be estimated with the application of the bed load formula proposed by Watanabe.
2) The predominant direction of sand drift could be recognized using the judgment equation which was derived from our previous study.
3) The distribution of wave height on the post-breaking region could be analyzed by use of a bore model. And the distribution in the pre-breaking region could be obtained form the third order approximate solution of Stokian wave on the coexistent system.
4) The prediction of the bottom transformation in the two-dimensional coexistent system could be done by use of the equation of continuity of the sand transport rate.
5) The results obtained by this analytical model agreed well the experimental results of laboratory. That is, the model developed in this paper is capable of describing aspects of two-dimensional sand movement, wave deformation and beach transformation on a wave-current coexistent system.

Appendix

(1) *The friction law for the wave-current coexistent system*

The next equations indicate the friction law given by Tanaka and Shuto[5] on the assumption of rough bottom.

(i) Friction coefficient

$$f_{cw} = 2B\left(\frac{\bar{u}_c}{\hat{u}_w}\right)^2 + 4BC\frac{\bar{u}_c}{\hat{u}_w}\cos\alpha + 2C^2$$

$$B = \frac{\kappa}{ln(h/z_0) - 1}$$

$$C = \begin{cases} \dfrac{\kappa}{\pi}\left\{0.25 + 0.101\left(ln\dfrac{\sigma z_0}{\hat{u}_w} - \dfrac{1}{2}ln\, f_{cw} + 0.24\right)^2\right\}^{-1/2}, & \dfrac{\hat{u}_w}{\sigma z_0} \geq 50 \\ 0.738\, f_{cw}^{-1/4}\left(\dfrac{\hat{u}_w}{\sigma z_0}\right)^{-0.408}, & 50 \geq \dfrac{\hat{u}_w}{\sigma z_0} \geq 10 \end{cases}$$

where f_{cw} is the friction coefficient, z_0 is the roughness length, $\bar{u}_c$ is the velocity averaged over the section of steady flow, $\hat{u}_w$ is the bottom velocity amplitude of orbital motion induced by wave and σ is the angular frequency.

(ii) Maximum Shear stress on the coexistent system

$$\tau = \frac{1}{2}\rho f_{cw}\hat{u}_w^2$$

where τ is the maximum shear stress acting to the bottom.

2) *The third order approximate solution of Stokian wave on the wave-current coexistent system.*

The next equations indicate the third order approximate solution of Stokian wave given by Sakai, Saeki and Ozaki[8].

(i) Velocity potential

$$\Phi = Ux + \frac{C-U}{k}\{(\lambda A_{11}+\lambda^3 A_{13}) \times \cosh k(h+z) \sin\theta$$
$$+\lambda^2 A_2 \cosh 2k(h+z) \sin 2\theta$$
$$+\lambda^3 A_3 \cosh 3k(h+z) \sin 3\theta\}$$

$$\zeta = \frac{1}{k}\left(\lambda f_1 \cos\theta + \lambda^2 f_2 \cos 2\theta + \lambda^3 f_3 \cos 3\theta\right)$$

$$B = B_0 + \lambda B_1 + \lambda^2 B_2 + \lambda^3 B_3$$

$$C = C_0 + \lambda C_1 + \lambda^2 C_2 + \lambda^3 C_3$$

$$k = 2\pi/L,\ \lambda = ka,\ \theta = k(x-Ct)$$

$$A_{11} = \frac{1}{\sinh\beta}, \qquad f_2 = \frac{1}{4}\frac{\cosh\beta}{\sinh^3\beta}(\cosh 2\beta + 2)$$

$$A_{13} = -\frac{\cosh^2\beta}{8\sinh^5\beta}(5\cosh^2\beta + 1), \qquad f_3 = \frac{3}{64}\frac{8\cosh^6\beta + 1}{\sinh^6\beta}$$

$$A_2 = \frac{3}{8}\frac{1}{\sinh^4\beta}, \qquad B = \frac{U^2}{2} + \frac{a^2 g}{2} k \frac{1}{\sinh 2\beta}$$

$$A_3 = \frac{1}{64}\frac{11 - 2\cosh 2\beta}{\sinh^7\beta}, \qquad (C-U)^2 = \frac{g}{k}\tanh\beta\left(1 + \frac{\lambda^2}{8}\frac{\cosh 4\beta + 8}{\sinh^4\beta}\right)$$

$$f_1 = 1, \qquad \pi\frac{H}{L} = \left(\lambda + \frac{3}{64}\lambda^3 \frac{8\cosh^6\beta + 1}{\sinh^6\beta}\right)$$

$$\beta = kh$$

where Φ is the velocity potential, ζ is the surface level variation, L is the wave length and H is the wave height.

(ii) Energy flux

$$W = \rho\frac{C}{k}\{U(C-U)Q_1 + (C-U)^2 Q_2\}$$

$$Q_1 = \lambda^2 \frac{1}{\sinh^2\beta}\left[\frac{1}{2}\sinh\beta\cosh\beta + \frac{\lambda^2}{32}\frac{\cosh\beta}{\sinh^5\beta}\{-4\cosh^6\beta + 20\cosh^4\beta - 16\cosh^2\beta + 9\}\right]$$

$$Q_2 = \lambda^2 \frac{1}{\sinh^2\beta}\left[\frac{1}{4}(\sinh\beta\cosh\beta + \beta) + \frac{\lambda^2}{8}\{\frac{1}{8\sinh^6\beta}(\sinh\beta\cosh\beta + \beta) \times (-20\cosh^6\beta + 16\cosh^4\beta + 4\cosh^2\beta + 9) + \frac{\cosh\beta}{4\sinh^3\beta} \times (16\cosh^4\beta + 2\cosh^2\beta + 9)\}\right]$$

where W is the energy flux, U is the velocity of steady flow and C is the relative wave celerity.

References

(1) Inman, D. L. and A. J. Bowen: Flume experiments on sand transport by waves and Currents, Proc. 8th Conf. on Coastal Engng., pp.137-150, 1963.
(2) Abou-Seida, M. M: Sediment transport by waves and current, Tech. Rep. No.HEL-2-7, Univ. of California, Berkley, 1964.
(3) Tanaka, H., H. Ozasa and A. Ogasawara: Experiments on sand movement by waves and current, Rep. Port & Harbor Res. Inst., Vol.12, No.4, 1973 (in Japanese).
(4) Watanabe, A., K. Maruyama, T. Shimuzu and T. Sakakiyama: Numerical prediction model of three-dimensional beach deformation around coastal structures, Proc. of Japanese Conf. on Coastal Engng., pp.406-410, 1984(in Japanese).
(5) Tanaka, H. and N. Shuto: Friction coefficient for a wave-current coexistent system, Coastal Engng. in Japan. Vol.24, pp.105-128, 1981.
(6) Sunamura, T: Prediction of on-offshore sediment transport rate in the surf zone including swash zone, Proc. of Japanese Conf. of Coastal Engng., pp.316-320, 1984(in Japanese).
(7) Kinose, K. and S. Okushima: Sand transport near sand bar on two-dimensional wave current coexistent system, Proc of '87 Annual Meeting of Japan Society of Irrigation, drainage and Reclamation Engng., pp.68-69, 1987(in Japanese).
(8) Sakai, S. and A. Ozaki: Effect of Current on wave Shoaling, Proc. of 29th Japanese Conf. on Coastal Engng., pp.70-74, 1982(in Japanese)
(9) Iwagaki, Y, T. Asano, T. Yamanaka and F. Nagai: Fundamental study on breaking of waves due to currents, Proc. of 27th Japanese Conf. on Coastal Engng., pp.30-34, 1980(in Japanese).
(10) Sakai, S., K. Ota, H. Oba, T. Kuribayashi and H. Saeki: Shoaling of irregular waves affected by opposing current, Proc. of Japanese Conf. on Coastal Engng., pp.224-228, 1985(in Japanese).
(11) Mase, H., A. Matsumoto and Y. Iwagaki: Calculation of random wave transformation in shallow water, Proc. of JSCE, No.375 /II-6, pp.221-230, 1986(in Japanese).
(12) Jonsson, I.G and J.D. Wang: Interaction between wave and current, Proc. of 12th Conf. on Coastal Engng., pp.489-507, 1970.
(13) Mase H. and Y. Iwagaki: Wave height distributions and wave grouping in surf zone, Proc. of 18th Conf. on Coastal Engng., pp.58-69, 1982.
(14) Sunamura, T. and K. Horikawa: Two-dimensional beach transformation due to waves, Proc. of 14th Conf. on Coastal Engng., pp.920-938, 1974.

CHAPTER 91

MOVABLE-BED EXPERIMENTS OF SHANTOU HARBOR

Huang Jianwei*

ABSTRACT

The paper, based on the macroscopic analysis of the sediment movement, deals with the movable-bed model experiments of the coast under the combined action of the wave and tidal current. By these experiments we mainly desire to solve the problem of the way how to regulate the outer sandbar, including the rational trend of the guide dike, its length and the expected siltation amount in the navigation channel.

1. INTRODUCTION

Shantou Harbor is an important commercial port in the north-east of Guandong Province and lies at the estuary which is the junction of two rivers, i.e. Rongjiang and Hanjiang.

As the result of the sediments transported from the upstream, i.e. Rongjiang and Hanjiang, the tide and wave actions in the seaboard region as well as the influence of coastal reclamation, the sediment deposits in the harbor basin in the inner navigation channel of Shantou Harbor have grown up. Especially, the development of the sandbar in the outer navigation channel not only affects the navigable depth of the passage of Shantou Harbor, but will obstructs the further expansion of the deep-water berth of the harbor as well.

The navigation channel outside of Shantou Harbor was dredged three times respectively in 1977, 1979 and 1980. However, the dredged channels were still silted up to the original states between six and nine months after dredging them. The obstruction of the further development of Shantou Harbor by the outer sandbar and the severe silting conditions after dredging of the outer navigation channel have attracted much attention of the production and construction departments concerned as well as scientific research institutions.

*Senior Engineer, Nanjing Hydraulic Research Institute, China

2. PROBLEMS ENCOUNTERED IN THE DESIGN OF COASTAL MOVABLE-BED MODEL UNDER THE COMBINED ACTION OF WAVE AND TIDE

2.1 About in-shore and off-shore models

French scholar C. Migniot, after summing up the experiences of movable-bed model for the investigation of the French coast, pointed out that with our gradual understanding of sediment transport process, the laws governing these natural phenomena have been found to be more and more complicated and the similarity of all conditions to the prototype can hardly be obtained in the same model for sediment study. Therefore, making selections from reproduction phenomena are required and the model scales specially used for the selected experiment type would be adopted. Portugal scholar Castanho, Canadian scholar J.W. Kamphuis and Chinese scholar Chen Zixia have made some specialized discussions in this respect. Especially Kamphuis explicitly divides a coastal movable-bed model into two kinds, i.e. inshore model and offshore model. Either of these two kinds of model, however, lays respective emphasis on its reproduction. They have different condition of similarity and different sediment transport scale and scale of scouring and silting time.

Basing upon the findings in our experiments as well as our specific purpose of experiment and specific district, we use the movable-bed model for regulation of the outer sandbar ---- the offshore model, and movable-bed model for sand catching effect by sediment retention dike ---- inshore model.

2.2 About rate of distortion of the coastal movable-bed model under combined action of wave and tide

In the experiments on the movable-bed model for sandbar regulation, it is necessary to stimulate the combined action of two dynamic effects by the wave and tide at the same time. A distorted model should also be made, because both the similarity requirements of the sediment under the combined action of the wave and tide and the problems of the experimental equipment and sand material should almost be considered.

According to the few examples that can be found for the time being the rate of distortion of coastal movable-bed models under the combined action of wave and tide is from 3 to 8.

It may be seen from the description above-mentioned that, if the wave diffraction is not selected as the key conjunction in the model reproduction and proper material taken as the model sand to satisfy the starting similarity and the falling similarity of the sediment movement, some model distortion should be allowable.

3. MOVABLE-BED MODEL TESTS FOR THE SANDBAR REGULATION

3.1 Analyses of evolution and grain size of the sediment

The sandbar in the navigation channel outside of Shantou Harbor was dredged four times respectively in September of 1977, July of 1979, October of 1980 and September of 1981. The cross sections of the excavated channel after silting are shown in Fig.1.

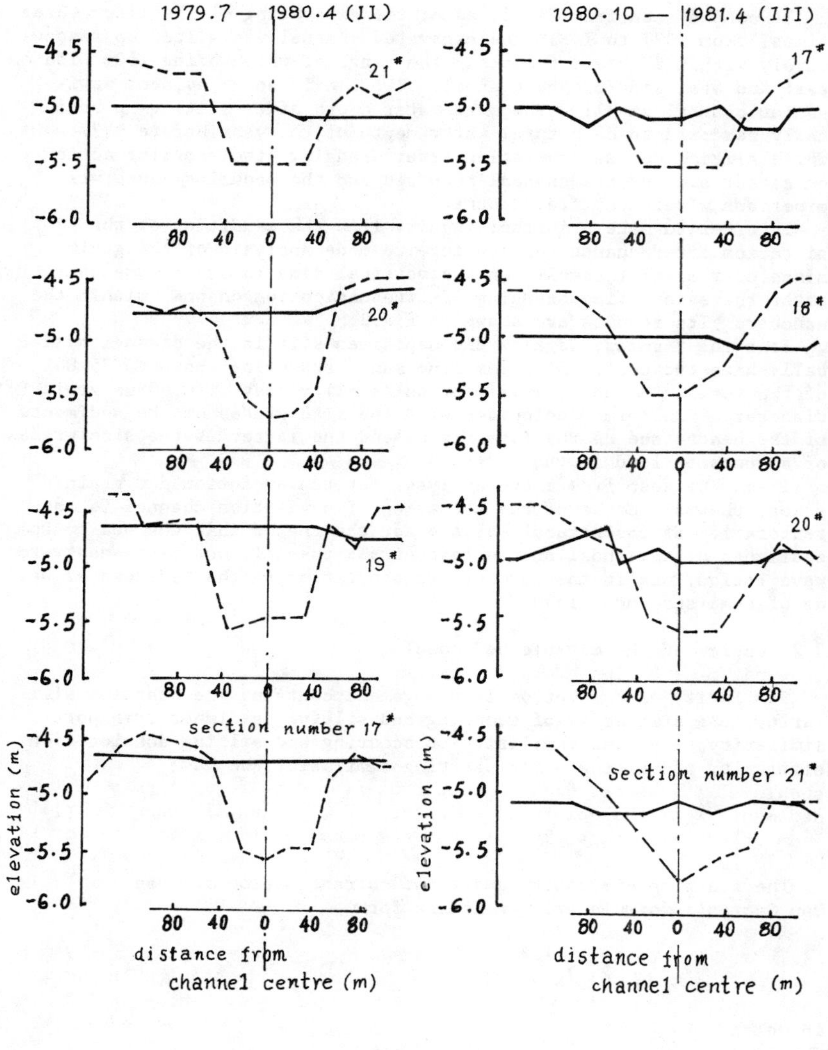

Fig.1 Variations in cross sections of the outer navigation channel after the first and second time's excavating

It can be seen from Fig.1, after the excavating of the first three times, from 1977 to 1981, the excavated channel was silted up respectively within 8, 9 and 6 months, the range of which being 150m wide on east and west side of the channel. There existed no evident navigation channel at all. The mean water depth after silting-up basically restored to be natural water depth of the year before 1977. While silting-up the navigation after dredging, the scouring action on either side of the channel occurred and the scouring range was wider and wider one after another.

In order to make a further inquiry into the condition of the siltation after channel excavating, we made analysis of the grain sizes of four silt samples on the vertical line taking in one year after the second time dredging, in the navigation channel within the sandbar. Its results are shown in Fig.2.

From the figures, we know the deposited silt in the channel basically have the grains of 0.1mm fine sand, amounting about to 75-80% of the total. Besides, 20-25% or so is silty clay of 0.009mm grain diameter. The former coincides with the size grades of the sediments of the beach face nearby (see Fig.2) and the latter is the size grades of suspension load in the river.

It can be seen from above analyses for the evolution and grain sizes, the main cause of the siltation of navigation channel is the restoration of the natural balance for the reason that the sea-bottom sediments of the shoal on the east of the channel, under the eastward wave action, was in the form of transportation on the bed load as well as of semi-suspended load.

3.2 Design of the movable-bed model

Similarity of the bottom load movement contains the starting similarity, the similarity of scouring and silting, sediment transport similarity, the time similarity of scouring and silting and location similarity of silting. For starting similarity, we have

$$\lambda_v = \lambda_{u_o} = \lambda_{v_k} \tag{1}$$

The starting similarity under the current action may use Dou Guoren's formula written in the form

$$\lambda_{v_k} = \lambda_\phi \, \lambda_{\gamma_{s-\gamma}}{}^{1/2} \, \lambda_d{}^{1/2} \tag{2}$$

in which

$$\lambda_\phi = \frac{\left[\ln 11 \frac{H}{\Delta} \left(1 + 0.19 \frac{\gamma}{\gamma_s - \gamma} \frac{\epsilon_\kappa + gh\delta}{gd^2}\right)^{1/2}\right]_p}{\left[\ln 11 \frac{H}{\Delta} \left(1 + 0.19 \frac{\gamma}{\gamma_s - \gamma} \frac{\epsilon_\kappa + gh\delta}{gd^2}\right)^{1/2}\right]_m} \tag{3}$$

For computing the bottom sediment discharge under tidal current action, Dou Guoren's formula may be used:

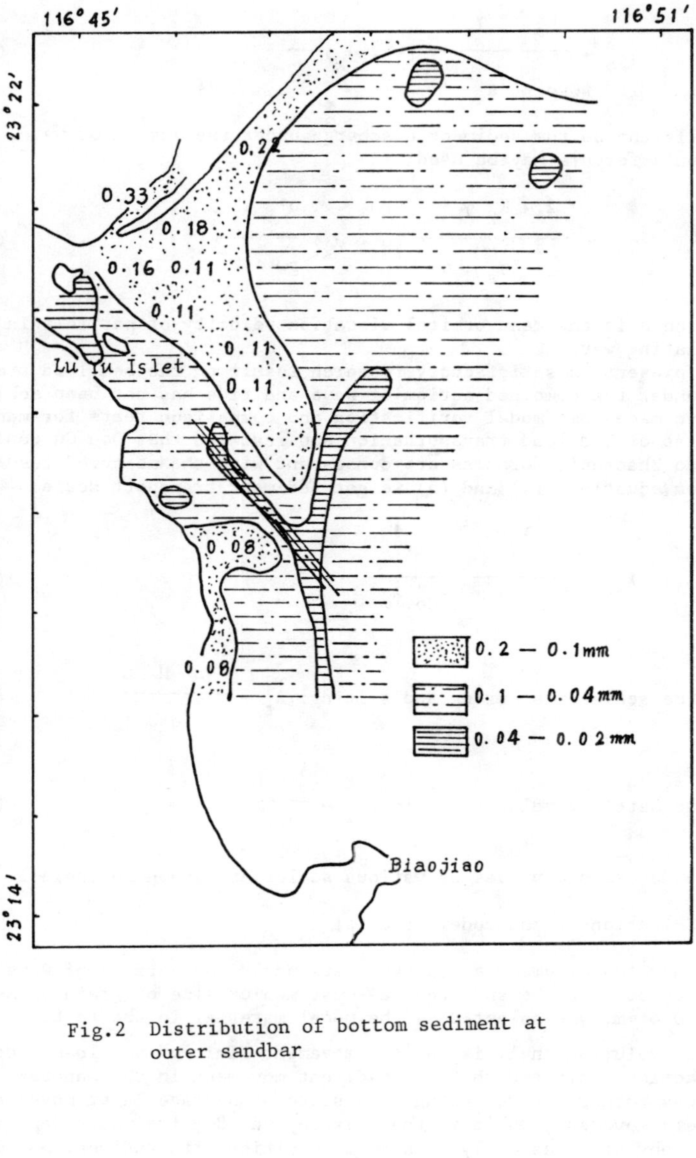

Fig.2 Distribution of bottom sediment at outer sandbar

$$q_{sb} = \frac{K_o}{C_o^2} \frac{\gamma_s \gamma}{\gamma_s - \gamma} (v - v_k) \frac{v^3}{g\omega} \tag{4}$$

While the bottom sediment discharge under the action of wave, Lua Zhaosen's formula may be used:

$$q_{sb} = \frac{2.4 K_2 \gamma \gamma_s}{C_o^2 (\gamma_s - \gamma)} (u - u_k) \frac{u^4}{g\omega} \frac{T}{L} \tag{5}$$

in which u is the mean orbital vibration velocity of particle in oscillating wave.

At present, a satisfactory research result on the bed load transport under the combined action of wave and tide has not been achieved. We have made some model verification and comparison tests for many formulas of bed load transportation and realized that Dou Guoren's and Luo Zhaosen's formulas are consistant with the measured results.

From equations (4) and (5) we get sediment transport scale:

$$\lambda_{qsb} = \frac{\lambda_{\gamma_s}}{\lambda_{\gamma_s} - \gamma} \frac{\lambda_v^4}{\lambda_{Co}^2 \lambda_\omega} \tag{6}$$

the time scale of scouring and silting: $\lambda_t = \dfrac{\lambda_{\gamma o} \lambda_H \lambda_L}{\lambda_{qsb}}$ (7)

and the settling velocity scale: $\lambda_\omega = \dfrac{\lambda_v \lambda_H}{\lambda_L}$ (8)

The determined values of various scales are shown in Table 1.

3.3 Selection of the model material

We have made some preparative tests under the actions of wave and tidal current in the end, the saw-dust medium-size of grain diameter D_{50} = 0.4 mm, was selected as the model material in the test.

The evolution analysis and the movement limit of bed load under wave action indicated that the sediment movement in the sandbar area was rather active, being in a state of surface layer movement and complete movement, while in the area beyond -8m, its water depths and topography were basically in stable condition, its sediment movement being in a state of surface layer movement and more and more weaker as it moves offshore. As stated above, if the wave height and wave period are regulated to a proper value, it would be easy to realize the similarity of the sediment movement regime in the above-mentioned area.

Table 1 : Scale for movable-bed model

Name of scale	Symbol	Computed value	Adopted value
Horizontal scale	λ_l		500
Vertical scale	λ_h		80
Current velocity scale	λ_v	8.944	8.944
Wave height scale	λ_H	80	80
Wave length scale	λ_L	80	80
Wave period scale	λ_T	8.944	8.944
Bottom orbit velocity scale	λ_{u_o}	8.944	8.944
Mass-transport scale	λ_v	8.944	8.944
Starting velocity scale	λ_{v_k}	8.944	8.944
Chezy coefficient scale	λ_c	2.50	2.10
Settling velocity scale	λ_ω	1.43	1.01
Grain diameter scale	λ_d		0.275
Dry density scale	λ_{γ_o}		2.55
Sediment discharge scale	λ_{qsb}	150	87
Scouring and silting time scale	λ_t	667	1168

Basing on the relation between the navigation channel water depth after dredging and the silting speed, it indicates that the mean water depth in the dredged channel less than 5 months after dredging could restore to its original water level. Three hours after the beginning of the experiment the longitudinal profile of the excavated channel in the model has a better coincidance in silting regime with that of the prototype. Thus obtaining the actual model scale of scouring and silting time λ_{t2} = 1168.

3.4 The experiments of regulation schemes

Though the comparison and selection of various options, the movable-bed model experiments finally gives the proper trend and length of the guide dike and the expected siltation in the dredged channel. Its results are shown in Fig.3 and Table 2.

The test results indicate: The option of constructing a 5.2km sediment retention dike attaching 2.75km submerged dike is rather adaptable.

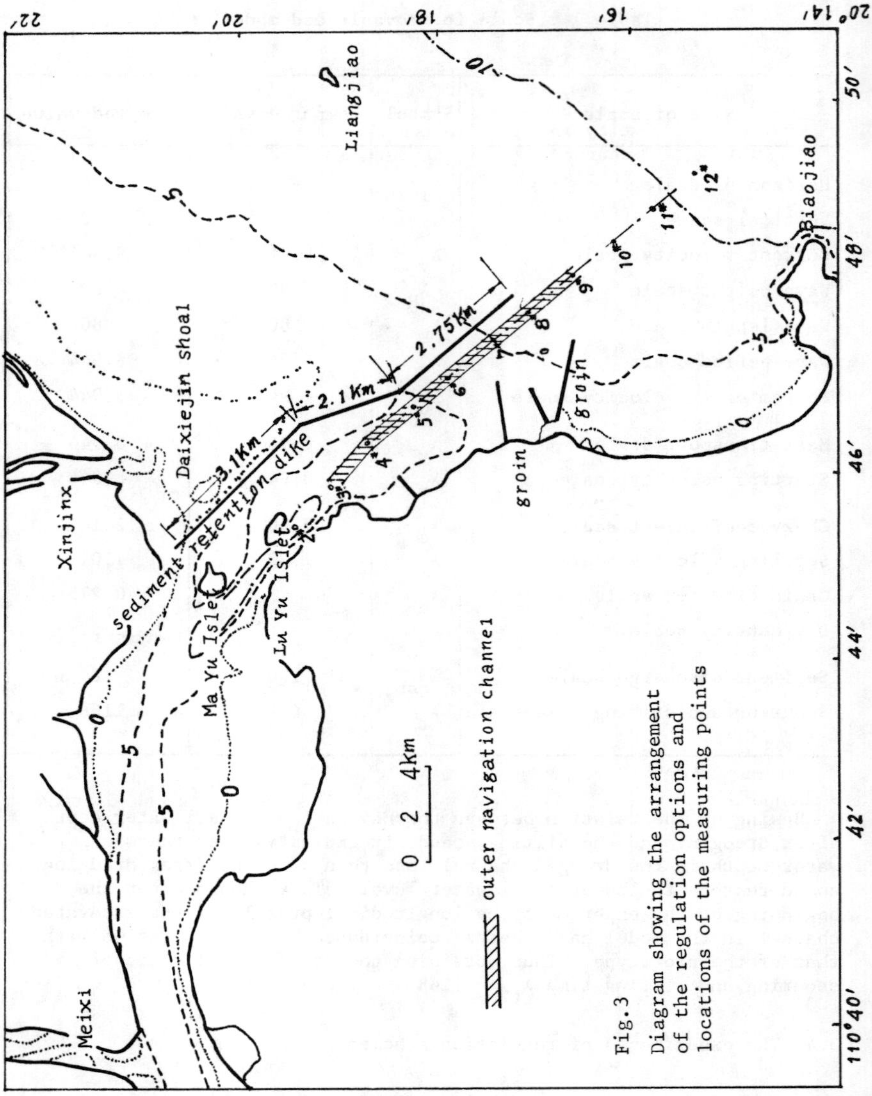

Fig.3

Diagram showing the arrangement of the regulation options and locations of the measuring points

Table 2 : Comparison of the options

Options	Silting quantity in the channel at initial stage ($x10^4 m^3$/year)	Silting trend in the channel in further	Resistance to typhoon's silting	Scouring at the dike head
1. To excavate the navigation channel	160	a little reduction	not possessing	
2. To excavate the navigation channel +5.2km sediment retention dike	70	a little reduction	not possessing	greater transverse current and scouring
3. To excavate the navigation channel +5.2km sediment retention dike+2.75km submerged dike	50 - 55	decreased year by year	possessing	basically has no dike head transverse current and scouring
4 To excavate the navigation channel +5.2km sediment retention dike+2.75km submerged dike + two southwest groins	50	decreased year by year	possessing	basically has no dike head transverse current and scouring
5. To excavate the navigation channel +5.2km sediment retention dike+2.75km overflow dike	35 - 40	decreased year by year	possessing	basically has no dike head transverse current and scouring

4. CONCLUSIONS

1. The paper discusses the theory and practice of the coastal movable-bed model tests under the combined action of the tide and wave for the regulation of the outer sandbar at Shantou Harbor on the test results. Our research shows that two kinds of the coastal movable-bed model should be devided according to the experimental purposes and the practice. That is off-shore model and in-shore model; the movable-bed model for the regulation of the outer sandbar at Shantou Harbor belongs to the off-shore model.
2. Our research also indicates that, if we do not select the diffraction phenomenon as the key conjunction in the model reproduction and select fit material as the model sand, which meet the falling similarity of the sediment movement, some model distortion would be allowable.
3. The key point of the determination of the scale of scouring and silting time in the coastal movable-bed model experiment lies in the selection of a better sediment transport formula and sediment transport scale.
4. The macroscopic analysis of the sediment movement in the current of outside and inside of Shantou Harbor indicates Shantou Bay is a type of estuary with the tide action as a dominant factor. The sediment deposit of the outer sandbar mainly comes from Xinjinxi and Wai Sha river and transported from northeast to southwest under the action of the tide and wave. Generally speaking, there is small quantity of sediment deposit in the whole seas and fundamentally in the state of dynamic equilibrium. The main reason of the siltation in the navigation channel is the bed load in the sea bottom transport on the spot and in semi-suspension condition to cause the restoration of natural balance of the sediment and then the siltation of the excavated navigation channel.
5. By way of the fixed-bed model tests under tidal current action as well as the movable-bed model tests under the combined action of the tide and wave, we have made some comparison tests for different option for the outer sandbar regulation. The test results indicate:
A 5.2km sediment retention dike attaching a 2.75km submerged dike is rather adaptable, possessing the effects of restricting the water way to scour the sand as well as the retention of the sediment.

5. REFERENCES

Bijker, E.W. Some considerations about scales for coastal models with movable bed. Delft Hydraulics Laboratory, Publ. No.50, 1967.

Chen Zixia et al. Model test on the sedimentation accumulation at Island Terminal. Proc. of the First International Symposium on River Sedimentation, Beijing, 1980

Dou Guoren. Law of all-sand model similarity and its designing examples. Water Conservancy & Transportation Science Information, March, 1977.

Huang Jianwei et al. The model experimental report on the regulation of the sandbar outside of Old Harbor District of Shantou Harbor. Nanjing Hydraulic Research Institute, 1986.

Huang Jianwei et al. The sediment transporting regularities of the sandbar outside of Shantou Harbor and a preliminary analysis of its regulatory prospect. Nanjing Hydraulic Research Institute, 1986.

Kamphuis, J.W. Practical scaling of coastal models. Proc. 14th International Conference on Coastal Engineering, Copenhagen, 1974, Vol.III, Chapter 121.

Lou Zhaosen and Xin Wenjie. Research report on the conditions of the harbor construction at Shuidong Bay and the practicability of the Navigation Channel Development at the Sandbar. Nanjing Hydraulic Research Institute, June, 1987.

Mignot, C. Représentation des transports et des dépôts de sédiments dans les modèles réduits maritimes, 1972.

You Fanghu et al. A. finding and research report on the sediment silting in the Navigation Channel through the Sandbar outside of Shantou Harbor. Institute of Oceanography of the Chinese Academy of Sciences, 1981.

CHAPTER 92

ENVIRONMENTAL CONTROLS ON LITTORAL SAND TRANSPORT

Paul D. Komar[1]

ABSTRACT

Quantities of sand transported along beaches are generally related to the "longshore component of wave power", P_ℓ, through the proportionality $I_s = KP_\ell$ where I_s is the immersed-weight sand transport rate and K is a dimensionless proportionality factor. A more-generally applicable relationship is that of Bagnold, $I_s = K'(ECn)_b \bar{v}_\ell / u_m$ where $(ECn)_b$ is the energy flux or total power of the breaking waves, $\bar{v}_\ell$ is the longshore current, u_m is the mean orbital velocity under the waves, and K' is another dimensionless coefficient. It is apparent that sediment transport rates on beaches should depend on environmental factors such as the grain diameter or settling velocity, and possibly on factors such as the beach slope or wave steepness. However, examinations of such dependencies for K and K' within the field data are hampered by problems with large random scatter within any one data set, and by systematic differences between separate studies which have employed diverse measurement techniques. Examinations of the field data for K and K' variations indicate that meaningful dependencies on sediment grain diameters and other factors cannot be established with confidence in the sand-size range. Limited data available from gravel beaches support the expected decreases in K and K' with increasing grain sizes. These data are too few in numbers to establish firm trends, but do suggest that future investigations to establish dependencies on environmental factors would be most profitably undertaken on gravel beaches. The measurements collected in recent years from sand beaches suggest revisions in average K and K' coefficients to be used in transport evaluations, but such revisions must be coordinated such that K/K' = 2.7 so as to maintain agreement with the longshore current data.

INTRODUCTION

It has been 35 years since George Watts collected the first field measurements of sand transport rates along beaches which permitted correlations with the causative wave conditions (Watts, 1953). Since that time, a number of studies have contributed to our data base, investigations which have obtained measurements from beaches in lakes, bays, and along ocean shores. The list of published field studies now includes 16 beach locations, yielding data for various types of correlations between sand transport rates, the wave conditions, and longshore current magnitudes. Given the considerable amount of accumulated data from a variety of beach locations, the question arises as to whether we can now detect environmental controls on the resulting sediment transport rates, that is, factors beyond the direct dependence on the wave power and

---

[1]College of Oceanography, Oregon State University, Oceanography Admin. Bldg. 104, Corvallis, OR 97331-5503

angle of wave breaking. Various dependencies have been suggested including a simple correlation with the beach sediment grain size (or alternately the grain settling velocity), the beach slope, or more complex combinations of various environmental parameters. The objective of this study is to reexamine those suggested correlations to determine whether improved predictions of longshore sediment transport rates are possible if we include considerations of such environmental factors. This examination of environmental factors necessitates a compilation and general review of the existing data, which will be undertaken in the first section of this paper. The second half of the paper will then be devoted to examining possible dependencies on the various environmental factors. This examination will be limited to field data, even though many of the proposed correlations with environmental parameters are based on measurements from laboratory wave basins. The laboratory data are not included here due to the well-known uncertainties in the scaling of the laboratory sand-transport measurements, and because the ultimate goal is to establish predictive relationships for prototype beaches.

FIELD DATA AND EMPIRICAL CORRELATIONS

The initial motivation for studying the longshore transport of beach sediments came in response to adverse impacts due to jetty construction which interrupts this transport. It is not surprising then that the first quantitative investigation of this process, the study by Watts (1953), made use of sand blockage by jetties, specifically those at South Lake Worth Inlet, Florida. This approach has continued to be employed by various investigations, including the studies of Caldwell (1956), Bruno and Gable (1976), Bruno et al. (1980, 1981), and most recently those by Dean et al. (1982) and Dean et al. (1987). The locations of these various studies are given in Table 1 together with relevent environmental parameters. There are obvious problems associated with the use of jetties or breakwaters to measure longshore sediment transport rates, the foremost being the local effects of the structures on waves and currents, and the long-term nature of the measurements. In some cases it takes a month or longer for sufficient quantities of sand to accumulate in order to make the volume determinations meaningful, an interval during which waves and currents are continuously changing.

Beginning in the 1960's, sand tracers came into use to determine "instantaneous" littoral transport rates that could be related to relatively constant conditions of waves and currents. There are three potential approaches in the use of tracers to measure sand transport rates; (1) the spatial-integration method, (2) the time-integration method, and (3) the dilution method. These approaches have been reviewed by Komar (in press). Investigations that have employed this technique for determining correlations between sand transport rates and the wave conditions include Komar and Inman (1970), Knoth and Nummedal (1977), Inman et al. (1980), Duane and James (1980) and Kraus et al. (1982); Table 1 gives the locations of these several studies.

The data listed in Table 1 are not of uniformly high quality, so there is a temptation to eliminate some. Most obvious in this respect are the early measurements of Watts (1953) and Caldwell (1956), which involved long-term averaging of the sand transport determinations, and had basic weaknesses in their measurements of the wave conditions. Rather than a whole-scale casting out of certain data sets at the start, all will be retained in the analyses undertaken here. In that way we can focus on potential errors and systematic differences between the various studies and measurement techniques, an approach that will be important in discerning whether there are environmental controls on longshore sediment transport rates.

TABLE 1: Field Data for Longshore Sediment Transport Rates

Source	Location	D_{50} (mm)	# of points	K	K'
Watts (1953a)[1]	Ft. Lake Worth Inlet, Fla.	0.40	4	0.89(.73-1.03)	
Caldwell (1956)	Anaheim, Calif.	0.40	6	0.63(.16-1.65)	
Moore & Cole (1960)	Cape Thompson, Alaska	1.0	1	0.18	0.18
Komar and Inman (1970)	El Moreno, Mexico	0.60	8	0.82(.48-1.15)	0.35(.22-.53)
	Silver Strand, Calif.	0.18	4	0.77(.52-.92)	0.22(.06-.36)
Lee (1975)	Lake Michigan	?	8	0.42(.24-0.72)	
Knoth & Nummedal (1977)	Bull Island, S.C.	0.18	5	0.62(.23-1.0)	
Inman, et al. (1980)	Torrey Pines, Calif.	0.20	2	0.69(.26-1.34)	
Duane & James (1980)	Pt. Mugu, Calif.	0.15	1	0.81	
Bruno, et al. (1981)[2]	Channel Islands Harbor, Calif.	0.2	7	0.87(.42-1.5)	
Kraus, et al. (1982)	Ajigaura, Japan	0.25	3		0.19(.16-.22)
	Shimokita	0.18	2		0.32(.28-.36)
	Hirono	0.59	2		0.091(.08-.10)
	Oarai	0.29	4		0.18(.16-.19)
Dean, et al. (1982)	Santa Babara, Calif.	0.22	7	1.15(.32-1.63)	
Dean et al. (1987)	Rudee Inlet, Va.	0.3	3	1.00(.84-1.09)	

[1] Only the monthly-averged data of Watts (1953a) are used in the analysis.

[2] Includes only the data where the wave statistics are based on measurements by gauges, not those based on LEO visual observations.

Data from the studies listed in Table 1 are plotted in Figure 1 as the transport rate versus

$$P_\ell = (ECn)_b \sin\alpha_b \cos\alpha_b \qquad (1)$$

where $(ECn)_b$ is the wave energy flux or power evaluated at the breaker zone, and α_b is the wave breaker angle. P_ℓ is often referred to as the "longshore component of wave energy flux or power." Longuet-Higgins (1972, p. 210) has taken exception of this terminology, but P_ℓ has continued to be employed in littoral sand transport evaluations, although now it is sometimes written as C_bS_{xy} where S_{xy}= En sinα cosα is the longshore-directed component of the radiation stress.

The sand-transport rate can be expressed either as the volume transport rate, Q_s, or as an immersed-weight transport rate, I_s , defined as

$$I_s = (\rho_s - \rho)ga'Q_s \qquad (2)$$

where ρ_s and ρ are respectively the sand and water densities, and a' is a pore-space factor such that $a'Q_s$ is the volume of solid sand alone, eliminating pore spaces included in the Q_s volume transport rate (a' is usually taken as 0.6). One advantage of using I_s is that this immersed-weight transport rate accounts for the density of the sediment grains. Also important is that I_s and P_ℓ have the same units, so that the relationship

$$I_s = KP_\ell \tag{3}$$

is homogeneous, that is, the K proportionality coefficient is dimensionless. In that Inman and Bagnold (1963) were the first to employ a littoral sand transport relationship having the form of equation (3), I will refer to this correlation as the "Inman equation."

The available field data are plotted in Figure 1, and it can be seen that the measurements are reasonably consistent with the direct proportionality given by equation (3). The solid line yields the proportionality coefficient K = 0.77, a value obtained by Komar and Inman (1970). The dashed line for K = 0.57 will be discussed later. By using K = 0.77 and taking ρ_s as the density of quartz sediments, one obtains the derivative relationship

$$Q_s = 6.8\, P_\ell \tag{4}$$

where Q_s has units m^3/day and P_ℓ is Watts/meter. Equation (4) applies only to quartz-density sands, and the value of the proportionality coefficient depends on the units employed; for example, the coefficient becomes 1.5×10^4 if Q_s is yds^3/yr and P_ℓ is ft-lbs/sec·ft. These latter units correspond to the comparable formula presented in the *Shore Protection Manual* (CERC, 1984), but the value obtained here for the proportionality coefficient is almost exactly twice that given in SPM. The reason for this difference is that in the SPM formula, the calculation of the wave energy and P are based on the significant wave height, $H_{1/3}$, whereas in the present analyses wave energies are based on the root-mean-square wave height, H_{rms}, the height which corresponds to the correct assessment of the wave energy as evaluated from complete spectra. In that $H_{1/3}/H_{rms} \approx 1.42$ (Longuet-Higgins, 1952), the calculated wave energies and powers would differ by a factor $(1.42)^2 \approx 2$. From this it is apparent that the SPM formula is equivalent to equation (3) and the results obtained earlier by Komar and Inman (1970). Of particular importance, this points out the need for recognizing that in using these relationships, one must be aware of whether they are based on significant wave heights or rms wave conditions.

Comparisons between the longshore sand transport rate and P_ℓ as undertaken above are empirical with little thought given to the physical processes. Early workers such as Grant (1943) stressed that sand transport in the nearshore results from the combined effects of waves and currents, the waves placing sand in motion and the longshore currents producing a net sand advection. Such a model was given a mathematical framework by Bagnold (1963), and applied specifically to the evaluation of sediment transport on beaches by Inman and Bagnold (1963). Their analysis yielded the "Bagnold equation",

$$I_s = K'(ECn)_b \frac{\bar{v}_\ell}{u_m} \tag{5}$$

where $\bar{v}_\ell$ is the mean longshore-current velocity, in practice measured at the mid-surf position, and u_m is the maximum horizontal orbital velocity of the waves evaluated at the breaker zone. K' is a dimensionless coefficient which again must be

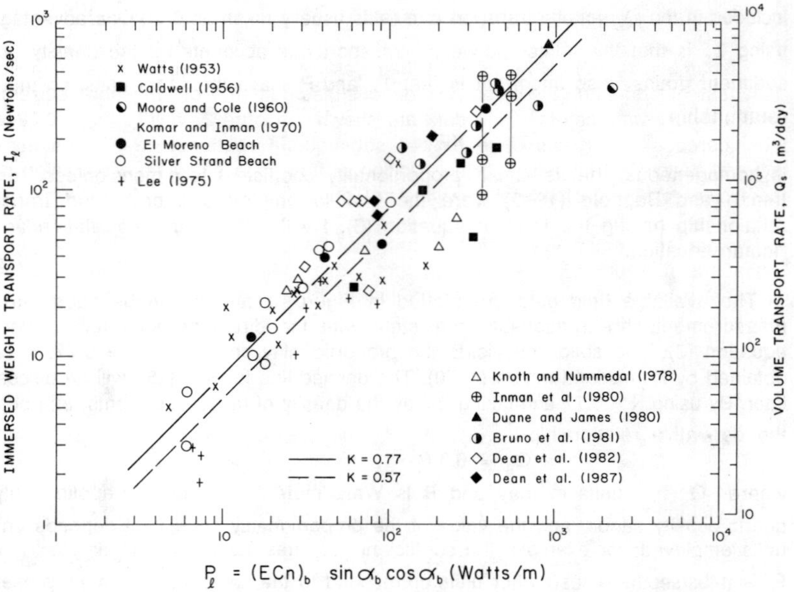

Fig. 1: Test of the Inman relationship, equation (3), for the longshore sand transport rate with the field data listed in Table 1.

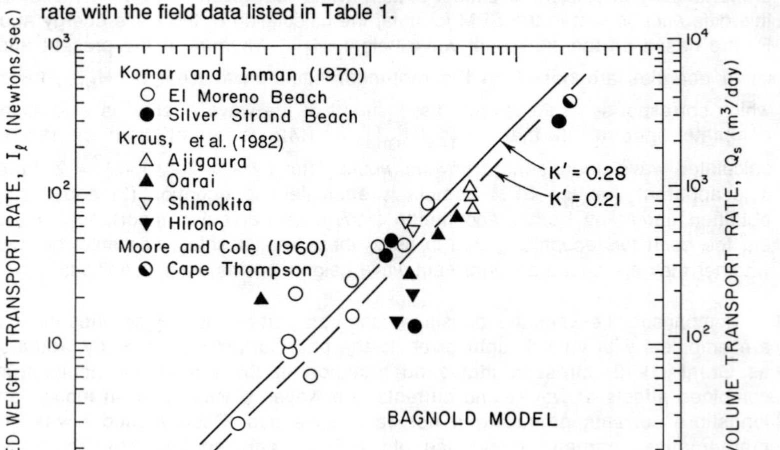

Fig. 2: Test of the Bagnold relationship, equation (5), for the longshore sand transport rate.

based on sediment-transport measurements.

Komar and Inman (1970) utilized their data on sand transport rates to make the first test of equation (5), employing direct measurements of longshore currents as well as wave parameters. Their data are shown in Figure 2, yielding K' = 0.28 when considered alone. Kraus et al. (1982) subsequently obtained measurements from several beaches in Japan, and the addition of their data suggests a reduction of the coefficient to K' = 0.21, the dashed line in Figure 2.

Komar and Inman (1970) found that their sand-transport data agreed with both equations (3) and (5), even though those relationships were seemingly based on different models. However, it was demonstrated that a simultaneous solution of these two equations yields

$$\bar{v}_\ell = (K/K')u_m \sin\alpha_b \cos\alpha_b \tag{6a}$$

or

$$\bar{v}_\ell = 2.7\, u_m \sin\alpha_b \cos\alpha_b \tag{6b}$$

if one takes K/K' = 0.77/0.28 = 2.7 from the sand-transport correlations. This relationship, based initially on sand-transport considerations, was found to agree closely with available measurements of longshore currents, and is similar in form to the relationship derived theoretically by Longuet-Higgins (1970) based on radiation stress evaluations. Substituting $u_m \propto \sqrt{gh_b} \propto \sqrt{gH_b/\gamma}$, equation (6b) can also be expressed as

$$\bar{v}_\ell = 1.17 \sqrt{gH_b} \sin\alpha_b \cos\alpha_b \tag{6c}$$

A comparison between this form of the relationship and available field and laboratory data sets for longshore currents again demonstrates that agreement is excellent (Komar, 1979; Komar and Oltman-Shay, in press).

One significance of this connection between the two relationships for the sand transport through the longshore-current equation is that the Bagnold relationship, equation (5), derived on the basis of considerations of processes of sand transport, should be viewed as more fundamental than the Inman equation which empirically correlates the sand transport with P_ℓ . Equation (3) or other correlations with P_ℓ should be applied only on beaches where one can be certain that the longshore current is produced solely by an oblique wave approach and hence given by equation (6). If the nearshore currents are affected by tides, cell circulation with rip currents, or by local winds, then the more basic equation (5) must be used with direct measurements of $\bar{v}_\ell$. This is confirmed by the sand-transport measurements of Kraus et al. (1982) at Oarai Beach. At that site the data were obtained in the sheltered region of a breakwater where the longshore current results from the combined effects of obliquely-incident waves and a longshore variation in wave heights. Under such conditions an evaluation of the sand transport from P_ℓ alone would be erroneous. This is especially illustrated by one measurement series where the direction of the longshore current and sediment transport was opposite to that expected from oblique-wave incidence. Even when the longshore currents and sand transport are produced entirely from waves breaking obliquely to the beach, the use of the Bagnold model, equation (5), may still be preferable to correlations with P_ℓ in that it is usually easier and more accurate to measure $\bar{v}_\ell$ than to measure breaker angles. Coastal engineering evaluations of sand-transport rates on beaches should be based more often on equation (5) than on the P_ℓ wave-power approach.

The connection between equations (3) and (5) for sand transport evaluations via equation (6) for the longshore current also places contraints on the respective K and K' coefficients since we must have K/K' = 2.7 for agreement with the longshore current data. In that longshore currents can be measured more accurately than sand transport rates, the 2.7 value is better established than the individual values of K and K'. It also requires that if we reduce K' from 0.28, as first established by Komar and Inman (1970), to K' = 0.21 given by Kraus et al. (1982), then we must also reduce the proportionality between I_s and P_ℓ to K = 0.57. This value yields the dashed line in Figure 1 and, given the large scatter of the data, the result is still a reasonable fit.

VARIATIONS IN THE K AND K' COEFFICIENTS

Considerable scatter is seen in Figures 1 and 2, and this is reflected in the K and K' proportionality coefficients in equations (3) and (5) for the individual data sets. These coefficients are compiled in Table 1 as averages and total ranges. The average K-values for the data sets are as high as 1.15 from Dean et al. (1982), and a low of 0.18 for the single measurement of Moore and Cole (1960). The question arises as to whether these variations in K and K' reflect environmental conditions such as sediment grain sizes, or whether they result from random scatter and systematic differences in procedures between the several studies.

Dependencies on Grain Sizes and Settling Velocities

The first investigation to use field data to examine possible environmental controls on the K coefficient of equation (3) was the study of Komar and Inman (1970). That study included two beaches of contrasting grain sizes (0.60 versus 0.18 mm), and the near equality of the average K coefficients (0.82 versus 0.77) from those two sites was taken as an indication of a lack of dependence on sediment diameter, at least within the limited range of sand sizes. This was further explored by Komar (1978) utilizing the then available field and laboratory data; it was again concluded that no relationship could be established between K and the beach sand-grain diameter or settling velocity.

With the inclusion of additional measurements, later investigators have reached the opposite conclusion; i.e., that the field measurements of transport rates do reveal a dependence on the grain size of the beach sediment (Bruno, et al., 1980, 1981; Dean et al., 1982; Dean et al., 1987). Figure 3 (left) is the plot of K versus the median grain diameter D_{50} from Dean et al. (1987), the plot and curve being the same as that

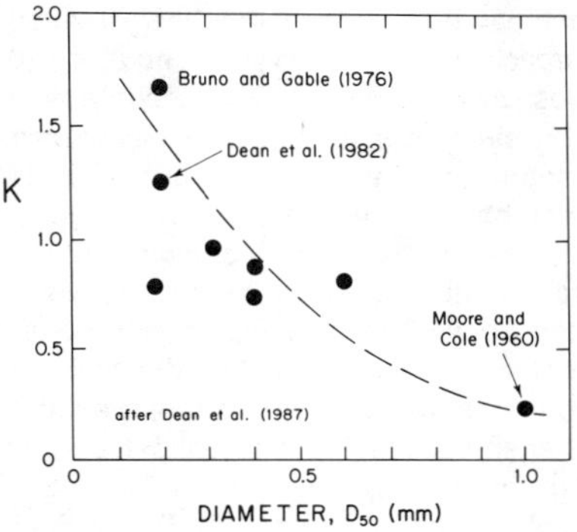

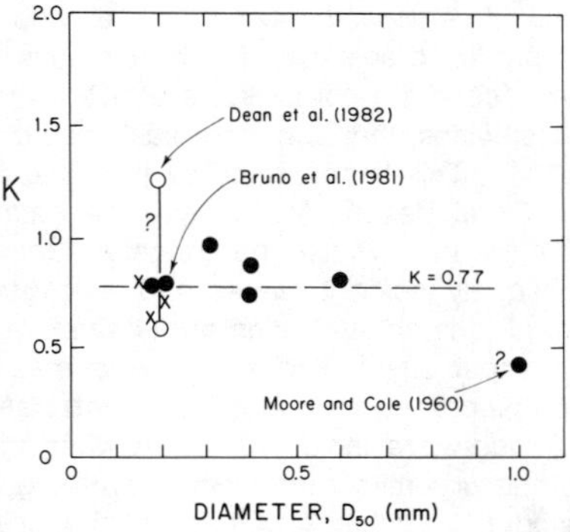

Fig. 3: K of eq. (3) versus the median grain diameter of the beach sand: (left) as analyzed by Dean et al. (1987), and (right) as re-analyzed here.

originally given in Bruno et al. (1981) excepting for the addition to two data points. The trend of decreasing K with increasing D_{50} is seen to be determined mainly by the K values obtained in the studies of Bruno and Gable (1976), Dean et al. (1982) and by Moore and Cole (1960). Unfortunately, the values from those three studies are either uncertain or clearly erroneous. The K = 1.61 plotted value from Bruno and Gable (1976) for the Channel Islands Harbor, California, was based on wave measurements collected by the Littoral Environment Observations (LEO) program. Bruno et al. (1980, 1981) later demonstrated the inadequacy of those visual data, and revised the coefficient downward to K = 0.87 on the basis of more reliable measurements from wave gauges. Unfortunately, the 1.61 value was mistakenly retained in their comparison with D_{50}, and this was perpetuated by Dean et al. (1987) in the plot of Figure 3 (left). The revised K = 0.87 coefficient was based on an assumed value of a' = 0.65 for the pore-space factor; other studies have assumed a' = 0.6, and if this value is employed, the coefficient is further reduced to K = 0.80, effectively the same as the 0.77 value proposed earlier by Komar and Inman (1970). This reduced K value is that plotted in the revised graph in Figure 3 (right).

Also uncertain is the K = 1.23 average coefficient obtained by Dean et al. (1982, 1987) at Santa Barbara, California, as part of the Nearshore Sediment Transport Study. Waves were measured by two S_{xy} gauges located offshore in approximately 9 m depth. It was found in the analysis that the west gauge yielded S_{xy} values which are approximately a factor 2 larger than those based on the east gauge. In their correlations with the sand-transport measurements, Dean et al. used the results from the east gauge. Had they instead used the west gauge, the $P_\ell = C_b S_{xy}$ wave powers would have been approximately 2 times larger and the resulting proportionality coefficient with the sand-transport data would have been reduced to K ≈ 1.23/2 = 0.61. This potential reduction is illustrated in Figure 3 (right), but it is not actually possible to determine which value, 1.23 or 0.61, should be employed in the correlation with D_{50}. It is clear, however, that too much weight should not be given to the average K value determined from the Santa Barbara data.

The single measurement of Moore and Cole (1960) is particularly important in the Dean et al. (1987) correlation between K and sediment grain size, Figure 3 (left), since it was derived from a coarse-sand beach where D_{50} = 1 mm. Unfortunately, this K-value from Moore and Cole is particularly questionable. The sand transport rate was determined from beach accretion within a breached area, so there is some uncertainty whether the total transport was measured. More important, the wave parameters were estimated visually. The velocity of the longshore current was measured, so it is possible to test the data of Moore and Cole against the Bagnold model of equation (5) to determine a K' coefficient. This yields K' = 0.18 which is effectively the same as the 0.21 value based on the combined data of Komar and Inman (1970) and Kraus et al. (1982). The Moore and Cole data point is seen in Figure 2 to agree with the trend established by the other measurements. This suggests that the mean breaker angle (25°) determined visually by Moore and Cole is too large, and this in turn makes P too large and hence K too small. Longshore currents generally can be measured more accurately than breaker angles, and this probably accounts for their data being more consistent with the Bagnold model of equation (5) than with the wave-power relationship of equation (3). This conclusion is further supported by comparisons of their data with equation (6) for the mean longshore-current velocity; their reported significant-wave breaker height [5.5 ft = 1.7 m (rms-H_b = 1.2 m)] and angle [α_b = 25°] yield a predicted longshore current $\bar{v}_\ell$ = 1.8 m/sec from equation (6c), whereas the measured current was only 0.64 m/sec. Working in the opposite direction, assuming the measured breaker height and longshore current are correct, then the breaker angle according to equation (6c) would have been approximately α_b = 9.5°. This angle reduces the corresponding P of equation (1), and the K coefficient of the Inman equation is accordingly increased to 0.43. This value is plotted in Figure 3 (right), and even though higher than the value used by Bruno et al. (1981) and Dean et al. (1987), it still possibly suggests a decrease in K with increasing D_{50}. However, given the uncertainties of the visual wave and current measurements of Moore and Cole, and that the measurement of sand accretion may not have accounted for the total transport, too much reliance should not be based on the resulting K value being lower than those found in other studies.

A similar analysis has been undertaken for K' in the Bagnold equation (5), and the results are given in Figure 4. It is apparent that there is no acceptable dependence of K' on D_{50}. The measurement of Moore and Cole (1960) fits within the data scatter rather than suggesting a decrease in K' at large D_{50}.

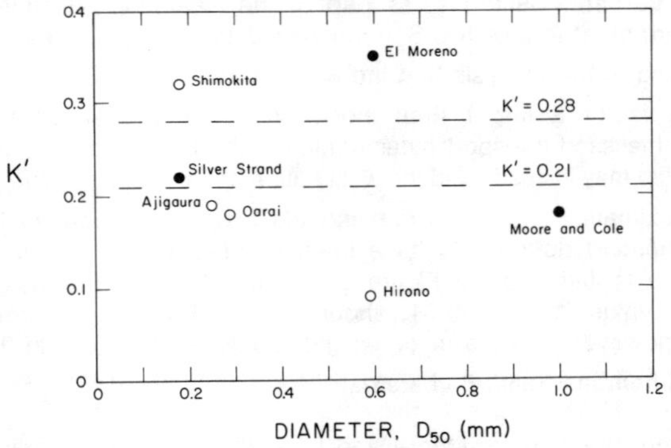

Fig. 4: K' of eq. (5) versus the median grain diameter of the beach sand. The data include those of Komar and Inman (1970), Kraus et al. (1982) and Moore and Cole (1960) [Table 1].

Grain-settling velocities of the beach sediments depend primarily on their sizes. As expected, attempts at correlations between K or K' and settling velocities are no better than found above for the median diameters.

It is expected that both K and K' should decrease as the sand size or settling velocity increase. The absence of such trends undoubtedly results from the considerable scatter in the field measurements, scatter that results from both random and systematic errors. The several studies that provide data for sand transport rates on beaches contrast markely in their techniques, involving both tracer experiments and blockage by jetties. It should come as no surprise that there are systematic differences in the results and that it would be difficult to establish dependencies on environmental factors such as D_{50}. Another problem is that essentially all of the data used above come from beaches with D_{50} between 0.15 and 0.60 mm, the one exception being the uncertain measurement of Moore and Cole (1960) at D_{50} = 1 mm.

If dependencies between K and D_{50} are to be established from field data, it will probably require that the measurements come from gravel beaches. Only with such a considerable increase in D_{50} might we find sufficiently large changes in K and K' that they will not be masked by systematic differences in measurement techniques. This is indicated by the study of Hattori and Suzuki (1979) on a gravel beach in Japan where $D_{50} \approx 2$ cm. Using tracer gravels, they measured longshore movements of 2 to 3 m/day under normal sea conditions, but as much as 400 m/day during storms. They found a good correlation between the mean advection rates of the gravel particles and P . If it is assumed that the movement has a thickness of one clast diameter, then their correlation is equivalent to $I_s \approx 0.2P$. Robert Nicholls (pers. comm.) has provided me with K values for their experiments on shingle beaches in England which involved the use of aluminum pebbles as tracers; Nicholls and Webber (1987) discuss the experiments with respect to sorting patterns of the pebbles according to their sizes and shapes. Three measurement series yield K = 0.011, 0.012 and 0.043 on two beaches where $D_{50} \approx 4$ cm. These limited results from gravel beaches do demonstrate the expected decrease in K from its value for sandy beaches. The results indicate that additional studies on gravel beaches could yield trends of K and K' versus D_{50}. However, it is also apparent that such analyses of coarse-sediment transport should include considerations of threshold criteria, modifying equations (3) and (5) to account for limits on particle movement.

Dependence on Beach Slope

There is a strong relationship between the beach slope and sediment grain size, so it might be anticipated that attempts to correlate K and K' with the beach slope would fare no better than the above correlations with diameters. There is an additional problem in that several of the studies do not report average beach slopes, so a smaller quantity of data is available to test such a correlation.

No correlation could be found between K and the beach slope within the sand-size range. The correlation between K' and beach slope is shown in Figure 5. Based on their data, Kraus et al. (1982) suggested that K' decreases with increasing beach slope. It is seen in the graph that such a trend is indicated by the low K' = 0.09 value obtained at Hirono beach, based on two individual sand-transport measurements (Table 1). However, the K' = 0.36 value for El Moreno Beach in the study of Komar and Inman (1970) diverges markedly from that trend. Therefore, it is too early to establish any

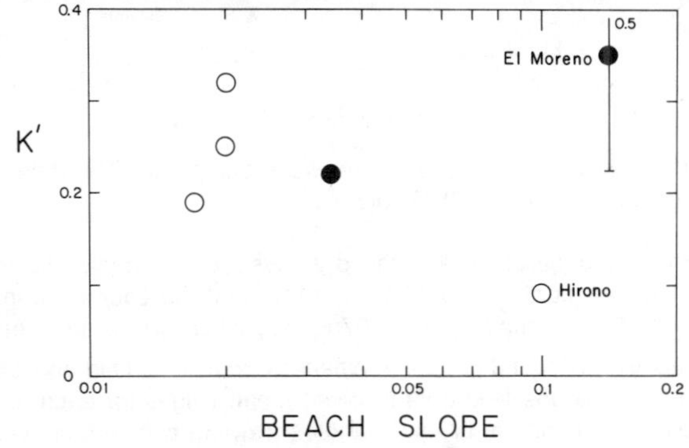

Fig. 5: K' of eq. (5) versus the beach slope. The data include those of Komar and Inman (1970) and Kraus et al. (1982) [Table 1].

dependence of K or K' on the beach slope for sand beaches. I have not yet attempted to extend the correlations to gravel beaches as undertaken above for mean grain diameters.

It of interest that the magnitude of the longshore current, as given by equation (6), does not depend on the beach slope. The *Shore Protection Manual* (CERC, 1984) and Galvin (1987) have proposed longshore-current relationships which contain a direct dependence on the beach slope, but these have been shown to yield serious errors when adequately tested with the available data (Komar, 1979; Komar and Oltman-Shay, in press). The implication to the longshore sediment transport is that any dependence on the beach slope will have to enter primarily through changes in concentrations of suspended sediments or thicknesses of the moving bedload carpet.

Dependencies on Wave Steepness and Other Wave Parameters

Laboratory studies such as that by Özhan (1982) have found a strong correlation between K and the wave steepness; the dependence was explained in terms of the control of the wave steepness on the breaker type (plunging, etc.). It is difficult to utilize field data to test such a relationship. The studies that evaluated sand transport rates by blockage at jetties involve very long time spans during which wave steepnesses and other wave parameters continuously change. Such correlations with field data will necessarily have to be limited to investigations where sand tracers were employed since these involve shorter intervals of time. Accordingly, I have tested the correlation between K and the deep-water wave steepness with the measurements of Komar and Inman (1970); the results, Figure 6, do not establish any reasonable trend.

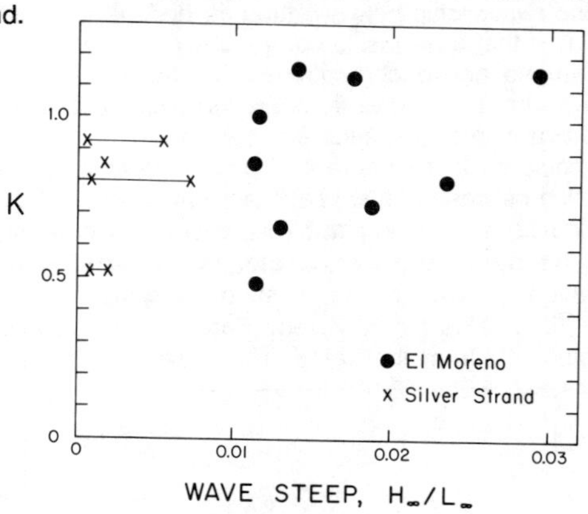

Fig. 6: K of eq. (3) versus the deep-water wave steepness. The data are those of Komar and Inman (1970) [Table 1].

Bailard (1984) has undertaken a detailed analysis of the bedload and suspension components of the transport. His analysis predicts that the coefficient in equation (3) is given by $K = 0.05 + 2.6\sin^2 2\alpha_b + 0.007u_m/w_s$, where w_s is the settling velocity of the beach sand. Bailard found good agreement by combining field and laboratory data. However, the inclusion of the field data involved mean values for each beach, ignoring the considerable variations in breaker angles (α_b) and in orbital velocities (u_m) which depend on the wave-breaker heights. I have tested this proposed relationship for K with the individual measurements contained within the data set of Komar and

Inman (1970), and could find no agreement.

Dependencies on Combined Wave and Beach Parameters

Some success has been achieved in analyzing surf-zone processes in terms of the Iribarren number, $\xi = m/(H_\infty/L_\infty)^{1/2}$, which combines the beach slope m with the wave steepness. Any test of dependencies of K and K' on ξ requires data from different beaches having a significant range of beach slopes, but where the sand-transport determinations involved only short time periods so that H_∞/L_∞ is nearly constant. The data of Table 1 are inadequate to meet these requirements. However, we already have established that K and K' do not depend on the beach slope for sandy beaches, although such a correlation will likely be found once the analysis is extended to gravel beaches and provides a larger range of slopes. Analyses with individual data sets show no dependence on H_∞/L_∞. Given these two findings, we can conclude that the present data cannot be used to establish variations of K and K' with the Iribarren number.

The heuristic model of Dean (1973) presents rational analyses of cross-shore and longshore transports of sediments. According to this model, the K coefficient in the Inman relationship, equation (3), is dependent on the wave-breaker height (H_b), breaker angle, grain settling velocity and beach slope (m) according to

$$K \propto m\sqrt{H_b}\cos\alpha_b/C_f w_s$$

where C_f is a drag coefficient. The principal new parameter introduced here that has not been tested is the proposed dependence on H_b. This again must be examined with data from short-term sand-tracer experiments so that breaker heights are relatively constant. Accordingly, I again used the data of Komar and Inman (1970); no trend was apparent between either K or K' and H_b.

Through a multi-stage analysis, Kamphuis et al. (1986) derived the empirical correlation $K \propto mH_b/D_{50}$ for the coefficient in the Inman relationship. The dependencies on the beach slope m and D_{50} would tend to be offsetting due to their positive correlation. Based on the empirical relationship $m = 1.8(H_b/D_{50})^{-0.5}$ established in their analyses, Kamphuis et al. further concluded that $K \propto (H_b/D_{50})^{0.5}$. Since we have been unable to find individual correlations between K and D_{50}, m or H_b , it can be concluded that this combined dependence proposed by Kamphuis et al. is not supported by the field data from sandy beaches.

SUMMARY OF CONCLUSIONS

The large quantity of field data available from a number of studies, Table 1, now makes it possible to re-examine the values of the K and K' proportionality coefficients respectively in equations (3) and (5). The connection between the two sand-transport relationships through the longshore current, equation (6), requires that K/K' = 2.7. The data of Komar and Inman (1970) yielded K' = 0.28 for the Bagnold equation, but subsequent measurements by Kraus et al. (1982) from several beaches suggest that the coefficient be reduced to K' = 0.21. This requires that K in the Inman equation (3) be reduced from 0.77 to 0.57; given the appreciable scatter of the data, Figure 1, such a modification in the mean value for K is acceptable.

Of particular interest to this study has been whether, by using the combined field data in Table 1, we can discern environmental controls on variations in K and K'. A number of such dependencies have been proposed in the literature. Here we initially explored direct correlations with environmental parameters such as beach-sand grain sizes, beach slopes and breaker heights, rather than using dimensionless combinations which would obscure the dependencies. Several studies have suggested that K decreases with increasing D_{50}, the median diameter of the beach sand. However, the analyses here lead to the conclusion that those previous correlations resulted from an erroneous K value for the Bruno and Gable (1976) data, and the likelihood that the K value employed for the Moore and Cole (1960) data was too large. The revised diagram of K versus D_{50} (Fig. 3 - right) provides little confidence for any dependence within the range of sand sizes. However, the limited data from gravel beaches do support the expected decrease in K with increasing D_{50}, but the measurements are too limited to establish a reasonable correlation. The results were similarly negative in attempts to correlate K and K' with beach slopes, wave-breaker heights, and finally with combinations of parameters such as the wave steepness and the Iribarren number.

It is apparent that K and K' should depend on basic environmental paramters such as sediment grain sizes. Therefore, the absence of such trends must result from the quality of the data. This should come as no surprise in view of the various techniques that have been used to measure sand transport rates on beaches and to collect the data on waves and currents. This has introduced appreciable systematic differences in the results from the several studies, along with considerable random uncertainties within individual data sets. Given the considerable difficulties in collecting data on waves, currents and sediment transport in nearshore field studies, and the general uncertainty in the scaling of laboratory results, it is hard to envision how this situation can be rectified.

ACKNOWLEDGEMENTS

I would like to thank Robert Nicholls for providing me with his unpublished data for gravel transport on beaches. This paper is a result of research sponsored by the Oregon State University Sea Grant Program, supported by NOAA Office of Sea Grant, Department of Commerce, under grant number NA85AA-D-SG0095.

REFERENCES

Bagnold, R.A. (1963) Mechanics of marine sedimentation: In *The Sea* , M.N. Hill (Editor), Wiley-Interscience, New York, p. 507-582.

Bailard, J.A. (1984) A simplified model for longshore sediment transport: Proc. 19th Coastal Engr. Conf., p. 1454-1470.

Bailard, J.A. and D.L. Inman (1981) An energetics bedload model for a plane sloping beach: local transport: Jour. Geophysical Res., v. 86, p. 2035-2043.

Bruno, R.O. and C.G. Gable (1976) Longshore transport at a total littoral barrier: Proc. 15th Coastal Engr. Conf., ASCE, p. 1203-1222.

Bruno, R.O., R.G. Dean and C.G. Gable (1980) Littoral transport evaluations at a detached breakwater: Proc. 17th Coastal Engr. Conf., ASCE, p. 1453-1475.

Bruno, R.O., R.G. Dean, C.G. Gable, and T.L. Walton (1981) Longshore sand transport study at Channel Island Harbor, California: U.S. Army Corps of Engrs., Coastal Engr. Res. Center., Tech. Paper No. 81-2, 48 pp.

Caldwell, J. (1956) Wave action and sand movement near Anaheim Bay, California: U.S. Army Corps of Engrs., Beach Erosion Board, Tech. Memo. No. 68, 21 pp.

CERC (1984) Shore Protection Manual: U.S. Army Corps of Engr., Coastal Engr. Res. Center, U.S. Govt. Printing Office, Washington, D.C.

Dean, R.G. (1973) Heuristic models of sand trans in the surf zone: Proc. 1st

Australian Conf. on Coastal Engr., Engineering Dynamics in the Surf Zone, Sydney, p. 208-214.

Dean, R.G., E.P. Berek, C.G. Gable, and R.J. Seymour (1982) Longshore transport determined by an efficient trap: Proc. 18th Coastal Engr. Conf., ASCE, p. 954-968.

Dean, R.G., E.P. Berek, K.R. Bodge, and C.G. Gable (1987) NSTS measurements of total longshore transport: Coastal Sediments '87, ASCE, p. 652-667.

Duane, D.B. and W.R. James (1980) Littoral transport in the surf zone elucidated by an Eulerian sediment tracer experiment: Jour. Sedimentary Petrology, v. 50, p. 929-942.

Galvin, C.J. (1987) The continuity equation for longshore current velocity with breaker angle adjusted for a wave-current interaction: Coastal Engr., v. 11, p. 115-129.

Grant, U.S. (1943) Waves as a transporting agent: Amer. Jour. Sci., v. 241, p. 117-123.

Hattori, M. and T. Suzuki (1979) Field experiment on beach gravel transport: Proc. 16th Coastal Engr. Conf., ASCE, p. 1688-1704.

Inman, D.L. and R.A. Bagnold (1963) Littoral processes: In *The Sea*, M.N. Hill (Editor), Wiley-Interscience, New York, p. 529-553.

Inman, D.L., J.A. Zampol, T.E. White, D.M. Hanes, B.W. Waldorf, and K.A. Kastens (1980) Field measurements of sand motion in the surf zone: Proc. 17th Coastal Engr. Conf., ASCE, p. 1215-1234.

Kamphuis, J.W., M.H. Davies, R.B. Nairn, and O.J. Sayao (1986) Calculation of littoral sand transport rate: Coastal Engineering, v. 10, p. 1-21.

Knoth, J.S. and D. Nummedal (1977) Longshore sediment transport using fluorescent tracer: Coastal Sediments '77, ASCE, p. 383-398.

Komar, P.D. (1978) The relative significance of suspension versus bed-load on beaches: Jour. Sedimentary Petrology, v. 48, p. 921-932.

Komar, P.D. (1979) Beach-slope dependence of longshore currents: Jour. Waterway, Port, Coastal and Ocean Div., ASCE, v. 105, WW4, p. 460-464.

Komar, P.D. (in press) Littoral sediment transport: In *Handbook of Coastal and Ocean Engineering*, J.B. Herbich (Editor), Gulf Publishing Co.

Komar, P.D. and D.L. Inman (1970) Longshore sand transport on beaches: Jour. Geophysical Res., v. 75, n. 30, p. 5514-5527.

Komar, P.D., and J. Oltman-Shay (in press) Nearshore currents: In *Handbook of Coastal and Ocean Engineering*, J.B. Herbich (Editor), Gulf Publishing Co.

Kraus, N.C., M. Isobe, H. Igarashi, T.O. Sasaki, and K. Horikawa (1982) Field experiments on longshore transport in the surf zone: Proc. 18th Coastal Engr. Conf., ASCE, p. 969-988.

Lee, K.K. (1975) Longshore currents and sediment transport in west shore of Lake Michigan: Water Resources Res., v. 11, p. 1029-1032.

Longuet-Higgins, M.S. (1952) On the statistical distribution of the height of sea waves: Jour. Marine Res., v. 11, p. 245-266.

Longuet-Higgins, M.S. (1970) Longshore currents generated by obliquely incident waves: Jour. Geophysical Research, v. 75, p. 6778-6789 and 6790-6801.

Longuet-Higgins, M.S. (1972) Recent progress in the study of longshore currents: *In* Waves on Beaches, R.E. Meyer (editor), Academic Press, N.Y., p. 203-248.

Moore, G.W. and J.Y. Cole (1960) Coastal processes in the vicinity of Cape Thompson, Alaska: *In* Geologic Investigations in Support of Project Chariot in the Vicinity of Cape Thompson, Northwestern Alaska – Preliminary Report, Kachadoorian et al. (editors), U.S. Geological Survey Trace Elements Investigation Report 753, p. 41-55.

Nicholls, R.J., and N.B. Webber (1987) Aluminum pebble tracer experiments on Hurst Castel Spit: Coastal Sediments '87, ASCE, p. 1563-1577.

Özhan, E. (1982) Laboratory study of breaker type effect on longshore sand transport: *In* Mechanics of Sediment Transport, B.M. Sumer and A. Müller (editors), Proc. Euromech 156, A.A. Balkema, Rotterdam, p. 265-274.
Watts, G.M. (1953) A study of sand movement at South Lake Worth Inlet, Florida: U.S. Army Corps of Engrs., Beach Erosion Board, Tech. Memo. No. 42., 24 pp.

CHAPTER 93

Predictive model for daily changes of shoreline

Kazumasa Katoh[1] and Shin-ichi Yanagishima[2]

Abstract

Beach profiles have been being measured every day on sandy beach at the Hazaki Oceanographical Research Facility, facing to the Pacific Ocean. Based on the data obtained during the period from March 12, 1986 to September 11, the relation between the daily changes of shoreline position and energy flux of incident waves is analyzed. A tentative predictive model of the short-term shoreline changes is proposed. A combination of this model and the one-line theory is examined.

1. Introduction

A shoreline changes due to the longshore sand transport in a long-term. A practical numerical simulation model, the one-line theory can predict the long-term shoreline changes. The shoreline also changes due to the cross-shore sand transport in a short-term. There is, however, no reliable method to calculate the short-term ones during and after a storm, mainly because of lack of information on the actual changes of shoreline.

The authors have measured the beach profile every day for 184 days. Based on the data obtained, the relation between the short-term shoreline changes and the energy flux of incident waves are examined. Furthermore, a black box model for predicting the short-term changes are developed.

2. Field observation (on beach profile)

The site of field observation is a entirely natural sandy beach. It is facing to the full wave energy of the Pacific Ocean (see Figure 1). A foreshore slope is mild, about 1/50, and the mean diameter of sediments on a beach is 0.18mm. According to eleven aerial photographs taken

---

1) Chief of Littoral Drift Lab., 2) Member of Littoral Drift Lab., Port & Harbour Research Institute, Ministry of Transport, 3-1-1, Nagase, Yokosuka, Japan.

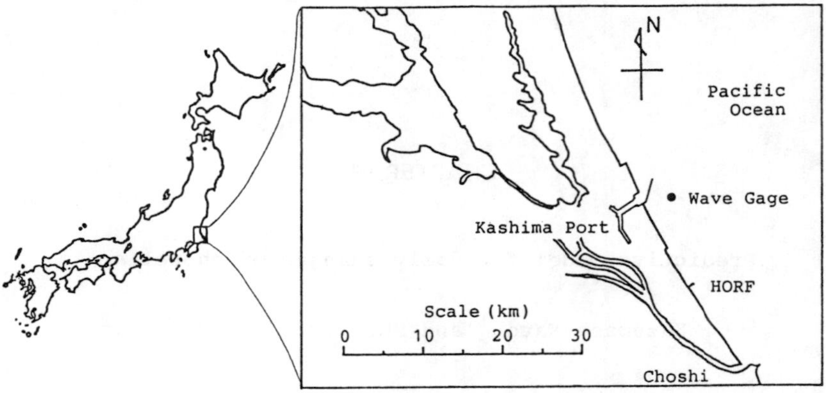

Figure 1 Site of field observation

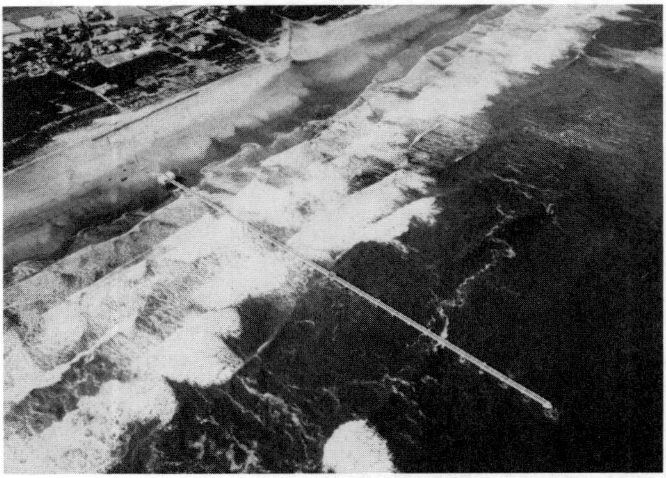

Photo. 1 Hazaki Oceanographical Research Facility

during a period from 1947 to 1984, the location of shoreline has been stable since 1979. On this beach, Port and Harbour Research Institute, Ministry of Transport, has constructed the Hazaki Oceanographical Research Facility (HORF) in 1986 for carrying out field observation in the surf zone even under sever sea conditions. The research pier is a 427 meters long concrete structure supported by 0.8 meter diameter concrete-filled steel piles in a single line, at 15 meters interval. The pier deck is 2.5 meters wide and 7 meters above L.W.L.(Photo. 1).

Beach profiles of 500 meters from the tip of pier to the backshore were measured with an interval of 5 meters once a day during 184 days from March 12 to September 11, 1986. The sea bottom profile was surveyed with a sounding lead

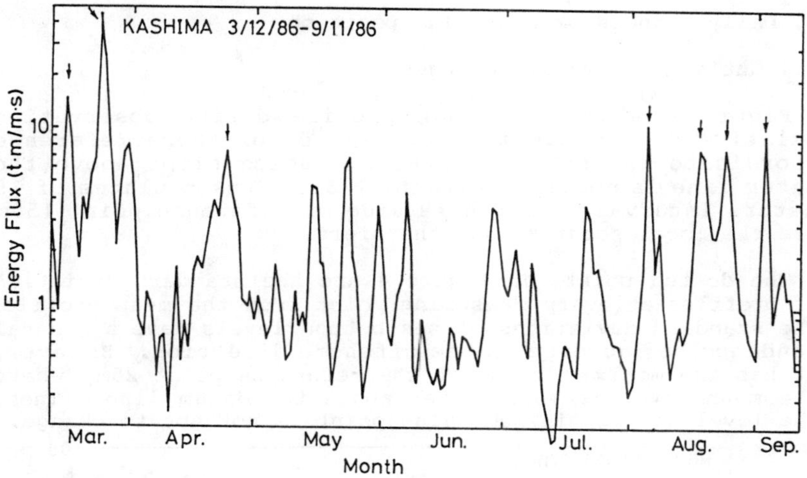

Figure 2 Changes of daily mean wave energy flux

Table 1
Representative high waves

Month :Day	Energy Flux (tm/ms)	Maximum waves	
		$H_{1/3}$ (m)	$T_{1/3}$ (s)
3:16	15.4	4.04	9.3
3:24	45.9	6.76	10.7
4:24	7.8	3.06	10.1
8: 5	11.0	4.12	9.1
8:18	8.0	3.12	11.9
8:24	8.2	3.10	9.2
9: 3	9.6	3.35	10.9

from the pier deck. The land survey of the backshore and foreshore was done by using a transit and surveyor's staff.

The offshore waves have been measured near Kashima Port (see Figure 1) during twenty minutes of every two hours at the depth of 22 meters below the datum line. The daily mean energy flux has been calculated with the significant wave data of every two hours.

Figure 2 shows the daily mean energy flux of incident waves during the period from May 12 to September 11, 1986. The energy flux was small in the months from May to July, while it was large in the other months. Storms larger than 3.0 meters in the significant wave height occurred seven times on days denoted by arrows in Figure 2. The daily mean energy flux, the maximum significant wave height and period on the storm day are listed in Table 1. The tide level is being observed every hour in Kashima Port. Figure 3 shows the distribution of frequency of tide level.

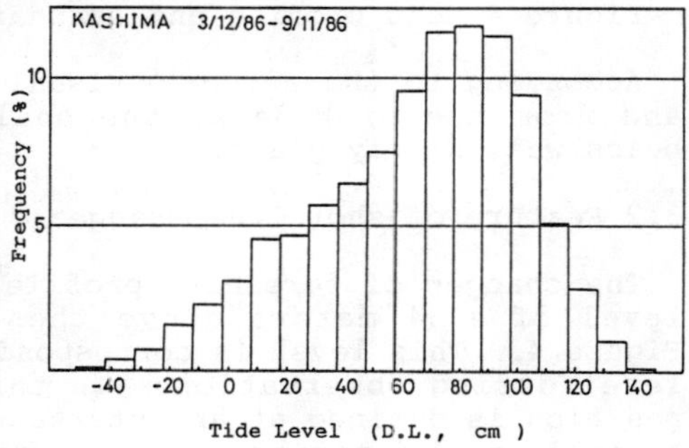

Figure 3 Frequency of tide level

3. Daily changes of shoreline position

3.1 Outline of profile changes

Figure 4 shows the mean profile during observation relative to the datum line. The on-offshore reference coordinate is inherent in HORF. Beach slope above the datum line is nearly constant, 1/50. Sunken places of 15 meters interval in offshore side of reference point 150m are the local scour around the piers.

The dotted points in Figure 4 are the standard deviation of profiles at every measuring point from the mean profile. The standard deviations of sea bottom levels have a general tendancy to increase in the offshore direction. However, it has the minimum value at the reference point 25m, where the mean level is +0.28 meter above the datum line. Then, the level at the minimum value point is not apt to change.

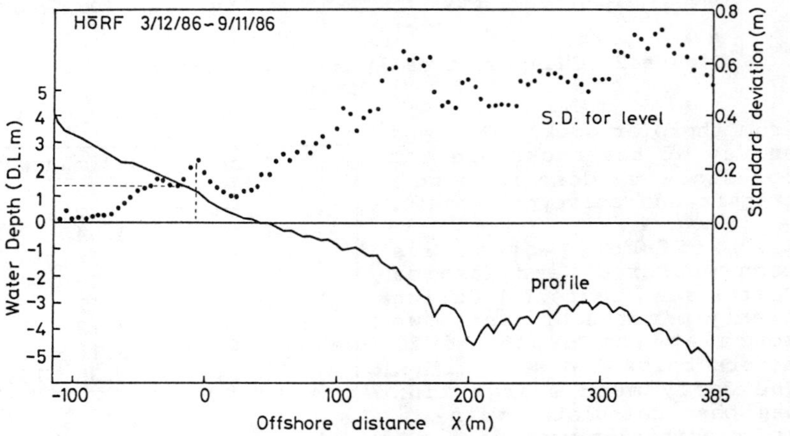

Figure 4 Mean profile and standard deviation of profiles

According to the authors' visual observation on the beach and from the pier deck, the shallow water area and the beach was nearly plane.

3.2 Feature of shoreline changes

The changes of foreshore profile was maximum at the mean level of +1.4 meters above the datum line, as seen in Figure 4. This level is corresponding to the maximum water level during observation. In this paper, the shoreline position is defined at an intersection of beach profile and a level of +1.4 meters above the datum line. The shoreline positions was calculated by interpolating the beach profile data of 5 meters interval.

Figure 5 shows the daily on-offshore changes of shoreline position. The shoreline was in onshore-side in the high energy flux months of March, April and August, while it was in the offshore-side during the low energy flux period from May to July. Then, a whole trend of shoreline changes are

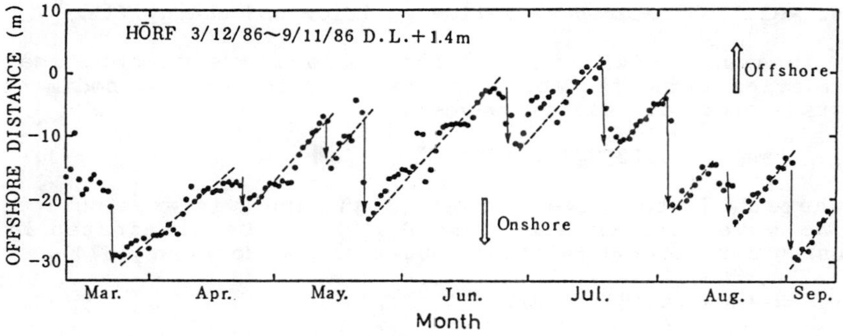

Figure 5 On-offshore changes of shoreline position

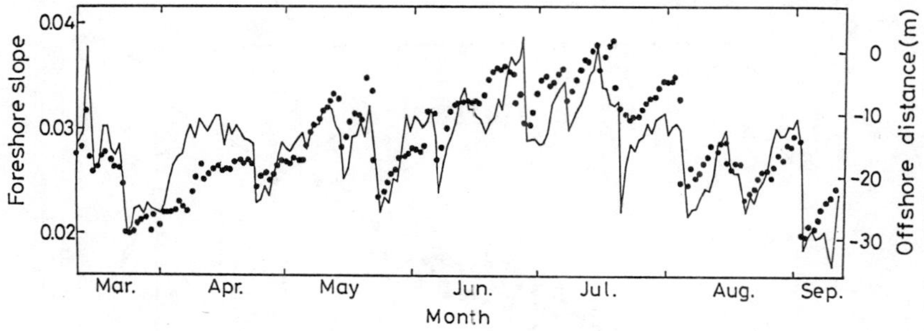

Figure 6 Changes of foreshore beach slope

roughly corresponding to the changes of incident energy flux.

The shoreline positions shifted in the offshore direction, independent of wave energy flux, in the accretionary process. Nine broken lines in Figure 5 approximate the accretionary process. The broken lines are nearly parallel each others, which means that the shoreline advanced with almost constant speed. The inclinations of the broken lines give the value of 0.68 meter/day for the speed of progression in average. On the other hand, the shoreline rapidly recessed in one or two days in erosional process as shown by arrows in Figure 5. The rapid recession have occurred even in the low energy flux months such as May, June, and July.

The solid line in Figure 6 shows the changes of mean foreshore slope calculated by the least square method with the data from the reference point -20m to +25m. The shoreline position are also plotted in this figure. The changes of the slope and the shoreline position have a same tendancy, although there is a some discrepancy between them. The foreshore slope suddenly became to be mild when the shoreline rapidly recessed in the erosional process.

4. Relation between shoreline position and energy flux

Sunamura(1984) proposed the following semi-empirical relation among the foreshore beach slope ($\tan\beta_f$), sediment grain size (d), and the waves:

$$\tan\beta_f = 0.12[H_B/(g^{\frac{1}{2}}Td^{\frac{1}{2}})]^{-\frac{1}{2}}, \quad (1)$$

where H_B is the wave breaker height and T is wave period. The wave breaker height in Eq.(1) can be eliminated by using a following relation (Sunamura and Horikawa,1974),

$$H_B/H = (\tan\beta)^{0.2}(H/L)^{-0.25}, \quad (2)$$

where $\tan\beta$ is mean bottom slope, H and L are offshore wave height and wave length, respectively. Then, we will have Eq.(3) instead of Eq.(1),

$$\tan\beta_f = 0.19d^{1/4}(\tan\beta)^{-0.1}(L^{1/3}/H)^{3/8}. \quad (3)$$

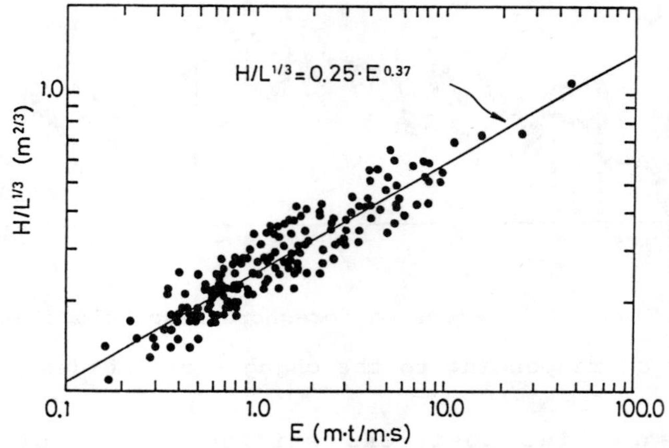

Figure 7 Relation between $H/L^{1/3}$ and energy flux

Figure 7 shows the relation between $H/L^{1/3}$ and the daily mean energy flux during the observation. The wave length was calculated by the small amplitude wave theory with the daily mean significant wave period. The height was reversely calculated based on the daily mean energy flux and mean wave period. As seen in Figure 7, there is a strong relation between them, a solid line in the figure, which is expressed as

$$H/L^{1/3} = 0.25E^{0.37}, \quad (4)$$

where E is the daily mean energy flux. By substituting Eq.(4) into Eq.(3), we have

$$E = [0.32d^{0.25}(\tan\beta)^{-0.1}]^{7.19}(\tan\beta_f)^{-7.19}. \quad (5)$$

Next, a simple relation between the foreshore slope and the shoreline position will be introduced. Let the

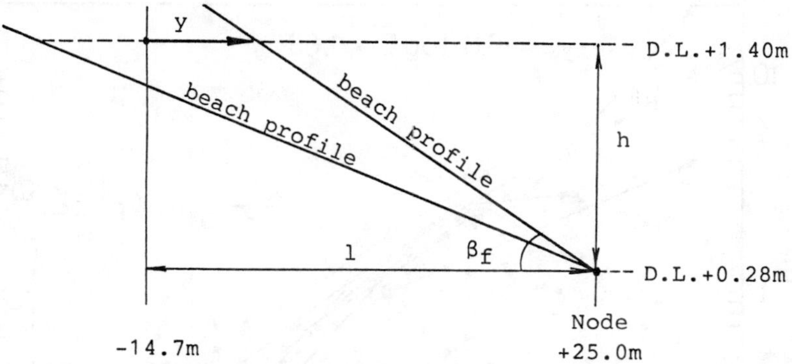

Figure 8 Relation between foreshore slope and shoreline position

reference point 25m be a node since the standard deviation of level changes at this point was minimum as seen in Figure 4. The mean level of reference point 25m is +0.28 meter above the datum line. Figure 8 illustrates the relation between the foreshore slope and the shoreline position. A symbol l is the horizontal distance from the node to the time averaged shoreline position, 39.7 meters. A symbol h is the vertical distance from the node to the level of shoreline, 1.12 meters. A symbol y is the location of shoreline, taking the origin at the mean shoreline position and being positive in the offshore direction. From this concept figure, the foreshore slope is represented by

$$\tan\beta_f = h/(l-y). \tag{6}$$

By substituting Eq.(6) into Eq.(5), and by introducing the actual values of d=0.18mm and tanβ=1/60 in the study site, and by normalizing the daily mean energy flux with the mean value of energy flux during the observation, 2.48tm/ms, we have

$$\hat{E} = 4.19\times10^{-10}[(l-y)/h]^{7.19}, \tag{7}$$

where $\hat{E}$ is the dimensionless energy flux.

Figure 9 shows a plot of the dimensionless wave energy flux ($\hat{E}$) versus the shoreline position (y) for 184 cases (days). Essentially speaking, the cross-shore component of energy flux must be considered. The measurement of wave direction, however, is not being done in Kashima Port. Then, the absolute energy flux are utilized without any further calculation. This approximation will introduce a little error of less than 3% when an angle of wave incidence to beach is less than ±10 degrees. The curved line in Figure 9 is Eq.(7) with the slightly modified constant of 7.15×10^{-10} and an exponential index of 6.0, which can be approximated on the figure by a straight broken line as

$$y = -16.0\,\log\hat{E} + 1.63, \tag{8}$$

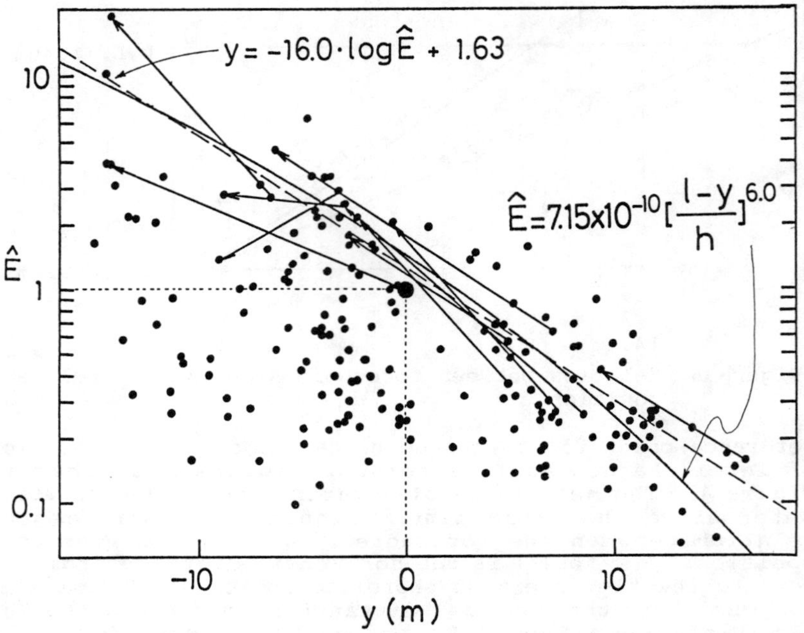

Figure 9 Relation between energy flux and shoreline position

where log is a common logarithm and a unit of y is in meter.

In Figure 9, the rapid recessions more than 5 meters in a day are connected with arrows. The direction of arrows are the same as those of shoreline changes, that is leftward in the figure. It is recognized that the arrows lie parallel near the straight broken line of Eq.(8). According to Figure 9, the shoreline position rapidly recesses by the lower energy flux when the more it is being advanced.

On the other hand, almost all data in the accretionary processes are plotted in the lower side of the broken line in Figure 9.

Furthermore, a relation between the averaged shoreline position ($y=0$) and the mean energy flux ($\hat{E}$) is just near the broken line, a larger closed circle. This situation is interesting and will be discussed later.

5. Tentative predictive model for daily shoreline changes

The shoreline advances with the constant speed in the accretionary process, while it rapidly recesses in one or two days in the erosional process. Then, without taking into account the time lag of shoreline changes behind the external force, the authors have developed an empirical predictive model for the daily shoreline changes as follow.

The shoreline position and the dimensionless energy flux on the i-th day are denoted by y_i and $\hat{E}_i$, respectively.

[Step 1] By assuming the accretionary process, the shoreline position on next day can be predicted by a following equation:

$$y_{i+1} = y_i + 0.68 \quad \text{(meter)}. \tag{9}$$

[Step 2] The assumption in Step 1 must be confirmed by plotting the data of (y_{i+1}, $\hat{E}_{i+1}$) in Figure 9. If it will be plotted in the lower side of the broken line, that is,

$$y_{i+1} < -16.0 \log\hat{E}_{i+1} + 1.63 \quad \text{(meter)} \tag{10}$$

y_{i+1} predicted by Eq.(9) is decided to be the shoreline position.

[Step 3] If it will be in the upper side of the broken line, it will be in the erosional process. Then, the shoreline position rapidly recesses along the broken line. A new position of y_{i+1} depends on $\hat{E}_{i+1}$ according to the following equation:

$$y_{i+1} = -16.0 \log\hat{E}_{i+1} + 1.63 \quad \text{(meter)}. \tag{11}$$

The above procedure from Eq.(9) to (11) is expressed in one equation as

$$y_{i+1} = \min[\, y_i + 0.68,\ -16.0 \log\hat{E}_{i+1} + 1.63 \,], \tag{12}$$

where the symbol min[a,b] gives the smaller of a or b.

Figure 10(a) shows the relations between the dimensionless energy flux and the shoreline position during the

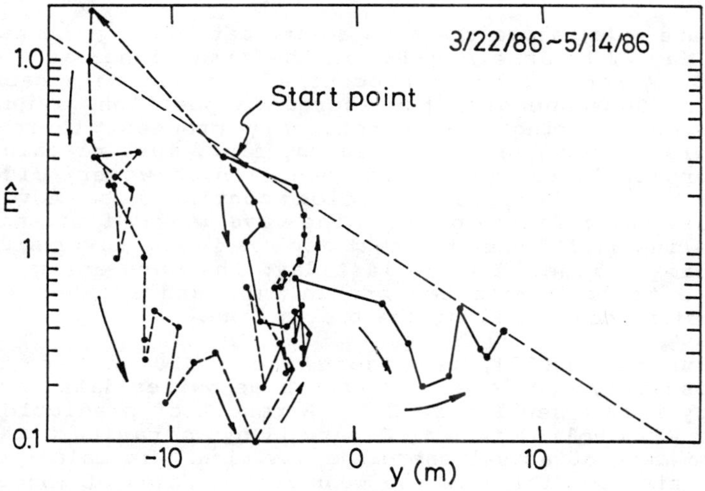

Figure 10(a) Relation between $\hat{E}$ and y (March 22 to May 14)

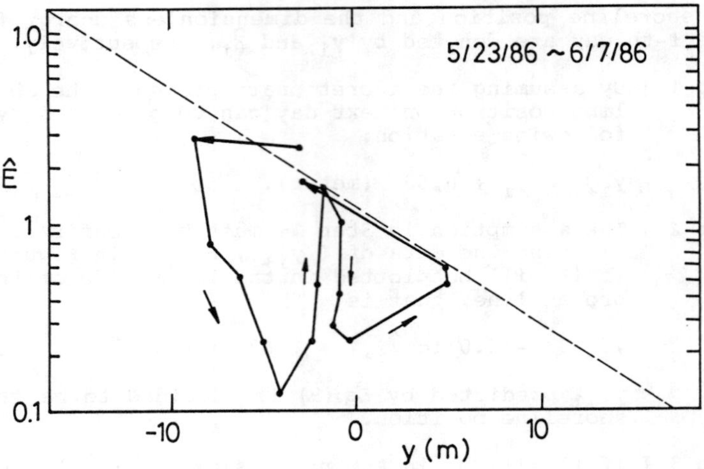

Figure 10(b) Relation between $\hat{E}$ and y (May 23 to June 7)

period from March 12 to May 14, 1986. Broken lines connect the time series data for the first half period and solid lines for second one. A broken straight line in this figure is corresponding to Eq.(8). At first, a rapid recession occurred along Eq.(8) when the value of energy flux was large. After that, the shoreline position gradually advanced with a decrease of wave energy flux up to be y=-5m. In turn, the wave energy flux became to be large, which eventually made the relation between the energy flux and the shoreline position to be on the broken line of Eq.(8). On that day, another rapid recession occurred again along the broken line. Next day, the shoreline position started to advance again under the lower energy flux.

Figure 10(a) shows the time series data during the period from May 22 to June 7, 1986, in the same manner as Figure 10(a). A rapid recession occurred also in this case at first. Subsequently, the shoreline position gradually advanced. During the accretionary process, there was temporary increase of wave energy flux, just reaching to the broken line. The rapid recession, however, did not occur. The shoreline position continued to advance, because the relation between the wave energy flux and the shoreline position on the next day was in the lower side of the broken line. On the last day, the wave energy flux became to be beyond the broken line and another rapid recession occurred along the broken line.

By using Eq.(12), the shoreline position have been successively predicted with the time series data of wave energy flux (see Figure 2). A result of prediction is shown with solid lines in Figure 11 by comparing with the dotted data of actual shoreline position. In calculation, an initial condition is the measured position of shoreline on May 12. As seen in Figure 11, the predicted shoreline

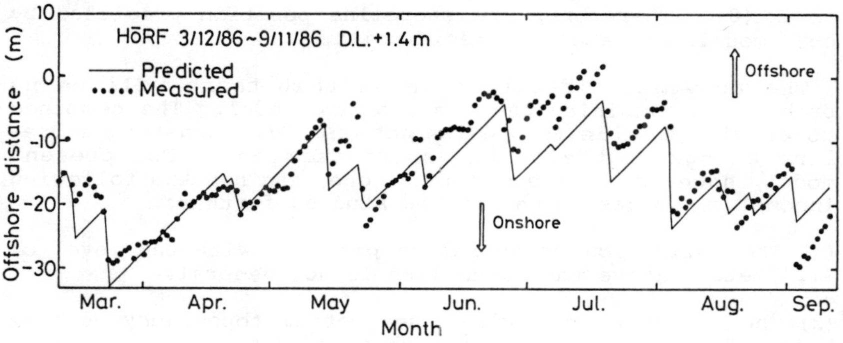

Figure 11 Comparison of changes of shoreline position predicted and those observed

position almost agree with the measured ones.

6. Possibility of combination of the present model and the one-line theory

For practical purpose of predicting the short-term shoreline changes, the present model must be combined with the one-line theory. For this purpose, any mutual contradiction between both models are not basically allowed. Concerning to this problem, the following four items are pointed out:

(a) Both models treat the mechanism of sand transport and the other phenomena in the surf zone in a black box. Therefore, both modeling accuracies of phenomena are almost the same each other.

(b) The external force in the present model is the cross-shore component of energy flux. That in the one-line theory is the longshore one. Therefore, there is no physical contradiction between the models.

(c) The time series data of energy flux are utilized in the present model, while the time averaged energy flux during a objective period is usually utilized in the one-line theory. Fundamentally speaking, the time series data have to be utilized also in the one-line theory. However, it will take much computer processing time for calculating the wave deformation. Usage of averaged energy flux in the one-line theory is only for convenience.

(d) The one-line theory can predict the long-term shoreline changes by inputting the mean energy flux. As shown in Figure 9, the relation between the time averaged energy flux (E=1) and the averaged position of shoreline (y=0) is just near the broken line, along which the rapid recession of shoreline position occurs. In short, the shoreline position predicted by the one-line theory with the averaged energy flux practically satisfies the relation of Eq.(8). Therefore, the shoreline position predicted by both models agree at the basic point.

The above consideration shows that there will be no problem in combining them into a new model. The combined model will be able to predict not only the long-term shoreline changes but also the short-term one. The present model, however, is the empirical one, and has the following incomplete points which must be studied further:

(1) The definition of shoreline position with the level of +1.4 meters above the datum line is not general.

(2) The effect of changes of sea bottom topography such as bars and troughs on shoreline changes is unknown.

(3) The changes of tide level is not taken into account in the present model.

(4) The mechanism of shoreline changes in the short-term has not been known.

7. Conclusions

The conclusions obtained in this study are as follows:

(a) The shoreline advanced with the constant speed of 0.68 meters/day in the accretionary process.

(b) In the erosional process, the shoreline rapidly recesses in one or two days. The rapid recession of shoreline depends not only on large energy flux of waves but also on the position of shoreline. In other words, the shoreline recesses by the lower energy flux when the shoreline has advanced, while it advances by the higher energy flux when it has recessed.

(c) The empirical criterion of an occurrence of rapid shoreline recession is obtained.

(d) Based on the results of data analyses, the authors have proposed a tentative predictive model of the short-term shoreline changes. This model can predict the shoreline position in order with the time series data of wave energy flux.

(e) The present model can be combined with the one-line theory without any contradictions.

References

(1) Sunamura,T. and K.Horikawa(1974): Two-dimensional beach transformation due to waves, Proc. of 14th ICCE, pp.920-938.
(2) Sunamura,T.(1984): Quantitative predictions of beach-face slopes, Geol. Soc. Am. Bull., Vol.95, pp.242-245.

CHAPTER 94

PROTOTYPE APPLICATIONS OF A GENERALIZED SHORELINE CHANGE NUMERICAL MODEL

by

Hans Hanson[1], Mark B. Gravens[2], and Nicholas C. Kraus[3], M.ASCE.

ABSTRACT

Three case studies are described in which long-term shoreline response was simulated using a numerical model. One project was along Kachemak Bay, Alaska and involved evaluation of design alternatives that included a detached breakwater, beach fill, and a revetment. The second project was on the north New Jersey coast and characterized by a long seawall and numerous groins. The third project was a model test of shoreline change behind three detached breakwaters Lake Erie. The shoreline response model used, called GENESIS, is demonstrated to have applicability to a wide range of commonly encountered shore protection situations.

INTRODUCTION

Numerical simulation models provide a powerful and unique capability for engineering studies of complex shoreline change occurring under realistic field conditions. Before the development of the modeling system GENESIS, each application of a numerical shoreline response model required extensive modification of an existing model and special refinements for the particular study. To allow application to an arbitrary prototype situation, considerable time was spent on devising a flexible and general internal structure of GENESIS. Through a simple interface, the user can simulate the effects on the shoreline of groins, jetties, detached breakwaters, seawalls, and beach fills. Almost arbitrary numbers, locations, and combinations of such structures can be represented, and user-specified operations can be introduced almost arbitrarily in space and time. The model is economical to run and, therefore, simulations can be performed for wide spatial extents and long time intervals. This paper presents results of applications of GENESIS to three prototype situations. The case studies demonstrate that GENESIS is capable of simulating long-term shoreline change in the field and assisting in the refinement of engineering analysis of shore protection alternatives involving beach fill and various types of structures.

---

(1) Associate Professor, Dept. of Water Resources Engineering, Lund University, Box 118, S-22100 Lund, Sweden; (2) Hydraulic Engineer; and (3) Senior Research Scientist, Coastal Engineering Research Center (CERC), U.S. Army Engineer Waterways Experiment Station, P.O. Box 631, Vicksburg, MS 39180-0631 USA.

GENESIS

GENESIS allows simulation of shoreline change occurring over a period of months to years as caused primarily by the action of breaking waves. The project horizontal length scale typically varies from one up to tens of kilometers. The model is generalized in the sense that the it can be used to simulate shoreline change under a wide variety of user specified beach and coastal structure configurations. In addition, the input wave conditions can be entered either (a) from an arbitrary depth as a single value specified by the user and involving a simplified wave refraction calculation, assuming parallel bottom contours, or (b) through interaction with a more rigorous wave refraction model RCPWAVE (Ebersole et al. 1985) allowing specification of an irregular bottom bathymetry (Figure 1).

GENESIS is based on the one-line (shoreline contour) theory of beach change (Pelnard-Considère, 1956). The same concept has been used in a number of previous studies (e.g., Price et al. 1973, Perlin 1979, LeMéhauté and Soldate 1980). In particular, Kraus et al. (1985), Kraus et al. (1988), and Hanson and Kraus (1986a) present applications of the model employed as an engineering tool for making shoreline change forecasts for a real beach. Based upon the results of these studies, recommendations for remedial measures were given.

BASIC EQUATIONS

As an extensive description of GENESIS, discussing basic assumptions, limitations, and governing equations is given in Hanson (1987), only a short review will be presented here.

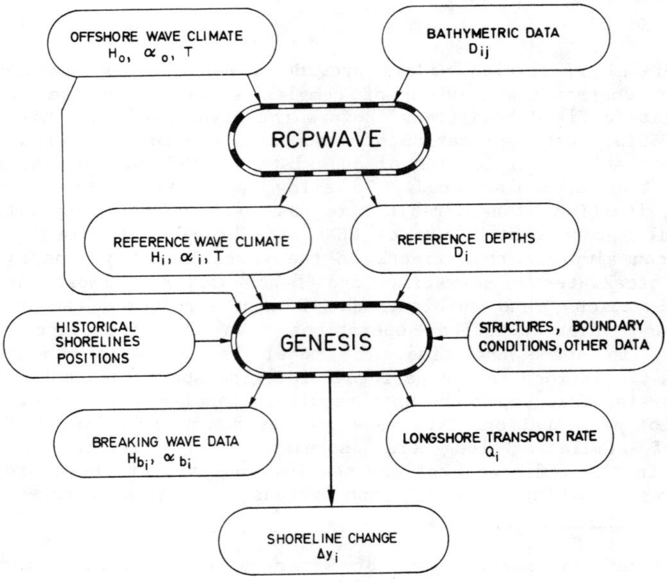

Figure 1. Relation between the wave model RCPWAVE and the shoreline change model GENESIS.

Mass conservation

Following the assumption that the bottom profile moves in parallel to itself to the depth of closure, conservation of sand for an infinitely small length, dx, of shoreline can be formulated as:

$$\frac{\partial y}{\partial t} + \frac{1}{D_B + D_C}\frac{\partial Q}{\partial x} = 0 \qquad (1)$$

where y is the shoreline position (m), x is the longshore coordinate (m), t is the time (s), D_B is the average berm height above the mean water level (m), D_C is the depth of closure (m), and Q is the longshore sand transport rate (m^3/s). In order to solve Eq. (1), expressions for the two quantities D_C and Q must be formulated. The berm height, D_B, is taken from the measured or assumed profile.

Depth of closure

For applications involving bypassing of sand at structures, knowledge of the depth to which sand is actively transported alongshore is required. Studies of beach change taking place over a long period of time (years) indicate that the profile varies to the depth of closure, D_C, associated with the wave climate over this long time period. Various values have been suggested for this depth (e.g. Willis and Price, 1975; Kraus and Harikai, 1983; Hands, 1983). These are all of the same order as the formulation of Hallermeier (1983), giving the annual depth of closure as slightly more than twice the extreme annual significant wave height. In the light of these formulations, and keeping the potential errors involved in determining these relations in mind, GENESIS uses a simple relation:

$$D_C = 2\,H_{mas} \qquad (2)$$

where H_{mas} is the maximum annual significant wave height (m) for the existing site. The value of H_{mas} for a given site must be specified by the user operating GENESIS. Alternatively, measured profiles can be compared and a closure depth estimated. H_{mas} would then be back-calculated by Eq. (2). It is also assumed that the dry portion of the beach profile, from the shoreline to the berm crest, moves with the wet part of the profile while maintaining its shape.

Longshore sand transport

Kraus et al. (1981) and Kraus and Harikai (1983) showed that the simulation of shoreline evolution, especially in the diffraction shadow zone near structures, is greatly promoted by taking the longshore gradient of breaking wave heights into account. For this reason, in GENESIS, the longshore sand transport volume rate, Q, is calculated as:

$$Q = (H^2 C_g)_b \left(a_1 \sin 2\alpha_{bs} - a_2 \cos\alpha_{bs} \frac{\partial H}{\partial x}\right)_b \qquad (3)$$

where C_g is the wave group velocity (m/s), α_{bs} is the angle of wave crests to the shoreline, the subscript b denotes the breaking condition, and the non-dimensional parameters a_1 and a_2 are given by:

$$a_1 = \frac{K_1}{16\,(\rho_s/\rho - 1)\,(1 - p)\,1.416^{5/2}} \qquad (4)$$

$$a_2 = \frac{K_2}{8\ (\rho_s/\rho\ -\ 1)\ (1\ -\ p)\ \tan\beta\ 1.416^{5/2}} \tag{5}$$

where K_1 and K_2 are calibration parameters, ρ_s and ρ are the densities of the sediment (quartz sand) and water (kg/m^3), p is the sediment porosity, and $\tan\beta$ is the average bottom slope. The factor 1.416 is used to convert from significant to RMS wave height. The first term in Eq. (3) expresses the longshore transport rate due to obliquely incident waves and is commonly known as the CERC-formula (SPM 1984). The second term, introduced by Ozasa and Brampton (1980), accounts for the longshore sand transport rate caused by the longshore variation in breaking wave height and is especially important in the diffraction zones near structures where the longshore wave height gradient often is strong (Kraus 1983).

HOMER SPIT SIMULATION

Homer Spit, a narrow peninsula southeast of the City of Homer, is between 90 and 450 m wide and extends approximately 7 km into Kachemak Bay in southcentral Alaska (Figure 2). At the tip of the spit are a small-boat harbor and a city dock that is used for year-round shipping. Apart from its commercial importance Homer Spit is also a cultural and social asset, being the only recreational beach in this part of Alaska.

A two-lane road leads from Homer to these developments following the southwestern shores of the peninsula (Figure 3). Since its construction in 1927, the inshore half of the roadway has experienced maintenance problems. Severe storms causing high water levels in combination with intense wave action have overtopped and washed out stretches of the roadway. Various means, including the installation of groins, revetments, and bulkheads have been attempted to control the erosion at the

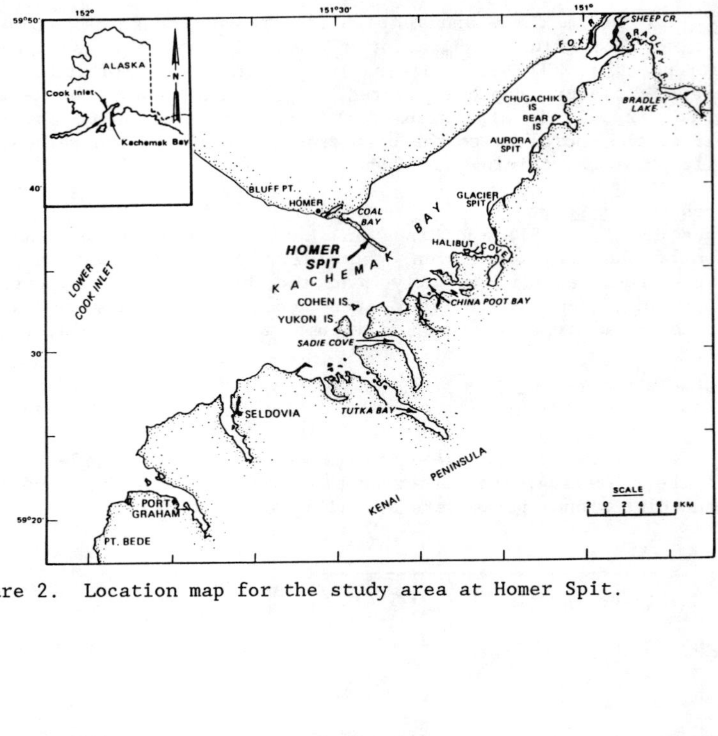

Figure 2. Location map for the study area at Homer Spit.

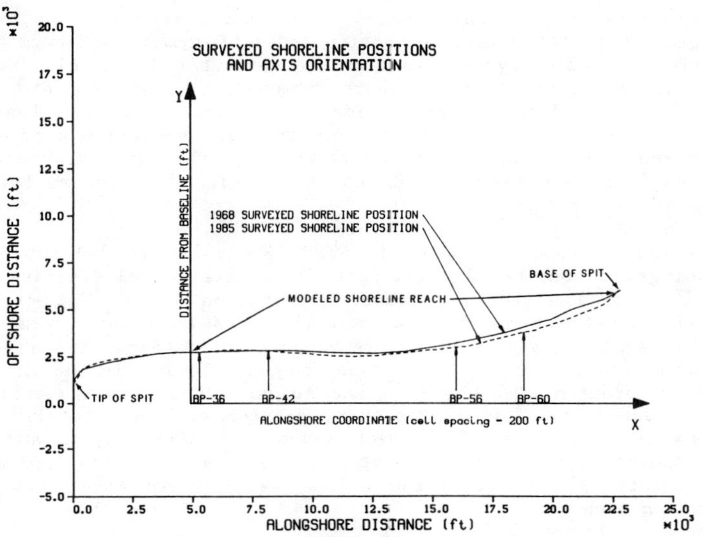

Figure 3. Map of Homer Spit, Alaska showing the extent of the calculation area.

southwestern beach facing outer Kachemak Bay and to mitigate damages to the roadway during extreme storm events. Results of these efforts have not been satisfactory.

The analyses performed by the Coastal Engineering Research Center (CERC) for the Alaska District of the Corps of Engineers (Smith et al. 1985, Chu et al. 1987) included: wind data analysis, deep-water wave hindcast, wave transformation modeling from deep to shallow water , tidal circulation analysis, geological history and sediment sample analysis, beach profile analysis, and longshore sediment transport estimate. By means of wave transformation modeling, a representative 1-year time-varying series of nearshore wave height, period, and direction specified at a 6-hr interval for each of the 79 model calculation grid cells was established.

Modeling Conditions

The input wave data set was prepared at 6-hr intervals, and the longshore sediment transport rates and associated shoreline change were also calculated with a simulated 6-hr time-step. Due to the large tidal range and the significant differences in the average sediment grain size at low, mean, and high tidal levels, the standard version of GENESIS was modified. For compatibility with the shoreline change model, the tide level was simulated at successive time-steps in a cyclical fashion through four representative stages: mean tide, mean high, mean, and mean low. This procedure, in effect, results in a semi-diurnal representation of the tidal cycle. The tides at Homer Spit are indeed semi-diurnal but do have a pronounced diurnal inequality. However, simulation of the tides in the stated manner is consistent with the accuracy of the shoreline change model. Representative grain sizes associated with the respective tidal stage were as determined from the sediment sample analyses: mean low tide, 0.25 mm; mean tide level, 10.23 mm; and mean high tide, 8.13 mm.

The inclusion of tidal changes necessitated input of the average water depth (controlling wave refraction and diffraction) and berm height (affecting the continuity relation) at the various tidal levels. Consistent with the one-line theory of shoreline change, only one contour line was modeled (shoreline at mean tide elevation), but the longshore sediment transport rate (Eq. 3) and corresponding shoreline change (Eq. 1) were calculated at three tide levels based on the particular physical properties (transport parameter K_1 and berm height) estimated for each elevation of the profile.

The longshore grid spacing in GENESIS was set to 200 ft (61 m). This spacing was considered sufficient for evaluating alternative plans but still allowed economical computer execution times. The grid was extended beyond the project area on both sides to obtain termination points that would provide appropriate boundary conditions. Two shoreline surveys of Homer Spit were judged to be adequate for use in the shoreline modeling calibration: October 1968 and August 1985. From inspection of the survey data, two nearly stationary sections of the shoreline were identified. Therefore, a fixed-beach boundary condition, in which the boundary shoreline position is constrained not to move, was implemented at each location. This condition allows sediment to move across the boundary from each side.

Model Calibration

The calibration procedure for GENESIS is to determine the transport parameters K_1 and K_2 of Eqs. (4) and (5) by reproducing measured shoreline change that occurred at the target site between two surveys (here, the 1968 and 1985 surveys). The simulation of shoreline change for this 17-year period was accomplished with the initial shoreline position given by the 1968 measured shoreline. The calibration constants K_1 and K_2 were systematically varied in successive runs of the model, and comparisons were made between the calculated and the measured shorelines of 1985. In addition to visual comparisons of plots, a measure of the calibration error, denoted as Yerr, was calculated to obtain an objective fitting criterion:

$$\text{Yerr} = \sum_{i=1}^{N} \left| \frac{\text{Ycalc85} - \text{Ymeas85}}{\text{Ymeas68} - \text{Ymeas85}} \right|_i \qquad (6)$$

where Ycalc85 = calculated shoreline position 1985; Ymeas85 = measured shoreline position 1985; Ymeas68 = measured shoreline position 1968; N = total number of calculation cells; and i = cell number.

As Yerr approaches zero, the accuracy of the calibration increases. Although Yerr was used as a numerical indicator of the relative accuracy of the calibration runs (on the order of 100 runs), the final judgement was based on inspection of plots of the full two-dimensional features of the shoreline planform. The model run chosen as the best fit is shown in Figure 4, giving the values of the calibration constants K_1 (mean low tide, 0.77; mean tide, 0.50; mean high tide, 0.55) and K_2 (0.15). The measured and calculated shorelines agree well.

If sufficient shoreline survey and wave data are available, the calibrated model is run to simulate observed shoreline change in a time interval not spanned by the calibration to verify that the calibration constants are independent of the time interval. Because of a severe earthquake in 1964, only the two previously mentioned shoreline surveys

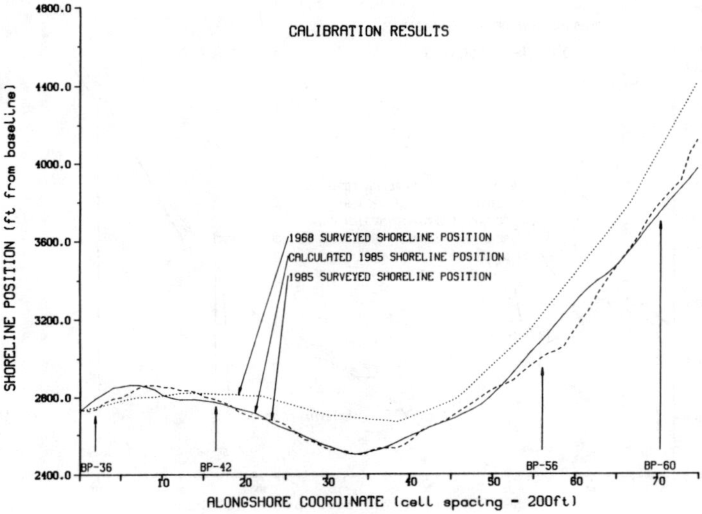

Figure 4. Results of shoreline change model calibration for Homer Spit.

were found suitable. Hence, verification could not be performed, thereby necessitating scrutiny of the sensitivity of the calibrated model to quantify variations in predicted results. A full description of this procedure is given in Chu et al. (1987).

Evaluation of Alternative Plans

Five design alternatives for erosion control (some alternatives including several variations) were analyzed using GENESIS. The alternatives simulated were: a. without-project, b. revetment extension, c. revetment extension with beach fill, d. beach fill, and e. offshore breakwater. A total of 19 alternative erosion control measures were modeled and studied. One best alternative from each of the general design options was selected for a comparative study (see Figure 5). Three of the alternatives shown (1A, 2A, and 5F) indicated considerable erosion and were not recommended for implementation. The remaining two alternatives were (3C) the extension of the existing revetment 610 m towards the tip of the spit combined with a beach fill (30 m added berm width and 610 m alongshore) and (4C) a major beach fill (23 m added berm width and 2,320 m alongshore).

It was difficult to determine which of these two alternatives would best solve the erosion problems at Homer Spit in that the extent of the specified beach fill will ultimately determine the shoreline position. However, model results clearly indicated that nourishment of the existing beach could control coastal erosion problems at Homer Spit. Revetment extension is required to protect the roadway during periods of high tide and storms. A structural approach without beach nourishment may resolve a local problem, but the area of erosion will migrate further down-drift.

Thus, as a conclusion, beach fill, along with extended revetment was considered as the most effective means for erosion control and storm damage reduction at the project area.

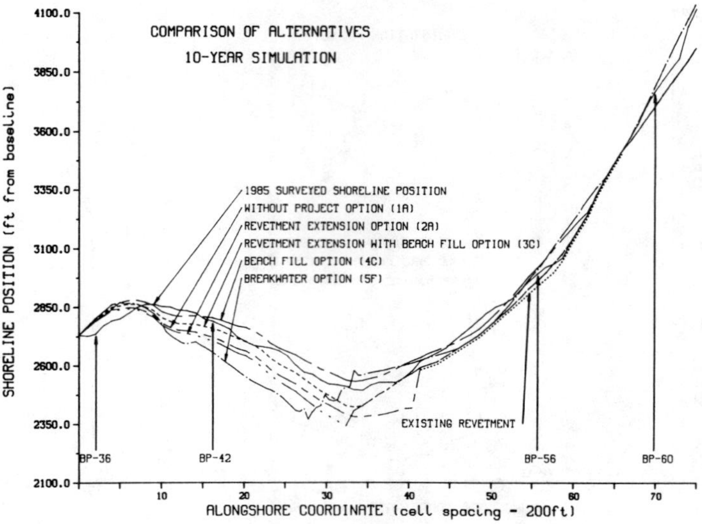

Figure 5. Ten-year simulations of alternative plans for Homer Spit.

SANDY HOOK SIMULATION

This study area constitutes 30 km of the New Jersey barrier island chain from Sandy Hook to Shark River (Figure 6). The coast has been suffering from severe erosion for a century. As an attempt to restore the beach and to secure the integrity of the shore protection structures, a shore protection plan was prepared by the New York District of the Corps of Engineers. The scope of the CERC study was to interpret data and to provide a predictive shoreline change model to assist in the implementation of this plan.

The northern part of the study area consists of Sandy Hook, one of the most famous and most investigated spits in the world. The spit, extending about 16 km into the New York Bight, was originally formed by sand eroding from the adjacent 30 km of the barrier island shores. Ever since the erosion became an increasing problem in the late 19th century, especially from Sea Bright to Monmouth Beach, an increasing portion of the beach has been protected with seawalls, revetments, and groins. Partially as a result of blocking off this sediment supply, sediment transport to the north decreased, causing erosion at the southern part of Sandy Hook, called the "critical zone". This critical zone was of particular interest for the present study as GENESIS was calibrated for this area.

Modeling Conditions

A large portion of the beach south of Sandy Hook is protected by an almost continuous rubblemound seawall and numerous groins. The seawall was modeled according to the principles discussed in Hanson and Kraus (1986b). This seawall constraint is imposed at the same level of approximation as the assumptions used to derive the shoreline change model. Thus, wave reflection, scouring, and flanking are not simulated. The lengths and locations of 91 operating groins were taken from aerial photographs and maps. Very short groins and non-functioning groin remnants were not included. The longshore transport of sand around the end of a groin is called bypassing, and the transport of sand over and through a groin is called transmission. In this study, there were no data sets available to estimate groin bypassing or transmission. Since the groins in the modeled area are mainly built of heavy, grouted stone, the transmission was set to zero. It is recognized that a limited amount of overtopping does occur during high tide and high wave conditions, but specification of that effect is not possible at the present time. Sand

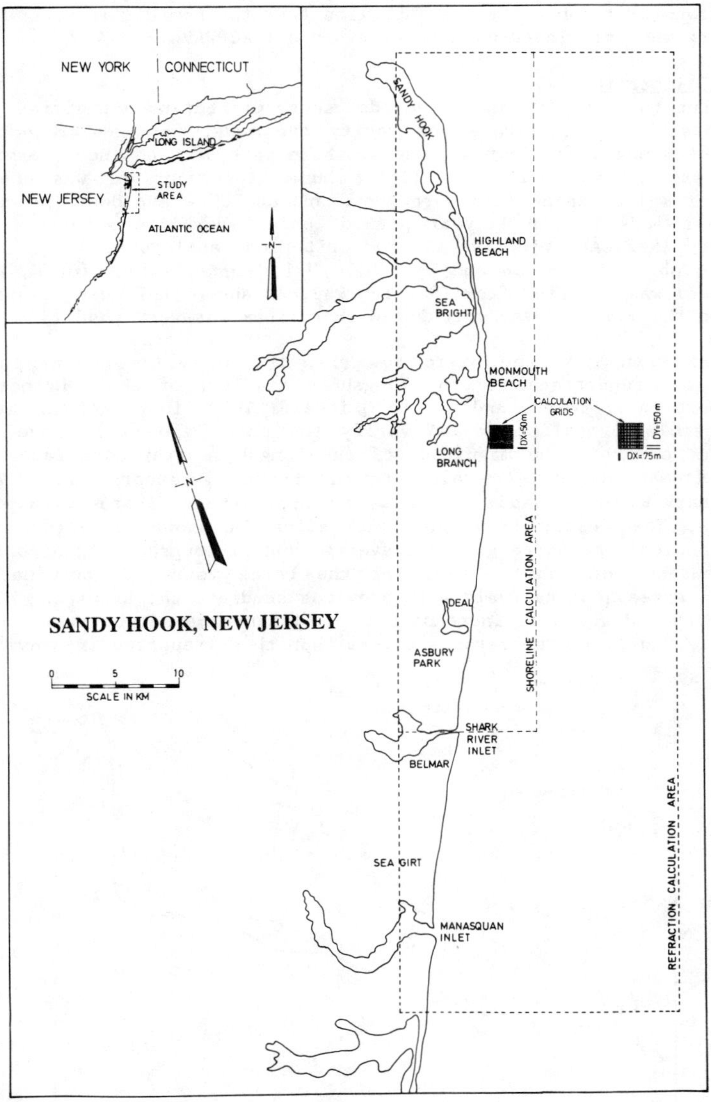

Figure 6. Location map for the study area, showing the extent of the calculation areas and grids for north New Jersey. (Modified from Kraus et al. 1988).

bypassing was calculated according to algorithms incorporated in GENESIS (Hanson 1987).

As no long-term wave measurements were available for the study area, the Wave Information Study (WIS) technique (Jensen, 1983) was used to generate a 20-year hindcast time series of wave height, period and direction at 18 m depth. To keep the data files to a reasonable size, three representative years were selected from the 20-year time series. Wave transformation along the shoreline from the nominal 18 m depth up to breaking was calculated using the wave model RCPWAVE.

Model Calibration

Due to extensive and unrecorded shore protection activities, it was not possible to calibrate and verify the model for the project coast south of Sandy Hook. Instead, the southern part of Sandy Hook, exhibiting beach erosion and with very little human intervention, was used. The dates of major beach fills were known and could be avoided. The calibration was made for an 11-year period, using shoreline surveys taken in 1971 and 1982. An objective fitting criterion, analogous to Eq. (6), was used to obtain $K_1 = 0.4$ and $K_2 = 0.1$. Using these values for K_1 and K_2, the model was verified from 1932 to 1953. As shown in Figure 7, agreement between the measured and calculated shorelines was very good.

As seen from the basic Eq. (1), the shoreline change in time, $\partial y/\partial t$, is proportional to the longshore gradient of the longshore sand transport rate, $\partial Q/\partial x$, and not to Q itself. Thus, in principle, it would be possible to calibrate and verify the model and still have a significant error in the magnitude of the longshore transport rate. Therefore, it was decided to calculate the annual longshore sand transport from Shark River to Sandy Hook and to compare these figures with previous studies. The result from the calculation is shown in Figure 8. The numbers along the shore give the average annual longshore transport rates in thousands of cubic meters for the three years of wave data. The results agree qualitatively with previous studies: the transport being to the north and somewhat increasing with distance moved northward. Quantitatively, however, the rates are less than those reported in previous

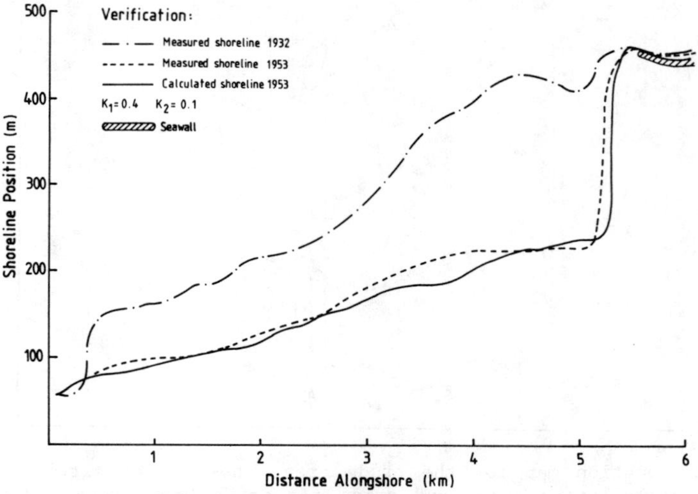

Figure 7. Result of model verification for Sandy Hook, New Jersey.

studies. Therefore, a reanalysis of the wave data, explicitly taking wave sheltering from Long Island into account, is presently being performed.

LAKEVIEW PARK SIMULATIONS

In 1977, three rubble-mound detached breakwaters were constructed at Lakeview Park, Lorain, Ohio, located on Lake Erie. These were the first breakwaters in the United States intended specifically to protect and stabilize a bathing beach (Pope and Rowen, 1983), in this case created by a beach fill (Figure 9). The purpose of the fill was to protect the park and serve as a recreational beach at the same time. In addition to the breakwaters, the beach fill was held in place by one groin on each side.

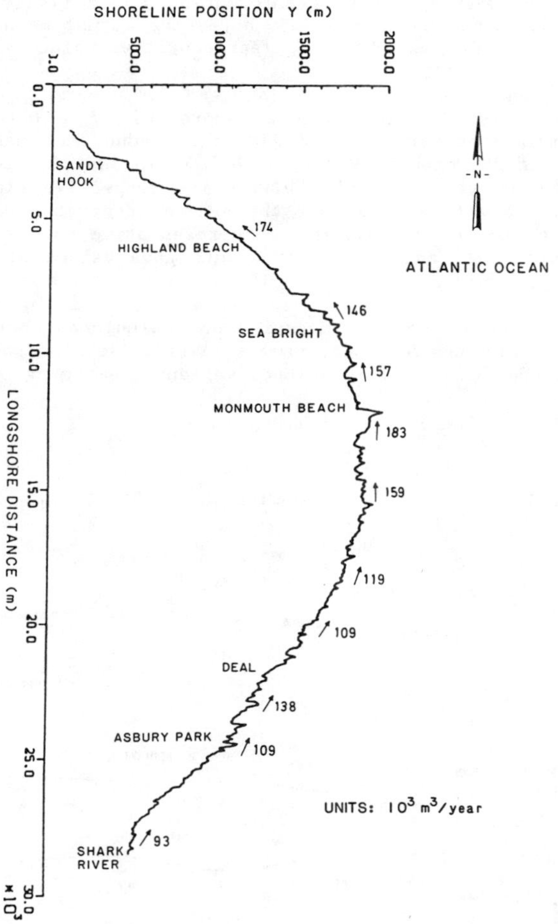

Figure 8. Calculated sand transport rates for northern New Jersey coast. (Modified from Kraus et al. 1988)

Modeling Conditions

The shoreline position and bottom contours were monitored by the Corps of Engineers, both before and after the fill, providing excellent data for a numerical model simulation. The wave data immediately available was limited, however, including only representative wave heights and periods from five different directions and their percentage distribution in time. Little information existed on the actual wave climate (height, period, and direction) between shoreline surveys. A wave time series, prepared at 6-hr intervals to be used in the model calibration/verification procedure, was synthesized for application of GENESIS. All necessary shoreline and structure configuration data were taken from survey charts on a 25-ft (7.6 m) interval. The total distance between the two groins was 1,200 ft (366 m).

Using this limited amount of data, the shoreline change was simulated for the first 24 days after the fill was completed. As the wave input was established on limited information it is likely that, for a short-term simulation as the one made here, the actual mean wave climate could deviate significantly from the representative values.

Model Calibration

Starting with the initial fill shoreline of 1 October 1977, a series of simulations were carried out to reproduce the measured shoreline position of 24 October 1977 (Figure 10). In addition to varying the calibration parameters K_1 and K_2 between the respective simulations, it was found necessary to assume that the average deepwater wave direction deviated 20 deg to the east from the representative values given by the input wave data. This calibration procedure gave values of $K_1 = 0.3$ and $K_2 = 0.3$.

A comparison between the measured and calculated shorelines of 24 October shows that the agreement, from a qualitative standpoint, is good. The model produces three well-developed salients (emerging tombolos) at

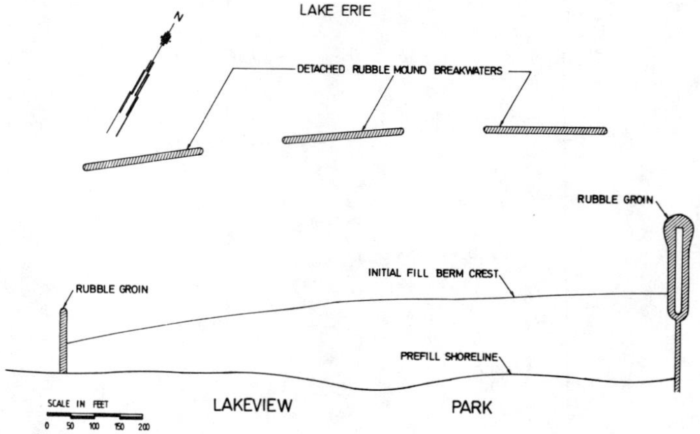

Figure 9. Shoreline and structure configuration at Lakeview Park, Lorain, Ohio.

the proper locations. However, the left-most calculated salient is somewhat too large whereas the other two are too small. An explanation for these discrepancies could be the simplified description of the bathymetry in the area. Due to the limited available wave data, it was decided not to use the wave model RCPWAVE. Instead, all wave calculations were made within GENESIS, assuming bottom contours were parallel to the calculated representative offshore contour line.

In a beach fill project of this type, the volumetric changes can be as informative as the shape of the shoreline. In terms of this volumetric change, the computational results were successful: the measured gain was 59,000 ft^3 (1,670 m^3) and the calculated gain was 53,000 ft^3 (1,500 m^3). Thus, the model accounted for 90 per cent of the volumetric change.

Evaluation

Being a small and well documented area, Lakeview Park serves well as a test case for a simulation model. Extensive monitoring of the bathymetry was not balanced with a similar wave documentation. The success of a model application is, to a large extent, limited by the degree to which the true wave climate can be reproduced. However, the lack of reliable wave data at the same time makes the area representative of most coastal projects. The site was therefore considered as an interesting and realistic test of GENESIS. Under the circumstances, and considering the limited effort spent on calibrating the model, the results were very encouraging.

In order to make more accurate predictions of shoreline evolution at the site, the following improvements would have to be made. The wave refraction pattern should be analyzed using the wave model RCPWAVE and the true bottom topography (this was not possible at the time). The breakwaters may also be overtopped, implying that wave transmission will contribute to mould the beach plan form. In addition, wave transmission through the detached breakwaters is believed to have a significant influence on shoreline change and should be represented.

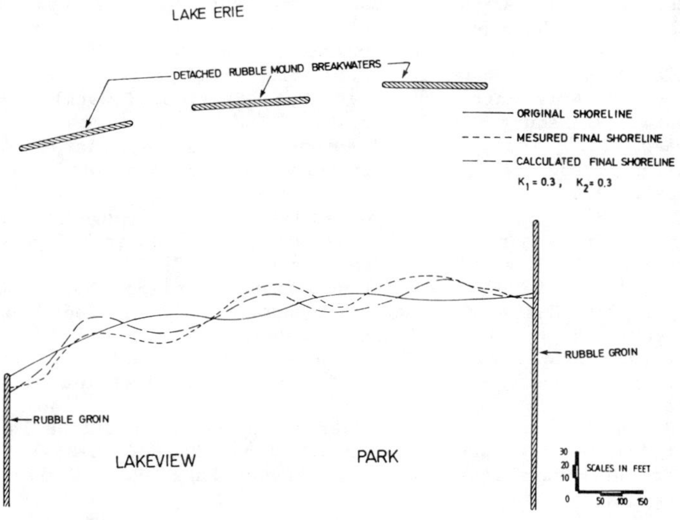

Figure 10. Measured and calculated shorelines at Lakeview Park.

SUMMARY

A numerical modeling system called GENESIS was developed to simulate the interaction between waves, longshore sediment transport, coastal structures, and other engineering activities in the nearshore area. The purpose of the model is to simulate shoreline change on a regional scale and in a long-term perspective ranging from a few months to several years. As opposed to previous models, GENESIS is generalized in the sense that it can be easily applied to almost any open coast and simulate the effects of almost arbitrary numbers, locations, and combinations of groins, jetties, breakwaters, seawalls, and beachfills. The capabilities of GENESIS were demonstrated through three prototype applications: Homer Spit, Alaska; Sea Bright, New Jersey; and Lorain, Ohio.

The calculated examples show that GENESIS is easily applied to quite complex prototype conditions and that the model is capable of simulating long-term shoreline change along coasts controlled by structures as well as open-coast natural beaches.

ACKNOWLEDGEMENTS

This study was partially supported by the work unit "Surf Zone Sediment Transport Processes," under the Shore Protection and Restoration Program of the U.S. Army Corps of Engineers, Waterways Experiment Station, Coastal Engineering Research Center. MG and NCK appreciate the permission granted by the Chief of Engineers to publish this information. The work by HH was carried out under contract No. DAJA45-85-C-0032 through the U.S. Army Research, Development and Standardization Group-UK. This support is gratefully acknowledged.

REFERENCES

Chu, Y., Gravens, M.B., Smith, J.M., Gorman, L.T., and Chen, H.S. 1987. "Beach Erosion Control Study, Homer Spit, Alaska," Miscel. Paper CERC-87-15, US Army Engr. Waterways Expt. Station, Coastal Engg. Res. Center, Vicksburg, MS.

Ebersole, B., Cialone, M.A., and Prater, M.D. 1985. "RCPWAVE - A Linear Wave Propagation Model for Engineering Use," Technical Report CERC-86-4. US Army Engr. Waterways Expt. Station, Coastal Engg. Res. Center, Vicksburg, MS, 160 pp.

Hallermeier, R.J. 1983. "Sand Transport Limits in Coastal Structure Design," Proc. of Coastal Structures '83, Am. Soc. of Civil Engrs., pp. 703-716.

Hands, E.B. 1983. "Erosion of the Great Lakes Due to Changes in the Water Level," In: Komar, P.D. ed., CRC Handbook of Coastal Processes and Erosion, CRC Press, Boca Raton, Fla., pp. 167-190.

Hanson, H. 1987. "GENESIS - A Generalized Shoreline Change Numerical Model for Engineering Use," Report No. 1007, Dept. of Water Resources Engg., Univ. of Lund, Sweden.

Hanson, H. and Kraus, N.C. 1986a. "Forecast of Shoreline Change Behind Multiple Coastal Structures," Coastal Engineering in Japan, Vol. 29, pp. 195-213.

Hanson, H. and Kraus, N.C. 1986b. "Seawall Boundary Condition in Numerical Models of Shoreline Change," Tech. Rep. CERC-86-3, US Army Engr. Waterways Expt. Station, Coastal Engg. Res. Center, Vicksburg, MS, 59 pp.

Jensen, R.E. 1983. "Methodology for the Calculation of a Shallow Water Wave Climate," Wave Information Study Report, No. 8, US Army Engr. Waterways Expt. Station, Coastal Engg. Res. Center, Vicksburg, MS.

Kraus, N.C. 1981. "One-Line Development and Simulation for Oarai Beach," NERC Report No. 13, Cooperative Research of Surf Zone Dynamics, Part 3, Beaches Near Breakwaters and Rocky Coasts, Nearshore Environment Research Center, pp. 155-192.

Kraus, N.C. 1983. "Applications of a Shoreline Prediction Model," Proc. of Coastal Structures '83, Am. Soc. Civil Engrs., pp. 632-645.

Kraus, N.C. and Harikai, S. 1983. "Numerical Model of the Shoreline Change at Oarai Beach," Coastal Engg., Vol. 7, No. 1, pp. 1-28.

Kraus, N.C., Hanson, H., and Harikai, S. 1985. "Shoreline Change at Oarai Beach - Past, Present and Future," Proc. 19th Coastal Engg. Conf. Am. Soc. Civil Engrs., pp. 2107-2123.

Kraus, N.C., Hanson, H., Scheffner, N.W., Chou, L.W., Cialone, M.A., Smith, J.M., and Hardy, T.A. 1988. "Coastal Processes at Sea Bright to Ocean Township, New Jersey," Miscell. Paper CERC-00, US Army Engr. Waterways Expt. Station, Coastal Engg. Res. Center, Vicksburg, MS.

LeMéhauté, B. and Soldate, M. 1980. "A Numerical Model for Predicting Shoreline Changes," Miscell. Rep. 80-6, US Army Engr. Waterways Expt. Station, Coastal Engg. Res. Center, Vicksburg, MS, 72 pp.

Ozasa, H. and Brampton, A.H. 1980. "Mathematical Modeling of Beaches Backed by Seawalls," Coastal Engg., Vol. 4, No. 1, pp. 47-64.

Pelnard-Considère, R. 1956. "Essai de Theorie de l'Evolution des Forms de Rivage en Plage de Sable et de Galets," 4th Journees de l'Hydraulique, Les Energies de la Mer, Question III, Rapport No. 1, pp. 289-298.

Perlin, M. 1979. "Predicting Beach Planforms in the Lee of a Breakwater," Proc. Coastal Structures '79, Am. Soc. Civil Engrs., pp. 792-808.

Pope, J. and Rowen, D.D. 1983. "Breakwaters for Beach Protection at Lorain, OH," Proc. Coastal Structures '83, Am. Soc. Civil Engrs., pp. 753-768.

Price, W.A., Tomlinson, D.W., and Willis, D.H. 1973. "Predicting Changes in the Plan Shape of Beaches," Proc. 13th Coastal Engg. Conf., Am. Soc. of Civil Engrs., pp. 1321-1329.

Smith, O.P., Smith, J.M., Cialone, M.A., Pope, J., and Walton, T.L. 1985. "Engineering Analysis of Beach Erosion at Homer Spit, Alaska," Miscell. Paper CERC-85-13, US Army Engr. Waterways Expt. Station, Coastal Engg. Res. Center, Vicksburg, MS.

SPM. 1984. "Shore Protection Manual," 4th ed., 2 vols., US Army Engr. Waterways Expt. Station, Coastal Engg. Res. Center, US Govt. Printing Office, Washington, DC.

Willis, D.H. and Price, W.A. 1975. "Trends in the Application of Research to Solve Coastal Engineering Problems," In. Hails, J. and Carr, A. (eds.), Nearshore Sediment Dynamics and Sedimentation, Wiley, N.Y., pp 111-121.

CHAPTER 95

Inception of Sand Motion around a Large Obstacle

by

Hidehiro Katsui* , Takao Toue**

ABSTRACT

To select a material for the top layer of the scour protection in the vicinity of a large-scale offshore structure, it is required to know when the material starts to move as a result of waves and currents. But, inception of the motion of sand particles in a region, where diffraction wave superimposes on incident wave and the water particles depict a resulting orbit of ellipse, differs from that calculated by formulae established in the two dimensional wave field.

This paper reveals the discrepancy in calculation when the conventional formulae were used and the data obtained by model experiment. It makes reference, also, to a possible method for a more accurate evaluation of the inception of motion.

1. Introduction

This paper describes the basic study of the local inception of motion of the sand particles which should provide important information in the design of the scour protection for large scale offshore structures.

Scour prevention work has a long history and has progressed rapidly in couple of decades as result of the demands and encouragement from the field of oil production in the North Sea and similar in the world. Before any further discussion of the subject, let us focus on the involved in the rip-rap type of protection as it has a popular practical application and the results of a study of this particular

---

* Senior Research Engineer, Technical Research Center, Taisei Corporation, 344-1 Nasemachi, Totsukaku, Yokohama, 245, Japan

** Research Engineer, ditto(present: Student of Post Graduate Course at University of Florida, U.S.A.)

type can be applied to a wide range of other types.

In the design of a scour protection, generally the first to be investigated are the external environmenal forces which cause the bottom sand or material of the protection, usually made of coarse shingles, to move. Then the design features of the protection such as, area, resistance, thickness, durability etc. are to be decided on.

For river structures and jacket members, which are subjected to a steady flow or are small in size compared to the wave length, a considerable amount of expertise concerning the mechanism of local scour and the protection work involved has been collected up to now. However, for a large scale offshore structure, where the diameter cf which reaches an order of a wave length or KC number in question is smaller than 0.5, sufficient expertise is not available concerning the protection work nor even the external forces.

Lets confine the problem to the condition of regular waves for the sake of simplicity. The wave field around a large scale structure can be divided roughly into three sections as shown in Fig.1. Firstly, in the front area where the standing wave is dominant, knowledge of the breakwater can be applied to the scouring phenomenon, what Irie(1984) has contributed for example. Secondly, in the sheltered area, there are no problems except for sand deposition. Thirdly, a considerably wide area is covered by the field where the diffraction wave is superimposed on the incident wave. Here on the bottom, the water particle depicts the orbit of ellipse as is shown later in Section 5(see Fig.14). The pure incident wave which is a single plane wave exists only in remote area away from the structure.

The question now remains whether the force which puts the particles in motion at this area is the same as that under the plane wave where the oscillatory flow is just in a single direction. To put the question more concretely, "Is the bottom shear stress the same as predicted by the conventional formulae established in the field of pure sinusoidal wave motion in one direction under two dimensional boundary layers? "

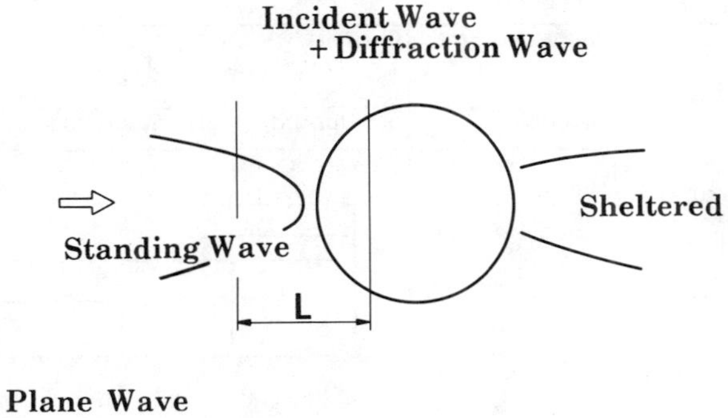

Fig.1 Wave field around a structure

As it is difficult to measure bottom shear directly in the diffraction wave field, so it is difficult to calculate the three dimensional boundary layer equation. Thus the research has begun where a series of model experiments are conducted to see what really happens around a large scale structure under waves. This paper presents a comparison from the observation of the movement of sand particles under plane waves and their movement under diffraction waves.

2. Experiment

The sand movement was investigated where the water depth was constant and the bottom was flat without any ripples. These conditions are plausible for the purpose of looking into the very early stage on particle motion on the top layer of scour protection, which is normally designed and executed flat.

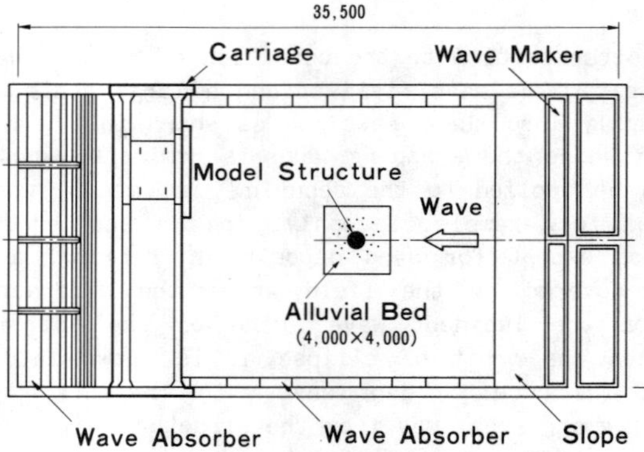

Fig.2 Experimental faciity

Table 1 Conditions of experiment

Sand	Grain Size	D_{50} = 0.58 mm
Bathimetry	Water Depth	h = 30 cm
Waves	Period	T = 0.8, 1.0, 1.5, 2.0 s
	Height	H = 6.0~9.0 cm
	Direction	θ = 0°, 45°
Structures	Shape and Size	Circular ϕ = 117 cm; Square D = 117 cm
	Height	I: Upright II: Submerged
	Sketch	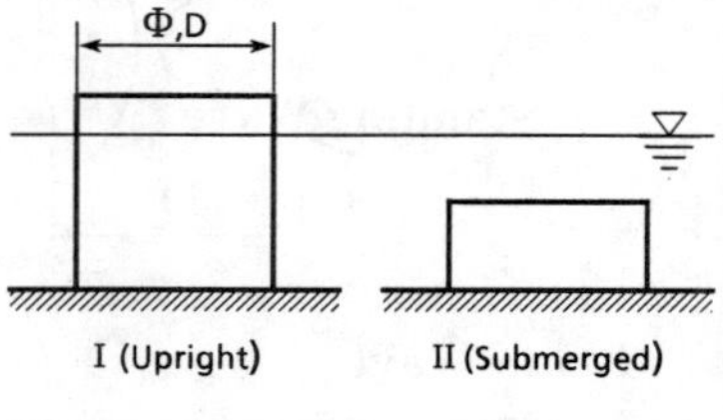

The experiment was carried out in a wave basin 35m×17m(see Fig.2) with the water depth kept a 30cm. The model offshore structures, measuring 117cm, were installed on a sand bed 4m×4m×1cm. The grain size of the sand was chosen as large as D_{50}= 0.58 mm with quite a uniform distribution, with the shingle material to be used in the scour protection in mind, and also giving the sand function as a tracer.

Gravity type oil production platforms being kept in mind, the shapes of the structure were made basically circular, and square to investigate the effect of flow separation at the corner of the square type. A submerged type, as a model of a submerged storage tank, was also investigated .

The water depth and the structure size were scaled down to the 1/66 model from the prototype planned somewhere offshore east of China. The waves were chosen so that the threshold of the inception of motion could be well resolved by observation. The wave condition together with the types of structures are listed in Table 1.

The movement of the sand was observed by a periscope and video camera. The periscope, a simple combination of a plastic pipe, 80 mm in diameter, with a plate of glass at the bottom, was put into the water to a depth of 10 cm to enable the observation of the movement of the bottom sand without the view being disturbed by the up and down movement of the water surface. Fig.3 shows how the observation device was set up. The observation points around a structure were so disposed by 10 cm pitch mostly to cover the region of 1 diameter cylinder by half of it as is displayed in Fig.4. There were a total of 268 observation points.

The observation procedure was as follows; As soon as the wave field became stationary, a computer control in the carriage mounting the observation device started to repeat a stop and go motion to follow the pre-programed paths of the observation points. Observation for one location took only half a minute and the total time took about three hours, during which it was found that the bottom remained almost flat even where the movement of the sand was greatest.

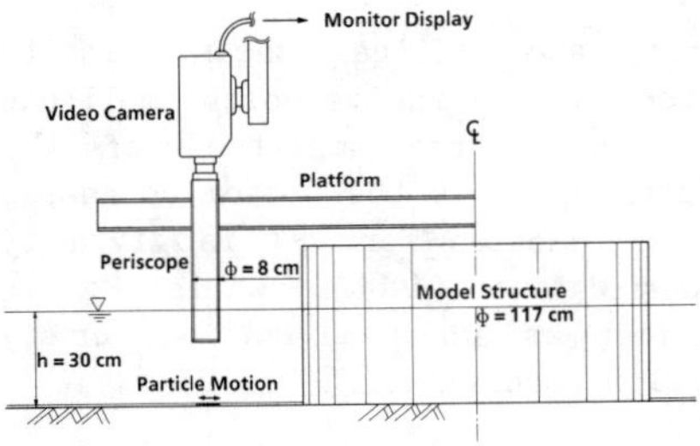

Fig.3 Observation device

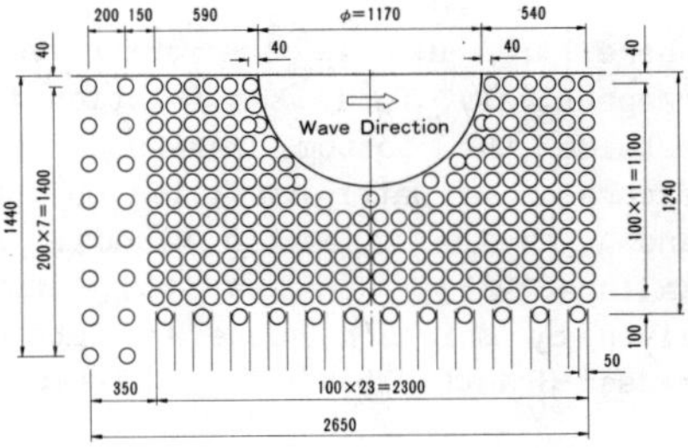

Fig.4 Observation points

3. Theoretical back ground

3.1 Inception of motion of particles under plane wave field

There have been already extensive research contributions in the area of prediction of inception of motion of particles under plane wave field. They are categorized into three approaches which follows;

(1) Critical velocity u_c is evaluated from Eq.(1), originally presented to give critical water depth, with the aid of linear (or nonlinear) wave theory.

$$(H/H_0)^{-1}(\sinh 2\pi h/L)(H_0/L_0) = \alpha(L_0/D)^n \quad , \qquad (1)$$

where H and L are wave height and wave length respectively at water depth h, suffix $_0$ indicates those parameters in the deep sea and D is diameter of sand particles. α and n are constants summarized in a table presented by Noda(1981), which vary with the condition of the boundary layer originally given by, Ishihara·Sawaragi(1960), Sato·Tanaka(1962) Horikawa·Watanabe(1966).

(2) The more direct method of obtaining critical velocity is to use Eq.(2).

$$u_c = \beta \cdot \Delta^{\gamma} \cdot D_{50}^{\delta} \cdot T^{\varepsilon} \, , \qquad (2)$$

where Δ is specific gravity of sand particle in water and T is wave period. β, γ, δ, ε are constants given by many researchers such as Bagnold(1946), Goddet(1960), Komar·Miller(1974). The merit of this formula is that it reflects the direct results of the experiments, but the demerit is that those constants vary with the different experiments.

(3) The third approach to calculating the threshold of the movement of sand particles is to use Eq.(3) and Eq.(4).

$$u_*/u_{b0} = \sqrt{f_w/2} = F_1(\, a_b/k_s,\ T\,) \quad , \qquad (3)$$

$$u_{*c} = F_2(u_* D_{50}/\nu) \quad , \qquad (4)$$

where, u_*, u_{*c} is the shear velocity and critical shear velocity respectively, f_w is the friction factor. u_{b0} is the velocity amplitude outside the bottom boundary layer, a_b is the amplitude of the excursion of water particles on the bottom, k_s is the bottom roughness and ν is the kinematic viscosity. The function form F_1 is given by Kajiura(1965), Jonsson(1966), Horikawa·Watanabe(1966), while F_2 is given by Shield's curve for steady flow as an approximation, or by Madsen·Grant(1976) in the oscilatory flow condition.

Eq.(3) gives shear velocity based on the consideration of the boundary layer, while Eq.(4) gives critical shear velocity, which is an experimental parameter. However, it is not certain at present whether these parameters can be applicable under the diffraction wave field where water particle motion is more complicated than with uni-directional oscillation.

3.2 Boundary layer equation under diffraction wave field

So far, the boundary layer equation of the oscillatory flow has been treated on the x-z plane, with x cordinate taken in the direction of the horizontal motion of the water particle induced by single directional wave (plane wave) motion and z in the vertical direction in which the boundary layer develops.

Here, to investigate the bottom shear caused by the multi-directional flow on the bottom where the path is elliptical, the boundary layer should be modeled in the x-y-z region, of which the x-y plane is taken on the bottom. In this respect, Tanaka·Shuto(1981), Fredsφe(1984) developed the models.

Tanaka(1981) treated the flow inside the boundary layer as a wave-current coexistent system. Let $\mathbf{u} = (u_x, u_y)$: the combined velocity inside the boundary layer on the x-y plane, p: water pressure, $\tau = (\tau_x, \tau_y)$: bottom shear stress, ρ: density of water, t : time, then,

$$\partial \mathbf{u} / \partial t = -(1/\rho)\, \nabla p + \partial(\tau/\rho)/\partial z \quad . \qquad (5)$$

It is assumed that the shear stress can be estimated by the velocity of the oscillatory component u_w by

$$\tau/\rho = K_z(\partial u_w/\partial z) \quad , \qquad (6)$$

where K_z is given by

$$K_z = \kappa\, u_w^* z \quad . \qquad (7)$$

u_w^* is the shear velocity under the wave-current coexistent system. With the boundary condition at outside edge of the boundary layer, Tananka derived Bessel's equation and thus obtained the solution for the bottom shear in the wave-current coexistent system.

However, the attempt by the authors to apply his theory to the field where the orbital motion of water is elliptical (without current) failed to bear anything new in evaluating bottom shear. The reason for this was because, with the assumptions of Eqs.(6) and (7), the solution of the bottom shear stress τ_b depicts a complete similar Lissajous figure(ellipse) as the velocity u_b at the outside of the boundary layer, with a constant phase lag. This means there was no change in the maximum shear as is predicted by the two dimensional treatment.

Fredsφe(1984) has started with the same basic equations as Eqs.(5),(6),(7), and has derived the more direct equation to obtain shear stress but with the more simple assumption that the orbital motion is circle and the phase lag between a flow outside the boundary layer and that inside the layer is negligibly small, which is not always the case. His conclusion is that the shear stress with the circle orbit increases by 10-15% more than with the simple oscillatory flow. However, the assumption that the phase lag can be neglected is rather difficult to accept.

4. Experimental result

4.1 Verification for method of experiment

There are some uncertainties when it comes to defining sand particle movement, which has caused some deviations in the data obtained by various researches so far. There was a fear that the presence of the periscope disturbed the bottom flow. To judge whether the observation was correctly carried out and to compare the criteria of movement defined by the authors with that of established results, the verification experiment was carried out for the plane wave.

Fig.5 shows the result. The symbols in the figure show the rank of movement. The curves are also drawn using Eq.(1) to Eq.(4) for the sake of comparison. Dashed lines represent the initial movement obtained by Isihara·Sawaragi and Jonnson+Shields. The solid lines indicate the severe movement shown by Horikawa-Watanabe and Sato-Tanaka. The lines of Komar-Miller and Goddet fall firmly on these solid lines. The result of the experiment is in good agreement with the calculation lines, except for the small deviations. This means that the observation method was correct.

4.2 Inception of sand motion around a large structure

In Fig.6 shows the result of the observation on the movement of sand

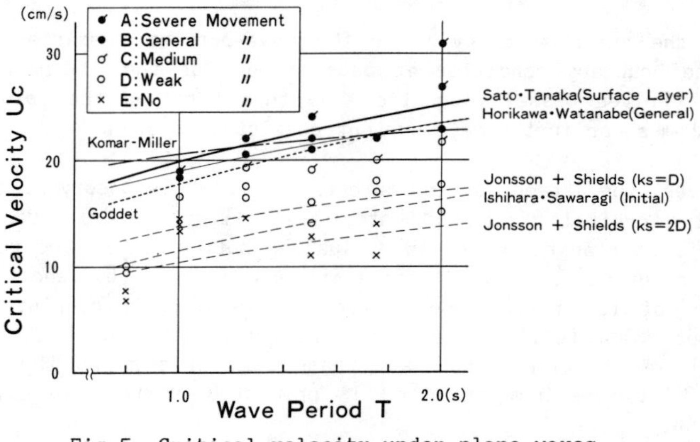

Fig.5 Critical velocity under plane waves

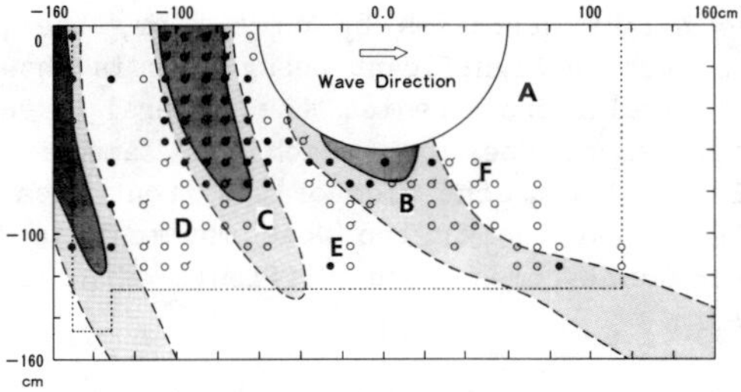

Fig.6 Particle movement(Circular-I, T=1.0s, H=8.0cm)

particles around a large scale circular cylinder. The observation area is enclosed by dotted lines and is somewhat rectangular in shape. The symbols have the same meanings as in Fig.5, while, for simplicity no symbols inside the observation area indicate no movement. The curves in Fig.6 are drawn by the method explained in Section 3.1. First, the bottom velocity field around a cylinder was calculated using potential wave theory, then it was compared with the threshold data obtained from the experiment for verification under plane waves explained in Section 4.1(see Fig.5). The data read from Fig.5 as follows; u_c= 0.15 m/s for the initial movement for T=1.0-2.0s, u_c=0.175,0.205,0.225 m/s for the general movement for T= 1.0,1.5,2.0s respectively. The light gray zones, enclosed by dashed lines, are for the initial movement, while dark zones, enclosed by solid lines, are for the severe movement. No color indicates no movement theoretically.

The general impression of the comparison of the state of movement between the experiment and the calculation is that they agree quite well. As for the initial movement, the sheltered area noted by character A in the figure agrees well, while in B and C where the particles were expected to move, they actually didn't move, and in D, E and F where no motion was expected some movement in fact occurred. As for the general movement, the agreement is fairly good.

The pattern of the sand movement in front of the cylinder ("front" here means the area facing the direct wave attack, and "lee" means the sheltered area, the opposite side of the front area), generally corresponds to that of the diffraction wave height. Severe movement occurred at the node, no movement or weak movement occurred at the antinode.

To summarize, even in the region of diffractive wave, the front side of the cylinder where the long crested standing wave was dominant, hence wave motion was rather two dimensional, as well as in the sheltered lee side where only small amplitude diffraction waves with their long crests penetrated, inception of motion agreed well with the conventional motion criteria. In other regions, such as to the side of the front of the cylinder where the orbit of the bottom water particles was circle or ellipse, the conventional formulae for inception were no longer valid, in the strict sense.

When the wave height decreases by only 1 cm, the pattern of the movement exhibits quite a significant change, as is shown in Fig.7. In this case, the calculations predict the general trend of movement well, but, the agreement does not reach the same extent as in the previous case. Especially to the side of the front area, the agreement becomes less. The prediction of the weak movement in this region by conventional formulae becomes more difficult than in the case of single plane waves.

The change of the wave period also changes the pattern of the movement. Fig.8 and Fig.9 show the case of T=1.5s and 2.0s respectively. The same comment as the above is drawn about the agreement. The longer the wave period, the more lee side the severe

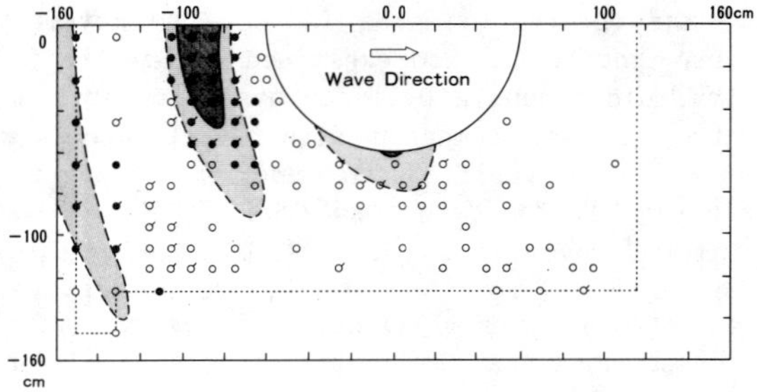

Fig.7 Particle movement(Circular-I, T=1.0s, H=7.0cm)

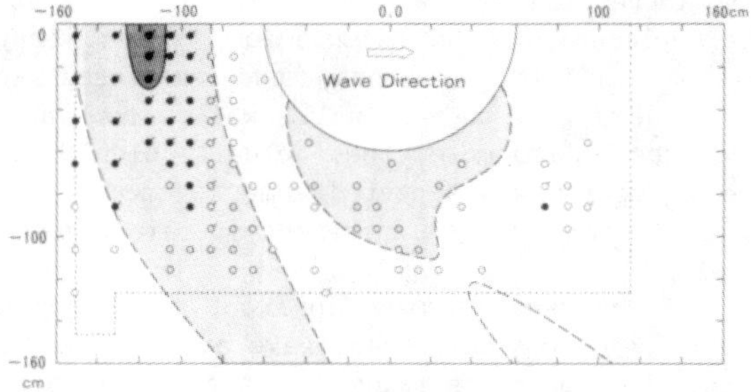

Fig.8 Particle movement(Circular-I, T=1.5s, H=6.0cm)

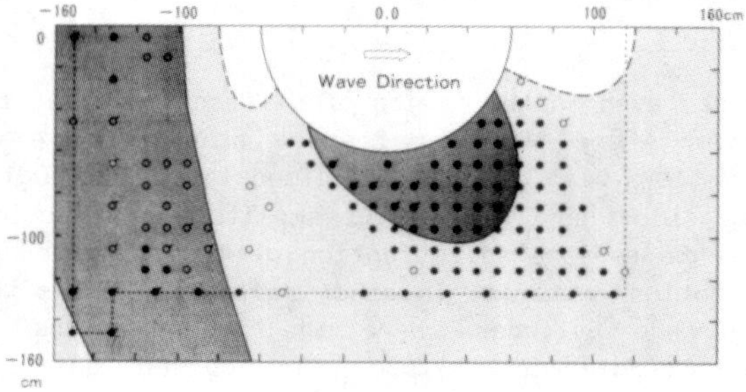

Fig.9 Particle movement(Circular-I, T=2.0s, H=8.0cm)

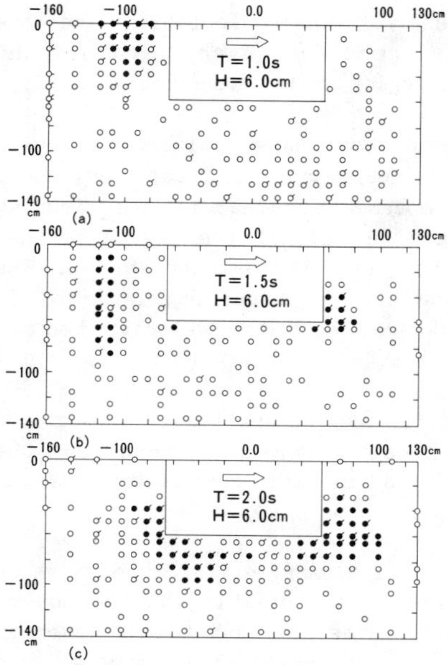

Fig.10 Particle movement (Square-I, θ=0°)

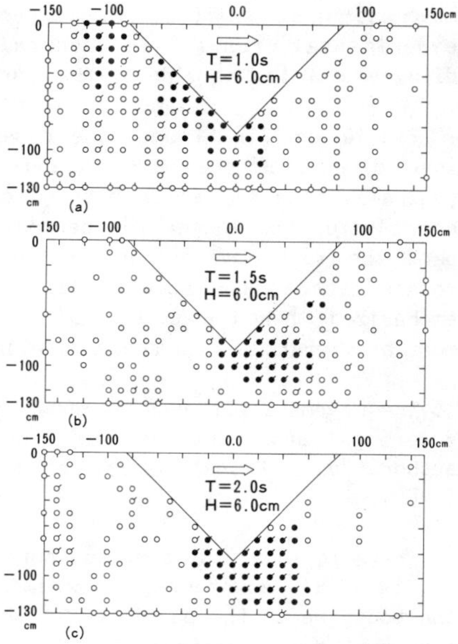

Fig.11 Particle movement (Square-I, θ= 45°)

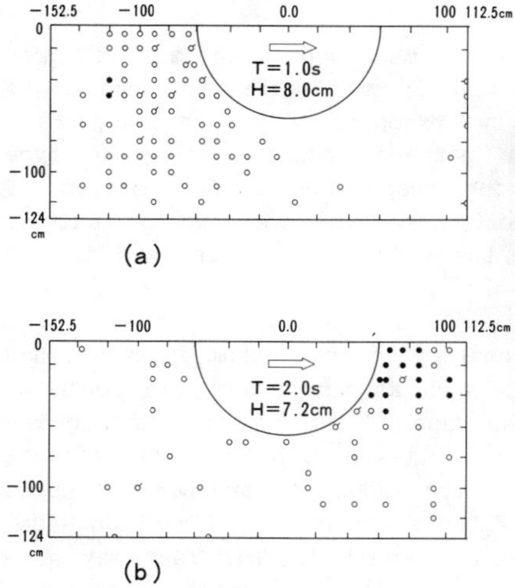

Fig.12 Particle movement (Circular -II)

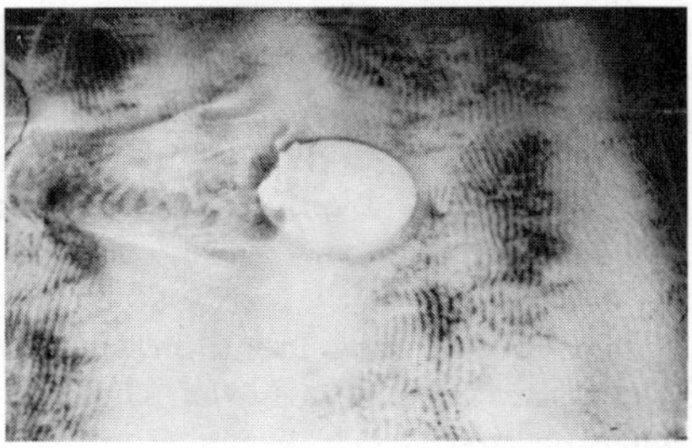

Photograph 1 Wave breaking on the top of a submerged cylinder (Circular -II)

movement area shifted. The agreement of the calculation and the experimental result is generally good. However, again, the disagreement to the side of the front region is not negligibly small.

Fig.10 and Fig.11 show the cases of the square cylinder, in which wave directions are θ = 0° and 45° respectively. Square cylinders generally give a similar pattern of movement to that of the circular one. Again, the longer the period, at the side of the region further back the more severe the movement becomes. A characteristic of the square type is the accelerated motion at the corners. This trend is emphasized when the corner locates where the water velocity becomes most accelerated as in the case where θ = 45 °, shown in Fig.11.

The movement patterns of shown so far are generally similar to the scour patterns obtained in the model experiment done by the authors(Toue ·Katsui 1985), as well as in one conducted by Rance (1980).

There is less impact at the bottom of a submerged structure as shown in Fig.12(a). However, it sometimes helps wave breaking at the top of the body, hence the particles in the lee side are easy to move even in the sheltered area(see Fig.12(b)). Photograph-1 shows how a wave breaks on the top of a submerged circular cylinder.

4.3 Ripple formation around a structure

The bottom configuration was also measured with the thicker layer(depth:30cm) of finer sand(D_{50}=0.15 mm), which was easier to move. Photograph-2 and Fig.13 show two examples of the results, which give a general view of the scour pattern around a circular type cylinder and a square type cylinder respectively. The pattern of erosion and deposition in the front area corresponded to that of severe motion. Erosion occurred to the side at the front and in the lee side.

Laboratory experiment on the morphology of the bottom, however, has its limitations. Whenever the data, such as the scour hole depth or scouring area, are translated to the case of the prototype or to the design procedure, special care should be taken with the scale effect. But, the pattern of the ripple formation around a structure is quite typical in the region to the side of the front where diffraction waves dominate as can be clearly seen in Photograph-3. This fact may give some clues in understanding the mechanism of bottom shear or presumable special flow.

5. Possible approaches to understanding the phenomena

It was confirmed that the inception of sand particle motion in the region where waves from different directions are superimposed on each other differs from what is calculated by conventional two dimensional formulae. This phenomenon could be considered inherent, too, in front

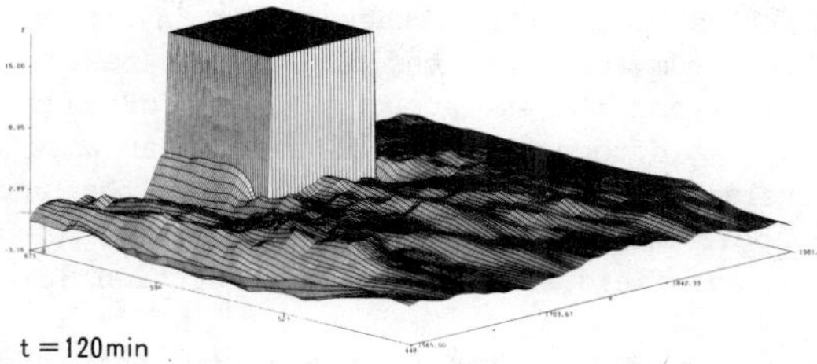

Fig.13 Bird's eye view of measured bottom change
(Square-I, θ=0°, T=1.0s, H=10.0cm, t=120min)

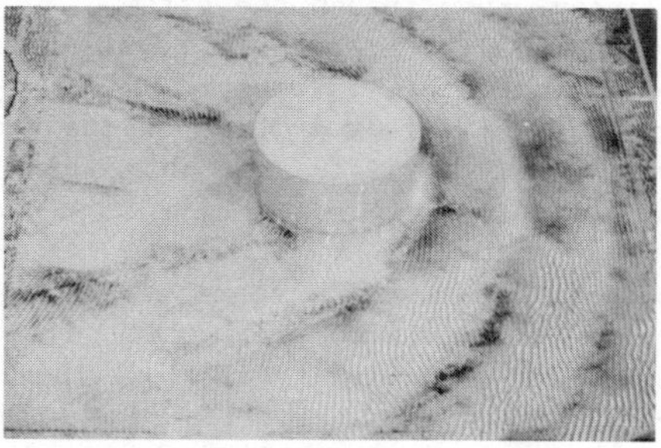

Photograph 2 Bottom change around a cylinder
(Cicular-I, T=1.0s, H=11.0cm, t=120min)

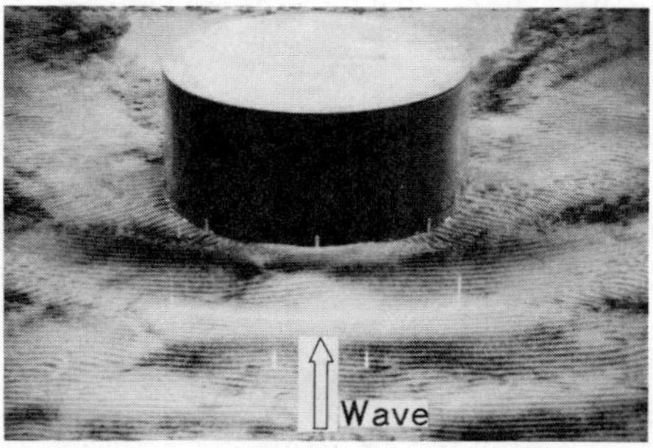

Photograph 3 Ripple pattern to the side of the front of a cylinder

of a breakwater when waves come obliquely to the normal direction of the breakwater and thus making a short crested standing wave field.

The phenomenon should be fully understood for the proper design of the scour protection around a structure. The reasonings can be accounted for in several possible ways; the measurement error, the nonlinearity of the bottom velocity (the velocity was estimated by linear wave theory), the special transformation of the bottom shear caused by the the elliptical orbital motion of the water particles or the induced flow around a structure.

As for the first possibility, there might have been, to some extent, wave reflection from the other end of the wave basin, inhomogeneity of the incident wave and the ambiguous judgement of sand movement despite the efforts to avoid them. The use of the linear wave theory to obtain the bottom velocity proved not to influence the accuracy. However, most of the above suspicions are cleared up with the veification test for the plane wave field already stated at Section 3.1.

Secondly, it would be natural to consider that the development of the boundary layer is influenced by the elliptical orbital motion of the water particles. Fig.14 shows the elliptical Lissarjous figure of the velocity vector, which is typical in the region to the side of the front of the cylinder. Even though the attempt by the authors failed to introduce theoretically the special development of the boundary layer there, the bottom shear, hypothetically, may be distorted both in amplitude and in phase lag from those in the two dimensional flow field, as Fredsφe(1984) partially has introduced the bottom shear where the orbital motion was circle.

The third possibility is the formation of the flow. In the experiment, the flow was also observed using dye from the point of view of directional flow not quantity of flow. There was a steady flow from the side of the cylinder to its lee side with a magnitude of a few cm's per second. It can be understood to a certain degree with the help of Fig.15, which shows a flow pattern of mass transport velocity around a circular cylinder inside the boundary layer. The figure was calculated after Mei(1983). But, in front of the cylinder, the flow

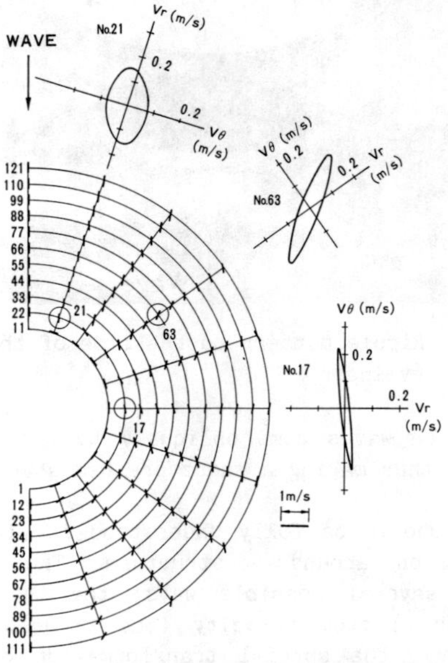

Fig.14 Lissajous figures of bottom velocity vector around a cylinder(Circular -I)

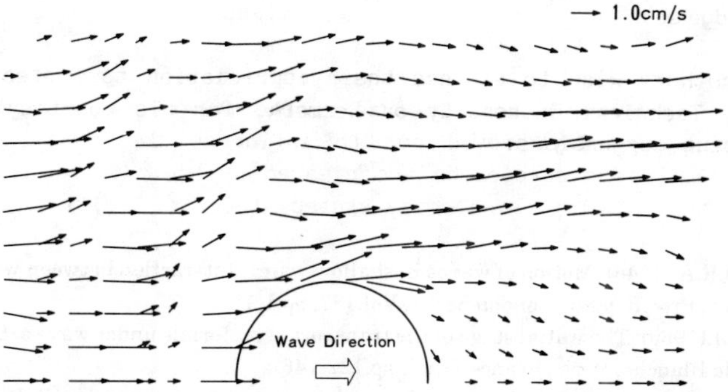

Fig.15 Mass transport velocity around a circular cylinder (Circular -I, T=1.0cm, H=10cm)

was observed to be in the offshore direction (against the direction of the wave incidence), with a magnitude of the same or a little stronger than that of the side flow. Thus in the area to the side of the front of the cylinder there was a kind of dead space with respect to the steady flow, in which the dye stayed for a considerably long time creating a circulation with the concentration of the dye color being faded at this particular place. Fig.15 cannot account for this kind of flow pattern. This fact indicates that there should be some other mechanism which creates the flow around the structure, something which may be caused by the gradient of radiation stress as is the case in the surf zone.

6. Conclusion

The basic, or preliminary, study of the inception of sand particle motion around a structure using the model experiment was carried out, as well as a simple application of the formulae established in plane wave motion. Through this study, it was confirmed that at the area where the wave motion is more two dimensional, areas like front area or sheltered area, the conventional formulae are applicable, but in the area where the wave field is more complex that the orbital motion of water particles depict circlular or elliptical path, there are discrepancies between such formulae and the experimental results.

The reasons of the discrepancies can be ascribed to two kinds of phenomena. One possibility is the flow induced by mass transport inside the boundary layer and/or by a flow which is presumably caused by the gradient of radiation stress around a structure. The second reason could be that the boundary layer, hypothetically, may develop in a more complex manner such as with some distortion in three dimensional space in the area where water particle figures the elliptical Lissajous.

Further research is required both on the theoretical treatment and to obtain a more precise measurement data concerning the flow in the region where multi-directonal waves are superimposed themselves.

Acknowledgement

The authors wish to express their appreciation to Professor Jφrgen Fredsφe, Technical University of Denmark, for his courtesy of giving significant suggestions.

Reference

1)Bagnold,R.A.(1946): Motion of waves in shallow water; Interaction between waves and sand bottoms, Proc.Roy.Soc.London Ser.A, Vol.187, pp.1-18

2)Goddet, J.(1960): The initial stage of the transport of materials under wave action, La Houille Blanche, Paris, France, No.2, pp.122-146

3)Fredsφe,J.(1984): The turbulent wave boundary layer along a vertical wall, Prog. Rep. 61, Inst. Hydrodyn. and Hydraulic Eng.Tech. Univ. Denmark, Sept. pp.23-33

4)Horikawa, K. and A.Watanabe(1966): A concideration of the movement of the sand under wave action, Proc.13th Japanese Conf. on Coastal Eng., pp.126-134(in Japanese)

5)Irie, I., K.Nadaoka, T.Kondo and K.Terasaki(1984): Two dimensional sea bed scour in front of breakwaters by standing waves, Report of Port and Harbour Research Institute, Vol.23, No.1, pp3-52

6)Ishihara, T. and T.Sawaragi(1960): Critical velocity for sand drift; Critical water depth and transport-rate, Proc.7th Japanese Conf. on Coastal Eng., pp.47-58(in Japanese)

7)Jonsson, I.G.(1966): Wave boundary layers and friction factors, Proc.10th International Conf. on Coastal Eng., Vol.I, pp.127-148.

8)Kajiura, K.(1965): On the bottom friction in an oscillatory current, Bull. Eathq. Res. Inst. Vol.42, No.1, pp.147-173

9)Komar, P.D. and M.C.Miller(1974): Sediment threshold under oscillatory waves, Pcoc.14th International Conf. on Coastal Eng., Vol.II, pp.756-775

10)Madsen, O.S. and W.D.Grant(1976): Sediment transport in the coastal environment, Ralph M.Parsons Lab.for Water Rersource and Hydrodynamics, M.I.T. No.209

11)Mei, C.C.(1983): The applied dynamics of ocean surface waves, John Willey & Sons, New York, pp.419-434

12)Noda, H.and Hashimoto(1981): Sand drift and shore protection facilities, Gihodo Press, p.36.(in Japanese)

13)Rance, P.J.(1980): The potencial for scour around large object; Scour protection Techniques around offshore structure, The Society of Underwater Technology, One-day seminar, pp.41-54

14)Sato, S. and N.Tanaka(1962): Sand motion on the horizontal bottom under waves, Proc.9th Japanese Conf. on Coastal Eng., pp.95-100(in Japanese)

15)Tanaka, H. and N.Shuto(1981): Friction coefficient for wave-current coexistent system, Coastal Engineering in Japan, Vol.24, pp.105-128

16)Toue, T. and H.Katsui(1985): On a scour around a large circular cylinder, Proc.32nd Japanese Conf. on Coastal Eng., pp.425-429(in Japanese)

CHAPTER 96

BEACH PROFILE CHANGE: MORPHOLOGY, TRANSPORT RATE, AND NUMERICAL SIMULATION

Magnus Larson[1], Nicholas C. Kraus[2], M.ASCE, and Tsuguo Sunamura[3]

ABSTRACT

An empirically based engineering numerical model is presented for simulating beach profile change in the surf zone produced by wave-induced cross-shore sand transport. The model simulates the dynamics of macroscale profile change, such as the growth and movement of berms and breakpoint bars. Model development was founded on two data sets from large wave tank experiments consisting of 42 cases with different incident wave conditions, median grain size, and initial beach shape. Model predictions are tested with field data, and reasonable agreement is found.

INTRODUCTION

Techniques to predict shore-normal evolution of the profile to changes in incident waves and water level are needed for design, maintenance, and evaluation of beach stabilization and shore protection projects. Time scales involved range from days for storm-induced beach and dune erosion to months for adjustment of beach fill to equilibrium. This paper presents a practical empirically based numerical model of beach profile change developed for engineering design which simulates the formation and movement of bars and berms. The model is applicable to describing profile change resulting from cross-shore sand transport produced by varying water level and short-period breaking waves in the range of approximately 3-20 sec.

The numerical model was developed through extensive investigation of beach profile change obtained in large wave tanks (LWTs) under controlled conditions, in order to understand the fundamental processes of macroscale profile change and cross-shore sand transport. Quantitative relationships were derived from the LWT data set to identify the important parameters of profile change and to establish cause and effect relationships between the incident waves and profile response.

---

(1) Assistant Professor, Department of Water Resources Engineering, Institute of Science and Technology, University of Lund, Box 118, S-221 00 Lund, Sweden.
(2) Senior Research Scientist, Coastal Engineering Research Center (CERC), U.S. Army Engineer Waterways Experiment Station, P.O. Box 631, Vicksburg, MS 39180-0631, USA.
(3) Professor, Institute of Geoscience, University of Tsukuba, Ibaraki 305, Japan.

The numerical model was then tested by reproducing LWT measurements and, when satisfactory agreement was obtained, it was applied to predict measured profile change in the field produced under time varying waves and water level. This paper summarizes and integrates results presented in more detail in four publications by the authors (Larson 1988, Larson and Kraus 1988a,b,c).

PROCEDURE

Field data sets useful for developing models of beach profile change are lacking because of the required high resolution measurement in time and space of the profile and the waves and water level that produced the change. Due to the great spatial and temporal variability of waves and the three-dimensional character of nearshore bathymetry, it is also difficult to extract cause and effect relationships between waves and profile change resulting solely from the wave-induced, cross-shore component of sand transport. However, any engineering simulation model must ultimately be tested with field data prior to actual application.

The most fruitful approach for empirical investigation of beach profile change appears to be use of data obtained with LWTs. Such facilities enable controlled reproduction of near-prototype conditions of beach slope, wave height and period, turbulence induced by wave breaking, and resultant sand transport and beach change. The problem of scaling encountered with small laboratory tanks is eliminated, and the required high resolution measurement of the profile can also be attained. Recently, two extensive independent data sets on beach profile change have become available from experiments performed using LWT and monochromatic waves. These experiments involved combinations of waves, water levels, beach slopes, and sand sizes that exist in the field, but with the advantages of true two-dimensionality, control of the external (wave) force, and an optimized measurement schedule. These data sets were selected for use in the present investigation and intensively analyzed.

One data set was obtained in experiments performed by the U.S. Army Corps of Engineers (CE) in the years 1956-1957 (Saville 1957) and 1962, recently compiled by Kraus and Larson (1988a). In the CE experiments 18 cases have been documented of which most were started from a plane slope of 1/15. The wave parameters ranged from periods between 3.75 and 16.0 sec and generated wave heights between 0.55 and 1.68 m in the horizontal section of the tank. The water depth in the horizontal section was in the range of 3.5-4.6 m in the different cases, and two grain sizes were employed with median diameters of 0.22 and 0.40 mm.

The second data set derives from experiments performed at the Central Research Institute of Electric Power Industry (CRIEPI) in Japan (Kajima et al. 1983a, 1983b). The CRIEPI experiments followed the pattern of the CE experiments. Quartz sand of median grain sizes 0.27 and 0.47 mm was employed in 24 cases of which 17 cases involved initial uniform beach slopes of 1/10, 1/20, 3/100, or 1/50. The cross-shore distribution of wave height was also measured between profile surveys.

RESULTS

Quantification of Morphologic Features

Profile morphologic features of interest are formations created by wave action, directly or indirectly, during time scales of several tens of the wave period. These features, particularly bars and berms, were defined with respect to the initial profile, which provided an unambiguous characterization for relating profile response to wave and beach properties.

The equilibrium profile is the shape achieved by a beach exposed to waves of constant characteristics for a long duration. In the equilibrium state sand particles may move, but there is no net transport along the profile. Despite the stochastic character of microscale fluid and sediment motion, as well as small variations in applied forces moving the sand, macroscale changes of physical and derived quantities from the experiments approached an apparent equilibrium value in a remarkably smooth manner.

If the profile is not in equilibrium with the waves passing over it, net transport of material occurs. It is of considerable scientific and engineering significance to predict whether a beach will erode or accrete, i.e., form a bar or a berm under given wave and beach conditions. In the present study, a new criterion was developed based on the deepwater wave steepness H_o/L_o and the dimensionless fall speed parameter H_o/wT using only LWT data (H_o=deepwater wave height; L_o=deepwater wavelength; w=sand fall speed; T=wave period). Fig. 1 is a plot of the prototype-scale data together with the developed empirical criterion to distinguish bar/berm profiles. The criterion is

$$\frac{H_o}{L_o} = 0.00070 \left(\frac{H_o}{wT}\right)^3 \qquad (1)$$

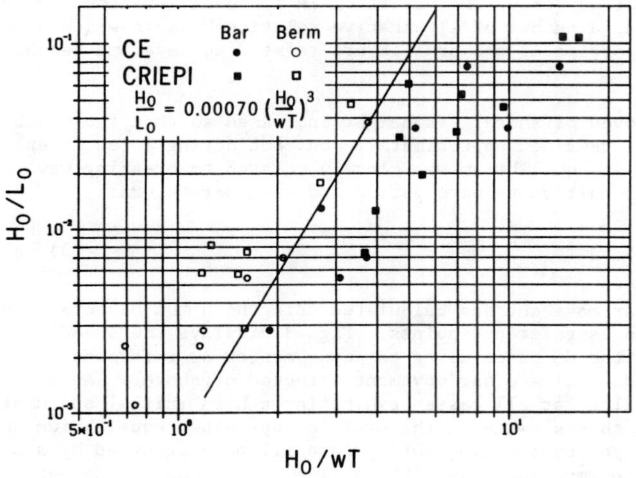

Fig. 1 Criterion for distinguishing bar and berm profiles.

Extensive correlation and regression analyses were conducted to investigate relations between geometric properties of the different profile morphologic features and wave and beach characteristics. The primary parameters used were: H_o, L_o, T, w, breaking wave height H_b, water depth h, median grain size D, and beach slope $\tan\beta$. Also, non-dimensional quantities were formed, both for deepwater and breaking conditions, such as H/L, H/wT, $\tan\beta/\sqrt{H/L}$, D/H, and D/L, in which L is the local wavelength. These quantities were related to geometric properties of the profile, presented in selected examples next.

Under steady waves and constant or slowly varying water level, the evolution of bars and berms was found to be regular, exhibiting clear growth and equilibrium properties that were readily described by simple regression expressions. The dimensionless fall speed H_o/wT emerged as an important parameter in predicting both profile response and geometric properties of various morphologic features. The strong relationship between wave and sand characteristics and morphologic features reinforced the possibility of quantitatively predicting the evolution of macroscale features of the profile in an empirical formulation.

As a bar moves offshore it simultaneously increases in volume to approach an equilibrium size. Fig. 2 shows evolution of bar volume for the main breakpoint bar for the CE experiments. Correlation analysis involving wave and beach profile parameters showed that equilibrium bar volume V_{eq} was most closely related to wave height and sand fall speed (or grain size). A larger wave height implied a larger bar volume, and a higher fall speed (or larger grain size) produced a smaller bar volume. A regression relationship was derived relating the non-dimensional equilibrium bar volume to the quantities H_o/wT and H_o/L_o according to

$$\frac{V_{eq}}{L_o^2} = 0.028 \left(\frac{H_o}{wT}\right)^{1.32} \left(\frac{H_o}{L_o}\right)^{1.05} \tag{2}$$

Eq. 2 explained 70% of the variation in the data. Larson and Kraus (1988a) present a number of alternative relationships in which other geometric properties of the profile are related to wave and beach parameters.

As a bar moved offshore, its height increased so that the depth to the crest h_c remained approximately constant during a run. Depth at the bar crest at equilibrium was closely related to breaking wave height and insensitive to wave period and sand parameters:

$$h_c = 0.66\ H_b \tag{3}$$

Speed of bar movement was calculated using both the bar crest and center of mass as reference points. Fig. 3 displays the speed of bar migration for the CE cases using the mass center as reference. Positive speeds indicate bar movement directed offshore. The main trend was similar for all cases, exhibiting a high initial speed of bar migration and then slowing as the profile approached equilibrium shape. The negative speeds occurring during Case 911 were produced by a water level variation imposed to simulate a tide, which caused the bar to move onshore during phases of increased water level.

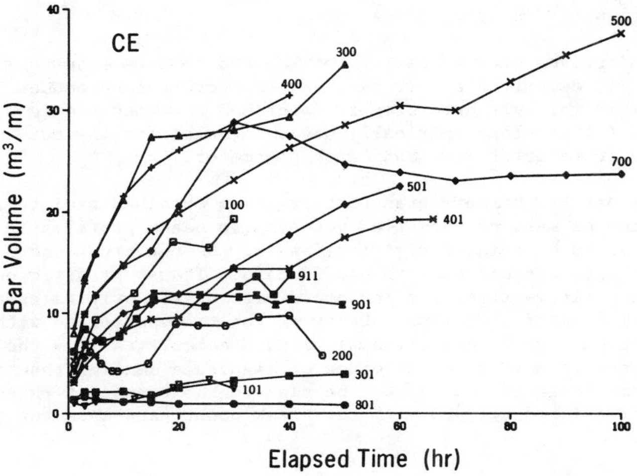

Fig. 2 Growth of bar volume with elapsed time for the CE data.

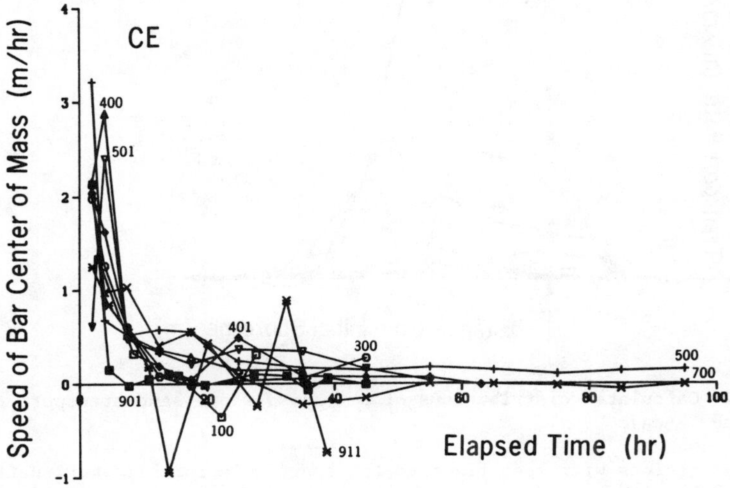

Fig. 3 Speed of bar center of mass (CE data).

Values of bar slopes formed in the LWTs under monochromatic waves are typically greater than those observed in the field. The difference between wave tank and field results is attributed to the action of random waves and varying water level, which would widen the breaker zone and smooth profile features in the field. Another factor is that steady wave conditions usually are not of sufficient duration in the field for the profile to reach equilibrium form.

Net Cross-Shore Sand Transport Rate

The main objectives of the analysis of the cross-shore transport rate were to: (1) demonstrate that the net transport rate produced by breaking waves is reliably predicted if described in a macroscale framework, and (2) develop empirically based formulas for the net transport rate in terms of wave and beach parameters.

The average net cross-shore transport rate was obtained by integrating the equation of mass conservation between two beach profiles in time. Fig. 4 shows calculated distributions of the net cross-shore sand transport rate associated with one CE case. Transport directed offshore has a positive sign, and the coordinate system originates at the initial still-water shoreline. Decay of the transport rate with time is clear from Fig. 4, and the maximum rate calculated from the final two surveys is more than one order of magnitude smaller than the maximum from the first two surveys. The peak in the transport rate distribution translated seaward with the break point, and thus the bar moved seaward.

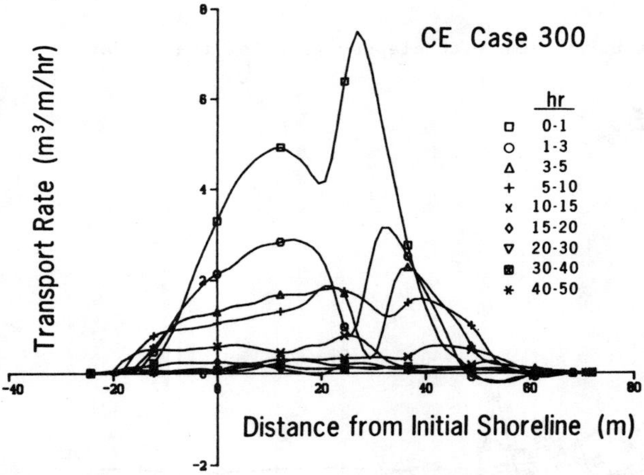

Fig. 4 Calculated distributions of net cross-shore sand transport rate for one CE case.

Four regions with specific sand transport relationships were defined in analogy with zonations of nearshore wave dynamics (for example, Svendsen et al. 1979; Basco and Yamashita 1987), as shown in Fig. 5. One region extends from the seaward limit of significant profile change to the break point, called the pre-breaking region (Zone I). In this region the transport rate is influenced by transport in the zone of wave breaking through the sand flux at its shoreward boundary, but the governing transport processes on either side of the boundary are different. Zone II corresponds to the breaker transition region and is located between the break point and the plunge point. From the location of the plunge point to the point of wave reformation one specific region, Zone III, is defined where the waves are fully broken and gradually decay (inner region in hydrodynamic terms). In this region the energy dissipation of the waves due to breaking becomes fully developed. Transport conditions in the swash zone differ from those prevalent in the surf zone, making it logical to define a fourth transport region, Zone IV.

The net transport rate in the zone of broken waves, where the most active transport is expected to occur, showed good correlation with the wave energy dissipation per unit water volume (Dean 1977). The net transport rate in the pre-breaking zone and in the breaker transition zone decayed exponentially with distance offshore both for erosional and accretionary cases. The exponential decay coefficient was related to the grain size and the breaking wave height (see Eq. 5). On the foreshore the net transport rate showed an approximately linear behavior, decreasing in the shoreward direction from the end of the surf zone.

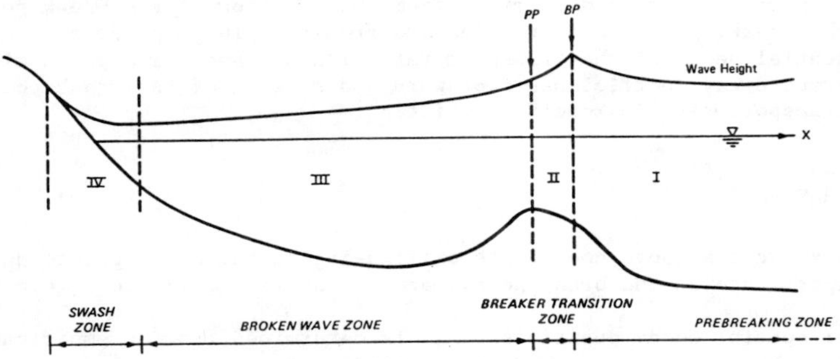

Fig. 5 Four principal zones of cross-shore sand transport.

Distributions of the net cross-shore transport rate calculated from measured profile change over intervals on the order of hours displayed regular and smooth properties, despite the random character of the grain-by-grain movement that actually took place. Consequently, it appears possible to estimate the net cross-shore sand transport rate with sufficient reliability to predict the development of main morphologic features of the beach profile.

Numerical Model of Beach Profile Change

A deterministic numerical model was developed to predict beach profile change resulting from cross-shore sand transport (Larson and Kraus 1988c), focusing on the main morphologic features of bars and berms. Many of the assumptions and relationships used in development of the model are founded on observations made from the LWT data. Changes in the beach profile are assumed to be produced by breaking waves; therefore, the cross-shore transport rate is determined from the local wave, water level, and beach profile properties, and the equation describing conservation of beach material is solved to compute profile change as a function of time.

The wave height distribution across-shore is calculated by applying small-amplitude wave theory to the point of breaking, and then the breaker decay model of Dally et al. (1985) is used to provide the wave height in regions of breaking waves. The profile is divided into specific regions (Zones I-IV) according to the wave characteristics at the given time step for specification of transport properties. The distribution of the cross-shore transport rate is then calculated from semi-empirical relationships applicable to the four regions. At the shoreward end of the profile the runup limit constitutes a boundary with no transport across it, whereas the seaward boundary is determined by the depth at which no significant sand transport occurs. Once the distribution of the transport rate is known, profile change is calculated from the mass conservation equation. This procedure is performed with the incident wave conditions and water level pertaining to the given time step.

The direction of the net cross-shore transport rate is determined in the model by Eq. 1, whereas the magnitude is a function of the transport rate in zones of fully broken waves. The waves are considered to be fully broken in the model from the plunge point to the end of the surf zone or to the point where wave reformation occurs. The location of the plunge point is determined from the location of the break point and the breaking wave height. Seaward from the plunge point an exponential decay of the transport rate with distance is applied with different decay coefficients shoreward and seaward of the break point. The transport rate in Zone I is written

$$q = q_b \, e^{-\lambda x} \qquad (4)$$

where x = cross-shore coordinate originating at the break point; q_b = transport rate at the break point; and λ = spatial decay coefficient.

The spatial decay coefficient λ is determined from an empirical relationship given by the LWT data according to

$$\lambda = 10.3 \left(\frac{D}{H_b}\right)^{0.47} \qquad (5)$$

The transport rate in Zone II is also described by an exponential decay but with a smaller value of the decay coefficient, 0.2λ.

A transport relationship similar to that used by Kriebel and Dean (1985) is applied in a region of fully broken waves (Zone III) with a term added to account for the effect of local slope. A steeper slope is expected to increase the transport rate down the slope. The transport relationship is

$$q = \begin{cases} K \left(D - D_{eq} + \dfrac{\epsilon}{K}\dfrac{dh}{dx}\right) & D > D_{eq} - \dfrac{\epsilon}{K}\dfrac{dh}{dx} \\ 0 & D < D_{eq} - \dfrac{\epsilon}{K}\dfrac{dh}{dx} \end{cases} \qquad (6)$$

where q = net cross-shore transport rate; K = empirical coefficient; D = wave energy dissipation per unit volume; D_{eq} = equilibrium energy dissipation per unit volume; and ϵ = empirical coefficient for the slope-dependent term.

The transport rate distribution in Zone IV is specified as a linear decrease from the end of the surf zone to the runup limit. Avalanching is initiated in the model if the local slope exceeds 28 deg at any point on the grid, and the process continues until an angle of 18 deg is reached. Larson (1988) describes the procedure for redistributing the sand in the model to simulate avalanching, and the limiting slopes are based on the LWT data analysis.

The numerical model was applied to simulate beach profile evolution for nine erosional cases from the LWT experiments. Values of model parameters were varied to minimize the sum of squares of the difference of measured and calculated depths for all profile surveys for a given case. After some experimentation, the empirical coefficient in the slope term in Eq. 6 was set as $\epsilon = 0.001\ m^2/sec$, leaving K and D_{eq} free to be varied in the calibration procedure. In calibration simulations, optimal values of D_{eq} were found to be on the order of 75% of those predicted by the design curve of Moore (1982), who used an energy dissipation approach with no slope dependence.

Seven cases from the LWT data were used for the calibration, and two independent cases were used to verify the generality of the optimal parameter values. Fig. 6 compares the calibration result for one of the CRIEPI cases, showing the profile evolution predicted by the model and the measured final profile. Also, the measured wave heights at the end of the run are shown together with the calculated wave height distribution. Values of K obtained in individual calibrations showed no strong dependence on wave and sand parameters. Qualitatively, K tended to decrease with increasing grain size and to increase with decreasing wave period. For predictive use, a single optimal value of K was determined by minimization of the total sum of squares for all profiles surveyed for the studied cases. A minimum occurred at $K = 1.6\ 10^{-6}\ m^4/N$, and this value was used in the verification.

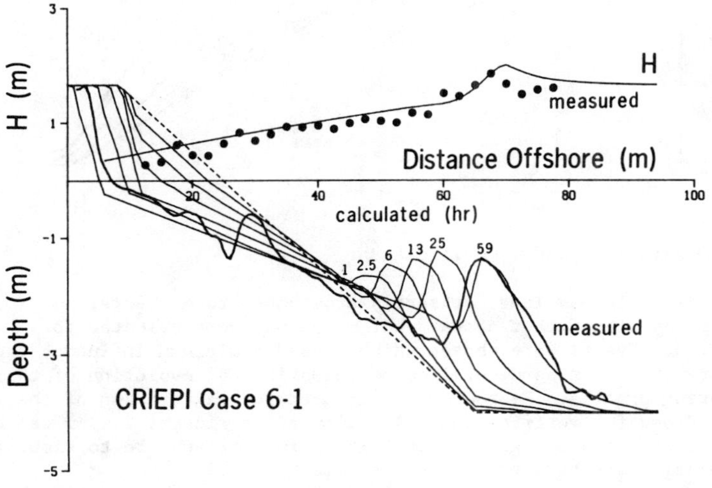

Fig. 6 Example of model calibration.

Fig. 7 displays the simulation result for one CE verification case. The volume of the main breakpoint bar, and the amount of erosion on the foreshore are well predicted. However, the crest of the bar is located somewhat too seaward, whereas the trough is not sufficiently deep. Details of the inshore features are not reproduced in the numerical model because secondary breaking was not allowed to operate, as reproduction of the main breakpoint bar was the primary goal of the calibration. Larson (1988) describes model tests performed to reproduce multiple bars.

Field Test

The numerical model was used to simulate beach profile change measured at the CERC's Field Research Facility (FRF) at Duck, North Carolina. This test represents a generalization of the model from monochromatic to random waves, the latter approximated by statistically representative waves allowed to vary in time together with the water level. Howd and Birkemeier (1987) compiled four years of profile change data taken at the FRF along two pairs of transects surveyed biweekly to approximately 10-m depth and more frequently surveyed during storms. Statistically representative wave height and period were available at 6-hr intervals at a depth of 18 m, and water levels at 1-hr intervals.

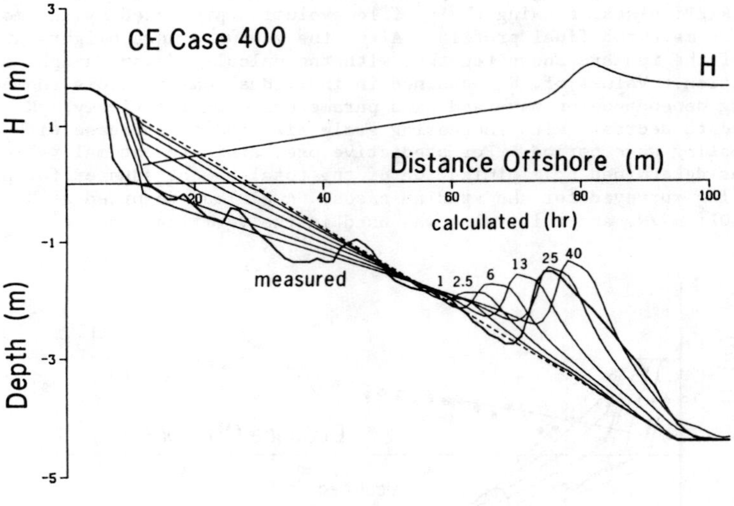

Fig. 7 Example of model verification.

Five 3- to 12-day long, primarily erosional storm events, as indicated by significant seaward bar movement, were selected for simulation. Events were chosen which showed a minimum influence from longshore transport as determined by comparison of evolution of the neighboring profile. Four events were used for calibration of the model and one for verification. A median grain size of 2.0 mm was used on the foreshore and a grain size 0.15 mm further seaward to simulate the bimodal character of the beach sediment.

At first, calibration parameters determined from the LWT comparisons were used in trial model runs. However, time rate of change of simulated profile development proved too rapid and bar development too pronounced, necessitating recalibration of the transport coefficient for the four storm events. The transport coefficient found applicable to the field profile change had a smaller value than that pertaining to the LWT calibrations (field average: $0.7\ 10^{-6}\ m^4/N$). Because of the fixed shoreline position at the FRF, which makes the foreshore act as a seawall to some extent, it is not clear whether the field-determined value of K has general applicability. Some decrease was expected since K qualitatively showed a weak inverse dependence on wave period in the LWT experiments, and the wave periods in the field were somewhat longer than in the laboratory experiments. Values of other empirical transport-related parameters in the model were the same as determined in the LWT calibration. Fig. 8a displays the wave height, wave period, and water level variation during one event, and Fig. 8b shows a typical calibration result, including the measured initial and final profile and the simulated final profile. Movement of the bar was fairly well predicted by the model, although the amount of material moved was underestimated, and the trough was not sufficiently pronounced.

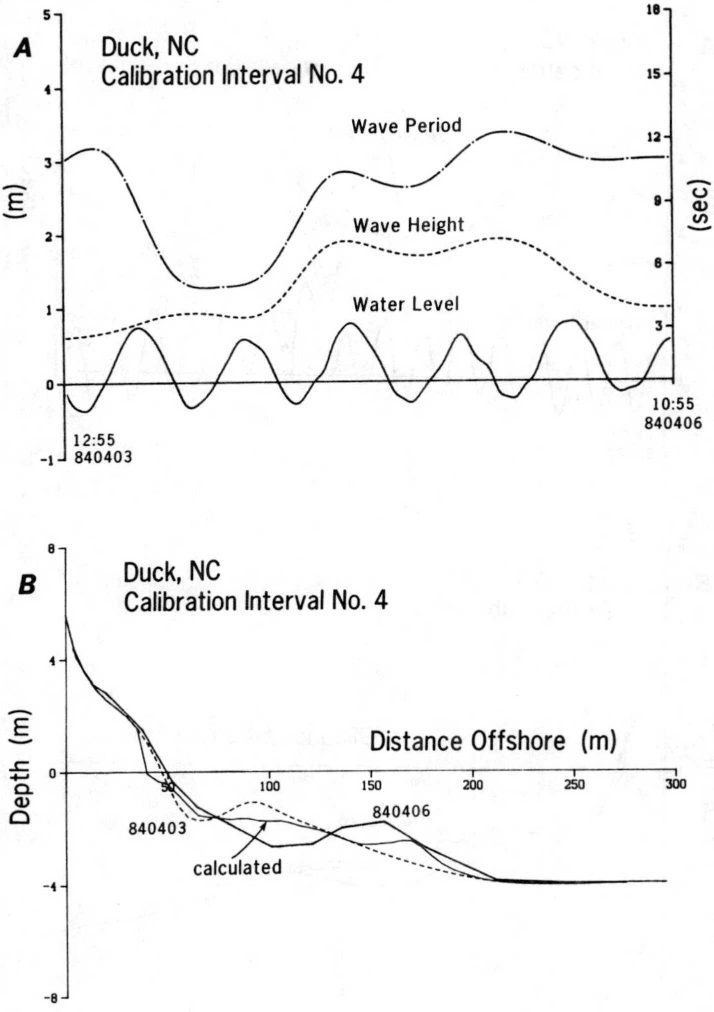

Figure 8. Calibration of numerical model against field data from Duck for one event. Variation with time of: (a) H, T, and water level; and (b) profile simulation result.

The field-calibrated model was used to simulate profile change that took place during another storm event to verify generality of the prediction. The wave and water level input are shown in Fig. 9a and the simulation result in Fig. 9b. The model reproduced main changes in the beach profile that occurred during the storm, with both bars moving offshore. Movement of the inner bar was overestimated, whereas the outer bar was located correctly but had a smaller volume than measured. In Fig. 9b the difference in measured beach volume was 45 m^3/m (a loss in beach volume, constituting 25% of the total absolute volume moved across the profile). This difference is attributed mainly to differentials in longshore sand transport.

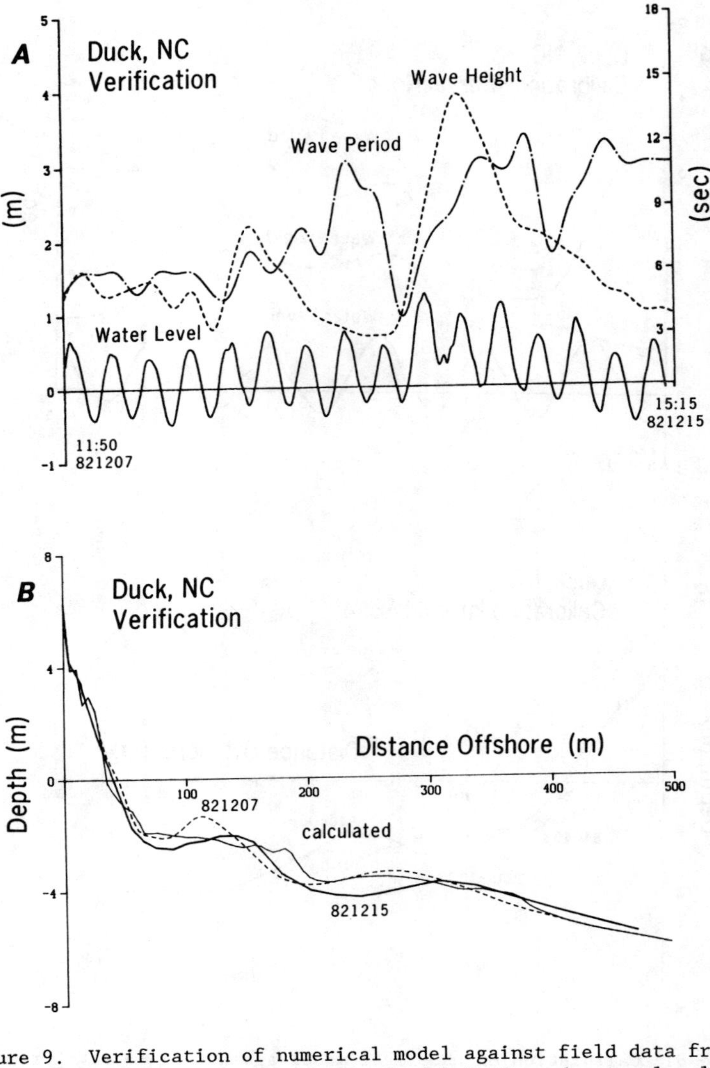

Figure 9. Verification of numerical model against field data from Duck for one event. Variation with time of: (a) H, T, and water level; and, (b) profile simulation result.

The importance of using varying forcing conditions in field simulations is illustrated in Figs. 10a, b, which show the verification condition calculated with (a) no change in water level and varying measured waves, and (b) average waves and no water level variation, respectively. The two calculated bars appearing in Fig. 10a correspond to the two peaks in the wave height time history. The bar shown in Fig. 10b was generated under five days of constant waves and water level, and is also unrealistic. The simulated results with constant forcing conditions shown in Figs. 10a, b are considerably inferior to the result obtained with varying waves and water level (Fig. 9).

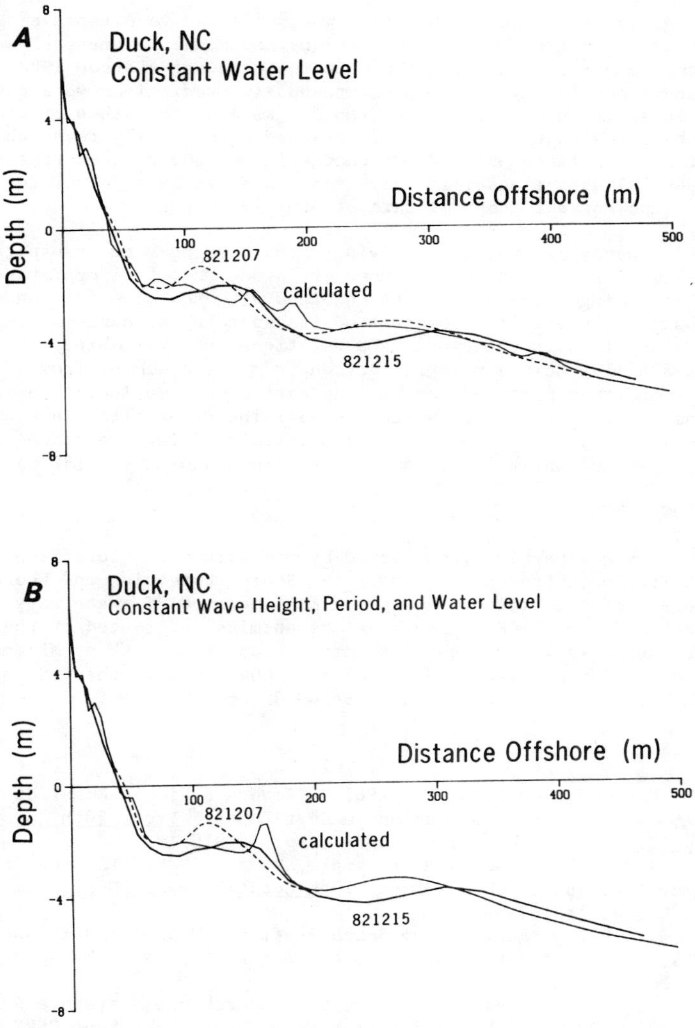

Fig. 10 Model prediction and field measurement: (a) omitting water level variation; and (b) omitting variation in H, T, and water level.

CONCLUDING DISCUSSION

A large data set comprised of 42 cases from two independent prototype-scale tank experiments was used to develop an empirical model of net cross-shore sand transport and beach profile change. The model reproduced bar formation and growth in the tank experiments, which involved monochromatic waves, and it performed well in a severe test to reproduce measured bar movement in the field over five separate simulations, each encompassing events of 3- to 12-day duration. The field comparisons were severe as all profile change, wave, and water level data were used directly as measured, and only one model parameter was adjusted in calibration.

Although not shown here, the model was subjected to extensive sensitivity analysis with input conditions and model parameters varied beyond the range of values available in the data set (Larson 1988, Larson and Kraus 1988c). Reasonable trends in predictions were always found. In addition, the model was run for several thousands of time steps; the calculated profile always reached a physically reasonable shape at earlier times and did not change in subsequent thousands of time steps. Thus the model is very stable and can be expected to reproduce the correct temporal rate of profile change.

Breaking waves are the sole driving force causing sand transport in the model in its present state. However, since the model operates in a general way using a transport rate combined with the material conservation equation, in principle other sand transporting mechanisms could be incorporated if their transport rate relations are available. These mechanisms could include wave reflection from the beach or from seawalls, and transport induced by long-period wave motion. The model is economical to run and has performed well in test applications to simulate month-long adjustment of beach fill involving storm and recovery wave and water level conditions (Kraus and Larson 1988b).

ACKNOWLEDGEMENTS

This study was partially supported by the work unit "Surf Zone Sediment Transport Processes," under the Shore Protection and Restoration Program of the U.S. Army Corps of Engineers, CERC, Waterways Experiment Station. NCK appreciates the permission granted by the Chief of Engineers to publish this information. The work by ML was carried out under contract No. DAJA45-86-C-0046 through the U.S. Army Research, Development and Standardization Group - UK.

REFERENCES

Basco, D. R., and Yamashita, T. 1987. "Toward a Simple Model of the Wave Breaking Transition Region in Surf Zones," Proc. 20th Coastal Engrg. Conf., Am. Soc. Civil Engrs., pp 955-970.

Dally, W. R., Dean, R. G., and Dalrymple, R. A. 1985. "A Model for Breaker Decay on Beaches," Proc. 19th Coastal Engrg. Conf., Am. Soc. Civil Engrs., pp 82-98.

Dean, R. G. 1977. "Equilibrium Beach Profiles: U.S. Atlantic and Gulf Coasts," Dept. of Civil Engrg., Ocean Engrg. Rep. No. 12, Univ. of Del., Newark, DE.

Howd, P. A., and Birkemeier, W. A. 1987. "Beach and Nearshore Survey Data: 1981-1984 CERC Field Research Facility," Tech. Rep. CERC-87-9, Coastal Engrg. Res. Center, U.S. Army Eng. Waterways Experiment Station, Vicksburg, MS.

Kajima, R., Shimizu, T., Maruyama, K., and Saito, S. 1983a. "Experiments of Beach Profile Change with a Large Wave Flume," Proc. 18th Coastal Engrg. Conf., Am. Soc. of Civil Engrs., pp 1385-1404.

Kajima, R., Saito, S., Shimizu, T., Maruyama, K., Hasegawa, H., and Sakakiyama, T. 1983b. "Sand Transport Experiments Performed by Using a Large Water Wave Tank," Data Rep. No. 4-1, Central Res. Inst. for Electric Power Industry, Civil Engrg. Div. (in Japanese)

Kraus, N. C., and Larson, M. 1988a. "Beach Profile Change Measured in the Tank for Large Waves, 1956-1957 and 1962," Tech. Rep. CERC-88-6, Coastal Engrg. Res. Center, U.S. Army Eng. Waterways Experiment Station, Vicksburg, MS.

Kraus, N. C., and Larson, M. 1988b. "Prediction of Initial Profile Adjustment of Nourished Beaches to Wave Action," Proc. Beach Preserv. Techn. '88, Florida Shore and Beach Preserv. Assoc., in press.

Kriebel, D. L., and Dean, R. G. 1985. "Numerical Simulation of Time-Dependent Beach and Dune Erosion," Coastal Engrg., Vol 9, pp 221-245.

Larson, M. 1988. "Quantification of Beach Profile Change," Rep. No. 1008, Dep. of Water Resources Engrg., Inst. of Science and Technology, Univ. of Lund, Lund, Sweden.

Larson, M., and Kraus, N. C. 1988a. "Beach Profile Change, 1: Morphology," submitted for publication.

Larson, M., and Kraus, N. C. 1988b. "Beach Profile Change, 2: Net Cross-Shore Sand Transport Rate," submitted for publication.

Larson, M., and Kraus, N. C. 1988c. "Beach Profile Change, 3: Numerical Model," submitted for publication.

Moore, B. D., 1982. "Beach Profile Evolution in Response to Changes in Water Level and Wave Height," Unpubl. M.S. Thesis, Univ. of Del., Newark, DE.

Saville, T. 1957. "Scale Effects in Two Dimensional Beach Studies," Trans. from the 7th General Meeting of the Intern. Assoc. of Hydraulic Res., Vol 1, pp (A3)1-10.

Svendsen, I. A., Madsen, P. A., and Buhr Hansen, J. 1979. "Wave Characteristics in the Surf Zone," Proc. 14th Coastal Engrg. Conf., Am. Soc. of Civil Engrs., pp 520-539.

CHAPTER 97

LOCAL SCOUR AROUND CYLINDRICAL PILES DUE TO WAVES AND CURRENTS COMBINED

by
Yoshiaki KAWATA* M.ASCE and Yoshito TSUCHIYA** M.ASCE

ABSTRACT

Local scour depth around a pile due to waves only is small, but under waves and weak currents combined it became large. If the flow velocity increases, the maximum scour depth increases and may approach to that by running water. In the processes of the scouring, Kármán vortex play an important role in deciding characteristics of a scour hole. In the case of the horseshoe vortex predominant, the maximum scour depth is found at the upstream end of a pile, but in the coupled field of waves and currents, the initial scouring position is a little apart from a pile in the offshore side and make an angle of about 45 degree in the shoreward direction. This depends on a vortex shedding from a pile. By using light weight aggregate, the effect of ripples on local scour depth is made clear.

1. INTRODUCTION

In recent year, we have a lot of various coastal structures in the shallow water, such as artificial reclaimed islands, access bridges, breakwaters and piers for fishing and field observation. When these structures are designed, many factors for desirable construction must be taken account. One of them is local scour around cylindrical piles of pier type structures.

Some studies on local scour due to waves have been made in hydraulic experiments and in field observations, as reviewed by Herbich et al.(1984). Local scour depth due to waves is not so large in comparison with that due to uni-directional flow. As to the experimental results obtained by Herbich et al. and through our preliminary tests, conditions of waves and weak currents combined makes scour depth large in comparison with that only by waves. The conditions can be usually seen in the nearshore zone in which several currents such as longshore currents, rip currents and undertow coexist with waves. The scouring mechanism is not so clear, therefore, it is very difficult to predict the scour depth and area around a cylindrical pile in the nearshore environment.

* Assoc. Prof., Disaster Prevention Research Institute, Kyoto University, Goka-sho, Uji, Kyoto 611, Japan
** Prof.,DPRI, Kyoto University

Studies on local scour due to currents only have been made by Shen et al.(1969), Raudkive et al.(1983) and other many investigators in river engineering. The works show that the horseshoe vortex is a dominant factor to determine the characteristics of local scour around a pile in uni-directional flow. Moreover, two types of scour may be identified as follows: (1) Clear water scour where sediment is removed from the scour hole and not replenished by the approach flow; and (2) Live-bed scour where sediment is continuously supplied into the scour hole by the approach flow. Their experiments also suggested that the scour depth increases with increase in mean water velocity in the clear water regime and reaches a maximum at a velocity approximately equal to the threshold velocity with which bed sediment is no longer removable from the hole. With live-bed scour, an equilibrium is attained when the rate of sediment transport into the scour hole by the approach flow is equal to the rate of sediment removed from the hole. The equilibrium scour depth in the live-bed scour regime is usually smaller than the maximum scour depth in the clear water scour regime. These characteristics are not clear yet under the condition of waves and currents combined. Moreover, the enlargement of the scour hole with time have not studied in detail, so that the limitation of the data keeps the physical modelling of scouring incomplete.

The objectives of this investigation are to examine processes of local scour around a cylindrical pile due to waves and currents combined in the regimes of clear water scour and live-bed scour and to make clear the effect of ripples on the characteristics of scouring.

2. EXPERIMENTS ON LOCAL SCOUR

2.1 Experimental apparatus and methods

Hydraulic experiments on local scour around a cylindrical pile were carried out in a recirculating wave flume at the Ujigawa Hydraulics Laboratory, Disaster Prevention Research Institute, Kyoto University. The experimental apparatus is shown in Fig.1 . The flume used was 40m long , 50cm wide and 65cm deep with a piston type wave generator at the end. This is a double-deck type flume in which rate of mass transport in waves in the upper deck can return in the offshore

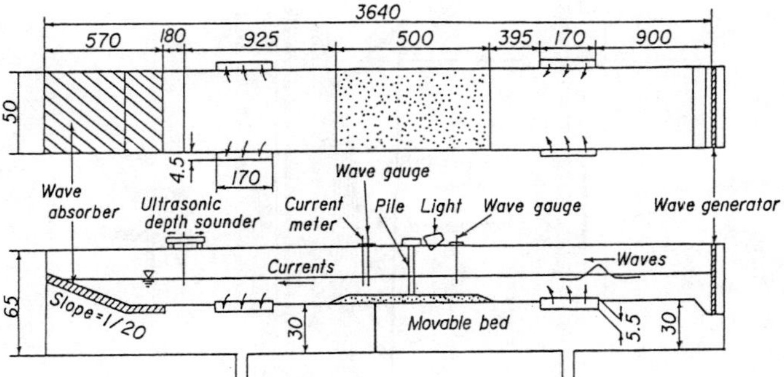

Fig. 1 Experimental apparatus and wave flume.

direction in the lower one. Pump system equipped in the flume can generate recirculating currents. We set a cylindrical pile in the midst of the movable bed of 5m long. Two kinds of sediment were used, i.e., one is very fine sand with a mean diameter of 0.11mm and its density of 2.65gr/cm³, and another is light weight aggregate whose diameter and density are 0.85mm and 1.81gr/cm³ respectively. Under the usual experimental wave condition, ripples are not formed on the movable bed of the light weight aggregate. Two capacitance type wave gauges and an electro-magnetic current meter were used.

All experiments were run at a depth of 10cm on the movable bed and with a constant wave period of 1.4s. The wave heights were changed 0 to 6cm systematically and the mean flow velocities were 13.4, 12.1, 6.71, 0 and -12.0cm/s, in which sign minus means reverse flow. Cylindrical piles of diameter 1.6, 2.0, 2.4, and 3cm were used. The depth of the local scour was measured continuously with a hundred plastic optical fibers of diameter 0.5mm and a high speed VTR as shown in Fig.2 . The number of the lights which pass through the optical fibers is equal to the depth of the local scour in real time. We measured the local scour depth both at the offshore and onshore sides of a pile. When ripples were formed on the movable bed, their heights and lengths were measured with a ultra-sonic depth sounder.

2.2. SCOUR PROCESSES

At first, the scour processes due to waves in our experiments are as follows: The local scour did not occur when the wave height was smaller than about 2cm in every piles used. As the wave height increases, scour holes were formed in the clear water scour regime. The scour due to Karman vortexes was observed at the bed nearly 2cm away from a pile and at an angle of nearly 45 degrees in the direction of wave propagation. We observed sediment suspension due to formation of the vortexes which were formed along the surface of the pile, and after that shed behind it. The vortex moved upward and downward in accompany with wave motion. The sediment movement is not so active at first and thus the scour depth is very shallow. The scour depth at the onshore side was generally larger than that at the

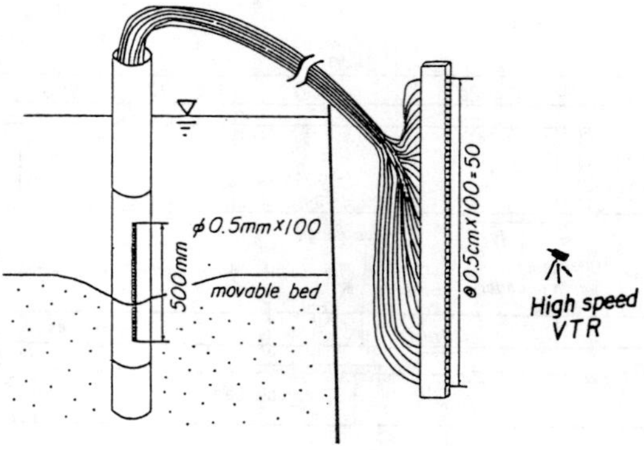

Fig. 2 Schematic diagram of a test pile with one hundred optical fibers and a high speed VTR.

offshore side because shoreward flow due to mass transport may strengthen a vorticity at the onshore side.

When the wave height was more than about 4cm, ripples were formed all over the surface of the movable bed. The depth of live-bed scour was nearly equivalent to the ripple height and the ripple length was usually of the same order of a pile diameter. Therefore, it was very difficult to estimate the scour depth. In this case, the sediment concentration near the bed generated by the vortex was very dense, but the scour depth is not so large.

Secondly, the scour processes due to currents are briefly explained. When the mean flow velocity was 6.71cm/s, local scour did not occur. When it was 13.4cm/s, scouring hole whose plan shape looked like a horseshoe was dug quickly. At the lee side of the pile, a slight accretion of sediment was recognized. When flow velocity was -12.0cm/s(reverse flow), the scour processes, of course, were very similar to the cases in which flow velocity was 13.4cm/s.

In the coupled field of waves and currents (u=6.71 cm/s), much sediment was easily swirled up with the vortexes generated behind a pile (onshore side), and the initial scouring position is slightly apart from a pile in the onshore side and make an angle of 45 degrees in the onshore direction as shown in Fig.3. This depends on the vortex shedding from a pile. The size and the depth of the hole were much larger than those due to waves only. As time went by, the scouring also occurred at the offshore side. After this stage, sediment which was swirled up at the offshore side of the pile was carried away due to currents, but some suspended sediment inside a scour hole can not be carried out with currents. Therefore, the former leads to net loss of sediment from the hole. This is a typical process of local scour enlargement at the offshore side of the pile. At the onshore side, the process of scouring is similar to that at the offshore side. However, some sediment which was flew up over the hole at the onshore side deposited on the edge of the scour hole at the onshore side. Behind the holes (onshore side), small ripples were formed.

In an equilibrium stage of live-bed scour, the shape of the scour hole was semicircular at the offshore side, and trapezoidal at the onshore side. When the wave height becomes larger than that at the beginning of ripple formation, it was very difficult to distinguish

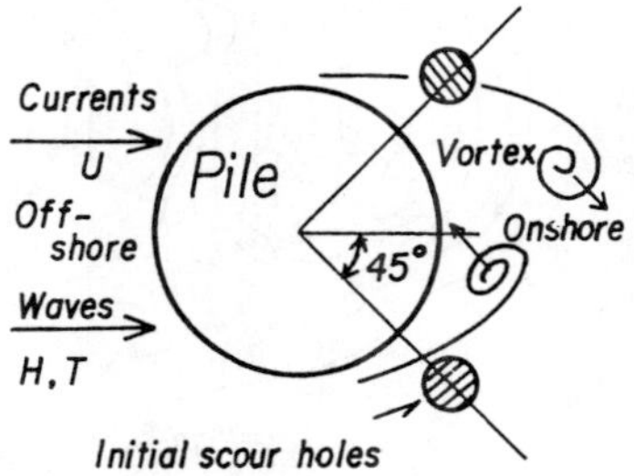

Fig. 3 Vortex shedding with a vertical axis and location of initial scouring.

between the trough of the ripple and the scour hole. The scouring processes in u=13.4cm/s resembled those in the cases of u=6.71cm/s, but the horizontal scales and the depth of the scour holes were much larger than the latter case. In the first stage, the vortex which looks like a horseshoe one contributes to local scour as well as the scouring process due to currents only. After this stage, the scouring advanced as same as that in the case of u=6.71cm/s. In some experiments, small ripples were formed at the onshore side of the scour hole. When wave height was larger, ripples were formed all over the bed. With the reverse flow (u= -12.0cm/s), the scour processes were very similar to the cases of those at the velocity of 13.4cm/s. The the shape of the scour hole in the former case, of course, was almost opposite to that in the latter case.

3. CHARACTERISTICS OF LOCAL SCOUR

Characteristics of local scour around a pile due to waves and currents combined are very different from those due to currents only. Using the data obtained in our experiments, we analyzed them with traditional method of dimensionless expressions.

3.1 Bottom topography of scour hole

Figure 4 shows the bottom topography of the scour hole given under the condition of wave period of 1.4s, wave height of 3.0cm, pile diameter of 3.0cm and current velocity of 13.4cm/s in the live-bed scour regime. At the offshore side of the pile, a hole like a reverse-circular corn is formed, and the offshore slope of the hole is an angle of 43 degrees and very uniform. This value is a little smaller than the angle of repose of the sand. At the onshore side, however, the plan shape of the scour hole looks like a trapezoid, and a narrow and flat zone of accretion is formed behind the pile (in the onshore side). In this area the shedding vortexes are being developed and bottom sediment are easily flown up. Mild bottom slope continues from this area to onshore side and at further onshore area, deposition of some sediment makes the bed flat and on the surface, small ripples are

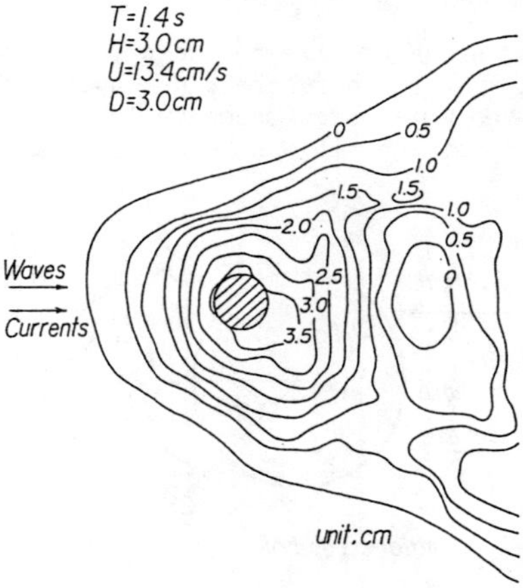

Fig. 4 Bottom topography of a typical scour hole.

generated. The lateral width of the scour hole increases shorewards with the distance from the pile. These characteristics depend on moving trajectories of the vortexes which shed from the pile and moves to onshore at an angle of 45 degrees in the onshore direction.

Figure 5 is a cross section of the scour hole along the longitudinal axis of the flume. The experimental conditions are as same as in the case of those in Fig. 4. At the offshore side of the pile , the angle of the bed slope is about 45 degrees. At the onshore side, the slope close to the pile is gentle, and with increase of the distance from the pile, it becomes steeper and approaches to 45 degrees. Fig. 6 shows a cross section of the scour hole along the transverse direction. The mean bed slopes both on the right and left sides of the pile make an angle of about 45 degrees to the horizontal bed and their shapes are nearly symmetric. It is found from our experiments that the bed slope of the scour hole is nearly equal to an angle of repose (it is about 45 degrees), but along the lines which make an angle of about 45 degrees in the shoreward direction as already shown in Fig.3, the slope is a little small due to active movement of sediment by vortexes.

3.2 Time development of scouring

Figure 6 illustrates some examples of the time development of scour depth z with time t in the case of a 3cm diameter pile with the parameter of wave height. The data may be each composed of three straight-line segments on a semilog graph paper. The first segment is associated with the slow scouring by the shedding vortexes. The middle segment describes the development of the scour hole as the vortexes move away from the pile and grow in strength. The last segment indicates an equilibrium stage. The development of scour in clear water scour is essentially as same as that in live-bed scour.

In the case of scour due to waves only, accretion of sediment is found at the offshore side of the pile. In the case of scour due to waves and currents at the velocity of 6.71cm/s, local scouring around a pile is so small that the effect of currents on local scour is very weak. This is due to a small amount of suspended sediment which is picked up by small waves. At a flow velocity of 13.4cm/s, local scour advanced much more in comparison with the other cases. This means that under the waves and currents combined, the local scour depth around a pile chiefly depends on the flow velocity.

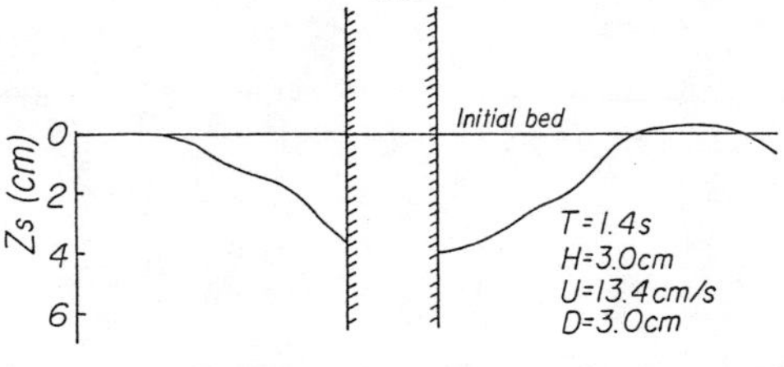

Fig. 5 Cross section of a scour hole in the longitudinal direction of the flume.

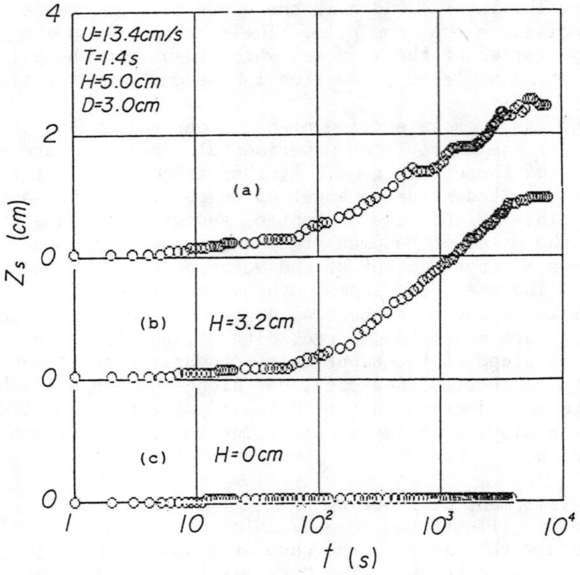

Fig. 6 Some examples of time development of clear water scour.

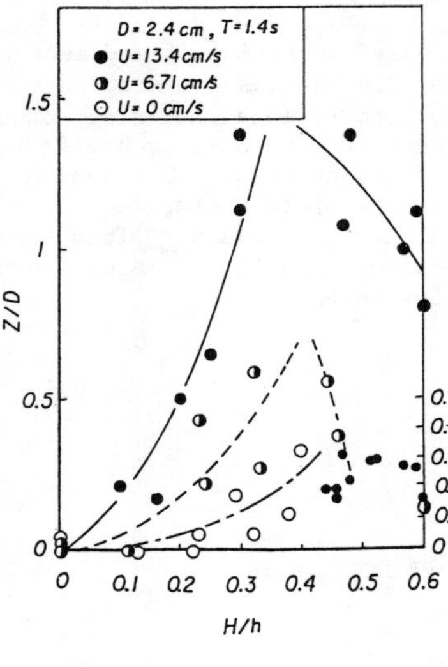

(a) D=2.4cm

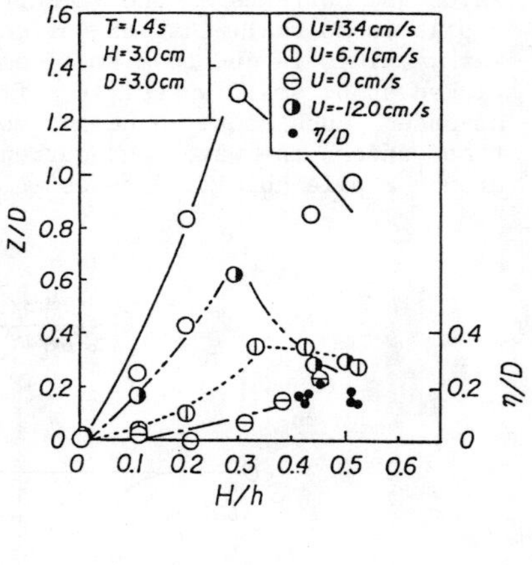

(b) D=3.0cm

Fig. 7 Maximum scour depth and ripple height vs. wave height.

3.3 Equilibrium maximum scour depth

Figure 7 shows the relationship between the dimensionless maximum scour depth and the dimensionless wave height in the cases of a pile diameter of 2.4 and 3.0cm respectively. Measuring point of the maximum scour depth is onshore side of a pile. In this figure, the dimensionless height of ripples which are formed in the dimensionless wave height of more than about 0.3 to 0.4 is also shown with small circles. With our experiments, the maximum scour depth is the largest under the conditions of the limit of beginning of ripple formation, i.e., the initial stage of live-bed scour regime. In clear water scour, the scour depth increases with increase of H/h. When H/h is more than about 0.3 (In live-bed scour regime in which movable bed is covered with ripples.), the maximum scour depth is as large as or slightly larger than the height of ripples except for the case of u=13.4cm/s. Our experiments shows that the length of ripples are larger than a diameter of a pile. Therefore, it is difficult to take away the local depression corresponding to trough of ripples from local scour depth. In the case of u=13.4cm/s, the scour process quickly advances due to large flow velocity, and thus the outflow rate of sediment transport from the scour hole is larger than those in other cases. Therefore, the maximum scour depth becomes large.

In Fig.8, the relationship between dimensionless radii of scour hole both in the onshore and offshore directions. The line in this figure indicates empirical relationship corresponding to each flow velocity. From Fig.8, the dimensionless radius of scour hole in the onshore direction is generally greater than that in the offshore direction. As the flow velocity increases, the horizontal size of the scour hole becomes large, as shown in Fig.4 and also the asymmetry of the plan shape of the scour hole in the longitudinal direction advances. This phenomenon can be explained with the characteristics of a vortex around a pile. The vortex which firstly sheds at the offshore side of the pile in a half cycle of wave period, moves to the onshore

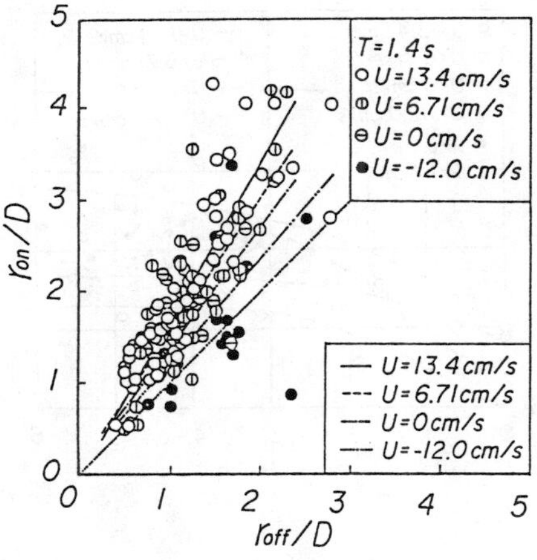

Fig. 8 Relationships between onshore and offshore radii of a scour hole.

side with growth in the next half cycle. Therefore, pick up rate of sediment at the onshore side of a pile is larger than that in the offshore side. In the case of waves only as shown in Fig.7, the onshore radius of a scour hole is also larger than the offshore one, though the scour hole is smaller than that under waves and currents combined. In the case of reverse flow at a velocity of -12.0cm/s, the data are much scattered in comparison with other cases. Experimental observation shows that difference between shoreward water particle velocity and seaward one under wave motion (nonlinear waves) contributes to this small asymmetry of the scour hole.

Figure 9 reveals the relationship between the dimensionless mean radius of the scour hole and the dimensionless wave height with the value of flow velocity as a parameter. The mean radius is defined as a half of the sum of the onshore and offshore radii of the scour hole. In this figure, the dimensionless ripple length is also shown. From the figure, it is found that the radius of the scour hole increases with increase of the wave height and the flow velocity. When wave height becomes larger and ripples are formed, the ripple length is as large as the mean diameter of a scour hole, and changes of ripple length with H/h is approximately equal to those in a scour hole. Therefore, in any case, the ripple length is larger than a diameter of a pile. Under these conditions, it is very difficult to estimate quantitatively a real scour depth. Taking scales of ripple length and a pile diameter in the field into account, their relationship is quite different from the experimental one, even if a ripple length depends on sediment diameter and an amplitude of water particle velocity. When it is necessary to predict a local scour depth around a pile in the field, the effect of ripples on local scouring process in the experiment should be made clear.

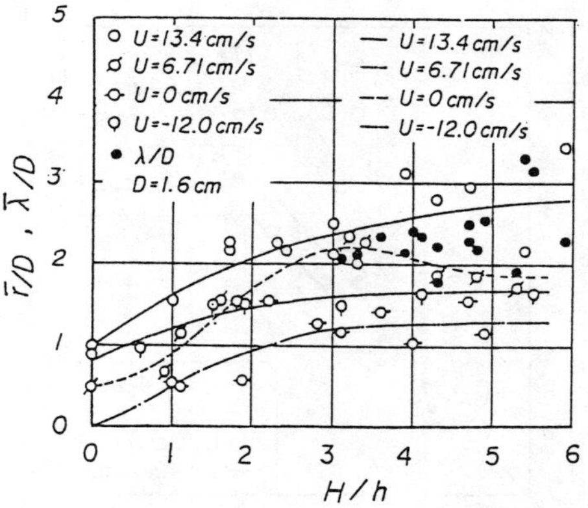

Fig. 9 Radius of scour hole and ripple length vs. wave height.

4. EFFECT OF RIPPLES ON LOCAL SCOUR

In our experiment with sand, ripples are always formed in live-bed scour regime. In order to obtain real local scour depth, it is necessary to exclude the contribution of ripple height. In the field, the pile diameter of a pier-type coastal structure is usually larger than a length of ripples which are appeared on the sea bottom, so that a local scour depth is larger than a ripple height. Therefore, there is a scale effect in the experimental data. In this chapter, we discuss the effect of ripples on local scour with the experimental data in no-ripple and live-bed scour regime which is obtained by using light weight aggregate.

Figure 10 shows the relationship between dimensionless local scour depth and dimensionless wave height at a velocity of 13.4cm/s in no-ripple regime. The critical dimensionless wave height between clear water scour and live-bed scour is around 0.35. After over this value, the scour depth also increases and rapidly decreases in the range of more than 0.42. In this figure it is not recognized the effect of wave period on local scour depth.

Figure 11 also shows the local scour depth in the two cases of no-ripple and ripple regime. In ripple regime, the data(open circle) are scattered within the area which is get between two dotted curves. The changes of local scour depth with wave height is similar to those of ripple height. Therefore, the scattering of the data might be not due to experimental error but due to existence of ripples. If the crest of ripples coincides with the location of a pile, local scour depth nominally becomes small. On the contrary, the trough comes, it increases. On the basis of our experiment, the effect of ripples on local scour is appeared due to following two mechanisms:

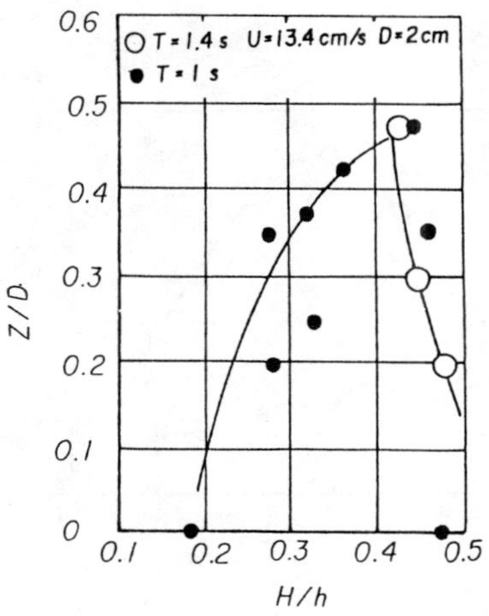

Fig. 10 Maximum scour depth vs. wave height in no-ripple regime.

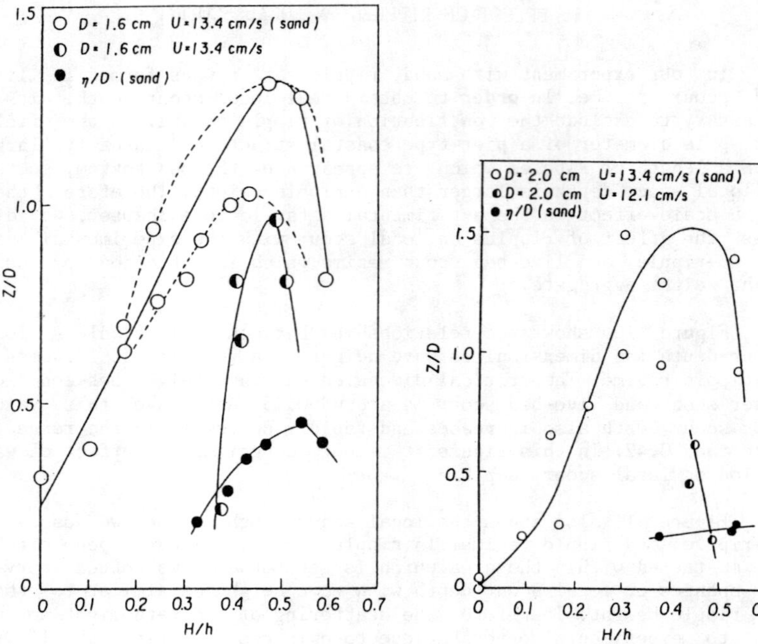

(a) D=1.6cm (b) D=2cm

Fig. 11 Comparison of maximum scour depth in ripple and no-ripple regime.

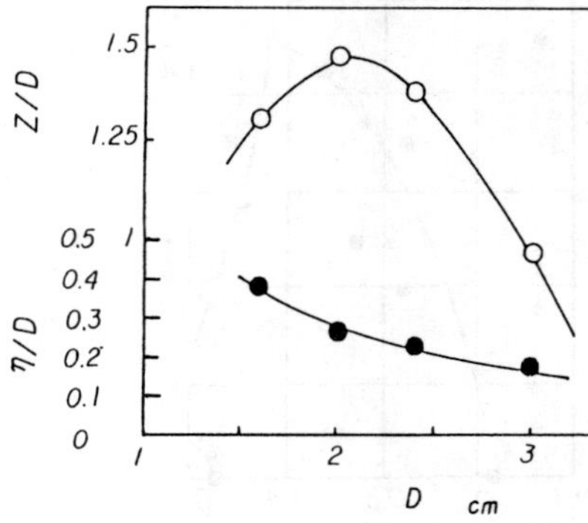

Fig. 12 Changes of scour depth and ripple height with a pile diameter.

(1) When ripples are generated on the bottom, the inflow rate of sediment transport into the scour hole reduces in comparison with that on a plane bed, because ripples spend certain energy of waves and currents.
(2) Suspended sediment cloud over ripples is easily spread out in the downstream area, therefore, transported sediment can not always be trapped with the scour hole.

From Fig.11, it is found that the dimensionless maximum local scour depth increases with increase of a pile diameter in ripple regime, but in no-ripple regime, this tendency become opposite. Moreover, the formation and characteristics of shedding vortexes are much affected with a pile diameter. Figure 12 show the changes of maximum scour depth and ripple height with a pile diameter.

5. MODELING OF LOCAL SCOUR

The most successive expression to explain mechanism of local scour around a cylindrical pile by running water was given by Shen et al.(1969) who introduced the concept of circulation. According to the conservation law of circulation, changes of bed shear stress in horse-shoe vortex region were obtained with development of scouring. This model have been modified by many researchers and some formulae are proposed to estimate the maximum scour depth under clear water scour. The horseshoe vortex with a almost horizontal axis is continuously generated along the sides of a pier.

Under wave motion, however, the horseshoe vortex is not clearly formed around a pile and the scour depth is not so large in comparison with that by running water. In the research field of wave forces, effects of a vortex formed with a vertical axis on lift force have been investigated. Through our experimental observation, Kármán vortex is a dominant factor to decide mechanism of local scour around a pile. Bottom sediment is easily picked up by the vortex and scoring process advances. Our physical model of local scour due to waves and currents combined introduces the effects of the vortex on sediment transport from scour hole by us(1987). Numerical simulation can roughly estimate the maximum scour depth. However, our model introduces many assumptions to predict the maximum scour depth due to incomplete knowledges about circulation of karman vortex and the formula of rate of sediment transport in the coupled field of waves and currents. We will reserve it for another occasion.

6. CONCLUSIONS

We investigated experimentally the characteristics of local scour around a cylindrical pile due to waves and currents combined, and made clear the effect of ripples in both clear water scour and live-bed scour regimes.

The main conclusions in our study can be summerized as follows:

1. Local scour depth around a pile due to waves only is small, but under waves and currents combined it became larger. If the flow velocity increases, the maximum scour depth increases and may come up to that in the steady flow. In the processes of the scouring, a vor-

tex with a vertical axis (Karman vortex) is dominant to decide characteristics of a scour hole. In the case of the horseshoe vortex in the steady flow, the maximum scour depth is found at upstream end of a pile, but in the coupled field of waves and currents, the initial scouring position is a little apart from a pile in the offshore side and make an angle of about 45 degree in the shoreward direction. This depends on a vortex shedding from a pile. It is found that the equilibrium scour depth around a pile in clear water scour is generally larger than that in live-bed scour.

2. By using light weight aggregate, the effect of ripples on local scour depth is made clear. In no-ripple regime, the scour depth suddenly decreases over a critical wave height due to increase of inflow rate of sediment transport. On the contrary, in ripple regime, suspended sediment cloud is not always trapped by a scour hole, so that its depth remains to be a certain value. The scattering of the data about scour hole depth might depend on the existence of crest and trough of ripples.

REFERENCES

Herbich, J.B., R.E. Schiller, R.K. Watanabe and W.A. Dunlap(1984). Sea Floor Scour, Design Guidelines for Ocean-Founded Structures, Ocean Engineering, Marcel Dekker, Inc., pp.145-201.

Raudkive, A.J., and R. Ettema(1983). Clear-water scour at cylindrical piers, Journal of the Hydraulic Engineering, Proceeding of ASCE, Vol.109, No.HY3, pp.338-350.

Shen, H.W., V.R. Schnider and S. Karaki(1969). Local scour around bridge piers, Journal of the Hydraulics Division, Proceedings of ASCE, Vol.95, No.HY6, pp.1919-1940.

Tsuchiya, Y., Y. Kawata and M. Nishikori(1987). Local scour around a cylindrical pile due to waves and currents combined, Proc. 34th Japanese Conf. on Coastal Eng., JSCE, pp. 402-406(in Japanese).

CHAPTER 98

STUDIES FOR VILLAJOYOSA'S BEACH REGENERATION (SPAIN)

José L. Campello Chorro *
Vicent J. de Esteban Chapapría **
José Aguilar Herrando ***
J. Javier Díez González **

ABSTRACT

In the last two years a whole of studies was realized in order to determine precise solutions to the regeneration of Villajoyosa's beach, in the spanish mediterranean coast. Investigations were carried out to the surrounding coastal areas based in field investigations and laboratory analyses of the beaches materials.

INTRODUCTION

In the last years, the spanish mediterranean coast has been subjected to a great deal of tourists. This supposed an increase in the population seasonal and permanent, wich has been solved with a lot of urban development and an intensification in uses of all types, wich have produced, and will continue producing, negative results on the coast. In many cases, these are superior than some produced in portual works. In must cases, the human activity developed in the coastal area goes well beyond geomorphologic processes. Alicante province is one of the coastal area in wich this demand has been present because its special climatologic and landscape conditions.

SITUATION AND CHARACTERISTICS

Villajoyosa village is placed between Alicante and Benidorm (fig. 1), finding a position around Amadorio river's mouth. The existence of a locked delta with a certain disymmetry is checked in this mouth. This river, with a tipical mediterranean regime, is today regulated by Amadorio dam, placed about four kilometers upstream from the mouth. Between this and the port, esentially fishing and sporting, takes place the Villajoyosa's

- - - - - - - - - - - - -
*Ingeniero Director Grupo de Ptos. Sur. ALICANTE (SPAIN).
**Profs. Dto. Ocean. e Ing. Costas. Un. Pol. MADRID (SPAIN).
*** Prof. Dto. Puertos. Univ. Pol. Valencia (Spain).

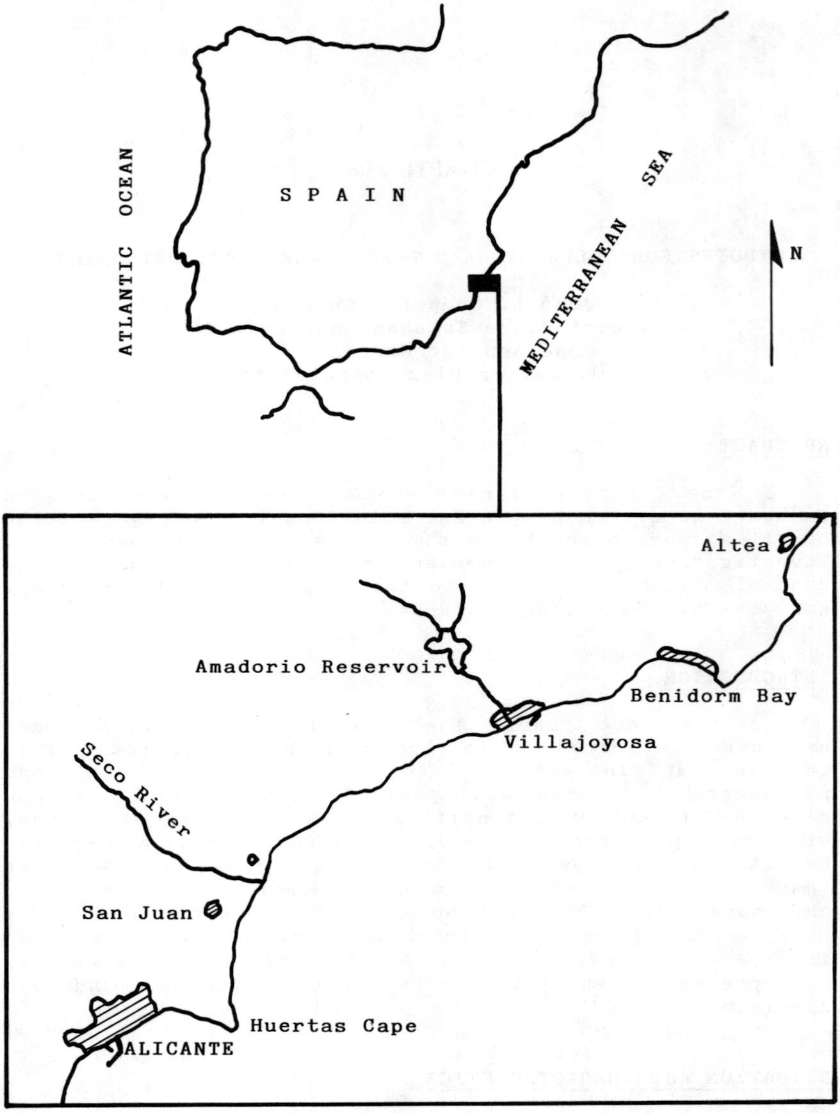

FIG. 1

beach integrated by hard grainsgravel, crashed stone and coble in most cases calcareus and ofitic. Its longitude is about 1400 m., and it was used to strand ships in old ages, according to a picture pated in the first years of the 19th Century, and wich is today found in the Naval Museum (Madrid), being used Alcocó Cala to search.

The line of this beach is not rectilinean, since, as well as its general curvature shoped like a big cave, the existence of reef and rocky outcroops have induced small hemitombol.

The beach has suffered a continous erosion in the last years, getting important damages in several works which have been done recently, specially in a seaside promenade with wich an important part of beach was occupied and in the way that merger the old maritime neighbourhood, in the old inner part of Villajoyosa with the port.

GEOMORPHOLOGIC FRAME

Separating Altea and Benidorm Bays we find Helada Mountain Range. It is a classic barrier of NE-SW direction formed in many cases by materials wich belong to Cretaceous period, with a submerged extension appearing in the small Benidorm Island. In the South there are Quaternary sandy sediments, especially the beaches of "levante" and "poniente" of Benidorm. Further away the change to limestone along the length of the coast which is developed to the South. Around Villajoyosa there is a whole, with more than 500 m. in power, of flischoid type, composed by claylу marls, levels of clay loams, bio-sandy and fosilifers limestones. All this are completed by a length of a rocky coast which gives way further away to the South to alow, backer coast, which have produced the presence of some deltas, such as Amadorio's, as we mentioned before, or gully Torre's one. Along the area there are small caves in which hard materials are found.

VILLAJOYOSA HARBOUR

The port was built in the place pointed in the mentioned picture of 1800. Villajoyosa harbour has been investigated because it can mean some disturbances to the surrounding area. The harbour was planned in 1917, June. Works, consisting in the breakwater construction starting at the Punta de L'Alcocó, began in 1923, March. After many paralysations and vicissitudes, works were nor finished to 1940 refering to the basic part. Beach has been gradually eroded after the breakwater construction, going materials into the harbour. Therefore a counter-dike was constructed.

In 1944 was planned a breakwater extension, and another one in 1952. The first one was disapproved, but the second was carried out, finishing the works in 1963 : 183 m. of new breakwater was constructed. Several storms (1943, February and December, 1947,January, 1965, January) pushed to make maintenance works. In 1984, January was planned a new extension of the harbour (Fig. 2) still not approved. Therefore predictions show negative influences to the near shoreline.

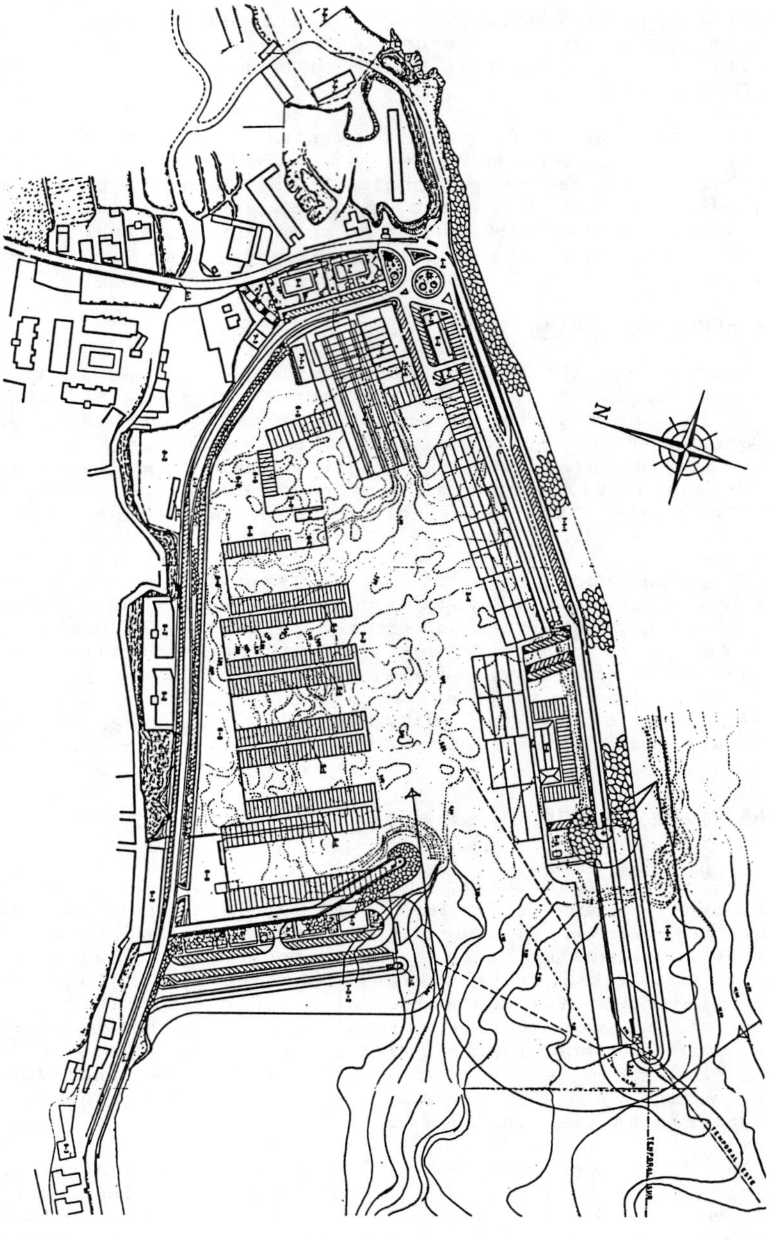

FIG. 2

STUDIES DONE AND CONCLUSIONS

With views to determine precise solutions to the regeneration of Villajoyosa's beach a whole of studies was realized referied to the determination of the coastal movement in our days, as well as in sedimentologic and batimetric analyses (Fig. 2, 3). On the other hand, we tried to study materials retained in Amadorio's Dam, but it was not possible get a quantification. It was investigated the influency in the beach erosion of the works realized that have been mentioned above. At last the possible sourcer of the precise materials to the beach regeneration were located, studying in this way the registered and enviromental impact in different means for the different alternatives.

Some of the studies developed were included in a general one of the Alicante coast. Studies show the independence of the coastal Villajoyosa area from near, especially with the north area. Works has investigated the clasification of the coastal zone in different "Morphodinamic coast units" of several orders, as a function considerating littoral dynamics, materials and geomorphologic frame.

In this case all that shows the Benidorm Bay connected to the Altea one as a result of the granulometric and mineralogic analyses. Nevertheless Villajoyosa area shows very different mineralogic compositions in the beach materials. That clearly mean another "Morphodinamic coast unit".It was been possible to demostrate have the materials of Villajoyosa's beach are inmerged with the contributions of Amadorio's River and also with streams which ran into this area which belongs to the second order "Morphodinamic coast unit" "Punta Plana Punta dels Banyets".

Coastal dynamic, with a clear tendency to the N-S, presents important partial components to the other direction, although they are inferior also the materials like autochtonous. Presence of high contents of anfibols and piroxen, and the longitudinal variation of the roundness grade in white quartz let confirm existence of real means of littoral transport towards the South around Amadorio's River mouth. However influence of its materials is losen quickly and this coastal area shows very clearly an important and general isolation as it has been studied widely considering different methods to determine littoral transport rates.

FRACCION RETENIDA TAMIZ ASTM Nº	50		
TAMANO DE PARTICULA(mm):	0,30-0,42	%RETENIDO:	49,4
FRACCION CARBONATADA :	97,1		
COMPOSICION MINERALOGICA EN % DE LA FRACCION NO CARBONATADA			
CUARZO BLANCO	63	TURMALINA	
CUARZO ROJO	9	EPIDOTA	
CUARZO AZUL		CIRCON	
CUARZO NEGRO		ANFIBOLES	8
MIXTOS CUARZO		PIROXENOS	
MOSCOVITA		OXIDOS	3
BIOTITA		OPACOS	
		RESTOS BIOLOG	17
MORFOLOGIA CUARZO BLANCO EN%		GRADO DE REDONDEO	
REDOND. 3	POCO ROD. 48	49	
MUY ROD. 46	NO ROD. 3		

FRACCION RETENIDA TAMIZ ASTM Nº	80		
TAMANO DE PARTICULA(mm):	0,18-0,30	%RETENIDO:	19,2
FRACCION CARBONATADA :	95,1		
COMPOSICION MINERALOGICA EN % DE LA FRACCION NO CARBONATADA			
CUARZO BLANCO	50	TURMALINA	
CUARZO ROJO	10	EPIDOTA	
CUARZO AZUL		CIRCON	
CUARZO NEGRO		ANFIBOLES	12
MIXTOS CUARZO		PIROXENOS	
MOSCOVITA		OXIDOS	4
BIOTITA		OPACOS	1
		MAGNETITA	*
REST.BIOLOG.	23	ILMENITA	*
MORFOLOGIA CUARZO BLANCO EN%		GRADO DE REDONDEO	
REDOND. 6	POCO ROD. 32	48	
MUY ROD. 48	NO ROD. 14		

FIG. 2 MINERALOGIC ANALYSES

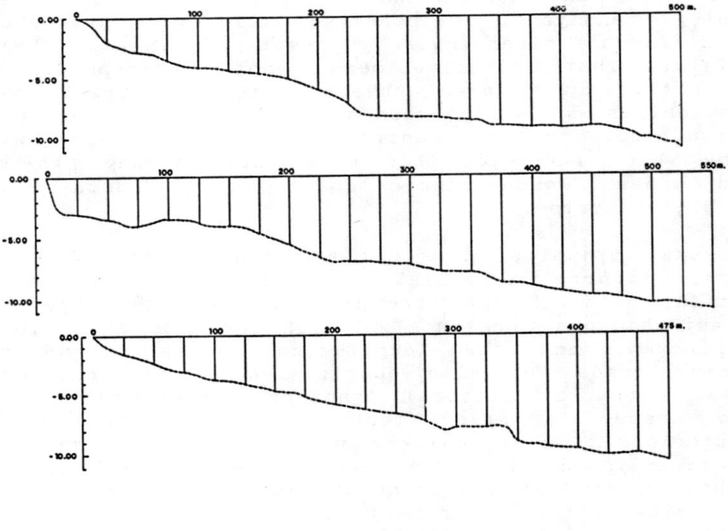

FIG. 3 VILLAJOYOSA'S BEACH PROFILES

REFERENCES

* ASENSIO AMOR,I. et al. (1973) : "Aspectos sedimentológicos del litoral de Alicante". Estudios Geológicos, Vol.XXIX, nº 3, pp. 253-264.

* DIEZ GONZALEZ, J.J. (1982) : "Bases para una aproximación a la dinámica litoral del País Valenciano". R.O.P. (Madrid).

* ESTEBAN CHAPAPRIA, V. (1987) : "Procesos litorales en las costas valencianas al Sur del Cabo de San Antonio". Tesis Doctoral. Univ. Politécnica de Valencia.

*LLORCA BAUS, C. et al. (1985) : "La vila del mar i el mar de la Vila". Exmo. Ayto. de Villajoyosa.

CHAPTER 99

COASTAL PROCESSES IN GUARDAMAR BAY (SPAIN)

Vicent J. de Esteban Chapapría *
J. Javier Díez González *
Miguel Arenillas Parra **
Rafael Cortés Gimeno ***

ABSTRACT

The surrounding coastal area of the Segura's River mouth has registered in the last years many erosion problems. It is one of the biggest coastal conjunct in the spanish mediterranean coast. A whole of studies was carried out to determine littoral and coastal processes in this area.

INTRODUCTION

As regard to the complete characterization of the conjunct coastal-littoral existing in Guardamar Bay some studies were realized some years ago to determine actuals and subactuals littoral processesin this area of the spanish mediterranean coast in order to the resolution of determined problems created around the Segura's River mouth.

Recently general occupation of the coastal area has generated deep disfunctions in many of them. Since that coast has experimented a deep revaluation due to tourism some works and actions have been got in general, which have not considered secondary effects that werw going to be done, producing erosive problems in a lot of coastal formations. Environmental degradation needs quick actions. However coastal arrangement requires exhaustive knowledge of coastal processes because its consideration lets have the reliability needed to undertake any action.

SITUATION. GEOMORPHOLOGIC

Guardamar Bay is placed in the great morphodinamic coastal unit which is between Santa Pola and Cervera Capes, in Alicante province. It constitute one of

*Depto. Ocean. e Ing. Costas. Univ. Pol. MADRID (Spain)
**Cát. Geol. Apl. Univ. Pol. MADRID (Spain)
***Cát. Geol. Apl. Univ. Pol. VALENCIA (Spain)

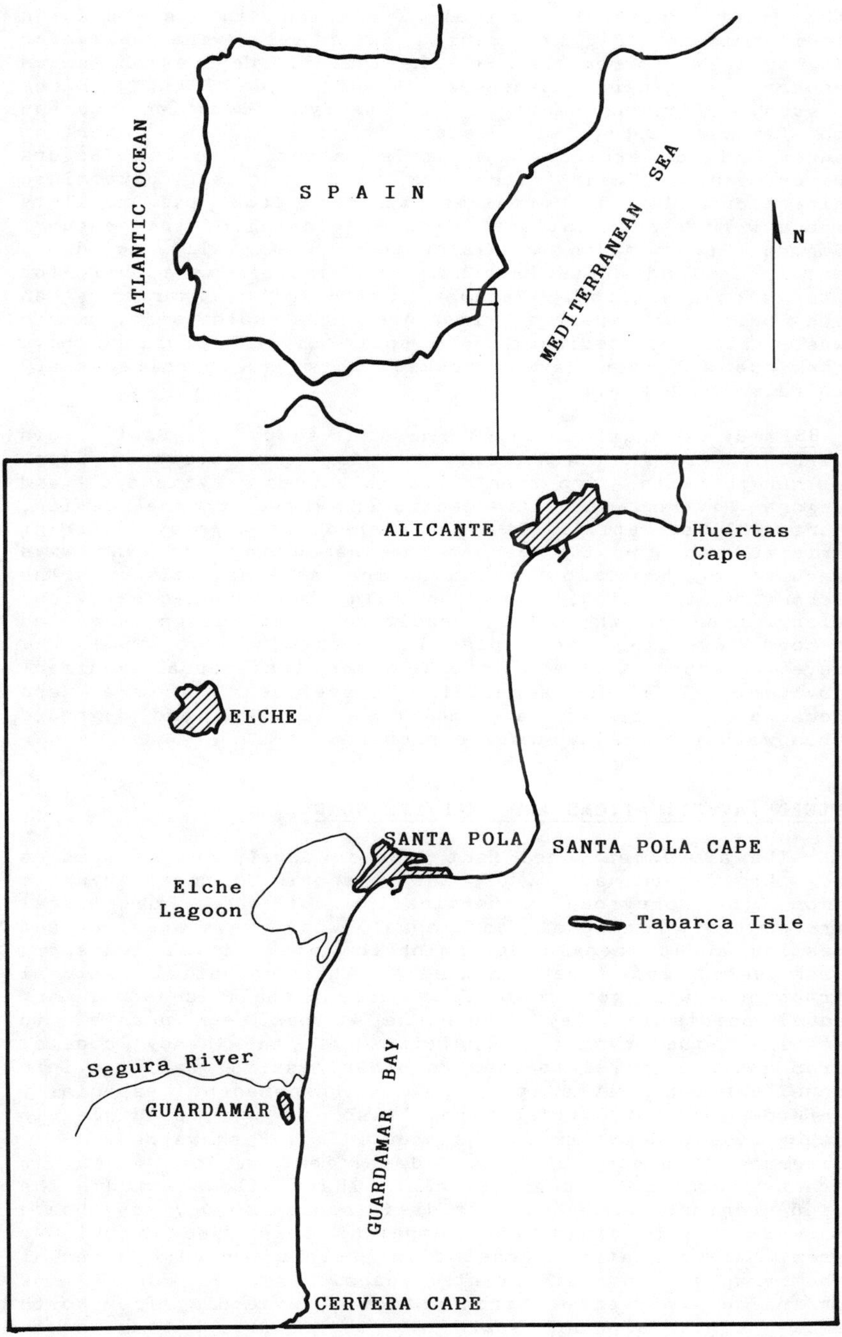
ATLANTIC OCEAN
S P A I N
MEDITERRANEAN SEA
N
ALICANTE
Huertas Cape
ELCHE
SANTA POLA
SANTA POLA CAPE
Elche Lagoon
Tabarca Isle
Segura River
GUARDAMAR
GUARDAMAR BAY
CERVERA CAPE

FIG. 1

the most relevant coastal formation in the spanish mediterranean (Fig.1). Behind it a very wide saltwater lagoonplace formed in Segura-Vinalopó depression known under the generic name of "Albufera de Elche" (Elche Lagoon). Uninterrumptly the sandy formation spread to Cervera Cape, orientated to the E. An important duner field expandes in this stretch around Segura River mouth. Great dunes are uned up along determined directions due to problems derived from its mobility there were reforestations in the beginning of the century. Segura River mouth was situated artifitially to drain easily lowland areas. Recently two jetties were built for its channelization. Because building a shorter than the built at present time over the soth side, mouth was filled of sediments. In spite of all problems have continue and even have increase, this have been necessary to raise it again.

Sandy formation start in the south of Santa Pola harbour almost uninterrumpted to the Cervera Cape, although with important singularness (Vinalopó and Segura Rivers, and its mentioned above channelization, Punta Pilas, etc.). It is refered to a coastal front oriented to the E. Beaches are named in different ways because of historic and situation seasons, although the formation is unique and with maintained characteristics. Along then a whole of traditional buildings used as second dwelling are placed longitudinally and they have supposed to be a problem for transversal sediment movement. This is especially grave near Cervera Cape because it makes up a completely built-up and degraded area with increasing erosive problems.

FIELD INVESTIGATIONS AND ANALYSES DONE

The above mentioned fact induced develpment of studies to the determination of the present littoral dynamic from the previous determination of the directional frecuential, seasonal and anual wind wave regime and results about theoric determination of littoral transport from were also taked initially. This potential littoral transport was got also from directional, seasonal and anual points of view. Therefore it has been possible to stablish the dynamic conditions of the sandy coastal from and the correlate them to other results, taking into consideration waves propagation phenomenon, especially refered to wave refraction. Also it was possible to study eolian deposits in the area : Fig. 2 shows rose diagram obtained by the classification of dunes in transverses and longitudinals. That allows study the wind regime. Also was studied sedimentology of beach materials and bathymetric campaigns were done in all the area, investigating coast line evolution from aerial photographs from differents years. Fig. 3 and 4 show grainsize analyses of beach material proceeding from north area (Fig. 3a) and south of the Segura River mouth (Fig. 3b). Mineralogic results (Fig. 4) are of sand strand of the second place mentioned above.

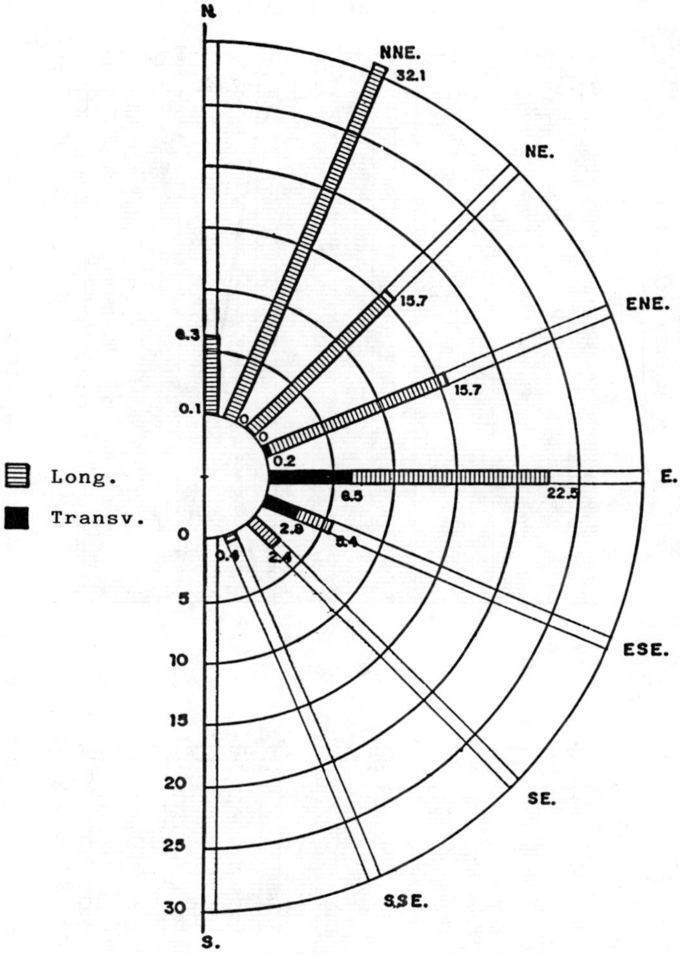

Fig. 2 Rose diagram of Guardamar dunes

GRAINSIZE ANALYSIS

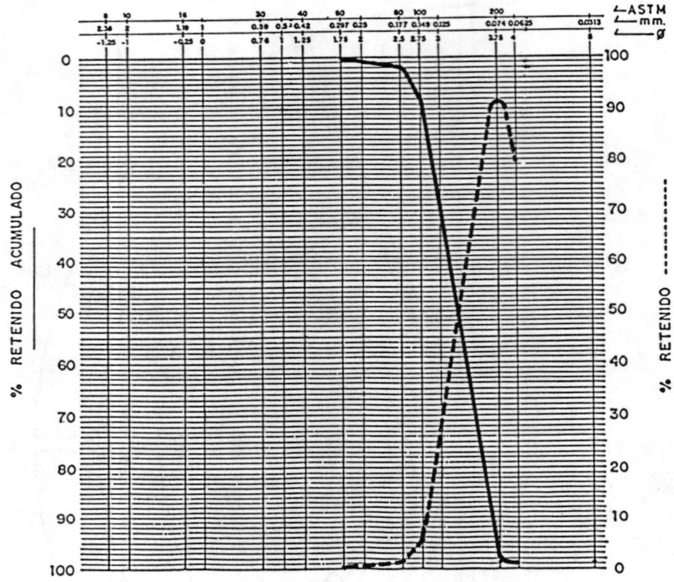

FIG. 3a

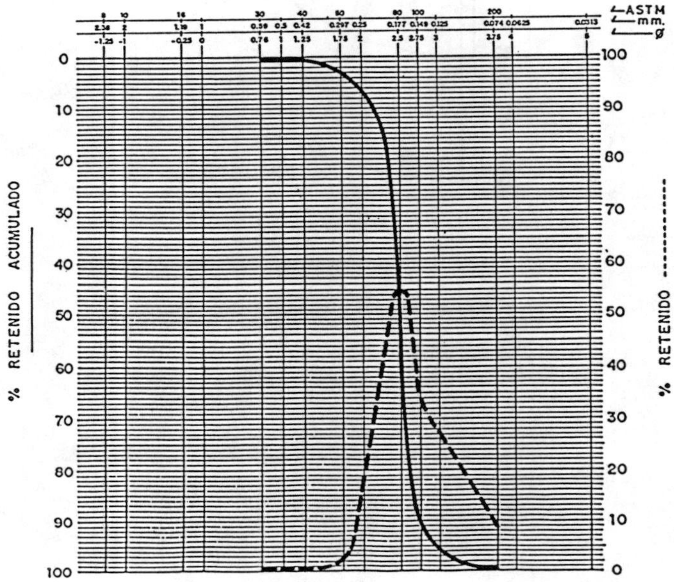

FIG. 3b

FRACCION RETENIDA TAMIZ ASTM Nº 50			
TAMANO DE PARTICULA(mm): 0,30-0,42		%RETENIDO: 3,39	
FRACCION CARBONATADA : 57,6 %			
COMPOSICION MINERALOGICA EN % DE LA FRACCION NO CARBONATADA			
CUARZO BLANCO	85	TURMALINA	
CUARZO ROJO	8	EPIDOTA	
CUARZO AZUL	5	CIRCON	
CUARZO NEGRO	2	ANFIBOLES	
MIXTOS CUARZO		PIROXENOS	
MOSCOVITA		OXIDOS	
BIOTITA		OPACOS	
MORFOLOGIA CUARZO BLANCO EN %		GRADO DE REDONDEO	
REDOND. 6	POCO ROD. 31	52	
MUY ROD. 53	NO ROD. 10		

FRACCION RETENIDA TAMIZ ASTM Nº 80			
TAMANO DE PARTICULA(mm): 0,18-0,30		%RETENIDO: 52,26	
FRACCION CARBONATADA : 53,4 %			
COMPOSICION MINERALOGICA EN % DE LA FRACCION NO CARBONATADA			
CUARZO BLANCO	84	TURMALINA	
CUARZO ROJO	10	EPIDOTA	
CUARZO AZUL	2	CIRCON	
CUARZO NEGRO	2	ANFIBOLES	*
MIXTOS CUARZO		PIROXENOS	
MOSCOVITA		OXIDOS	2
BIOTITA		OPACOS	
MORFOLOGIA CUARZO BLANCO EN %		GRADO DE REDONDEO	
REDOND. 11	POCO ROD. 40	48	
MUY ROD. 36	NO ROD. 13		

FRACCION RETENIDA TAMIZ ASTM Nº 100			
TAMANO DE PARTICULA(mm): 0,15-0,18		%RETENIDO: 32,31	
FRACCION CARBONATADA : 55,8 %			
COMPOSICION MINERALOGICA EN % DE LA FRACCION NO CARBONATADA			
CUARZO BLANCO	82	TURMALINA	1
CUARZO ROJO	13	EPIDOTA	
CUARZO AZUL	2	CIRCON	
CUARZO NEGRO		ANFIBOLES	1
MIXTOS CUARZO		PIROXENOS	
MOSCOVITA		OXIDOS	1
BIOTITA		OPACOS	
MORFOLOGIA CUARZO BLANCO EN %		GRADO DE REDONDEO	
REDOND. 13	POCO ROD. 26	54	
MUY ROD. 49	NO ROD. 12		

FRACCION RETENIDA TAMIZ ASTM Nº 200			
TAMANO DE PARTICULA(mm): 0,07-0,15		%RETENIDO: 11,95	
FRACCION CARBONATADA : 60 %			
COMPOSICION MINERALOGICA EN % DE LA FRACCION NO CARBONATADA			
CUARZO BLANCO	64	TURMALINA ANG.yR.	8
CUARZO ROJO	16	EPIDOTA	
CUARZO AZUL	5	CIRCON	
CUARZO NEGRO	3	ANFIBOLES	1
MIXTOS CUARZO		PIROXENOS	
MOSCOVITA		OXIDOS	3
BIOTITA		OPACOS	
MORFOLOGIA CUARZO BLANCO EN %		GRADO DE REDONDEO	
REDOND. 33	POCO ROD. 41	62	
MUY ROD. 23	NO ROD. 3		

FIG. 4 Mineralogic Analysis

WAVE REFRACTION AND LITTORAL TRANSPORT RATES

In order to understand the effect and influence of wave refraction it has been established six point along the area to determine results. Wave periods and heigths of 6, 8 and 10 sec., and 2, 6 and 8 m. were fixed respectively. A computer program was applied to find incident waves in deep water from S, SE, E and NE, and determine refraction coefficients and wave angle of incidence. Fig. 5 shows results in central point of the area.

Littoral transport rates were after calculated utilizing CERC methods :

$$Q = \sum \left[p_V x \left\{ 3.59 \times 10^6 \sum_H \left[(H_o \times K_r)^{5/2} p_{H_V} \text{sen } 2\alpha_b \right] \right\} \right]$$

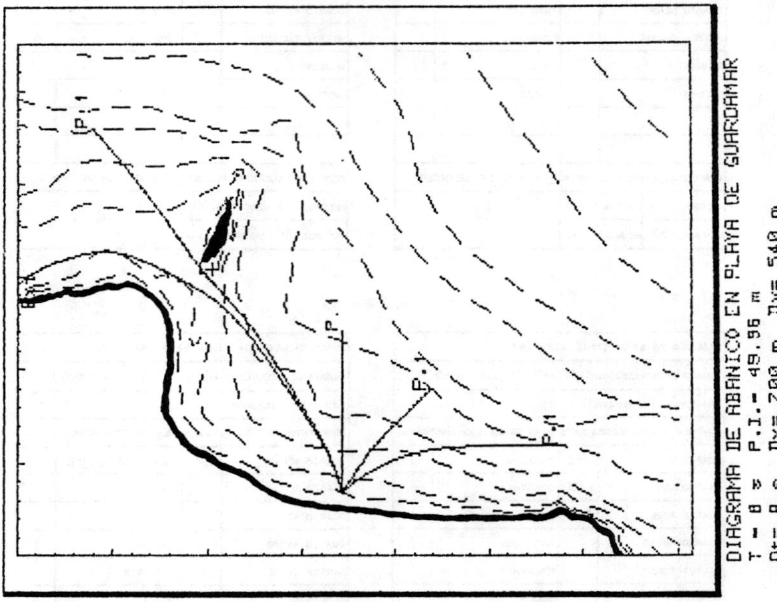

Ortogonal	Punto	Coord. X (m)	Coord. Y (m)	Prof. (m)	Ang. con DX+ (DEG)
1	1	17070.0	3405.0	4.8	45.0
1	142	28773.0	5932.3	50.0	-.7
2	1	17070.0	3405.0	4.8	58.0
2	87	21780.6	8945.6	50.0	45.3
3	1	17070.0	3405.0	4.8	86.0
3	100	17109.9	12074.7	50.0	90.8
4	1	17070.0	3405.0	4.8	99.0
4	274	3948.2	23006.1	50.0	130.5
5	1	17070.0	3405.0	4.8	99.5
5	300	44.7	14610.4	8.8	220.3

FIG. 5 Wave refraction diagram result

In addition, for the three southest points refered

it has been considered littoral transport induced by wave generation in Guardamar Bay.

Waves from E,SE and S reach the shoreline. Logically refraction is bigger in waves from SE and S. Waves from NE have a different way as a function of its characteristics : biggest reach the north area across seasurface between Santa Pola Cape and Tabarca Isle, but do not reach southest areas, except its extreme. It is due to the island. So there is a large stretch that it is not reached by waves from NE, these who would induce bigger littoral transport rates. Smaller waves reach the half north area across channel and in the south arrive waves from the Guardamar Bay. Estimated littoral transport rates are shown in Table 1.

TABLE 1

POINT	Q_1	Q_2	$Q(m^3/y)$
1. PLAYA LISSA	-281.800	-	-281.800
2. PLAYA PINET	-35.300	-	-35.300
3. PLAYA PINET (*)	+168.900	-	+168.900
4. PLAYA PESQUERAS	+442.400	-	+442.400
5. PLAYA GUARDAMAR	+221.400	+29.500	+250.900
6. PLAYA LA MATA	+199.000	+35.400	+234.400
7. CABO CERVERA	+209.000	+35.000	+244.000

(*) Were considered two orientations of the shoreline
Sign + indicates direction to the south.

REFERENCES

* AGUILAR HERRANDO, J. (1984) : "Naturaleza y distribución de las corrientes producidas por la rotura del oleaje". Tesis Doctoral. Universidad Politécnica de Valencia. Spain.

* DIEZ GONZALEZ, J.J. (1982) : "Bases para una aproximación a la dinámica litoral del País Valenciano". R.O.P. Madrid. Spain.

* DIEZ GONZALEZ, J.J. et al. (1982) : "Shore management in Santa Pola Bay (Alicante, Spain)". Proc. of Congress Int. Assoc. of Eng. Geol. New Delhi.

* ESTEBAN CHAPAPRIA, V. (1987) : "Procesos litorales en las costas valencianas al Sur del Cabo de San Antonio". Tesis Doctoral. Universidad Politécnica de Valencia. Spain.

CHAPTER 100

TRACERS OF SAND MOVEMENT ON THE OREGON COAST

KAREN E. CLEMENS[1] AND PAUL D. KOMAR[2]

ABSTRACT

The study of sand mineralogy and grain rounding can help answer many questions of immediate concern to coastal engineers or to broader issues of beach preservation. The heavy-mineral contents of sands, together with statistical techniques such as factor analysis, can be used to delineate sediment sources, trace transport paths, and map out patterns of mixing during sediment dispersal. Variations in the degree of grain rounding can similarly be used to trace sand movements, or to obtain additional information concerning the history of the sediment particles.

The techniques of studying sand mineralogies and grain rounding, and the types of problems they can address, are illustrated by research on the Oregon coast. Heavy mineral compositions of Oregon beach sands are the products of mixing contributions from four sources; the Columbia River on the north, the smaller rivers draining the Coast Range, the Umpqua River on the southern Oregon coast, and the Klamath Mountains of southern Oregon and northern California. Numerous headlands now prevent the longshore transport and mixing of sands from these multiple sources. The beach-sand compositions instead reflect along-coast mixing during Pleistocene lowered sea levels when blockage by headlands was absent. At that time there was a net littoral sand transport to the north, evident from the dispersal of Klamath-derived sands. With a rise in sea level and accompanying migrations of the beaches, headlands eventually interrupted the along-coast mixing of nearshore sands. Therefore, the north to south variation in compositions of beach sands is in part a relict pattern inherited from mixing during lowered sea levels. This has been modified during the past several thousand years by some additions of sand to the beaches from sea-cliff erosion and from rivers. However, studies of sediment mineralogy and grain rounding indicate that sands derived from most rivers draining the Coast Range are presently trapped in estuaries and so are not significant sources of beach sand. The Columbia River now supplies sand to Oregon beaches only to the first headland, Tillamook Head. At that headland there is a marked change in mineralogy and grain rounding with angular, recently supplied Columbia River sand to the north and rounded relict sand to the south.

---

[1]Greenhorn & O'Mara, Maryland Trade Center, 7500 Greenway Center Dr., Suite 700, Greenbelt, MD 20770

[2]College of Oceanography, Oregon State University, Corvallis, OR 97331

INTRODUCTION

The mineral compositions of sands and properties of individual particles such as their degrees of rounding can be utilized to trace movements of sediments in the coastal zone. This was illustrated many years ago by the pioneering studies of Parker Trask who investigated the movement of beach sand along the southern California coast (Trask, 1952). Trask was able to demonstrate that the sand filling the harbor at Santa Barbara comes from a distance of more than 160 km up the coast. He concluded this by using the heavy mineral augite as a tracer, a mineral whose source is ancient volcanic rocks near Morro Bay to the north of Santa Barbara. This mineral, together with the lighter quartz and feldspar, moves to the south along the beaches as littoral drift, until it is trapped by the Santa Barbara breakwater.

This simple example from the work of Trask illustrates the potential of applications in the use of the mineralogy of sediments to determine their sources and transport paths. In this example there was a clear application to an engineering problem – shoaling in the Santa Barbara boat basin. However, such techniques can be used to examine a range of questions concerning sources and losses of beach sands: Will dam construction on a river cut off a major source of beach sand?; Are river sands able to pass through estuaries to reach ocean beaches?; Is erosion of cliffs backing the beach a major source of nearshore sands?; Is the sand on the beach able to pass around headlands? Answers to such questions are often important to the proper design of coastal engineering structures, and to the preservation of beaches.

It is unfortunate that coastal engineers seldom draw upon this type of information. Considerable advances have been made in the techniques of studying sand mineralogies and grain rounding since the study of Trask (1952). These advances now permit more refined analyses and interpretations. A general objective of this paper is to illustrate the types of studies that can be undertaken, utilizing examples from the Oregon coast. These examples include the findings of studies which have attempted to determine the sources of beach sands on the central-Oregon coast, and examinations of the exchange of sand between estuaries and adjacent beaches. The Oregon coast is ideal to illustrate the types of investigations that can be undertaken in that all of the questions cited above are relevant to an understanding of beach-sand sources and transport paths.

TECHNIQUES OF STUDY

Heavy minerals such as augite contained within sands are distinguished from the light minerals, principally quartz and feldspars, by their contrasting densities. Laboratory techniques to separate the light and heavy minerals have in the past employed the heavy liquids bromaform (density 2.88 g/cm^3), tetrabromoethane (2.95) and methylene iodine (3.30). These heavy liquids are highly toxic, and so considerable care must be taken in the mineral separations. Fortunately, the use of the much safer sodium polytungstate (3.00) has recently been introduced (Callahan, 1987). Identification of the individual heavy minerals is accomplished with a microscope using standard petrographic techniques. The heavy-mineral separates are mounted on slides in a liquid such as Canada Balsam which has a known index of refraction, and usually some 100 to 300 non-opaque heavy minerals are identified and counted. This yields the percentages of the various non-opaque heavy minerals.

Variations in the degree of grain rounding can also be used to trace sand movements, or to obtain additional information concerning the history of the sand. Roundness is a measure of the sharpness of the grain's corners and edges, and is distinguished from sphericity which is a measure of the degree to which the grain deviates from a sphere. When grains are transported, collisions abrade their sharp edges so that the average degree of rounding progressively increases. In the case of sand-size grains, it has been shown that transport in rivers causes very little abrasion and rounding, and would produce minimal changes over typical river lengths (Kuenen, 1959). On the other hand, the long-term action of surf on beaches can eventually lead to grain rounding. It is difficult to evaluate the time periods required for rounding sand grains on beaches, but most investigators qualitatively assess it as involving thousands of years.

A major problem in studies of grain rounding has been in its quantification. Powers (1953) and Shepard and Young (1961) provide standardized series of grain images ranging in degrees of roundness. Comparisons between grains of the sample and these images allows one to place them in one of six categories ranging from "very angular" to "well rounded", or to assign corresponding numbers 1 through 6. However, this approach is subjective and can be viewed as only semi-quantitative. We have been able to use this approach in studies of Oregon coast sands due to the significant differences in angularity between the fresh river sands and the beach sands which have suffered appreciable abrasion. A more quantitative and reproduceable approach has been introduced by Schwartz and Shane (1969) and Ehrlich and Weinberg (1970) which involves Fourier analyses of the digitized outlines of the grains. The lower harmonics in such analyses relate to the overall shape of the grain, being more of a measure of the grain's sphericity, while the higher harmonics in part reflect the grain's degree of rounding. Studies such as that by Mazzullo et al. (1984) demonstrate that analyzing grain shapes by this approach can yield information on sand sources and transport paths. Techniques have been devised which automatically digitize grain outlines viewed under a microscope, with subsequent derivations of the harmonics of the Fourier analyses. This is necessary in that when the application is to trace sand movements, measurements are required on hundreds of grains in any given sample.

OREGON-COAST MINERAL STUDIES

The Oregon coast provides an ideal location to illustrate the usefulness of studies of sand mineralogy and grain rounding to obtain a better understanding of sources and transport paths of coastal sands. There are multiple sources having reasonably distinctive mineralogies which supply sands to the nearshore. The rivers contain angular sand grains which are progressively rounded during transport on beaches. It will be seen that combined studies of mineralogy and grain rounding indicate that Oregon beach sands have had a long and complex history. Contrasts in mineralogies and grain rounding between beach and river sands have also permitted detailed examinations of transport paths within Oregon estuaries.

Coastal Morphology and Potential Sand Sources

Much of the Oregon coast consists of a series of large rocky headlands which separate and isolate stretches of beaches. The headlands are composed of highly resistant volcanic rocks, while the beaches exist in areas backed by more easily eroded sedimentary rocks. This segments the coast into a series of pocket beaches where the lengths of the stretches of beach vary from about 5 to 100 km depending on the spacings of the volcanic rocks which form headlands. The northern half of the Oregon coast is shown in Figure 1, which illustrates this interplay of headlands and pocket beaches. The headlands extend well offshore from the beaches, and generally have considerable along-coast lengths. This implies that they prevent or severely limit any bypassing of beach sands. As will be seen, this view is supported by studies

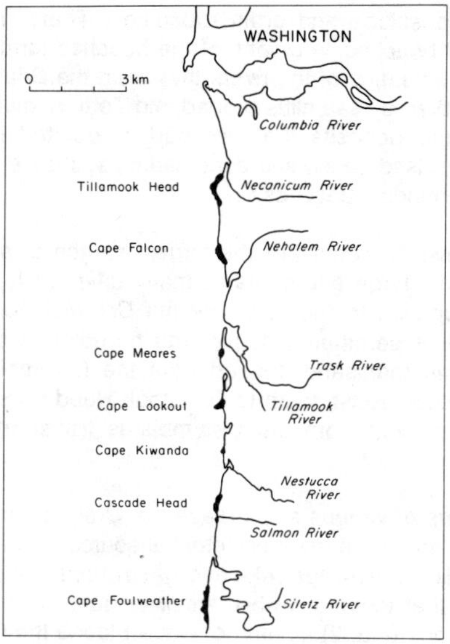

Fig. 1: The northern coast of Oregon, showing the interplay between headlands and beaches.

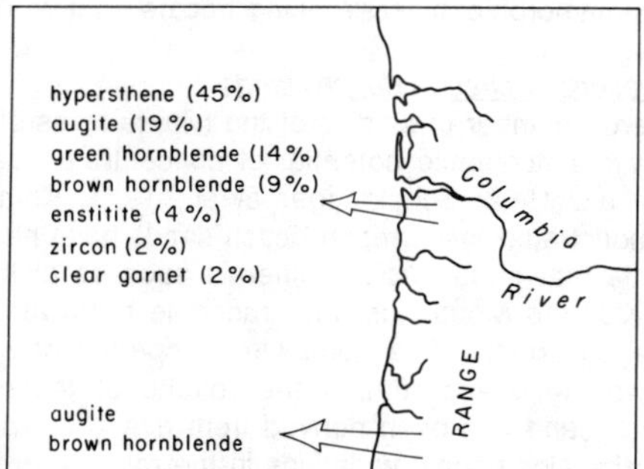

Fig. 2: Schematic of the principal sources of heavy minerals used to trace sand movements on the Oregon coast.

of beach-sand compositions and grain rounding. Therefore, the headlands act to isolate the individual beaches. Portions of the beaches form sand spits behind which are found bays and estuaries of the rivers that drain the Coast Range. Other portions of the beach are backed by sea cliffs eroded into Tertiary mudstones and siltstones, or into Pleistocene sand deposits that are part of uplifted marine terraces. These terrace deposits are raised beach and dune sands, so their erosion can be a significant source of sand to the modern beaches.

The Columbia River is potentially the largest source of sand to the nearshore. Its drainage basin is very large and contains many different types of rocks. A result of this is that sand supplied to the coast by the Columbia contains a wide variety of heavy minerals with hypersthene, augite and hornblende being the most abundant (Fig. 2). It is known that sand derived from the Columbia has resulted in beach accretion on the Oregon coast south to Tillamook Head (Fig. 1). However, at present most of the sand derived from the Columbia is transported northward along the Washington coast.

A number of rivers of various sizes reach the Oregon coast from the nearby Coast Range (Fig. 1), and these are another potential source of beach sands. This range of low mountains (450-m average elevation) contains volcanic rocks which yield augite, the mineral that together with a small amount of brown hornblende dominate river-sand compositions (Fig. 2). South of Cape Blanco the Coast Range gives way to the older but higher Klamath Mountains. The Klamaths contain complexly folded metamorphic rocks and intrusions of serpentinized ultrabasic and granitic rocks; these yield a wide spectrum of heavy minerals (Fig. 2), with glaucophane, pink garnet and other metamorphic minerals being tracers unique to this source.

Compositional Variations of Oregon Beach Sands

There have been a number of studies of the mineralogies of Oregon beach sands, largely induced by the economic potential of concentrated deposits of the opaque minerals (Kulm et al., 1968; Scheidegger et al., 1971; Komar and Wang, 1984). These studies demonstrated that Oregon beach sands have had multiple sources. Of particular interest is that metamorphic minerals such as pink garnet, which clearly originated in the Klamath Mountains, are traceable northward along nearly the full length of the Oregon coast. The southward dispersal of sand derived from the Columbia River was less apparent in the results of those studies, and it was uncertain how much sand has been derived from rivers that drain the Coast Range. There appeared to be along-coast variations in the relative mixtures of sands derived from these potential sources, but this had not been clearly delineated by the early investigations. In addition, an important question involved whether the rocky headlands are effective in blocking along-coast sand movements. Does the finding of Klamath Mountain sands in the beaches along nearly the entire Oregon coast imply that there is bypassing in spite of the large sizes of most of the headlands?

In order to answer such questions, we undertook a detailed study of the mineralogy and grain rounding of Oregon beach sands (Clemens, 1987; Clemens and Komar, 1988). Sand samples were collected from 26 beaches extending from immediately south of the Columbia River to Cape Blanco on the south-Oregon coast. Sand samples were also obtained from rivers and sea cliffs that are potential sources of beach sand. In total 18 different minerals were identified in microscopic analyses of the samples, but the major constituents are augite, hornblende, hypersthene and garnet. The percentages of those four major heavy minerals are plotted in Figure 3, revealing how they vary with along-coast distance south of the Columbia River. The most noticeable variation is in the percent of augite, which dominates beach-sand compositions between Tillamook Head, the first headland south of the Columbia River, and Cape Foulweather midway through the study area. These augite-rich beach sands appear to reflect contributions by the Coast Range volcanics directly landward from this stretch of coast. North of Tillamook Head the percentage of augite abruptly drops as the beach is richer in hornblende and hypersthene, minerals that are clearly derived from the Columbia River. The content of garnet in the beach sands systematically increases toward the south (Fig. 3), an expected distribution in view of its source in the Klamath Mountains of southern Oregon. Of interest is that garnet is found in beach sands all the way north to Tillamook Head, some 500 km north of its source.

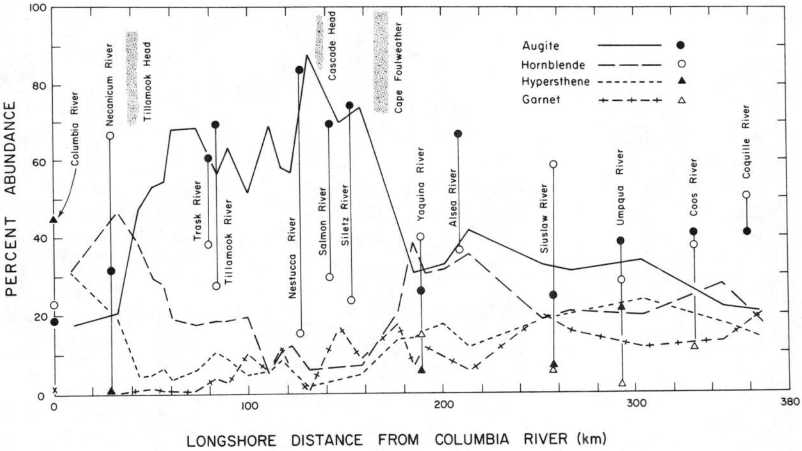

Fig. 3: Along-coast variations in abundances of the principal heavy minerals found in Oregon beach and river sands. [*from* Clemens and Komar (1988)]

The along-coast distributions of individual minerals, Figure 3, has already permitted tentative interpretations of sources and dispersal routes of beach sands on the Oregon coast. We have identified minerals that are suggestive of various sources and have qualitatively assessed their relative contributions along the length of the coast. The abrupt compositional changes at Tillamook Head and Cape Foulweather indicate that at least those headlands are effective in blocking longshore sand movements on the beaches.

Factor analysis is in general a more powerful approach to interpreting mineral assemblages than relying on individual minerals. Factor analysis examines natural groupings of minerals found within the assemblages contained in the series of individual samples (Imbrie and van Andel, 1964). The technique searches for end-member vectors or factors which can be summed in varying proportions to yield the mineralogies of the samples. Such an analysis (Q-mode factor analysis) was undertaken on the heavy-mineral percentages found in the Oregon beach-sand samples (Clemens and Komar, 1988). Three dominant factors were derived which account for 89% of the compositional variations in the samples. The mineral compositions of these end-member factors are given in Figure 4, and their

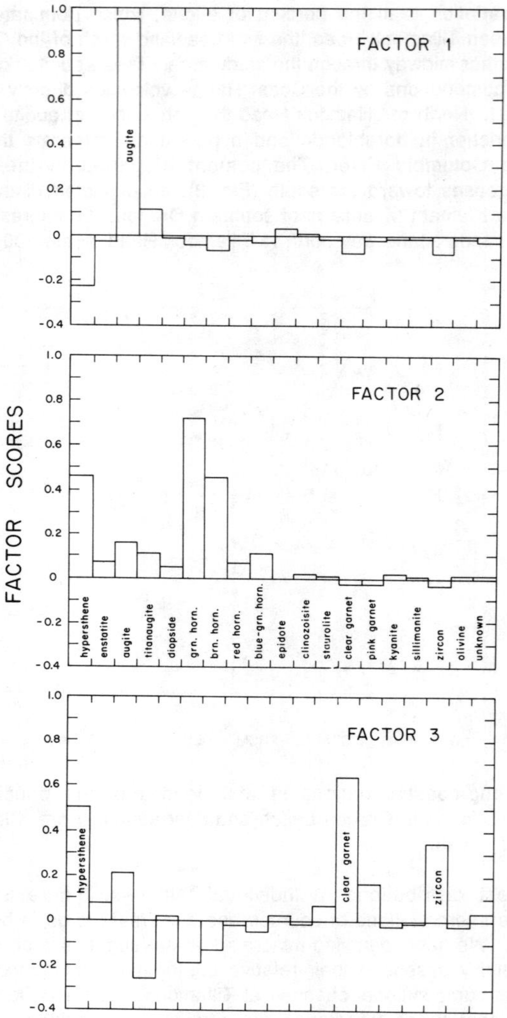

Fig. 4: Compositions of the factors obtained in factor analyses of the heavy mineral compositions of Oregon beach-sand samples. [*from* Clemens and Komar (1988)]

along-coast variations are graphed in Figure 5. Factor 1 consists almost entirely of augite, and it is seen that this factor dominates the beach-sand compositions between Tillamook Head and Cape Foulweather. In this case, since there is a source consisting of essentially one mineral, the results from the factor analysis correspond closely with our analysis based on individual minerals (Fig. 3). Factor 2 is rich in hypersthene and green and brown hornblende (Fig. 4). This factor has a bimodal distribution along the coast (Fig. 5). Its main loading, approaching 1, occurs on the beach north of Tillamook Head and is clearly associated with the Columbia River which is rich in those minerals. The more southerly increase in Factor 2, extending from 190 km south of the Columbia to the end of our sample range at 365 km south, is most likely contributed by the Umpqua River. Sand derived from that river has a high content of both hypersthene and hornblende, as well as augite, and with its large drainage basin the Umpqua could have been a major source of beach sand (more so than the other rivers draining the Coast Range). Therefore, Factor 2 appears to be associated with two major sources that are similar in compositions, the Columbia and Umpqua Rivers. This interpretation is verified if the river-sand samples are included in the factor analysis together with the beach samples. The Columbia River is then identified as 100% Factor 2, and the Umpqua River sand is found to consist of nearly equal portions of Factors 1 and 2.

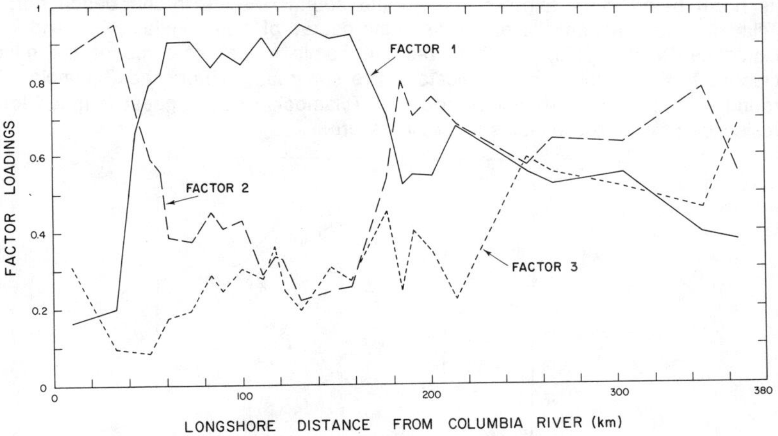

Fig. 5: Along-coast loadings of the three factors (fig. 4) obtained in the analyses of the beach-sand compositions. [*from* Clemens and Komar (1988)]

Factor 3 is rich in garnet and zircon (Fig. 4), and also includes the metamorphic minerals staurolite and epidote. This composition, together with its northward, along-coast decrease (Fig. 5), clearly points to a source in the Klamath Mountains. This interpretation is the same as in Figure 3 which was based on garnet alone, but is more firmly established here by a complete assemblage of metamorphic minerals.

It is apparent that more information about sand sources and dispersal paths can be obtained from factor analyses than from distributions of individual minerals. Considerable interpretation is still required, but the end-member factors can generally be traced back to specific sources. These sources may not necessarily be individual rivers, but instead could be rock terrains. This is the case on the Oregon coast where it is better to think in terms of the sources being the metamorphic rocks of the Klamaths, the augite-rich volcanics of the Coast Range, and the hypersthene and hornblende-rich rocks of the Cascade Mountains. We have seen that the Umpqua River combines two of these terrain sources.

The switch from Factor 1 (Columbia River sand) to Factor 2 (Coast Range) is seen in Figure 5 to be abrupt at Tillamook Head, more so than the change in augite alone (Fig. 3). This further substantiates the ability of Tillamook Head to block longshore movements of beach sands. While determining the heavy-mineral compositions of the samples, differences in grain rounding were also noted in the sands on opposite sides of Tillamook Head. North of this headland the heavy-mineral grains are highly angular and many delicate crystals are present, while to the south they are noticeably more rounded. This change was further established by using the photo-comparison roundness scale of Shepard and Young (1961). Roundness was evaluated for about 50 grains of each mineral in a sample, yielding a distribution of roundness values for a mineral. Such analyses were primarily performed on augite, but some evaluations were also made for hypersthene, hornblende and quartz. The results are shown in Figure 6 for samples from the Columbia River, and from the beaches to the south of the Columbia but on opposite sides of Tillamook Head. A significant change in rounding is seen to occur at Tillamook Head, the beach sands to its north being more angular than to the south. Augite in the beach north of Tillamook Head appears to retain the same degree of high angularity as sand in the Columbia River. This, together with the fragile nature of some of the mineral crystals in the beach sands, indicate active contributions from the Columbia. The rounder grains on the beaches south of Tillamook Head suggest a much longer residence time for those sands in the nearshore.

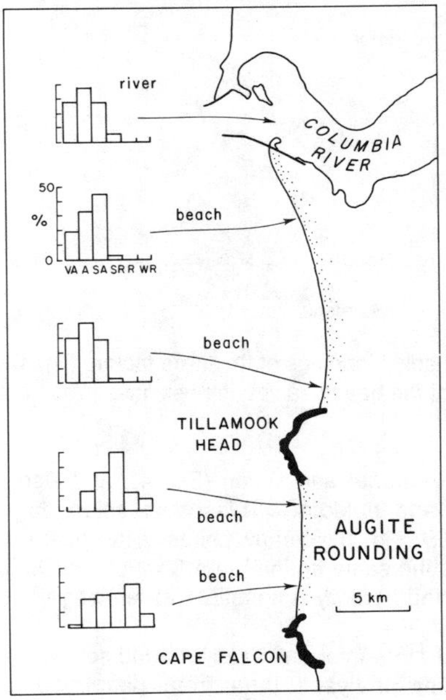

Fig. 6: Histograms of grain rounding for augite in river and beach sands: VA - very angular; A - angular; SA - subangular; SR - subrounded; R - rounded; WR - well rounded.

The only other headland which demonstrates a marked compositional change in beach-sand mineralogies on its north and south sides is Cape Foulweather (Figs. 3 and 5). This change can be attributed to differences in materials derived primarily from sea-cliff erosion. North of Cape Foulweather most of the cliff erosion has been in alluvium derived from the Coast Range and therefore is rich in augite (like the rivers draining the Coast Range). Pleistocene marine terraces are more important south of Cape Foulweather, and these terraces contain uplifted beach and dune sands. These old deposits contain a full spectrum of minerals from the various sources including the Columbia River and the Klamath Mountain metamorphics.

It might mistakenly be concluded from our results that Tillamook Head and Cape Foulweather are the only headlands that are effective in blocking movements of sand on the modern beaches. However, blockage by those headlands is made apparent by differences in sand sources on their opposite sides. If the sand sources to the north and south are much the same and the long-term history of the sands do not differ, then there will be no compositional or textural differences. Tillamook Head and Cape Foulweather are not unusually large, nor do they extend to deeper water than most of the other headlands. Therefore, it is probable that the other headlands are also effective barriers to longshore sand movements, even though this could not be established directly by our analyses of beach sands.

The results of our investigations of Oregon beach-sand mineralogies and grain rounding have confirmed that the compositions are derived from the mixing of sands from multiple sources. All beaches, excepting those south of the Columbia River to Tillamook Head, contain significant quantities of metamorphic minerals derived from the Klamath Mountains of southern Oregon and northern California. This indicates a northward transport of sand along the coast, but this transport cannot occur now due to the presence of headlands. Instead, it must have taken place during Pleistocene lower stands of sea level when headlands did not interrupt such sand movements. With a rise in sea level at the end of the ice ages and accompanying migrations of the beaches, the formerly continuous beach became segmented and isolated between headlands. Therefore, much of the along-coast variation in beach-sand mineralogies on the Oregon coast is relict, preserving the longshore patterns established thousands of years ago when sea levels were lower. The beach-sand compositions have been altered somewhat during the last few thousand years following their isolation between headlands. Most noteworthy has been the growth of the beach north of Tillamook Head due to large sand contributions by the Columbia River. However, elsewhere the additions must have been comparatively small since the volumes have not been sufficiently large to obscure the relict pattern established thousands of years ago. This implies that modern-day inputs of sand to most Oregon beaches from sea cliff erosion and from rivers draining the Coast Range are relatively small.

Exchanges of Sand Between Beaches and Estuaries

A primary reason for the small contributions of sands to beaches from rivers draining the Coast Range is that most of these rivers are separated from the ocean by estuaries. To varying degrees, these estuaries trap the river sands. This has been established by Kulm and Byrne (1966) and Byrne and Kulm (1967) in Yaquina Bay, by Scheidegger and Phipps (1976) in Grays Harbor, Washington, and by Peterson et al. (1982, 1984) in several other Oregon-coast estuaries.

In their investigations of Yaquina Bay, the estuary of the Yaquina River, Kulm and Byrne (1966) were able to establish the three realms of deposition diagramed in

Figure 7; a marine realm which extends approximately 2.5 km into the bay from the ocean inlet, a fluviatile realm in the upper reaches of the bay, and a broad "mixed" realm which is a zone where the two sand sources mix in varying proportions. These were delineated on the basis of sediment grain sizes, compositions, and particle rounding. The compositional distinctions included not only differences in heavy minerals found in the marine versus fluviatile sands, but also in the mineralogy of the feldspars and the distribution of "yellow grains". Yellow grains are diagnostic of marine sands where they constitute about 10% of the light-mineral fraction; they consist primarily of weathered feldspars and chert, and are derived from erosion of terrace sands in the sea cliffs. Because they are distinctive and easily recognized, yellow grains can serve as a simple tracer of marine-sand movements into Yaquina Bay (Fig. 8). Heavy minerals such as staurolite and kyanite, found in the beach sands but not in the Yaquina River, were also used by Kulm and Byrne as tracers of marine-sand movement into Yaquina Bay.

Peterson et al. (1982) has demonstrated a similar pattern of mixing of marine and fluviatile sands in Alsea Bay. However, the use of factor analysis of heavy-mineral assemblages, together with quantitative measurements of grain rounding, permitted a more detailed mapping of the transport paths of sands in this estuary. Based on the abundances of 15 non-opaque heavy-mineral varieties in surface sand samples, two factors were obtained in a factor analysis which account for 95% of the sample variance. One factor consisted almost exclusively of augite, and this end member represents contributions by the Alsea River. In contrast, the marine-related factor was composed of a number of minerals including hyperthene, hornblende and garnet. The weightings assigned to these two factors for each sand sample in Alsea Bay represented the respective contributions by the river versus the marine beach. Peterson et al. also noted that the augite and hornblende grains of fluvial origin are angular, while those of marine origin are rounded. This provided another means to evaluate sources and trace sand movements in Alsea Bay. The delineation of sand mixing in Alsea Bay by the two methods, factor analysis of heavy-mineral compositions and grain rounding, yielded essentially the same results. The general patterns are similar to those found by Kulm and Byrne (1966) in Yaquina Bay (Fig. 7), with marine sand dominating near the mouth, a broad zone of marine-fluviatile mixing, and pure riverine sands occurring only in the upper reaches of the estuary. However, more details of the mixing patterns were established in Alsea Bay, the tidal channels clearly serving as conduits of sand transport throughout the central estuary with an interfingering of sands from the two sources. Repeated sampling in different seasons revealed that down-channel transport of fluviatile sands occurs during winter periods of high river discharge, while beach-sand intrusions under tidal currents occur during the summer months of low river discharges.

Similar analyses have been conducted in other Northwest estuaries: Grays Harbor, Washington (Scheidegger and Phipps, 1976), the Sixes River estuary, Oregon (Boggs and Jones, 1976), Tillamook Bay, Siletz Bay, and the estuaries of the Siluslaw and Salmon Rivers (Peterson et al., 1984). Although these investigations demonstrated that most Northwest estuaries act as traps for both fluviatile and marine sands, it is more difficult to establish whether some river sand is able to pass through an estuary to become a source of beach sand. Small rivers such as the Sixes River studied by Boggs and Jones (1976), have essentially no estuary, and during winter floods the riverine sands are flushed out onto the neighboring beaches. However, since the rivers are small, the volumes of sand contributed to the beach are small. More uncertain is whether the larger rivers, such as the Yaquina and Alsea,

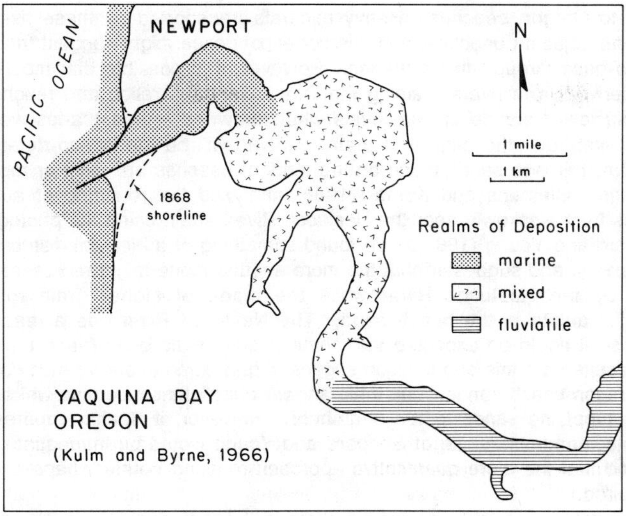

Fig. 7: Realms of sand deposition in Yaquina Bay. [*after* Kulm and Byrne (1966)]

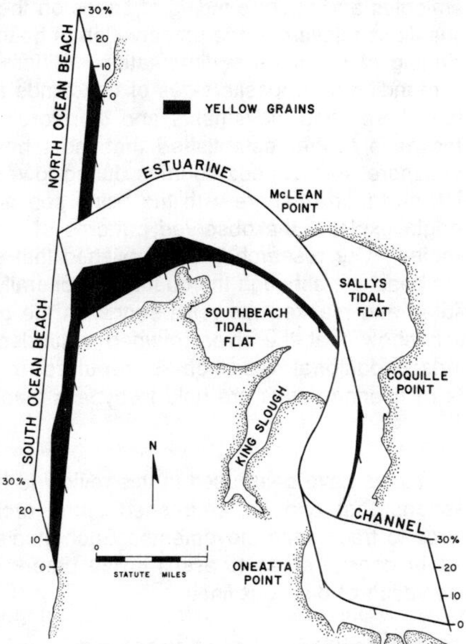

Fig. 8: Distribution of "yellow grains" in Yaquina Bay, used as a tracer of marine-sand movements into the estuary. [*from* Kulm and Byrne (1966)]

supply sands to Oregon beaches. Heavy minerals transported by these rivers are abundant on the adjacent beaches, and this correspondence might suggest that some river sands do pass through the estuaries. However, it is possible that most of the beach sand derived from rivers draining the Coast Range is relict, and reached the beaches during lower stands of sea levels when drowned-river estuaries were not present. It is likely that the degree of grain rounding will be the key to determining whether the Coast Range sands found on the modern beaches are relict or represent recent additions. Clemens and Komar (1988) analyzed the rounding of augite in beach samples from Neskowin near the Nestucca River, again using the photographic chart of Shepard and Young (1961). We found something of a bimodal distribution in the sand angularity, and suggested that the more angular mode might represent recent contributions by the Nestucca River while the mode of higher grain rounding represents relict augite in the beach sand. The Nestucca River has a reasonably small estuary so it could be expected that some riverine sand does reach the ocean beach. The results from this one location suggest that a comprehensive study of grain rounding of Oregon beach sands would help answer questions concerning which rivers are presently supplying sands to the nearshore. However, it is also apparent that using the photo roundness scale of Shepard and Young would be inadequate in this application, and that the more quantitative approach of using Fourier shape analyses would be required.

SUMMARY OF CONCLUSIONS

Studies of the mineralogies and grain rounding of sands on the Oregon coast have helped answer many questions relevant to the preservation of beaches and have led to an increased understanding of estuarine sedimentation. Differences in beach-sand mineralogies and grain rounding on opposite sides of headlands confirm that they are effective in blocking longshore sand movements and therefore isolate the beaches. The mineralogy studies have further established that most beaches contain relict sands, transported alongshore from various sources during lowered sea levels when headlands did not exist, and then onshore with the rising sea at the end of the last glacial period. This origin explains the observed patterns of along-coast variations in beach-sand mineralogies. Our research has established that sea-cliff erosion is a source of sands to some beaches, although the quantities generally are not so large as to have completely diluted and masked the relict sands on the beaches. Studies of estuarine sedimentation show that the river-drowned estuaries are sinks of both river and beach sands; additional research is required to establish whether measureable quantities of riverine sands are able to bypass these estuaries to reach the ocean beaches.

These Oregon-coast studies have been used in this review to illustrate the types of applications and questions that can be addressed using techniques of mineral compositions and textures to trace sand movements. Such studies could be used by engineers working in other coastal areas to assist in the proper design of structures and to help in the preservation of our coastlines.

ACKNOWLEDGEMENTS

We would like to thank L. D. Kulm for his helpful comments in reviewing this paper. This paper is a result of research sponsored by the Oregon State University Sea Grant Program, supported by NOAA Office of Sea Grant, Department of Commerce, under grant number NA85AA-D-SG0095.

REFERENCES

BOGGS, S., and JONES, C.A. (1976) Seasonal reversal of flood-tide dominated sediment transport in a small Oregon estuary: Geol. Soc. Amer. Bull., v. 87, p. 419-426.

BYRNE, J. V., and KULM, L. D. (1967) Natural indicators of estuarine sediment movement: *Jour. Waterways and Harbors Div., Proc. Amer. Soc. Civil Engrs.*, v. 93, no. WW2, p. 181-194.

CALLAHAN, J. (1987) A nontoxic heavy liquid and cheaper filters for separation of mineral grains: *Jour. Sedimentary Petrology*, v. 57, p. 765-766.

CLEMENS, K. E. (1987) Along-coast variations of Oregon beach-sand compositions produced by the mixing of sediments from multiple sources under a transgressing sea: Unpubl. Master's Thesis, Oregon State Univ., Corvallis, 75 p.

CLEMENS, K. E., and KOMAR, P. D. (1988) Oregon beach-sand compositions produced by the mixing of sediments under a transgressing sea: *Jour. Sedimentary Petrology*, v. 58, p. 519-529.

EHRLICH, R., and WEINBERG, B. (1970) An exact method for characterization of grain shape: *Jour. Sedimentary Petrology*, v. 40, p. 205-212.

IMBRIE, J., and VAN ANDEL, TJ. H. (1964) Vector analysis of heavy mineral data: *Geol. Soc. America Bull.*, v. 75, p. 1131-1156.

KOMAR, P. D., and WANG, C. (1984) Processes of selective grain transport and the formation of placers on beaches: *Jour. of Geology*, v. 92, p. 637-655.

KUENEN, Ph. H. (1959) Experimental abrasion 3. Fluvial action on sand: *Amer. Jour. Science*, v. 172-190.

KULM, L. D., and BYRNE, J. V. (1966) Sedimentary response to hydrography in an Oregon estuary: *Marine Geology*, v. 4, p. 85-118.

KULM, L. D., SCHEIDEGGER, K. F., BYRNE, J. V., and SPIGAI, J. J. (1968) A preliminary investigation of the heavy mineral suites of the coast rivers and beaches of Oregon and northern California: *The Ore Bin*, v. 30, p. 165-180.

MAZZULLO, J., EHRLICH, R., and HEMMING, M.A. (1984) Provenance and areal distribution of Late Pleistocene and Holocene quartz sand on the southern New England continental shelf: *Jour. Sedimentary Petrology*, v. 54, p. 1335-1348.

PETERSON, C., SCHEIDEGGER, K. F., and KOMAR, P. D. (1982) Sand-dispersal patterns in an active-margin estuary of the northwestern United States as indicated by sand composition, texture and bedforms: *Marine Geology*, v. 50, p. 77-96.

PETERSON, C., SCHEIDEGGER, K. F., KOMAR, P. D., and NIEM, W. (1984) Sediment composition and hydrography in six high-gradient estuaries of the northwestern United States: *Jour. Sedimentary Petrology*, v. 54, p. 86-97.

POWERS, M.C. (1953) A new roundness scale for sedimentary particles: *Jour. Sedimentary Petrology*, v. 23, p. 117-119.

SCHEIDEGGER, K. F., KULM, L. D., and RUNGE, E. J. (1971) Sediment sources and dispersal patterns on Oregon continental shelf sands: *Jour. Sedimentary Petrology*, v. 41, p. 1112-1120.

SCHEIDEGGER, K. F., and PHIPPS, J. P. (1976) Dispersal patterns of sand in Grays Harbour estuary, Washington: *Jour. Sedimentary Petrology*, v. 46, p. 163-166.

SCHWARTZ, H.P., and SHANE, K.C. (1969) Measurements of particle shape by Fourier analysis: *Sedimentology*, v. 13, p. 179-212.

SHEPARD, F.P., and YOUNG, R. (1961) Distinguishing between beach and dune sands: *Jour. Sedimentary Petrology*, v. 31, p. 196-214.

TRASK, P. D. (1952) Sources of beach sand at Santa Barbara, California, as indicated by mineral grain studies: *Beach Erosion Board Tech. Memo. No. 28*, U.S. Army Corps of Engrs., 24 p.

CHAPTER 101

PROBABILISTIC ANALYSIS OF SEAFLOOR LIQUEFACTION

Rafael Blázquez *
Felipe M.Martínez **

To investigate the reliability of a sandy soil layer in an ocean wave environment a liquefaction model is used in conjunction with a first order reliability method. Thus, sensitivity indices of the soil-water system with respoect to the uncertain strength and input variables are computed, and the relative importance of the various factors defining the problem can be determined. The relationship of this approach with more conventional design methods (deterministic models, risk models) is discussed along with the range of applicability of the different safety measurements.

INTRODUCTION

When ocean waves propagate over a cohesionless oceanfloor, hydrodynamic pressures are continuously being generated at a rate which depends mainly on the permeability of the material. Sandy soils with low permeability coefficients are unable to dissipate the excess pore pressures, which eventually may become equal to the total stresses acting on the soil element. The result is the complete loss of all intergranular effective stress and bearing capacity of the material, a cyclic instability phenomenon known as liquefaction of the sand.

In this paper the liquefaction behavior of seafloor soils is evaluated by means of an analytical model-proposed earlier in the literature-which takes into account the physics of the phenomenon. Based on that model the following steps are taken:

a) Carry out a parametric study to rank the relative importance of the factors governing liquefaction processes (deterministic approach).

---

* Research Engineer and Physicist, PhD; Head of the Section of Stochastic Analysis, CEPYC (Madrid-SPAIN), Formely Visiting Lecturer, Civil Engineering Department, Department, University of California (Berkeley, USA).

* Director Centro de Estudios de Puertos y Costas, (CEPYC)-CEDEX. Antonio López, 81, 28026 Madrid, SPAIN

b) Incorporate uncertainties in input/output variables into a fully probabilistic analysis of the problem.

c) Use reliability techniques to compute sensitivity factors and critical values of the random variables.

d) Compare the above methodology with classical risk methods.

DETERMINISTIC APPROACH

For the purposes of this study, a simplified pore pressure analysis of the liquefaction phenomenon, first described by Layas (1982) and Rahman and Jaber (1986), has been chosen. In this model harmonic linear waves are used, with fixed periods, T, and wave lengths, L, related by the formula:

$$T = \sqrt{\frac{2\pi L}{g \tanh(kd)}} \qquad (1)$$

where $k=\frac{2\pi}{L}$ is the wave number and d is the water depth.
Two cases are considered:

a) Undrained model (infinite layer)
b) Drained model (finite layer)

Undrained model

If the permeability of the soil is zero, the pore pressure, u_g, generated at the sand layer (semi-infinite half space) at depth z is derived, according to the following algorithm.

1. Assume a linear pore pressure generation curve, $u_g(t)$:

$$\frac{u_g}{\sigma'_{vc}} \simeq \frac{N}{N_l} = \frac{t}{t_l} \qquad (2)$$

(σ'_{vo}=initial effective stress; N_l=no. of cycles to liquefaction)

2. Assume a liquefaction cyclic strength curve, $\tau_l(N_l)$, of fatigue type:

$$\frac{\tau_l}{\sigma'_{vc} D_r} = a N_l^{-b} \qquad (3)$$

(a,b=constants; D_r=relative density)

3. Compute the hydrodynamic pressure on the sea bed by linear wave theory:

$$p_o(x,t) = \frac{\gamma_w H}{2\cosh(kd)} \cos(kx - \omega t) \qquad (4)$$

(k=wave number; w=wave frequency; d=water depth)

4. Compute the wave-induced shear stresses in the soil by elasticity theory:

$$\tau(x,z,t)=\frac{\gamma_w HKz}{2\cosh(Kd)}e^{-Kz}\cos(Kx-\omega t-\varphi) \qquad (5)$$

5. Evaluate the liquefaction time at depth z, t_l, from the conditions at the onset of liquefaction (equations 2, 3 and 4) written for $z=z_l$, $N=N_l$, and $t=t_l$)=

$$t_l=T\cdot N_l=T\cdot\left[\frac{\gamma' 2a\cdot D_r\cosh(Kd)}{\gamma_w HK}\right]^{\frac{1}{b}}\times e^{Kz_l/b} \qquad (6)$$

Then, substituting in eq. (2), yields:

$$u_g=\frac{\gamma' z t}{t_l} \qquad (7)$$

where γ' is the buoyant unit weight of the soil.

Notice that the amplitude of the harmonic expressions 4 and 5 is independent of x, meaning that at the same depth the induced stresses at different horizontal positions have the same maximum values. Therefore neglecting the differences in phase, only one-dimensional responses need to be considered (Seed and Rahman, 1978).

Drained model

For a permeable stratum the equation governing the pore pressure rresponse of the soil, u, becomes:

$$c_v\frac{\partial^2 u}{\partial z^2}=\frac{\partial u}{\partial t}-\frac{\partial u_g}{\partial t} \qquad (8)$$

where c_v is the consolidation coefficient of the sand.

Equation 8 quantifies the balance of the simultaneous generation and dissipation of pore pressures that continuously occurs inside the soil layer. By inserting in eq. 8 the correspondig value of the generation rate term:

$$\frac{\partial u_g}{\partial t}=Cze^{\tau z}\frac{\gamma'}{T}=\left(\frac{\gamma_w\pi}{\gamma' a D_r}\right)^{\frac{1}{b}}\left(\frac{H/L}{\cosh\left(\frac{2\pi d}{L}\right)}\right)^{\frac{1}{b}}z\,e^{\tau z}\frac{\gamma'}{T} \qquad (9)$$

and solving for the appropiate boundary and initial conditions, namely (Fig. 1):

B.C. { previous top $\longrightarrow$ $\left[u(z,t)\right]_{z=0}=0$ (10-1)

impervious bottom $\longrightarrow$ $\left[\frac{\partial u}{\partial z}(z,t)\right]_{z=D}=0$ (10-2)

I.C. $\longrightarrow$ $\left[u(z,t)\right]_{t=0}=0$ (10-3)

the following expression for the excess pore-pressure curves is found:

$$u(z,t)=\frac{C}{c_v}\frac{\gamma'}{T}\left\{W(z)+V(z,t)\right\} \qquad (11)$$

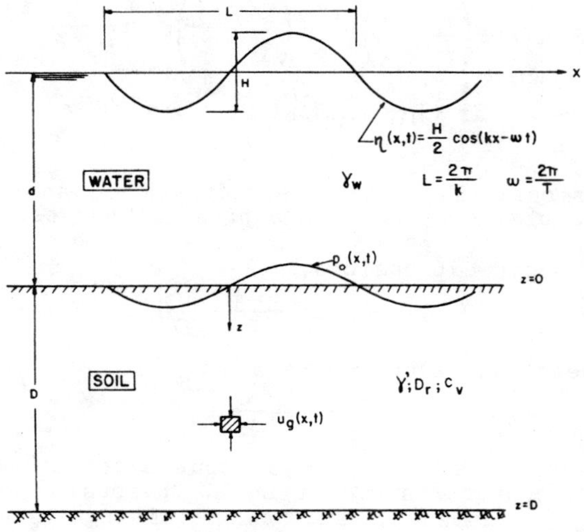

Fig.1. Scheme of the problem.

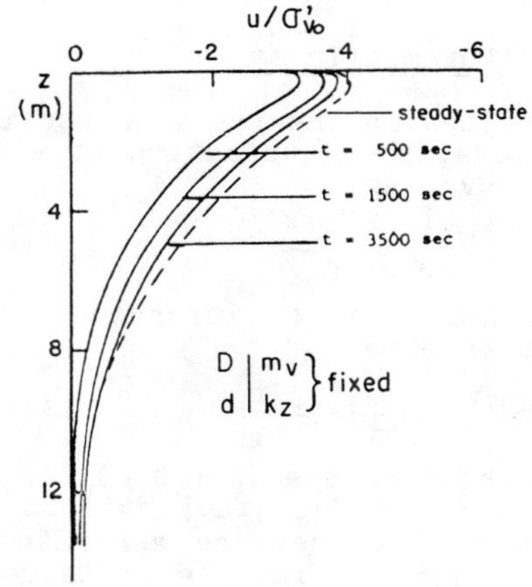

Fig.2. Dynamic isochrones ata different stages of the storm. Drained layer (after Layas,1982)

in which:

$$w(z)=\left(\frac{2}{\eta^3}-\frac{z}{\eta^2}\right)e^{\eta z}+\left(\frac{D}{\eta}-\frac{1}{\eta^2}\right)e^{\eta D}z-\frac{2}{\eta^3} \qquad (12\text{-}1)$$

and

$$v(z,t)=-\frac{2}{D}\sum_{n=0}^{\infty}\left\{h(\alpha_n,D,\eta)\right\}e^{-c_v\alpha_n^2 t}\sin(\alpha_n z) \qquad (12\text{-}2)$$

are, respectively, the steady-state and transient components of the dynamic pore pressure curves.

In the above expressions,

$$\alpha_n=\frac{2n+1}{2}\frac{\pi}{D} \qquad (13\text{-}1)$$

is a geometrical value, whereas

$$\eta=-\frac{K}{b} \qquad (13\text{-}2)$$

is a parameter related to the input storm as well as the mechanical properties of the soil. Obviously, as $t\rightarrow\infty$ the term v(z,t) vanishes, and the limiting (steady-state) isochrone becomes fixed in shape:

$$u(z)\simeq\frac{C\gamma'}{c_vT}\,w(z) \qquad (14)$$

Figure 2 exemplifies this behavior for different times of evolution of a given storm at a liquefiable site.

Factor of safety. Parametric study

The model described in the previous section defines the factor of safety against liquefaction at a given time and depth within the layer as:

$$(SF)_d=\frac{u}{\sigma'_{vc}} \qquad (15)$$

For the particular case of impervious soil, $u=u_g$, and the safety factor becomes:

$$(SF)_u=\frac{u_g}{\sigma'_{vc}}=\frac{t}{t_l} \qquad (16)$$

Where t_l is a known parameter (eq.6). In both cases the onset of liquefaction (safety factor=1) is defined as the state in which the pore pressure generated by the ocean waves equals the initial effective pressure at some point in the deposit (soil resistance).

Since the parameter SF constitutes a global index of the safety of the deposit against liquefaction under various drainage conditions, its sensitivity to the variables involved is of practical importance.

To investigate further this point a classical parametric study hass been carried out on a digital computer. Table I shows the mean values assumed for all variables (reference state), whereas the corresponding ranges of variation and their effects on $(SF)_u$ and $(SF)_d$ are represented, side by side, on Figures 3 to 11 for a storm of 10 minutes duration. From a comparative analysis of these figures it can be concluded that:

a) The safety factor increases with the wave length and all soil parameters but b (when H and/or b increases the safety factor decreases).

b) Wave heigth and consolidation coefficient affect significantly the liquefaction behavior of the soil.

c) The liquefaction resistance of drained deposits is roughly one order of magnitude (ten times) greater than that of undrained ones.

d) Monotonic variations of the geometrical parameters (d,D) lead to non-monotonic trends for the safety factor. For this reason these parameters are assumed fixed hereafter.

PROBABILISTIC APPROACH

Basic Concepts

In order to evaluate realistically the liquefaction potential of the stratum of sand, the method employed should take into account the probabilistic nature of the problem. Uncertainties in the dynamic bahavior of the soil and/or in its geotechnical parameters are important and cannot be neglected. Furthermore, very often the input load is defined statistically, and so the wave parameters are random in nature.

Again, according to the time span used in the study, two types of models are considered:

a) Reliability models (short term conditions)
b) Risk models (long term conditions)

Reliability model

In this case the wave parameters are those associated to storm conditions at the sea, which in turn depend on the location of the site relative to the zone of wave generation. Then, for a given storm (fixed time), a reliability technique is used to compute the probability of liquefaction of the deposit for the assumed sea states.

A mean-centered, first-order, second-moment reliability technique (FOSM model) has been proposed recently by Layas (1982), but no quantitative results were reported. In this paper a first-order marginal distribution method (Madsen

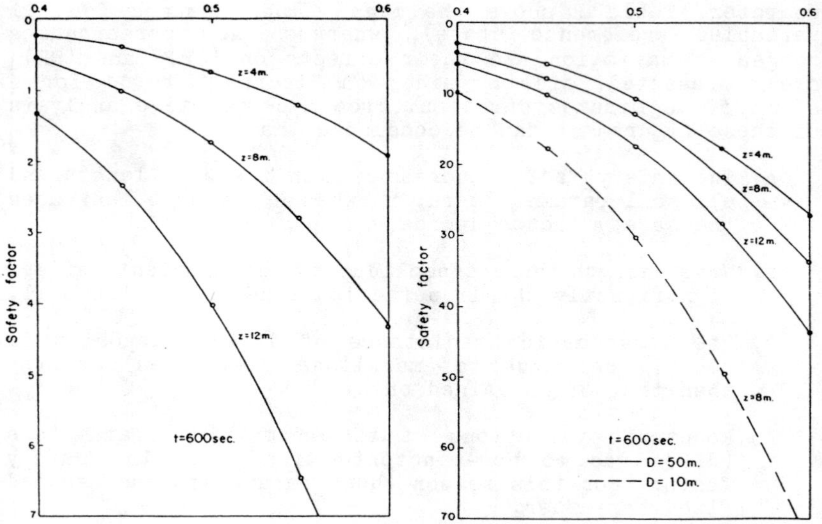

Fig.3. Parametric study.

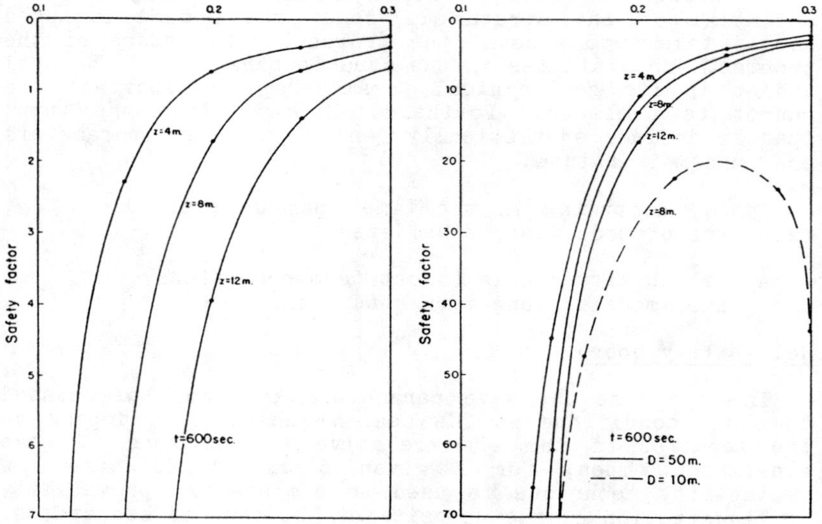

Fig.4. Parametric study (cont.)

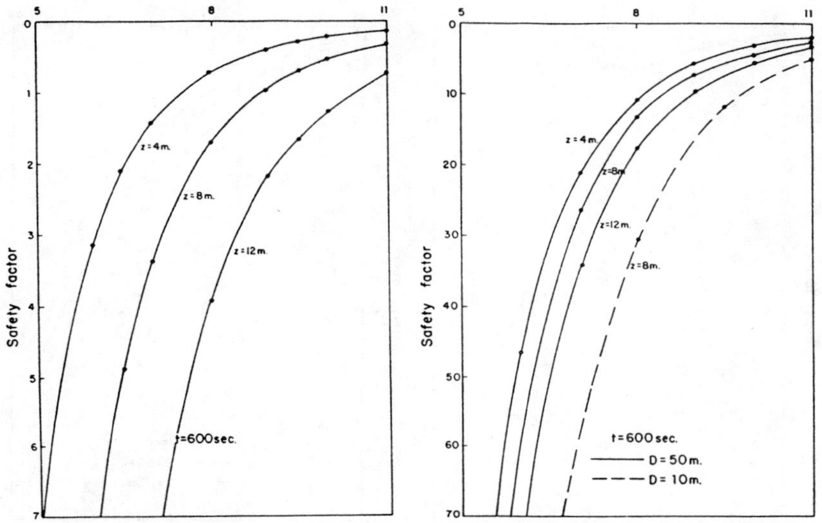

Fig.5. Parametric study (cont.)

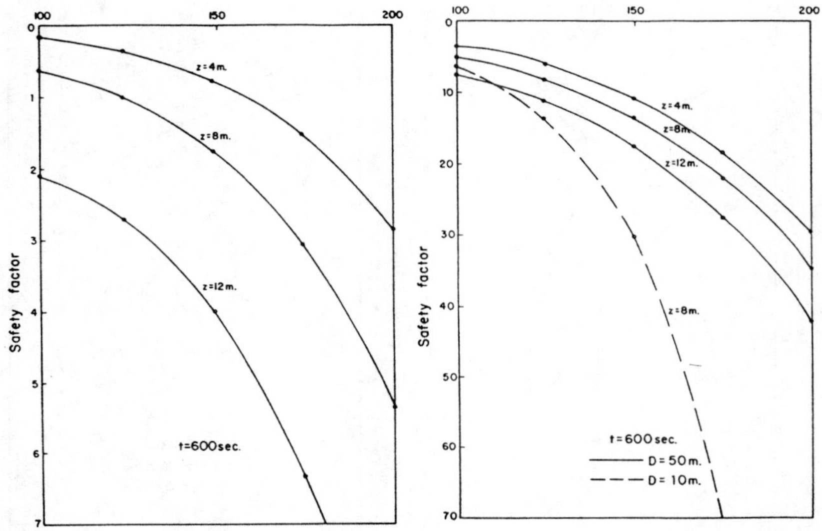

Fig.6. Parametric study (cont.)

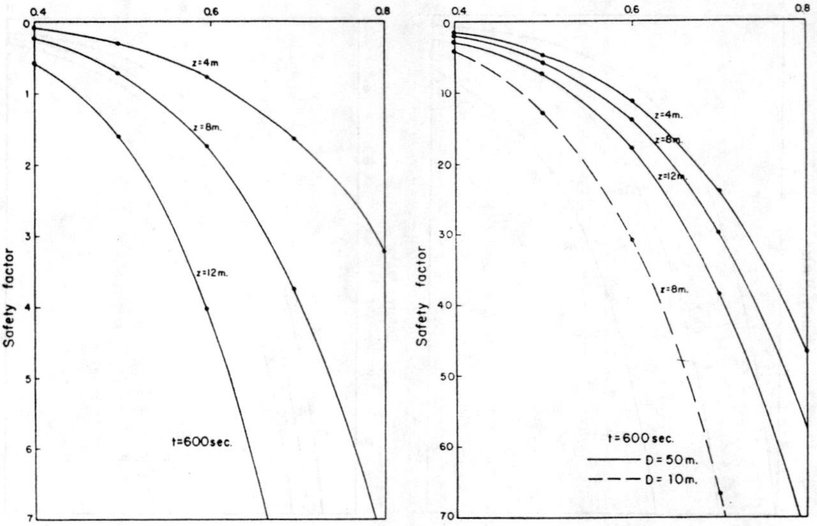

Fig.7. Parametric study (cont.)

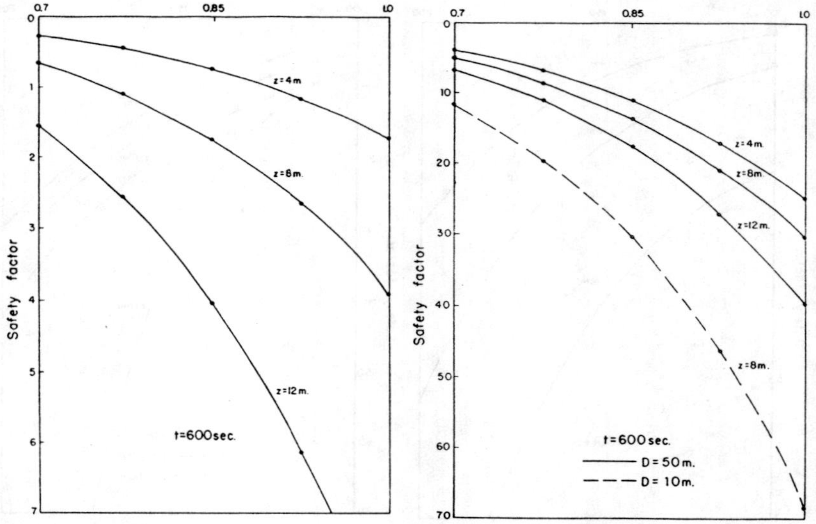

Fig.8. Parametric study (cont.)

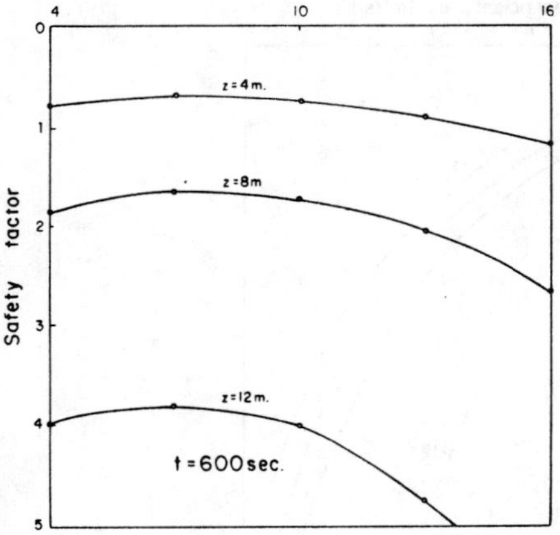

d

Fig.9 Parametric study (cont.)

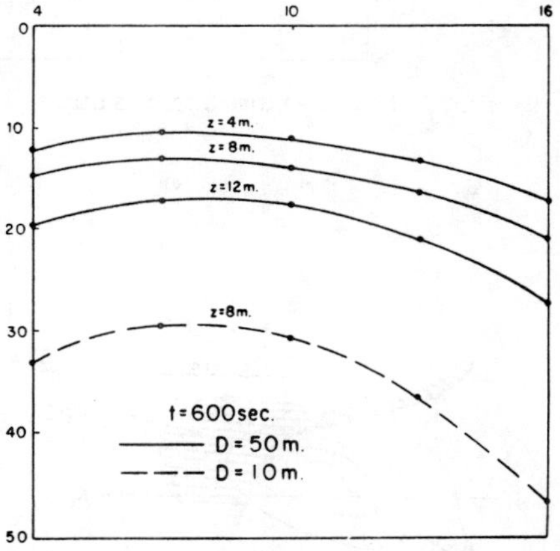

d

Fig.9 Parametric study (cont.)

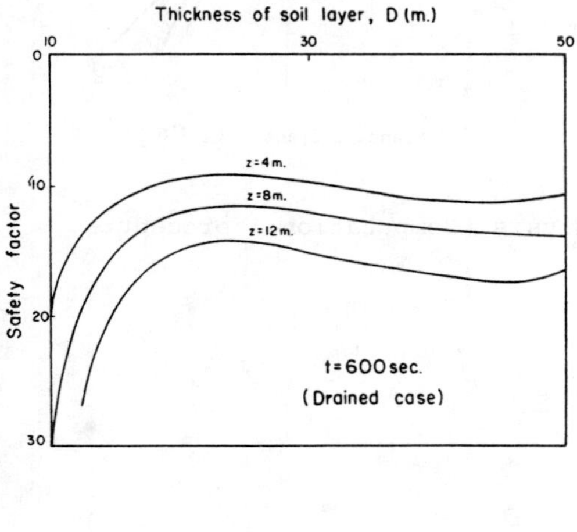

Fig. 10. Parametric study (cont.)

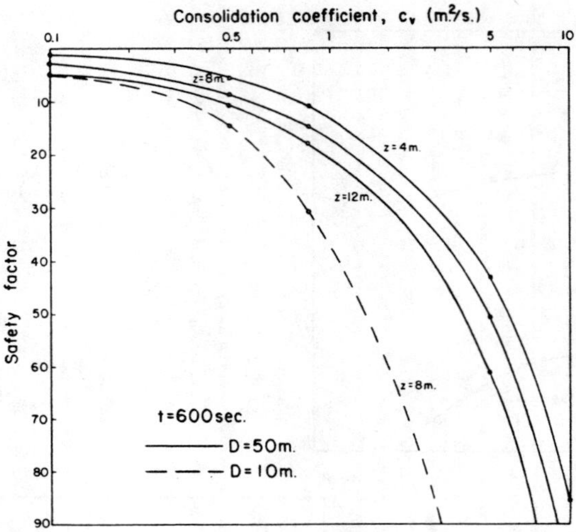

Fig. 11. Parametric study (cont.)

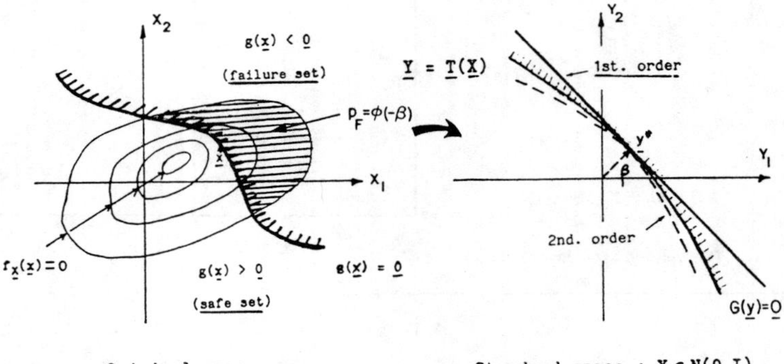

Fig.12. Reliability analysis. Computational procedure.

et al, 1986; Blázquez and Der Kiureghian, 1987) has been applied to compute the liquefaction susceptibility of the layer (unimodal system). Within this contest, the value of the actual short term probability of failure, p_l, is computed as:

$$p_\ell = \int_{g(\underline{x})<\underline{0}} f_{\underline{x}}(\underline{x})\, d\underline{x} \qquad (17)$$

in which $g(\underline{x})=0$ is the limit state function of the layer in the liquefaction mode of failure and $f_{\underline{x}}(\underline{x})$ denotes the pdf of the set of joint random variables, $\underline{x}$, describing the state fo the layer. Figure 12 depicts schematically the way to proceed in order to compute p_l for a specific case, namely $\underline{x}=(x_1, x_2)$. Details on the computation procedure may become quite involved and are given elsewhere (Der Kiureghian and Liu, 1986).

For the liquefaction problem, the limit-state functions are formulated as follows:

Undrained case

$$\{g_{\underline{x}}(\underline{x})=\underline{0}\} \equiv \{u_g-\sigma'_{v_0}=0\} \equiv \left\{\gamma' z\left[1-\frac{t}{T}\left(\frac{\gamma' 2aD_r\cosh(Kd)}{\gamma_w HK}\right)^{-\frac{1}{b}} e^{-\frac{Kz}{b}}\right]=0\right\} \qquad (18\text{-}1)$$

Drained case

$$\{g_{\underline{x}}(\underline{x})=\underline{0}\} \equiv \{u-\sigma'_{v_0}=0\} \equiv \left\{\gamma'\left(z-\frac{c}{c_v T}\left[w(z)+v(z,t)\right]\right)=0\right\} \qquad (18\text{-}2)$$

where z,t are parameters and the other symbols stand for random variables. The statistical distribution and first two moments of these variables is given in Table II. Only first order reliability analysis is considered in the present work.

Figure 13 illustrates the results of such an analysis. The probability of liquefaction of either a drained or undrained stratum at a given time and location is computed for a fixed geometry of the problem, specified in terms of water depth and thickness of the layer. A sample calculation (point P; Fig 13) is shown below:

Sample Calculation (point P)

Variables	x_i^*	y_i	Parameters
H	9.61	0.737	t = 600 sec.
			z = 6 m.
L	164	-0.301	D = 50 m.
D_r	0.50	-0.398	d = 4 m.
γ'	0.75	-0.338	β (First order) = 1.71
a	0.47	-0.298	
b	0.20	0.058	P_l (First order) = 0.0438
c_v	1.34	-0.023	

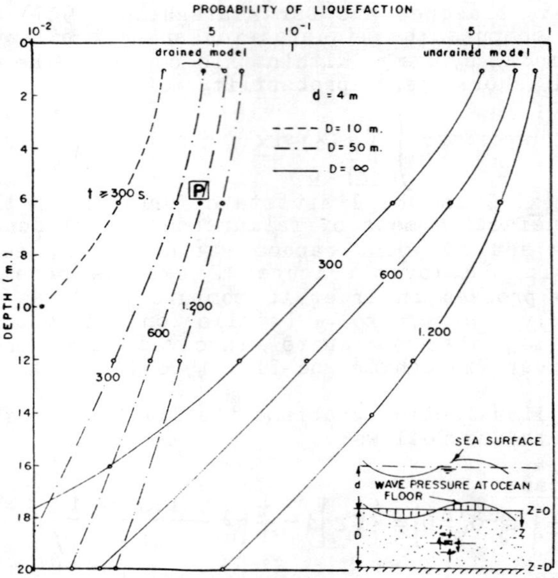

Fig.13. Probability of liquefaction. First-order reliability analysis.

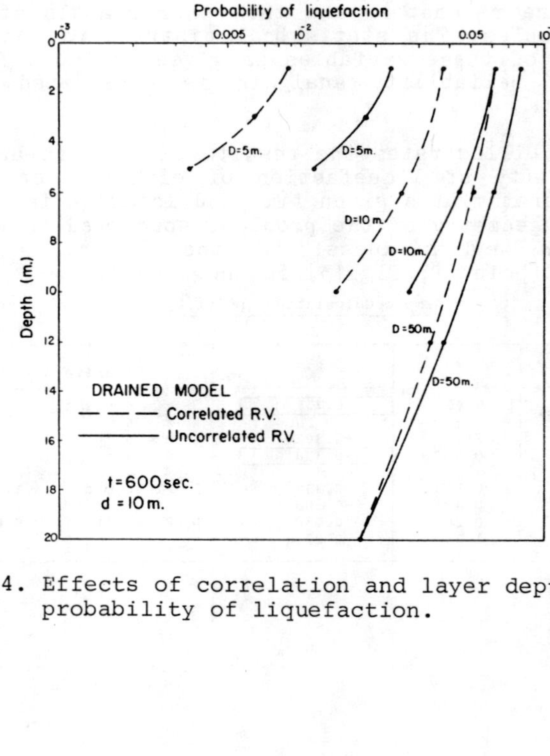

Fig.14. Effects of correlation and layer depth on the probability of liquefaction.

This calculation indicates that, for drained soils, the liquefaction behavior of the soil is mostly influenced (γ_i high) by the wave height and drainage conditions and, to a less degree and with opposite sign, by the relative density of the sand. The same conclusions apply to undrained deposits which, other conditions being equal, are roughly one order of magnitude unsafer than drained ones.

The combined effects of correlations among random variables and layer depth have also been investigated and are displayed graphically in Figure 14. A quick inspection of this figure reveals the probability of liquefaction becomes higher as the degree of correlation decreases and/or the length of the drainage path increases. Also, as could be expected, deep-sea deposits are less susceptible to liquefaction than surficial seabeds. This effect is more dramatic for thin permeable layers (Figure 15).

Risk model

In this case the joint distribucion function of the set $\underline{X}$ of random variables (eq. 17) is no longer employed. Wave input and liquefaction response random variables are split up, and the annual (long term) probability of liquefaction of a given soil layer is written as:

$$(p_\ell)_{D_r} = \int_0^H (p_\ell \mid H, D_r)\, f_H(H)\, dH \qquad (19)$$

Where $(p_l \mid H,D_r)$ = the probability of liquefaction at the site for specified values of the significant wave height and the relative density of the soil, and $f_H(H)$ = the density function of the sea state H. The last function is independent of the geomechanical properties of the site and can be derived from the long-term statistics of the severity of the sea, a topic that has been throuroughly discussed in the literature (Ochi, 1982). On the contrary, the conditional local probability of liquefaction for a given wave climate at the site, $(p_l \mid H,D_r)$, is not well known yet and depends solely on the uncertainties in the liquefaction model.

As an example, Figure 16 shows the risk of liquefaction at a depth of 1 m for loose and dense sand deposits located at two well different sites in the spanish shoreline, namely Bilbao (Biscay Gulf) and Alicante (Mediterranean Sea). The dramatic effect of the wave climate in the expected liquefaction behavior of the soil is self-evident.

Conclusions

Application of probabilistic techniques to the liquefaction analysis of sandy soil layers in an ocean

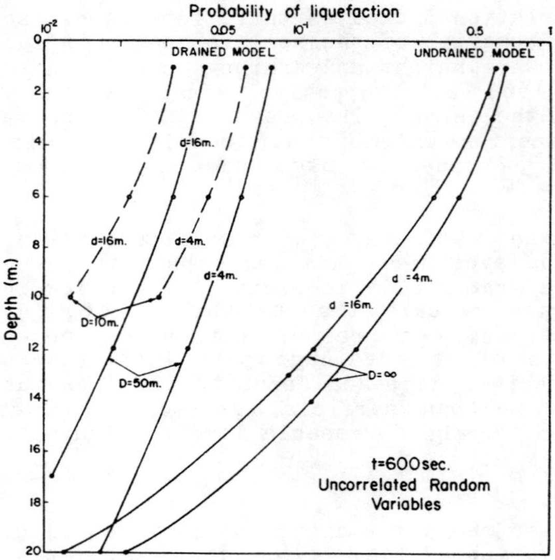

Fig. 15. Effect of water depth on the probability of liquefaction.

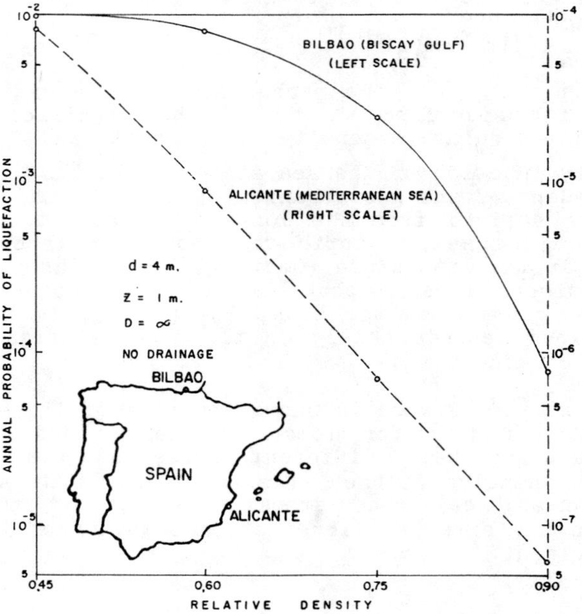

Fig.16. Risk of liquefaction at two sites on the Spanish coast.

wave environment is reviewed. The following conclusions can be drawn:

* Seafloor liquefaction failure is mostly influenced by drainage conditions, and to a less degree by wave height and relative density of the sand.

* Drained soil deposits are roughly one order of magnitude safer than undrained ones.

* Uncorrelated models of liquefaction reliability are overconservative, especially for thin soil layers.

* The probability of liquefaction increases with the thickness of the layer (⇒slower rate of consolidation) and shallow water depth (⇒less confinement).

* Risk liquefaction models-in which wave action and liquefaction resistance are uncoupled-can be easily derived from reliability models for long-term wave conditions.

REFERENCES

- Blázquez, A. and A.Der Kiureghian (1987), "Seismic Reliability of Retaining Walls", Fifth ICASP, Vancouver, Canada, Vol 2, pp. 1149-156.

- Der Kiureghian, a. and P.L. Liu (1986), "Structural Reliability under Incomplete Probability Information", Journal Engineering Mechanics, ASCE, 112:1, pp. 85-104.

- Layas, F.M. (1982), "Response and Stability of Ocean Floor Soils under Random Waves", PhD thesis, North Carolina State University, Raleigh, N.C., USA.

- Madsen, H.O, S. Krenk and N.C. Lind. (1986), "Methods of Structural Safety", Prentice-Hall, Englewood Cliffs, N.J., USA.

- Ochi M.K. (1982), "Stochastic Analysis and Probabilistic Prediction of Random Seas", Advances in Hydroscience, Vol. 13, pp. 218-375.

- Rahman, M.S. and W.Y. Jaber (1986), "A Simplified Drained Analysis for Wave-Induced Liquefaction in Ocean Floor Sands", Soils and Foundations, Vol. 26, Nov. 3, pp. 37-68, Sept.

- Seed, H.B. and Rahman, M.S. (1978), "Wave-induced pore pressure in relation to ocean floor stability of cohesionless soils", Marine Geotechnology, Vol. 3, No.2, pp. 123-150

CHAPTER 102

EROSION AROUND A PILE DUE TO CURRENT AND BREAKING WAVES

E.W. Bijker[1], M. ASCE, and C.A. de Bruyn[2]

ABSTRACT

Tests have been performed on a vertical pile subject to current only and to a combination of current with normal waves and current with breaking waves. The scour around the pile produced by current only is decreased by normal short waves superimposed upon that current and increased when breaking waves are superimposed upon the current. After analysis of the velocity profiles in the undisturbed area upstream of the pile and next to the pile, the following explanation is found for this phenomenon. When normal short waves are superimposed upon a current, the bottom shear stress of the combination of current with waves is increased more in the undisturbed area than next to the pile in the scour area. This results in a decrease of the scour around the pile. Due to the large values of the orbital velocity under breaking waves this effect is reversed for the combination of a current with breaking and relatively long waves. This results in an increase of the scour around the pile.

1. INTRODUCTION

Around piles of jack-up platforms along the Dutch coast in a zone where breaking waves occur occasionally, scour depths have been observed which are significantly more than the normally occurring 1.0 to 1.5 times the pile diameter. Normally the scour around a structure due to a combination of waves with current is less than the scour as a result of current only (Bijker, 1986). In order to try to find an explanation for the unexpected deep scour with a combination of current and breaking waves, tests have been performed in a flume and in a basin of the Laboratory of Fluid Mechanics of the Delft University of Technology. Since there are no bracings between the legs of a jack-up platform and the distance between the legs is large as compared with their diameter, one single pile is studied.

2. TESTS

2.1 General

Three test-series have been performed, with an as much as possible constant mean current velocity, $v = 0.40$ m/s, a water depth, $h = 0.285$ m, and a mean grain size diameter $D_{50} = 0.2$ mm.

Tests of series I have been performed in a flume of 0.8 m width and with a pile diameter $\phi = 0.048$ m.

---

[1] Professor of Coastal Engineering and [2] Graduate Student, Faculty of Civil Engineering, Delft University of Technology, Stevinweg 1, 2628 CN Delft, The Netherlands.

Tests of series II have been performed in a wave basin with the wave direction perpendicular to the current direction. The pile diameter was again 0.048 m.

Tests of series III have also been performed in the wave basin, but with a pile diameter ϕ = 0.09 m.

2.2 Specific conditions

All series have been performed with three conditions, viz.:

A. Only current, no waves. The water depth varied between 0.283 and 0.296 m and the current velocity between 0.391 and 0.461 m/s.
B. Current with normal short waves. Current velocity and water depth were as much as possible equal to that for current only. The wave height varied between 0.034 and 0.108 m and the period between 0.97 and 1.23 s.
C. Current with breaking waves. Current velocity and water depth were again almost equal to those for conditions A and B. The height of the breaking waves varied between 0.190 and 0.214 m with periods between 2.83 and 3.11 s.

In all tests rather uniform sand with a mean grain size D = 0.2 mm was used.

All tests are summarized in Table 1. Velocity profiles have been measured in most tests by means of an electromagnetic current meter (made available by Delft Hydraulics) in the undisturbed area upstream of the pile and next to the pile at a distance of 0.02 m from the wall of the pile. The bed form has been measured with a profile-follower.

The tests of series I have been performed in a flume of 25 m length with a width of 0.8 m which at the inflow side was equipped with a generator for regular waves. The sand bed had a thickness of about 0.1 m. The pile was placed at a distance of 10 m downstream of the inflow of the flume. It is assumed that at this place the flow will be completely adjusted. A sketch of the flume is given in Figure 1. From comparison with the tests in the basin it is concluded that the current pattern around the pile is somewhat influenced by the limited width of the flume. The influence is, however, so small that the tests will still demonstrate the differences between the various conditions.

During the tests of this series I the direction of propagation of the waves has been in the flow direction.

The tests of series II and III have been performed in a wave basin which is shown in Figure 2. The current was guided towards the test section by two training walls. In the tests of series II and III the wave direction has been perpendicular to the current direction and the -regular- waves were guided from the wave generator by training walls to the test section. At the end of these training walls the waves will diffract. The influence of this diffraction is, however, negligible at the location of the pile. Opposite to the wave generator a wave damping slope has been installed.

Since the distance between the location of the pile and the upstream side of the actual test section is rather short, sand has been applied on the slope upstream of the test section. This sand has avoided excessive scour of the 0.2 m thick sand layer at the upstream side of the test section and by the roughness it assisted in establishing the equilibrium velocity profile. Measurements of the velocity profile near the pile have indicated that this goal has been reached.

In Figure 3 typical examples of normal short waves and relatively long breaking waves are given. The orbital velocities at the bottom corresponding to short and breaking waves are given in Figure 4. Figure 5 shows the various measuring lines in the wave basin along which depths are measured. The values of the maximum scour are indicated

Table 1

Date		u* cm/s	u* norm. cm/s	r' cm	r'' cm	ug cm/s	h cm	cz cm	H cm	T sec	Ûb cm/s	ab cm	Tcw N/m2	Sb 10↑-2 cm2/s	Ss 10↑-2 cm2/s
I A															
	V	4.43	3.85	5.92	3.7	46.1	29.5	6.40					1.88	4.06	4.34
	N	6.56	5.69	21.12		49.5	35.6						4.88	9.70	
I B															
18-11	V	4.42	3.87	5.40	2.4	45.7	28.3	5.80	3.40	1.23	7.46	1.46	2.33	5.34	6.32
	N	5.41	4.73	10.56		50.2	35.6				6.07	1.19	5.32	10.73	
26-11	V	4.43	3.87	5.91	4.5	45.8	30.3	6.30	6.40	1.23	13.20	2.60	3.24	7.41	11.84
	N	4.93	4.31	7.56		49.9	35.7				11.40	2.20	5.99	11.89	
I C															
	V	2.65	2.41	1.09	5.8	43.8	29.6	7.00	19.20	3.11	53.00	26.20	9.50	12.34	36.75
	N	3.01	2.75	1.23		45.0	35.9				47.00	23.65	22.46	20.04	
II A															
	V	4.17	4.15	7.59	3.8	40.2	29.6	6.55					1.69	2.66	2.24
	N	4.91	4.79	19.88		37.4	34.4						2.89	3.49	
II B															
21- 1	V	3.15	3.34	2.77	2.2	37.7	27.3	5.80	6.30	1.16	13.70	2.50	3.04	5.60	5.95
	N	5.01	5.31	23.60		35.4	32.9				9.12	1.76	3.32	4.14	
19/23- 1	V	3.45	3.76	4.77	3.4	36.7	27.8	3.80	10.75	1.16	23.00	4.25	5.72	8.75	12.81
	N	3.21	3.48	4.21		36.0	30.6				16.88	3.12	5.17	7.43	
II C															
26- 1	V	1.77	1.54	0.01	2.8	46.1	27.4	7.00	19.20	2.83	54.00	24.70	13.58	15.24	35.03
	N	4.32	3.75	4.09		49.4	32.6				39.36	17.76	17.85	21.35	
28/29- 1	V	2.95	2.63	0.71	2.5	44.9	29.8	8.10	21.40	2.83	59.30	26.70	14.55	14.89	37.70
	N	5.37	4.79	15.35		44.5	33.7				43.88	19.42	19.27	19.47	
III A															
	V	3.89	3.98	6.57	4.0	39.1	28.3	10.15					1.63	2.40	1.00
	N	5.02	5.14	26.99		35.6	37.8						2.42	2.45	
III B															
12- 5	V	4.27	4.39	9.55	2.8	38.9	28.2	8.00	9.20	0.97	16.13	2.47	3.69	6.93	8.40
	N	4.60	4.75	12.59		41.0	35.1				9.84	1.51	4.15	6.66	
14/15- 5	V	3.60	3.62	4.15	3.4	39.8	28.3	8.95	9.20	1.16	19.35	3.55	4.67	8.43	12.00
	N	5.44	5.48	25.16		39.3	37.2				11.84	2.16	4.11	6.33	
III C															
14/15- 5	V	2.53	2.23	0.30	2.8	45.4	31.8	11.00	15.20	1.88	37.10	11.10	12.32	14.23	36.92
	N	6.16	5.43	19.24		49.6	39.7				20.00	8.67	11.43	17.32	
13- 5	V	1.77	1.81	0.05	3.2	39.2	28.5	10.40	21.00	2.81	58.60	26.20	13.94	12.93	31.34
	N	4.35	4.44	7.92		44.7	39.9				38.40	17.28	15.72	17.70	

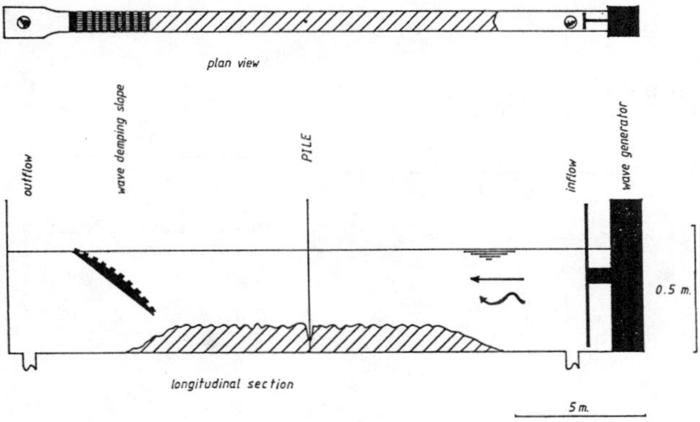

Figure 1
Flume

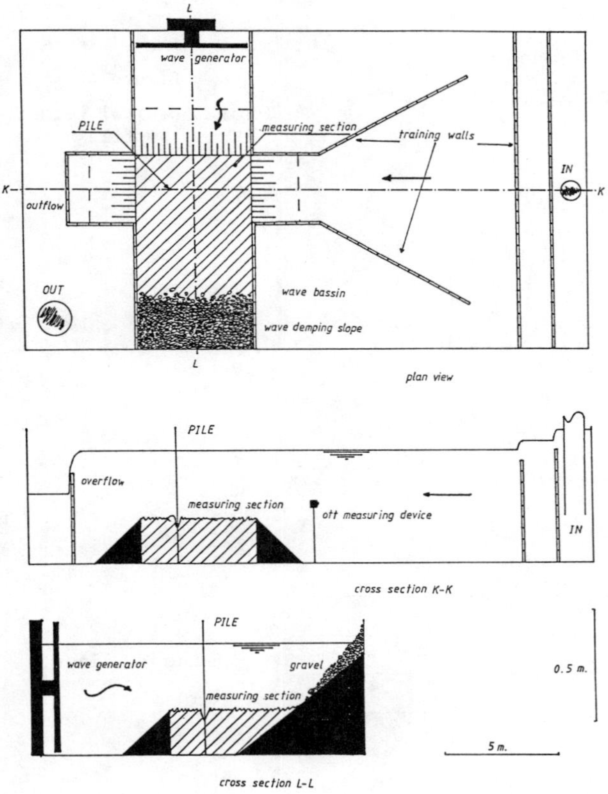

Figure 2
Wave basin

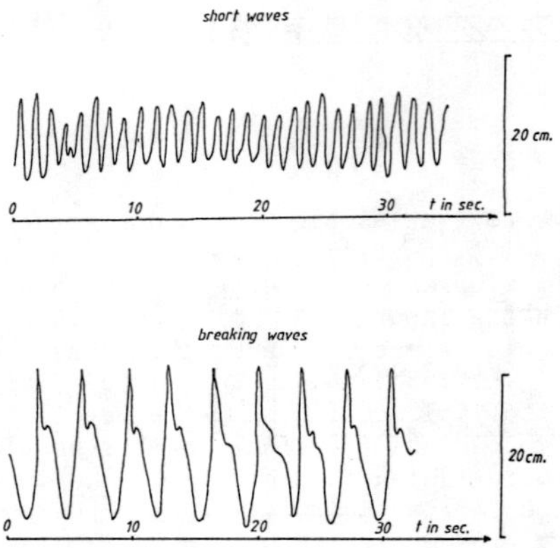

Figure 3
Typical wave profiles

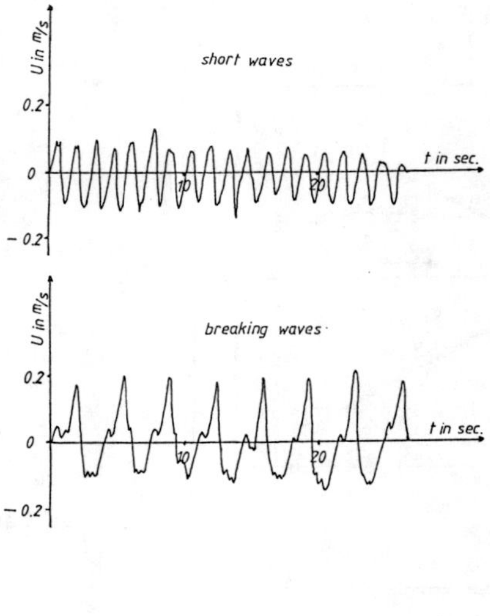

Figure 4
Typical orbital velocity profiles

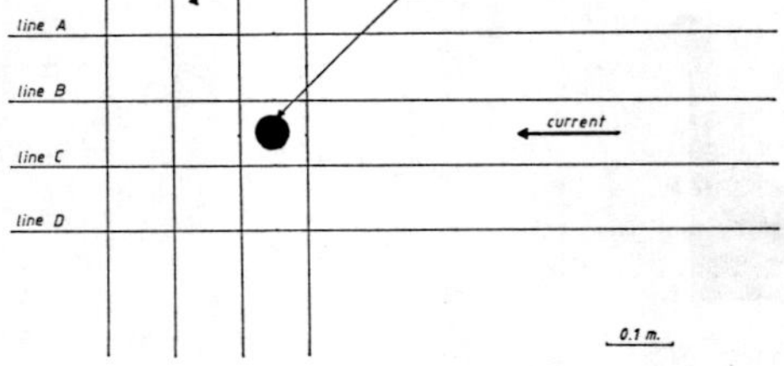

Figure 5
Measuring lines

in Table 1 with the line indication and an asterisk (so deepest point in line C is indicated by C*).

3. TEST RESULTS

3.1 Principles

The principle reason for any scour development around a structure is the difference in transport capacity upstream of the structure and around it. The possible relative increase -or decrease- of suspended load with respect to bed load in the scour area will result in extra -or less- erosion than would result from the change of the value of the total transport (Leeuwestein et al., 1985, and Bijker, 1986). Since, however, in this case the length of the scour area is rather limited, this effect can probably be neglected and the change of the bed load transport, dS_b/dx, will be the determining parameter.

Although in principle a method for the computation of the transport and subsequent scour development around a structure is available (Leeuwestein and Wind, 1984), it was not considered feasible to apply this method here since the transport module in this procedure is not yet fully developed. Also a more general computation of transports upstream and around the structure is not considered for the following reasons.

a. The development of the velocity distribution should be measured very detailed and the actual scour computation would require extensive transport computations.
b. The current around the pile is accelerating and the effect of an accelerating current on the transport is not yet fully understood.

The main goal in this research is, therefore, to explain the difference in scour development under the various conditions and to define the main parameters which determine this development.

3.2 Scour data

In order to give an impression about the order of magnitude of the transport, the quantities as measured in the wave basin and as calculated for the undisturbed conditions are given in Table 2.

Table 2: Measured and calculated transports

Condition	Measured S_{tot} in m^3/h	Calculated S_{tot} in m^3/h
A	0.05 - 0.10	0.05
B	0.10 - 0.20	0.19
C	0.50 - 1.00	0.54

The transport is calculated by the method suggested by Bijker for the combination of waves and current (Bijker, 1971).

In the ultimate (equilibrium) situation of the scour development the transport upstream of the pile and around the pile must be equal. Since, however, the flow lines around the pile are closer together than upstream of the pile, the transport per unit of width around the pile will be, also in the equilibrium situation, somewhat bigger than in front of the pile in the undisturbed area. The bottom shear stress around the pile must be, therefore, also in the equilibrium situation, bigger than in the undisturbed area.

In the Figures 6 through 8 some typical profiles along line B are shown. Since these profiles resulted from momentary surveys, the maximum scour depths are not necessarily equal to those given in Table 1.

4. DISCUSSION

4.1 Bottom shear stresses

The final scour around a structure will depend, as demonstrated in Chapter 4, on the ratio between the transport upstream of the structure and around it. It is difficult to calculate this transport, but at any rate it will be a function of the bottom shear stress. Therefore, these bottom shear stress values will be compared. As it is not well possible to measure these bottom shear stresses directly, they will be determined from the velocity profiles.

In the Figures 9 through 11 the velocity profiles for the undisturbed flow are given. The bottom shear stress can be calculated from the velocity profile under the assumption that it is that of a completely developed boundary layer. Such a profile can be described according to Prandtl (1926) and von Karman (1930) by

$$v(z) = \frac{v_*}{\kappa} \ln \frac{z}{z_o} ,$$

with $v_* = \sqrt{\tau/\rho}$ = shear stress velocity, z = height above the bottom, z_o = height above the bottom, where v(z) is theoretically zero and κ is the constant of von Karman, equal to 0.4.

Since it is difficult to determine exactly the height of a measuring point above a rippled sand bed, the formula is written as

$$v(z) = \frac{v_*}{\kappa} \ln \frac{z' + \Delta z}{z_o' + \Delta z} ,$$

in which z' is the height above the assumed bottom and Δz is the difference between the assumed and real bottom.

With a least square procedure the value of Δz is chosen in such a way that the difference between the actual profile and the logarithmic profile is minimal. In this way the most reliable value of v_* can be determined.

4.2 Combination of waves and current

The above described method cannot be used in case of a combination of waves and current because the bottom shear stress is determined by the actual velocity, which in this case is fluctuating in magnitude and direction through the orbital motion. This velocity differs from the velocity component in the main flow direction which is measured by the electromagnetic current meter (Bijker, 1968, 1971). The resultant bottom shear stress can be written as: (Bijker, 1986)

$$\tau_{cw} = \tau_c + \tfrac{1}{2} \tau_w ,$$

in which $\tau_w = \frac{1}{2} f_w \rho \hat{u}_b^2$, with f_w = bed friction coefficient according to Jonsson (1966). Swart (1976) has written this factor as

$$f_w = \exp \left[- 5.977 + 5.21 \, (a_b/r)^{-0.194} \right] ,$$

with a_b = amplitude of the orbital motion at the bottom and r = bottom roughness.

This formula holds for $a_b/r > 1.47$. For $a_b/r < 1.47$, f_w remains constant and equal to 0.32.

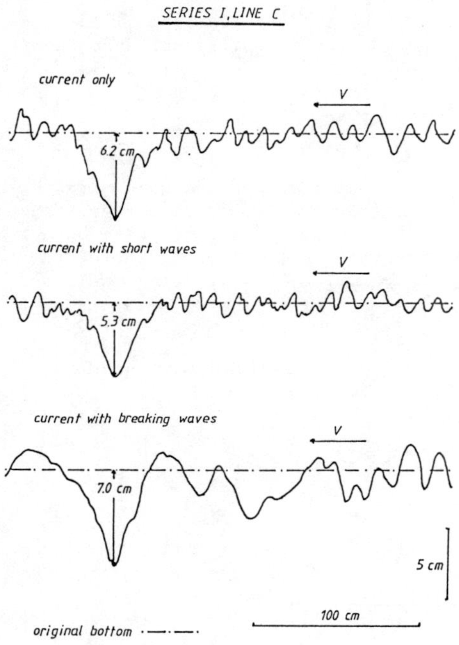

Figure 6
Typical scour profile

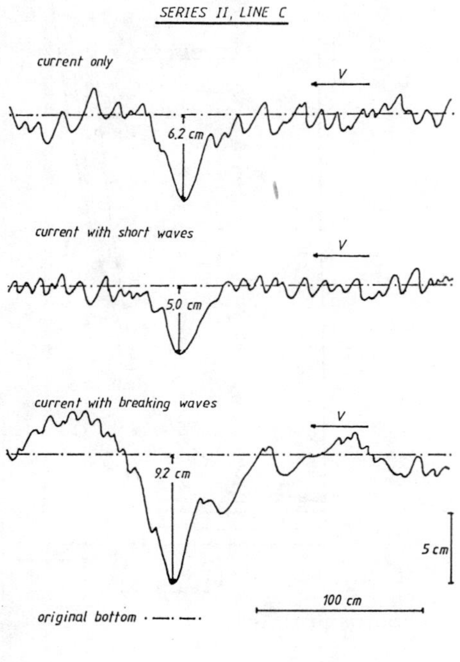

Figure 7
Typical scour profile

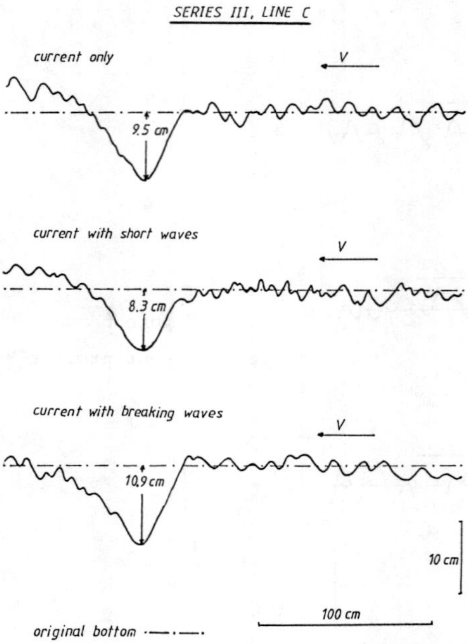

Figure 8
Typical scour profile

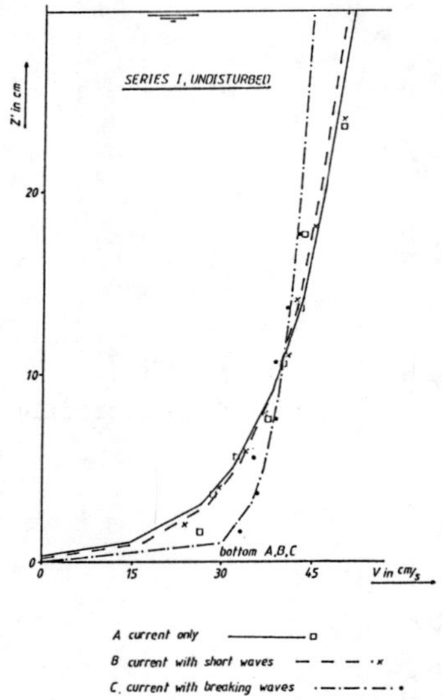

Figure 9

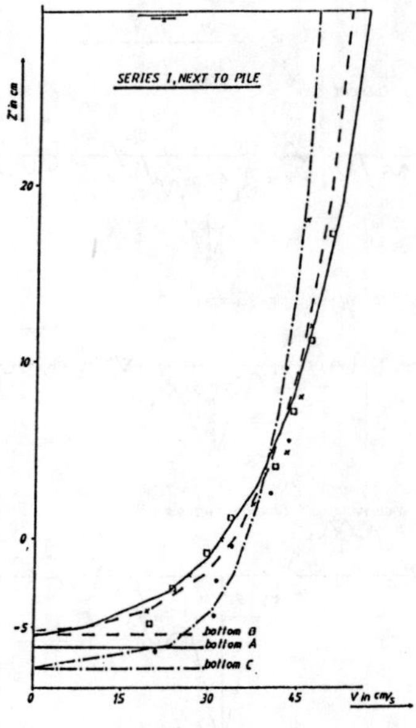

Figure 12

Typical velocity profiles

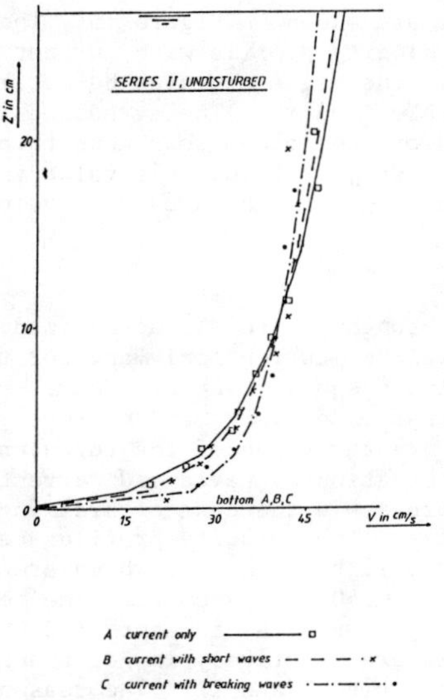

Figure 10

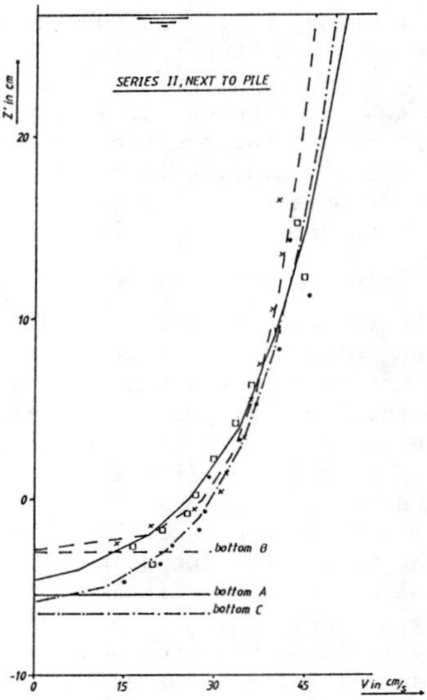

Figure 13

Typical velocity profiles

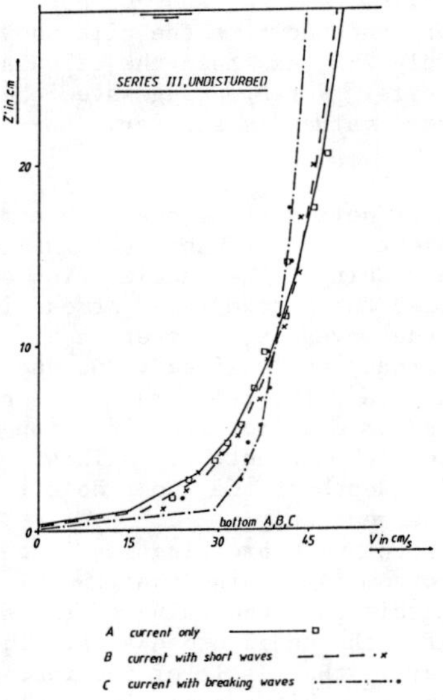

Figure 11

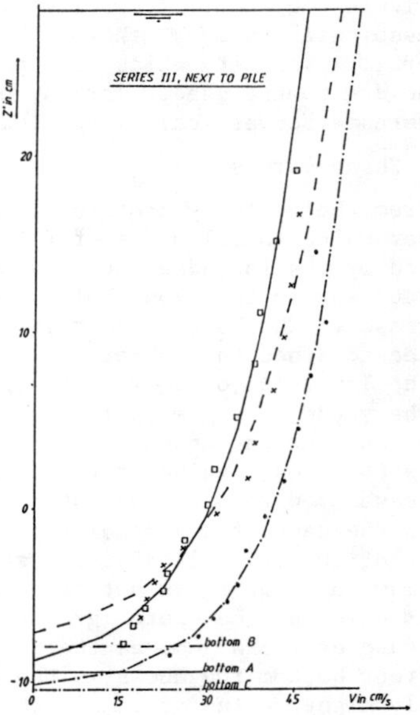

Figure 14

Typical velocity profiles

The values of f_w as function of a_b/r are shown in Figure 15. The value of r can be determined from the velocity profile with current only as 33 z_o and is given as r' in Table 1. The value can also be estimated from the ripple size and is given in Table 1 as r". The agreement is acceptable, but since the determination of the bottom roughness through the velocity profile is physically the most justified, this value is used as reference. In the flume the average of the calculated values is 0.05 m and in the basin 0.07 m.

4.3 Velocity profiles for the current

The velocity profiles of Figures 9 through 11 are all normalized to a mean velocity of 0.4 m/s. The profiles for current only and for the combination of waves and current are almost equal. There is some steepening of the profile for the combination of waves and current which is most likely the result of the greater value of the turbulent mixing coefficient in the case of a combination of waves and current (Van de Graaff, 1988). This is even more so in the case of breaking waves with much higher orbital velocities. The velocity profiles next to the pile above the scour hole (Figures 12 through 14), which are also normalized to the undisturbed velocity of 0.40 m/s, show the same tendency. However, in series III with the larger pile diameter of 0.09 m, the velocities measured next to the pile are relatively larger than for the smaller pile diameter of 0.048 m in series I and II. The reason is that the velocities are always measured at equal distances (0.02 m) from the pile wall. For the large pile the velocities are measured, therefore, relatively closer to the pile and so they are higher.

4.4 Velocity profiles for orbital motion

The orbital velocities are calculated by the first order linear wave theory and the results are shown, together with the measured values, in Figures 16 and 17 for series II and III with waves and current perpendicular to each other. The measurements next to the pile show an orbital velocity which is approximately 20% less than the calculated value due to the shadow working of the pile. For breaking waves the difference between calculated and measured values is greater.

4.5 Shear stress

From the velocity profiles in the scour hole next to the pile a much higher value is calculated for the apparent bottom roughness. This is caused by the increased turbulence as a result of the deceleration of the current in the scour hole. Therefore, the bottom shear stress is increased. The bottom shear stress of the waves is, however, not increased since this shear stress has already the maximum value due to the low value of a_b/r. This results in a relatively smaller increase of the value of $\tau_{cw} = \tau_c + \frac{1}{2}\tau_w$ in the scour area for the situation with waves and current than for the situation with current only. This leads to a lower value of the equilibrium depth of the scour hole in the case of waves and current than for current only.

In the case of the combination of current with breaking waves, the value of the orbital excursion at the bottom is so high that the maximum value of f_w is not reached. In this case the value of f_w in the scour hole is, therefore, greater than for the undisturbed area. This increase of f_w is just as the increase of τ_c the result of the increased apparent bottom roughness in the scour hole. This results in a relatively higher value of τ_{cw} in the scour hole for the combination of breaking waves and current than for current only.

The various values of the ratio between the bottom shear stress in the scour hole (τ_{cw}/N) and in the undisturbed area (τ_{cw}/V) are shown in Figure 18.

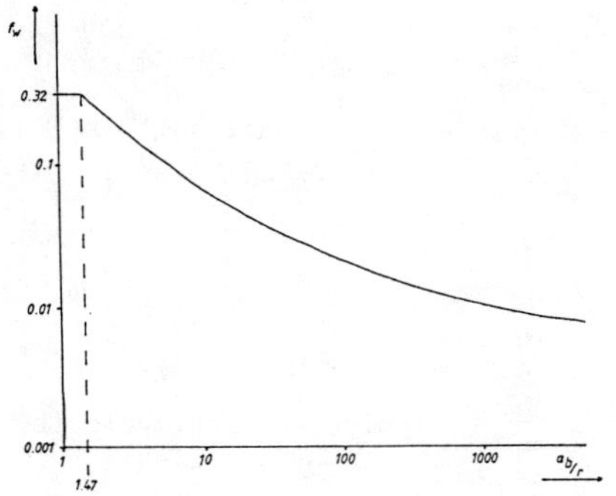

Figure 15

Relation f_w versus ab/r

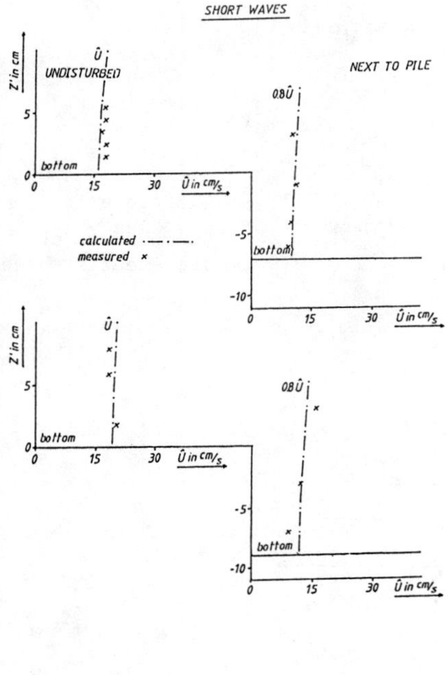

Series II

Figure 16

Typical orbital velocity profiles

Series III

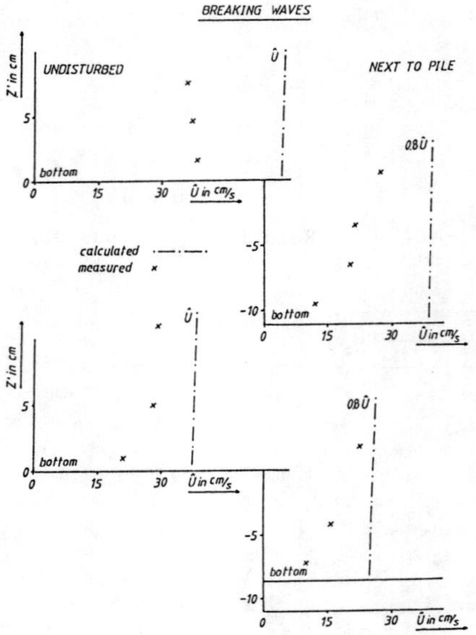

Series II

Figure 17
Typical orbital velocity profiles

Series III

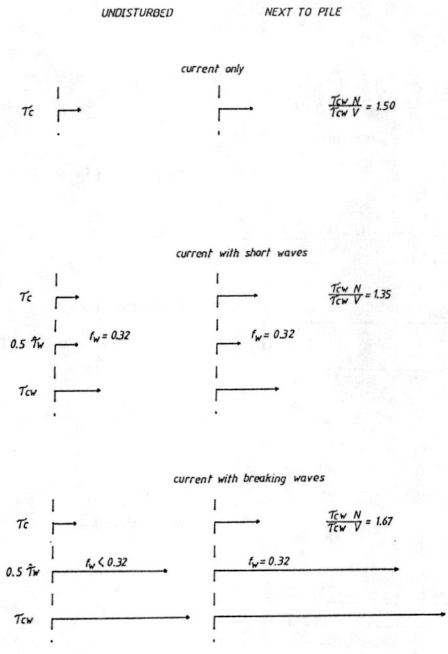

Figure 18
Comparison of
bottom shear stresses

The equilibrium depths of the scour hole have been explained by the effect of the apparent bottom roughness in the scour hole on the bottom shear stress for the situation with current only and for the situation with a combination of current with breaking and non-breaking waves. When the equilibrium depth is not yet reached, the scour is simply the result of the increased velocities around the pile. Only the extent to which the equilibrium scour developes will differ for the various circumstances.

4.6 Comparison with prototype situations

This research has been started because of the observed strong scour around the piles of a jack-up platform placed in -sometimes- breaking waves. The difference between the prototype situation and these tests is caused by the difference in bottom roughness. In front of the pile, in the undisturbed flow, the roughness in the prototype is probably not more than 2 to 3 times the roughness in the model. However, in the scour hole, around the pile, the roughness depends on the depth of the scour hole which is a function of the pile diameter. This diameter will be in the prototype 10 or 20 times that in the tests and the relative increase of the bottom roughness in the scour hole will be, therefore, more in the prototype than in the model. This results in a greater increase of f_w in prototype conditions with breaking waves than in the model. This leads to a greater value of the equilibrium scour depth. In prototype conditions indeed scour depths of 1 to 3 times the pile diameter have been found.

4.7 Conclusions

Normally waves will not increase, but even decrease the scour around a structure as compared with that by current only. The depth of this scour is in the order of 1.5 times the pile diameter. In the case of breaking waves this value can be, however, considerably higher. This may necessitate protective measures where otherwise the scour could be accepted.

5. REFERENCES

Bijker, E.W., 1967. Some considerations about scales for coastal models with movable bed. Delft Hydraulics Laboratory, Publ. No.50.

Bijker, E.W., 1971. Littoral drift as function of waves and current. Proc. A.S.C.E., Journ. of Waterw., Harbors and Coastal Eng. Div., WW4, November 1971.

Bijker, E.W., 1986. Scour around structures. Proc. on the 20th Intern. Conf. on Coastal Engineering, Taipei, Taiwan, 1986, Vol.II, pp.1754-1768.

Graaff, van de J., 1988. Sediment concentration due to wave action (in preparation).

Karman, von Th., 1930. Mechanical similarity and turbulence. Proc. of the 3rd Intern. Congress of Applied Mechanics, Stockholm, Sweden, 1930, pp.85-92.

Leeuwestein, W. and H.G. Wind, 1984. The computation of bed shear in a numerical model. Proc. of the 19th Intern. Conf. on Coastal Engineering, Houston, Texas, 1984, Vol.II, pp.1685-1702.

Leeuwestein, W., E.W. Bijker, E.B. Peerbolte and H.G. Wind, 1985. The natural self-burial of submarine pipelines. Proc. of the 4th Intern. Conf. on Behaviour of Offshore Structures (BOSS'85), Delft, The Netherlands, 1985, pp.717-728.

Prandtl, L., 1926. On fully developed turbulence, Zürich, Switzerland, 1926, pp.72-74.

Swart, D.H., 1974. Offshore sediment transport and equilibrium beach profiles, Delft Hydraulics Laboratory, Publ. No.131.

CHAPTER 103

IMPROVED LONGSHORE SAND TRANSPORT EVALUATION

A.S.- Arcilla, Professor*
A. Vidaor, Research Ass.*
J. Pous, Research Ass.*

Abstract

In this paper an improved bulk formulation for the longshore sediment transport rate is presented. It is based on a simplified hydrodynamic analysis of surf zone flow and supplemented by an exhaustive dimensional analysis. The proposed model includes the effect of the surf zone dynamic state (e.g. variation of longshore sand transport, Il, with breaker type) and it is now being used in the one- and two-line shoreline evolution models developed by the Maritime Engineering Laboratory in Barcelona.

1. Introduction

The interest of engineers in estimating longshore sand transport comes from the need to study coastal evolution in order to solve existing coastal problems.

General longshore sand transport models have been developed in recent years, all of them using stringent simplifying hypotheses, due to the poor present understanding of the physics of the problem. These models are usually based on extensive data measurement campaigns without providing the level of accuracy their cost would suggest.

Because of these drawbacks in the research stage, most of the models are still under development (analyses of basic hypothes, comparison with other data sets, etc.). In this context, the bulk formulations, basically because of their simplicity, have been the most extensively calibrated and used for coastal engineering analyses.

2. Background

In nearly all approaches to longshore sand transport evaluation, no matter whether they are energetics based (Komar and Inman, 1970)(Bailard, 1981) or shear-stress based (Bijker, 1971)(Swart, 1976), the influence of the surf-zone dynamic state (e.g. breaker type) on the Il model or formulation is seldom considered. This influence is slight-

*Maritime Engineering Laboratory
Catalonia University of Technology in Barcelona, SPAIN.

ly better understood for surf zone hydrodynamics, which are the driving factor for the sediment transport. As it was proposed in earlier works (S.-Arcilla et al.,1986)the mean rate of wave energy dissipation per unit area,D (accross-shore integrated value) can be related to the Iribarren parameter (Iribarren, 1949)(Battjes, 1974):

$$Ir = \frac{m}{(Hb/Lo)^{1/2}} \qquad (2.1)$$

where:
m: bottom slope (average for the surf-zone).
Hb: breaker wave height.
Lo: deep water wave length.

This relationship can be derived from bore theory, energetic balance and dimensional analysis, and has been tested with an extensive data set, including field and laboratory measurements.

Similarly, and due to the relationship between the driving term for longshore currents and D (established from the alongshore momentum balance equation) the longshore current velocity, Vl, was also shown to be related to the surf zone dynamic state, via iribarren's parameter (S.-Arcilla et al. 1986)

Both fits showed a bell-shaped trend for the relationship bettween the non-dimensional D and Vl and Ir, relating thus the breaker type with the amount of dissipated energy and the value of the longshore current. As an intermediate result of the research,the bottom friction coefficient, Cf, was also shown to be a function of the Iribarren parameter. The remaining step of this research is to analyse the influence of the surf zone dynamic state on the longshore transport rate, starting from the relationships obtained for D and Vl. This point is developed in the following sections.

3. Predictive Formulation

3.1. General

The sediment transport rate expressed in units of submerged weight per unit time, shall be denoted by I. This variable, as said before, has been evaluated in various ways during the last decades. One of the most physically-based approaches consists in assuming I to be proportional to a stirring term times a transporting term. This approach is shown in (Bagnold, 1966)(Bijker, 1968) and (Soulsby,1986). The significant differences in the details illustrate the difficulties and uncertainties associated to this problem. In this paper, the alongshore component of I,across-shore integrated, is expressed by:

$$Il = Kl\ Vl\ lb\ (\tau b - \tau bcr) \qquad (3.1)$$

in which:

Il: alongshore component of the sediment transport rate,expressed in Newtons/second and corresponding to the submerged weight of sediment passing throuhg a section (active profile) of the surf zone per unit time.

Kl: dimensionless proporcionality parameter.

Vl: across-shore averaged longshore current velocity.

lb: some measure of the width of the surf zone (e.g.horizontal distance from breaker- to shore-line).

τb: bottom shear stress.

In equation (3.1) the transporting term is (Vl.lb) and the stirring term is (τb- τbcr).This last term includes a threshold condition in the form of a critical value, τbcr, for the bottom shear stress. This means that for $\tau b \leq \tau bcr$, there is no Il. For $\tau b > \tau bcr$ the sediment transport rate is given by (3.1). The transporting term requires a longshore current and, thus, a non-zero value of D (mean rate of wave energy dissipation per unit time)to act. The stirring term, on the other hand, is associated to τb and depends on the dynamics of the wave/current boundary-layer.

The transporting term associated to Vl, will be mainly due to the incident energy flux, Efi, because these incident waves (Hbi) will generate a significant dissipation, D, and, thus, Vl associated to the breaking process. The reflected waves (Hr), in an across-shore integrated sense, will generate smaller D and Vl.

The stirring term, on the other hand, will be due to both wave trains (incident and reflected) because both of them generate a boundary layer and, thus, contribute to τb. In other words, the reflected waves will increase the roughness felt by the current and τb is, therefore, estimated based on a total wave-height, H_T:

$$H_T = Hbi + \beta \, Hr \qquad (3.2)$$

in which $\beta = \beta$ (Ir) is a parameter between 0 and 1. We can now consider the effect of increased reflection (due,e.g., to some new structure built in the surf zone) on the across-shore integrated sediment transport rate, Il. Increased reflection means that H_T increases from which both τb and the stirring term increase. On the other hand, for a given incident energy flux (Efi) and width of the surf zone (lb),increased reflection (Efr) implies a smaller D, according to the equation (S.-Arcilla et al., 1986):

$$Efi = Efr + D \, lb \qquad (3.3)$$

Therefore, the effect of increased reflection is to reduce D and Vl and, consequently, the transporting term. The global effect on Il, product of a stirring term that increases and a transporting term that decreases, is not easy to estimate a priori and must be obtained from the Il formulation.

3.2. Il Formulation

This formulation requires, according to equation (3.1), estimates for both Vl and τb. They will be derived from simplified hydrodynamic arguments. On the other hand, Kl and τbcr will be empirically calculated,by comparing this model with an alternative formulation derived from dimensional analysis.

i) V1 ESTIMATION

The alongshore component of the vertically integrated momentum conservation law reads for stationary conditions (e.g.(Mei, 1983)):

$$\frac{\partial S'_{xy}}{\partial x} + \frac{\partial S''_{xy}}{\partial x} + \tau_{yb} = 0 \qquad (3.6)$$

in which:

S'_{xy} : excess momentum flux tensor due to wave motion.
S''_{xy} : excess momentum flux tensor due to turbulent motion.
τ_{yb}: bottom shear-stress (y-component).

We can now evaluate the wave thrust in terms of D(Longuet-Higgins, 1970):

$$\frac{\partial S'_{xy}}{\partial x} = D\,\frac{\sin\alpha_b}{C_b} \qquad (3.5)$$

and D can be written (including reflection effects) as (S.-Arcilla et al.,1986):

$$D = \rho_w g \,\frac{Hbi^2}{T}\,\frac{\sqrt{2\pi}}{8}\,\gamma^{1/2} Ir \cos\alpha_b (1 - Kr^2) \qquad (3.6)$$

The bottom shear stress is usually expressed by the following linearised equation (obtained assuming small V1 and αb):

$$\tau_{yb} = \rho_w\, Cf\, Uorb\, V1 \qquad (3.7)$$

Substituting all these equations into (3.4), we can obtain (after neglecting turbulent mixing):

$$V1 = F(Ir)\, m\, (gHbi)^{1/2} \sin 2\alpha_b \qquad (3.8)$$

in which F(Ir) is a function of Iribarren's parameter.

ii) τb ESTIMATION

The τb estimation is based on the eddy-viscosity concept which allows us to write:

$$\tau_b = \rho_w A_H \frac{\partial U_c}{\partial Z} \qquad (3.9)$$

in which:

ρ_w: mass density of sea-water
A_H : eddy viscosity coefficient
U_c : characteristic velocity in the boundary layer
Z : vertical coordinate, positive upwards

The eddy viscosity coefficient, A_H, is estimated using the model proposed in (Battjes, 1975).

$$A_H = M.\ h_{Turb}.\ U_{Turb} \tag{3.10}$$

in which M is a parameter of order O(1) and h_{Turb} and U_{Turb} are respectively, the length and velocity scales of the turbulent motion. By writing $U_{Turb} = (D/\rho_w)^{1/3}$ it is easy to obtain:

$$U_{Turb} = f_1\ (Ir)\ Uorb \tag{3.11}$$

with Uorb being the wave orbital velocity and $f_1(Ir)$ a function of Iribarren's parameter.

The vertical velocity gradient is expressed by:

$$\frac{\partial U_c}{\partial Z} \propto \frac{Uorb}{\delta} \tag{3.12}$$

in which δ is the thickness of the boundary layer and U_c has been approximated by Uorb because the wave boundary layer is expected to be much thinner than that of the current and, thus, more relevant for the calculation of τb. If we express δ by (e.g.(Nielsen, 1986))

$$\delta = (A_H.T)^{1/2} \tag{3.13}$$

and substitute all previous equations into (3.9) it is easy to obtain:

$$\tau b = \rho_w\ f\ (Ir)\ G(\theta)\ Uo^2rb \tag{3.14}$$

with θ the mobility parameter.

iii) Il FORMULATION

Substituting (3.8) and (3.14) into (3.1) we get:

$$Il = \rho_w\ \hat{F}(Ir)(gHbi)^{3/2}\ Hbi\ \sin 2\alpha b(\hat{G}(\theta) - \hat{G}(\theta cr)) \tag{3.15}$$

The obtained formulation, apart from including a threshold level for incipient motion, presents an explicit variation of the transport rate with the breaker type or, more generally, with the surf zone (S.Z.) dynamic state (via Ir). This allows to consider (albeit in an across-shore integrated manner and somewhat simplistically) the effect of increased reflection or other changes in the S.Z. dynamic state on the transport rate.

The functions F(Ir) and G(θ) will be calibrated in section 5, making use of some of the results obtained from the dimensional analysis presented in section 4.

4. Dimensional Analysis

4.1. General

The general expression of the dimensional formula for longshore sand transport (in a longshore uniform coast with a plane shore) following (Kamphuis et al.,1986) can be written as:

$$Il = f(\mu, \rho_w, g, \rho_s, D, \psi, F(D), x_c, y_c, h_b, H_b, T, \alpha_b) \quad (4.1)$$

where: μ (dynamic viscosity), ρ_w (water density) and g(gravity acceleration) are the fluid parameters; ρ_s (sediment density), D(characteristic grain diameter) ψ (grain shape factor) and F(D) (granulometric distribution function) are the sediment parameters; x_c (across-shore characteristic scale) and h_b (water depth at breaking point) are the domain parameters; H_b (breaking wave height), T (wave period) and α_b (angle of wave incidence at breaker line) are the wave parameters.

Stationary conditions are assumed (time is not included) and only wave parameters are included since currents are considered to be wave induced. Other variables usually related to longshore transport conditions (e.g., wave orbital velocity, u_{orb}, or the sediment fall velocity, w), turn out to be, in this context, not-independent parameters and will, thus, appear only by exchange. Shape and granulometric distribution factors are not included (in general) in the data set used for calibration, and they will not be further considered.

For the development of the dimensional analysis four sets of repeaters have been employed: $\{\rho_w, g, H_b\}$, $\{\rho_w, h_b, T\}$, $\{\rho_s, g, D_{50}\}$, $\{\rho_w, g, H_b, T\}$. The main conclusion of these analyses was that the first set provided a better fit while giving the more physically meaningful parameters. This set was, therefore, chosen for the development of the model and is the only one here presented. Applying the Π theorem, and noticing that 8 non-dimensional parameters are expected to appear (13 independent variables, 2 or which are neglected and three of which are taken as repeaters, i.e. associated to the measurement system), the following expressions are obtained:

$$\Pi_{Il} = \frac{Il}{\rho_w g\, H_b^2\, (gH_b)^{1/2}} \quad (4.2)$$

$$\Pi_{Il} = \emptyset\left(\frac{\mu}{\rho_w H_b (gH_b)^{1/2}}, \frac{\rho_s - \rho_w}{\rho_w}, \frac{\rho_w H_b}{\rho_s D}, \frac{X_c}{H_b}, \frac{Y_c}{H_b}, \frac{h_b}{H_b}, \frac{g^{1/2} T}{H_b^{1/2}}, \alpha_b\right) \quad (4.3)$$

Combining these expressions Π_{Il} can be approximately written as:

$$\Pi_{Il} = \emptyset\,(Re, Ir, Id, \theta, m) \sin 2\alpha_b \quad (4.4)$$

in which

Re: Reynolds number (ratio between inertial and viscous forces).

θ : Mobility or shields Number (ratio between disturbing and resisting forces).

I_D: Dean number (ratio between fall velocity and a characteristic wave velocity)(Dean, 1973)(Dalrymple, 1976).

and assuming that the relationship between the angle of wave incidence (αb) and the longshore sand transport is (Komar and Inman, 1970)(Longuet-Higgins, 1072):

$$Il <> \sin 2\alpha b \qquad (4.5)$$

The expression for the Ø function has been thoroughly calibrated with the following field and laboratory data:

- field data : (Watts, 1953)(Komar, 1969)(Kana, 1977)(Kana, 1980)(Inman et al., 1980)(Gable, 1981)(Bruno et al., 1981)(Kraus et al., 1982).
- laboratory data:(Kamphuis et al., 1978)(Vitale, 1981)(Sayao and Kamphuis, 1983).

The calibration has been based mainly on field values due to scale effects. Exhaustive analyses have been performed in order to select the most adequate data from each set, deleting those data not fulfilling the imposed quality checks. The breaking depth, h_b, was obtained via Goda's criterion when not measured. Curve adjustement and fitting was obtained both visually and by a least-squares technique.

4.2. Relationships among parameters

4.2.1. Iribarren parameter/Dean number

Both parameters refer to onshore/offshore transport and beach profile variations. It is easy to show, from physical arguments, that a relationship as the one shown in figure 4.1 (corresponding to field data) must exist between the two parameters, defined by (2.1) and $I_D = H/wT$.

The proposed relation can be derived from the influence of grain size, D_{50}, on both bottom slope, m and fall velocity, w. This results in a somewhat equivalent use of the two parameters for across-shore-integrated analyses (e.g. Il). As a consequence, only Ir will be employed in what follows.

4.2.2. Reynolds number/Mobility number

Both parameters refer to the stage of flow development and, thus, to the stage of transport. They are related for initiation of motion conditions (Shields threshold curve). By combining their expressions, they can also be related for actual transport conditions:

$$Re = \theta^{2/3} f(k)$$

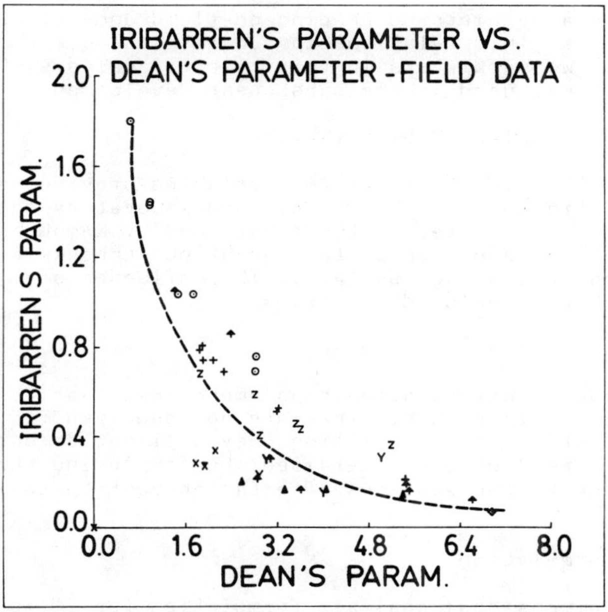

Figure 4.1: Iribarren parameter versus Dean's number (field data).

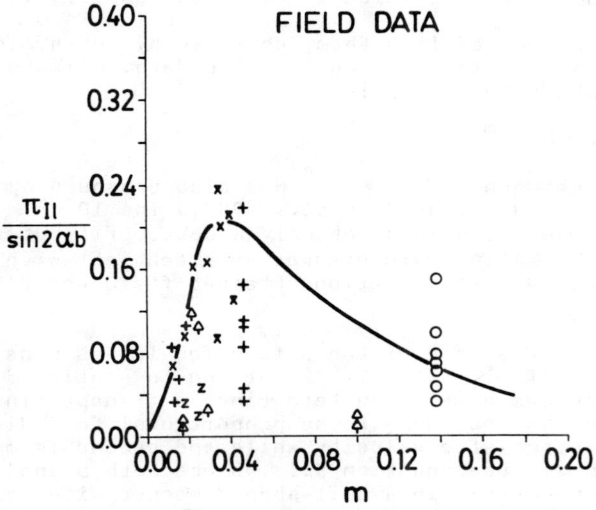

Figure 4.2: Π_{Il} versus m (beach slope) and the corresponding least-squares fit.

where k is a parameter depending on fluid and sediment characteristics. This relationship suggests the use of just one of the two parameters in this type of analysis. Only θ will be, thus, used in the subsequent development.

4.2.3. Bottom slope/Mobility number

A simple plot of these two variables provides an acceptable relationship for both field and laboratory data. These results are similar to those obtained by Kamphuis et al., 1982 and 1986) and suggest to discontinue the explicit use of m in equations such as (4.4). The influence of m on Il is, therefore, included via Ir.

4.2.4. Other relationships

Various other relationships among these parameters can be easily obtained. Since they are not required for the development of the Il formulation they will not be here included. Due care should be exercised when employing these relations due to the spurious effects inherent to this type of analysis.

4.3. Il formulation

The dimensional analysis formulation for Il is obtained by substituting the relationships derived in sections 4.2.1, 4.2.2 and 4.2.3 into equation (4.4):

$$Il = \sin 2\alpha_b \cdot \emptyset(Ir,\theta) \qquad (4.7)$$

This simplified expression is functionally identical to the one derived in section 3 (see equation 3.15).

The variation of Il with m, obtained by performing a least-squares fit to measured Il field data, turned out to be of the form (figure 4.2):

$$Il \leftrightarrow m^{1/2} \cdot 10^{-8m} \qquad (4.8)$$

The relationship $\Pi_{Il} \leftrightarrow m^{1/2}$ was also proposed by (Bruun et al., 1986) and (Deigaard et al., 1986). The 10^{-8m} factor included here provides not only a better fit but also a bell shaped variation with breaker type coherent with the corresponding laws of evolution obtained for D and Vl (see section 2).

To derive the form of the $\emptyset$ function let us consider the variation of Π_{Il} with Ir. According to section 3, Il can be written as a stirring term times a transporting term. The first one can be shown to be proportional to D divided by some characteristic velocity while the seocnd is mainly dependent on Vl (see equation 3.1). Since both D and Vl have been shown to vary in a bell-shaped manner with Ir, this same behaviour should be expected for Π_{Il}. When plotting Π_{Il} versus Ir (figure 4.3) a bell-shaped trend is in fact found. The mobility number θ appears to act as a parameter of the obtained family of curves, controlling the scale of the process. This type of variation with Ir is fully cohe-

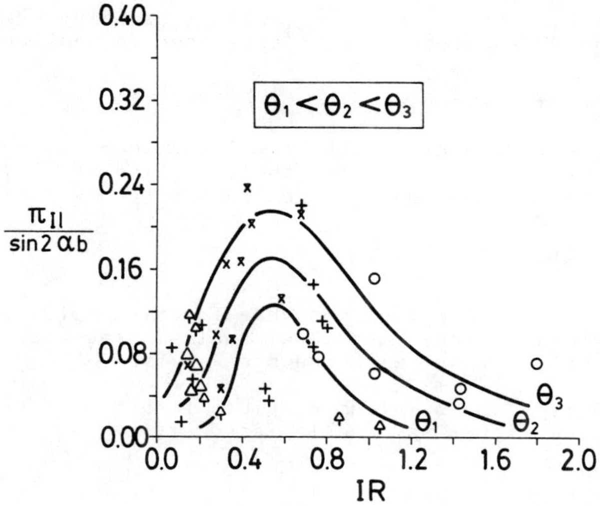

Figure 4.3: Π_{Il} versus Iribarren parameter (Ir) with Θ (mobility number) as family parameter.

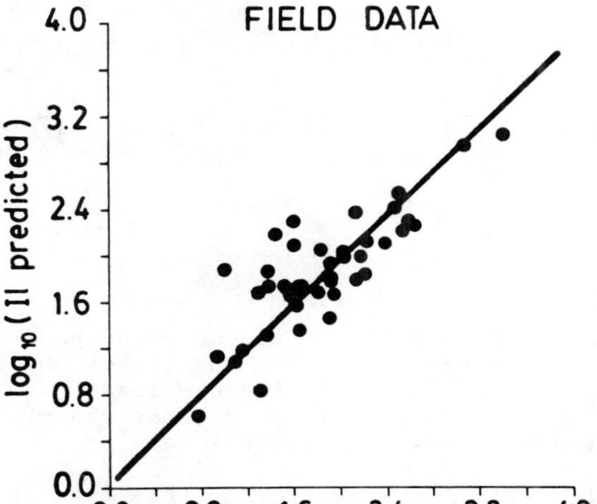

Figure 5.1: Il predicted versus Il measured (field values).

rent with the variation of π_{Il} with m given by equation (4.8) and it shall be further exploited in section 5.

5. Calibration of the Model

The variation of Il with m in equation (3.15) must be via the F(Ir) function. Using the variation of π_{Il} with m obtained in the previous section we may then write:

$$\hat{F}\,(Ir) = A\ Ir^{1/2}\ 10^{-8\ Ir} \qquad (5.1)$$

This is the simplest possible function varying with m as prescribed by the dimensional analysis and possessing a bell-shaped behaviour with the breaker type.

Likewise, the simplest possible function of θ including this parameter as a scale factor is:

$$\hat{G}(\theta) = \left(\frac{\rho_s - \rho_w}{\rho_w}\right)\ \theta^c \qquad (5.2)$$

in which the first term has been included to express Il as a function of $(\rho_s - \rho_w)$ instead of (ρ_w).

The resulting final equation for Il is obtained by substituting (5.1) and (5.2) into (3.15):

$$Il = (\rho_s - \rho_w) A\ Ir^{1/2}\ 10^{-8.00 Ir} g^{3/2} Hbi^{5/2} \sin 2\alpha\, b.(\theta^c - \theta^c_{cr}) \qquad (5.3)$$

By performing a least squares fit to measured Il values (field data only) it is easy to obtain:

$A = 4.8 \times 10^{-3}$

$c = 0.44$

$\theta_{cr} = 12.77$

in which the factor 10^{-3} in A can be obtained from dimensional analysis considerations. The quality of the obtained fit was not very sensitive to the precise θ_{cr} value, probably indicating that threshold conditions cannot be accurately established with the available data base.

To illustrate the structure of this formulation it is compared in Table I with three other well known models for Il. This comparison is by no means exhaustive since many other significant Il models have not been included (e.g. (Dean, 1973)(Watanabe et al., 1988)). Its main purpose is to give a feeling of the functional structure obtained with various formulations. A quantitative measure of the corresponding goodness of fit appears in Table II. The agreement obtained with the proposed formulation is shown in figure 5.1. It should be, however, mentioned that the number of free parameters for the fit of the proposed model is higher

Il model	PARAMETER DEPENDENCY						
	Hb	T	m	D_{50}	K_R (via Ir)	αb	ρ_s
CERC, 1970	$Hb^{2.5}$	-	-	-	-	$\sin 2\alpha b$	-
Kamphuis et al. 1986	$Hb^{3.5}$	-	m^{1}	D_{50}^{-1}	-	$\sin 2\alpha b$	$\rho_s - \rho_w$
Kraus et al. 1986	$Hb^{2.5}$	-	m^{-1}	-	-	$\sin 2\alpha b$	$\rho_s - \rho_w$
Arcilla et al. 1986	$Hb^{2.69}$	$T^{1/2}$	$m^{1/2}$	$D_{50}^{-0.44}$	$10^{-0.201r}(1-K_R^{*})^{7/6}$	$\sin 2\alpha b$	$(\rho_s-\rho_w)(\frac{\rho_w}{\rho_s})^{0.44}$

Table 1: Parameter dependency of various Il models compared to the formulation proposed in this paper.

Il model	S_{RSE}
CERC, 1970	0.6058
Kamphuis et al. 1986	0.3679
Kraus et al. 1988	3.7216
Arcilla et al. 1988	0.3308

Table II: Fit obtained with various Il formulations using the following measure (standard error) of the scatter:

$$S_{RSE} = \left[\sum (\log_{10}(Il) \text{ predicted} - \log_{10}(Il) \text{ measured})^2/(N-1) \right]^{1/2}$$

than for other formulations. This could be associated to the obtained reduction in dispersion. It should also be pointed out that the high scatter obtained for (Kraus et al., 1988) model is probably due to the fact that apparently there were some misprints in the coefficients reported in the original paper and used here for the fit (Kraus personal communication).

6. Conclusions

1. A bell-shaped variation with Ir (and thus, with the breaker type) has been obtained from physical arguments and field data for the following variables: D (mean rate of wave energy dissipation per unit area), Vl (mean longshore current velocity) and Il (mean alongshore sand

transport rate).

2. A reasonable correlation has been found between the following pairs of dimensionless parameters:

   - Iribarren's and Dean's parameters
   - Reynolds and mobility parameters
   - Beach slope and mobility parameter

   This means that these parameters can be used somewhat equivalently in the context of across-shore integrated transport evaluation. It should also be remarked that this conclusion is limited to the employed data base and could also be affected by some spurious effects, always present when dealing with dimensionless parameters obtained from dimensional analysis. Because of this the predictive formulation proposed has been directly calibrated from equations (4. 8) and (5.3), therefore avoiding the use of repeated variables in the two axes of the fitting diagram. The correlation among dimensionless parameters can illustrate the dangers of using models exclusively based on dimensional analysis.

3. The proposed Il formulation calculates this variable as a transporting term (proportional to Vl) times a stirring term (proportional to τ b). The main advantages of this model are:

   - It allows using different formulations for any of the two terms involved.
   - It explicitely shows the variation of Il with Ir and θ parameters and, in general, with the surf-zone dynamic state. In particular, this model includes the effect of varying reflection on Il.
   - It provides a slightly better fit to the employed data base than other available formulations.

The proposed model represents, thus, an attempt to provide a more precise and physically based Il formulation. This conclusion must be, however, interpreted in the context of the limitations (in quantity and quality) associated to the available Il field data.

7. References

- Bailard, J.A. (1984) "A simplified model for longshore sediment transport". Proc. Int. Conf. Coastal Eng. 1984.
- Bijker, E.W. (1968) "Litoral drift as a function of waves and current". Delft Hyd. Lab. Pub. nº 58.
- Dean, R.G. (1973) "Heuristic models of sand transport in the surf zone". Proc. Conf. on Eng. Dyn. in the Surf Zone, Sidney, Australia, p. 208-214.
- Kamphuis, J.W. and Readshaw, J.S. (1978) "A model study of alongshore sediment transport rate". Int. Conf. on Coastal Eng. ASCE.

- Kamphuis, J.W., Davies, M.H., Nairm, R.B. and Sayao, O.J. (1986) "Calculation of littoral sand transport rate".Coastal Eng., 10, 1-21.
- Komar, P. and Inman, D.L. (1970) "Longshore sand transport on beaches", Journal Geophys. Res., Vol. 75 nº 30 pp.5914-5927.
- Kraus, N.C., Hanson, H. and Larson, M. (1988) "Threshold for longshore sand transport and application to a shoreline change simulation model". IAHR. Symp. on Math. Modelling of Sediment Transport in the Coastal Zone.
- S.- Arcilla, A., Vidaor, A. and Monsó, J.L. (1986) "Influence of breaker type in surf zone dynamics". Proc. Int. Conf. Coastal Eng. 1986.
- Sayao, O.S. and Kamphuis, J.W. (1983) "Littoral sand transport rate: Model tests 1977-1982" CE, Res. Rep. 79,Civil Eng. Dep., Queen's Univ., Kingston, Canada.
- Soulsby, R.L. (1988) "The structure of suspended sediment transport formulae for unidirectional and wave-plus- current flows". IAHR Symp. on Math. Modelling of Sediment transport in the Coastal Zone.
- Vitale, P (1981) "Movable bed laboratory experiments comparing radiation stresses and Energy Flux factor as predictors of longshore transport rate", res. nº 81-4, C.E.R.C.

CHAPTER 104

LONGSHORE CURRENT AND TRANSPORT ACROSS NON-SINGULAR EQUILIBRIUM BEACH PROFILES

Kevin R. Bodge[1] (Associate Member, ASCE)

ABSTRACT

The longshore current and longshore sediment transport distributions are described across an equilibrium beach profile comprised of an intersecting planar foreshore and a concave-up profile. Such a profile shape avoids the singularity associated with the infinite-slope at the shoreline described by traditional equilibrium profile forms and allows prediction of beach processes at and above the shoreline. The mathematical expressions which describe the distributions are simplified and can be more readily applied relative to expressions previously presented in the literature. The findings are in general agreement with similar previous analytic studies and indicate that the current and transport maxima are generally located at about the intersection of the planar and concave-up portions of the profile.

INTRODUCTION

Recent analytical solutions and data indicate that the longshore current and transport distributions across non-planar beaches can be markedly different than that predicted for planar beaches (see Sawaragi and Deguchi, 1978; Berek and Dean, 1982; Sternberg et al., 1984, Bodge and Dean, 1987; Fulford, 1987; Kraus, Rosati and Gingerich, 1988; Bodge, in press; among others). Since few natural beaches are planar in profile, the solution of the longshore current and transport distributions across non-planar beaches is relevant to improved understanding and modelling of littoral processes for physically realistic beach profiles.

BACKGROUND

Bruun (1954) and Dean (1977) have suggested that equilibrium beach profiles are described by the form

$$h = a\ x^{2/3} \qquad (1)$$

[1]Senior Engineer, Olsen Associates, Inc., 4438 Herschel Street, Jacksonville, FL 32210.

where h=depth, x=distance offshore, and a is a dimensional coefficient. A disadvantage of the concave-up profile described by this equation is the infinite beach slope at the shoreline (x=0) which leads to (i) a singularity in the description of the longshore current and transport at the shoreline, and (ii) no description of the swash zone. In contrast, most equilibrium profiles in nature are characterized by a relatively planar beachface which extends from the berm to just below the shoreline.

McDougal and Hudspeth (1984) analytically considered the wave-induced set-up and longshore current and transport across beach profiles described by a planar foreshore and an $ax^{2/3}$-type profile seaward of the foreshore. These authors considered both "stress" and "energetics" longshore transport models after Komar (1977) and Bagnold (1963), respectively.

In the present study, the set-up, longshore current and longshore transport across such a combined planar-foreshore/concave-up profile is analytically described using (1) the longshore current solutions described by Longuet-Higgins (1970) and in part by McDougal and Hudspeth (1984), respectively, across the planar and concave-up portions of the profile; (2) a simplistic assumption that the local wave height is proportional to the local set-up water depth, and (3) a distributed longshore transport model described by Bodge and Dean (1987).

THE NON-SINGULAR EQUILIBRIUM BEACH PROFILE

Consider a beach profile such as shown in Figure 1 with x-axis directed offshore and origin x=0 at the still water shoreline. The profile is composed of two mathematically described segments: (1) a planar foreshore of slope m_f, and (2) a concave-up profile of the form $h=ax^{2/3}$, where h is the still-water depth and a is a shape parameter. The profiles intersect at the "match-point" $x=x_m$. The usual origin of the concave-up profile (i.e., at depth h=0) is displaced seaward of the x-axis origin by the distance δ. Specifically,

$$h = \begin{cases} m_f\, x & x \le x_m \\ a(x-\delta)^{2/3} & x \ge x_m \end{cases} \tag{1}$$

Requiring that the bottom depths and slopes are identical at the match-point x_m yields

$$x_m = \frac{4}{9}\frac{a^3}{m_f^3} \tag{2}$$

and

$$\delta = \frac{x_m}{3} \tag{3}$$

That is, the location of the match-point can be simply

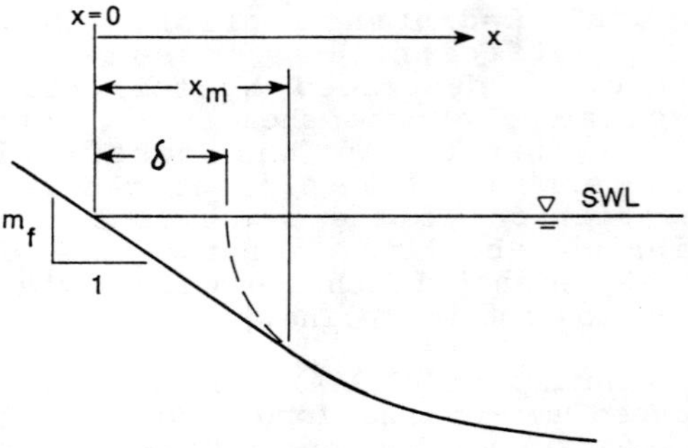

Figure 1: The non-singular equilibrium beach profile.

described by the foreshore slope and the shape factor of the concave-up part of the profile.

WAVE-INDUCED SET-UP

The variation in mean water surface associated with change in onshore radiation stress, or wave-induced set-up $\bar{\eta}$, has been shown to be equal to

$$\bar{\eta} = \begin{cases} \nu(h_b - h) - \dfrac{\kappa^2}{16} h_b & x \leq x_b \\ \dfrac{-H^2 k}{8 \sinh 2kh} & x \geq x_b \end{cases} \tag{4}$$

after Longuet-Higgins and Stewart (1964), and Dean and Dalrymple (1984), among others, where h_b and x_b are the still-water depth and location of the wave breakpoint, respectively, H is the local wave height, and k is the wavenumber. Shoreward of the breakpoint, the wave height H is assumed proportional to the total water depth d, where

$$\kappa = H/d \tag{5}$$

and

$$\nu = \frac{3\kappa^2}{8 + 3\kappa^2} \tag{6}$$

Across the planar foreshore part of the profile, the slope of the set-up water surface is uniform and equal to $-\nu\, m_f$. The point of intersection of the set-up water surface and the foreshore, x_s, is

$$x_s = \frac{(\nu - \kappa^2/16)\, h_b}{(\nu - 1)\, m_f} \tag{7}$$

This location corresponds to the "set-up still water line", or more precisely, the mean water line in the influence of static wave set-up.

The total water depth d across the profile is

$$d = h + \bar{\eta} \tag{8}$$

Combining Eqs. (1), (4), and (8), the total water depth across the non-signular equilibrium profile is

$$d = \begin{cases} (1-\nu)\, m_f\, x + (\nu - \frac{\kappa^2}{16})\, h_b & x_s \le x \le x_m \\ (1-\nu)\, a\, (x-\delta)^{2/3} + (\nu - \frac{\kappa^2}{16})\, h_b & x \ge x_m \end{cases} \tag{9}$$

For purposes of notation, observe that the set-up water surface slope across the planar foreshore part of the profile is

$$s_f = (1-\nu)\, m_f \tag{10}$$

McDougal and Hudspeth (1981, 1984) showed that the total water depth across the concave-up $h=ax^{2/3}$ profile can be well-approximated by the form

$$d = \tilde{a}\, x^{1/2} \tag{11}$$

This expression models the set-up water surface well except as x approaches 0 (since d does not go to zero as $x \to 0$). However, the poorer fit near x=0 is essentially inconsequential in the present case if the exact form of the set-up water surface is used across the planar foreshore part of the profile; i.e., in the region x=0. Accordingly, the following expressions are adopted for the total water depth:

$$d = \begin{cases} s_f\, x + (\nu - \frac{\kappa^2}{16})\, h_b & x_s \le x \le x_m \\ \hat{a}\, (x-\hat{\delta})^{1/2} & x \ge x_m \end{cases} \tag{12}$$

Requiring continuity of total water depth and water surface slope at the matchpoint x_m yields

$$\hat{a} = s_f \sqrt{2(x_m - x_s)} \tag{13a}$$

$$\hat{\delta} = \frac{1}{2}(x_m + x_s) \tag{13b}$$

Non-dimensionalizing all length scales by the distance to the breakpoint x_b yields

$$D = \begin{cases} s_f\, X + (\nu - \frac{\kappa^2}{16})\, s_b & X_s \le X \le X_m \\ \hat{A}\, (X - \hat{\Delta})^{1/2} & X \ge X_m \end{cases} \tag{14}$$

$$\hat{A} = s_f \sqrt{2(X_m - X_s)} \tag{15a}$$

$$\hat{\Delta} = \frac{1}{2}\,(X_m + X_s) \tag{15b}$$

where upper-case denotes non-dimensional terms such as

$$X = x \,/\, x_b \tag{16}$$

and where the term s_b has been introduced as

$$s_b = h_b \,/\, x_b \tag{17}$$

LONGSHORE CURRENT

In dimensional form, wave-induced longshore current can be described through the depth- and time-averaged equation of motion in the alongshore direction (Bowen, 1969; Longuet-Higgins 1970),

$$-\frac{\partial}{\partial x} S_{xy} + \tau_{by} + \frac{\partial}{\partial x}(\mu_e \, d \, \frac{\partial}{\partial x} v) = 0 \tag{18}$$

where S_{xy} is the onshore-directed flux of the alongshore component of momentum, τ_{by} is the alongshore component of bottom stress, v is the depth- and time-averaged longshore current, and μ_e is the turbulent eddy viscosity. Following the classical approximations of Longuet-Higgins (1970) and others,

$$\tau_{by} = -\frac{\kappa}{\pi}\, C_f \,\rho\, \sqrt{gd}\;\; v \tag{19}$$

where C_f is an empirically determined friction factor of order 0.01. The turbulent viscosity is assumed to vary linearly with distance offshore

$$\mu_e = N \,\rho\, x\, \sqrt{gd} \tag{20}$$

where N is a numerical constant.

<u>Non-dimensionalizing</u> all length-scales by the breaker distance x_b and velocity by the classical Longuet-Higgins solution of longshore current at the breaker line with no mixing, i.e.,

$$v_o = \frac{5}{16}\frac{\kappa\pi}{C_f}\, \sin\Theta_b \,\sqrt{gd_b}\;\; s_b \tag{21}$$

where Θ is the wave angle and the "b" subscript refers to breaking conditions, the alongshore equation of motion (Eq. 18) becomes

$$\frac{N\,\pi}{\kappa C_f}\,\frac{\partial}{\partial X}(XD^{3/2}\,\frac{\partial V}{\partial X}) - D^{1/2}V \;=\; -\,D^{3/2}\,\frac{\partial D}{\partial X}\,\frac{1}{{s_b}^2} \tag{22}$$

where upper-case denotes non-dimensional terms once again such as

$$V = v/v_o \qquad \text{and} \qquad X = x/x_b \tag{23}$$

To determine the longshore current, Eq. (22) must be solved across each of three regions across the non-singular equilibrium profile: (i) the planar foreshore, (ii) the concave-up segment landward of the breakpoint, and (iii) the concave-up segment seaward of the breakpoint. Both the magnitude and gradient of the velocity must be matched between each region. McDougal and Hudspeth (1984) demonstrated these solutions but the resultant expressions are somewhat cumbersome and tedious to apply.

In the present study, Eq. (22) is solved for the longshore current across each of the three regions; both the magnitude and gradient are matched at the planar/concave-up matchpoint x_m but only the magnitude is matched at the breakpoint x_b. The resulting discontinuity in the velocity gradient at the breakpoint is small for the physically-typical case where x_m is small compared to x_b (i.e., where the toe of the foreshore is located close to shore relative to the breakpoint).

Incorporating the non-dimensional expressions for total water depth D (Eq. 19) to Eq. (22), and solving for velocity subject to the above-described matching conditions, the non-dimensional longshore current is given by

$$V = \begin{cases} A_1 \, (X-X_s)^{p_1} + \left(\frac{s_f}{s_b}\right)^2 \frac{1}{1-\frac{5}{2}P} \, (X-X_s) & X_s \le X \le X_m \\ C_1 \left[\frac{\cosh(\lambda \tilde{X}^{1/4})}{\lambda \, \tilde{X}^{1/2}} - \frac{\sinh(\lambda \tilde{X}^{1/4})}{\lambda^2 \, \tilde{X}^{3/4}}\right] + \frac{\hat{A}^2}{2 s_b^2} & X_m \le X \le 1 \\ C_2 \left[\frac{1}{\tilde{X}^{1/2}} + \frac{1}{\lambda \, \tilde{X}^{3/4}}\right] \exp(-\lambda \tilde{X}^{1/4}) & X \ge 1 \end{cases} \qquad (24)$$

where

$$p_1 = -\frac{3}{4} + \left(\frac{9}{16} + \frac{1}{P}\right)^{1/2} \qquad (25)$$

$$P = \frac{N \, \pi \, s_f}{\kappa \, C_f} \; ; \quad P \neq 2/5 \qquad (26)$$

$$\lambda = 4/\sqrt{P_2} \qquad (27)$$

$$P_2 = \frac{N \, \pi \, \hat{A}}{C_f} = P \, \sqrt{2(X_m - X_s)} \qquad (28)$$

$$\tilde{X} = (X - \hat{\Delta}) \qquad (29)$$

The coefficients are given by

$$A_1 = (X_m - X_s)^{-p_1} \left[C_1 \, \Phi - \alpha(X_m - X_s) + \frac{\hat{A}^2}{2 s_b^2}\right] \qquad (30)$$

$$C_1 = -\frac{\hat{A}^2}{2 s_b^2} \left(3Q + \lambda Q^2 + \frac{3}{\lambda}\right) \exp(-\lambda Q) \qquad (31)$$

$$C_2 = \frac{\hat{A}^2}{2s_b^2}\left[-\frac{3Q}{\lambda}\cosh\lambda Q + \left(Q^2 + \frac{3}{\lambda^2}\right)\sinh\lambda Q\right] \qquad (32)$$

and where

$$\alpha = \left(\frac{s_f}{s_b}\right)^2 \frac{1}{1-\frac{5}{2}P} \quad ; \quad P \neq 2/5 \qquad (33)$$

$$\Phi = \frac{1}{\lambda R^2}\cosh\lambda R - \frac{1}{\lambda^2 R^3}\sinh\lambda R \qquad (34)$$

$$R = (X_m - \hat{\Delta})^{1/4} \qquad (35)$$

$$Q = (1 - \hat{\Delta})^{1/4} \qquad (36)$$

Application of Eqs. (24) through (36) can be cumbersome. However, each of the terms in Eq. (24) which describe the longshore current can be expressed solely as a function of the non-dimensional matchpoint location X_m and the mixing parameter P. Recall that X_m is simply described by

$$X_m = x_m / x_b \qquad (37)$$

where

$$x_m = \frac{4}{9}\left(\frac{a}{m_f}\right)^3 \qquad (38)$$

and

$$x_b = \left(\frac{H_b}{\kappa a}\right)^{3/2} + \frac{4}{27}\left(\frac{a}{m_f}\right)^3 \qquad (39)$$

Hence, the non-dimensional longshore current profile can be readily calculated from the (i) breaking wave height H_b, (ii) breaker index κ, (iii) concave-up beach shape parameter a (commonly denoted by "A" in the literature), (iv) foreshore slope m_f, and (v) selection of a mixing parameter P, where $P \neq 0.4$.

Figures 2 through 9 depict each of the terms necessary to evaluate the longshore current distribution from Eq. (24) for a breaker index $\kappa = 0.8$, for various chosen values of the mixing parameter P and for calculated values of the non-dimensional planar-foreshore/concave-up matchpoint location X_m.

Figure 10 illustrates the non-dimensional longshore current across the non-singular equilibrium beach profile for various locations of the profile matchpoint X_m. Note that the current maxima approximately corresponds to the location of the profile matchpoint.

Figure 11 illustrates longshore current distributions observed across beaches with non-singular equilibrium shaped profiles -- compared to the distributions predicted for each case by Eq. (24) for several values of the mixing parameter. The first five cases represent dye observations from a moveable-bed laboratory model for five different breaking/surf conditions; the last case is from current meter data collected at Duck, N.C., (Bodge and Dean, 1987). The agreement is fairly good, although the longshore current distributions observed in the laboratory appear displaced landward by about 2/10'ths of a surf zone width relative to the predicted distributions.

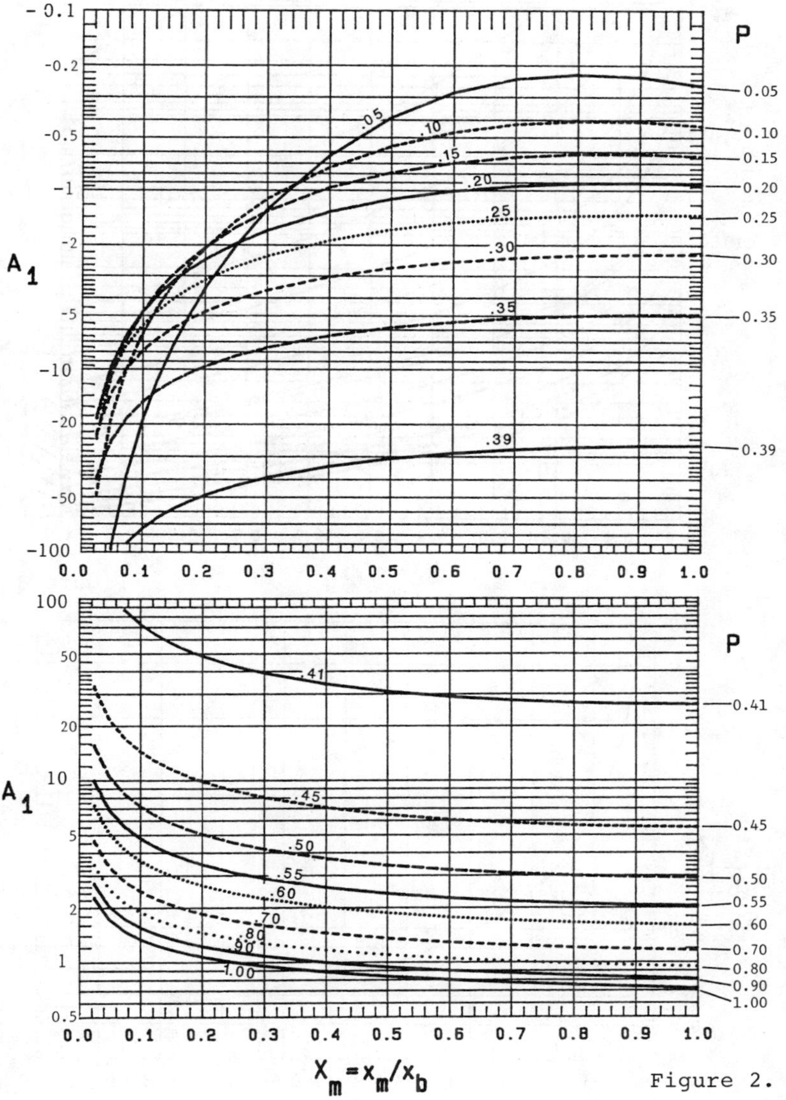

Figure 2.

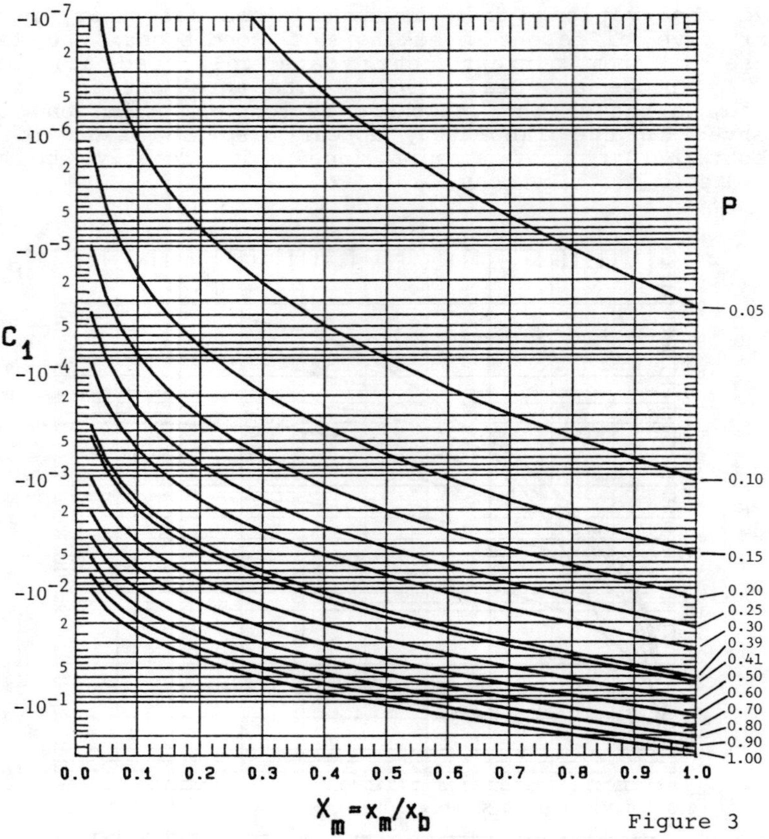

Figure 3

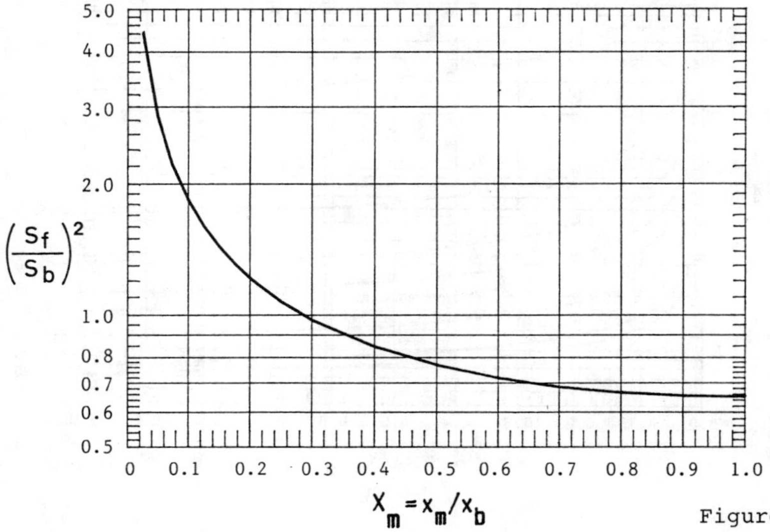

Figure 4

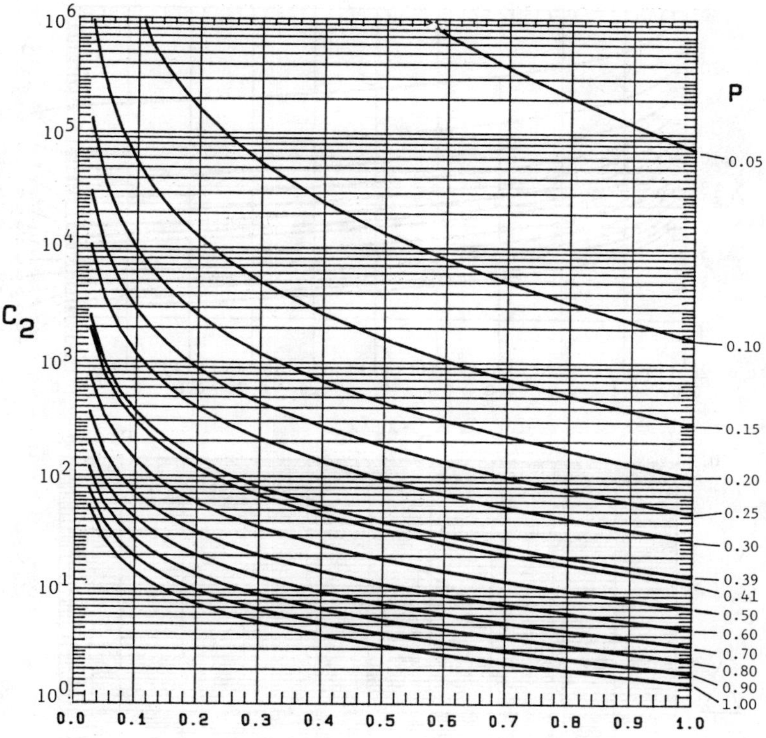

Figure 5

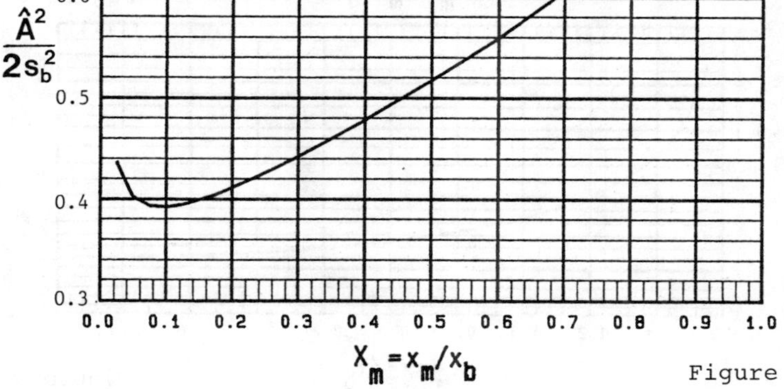

Figure 6

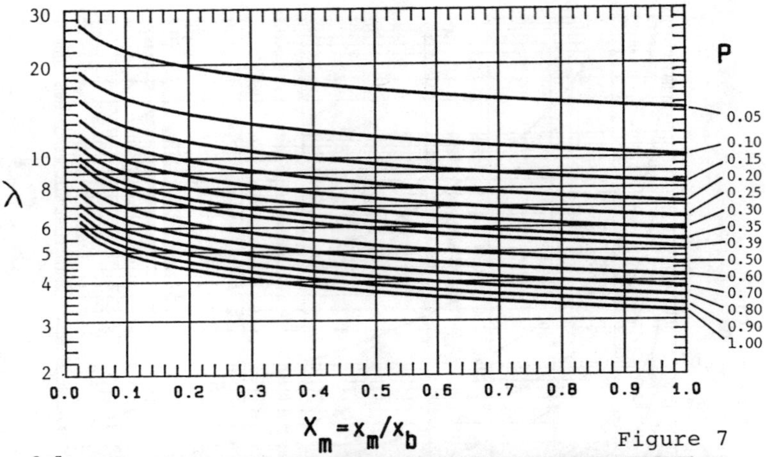

Figure 7

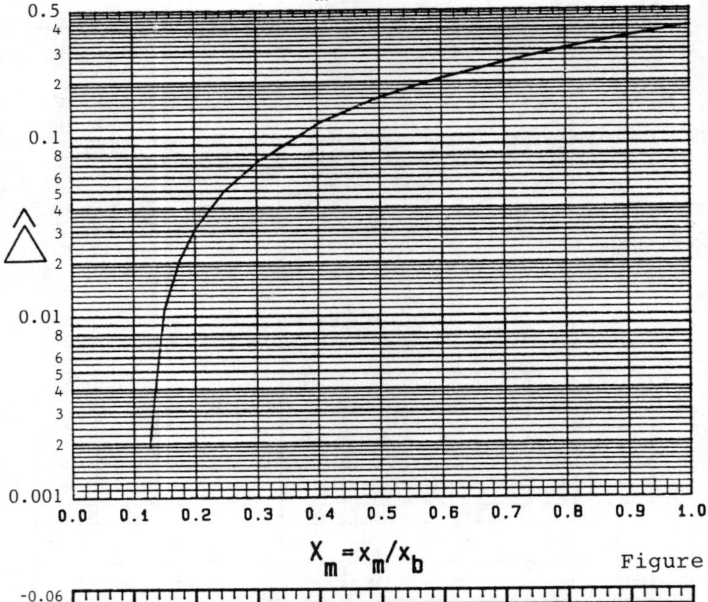

Figure 8

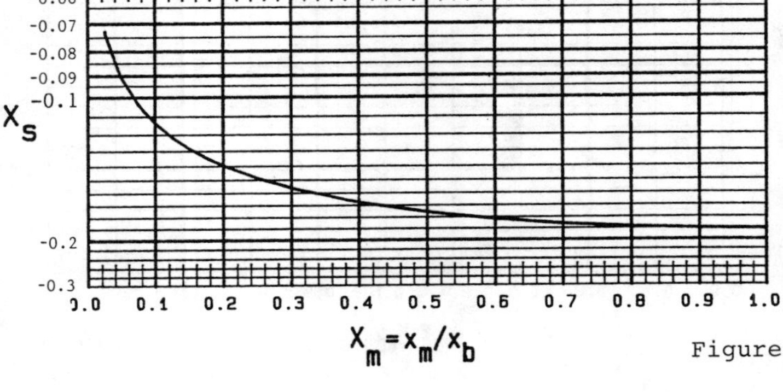

Figure 9

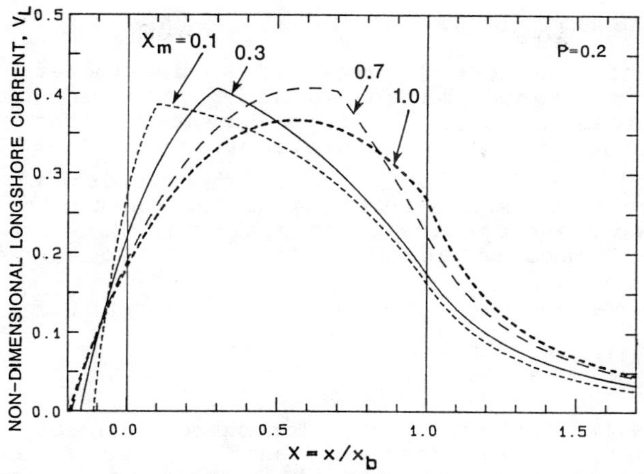

Figure 10: Longshore current predicted across a non-singular equilibrium profile for various profile matchpoints.

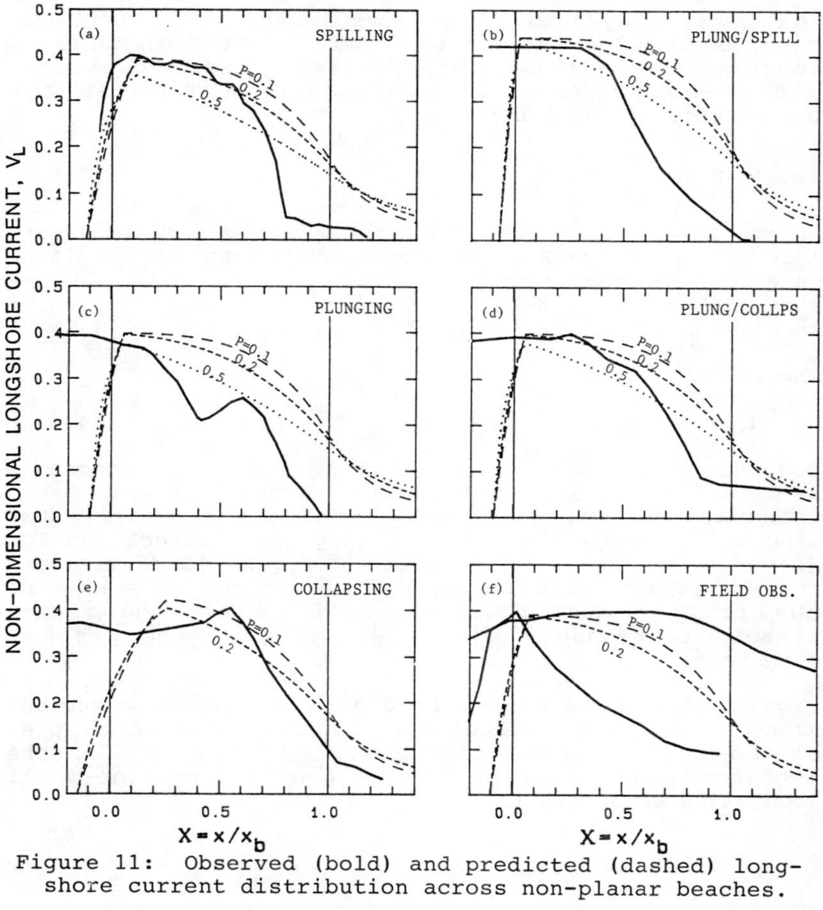

Figure 11: Observed (bold) and predicted (dashed) longshore current distribution across non-planar beaches.

LONGSHORE SEDIMENT TRANSPORT

Bodge and Dean (1987) presented a distributed longshore transport model based upon results of short-term impoundment of longshore sediment transport across rapidly-deployed, low-profile groins in the field and laboratory environments. Their expression suggests that sediment is mobilized in proportion to the rate of local wave energy dissipation per unit surf zone volume and is transported alongshore by local longshore current; viz.,

$$q_\ell(x) = k_\ell \frac{1}{h} \frac{\partial}{\partial x}(e\ c_g)\ v \tag{40}$$

where $q_\ell(x)$ is the local longshore sediment transport rate per unit offshore distance, k_ℓ is a dimensional coefficient of proportionality, and ec_g is the local wave energy flux.

Again assuming that local wave height is proportional to the local water depth, and non-dimensionalizing lengths by the breaker distance and velocities by the planar-beach no-mixing longshore current at the breakpoint, Eq. (40) becomes

$$Q_\ell = D^{1/2}\ s_b^{-3/2}\ V \frac{dD}{dX} \tag{41}$$

where Q_ℓ is the local (not total) longshore transport rate per unit distance offshore normalized by the local rate at the breakpoint; i.e.,

$$Q_\ell = q_\ell \ /\ q_{\ell b} \tag{42}$$

where

$$q_{\ell b} = k_\ell \frac{5}{16} \rho\, g^{3/2}\, \kappa^2\, d_b^{1/2}\, v_o\, s_b \tag{43}$$

Equation (41) can be evaluated to yield the longshore sediment transport distribution across the surf zone (landward to the location of the set-up still water line x_s) via Eq. (24) for the longshore current and Eq. (14) for the total water depth. Figure 12 illustrates the transport distributions for various locations of the profile matchpoint X_m. Note that the maxima closely corresponds to the location of the profile matchpoint.

Figure 13 compares the longshore transport distributions across a non-singular equilibrium profile as predicted by the "energetics" model (Bagnold, 1963), the "stress" model (Komar, 1977), and the present model (Bodge and Dean, 1987).

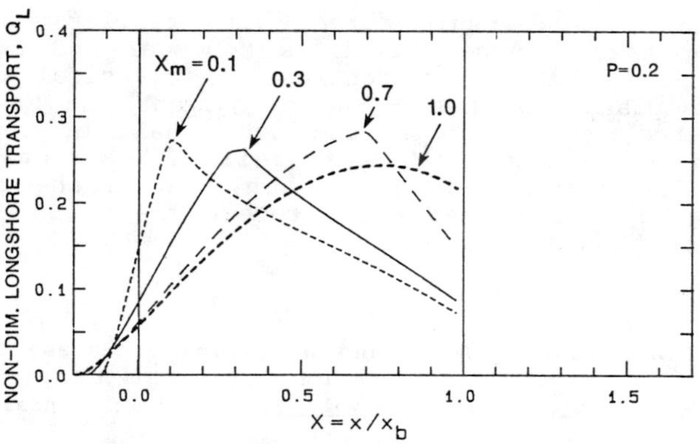

Figure 12: Longshore transport across a non-singular equilibrium profile for various profile matchpoint locations.

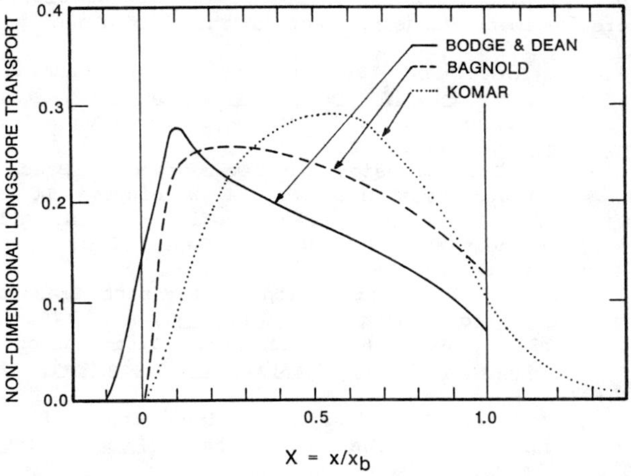

Figure 13: Comparison of longshore transport distribution models evaluated for a non-singular equilibrium profile with matchpoint X_m=0.1 and mixing parameter P=0.1.

SUMMARY AND CONCLUSIONS

The non-singular equilibrium beach profile consists of a planar foreshore "matched" to a concave-up profile across the surf zone. The longshore current and sediment transport distributions across such a profile can be readily developed through knowledge of the foreshore slope, concave-up profile shape parameter, breaker height, and selection of a Longuet-Higgins type mixing parameter. The current and transport maxima are predicted at or about the location of the planar-foreshore/concave-up profile matchpoint. Since this point is typically located fairly close to shore relative to the breakpoint, this suggests that the peak current and transport are more closely located to shore than to the breakpoint. This result agrees with similar solutions for non-planar beaches, such as McDougal and Hudspeth (1981, 1984), but is in contrast to classical predictions for planar beaches -- where current and transport is predicted as maximum at the outer half of the surf zone and vanishing at the shoreline. Moreover, the results for the non-planar beach, as in the present study, agree more closely with recent field and laboratory observations.

REFERENCES

Bagnold, R.A., 1963. Beach and Nearshore Processes Part I: Mechanics of Marine Sedimentation. In: M.N. Hill (ed.), The Sea: Ideas and Observations, Vol. 3. New York: Interscience, 507-528.

Berek, E.P. and R.G. Dean, 1982. Field Investigation of Longshore Transport Distribution. Proc., 18th Int. Conf. on Coastal Eng., ASCE, 1620-1639.

Bodge, K.B., in press. A Literature Review of the Distribution of Longshore Sediment Transport Across the Surf Zone. J. Coastal Research.

Bodge, K.B. and R.G. Dean, 1987. Short-Term Impoundment of Longshore Transport. Proc., Coastal Sediments '87, ASCE, 468-483.

Bowen, A.J. The Generation of Longshore Currents on a Plane Beach. J. Marine Research, 27, 206-215.

Bruun, P., 1954. Coast Erosion and Development of Beach Profiles. Beach Erosion Board, Tech. Memo. No. 44, Washington, DC.

Dean, R.G., 1977. Equilibrium Beach Profiles: U.S. Atlantic and Gulf Coasts. Ocean Eng. Rpt. No. 12, Dept. of Civil Eng., Univ. of Delaware, Newark, DE; 45 pp.

Fulford, E.T., 1987. Distribution of Sediment Transport Across the Surf Zone. Proc., Coastal Sediments '87, ASCE, 452-467.

Komar, P.D., 1977. Beach Sand Transport: Distribution and Total Drift. J. Waterway, Port, Coastal and Ocean Eng., ASCE, 103, WW2, 225-239.

Kraus, N.C., Rosati, J.D. and K.J. Gingerich, 1988. Longshore Sediment Transport in the Surf Zone. Proc., 21st Conf. on Coastal Eng., ASCE.

Longuet-Higgins, M.S., 1970. Longshore Currents Generated by Obliquely Incident Sea Waves, 1 and 2. J. Geophysical Research, 75(33), 6778-6801.

McDougal, W.G. and R.T. Hudspeth, 1981. Wave Induced Setup/Setdown and Longshore Current; Non-Planar Beaches: Sediment Transport. Oceans, IEEE, 834-846.

McDougal, W.G. and R.T. Hudspeth, 1984. Longshore Sediment Transport on Dean Beach Profiles. Proc., 19th Int. Conf. on Coastal Eng., ASCE, 1488-1506.

Sawaragi, T. and I. Deguchi, 1978. Distribution of Sand Transport Rate Across a Surf Zone. Proc., 16th Int. Conf. on Coastal Eng., ASCE, 1596-1613.

Sternberg, R.W., N.C. Shi and J.P. Downing, 1984. Field Investigations of Suspended Sediment Transport in the Nearshore Zone. Proc., 19th Int. Conf. on Coastal Eng., ASCE, 1782-1798.

CHAPTER 105

DESIGN OF POCKET BEACHES. THE SPANISH CASE

José Mª Berenguer *
Javier Enríquez **

Pocket beach is a usual method to restore an eroded or regressive coast without natural sand supply. Nowadays, there are more than 40 of these beaches on the Spanish coasts and a lot of them all over the world.

This paper deals with the design parameters of pocket beaches, based on the analysis of data collected from 24 existing beaches on the Mediterranean coast of Spain. Fourteen of these beaches have been studied in detail; nine of them located on the Alboran Sea and the other five on the coast of Cataluña. Data from 10 additional beaches have also been considered.

Different parameters such as structural design, location of breakwaters, beach planform, beach profile, wave conditions at the site and sedimentary conditions have been analyzed. A mathematical model has been used for the study of the shoreline equilibrium planform.

Practical coastal criteria design considerations have been drawn from the observed behaviour of the aforementioned beaches.

1.- INTRODUCTION

The term "pocket beach" is going to be used for all those beaches which have been artificially created using constructions that limit them laterally, and also in certain cases frontally, either totally or partially. By definition, the term suggests that the dimensions are limited and generally reduced.

* Civil Engineer, Deputy Director of the Harbour Division at the "Centro de Estudios de Puertos y Costas"(CEDEX)
Antonio López, 81, 28026 Madrid, SPAIN
** Civil Engineer, Head of the Coastal Defence Division at the "Centro de Estudios de Puertos y Costas"(CEDEX)

This work is concerned with those pocket beaches protected by works parallel to the coast. Beaches which are bounded by groynes that are straight and perpendicular to the coast are thus excluded. On the other hand, constructions that are of the detached breakwater kind, L-shaped or T-shaped or curved groynes and platform islands are those which will be studied.

The main purpose of these constructions is to obtain a stable and safe beach. Pocket beaches undoubtedly have their advantages and drawbacks. Each and every coastal problem has its own specific characteristics, (purpose, wave climate, environment, etc.), and this makes the indiscriminate use of theoretical plans and the extrapolation of practical solutions unadvisable even if they have been successful in other cases.

The methods that have been developed in coastal engineering for the control of sedimentary processes which are of an erosive nature, are basically three.

1. To protect the coast with works that prevent or limit the direct action of the waves on sedimentary deposits.

2. To reduce or alter the rate of littoral transport.

3. To supply new sediment to the beaches in order to replace the losses or to increase size.

The pocket beaches are a compendium of all these systems and as such can act in any of these three capacities. Their projection parallel to the coast reduces the average energy that reaches the shoreline. The differential way in which this effect is brought about, causes a considerable change in the littoral transport of sand. Finally the reduced dimensions of the physiographical micro-unity created make it possible to gain stable and economical beach by artificial nourishment or natural trapping of new sand.

In recent years, the increase of pocket beaches has been remarkable. In Mediterranean countries where tidal effects are insignificant, especially in Spain, Italy, France and Israel, this system of beach regeneration has been widely used because of its rapid results and its economical nature. Japan also has a lot of experience in this kind of work.

In Spain, there are over 40 beaches of this kind. They are usually situated on stretches of coastline where there is a lot of tourism and in which there is also a fairly rapid erosional process as a consequence of either causes exogenous to coastal use (dams, wells), or endogenous causes (marinas, coastal works, dredging, etc.).

2.- PREVIOUS RESEARCH

The influence of detached breakwaters on the coast and the formation of tombolos has been studied by several researchers. Sawaragi (1957), using movable-bed model, analyzed the variation induced on the longshore current by this kind of coastal structures. Shinohara and Tsubaki (1966) developed a new physical model for the analysis of the role of detached breakwaters in the formation of tombolos. Toyoshima (1974) carried out a statistical survey of 308 detached breakwaters constructed in Japan, looking for a series of relationsships between the lengths of the works, their distance from the coast and the accumulations caused in the sheltered area.

The planform of tombolos in the lee of breakwaters has been studied by Perlin (1979) through the use of a numerical model. One of the conclusions points to the considerable influence that the parameter L_o (wave length) has in the shape of the resulting tombolo. Noble (1978) carried out an analysis of the behaviour of diverse coastal protection structures and their effects on the beaches, reaching the conclusion that an off-shore breakwater has virtually no influence on the coast when its distance from the coast is over six times its length. Berenguer (1986) presents graphs which are similar to Toyoshima's; these are drawn up from different artificial beaches on the Spanish Mediterranean Coast.

The study of the planform of the beaches formed behind offshore structures, whether these be natural or artificial has been the subject of considerable Coastal Engineering work. Sauvage (1954) was one of the first authors to analyze the equilibrium of these coastal formations, showing that the shape of the planform was markedly elliptical. Yasso (1965) checked, by analyzing real cases, of adjustment of the planforms of the beaches to logarithmic spirals of equation $r = e^{\theta ctg\alpha}$, where r is the radius from the pole of the spiral, θ is the horizontal angle from the origen and angle α is the constant angle of the radii to the tangents of the curve.

Silvester et al. (1970), (1972), using prototype data, concludes that the beach planforms between headlands consist of three distinct zones:

- an arc behind the headland which is located upstream in the direction of littoral transport.

- a stretch which is logarithmic spiral in shape.

- a straight stretch which extends downstream towards the next headland.

Applying these conclusions, Silvester and Ho (1972) have used the logarithmic spiral for the stabilization of artificial reclamations through the concept of headland control.

Garau (1973), using observations made at various beaches of the Spanish Mediterranean Coast, defines the planforms of the diffraction beaches through their sediment filling and emptying processes. His most important conclusions are that the maximum emptying curve would be a logarithmic spiral whose centre is in the diffraction pole and that the characteristic angle of the family of spirals depends basically on the friction angle of the sand in the trough of water, and which can be generalized for a value of $\alpha = 60^{\circ}$.

Rea and Komar (1975) developed a numerical model of evolution of spiral beaches behind a rocky headland. Font, Sanabria and Silva (1976) include the effects of the gradient at the height of the breaking wave and the associated current. The longitudinal current is estimated as a consequence of the obliqueness of the incident wave and of the set-up gradient caused by this. Dean (1978) presents a method for calculating the equilibrium shape of pocket beaches. In this method the bathymetry is considered stable when the wave front is tangent everywhere to the local bottom contours. The calculation is carried out for relatively narrow openings with normal and oblique waves. For high wave-periods the planform obtained is markedly semi-circular.

Another of the aspects dealt with in the question of beaches situated in diffraction zones, is the hydrodynamic conditions required in the sheltered area for the systems to be equilibrium in the face of a specific wave condition. This question has caused different authors to try and clarify the processes associated with the propagation of waves in the diffraction zone (currents, set-up, etc.). The works of O'Rourke and Le Blond (1972), and Gourlay (1974) (1976), are to be found in this section. The presence of longshore currents in the sheltered zone when the beach is stable, phenomenon confirmed in tests by Gourlay, contrasts with the assumptions made by other authors with regard to the stability situation on a protected beach.

Walton (1977), after analyzing the planform of spiral beaches, concludes that their position is at all points perpendicular to the energy vector of the incident waves. Rosen and Vajda (1982) deduce that the stability of a tombolo is achieved when the bathymetric lines are such that the diffracted waves have a component of movement quantity opposite to the gradient of average level induced by the radiation stress.

3.- METHODOLOGY

The basis of the present study is the experience achieved in cases where constructions are already present on the Spanish Coast. This is why a detailed study of formal and sedimentological characteristics of 14 "pocket beaches" on the Mediterranean Coast have been carried out. Nine of these beaches are to be found in a limited marine area, which is the Alborán Sea, that reaches the coast of the province of Málaga. The other five are in Catalonia: two in the province of Gerona and three in Tarragona. In addition to these beaches, some geometrical parametres of another ten beaches are analyzed, all of these also on the Mediterranean Coast.

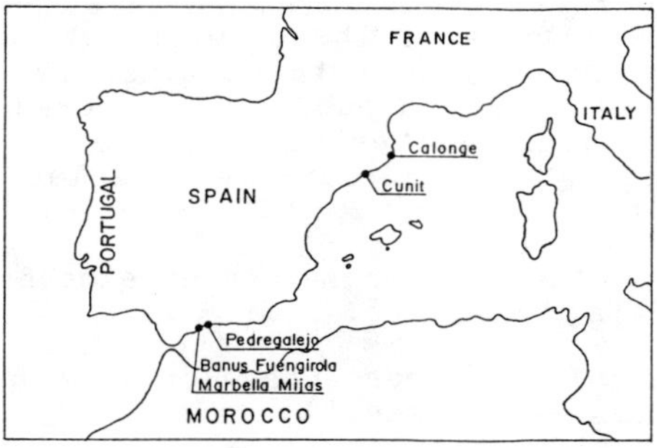

Fig. 1.- Site Location.

4.- DATA COLLECTION

Table 1 shows the main information collected from each of the beaches studied. In 10 of these, the information has been obtained through special field studies whereas in the rest, the data has been got from plans and aerial photography.

The information includes the following aspects:

- Geometrical characteristics: The measurements have been carried out using topographical techniques. The criterion used for the determining of distances between points of contact between water and land has been to adjusted to situations of mean sea level.

- Bathymetry: The survey has been extended to the inside of the cell, the gap and the outer area in front of the gap. The depth of the bottom within each cell has been determined at base, with at least 5 profiles.

- Slope: The slope measurements refer to those carried out at different points throughout the length of the beach face.

- Granulometric distribution: On 12 beaches, sediment samples have been collected on both the subaerial and submerged parts of the beach. These samples have been dried, weighed and sifted by the ASTM series.

The parameters used for defining the characteristics of pocket beaches are sef out in fig. 2.

5.- WAVE CLIMATE

Tidal and wave conditions on the Spanish Mediterranean Coast are fairly uniform. The maximum range of astronomical tides varies between 0.30 m. (Catalonia) and 0.85 m. (Málaga). Wave energy in the whole area is moderate. The significant wave height of the one year return period storm, is in Málaga about 3.0 metres, and slightly higher in Catalonia. For calculation purposes, the characteristic value of the parameter "wave length", L, has been estimated from data obtained by wave buoys located in these two areas. The average of the zero-crossing significant period, $T_{z,s}$, of all the wave records in which $H_s \geqslant 1.0$ m. during a complete year, has been calculated:

Catalonia: $H_s \geqslant 1.0$ m. (percentage of exceedence: 8'9%)
$T_{z,s} = 8.62$ sec.

Málaga: $H_s \geqslant 1.0$ m. (percentage of exceedence: 3'8%)
$T_{z,s} = 7.37$ sec.

6.- RESULTS AND DISCUSSIONS

6.1.- Beach planform

From the analysis of the principle geometrical parameters, it is found that the planform of pocket beaches follows a pattern which is very similar in all the cases examined. In fig. 3, values of A_o and A_1 have been drawn for each beach. It can be observed how the points obtained generally adjust to the following straight line:

$$A_o = 2 A_1 \qquad (1)$$

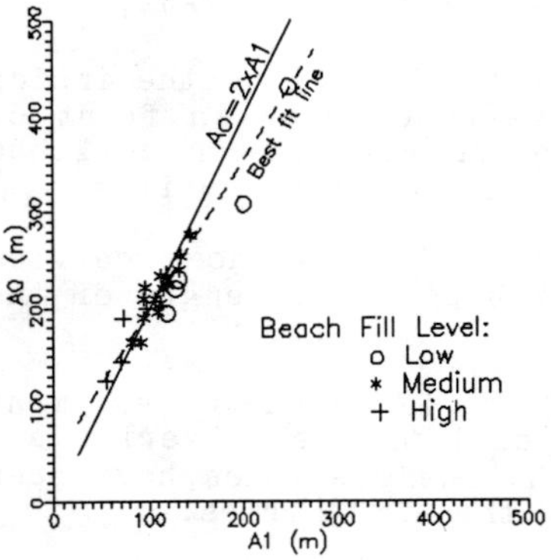

Figure 3.- Relation between Ao and A1

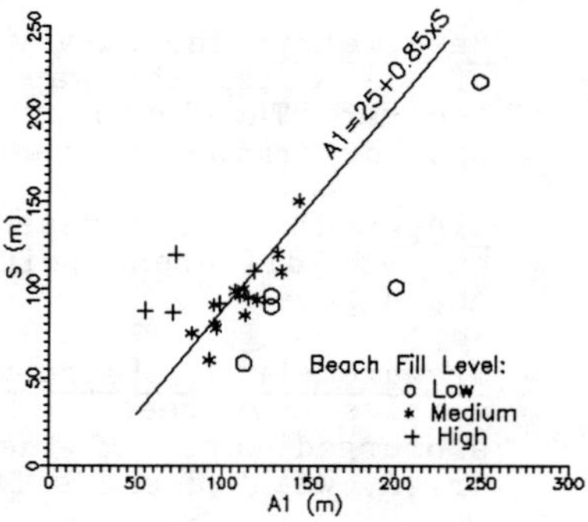

Figure 4.- Relation between A1 and S

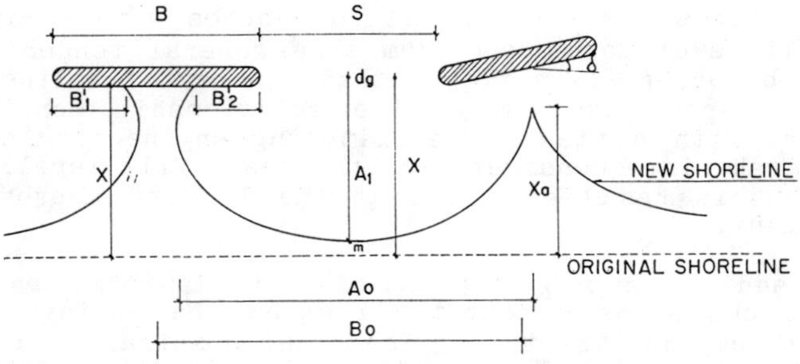

Figure 2.- Definition of pocket beach parameters.

Beach	X(m)	S(m)	A1(m)	Ao(m)	Dg(m)	Bo(m)	Dm(mm)
BANUS 1	150	90	128	230	3.40	247	0.29
BANUS 2	150	94	120	224	-	250	-
BANUS 3	156	95	115	220	4.10	222	0.84
BANUS 4	144	91	95	210	-	190	-
DON PEPE	108	78	96	222	-	192	-
MARBELLA-1	130	110	134	254	-	260	-
MARBELLA-2	-	99	107	210	2.70	-	0.25
MARBELLA-3	-	75	82	165	2.80	-	0.25
MARBELLA-4	160	150	144	276	-	-	-
MARBELLA-5	-	88	55	125	1.50	-	0.25
MARBELLA-6	-	87	71	145	2.50	-	0.25
MARBELLA-7	-	120	73	190	-	-	-
FUENGIROLA	120	60	92	165	2.10	210	0.31
MIJAS	108	120	132	240	-	-	-
PEDREGALEJO-1	110	80	95	190	3.50	220	1.86
PEDREGALEJO-2	122	92	98	200	3.70	225	1.12
PEDREGALEJO-3	124	110	118	235	3.50	268	1.86
PEDREGALEJO-4	148	58	112	195	3.50	202	1.86
CUNIT 1	120	96	110	196	-	230	-
CUNIT 2	150	96	128	220	2.70	249	0.15
CUNIT 3	210	101	200	308	3.50	313	0.14
CUNIT 4	270	218	248	430	3.55	520	0.13
CALONGE-1	125	85	113	205	5.70	247	1.89
CALONGE-2	128	100	112	234	6.40	255	1.42

Table 1.- Summary of Basic Data.

The points that correspond to beaches with a high or low fill level move away from this general tendency. It should be observed in fig. 3 that the best fit line does not correspond exactly to the relationship mentioned. However, with a view to facilitating engineering design and to take into consideration the beach fill level, this relationship has been taken as the line of approximate adjustment.

In addition, an analysis of the planform has been carried out using a computer program that brings about the optimum adjustment of a logarithmic spiral to a given shoreline, in accordance with Yasso's method. Previous studies that have taken place at the CEPYC (J.M.Berenguer (1981), J.Almenar (1983), J.Enríquez (1984)) showed how, considering the two halves of the cell separately, the planform of these stretches adjusted reasonably well to the logarithmic spirals with a pole at the head of the detached breakwaters, that in each case caused diffraction, and characteristic angles (α) of around 60° - 65°. Taking the pocket beach as a whole, the optimum adjustment is obtained using the logarithmic spiral, whose characteristic angle (α) is situated between 80° and 90°, the pole of the curve being close to the average point of the gap. With a view to design, a theoretical central circumference at the centre point of the gap adjusts very closely to the aforementioned beach line. When there is a low-fill level the beach line will lag behind with respect to the circumference mentioned, and when this level of filling is high, the shore line will stand out with respect to this theoretical curve.

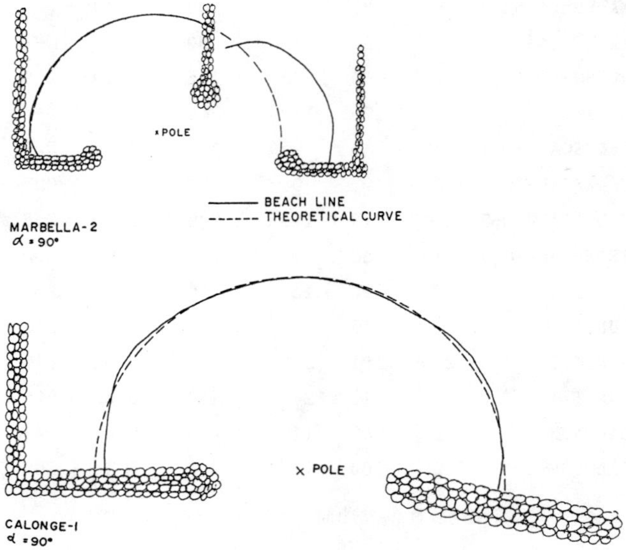

Figure 5.- Examples of adjust of beach planform.

The determining of the radius of the theoretical circumference that defines the shoreline of the cell in each case, can be found by using fig. 4. This figure represents the relationship between the opening of the gap (S) and the distance from the shore (A_1). A good adjustment of these points is obtained with line of equation.

$$A_1 = 25 + 0.85xS \qquad (2)$$

The points corresponding to cells with a high or low level of filling deviate considerably from this straight line.

Another interesting result obtained in the analysis is the one shown in fig. 6. The product (XB_o) represents the surface area of water in the lee of the new works, whileA_1^2 is proportional to the surface of water that will remain after the filling of the cell with sand, given that this surface area is assumed to be a semicircle with an area ($\pi A_1^2/2$). The relationship obtained is that expressed by the equation

$$XB_o = 2.5A_1^2 \qquad (3)$$

Using this relationship as a basis, the maximum surface area (S_p) of beach that could be stable can be obtained

$$S_p = XB_o - (\pi A_1^2/2) \qquad (4)$$

Thus: $$S_p = 0.37\ X.B_o \qquad (5)$$

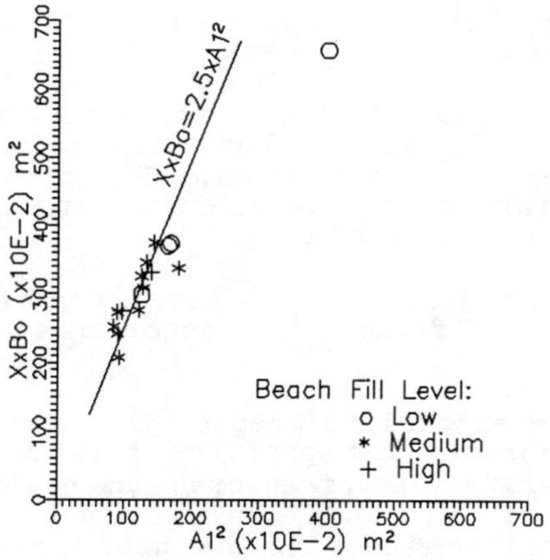

Figure 6.- Relation between A1² and XxBo

6.2.- Beach profiles.

The general bottom configuration of a pocket beach presents a characteristic that makes it different from the typical pattern of an open beach. The groynes or breakwaters that make up the pocket beach, act as wave filters to a greater or lesser extent. At any point on the beach, the range of variation of the wave height is considerably reduced and as a consequence, so is its wave steepness. The direction of the wave as it propagates towards the shallower zones of the beach becomes more independent with respect to the deep-water direction, when the breakwater gap is narrower. Restrictions of both energy and direction, cause marked differences to the profiles of the submerged beach as regards its position along the shoreline, but with great annual stability.

In most of the theoretical studies carried out up to the present on pocket beaches, the simplification of assuming a symmetry of actions and responses along a central axis, has been adopted.

This hypothesis is only borne out in two cases:

a) When the directional distribution of the wave energy is more or less constant in time and has an incidence near to the normal direction of the gap.

b) When the relationship between the width of the gap (S) and length of wave (L) is small enough, in such a way that the action of the filter caused by the breakwaters through diffraction is considerable.

In all other cases, the assymetry shows itself clearly in the form of the profiles of the different zones of the beach, and also, but in a less clear way, in the planform of the shoreline.

The homologous profiles have been compared by grouping homogeneous beaches together according to their climatic and sedimentary characteristics. The most significant conclusions obtained in this analysis are summarized in the following line (see fig. 7).

- Within one single beach the diverse profiles take on characteristic forms in accordance with their position.

- In the areas exposed to greater wave energy, (profile num. 3, central), the profiles take on a parabolic shape similar to that of an open beach without bar.

- The end profiles (num. 1 and num. 5) of the cells present two clearly differentiated zones. The upper part of the profile where the residual energy of the wave after diffraction, is capable of moving the sediment by drag and suspension, and the lower part with a gentle slope, where the wave acts dragging the material along the bottom. A marked step is produced as a transition of the two, and its slope depends on the angle of internal friction of the sediment and its height from the depth of the gap. The lower limit of the highest zone is a function of the height of the residual wave of the most severe storms that reach the cell every year.

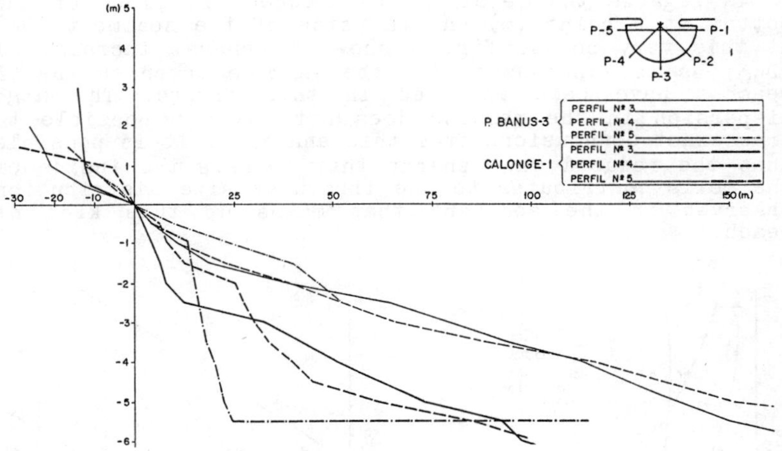

Figure 7. Beach profiles from two selected pocket beaches

Within one cell the transition between the central profile and the ends (profiles num. 2 and num. 4) takes place gradually and in accordance with the spatial variation of the resulting energy from the process of double diffraction. The progressive reduction of wave height as the sheltering increases, implies a change in the steepness of the wave, and with this a different response on the beach profile.

6.3.- Sediment-size distribution

This analysis has been carried out on 12 beaches from which over 25 samples were taken from each.

With a view to design, it is considered essential to carry out a survey of two relationships: the distribution of the average grain-size according to depth and the relationship between the average grain-size and the slope of the bottom.

In fig. 8 the relationship between the relative diameter (D_i/D_m) and the relative depth (d_i/d_g) of the samples corresponding to profiles num. 2, 3 and 4 of three beaches is represented. D_i is the mean size of each sample, D_m is the average of the mean sizeof all the samples in the cell, d_i is the depth where the sample is collected and d_g is the depth at the centre of the gap.

As can be observed, in beaches with fine and medium sand, (Cunit-4 and Banús-1) the dispersion of the relative diametre becomes progressively smaller with the relative depth. In beaches with coarse sand, (Calonge-2) a considerable discontinuity is caused, together with a greater dispersion in the samples situated near the shore.

As regards the relationship between the slope of the bottom at a point (m) and the size of the sediment (D_5) at this very point, fig. 9 shows a general increase of (D_{50}) as (m) increases. All the samples taken on the 12 beaches have been included in this figure. The high dispersion in the results does not make it possible to draw exact conclusions from this analysis. It is possible that the reduced wave energy that reaches the inside of the cells contributes to the irregular size distribution observed in the sediment that makes up this kind of beach.

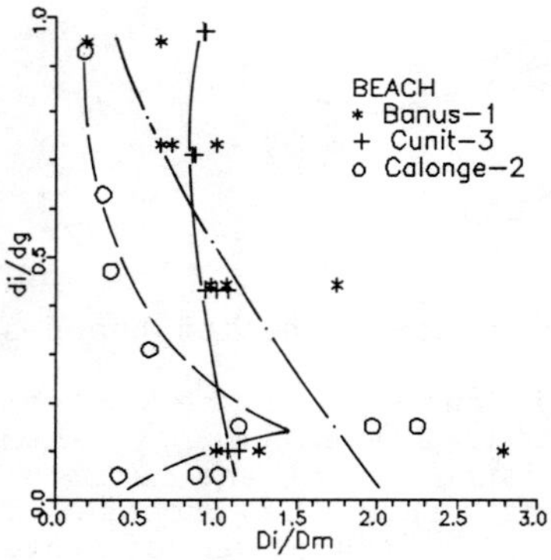

Figure 8.- Relation between Di/Dm and di/dg

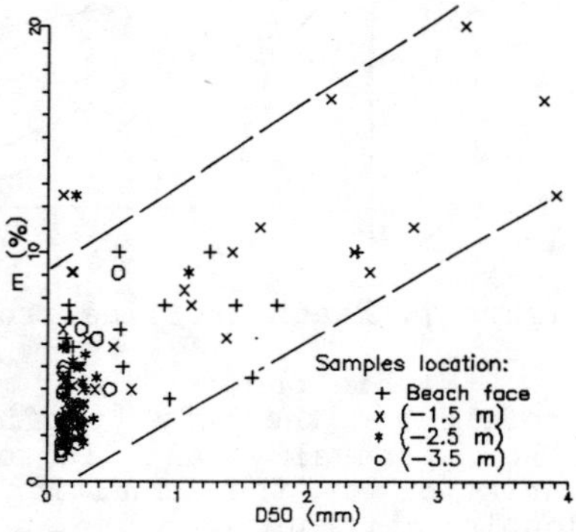

Figure 9.- Relation between D50 and m

6.4.- Other factors

In addition to the relationship between the various pocket beach parametres previously studied, others have been sought that help to define relationships between them that are useful in designing. In particular, an attempt has been made to relate the surface of the gap (Sd_g) and the dimensions of the resulting beach ($A_1{}^2D_m$). As a consequence of the relationship that exists between the mean size of the sediment (D_m) and the general slope of the submerged beach, this parameter can be adopted as a measure of the volume of water contained in the cell.

The results are presented in fig. 10, in which it can be seen just as expected that the greater the gap surface, the larger the cell. For values of (Sd_g) situated at the interval $1.5\ 10^2 < Sd_g < 5.10^2$ the fitting to a straight line such as that shown is reasonably good, this being a very useful tool for the evaluation of the balance between the different parameters, both geometrical and granulometrical, that define a pocket beach. Below a value of $1.5\ 10^2$, it is logical that the adjustment line undergoes inflection and shows a tendency towards the origin of coordinates as the size of the cell becomes smaller. For the adjustment of the upper zone of the graph it will be necessary to rely on additional data in the future, that will make it possible to increase it for values of (Sd_g) above 5.10^2.

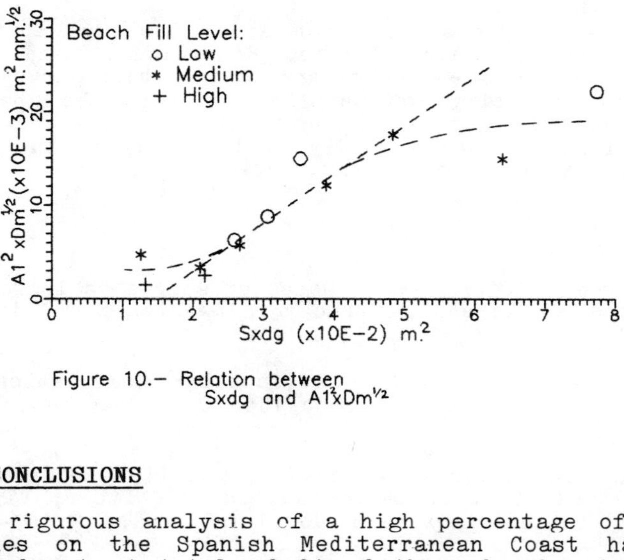

Figure 10.– Relation between Sxdg and A1²xDm½

7.- CONCLUSIONS

A rigurous analysis of a high percentage of pocket beaches on the Spanish Mediterranean Coast has been carried out. A total of 24 of these beaches have been studied from geometry, bathimetry, and granulometric characteristics of sediment perspectives.

The limited amount of wave energy that reaches the inside of the cell, causes specific areas of the beach to show profiles that are hardly shaped by the waves, with slopes close to the natural slope of the sediment. This phenomenon is clearly visible at the end profiles of virtually all the cells that have been studied, where swift changes in the profile occur as a result of the lack of there being sufficient energy to form normal beach profiles.

Another consequence of the low and irregular exposure to the wave energy that occurs on these beaches is the great irregularity observed in the granulometric distribution of the sediments that it is made up of.

It has been demonstrated that the planform of the shoreline can be predicted with sufficient accuracy, by means of different kinds of theoretical curves. In beaches where the variables S,A_1 an L are within the range of those analyzed (1 S/L 5 and 1.5 A_1/L 5, being L the wave length at the gap) it has been possible to get a good adjustment using arcs of circumference whose centre is in the middle of the gap of the cell. The radius of the theoretical circunference that defines the shoreline can be estimated by the equation that relates the parameter (S) (to be defined in the design) with (A_1).

Knowing the characteristics of the sand that is going to form the new beach, fig. 10 provides a quick method of assuring that the diverse design parameters established adjust to a balanced relationship between them.

It should be observed that given the similarity of characteristics (wave climate, dimensions, sand size, etc), of most of the beaches analyzed in this survey, the range of validity of application of the partial conclusions is, to a certain extent, limited. Its application to other markedly different types must take place with caution and by engineering experience.

REFERENCES

- Berenguer, J.M., (1986) "Caracteristicas de los diques exentos en España", Curso de Ingeniería de Costas, CEPYC.

- Dean, R.G., (1978) "Diffraction Calculation of Shoreline Planforms", Proc 15th Conf.Coastal Eng. 2, 1903-1917.

- Font, J.B.,P. Sanabria and A.Silva, (1974) "Geometria des Playas Protegidas en Equilibrio", Proc., VI Congreso Latino Americano de Hidráulica, p. C3-1 to C3-12, July.

- Garau Sacrista, C (1973), "El rozamiento y la estabilidad de las playas", Revista de Obras Públicas Num. 3099 y 3100, Julio, Agosto.

- Gourlay, M.R. (1974), "Wave Set-up and Wave Generated Currents in the Lee of a Breakwater or Headland". Proc. 14th Conf. on Coastal Eng.

- Gourlay, M.R. (1976), "Non Uniform Alongshore Currents" 15th Conference on Coastal Engineering. ch. 40, pp. 701-720.

- Noble, R.M. (1978), "Coastal Structures' Effects on Shorelines" Proc. 16 Conf. on Coastal Eng. Vol. III, ch. 125, pp. 2069-2085.

- O'Rourke, J.G. and P.H. LeBlond, (1972) "Longshore Currents in a Semi-Circular Bay", J.Geophy. Res. Vol. 77, p. 444-452.

- Perlin, M., (1979), "Predicting Beach Planforms in the Lee of Breakwater" Symp. on Coastal Structures 79, ASCE, vol. 2, pp. 792-808.

- Rea, C.C. and Kohar, P.D., (1975) "Computer Simulation Models of a Hooked Beach Shoreline Configuration" Jour. Sed. Petrology vol. 45, nº 4, p. 866-872.

- Rosen D.S. and Vajda. M., (1982) "Sedimentological Influences of Detached Breakwaters". Proc. Conference on Coastal Engineering. ASCE.

- Sauvage de St. Marc M.G., and Vincent, M.G., (1954), "Transport littoral, formation de fleches et de tombolo", Proc. of 5th Conf. on Coastal Eng.

- Sawaragi, T., (1957), "Effects of Shore Structures on Coastal Sand Drift (Report Nº. 1), "4th Conf. on Coastal Engineering in Japan.

- Shinohara, K. and Tsubaki, T., (1966), "Model Study on the Cange of Shoreline of Sandy Beach by the Offshore Breakwater", Proc. 10th Conf. on Coastal Engineering, pp. 550-563.

- Silvester, R., (1970), "Growth of Crenulate-Shaped Bays to Equilibrium", Journal of the Waterway and Harbours Division, May 1970, WW2, pp. 275-287.

- Silvester, R. and HO, S.K., (1972), "Use of Crenulate Shaped Bays to Stabilize Coasts", Proc. 13th Conf. on Coastal Eng., Vol w, pp. 1347-1366.

- Toyoshima. O. (1974), "Design of a Detached Breakwater System". Proc. Conference on Coastal Engineering. ASCE.

- Toyoshima. O. (1976), "Changes of Sea Bed Due to Detached Breakwaters". Proc. Conference on Coastal Engineering. ASCE.

- Toyoshima. O., (1982), "Variation of Foreshore Due to Detached Breakwaters". Proc. Conference on Coastal Engineering. ASCE.

- Walton, T.L. (1977), "Equilibrium Shores and Coastal Design", Proc., ASCE Speciality Conference on Coastal Structures, p. 1-16, Nov.

- Yasso, W.E. (1965), "Plan Geometry of headland bay beaches" Jour. Geology: v. 73, p. 702-714

CHAPTER 106

DUNE EROSION AND SEDIMENT PROFILE DUE TO WAVE UPRUSH

T. A. Fenaish*
M. F. Overton, Assoc. M. ASCE**
J. S. Fisher, M. ASCE***

Introduction

Shorelines are continuously adjusting in response to the changing hydraulic and meteorological conditions. Storms that generate large waves and surge conditions can alter the nearshore topography and relocate the beach shorelines, often with substantial amounts of beach and dune erosion. Such storms pose a major threat to coastal developments for which the economic impact can be significant. The ability to predict the rate of erosion and, consequently the shoreline change, is important in making decisions regarding the planning and managing of the coastal regions.

In general, the available methods for the prediction of beach and dune erosion are based on the assumption of post-storm equilibrium profile. In this approach it is assumed that, for a given set of wave and surge conditions, the entire beach reaches a steady-state, and that the volume of sand released from the dune is equal to the volume of sand required to establish this profile. Existing methods that are based on this concept include those developed by Edelman (1968, 1972), Vellinga (1982, 1983, 1986), Kriebel and Dean (1984), Sargent and Birkemeier (1985), and Kobayashi (1987). The reliance of these methods on the assumption of steady-state condition limits their application to extreme events generated by severe storms. Generally, storms do not have sufficient duration or intensity, such that the beach profile attains equilibrium during the storm.

A numerical model has been developed to predict beach adjustment due to dune erosion during a single storm event. In this model, the estimation of the shoreline change is based on the mechanics of sediment-

* Research Assistant, Dept. of Civil Engineering, North Carolina State University. Raleigh,NC.
** Associate Professor, Dept. of Civil Engineering, NCSU
***Professor, Dept. of Civil Engineering, NCSU.

swash interaction at the dune face, Overton, et al. (1988). This approach estimates the magnitude of dune erosion and beach adjustment associated with each swash, with no dependency on an equilibrium beach profile. Therefore, for a given storm condition, the total change in the beach profile is determined by a series of swash attacks. This technique has the advantage of considering small events in which post-storm equilibrium conditions can not be reached, as well as extreme events. In the numerical scheme presented herein, attention is restricted to the prediction of dune erosion and subsequent beach adjustment due to a single swash.

Methodology

The dune erosion and sediment profile model consists of three distinct components - the hydrodynamics of a bore impacting a vertical barrier, the volume of sand eroded from the dune due to that impact, and the redistribution of that volume of sand on the beach face during the subsidence of the bore. The first two components have been discussed elsewhere in the literature and will be discussed herein only briefly. Attention is focused on the development of the third component of the model - the adjustment of the beach face due to the erosion event.

The wave uprush is idealized as a bore propagating over a sloping beach. Input data include the depth of the bore at the seaward edge, the slope of the beach, the location of the dune relative to the mean water level, and the sediment characteristics. The bore motion on the beach is described using a two-dimensional hydrodynamic model. This model solves the momentum and continuity equations using a finite-difference technique. The finite-difference mesh is placed at right angles to the beach face in order to simplify computations. The model determines the bore height and velocity as it travels up the sloping beach, impacts with, and retreats from the onshore barrier (dune). The solution algorithm of the velocity and pressure fields are based on a modified SOLA-SURF code, Hirt et al. (1975). Modifications include those which incorporate the sloping beach and the moving boundary of the borefront. The hydrodynamics of the bore model are verified using a simplified analytical solution for the horizontal bed case, as well as a series of laboratory experiments for the sloping bed case.

The volume of sand released from the dune by the impacting bore is computed in terms of a force parameter which in turn is a function of a the flow characteristics at the dune face. This functional relationship is derived from laboratory tests designed to isolate a single bore and determine its erosive power, Overton, et al. (1988). This relationship has been modified for use in the hydrodynamic model, Overton, et al. (1987).

The amount of sand extracted from the dune is redistributed in front of the dune by the retreating bore. The resulting sediment profile is computed as a function of the backwash flow parameters and the sediment characteristics. The third component of the model which predicts the deposition pattern in the dune vicinity is based upon both theoretical considerations as well as laboratory results. Because of the nature of the flow pattern at the wall and the importance of turbulence in the mechanics of sediment transport, laboratory experiments were conducted for the development, calibration and independent verification of the model.

Experimental Study

Wave Flume:

The experiments were performed in the Civil Engineering wave flume at North Carolina State University. The glass walled flume, is 15.0 m long, 0.6 m wide, and 0.9 m deep. Young (1986), in his investigation of dune erosion due to wave uprush, modified the flume to isolate a single bore. The flume was divided by a sheet of plywood into two portions. The lower portion was used as a reservoir, and the upper portion as the main channel with plywood surface bottom as shown in Figure 1. A head tank, equipped with back and front rotating gates, was installed inside the flume. The head tank was 1.8 m long, 0.6 m wide, and 0.6 m deep. The front gate was equipped with a cantilever weight and springs allowing a sudden release of the gate and, consequently, the water in the head tank. The water released from the head tank formed a

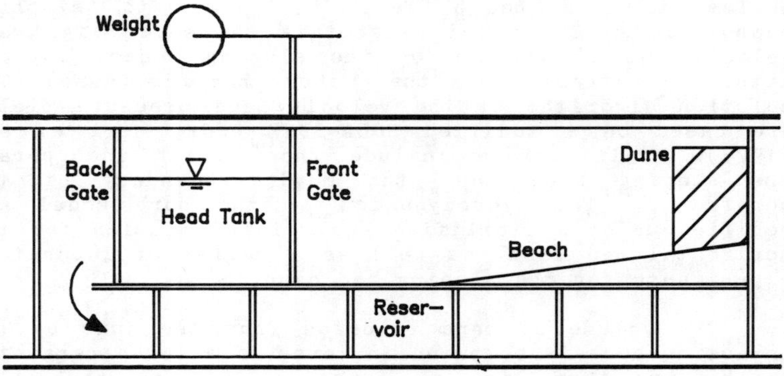

Figure 1. Schematic Diagram of Experimental Setup.

bore propagating on the sloping beach. As the bore impacted with the upstream boundary and retreated, the back gate was opened, using a release spring, allowing the backwash to flow back to the reservoir and thus preventing reflection. A plywood beach, 2.4 m long, was installed at an angle of three degrees to the horizontal bottom of the channel and 1.2 m in front of the head tank. A vertical barrier of plywood was erected at the end of the sloping beach to serve as the dune boundary.

Dune Model:

The sand dune model was constructed in a box inside the flume. The box extended the full width of the flume and had a fixed back board and removable front. The sand used to build the dune had a mean grain size, d_{50}, of 0.68 mm and a grain size distribution as shown in Figure 2. The oven dried sand was then rained into the

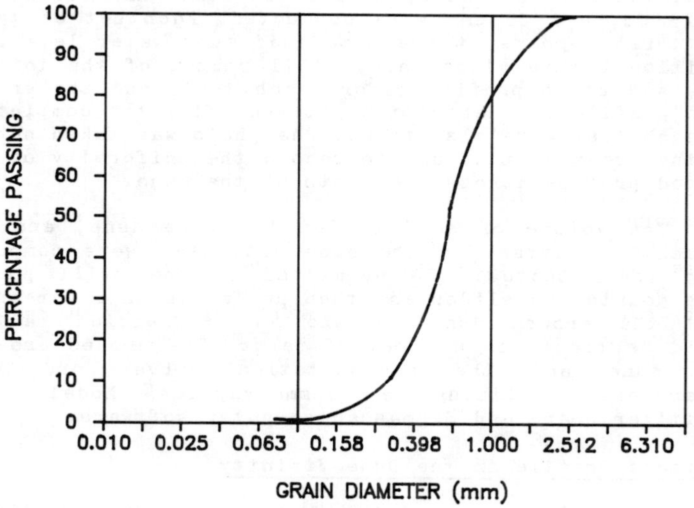

Figure 2. Grain Size Distribution Curve.

box from a constant head of 0.3 m so that a uniform degree of compaction was attained. The sand was rained in using a shallow plastic tray containing two rows of 8 mm diameter holes, three holes in each row. After the sand set in place, 7.5 l of water were poured at the top of the dune at a slow, constant rate. The dune was then allowed to drain for 30 min through holes provided in the front board. The amount of water and the drainage time were determined such that a sufficient strength was developed in the dune to maintain a vertical face when the front was removed. After removing the front of the box the first 6 cm of the dune front were scraped away to eliminate any cracks or instabilities caused by the sand leaning on the front board.

A video cassette recording system equipped with time display and single frame advancement was used to record the impact of the bore with the upstream boundary in all tests. A transparent plastic sheet containing a scale, used to depict the initial dune and beach profiles before each run, was mounted on the glass side of the flume, covering the dune area and extending in front of the dune about 100 cm. In addition, a 35 mm camera was utilized to record the dune and beach profiles immediately before and after each test.

Experimental Procedure:

The dune was constructed at the upstream boundary of the flume and the entire sloping beach was covered with a uniform layer of sand. The sand layer was 4.0 cm thick and had the same characteristics as that of the dune. Two sets of tests, at initial head tank elevations of 23 cm, and 26 cm, were conducted. Each set of tests had three repeats at the same head tank water level. In addition to the video tape, still photos of the initial dune and beach profile before each test, and the eroded dune profile and deposition pattern after the completion of each test were also taken. One photo was taken normal to the front of the dune to record the uniformity of the eroded profile across the width of the dune.

The volume of sand eroded from the dune, and the deposition pattern in the dune vicinity were computed from still photos. The negatives of the still photos were mounted on slides and then projected on a screen so that the eroded dune profile and deposition pattern could be traced on a sheet of paper. The eroded area of the dune and the sedimentation curve were then determined utilizing a Summagraphics Model 1201 digitizer board and Sigmascan computer software.

Sediment Profile in the Dune Vicinity

Generally, the eroded volume of dune sediment deposited on the beach is a function of the retreating bore flow variables and sediment characteristics. In order to determine the backwash flow variables that best represent the sediment profile, it is important to analyze: (1) the laboratory results of the deposition pattern and (2) the flow behavior as described by the verified numerical scheme. Based on visual observations of several flume tests, the volume of sand eroded is agitated and kept in suspension by the highly turbulent flow. As the bore begins to retreat, the flow becomes more stable and most of the sediment load is observed to quickly settle with the exception of the smaller size particles. In other words, the deposition pattern is mostly controlled by the flow parameters of the less turbulent backwash and not the flow field at the time of impact. Laboratory results also show that the sediment profile extended approximately 25 cm in front of the dune. A significant volume of the sediment load, consisting mainly of large to medium size particles, is deposited in the first 10 to 15 cm.

The numerical hydrodynamic results reflect qualitatively what was seen in the laboratory. During the period of impact of the bore with the dune, the flow field (velocity and depth) is highly unstable, Figures 3 and 4. The depth increases rapidly while the velocity

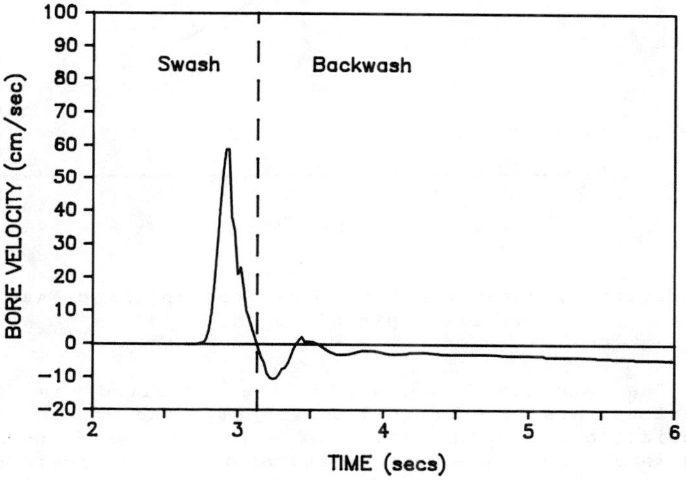

Figure 3. Bore Velocity Averaged over Depth at 6 cm from the Dune Face, Initial Depth of 26 cm.

goes to zero. As the backwash process begins, the depth begins to decrease and the velocity reverses direction. Both velocity and depth then stabilize changing almost linearly with time. The flow field in the 25 cm in front of the dune varies predictably with distance from the dune. The maximum depths and velocities are reduced and there is a shift in time for the maximum but the general shape of the curves is the same as that in Figures 3 and 4.

Based on the above analysis, the sediment profile in front of the dune is computed in terms of the backwash flow height and velocity in the 25 cm adjacent to the dune face and the sediment properties. The flow height is defined as the average height in this regime. The water height is obtained by computing the water height in each time increment and then averaging over the time period that the bore is in contact with the dune and propagating down slope. The flow velocity is defined as the depth averaged velocity in the same 25 cm in front of the dune and over the same period of time.

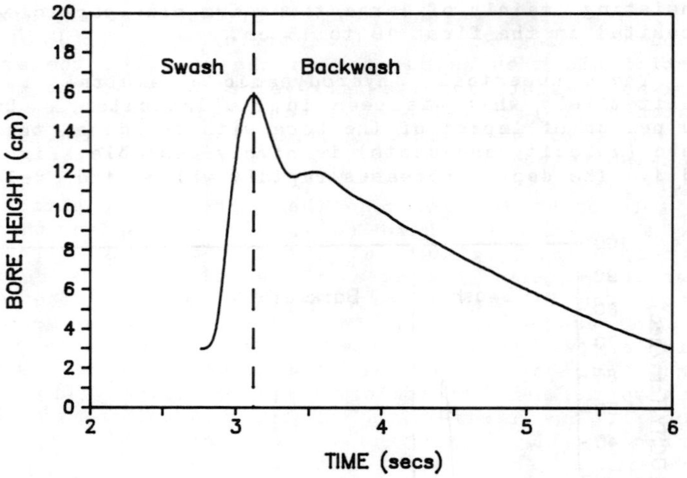

Figure 4. Bore Height at 3 cm from the Dune Face, Initial Depth of 26 cm.

The sediment characteristics utilized in this algorithm are the specific weight, the grain size distribution, and the fall velocity. The sand used in this model has a specific weight of 2.65, and grain size distribution as shown in Figure 2. The fall velocity of the sand particles is computed using the following equation, Henderson (1966)

$$W_s = C\,[(SS - 1)gd_s]^{1/2} \tag{1}$$

and,

$$C = \left[\frac{2}{3} + \frac{36\,n^2}{gd_s^3(SS-1)}\right]^{1/2} - \left[\frac{36\,n^2}{gd_s^3(SS-1)}\right]^{1/2} \tag{2}$$

where n is the kinematic viscosity of the fluid, d_s is the grain size diameter, and SS is the specific gravity of the sediment.

Having determined the retreating flow height and velocity, and sediment properties, the sedimentation profile in the dune vicinity is computed as outlined below.

(1) The grain size distribution curve is subdivided into representative intervals. The sediment volume ratio (V_r) and the average grain size diameter (d_s) for each interval are determined.

(2) The fall velocity (W_s) of each grain size is computed using Eq. 1. The settling time (t_s) of each particle is then calculated as the ratio of the average backwash height, and its fall velocity (W_s). The maximum distance (X_s) travelled by each grain size is computed in terms of its settling time (t_s) and the average backwash velocity.

(3) In order to determine the depth of deposition (Y_s) for a given grain size (d_s), it is assumed that its corresponding volume of sand (V_r) is uniformly distributed over the area of deposition. The depth of deposition is then computed as the ratio of the volume of sand eroded over the deposition area. The area over which a given grain size may be distributed is estimated based on the study of the mechanism of sediment transport and laboratory visual observations. In general, the smaller the particle size, the longer a particle will stay in suspension. Therefore for a given sand sample which has a wide range of particle sizes, the small sand particles at the lower end of the curve have a relatively longer settling time than those at its upper end. Thus, based on the settling time of each grain size, most of the small grains do not settle in the immediate vicinity of the dune as do those of large grain sizes. In order to determine the deposition area of the small particles, the point where they begin to settle should be estimated first. This is done by arbitrarily dividing the grain size distribution curve into three zones: lower, middle, and upper. The lower and upper zones consist of the relatively small and large grain sizes, $d_s<0.5$ and $d_s>1.0$, respectively, while the middle zone consists of the rest of the sand particles, $0.5 \leq d_s \leq 1.0$. In order to prevent the smaller size particles, ($d_s<0.5$) from settling too soon, it is required that the deposition area for the lower zone material begins at a distance in front of the dune equal to the travel distance of the smallest particle in the upper zone. The deposition depth (Y_s) is computed as the ratio (V_r/X_d).

(4) Finally the individual deposition patterns of each distribution interval are superimposed to yield the total sedimentation profile due to a single wave swash in front of the dune.

Results

Applying the above methodology as the third component of the erosion and sediment profile model, the redistribution of sediment in front of the dune after the retreat of a single swash event can be computed numerically. Two tests at different head tank levels were conducted in the laboratory. The first of these, with a head tank level of 26 cm, was used to qualitatively study the process and to aid in the development of the model. The numerical versus the experimental results are shown in Figure 5. Note that

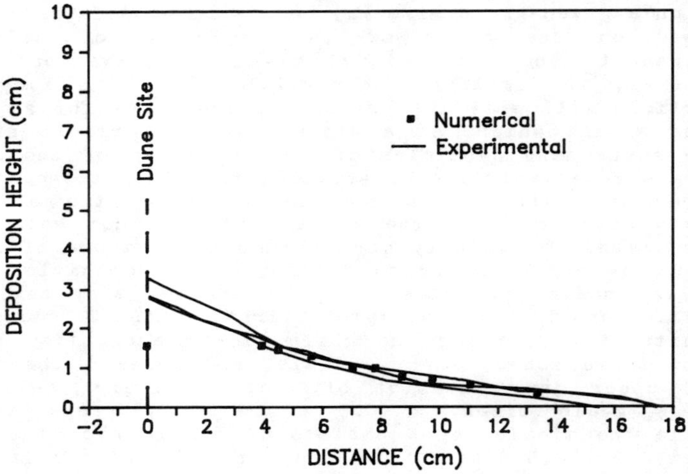

Figure 5. Comparison of Numerical and Three Replicate Experimental Sediment Profiles, Head Tank Level, 26 cm.

the high degree of correlation between the numerical and the experimental results is because the data were used to calibrate the model. In order to further test the ability of the model to reproduce deposition patterns, an independent set of data at 23 cm initial head tank elevation was used. The numerical model was run with head tank elevation set at 23 cm with no additional adjustments to the model. Figure 6 shows comparison of numerical and experimental beach profiles obtained. The model closely predicts the experimental results though with a slight over estimation of the depth of deposition along the beach face. Based on the limited number of cases investigated to date, it is shown that this approach is a viable alternative to use in the prediction of beach and dune erosion due to a storm event.

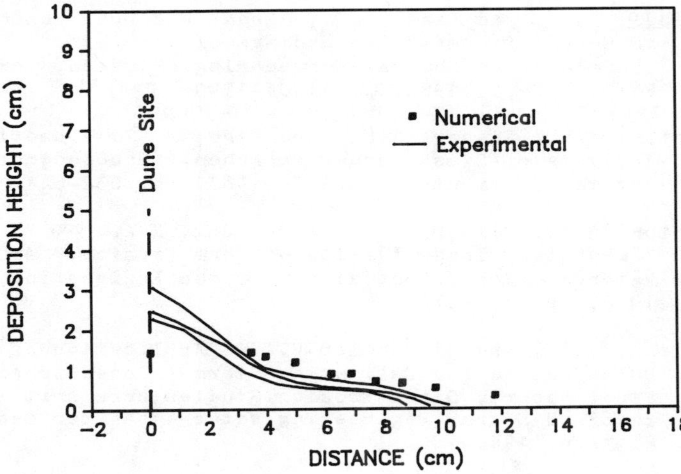

Figure 6. Comparison of Numerical and Three Replicate Experimental Sediment Profiles, Head Tank Level, 23 cm.

References

Edelman, T., "Dune Erosion and Storm Conditions," Proceedings of the 11th Conference on Coastal Engineering, London, England, Vol. 1, Sept. 1968, pp. 719-722.

Edelman, T., "Dune Erosion During Storm conditions," Proceedings of the 13th Conference on Coastal Engineering, Vancouver, Vol. 2, July 1972, pp. 1305-1311.

Henderson, F.M., "Open Channel Flow, " McMillan Co., New York, New York, 1966.

Hirt, C.W., Nichols, B.D., and Romero, N.C., "SOLA - A Numerical Solution Algorithm for Transient Fluid Flows," Los Alamos Scientific Laboratory Report LA-5852, 1975.

Kobayashi, N., "An Analytical Solution for Dune Erosion by Storms," Journal of Waterway, Port, Coastal, and Ocean Engineering, ASCE, Vol. 113, No. 4, July 1987.

Kriebel, D.L., and Dean, R.G., "Beach and Dune Response to Severe Storms," Proceedings of the 19th Conference on Coastal Engineering, Houston, Texas, Vol. 2, Sept. 1984, pp. 1584-1599.

Overton, M.F., Fisher, J.S., and Fenaish, T. "Numerical Analysis of Swash Forces on Dunes," Proceedings Coastal Sediments'87, ASCE, 1987, pp. 632-641.

Overton, M.F., Fisher, J.S., and Young, M.A., "Laboratory Investigation of Dune Erosion," ASCE J. Waterway, Port, Coastal, and Ocean Engineering,May, 1988, pp. 367-373.

Sargent, F.E., and Birkemeier W.A., "Application of the Dutch Method for Estimating Storm-Induced Erosion," Final Report, Storm Erosion Studies Work Unit 31467, U S Army Engineering Waterways Experiment Station, 1985.

Vellinga, P., "Beach and Dune Erosion During Storm Surges," Ph.D. Thesis, Delft Technical University, 1986.

Vellinga, P., "Predictive Computational Model for Beach and Dune Erosion During Storm Surges," Coastal Structures, 1983, pp. 806-819.

Vellinga, P., "Beach and Dune Erosion During Storm Surges," Coastal Engineering, Vol. 6, No. 4, 1982.

Young, M.A., "Modeling of Storm-Induced Dune Erosion Due to Wave Uprush" M.S. Thesis, Department of Civil Engineering, North Carolina State University, 1986.

CHAPTER 107

Barcelona's littoral regeneration looking forward to the Olimpic Games. Numerical model

Vicente Negro Valdecantos *

Manuel Quecedo Gutierrez **

* Civil Engineer. Research Assistant. University of Madrid. Spain.

** Civil Engineer. M. Sc.

1.- ABSTRACT

The modern technique of Coastal Engineering has enabled the succesful modification of the environment, the recuperation and the stabilization of those sandy areas which due to human action, were in a process of regression and erosion. This work developes a mathematical and numerical model to adjust and analyze the variations between the dissipative profiles and the reflectives ones. The research also emphasizes the enveloping profile to calculate the stable behaviour of the beach.

2.- INTRODUCTION

After several series of observations to get data about bathimetry topography, location of sand, the littoral process, sea, swell, tides, the longitudinal and onshore-offshore current, the authors analyzed the conditions of the progressive erosion of our shoreline during the last decades. Considering these series of initial conditioners, the design criteria were developed in a mathematical and numerical model divided in the following chapters:

a) Variations between the dissipative and reflective profiles.

b) Incipient motion on a granular bed in water waves.

c) The depth at breaking and the signigicant wave height.

d) Beach profiles.

e) Numerical model to adjust. Enveloping profile in

$x/x_{máx}$ d/d_{max}

3.- VARIATIONS BETWEEN THE DISSIPATIVE AND REFLECTIVE PROFILES

Equilibrium profiles have been studied extensively through the history of coastal engineering. For many years it has been held that the steepness H_o/L_o, the ratio of the deep water waves of the wave height and wave length divided the bar and berm profiles; the dimensionless fall velocity H_o/Tw, the erosion parameter Q/H_{odi}, the grain size D_{50}, the depth d, and the relations between the steepness and the fall velocity separate the dissipative and the reflective profile.

The numerical model calculates the beach profile shape depending of wave climate and the variations with H_o, L_o, γ_s, γ_w, D_{50} data.

In Table 1, the authors include the initial variables and the design criteria.

5.- THE DEPTH AT BREAKING AND THE SIFNIFICANT WAVE HEIGHT

A large number of factors influence the shape of beach profiles in nature. The series of observations to get data about bathimethy are considered the design criteria to establish the refraction, difraction and the breakpoint.

The depth at breaking and the relationship with the significant wave height, type of breaker, broken wave and the accreting, transition and eroding beaches correspond with the occurence of surging, plinging and spilling breakers respectively.

The authors analyze the different criteria and adjust several ones to compute the break point and the type of breaker.

In table 2, they include the variables and the design criteria. Numerical model uses the following formulae:

- Munk
- Weggel
- Gumbak
- Goda

TABLE 2. MODEL OF WAVE BREAKING

- Boussinesq	(1881)	H/d =	0,73
- McGowan	(1881)	H/d =	0,781
- Michell	(1893)	H/L =	0,127
- Gwyther	(1900)	H/d =	0,83
- Miche	(1944)	H/d =	2 H/L/arcth (7,04 H/L)
- Davies	(1951)	H/d =	0,83
- Packham	(1952)	H/d =	1,03
- Yamada	(1957)	H/d =	0,828
- Laitone	(1962)	H/d =	0,7273
- Lenau	(1966)	H/d =	0,827
- Horikawa-Kuo	(1966)	Grafico	
- Kishi-Saeki	(1966)	H/d =	$5{,}618\ m^{0{,}40}$
- Camfield-Street	(1967)	H/d =	$0{,}75 + 25\ m - 111\ m^2 + 3.870\ m^3$
		$0 \leq m \leq 0{,}045$	
- Galvin	(1969)	$\frac{H}{d} = \frac{1}{B - \tau i K}$	
		B =	1,40 - 6,85
		τ =	4 - 9,25 i
		$0 \leq K \leq 2 \; ; K \simeq 1$	
- Collins	(1969)	H/d =	0,72 + 5,60 m
- Bryar-Smith	(1970)	H/d =	0,86
- Grimshaw	(1971)	H/d =	1,21
- Strel Koff	(1971)	H/d =	0,85

Weggel	(1972)	$\frac{H}{d} = \frac{b}{1 + a/gT^2}$
		$b = \frac{1,56}{1 + e^{-19,5i}}$
		$a = 43,70\ (1-e^{-19i})$
Goda	(1975) $H_{1/3}$	$K_s\, H_o;\ D/L_o \geq 0,20$
		$\min\ (B_o\, H_o + B_1 d),\ B_{max}\, H_o,\ K_s\, H_o$
Gumbak	(1977)	$\frac{H}{d} = 0,80;\quad I_r < 0,20$
		$\frac{H}{d} = 0,87\ I_r + 0,63;\quad 0,20 \leq I_r \leq 0,66$
		$\frac{H}{d} = 1,20;\quad 0,66 \leq I_r \leq 2,20$
Ostendorf-Madsen	(1979)	$H/d = 0,829$
Ostendorf-Madsen	(1979)	$H/d = (0,80+5m)\ 2\ \ H/L\ /\ \text{argth}\ (7,14\ H/L)$ $m < 0,10$
		$H/d = 0,13\ 2\ \ H/L\ /\ \text{argth}\ (7,14\ H/L)$ $m > 0,10$

6.- BEACH PROFILES

To find a connection between the depth, the wave height and length, the grain diameter, the dimensionless fall velocity and to evaluate the response of the beach, the bar or berm profiles the researches have been studied the problem through the history of the Coastal Engineering. The lineal relation, the adjustment of storm profiles by cubic parabolas, the location of the inflection point and the shoreline process of regression are the problems to analyze and resolve.

In table 3, the authors include the state of art and they develope a mathematical and numerical model to determine the variations between bar and berm profiles. Considering the initial data:

- Beach shape
- Motion initiation by oscillatory flow
- Break point
- Beach profile

The numerical model evaluate in 600 points in x and depth the value of each profile, the risk of failure depending of the grain size and wave climate, and the enveloping profile of the stable beach.

The mathematical curve adjusts the stable beach, the validity of each one (Bruun, Larras, ...) and with the initial data the defensive barrier that we need is created an artifitial beach. We shall emphasize the enveloping profiles in dimensionless parameters:

$x/x_{máx}$

$d/d_{máx}$

TABLE 3. BEACH PROFILES

- Bruun (1954, 1956, 1978, 1986) $x = L_o \sqrt{2\pi y}\left(2\left(\frac{2\pi y}{L_o}\right) + \frac{1}{3}\left(\frac{2\pi y}{L_o}\right)^2 + \ldots\right.$

  $y < L_o/8 \quad y^{3/2} = px$

  $y^{5/4} = px$

  $y^{3/2} = px$

- Keulegan y Krumbein (1949) $y = px^{4/7}$

- Inmann y Bagnold (1963) $y = px$

- Larras $\frac{y}{L_o} = k\left(\frac{x}{L_o}\right)^m$

  $k = f(H_o, L_o, D, \rho, \rho_o)$

  $m = g(H_o, L_o, \rho, \rho_o)$

- Sitarz (1969) $x = ay^2 + x_o$

  $a = f(D, H_o, \rho, \rho_o)$

  $x_o = g(A_1, H_o, D, \rho, \rho_o)$

A_1	43,50	Berm
	66	Bar
	75	Nature

- Dean (1973)

$$y = A_D\, x^{2/3}$$

$$A_D = \left[\frac{24}{5}\,\frac{f(D)}{f\,g\,\sqrt{g}\,\gamma_b^{2}}\right]^{2/3}$$

f (D) = Energy wave regime

γ_b = H/d breaker point

- Vellinga (1982,19854)

$$y\,\frac{7{,}60}{H_{os}} = 0{,}47\ (7{,}6/H_{os})^{1{,}28}\ x$$

$$x\,(W_f/0{,}0268)^{0{,}56}\ x + 18\ \tfrac{1}{2} - 2{,}00$$

$$y = 0{,}70\,\left(\frac{H_o}{L_o}\right)^{0{,}17}\ w_f^{0{,}44}\ x^{0{,}78}$$

H_o = 7,6 m

T = 12 sg

D_{50} = 0,225 mm

- Sunamura (1980, 1984) $y = 0{,}75\,\left(\frac{D}{H}\right)^{0{,}25}\left(\frac{L}{H}\right)^{0{,}15}\ x$

- Horikawa (1980, 1984) $y = 0{,}21237\ H^{-0{,}50}\ D^{-0{,}25}\ T^{-0{,}50}\ x$

- Sayao (1982) $y = 0{,}082\ x^{0{,}633}$ D_{50} = 0,225 mm

- Hattori y Kawamata (1982) $y = p\ x^{2/3}$

$$\frac{H_o}{L_o}\ \mathrm{tg}\ B/\frac{w_f}{gT} = k$$

Beach profile shape

TABLE 1. Dissipative and reflective profile

-	Jonson (1952)	$0,025 < \frac{H_o}{L_o} < 0,030$		
-	Saville (1957)	$0,025 \simeq \frac{H_o}{L_o}$		
-	Iwagaki-Noda (1962)	$\frac{H_o}{L_o}, \frac{H_o}{D_{50}}$		
-	Sitarz (1963)		< 0,60	Berm
		$\frac{H}{T\left(g(s-1)D\right)^{1/2}}$		
			> 0,60	Bar
-	Nayak (1970)	$\frac{H_o}{L_o}, \frac{H_o}{D_{50}}, \frac{\gamma_s}{\gamma_w}$		
-	Dean (1973)		< 0,85	Berm
		$\frac{H_o}{T_w}$		
			> 0,85	Bar
			> T/2	Bar
		tf		
			< T/2	Berm
		$\frac{H_o}{L_o} > 1,70 \frac{\pi w_f}{gT}$	Bar	

- Van Hijum (1974) $\frac{H_o}{L_o} > 2{,}50 \frac{D_{90}}{H_o}$ Bar

  < 0,54 Berm

  $\frac{H_o}{Tw\, D_{50}}$

  > 0,54 Bar

- Sunamura y Horikawa (1974) $\frac{Ho}{L_o} > 4\ tg\,\alpha^{-0{,}27} \left(\frac{D}{L_o}\right)^{0{,}67}$ Bar

- Dalrymple y Tompson (1976) $\frac{Q}{H_o\, di}$ erosion parameter

- Hattori y Kawamata (1980) < 0,50 Berm.

  $$\frac{\frac{H_o}{L_o}\ tg\, B}{\frac{w}{gT}}$$

  > 0,50 Bar.

- Gourlay (1980) <1,55 Berm

  $\frac{H_o}{T_w}$

  >1,55 Bar

  Marine sand, D_{50} 0,20 a 0,30 mm.

$$\frac{H_o}{T_w} \quad \begin{cases} < 0{,}40 & \text{Berm} \\ > 0{,}40 & \text{Bar} \end{cases}$$

$D_{50} \simeq 2$ mm.

Wang (1981)

$$I_w = \frac{H_b^{\frac{1}{2}}}{g^{\frac{1}{2}}\ T\ tg\alpha}$$

Iw ↓ Berm

Iw ↑ Bar

4.- INCIPIENT MOTION ON A GRANULAR BED IN WATER WAVES

Motion initiation on a granular bed by oscillatory flow is of great importance to coastal engineers, it is a necessary condition for bed erosion and to determine the changes onshore-offshore in beaches, to establish the failure and to use in different beach profiles.

The numerical model adjusts three formulae in this case:

- Swart $u = 4{,}58\ D_{50}^{0{,}38}\ T^{0{,}043}$
- Hallermeier $u = \left(8\ (\frac{\gamma s}{\gamma w} - 1)\ g\ D_{50}\right)^{\frac{1}{2}}$
- Vellinga $\log\ (\frac{1}{w}) = 0{,}476\ (\log\ D)^2 + 2{,}18\ \log\ D + 3{,}190$

- Readshaw (1982)	$y = 0{,}112\ x^{0{,}98}$ $D_{50} = 0{,}56$ mm
- Bailard (1981)	$y = p\ x^{\frac{1}{2}}$
	$y = p\ x^{2/7}$
- Sayao (1982)	$y = A\ x^{n}$ $n > 1,\ D_{50} > 0{,}60$ mm
	$n < 1,\ D_{50} < 0{,}60$ mm
- Garau (1974-1984)	$m_o = 0{,}075 + 0{,}035\ L_n\ D_{50}$
	$m_s = 0{,}025 + 0{,}010\ L_n\ D_{50}$
	$m_p = 0{,}0223 + 0{,}005\ L_n\ D_{50}$
	$tg \quad = m = 0{,}138 + 0{,}056\ L_n\ D_{50}$
- Garau-Friedman (stable beachline)	$m_{sa} = (0{,}0255 + 0{,}010\ L_n\ D_{50})\ \frac{1}{Ka^{1{,}11}}$
	$m_{oa} = (0{,}075 + 0{,}035\ L_n\ D_{50})\ \frac{1}{Ka^{1{,}22}}$

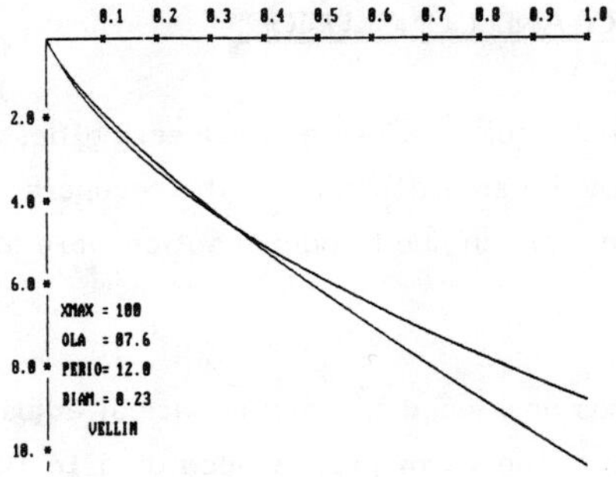
0.1 0.2 0.3 0.4 0.5 0.6 0.7 0.8 0.9 1.0
2.0
4.0
6.0
8.0
10.
XMAX = 100
OLA = 07.6
PERIO= 12.0
DIAM.= 0.23
VELLIN

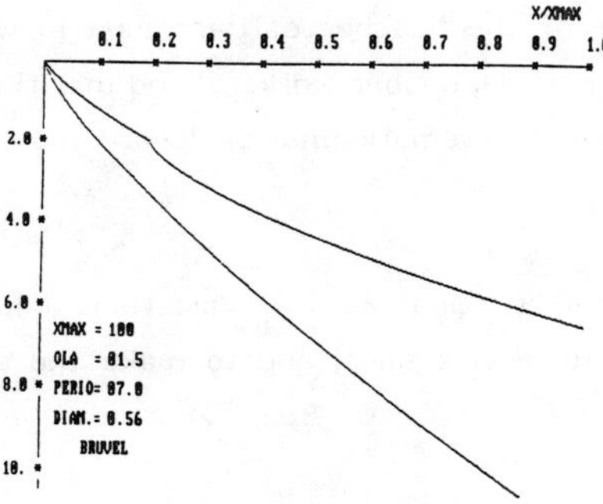
X/XMAX
0.1 0.2 0.3 0.4 0.5 0.6 0.7 0.8 0.9 1.0
2.0
4.0
6.0
8.0
10.
XMAX = 100
OLA = 01.5
PERIO= 07.0
DIAM.= 0.56
BRUVEL

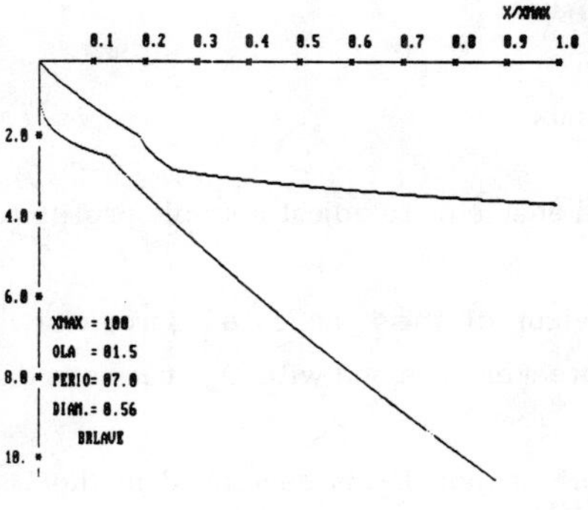
X/XMAX
0.1 0.2 0.3 0.4 0.5 0.6 0.7 0.8 0.9 1.0
2.0
4.0
6.0
8.0
10.
XMAX = 100
OLA = 01.5
PERIO= 07.0
DIAM.= 0.56
BRLAVE

7.- SUMMARY AND CONCLUSIONS

The modern Technique of Coastal Engineering has enabled the successful modifications of the environment and the recuperation and stabilization of those sandy areas which due to human action were in a process of regresion and erosion.

The research has developed the mathematical adjustment of curves to the storm profile and the berm profile according to Bruun, Vellinga, Larras, Sitarz ... and the behaviour of the beaches of Barcelona in which both human action (growing, construction, habour works...) and maritime actions (sea movements) have produced a remarkabled backward movement and regression.

The numerical model is able to analyze the variations between the dissipative profiles and the reflective ones, and to make the enveloping profile in

x/x_{max}

d/d_{max}

Which will enable us to adjust a stable profile to the beach.

The behaviour of the beaches of Barcelona may consider as fossil ones, near the breaker zone and with D_{50} between $0{,}16 \text{ mm} \leq D_{50} \leq 2 \text{ mm}$.

The numerical model was developed in the University of Madrid and the results were analyzed by experimental data for the beach profile tests in the Spanish Center of Investigation (CEDEX, CEPYC. Madrid).

8.- ACKNOWLEDGEMENTS

This research was supported by the Madrid University the author wish to thank Professor Suarez Bores and Mr. Victoriano Fernández. Both of them provided valuable comments.

The author is grateful for the efforts of Mr. Quecedo, civil Engineer and M.Sc. He was dedicated to programme the numerical model.

CHAPTER 108

Wave–Induced Flow And Nearshore Suspended Sediment

J.C. Doering* and A.J. Bowen

1. Introduction

It has been realized for nearly one hundred years that the transport of sediment is related to the characteristics of a wave, in particular its shape. Cornish (1898) noticed that the shoreward velocity associated with a wave crest was more effective at moving coarse sediment than was the seaward velocity associated with the wave trough. Cornish's observation was consistent with the theory of Stokes (1847), which predicts the onshore velocity associated with the wave crest is stronger and of shorter duration than the offshore velocity associated with the wave trough. This horizontal asymmetry of the cross–shore flow, which is a reflection of the wave shape, is known as velocity skewness. It has been suggested that "the existence of the beach depends on small departures from symmetry in the velocity field balancing the tendency for gravity to move material offshore"(Bowen, 1980). Although the concept of velocity skewness has been incorporated into detailed predictors of sediment transport (Bowen, 1980; Bailard and Inman, 1981) it is only one of many facets that needs to be understood in order to make the accurate prediction of sediment transport realizable. A comprehension of sediment transport is hampered by both an incomplete knowledge of the hydrodynamics and a lack of instrumentation to directly measure instantaneous sediment concentration and the accurate prediction of sediment transport is probably the most enigmatic problem in coastal engineering.

Occasionally, suspended sediment concentration has been inferred from *in situ* pumps and hand–held tubes, but these methods lack the temporal and spatial resolution necessary to elucidate the details of the interaction between the wave–induced flow and the sediment. Recently, a miniature optical backscatter sensor (MOBS), which provides a time series of suspended sediment concentration at a "point", was developed by Downing *et al.* (1981). During a recent field experiment a vertical array of 5 of these optical backscatter sensors and a colocated flow meter was deployed close to the sea bed. These colocated measurements provide a unique opportunity to investigate the response of near–bed suspended sediment concentration to the wave–induced flow.

Department of Oceanography, Dalhousie University, Halifax, Nova Scotia B3H 4J1 Canada

* Presently at Hydraulics Lab., National Water Research Institute, P.O. Box 5050, Burlington, ON L7R 4A6

2. Field Measurements

Data were collected at Pte. Sapin, New Brunswick, as part of the Canadian Coastal Sediment Study. This field experiment was conducted during October–November, 1983. Data from a Marsh–McBirney flow meter, measuring the horizontal components of the flow, and a colocated miniature optical backscatter sensor (MOBS) were used to examine the temporal response of suspended sediment to the wave–induced flow field. The MOBS and flow meter were located at 0.02 m and 0.12 m above the sea bed, respectively. Note that only the lowest MOBS sensor data is used in this paper; a description of the data from all five sensors can be found in Hanes and Huntley (1986).

Data for this paper is confined to a run collected on the 20th of October (run FM). The significant wave height for run FM was 0.20 m. The instrument array was located well seawards of wave breaking in about 1.1 meters of water. Figures 1a and 1b show the first twelve minutes of cross–shore velocity (u) and near–bed sediment concentration (c) for the 20th of October at Pte. Sapin, respectively. Since the alongshore velocity is much smaller than the cross–shore velocity ($\frac{\langle v^2 \rangle}{\langle u^2 \rangle} \sim 0.06$), the alongshore component of the flow is neglected from this analysis and discussion. It is obvious from figure 1 that the concentration of near–bed suspended sediment responds to both the individual waves and the wave groupiness. The response of the concentration to the flow associated with the passage of each wave is intriguing because the concentration seems to respond strongly and rapidly to the onshore flow associated with a wave crest, but very weakly to the offshore flow associated with a wave trough. A close inspection of figure 1a suggests that the cross–shore flow is strongly skewed; in particular, the onshore velocities are considerably stronger and of shorter duration than the offshore velocities (this can be readily seen with respect to the horizontal line denoting the mean flow). The suggestion from figure 1 is the stronger onshore flows associated with the skewed wave crests exceed the threshold velocity for the mobilization and subsequent suspension of sediment, while the weaker offshore flows associated with the "flattish" wave troughs do not. In addition, this suggestion implies that there should be a net onshore transport of sediment by the wind–waves at this height because the onshore flow will advect a high concentration of sediment shoreward whereas the offshore flow will advect a low concentration seaward. The cross–shore transport of suspended sediment at this height is given by the product $u \cdot c$, and is shown in figure 1c. The shoreward transport of suspended sediment associated with a wave crest is clearly larger than the seaward transport associated with a wave trough, as anticipated. For Bagnold–type models of sediment transport the instantaneous cross–shore sediment transport varies with the velocity moment $u^3|u|$, which is shown in figure 1e. However, $u^3|u|$ is in general very strongly correlated with u^3, the dimensional skewness of the flow (figures 1e and 1d, respectively). For the present case $r^2 = .93$. Thus, the transport of suspended sediment is intimately related to the skewness of the flow; the covariance of $u \cdot c$ and u^3 is readily apparent from figure 1.

Particularly interesting is the way the transport tracks u^3 even though the effect of a critical stress (threshold velocity) has not been included, nor has any representation of the settling characteristics of the sediment. Obviously both effects need to be included in a realistic model, the critical stress for suspension must

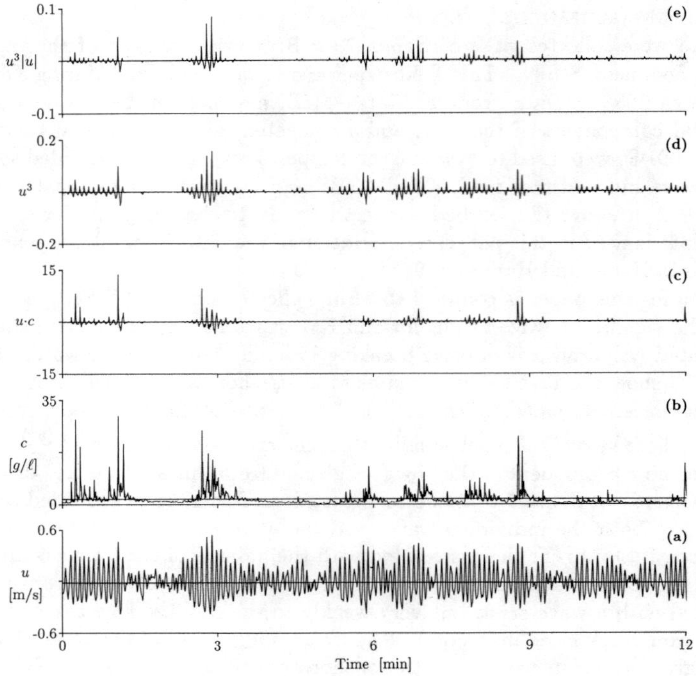

Figure 1 Time series of cross–shore velocity **(a)**, near–bed sediment concentration **(b)**, instantaneous suspended load transport **(c)**, and two moments of the cross–shore velocity field, u^3 **(d)** and $u^3|u|$ **(e)** for the first twelve minutes of run FM at Pte. Sapin. A positive velocity denotes an onshore flow. The horizontal line through each record denotes the mean.

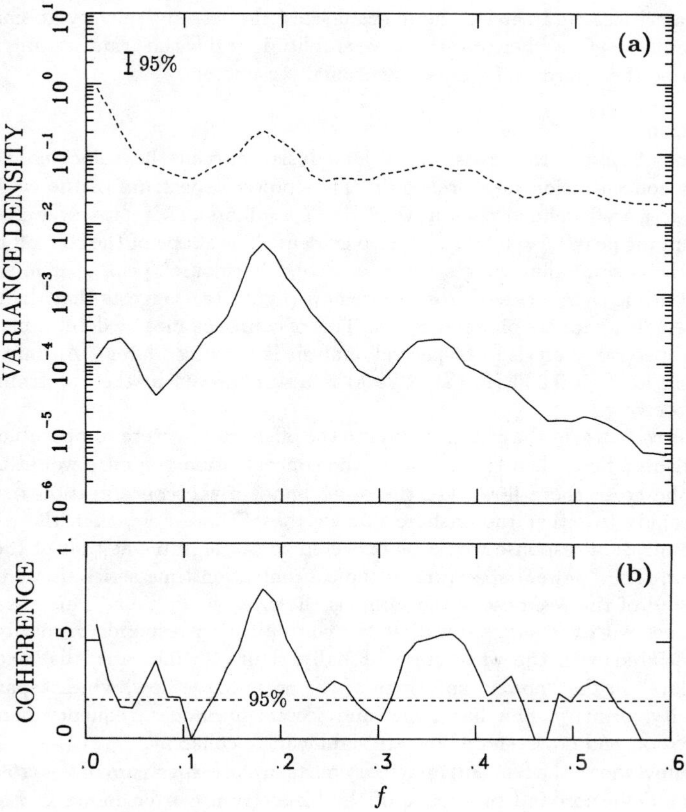

Figure 2 **(a)** Spectra of cross–shore velocity (——) and MOBS (- - -) for run FM at Pte. Sapin. Error bars show the 95% confidence limits. **(b)** Coherence between the spectra in (a). The 95% confidence limit for zero coherence is given by - - -. There are 100 degrees of freedom and $\delta f = 0.0121$ Hz.

become a critical parameter at large grain sizes, the settling velocity at small sizes (the parameter af/w where a is the wave amplitude and w the grain settling velocity seems to be the appropriate non–dimensional parameter).

3. Analysis

Figure 2 shows the cross–spectral analysis between the colocated flow and sediment concentration measurements. The "power" spectrum of the cross–shore flow shows a well–defined peak at 0.18 Hz ($T_p = 5.5$ s). A relatively well–defined first harmonic peak ($f = 0.36$ Hz) is also evident. The shape of the roll–off between 0.4–0.6 Hz is somewhat suggestive of a second harmonic "peak". The skewness observed in the time series of cross–shore flow (figure 1a) suggests that the primary and these harmonics are phase–coupled. This of course cannot be determined from a "power" spectral analysis (a bispectral analysis is discussed later). An infragravity wave peak at $f = 0.036$ Hz ($T \sim 28$ s) is also observed in the spectrum of the cross–shore flow.

If the response of the concentration to the offshore flow were comparable to that of the onshore flow, then the period of the concentration response would be twice that of the cross–shore flow. On the other hand, if the concentration responded more strongly to either the onshore flow or the offshore flow, then the period of the concentration response would be expected to be the same as that of the cross–shore flow. The "power" spectrum of the concentration time series shows that the peak period of the response is the same as the cross–shore flow. This observation is consistent with the suggestion that the concentration responds primarily to the flow associated with the wave crest. Finally, figure 2b indicates that except for the "valleys" in the "power" spectrum of the cross–shore flow, which separate the infragravity, primary, first harmonic, and second harmonic frequency bands, the concentration and cross–shore flow are significantly coherent.

Figure 3 shows the real and imaginary parts of the bispectrum of the cross–shore flow. The definition and properties of the bispectrum can be found in Elgar and Guza (1985), Doering and Bowen (1986), and many others. In a few words though, the bispectrum is used to identify triad(s) (*i.e.*, three phase–coupled frequencies) and to determine the relative contribution of a triad to the total skewness and asymmetry (a measure of the lack of vertical, as opposed to horizontal, symmetry) of the waves in a record. The peak centered at (0.18 Hz, 0.18 Hz) denotes phase–coupling between primary and first harmonic frequencies, and is suggestive of a self–self sum interaction between primary wavetrains. The smaller peak located at (0.36 Hz, 0.18 Hz) indicates phase–coupling between first, primary, and second harmonic frequencies. These two peaks clearly indicate that harmonic wavetrains are phase–coupled to the primary. The wind–wave skewness (S_{ww}) arising from these two interactions is relatively large, $S_{ww} = 0.676$. This confirms that the cross–shore flow is strongly skewed. However, the wind–wave asymmetry (A_{ww}) arising from these two interactions is relatively small, $A_{ww} = -0.159$; that is, the waves are almost vertically symmetric. The depression located at (0.16 Hz, 0.02 Hz) indicates that negative skewness arises from phase–coupling between primary wavetrains and long waves. The infragravity wave skewness and asymmetry arising from this interaction are -.176 and 0.017, respectively; hence, the biphase of this interaction

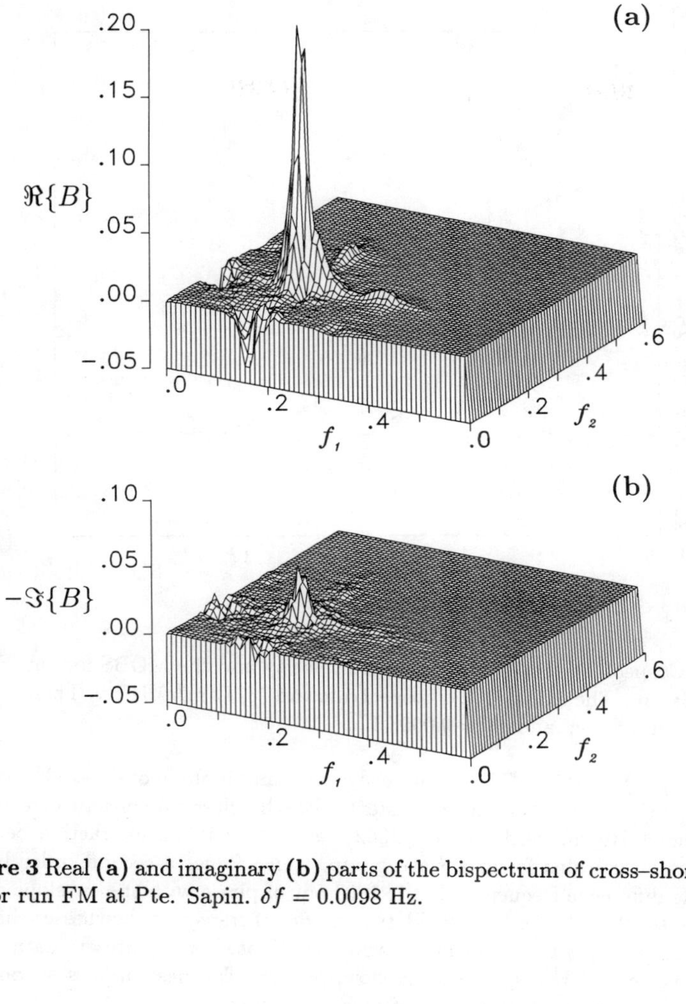

Figure 3 Real **(a)** and imaginary **(b)** parts of the bispectrum of cross–shore velocity for run FM at Pte. Sapin. $\delta f = 0.0098$ Hz.

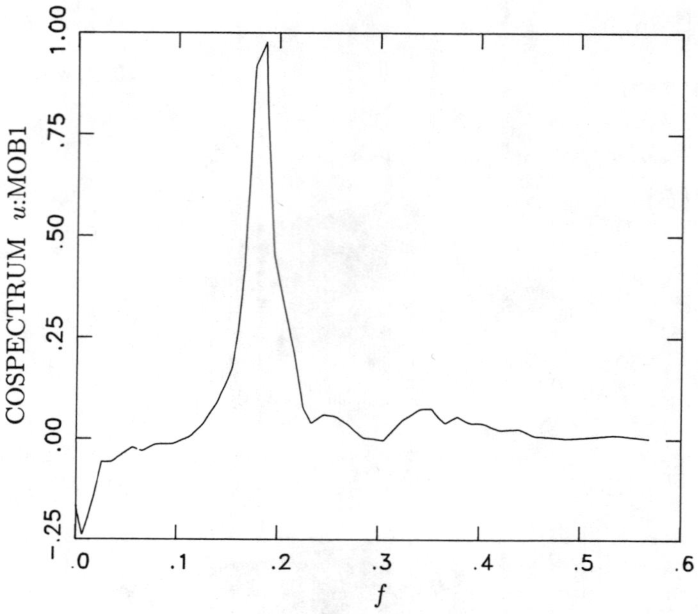

Figure 4 Cospectrum between cross–shore velocity and the MOBS for run FM at Pte. Sapin. The units of the cospectrum are $(g/\ell)(\text{m/s})\text{Hz}^{-1}$. There are 58 degrees of freedom and $\delta f = 0.0098$ Hz.

$(= \tan^{-1}\{A/S\})$ is $-174°$. The negative skewness and biphase of $\sim -180°$ that are observed for this interaction are consistent with the classical concept of a bound wave (Longuet–Higgins and Stewart, 1962, 1964); that is, an interaction between two primary wavetrains forms a wave group, which forces a second–order bound wave at the difference frequency that is 180° out of phase with the envelope of the wave group (§2.2–2.3). For a Bagnold–type model of transport skewnesses that are opposite in sign imply transport in opposite directions. For the present data where a positive velocity denotes an onshore flow, positive skewness implies an onshore transport and negative skewness, an offshore transport.

To determine the net transport in the cross–shore direction at some height above the sea bed, the cospectrum of the cross–shore flow and concentration c at that height is formed, *i.e.*, $\langle u \cdot c \rangle$. Figure 4, reproduced from Huntley and Hanes (1987), shows the cospectrum between the MOBS sensor and the cross–shore velocity. The most striking feature is the large positive or onshore flux of sediment by the primary waves. The indication here is the strong onshore flow associated with the passage of a skewed wave crest leads not only to the mobilization and suspension of sediment, but also to an onshore transport. This implies that the bulk of the sediment at this height (of ~ 2 cm above the sea bed) settled before the offshore flow associated

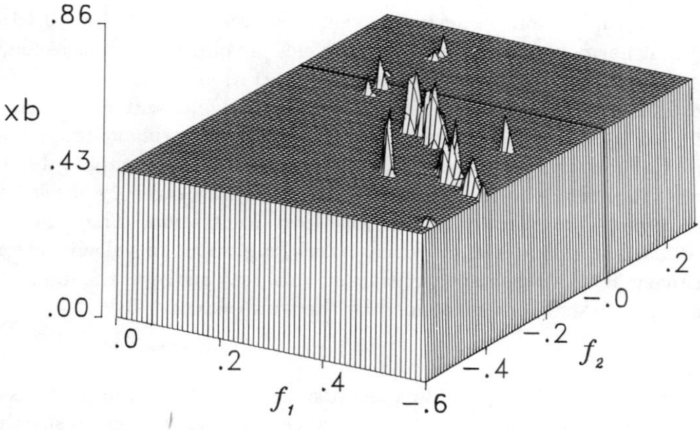

Figure 5 Cross–bispectrum between cross–shore velocity and the MOBS signal for run FM at Pte. Sapin. There are 32 degrees of freedom and $\delta f = 0.0098$ Hz.

with the wave trough could transport it back, *i.e.*, the net flux of sediment was onshore, which is consistent with figure 1 where a much higher concentration was observed during the onshore flow. The offshore transport of sediment by infragravity waves is probably due to the offshore skewed flow associated with the bound long wave forced by a wave group (Longuet–Higgins and Stewart, 1962; Wells, 1967); the general tendency for stronger offshore flow to occur when the waves are large, provides a perturbation which tends to move sediment seaward (Shi and Larsen, 1984). This cospectrum does not necessarily imply that the beach was locally accreting, as neither the transport by the alongshore current nor the mean flow have been considered.

The evidence presented so far is consistent with the idea that the wind–waves result in an onshore transport of sediment because the flow associated with them is skewed onshore (*i.e.*, the strong onshore velocities exceed the threshold velocity required to mobilize, suspend, and transport the sediment, whereas the offshore velocities generally do not), while the interaction of the infragravity waves with the sediment suspended by the wind–waves leads to an offshore transport of sediment because the flow associated with these long waves is skewed offshore. In other words, the suggestion is the response of concentration should be phase–coupled to the wave–induced flow. One can test this idea using the cross–bispectrum.

Figure 5 shows the unique part of the cross–bicoherence spectrum between velocity and concentration. The peaks shown are above the 95% significance level for zero cross–bicoherence. The convention adopted is that $u(f_1) \pm u(f_2) \rightarrow c(f_3)$, where $u(f_1)$, $u(f_2)$, and $c(f_3)$ denote frequencies in the velocity and concentration series, respectively, and $f_1 + f_2 = f_3$. This convention assumes that coupling between the spectral components of the wave–induced flow are phase–coupled to the variation in concentration. The broad peak centered at (0.36 Hz, -0.18 Hz) indicates that first harmonic and primary frequencies in velocity are phase–coupled to 0.36+(-0.18)=0.18 Hz, or the primary in concentration. The peak at (0.54 Hz, -0.36 Hz) indicates that phase–coupling between second and first harmonic frequencies in velocity are also coupled to fluctuations at the primary frequency in concentration. Collectively, these two peaks indicate that phase–coupling between the primary and its harmonics in velocity, which result in a strongly skewed flow (figure 3a), are coupled to primary fluctuations in concentration. The small peak at (0.19 Hz, -0.03 Hz) indicates that primary and long waves in velocity are also coupled to primary fluctuations in concentration; this peak suggests the interaction of long waves with the sediment suspended by the wind–waves.

4. Discussion

The observations in the preceding section suggest that the response of concentration is strongly coupled to the wave–induced flow; for this grain size range the complexities of the critical stress for suspension and the fallout of sediment play a minor role and might be adequately modelled in a relatively simple way. The interesting part is that the sedimentary response depends critically on the instantaneous velocity, clearly reflecting the pattern of both the individual waves and the wave groups.

Historically, the predictors used to model nearshore sediment transport, assume that the total rate of sediment transport can be empirically related to simple, time-averaged characteristics of the incident wave field, *e.g.* breaker height, angle of incidence with respect to shore–normal, and the depth of water at the location of breaking. More detailed predictors assume that the sediment is suspended by the waves and then transported by the superposition of a mean flow. Most bulk predictors of sediment transport do not explicitly incorporate any of the effects of velocity skewness, wave groups, settling velocity, or a critical velocity. In light of this and the many other sweeping assumptions that are implicit in these models, it is not not surprising that they are not accurate predictors of sediment transport (Fleming *et al.*, 1986; Baird *et al.*, 1986).

If the skewness of the flow is linked to the mobilization, suspension, and transport of sediment as suggested by the observations in this paper, then the implication is the cross–shore variation of skewness should lead to convergences and divergences of sediment. For example, if the flow is skewed onshore through the shoaling region, reaches a maximum seaward of wave breaking, and then becomes skewed offshore in the surf zone due to a dominance of infragravity wave skewness, then this implies a convergence of sediment and thus the formation of a breakpoint bar. This idea for bar formation is consistent with the skewness observations on a barred beach by Greenwood and Sherman (1984), who observed that velocity skewness tended to be positive on the lakeward side of a bar and negative on the landward side of a bar. However, whether or not a bar actually forms and where it forms will also depend on the spatial variation of the mean flow; that is, it is the skewness of the total flow (*i.e.*, including the mean) that must be considered, not just the skewness of the the oscillatory component. Bowen (1980) has suggested that an onshore skewed flow is required to balance the downslope component of gravity; therefore, the bar would be expected to form seaward of the location of zero skewness. However, the bar could form closer to the location of zero skewness if the downslope component of gravity is partly balanced by an onshore mean flow in the bottom boundary layer (Longuet–Higgins, 1953).

The role of the wave skewness is the longshore direction is probably much less important due to the existence of strong longshore currents in the mean flow. However, the role of wave groups in mobilising the sediment may well be important and it is far from clear that this groupiness is well parameterised by a mean quantity such as incident wave energy.

5. Conclusions

Colocated measurements of near–bed suspended sediment concentration and velocity were used to investigate the variation of sediment concentration with respect to the wave–induced flow. The data clearly showed that sediment responds to both the individual waves and to wave groups. The response of sediment concentration to the individual waves within a wave group is particularly interesting because the concentration responds strongly to the onshore flow associated with wave crests, but very weakly to the offshore flow associated with a wave trough. These field measurements clearly show the behavior of sediment concentration and sediment transport depends on some high power of the velocity with very little time lag. They suggest that the traditional assumption that bulk transport is simply related to basic characteristics of the incident waves, such as peak period, breaker height, and angle of breaking, is tenuous.

The cospectrum of cross–shore velocity·concentration, which was used to examine the wave–induced transport of suspended sediment by the cross-shore velocity, showed that there was a strong onshore transport of sediment due to wind–wave frequencies and a weaker offshore transport due to infragravity waves. The onshore transport observed at the primary frequency suggests that the strong onshore flow associated with the passage of each skewed wave crest leads not only to the mobilization and suspension of sediment, but also to an onshore transport of the sediment. Moreover, the implication is the bulk of the sediment suspended by the passage of a wave crest (at this height of ~ 2 cm above the sea bed) settles before the offshore flow associated with the following wave trough can transport it back seaward; the net transport of suspended sediment at this height is strongly shoreward. The offshore transport by infragravity waves is apparently due to the offshore skewed flow associated with the bound long wave forced by a wave group (Longuet–Higgins and Stewart, 1962, 1964).

Shi and Larsen (1984) suggested that the offshore transport of fine silt and sand on the continental shelf might be due to the offshore flow associated with the long wave forced by a wave group. The cospectrum observation tends to confirm this suggestion. Moreover, this cospectrum observation underlines the sedimentary importance of infragravity wave energy in the nearshore, and again undermines the traditional assumption that the transport of sediment is strictly a wind–wave related phenomenon.

All the observations shown in this paper collectively suggest that sediment dynamics is fundamentally related to the time variability of the velocity field. The sediment is not simply stirred up by the waves and moved by the mean flow. The important parameters include the mean value of the high movements of the flow field such as skewness, as theoretically suggested. In light of this relation between sediment transport and velocity skewness, it was suggested that the cross–shore variation of skewness should lead to convergences and divergences of sediment. For example, if the flow is skewed onshore in the shoaling region due to a dominance of wind–wave skewness and offshore in the surf zone due to a dominance of infragravity wave skewness, then this leads to the possibility of a convergence of sediment and the formation of a break point bar. However, a complete model for transport should also include: the contribution to moments from the mean flow; the downslope component of gravity, *i.e.*, a beach slope dependence; higher order velocity moments such as those that appear in Bagnold–based models for bed and suspended load transport; a threshold velocity for the initiation of transport and a representation of the settling behavior of the grains.

REFERENCES

Baird, W. F., J. S. Readshaw, and O. J. Sayao, 1986. Nearshore sediment transport predictions Stanhope, Lane, Prince Edward Island, National Research Council Canada, C^2S^2 report **21**.

Bailard, J. A. and D. L. Inman, 1981. An energetics bedload model for a plane sloping beach: Local transport, *J. Geophys. Res.*, **86**, 2035-2043.

Bowen, A. J., 1980. Simple models of nearshore sedimentation; beach profiles and longshore bars, in *The Coastline of Canada*, S. B. McCann, editor, Geological Survey of Canada, Paper **80-10**, 1-11.

Cornish, V., 1898. On sea beaches and sand banks, *Geol.*, **2**, 628-674.

Doering, J. C. and A. J. Bowen, 1986. Shoaling surface gravity waves: A bispectral analysis, *Proc. of the* 20th *Conf. on Coastal Eng.*, ASCE, 150-162.

Downing, J. P., R. W. Sternberg, and C. R. B. Lister, 1981. New instrumentation for the investigation of sediment suspension processes in shallow marine environments, *Mar. Geol.*, **42**, 14-34.

Elgar, S. L. and R. T. Guza, 1985. Observations of bispectra of shoaling surface gravity waves, *J. Fluid Mech.*, **161**, 425-448.

Fleming, C. A., B. M. Pinchin, and R. B. Nairn, 1986. Evaluation of coastal sediment transport estimation techniques, National Research Council Canada, C^2S^2 report **19**.

Greenwood, B. and D. J. Sherman, 1984. Wave, currents, sediment flux, and morphological response in a barred nearshore system, *Mar. Geol.*, **60**, 31-61.

Hanes, D. M. and D. A. Huntley, 1986. Continuous measurements of suspended sand concentration in a wave dominated nearshore environment, *Cont. Shelf Res.*, **6**, 585-596.

Huntley, D. A. and D. M. Hanes, 1987. Direct measurements of suspended sediment transport, *Proc. of Coastal Sediments 87*, ASCE, 723-737.

Longuet-Higgins, M. S., 1953. Mass transport in water waves, *Proc. Roy. Soc.*, **A245**, 535-581.

Longuet-Higgins, M. S. and R. W. Stewart, 1962. Radiation stress and mass transport in gravity waves, with application to "surf beats", *J. Fluid Mech.*, **13**, 481-504.

Longuet-Higgins, M. S. and R. W. Stewart, 1964. Radiation stresses in water waves; a physical discussion with applications, *Deep-Sea Res.*, **11**, 529-562.

Shi, N. C. and L. H. Larsen, 1984. Reverse sediment transport induced by amplitude–modulated waves, *Mar. Geol.*, **54**, 181-200.

Stokes, G. G., 1847. On the theory of oscillatory waves, *Trans. Cambridge Philos. Soc.*, **8**, 441-455.

Wells, D. R., 1967. Beach equilibrium and second–order wave theory, *J. Geophys. Res.*, **72**, 497-504.

CHAPTER 109

WAVE-INDUCED BREAKOUT OF HALF-BURIED MARINE PIPES

Mostafa A. Foda[1]
Adrian W-K Law[2]
Jo Y-H Chang[2]

ABSTRACT

An experimental study is conducted in order to identify the major physical processes leading to the breakout of half-buried submarine pipelines from the seafloor under ocean wave action. Both the hydrodynamic loading on the exposed surface of the pipe, as well as the resulting displacement history of the pipe, were recorded and analyzed in order to identify the critical pipe-soil-wave conditions for the detachment of the pipe from the seabed. The paper also examines the balance of the pipe under the combined lift and drag loading from the water wave. An experimental breakout force-time power law is obtained and compared to available theoretical breakout models.

1. INTRODUCTION

During severe storm or hurricane conditions, wave-induced loading on submarine pipelines can be large enough to detach some of the placed pipes from the sea floor, causing them to move or float to great distances with the ocean currents. Numerous reports of this type of failure to submarine pipes are documented in the literature (e.g. Blumburg 1964, Grace 1978, Christian *et al*. 1974, Nataraja and Gill 1983). With the repeated storm-induced failures of marine pipes, the cost of maintaining their operational integrity can be quite high (e.g. Gerwick 1986). The needed improvement of the cost-effectiveness of the present design and maintenance practice for marine pipelines can clearly be established by gaining more insight into the various processes that constitute the failure mechanism.

This paper aims at studying the mechanical stability of the pipe-soil system under wave action. In particular, we focus on the conditions that lead to pipeline detachment or *'breakout'* from the sea floor under hydrodynamic wave loading. We attempt to do that by simulating the breakout process in the laboratory, and further identifying the critical wave-soil-pipe conditions for pipeline breakout. The experimental study is conducted for the configuration of partially buried pipe in a sandy bed under shallow water wave forcing. Partial burial is a commonly observed configuration of submarine pipelines on the sea floor. This may differ from the initial set-up of the pipeline, which may simply be through the direct laying of the pipe on top of the seabed if the water depth is large, or the full or the partial burial of the pipe if the water depth is small. Subsequent interaction between the pipe and the soil will most likely change such initial setting. A marine pipe would then enter into cycles of partial exposure and reburial under the usually variable wave climate conditions. For example, the pipeline may be totally buried initially, but becomes partially

[1]Associate Professor, Department of Civil Engineering, University of California, Berkeley, California 94720.
[2]Graduate Student, Department of Civil Engineering, University of California, Berkeley, California 94720.

exposed due to the erosion of the cover soil by waves. The opposite process of self-burial of the pipe may occur instead due to the erosion of soil particles from underneath the fully-exposed pipe, resulting in the gradual sagging of the pipe into the seabed (e.g. Leeuwestein *et al.* 1985, Hulsbergen 1984, Mao 1986). Another process that results in a partial-burial situation occurs when the pipe is initially laid in an open trench in the seabed. After placing the pipe, local waves and currents at the site will work the spoils back to the artificially-dug trench, leading to a partial-burial configuration.

The first part of this study will focus on the hydrodynamic loading of water waves on partially-buried pipes. This will be done by measuring, in the wave flume, the wave-induced pressure over the exposed surface of the pipe. Then, by integrating the pressure distribution we calculate the resulting forces; drag and lift, and moments on the pipe. Most of the previous studies on wave forcing on submarine pipes were concerned with unburied pipes, with or without close proximity to the seabed (e.g. Sarpakaya 1976, Verley *et al.* 1987). Ismail *et al.* (1986) reviewed the very limited data available on wave loading on partially-buried pipes and stressed the need for detailed measurements of both the hydrodynamic forces as well as the seepage-flow loading on partially-buried marine pipes. In the present study, the measured wave forces will be correlated with wave parameters in order to obtain nondimensional force coefficients for both the drag and lift forces. The widely used Morison Equation will be used to curve-fit the measured drag force on the exposed surface of the pipe, and best-fit coefficients of drag C_D and added mass C_M will be obtained for the range of Keulegan-Carpenter numbers of our experimental set-up. A nondimensional lift-force coefficient C_L will be obtained following a similar procedure.

Then, the study will concentrate on the response of the pipe-sand system to the applied wave forcing. In particular, we examine how will the pipe react to the applied forcing, and how does the detachment process take place. Earlier body breakout studies were concerned with the detachment process under constant applied lift force (e.g. Liu 1969, Muga 1969, DeHart and Ursell 1967). A complete detachment was observed to take place only after the lift force has been applied for a certain period of time, normally called 'the breakout *time*'. The process is usually very slow in the beginning, concluded by a rather sharp release at the breakout time. It was concluded that this almost yield-type release is related to the strong coupling between the body and the saturated bed. Foda (1982, 1983) developed a theory that describes the quasistatic detachment of a large semiburied object from a poroelastic bed. In the theory, this initial resistance to body movement is attributed to the development of negative pore-pressure, or 'mud-*suction*' at the body-soil interface, which gives rise to a certain seabed holding strength against body breakout. The nature of this mud-suction resistance can be illustrated by considering the behavior at the body-soil interface. Initially, as the body moves upward in response to the applied lift force, some pore water will have to be moved, or sucked, upward in order to fill in the resulting gap between the body and the soil. Such movement of pore water requires pore-pressure gradient in order to push the flow through the pores against viscous resistance stresses (Darcy's law), giving rise to the negative pore-pressure at the body-soil gap. The yield-type release will occur late into the lift-off process, when the body-soil gap is wide enough to allow ambient water from above the bed to flow laterally into the gap through its periphery. The resistance to such lateral flow into the gap is far less than that for the upward pumping of pore water into the gap, and at this stage there is virtually no resistance to the further expansion of the body-soil gap, and hence the quick release. The analysis resulted in a general breakout time-force power law of the form:

$$F_b = C\, t_b^{-n} \quad ; \quad n = 1.5 \tag{1}$$

where t_b is the breakout time, F_b is the breakout force, or the applied lift force in excess of the submerged weight of the body, and C is a material constant that depends on the permeability, stiffness and porosity of the saturated bed, as well as the size of the body (Foda 1983). However, the above law is limited to the conditions of the boundary-layer theory of Mei and Foda (1981), which requires that the pipe diameter D is much larger than the seabed boundary-layer thickness δ which is roughly proportional to $\sqrt{k/\mu\omega}$ where k is the soil permeability, μ is water viscosity, and ω is wave frequency. Mei *et al.* (1985) developed another breakout law for bodies on a porous but rigid bed that has the same form as (1), but with a different material constant C and power n ($n = 1$). This law would be more applicable to pipes with a small enough diameter.

Foda (1985) extended his theory to the present case of a half-buried marine pipe under the oscillatory loading of water waves. The hydrodynamic loading on the pipe is calculated assuming a potential flow above the seabed. Furthermore, it was assumed that the pipe diameter D is much smaller than both the wave length L and the water depth h, so that wave scattering by the pipe was ignored. Only the vertical equilibrium of the pipeline was considered under the the wave-induced lifting force, which is composed in this case of an average lift force plus an oscillatory component. The analysis has shown that for typical physical parameters, the wave period $T \simeq 10$ sec is usually much shorter than the quasistatic breakout time t_b, and hence one expects that the oscillatory wave-induced force component will not influence, to leading order, the breakout process of the pipe.

The possibility of resonating the system by the oscillatory force was also examined. The obtained response function of the pipe-seabed system was shown to exhibit a weak nonlinearity of the order $\epsilon = T/t_b \ll 1$. This should lead to the slow generation of higher harmonics and hence raises the possibility of resonating the system when one of the generated harmonics approaches the system's natural frequency. However, the analysis has shown that the natural frequency of a *'typical'* pipe-soil system would be much higher than water wave frequency, so that a long chain of generated higher harmonics would be needed in order to approach the resonance condition.

The present experimental study examines many of the features of the above mentioned breakout models, along with the associated assumptions and limitations. This includes investigating the nature of the pipe breakout (quasistatic or resonant), the relative role of pore-pressure *vs.* soil effective stresses in holding the pipe in place, and the ultimate mode of failure for the supporting soil at breakout.

2. EXPERIMENTAL SETUP

The experimental work was conducted in the 8 *ft* (*W*) *x* 5 *ft* (*H*) *x* 180 *ft* (*L*) wave flume at the Richmond Field Station of the University of California at Berkeley. The relatively large cross-sectional area of the flume allowed the building of a reasonable-size sand basin inside the flume to model the seabed. A layout of the experimental setup is shown in Figure *1a.* The flume is fitted at one end with a mechanical flap-type wave maker that generates monochromatic water waves, with a range of possible wave amplitudes and frequencies. At the other end of the flume, a 20° sloping beach with four layers of horse-hair mats is built to absorb the incoming wave energy. The test section is built about 50 *ft* from the wave maker where there is a side-viewing glass window in the flume's concrete wall. As shown in Figure 1, the test section is made of a rectangular sand basin 2 *ft* high, 8 *ft* wide and 4.5 *ft* long. Two 20°-plywood ramps are fitted at both ends of the basin in order to provide smooth transitions for the water wave as it passes over the test section. The sand used in the experiments was a well sorted medium-size sand of $D_{50} = 0.3$ *mm.* The basin was filled with the sand, manually compacted in the dry, then an 8" PVC pipe was placed across the flume, half-buried in the sand bed. The length of the pipe was cut slightly shorter than the width of the flume (8 *ft*) to prevent friction between the pipe and the flume's concrete walls. The pipe would then be connected to the various sensors of the experiment, feeding into a data-acquisition system whose details are discussed in the following section. Then, the flume would be filled with water to a total depth of 4 *ft* (2 *ft* of water above the sand bed).

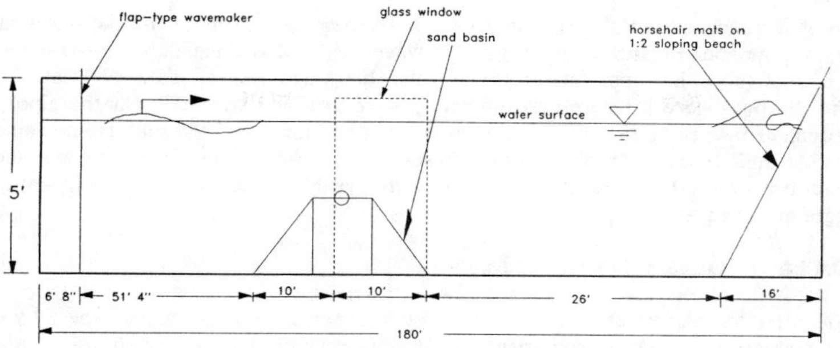

Figure 1a: A sketch of the flume dimensions

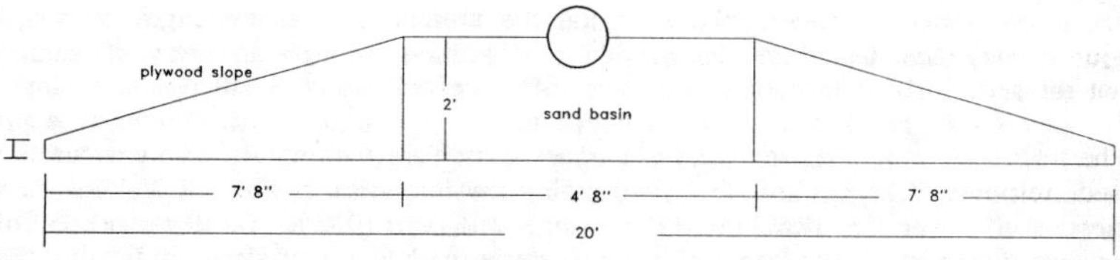

Figure 1b: The sand basin dimensions

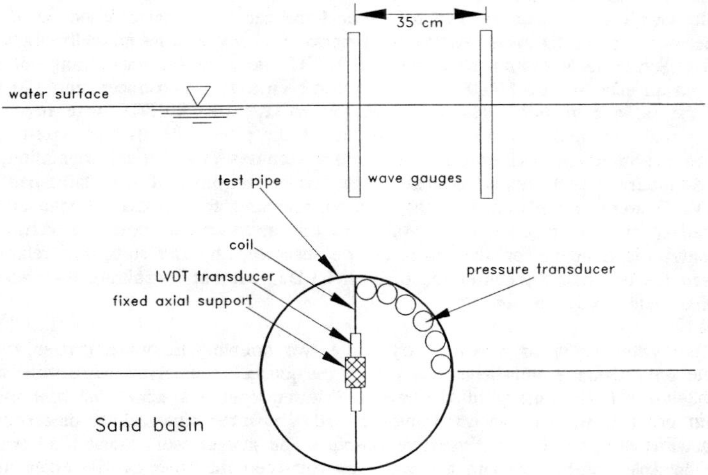

Figure 2: Experimental Setup

A typical experimental run would start by turning on the motor of the wave maker to generate a monochromatic wave of pre-set wave height and frequency (typically, 6" and 3 sec, respectively), and continuously recording the measurement data on water pressure around the pipe, pore pressure measurements, wave profile above and near the pipe, and the displacement history of the pipe, until failure, or detachment, of the pipe from the sand bed occurs (or until termination of the run if breakout does not occur). In many runs, the whole test section would be videotaped to record the visible behavior of the pipe-soil system throughout the experiment.

3. DATA-ACQUISITION SYSTEM:

In order to measure the wave-induced hydrodynamic loading on the pipe a total of six pressure transducers (Data Instrument Model AB High Performance 0-6 psi) was used to measure pressure distribution over one quarter of the exposed pipe surface (Figure 2). The pipe would simply be rotated 90^0 to measure water pressure over the other quarter surface. The pressure transducers were placed inside the 8"-pipe at the midlength section. The transducers were connected to the outside of the pipe through an array of small holes 15-degrees apart drilled through one quarter of the pipe's wall. Small plastic tubings (1/8" inner diameter, 4" long) were used to connect these holes to the diaphragm-type pressure transducers. The output signals from the transducers were then amplified by accurate strain gauge conditioners (Daytronic 9178*A*), enhancing the resolution to about 1 N/m^2. Although this accuracy is over the manufacturer recommended limit (0.25% of full range), calibration of the transducers showed a repeatable linear curve with constant slope, indicating that the extrapolation of accuracy is valid.

The measurement of pore-pressure, or mud-suction stress, along the pipe-soil interface was made following the same procedure by rotating the pipe so that the measurement holes would be on the buried half of the pipe surface.

As for the movement of the pipe, two Linear Variable Differential Transformers, LVDT, were used to measure both the vertical and the horizontal components of the pipe's displacement. Since the pipe was totally immersed into water, a hermitically-sealed type was selected (Schaevitz 500HPA). The selected LVDT has a measurement range of about one inch and an accuracy of 1/4000 of an inch using an analog transducer amplifier (ATA101) from the same company. The magnetic coils of the two LVDTs were mounted on an aluminum frame that runs through the center of the pipe, with its ends fixed on the side walls of the wave flume (Figure 2). One coil is mounted in a vertical orientation, while the other is mounted horizontally, normal to the axis of the pipe. The metallic center rods of the LVDTs were then placed through, but not touching, their respective magnetic coils and attached to the the pipe's interior wall. In this arrangement, both the vertical and the horizontal displacement of the pipe can be measured by detecting the relative motion between the coil and the center rod of each LVDT, without restraining the free motion of the pipe under wave loading.

The wave profile is measured by using two double-wire wave gauges which relate changes in resistance and hence voltage of the partially-immersed wires with changes in water level. The accuracy of the recorded water level was about 0.3 *mm* with careful calibration. One wave gauge was located directly above the pipe and the other was about 35 cm downwave towards the dissipative beach. The gauges were isolated electronically to avoid possible interference and together they measured the wave profile at the test section and also the amount of reflection from the beach. They are excited and their signals amplified by strain gauge conditioners (Daytronic 9178*A*) similar to those used for the pressure transducers.

The data-processing unit was an IBM *PC/AT* equipped with two data acquisition cards (IBM Data Adapter and Datatranslation Data Board DT2801). This provides for 12 channels with 12-bit accuracy. A program is written in *C*-language to allow the start or termination of the recording of any signal at any time during the experiment independently or synchronically with other signals. In addition, during the recording, the program can still report values of any desired signal. These features add more flexibility on memory storage and early detection of errors.

4. THE BREAKOUT EXPERIMENT:

In all of the conducted experiments, the observed behavior was very similar, consisting primarily of two distinct phases. First, there is the build-up phase, which covers more than 90% of the duration of the experiment. Then, there is the short phase of actual pipe-breakout from the sand bed. During the build-up phase, there is hardly any discernible movement of the half-buried pipe. The only visible movement was that of some of the sand grains on the surface of the sand bed, which quickly resulted in the establishment of a regular ripple pattern on both sides of the pipe. However, in the near proximity of the pipe (about one pipe diameter distance on either side), no motion of sand grains was visible and the sand surface there remained flat throughout this phase. In other words, one can say that in our set-up soil erosion did not play any role in the ensuing behavior of the pipe-sand system.

The duration of this build-up phase ranged from about half an hour to a little less than 5 minutes. During this phase, the pipe was clearly subjected to the hydrodynamic lift and drag due to the passing of water waves above it. Flow separation was observed to take place in all the experiments. Separation was made visible by injecting liquid dye near the exposed surface of the pipe. The dye would quickly encounter and make visible one or more of the shedded vortices from the surface of the pipe. Such viscous separation would result in an asymmetric distribution of pressure around the pipe, and hence a net loading on it. Another cause for pressure asymmetry is the added-mass effect associated with the unsteadiness of the flow. Furthermore, there is the lift-force component associated with the Bernoulli's effect, where the increase in velocity at the summit of the pipe is accompanied by a decrease in pressure, i.e. *a* lift. The details of the measured hydrodynamic loading on the pipe will be given in the following Section.

By examining the displacement record of the pipe during this initial build-up phase, we observe that the response of the pipe to these applied hydrodynamic loadings is in the form of a very small oscillatory motion, with a very slow net movement in the upward direction. Figure 3 shows the vertical displacement history of the pipe center during a typical run. In this Figure, the amplitude of the pipe oscillation is about .008 inch, and the average rate of rise of the pipe during this build-up phase is of the order of .10 inch/minute, which can hardly be detected by the naked eye. Superimposed on this gradual rise, there was an even slower rotational movement of the pipe around its longitudinal axis. This suggests that the pipe is actually rolling out of the sand bed, as opposed to being uniformly lifted off from the bed. With the wave coming from left-to-right the observed net rotation was always anticlockwise. In other words, the combined rise and roll motion of the pipe did have an effective pivot point (point of zero net displacement) which was always at the wave-maker side from the pipe center. The rotational displacement was measured by placing markers at equal distances on the pipe wall and visually recording the difference between the reading at the two pipe-soil contact lines. This would typically reach a maximum value, at pipe breakout, of about 10-15 degrees.

It is important to note that throughout this phase where the pipe is slowly building up towards the ultimate breakout, the soil mass surrounding the pipe remained intact, with no sign of any soil failure, such as soil sliding, erosion or soil liquefaction. Finally, near the end of this build-up phase, the pulsating motion of the pipe becomes easier to detect by the naked eye, and this signals the start of the very short second phase of pipe breakout.

As seen in Figure 3, the very slow, almost unnoticeable rise of the pipe during this long build-up phase is followed by the rather sudden and violent release of the pipe from the sand bed in the breakout phase. The release is so sudden that it is relatively easy to identify a specific time when that happens, and we may term that the 'breakout *time*'. From Figure 3 we see that approaching the breakout time there is no significant increase in the amplitude of the oscillatory motion, and that the breakout is essentially associated with the increase in the net rise component of the motion. This should quite reasonably exclude resonance as the possible mechanism for such a breakout. Instead, it is a quasistatic process through which the pipe gradually detaches itself from the sand bed, against the so-called 'mud- *suction*' resistance force. The eventual breakout occurs due to the dissipation of this mud-suction force at breakout time. The details of the breakout mechanism will be discussed later in this paper. But first, we discuss the hydrodynamic loading that forces such breakout.

5. HYDRODYNAMIC LOADING:

Figure 4 shows the distribution of dynamic water pressure on the exposed half-surface of the pipe during one wave cycle (wave period of 2.9 seconds and wave height of 6"). This is obtained by combining two 1/4-surface data, under identical wave profiles. For any point on the exposed pipe surface, it is seen that pressure is nearly in-phase with the wave profile, i.e., the maximum and minimum pressure occur almost under the crest and trough of the wave, respectively. From the Figure, it is seen also that under either the crest or the trough, the pressure distribution has a minimum near the summit of the pipe (near points 6 and 7), where the kinetic energy is a maximum due to the pipe-induced contraction of the flow, (the Bernoulli's effect). This will clearly result in a net lift force on the pipe. Aside from that, there is some asymmetry in crest and trough pressure distribution around the summit point. This may be attributed to flow separation effect. A more pronounced asymmetric distribution of pressure is however due to the added-mass effect and is evident in the Figure near the wave's nodal points (time = 1.0 *s* and 2.0 *s*), where flow acceleration is maximum. This asymmetry and the resulting loading on the pipe are due to the pipe-induced deceleration of a certain amount of fluid mass in the pipe's vicinity.

By integrating the measured pressure distribution over the pipe surface, the wave-induced drag and lift can be found. Figure (5) shows sample results for the measured drag force on the half-buried pipe. The Figure also shows the least-square fitting of the data in the form of a Morison Equation:

$$F_D = \frac{1}{2} C_D \, \rho(D/2) \, u|u| + C_M \, \rho(\pi D^2/2) \frac{du}{dt} \tag{2}$$

where $u(t)$ is the unsteady ambient water velocity near the pipe, D is the pipe diameter, ρ is the water density, C_D is the drag coefficient, C_M is the added-mass coefficient. The ambient water velocity was estimated using Dean's stream-function calculation. First, the stream-function was tuned to reproduce the measured water surface elevation $\eta(t)$, and then the corresponding bottom water velocity was calculated. Figure 6 shows a sample of the calculated bottom velocity for a measured $\eta(t)$. Over many selected wave cycles, the best-fit values for the drag coefficient ranged between -0.188 to +0.861, while the added-mass coefficient ranged from 1.51-2.08. From potential theory, assuming no flow-separation, the added-mass coefficient (which should be the same for either a half or a full cylinder) is predicted to have a value of 2 (Lamb 1932). For real fluids, the values of C_M and C_D would depend on the Keulegan-Carpenter Number U_0T/D, where U_0 is the amplitude of the velocity oscillation, T is the wave period, as well as on Reynolds Number U_0D/ν, where ν is the kinematic viscosity (e.g. see Sarpakaya 1976 for full cylinders near flat bed). In our experiments, the Keulegan-Carpenter Number was around the value of 6, and the Reynolds Number was in the order of $(0.8 - 1.0) \times 10^5$.

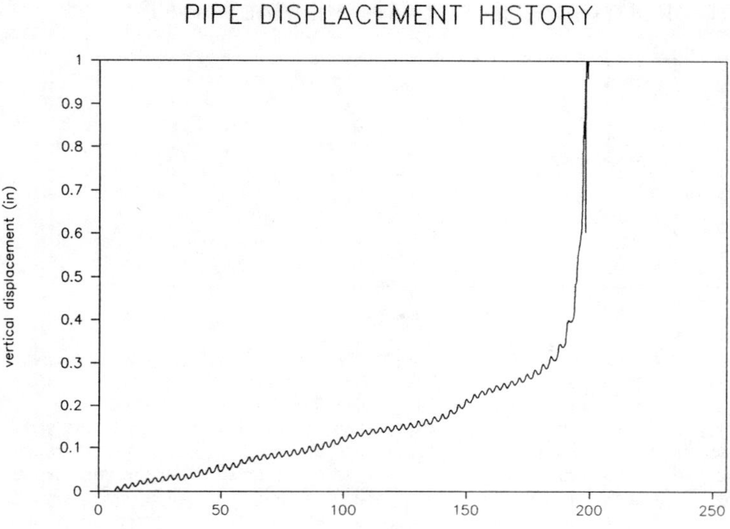

Figure 3: Vertical displacement history of the half-buried pipe under 6"-2.9sec water wave

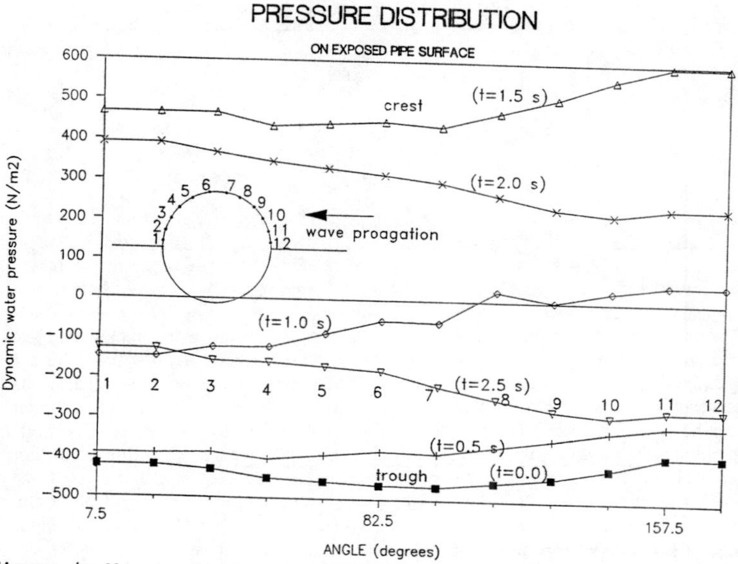

Figure 4: Measured water pressure on the exposed half-surface of the pipe over one full wave cycle (wave height 6", wave period 2.9 sec).

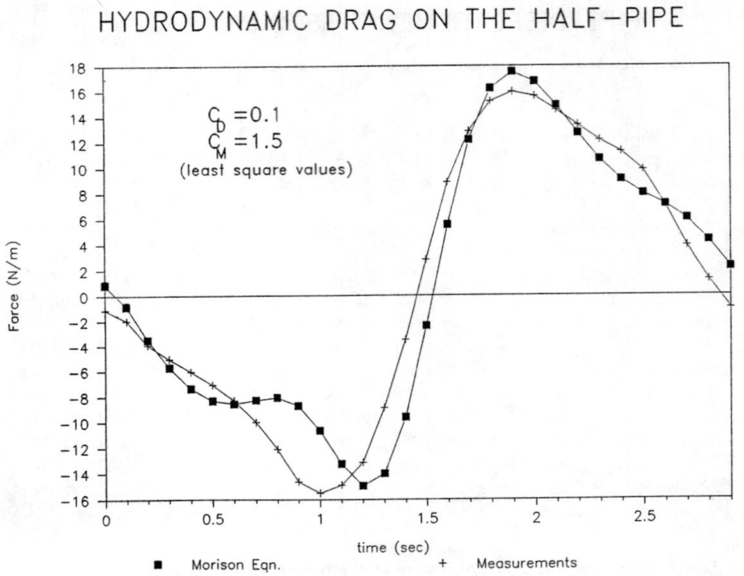

Figure 5: Curve fitting the measured drag force on the half-surface of the pipe with Morison Equation.

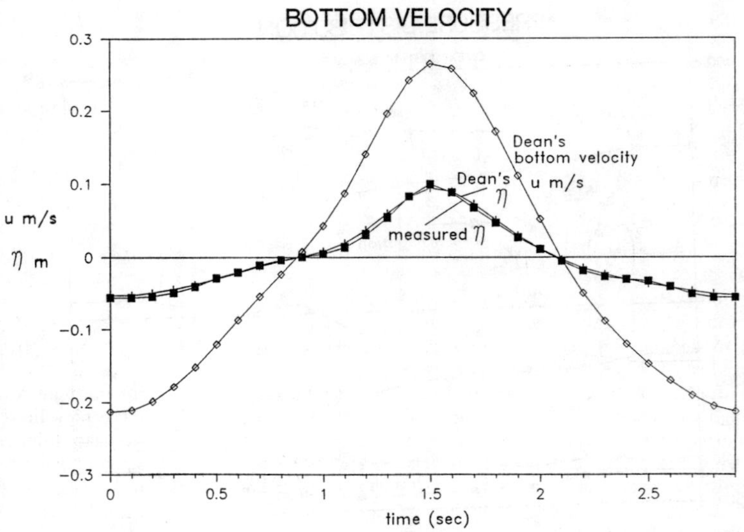

Figure 6: Comparison between measured vs. predicted water wave profile, and the resulting bottom velocity u using Dean's stream function.

As for the lift force F_L, we used the following expression to curve-fit the data

$$F_L = \frac{1}{2} C_L \rho D u^2 \tag{3}$$

where C_L is the nondimensional lift coefficient. The best-fit value for C_L, averaged over many wave cycles, was found to be 2.556, which is close to the theoretical value of 2.66 from potential flow theory (e.g. Milne-Thomson 1967).

Figure 7 shows sample results for the measured resultant hydrodynamic force on the pipe during one wave cycle. Notice that the resultant pressure force is passing, by definition, through the center of the pipe. With the wave propagating from right-to-left, the resultant force is rotating in an anticlockwise sense with a non-constant angular velocity, as is seen from the variation of the force angle θ with time. Clearly, the average angular velocity of the resultant force is equal to the wave frequency ω, but as seen from $\theta(t)$, the resultant force tends to slow down (below ω) when the force is near vertical, upward or downward, with a fast transition (above ω) between them. Notice also that the wave average drag and lift forces are much smaller than the respective amplitudes of the oscillatory forces. The wave-average lift is always directed upward, and its magnitude is found to be very close to the potential theory calculation, i.e., by wave-averaging (3) with $C_L = 2.66$. However, the far smaller net drag is observed to fluctuate in magnitude and even in direction; sometimes in the wave direction (Fig *7a*) and sometimes opposite to the wave direction (Fig *7b*).

6. THE BREAKOUT MECHANISM:

If the wave-averaged lift force exceeds the buoyant weight of the pipe, then the pipe will try to move upward in response. This will initiate two types of resistance forces along the pipe-sand interface, both acting to keep the pipe in place. One is the shear stress between the pipe and the sand grains, and the other is the negative pressure in the pore water below. That negative pore-pressure would be developed as some pore water will be forced to move upward along with the upward-moving pipe. For typical, low-permeability soils, the developed negative pore-pressure can be quite appreciable even for a very slight initial motion of the pipe. For example, a 1-*m* diameter pipe with an initial motion of only one hundredth of an inch above a fine-sand bed (permeability $k = 10^{-12}\ m^2$) will produce, according to Darcy's law, a negative pore-pressure impact force of the order of 10^5 *N-s/m*. On the other hand, the value of the sand shear stress has an upper limit given by the relatively low Coulomb sliding shear for this unconfined, near-surface sand around the pipe.

The rate of vertical movement of the pipe will depend on the rate of water penetration into the developing gap between the pipe and the soil below. This water penetration can come from two possible sources: (*a*) the seepage flow of water into the gap across the soil surface, and this depends on the soil properties, especially its permeability and stiffness, and (*b*) the ambient water flow above the seabed through the gap periphery, which according to the lubrication theory is proportional to the gap thickness to the third power. In the initial stages of lift-off, the gap thickness is very small and thus its expansion will be essentially due to the pentration of pore water through the soil matrix. Since the pore water flow through the soil is typically very slow (governed by soil permeability), the pipe lift-off velocity during this stage will be very small (see Fig 3). With time, the gap thickness becomes large enough so that the lateral flux of ambient water through the gap periphery gains dominance over the pore water flux. As a result, the resistance to further gap expansion becomes much smaller than the pull-up force (e.g. lubrication theory), and hence the quick release at breakout time.

This is the basis of the theoretical breakout model of Foda (1985). In reality, however, there are a number of other factors that may be contributing to the breakout process, but are

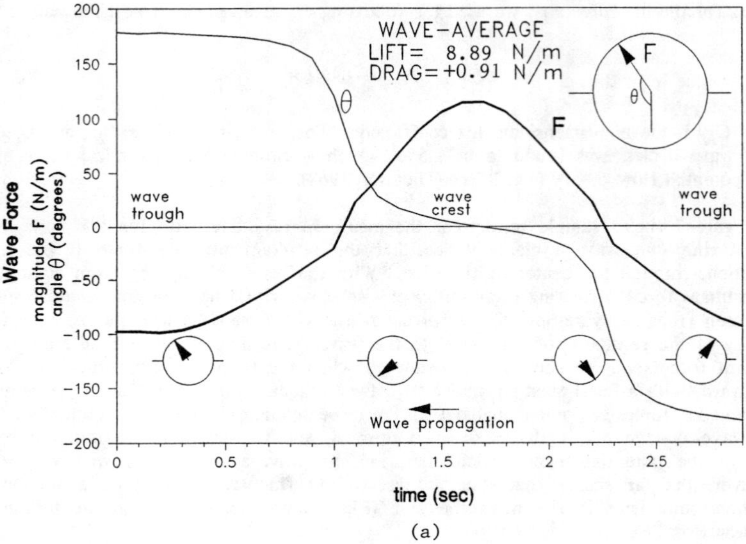

(a)

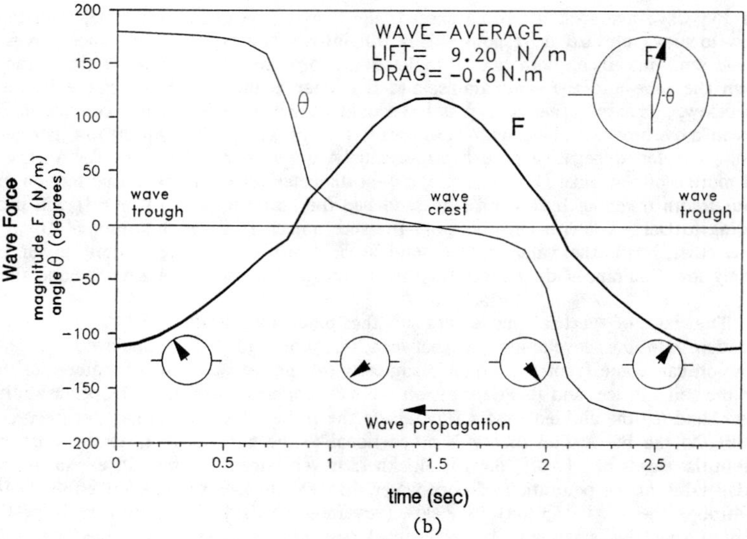

(b)

Figure 7: The variation of the total wave force magnitude and direction over two selected wave cycles.

not considered in this idealized model of Foda. For example, there is the question of the role of wave-induced drag force during the breakout process. In the above discussion, we focused on the response due to the applied lift force, which if acting alone on the pipe would result in a uniform lift-off of the pipe away from the underlying soil. However, in reality, the pipe is responding to the combination of lift and drag forcings. Furthermore, the model assumes that the pipe resistance against breakout is entirely provided by the negative pore-pressure or mud-suction stress at the pipe buried surface. In general, however, there will be some direct contact between the pipe surface and the sand solid skeleton, and hence there is the combination of pore pressure and effective (solid) soil stress that will be acting there, and it is of interest to compare the relative magnitudes of the solid *vs.* pore-fluid stresses that are actually acting on the buried pipe surface.

By integrating the pressure over either the upper, exposed surface l_e, or the lower, buried surface l_b, we get the corresponding fluid loadings on the pipe as follows:

$$\{F_L \, , F_D\}_{h,p} = \int_{l_e, l_b} p(\theta,t) \, \{\sin\theta \, , \cos\theta\} \, d\theta \qquad (5)^{+}$$

where F_{Lh}, F_{Dh} are the hydrodynamic lift and drag forces, and F_{Lp}, F_{Dp} are the pore-water lift and drag forces, respectively. It is further seen from Figure 4 that the pipe displacement is very small so that we may invoke the static balance equations in both the horizontal and vertical directions:

$$F_{Dh}(t) = -[\, F_{Dp}(t) + F_{Ds}(t) \,] \qquad (6a)$$

$$F_{Lh}(t) = -[\, W + F_{Lp}(t) + F_{Ls}(t) \,] \qquad (6b)$$

where F_{Ls} and F_{Ds} are the remaining soil lift and drag forces on the pipe due to the direct contact between the pipe and the soil's solid skeleton, and W is the submerged weight per unit length of the pipe.

A Fast Fourier Transform (FFT) is performed on Equations *6a,b* in order to separate out the balances at the wave-averaged or zeroth-harmonic mode from the first and higher harmonic balances. The results for the fluid loadings are shown in Figures *8a,b*. The discrete Fourier transform is defined as:

$$\overline{F}(m.\Delta\omega) = \frac{T_0}{N} \sum_{n=0}^{N-1} F(n.\Delta t)\, e^{-i2\pi mn/N} \; ; \quad m=0,1,...,N/2 \qquad (7)$$

where the overbar denotes the Fourier transform, Δt is the sampling time increment which is taken here to be equal to 0.1 sec, N is the total number of data points, T_0=N.Δt is the duration of sampling, taken equal to 102.4 sec, and the incremental frequency $\Delta\omega = 2\pi/T_0$. The results in Figures *8a,b* correspond to the initial period (i.e. the first 102.4 *s*) of a selected breakout experiment with the same conditions as those for Figure 4.

Let us first examine the zeroth-harmonic (m=0) vertical response, which requires from (*6b*) that,

$$\overline{F}_{Lh}(0) \,/\, T_0 = -\,[\, W + (\, \overline{F}_{Lp}(0) + \overline{F}_{Ls}(0) \,) \,/\, T_0 \,] \qquad (8)$$

By accurately measuring the dry weight of the pipe and subtracting the water buoyancy we obtained a value of W = 7.796 N/m. Furthermore, as seen from Figure *8a*, we find that the

---

+Equation (4) is missing due to an error in numbering the equations.

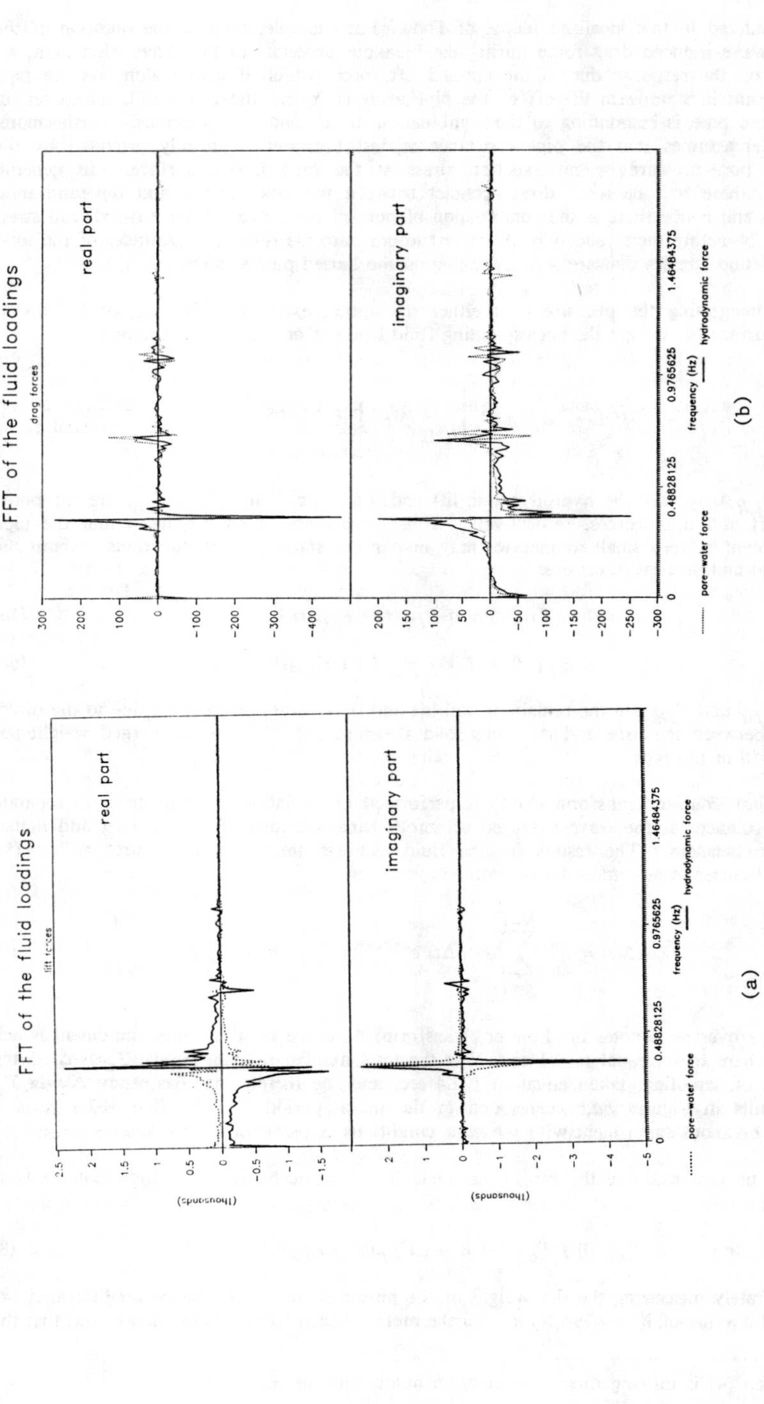

Figure 8: The Fast Fourier Transform of the hydrodynamic (solid lines) and pore-water (dotted lines) loadings on the half-buried pipe; (a) lift forces, (b) drag forces.

wave averaged hydrodynamic lift $-\overline{F}_{Lh}(0)/T_0$ is equal to 1021.8/102.4 = 9.98 N/m, while the wave-averaged pore-water suction force $\overline{F}_{Lp}(0)/T_0$ is equal to 230.8/102.4 = 2.25 N/m, which virtually satisfies the vertical balance in (8) with negligible contribution from the soil effective stress forcing on the pipe; a mere -0.066 N/m according to (8). Turning to the zeroth-harmonic horizontal balance (see Figure *8b*), we find a steady-state hydrodynamic drag on the pipe of a magnitude of 0.73 N/m, in the direction of wave propagation. We also find that the steady-state pore-water reaction force is equal to 0.817 N/m acting in the opposite direction. Again, there is an almost complete balance between the hydrodynamic and pore-water forces, with only a small contribution from the soil effective-stress loading, equal to the difference 0.087 N/m. It is curious, however, to note that these small soil loadings in both the horizontal and vertical directions are acting in the same direction as the the forcing hydrodynamic loadings. A partial explanation of this observation will be offered shortly.

We turn next to examine the balance of forces at the primary first harmonic, @ the wavemaker frequency $1/T$ = 0.345 Hz; T = 2.9 sec. First, from Figure *8a* we see that there is a 180° phase-shift between the hydrodynamic and pore-water vertical loadings almost throughout the frequency spectrum. But, we also see that the pore-water vertical force balances out only a proportion α of the forcing hydrodynamic loading, i.e.

$$\overline{F}_{Lp} = -\alpha \overline{F}_{Lh} \tag{9}$$

Except for a few very low frequencies near m=0 (giving higher values for α), the value of α is found to be relatively constant around α = 0.55. That is to say that roughly 55% of the total soil reaction force to the unsteady hydrodynamic lift on the pipe is taken up by the pore-pressure. The remaining 45% of the soil reaction has to come from the soil-effective stress loading on the pipe. This contrasts drastically with the above-mentioned steady-state (m=0) results, where there is negligible effective stress contribution to the balance of forces.

The role of effective stresses is even more prominant in the horizontal forces balance. From Figure *8b* it is seen that the unsteady ($m \neq 0$) hydrodynamic and pore-water drag forces on the pipe are in fact in-phase. This means that the pore-water unsteady drag is reinforcing, not balancing, the hydrodynamic unsteady drag. Therefore, it is the role of the soil solid skeleton to provide the effective stress loading necessary to balance the combined fluid drag force (hydrodynamic + pore-water). On the other hand, it should be noted that even this combined fluid drag on the pipe is an order of magnitude smaller than the hydrodynamic lift, as can be seen by comparing the ordinate scales in Figures *8a vs. 8b*. Furthermore, we note that the drag-induced effective stress is not additive to the lift-induced effective stress, since they are almost 90° out-of-phase. This is evident from Figures *8a,b* where we have two almost real lift spectra, and two almost imaginary drag spectra. It is also a consequence of the dominance of the added-mass drag component, as discussed before and as evident in Figure 5.

From the above, we may conclude that the pipe is in constant contact with the underlying soil, and the resulting effective stress loading on the pipe is, in part, balancing a proportion (~45%) of the wave lift force, and in part balancing the total fluid drag on the pipe. The elastic pipe displacement should be in-phase with the effective-stress loading, and thus should follow the same pattern as that of the force F in Figure 7, with a resulting anticlockwise orbital path of the pipe center over one wave cycle.

Superimposed on this oscillatory motion, there is the zeroth-harmonic response due to wave-averaged lift and drag forces. As predicted by Foda (1985), this response is entirely in the form of a mud-suction force or negative pore-pressure along the pipe-soil interface. The resulting motion is a gradual lift-off of the pipe as shown in Figure 3. The much smaller steady-state drag force is causing a slight antisymmetry in the mud-suction stresses and hence a very small deviation from the uniform lift-off condition. In other words, the slow pipe motion will have both a net vertical upward component as well as a much smaller horizontal component in the same direction as the fluid drag, i.e. in the direction of wave propagation.

Next, we briefly discuss some of the possible dynamic interaction between the first and zeroth harmonic responses (e.g. Foda 1985). In particular, we heuristically discuss one specific possibility, namely, the role of the first-harmonic drag force on the net, or zeroth-harmonic pipe motion. First, we note that due to the oscillatory fluid drag, the pipe will be bouncing back-and-forth against the sides of the soil circular interface. The problem can be modelled as a simple elastic bouncing of a body off an elastic surface with an oblique impact angle. Therefore, as the pipe impacts on this nonvertical soil surface it will clearly bounce back from the surface acquiring a small additional vertical velocity in the upward direction. This is because the elastic strains in the soil due to this horizontal pipe impact will include vertical compressional strains along the circular soil interface. This small upward bounce, or roll-out, due to wave drag would clearly occur regardless of the direction of the drag. Therefore, there should be a net, or a wave-averaged (zeroth-harmonic) upward motion, as well as an upward effective-stress force on the pipe due to such mechanism. An indication of such upward effective-stress force was observed above when considering the balance of forces according to Equation (8). However, the obtained magnitude of this force, from the analysis that follows Equation (8), is seen to be negligibly small so that, at least in our case, such roll-out mode of motion is deemed to be of little significance to the total breakout process.

Finally, Figure 9 shows the experimentally obtained data on the breakout force-time relation. The breakout force F_b is defined here as the net uplift force acting on the pipe, which is given by

$$F_b = \overline{F}_{Lh}(0)/T_0 - W \tag{10}$$

and the breakout time t_b is determined from the recorded displacement histories (similar to Figure 4) when the slope of the curve reaches 90% of vertical. The best-fit power law that runs through the experimental data was obtained using a least-square routine, and is given in SI-units by

$$t_b = A\, F_b^{-0.93} \quad ; \quad A = 1477 \tag{11}$$

The exponent is very close to Mei *et al.* (1985) power law for breakout of flat-bottomed bodies from a porous rigid bed. Their law is given by

$$t_b = \tau(\mu\, L^{7/3} k^{-2/3})\, F_b^{-1} \tag{12}$$

where μ is the water viscosity, k is the soil permeability, L is contact width of the two-dimensional body, and τ is a nondimensional coefficient that ranges between 0.4-0.62 depending on the assumed slip boundary condition at the bed surface. Using conformal mapping, the above analytical relation was extended to our configuration of a half-buried pipe and the details are given in Law *et al.* (1988). The result is a modification in the value of τ from that for flat-bottomed bodies. For example, for a no-slip boundary condition at the bed surface, we get

$$\tau = 1.9 \tag{13}$$

The increase in the value of τ, and hence the breakout time t_b, is due to the nonuniform shape of the resulting gap between the pipe and the bed, with the smallest gap thickness being at the gap periphery. This will clearly reduce the lateral flux of ambient water through the gap periphery, as compared to the case of a flat-bottomed body, and hence delay the breakout time.

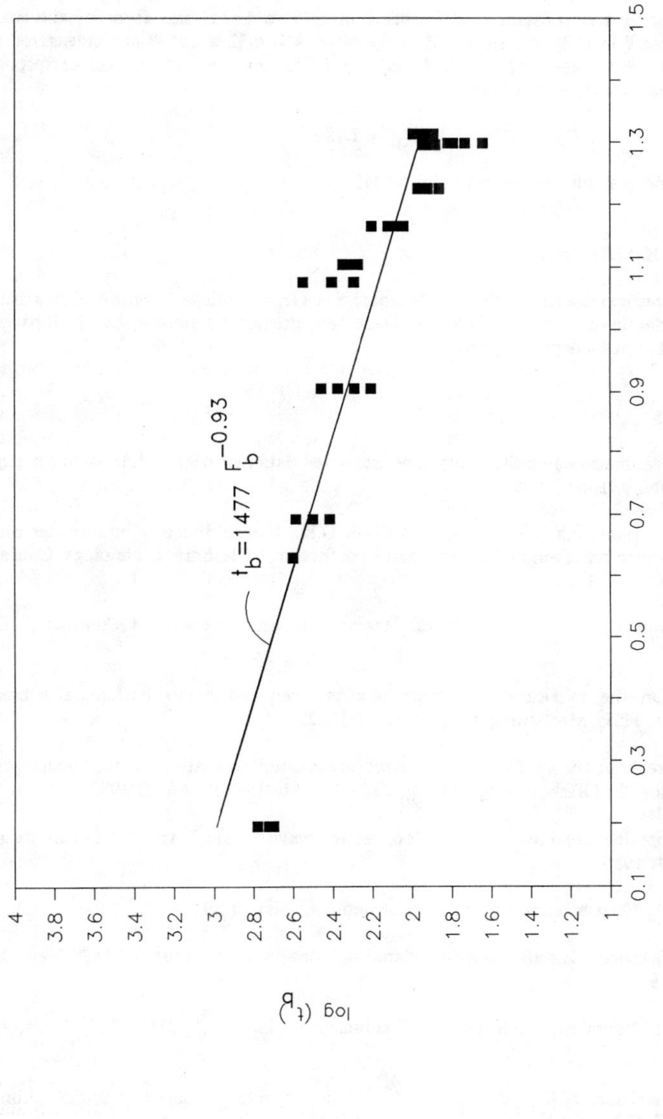

Figure 9: The experimentally-obtained breakout force–time relation

Now, recasting the constant coefficient A in (11) to be of the form of the material constant of the law in (12), where $\mu = 10^{-3}\ kg/m\text{-}s$, $k = 0.32 \times 10^{-10}\ m^2$ (measured using a constant head permeometer), $L = \pi D/2$, and $D = 0.2191\ m$, we find that the corresponding value of the free parameter τ is equal to

$$\tau_{\text{exp}} = 1.78 \tag{14}$$

which is very close to the theoretical value in (13).

ACKNOWLEDGMENTS

This research was sponsored by NOAA, National Sea Grant College Program, Department of Commerce, under grant number NA85AA-*D-SG140*, project number *R/OE*-2, through the California Sea Grant College Program.

REFERENCES

Blumberg, R., "Hurricane winds, waves and currents test, marine pipeline design," Pipeline Industry, June-Nov. 1964.

Christian, J.T., Taylor, P.K., Yen, J.K. and Erali, D.R., "Large diameter underwater pipeline for nuclear power plant designed against soil liquefaction," Offshore Technology Conference Proc., 2, 597-602, 1974.

DeHart, R.S. and Ursell, C.R., Southwest Research Institute Report on Contract *N*-ONR-336300, 1967.

Foda, M.A. "On the extrication of large objects from the ocean bottom, the breakout phenomenon," J. Fluid Mechanics, 117, 211-231, 1982.

Foda, M.A. "Breakout theory for offshore structures seated on seabed," Proc. Specialty Conf. Geotech. Practice In Offshore Engineering, 288-299, Austin-TX, ASCE 1983.

Foda, M.A. "Pipeline breakout from seafloor under wave action," Applied Ocean Research, Vol 7(2), 79-84, 1985.

Gerwick, B.C.Jr. "Construction of Offshore Structures," New York: Wiley, 1986.

Grace, R.A. "Marine Outfall Systems: Planning, Design and Construction," New Jersey: Prentice-Hall 1978.

Husbergen, C.J. "Stimulated self-burial of submarine pipelines," OTC-2967, Houston-TX, May 1984.

Ismail, N.M., Wallace, N.R. and Nielsen, R., " Wave forces on partially buried submarine pipelines," OTC-5285, 27-36, Houston-TX, May 1986.

Law, A. and Foda, M.A., "Breakout of half-buried pipe from a rigid porous bed," Technical Report HEL- ,(in-press),*University* of California, Berkeley, 1988.

Leeuwestein, N., Bijker, E.W., Peerbolte, E.B. and Wind, H.G., "The natural self-burial of submarine pipelines," BOSS'85, Delft July 1985, Amsterdam: Elsevier, Dev. Mar. Technol., *Vol.2*, Paper *C18*, 717-728, 1985.

Liu, C.L., "Ocean sediment holding strength against body breakout," U.S. Naval Civil Eng. Lab., Port Hueneme-CA, Tech. Rep. *R635*, 1969.

Mao, Y., "The interaction of a pipeline with an erodible bed," Institute of Hydrodynamics and Hydraulic Engineering, Technical University of Denmark, Research Series Paper No. 39, 1986.

Mei, C.C., Yeung, R.W. and Liu, K.P., "Lifting of a large body from a porous seabed," J. Fluid Mechanics, 152, 203-215, 1985.

Mei, C.C. and Foda, M.A., Wave-induced response in a fluid-filled poro-elastic solid with a free surface - *A* boundary-layer theory," Geophy. J. R. Astro. Soc., 66, 597-637, 1981.

Milne-Thomson, M.N. "Theoretical Hydrodynamics," 5th *ed.*, New York:MacMillan.

Muga, B.J., "Ocean bottom breakout forces, including field test data and the development of an analytical method, Tech. Rep. *R591*, U.S. Naval Civil Eng. Lab., Port Hueneme-CA., 1969.

Nataraja, M.S., Gill, H.S., "Ocean wave-induced liquefaction, J. Geotech. Eng., 109(4), 573-590, 1983.

Sarpakaya, T., "Forces on cylinders near a plane boundary in a sinusoidally oscillating fluid," Trans. ASME J. Fluid Eng., 98, 499-505, 1976.

Verley, R.L. and Reed, K., "Prediction of hydrodynamic forces on seabed piplines," OTC-5503, 171-180, Houston-TX, April 1987.

CHAPTER 110

Perched Beach Profile Response to Wave Action

R. M. Sorensen* F.ASCE
N. J. Beil**

Abstract

Wave tank experiments were conducted to investigate the response of a perched beach profile to storm wave attack. Irregular waves having a JONSWAP spectrum were used for all of the test runs. Beach profile erosion was measured initially for a 1:20 slope nourished profile without a toe structure in place. This established base test profiles for comparison with subsequent erosion profiles with a toe structure or sill in place. The subsequent tests with a sill were conducted with the sill placed at a range of depths along the nourished profile slope.

Each of the test conditions was subjected to erosive wave attack for 42 hours with periodic interruptions for measurement of resulting beach profiles. Test results are presented and interpreted to evaluate the effect of the sill structure on beach profile response and the effectiveness of the structure in retaining the beach. Scaling of the results to prototype conditions is also discussed.

Introduction

A common solution to beach erosion problems is to periodically nourish the beach by the placement of sand fill. Often a structure is constructed to compliment nourishment - the structure increasing the stability of the fill to storm wave attack and/or deficiencies in longshore sediment transport. Common types of beach stabilization structures include groins and offshore breakwaters. The perched beach concept is a variation on the use of offshore breakwaters for stabilizing nourished beaches and protecting shore facilities.

A low submerged sill is constructed offshore and parallel to shore. Sand fill is placed landward of the sill, creating the perched beach (see Figure 1). The toe structure retains the perched beach, greatly reducing the

----

* Professor, Department of Civil Engineering, Lehigh University, Bethlehem, PA 18015
** Senior Project Engineer, Kidde Consultants Inc., Baltimore, MD 21204

volume of sand required to nourish the beach. The sill also triggers breaking of the larger waves thereby providing a measure of protection to the shore even if the fill were not in place.

Most commonly, the sill structure would be constructed of a homogeneous mound of stones. A preceeding study (Givler and Sorensen, 1986 and Sorensen, 1987) investigated stability requirements for a submerged homogeneous stone sill structure. This investigation is concerned with the response of a nourished perched beach profile to storm wave attack. It involves two-dimensional wave tank experiments using irregular waves and a sand beach with and without a sill structure in place. Additional information on several aspects of this study are given in Beil and Sorensen (1988).

Research Facility

The experiments were conducted in a 32.7 m long, 0.91 m wide, 0.91 m high wave tank with a programmable spectral wave generator. The wave generator has an absorption capability, i.e. reflected waves are sensed at the generator paddle and subtracted from the primary wave being generated. This capability allows for long term generator operation without the buildup of troublesome reflected waves.

A parallel-wire resistance wave gage was installed 8.23 m from the wave generator and seaward of the toe of the sand beach. Output from the wave gage was recorded in analog form on chart paper and in digital form (0.12 hz) on a portable PC.

Beach profiles were measured along the tank centerline by a point gage mounted on a carriage that moved along the wave tank rails. Gage readings were made at 0.03 m horizontal intervals along the active portion of the beach profile.

The beach was made of a fine, uniformly graded sand having a median diameter of 0.145 mm and a corresponding median diameter fall velocity (Vf) of 0.018 m/s (calculated for 20 degrees C, S.G. = 2.65).

Experimental Set-up and Procedure

The experimental set-up is summarized schematically in Figure 2. All tests were run with a tank water depth of 0.483 m. A bulkhead was installed at a distance of 24.7 m from the wave generator. The sand beach was placed in front of the bulkhead with a berm crest elevation of 0.57 m above the tank floor and extending 2.44 seaward from the bulkhead. The nourished beach face extended sea-

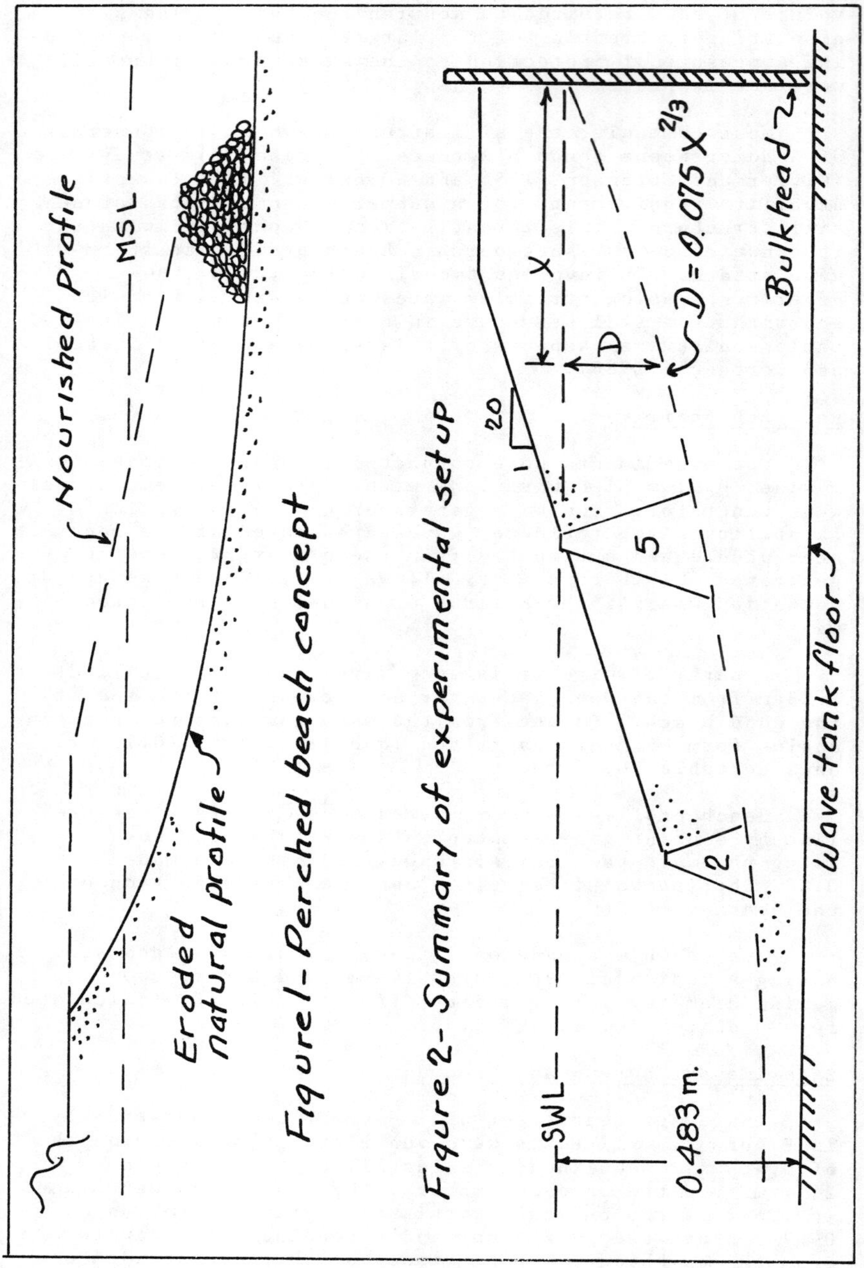

Figure1- Perched beach concept

Figure 2- Summary of experimental set up

ward from the berm at a 1:20 slope for all of the tests with and without the sill structure. The 1:20 slope was selected based on a review of recent beach nourishment projects in the U.S. (Hobson, 1981).

The same wave conditions were used for all of the tests. They were JONSWAP spectrum waves with a significant height (Hs) of 0.09 m and a peak period (Tp) of 1.6 sec measured at the toe of the beach slope. These waves yield a fall time parameter (Hs/Vf*Tp) equal to 3.1. Kriebel et al. (1986) show that waves having a fall time parameter in excess of 2.0 to 2.5 should produce an erosive or storm type beach profile. This was confirmed by the resulting beach profile behavior.

Five test cases were investigated (Figure 2). The first consisted of a 1:20 nourished profile without a toe structure. The remaining four cases (2 through 5) were perched beach conditions with a sill located at various depths below the SWL and along the 1:20 beach face slope. The four sill crest depths were 1.0, 0.5, 0.25 and 0 times the incident significant wave height (i.e. 0.091, 0.046, 0.023, and 0 meters below SWL). For the four perched beach conditions, the beach profile seaward of the sill was a typical natural profile (Dean, 1977) as shown in Figure 2.

The sill structure was constructed of plywood with a crest width of 7.62 cm and 1:1.5 front and rear slopes. The seaward slope of the sill was covered with a plastic artificial turf material to reduce wave agitation in front of the sill.

Waves were run for a total of 42 hours for each of the five test cases. After the first six hours of waves the generator was stopped and the beach profile was measured. The profile was measured again at 12 hours, 18 hours, 24 hours, and 42 hours. At the 42 hour time period profile change was still occurring for each test case but the beach had approximately reached an equilibrium condition.

Experimental Results

The experimental data for the five test cases consists of beach profiles for the 0, 6, 12, 18, 24, and 42 hours of wave attack for the selected storm wave condition. Those portions of the measured profiles that are of primary interest are shown in Figures 3A to 3E. The profile for the first case (without a structure) is truncated at the seaward extreme of profile erosion. The profiles for the other cases are truncated a short distance seaward of the structure. These give sufficient scale in the figures to see what changes occurred from the structure to the beach face, the area of greatest interest. The regions seaward of the point of truncation just show successive accumulations of sand tapering seaward.

Table 1 is a tabulation of horizontal retreat distances for the point on the beach profile at the still water line. The distances are measured from the original filled profile position before the start of wave attack and are for the given time intervals. Note that initially (0 - 6 hrs) the retreat is the least for the nonperched profile and increases as the structure crest moves up the profile to the SWL. Also note that the total retreat after 42 hours follows the same pattern, primarily owing to the large early retreat rates. However, after the initial 6 hour period, the retreat rate is reversed for the four perched cases with the retreat rate being minimum from hour 6 to hour 42 for the case with the structure crest at the SWL. It appears that, after the initial adjustment (which could be allowed for in the initial fill placement) location of the sill crest at the still water line results in the least recession of the beach face. Particularly in comparison to the nonperched profile, there is a significant reduction of later retreat with a sill installed near the SWL.

Figure 4 shows the initial fill profile, the 42-hour nonperched profile and the 42-hour perched profile with the sill crest at the SWL. Comparison of these profiles largely summarizes the effect of the sill structure on the resulting beach response and demonstrates the greatly reduced volume of fill required for the perched beach. A larger net volume of sand is transported further seaward for the nonperched condition.

Figure 5 is a summary of the 42 hour profiles landward of the sill structure for each of the four perched beach test cases. The beach face had a consistent slope of 1:5. There was a platform fronting the beach face that had a less consistent slope (but one that did not change with a noticeable trend) of about 1:50. As discussed above, the distance from the sill to the beach face progressively decreased for decreasing submergence of the sill crest as did the depth to the break between the beach face slope and the platform slope.

Scale Effects

The storm wave characteristics employed in these tests were limited largely by the wave amplitude generating capability of the wave generator. This in turn largely dictated the spectral peak period employed so that the resulting wave steepness was similar to typical prototype storm waves. An indication of the scaled prototype conditions represented by these tests can be obtained from

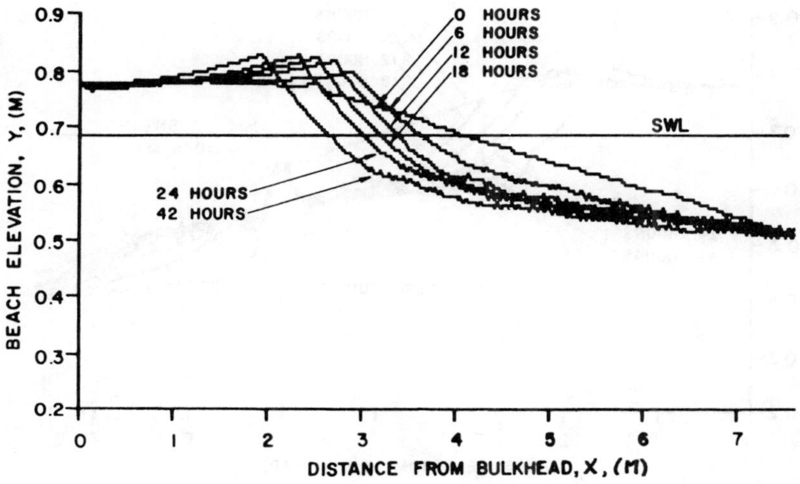

Figure 3A - Nourished beach, no structure.

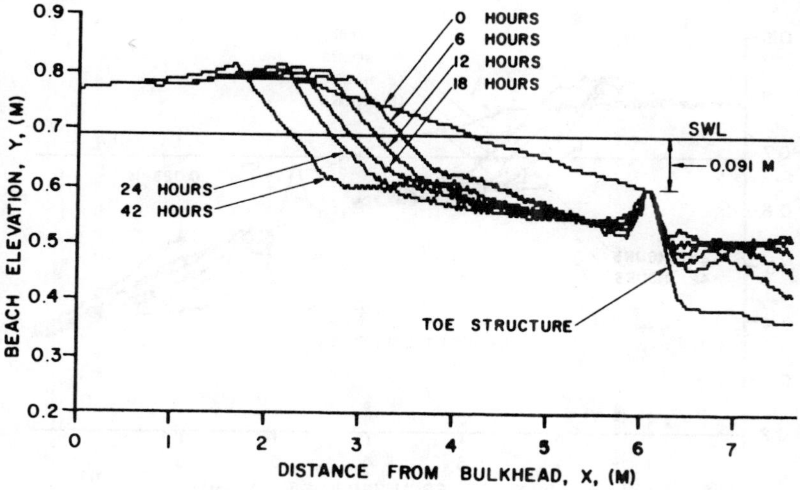

Figure 3B - Perched beach, structure crest depth 0.091 meters.

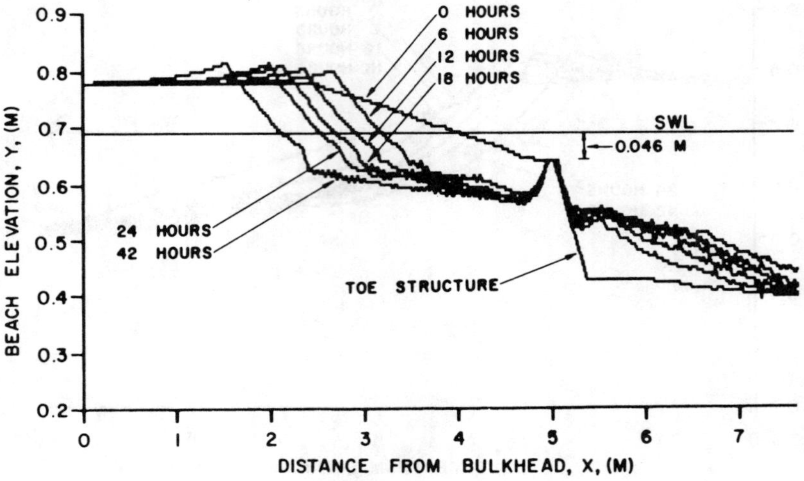

Figure 3C - Perched beach, structure crest depth 0.046 meters.

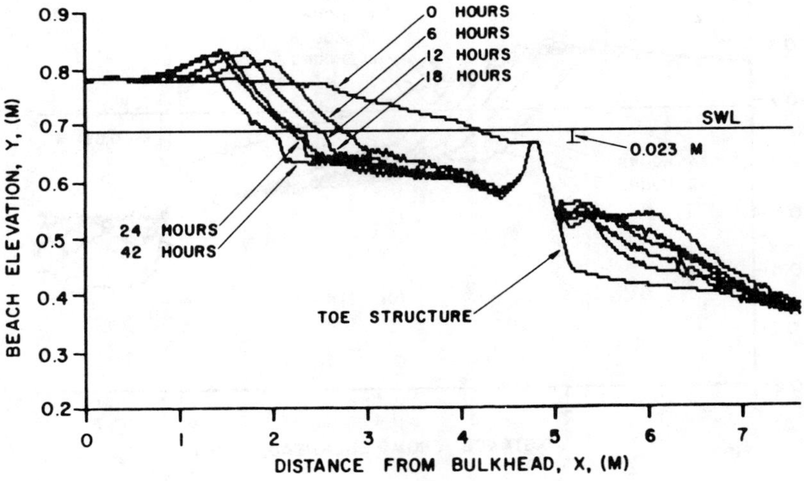

Figure 3D - Perched beach, structure crest depth 0.023 meters.

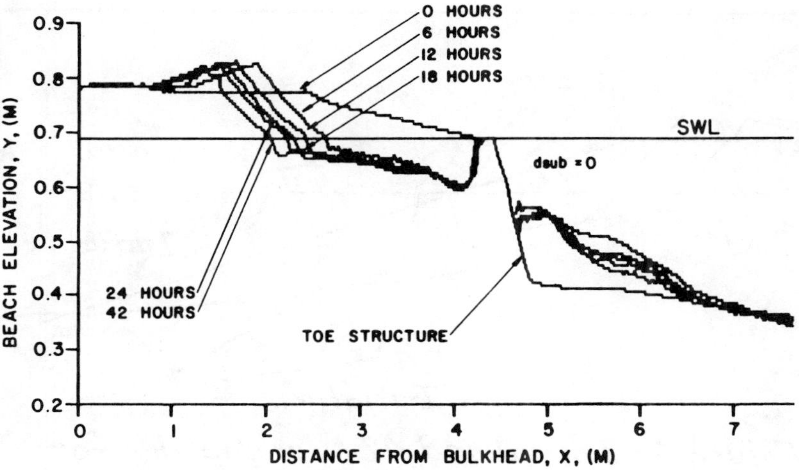

Figure 3E - Perched beach, structure crest depth 0.0 meters.

Test	$\frac{dsub}{Hs}$	0-6 hrs	6-12 hrs	12-18 hrs	18-24 hrs	24-42 hrs	Total
1	-	.498	.234	.186	.198	.480	1.596
2	1.0	.786	.264	.270	.216	.552	2.088
3	0.5	.816	.258	.222	.162	.552	2.010
4	0.25	1.428	.246	.216	.072	.336	2.298
5	0	1.662	.216	.120	.066	.168	2.232

Table 1 - Retreat of SWL point (meters) for time intervals shown.

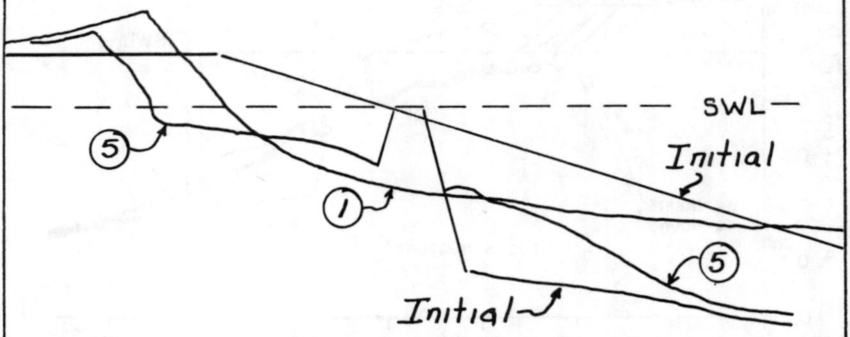

Figure 4 - Initial and 42 hour profiles for Cases 1 and 5

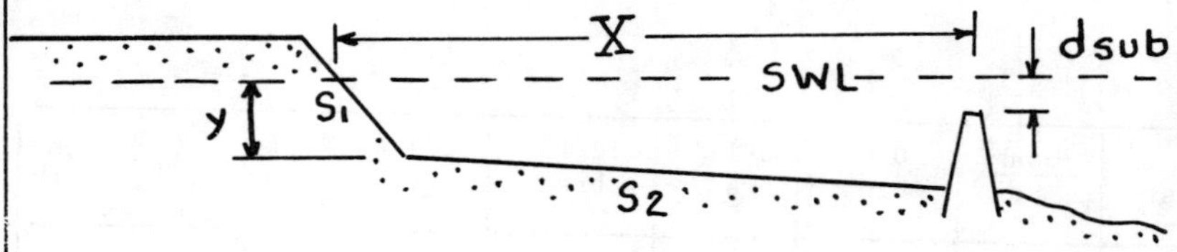

Test	dsub (m)	S_1	S_2	X (m)	Y (m)
2	0.091	0.19	0.015	3.79	0.09
3	0.046	0.19	0.023	3.06	0.06
4	0.023	0.20	0.014	2.78	0.045
5	0.0	0.19	0.023	2.47	0.015

Figure 5- 42 hour profile geometries, for Cases 2 through 5

the recent work of Kriebel et al. (1986). They showed that erosive beach profiles can be scaled using Froude similarity (equal model and prototype Froude Numbers) and equal fall time parameters in model and prototype. With these two criteria, the model profile can be scaled up to prototype dimensions without distortion of profile geometry.

Assuming a prototype median sand diameter of 0.45 mm (Vf = 0.065 m/s) which is typical for beaches on the north shore in New Jersey and the model fall time parameter value of 3.1, along with Froude similarity requirements, yields an equivalent prototype significant wave height of 1.2 meters and a spectral peak period of 5.8 seconds (i.e. length ratio = 13.04, time ratio = 3.61). These values are somewhat low for typical storm wave conditions on the New Jersey shore which would be more in the order of 7 seconds period and 2 meters or more in height. Higher and longer waves than those used in the experiments would likely have resulted in a somewhat flatter beach face slope and more sand loss, in turn resulting in a greater distance between the sill and the beach face.

Discussion and Conclusion

The experiments presented herein are for only one wave condition and thus do not provide the range of data needed for the design of a perched beach under various conditions. However, the results suggest that a perched beach with the sill crest near the still water level can be an effective concept for beach nourishment under appropriate conditions - for example, where there is a steep offshore slope or a narrow shelf width. Significantly less fill volume is required with a sill in place and offshore losses should be reduced. So, a perched beach may be cost effective for many locations. Additional data, covering a much wider range of conditions, are needed.

References

Beil, N. J. and R. M. Sorensen (1988) "An Investigation of Perched Beach Profile Response to Wave Action" H. R. Imbt Hydraulics Lab Report 115-88, Lehigh University, Bethlehem, PA.

Dean, R. G. (1977) "Equilibrium Beach Profiles, U.S. Atlantic and Gulf Coasts", Ocean Engineering and Technical Report No. 12, University of Delaware, Newark, DE.

Givler, L. D. and R. M. Sorensen (1986) "An Investigation of the Stability of Submerged Homogeneous Rubble-mound Structures Under Wave Attack", H. R. Imbt Hydraulics Lab Report 110-86, Lehigh University, Bethlehem, PA.

Hobson, R. D. (1981) "Beach Nourishment Techniques" Report 3 - Typical U.S. Beach Nourishment Projects Using Offshore Sand Deposits, U.S. Army Coastal Engineering Research Center, Ft. Belvoir, VA.

Kreibel, D. L., W. R. Dally and R. G. Dean (1986) "Undistorted Froude Model for Surf Zone Sediment Transport" Proceedings, 20th International Conference on Coastal Engineering, Taipei, pp. 1296-1310.

Sorensen, R. M. (1987) "Evaluation of Perched Beach Concept for Shore Stabilization" Proceedings, Coastal & Port Engineering in Developing Countries, Beijing, pp. 318-326.

CHAPTER 111

EFFECTS OF A VERTICAL SEAWALL ON PROFILE RESPONSE

Michael R. Barnett[1], A. M., ASCE, and Hsiang Wang[2]

ABSTRACT

An attempt is made to determine beach profile response due to the presence of a vertical seawall placed in various cross-shore positions, and to examine the differences between natural beaches and seawall-backed beaches in response to normally incident wave attack. The investigation was mainly restricted to two-dimensional profile response under erosive wave conditions, with beach recovery response monitored to a limited extent. Spatial and temporal profile response was investigated by examining time-series profile configuration, volumetric changes, sediment transport patterns, and quasi-equilibrium profile configuration. Additionally, dominant profile features such as the break point and reflection bars (as well as scour at the toe of the seawall-backed profiles) were observed and quantified.

INTRODUCTION

An emerging concern by coastal communities and engineers is the effect of seawall placement along beach-dune systems. Since seawalls are most likely to be placed along eroding coastlines, any added erosional pressure due to the presence of the structure is deemed serious.

Important questions raised by interrupting a natural system with a seawall are

1) will the presence of the seawall accelerate the erosion process of the beach fronting it;
2) what effect does cross-shore placement of the seawall have on profile response;
3) what is the resulting profile configuration under erosive wave conditions;
4) will profile recovery be impeded by the structure.

[1]Coastal Engineering Specialist, Florida Sea Grant Extension Program, 336 Weil Hall, University of Florida, Gainesville, Florida 32611 USA

[2]Professor, Coastal & Oceanographic Engineering Department, 336 Weil Hall, University of Florida, Gainesville, Florida 32611 USA

These concerns regarding construction and placement of a seawall necessitate that quantitative and qualitative relationships and trends be investigated. The focus of this study was to examine a fundamental aspect of a seawall-beach system by assessing the influence of a vertical seawall on beach profile response through laboratory experimentation.

Numerous field and laboratory investigations documenting seawall-beach interaction exist, but very few tests have focused on structural influence on profile response. Kraus (1987) provides a detailed literature review of over 70 laboratory, field, theoretical, and conceptual studies; due to space limitations here, the reader is referred to this thorough treatment for a more detailed background on the subject.

OBJECTIVES AND SCOPE

The main objectives of this investigation were to determine beach profile response due to the presence of seawalls, and to examine the differences between natural beaches and beaches fronting seawalls in response to normally incident wave attack. Seawall location was varied throughout the experiments, with each location subjected to the same wave conditions as the "no-seawall" tests. The present investigation differs from previous investigations in that the initial profile shape of $Ax^{2/3}$ has not been examined in detail with a seawall placed along the profile.

The experiments involved monitoring temporal and spatial changes of beach forms on natural and seawalled beaches under various wave conditions and seawall positions. The results were analyzed toward an understanding of:

1) characteristic profiles;
2) volumetric changes;
3) patterns and mechanisms of sediment movement; and
4) equilibrium profile configurations.

Profile features of special interest include break point bar formation and the presence of reflection bars. For seawall-backed beaches, the scour trough at the structure toe, the eroded and accreted volumes, and the rates of erosion and recovery were also examined.

DESIGN OF THE EXPERIMENTS

Selection of the initial profile geometry for all tests is based on an empirical relationship proposed by Dean (1977); the same criterion was employed by Kriebel et al. (1986) for tests without a seawall. To simulate natural conditions, similarity criteria are established on the assumption that the energy dissipation per unit volume along a beach profile is uniform, and that the wave properties can be properly scaled by Froude criteria.

Beach Profile Geometry. Each of the model tests conducted in the present study was molded to an initial profile of $Ax^{2/3}$; this profile geometry was determined by Dean (1977). Dean postulated that if the energy dissipation per unit volume in the surf zone is uniform and is a function of sand grain size, D, only, then the following relationship exists:

$$\frac{1}{h} \frac{\partial F}{\partial x} = \zeta_*(D) \qquad (1)$$

where F is the energy flux per unit width, h is the water depth, x is a shore-normal coordinate, and $\zeta_*(D)$ is a constant depending on D only. By applying linear wave theory assumptions and shallow water approximations, the following relationship is derived:

$$h(x) = \{[\frac{\zeta_*(D)}{\frac{24}{5} \rho g\, g^{1/2} \kappa^2}]^{2/3}\} x^{2/3} \qquad (2)$$

The quantity enclosed by { } is defined as A(D), leaving

$$h(x) = A(D) x^{2/3} \qquad (3)$$

This profile shape was examined by Dean and compared to over 500 natural beach profiles along the U.S. Atlantic and Gulf coasts. All initial profiles in the present study utilized the geometry represented by (3).

Similarity Criteria. In order for the laboratory model to faithfully represent prototype conditions, similarity criteria should be established and observed. Constructing a moveable bed model requires the horizontal scale, vertical scale, and grain size and specific gravity of the bed material to be specified.

To maintain proper similarity between model and prototype requires the geometry conditions, flow conditions, and sediment transport processes to be similar. By utilizing an undistorted model with sediment identical to the prototype, the transport process can be approximately simulated if

1) the Froude criterion is fulfilled for the flow field;
2) the sediment fall velocity scale ratio behaves as the square root of the length scale; and
3) the sediment is large enough to ensure a turbulent boundary layer and to ensure that the properties of the granular material are maintained.

Empirical Orthogonal Functions. By utilizing empirical orthogonal functions (eigenfunctions) to represent a series of beach profiles, all the profile data in that series may be represented in a compact form by showing spatial and temporal variations of dominant profile features. Profile variations are taken from an elevation, h, measured a distance, x, from a baseline over time intervals of variable spacing. Profile elevation at a given time may be represented by h_{xt}, where x represents an index for the spatial profile coordinate and t represents the temporal index.

The empirical eigenfunction analysis seeks to represent h_{xt} in terms of a linear expansion of the product sum of the spatial and temporal eigenvectors in the form

$$h_{xt} = \sum_{n=1}^{N} w_n c_{nt} e_{nx} \qquad (4)$$

where w_n is the weighting coefficient for each of the N eigenvectors, c_{nt} is the temporal eigenvector, and e_{nx} is the spatial eigenvector.

The data set input into such a form results in a set of empirical eigenfunctions which best fit the data in a least-squares sense. The first set of eigenvectors, for the mode n = 1, represents the largest percentage of the total variance of the data set, or the dominant mode of the transport process in this case, while successively higher modes represent successively higher order perturbations. This method has been applied to beach profile field data by Winant et al. (1975), Aubrey (1979), and Kriebel et al. (1986).

As each eigenvector exhibits the property of orthogonality, individual eigenfunctions represent a certain percentage of the variance of the mean square of the data; therefore, each solution is unique in that it is not correlated to any other solution, allowing for an explanation of the variance from the input data set (Kriebel et al., 1986). Hence, large numbers of variables may be expressed by those few empirical functions which describe the major percentage of the mean square value of the data.

APPARATUS, PROCEDURE, AND CONDITIONS

System Components

Wave tank. The laboratory tests were conducted in the "air-sea" tank at the University of Florida Coastal Engineering Laboratory. The usable portion of the tank is 37 m long, 1.2 m high, and partitioned by a 1.9 m high concrete block wall into two sections of equal width, each approximately 0.87 m wide. The section of the tank in which the laboratory tests were conducted has one wall constructed of glass panels to permit visual observation of profile changes. The beach was constructed with the toe 18 m from the wave generator and with an approximate profile length of 17 m. The tank dimensions, along with the location of the profile in the tank, are shown in Figure 1.

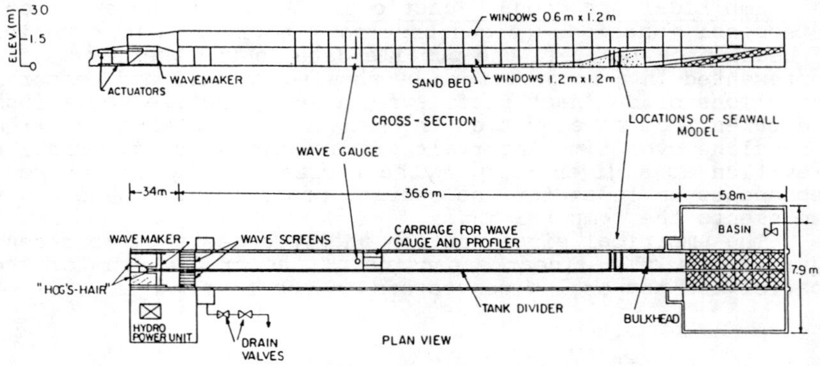

Figure 1. Schematic of Test Facility.

Sediment. The bed material used in the model tests was a fine quartz sand with a median grain diameter of 0.15 mm as determined by sieve analysis of samples obtained at four locations along an eroded equilibrium profile; a settling tube analysis was also performed. Two independent tests were performed on each sample, and the median fall velocity was determined to be 1.77 cm/sec.

Seawall model. The model seawall was constructed of 13 mm plywood 0.85 m wide and 0.88 m high coated with fiberglass resin to protect against both warping and abrasion. All experiments which were conducted with a seawall had the model installed vertically and flush with the tank bottom.

Instrumentation. Data acquisition was accomplished by utilizing a capacitance-type wave gauge and a Mark-V Electronic Profile Indicator (capacitance-sensing bottom profiler) mounted on a motorized cart which ran along parallel rails located above the test section.

Procedure

Initial Conditions. In each test, the sand bed was initially molded to an $Ax^{2/3}$ profile shape, with an A value of 0.075 $m^{1/3}$ and a beach face slope of 1/5 from the point of tangency to an elevation above the expected runup limit (Kriebel, Dally, and Dean, 1986). At this elevation, a 30 cm wide berm was constructed, the landward extent of which served as a control point for all profiles. The wave tank tests conducted by Saville (1957) were selected as the prototype tests; based on calculation of the sediment fall velocities, Kriebel et al. (1986) found the prototype to model time scale to be 3.09, and the length ratio to be 9.6. These values were also adopted for the present study. Once the profile was graded, the tank was filled with fresh water and the sand bed was allowed to soak for 10 to 12 hours. Before beginning a test, monochromatic waves were run against the initial profile for one to two minutes to allow for further bed consolidation; the initial profiles were then recorded by moving the profiler along the entire bed.

Data Acquisition. Each experiment was run a sufficient duration to reach a quasi-equilibrium condition (that is, no appreciable changes in profile configuration over time). Wave data were collected intermittently, and profile data were acquired as a time series to monitor profile evolution.

Further Profile Data Conditioning. Two independent profiles were taken at approximately one-third the cross-tank distance from each of the flume walls at each time interval in an attempt to reduce the error introduced by cross-tank variations. Such variations were also noted by Kriebel et al. (1986); possible causes are: wall boundary layer producing wave refraction, uneven bed compaction, and reflected wave interaction with incident waves.

A short series of experiments was conducted to test the profile recovery characteristics of a seawall-protected beach versus a "natural" (unprotected) beach. To accomplish this, the profiles were first subjected to erosive wave conditions accompanied by a storm surge; after eroding the profile for a fixed length of time, the water level was reduced, and lower steepness waves were produced.

Test Conditions

Figure 2 illustrates the seawall locations, water depths, and initial profile geometry used in the present investigation. Table I lists pertinent test parameters such as water depth, wave height, wave period, and seawall location. The letter designations for each Test Classification listed (A, B, C, and D) refer to tests with identical input wave conditions and water depths. Tests 1-13 were performed under erosive wave conditions, both with and without a 10 cm model scale storm surge (0.96 m full scale). This value was chosen to prevent overtopping of the seawall and, for the natural beach test cases, to prevent overtopping of the berm. Tests 14 and 15 were conducted to determine the rate and extent of profile recovery on one natural and one seawall-backed beach, respectively.

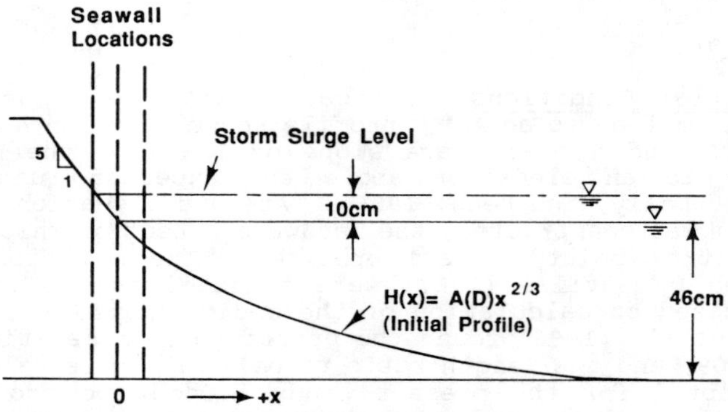

Figure 2. Schematic of Test Conditions.

Table I. Pertinent Test Parameters.

Test Number and Classification		Period (sec)	Water depth (cm)	Wave height (cm)	Seawall location, (m from 46 cm SWL)
1	A	1.81	46	11.75	N/A
2	A	1.81	46	11.75	-0.3
3	A	1.81	46	11.75	0.0
4	A	1.81	46	11.75	+0.3
5	B	1.30	46	8.80	N/A
6	B	1.30	46	8.80	-0.3
7	B	1.30	46	8.80	0.0
8	B	1.30	46	8.80	+0.3
9	C	1.81	56	11.75	N/A
10	C	1.81	56	11.75	N/A
11	C	1.81	56	11.75	-0.3
12	C	1.81	56	11.75	0.0
13	C	1.81	56	11.75	+0.3
14	D	1.81	46	4.00	N/A
15	D	1.81	46	4.00	-0.3

RESULTS AND DISCUSSION

Composite Profiles

Wave data were obtained periodically between profile data-taking intervals; wave height records were acquired by sampling with the capacitance gauge immediately seaward of the profile toe. Reflection envelopes were obtained by moving the gauge horizontally in the direction of wave propagation at a fixed rate of speed.

For tests with a 46 cm water depth, profile configurations both with and without seawalls were remarkably similar in overall planform. With the exception of local effects due to the presence of the seawall, the main bar-trough features existed for all profiles. The presence of the seawall on the profile shifted the location of the break point bar and also created a scour trough at the toe of the structure. For an elevated water level (10 cm storm surge), on the other hand, the profiles backed by a vertical wall were markedly dissimilar in final planform to those profiles without a seawall. This is apparently due to the artificially reduced surf zone width and increased wave reflection created by the wall. Reflection bars were more evident for the walled profiles than the "natural" beach profiles for the storm surge conditions; reflection coefficients were higher for the seawall-backed profiles.

Test Classification A experiments exhibited a similar trend in the formation and migration of the break point bar. Breaking waves rapidly formed a small break point bar formation at a distance of 4.3 m to 5.3 m from the baseline (intersection of 46 cm SWL with the initial profile). Bar migration in all tests in this category ($H = 11.75$ cm, $T = 1.81$ s) was observed to be offshore, with the break point location becoming fairly well-stabilized between 3.23 hours and 4.2 hours model time, as seen in Figures 3 and 4; once this occurred, bar width decreased and bar height increased. Sediment transport seaward of the break point bar then became predominantly offshore, forming a large trough 3.1 m to 3.5 m in length, with an offshore bar at the profile toe.

Test Classification B experiments also exhibited a consistent unidirectional mode of offshore sediment transport. Break point location ranged from 1.7 m (Test 7, seawall at SWL) to 2.8 m ("natural" beach) from the baseline.

Laboratory experiments conducted to simulate a 10 cm storm surge with the same wave conditions as Tests 1 through 4 displayed rapid bar formation. Bar migration for the "natural beach" tests was observed to progress offshore, with bar location stabilizing at 3.5 m from the baseline (see Figure 5). Test 10 was conducted as both a confirmation of the repeatability of the test program (Test 9) and as a means to establish initial conditions for the "natural" recovery test (Test 14).

Classification C experiments which investigated the effects of varying seawall location (Tests 11 through 13) established break point bar formation within the first 0.5 hours of testing. Bar migration for all tests was minimal due to the "fixed" shoreline created by the wall; as shown by figure 6, this, coupled with a comparatively high reflection

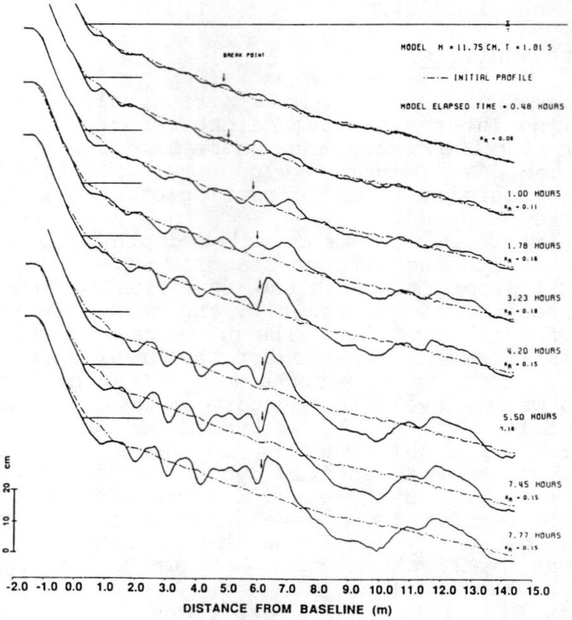

Figure 3. Time-series Profile Evolution, Test 1.

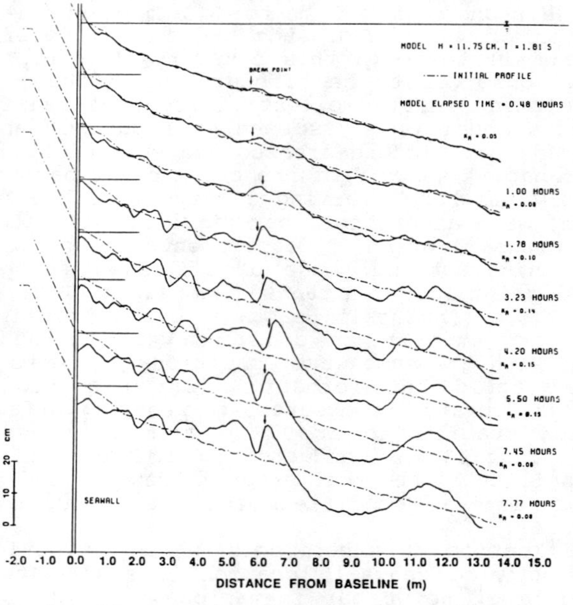

Figure 4. Time-series Profile Evolution, Test 3.

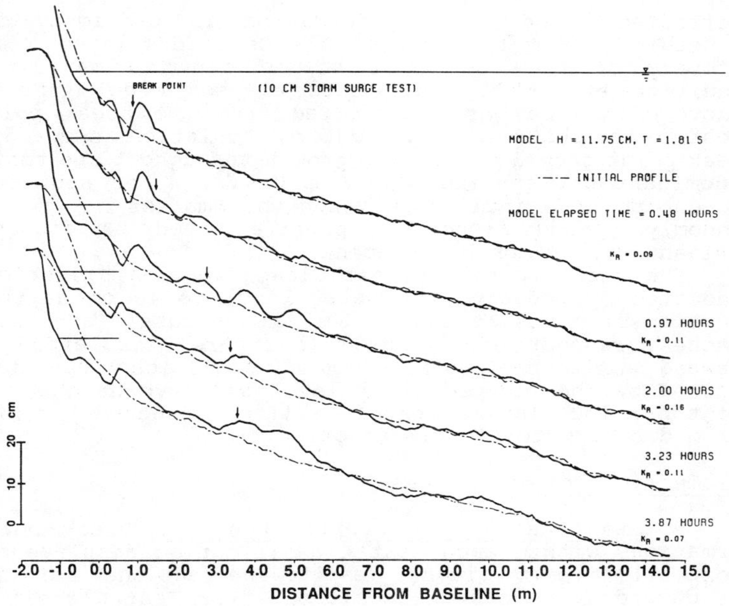

Figure 5. Time-series Profile Evolution, Test 9.

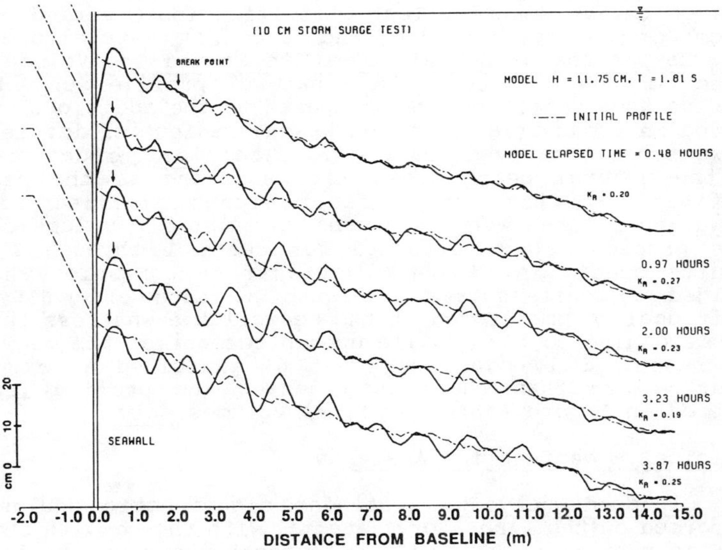

Figure 6. Time-series Profile Evolution, Test 12.
(Same wave parameters as Test 9).

coefficient, led to bar formation in proximity to the structure (0.5 m to 1.0 m). All tests displayed a partial recovery of the toe scour volume after a maximum scour condition at roughly 2.0 hours elapsed model time. This recovered volume was transferred from the break point bar crest to the scour trough. Bedform undulations seaward of the break point showed continuous growth throughout the test, with a dominant bar spacing of 1.2 m to 1.3 m, or approximately one-quarter the incident wavelength. Smaller bar formations randomly spaced along the profile acted to disrupt the dominant reflection bar spacing.

The recovery tests (Test Classification D), which were conducted by reducing the water level to 46 cm at the termination of Tests 10 and 11 for the "natural" and seawalled beaches, respectively, showed an apparent isolation of the seaward half of the profile from any significant bar migration (caused by the reduced water level and wave height). Break point bar location in Test 14 shifted onshore approximately 1.1 m due to water level reduction.

Volumetric Profile Changes

Volume Changes over Profile Length. Break point bar formations within each test classification displayed nearly identical maximum volumes, despite the fact that bar location was dependent on seawall position. For Test Classification A, bar volumes ranged from 110 cm^3/cm (seawall located -0.3 m (shoreward) of 46 cm SWL) to 123 cm^3/cm for the natural profile. Test Classification B bar volumes ranged from 60 cm^3/cm (natural profile) to 74 cm^3/cm (seawall +0.3 m). Maximum bar volumes for Test Classification C varied between 62 cm^3/cm (seawall $\pm$ 0.3 m) and 70 cm^3/cm (natural profile).

Comparison of Denial Volume to Eroded Toe Volume. The volume change obtained with a natural profile subjected to erosive wave conditions was compared to the additional volume eroded in front of a seawalled beach; regions of interest are shown schematically in Figure 7. The final eroded profiles of the natural beach cases were selected as the baseline profiles for this study; final eroded profiles for the seawalled beaches were then used to calculate volume change. This procedure was performed for tests 1 through 13; the results appear in Figure 8. A linear regression analysis yielded a least-squares fit with a slope of 0.616. The additional volume eroded at the seawall toe was less than the volume denied to the profile upon placement of the seawall on the beach. Only one case (Test 6) exhibited a larger toe scour volume than the volume denied the profile (results plotted in Figure 8 are for final volumes only).

Empirical Eigenfunction Analysis

Standard Analysis. A standard eigenfunction analysis was performed on the laboratory data; as with the results obtained by Winant et al. (1975), the mean beach function was found to represent over 99 percent of the mean square value of the data. Laboratory test results also showed similarity to the second and third eigenfunction data obtained at Clearwater Beach, Florida by Kriebel et al. (1986).

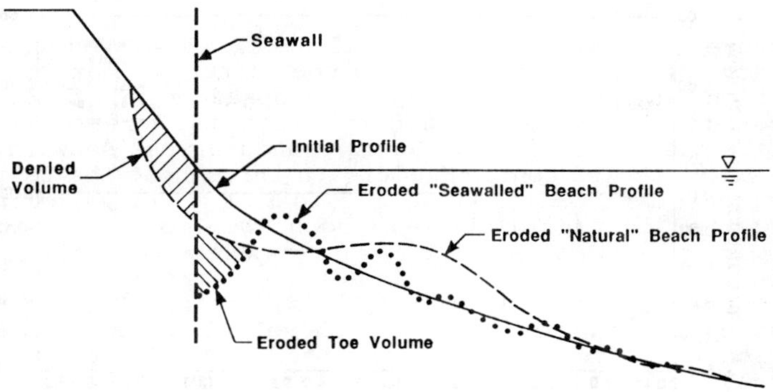

Figure 7. Schematic of Profile Features of Interest.

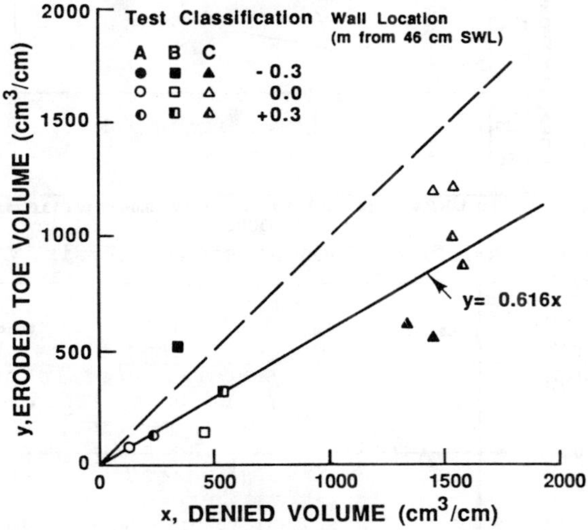

Figure 8. Denied Volume versus Eroded Toe Volume.

Modified Eigenfunction Analysis. Based on further interpretation of the remaining data, the influence of the initial profile on the mean profile configuration (first eigenfunction) was found to be highly dominant. Since the field studies cited previously dealt with spatial and temporal changes over long periods of time, no true "initial" profile existed in these investigations. Hence, a modified eigenfunction analysis was undertaken in which the method of analysis was unchanged, but the input data set was reformatted such that the initial profile acted as a baseline configuration. All profiles within a test set were subtracted from the initial profile, and the eigenfunction analysis performed on this modified data; typical results are illustrated in Figures 9 and 10. Additionally, the temporal

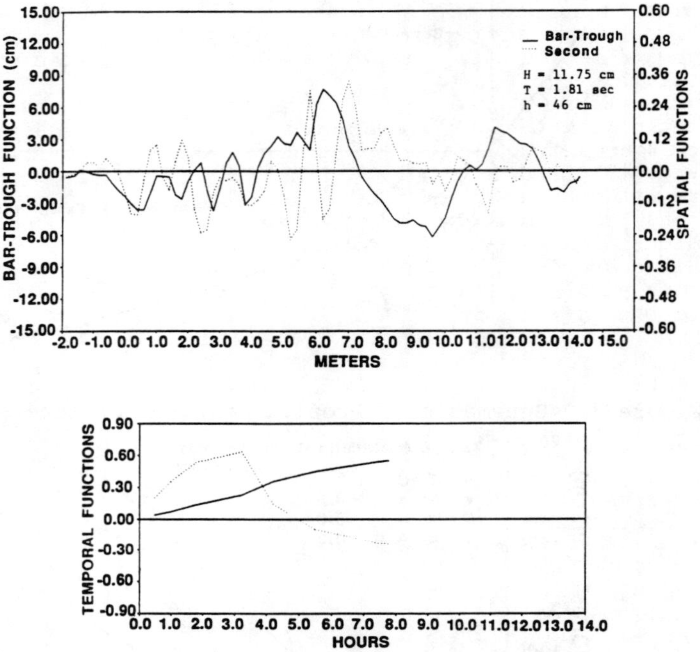

Figure 9. Modified Eigenfunction Analysis, Test 1.

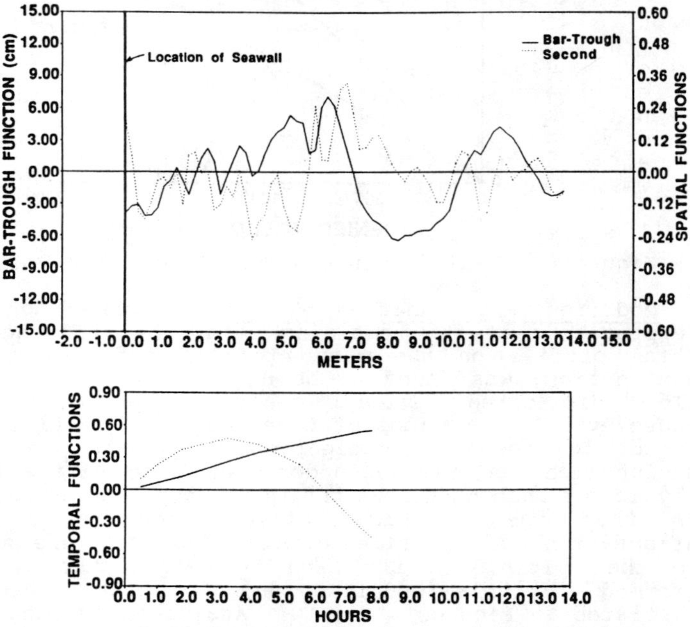

Figure 10. Modified Eigenfunction Analysis, Test 3.

functions were unified such that all values originated as positive, with an erosion from the modified mean profile assigned a negative spatial value, and an accretion a positive spatial value.

The first three eigenfunctions represent over 97.5 percent of the variance from the mean in all test cases. The first eigenfunction, referred to as the "bar-trough" function, explains 79.88 to 95.91 percent of the variance from the mean square value of the data. The second eigenfunction represents 60.0 to 86.8 percent of the variation from the first eigenfunction, with 6.2 to 29.0 percent of the variance explained by the third eigenfunction. Due to the complex behavior exhibited by the third eigenfunction, only the first and second eigenfunctions were analyzed.

The first eigenfunction appears to represent a "bar-trough" behavior along the profile length. The solid line for the spatial function (shown in Figures 9 and 10) displays the location and magnitude of the bar and trough features present in each profile as a weighted average of the input data; by coupling the first temporal function with the first spatial function in each plot, the time rate of change of the profile features could be interpreted.

Tests 1 through 3 showed maxima on either side of the break point bar crest which dominated the first 3.23 hours of the test, then diminished in effect with further profile evolution. By comparing these features with Figures 3 and 4, the dual-crested bar was observed to initially steepen, then become subdominant upon emergence of single-peaked, highly skewed break point bars. The maxima which appear shoreward of the break point correspond to the skewness of the berm erosion or toe scour trough and bar feature (depending on whether a seawall was located on the profile).

For the storm surge tests, the seawalled profiles revealed numerous reflection bars. The main incongruity between the storm surge and normal water level tests with erosive wave conditions was the behavior of the second temporal function. A rapid decrease from an initial peak value was noted in all storm surge tests, which corresponds to a temporal decrease in the scour trough skewness. The no-seawall tests showed a peak spatial function value at the shoreward dune erosion limit, which corresponded to the temporal steepening of the dune face.

The recovery tests displayed extremely "noisy" second spatial functions due to the presence of well-established reflection bars which exhibited minimal spatial change over the seaward half of the profile (owing primarily to a decreased wave height); numerous maxima and minima over the shoreward portions of the profile were also revealed. Temporal changes differed greatly between the two test cases, with the seawalled profile demonstrating a more stable trend of the skewness of the surf zone bar formations in the latter 4.0 hours of testing than that exhibited by the no-wall case. The seawalled profile did not exhibit a relative minimum value of the bar skewness in the vicinity of the break point bar; however, both profiles exhibited a distinct break point bar-trough feature (due primarily to the reduction in water level) which caused the bed to scour shoreward of the break point.

SUMMARY AND CONCLUSIONS

Summary of Investigation. Due to the recent attention being focused on the effects of seawall placement on eroding coastlines, there exists a need to examine the resulting profile configuration along with the erosion or recovery rates to ascertain whether the presence of the structure is detrimental to the beach-dune system. A laboratory test series was developed which examined the cross-shore effects of seawall placement on profile response. Erosive wave conditions comprised the majority of the experiments; a limited study of beach recovery characteristics was also undertaken. A substantial set of laboratory data was obtained, and the results have been presented in a comparative sense to further emphasize the structure effects. The complete data set and analysis are presented in Barnett (1987).

Important Findings. The test results were examined by time-series profile evolution, volume change over profile length, and empirical eigenfunction analysis. For all cases tested, profile configurations with and without a seawall were remarkably similar in overall planform; this suggests that the major transport process is not significantly influenced by the presence of the seawall.

Under erosive wave conditions, the dominant spatial feature is a bar-trough system, with the bar forming in proximity to the wave breaking point and the trough occurring near the still water shoreline. The presence of the seawall accentuates the trough into a scour hole instead of spanning over the swash zone, as is the case with a natural beach. However, while local scour was noted to be severe in many of the seawalled profiles, the volume of sand retained upland of the structure (which would otherwise be eroded under identical wave conditions without a seawall) was found experimentally to be approximately 60% greater than the additional volume eroded at the toe of the structure.

Wave reflection, often considered to be a major adverse influence on scour in front of a seawall, did not appear to play a significant role. Reflection bars of varying quantity and spacing did occur, but were usually secondary features of a predominantly bar-trough system. Water depth, on the other hand, appeared to be a dominant factor affecting erosion for all cases tested, with higher water levels causing a significant increase in erosion. For the seawalled cases, the break point bar was in proximity to the wall; the natural profiles displayed a more seaward bar location. This affected the recovery process when the water level returned to normal.

The empirical eigenfunction analysis is a useful tool for examining profile evolution. By conducting a modified eigenfunction analysis, the spatial and temporal behavior of the profile evolution process was more readily facilitated. The primary bar-trough system and the secondary reflection bar system were clearly revealed in spatial plots for all erosive test cases. The primary system accounted for more than 80 percent of the variation from the initial profile. A relatively smooth spatial function coupled with a slowly-varying temporal function revealed the stable nature and steady evolution of the bar-trough system.

The "noisy" behavior exhibited by the second spatial function of the recovery tests showed the recovery process to be more unstable. The primary recovery process was the removal of the break point bar and the redistribution of sand both onshore and offshore. The more prominent primary spatial function in the seawalled case revealed a more efficient recovery.

The seawalled beach exhibited a more substantial recovery volume in the vicinity of the structure toe than that observed for the natural profile, which showed only a small berm growth. This is not sufficient evidence to construe that placement of a seawall on an eroded beach will promote recovery; rather, the recovery process appears possible on a seawalled beach provided the water level and wave conditions are capable of transporting sediment to the scour trough created under erosive conditions. Further recovery tests should be conducted to confirm this assumption.

REFERENCES

Aubrey, D.G., "Seasonal Patterns of Onshore/Offshore Sediment Movement," Journal of Geophysical Research, AGU, Vol. 84, No. C10, October, 1979, pp. 6347-6354.

Barnett, M.R., "Laboratory Study of the Effects of a Vertical Seawall on Beach Profile Response," UFL/COEL-87/005, University of Florida, Coastal & Oceanographic Engineering Department, Gainesville, FL, May, 1987.

Dean, R.G., "Equilibrium Beach Profiles: U.S. Atlantic and Gulf Coasts," Ocean Engineering Report 12, Department of Civil Engineering, University of Delaware, Newark, DE, 1977.

Dean, R.G., "Physical Modelling of Littoral Processes," In: R.A. Dalrymple, Physical Modelling in Coastal Engineering, Univ. of Delaware, Newark, DE, 1985, pp. 119-139.

Kraus, N.C., "The Effects of Seawalls on the Beach: A Literature Review," Coastal Sediments '87, ASCE, Vol. 1, 1987, pp. 945-960.

Kriebel, D.L., Dally, W.R., and Dean, R.G., "Beach Profile Response Following Severe Erosion Events," UF/COEL-86/016, Coastal & Oceanographic Engineering Department, University of Florida, Gainesville, FL, December, 1986.

Saville, T., "Scale Effects in Two-Dimensional Beach Studies," Proceedings of the Seventh General Meeting of the International Association of Hydraulic Research, LNEC, Lisbon, Portugal, Vol. II, 1957, pp. A3.1-A3.8.

Winant, C.D., Inman, D.L., and Nordstrom, C.E., "Description of Seasonal Beach Changes Using Empirical Eigenfunctions," Journal of Geophysical Research, AGU, Vol. 80, No. 15, May, 1975, pp. 1979-1984.

CHAPTER 112

BED RESPONSE TO FAIRWEATHER AND STORM FLOW ON THE SHOREFACE

Malcolm O. Green[1], John D. Boon[2], Jeffrey H. List[2]

and L. Don Wright[2]

Abstract

Bailard's (1981) model of combined-flow, total-load sediment transport was used to calculate sediment flux at 8-m depth on a wave-dominated shoreface during fairweather and during a storm. Waves were skewed onshore during fairweather, however it was the reversing tidal current that controlled predicted transport direction. Transport direction during the early phase of the storm was also controlled by the mean flow, this time a wind-driven jet-like flow with an offshore component. As the storm progressed, waves became more organized and highly skewed, and by the end of the storm, predicted sediment transport was turned onshore by the shoreward-skewed waves against the mean flow. Measurements of changes in relative bed elevation at the 8-m depth site were used to verify the transport predictions. A total of 6 cm of accretion occurred over 4.5 days of low-energy flow. It was found that predicted onshore transport was strongly correlated with erosion at the 8-m depth site, and predicted offshore transport was strongly correlated with accretion. Five cm of scour occurred during the initial phase of the storm, followed by 15 cm of rapid accretion. Onset of accretion was coincident with the organization of surface waves into long-period swell, and the maximum accretion rate was coincident with the most highly-skewed waves.

----

[1]Department of Earth Sciences, University of Cambridge, Downing Street, Cambridge, CB2 3EQ, United Kingdom.

[2]Virginia Institute of Marine Science, School of Marine Science, College of William and Mary, Gloucester Point, VA 23062, U.S.A.

Introduction

Direct measurement of sediment transport by combined wave and current flow has not kept pace with measurement of the velocity fields that drive transport. Consequently, although the properties of the velocity field that are theoretically germane to sediment transport are generally recognized, and in some cases well understood, models that link bed response to flow properties remain largely untested.

Properties of the combined flow relevant to sediment transport include, in one form or another, the mean, variance, and skewness. The mean flow advects sediment suspended in the water column; the variance is a measure of the kinetic energy available for entraining sediment from the bed; and velocity skewness translates into net transport of that component of the sediment load that fluctuates with the instantaneous velocity. Rigorous extensions of Bagnold's (1963) unidirectional transport model by Bowen (1980) and Bailard (1981) have given specific form to the relevant flow properties and subsequently stimulated much research. Bailard (1981) gave the following equation for total-load (sum of bedload and suspended load) time-averaged sediment transport:

$$\langle \mathbf{i}_t \rangle = (\rho C_f \varepsilon_b / \tan\phi) \, [\langle \mathbf{U}_t^3 \rangle + (\tan\beta/\tan\phi) \langle |\mathbf{U}_t|^3 \rangle \hat{\mathbf{i}}]$$

$$+ (\rho C_f \varepsilon_s / w) \, [\langle |\mathbf{U}_t|^3 \mathbf{U}_t \rangle + (\varepsilon_s \tan\beta / w) \langle |\mathbf{U}_t|^5 \rangle \hat{\mathbf{i}}] \quad \ldots(1)$$

where $\mathbf{U}_t$ is the instantaneous velocity, β is the local bed slope, ϕ is the coefficient of internal friction of the bed, w is the sediment settling velocity, ρ is the fluid density, C_f is a drag coefficient, ε_b and ε_s are the suspended-load and bedload efficiencies respectively, and $\langle \mathbf{i}_t \rangle$ is an immersed-weight transport rate (units of weight per time per unit width, i.e. mass/time^3). The angle brackets denote time-averaging, and boldface characters indicate vector quantities. $\hat{\mathbf{i}}$ is the unit vector in the cross-slope direction, positive downslope, and for the purposes of the following analysis, the unit vector $\hat{\mathbf{i}}$ also corresponds to the shore-normal direction (positive offshore).

The two terms in equation (1) that involve time-averages of odd or signed powers of the instantaneous velocity, i.e. $\langle U_t^3 \rangle$ in the bedload term and $\langle |U_t|^3 U_t \rangle$ in the suspended-load term, reflect the skewness of the velocity distribution and denote the bedload and suspended-load fluxes that are driven by and parallel to the instantaneous velocity. Both terms are identically zero for a symmetrical, zero-mean velocity distribution. The two terms that involve time-averages of powers of absolute values of the instantaneous velocity, i.e. $\langle |U_t|^3 \rangle$ in the bedload term and $\langle |U_t|^5 \rangle$ in the suspended-load term, reflect the flow energy and are considered to govern the bedload and suspended-load fluxes that fall downhill under the influence of gravity. Unlike the former terms, both of these terms are non-zero for a symmetrical, zero-mean velocity distribution.

Guza and Thornton (1985) decomposed the total velocity into mean ($\bar{\ }$) and oscillatory ($\tilde{\ }$) components:

$$\mathbf{U}_t = (\tilde{u} + \bar{u})\hat{\mathbf{i}} + (\tilde{v} + \bar{v})\hat{\mathbf{j}} \quad \text{.............................(2)}$$

where $\hat{\mathbf{j}}$ is the unit vector in the alongslope direction, and showed how each component theoretically contributes to each of the terms in the transport equation. The terms involving velocities in equation (1) are frequently referred to as "velocity moments" although this is technically inaccurate, firstly because they are not dimensionless, and secondly, and more importantly, because they are not computed from a zero-mean velocity distribution. The mean flow skews the velocity distribution and adds kinetic energy; the mean flow cannot be presumed **a priori** to be subordinate, in terms of sediment transport, to the oscillating component of the total velocity.

The objective of this paper is to test the theoretical links between the properties of the velocity distribution and sediment transport as expressed in equation (1). The data set, obtained from 8-m depth on a wave-dominated, microtidal shoreface, consists of simultaneous measurements of near-bed velocities and changes in relative bed elevation at a single point. Sediment flux thus was not measured directly, however the record of local sedimentation is useful for the stated objective if it is viewed as the product, and therefore the close analog, of sediment transport.

Observed velocity distributions are described, and the contributions made by the mean flow and oscillating flow to each of the velocity "moments" in the transport equation are compared. Equation (1) is used to calculate sediment transport, and the transport predictions are tested by seeking purely qualitative correlations between terms in the transport equation and observed changes in relative bed elevation.

Methods

Observations were made over a period of 8 days from 6 to 14 September, 1985, at 8-m depth on the shoreface that fronts the barrier-island coast at Duck, North Carolina, on the U.S. Atlantic coast (Figure 1). The experiment was one component of a larger nearshore processes experiment, DUCK85, which was conducted at the U.S. Army Corps of Engineers' Field Research Facility; details of the DUCK85 experiment can be found in Mason et al. (1987).

Currents were measured at 20 cm above the bed with a 4.0-cm diameter Marsh-McBirney electromagnetic current meter supported on a bottom-mounted tripod and controlled by a Sea-Data Model 635-9RS logger. Currents were sampled at 1 s intervals for 2048 s every 2 hours . Relative bed elevation was measured using a Datasonics Model ASA-920 digital sonar altimeter (DSA) mounted 4 m from the main tripod on a platform supported by pipes driven 50 cm into the seabed. The DSA is a high-resolution (0.15 cm) sonar device that converts reflected travel time of a high-frequency (1 MHz) acoustic pulse into a digital number that is proportional to the distance between the transducer and the seabed. Green and Boon (1988) described the calibration of the unit, its zero stability, and its performance over a rippled bed and when suspended sediment concentration is high, and concluded that the altimeter performed reliably and predictably in the field. The DSA was polled every 8 s by the Sea-Data logger, for a total of 256 observations per 2048-s burst.

Green (1987), drawing from bathymetric and side-scan sonar data, demonstrated that the Duck shoreface exhibits scant longshore variability, and is well represented by a simple two-dimensional concave-upwards profile. If it is assumed that sediment transport along the straight and parallel contours of the Duck shoreface is uniform, then it is the cross-shore component of the instantaneous velocity that controls erosion and sedimentation seaward of the surfzone. Thus, the longshore component of velocity can be neglected; for the purposes of the following analysis it is assumed that $\mathbf{U}_t = (\tilde{u} + \bar{u})\hat{\mathbf{i}}$, and all sediment fluxes are thus cross-shore fluxes.

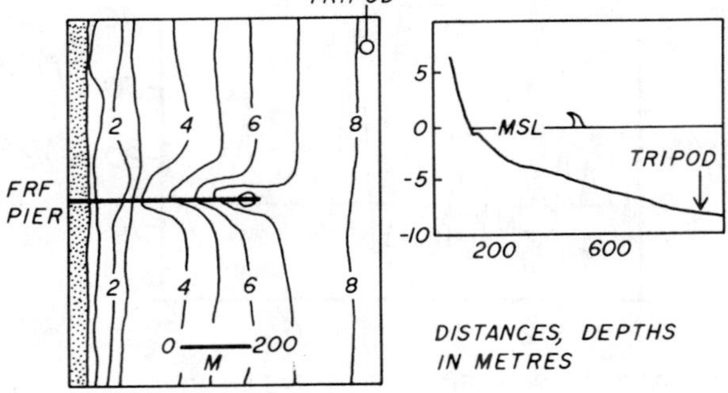

Figure 1. Experiments were conducted at 8-m depth seaward of the surfzone. The Field Research Facility (FRF) is situated in the southern Mid-Atlantic Bight, between Cape Hatteras to the south and Chesapeake Bay to the north. Bathymetry based on surveys by personnel of the FRF.

Predicted Sediment Transport

During the first 4.5 days of the deployment, waves were low (<60 cm) and long (10 s) and the wind light (<6 m/s) and directed offshore. A shift of the wind to the northeast (onshore) at 0700 hours on September 11 signalled the end of fairweather conditions; for the next 3.5 days, the wind blew from the northeast at speeds generally in excess of 10 m/s. See Wright et al. (1986) for a detailed description of the evolution of the storm.

Prior to the storm, the burst standard deviation of the cross-shore component of velocity measured at z = 20 cm averaged 10 cm/s (Figure 2), and most of the variance was due to surface gravity waves: infragravity band (0.01 to 0.05 Hz) energy in the velocity record was two orders of magnitude smaller than energy in the incident band (0.05 to 0.5 Hz). The burst-averaged cross-shore current attained maximum speeds of ~5 cm/s, and the sign of $\bar{u}$ reversed at approximately the semi-diurnal frequency (Figure 2), indicating a strong contribution by the astronomical tide.

Current energy in the incident and infragravity bands increased through the storm at approximately equal rates. The burst standard deviation of the cross-shore velocity reached a peak of nearly 40 cm/s 2.2 days after the windshift (Figure 2), at which time the significant wave height was 2.10 m and individual maxima in the cross-shore velocity exceeded 150 cm/s. The mean flow during the storm resembled a jet-like coastal flow, (Wright et al., 1986), setting alongshore to the south at an average of 33 cm/s and offshore at a maximum of 10 cm/s.

Those terms in equation (1) that denote sediment flux parallel to the instantaneous velocity are governed themselves by both the mean and skewness of the velocity distribution. Shown in Figure 3 are observed values of

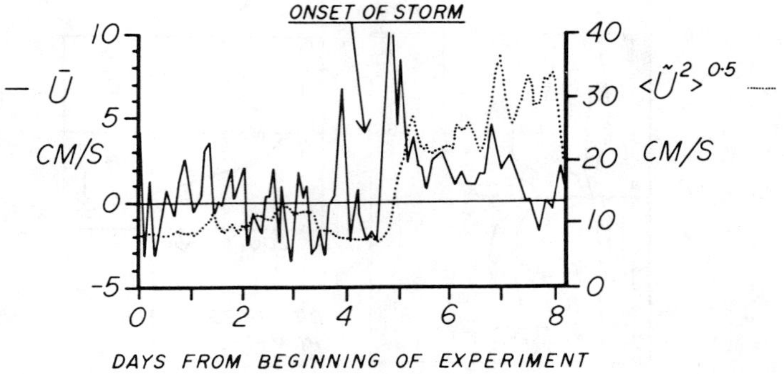

Figure 2. Burst mean cross-shore current speed ($\bar{u}$) and burst standard deviation of the cross-shore component of velocity ($<\tilde{u}^2>^{0.5}$), both measured at z = 20 cm, 8-m depth. The onset of the northeaster storm is indicated.

$\langle \tilde{u}^3 \rangle$, which reflects the wave-orbital velocity asymmetry only since the mean flow is removed; and $\langle U_t^3 \rangle$, which reflects the asymmetry of the total flow. The averaging period corresponds to the burst duration. Observed values of $\langle \tilde{u}^3 \rangle$ were negative throughout the fairweather period, which indicates a higher than Gaussian occurrence of large negative (i.e. onshore) velocities and which is consistent with the expectation from theory of relatively long (10 s) waves in relatively shallow water (~10 m) being skewed in the direction of propagation. In contrast, during the early stages of the storm there was no consistent sign to the wave skewness, which reflects the stage of developing wind-waves at that time with surface waves driven locally before the strong northeast wind. Approximately 1.3 days after the onset of the storm, $\langle \tilde{u}^3 \rangle$ again became consistently negative, and thereafter increased in magnitude. Towards the end of the experiment, 3.5 days from the onset of the storm, $\langle \tilde{u}^3 \rangle$ attained its largest negative value, reflecting the marked asymmetry of the organized long-period and large-amplitude swell that was the legacy of the storm.

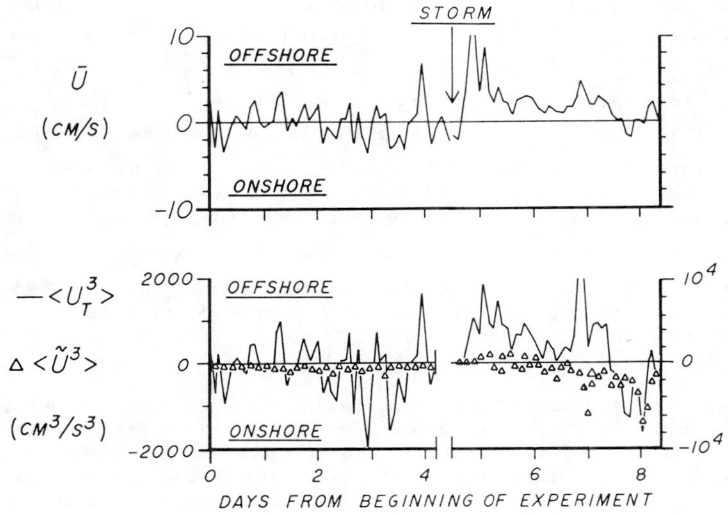

Figure 3. Wave-orbital velocity skewness, $\langle \tilde{u}^3 \rangle$, and total velocity skewness, $\langle U_t^3 \rangle$, measured at the 8-m depth site, z = 20 cm. Horizontal axis is time in days from the beginning of the experiment; note the break of scale at 4.5 days which corresponds to the onset of the storm. The vertical scale on the left refers to the fairweather period, and the vertical scale on the right to the storm period. Also shown is the mean cross-shore current speed flow at z = 20 cm.

The oscillatory velocities, however, did not usually make the major contribution to the skewness of the **total** flow. Except at the end of the storm, when wave-orbital velocities were most highly skewed, the sign of the total-flow skewness (i.e. $\langle U_t^3 \rangle$) was the same as the sign of the mean cross-shore flow, regardless of the wave asymmetry (Figure 3). Thus, the skewness of the observed velocity distributions was controlled by the tidal mean flow during fairweather, by the wind-driven mean flow for much of the storm, and by the oscillatory component of the total flow only towards the end of the storm. The higher-order signed velocity moment that appears in the suspended-flux term (i.e. $\langle |U_t|^3 U_t \rangle$) behaved in essentially the same fashion as the lower-order moment, and need not be examined further here.

In contrast, the mean flow was less important to those velocity terms that reflect the kinetic energy of the flow and that govern the downhill sediment flux. The third-order zero-mean term, $\langle |\tilde{u}|^3 \rangle$, averaged $1.56\langle \tilde{u}^2 \rangle^{3/2}$ and $1.59\langle \tilde{u}^2 \rangle^{3/2}$ over the fairweather and storm periods respectively, where the theoretical value is $1.20\langle \tilde{u}^2 \rangle^{3/2}$ assuming a monochromatic sea and $1.60\langle \tilde{u}^2 \rangle^{3/2}$ assuming a linear random sea (Guza and Thornton, 1985). The fifth-order zero-mean term, $\langle |\tilde{u}|^5 \rangle$, averaged $5.76\langle \tilde{u}^2 \rangle^{5/2}$ and $6.26\langle \tilde{u}^2 \rangle^{5/2}$ over the same periods, where the theoretical value is $1.92\langle \tilde{u}^2 \rangle^{5/2}$ and $6.38\langle \tilde{u}^2 \rangle^{5/2}$ assuming monochromatic and linear random seas respectively. Over the fairweather period, the mean flow increased the value of the third-order zero-mean term by 8% on average (i.e. $\langle |U_t|^3 \rangle / \langle |\tilde{u}|^3 \rangle = 1.08$) and the fifth-order zero-mean term by 13% on average. During the storm, 10% and 19% on average were added to the third- and fifth-order zero-mean terms respectively by the mean flow.

Sediment flux at the 8-m depth site was predicted from the velocity data using equation (1) and averaging over the burst length (34 minutes). The predicted burst-averaged bedload flux, suspended-load flux, and total-load flux are shown in Figure 4 for both the fairweather and storm periods. All fluxes are cross-shore components and the following constants were assumed: $C_f = 0.003$ (Sternberg, 1972), adjusted to apply to the velocity at z = 20cm; $\varepsilon_b = 0.21$, $\varepsilon_s = 0.025$ (Bailard, 1981); w = 1.5 cm/s (determined in the laboratory); $\tan \phi = 0.63$ (Guza and Thornton, 1985); and $\tan \beta = 0.0055$ (measured).

During the fairweather period, the bedload and suspended-load components of the predicted total load were approximately equal in magnitude, while during the storm, suspended load accounted for approximately three-quarters of the predicted total load (Figure 4). Since the factor $\rho C_f \varepsilon_b / \tan\phi$ is three orders of magnitude greater than the factor $\rho C_f \varepsilon_b \tan\beta / \tan^2\phi$, for the gravity component of the bedload flux to be comparable in magnitude to the velocity-driven component, the total velocity must be both energetic ($\langle|\mathbf{U}_t|^3\rangle$ large) and nearly symmetrical about zero ($\langle\mathbf{U}_t^3\rangle$ small), which was generally never the case in either fairweather or storm conditions. Similarly, $\rho C_f \varepsilon_s / w$ is four orders of magnitude greater than $\rho C_f \varepsilon_s^2 \tan\beta / w^2$, and thus the gravity component of the suspended-load flux was subordinate to the velocity-driven component. Thus, during fairweather, since the mean flow controlled the total-flow skewness and the sediment flux was predominantly driven by the instantaneous velocity, the cross-shore components of the predicted bedload and suspended-load fluxes reversed sign in phase with the mean (tidal) flow (Figure 4). During the early stage of the storm, when the total-flow skewness was controlled by the wind-driven mean flow, predicted net sediment flux was directed offshore, and during the later stage of the storm, when the total-flow skewness was controlled by the wave asymmetry, predicted sediment flux was directed onshore (Figure 4).

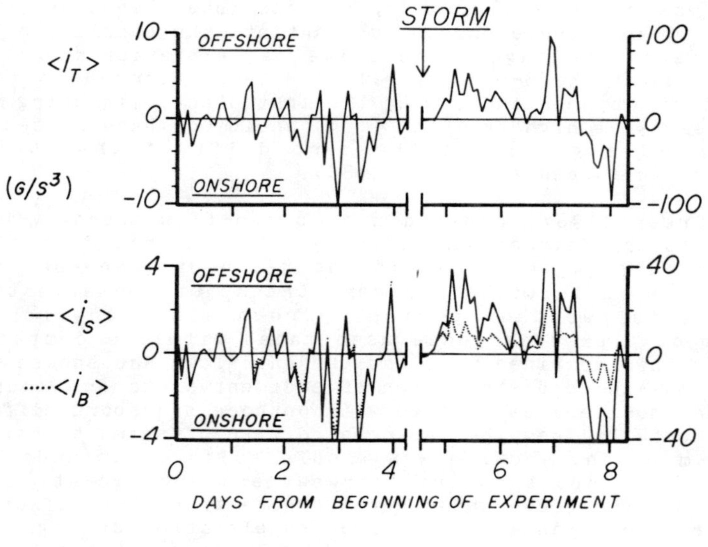

Figure 4. Predicted cross-shore total-load, suspended-load and bedload sediment flux at the 8-m depth site. The horizontal axis is time in days from the beginning of the experiment; note the break of scale at 4.5 days which corresponds to the onset of the storm. The vertical scale on the left refers to the fairweather period, and the vertical scale on the right to the storm period.

Observed Bed Response to Fairweather and Storm Flow

Fairweather

The DSA record was burst-averaged after all wild points in the data were removed. Generally, during low-energy conditions, the DSA signal was quite flat, however, when spikes did occur, they were easily recognized and removed. Green and Boon (1988) provide a discussion of the editing procedure, and show actual time-series of instantaneous relative bed elevation.

In a series of laboratory tests, Green and Boon (1988) also demonstrated that the sonic beamwidth was finite, and that positioning the instrument at a height above the bed greater than eight times the height of the wave ripples guaranteed that only the tops of the wave ripples were sensed. Since bedforms at the 8-m depth site were symmetrical wave ripples 3 cm high by 15 cm long, and the DSA was located approximately 50 cm above the bed, the DSA record should therefore not be confounded by the local topography.

The burst-averaged relative bed elevation over the fairweather segment of the experiment, expressed as cm below the fixed elevation of the transducer, is shown in Figure 5. Over the 4.5 days of low-energy preceding the storm, a total decrease in relative bed elevation of nearly 6 cm was recorded. Assuming the absolute elevation of the DSA is fixed, three points substantiate the conclusion that the observed change in relative bed elevation represents accretion at the 8-m depth site: divers observed no local bed disturbances induced by the instrument; migrating wave ripples (3 cm high by 15 cm long) could not have caused the observed signal; and the zero drift of the DSA was negligible (Green and Boon, 1988).

Green (1987) applied a combined-flow boundary-layer model to the fairweather velocity data and showed that the combined-flow skin friction at 8-m depth exceeded that necessary to initiate sediment transport for almost the entire fairweather period. Green also modelled the sediment-flux divergence associated with the competent fairweather combined wave and current flow, and showed that the cross-shore distribution of sediment sinks and sources on the shoreface associated with onshore transport differed fundamentally from that associated with offshore transport. At 8-m depth, erosion was predicted to occur under net onshore transport during fairweather, and accretion was predicted to occur under net offshore transport. Thus, if the observed change in relative bed elevation during times of predicted onshore transport differs consistently from that observed during times of predicted offshore transport, then this would be at least qualitative verification of the transport predictions provided by equation (1).

Inspection of the DSA record in Figure 5 reveals a connection between predicted transport direction and

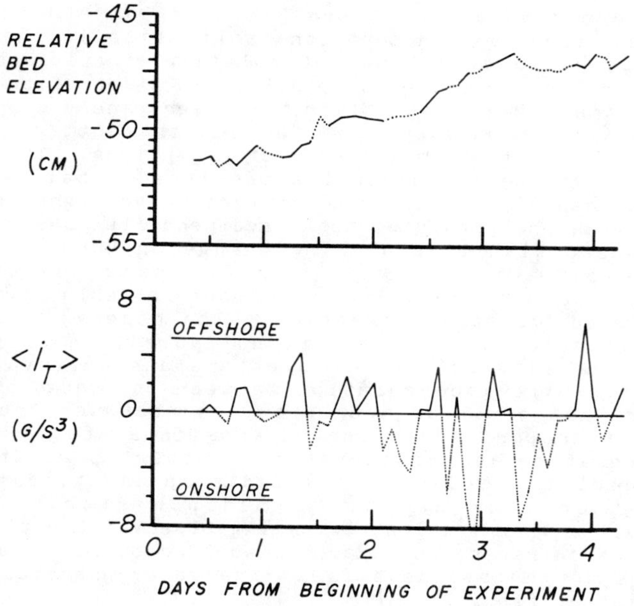

Figure 5. Burst-averaged relative bed elevation, expressed as cm below the fixed elevation of the DSA transducer, and predicted total-load sediment flux at 8-m depth during the fairweather period. Both records are marked with a solid line when the predicted transport was offshore and with a dotted line when the predicted transport was onshore.

PREDICTED TRANSPORT DIRECTION

	Offshore	Onshore
	+0.00388	-0.00218
	+0.00221	-0.00267
dž/dt	+0.00083	+0.00094
(cm/min)	+0.00017	+0.00154
	+0.00375	-0.00065
	+0.00204	-0.00267
	+0.00171	-0.00592
	+0.00317	

Table 1. Observed time-rate-of-change of relative bed elevation at 8-m depth during fairweather, calculated over consecutive bursts in which predicted transport direction (onshore or offshore) remained constant.

observed bed response at the 8-m depth site: times of predicted offshore transport (the solid line in the lower panel) correspond to times of relatively rapid accretion (the solid line in the upper panel), and times of predicted onshore transport (dotted line in lower panel) correspond to erosion or relatively slower accretion (dotted line in upper panel), with only a couple of exceptions. There were seven separate periods when the predicted total sediment flux was onshore over consecutive bursts and eight separate periods when the predicted total sediment flux was offshore over consecutive bursts. The change in burst-averaged relative bed elevation over each of these periods was calculated and these values are shown in Table 1, grouped by predicted transport direction. The observed rates of change in relative bed elevation (Table 1 and Figure 5) confirm the link between predictions and observations: there is a strong correlation between predicted offshore transport and observed accretion, and between predicted onshore transport and observed erosion at the 8-m depth site. Equation (1) is therefore a useful predictor, at least qualitatively, of the bed response to competent fairweather flow at this 8-m depth site. A more quantitative verification, comparing observed and predicted sedimentation rates, for example, would require a detailed analysis of the sediment-flux divergence associated with the competent flow.

The correlation between mean flow, which was driving transport direction at 8-m depth, and bed response highlights a fundamental difficulty with equation (1). Although the vertical variation of the mean flow in a combined-flow boundary layer is known theoretically (e.g. Grant and Madsen, 1979), it is not at all clear at which elevation the mean flow must be measured for use in the transport model. This problem precludes any further analysis of transport rates here, even though transport directions predicted on the basis of the velocity measured at $z = 20$ cm appear justified. Green (1987) circumvented this difficulty by using a traction model of transport, in which the combined-flow bed shear stress drives the sediment flux. The importance of the mean flow also emphasizes a potential deficiency of the electromagnetic current meter, which is presently in widespread use. Although Bowen and Doering (1984) found that the measurement of pure-wave skewness using electromagnetic current meters was rather more consistent than expected, and indeed the skewness measurements reported herein appear to be sensible, recent detailed laboratory tests by Aubrey and Trowbridge (1985) have shown that it may be difficult to obtain accurate estimates of the mean flow using these instruments.

Storm

The DSA data obtained during the storm were significantly noisier than the fairweather data; high concentrations of suspended sediment in the line-of-sight of the transducer caused premature echoes and signal scattering that occasionally resulted in complete loss of echo detection (Green and Boon, 1988). Even after removal of the primary spikes, significant noise remained in the record, which may have biased the estimates of burst-averaged relative bed elevation. Therefore, only gross trends can be examined with confidence. The major features of the storm record were: a negligible change in bed level in response to the initial impulse of the wind; ~5 cm of bed scour during the mid-phase of the storm; and ~15 cm of rapid accretion during the waning phase of the event (Figure 6).

A significant feature of the storm record is the change from erosion to rapid accretion of the bed that occurred approximately two days after the onset of the storm. This corresponded to a sudden increase in the magnitude of the predicted total-load transport (Figure 6), however there was no fundamental change in the sedimentation rate corresponding to the predicted change in transport direction shortly afterwards. The initiation of accretion was also roughly coincident with the beginning of the period of wave organization following the initial phase of the storm, as reflected by the wave skewness (Figure 6); furthermore, as the wave-orbital velocities became increasingly skewed, the accretion rate increased, culminating in a maximum in both towards the end of the experiment.

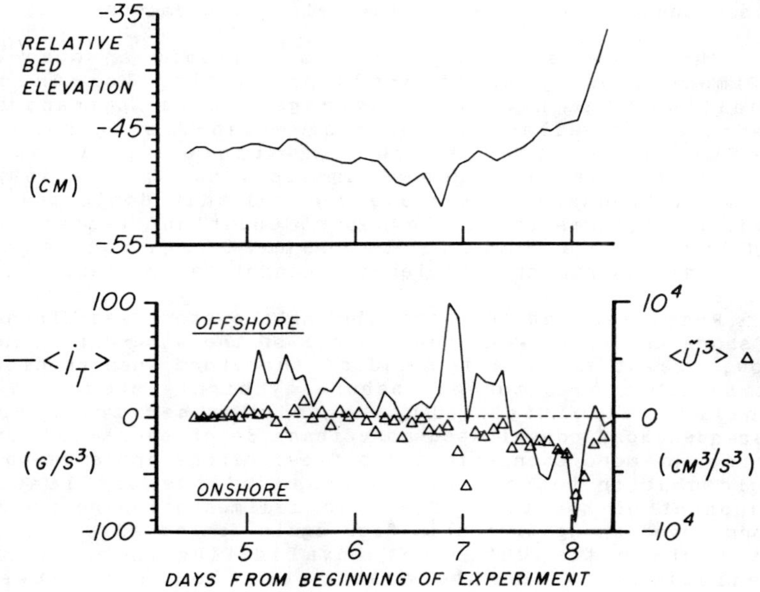

Figure 6. Burst-averaged relative bed elevation, expressed as cm below the fixed elevation of the transducer, predicted total-load sediment flux, and wave-orbital velocity skewness at 8-m depth during the storm.

Storm sedimentation thus differed fundamentally from fairweather sedimentation, and equation (1) failed as an indicator of bed response under the storm conditions. This test, however, does not necessarily constitute invalidation of the transport model; it may be that there is a complex relationship between sedimentation (measured by the sonar altimeter) and sediment transport (predicted by equation 1) during high-energy conditions. For example, it appeared that the bed responds primarily to the oscillatory component of the combined flow. Another possibility is that the measured accretion reflects the migration of a large-scale bedform under the sensor; Wright et al. (1986) argued that the rapidity of the accretion was suggestive of a pulse-like migration of a sediment front or large-scale bedform over the installation site, and showed side-scan sonar imagery to support that hypothesis.

Conclusions

During fairweather at the 8-m depth site, the predicted cross-shore burst-averaged bedload and suspended-load fluxes were approximately equal in magnitude and, since the near-bed velocity was both highly skewed and not very energetic, the predicted downslope flux of sediment under the influence of gravity was relatively insignificant. The mean flow dominated the asymmetry of the observed velocity distributions, indicating that transport direction was controlled by the tidal mean flow rather than by the wave-orbital velocity asymmetry.

There was a strong correlation between observed sedimentation rates at 8-m depth during low-energy conditions and net burst-averaged sediment transport predicted by Bailard's (1981) combined-flow transport model. This correlation constitutes qualitative verification of the transport predictions. Since it was the mean flow at 20 cm above the bed that dominated the significant terms in the transport equation, fundamental difficulties in the transport equation and potential instrument deficiencies cannot be ignored.

Predicted net transport during the storm was directed offshore under the dominant control of the wind-driven mean flow, except towards the end of the storm when predicted net transport was turned onshore by highly-skewed wave-orbital velocities. Unlike fairweather transport, suspended load constituted more than 75% of the total load during the more energetic storm flow. There appeared to be a correlation between the skewness of the oscillatory component of the total flow and sedimentation during the storm, even though the mean flow again dominated predicted net transport. While not invalidating the transport predictions, since the exact relationship between sedimentation and sediment transport is unknown, the oscillatory flow may control storm sedimentation, at least during certain phases of the storm.

Acknowledgements

This study was supported by the Virginia Institute of Marine Science and the National Science Foundation, Grant No. OCE-8610635. We thank Bob Gammisch for expert assistance in the field, and Captain Durand Ward of the R/V Captain John Smith. Thanks are also due to Curt Mason and Bill Birkemeier of the U.S. Army Corps of Engineers' Field Research Facility. Contribution No. 1459 from the Virginia Institute of Marine Science.

References

Aubrey, D.G. and Trowbridge, J.H., 1985. Kinematic and dynamic estimates from electromagnetic current meter data. J. Geophys. Res., 90: 9137-9146.

Bagnold, R.A., 1963. Mechanics of marine sedimentation. In: M.N. Hill (Editor), The Sea, Vol. 3. Wiley-Interscience, New York, pp. 507-528.

Bailard, J.A., 1981. An energetics total load sediment transport model for a plane sloping beach. J. Geophys. Res., 86: 10,938-10,954.

Bowen, A.J., 1980. Simple models of nearshore sedimentation; beach profiles and longshore bars. In: S.B. McCann (Editor), The Coastline of Canada. Paper 80-10, Geol. Surv. of Canada. p. 1-11.

Bowen, A.J. and Doering, J.C., 1984. Nearshore sediment transport: estimates from detailed measurements of the nearshore velocity field. Proc. 19th Int. Coastal Eng. Conf., Houston, A.S.C.E., pp. 1703-1714.

Grant, W.D. and Madsen, O.S., 1979. Combined wave and current interaction with a rough bottom. J. Geophys. Res., 84: 1779-1808

Green, M.O., 1987. Low-energy bedload transport by combined wave and current flow on a southern Mid-Atlantic Bight shoreface. Unpublished Ph.D. Dissertation, College of William and Mary, Williamsburg, Virginia, 160 pp.

Green, M.O. and Boon, J.D., 1988. Response characteristics of a short-range, high-resolution, digital sonar altimeter. Mar. Geol., in press.

Guza, R.T. and Thornton, E.B., 1985. Velocity moments in nearshore. Proc. A.S.C.E., J. Waterway, Port, Coastal and Ocean Eng., 111: 235-256.

Mason, C., Birkemeier, W.A. and Howd, P.A., 1987. Overview of DUCK85 nearshore processes experiment. Coastal Sediments '87, New Orleans, pp. 818-833.

Sternberg, R.W., 1972. Predicting initial motion and bedload transport of sediment particles in the shallow-marine environment. In: D.J.P. Swift, D.B. Duane and O.H. Pilkey (Editors), Shelf Sediment Transport: Process and Pattern. Dowden, Hutchinson and Ross, Stroudsburg, Pennsylvania, pp. 6-82.

Wright, L.D., Boon, J.D., III, Green, M.O. and List, J.H., 1986. Response of the mid shoreface of the southern mid-Atlantic Bight to a "northeaster". Geo-Marine Letters, 6: 153-160.

CHAPTER 113

Phase–Locking of Modes in the Nearshore: Field Evidence

J. W. Haines*
A. J. Bowen*

Field investigations have shown that infragravity frequencies may dominate nearshore fluid motions (Bowen and Huntley, 1984). Theoretical studies have suggested that the modal structure of the infragravity field, including the degree of modal coupling, is important in generating complicated beach morphologies (Holman and Bowen, 1982) but previous analyses of field data have not provided a description of the infragravity field which is sufficiently detailed to assess the importance of these ideas.

In this paper we present a statistical method which, when applied to data from any large sensor array, provides the desired description of the infragravity field. The method is applied to data from the NSTS (Nearshore Sediment Transport Study) experiment. The variance of the observed velocity field is described by a set of free wave modes. The results indicate that a complicated infragravity field composed of both edge waves and leaky modes is present. The modes modelled are also found to be significantly coupled over a range of modal pairs and frequencies.

Introduction

Results from the analysis of extensive sets of field data have convincingly demonstrated the importance of infragravity motions in the nearshore (Bowen and Huntley, 1984). Observations show that these lower frequency motions may dominate the velocity and elevation fields at the shoreline and across the inner surfzone (Guza and Thornton, 1982). Numerical and analytical models have suggested that these infragravity waves may be important agents of morphologic evolution and may significantly modify the incident wind–wave field. Evaluation of such models requires knowledge of the detailed structure of the infragravity wave field in natural systems.

Bowen and Inman (1971) and Guza and Inman (1975) considered simple models of phase–locked edge waves progressing in opposite directions alongshore. The second–order mean drift velocity and the alongshore variations in shoreline run–up due to these standing edge waves were shown to produce zones of convergence leading to the formation of shoreline cusps and crescentic bars. More complicated infragravity wave fields were modelled by Holman and Bowen (1982). They described the mean drift due to coupled modes of variable mode number and direction of propagation. The resulting time–averaged velocity field varied both across and along the beach, providing a mechanism for the generation of complicated topography varying alongshore on a variety of scales.

These theoretical studies suggest that the coupling or phase–locking of infragravity modes is necessary for the generation of three–dimensional topography. Field evidence for the existence of coupled modes is scarce. Holman and Bowen (1984) showed that standing edge waves were present on a beach bounded by reflective headlands. More complete descriptions of the infragravity field, utilizing large sensor arrays, have failed to addresss the possible coupling of modes. In this paper we shall describe a new method of describing a large data set in terms of prescribed modes (Haines, et al., in prep.). The method returns

* Department of Oceanography, Dalhousie University, Halifax, NS B3H 4J1 Canada

the time–averaged modal amplitudes, the degree of modal coupling and the relative phase between modes. The resulting detailed picture of the infragravity wave field should provide constraints for physical models of infragravity wave generation, interaction and decay.

Theory and Background

A variety of motions in the nearshore zone are driven by the flux of momentum due to wind–generated waves incident on the beach. The response to this forcing is spread across a broad band of frequencies. The higher frequencies, harmonics of the incident waves and turbulence at the boundaries may stir up and transport the underlying sediment. Larger scale spatially coherent responses to the forcing are predominantly confined to the infragravity frequencies; i.e. those frequencies below the incident wave band. Theoretical investigations have identified a number of possible infragravity responses to various types of forcing.

Both forced and free responses may occur in the nearshore. A broad–banded (in frequency and direction) incident wave field will result in a similarly complicated forcing function. Modification of the incident field due to shoaling and breaking may significantly alter the forcing (Symonds *et al.*, 1982). We are concerned here with the free wave response of the system as described by the linearised shallow–water equations of motion. These solutions include both trapped modes confined to the nearshore and leaky modes which may propagate into and out of the region.

The trapped modes, or edge waves, are a discrete set of modes described by the eigensolutions of the shallow–water equations (Eckart, 1951). For a planar beach with slope $\tan\beta$ the velocity potential and dispersion relation are given by

$$\Phi_n = \Re\{g\sigma^{-1}e^{-l_n x}L_n(2l_n x)\alpha_n e^{i(l_n y+\sigma t)}\} \tag{1}$$

and

$$\sigma^2 = gl_n(2n+1)\tan\beta. \tag{2}$$

In the above equations

Φ is the velocity potential,
σ is the radial frequency,
l_n is the wave number,
n is the mode number,
L_n is the Laguerre polynomial of order n,
α_n is the complex modal amplitude,
(x, y) are the offshore/alongshore coordinates and $\Re$ denotes the real part.

These shallow–water solutions are valid for low modes where $(2n+1)\tan\beta \ll 1$ and are appropriate for the modes and beach slope to be discussed further (Guza and Davis, 1974).

The leaky modes, including incident and reflected waves, form a continuum of modes in (σ, k) space. The velocity potential and deepwater dispersion relation are given approximately by

$$\begin{aligned}\Phi_i &= \Re\{g\sigma^{-1}(J_0(x), iY_0(x))\alpha_i e^{i(ly+\sigma t)}\},\\ \Phi_r &= \Re\{g\sigma^{-1}(J_0(x), -iY_0(x))\alpha_r e^{i(ly+\sigma t)}\},\end{aligned} \tag{3}$$

and

$$\sigma^2 = g|K|\tanh|K|h, \tag{4}$$

where J_0, Y_0 are the zeroth order Bessel functions, subscripts (i,r) denote incident and reflected waves, h is the water depth, and K is the vector wavenumber. These solutions are valid when $\sin\delta\tan\beta$ is small, where δ is the angle of incidence (Guza and Bowen, 1975).

The higher mode edge waves and the leaky modes are quite similar in structure. This, along with the large number of possible modes, complicates observational studies seeking to identify modes. An array capable of separating motions due to many similar modes would be prohibitively large. In addition the modes are likely to be coupled. The modes may arise from a common forcing, or from coherent forcing functions, and the total response must satisfy external boundary conditions. Thus it is likely that the modes are coherent or phase–locked, i.e. $\langle \alpha_n \alpha_m^* \rangle \neq 0$ (where * denotes the complex conjugate and $\langle \ \rangle$ denotes time averaging).

The modal structure of the infragravity field and the degree of modal coupling are important parameters in models of generation of beach morphology (Holman and Bowen, 1982). A detailed description of the modal response to forcing will also contribute to improved understanding of the dynamics of the nearshore. The large number of possible modes and the potential for coupled modes requires special treatment of observations made in the nearshore. We shall describe and apply a method which utilizes observations from a variety of sensor types and locations, and which explicitly considers modal coupling. Additionally the method may be used to address the relationship between the estimated modal response and observed forcing variables.

Previous investigators have shown that the bulk of observed infragravity energy may be described by free wave modes. Oltman–Shay and Guza (1987) applied maximum–likelihood 2–D spectral estimators to data collected on California beaches during the NSTS study. The method employed utilized only a single observed variable from sensors located along a shore parallel line. Additionally the method assumes the modes are random in phase. The results of the analysis showed that much of the observed energy fell along edge wave modal dispersion lines. The modal structure of the field, as suggested by this method, varied dramatically depending on the variable type (alongshore or cross–shore velocity) subjected to the analysis. This is consistent with the observation that higher modes (and leaky modes) have a relatively large cross–shore velocity component.

Huntley (1988) examined the same data using a qualitative method to assess the importance of modal coupling. He found strong evidence that, over the period of the observations, modes present were coupled. This coupling was associated with alongshore inhomogeneity in the observed velocity field. No quantitative assessment of the modal coupling or relative phases between modes was possible.

We shall present a method which allows simultaneous incorporation of a variety of variables from any sensor location. The method also explicitly estimates the coupling and relative phases of the modes modelled. We shall briefly discuss how measured indices of the forcing might be included to yield estimates of the gain between the forcing and the modal response as modelled.

The Method of Analysis

The observations will be modelled as the resultant of a linear superposition of modes at a single frequency, σ_0. The observations from the ith sensor, $\theta_i(t)$, are given by

$$\theta_i(t) = \alpha_j(t) f(\phi_{ij}(x,y) \exp i(\ell_j y_i + \sigma_0 t)) + \epsilon_i(t) \tag{5}$$

where i denotes the ith sensor and j the jth mode and $j_{max} < i_{max}$. $\epsilon_i(t)$ is the error term at the ith sensor which may contain unmodelled spatially coherent motions. The function $f()$ depends on the sensor location and the variable type. The model is complex and $\theta_i(t)$ may be thought of as the complex demodulate of the observed series at σ_0.

The model equation may be written in vector form as

$$\underset{\sim}{\theta} = \mathbf{E}\underset{\sim}{\alpha} + \underset{\sim}{\epsilon} \tag{6}$$

where $\mathbf{E}$ is the matrix of modal coefficients and $E_{i,j}$ defines the jth mode at the ith sensor. The cross–spectral matrix of observations, $\langle \underset{\sim}{\theta}\,\underset{\sim}{\theta}'^{*}\rangle$, (where $'$ is the transpose) at frequency σ_0 is given in terms of the cross–spectral matrix of modal amplitudes by

$$\langle \underset{\sim}{\theta}\,\underset{\sim}{\theta}^{*\prime}\rangle = \mathbf{E}\langle \underset{\sim}{\alpha}\,\underset{\sim}{\alpha}^{*\prime}\rangle\mathbf{E}^{*\prime} + \mathit{errors}. \tag{7}$$

It has been shown (Haines *et al.*, in prep.) that the least squares estimate of $\langle \underset{\sim}{\alpha}\underset{\sim}{\alpha}'^{*}\rangle$ is given by

$$\widehat{\langle \underset{\sim}{\alpha}\underset{\sim}{\alpha}^{*\prime}\rangle} = (\mathbf{E}^{*\prime}\mathbf{E})^{-1}\mathbf{E}^{*\prime}\langle \Theta\Theta^{*\prime}\rangle\mathbf{E}(\mathbf{E}^{*\prime}\mathbf{E})^{-1}. \tag{8}$$

The quantity minimised is the squared error between the predicted and observed series, summed over the entire cross–spectral matrix. This formulation guarantees that the predicted cross–spectral matrix and the cross–spectral matrix of modal amplitudes are physically realizable and so may be interpreted as cross–spectral matrices. This provides estimates of modal power, coherences and relative phases. As the solution is a linear tranformation of the cross–spectral matrix of observations, significance levels follow from standard formulae.

The method may be simply understood by solving (6) in the time domain. The least squares solution gives

$$\hat{a} = (\mathbf{E}^{*\prime}\mathbf{E})^{-1}\mathbf{E}^{*\prime}\underset{\sim}{\theta}. \tag{9}$$

Squaring and averaging, or forming $\langle \hat{a}\hat{a}^{*\prime}\rangle$, returns (8). Similarly from (9) we may define a cross–spectral matrix of residuals given by

$$\begin{aligned}\langle \underset{\sim}{\epsilon}\underset{\sim}{\epsilon}^{*\prime}\rangle = {} & \langle \underset{\sim}{\theta}\,\underset{\sim}{\theta}^{*\prime}\rangle + \mathbf{E}(\mathbf{E}^{*\prime}\mathbf{E})^{-1}\mathbf{E}^{*\prime}\langle \underset{\sim}{\theta}\,\underset{\sim}{\theta}^{*\prime}\rangle\mathbf{E}(\mathbf{E}^{*\prime}\mathbf{E})^{-1}\mathbf{E}^{*\prime} \\ & - \mathbf{E}(\mathbf{E}^{*\prime}\mathbf{E})^{-1}\mathbf{E}^{*\prime}\langle \underset{\sim}{\theta}\,\underset{\sim}{\theta}^{*\prime}\rangle - \langle \underset{\sim}{\theta}\,\underset{\sim}{\theta}^{*\prime}\rangle\mathbf{E}(\mathbf{E}^{*\prime}\mathbf{E})^{-1}\mathbf{E}^{*\prime}.\end{aligned} \tag{10}$$

This matrix is also physically realizable. Decomposition of the cross–spectral matrix of residuals will indicate whether spatially coherent motions remain which are not adequately described by the model fitted.

In practice the sensor array may be incapable of resolving the wide range of modes possible. The model must be limited to those modes which are well resolved. The degree to which modes modelled are resolved is indicated by the condition of the matrix $(\mathbf{E}'^{*}\mathbf{E})$, where the condition is the ratio of the smallest to largest eigenvalues of the matrix. This provides a useful tool for array design as the matrix may be formed, for given topography and frequency, prior to collecting data. The resolution of the array will vary with frequency as the modal shapes vary and frequency dependent models may be necessary. As a general rule of thumb the condition of the model matrix should exceed the estimated noise to signal ratio of the array.

The incorporation of a variety of sensors may necessitate weighting of the model. This will prevent single energetic observations from dominating the fit. The most straightforward approach is to weight the model by the standard deviation of the observations.

The modelling may also be extended to include forcing variables. In such a case the observed forcing time series are included as observation series and simply passed through the fitting step. The forcing series, $F(t)$, are modelled as

$$F_i(t) = E_{i,j}\alpha_n(t) + 0, \tag{11}$$

where $E_{i,j} = 1$ when $j = n$ and 0 otherwise. The result is that $F_i = \alpha_n$, and the final fitted cross–spectral matrix includes terms of the form $\langle F_i\alpha_m{}^{*}\rangle$.

Application

The method of analysis described has been applied to a single data run from the NSTS experiment. Measurements were made at Torrey Pines beach, San Diego in November of 1978. Torrey Pines is a 3 km long beach with slope approximately .02 below MSL. The incident wave field is directionally restricted by offshore islands and for the November 21 run examined 17 two-component Marsh–McBirney flowmeters were deployed. The data were sampled at $64Hz$, low-pass filtered and reduced to $2Hz$. The cross-spectral estimates have a bandwidth of .00049 Hz giving 10 degrees of freedom. The incident wave field was characterised by a peak frequency of $.067Hz$ and a significant wave height of $.86m$ (Guza and Thornton, 1985). The cross-spectral matrix is modelled as described and the results averaged across frequency to increase the degrees of freedom. This allows us to model the variation in modal shapes with frequency rather than averaging prior to fitting.

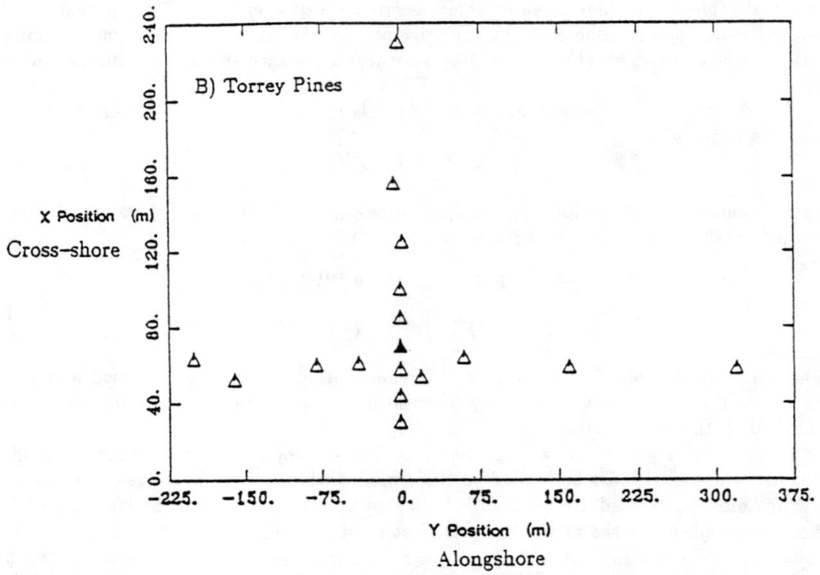

Figure 1. Configuration of flowmeters used in analysis. Symbols indicate position of Marsh–McBirney electromagnetic flowmeters. Shoreline is lower horizontal axis.

The array configuration is shown in Figure 1. The studies previously discussed (Oltman Shay and Guza, 1987, Huntley, 1988) used only the alongshore transect and treated the two velocity components separately. We simultaneously model all the available data. This is particularly helpful in avoiding the spectral gaps caused when an offshore position sees a node in a measured variable. Given the large number of modes possible, some care was taken in choosing a model, as the results of Oltman–Shay and Guza (1987) suggested a wide range of modes were present during the study. However, it is necessary to fit a restricted number of modes which will be representative of the entire field. After a number of trials we chose to fit edge wave modes 0,1,2,4 and 7 (progressive in either direction) and an incident and reflected pair. We chose to fit the same model across all frequencies in order to identify the frequency dependence of the modal field. At any particular frequency simpler models are as efficient at describing the data – but the complete model is superior across the entire frequency range.

The success of the model in describing the observations is shown in Figure 2. Across most of the range the bulk of the variance and the spectral peaks are well described. The more energetic alongshore velocities are better described than the cross–shore velocities, though the residual variance for the two sensor types is comparable. At low frequencies the model fails to describe the motions present. It is clear that no simple edge wave model will fit these motions. Eigenvector analysis of the cross–spectral matrix of residuals shows that the bulk of the unexplained variance is spatially coherent. This indicates that coherent motions incompatible with edge/leaky wave models are present.

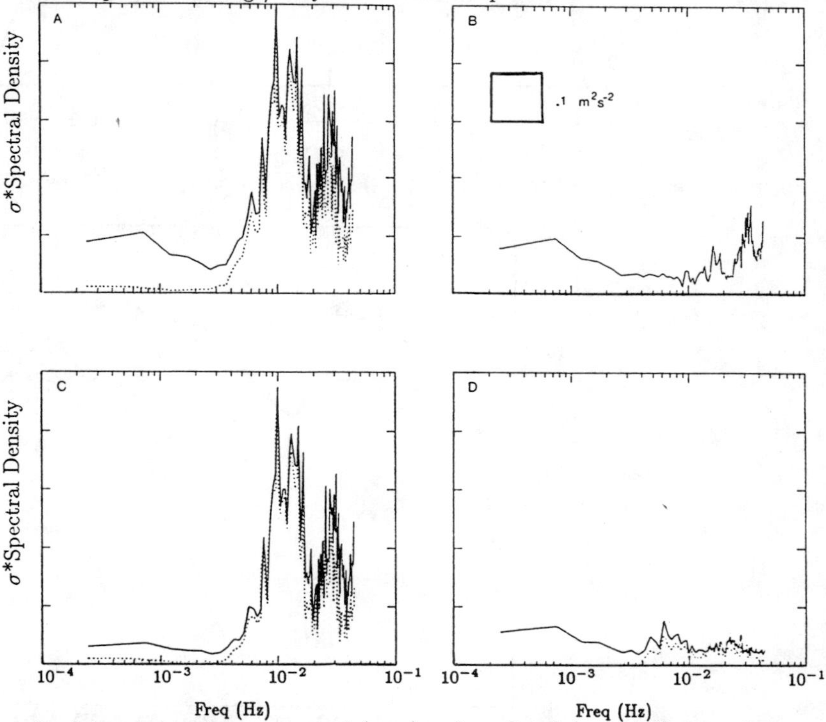

Figure 2. Integrated variance measured (——) and predicted (- - -). Variance is integrated across a) the total array, c) the cross–shore velocoity sensors and d) the alongshore velocity sensors. Panel b) shows the total residual variance.

As a rough but independent check of the results Figure 3 shows the predicted shoreline run–up for the model fitted. Measured run–up (Guza and Thornton, 1985) is also shown. The predicted run–up has been calculated by assuming that a standing wave of twice the reflected wave amplitude is seen at the shoreline in addition to the edge waves. Given the limitations of this model and the significant residual energy the agreement is encouraging. Figure 4 shows the predicted run–up for a number of alongshore positions. The substantial variation shows that the degree of modal coupling found is important. It also points out the dangers of assuming spatial homogeneity and the advantages of having shoreline data to verify and constrain modelling techniques.

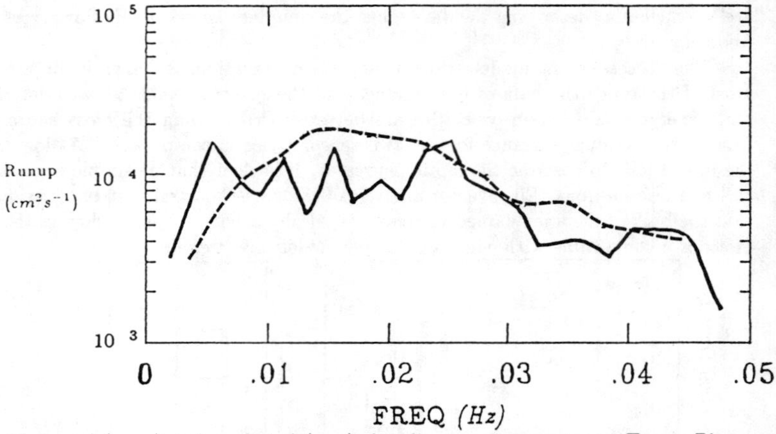

Figure 3. Measured (——) and predicted (- - -) shoreline run–up spectra at Torrey Pines.

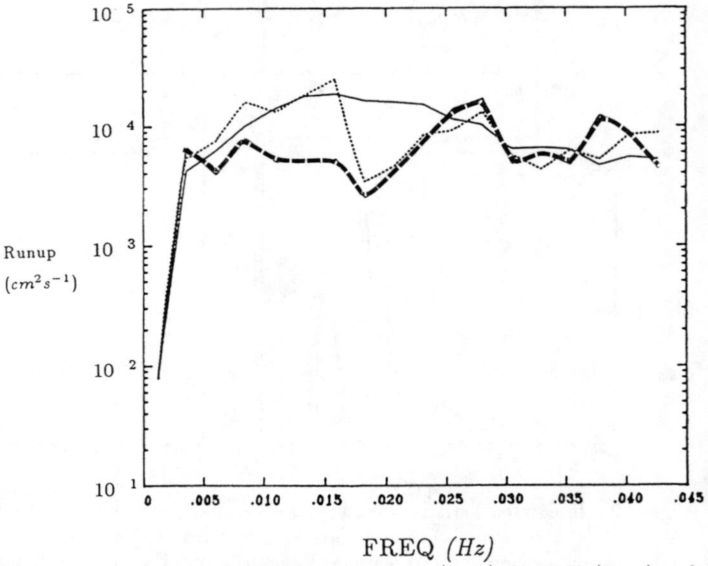

Figure 4. Predicted run–up spectra at Torrey Pines at $y = 0$ (——), $y = 250$ (· · ·) and $y = 500$ (- - -).

The predicted run–up spectra clearly suggest that the wave field is inhomogeneous. This has important implications for sediment transport and beach profile modelling. In particular it validates the modelling assumptions of Holman and Bowen (1982) that phase–locked modes are important.

The estimated modal amplitude spectra are shown in Figure 5. All modes show neglible amplitude at the lowest frequency. With the possible exception of mode zero, all modes show enhanced modal amplitudes around .015Hz. Modes higher than mode zero show a second peak between .025 and .030Hz. There is a tendency for northward progressive waves (dotted lines) to be slightly larger, the greatest differences appearing in the neighbourhood of the spectral peaks. Above .025Hz mode zero is forced to small amplitudes as it decays inshore of the array. In general the modal amplitude spectra indicate:

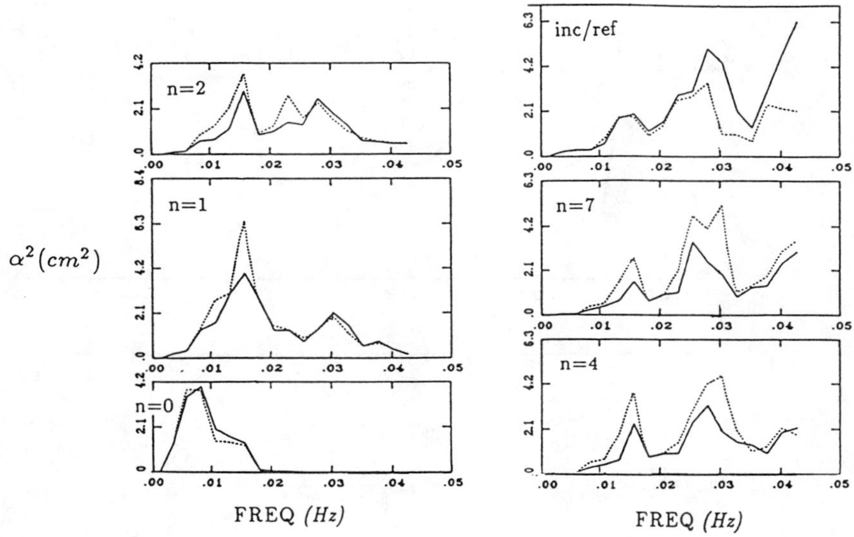

Figure 5. Shoreline amplitude spectra for modes fitted. Mode numbers are indicated in upper left hand corner of spectral plots. Solid (dashed) lines indicate positive (negative) alongshore progression for edge waves and the incident (reflected) wave in the upper right hand panel.

1) A wide range of modes, including high modes and leaky waves are present.
2) The two directions of edge wave propagation show roughly equivalent amplitudes, with a slight enhancement of the northward propagating waves.
3) There are two distinct peaks in the amplitude spectra. The peaks are present for all modes higher than zero. This frequency structure is also seen in the incident and reflected waves.
4) Modes higher than two show evidence of increasing amplitude above .035Hz.
5) The higher frequency peak, $\sim$ .025 – .030Hz, is dominated by higher modes with amplitude increasing with mode number.
6) The incident wave amplitude increases sharply at the upper frequency limit, and the incident wave is uniformly larger than the reflected wave.

Figures 6 and 7 show the gain and phase for the mode zero and incident and reflected wave pairs. Significantly non-zero gain is equivalent to significantly non-zero modal coupling. The mode zero pair shows significant coupling over a wide range of frequencies, although primarily outside the energetic region of the amplitude spectra. Phase shows no obvious σ^2 dependence which would indicate the presence of an alongshore reflector. The incident/reflected pair also shows significant coupling with the gain less than 1 and generally decreasing with frequency. This frequency dependence of the reflection coefficient has been

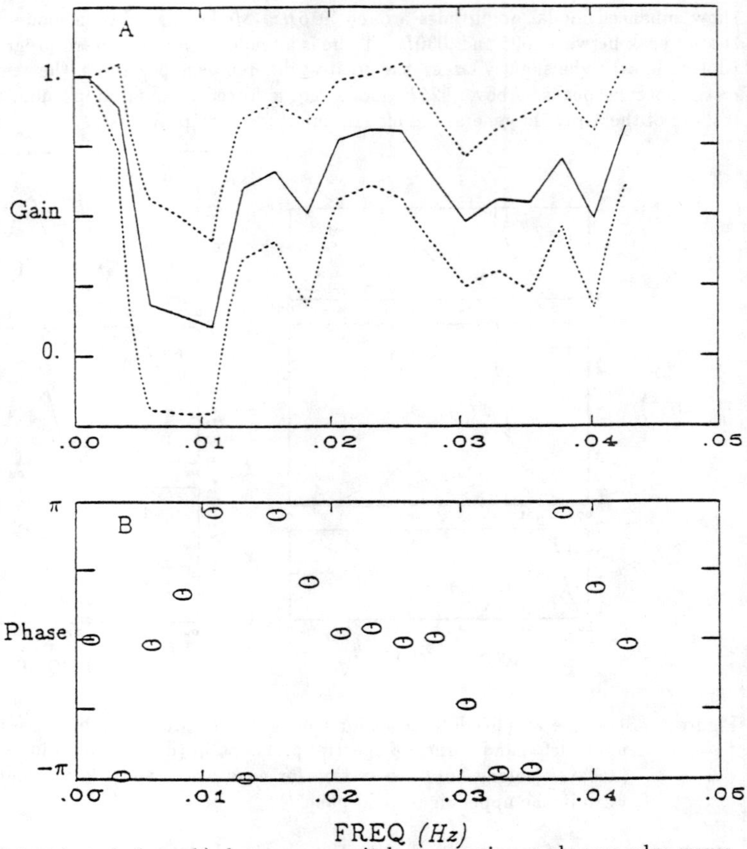

Figure 6. Gain, a), and phase, b), between oppositely progressing mode zero edge waves. Dashed lines on gain plot are upper and lower 95% confidence levels (Jenkins and Watts, 1968).

suggested by a number of authors (Elgar, 1985). The pronounced, quasi-linear trend to the phase is unexplained. Because the incident and reflected waves are largely nondispersive at these frequencies, a linear phase shift may suggest that the cross-shore origin is misplaced, i.e. reflection is occuring away from the shoreline. Determination of this reflection point would require a physical model for the process being described. Phase differences may also be attributable to nonlinearities in run-up or the effects of a porous beach face. These introduce what are presumably higher order modifications to the modelled waves.

Significant squared coherences are shown in Table 1 for all modal combinations at selected frequencies. The occurrence of significant coherences is substantially greater than that expected for a system composed of noise. Apart from a slight tendency for higher modes to be coupled to the incident wave, there appear to be no particular pairs which exhibit consistently high coupling. Significant coupling also seems unrelated to modal amplitude

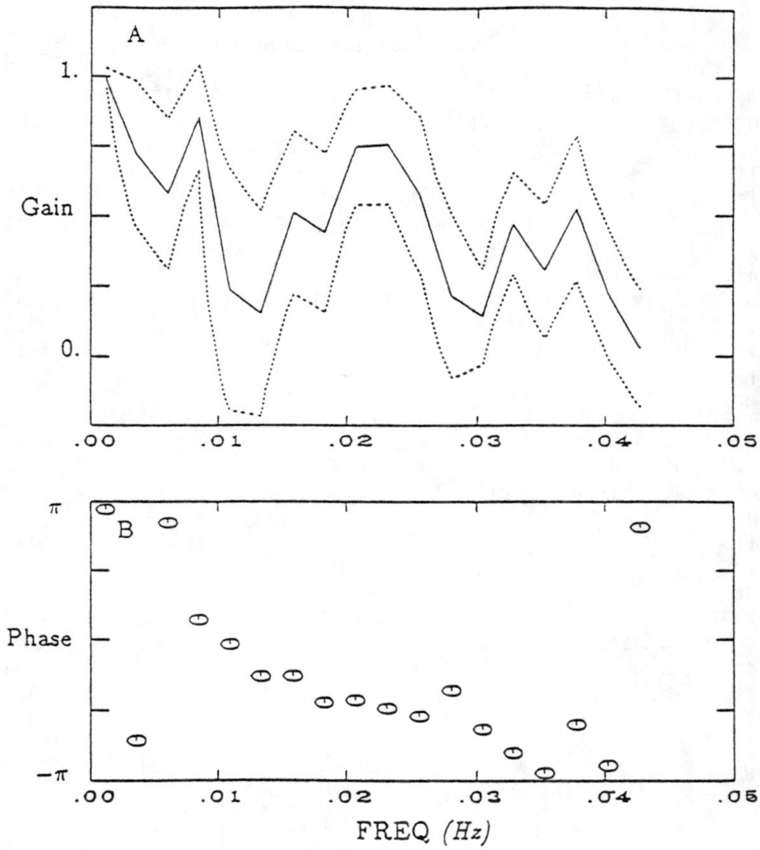

Figure 7. Gain, a), and phase, b), between incident and reflected waves. Dashed lines on gain plot are upper and lower 95% confidence intervals (Jenkins and Watts, 1968).

or frequency.

Conclusions: We have presented and applied a method of extracting modal information from observed dynamic variables. The method uses all available data and explicitly considers phase–locking of modes. The method may be extended to include forcing variables.

The data set examined comprised an unusually large sensor array. Even so it is not possible to consider all the modes which might be expected to be present. The results of the

Table 1
Modal Coupling – Torrey Pines

	0^+	0^-	1^+	1^-	2^+	2^-	4^+	4^-	7^+	7^-	I	R
0^+	-	-	.65	.14	.32	.26	-	.27	-	.21	.20	.20
0^-		-	.18	.69	.30	.36	.27	-	.19	-	-	-
1^+			-	.38	.87	.29	.55	-	.33	-	-	-
1^-				-	.28	.88	-	.58	-	.36	-	-
2^+					-	.12	.87	-	.67	-	.13	-
2^-						-	-	.88	-	.69	-	-
4^+	$\sigma = .0061 Hz$						-	-	.94	.25	.24	-
4^-								-	.25	.95	.23	.21
7^+									-	.47	.31	.19
7^-										-	.31	.26
I											-	.38
R												-

	0^+	0^-	1^+	1^-	2^+	2^-	4^+	4^-	7^+	7^-	I	R
0^+	-	.43	-	-	-	-	-	-	-	-	-	-
0^-		-	-	-	-	-	-	-	-	-	-	-
1^+			-	.18	.45	-	-	-	.19	-	.14	.20
1^-				-	-	.40	-	-	-	-	-	-
2^+					-	-	.45	.25	-	.27	-	-
2^-						-	-	.73	-	.49	-	.12
4^+	$\sigma = .0134 Hz$						-	-	.77	-	-	.13
4^-								-	-	.93	.36	.32
7^+									-	.22	.32	.38
7^-										-	.47	.44
I											-	-
R												-

	0^+	0^-	1^+	1^-	2^+	2^-	4^+	4^-	7^+	7^-	I	R
0^+	-	.64	-	-	-	.14	-	.19	.26	.49	.76	.76
0^-		-	-	-	-	-	.25	-	.49	.25	.75	.76
1^+			-	-	-	-	-	-	-	-	-	-
1^-				-	-	-	-	-	-	-	-	-
2^+					-	-	.37	-	-	-	-	-
2^-						-	-	.47	-	.14	-	-
4^+	$\sigma = .0208 Hz$						-	-	.74	.12	.19	.19
4^-								-	-	.74	.19	.16
7^+									-	.13	.49	.39
7^-										-	.46	.48
I											-	.63
R												-

Table 1 The coherence squared between modes. Significant values of coherence squared (Jenkins and Watts, 1968) are shown for all modal pairs. +/− indicates a wave progressing northward/southward.

Table 1 - continued
Modal Coupling – Torrey Pines

	0^+	0^-	1^+	1^-	2^+	2^-	4^+	4^-	7^+	7^-	I	R
0^+	-	.49	-	.13	-	.14	.19	.40	-	.14	.45	.16
0^-		-	.13	-	-	-	.24	.36	-	-	.38	.20
1^+			-	-	-	-	-	.17	-	-	-	-
1^-				-	-	-	-	-	-	-	-	-
2^+					-	-	.19	-	.12	.26	-	-
2^-						-	-	-	.13	-	-	-
4^+	$\sigma = .0281 Hz$						-	.12	.30	.13	.25	-
4^-								-	.15	.39	-	.25
7^+									-	.55	-	-
7^-										-	.13	.18
I											-	-
R												-

	0^+	0^-	1^+	1^-	2^+	2^-	4^+	4^-	7^+	7^-	I	R
0^+	-	.28	.83	.38	.12	-	-	-	-	-	-	-
0^-		-	.30	.77	-	.15	-	.13	-	-	.15	-
1^+			-	.43	-	-	-	-	-	-	-	-
1^-				-	-	-	-	-	-	-	-	-
2^+					-	-	-	-	-	-	-	-
2^-						-	-	-	-	-	-	-
4^+	$\sigma = .0354 Hz$						-	-	-	.13	.25	.20
4^-								-	-	-	.27	-
7^+									-	.38	.19	-
7^-										-	.14	-
I											-	.18
R												-

	0^+	0^-	1^+	1^-	2^+	2^-	4^+	4^-	7^+	7^-	I	R
0^+	-	.67	.92	.55	-	-	-	-	.21	.32	.53	.22
0^-		-	.80	.90	-	-	.15	-	.24	.14	.57	.18
1^+			-	.45	.12	-	-	-	.73	.37	.52	.25
1^-				-	-	-	.12	-	.25	.17	.69	.14
2^+					-	-	-	-	-	-	-	-
2^-						-	-	.14	-	-	-	-
4^+	$\sigma = .0427 Hz$						-	.14	.23	-	-	-
4^-								-	-	-	-	-
7^+									-	-	.15	-
7^-										-	.30	-
I											-	-
R												-

Table 1 The coherence squared between modes. Significant values of coherence squared (Jenkins and Watts, 1968) are shown for all modal pairs. +/− indicates a wave progressing northward/southward.

fit suggest a variety of modes are present and that significant phase–locking occurs. This has important implications for sediment transport, beach morphology and the modelling of edge wave dynamics.

References

Bowen, A.J., and D.L. Inman, Edge waves and crescentic bars, J. Geophys Res., 76, 862-8671, 1971.

Bowen, A.J., and D.A. Huntley, Waves, long waves and nearshore morphology, Marine Geology, 60, 1-13, 1984.

Eckart, C., Surface waves in water of variable depth. Wave report, SIO Ref. 51-12, Scripps Inst. Oceanogr., La Jolla, CA, 99pp, 1951.

Elgar, S., Shoaling surface gravity waves, Ph.D. thesis, Scripps Inst. Oceanogr., La Jolla, CA 123 pp., 1985.

Guza, R.T., and R.E. Davis, Excitation of edge waves by waves incident on a beach, J. Geophys. Res., 79, 1285-1292, 1974.

Guza, R.T., and A.J. Bowen, Finite amplitude edge waves, J. Mar. Res., 34, 269-293, 1976.

Guza, R.T., and Inman, D.L., Edge waves and beach cusps, J. Geophys. Res., 80, 2997-3012.

Guza, R.T., and E.B. Thornton, Observations of surf beat, J. Geophys. Res., 80, 2997-3172, 1985.

Haines, J.W., K.R. Thompson, and D. Wiens, Fitting dynamical modes to data. (in prep.)

Holman, R.A., and A.J. Bowen, Bars, bumps and holes: models for the generation of complex beach topography, J. Geophys. Res., 87, 457-468, 1982.

Holman, R.A., and A.J. Bowen, Longshore structure of infragravity wave motions, J. Geophys. Res., 89, 6446-6452, 1984.

Huntley, D.A., Evidence for phase–coupling between edge wave modes, (in prep.)

Jenkins, G.M., and D.G. Watts, Spectral analysis and its applications, Holden–Day, Oakland, CA, 525 pp., 1968

Oltman-Shay, J., and R.T. Guza, Infragravity edge wave observations on two California beaches, J. Phys. Oceanogr., 17, 644-663, 1987.

Symonds, G., D.A. Huntley and A.J. Bowen, Two–dimensional surf beat, long wave generation by a time–varying breakpoint, J. Geophys. Res., 87, 492-498, 1982.

CHAPTER 114

BEACH PROFILE CHANGE UNDER VARYING WAVE CLIMATES

Yen Kai[1] Ren Rushu[2] Wang Liang[3]

Abstract

In the present paper results of experimental study of two-dimensional transformation of sandy beach under varying wave climates are presented. The varying wave climates were composed of different systems of regular waves exerted one after another on the model beach. Through experiments it was found that sandy beach transformation within surf zone could be expressed by the changes of characteristic point A and characteristic slope tan ß, and that although the expression for beach erosion-accretion criteria is dimensionless, similitude scale effects should still be taken into consideration.

I. Introduction

Beach erosion-accretion process changes with wave climates. A new beach profile is always developed from the profile created by preceding wave climate. However, previous experimental studies on beach profile change were usually carried out by assuming an initial straight beach slope i_0, which is not in conformity with the natural situation. To avoid this a series of tests on beach transformation were conducted using varying wave climates, i.e. different systems of regular waves one after another. Part of experimental results has been published[3], in which the conception of characteristic point A and characteristic slope tanß was introduced and a beach erosion-accretion criterion was proposed. In the present paper some new experimental results and advancement in the authors' research works will be given.

II. Experimental Investigation

From 1980 to 1987, experiments were performed in a wave basin, 23.8m long, 8.7 m wide and 0.58 m deep. The model stretch is separated lengthwise into six wave flumes, each 17 m long, 0.48 m wide. Ordinary quartz sand of different sizes were used and four different initial beach slopes 1/5, 1/10, 1/15 and 1/20 were moulded in wave flumes. The ground water level in the model was controlled so as to maintain its original still water level in the wave basin.

---

(1) Professor and Hon. President, (2) Associate Professor, (3) Research Assistant, Hohai University, Nanjing, China

Three kinds of waves climates were used:

1st kind: A train of regular waves with constant wave height H_o and wave period T acting intermittently on model beach in varying durations. Through experiments the relations between initial, intermediate and equilibrium profiles can be observed. The wave climates used in 1st kind are shown in Table 1.

Table 1 Experimental conditions of 1st kind wave climates

Sand size	d_{50} = 0.30 mm				d_{50} = 0.69 mm			H_o	T	No. of
i_o / Flume No. / case	No.1	No.2	No.3	No.4		No.5	No.6	(cm)	(sec)	tests & durations(hr)
A-1	1/5	1/10	1/15	1/20		1/10		6.8	1.25	1×8
A-2	1/5	1/10	1/15	1/20		1/10		6.8	1.25	2×4
A-3	1/5	1/10	1/15	1/20		1/10		6.8	1.25	4×2
A-4	1/5	1/10	1/15	1/20		1/10		6.8	1.25	8×1
A-5	1/5	1/10	1/15	1/20		1/10		6-8	1.25	16×0.5
A-6	1/5	1/10	1/20		1/20	1/10	1/5	6.8	1.25	2
A-7	1/5	1/10	1/20		1/20	1/10	1/5	6.8	1.25	54
B-1	1/5	1/10	1/15	1/20		1/20	1/10	5.1	1.25	1×8
B-2	1/5	1/10	1/15	1/20		1/10	1/5	5.1	1.25	2×4
B-3	1/5	1/10	1/15	1/20		1/10	1/5	5.1	1.25	4×2
B-4	1/5	1/10	1/15	1/20		1/10	1/5	5.1	1.25	2
C-1	1/5	1/10	1/15	1/20		1/10	1/5	5.1	1.25	8
C-2	1/5	1/10	1/15	1/20		1/10	1/5	6.8	1.00	2×4
C-3	1/5	1/10	1/15	1/20		1/10	1/5	6.8	1.00	4×2
C-4	1/5	1/10	1/15	1/20		1/10	1/5	6.8	1.00	8×1
C-5	1/5	1/10	1/15	1/20		1/10	1/5	6.8	1.00	16×0.5
C-6	1/5	1/10	1/20		1/20	1/10	1/5	6.8	1.00	2
C-7	1/5	1/10	1/20		1/20	1/10	1/5	6.8	1.00	22
C-8	1/5	1/10	1/20		1/20	1/10	1/5	6.8	1.00	136

2nd kind: Wave climates consisting of constant regular waves and different systems of regular waves with varying wave height H_o and wave period T. Through experimental studies of the actions of constant waves and different systems of waves on straight as well as on transformed beachs, the transformation relations for different beach configuration in surf zone can be observed. The wave climates used in 2nd kind are shown in Table 2. In Table 2(a), E-K are constant wave cases. The experiments were continued until the beach profile reaches the state of equilibrium or quasi-equilibrium. In Table 2(b), groups A and BC were performed using wave cases F, H, I, J, K, arranged in the order and for the duration shown.

Table 2 Experimental conditions of 2nd kind wave climates
(a) Constant regular waves

Wave Case	E	F	G	H	I	J	K
H_o (cm)	6.80	2.96	2.32	4.96	4.28	7.00	9.42
T (sec)	1.25	2.20	2.70	1.42	1.80	1.24	0.87
Test Duration (hr)	21.0	60.0	32.5	24.0	28.0	27.0	35.0
Flume Number	1	2	3	4	5	6	
Initial Slope i_o	1/5	1/10	1/20	1/20	1/10	1/5	
d_{50} (mm)	0.30	0.30	0.30	0.69	0.39	0.69	

(b) Systems of regular waves

Wave Group	A					BC				
Case Order	K	J	I	F	H	F	I	J	K	H
Test Duration (hr)	34.5	3.0	5.0	8.0	13.0	34.5	10.5	13.0	12.5	13.0

3rd kind: Wave climates consisting of constant regular waves and different systems of regular waves with constant wave height H_o and varying wave period T, varying H_o and constant T as well as varying H_o and T (Table 3). Through experiments under this kind of wave climates, we can get some more knowledge of two-dimensional transformation of sandy beach under varying wave climates.

III. Erosion-Accretion Parameters and Beach Types within Surf Zone

1. Erosion-accretion parameters

Through experiments mentioned above, it can be noted that in the course of beach transformation under a certain wave action there is a point on the beach profile, located near the wave breaking point, undergoing very little change as shown in Figure 1 and Figure 2. The same point also existed in T. Sawaragi's model test (1980)[5]. We defined this point as the characteristic point A. Make use of its relative stable position, point A can be taken as a coordinate origin on the beach profile to facilitate the study of beach transformation processes. Slope $\tan\beta$ formed by connecting the characteristic point and shoreline point may be defined as the characteristic slope. Characteristics slope $\tan\beta$ changes simultaneously with advance and recession of the shoreline in the course of beach transformation, i.e. in the course of beach erosion $\tan\beta$ tends to become gentle with the recession of shoreline, conversely, in the course of beach accretion $\tan\beta$ tends to become steep with advance of shoreline. In the state of erosion-accretion equilibrium the characteristic slope will almost remain unchanged and may be denoted by $\tan\beta_c$. So by comparing the characteristic slopes before and after a certain series of wave action, we can judge whether the shoreline is advancing or receding and define the types of beach erosion-accretion accordingly. For this reason point A and $\tan\beta$ may be chosen as two important parameters to account for beach transformation within surf zone. Figure 3 shows that when a certain wave action is continued,

Table 3 Experimental conditions of 3rd kind wave climates

Flume Number	1	2	3	4	5	6	H_o (cm)	T (sec)	Test Duration (hr)
Initial slope i_o	1/5	1/10	1/20	1/20	1/10	1/5			
d_{50} (mm)	0.40	0.40	0.40	0.54	0.39	0.40			
Wave Group	Case Order								
T	T_3						5.85	1.2	30
	T_9						6.93	1.2	23
	T_{19}						8.90	1.2	30
	T_{14}						7.96	1.2	16.6
H	H_4						6.92	0.98	13.3
	H_{35}						7.31	1.70	14.3
	H_{25}						7.42	1.53	16.3
	H_{13}						6.92	1.34	16.6
H_o/L_o	H_o/L_o - 22						5.34	1.74	16.9
	H_o/L_o - 34						6.31	1.84	16.2
	H_o/L_o - 38						3.66	1.34	15.5
	H_o/L_o - 7						3.61	1.48	17.8
S_y	S_y - 16						2.71	2.63	15.8
	S_y - 12						3.91	1.77	20.4
	S_y - 30						8.16	1.54	15.9
	S_y - 15						10.60	0.98	15.8
	S_y - 4						6.92	0.98	17.5
	S_y - 17						3.44	1.90	20.4
B - 17							3.44	1.90	20.2

the characteristic slope gradually reduces its developing speed and finally attains a state of equilibrium or quasi-equilibrium, which can be denoted by $\tan\beta_c$.

2. Beach types within surf zone

Beach erosion-accretion within surf zone has great influence on the stability of shoreline. According to whether the shoreline is advancing or receding and the beach is under erosion or accretion within surf zone, the beach profile within surf zone could be classified into three types as shown in Figure2:

Type I: The shoreline recedes and surf zone is under erosion.

Type II: The beach profile and shoreline have reached quasi-equilibrium state.

Type III:The shoreline advances and surf zone is under accretion.

when the wave action is continued, the beach profile would always transform from type I or type III into type II as shown in Figure 3.

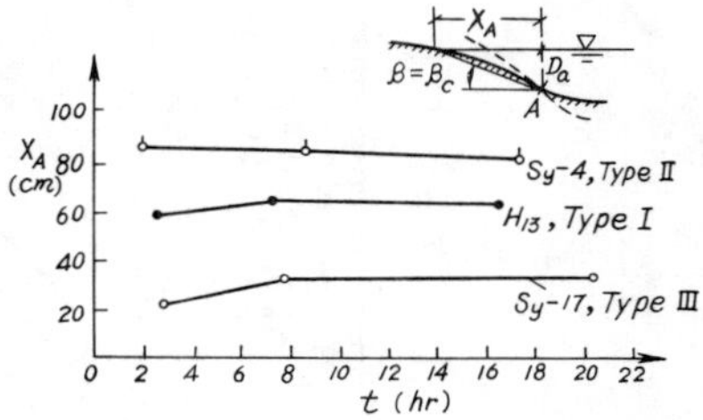

Fig. 1 Distance X_A between characteristic point A and equilibrium shoreline point versus time t

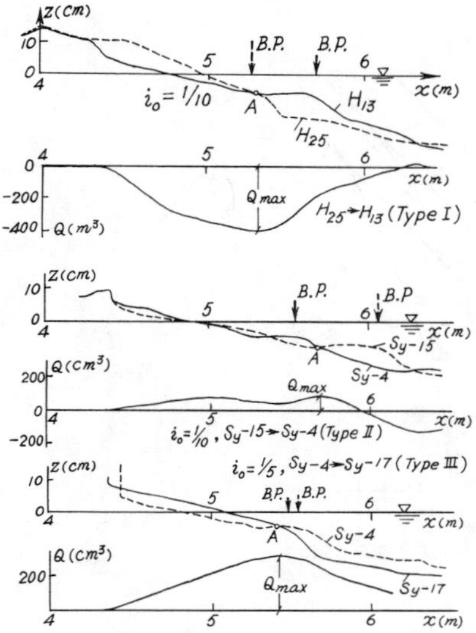

Fig. 2 Beach types within surf zone and distribution of net on-offshore sediment transport load

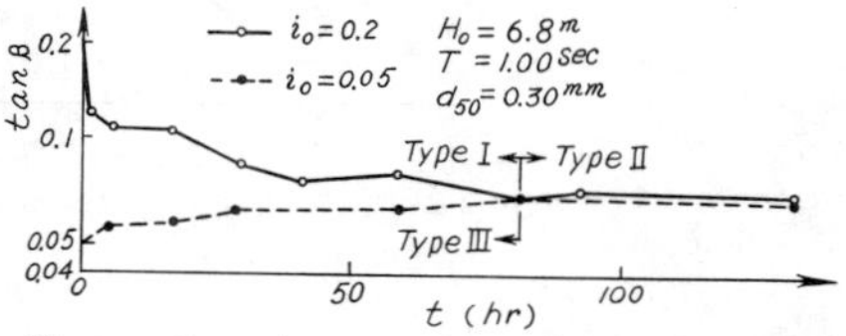

Fig. 3 Characteristic slopes $\tan\beta$ change with advance and retreat of the shoreline in the course of beach transformation and tend to attain equilibrium

IV. Beach Criterion Number

1. Model scales

The authors' experimental data are compared with Sunamura and Horikawa's criterion (1974)[8] as shown in Figure 4. The data points from 205m long large wave tank (R. Kashima et al, 1982)[2] are also shown in the figure. It shows that for Sunamura's small wave tank, the beach erosion-accretion criterion number K=4 - 8, but for the large wave tank K=18, and from the author's experiments the criterion number for the two different grain sizes are different. Although the expression for beach erosion-accretion criteria has been made dimensionless, yet the wave and grain size effects have not been eliminated. Obviously such criteria can not be used in practice.

Since beach erosion-accretion transformation is principally produced in the surf zone, a normal similarity of wave is also required in the model in order to satisfy the dynamic similarity of wave action within the surf zone, i.e. the following relations should be satisfied:

$$\left.\begin{aligned} \lambda_H &= \lambda_L = \lambda_D \\ \lambda_T &= \lambda_D^{0.5} \end{aligned}\right\} \qquad (1)$$

When prototype sand is used the beach model is a distorted model, so the following relations also should be satisfied:

$$\lambda_d = \lambda_\omega = 1 \qquad (2)$$

$$\lambda_l = \lambda_D^{\alpha} \qquad (3)$$

where λ_H, λ_L, λ_T, λ_D represent the scale of wave height, wave length, wave period and water depth respectively, λ_l is the horizontal scale: λ_d and λ_ω are the scale of sand grain size and settling veloci-

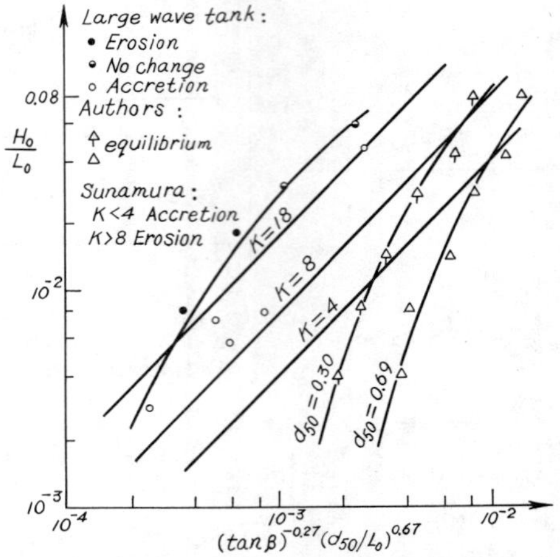

Fig. 4 Examination of similarity of T. Sumanura's erosion–accretion criterion

ty respectively. In order that the criterion number K obtained in model may be used in practice, the scale of number K must equal one in any situation, i.e.:

$$\lambda_K = 1 \tag{4}$$

If the relations in equation (1) and (3) are substituted into Sunamura's criterion, and adopt α =1.5, Sunamura's criterion can be rewritten as:

$$\lambda_K = \left(\frac{\lambda_H}{\lambda_d}\right)^{0.67}\left(\frac{\lambda_D}{\lambda_L}\right)^{0.27} = \frac{\lambda_H^{0.535}}{\lambda_d^{0.67}} \tag{5}$$

It is obvious that, the above equation is not simultaneously in accord with the prerequisites indicated in equation (2) and (4). If λ_d=1, then $\lambda_K \neq 1$, λ_K is directly proportional to $\lambda_H^{0.535}$, i.e. criterion number K will be increased with increase in wave force; while if λ_H=1, λ_K is inversely proportional to $\lambda_d^{0.67}$, i.e. the criterion number K will be increased with decrease in sand size. So that, Sunamura's criterion for large wave tank, small wave tank and different sand grain sizes exhibited non-identity as shown in Figure 4.

The criterion expression of CERC (1975)[9], $K=H_0/\omega T$, was vertified. The scale of the criterion can be written as

$$\lambda_K = \lambda_H^{1/2} / \lambda_\omega \tag{6}$$

when λ_ω=1, then $\lambda_K = \lambda_H^{1/2}$. Obviously, in CERC's criterion λ_K is a value affected by wave height or wave length, the value K increases with the wave height or wave length, so the criterion of CERC is rather limited for use.

2. Beach erosion-accretion criterion

A beach erosion-accretion criterion was presented in 1986 by authors, which is:

$$\frac{H_o}{L_o} = K\left(\frac{\omega}{gT}\right)^{1.3}\left(\frac{U_c}{\omega}\right)^{0.4}(\tan\beta)^{-1.5}, \quad K = Cf_w^{0.2} \tag{7}$$

In this formula several points are worthy to mention. First, the conception "characteristic slope" $\tan\beta$ is introduced instead of initial straight slope i_o, to make the criterion applicable to any stage of beach evolution and irregular beach profile under varying wave climates.

Second, the critical velocity for incipient movement of the sediment U_c is introduced which is an important factor in determining the net sediment transport on the beach under wave actions. Moreover, the ratio of bottom frictional velocity of wave motion to settling velocity U_*/ω indicates the direction of net sediment transport on the beach (T. Sawaragi, 1982)[6] and reflects what kind of sediment movement is dominant between suspended load and bed load. The tendency of beach to attain equilibrium state is shown by the value of U_* to approach $\sqrt{2f_w}\,U_c$. Thus, on an equilibrium beach $\sqrt{2f_w}\,U_c/\omega$ could be used instead of U_*/ω, where f_w is the frictional coefficient.

Third, the criterion formula satisfies the scale relations. After analysing over 100 data sets of the authors' as well as other researchers' experiments the following results were obtained: A beach will be in the state of quasi-equilibrium when $0.32 \leqslant K \leqslant 0.83$, with average value $K=0.5$, in erosion state when $K > 0.83$; and in accretion state when $K < 0.32$. The data extent is: $H_o=2.32\sim176^{cm}$, $T=0.87\sim12^{sec}$, $D_{50} = 0.2\sim6.1^{mm}$ [3] .

V. Characteristic Point and Equilibrium Characteristic Slope

1. Characteristic point A

In the paper of 1986[3], it has been indicated that on model beach composed of a certain kind of sand under the action of different systems of waves run in succession, a new characteristic point will be produced when one wave system is changed into another, whatever the previous beach transformation process may be. The position of point A (relative to the equilibrium shoreline) created by a certain wave case on any beach profile (irregular beach profile or simple straight beach) is generally the same. That is to say, the relative position of point A is closely related to wave parameters and sediment properties only and is almost irrespective of the previous processes of evolution of beach transformation.

Based on the above analysis a statistical formula for calculating water depth D_a at point A was obtained:

$$D_a/L_o = 0.52\,(H_o/L_o)^{0.96}\,(U_c/\omega)^{0.33} \tag{8}$$

the data extent is: $H_o=2.32\sim176^{cm}$, $T=0.87\sim12^{sec}$, $d_{50}=0.2\sim6.1^{mm}$.

For any type of beach profile, the point A corresponds also to the point where the net sediment transport rate or net sediment load reached a limiting value as shown in Figure 2.

Using net on-offshore sand transport formula of Watanabe (1981)[2,7], $\phi = 3(\psi - 0.12)\,\psi^{0.5}$, taking its limiting value, and then considering the correction for the effects of beach slope and sand size, another formula for calculating water-depth D_a at point A is obtained:

$$\frac{D_a}{L_o} = 0.192 \left(\frac{\rho_s - \rho}{\rho}\right)^{-0.8} \left(\frac{U_c^2}{g\, d_{50}}\right)^{8/15} \left(\frac{H_o}{L_o}\right)^{0.8} \qquad (9)$$

where ϕ is the non-dimensional transport rate and ψ is the Shields parameter. If the wave parameters H_o and L_o as well as beach sand size d_{50} (cm) are known, the value of D_a can be calculated.

Formula (9) is compared with formula (8) by 56 experimental data obtained from large and small wave flumes. The relative error for formula (9) is 0.28 and for formula (8) is 0.43. It is clear that formula (8) is simpler than formula (9), but the latter seems to be more accurate and rational in theory.

2. Equilibrium characteristic slope

Equilibrium characteristic slope $\tan\beta_c$ can be calculated from formula (7) using average value K=0.5.

It is interesting to note that in the course of beach transformation under varying wave climates, although the initial and intermediate beach profile or characteristic slope $\tan\beta$ are not the same, yet when a certain wave action is continued, the beach always tends to transform into the same equilibrium or quasi-equilibrium characteristic slope as shown by the black data points of group S_y - 12, S_y - 15, S_y - 4 and S_y - 17 in Figure 5.

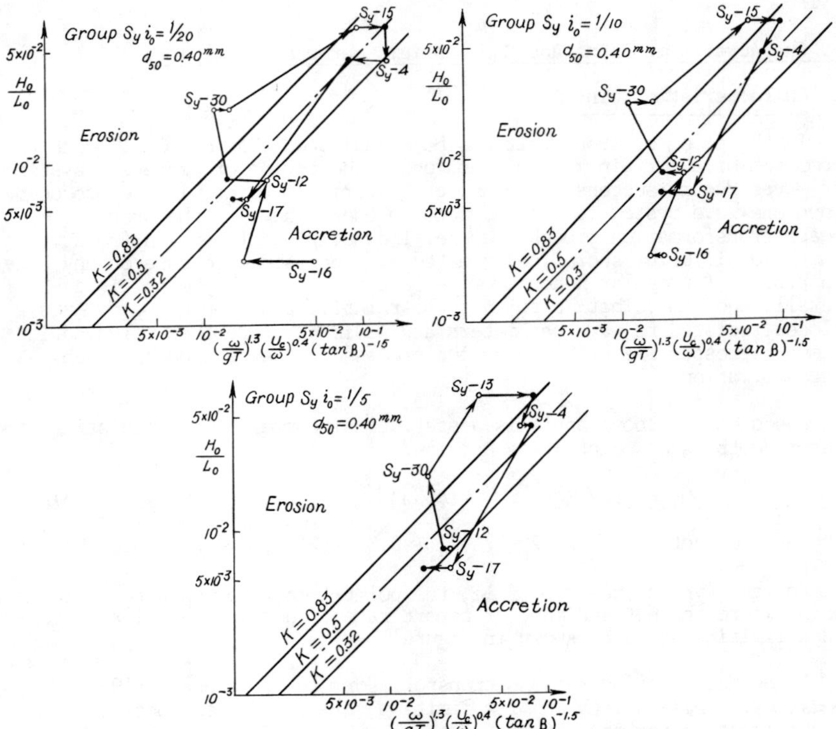

Fig. 5 H_o/L_o *versus* $(\omega/gT)^{1.3}(U_c/\omega)^{0.4}/(\tan\beta)^{1.5}$ *for wave climates of group* S_y

Conclusion

1. Based on the shoreline recession-advance and beach erosion-accretion, the beach profile within surf zone could be classified into three types.

2. Two-dimensional beach transformation under varying wave climates could be expressed by the changes of characteristic point A and characteristic slope tanß. Point A and slope tanß are closely related to wave characteristics and sediment properties only and could be obtained from formule (7) and (9).

3. Although the expression for beach erosion-accretion criteria is dimensionless, similitude scale effects should still be taken into consideration. Without similarity consideration the criterion may not be available for use.

References

1. Hallermeier, R. J., (1978) "Use for a calculated limit depth to beach erosion", Proc. of the 16th coastal Engineering Conf., pp.1493 - 1512.

2. Kashima, R., Shimizu, T., Maruyam, K., and Saito, S., (1982) "Experimental study of on-offshore sediment transport in a large wave tank", Proc. 29th Japanese Conf. Coastal Eng. pp. 125-129.(in Japanese).

3. Ren Rushu and Wang Liang, (1986) "Investigation on two-dimensional transformation of sand beach under wave action",Proc. of 3rd International Sympsoium on River Sedimentation, pp.378 - 387.

4. Saville, T., (1975) "Scale effects in two-dimensional beach study", Proc. 7th IAHR, pp. A3-1-A3-10.

5. Sawaragi, T. and Deguchi, I., (1980) "On-offshore sediment transport rate in the surf zone", Prc. of the 17th Coastal Engineering Conf., pp 1195 - 1214.

6. Sawaragi, T., (1982) Sediment trasnport and beach erosion, Morikita Publishing Company, Tokyo. (in Japanese).

7. Shimizu, T., Saito, S., Maruyam, K., Kashima, R., et al, (1985) "Experimental study of on-offshore sediment load distribution model in a large wave tank", Proc. 32th Japanese Conf. Coastal Eng. pp 311 - 314

8. Sunamura, T., Horikawa, K., (1974) "Two-dimensional beach transformation due to waves", Proc. of the 14th coastal Eng. Conf., pp 920-938.

9. U. S. army CERC, (1975) Shore Protection Manual, Vol I, pp 4-81-4-83.

10. Van Hijum, E., (1974) "Equilibrium profiles of coarse material under wave attack". Proc. of the 14th coastal Eng. Conf. pp.939 - 957.

11. Xu Xiao, (1981) Two-dimensional transformation of Sandy beach and on-offshore sediment transport, Master Thesis, East China Technical University of Water Resources, Nanjing, China, (in Chinese).

CHAPTER 115

REPRODUCTION MODELS OF BEACH CHANGE BY STORM WAVES

Masahiro Ito[1] and Yoshito Tsuchiya[2]

ABSTRACT

This paper presents a technique to reproduce, by a two-dimensional moveable-bed model, beach change due to the time-dependent storm waves which are generated by the passage of an atmospheric depression. In the model test, scaling conditions for sand grain-size, vertical and horizontal lengths, and wave height and period characteristics were established by applying the authors' scale-model relationship which was reported; and wave duration time also was decided. A method of employing regular waves in the model to represent irregular waves in the field is proposed. From the results, it was shown that the model can reproduce well the beach change in the field using the regular waves having the mean wave properties in the irregular waves.

1. INTRODUCTION

Coastal disasters, such as a beach erosion and seawall collapse which are caused by rapid sediment transport that occurs during storm conditions, continue to be frequently reported. One very important method used to approximate the coastal disaster, in hopes of preventing it, is the movable-bed model which can very closely approximate the beach change processes under the conditions of time-varying waves. However, because the scale-model relationship was not properly, we could not extrapolate the quantitative results which were obtained by the moveable-bed scale model test to the actual prototype, and therefore the results were applied qualitatively. It is necessary to establish the scale-model relationship and the experimental technique for the moveable-bed model test which is applicable for short-term beach change by unsteady time-dependent waves, as occur in the field.

As an example of a moveable-bed model investigation, Vellinga (1982) tried to reproduce, by a two-dimensional model, erosion of a coastal sand dune on the Dutch coast, the Netherlands due to wave action during an extreme storm surge. In the model test, vertical and

1 Associate Professor, Department of Civil Engineering, Meijo University Tempaku-ku, Nagoya 468 Japan
2 Professor, Disaster Prevention Research Institute, Kyoto University Kyoto, 611 Japan

horizontal scales were distorted, and the model sediment grain-size was determined by a new parameter which was composed of both the dimensionless sediment-settling parameter proposed by Dean (1973) and the vertical scale. Also, Hughes (1983) examined the application of Dean's dimensionless parameter, Froude's time scale of vertical scale, and the distorted model between the vertical and horizontal scales. The relationship which Hughes proposed was verified by results which reproduced the coastal dune erosion in the Florida Panhandle, USA caused by the 1975 Hurricane Eloise, using two-dimensional model test and regular waves. Tsuchiya et al. (1985) considered experimentally how to efficiently control beach erosion due to the T-type groin and the artificial headland works at Shirarahama, Wakayama, Japan using a three-dimensional movable-bed model and regular waves. The model test condition was established using the modified author's scale-model relationship for length and grain-size scales, Froude's scale for time scale, and an undistorted scale for vertical and horizontal scales.

In these previous experimental studies, the reproduction of filed beach profiles by model test are not only defined quantitatively, but also in how to determine set the optimum regular waves for the model. Therefore, as the first step, the approach due to regular waves is hoped strongly rather than use of the time-dependent irregular waves as are found in the field, due to the lack of an adequate irregular wave scaling relationship.

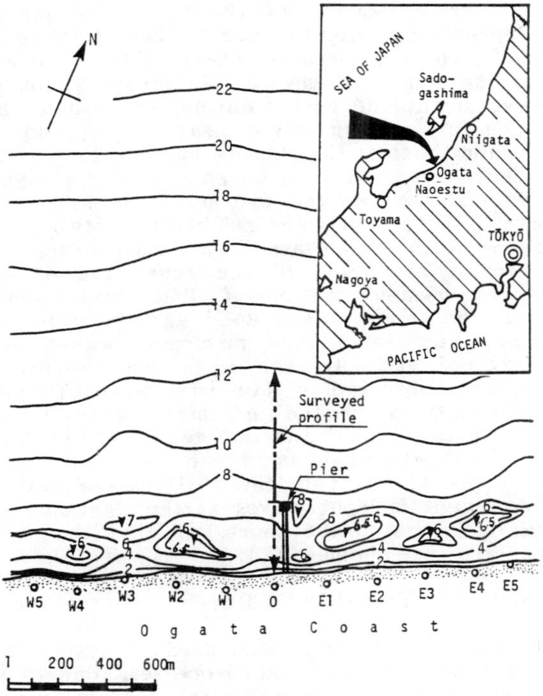

Figure 1 Bottom topograhpy of Ogata coast and study site.

2. CONDITIONS IN STUDY SITE

(1) Coastal morphology

The Ogata coast facing the Sea of Japan is about 100km southwest from Niigata-city, Japan and has a considerably straight shoreline as shown in Figure 1. The sediment grain-size of this coast becomes gradually fine in the offshore direction, and its medium grain-size diameter, d_{50}, is 1.5 to 0.2mm. The beach profiles ranging from shoreline to about 300m offshore had been surveyed at one week intervals using a rod along an observation pier known as the Ogata Wave Observatory of the Disaster Prevention Research Institute, Kyoto University. The beach profile father than 300m offshore, that is, deeper than a water depth of 8m, had been surveyed occasionally by using an ultrasonic sounding instrument. The sea bottom father than 300m offshore has an average slope of 1/100. A longshore exist bar within the zone from the shore to about 300m offshore, beyond this 300m offshore distance the seabed contours remain fairly strait and parallel to the shoreline. Therefore, the topography of this coast may be considered almost two-dimensional in the on-offshore direction.

(2) Wave climate

Wave conditions in this area are usually observed at the offshore end of the observation pier in a depth of 6m, and Naoetsu Port Office, about 10km distance from the Ogata, in a depth of about 21m.

The Naoetsu Port Office continues to perform daily wave observations. From March 14-18, 1981 a large-scale lower atmospheric depression passed through the Sea of Japan, off the Ogata coast. Figure 2 shows the weather chart on the 15th when the Ogata coast wave attack was most severe. In Figure 3, time histories for significant wave height and period during the 15th to the 17th at the observation pier are shown using heavy solid and broken lines, respectively, while those at Naoetsu harbor are shown using shin solid and broken lines, respectively. As shown in this figure, the wave period records observed at the two positions agree well with each other. Whereas, records of wave height at the two positions do not completely agree with each other. This discrepancy of wave height higher than 4m is attributed to the generation of wave breaking because the measurement station of observation pier is relatively shallows. However, there was a good agreement for waves smaller than 4m height. Because the time history of waves observed at the two positions agreed well with each other, we choose to use the wave records which were observed in Naoetsu instead of Ogata. Figure 4 shows the time history of the maximum wave height, H_{max}, the significant wave height, $H_{1/3}$, and the mean wave height, H_{mean}, for 1 hour, which were calculated using the wave records observed at Naoetsu during the storm from the 10th to the 18th. During this period, the dominant incident waves at the Ogata coast attacked at nearly normal incidence to the shore, so that the coastal condition could be considered two-dimensionally.

3. SCALE-MODEL RELATIONSHIP USED

The scale-model relationship was used to design the model such that the Ogata coast beach change processes, due to storm waves, could be reproduced. The scale-model relationship, which is indicated as the similitude zone in Figure 5, had been earlier defined by the authors (1984) by considering experimentally the similarity between the

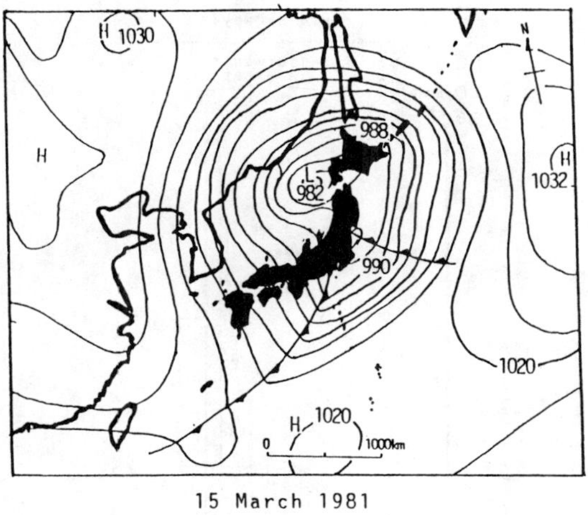

Figure 2 Weather chart of the storm at maximum intensity attacked to the Ogata coast on 15 March 1981.

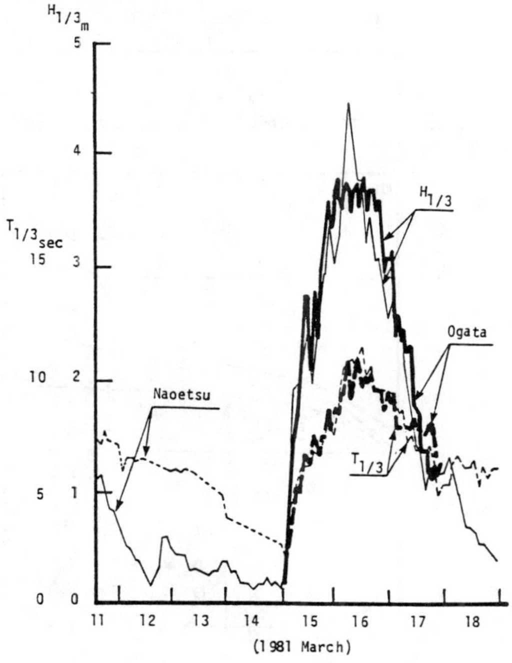

Figure 3 Time history of the significant wave height and period for 1 hour observation averaging at the tip of Ogata's pier and in Naoetsu harbour during 15 March 1981 to 17.

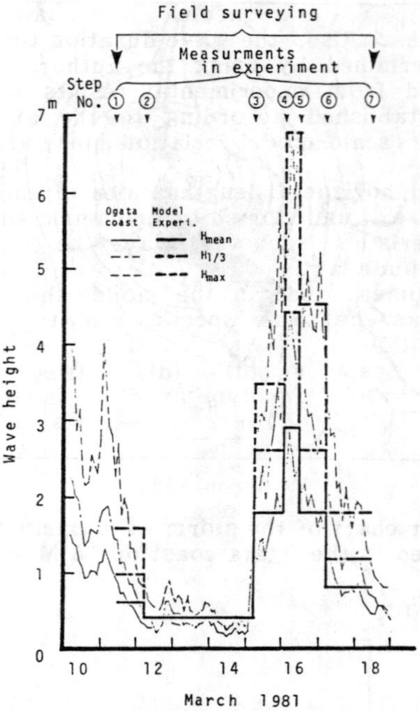

Figure 4 Variation of wave height during storm and experimental waves used.

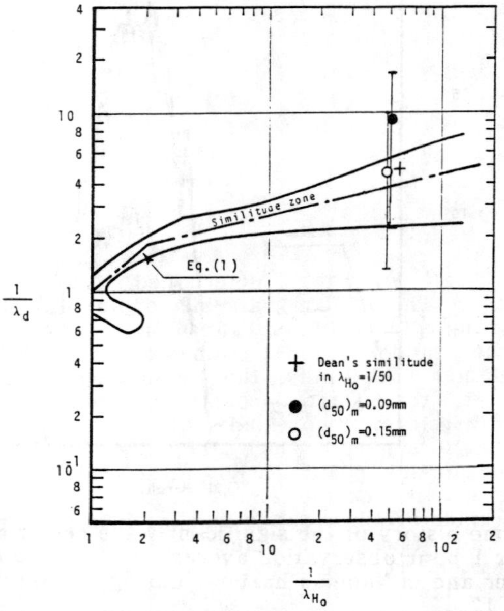

Figure 5 Scale-model relationship and experimental conditions set up in model.

large-scale beach profile and medium to small-scale ones under equilibrium conditions. Also, the wave duration time scale used in the model test was determined by using the authors (1986). This time scale was introduced from experimental results which were derived under conditions established according to the authors' scale-model relationship. These scale-model relationships are summarized as follows:

(1) The vertical and horizontal length scales of the beach profile are taken as the same (i. e., undistorted or geometric similarity).

(2) Wave characteristics such as wave height and period are determined by the Froude law.

(3) The beach sediment used in the model should be the sand or silica-sand which has the same specific gravity as the sea-bottom sediment of the prototype.

(4) The grain-size scale $\lambda_d = (d)_m / (d)_p$ between the grain-size of model sand, $(d)_m$, and prototype, $(d)_p$, is determined by the relationship indicated by the dot-dashed line in Figure 5. This relationship is given by

$$\frac{1}{\lambda_d} = 1.7^a \left(\frac{1}{\lambda_l}\right)^b \tag{1}$$

where a=1, b=0.87 for the range of experimental scale $1 \geq \lambda_l \geq 1/2.2$; a=0, b=0.2 for $\lambda_l < 1/2.2$, and λ_l is the scale of length (1/n), that is, experimental scale.

(5) Time scale, λ_t, is given by

$$\frac{1}{1.6}\sqrt{\lambda_l} \leq \lambda_t \leq \frac{1}{0.65}\sqrt{\lambda_l} \tag{2}$$

Then we use the time scale which is indicated the average tendency of Eq. (2), given as

$$\lambda_t = \sqrt{\lambda_l} = 1/\sqrt{n} \tag{3}$$

This relationship is none other than the Froude's time scale.

4. MODEL TEST CONDITION

The experimental scale of the model was set at 1/50, and the sand-grain size scale was then determined by Eq. (1) to be $\lambda_d = (d)_m / (d)_p = 1/3.7$. The median grain-size of on-offshore direction in the Ogata coast ranges from $(d_{50})_p$=1.5~0.2mm, therefore, according to Eq. (1), the model sand grain-size should range from $(d_{50})_m$=0.41~0.05mm. Especially, the sediment grain-size, where the beach profile at the Ogata coast changes remarkably, is $(d_{50})_p$=0.4mm, therefore, two kinds of silica-sands were used, $(d_{50})_m$=0.09 and 0.15mm, which have a uniform grain-size distribution. The sorting coefficient of these silica sands are 1.32, and the specific gravity in the air is 2.71. The grain-size scale between the two kinds of silica-sands and the field sediment are shown in Figure 5. The value of sediment grain-size in the field are scattered, so that their representative values and scattering ranges are shown in this figure. This figure shows the silica-sand, $(d_{50})_m$=0.15mm (open circle) sufficiently satisfies the scale-model relationship, however the $(d_{50})_m$=0.09mm silica-sand is shifted slightly upward from the scale-model relationship. The time dependency of the wave height and period during the storm were considered by separating the storm term into some time intervals which wave height and period wereevaluated, furthermore, three classification methods were employed for comparison purpose, mean significant and maximum, as shown in Figure 4. In the model test the three types of waves were then scaled down to 1/50 by the Froude law and regular waves employed. Therefore, wave height, period, and wave duration time were each taken as a variable in the model. Figure 6 shows that the mean tidal level in Naoetsu during the storm, 11th to 18th, varies from -2cm to 45cm. From this figure, it is clear that the passing of the lower depression generated a significant storm surge. Therefore, it is assumed that this tidal change influenced the Ogata coast as well. If the variation of tidal level is scaled down to 1/50, the range of variation in the model becomes 0. 94cm. Then, water level in the model test, that is the water depth of wave flume was held a constant. The model test conditions which were set by thus modeling technique, were arranged in Table 1.

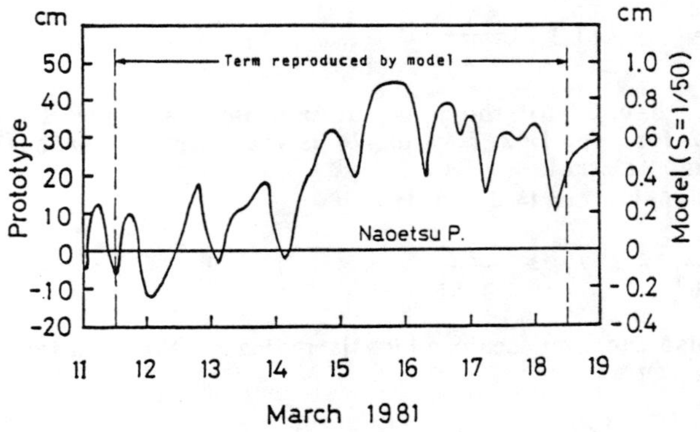

Figure 6 Variation of tidal level and storm surge in Naoestu.

Table 1 Experimental conditions to reproduce beach processes of Ogata coast.

Run No.	Grain size d_{50} mm	Wave character-listics		Step No. ①	② -1	-2	-3	③	④	⑤	⑥
				t=2h33m	3h	6h	9h37m	3h32m	1h16m	2h33m	4h14m
2	0.15	Hmean	H(cm)	1.2	0.8	0.8	0.8	3.5	5.8	3.5	1.6
4	0.09		T (s)	0.71	0.71	0.71	0.71	0.71	1.13	1.13	0.71
			H_0/L_0	0.0211	0.0141	0.0141	0.0141	0.0626	0.0335	0.0222	0.0283
1	0.15	$H_{1/3}$	H(cm)	2.0	0.8	0.8	0.8	5.2	8.8	5.2	2.4
5	0.09		T (S)	1.06	0.92	0.92	0.92	0.92	1.6	1.5	0.92
			H_0/L_0	0.0125	0.0079	0.0079	0.0079	0.0511	0.0234	0.0139	0.0235
3	0.15	Hmax	H(cm)	3.6	0.8	0.8	0.8	7.0	13.6	9.0	3.6
6	0.09		T (s)	1.06	0.92	0.92	0.92	0.92	1.7	1.98	1.13
			H_0/L_0	0.0257	0.0079	0.0079	0.0079	0.068	0.0309	0.0142	0.0222

5. REPRODUCTION BY MODEL TEST

Movable-bed model tests was carried out in a medium-sized wave flume of dimensions 28m long, 0.5m wide and 1m deep, with a flap-type regular wave generator. An initial model beach profile was formed in the on-offshore direction corresponding to a 900m offshore distance in the prototype. This initial profile corresponds to the location indicated by the survey profile line in Figure 1, taken on the 11th just prior to the storm. Wave properties are varied according to Step Nos. ①~⑦ in Table 1, while the wave duration intervals are specified by the Froude's time scale for the step like functions shown in Figure 4. The changes in beach profile were also measured at the times indicated in Step Nos. ①~⑦ in Table 1. We classified the measured beach profile data into the following three groups, and discussed each from the standpoint of the reproduction of field beach profile.

(1) Beach profile

Figure 7 shows the beach changes in the model due to regular waves. The beach profiles that changed more significantly than those in the step are only shown in this figure. In this figure, the initial beach profile marked by ① (Step No. ①), and final beach profile by ⑦ are shown overlapping their corresponding field beach profile. From this figure, it is found that sand dunes (reflection bars) form on the onshore and offshore sides of the longshore bar when the significant and maximum waves were employed. The influence of the partial standing waves is considered to contribute to bar formation.

The initial and final Ogata coast beach profiles, 11th and 18th, were used to examine and compare the reproductive capability of the based on variations in sediment grain-size and wave height classification, where the solid and dash-dotted lines represent grain-sizes of 0.09mm and 0.15mm, respectively as shown in Figure 8. From this figure, it is shown that the reproduction of the beach profile is best when using regular waves which correspond to the mean waves of the observation records, that the reproduction capability significantly decays as the order of the wave exceeds the significant. However, the sediment grain-size was found to produce little effect on model reproduction as compare to wave scaling.

(2) Shoreline change

The temporal changes in the shoreline of the model tests from the initial state on the 11th by the mean, significant and maximum waves are arranged in Figure 9, such that the profiles corresponding to the two of sediment grain-sizes are distinguishable. In this figure, the results are shown only for Step No. ⑦ (final experimental data), along with the corresponding range of experimental error. Also, a field shoreline measurement taken on the 18th is shown along with the tidal variation corresponding to that day. From the comparison of the shoreline change by the model tests with the field, it is found that, when mean waves are used in the experiment, the reproduction is much better than profile reproduction using the significant and maximum waves.

(3) Sand volume change

We examine the reproduction in term of the sand volume change which can be calculate from the beach profile change. The sand volume change per 1m width in the field was calculated over a 260m

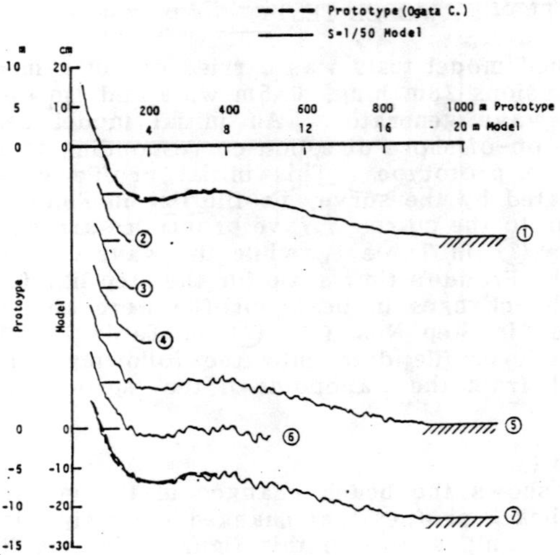

7 (a) mean wave height, H_{mean}, Run No. 2

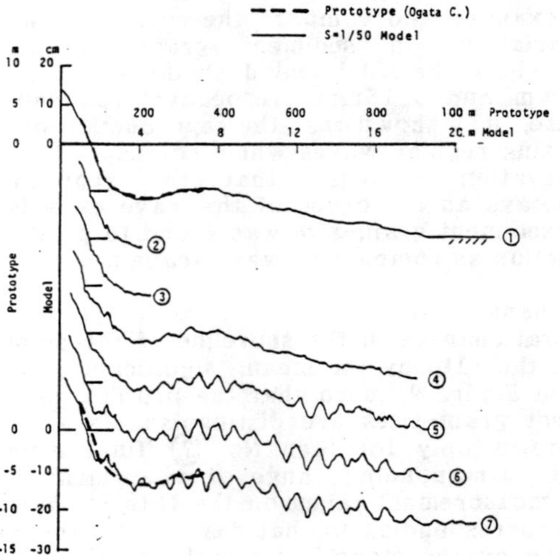

7 (b) significant wave height, $H_{1/3}$, Run No. 1

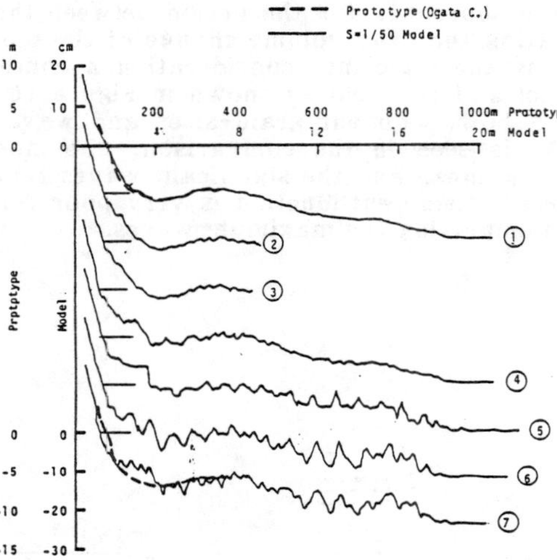

7 (c) maximum wave height, H_{max}, Run No. 3

Figure 7 Reproduction of beach change processes in the model dye to the regular waves.

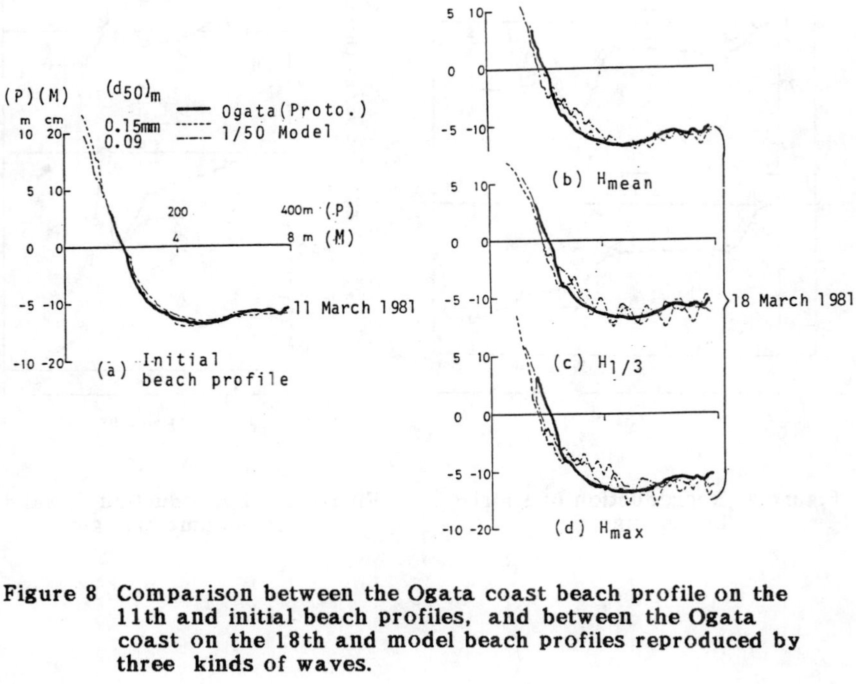

Figure 8 Comparison between the Ogata coast beach profile on the 11th and initial beach profiles, and between the Ogata coast on the 18th and model beach profiles reproduced by three kinds of waves.

range included the shoreline, for the period between the 11th and the 18th. By calculating the sand volume change of the model in the same manner and taking the scale into consideration, a comparison is made between the model and the field as shown in Figure 10. In this figure the differences in both sediment grain-sizes and wave properties are distinguished. As is seen in the comparison, the model test results which were used the mean and the significant waves reproduce the field very well, however, field reproduction is very poor for both sediment grain-sizes when employing the maximum waves.

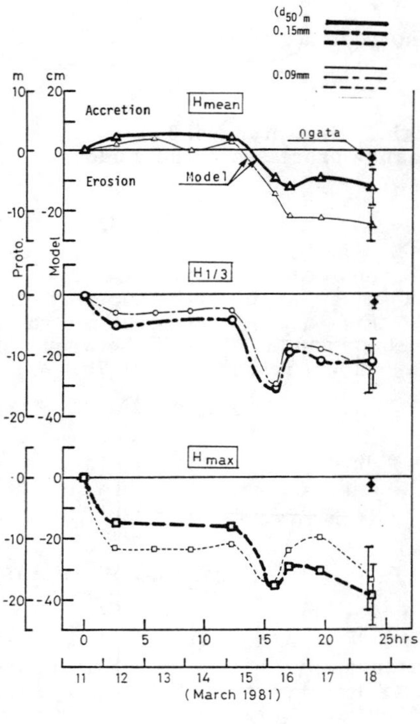

Figure 9 Reproduction of shore-line change.

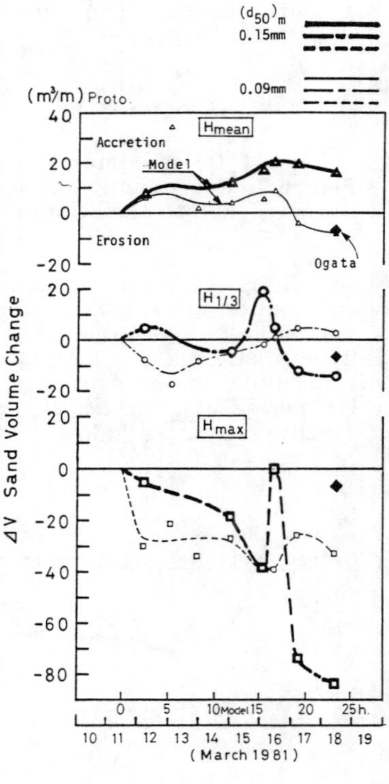

Figure 10 Reproduction of sand volume change.

6. DISCUSSION

Some scale-model relationships of beach change processes have reported till now, as reviewed by Hallermeier (1985). This model test study was performed applying the author's scale-model relationship which was described in the sections 3. (1)~3. (5). Therefore, quoting the scale-model relationship which is possible to compare to the author's one, we discuss about the relation between them. Especially, the fall time perameter which is introduced from the criterion separating on-offshore transport by Dean (1973) is frequently quoted in the study on the similitude of beach change, so that we examine the correspondence between the authors' and Dean's scale-model relationship.

Dean (1973) studied experimentally the onshore or offshore direction of sediment motion, and reported that the normal and storm beach profiles can be classified by:

$$\frac{H_0}{L_0} \gtrless 1.7 \frac{\pi w}{g T} \quad \begin{array}{l} \text{offshore (storm profile)} \\ \text{onshore (normal profile)} \end{array} \tag{4. a}$$

or

$$\frac{H_0}{w T} = c \tag{4. b}$$

where H_0 is the wave height in deep water, L_0 the wave length in deep water, g the gravity acceleration, w the settling velocity of sediment, T the wave period, and c the constant. Eq. (4. b) is described as tha fall time parameter. Letting the notation $\lambda_{parameter}$ represent the ratio of the model to the prototype and using Froude similitude between the wave period and the length, the scale-model relationship of Eq. (4. b) is finally written as follows:

$$\lambda_w = \lambda_l^{1/2} \tag{5}$$

Because the experimental scale of the model tests is $\lambda_l = 1/50$, the required settling velocity rate of model and prototype sediments by Eq. (5) becomes

$$\frac{(w)_m}{(w)_p} = \frac{1}{7.1} \tag{6}$$

As mentioned above, the sea-bed sediment grain-size at Ogata coast varies from $(d_{50})_p = 0.2$mm to 1.5mm in on-offshore direction and its typical grain-size is $(d_{50})_p = 0.4$mm. By using the relationship between the sediment grain-size and the settling velocity, that is, Eq. (6), the sediment grain-size to be used in model becomes $(d_{50})_m = 0.06 \sim 0.14$mm and the typical size is $(d_{50})_m = 0.086$mm. Thus when $(d_{50})_p = 0.4$mm, the grain-size scale by Dean's parameter becomes $\lambda_d = (d)_m / (d)_p \cong 1/4.7$, while author's similitude is $\lambda_d = 1/3.7$. The comparison between authors' similitude (dotted dash line) and Dean's grain-size scale is shown by the cross symbol "+" in Figure 5. From this comparison, it is interesting to note Dean's grain-size scale is slightly shifted from the authors, that is, the grain-size of the model obtained by Dean's grain-size scale relationship becomes slightly finer than the authors.

Therefore, one of the two kinds of sediments which were used in this model test approximately agrees with the grain-size scale by the authors and Dean.

Table 2 Synthetic evaluation of the model reproduced processes of Ogata coast.

Run No.	Wave characteristics	Beach profile	Shoreline change	Sand volume
2 4	Mean wave H_{mean}	○	○	△
1 5	Significant wave $H_{1/3}$	△	△	○
3 6	Maximum wave H_{max}	◆	◆	◆

○ : "very good" reproduction △ : "good" reproduction ◆ : "bad" reproduction

7. SUMMARY AND CONCLUSION

In this model study the Ogata coast beach profile changes were modeled according to the methods described in sections 3. (1), 3. (3), and 3. (4) for the initial beach profile, wave characteristics, sediment material, and its grain-size, respectively. The model eaves were varied according to the measured Ogata coast storm wave intensity, and the wave duration time of regular waves in the model test were scaled down using Eq. (3). Wave characteristics were determined by the method described by 3. (2). Beach change processes due to time-dependent storm waves were found to be reproduce by using model waves corresponding to a one hour mean wave representation of field observations.

The two different sediment grain-sizes used in model test did not have a clear effect upon the reproduction.

The degree of reproduction due to regular waves corresponding to H_{mean}, $H_{1/3}$ and H_{max} with regard to the beach profile, the shoreline change, and the sand volume change are summarized in Table 2. From this table, it is found that, when the beach profile was reproduced very well, the reproduction of shoreline change and sand volume change were also.

It is concluded that reproduction in the model tests is "very good" when the regular waves corresponded to the mean wave in the field was used; "good", when using significant waves; and "bad" when using maximum waves.

The sediment grain-size of model which was determined by the authors' scale-model relationship was nearly equal to the value derived by Dean's relationship where the typical sediment grain-size in the Ogata coast was $(d)_P$=0.4mm. It should be noted that sediment grain-size of model, determined by Deans' scale-model relationship, changes depending on the settling velocity of sediment of field (prototype).

8. REFERENCES

1) DEAN, R. G. (1973) Heuristic models of sand transport in the surf zone, Proc. Eng. Dyn. in Surfzone, Institution of Engineers, Australia, pp. 208-214.

2) HUGHES, S. A. (1983) Movable-bed modeling law for coastal dune erosion, Jour. Water Port Coast. and Ocean Eng., ASCE. pp. 164-179.

3) HALLERMEIER, R. J. (1985) Unified modeling guidance based on a sedimentation parameter for beach changes, Coastal Engineering, Vol. 1, No. 1, pp. 27-70.

4) ITO, M. and TSUCHIYA, Y. (1984) Scale-model relationship of beach profile, Proc., 19th ICCE, ASCE, pp. 1386-1402.

5) ITO, M. and TSUCHIYA, Y. (1986) Time scale for modeling beach change, Proc., 20th ICCE, ASCE, pp. 1196-1209.

6) TSUCHIYA, Y., KAWATA, Y. and SILVESTER, R. (1985) Beach processes of Sirarahama and its preservation (2), Disaster Prevention Research Institute Annuals, Kyoto Univ., Vol. 28, B-2, pp. 565-590 (in Japanese).

7) VELLINGA, P. (1982) Beach and dune erosion during storm surges, Coastal Engineering, Vol. 6, No. 4, pp. 361-387.

CHAPTER 116

REALISTIC ECONOMIC BENEFITS FROM BEACH NOURISHMENT

Robert G. Dean*, M.,ASCE

ABSTRACT

A method is presented and illustrated with examples to establish appropriate storm damage reduction and recreational benefits from beach nourishment projects. Unlike previous methods, benefits to project adjacent areas are recognized due to sand transport out of the project area and deposition on adjacent beaches. Assuming homogeniety along the shoreline, the character of storm damage reduction and recreational benefit relationships are such that sand transported from a project area and deposited on adjacent beaches always results in an increase rather than a reduction in benefits. A central element in calculating storm damage reduction benefits is the establishment of a proportional damage curve for upland structures as a function of beach width and storm return period. To illustrate the method, limiting cases are presented in which (A) all sediment remains within the area placed, and (B) all sediment spreads out immediately over a long segment of shoreline. Using Monte Carlo simulation to represent the random character of the storms, the method is applied to 15 realistic cases with varying project lengths, representative wave heights, added beach widths and interest rates. The present worth storm damage reduction and recreational benefits are calculated to demonstrate the effects of the various parameters. It is found that for short project lengths and relatively large wave heights, the benefits from project adjacent areas exceed those in the project area where the sand is placed. Although no littoral control structures, such as jetties are included in the present application, the method could be extended readily to include their effects.

INTRODUCTION

Policies and methodologies should evolve continuously to remain consistent with modern understanding of coastal processes and the true equities of those residing along the shoreline. Several changes have occurred in the last few decades that argue for an examination and modifications of present economic analysis procedures relating to beach

*Graduate Research Professor, Coastal and Oceanographic Engineering Department, University of Florida, Gainesville, FL 32611.

nourishment: (1) It is now clear that on a long, uninterrupted shoreline, good quality sand placed in a beach nourishment project will eventually be transported out of the region placed, but will remain within the active nearshore system, (2) Sand transported from a project area and deposited on project adjacent areas provides not only continuing damage reduction and recreational benefits, but provides enhanced benefits, and (3) With increasing concern over the use of "hard structures" as a means of shoreline control, beach nourishment will play an increasing future role.

This paper considers the economic consequences of sand eroded from a beach nourishment project area and deposited on project adjacent areas. Realistic damage reduction relationships and recreational benefits for a widened beach are utilized to demonstrate that this evolution process actually results in a net increase in project benefits. Benefits from simple limiting cases are examined in which (1) the sand remains in the area placed, and (2) the sand spreads out immediately. A direct procedure is presented to account for total present worth project benefits. The procedure utilizes Monte Carlo simulation to faithfully represent the probability of storm occurrences.

Although the methodology presented here is not applicable to shorelines which include features which would cause longshore sediment transport interruptions, the concepts could be extended readily for such cases.

CONCEPTS

There are two simple concepts which are critical to the methodology presented here:

(1) Good quality sand placed in a beach nourishment project will be eroded from the area placed but will remain indefinitely in the active nearshore region, and

(2) The greatest storm damage and recreational benefits are generally realized for the initially narrower beaches.

The first concept will be considered as valid without much discussion. Although "good quality sand" is a matter of degree, here it refers to sand that is greater than 0.14 mm or so in diameter and that is coarser than or as coarse as the material originally present on the beach. For those nourishment materials in which the above is not the case, this paper refers to that sand fraction which is compatible. Monitoring results from a number of beach nourishment projects have demonstrated the first concept, for example at Port Canaveral, FL (Dean, 1988) and Captiva Island, FL (Tackney and Associates, 1983).

The second concept is illustrated by Figure 1a which represents a survey (by Shows, 1978) of the structural damage caused by Hurricane Eloise (1975) in Bay County, FL as a function of proximity of the structures relative to a jurisdictional control line which is generally parallel to

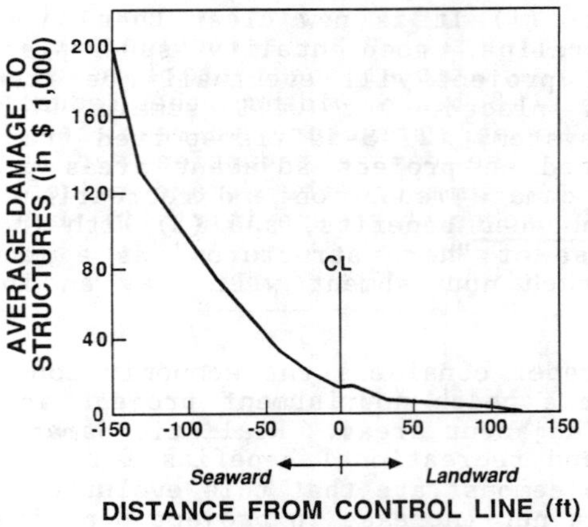

a) Damage to Structures in Relation to their Location with Control Line (Resulting from Study of 540 Structures in Bay County after Hurricane Eloise, by Shows, 1978).

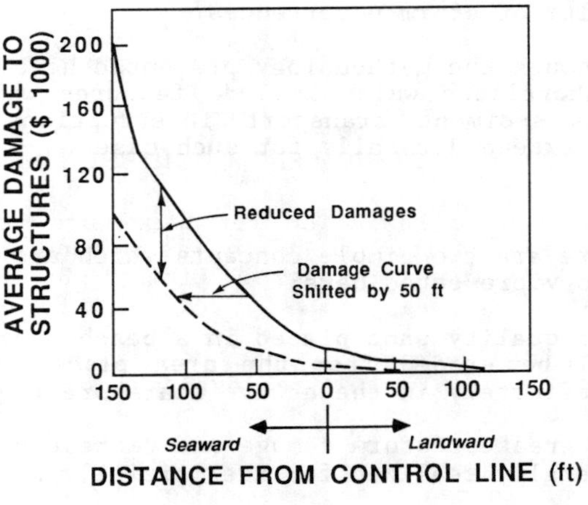

b) Damage Reduction Due to Beach Nourishment Advancing the Profile Fifty Feet Seaward.

Figure 1. Structural Damages Due to Hurricane Eloise (1975) and Example of Reduced Damages by Beach Nourishment Advancing the Shoreline Seaward by Fifty Feet.

the shoreline. Of particular significance in Figure 1a is the steeply sloped portion of the damage curve near its seaward end and the relatively mild slope near its landward end. It is instructive to consider the effect of a beach nourishment project which displaces the beach seaward by a certain amount such as 50 ft as shown in Figure 1b. It is seen that due to the slope characteristics discussed above, the greatest damage reductions occur for those structures which initially have very little beach in front of them. Figure 2 presents the damage reduction per structure associated with an additional one foot of beach width. For the narrower initial beach widths, the reduction is approximately $3,000 per structure whereas for greater initial beach widths, the damage reduction per structure is less than $500. In summary, the damage reduction benefits are greater for beaches which are initially much narrower.

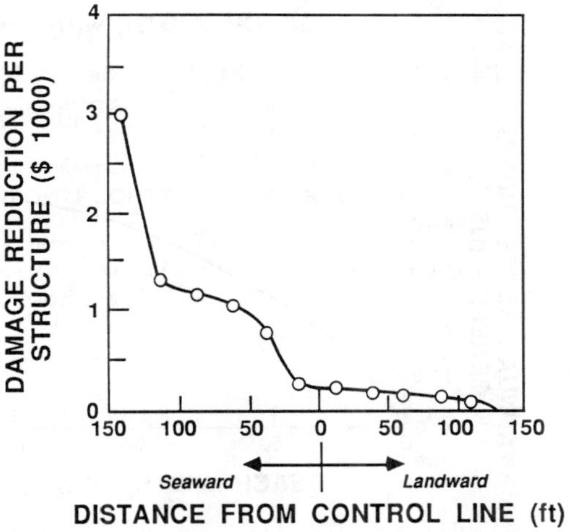

Figure 2. Damage Reduction Per Structure Resulting from a One Foot Wide Additional Beach, as a Function of Structure Location Relative to Control Line. Based on Hurricane Eloise Data.

The same concepts demonstrated above for damage reduction benefits apply for recreational benefits. Figure 3 presents the hypothetical usage and associated recreational benefits for beaches of varying widths. The number of people using the beach will increase with beach width; however, the rate of increase decreases for the greater widths. The results in Figure 3b are based on a visitation value of $6.00 per visitor per day and a plan area visitation requirement of 200 square feet. The annual recreational benefits associated with an additional foot of beach width versus initial beach width, based on Figure 3, are presented in Figure 4. As before, it is seen that the greatest benefits occur for the initially narrower beaches.

Referring to Figure 5, the significance of greater benefits for initially narrower beaches is that as a beach nourishment project evolves with the beach fronting the project area narrowing and the project adjacent beaches widening, benefits are lost in the initially wider project area. This loss of benefits is small compared to the gain of relatively large benefits in the initially narrow project adjacent areas. Assuming that the value of the upland structures protected by the project and the initial beach

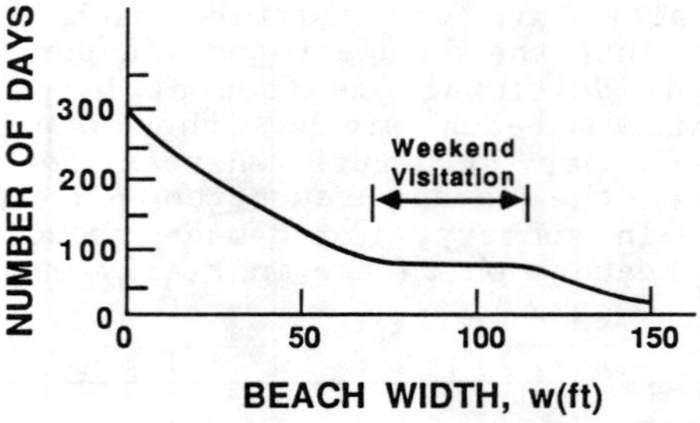

a) Number of Days Per Year that Full Beach Width is Used.

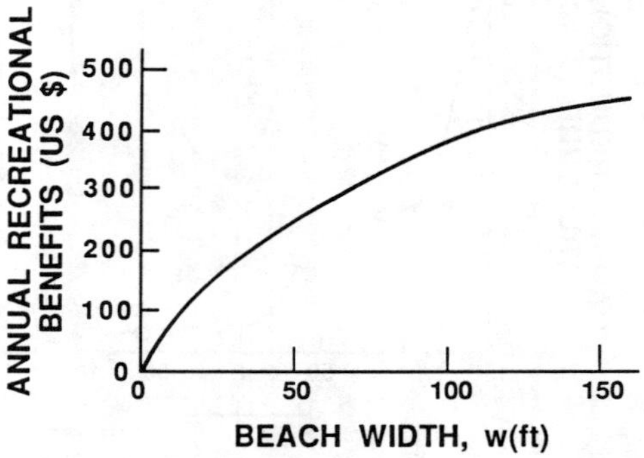

b) Annual Recreational Benefits vs Beach Width Per Foot of Beach Length.

Figure 3. Hypothetical Usage and Recreational Benefits of Sandy Beaches.

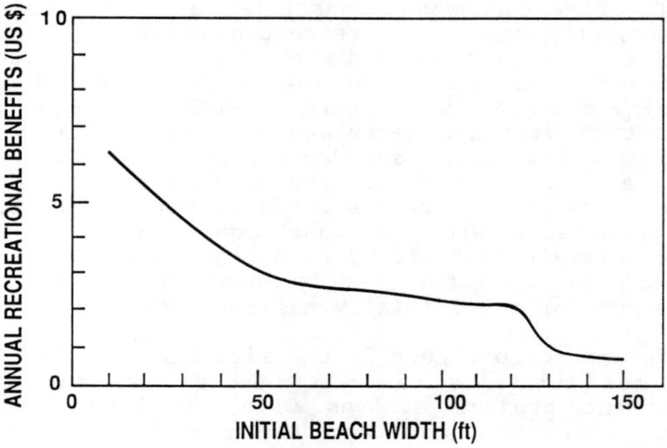

Figure 4. Annual Recreational Benefits Per Additional Foot of Beach Width as a Function of Initial Beach Width, Per Foot of Beach Length. Developed from Figure 3b.

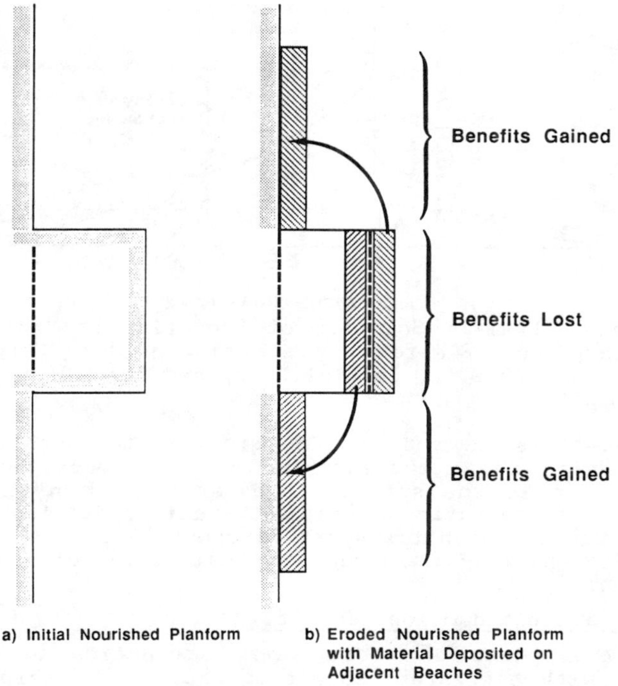

Figure 5. Schematic of Erosion of Nourished Area and Deposition in Project Adjacent Areas.

widths in project adjacent areas are uniform along the beach, there is always a net gain in storm reduction benefits as a result of project evolution. Similarly with respect to recreational benefits, assuming that the need for and access to recreational beaches are uniform, etc., the net effect of project evolution is a gain in recreational benefits.

METHODOLOGY

The methodology will be described and illustrated for idealized cases of no project evolution and rapid project evolution and general cases of benefits due to project evolution over realistic time frames.

Shoreline Evolution Model

The shoreline evolution model adopted here will be that due to Pelnard-Considerè for an initially rectangular planform as presented in Figure 6. The factor G is the so-called "longshore diffusivity" and for small angles of wave incidence is

$$G = \frac{K\, H_b^{5/2}\, \sqrt{g/\kappa}}{8(s-1)(1-p)(h_* + B)}$$

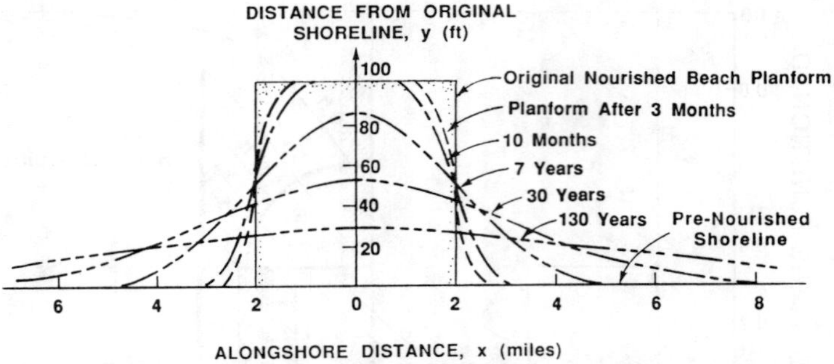

Figure 6. Example Solution of Evolution of Initially Rectangular Beach Planform. Pelnard Considere Method. Wave Height, H_b = 2.0 ft, Initial Nourished Beach Width = 100 ft, Fill Length, ℓ = 4 miles, t = time.

in which K is the sediment transport factor usually taken as 0.77, H_b is the representative breaking wave height, g is gravity, κ is the spilling breaker ratio (on the order of 0.8), s is the ratio of sediment specific gravity to that of the water in which transport is occurring, p is the in situ porosity and ($h_* + B$) is the vertical extent of beach profile response.

Storm Damage Reduction Benefits

Development of storm damage reduction benefits commences with the establishment of the relationship of a proportional storm damage factor, D, as a function of beach width fronting the structure, w, and storm return period, T_R. Figure 7 presents one example of such a relationship which has been used in the state of Florida Beach Management Plan. Development of this relationship is by no means trivial and should be based on an analysis of the expected damage to a range of representative structures as well as calibration with available storm results if such data are available. The proportional storm damage factor, D, depends on the foundation and structural and elevation characteristics of the buildings as well as the beach morphology, and presence and integrity of coastal protection structures, etc.

With the availability of $D(w,T_R)$ it is possible to predict the present worth damage reduction benefits PWDRB(N) during N years by the following equation

$$PWDRB(N) = \sum_{n=1}^{N} \frac{1}{(1+I)^n} \int_{\substack{\text{Project}\\ \text{Area}}} V(x,n)\left[D(w(x,n)T_R(n)) - D(w_o,T_R(n))\right]dx$$
$$+ \sum_{n=1}^{N} \frac{1}{(1+I)^n} \int_{\substack{\text{Project}\\ \text{Adjacent}\\ \text{Areas}}} V(x,n)\left[D(w(x,n),T_R(n)) - D(w_o,T_R(n))\right]dx \quad (1)$$

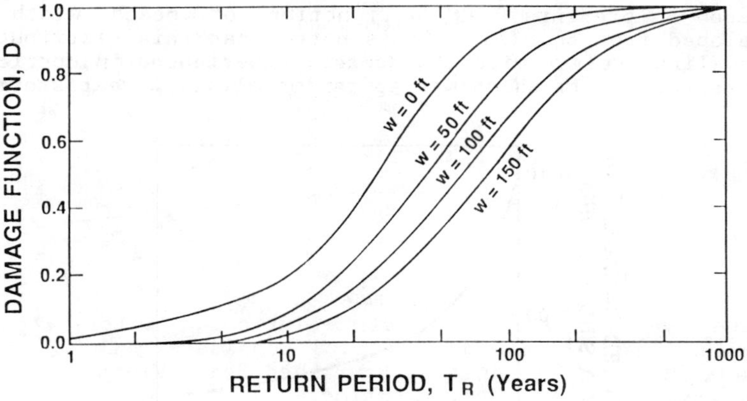

Figure 7. Hypothetical Proportional Storm Damage, D, as a Function of Storm Return Period, T_R, and Beach Width, w.

in which I is the interest rate and V(x,n) represents the structure value at a location, x, at a time n years into the future. $T_R(n)$ is the storm return period n years into the future. The two integrals differ only in their respective intervals of integration and are written separately here to illustrate the contributions from the two areas.

Eq. (1) accomplishes the objective of providing methodology for quantifying storm damage reduction. However, it is instructive to develop concepts further. Referring to Figure 7 which presents the proportional storm damage factor, D, the expected damage by a single storm $\bar{D}(w)$ as a function of beach width, w, is

$$\bar{D}(w) = \int_0^1 D(w,T_R)p(D)dD \tag{2}$$

in which p is the probability density function and is related to the cumulative probability distribution P by

$$p(D) = \frac{dP}{dD} \tag{3}$$

and noting that

$$T_R = \frac{1}{P} \tag{4}$$

Eq. (2) simplifies to

$$\bar{D}(w) = \int_0^1 D(w,T_R)\frac{dP}{dD}\,dD = \int_0^1 D(w,T_R)dP \tag{5}$$

Figure 8 presents $\bar{D}$ as a function of beach width as developed from Eq. (5). It is noted that this distribution is qualitatively similar to damages experienced in Hurricane Eloise (Figure 1) which was approximately a 70 year storm.

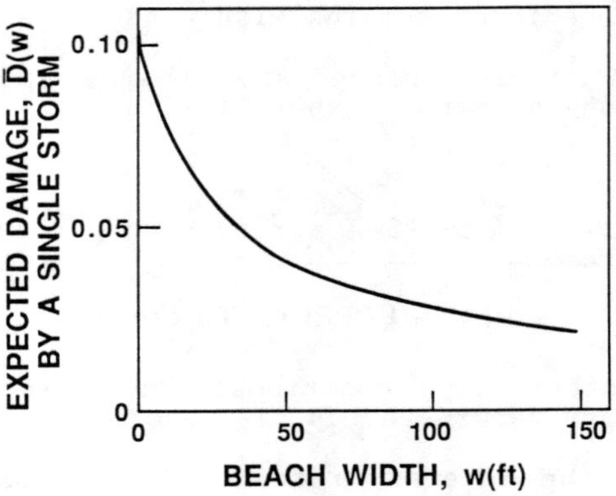

Figure 8. Expected Damage $\bar{D}(w)$ Due to a Single Storm as a Function of Beach Width, w.

Assuming that the value of the upland structures remain constant with time and that damaged structures are rebuilt to the same standards (both considerable assumptions), the present worth damage factor, PWDF(w) as a function of beach width for N years into the future is

$$PWDF(w,N) = \sum_{n=1}^{N} \frac{1}{(1+I)^n} \bar{D}(w) = \frac{1}{I} \left[1 - \frac{1}{(1+I)^N}\right] \bar{D}(w) \qquad (6)$$

and again, I is the interest. The bracketed factor in Eq. (6) approaches unity with large N. Table I presents values of PWDF(w,∞) for several beach widths interest rates. It is noted that the present worth damage factor can range as high as 1.31 for the case of zero beach width and an interest rate of 8%.

TABLE I

PRESENT WORTH DAMAGE FUNCTION, PWDF(w,∞)
VERSUS BEACH WIDTH, w, FOR ALL FUTURE DAMAGE

Interest Rate	Present Worth Damage Function, PWDF(w,∞) For Beach Width, w			
	0 ft	50 ft	100 ft	150 ft
6%	1.75	0.67	0.47	0.35
8%	1.31	0.50	0.35	0.26
12%	0.88	0.33	0.23	0.18

Idealized Cases

In contrasting project benefits realized within the project area to those outside the project area, it is instructive to consider two simple cases:

Case (A). All sediment remains within the area placed, and

Case (B). The sediment placed spreads out immediately over a long segment of shoreline.

Case (A).

The expected storm damage reduction benefits due to a single storm are

$$(SDRB)_A = [\bar{D}(w_o + \Delta w) - \bar{D}(w_o)]\ell \quad (7)$$

Case (B).

Denoting the (long) distance over which the sediment has been distributed as ℓ' and the associated additional width as $\Delta w'$, we have

$$(SDRB)_B = -\left(\frac{\partial \bar{D}}{\partial w}\right)_{w_o} \Delta w' \ell' \quad (8)$$

and since sediment is conserved $\Delta w \ell = \Delta w' \ell'$,

$$(SDRB)_B = -\left(\frac{\partial \bar{D}}{\partial w}\right)_{w_o} \Delta w \ell \quad (9)$$

The ratio, R_{SD}, of storm damage reduction benefits for the case of sand spreading out immediately to the case in which sand remains where placed is

$$R_{SD} = \frac{-\left(\frac{\partial \bar{D}}{\partial w}\right)_{w_o} \Delta w}{[\bar{D}(w_o + \Delta w) - \bar{D}(w_o)]} \quad (10)$$

It is noted that the ratio R_{SD} is always greater than unity. As shown in Figure 9, the interpretation is simple with the numerator representing the tangent of the damage curve at w_o and the denominator the secant slope between w_o and $w_o + \Delta w$. Due to the character of the curve, the ratio will always exceed unity. Figure 10 presents the ratio R_{SD} vs w_o for several values of Δw.

The same general discussion presented above applies to recreational benefits relationships. The ratio of benefits R_{RB} for a project that spreads out immediately to one that remains in place is

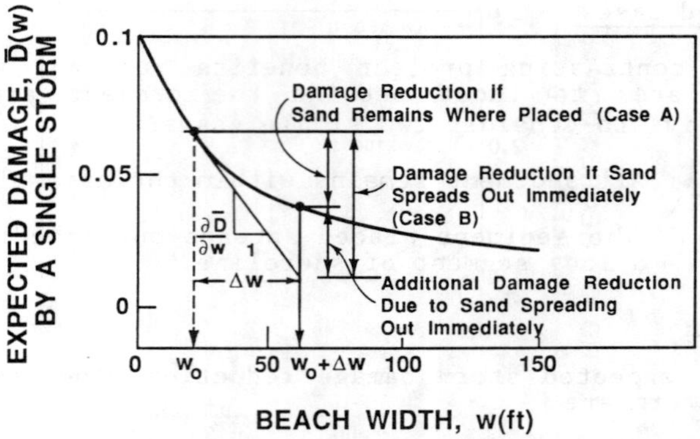

Figure 9. Interpretation of Damage Reduction Benefits if Sand Remains Where Placed (Case A) and if Sand Spreads Out Immediately (Case B).

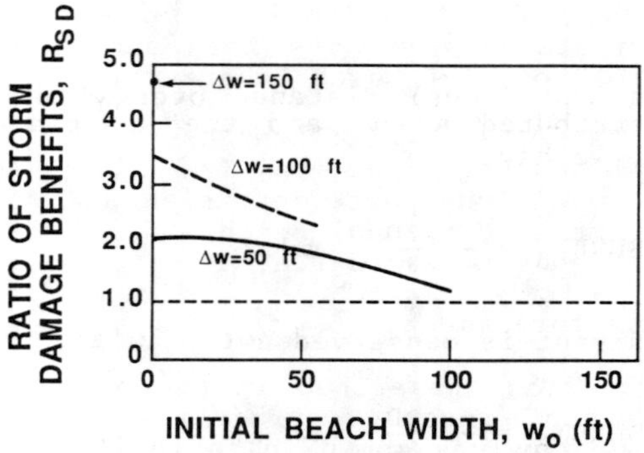

Figure 10. Ratio of Storm Damage Benefits, R_{SD}, vs Initial Beach Width, w_o, and Additional Beach Width, Δw. R_{SD} is Ratio of Storm Damage Benefits for Sand Which Spreads Out Immediately to Those for Which Sand Remains Where Placed.

$$R_{RB} = \frac{(\frac{\partial R}{\partial w})_{w_o} \Delta w}{[R(w_o + \Delta w) - R(w_o)]} \tag{11}$$

and this ratio will always exceed unity by the same argument as for the damage reduction benefits. For the recreational benefits shown in Figure 3, values of the ratio, R_{RB}, are presented in Figure 11.

RESULTS

Prior to presenting results for the general case, in which the beach planform evolves with time, it is worthwhile to consider the variables which will tend to favor Case A (sand remains in place) or Case B (sand spreads out immediately). Case A conditions would tend to dominate for:

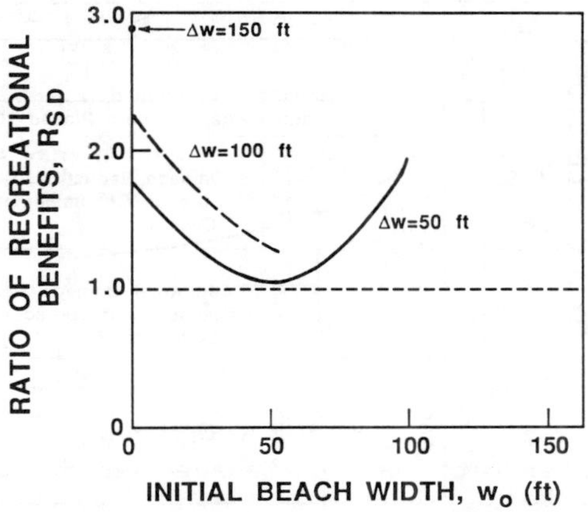

Figure 11. Ratio of Recreational Benefits, R_{RB}, vs Initial Beach Width, w_o, and Additional Beach Width, Δw. R_{RB} is Ratio of Recreational Benefits for Sand Which Spreads Out Immediately to Those for Which Sand Remains Where Placed.

Large Beach Fill Lengths, ℓ
Low Wave Height, H_b
Small Transport Coefficient, K
Small Additional Beach Widths, Δw
High Interest Rates, I

and vice versa for Case B.

The methodology described in the previous section was incorporated into a computer program which was "exercised" for the variable values shown in Table II. Results will be presented in two different forms.

Figure 12 presents variations of storm damage and recreational benefits with time for Run 5. The relatively large wave height and short beach fill associated with Figure 12 favors Case B conditions and it is seen that the dominant benefits occur within the adjacent project areas. It is also of interest to note that the benefits in the project adjacent areas lag those in the project area due to the time required for sediment transport to these adjacent areas. The longer project lengths and smaller wave heights would favor Case A conditions and the benefits inside the project area would dominate and commence quite early.

Table II summarizes results for all 15 runs conducted.

SUMMARY AND CONCLUSIONS

The methodology and results presented herein support the following statements.

Wider beaches seaward of structures perform as effective energy dissipators during storm conditions and, where

TABLE II. PRESENT WORTH STORM DAMAGE AND RECREATIONAL BENEFITS FOR VARIOUS WAVE AND PROJECT CONDITIONS

Run	Characteristics					Storm Damage Reduction*			Recreational Benefits**		
	Project Length, ℓ, (miles)	Wave Height H_b (ft)	Initial Beach Width w_o (ft)	Added Beach Width Δw (ft)	Interest Rate I	In Project Area	In Project Adjacent Area	Total	In Project Area	In Project Adjacent Area	Total
1	1.0	1.0	0	100	0.08	0.368	1.037	1.405	4.4	12.7	17.1
2	1.0	2.0	0	100	0.08	0.176	1.434	1.610	2.1	17.8	19.9
3	1.0	4.0	0	100	0.08	0.072	1.561	1.633	0.9	19.7	20.6
4	2.0	1.0	0	100	0.08	0.592	0.532	1.124	15.4	13.0	28.4
5	2.0	2.0	0	100	0.08	0.323	1.136	1.459	7.6	27.9	35.5
6	2.0	4.0	0	100	0.08	0.148	1.470	1.618	3.6	36.7	40.3
7	4.0	1.0	0	100	0.08	1.009	0.084	1.093	55.9	4.2	60.1
8	4.0	2.0	0	100	0.08	0.525	0.672	1.197	26.6	32.7	59.3
9	4.0	4.0	0	100	0.08	0.280	1.234	1.514	13.0	60.7	73.7
10	2.0	2.0	0	150	0.08	0.388	.679	2.067	10.1	38.5	48.6
11	2.0	2.0	50	100	0.08	0.066	0.178	0.244	4.9	14.5	19.4
12	2.0	2.0	100	150	0.08	0.057	0.144	0.201	7.4	21.7	29.1
13	2.0	2.0	0	100	0.12	0.240	0.697	0.937	5.7	17.0	22.7
14	2.0	2.0	0	100	0.04	0.506	2.413	2.919	11.7	60.0	71.7
15	16.0	2.0	0	100	0.08	1.733	0.002	1.735	392.4	0.5	392.9

*Relative to immediately adjacent upland property values within project area
**Expressed in millions of dollars

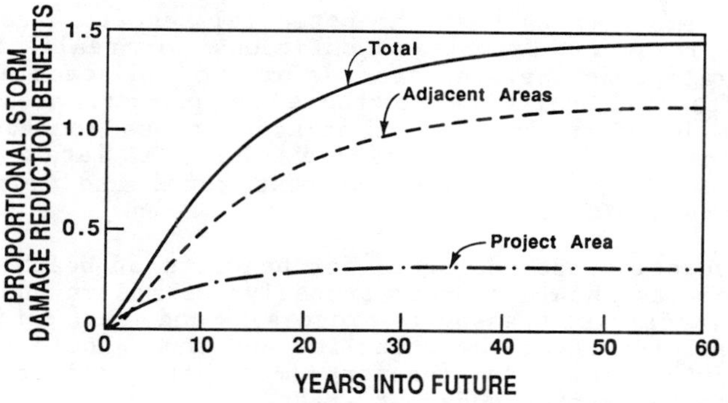

a) Proportional Present Worth Storm Damage Reduction Benefit Components vs. Years into Future.

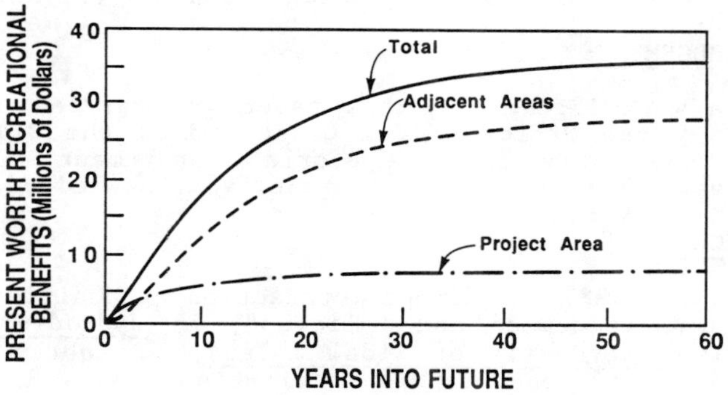

b) Present Worth Recreational Benefits vs. Years into Future.

Figure 12. Present Worth Storm Damage Reduction and Recreational Benefits. H_b = 2.0 ft, ℓ = 2.0 miles, w_o = 0.0, Δw = 100.0 ft., Interest Rate = 8%, Run No. 5.

the demand exists, also provide recreational benefits. These benefits can be enhanced through increasing beach widths by nourishment projects.

Beach nourishment projects conducted with good quality sand will evolve with erosion occurring within the project area and deposition in the project adjacent areas. Good quality sand will remain within the active nearshore region and provide continuing storm damge reduction and recreational benefits.

A simple method is presented for quantifying the benefits in and adjacent to beach nourishment project areas. Considering limiting cases in which (a) all sand stays within the area placed, or (b) all sand placed spreads out rapidly demonstrates that the potential benefits are greater for the latter. Example calculations for realistic cases demonstrate that the benefits for project adjacent areas can be substantial relative to those in project areas. The relative benefits in project adjacent areas increase with: short project length, large wave height, large sediment transport coefficient, low interest rate, and large additional beach width.

Accounting methodologies for benefits of beach nourishment projects should be representative of modern understanding of sediment transport processes and the equities of those residing along the shoreline and thus should recognize the benefits from sand transported from the project area and deposited in project adjacent areas.

Although the method presented here applies to the case of projects placed on long uninterrupted shorelines, similar procedures could be applied to situations where littoral controls exist, such as jetties at a channel entrance.

ACKNOWLEDGEMENTS

The work leading to this paper was carried out under funding by the Office of Sea Grant and by the Division of Beaches and Shores of the Florida Department of Natural Resources. This support is gratefully acknowledged.

REFERENCES

Dean, R.G. (1988) "Sediment Interaction at Modified Coastal Inlets: Processes and Policies", in Hydrodynamics and Sediment Dynamics of Tidal Inlets, D. Aubrey, Editor, Woods Hole Oceanographic Institution, Woods Hole, Mass. (In Press).

Pelnard Considere, R. (1956) "Essai de Theorie de l'Evolution des Formes de Rivate en Plages de Sable et de Galets", 4th Journees de l'Hydraulique, Les Energies de la Mar, Question III, Rapport No. 1.

Shows, E.W. (1978) "Florida's Coastal Setback Line - An Effort to Regulate Beachfront Development", Vol. 4, Nos. 1/2, Coastal Zone Management Journal, p. 151-164.

Tackney and Associates (1983) "Physical Monitoring Captiva Beach Restoration Project", Final Report, August.

CHAPTER 117

Effects of structure on deposition of discharged sediment around rivermouth

by
Ichiro Deguchi* and Toru Sawaragi**

ABSTRACT

The influence of coastal structures on a discharged flow and a depositional pattern of discharged sediment from a river are investigated experimentally. It is found that a pair of offshore detached breakwaters, as well as a pair of jetties, had a little influence on them. The offshore detached breakwater has also a function to prevent rivermouth from filling up with depositional sediment by waves. A numerical procedure for the prediction of depositional pattern of discharged sediment is developed based on the experimental results. The proposed numerical procedure is shown to reproduce depositional patterns around a rivermouth without structures and a jetty-protected rivermouth satisfactorily. However, the flow pattern around the offshore detached breakwater-protected rivermouth can not be reproduced.

INTRODUCTION

Until now, jetties have been widely used to protect a rivermouth from blockage. This is because a jetty possesses an effective function to trap sediment transported in the longshore direction. However, this function of jetty produces an abrupt discontinuity of sediment transport and consequently brings erosion of a coast around a rivermouth. The decrease of discharged sediment due to constructions of dams and improvements of river channels accelerate the erosion.

Therefore, to deal with rivermouthes by constructing coastal structures successfully, it is required that such structures have functions not only to prevent the rivermouth from blockage but also not to bring an extreme discontinuity of sediment transport in the longshore direction. It is also desired that the discharged sediment from the rivermouth should be fed back effectively to the beach around the rivermouth.

---

*Associ. Prof., Dept. of Civil Engineering, Osaka University
**Prof., Dept. of Civil Engineering, Osaka University
Yamada-oka, Suita-city, Osaka 565, Japan

The authors have been conducting a series of investigations to establish the optimum design method for rivermouth treatments. This study is a part of the investigations in which the effects of jetties and offshore detached breakwaters, which will be one alternative of jetties, on the behavior of discharged sediment from the river are discussed based on the experimental results. A numerical model for predicting the depositional pattern of discharged sediment is also proposed and the applicability of the model is verified through experiments.

EXPERIMENTS ON THE DEPOSITIONAL PATTERN OF DISCHARGED SEDIMENT

The effects of jetties and offshore detached breakwaters on the discharged flow from the river and depositional patterns of discharged sediment were first investigated by conducting three-dimensional experiments in a wave basin of 5m wide, 15m long and 0.6m deep. A sketch of the wave basin and the model river constructed in it is shown in Fig.1 together with the coordinate system and symbols used in the following descriptions.

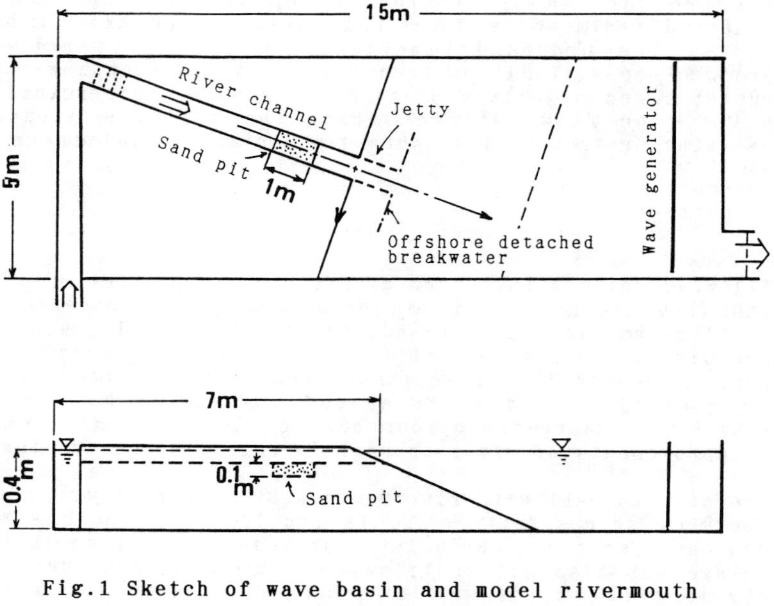

Fig.1 Sketch of wave basin and model rivermouth

A width of the model river channel, B, was 0.5m and a depth at the rivermouth , h_0 , defined at X=0 in the figure was 6.5cm. A river discharge, Q_r , in the experiment was kept constant to be 11700cm^3 /sec and a mean discharged velocity at the rivermouth, U_0, was 45-47cm/sec.

Two kinds of sand of mean grain size d_{50} =0.15mm(fine sand) and 0.35mm(median sand) were used as a bed material of the river and were discharged from a pit by feeding sand into the pit so that a volume of sand in the pit became almost constant. A depth and a length of the pit was 10cm and 1.0m. Fine sand was easily brought into suspension by currents and median sand was mainly transported as bed load.

Jetties and offshore detached breakwaters of different length, the dimensions of which are given in Table 1, were used. The locations of these structure are illustrated in Fig.1.

Table 1 Dimension of structure

Structure		Length:Ls(m)	Ls/B	Ls/Lo
Jetties	I	0.5	1.0	0.5
	II	1.0	2.0	1.0
Offshore detached	I	0.8	1.6	0.8
breakwater	II	1.2	2.4	1.2

Lo:the wave length in deep water

A mean water depth of the wave basin was 30cm and a bottom slopes of the river channel and beach were 1/100 and 1/10, respectively.

After measuring velocities and surface displacements around the rivermouth on a fixed bed, sand was discharged. Depositional patterns were measured after 2 hr's sediment discharge with a resistance type bottom profiler. The velocity was measured with a 2-component electromagnetic current meter in the middle layer and capacitance type wave gauges were utilized to measure surface displacements. Trajectories of tracers thrown into the river and around the rivermouth were also recorded by a video-camera to analyze flow patterns.

A deformation of depositional pattern caused by waves was measured by generating waves obliquely to the shore line for one hour. A height and a period of incident waves were 4cm and 0.8sec and an angle of wave incidence was 20°.

The effects of jetties and offshore detached breakwaters on the discharged flow and depositional patterns of discharged sediment were investigated based on the experimental result.

Figure 2 shows the effects of coastal structures on the velocity,U, of the discharged flow and the mean water level, E, along the center line of the river channel(Y=0).

The discharged velocity and the mean water level measured in the jetty-protected rivermouth shown by open circles do indicate almost the same values as the results obtained in the case where there was no coastal structure which are illustrated by closed circles.

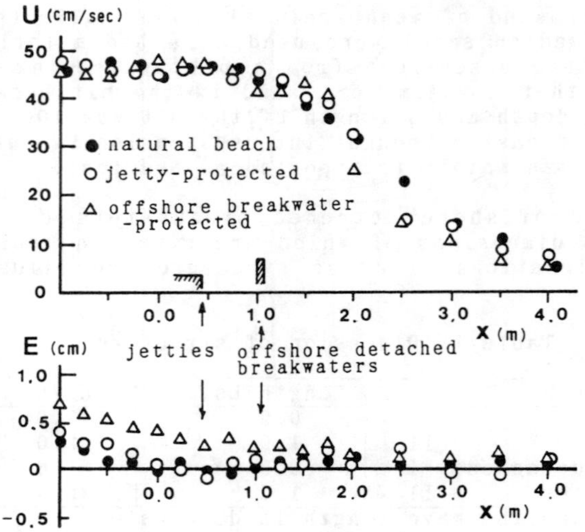

Fig.2 Effects of coastal structures on the discharged velocity and mean water level

When the offshore detached breakwaters were constructed, the rise of mean water level behind them became conspicuous when compared with other two cases shown in the figure. However, the corresponding difference in the discharged velocity in that region can not be found. Although the discharged velocity decreases a little in the region of 2m<X<3.5m (4<X/B<7), it recovers to the same velocity as other two cases.

Figure 3 shows examples of discharged flow patterns. Figures (a) is the case of natural rivermouth without structures. Figures (b) and (c) correspond to the cases of jetty-protected rivermouth and offshore detached breakwater-protected rivermouth.

In the case of natural rivermouth, the discharged flow did not spread in the longshore direction and an entrainment of surrounding water can be seen in the wide region.

In the case of jetty-protected rivermouth, the entrainment of surrounding water took place no sooner than the discharge flowed out of the jetties. Of course, strong divergence can not be found.

On the other hands, at the rivermouth where the offshore detached breakwaters were constructed, the remarkable entrainment took place within a narrow region near the shoreline and the discharged flow diverged outwards behind the breakwaters. However, in the offshore of the breakwater, a small volume of surrounding water was again entrained and the discharged flow did not diverged significantly.

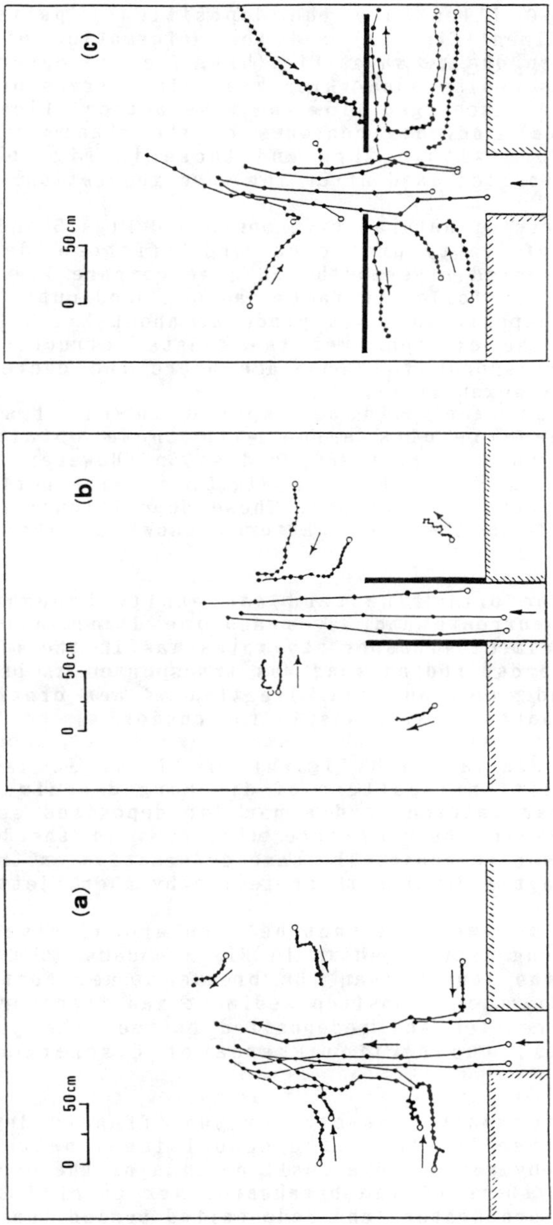

Fig.3 Flow patterns around rivermouth

Figures 4 to 6 illustrate the depositional pattern of discharged sediment(Fig.(a)) and the deformation of depositional pattern due to waves(Fig.(b)). The bed material of these cases was median sand which was mainly transported as bed load by the discharged flow and wave action. Fine solid lines in Fig.(a) indicate contours of the change of water depth took place within 2hrs and those in Fig.(b) give contours of deposited sand after 1hr wave generation.

Fig.4 is the case of natural rivermouth and Figs.5 and 6 are the results of jetty-protected and offshore detached breakwater-protected rivermouth. When we compare Fig.(a) of these figures, the following facts can be found out:
1)The maximum deposition takes place at about X=1.5m(X/B=3) regardless of the existence of the coastal structures and this point corresponds to the place where the decrease of discharged flow began in Fig.2.
2)Discharged sediment did not spread in the longshore direction. These results agree with those obtained by Butakov(1971), Suga et al.(1986) and so on. However, behind the breakwater which is shown in Fig.6, a small portion of discharged sediment was trapped. These depositional pattern correspond well to the flow patterns shown in the former figures.

As mentioned before, fine sand was easily brought into suspension by currents and waves and the direction of net cross-shore sediment transport by waves was in the offshore On the other hands, median sand was transported as bed load by currents and waves and the direction of net cross-shore sediment transport by waves was in the onshore.

When compared Fig.(a) with Fig.(b) of Fig.4, it is found that the depositional pattern of discharged sediment was flattened by wave action and a part of deposited sediment was transported in the longshore direction in the breaker zone around the rivermouth. The same deformation as this was observed around the rivermouth protected by short jetties.

Any longshore movement can not be seen around rivermouth protected by long jetties shown in Fig.5 because the length of the jetty was longer than the breaker zone. Further, a signifi- cant part of deposited sediment was transported in the onshore direction and redeposited between the jetties. As the results, the sectional area of discharged flow decreased.

Around the rivermouth protected by the offshore detached breakwaters given in Fig.6, the depositional pattern was also flattened by waves and a small portion of the deposited sand in the offshore of the breakwater was carried through the gap of the breakwaters and redeposited behind the breakwater. However, the sectional area of the discharged flow was not reduced by these sand movement.

In the cases of fine sand, decrease of sectional area by wave action did not take place even in the case where the long jetties were constructed.

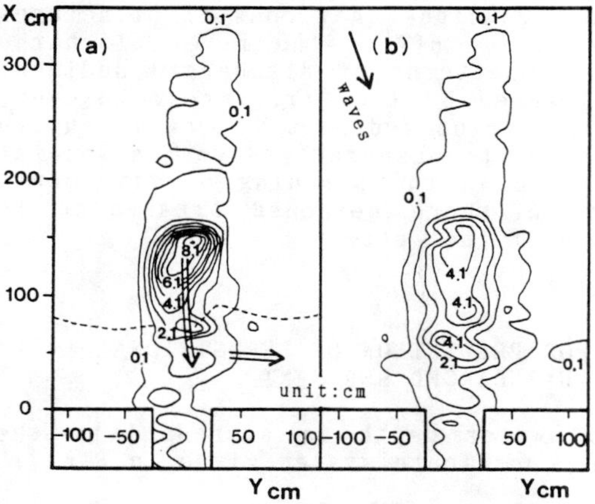

Fig.4
Depositional pattern of discharged sediment (without structure)

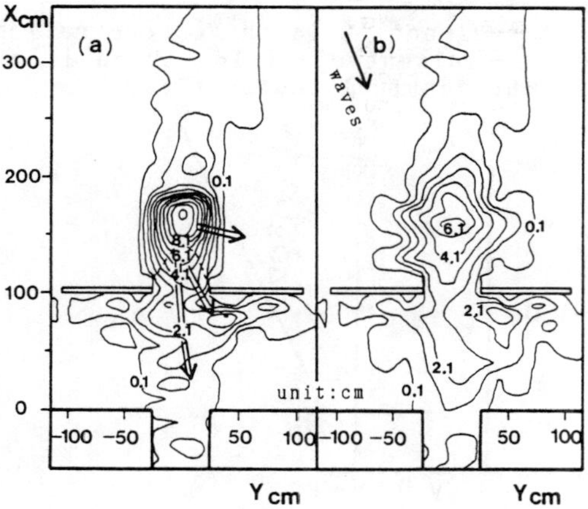

Fig.5
Depositional pattern of discharged sediment (with jetties)

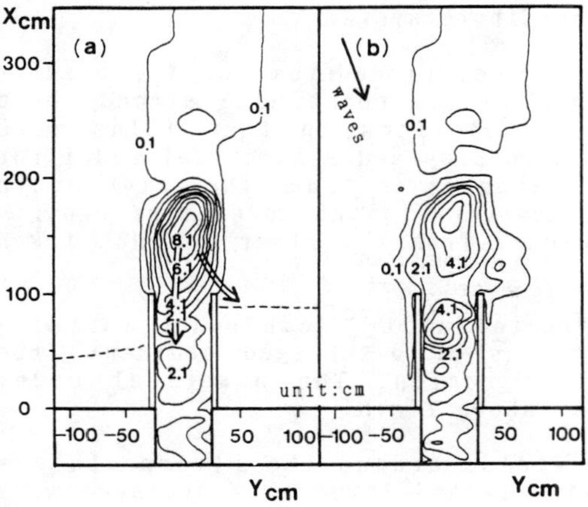

Fig.6
Depositional pattern of discharged sediment (with offshore detached breakwaters)

From these results, we can conclude that coastal structures used in the experiments do not affect the river discharge significantly. Depositional patterns of discharged sediment are also not deeply influenced by the structures. However, when the grain size of discharged sediment is coarse enough so that it is transported in the onshore by waves, a special consideration has to be paid in the planning of rivermouth treatments. Further, the offshore detached breakwater is shown to be an alternative of the jetty.

NUMERICAL PROCEDURE FOR PREDICTION OF DEPOSITIONAL PATTERNS OF DISCHARGED SEDIMENT

Generally, the change of water depth is expressed by the following equation in the coordinate system given in Fig.7:

$$\frac{\partial h}{\partial t} = \frac{1}{1-\lambda}\left(\frac{\partial}{\partial x}\int U_s \overline{C}\, dz + \frac{\partial}{\partial y}\int V_s \overline{C}\, dz + \Delta Q_s\right) \qquad (1)$$

where λ is the void ratio of sediment, $\overline{C}$ is the concentration of sediment, (U_s, V_s) are the sediment migration speed in x- and y-direction and ΔQ_s is the vertical sediment flux.

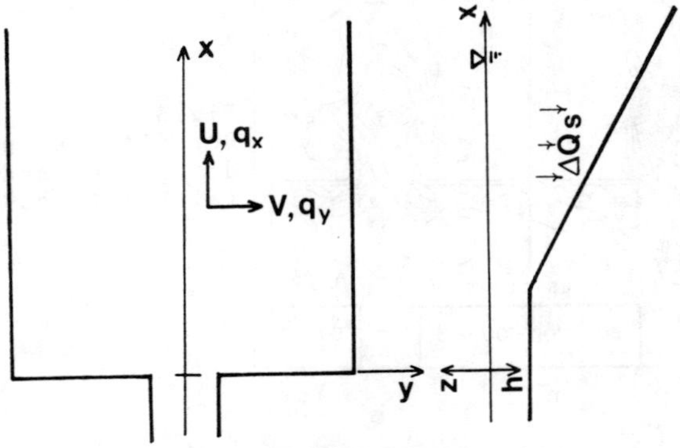

Fig.7 Coordinate system

Based on Eq.(1), some numerical procedures for the prediction of topographic change in the coast have already been proposed and practical simulations in the fields were carried out. The authors also proposed a numerical model for predicting topographic changes around the rivermouth (Sawaragi et al.(1985)). However, in the model, the contribution of discharged sediment from the river was not taken into account.

In this study, we constructed a numerical model for predicting depositional patterns of discharged sediment from the river in the coastal region. The numerical model consists of three parts as shown Fig.8.

First of all, the flow fields around the rivermouth was calculated based on vertically and temporally averaged mass and momentum conservation equations. Then, the horizontal flux and the vertical flux of sediment are estimated and finally, the change of water depth around the rivermouth is calculated from Eq.(1).

1)Calculation of the flow fields
U, V, E
↓
2)Estimation of sediment fluxes
horizontal fluxes : q_x and q_y
vertical flux : ΔQs
↓
3)Calculation of topographic change
Eq.(1)
↓
out put
↓

Fig.8 Flow of the numerical model

1) Calculation of the flow fields
Fundamental equations for the calculation of flow fields are

$$\frac{\partial E}{\partial t}+\frac{\partial}{\partial x}U(h+E)+\frac{\partial}{\partial y}V(h+E)=0 \qquad (2)$$

$$\frac{\partial U}{\partial t}+U\frac{\partial U}{\partial x}+V\frac{\partial U}{\partial y}=-g\frac{\partial E}{\partial x}-\frac{Tx}{\rho(h+E)}+L'\nabla^2 U \qquad (3)$$

$$\frac{\partial V}{\partial t}+U\frac{\partial V}{\partial x}+V\frac{\partial V}{\partial y}=-g\frac{\partial E}{\partial y}-\frac{Ty}{\rho(h+E)}+L'\nabla^2 V \qquad (4)$$

where U and V are the vertically averaged cross-shore and longshore velocities of discharged flow, E is the surface displacement from the still water, Tx and Ty are the time averaged bottom shear stresses in the x-(cross-shore) and y-(longshore) direction and L' is the horizontal mixing coefficients. In the present numerical model, following expressions for (Tx,Ty) and L' are used:

$$(Tx,Ty)=\rho f_w Fc(U,V)/2, \quad Fc^2=(U^2+V^2)$$
$$f_w=(2g/100)/\{18\log(12R/k_s)\}^2 \qquad (5)$$

$$L'=0.01(h+E)\sqrt{g(h+E)}/\tan\theta \qquad (6)$$

where R is the hydraulic radius, k_s is the equivalent roughness height and $\tan\theta$ is the bottom slope.

These equations are transformed into finite difference equations at homogeneous grid points and solved by so-called ADI method. A homogeneous grid system of a distance ΔS=20cm was used and a time increment Δt was 0.15sec.

A boundary condition for the discharge of the river was given as a surface elevation, E_0, at the upstream end of the river channel. By preliminary calculations, E_0 was determined to be 0.8cm to give the discharge of 11.7 l/sec at the rivermouth whose depth was about 6.5cm.

2) Horizontal and vertical flux of sediment

When the sediment concentration is in equilibrium, a formula of sediment transport rate in a steady state can be applied to the horizontal sediment flux, ie, the 1st(=q_x) and 2nd(=q_y) terms in the right hand side of Eq.(1). In this study, the authors apply the following Rijin's formula (Rijin(1985)) to estimate the horizontal sediment flux:

$$\vec{q} = (q_x, q_y) = \vec{q_b} + \vec{q_s}$$
$$\vec{q_b} = C_b \delta_b \vec{U_{sb}}, \quad \vec{q_s} = F d C_a \vec{U_{ss}} \qquad (7)$$

In these equations, C_b and C_a are the sediment concentration in the bed load layer and the reference level, δ_b is the thickness of bed load layer, U_b and U_s are the migration speeds of sand in the bed load layer and suspended load layer and F is the vertical distribution function of the concentration of suspended sediment. These are given by the following equations:

$$C_b = 0.18\, C_0 T/D_* \ (C_0 = 0.65), \quad C_a = 0.015(d_{50}/z_a)(T^{1.5}/D_*^{0.3})$$
$$|\vec{U_{sb}}| = 1.5T^{0.6}\{(\rho_s/\rho - 1)gd_{50}\}^{0.5}, \quad \vec{U_{ss}} = \vec{U}$$
$$\delta_b = 0.3d_{50}D_*^{0.7}T^{0.5},$$
$$F = \{(z_a/d)^{z'} - (z_a/d)^{1.2}\} / (1 - z_a/d)^{z'}(1.2 - z')$$

in which, $T = (u_*^2 - u_{*c}^2)/u_{*c}^2$, $u_*^2 = g/Cz^2(U^2 + V^2)$, $Cz = 18\log(12R/Ks)$, $d = h + E$, $D_* = d_{50}[(\rho_s/\rho - 1)g/\nu^2]^{1/3}$, $z' = W_f/\beta K u_* + 2.5(W_f/u_*)^{0.8}(C_a/C_0)^{0.4}$, and $\beta = 1 + 2(W_f/u_*)^2$.

W_f and K in the above expressions are the settling velocity of sand and Karman's constant and z_a is the height of the referent level of suspended sediment. In the calculation, z_a is assumed to be 0.01*d. u_{*c} is the critical shear velocity for the sand movement. The authors apply u_{*c} on the horizontal bottom proposed by Iwagaki(1956). The effect of bottom slope on u_{*c} is taken into account as the influence of the gravity at the critical stage of the sand movement.

In the Rijin's expression, the suspended load was formulated by giving the bed load as a boundary condition. Therefore, to reproduce depositional patterns observed in the experiments numerically, a special technique was used which will be mentioned afterward.

On the other hands, various expressions for the vertical flux ΔQs in Eq.(1) have been proposed(for example, Sawaragi et al.(1985), Hosokawa et al.(1986)). In this study, ΔQs was defined as the volume of settling sediment which entered the calculation region from the upstream boundary of the river channel and was determined from the convection diffusion equation:

$$\Delta Qs = +\bar{C} W_f \qquad (8)$$

$$\frac{\partial \bar{C}}{\partial t} + U\frac{\partial \bar{C}}{\partial x} + V\frac{\partial \bar{C}}{\partial y} = \frac{\partial}{\partial x}\left(Ksx\frac{\partial \bar{C}}{\partial x}\right) + \frac{\partial \bar{C}}{\partial y}\left(Ksy\frac{\partial \bar{C}}{\partial y}\right) + \frac{1}{d}\left(Ksz\frac{\partial \bar{C}}{\partial z} + W_f\bar{C}\right)_{z=-h} \qquad (9)$$

where Ksx and Ksy are the diffusion coefficient of suspended sediment and are estimated by the following expression (Murray(1968)):

$$Ksx = Ksy = 0.15\, Fc\, d \qquad (10)$$

Eq.(9) is solved numerically at the same grid points as the calculation of flow fields to determine the vertical flux of Eq.(8) with the boundary condition at X= Xstart as

$$C = Cor = \frac{1}{d}\int q_x \, dz \qquad (11)$$

3)Calculation of the change of water depth

Change of water depth, that is, the depositional patterns of discharged sediment from the rivermouth was calculated from Eq.(1).

The topographic change took place within a unit time Δt is expressed by the sum of those caused by vertical flux Δh_v and horizontal flux Δh_h as follows:

$$\frac{\Delta h}{\Delta t} = \frac{1}{\Delta t}(\Delta h_v + \Delta h_h)\ , \quad \frac{\Delta h_v}{\Delta t} = \frac{1}{1-\lambda}\Delta Qs\ ,$$

$$\frac{\Delta h_h}{\Delta t} = \frac{1}{1-\lambda}\left(\frac{\partial q_x}{\partial x} + \frac{\partial q_y}{\partial y}\right) \qquad (12)$$

The change of water depth caused by ΔQs is easily estimated provided that ΔQs is given. However, it is required a special treatment to reproduce depositional pattern brought about by the horizontal flux in the experiments. Because in the experiments, sediment was discharged from the sand pit to the river channel of the fixed bed where there is no source of suspended sediment.

Therefore, in the calculation of the change of water depth, the following procedure was employed:

For the simplicity, consider the phenomena in one-dimension and let Δx and Δt be the space and time increments as shown in Fig.9.

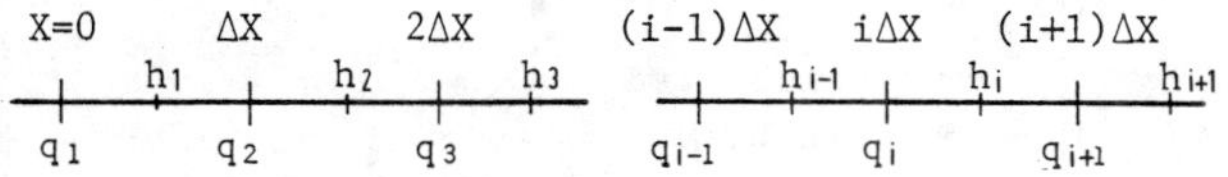

Fig.9 Descritization scheme for the calculation of topographic change due to the horizontal sediment flux

Then, during $t=0-\Delta t$, only q_1 at $x=0$ takes place. During time $t=\Delta t-2\Delta t$, q_1 and q_2 occur at $x=0$ and Δx. During $t=(n-1)\Delta t-n\Delta t$, sediment movement tale place in the region $x<(n-1)\Delta x$. At this time, the change of water depth Δh_i at $x=i\Delta x$, can be expressed by

$$\Delta h_i = \{(n-i+1)q_i - (n-i)q_{i+1}\}(\Delta t/\Delta X)/(1-\lambda) \qquad (13)$$

In the calculation of Δh we used this procedure in both x- and y-direction.

APPLICABILITY OF THE NUMERICAL MODEL

Figure 10 illustrates calculated and measured flow patterns in the case where there was no coastal structures. Calculated results is shown by the velocity vectors and measured result is illustrated as the trajectories of floats.

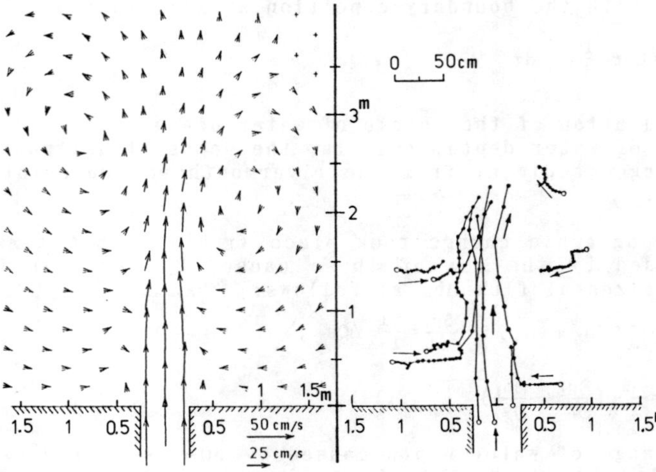

Fig.10 Calculated and measured flow pattern around river-mouth without structure

Figure 11 shows the comparison of measured and calculated discharge velocity U and mean surface displacement E along the center line of the river channel.

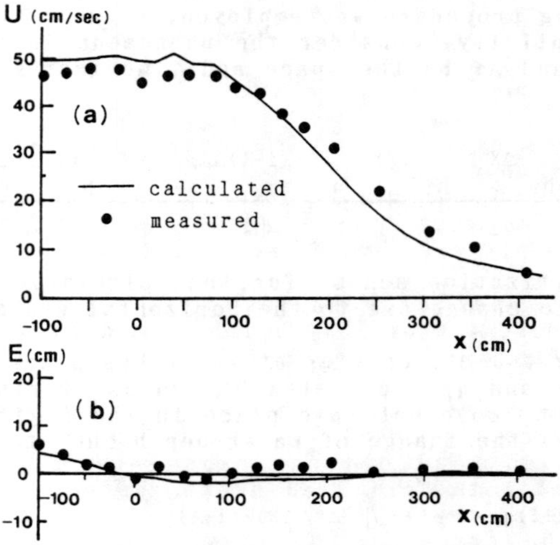

Fig.11 Comparison between calculated and measured discharged velocity and mean water level along a center line of the river channel

These results show that the present numerical procedure for predicting flow field is adequate.

Figure 12 illustrates the comparison of measured and calculated depositional profiles along the center line of the river channel. Fig.(a) is the case of fine sand and (b) is the case of median sand.

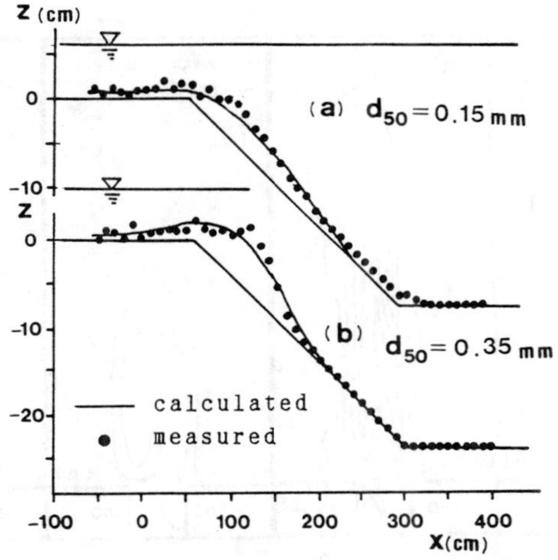

Fig.12 Comparison between calculated and measured depositional profile along a center line of the river channel

From the figure, it can be seen that the depositional profiles of discharged sediment can be predicted by the present model accurately regardless of the grain size of discharged sediment.

Figures 13(a) illustrates the depositional pattern of discharged sediment in the case of fine sand without structure which corresponds to Fig.12(a). Figure 13(b) shows the predicted depositional pattern. When compared with these figures, it is judged that depositional pattern can also be predicted fairly well by the present model.

In Figs.(c) and (d), changes of water depth caused by vertical sediment flux ΔQs, bed load and suspended load are shown separately.

Although the effects of ΔQs extended till X>3m, the amount of the topographic change is small. Change of water depth caused by bed load did not expand beyond X=2.0m which coincides with the critical depth where the bottom shear stress exceeded the critical shear of the sediment movement shown by the arrow in the figure. On the other hands, deposition due to suspended load extended far beyond this critical depth.

The same degree of agreements can be seen between measured and predicted depositional patterns in the cases of jetty-protected rivermouth. However, when the offshore detached breakwaters were constructed, discharged flow pattern in the experiments can not be reproduced by the present model. Therefore, there is some room for further improvements of the model.

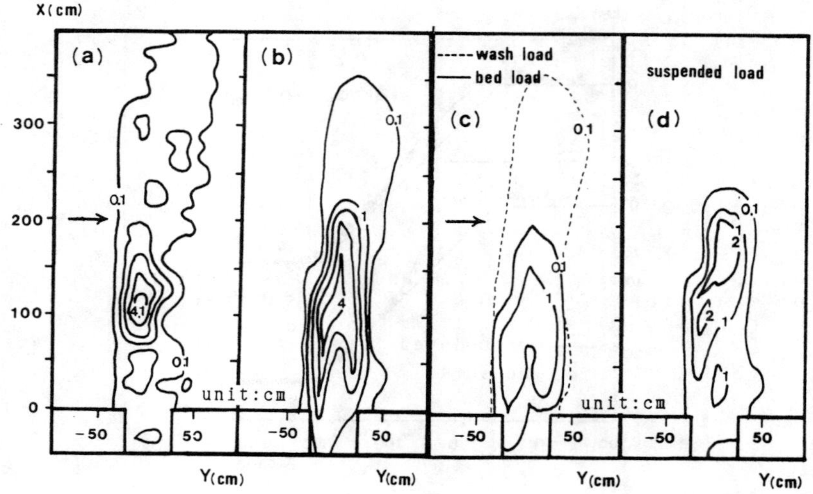

Fig.13 Depositional pattern of discharged sediment in the case of fine sand

CONCLUSIONS

The effects of coastal structures on the discharged flow and depositional patterns of discharged sediment from the river were investigated through experiments. A pair of offshore detached breakwaters is shown to be one alternative of a pair of jetties which have been commonly constructed as a countermeasure against a blockage of rivermouth.

A numerical procedure is also developed for predicting the depositional patterns of discharged sediment. The depositional patterns measured around the rivermouth without structures and around the jetty-protected rivermouth can be reproduced by the numerical procedure precisely. However, the flow pattern measured around rivermouth protected by the offshore detached breakwater can not be reproduced exactly.

REFERENCES

Butakov,A.N.(1971) : Study of development and deformation of mouth bar, Proc. 14th Conf. on IAHR, Paris, pp.95-102.

Hosokawa,Y., N.Tanaka, M.Kudaka and K.Sato(1986) : Method for forecasting quantity of sedimentation in trenches and its application in the field, Proc. 33rd Japanese Conf. on Coastal Engineering, pp.312-316(in Japanese).

Iwagaki,Y.(1956) : Basic study on the critical shear stress for sand movement in the open channels, Proc. JSCE, Vol. 41, (in Japanese).

Murray,S.P.(1968) : Simulation of horizontal turbulent diffusion of particle under waves, Proc. 10th ICCE, pp. 446-466.

Rijin,L.C.(1985) : Sediment transport, Delft Hydraulic Labo. Publication No.334.

Sawaragi,T., J.S.Lee and I.Deguchi(1985) : A new model for a prediction of beach deformation around a rivermouth, Proc. Int'l. Sympo. Ocean Space Utilization '85, Vol.2, pp.229-236.

Suga,K., T.Ishikawa, K.Nadaoka and H.Tanaka(1987): Formation of sand terrace in front of a river mouth and its decline, Technical Note in Proc. of the Japan Society of Civil Engineers, No.381/II-7, pp.227-230(in Japanese).

CHAPTER 118

DUNE STABILIZATION WITH A SAND/GEL COMPOSITE SYSTEM

M. H. Auerbach*, G. W. Borden*, and B. L. Edge, ASCE-M**

ABSTRACT

The effectiveness of a novel sand/gel composite system for the temporary stabilization of coastal dunes was demonstrated in wavetank and field testing. The composite consists of 97% beach sand and water, with a few percent of a biodegradable aqueous polymeric gel made from non-toxic ingredients. The gel binds the sand into a firm but resilient composite that is natural in appearance and resists erosion from waves and tides. The composite is applied by pneumatic gun to the front face of a dune to a few feet below the berm level for toe protection.

This paper describes the initial development and performance of the composite system in wavetank tests at the Oregon State University Wave Research Facility and field trials at Anastasia State Park, St. Augustine, Florida; Chuck's Steak House, Melbourne Beach, Florida; and Ocean Dunes Condominiums, Fort Fisher, North Carolina.

INTRODUCTION

Storm and tidal erosion threatens thousands of miles of shoreline and associated coastal property in the US and throughout the world. Current stabilization methods include permanent armoring, such as concrete seawalls, rock revetments, groins and jetties, and temporary measures such as sand-filled bags and replenishment with dredged sand. Several states in the US have prohibited hard erosion control structures. The cost-effectiveness and environmental consequences of these approaches remain uncertain.

The Chemical Division of Pfizer Inc. has produced water-soluble polysaccharide thickeners and gelants for numerous applications in the oilfield and other industries

---

* Assistant Director and Director, respectively, Specialty Chemicals R & D, Pfizer Central Research, Eastern Point Road, Groton, CT 06340

** President, Cubit Engineering division of Kimley-Horn, 207 East Bay Street, Charleston, SC 29401

for over ten years. The adaptation of this technology for coastal erosion control was investigated.

EXPERIMENTAL PROGRAM

Laboratory Testing

A wide range of gel/sand compositions was evaluated in the laboratory using a water impact test device. Successful compositions, based on a weight loss/hour limit under continual water impact, had unconfined compression strength values of 3000 - 30,000 lb/ft^2.

Environmental Exposure

The leading composition candidates were prepared in a small cement mixer, cast in 1 ft^3 wooden pans and anchored in the sand at a 45° angle on the eastern bank of the mouth of the Thames River in Groton, Connecticut. Although not subject to appreciable wave impact at this Long Island Sound location, the pans were totally submerged twice a day at high tide. Successful compositions endured intact for over seven months with only minor gouges from driftwood impacts, until the pans were uprooted in a storm.

Emplacement Testing

Pilot sand drum tests showed that ungelled polymer solution sprayed onto the sand surface did not penetrate significantly. Shank-injected ungelled polymer solution did not permeate the sand but left gel-filled voids. Large batch mixing of sand with ungelled polymer solution in a cement mixer and shoveling onto a dune was not possible because of rapid polymer gelation kinetics. Accordingly, the pneumatic gun approach, where sand and polymer gel component solutions are comixed with large amounts of air and sprayed onto the dune, was attempted.

Gun Trials

Five trials were conducted from 1985-87 with commercial gunite equipment at sand feed rates of 120-200 lb/min (1.6-2.7 ft^3/min composite) to test various spray nozzle designs, equipment configurations, and component combinations. Numerous beach pans were sprayed full and evaluated for pregel slump and drying rate. Several of these composite panels were placed on the Groton beach for confirmation of environmental endurance. These panels also remained intact without erosion for the several month exposure period.

OSU WAVETANK TRIALS

To evaluate the full-scale strength and erosion resistance of the composites under controlled conditions, tests were conducted at Oregon State University's Wave Research Facility in May, 1986. The 12-ft wide x 18-ft deep x 350-ft long tank has a computer-controlled hydraulically-driven wave generator.

Dune Preparation

A 1:12 sloping berm was prepared with movable concrete slabs covered with sand in the bottom of the tank. The berm was seasoned with gentle waves and three 8-ft dunes were constructed about 20 feet apart with a 1:1 sloping surface at the end of the tank with local sand. A thin non-woven geotextile was placed over the front surface of each dune to keep the sand base from blowing away during spraying and from washing out if the composite should crack during wave testing. The composites were emplaced with a pneumatic gun onto this filter fabric at a rate of 2.0-2.7 ft^3/min.

Each composite was about 2-ft thick at the bottom, 1-ft thick at the top, and 7 feet high (Figure 1). Different compositions and feed configurations were used for each dune. As the weather in Corvallis in May was cool and damp, a wooden frame and plastic tent were erected over the dune end of the wavetank and kerosene heaters were placed inside to speed composite drying. Colored horizontal stripes were painted on the composite surfaces to measure runup, and a swash gauge was built in the tank to measure wave velocity.

Wave Testing

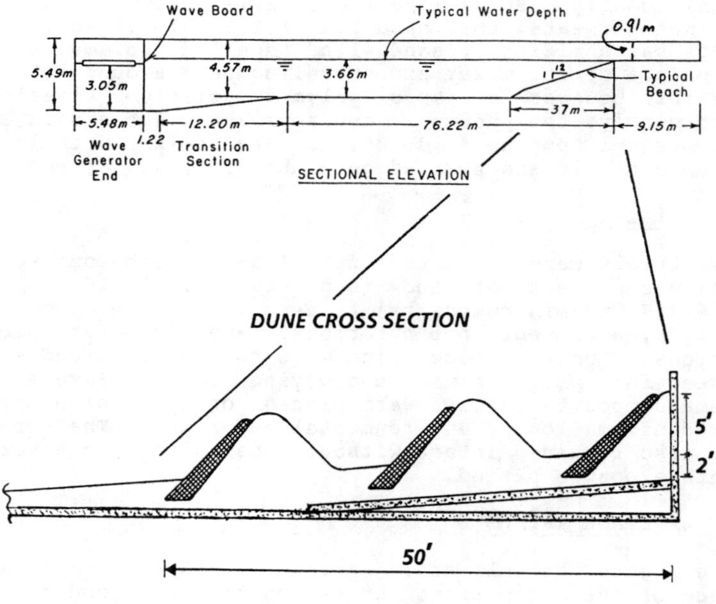

Figure 1. OREGON STATE UNIVERSITY WAVE RESEARCH FACILITY

The tank was filled with water to the base of the front dune. Monochromatic and Bretschneider spectrum random waves were generated with 7-11 ft/sec velocity, 1-11 second period, and 2-5 foot height (Figure 2). Conditions were controllable to allow berm sand to erode away from the dune face or accrete against it. They were chosen so that sand and gravel were thrown against the composite for lengthy periods to exacerbate erosion. The water level was adjusted so that the bores ran up to the top of the composite without constantly overtopping the dunes. In many cases, the waves broke just in front of the dune so that the swash generated was almost entirely dissipated on the dune itself. These conditions simulated average US East Coast annual storms, and likely represented a more severe 2-3 year event.

One dune was tested each day. Each dune was exposed to wave impacts for 4-6.5 hours. Aside from minor sloughing of a crusty skin from one of the composites and loss of a chunk from another because of a gunning defect, there was no appreciable erosion of the composites during testing (Figure 3).

At the end of the first dune test, a 1-ft hole was dug in the center of the composite with a shovel and wave

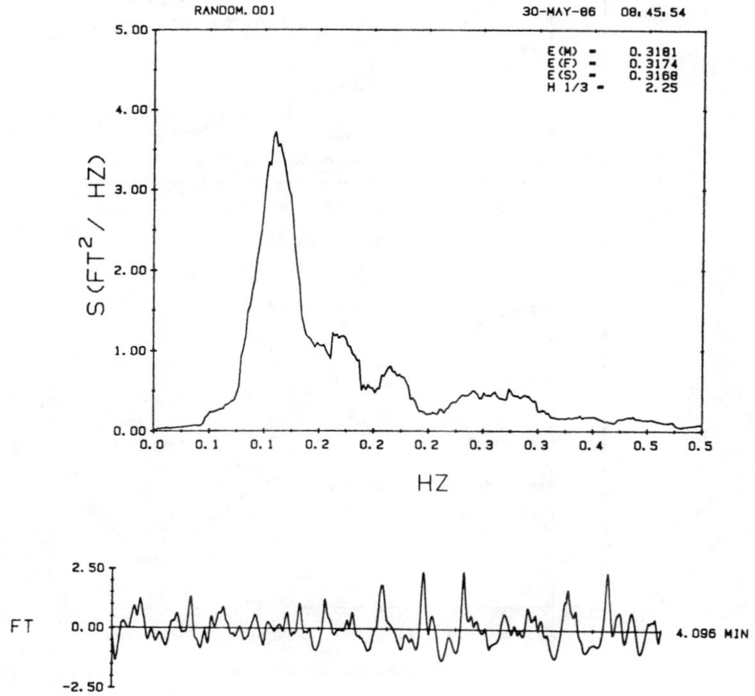

Figure 2. OSU BRETSCHNEIDER WAVE SPECTRUM

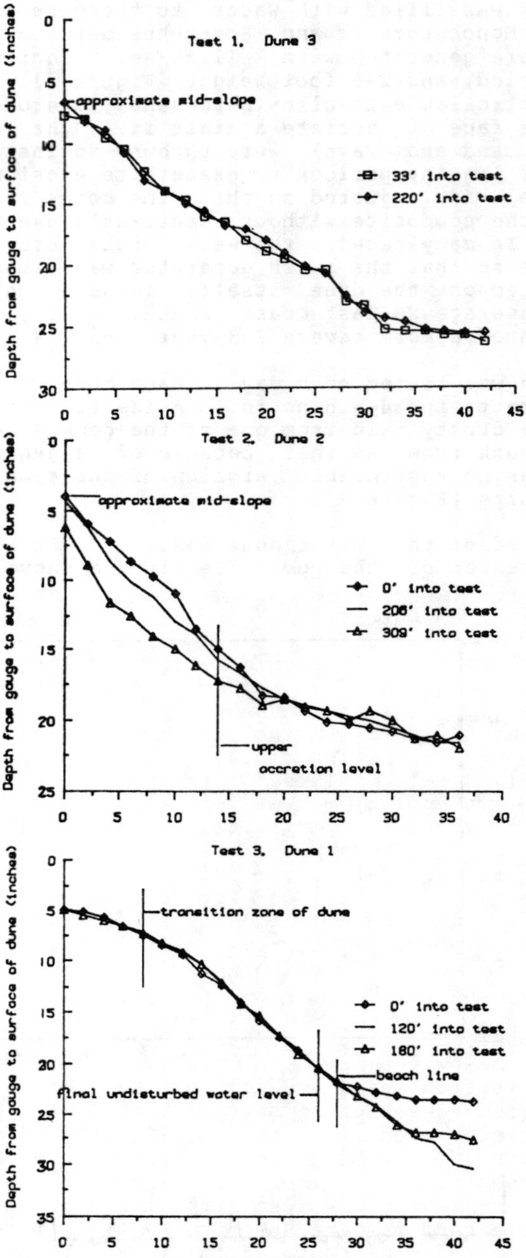

Figure 3. OSU Wavetank Tests
Surface Profiles of Test Dunes
Measured 2-ft from Right Side of Dune

uprush pumped water into the hole. With the exposed geotextile intact, no sand was lost from the dune base and the composite hole was not widened by additional wave impacts. After wave tests on the second dune were completed, a hole was dug in the composite and the geotextile was cut open. Additional waves for a half hour created a 2-ft diameter cavity in the dune base underneath without breaking the composite.

After each day's testing was completed, the intact dune was removed from the tank with a clamshell and the berm repaired for the next test. After all three composites were tested and removed, a fourth dune was built in the tank with untreated sand as a control. Full amplitude 8-second waves (~ 3-ft height) obliterated 80% of this dune within seven minutes, leaving a vertical escarpment.

SHORELINE FIELD TRIALS

Three field tests of the sand/gel composite system have been initiated to date. These are summarized in Table 1 and described below.

Florida Field Trial I - Anastasia

Permitting: Permission was granted by the Florida Department of Natural Resources, Division of Recreation and Parks, in November, 1986, for a field test at the south end of Anastasia State Recreation Area in St. Augustine. As the east coast of Florida is a prime nesting area for endangered marine turtles, the US Department of Interior Fish and Wildlife Service was also consulted about the test. No turtle nests had been observed at the south end of the park in several years, so no turtle monitoring was required. The USDI FWS also advised that the Anastasia Island beach mouse (*Peromyscus polionotus phasma*) was not listed as an endangered species but was a candidate for listing. *P. p. phasma*, a subspecies of the common beach mouse and closely related to the Santa Rosa Island, Perdido Key, Choctawhatchee, and Alabama beach mice, lives among vegetation behind the dune face (USDI FWS, 1986) and would not be affected by the test.

Emplacement: The composite was designed to extend several feet below the berm to provide toe protection and prevent it from being undercut during a storm that might lower beach elevation by several feet. A trench 3 feet deep was excavated in front of 90 feet of eroded dune just south of the beach access ramp. The dune surface was dressed to a 1:1 slope to the bottom of the trench and geotextile was unrolled along the dune face. Beach-compatible sand was imported from a local pit to build a 200-ft control dune south of the test section (sieve analyses in Table 2). The 90-ft test dune was gunned with composite in two configurations (Figure 4) using beach sand from the toe trench. After spraying was completed, a heated tent was erected over the composite to expedite

Table 1: SAND/GEL COMPOSITE FIELD TRIALS

Site	Anastasia State Park	Chuck's Steak House	Ocean Dunes Condos
Location	St. Augustine, FL	Melbourne Beach, FL	Fort Fisher, NC
Empl. Date	December 1986	April 1988	May 1988
Elevation			
Crest	+ 12.0 NGVD*	+ 17.0 NGVD	+ 17.0 NGVD
Toe	+ 5.0 NGVD	+ 8.0 NGVD	+ 8.0 NGVD
Height	7 feet	9 feet	9 feet
Length	90 feet	170 feet	100 ft 100 ft
Slope	1:1	1:1	1:1 1:1.5
Thickness			
Crest	12 inches	15 inches	16" 12"
Toe	24 inches	18 inches	22" 18"
Ft^3/lin-ft	15.0	18.0	20.5 21.0
Spray Rate	1.8-2.4 ft^3/min	1.7-4.5 ft^3/min	2.6-5.4 ft^3/min
Profile Monitoring (Number Profiles - Total Feet)			
N Control	5 - 75	5 - 180	4 - 300
Composite	6 - 90	5 - 170	6 - 200
S Control	5 - 75	6 - 250	8 - 700
Other Monitoring	Storms	Storms Vegetation Sea Turtles Waves/Tides	Storms Vegetation

* National Geodetic Vertical Datum

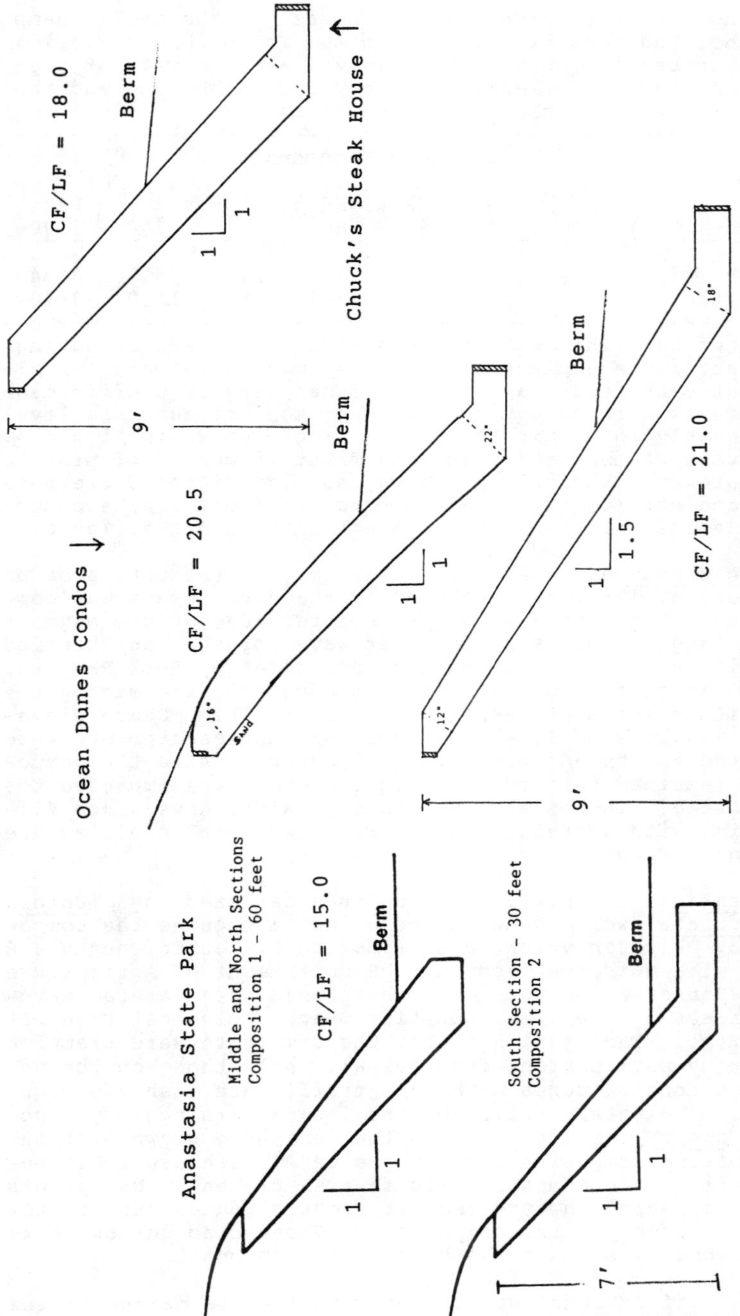

Figure 4. Sand/Gel Composite Cross-Sections

drying, a precaution taken because of the cool, damp weather and possibility of storms in northeast Florida in December. The tent was removed after a week of drying and the side flanks were covered with sand and the berm in front of the composite was replaced.

Table 2: Anastasia Sand Analyses

Sand	% H_2O	% Retained on - Mesh 40	60	80	100	200	Pass 200
Beach	9.4	0.0	5.0	61.6	24.4	8.8	0.24
Imported	9.6	0.0	5.5	53.6	29.6	11.0	0.30

Monitoring: The test area was videotaped before, during, and after the emplacement. Survey monuments were installed atop the test and control dunes and 16 profile measurements were made from the dune top to mean sea level at monthly intervals at the positions shown in Figure 5. Analysis of the data from the first 15 months of profile measurements indicate there was no significant difference in sand erosion or accretion rate in front of the composite vs. in front of the adjacent control dunes (Table 3).

Storm Exposure: A major northeaster hit the test area on January 5, 1987, two weeks after the emplacement was completed. The storm surge was not recorded at the Mayport tide gauge 60 miles north, but wave uprush was observed running up to the top of the composite by Park Rangers. This was confirmed by a storm hindcast performed by the Institute for Storm Research, Houston, TX. Beach elevation was lowered 1.5-2 feet and 2-5 foot escarpments were created in the adjacent control dunes, while the composite remained intact. No other storms have impacted the composite. The escarpments on both sides have since filled in with accreted sand. Selected dune profiles are shown in Figure 6.

Vegetation: In April, 1987, 600 sea oat seedlings (Uniola paniculata) were planted above and alongside the compoite by children volunteers from the St. Johns County 4-H Club with guidance from the US Department of Agriculture Soil Conservation Service. The plants were watered several times by the St. Augustine Beach Volunteer Fire Department. The seedlings atop the composite were trampled later by park patrons (see below), while those on the adjacent control dunes with less traffic are lush and vigorous. Voluntary railroad vine (beach morningglory, Ipomoea pes-caprae) from behind the dune have grown over the top of the composite down to the berm, with new roots penetrating the composite in several places. New plants also appeared in the rebuilt control dunes and in the berm in front of the composite. Ghost crab burrows were occasionally seen in the base of the composite.

Foot Traffic/Vandalism: The only noticeable damage to the composite has been the result of frequent foot traffic and occasional vandalism by park patrons. Some shovel

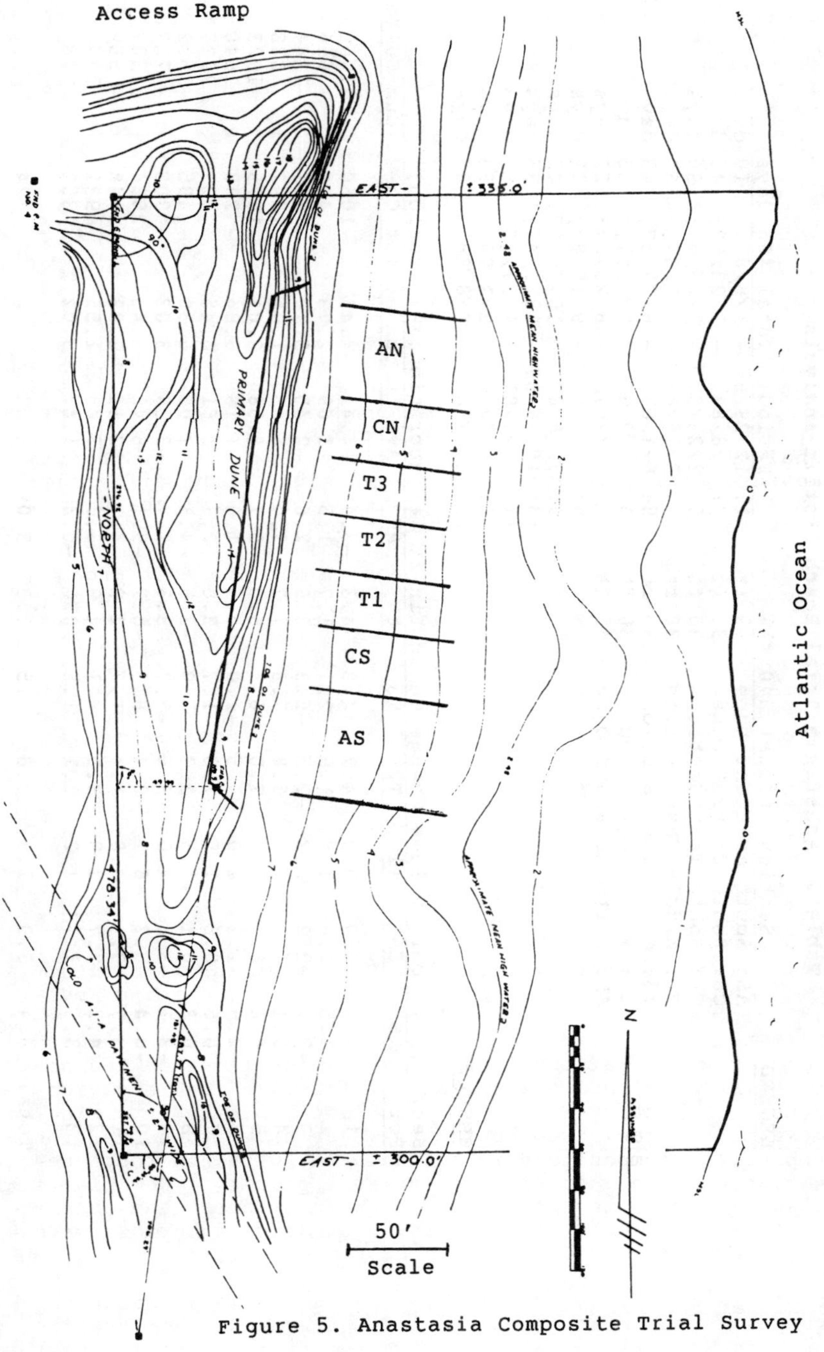

Figure 5. Anastasia Composite Trial Survey

Table 3: Anastasia Trial Beach Profile Analysis

LEGEND

Section (Total 240 ft)	
AS: South Adjacent Dune	45 ft
CS: South Control Dune	30 ft
T1: South Test Section	30 ft
T2: Middle Test Section	30 ft
T3: North Test Section	30 ft
CN: North Control Dune	30 ft
AN: North Adjacent Dune	45 ft

Period (Total 466 days)			
1:	12/29/86	1/8/87;	10 days
2:	1/8/87	2/23/87;	46 days
3:	2/23/87	3/25/87;	30 days
4:	3/25/87	6/1/87;	68 days
5:	6/1/87	6/29/87;	28 days
6:	6/29/87	8/3/87;	35 days
7:	8/3/87	8/25/87;	22 days
8:	8/25/87	9/16/87;	22 days
9:	9/16/87	10/26/87;	40 days
10:	10/26/87	12/1/87;	36 days
11:	12/1/87	1/27/88;	57 days
12:	1/27/88	3/1/88;	34 days
13:	3/1/88	4/8/88;	38 days

RESULTS

Sec/	Unit Volume Change Cu-Yd/Lin-ft (+: accretion; -: erosion)										
Per	AS	CS	T1	T2	T3	CN	AN	Avg/ft	Cum Av	Tot Cu-Yd	Cum Cu-Yd
1	-4.4	-4.9	-5.0	-4.7	-4.8	-4.3	-3.4	-4.42	-4.42	-1060	-1060
2	1.5	1.6	1.8	1.9	2.1	1.7	1.5	1.69	-2.73	405	- 656
3	-2.1	-2.2	-2.2	-2.5	-2.8	-2.9	-2.6	-2.45	-5.18	- 589	-1244
4	6.1	6.1	6.1	6.8	7.3	7.2	7.0	6.63	1.45	1592	348
5	6.8	6.9	7.3	7.0	6.4	6.0	5.7	6.56	8.01	1575	1923
6	-2.6	-2.2	-2.0	-1.5	-1.0	-0.4	1.0	-1.18	6.84	- 283	1640
7	0.6	0.8	1.5	1.0	1.7	3.0	2.7	1.61	8.45	387	2027
8	0.3	-0.1	-1.3	-2.3	-2.9	-3.6	-4.5	-2.06	6.38	- 495	1532
9	-3.1	-3.8	-4.0	-3.5	-3.7	-3.3	-2.7	-3.36	3.03	- 805	727
10	-8.4	-8.2	-8.1	-8.4	-8.3	-8.7	-9.1	-8.48	-5.45	-2034	-1307
11	3.7	3.9	3.8	1.4	1.7	2.2	4.3	3.12	-2.33	748	- 559
12	4.3	3.7	2.9	3.5	4.1	4.9	3.6	3.88	1.55	931	372
13	3.5	1.7	1.2	2.1	1.5	0.6	-0.4	1.47	3.02	352	724
Total	6.4	3.3	1.9	0.8	1.5	2.3	3.0	3.02		724	

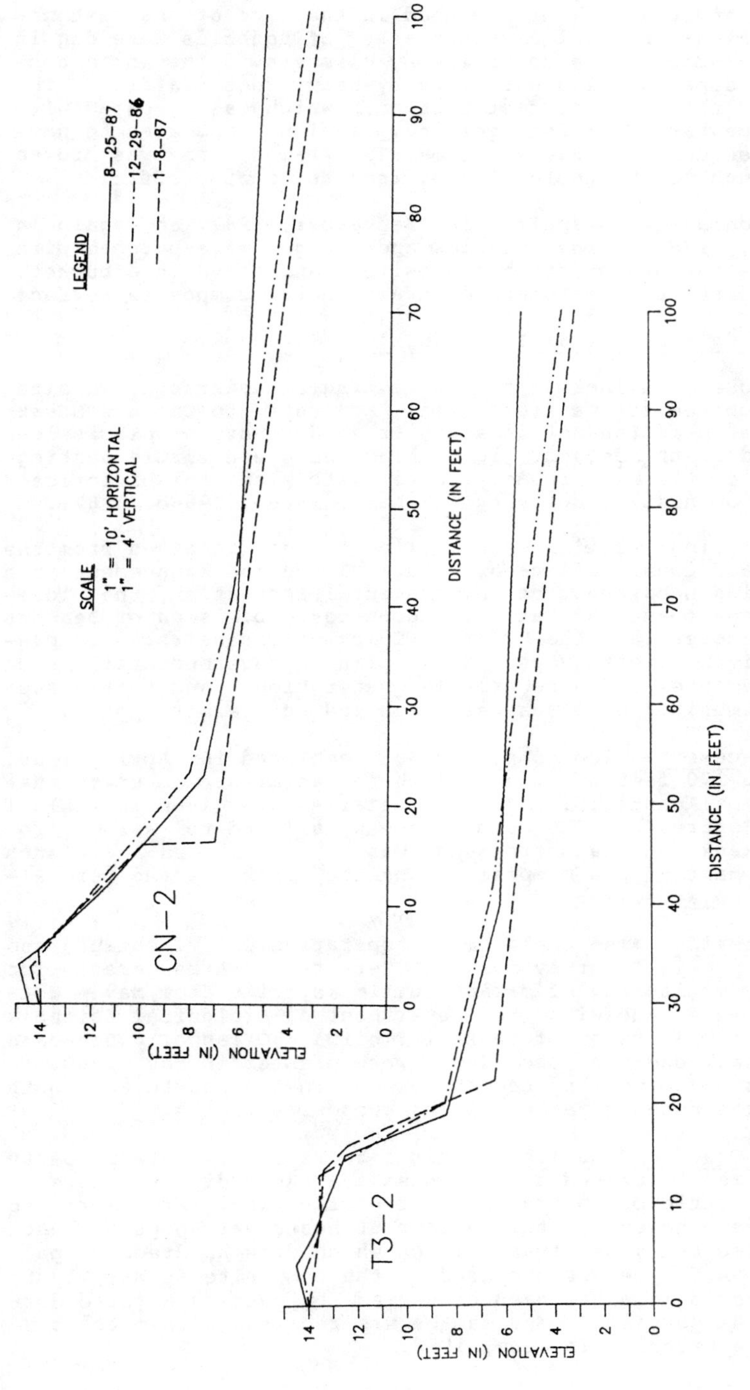

Figure 6. Anastasia Trial Dune Profiles

cuts and graffiti were gouged in the face of the test material and several distinct sets of toeholds were dug in the middle of the composite by climbers. The north control dune was also worn down by heavy foot traffic. The south third of the test section, which was prepared with a somewhat different gel composition, appeared to have softened in the heavy summer 1987 rains, and was broken through to the geotextile by repeated foot traffic.

The damage was repaired in September, 1987, and again in June, 1988. The toeholds and gouges were patched with small batches of fresh composite hand-mixed in a bucket. The patches have remained intact in the composite surface.

Florida Field Trial II - Chuck's

Because of a lack of turtle nesting at Anastasia, a site was chosen for a field trial adjacent to Chuck's Steak House, a restaurant in south Brevard County. This is located in the densest loggerhead and green turtle nesting area of the Western Hemisphere, with plentiful historical data on nesting activity (Ehrhart, 1986a, 1986b, 1987).

Permitting: Permits were applied for and received from the Brevard County Office of Natural Resources Management; the Florida Department of Environmental Regulation; the Florida Department of Natural Resources, Division of Beaches and Shores; and the US Army Corps of Engineers. Comprehensive site surveys, work plans, environmental impact statements, and dune profile, vegetation, and turtle nesting monitoring plans were required for permit approval.

Emplacement: The composite was emplaced in April, 1988, along 170 feet of dune surface in a similar manner to that used at Anastasia. Further details are given in Table 1 and Figure 4. The composite was allowed to air dry for one week and then berm sand was restored such that half of the composite's height was below the berm elevation at + 12.5 ft NGVD.

Monitoring: The emplacement operation was videotaped and dune profile survey markers were set. Arrangements for daily monitoring of marine turtle activity from May - October were made with the Department of Biological Sciences of the University of Central Florida, Orlando. Sea oats and railroad vine seedlings were planted in May, 1988, in the sand above the composite and in the north and south control dunes adjacent to the composite flanks.

Foot Traffic/Vandalism: This site is on private property much more isolated than Anastasia. Accordingly, there is very little beach traffic most of the time. From April to October, however, the restaurant hosts very popular beach parties every Sunday afternoon which draw hundreds of people. A few gouges appeared in the composite in May, 1988; further damage has been prevented by posting a guard during the parties. The damage was repaired with patch composite material in June, 1988.

North Carolina Trial - Ocean Dunes

Because of the lack of significant storm exposure at Anastasia, a site adjacent to the Ocean Dunes Condominiums, Fort Fisher, was chosen for a field trial. This area has endured several severe winter storms that caused property damage in the last few years. A permit for a test was obtained under the North Carolina Coastal Area Management Act from the Department of Natural Resources and Community Development.

Composite was emplaced along 200 feet of eroding dune in two configurations (Figure 4) in May, 1988. The composite was air-dried for one week and the berm was replaced as at Chuck's Steak House. Dune profile monitoring along 1200 feet of beach was initiated to measure longshore sand transport effects. As this is a private development, no significant damage from foot traffic has been observed.

CONCLUSIONS

The sand/gel composite system has been shown to be resistant to erosive forces, natural in appearance and compatible with beach vegetation. As intended, it appears to be resilient and environmentally compatible. Monitoring of the three field trials will continue for several years.

This technology is intended to stabilize threatened dunes against typical tides and storms long enough to allow natural vegetation to take hold, perhaps 3-5 years. In time, the composite should slowly disintegrate back to sand and water. Individual pieces of composite material, if broken free in a severe storm or hurricane, should break up in the surf and not accumulate on the beach. This novel system will hopefully provide a cost-effective, environmentally acceptable alternative to current approaches for coastal protection.

ACKNOWLEDGEMENTS

We gratefully appreciate the support and assistance of the following key people in carrying out this program: Merrill Lozanov and Dr. Irving Goldman, Pfizer Central Research; Sheldon Brandes, Pfizer Chemical Division; C. W. Hartsfield, Florida Department of Natural Resources, Division of Recreation and Parks; Ralph Clark and Neal Rogers, Florida Department of Natural Resources, Division of Beaches and Shores; Ross Witham and Alan Huff, Florida Department of Natural Resources, Division of Marine Resources; John Roche and Doug Carter, Florida Park Service; Derek Busby, Brevard County Office of Natural Resources Management; Earl Possardt, US Department of Interior, Fish and Wildlife Service; Calvin Hubbard, Don Hamer, and Keith Salvo, US Department of Agriculture, Soil Conservation Service; Cynthia Goodman, St. Johns County Extension Service; Henry Beeker, North Carolina Department of Natural Resources and Community Development; Dr. Charles Sollitt, Department of Civil Engineering, Oregon State University; Dr. Llewellyn Ehrhart, Department of Biological Sciences, University of Central Florida; Dr. Richard Seymour, Scripps Institution of Oceanography, University of California at San Diego; and Dr. John Fisher, Department of Civil Engineering, North Carolina State University.

APPENDIX - REFERENCES

1. L. M. Ehrhart and B. E. Witherington, "Human and Natural Causes of Marine Turtle Nest and Hatchling Mortality and Their Relationship to Hatchling Production on an Important Florida Nesting Beach," Florida Game and Fresh Water Fish Commission Project GFC-84-018, 12 June 1986.

2. L. M. Ehrhart and B. E. Witherington, "Studies of the Ecology of Indian River Marine Turtle Populations and Marine Turtle Nesting Beach Productivity in South Brevard County, Florida," 1986 Progress Report to Bureau of Marine Research, Florida DNR Contract C3848, 15 December 1986.

3. L. M. Ehrhart and P. W. Raymond, "Loggerhead and Green Turtle Nesting Densities in South Brevard County, Florida, 1981-84," in W. N. Witzell, ed., Ecology of East Florida Sea Turtles, NOAA Tech. Rep. NMFS 53, USDC, 1987.

4. US Department of Interior, Fish and Wildlife Service, "Beach Mouse Recovery Plan," Atlanta, GA (1986)

CHAPTER 119

Littoral Drift Model for Natural Environments

Rolf Deigaard [1], Jørgen Fredsøe [2], Ida Brøker Hedegaard [1]
Julio A. Zyserman [2] and Ole Holst Andersen [1]

ABSTRACT

The littoral drift model developed at DHI and ISVA, see Deigaard et al. (1986b) has been extended to include the effects of the irregularity of the waves, of a coastal current and a wind acting on the surf zone. Further, a mathematical model to simulate the near-shore current pattern along a barred coast with rip channels has been developed.

The influence on the littoral drift of the irregularity of waves, wind, coastal current, and rip channels is discussed. It is concluded that irregularity of waves and presence of rip channels must be considered while coastal current and wind action are of minor importance.

INTRODUCTION

The modelling of longshore sediment transport has been improved significantly during the last decade. From formulae based on 'longshore energy flux' originating from the incoming waves, a more detailed formulation of the problem is now appearing. However, it is common to most of the approaches that the description still relies on regular incoming waves which meet a coast with a constant slope, and with quasi-uniform longshore conditions.

The purpose of this paper is to discuss the influence on the littoral drift of the natural variability of waves and coastal profiles. Besides analysis is carried out of the effect on the littoral drift of a coastal current and a wind blowing over the surf zone.

The starting point of the present analysis is the littoral drift model, developed at DHI and ISVA, and described by Deigaard et al. (1986b). It is a mathematical model which consists of two main elements.

----

1) Danish Hydraulic Institute (DHI), Agern Allé 5, DK-2970, Hørsholm, Denmark.
2) Institute of Hydrodynamics and Hydraulic Eng. (ISVA), Technical University of Denmark, DK-2800 Lyngby, Denmark.

- A hydrodynamic module describing the wave height and direction in the coastal zone and the wave generated littoral current.

- A sediment transport module which calculates the local sediment transport rate as a function of the local hydraulic parameters. The sediment transport rate is integrated over the coastal profile.

The processes involved in wave transformation are refraction, shoaling and breaking. The associated variation in radiation stresses is used to calculate the shore normal component causing a wave set-up and the shear component of the radiation stress, S_{xy}. The shear component is driving the littoral current, it is constant outside the breaker line where no energy dissipation occurs. The littoral current is calculated from the balance between bed shear stress τ_b, cross-shore momentum exchange and the gradient in shear radiation stress:

$$\tau_b - \frac{d}{dx}\left(\rho E D \frac{dV}{dx}\right) = -\frac{dSxy}{dx} \qquad (1)$$

where x is the cross-shore coordinate, ρ is the density of water, E the momentum exchange coefficient, D the water depth, and V is the depth averaged littoral drift velocity.

The local sediment transport is calculated by the model described by Fredsøe (1984), Fredsøe et al. (1985), and Deigaard et al. (1986a). The bed load and suspended load transport is calculated separately. The suspended load is normally dominant in the surf zone; it is described by the vertical turbulent diffusion equation:

$$\frac{\partial c}{\partial t} = \frac{\partial}{\partial z}\left(\varepsilon_s \frac{\partial c}{\partial z}\right) + w \frac{\partial c}{\partial z} \qquad (2)$$

where z is the vertical coordinate, c the sediment concentration, t is time, ε_s turbulent diffusion coefficient, and w is the settling velocity. ε_s is calculated by taking the turbulent wave-current boundary layer as well as the turbulence generated by breaking or broken waves into account. The boundary conditions for Eq. (2) are zero flux of sediment through the water surface and a time-varying near-bed concentration calculated by the model of Engelund and Fredsøe (1976).

Fig. 1 gives an example of model results showing the profile of the littoral current and the littoral sediment transport rate across a coastal profile with one longshore bar.

The above described model has a number of limitations, the most important being:

. the incoming waves are unidirectional and regular
. the conditions along the coast are uniform
. only driving forces from the waves are included.

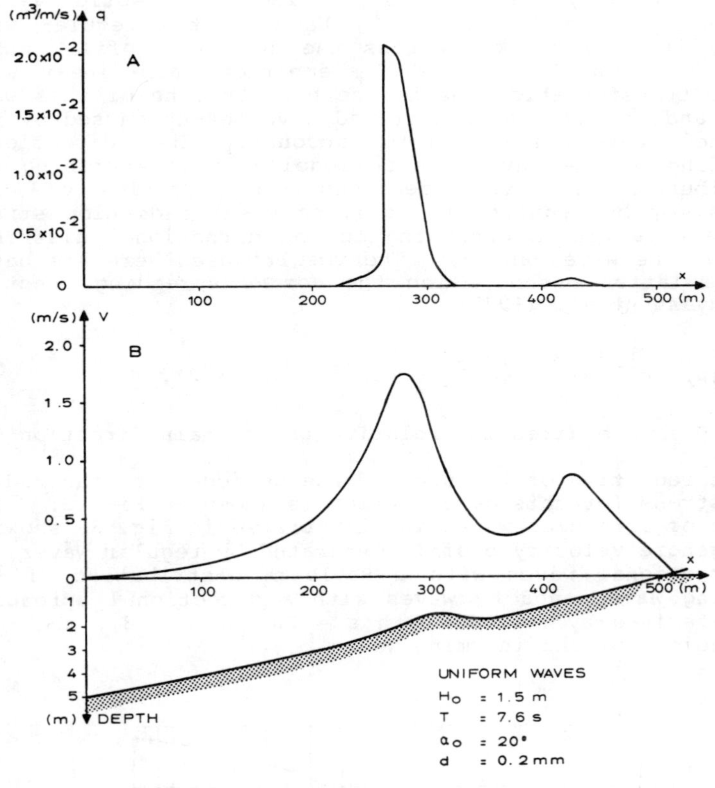

Fig. 1. Example of model results.
A: Littoral Sediment Transport Rates.
B: Profile of Littoral Current and Coastal Profile.

In the following it is described, how the model has been modified to overcome these limitations, and the effect of the various mechanisms treated is discussed.

IRREGULAR WAVES

The effect of irregular waves versus regular monochromatic waves has been included in the modelling of the longshore current as well as of the sediment transport rate.

The longshore current velocity profile driven by irregular waves with directional spreading was analysed in detail by Battjes (1974) and the present model conforms with his results. The present model describes the irregular wave situation as a series of regular wave trains each characterised by its height H_i, period T_i, direction α_i and frequency of occurrence, ϕ_i. Each of the regular wave trains is then tracked across the coastal profile, which makes it possible to use the empirical non-linear wave height transformation inside the breaker line of this wave train and to take the integrated wave set-up caused by the combined wave trains into account. The directional spreading of the waves can be modelled by prescribing the distribution of wave directions for a series of wave trains, or by reducing the driving shear radiation stress of the wave trains according to the directional distribution of the wave energy. The results used here are based on the latter method using the cosine spreading function (Mitsuyasu et al. (1975):

$$H(\Theta) = \frac{2^{2s-1}}{\pi} \frac{\Gamma^2(s+1)}{\Gamma(2s+1)} \cos^{2s}((\Theta-\Theta_o)/2), \tag{3}$$

where Θ is the direction relative to the main direction Θ_o.

The reduction of the shear force component of the radiation stress for this distribution is shown in Fig. 2. The effect of irregular waves is illustrated in Fig. 3, showing a longshore velocity profile generated by regular waves, by unidirectional waves with a Rayleigh distribution of the incoming waves, and by waves with a directional spreading characterised by Eq. (3) with s = 20 and s = 3, and regular height of the incoming waves.

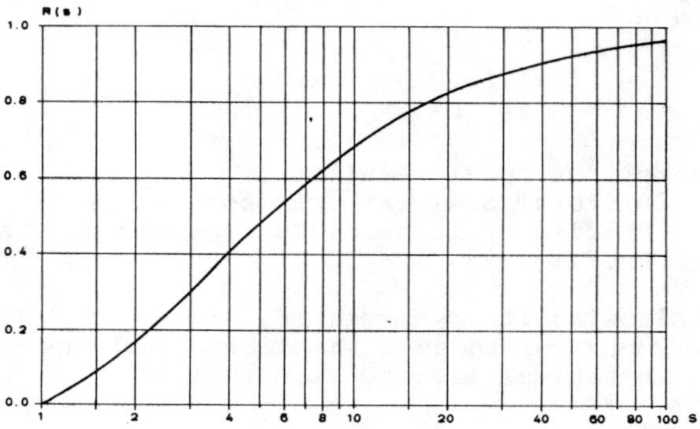

Fig. 2. The Reduction of the Longshore Shear Component of the Radiation Stress due to Directional Spreading.

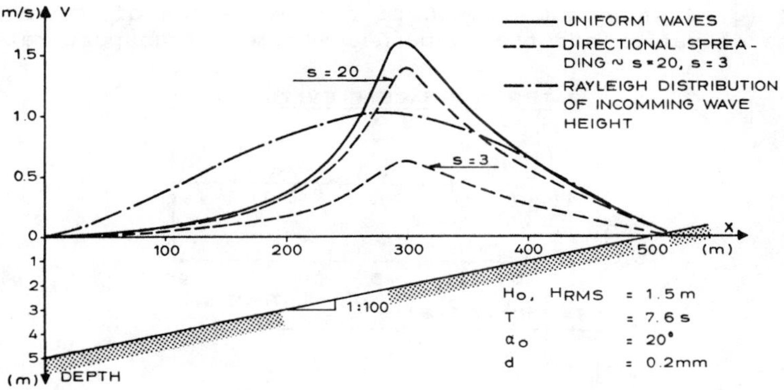

Fig. 3. Longshore Velocity Profiles.

The sediment transport rate is strongly influenced by the irregularity of the waves. The two most important effects are: A) The near bed orbital motion is irregular, which has to be taken into account when calculating the suspended sediment concentrations. B) In the surf zone only a fraction of the waves at a given point will be breaking/broken as there will be a fraction of the smaller waves which are breaking further inshore. The two effects are treated independently.

The irregular wave orbital motion has been analysed by using a time series of waves derived from a Pierson-Moskowitz spectrum. The nearbed wave orbital motion is calculated by a transfer function according to linear wave theory. The time series of the wave orbital velocity has been used in the model of Fredsøe et al. (1985) to calculate variation in time of the suspended sediment concentration and the sediment transport rate. The response of the mean current velocity profile to the irregular wave condition has not been analysed in detail, and two extreme situations have therefore been considered: 'full response' and 'no response'. By 'full response' it has been assumed that the mean current velocity profile outside the wave boundary layer adjusts immediately to the instantaneous wave conditions. In the case of 'no response' it has been assumed that the outer velocity and eddy viscosity profiles are determined from the mean bed shear stress integrated over the entire simulation period, and the outer velocity profile is thus kept constant.

Fig. 4 shows an example of a time series of water surface elevation and the suspended sediment transport rate.

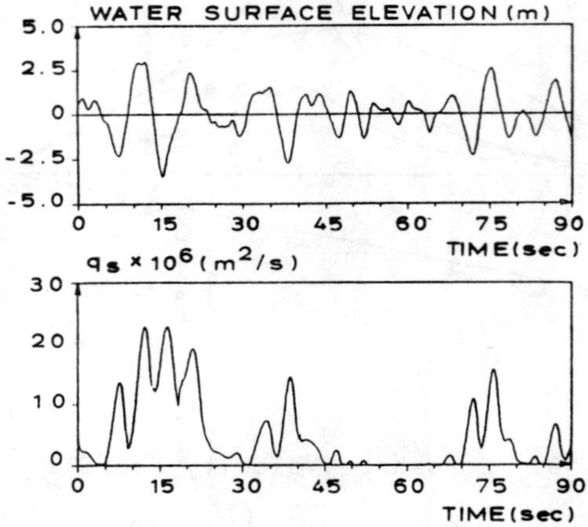

Fig. 4. Time Series of Surface Elevation, and Suspended Sediment Transport Rate, q_s, calculated for the following Parameters:
H_{rms} = 3.5 m, T_p = 10 s, D = 10 m, V = 0.1 m/s, d = 0.22 mm. 'Full Response'.

The purpose of these sediment transport calculations has been to determine the wave parameters which are best able to describe the time-averaged sediment transport under irregular waves.

In order to determine these wave parameters a series of numerical tests have been carried out. For all the tests, V was taken as $0.1U_{b,rms}$ of the corresponding series, where $U_{b,rms}$ is the root-mean-square of the orbital velocity amplitudes.

The value of q_s determined for each test was compared to that determined for a regular wave of height H and period T. From the analysis of the 16 combinations of H and T used for the sinusoidal motion it turns out that the best representation of the mean suspended sediment transport due to irregular waves over the whole range of values of $U_{b,rms}$ studied is given by the regular waves having $H = H_{rms}$ and $T = T_s$, as shown by Fig. 5.

Fig. 5 also illustrates the influence of changing the eddy viscosity profile outside the wave boundary layer from 'full response' to 'no response'. In general, the difference is of order 10% or less.

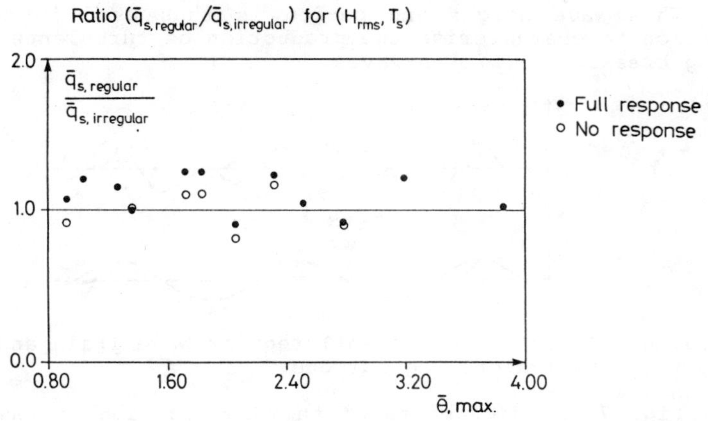

Fig. 5. Comparison between the Mean Suspended Sediment Transport due to Regular and Irregular Waves, and Illustration of the Influence of Current Response.

In the surf zone only a fraction of the waves is breaking due to the irregularity. The effect of breaking and broken waves is modelled according to Deigaard et al. (1986a) by describing the production, vertical spreading and decay of turbulence generated by the passage of the front of a spilling breaker or a broken wave. The input parameters, wave height H_b and period T_b, to this model are modified to reflect the conditions in the surf zone with irregular waves. First the wave period is increased to actual period of wave breaking, i.e. the non-broken waves are neglected, cf. Fig. 6.

$$T_b = \frac{\Sigma T_i \phi_i}{\Sigma \phi_{bi}} \tag{4}$$

where subscript 'bi' indicates that only breaking/broken waves are considered.

The representative wave height is determined from the energy dissipation, using a bore to characterise the dissipation in the passing wave fronts. The dissipation is thus proportional to the cube of wave height, giving:

$$H_b^3 = \frac{\Sigma H_{bi}^3 \phi_{bi}}{\Sigma \phi_{bi}} \tag{5}$$

This wave height and period is thus used in each location to characterise the production of turbulence by passing breaking or broken waves.

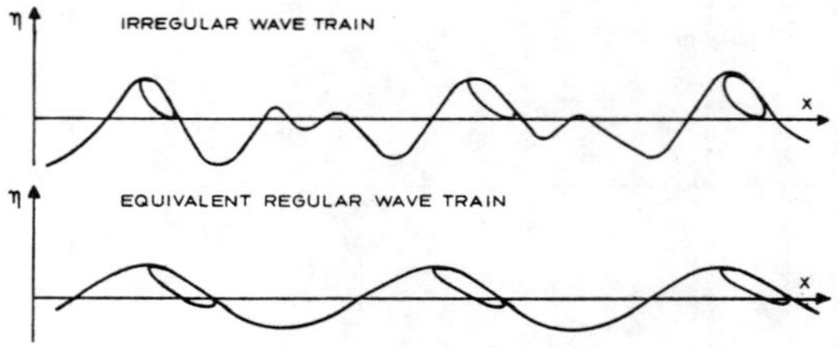

Fig. 6. Illustration of an Irregular Wave Train and the Equivalent Regular One.

Fig. 7 shows examples of the distribution of the transport in the coastal profile, calculated from regular and irregular waves. The main effect of including irregular waves is a much more smooth distribution of the transport and a reduction of the total transport by a factor of 0.3 in this particular case. The directional spreading of the waves s = 20 and s = 3 causes a reduction of the sediment transport in regular unidirectional waves by a factor 0.7 and 0.1, respectively.

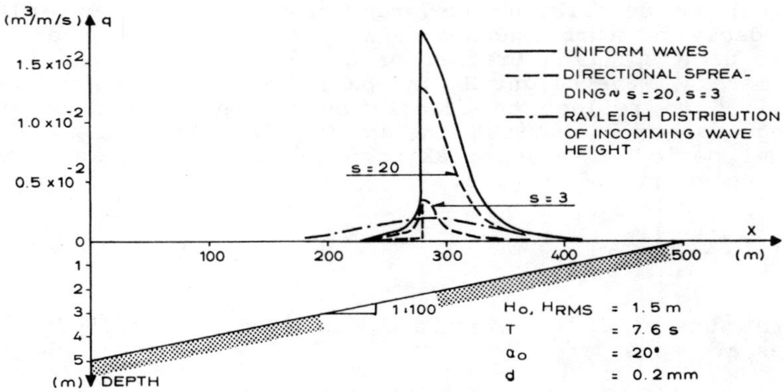

Fig. 7. Distributions of the Longshore Sediment Transport across a Plane Coast.

A comparison has been made between the calculated longshore sediment transport and the backfilling of a 1.600 m long trench (volume: 200.000 m^3) dredged through a three-bar coastal profile at the Danish North Sea Coast, cf. Mangor et al. (1984). The backfilling during a storm in the spring of 1982 is considered with measured wave heights up to H_s = 4.75 m. At this period the trench, width 90 m, depth 10 m, was only dredged at the outer bar. The total backfilling of the trench in this period was measured to be 90.000 m^3 (solid) grain volume. Fig. 8 shows the distribution of the backfilling and the calculated distribution of the longshore transport on the outer bar during the storm. The wave heights have been assumed to follow the Rayleigh distribution. The calculations have been made for two directional spreadings, s, of which the smallest one is the most likely on that specific location.

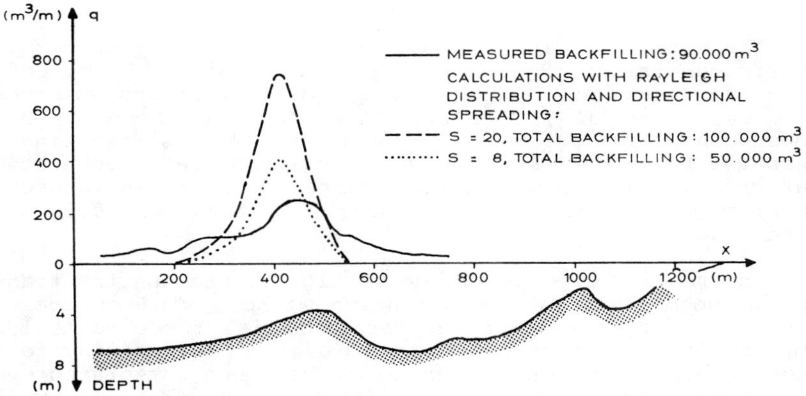

Fig. 8. Comparison between Measured Backfiling, cf. Mangor et al. (1984), and Calculated Backfilling.

WIND-DRIVEN AND COASTAL CURRENTS

In addition to radiation stress gradients, wind and coastal current can contribute to the forces driving the longshore current. The effect of wind shear stress and a coast parallel current has been included in the model. The wind shear stress is expressed through a friction factor (given the value: $f_w = 0.005$) and the wind speed, U_{10} at 10 m elevation:

$$\tau_w = \frac{\rho_a}{2} f_w U_{10}^2 \tag{6}$$

ρ_a is the density of air. τ_w is resolved into a shore normal and a shore parallel component according to the angle α_w between the wind direction and the shore normal direction. The shore normal component is included in the calculation of set-up and the shore parallel component is added to the driving forces of the littoral current.

The coastal current is assumed to be driven by a shore parallel gradient of the water surface, I, and the contribution to the driving force for the littoral current is ρgDI, where D is the local water depth.

The resulting equation for the littoral current thus becomes:

$$\tau_b - \frac{d}{dx}\left(\rho ED \frac{dV}{dx}\right) = -\frac{dS_{xy}}{dx} + \tau_w \sin\alpha_w + \rho gDI \qquad (7)$$

The effect of wind and current has been analysed for the basic conditions: $H_o = 1.5$ m, T = 7.6 s, d = 0.2 mm, and a coastal slope of 1:100. Three wave directions are considered: $\alpha_o = 10^o$, 20°, and 45°. Three wind speeds have been considered U_{10} = 5 m/s, 10 m/s and 20 m/s, the wind is assumed to be parallel to the direction of wave propagation at deep water. Three coastal current velocities, measured at 5 m water depth, are analysed: V = -0.25 m/s, 0.25 m/s, and 0.50 m/s.

The results are summarised in Fig. 9, showing the transport, normalised by the situation without wind or coastal current plotted against the total driving force normalised by the driving force due to waves alone. The driving force due to wind or water surface slope has been integrated over the area between the breaker line and the coastline, while the driving force due to the waves is the shear radiation stress, S_{xy} at the breaker line. The driving forces due to waves, wind and water slope can easily be calculated, and generalised plots similar to Fig. 9 can then be used to estimate the relative importance of the different contributions. The difference between the effect of wind and current, revealed in Fig. 9, is due to the difference in distribution across the coastal profile; wind shear stress is constant while current-induced bed shear stress decreases with decreasing depth. For the same integrated force the coastal current will thus give a larger contribution near the breaker line where the sediment concentration and transport are highest.

RIP CURRENT

Along natural coasts bars, interrupted by rip channels, often show up as a result of a complex current and sediment transport pattern.

Rip currents can be generated as follows: The incoming waves break on the foreslope of the bar. When the waves pass the crest of the bar and enter the deeper waters of the trough, they cease breaking and are reformed as non-breaking waves, and shoaled and refracted once again until breaking occurs again on the inner beach.

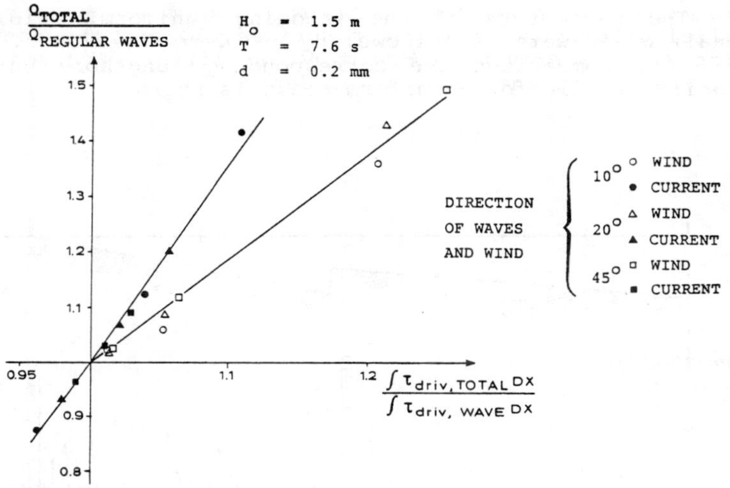

Fig. 9. The Sediment Transport including the Effect of Wind or Current normalised by the Transport due to Waves only versus the Total Driving Forces Normalised by Driving Forces due to Waves only.

The process of wave-breaking dissipates energy, producing a gradient in the components of the radiation stress tensor S. At the holes, the greater depths will cause that the waves are not breaking, but will instead continue propagation towards the shoreline. As a result of this, the wave set-up produced by the cross-shore component of the radiation stress S_{xx} will be small at the holes, and greater behind the bar.

This results in a net pressure gradient accelerating the water at the trough towards the holes. Afterwards this mass of water flows through the holes in the seaward direction in the form of rip-currents. The amount of water carried by the rips is compensated by the water transported over the crest of the bar. Under these conditions, a flow will exist at the trough behind the bar even for waves having an angle of incidence normal to the shoreline.

A mathematical model of flow in this situation has been developed. The model solves a simplified version of the equations describing the conservation of mass and momentum for time- and depth-averaged steady flow, see Zyserman and Fredsøe (1988).

Fig. 10 shows the flow field obtained using this model when the bar length was L = 180 m, the width of the holes was Y_b = 50 m, and the coastal profile was as shown on Fig.

10. The parameters of the incoming (uniform, unidirectional) waves were as follows: H_o = 1.2 m, T = 7.5 s, α_o = 45°. For comparison the corresponding longshore current velocity profile for a uniform coast is shown.

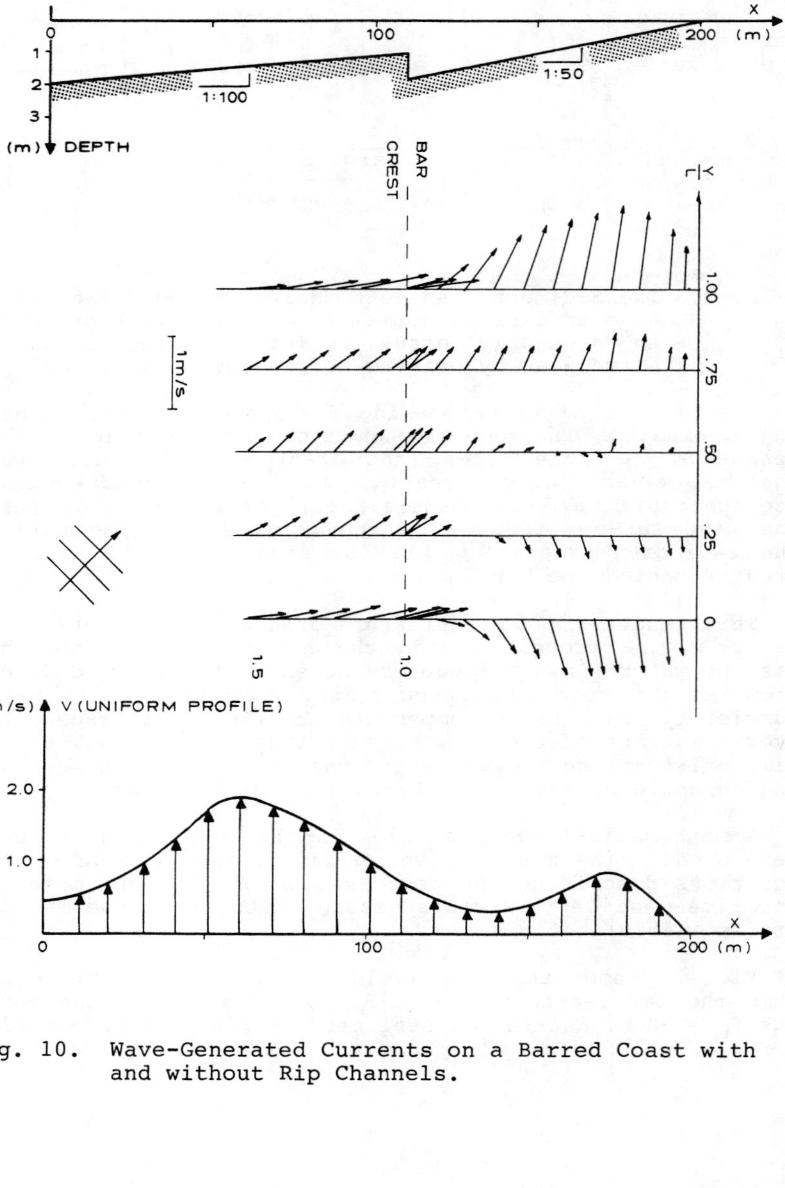

Fig. 10. Wave-Generated Currents on a Barred Coast with and without Rip Channels.

The following conclusions can be extracted from the results obtained for the wave-induced flow: a) The maximum velocity of the current on the bar is greatly reduced. The main cause for this is the cross-shore flow over the bar. The water carried in from offshore must continuously be accelerated up to the longshore velocity, which drains a considerable amount of longshore momentum. b) The flow reverses at some locations in the trough due to the pressure gradient which is opposite to the forcing term provided by the gradient of the shear component of the radiation stress tensor, S_{xy}. c) The velocity of the flow in the trough behind the bar crest becomes important, especially close to the holes.

Calculated current fields have been combined with the sediment transport model to describe the cross-shore variation of the longshore transport q_s as well as the total longshore transport Q_s for several transversal sections along the bar.

Due to the variability of the longshore current profile along the bar, the total transport varies from one section to another. In order to compare the sediment transport with that corresponding to the uniform situation, the average value of Q_s along the bar was calculated.

Fig. 11 shows the relation between the average longshore sediment transport when rip-currents are present $\bar{Q}_s$ and the transport for the uniform situation Q_o as a function of the ratio between Y_b and L , where Y_b has been kept constant equal to 50 m. In these examples the characteristics of the shore and of the waves were chosen as for the situation, shown in Fig. 10.

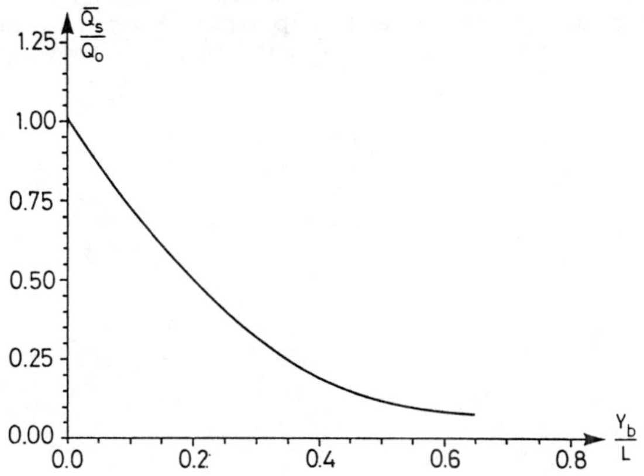

Fig. 11. Influence of Rip Currents on the Longshore Transport of Sediment. Coastal Profile and Wave conditions as Fig. 10.

It can be seen that the longshore sediment transport decreases for decreasing distance between rips, i.e. for shorter bars. This fact is easily explained by the reduction in the magnitude of the longshore current velocity when L decreases, due to an increased influence of the rip-currents.

CONCLUSION

A detailed description of littoral drift has been applied on a natural environment. It can be concluded that such a detailed model is rather sensitive to even small changes in the hydrodynamic description.

The following can be concluded from the test examples carried out with the extended littoral drift model including irregular waves, Fig. 7, coastal current and wind, Fig. 9, and the calculation of sediment transport along a barred coast with rip channels, Fig. 11:

. Both the irregularity of the waves and the presence of rip channels strongly reduce the total littoral drift compared to monochromatic waves and uniform conditions.

. The modifications of the littoral drift due to coastal current and wind action are less important.

The analysis rises a more general conclusion concerning the subject - modelling of littoral drift: Bearing in mind that all parameters in the models represent a physical phenomenon it must be concluded that the calculated littoral drift varies considerably by changing these parameters inside realistic ranges. The above means that modelling the littoral drift demands more information about the conditions (waves, coastal profiles etc.) than up till now has been established as basis for an investigation of littoral drift.

REFERENCES

Battjes, J.A. (1974) Computation of Set-up, Longshore Currents, Run up and Overtopping due to Wind-Generated Waves. Delft Technische Hogeshool.

Deigaard, R., Fredsøe, J. and Hedegaard, I.B. (1986a) Suspended sediment in the surf zone. Journal of the Waterway, Port, Coastal and Ocean Engineering, ASCE, Vol. 112, No. 1, pp. 115-128.

Deigaard, R., Fredsøe, J. and Hedegaard, I.B. (1986b) Mathematical model for littoral drift. Journal of the Waterway, Port, Coastal and Ocean Engineering, ASCE, Vol. 112, No. 3, pp. 351-369.

Engelund, F. and Fredsøe, J. (1976) A sediment transport model for straight alluvial channels, Nordic Hydrology, 7, pp. 293-306.

Fredsøe, J. (1984) The turbulent boundary layer in combined wave-current motion. Journal of Hydraulic Engineering, ASCE, Vol. 110, No. HY8, pp. 1103-1120.

Fredsøe, J., Andersen, O.H. and Silberg, S. (1985) Distribution of suspended sediment in large waves. Journal of the Waterway, Port, Coastal and Ocean Engineering, ASCE, Vol. 111, No. 6, pp. 1041-1059.

Mangor, K., Sørensen, T. and Navntoft, E. (1984) Shore approach at the Danish North Sea Coast, monitoring of sedimentation in a dredged trench. Proc. Coastal Engineering Conference, pp. 1818-1829.

Mitsuyasu, H., Tasau, F., Suhara, T., Mizuno, S., Ohkusu, M., Honda, T. and Rikiishi, K. (1975) Observations of the directional spectrum of ocean waves using a clower leaf buoy. Journ. Phys. Oceanography, Vol. 5, pp. 750-760.

Zyserman, J. and Fredsøe, J. (1988) The effect of ripcurrents on the longshore sediment transport. Second International Symposium on Wave Research and Coastal Engineering, Hannover, F.R. of Germany.

CHAPTER 120

CARIBBEAN BEACH-FACE SLOPES AND BEACH EQUILIBRIUM PROFILES

John D. Boon* and Malcolm O. Green**

ABSTRACT

Field measurements performed on two Caribbean islands revealed that two-dimensional nearshore bottom morphology is well represented by Dean's (1977) model of the beach equilibrium profile, $h = A x^m$, where h is depth below mean water level at a distance x offshore and A is a scale factor. For the curvature, m, we obtained an average value of approximately m = 1/2 through least squares curve fitting of observed profile data, yielding a more concave and therefore steeper profile inshore than m = 2/3, the average previously reported by Dean for quartz sand beaches in the United States. Furthermore, an objective measure of beach steepness was found to be $A^{1/m}$, a quantity which utilizes both of Dean's parameters and which may serve as a surrogate for the beach-face slope, $\tan \beta$, on highly concave beaches. Reasonable correlations were found between $A^{1/m}$ and the environmental parameter, H_b^2/gDT^2, where H_b is breaker height, D is sediment grain size, T is wave period and g is gravitational acceleration. Improved prediction of Caribbean beach slopes and beach equilibrium profiles is an important practical result.

INTRODUCTION

Beaches on the islands bordering the Caribbean Sea differ in at least two respects from beaches in temperate zones: (1) they consist mainly of skeletal calcium carbonate sands of marine origin rather than quartz and feldspar-rich detrital sands, (2) much of the time they experience fairweather waves of relatively low energy and steepness produced either locally by trade winds or received as swell from distant storms (Wilson, 1969; Wilson et al., 1973; Terwindt et al., 1984). Mean tidal range is also minimal (less than 20 cm) within the eastern Caribbean basin (Kjervfe, 1981) around which most of the Caribbean islands are located.

As a result of coastal processes dependent on these factors, Caribbean island beaches typically have a steep, concave-upward profile and lack an offshore bar and bar-related surf zone. Morphodynamically, they exemplify the reflective beach state (Wright and Short, 1984; Wright et al., 1985) wherein incident waves are strongly reflected and thus are conducive to nearshore standing wave motions at subharmonic frequencies. Exceptions do exist. Dissipative surf zones with offshore

* Virginia Institute of Marine Science, Gloucester Point, VA 23062 USA

** Department of Earth Sciences, University of Cambridge, Cambridge CB3 OEZ United Kingdom

bars and plunging breakers are occasionally seen and certain beaches contain an abundance of black sands rich in ferromagnesian minerals rather than pure calcium carbonate. Pocket beaches lying between rocky headlands are more common than long, straight beaches, however, and the former may have steeper slopes due in part to the absence of strong longshore currents (Bailard, 1981).

Our research interests throughout this study have focused on the question of predictability among various elements of Caribbean beach morphology and the extent to which this knowledge may prove useful in beach construction or restoration projects. Among these elements, none has appeared more promising than the quantitative description of the two-dimensional form of beach profiles in approximate equilibrium with existing environmental parameters including wave height and period, and sediment grain size.

DESCRIPTION OF THE STUDY ENVIRONMENT

Ten carbonate beaches were selected for study on the island of Sint Maarten/St. Martin (Netherlands Antilles/ French West Indies) located in the extreme northeast corner of the Caribbean Sea at the top of the Lesser Antillean arc system (Figure 1). Additional beaches were surveyed on the island of Curacao (Netherlands Antilles) located at the

Figure 1. Location of Study Area and Beaches Surveyed.

southern margin of the Caribbean Sea near Venezuela. Due to the extremely narrow shelf (150 to 200 meters wide) surrounding almost the entire coast of Curacao, only one of its beaches possessed a fully covered, natural sand bottom (Playa Abao) suitable for the purposes of this study. Sint Maarten/St.Martin lies at the western boundary of an elongate, raised shoal platform known as the Anguilla Bank, a feature approximately outlined by the 100-meter depth contour, beyond which depths in excess of 500 m are quickly reached.

Beaches along the southwest side of Sint Maarten are fully exposed to deep water waves (100-meter depth contour less than 2 kilometers from shore). Wave energy reaching this coast consists mostly of ocean swell from the Caribbean Sea, or the Atlantic via the Anegada Passage, mixed with only intermittent local wind waves due to ample protection from dominant northeast and easterly trade winds (Netherlands Antilles Meteorological Service, 1981).

The eastern side of the island is bordered by an extensive shallow platform wherein the 30-meter depth contour lies approximately 5 kilometers from shore. The coastline here is generally rocky or fronted by reefs. Numerous pocket beaches are found along the east coast behind reef platforms. Beaches on the north side of St. Martin are characterized by relatively low energy conditions since this side fronts the narrow and sheltered Anguilla Channel between Sint Maarten and the neighboring island of Anguilla. Deep water waves in this region have a mean height of approximately 1.5 - 2.0 meters; about half of all waves have a period of 6 seconds or less and about one-third have a period of between 6 and 9 seconds (U.S. Naval Weather Service Command, 1974).

Curacao differs from Sint Maarten/St. Martin in having little or no shelf. A submerged terrace about 60 meters deep occurs a short distance offshore, in some places only 125 meters from the shoreline. Consequently, very few natural beaches are able to retain sand cover and most consist of coral rubble at the foreshore and step. The only exceptions are found among a few small pocket beaches located at the heads of narrow limestone cliff re-entrants prevalent at the northwest corner of the island, the largest example of which is Playa Abao. Curacao lies within the trade wind belt at a latitude where the dominant winds are from the east. Deep water wave height averages about 1 meter with more than half of all waves having periods of 6 seconds or less (U.S. Naval Weather Service Command, 1974). Swell waves with periods of 9 seconds or more are rare.

DATA COLLECTION METHODS

Beach profiles from the backshore to the wave step were measured with an automatic level and metric surveyor's rod. The nearshore profile from the approximate still water line (depth zero) to an offshore distance of approximately 120 meters was surveyed by divers who carried the profile out to depths of 4 to 6 m where the sand cover on the offshore bottom typically begins to thin. At one beach (Playa Abao) the entire profile was surveyed using rod and level measurements. At all other beaches, depths were measured by divers using an electronic depth gauge. The depth sensing device consisted of a Senso-Metrics model SP91 pressure sensor (accurate to approximately $\pm$ 4 cm) mounted in a clear plastic tube containing batteries, amplifier circuitry and a digital voltmeter displaying the depth in centimeters. Wave filtering was done by visually averaging the readout over several wave periods.

Distances to marks on the bottom were measured using a 50-meter fiberglass tape and depths were recorded at each mark. Taped (slant) distances were later corrected to horizontal distances in a surveying program written for a portable computer. Divers also collected bottom samples in small plastic vials and recorded the sample numbers and depth-distances on an underwater slate together with descriptions and measures of bedforms. Sediment samples were later introduced into a rapid sand analyzer to determine size characteristics based on the formulation of Gibbs et al. (1971). The graphic mean size and Inclusive Graphic Standard Deviation of Folk and Ward (1957) were used to represent sediment size in this study.

Bulk density determinations were made using selected carbonate sand samples which revealed an average density of approximately 2.65 g/cm^3, the density of quartz. Although pure crystals of calcite and aragonite have a greater density than quartz, most of the carbonate sands examined consisted of porous coral fragments and broken plates of calcareous algae (Halimeda sp.), sediments similar to those described by Folk and Robles (1964).

Following each beach survey, a sample of breaking wave heights (H_b) and breaker periods (T) were recorded using the metric survey rod and a stop watch. Mean breaker period and root-mean-square breaker height were then calculated for each site. Water temperature was also recorded and found to be nearly constant at 28.5°C.

BEACH EQUILIBRIUM MODEL

Among the processes that drive the cross-shore exchange of sediment between the beach and the nearshore region are: asymmetric waves, rips, combined wave and steady current flow, downwelling and upwelling, gravity flows and "groupy" waves (Wright, 1987). The temporal and spatial expression of the nearshore seabed morphology is influenced simultaneously and at different times by at least all of these processes. Dean (1977) argued that the details of the individual processes could be neglected and the offshore profile of equilibrium modelled in terms of very simple propositions regarding "destructive" or sediment-mobilizing forces that mold the bed profile. Dean showed that the time-integrated, two-dimensional equilibrium profile over an unspecified distance seaward could be modelled by the power equation

$$h = A x^m \qquad (1)$$

in which h is the depth below still water level expressed as a function of x, the distance seaward from the shoreline. "A" is a scale parameter numerically equal to the depth at a unit distance from shore whereas "m" is a parameter representing the degree of profile concavity (m<1), convexity (m>1) or a plane profile (m=1). The theoretical values obtained for the parameters in equation (1) vary according to the way in which the destructive forces are expressed: assuming uniform energy dissipation per unit volume in the surf zone gives m = 2/3, whereas m = 2/5 results from assuming either uniform alongshore shear stress or uniform energy dissipation per unit surface area in the surf zone.

Bowen (1980), using an energetics model of suspended sediment transport based on symmetric wave orbital motion and a "perturbation" or drift velocity in the nearshore zone, also obtained an expression for the equilibrium beach profile similar to equation (1) with an m = 2/3 exponent. However, by including a higher harmonic term to simulate asymmetrical wave orbital motion, Bowen obtained a new approximation of $h \cong g(5.7wx/\sigma^2)^{2/5}$ where w is the grain settling velocity of the beach sediment and $\sigma = 2\pi/T$ is the wave radian frequency. This equation, which is valid for both the surf and nearshore zones, yields an m = 2/5 exponent and implies that the parameter A should be a function of the wave period in addition to sediment grain size or settling velocity.

Based on least squares fitting of equation (1) to some 502 profiles measured from the shoreline to approximately 365 meters offshore (3 to 5-meter depths) along the U.S. East and Gulf coasts, Dean (1977) adopted a mean value of m = 2/3 to be used as a functional constant although m for individual profiles ranged from about 0.2 to 1.2 in value. Justification for selecting m = 2/3 was based in part on the apparent Gaussian distribution and observed central tendency of the 502 m-values reported by Dean and partly for the convenience of examining variations in A with m held fixed. The latter variation was attributed to a functional relationship between A and the grain diameter, D (Moore, 1982; Dean, 1983) or to the equivalent settling velocity, w, of the bed sediment. Dean (1987) recently noted a close fitting of available A and w data from detrital (quartz) sand beaches by an empirical equation given as $A = 0.067\ w^{0.44}$, where A has units of $meters^{1/3}$, (since m was fixed at 2/3) and w is in cm/s.

The use of particle settling velocity rather than particle size would appear to be more in keeping with governing fluid dynamic processes but this assumes that w has been determined with the necessary adjustment for the in-situ water temperature, a correction that may show considerable variation from one beach (or season) to the next. Particle settling in the Stokes (viscous) range increases by more than 50 percent as water temperature increases from 5 °C to 20 °C, extremes not at all uncommon in the temperate zone. Water temperatures are, of course, higher and less variable in tropical regions (nominally 25 °C to 30 °C); Hydraulic-equivalent grain sizes should be larger there when compared with grains of the same physical size in colder, temperate regions. Meaningful grain settling velocity data should therefore be based upon stated field (not laboratory) water temperature, a practice not commonly evident at present in the literature.

Dean's theoretical arguments leading to the general expression for the equilibrium profile (equation (1) with m = 2/3) are strictly applicable to dissipative surf zones although the profiles used by Dean to verify the model and support the choice for m extend well beyond the surf zone into the nearshore region. Since Dean's field examples mainly represent beaches with dissipative surf zones, other surf-zone types may contain exceptions to the m = 2/3 "rule", and, if so, it is of interest to know under what conditions they may occur. The singular dependence of the A parameter on either particle size or sediment settling velocity may also require further examination; one intuitively expects the A parameter to be a function of wave characteristics as well as beach sediment properties.

The Equilibrium Profile on some Caribbean Beaches

The fitting of equation (1) to the observed profile data obtained at our study sites was, in general, quite successful. In a typical example from Mullet Bay, shown in Figure 2, the fitted profile is designated as FP3458 in which the parameters determined by least squares methods are A = 0.34 and m = 0.58. In Figure 3 (Baie Rouge), the fitted profile is FP6550 (A = 0.65, m = 0.50).

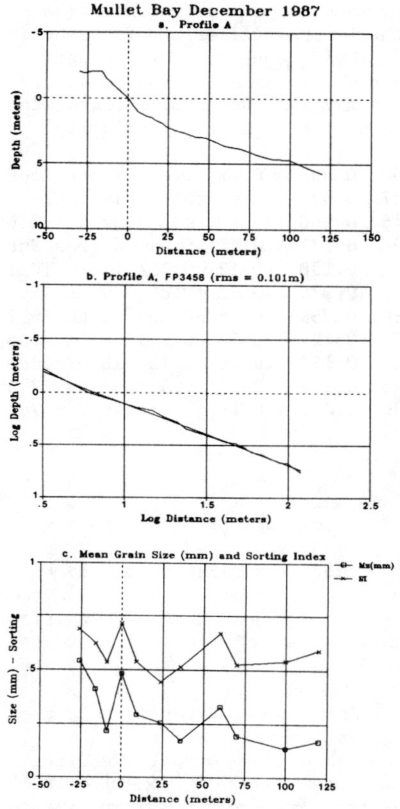

Fig. 2 Beach Profile Data, Mullet Bay

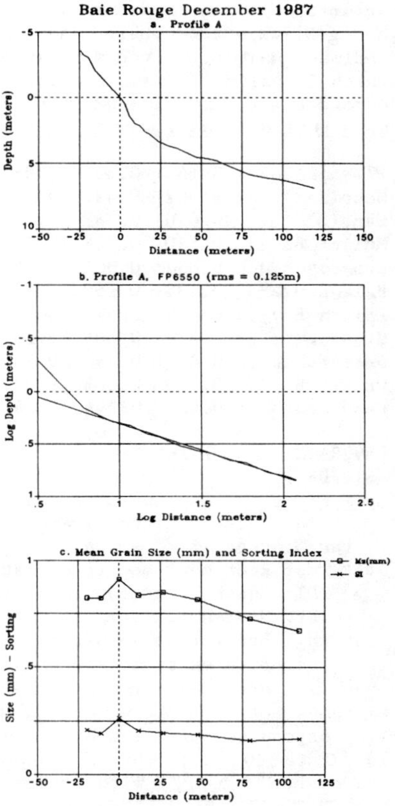

Fig. 3 Beach Profile Data, Baie Rouge

Only one measured beach profile, that of Guana Bay (Figure 4) on Sint Maarten, failed to achieve a reasonable fit. The reason for the deviant shape is presently unclear but the Guana Bay beach differed from all others in being fully dissipative and habitually displaying a wide surf zone with large, plunging breakers approximately 50 meters offshore; the profile, however, steepens markedly beyond this point and thereafter resembles the shape that other profiles have closer to shore. Guana Bay lies between headlands on the windward side of the island and is the only windward beach not protected by an offshore reef.

Table 1 contains the fitted values of the parameters A and m together with root mean square deviations of the observed profile points from the least squares line of best fit. It is noted that the mean value of m for the eleven fitted profiles (two profiles, A and B, were measured at Maho Beach) is approximately 0.55 with a standard deviation of $\pm$ 0.10. The present sample of Caribbean beach profiles is perhaps too small to warrant a definitive statement on the exact distribution of m-values for this population. However, if the assumption of normality is made, one can infer (using the Student's t statistic at the 0.995 level of confidence, one-tailed test) that the population mean is less than 0.67, the m-mean reported by Dean (1977).

Table 1. Summary of Parametric Values for Surveyed Beaches on Curacao, Sint Maarten, Netherlands Antilles and St. Martin, French West Indies.

BEACH	m	A	tan β	$A^{1/m}$	fit rms	D1	D2	H_b	T
Playa Abao	0.544	0.356	0.08	0.150	0.069	0.25	0.13	20.0	5.0
Maho(A)	0.502	0.431	0.12	0.187	0.071	..no data		..no data	
Maho(B)	0.470	0.487	0.13	0.216	0.080	0.40	0.31	32.5	9.0
Mullet Bay	0.580	0.338	0.11	0.154	0.101	0.41	0.21	44.2	9.8
Cupecoy Beach	0.637	0.301	0.14	0.152	0.130	0.53	0.35	51.8	10.1
Baie Rouge	0.498	0.650	0.13	0.421	0.125	0.85	0.78	39.2	12.2
Friars Bay	0.512	0.416	0.19	0.180	0.098	0.25	0.13	12.0	10.0
Grand Case	0.432	0.506	0.15	0.207	0.229	0.50	0.27	10.0	9.5
Anse Marcel	0.479	0.395	0.16	0.144	0.139	0.25	0.15	13.0	10.0
Orient Bay	0.773	0.135	0.09	0.075	0.212	0.22	0.47	32.4	7.3
Dawn Beach	0.610	0.246	0.16	0.100	0.297	0.29	0.35	49.2	7.3
AVERAGE	0.549								
STD.DEV.	0.097								

m, A: Least squares parameters for fitted model profile, $h = A x^m$.
tan β: Beach-face slope
fit rms: Root-mean-square deviation from model profile in meters.
D1: Beach-face mean sand size in mm.
D2: Nearshore mean sand size in mm.
H_b: Breaker height (rms) in cm.
T: Breaker period in sec.

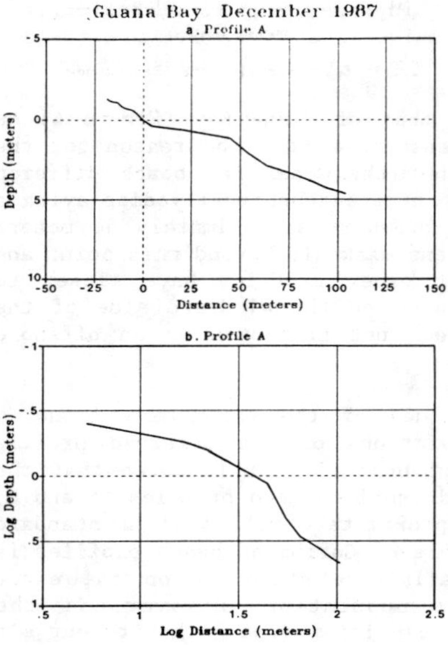

Fig. 4 Beach Profile Data, Guana Bay

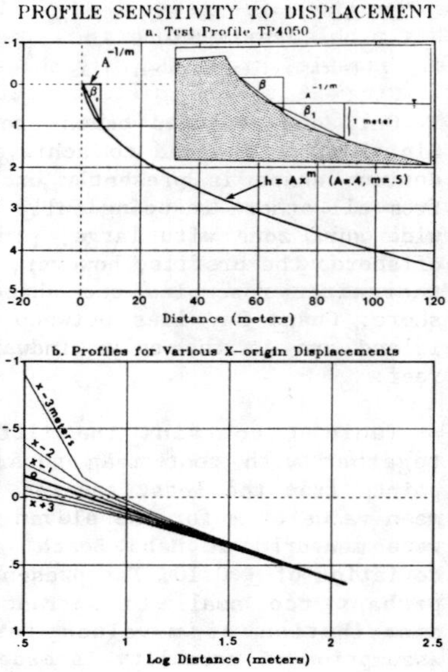

Fig. 5 Simulated Profiles with x-origin Displacements from shoreline zero

Profile Sensitivity to Measurement Error

Field profiles modelled by equation (1) require the precise location of the h,x reference or "zero" point. The vertical reference is taken to be the mean water level averaged over the diurnal tidal period and the horizontal reference is found as the intersection of mean water level at the shoreline. Due to the extremely small variation in tidal range (including "weather" tides) at the study locations, we considered that the major error in defining the profile reference would be the error associated with the elimination of gravity wave motion. Consequently, we found the point of sea level intersection on the beach foreshore by visual estimation, choosing a point just above the lower limit of the swash zone. Since depths below mean water level were obtained with a pressure gauge, we assumed these measures to be independent of horizontal distance. To test the sensitivity of the measured profiles to possible errors in reference location, a typical beach profile was simulated using representative model parameters ($A = 0.4$ and $m = 0.5$) to produce a test profile (Figure 5a). Following its construction, the test profile was re-measured (numerically) after each one of a series of fixed x-origin displacements landward and seaward of the reference point. The erroneous profiles, shown in Figure 5b, suggest that measures of real profiles will be most affected by x-origin measurement error occurring in the first 10 meters of the profile but that little change will be noted seaward of that point. Fitting of model profiles in the range x = 10 to 100 meters on log-log linear plots will be only slightly affected by a horizontal error of nearly 3 meters in x-origin location.

Beach Slope based on the Equilibrium Profile

The parameter A is a potentially useful representation of beach slope or steepness in the nearshore zone since, from equation (1), it is simply the depth at a unit distance from the shoreline. The parameter A in equation (1), however, has units of length^{1-m}. It is therefore impossible to compare A values in the same units when m is allowed to vary, as we believe it must in some situations. Basically, this problem arises due to a lack of unique spatial scale. One could, for example, make the variable x in equation (1) dimensionless by dividing it by the distance out to wave base (depth at which which waves initiate bottom sediment motion) but this distance clearly varies as a function of the local wave regime. To achieve standardization, we propose to make horizontal distances nondimensional using x_1, the distance seaward to a depth (h_1) of one meter, roughly the order of the usual depth at the surf zone limit. At this depth, equation (1) yields a "scaling" distance of $x_1 = A^{-1/m}$ and a new, dimensionless representation of beach slope is obtained as

$$\tan \beta_1 = h_1/x_1 = A^{1/m} \qquad (2)$$

in which both of the fitted curve parameters, A and m, are utilized. The slope angle, β_1, is given a subscript to distinquish it from β, the angle of the beach-face slope with which it overlaps (Fig. 5a). The former angle approaches 90° as m approaches 0 (vertical wall) and approaches 45° when m>>1 (highly convex profile). Therefore, assuming that a two-dimensional beach profile can be represented adequately by equation (1), we conclude that equation (2) is an objective means of expressing the beach slope and is probably intermediate between what is termed beach-face slope (tangent extending from about mid-foreshore to about the low tide mark) and various linear approximations of tan β taken across the nearshore region (e.g., Bowen, 1980, Fig. 1.1). For the steep, reflective-type profiles under discussion here, we consider that equation (2) may provide a close approximation to the beach-face slope, tan β, as discussed in the next section.

RESULTS AND DISCUSSION

Nearshore beach slopes, calculated as $A^{1/m}$, are presented in Table 1 along with the corresponding values of beach-face slope, tan β, obtained from leveling data on the beach-face of the eleven beach profiles of the present study. In several instances the two measures of slope are quite similar but, overall, it is clear that $A^{1/m}$ is more variable, showing greater range of slope values than tan β.

Values of the dimensionless parameter $H_b/g^{0.5}D^{0.5}T$ were determined for comparison with beach slope, where H_b is breaker height, D is beach sediment grain size, T is breaker period and g is the acceleration due to gravity. Sunamura (1984) has shown that beach-face slope (tan β) may be quantitatively predicted using

$$\tan\beta = [0.013/(H_b^2/gDT^2)] + 0.15 \qquad (3)$$

which was determined from existing laboratory data, or

$$\tan\beta = 0.12/(H_b/g^{0.5}D^{0.5}T)^{0.5} \qquad (4)$$

which was based on field data. Table 2 contains data from the present field experiment in which two sets of $H_b/g^{0.5}D^{0.5}T$ values were determined, one using the mean sand grain diameter from the beach-face and one using the mean grain diameter from the nearshore zone. Samples taken near the wave step were excluded as they usually contained very coarse material of limited extent (e.g., Figure 2c). Plots showing the four combinations of data (tan β versus $A^{1/m}$, beach-face versus nearshore sand size) are shown in Figure 6 with the curve for equation (4).

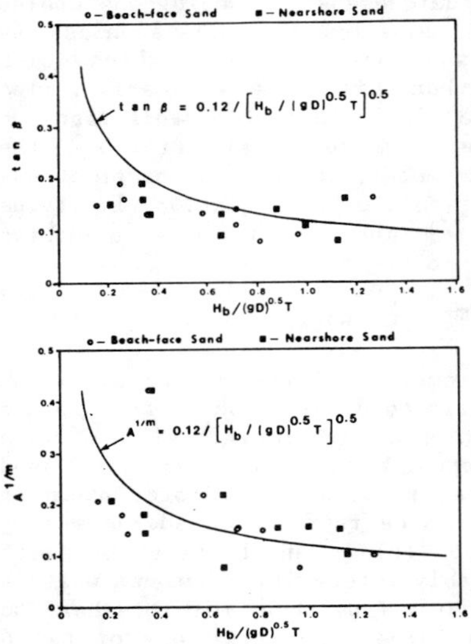

Fig. 6 Beach slope measured as tanβ and $A^{1/m}$ plotted against the dimensionless parameter $H_b/(gD)^{0.5}T$

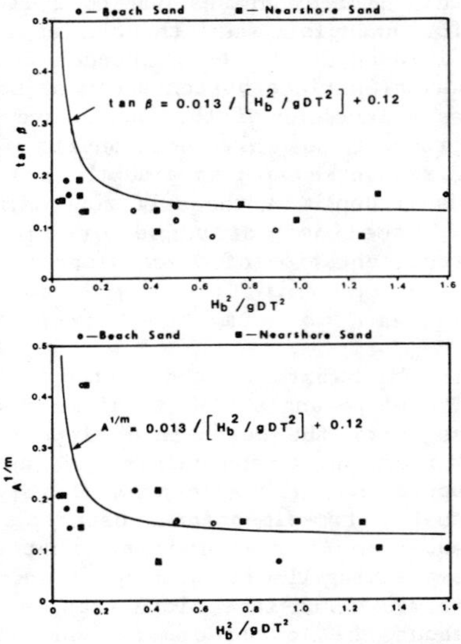

Fig. 7 Beach slope measured as tanβ and $A^{1/m}$ plotted against the dimensionless parameter H_b^2/gDT^2

Table 2. Comparison of Beach Slope with functions of $H_b/g^{0.5}D^{0.5}T$

f1(D1)	f1(D2)	f2(D1)	f2(D2)	tan β	$A^{1/m}$	Beach
0.150	0.204	0.023	0.042	0.15	0.207	Grand Case
0.242	0.336	0.059	0.113	0.19	0.180	Friars Bay
0.262	0.339	0.069	0.115	0.16	0.144	Anse Marcel
0.352	0.367	0.124	0.135	0.13	0.421	Baie Rouge
0.577	0.655	0.332	0.429	0.13	0.216	Maho(B)
0.711	0.994	0.506	0.988	0.11	0.154	Mullet Bay
0.711	0.875	0.506	0.766	0.14	0.152	Cupecoy Beach
0.808	1.120	0.652	1.254	0.08	0.150	Playa Abao
0.955	0.654	0.913	0.428	0.09	0.075	Orient Beach
1.264	1.150	1.597	1.322	0.16	0.100	Dawn Beach

D1: Beach-face mean sand size in cm.
D2: Nearshore mean sand size in cm.
$f1(D1) = H_b/g^{0.5}D1^{0.5}T$ $f2(D1) = (H_b/g^{0.5}D1^{0.5}T)^2$
$f1(D2) = H_b/g^{0.5}D2^{0.5}T$ $f2(D2) = (H_b/g^{0.5}D2^{0.5}T)^2$

In Figure 6, tan β shows little, if any, systematic variation with $H_b/g^{0.5}D^{0.5}T$. The beach slope calculated as $A^{1/m}$ does show the expected variation with the independent variable, allowing for scatter in the data similar to that seen in Sunamura's graphs (Sunamura, 1984, Figures 1 and 2). Although Sunamura attributed some of the scatter to inaccuracies in the measure of tan β, he pointed to 1) spatial and temporal change in D and 2) temporal changes in H_b and T as the leading sources of error. As expected, less error was associated with laboratory as opposed to field data plots.

The quantity $H_b/g^{0.5}D^{0.5}T$ is in part an index for wave steepness since the numerator contains wave height and the denominator furnishes a length based on joint wave-sediment properties. A similar quantity can be derived as a ratio of wave and sediment-related forces expressed as unit mass accelerations. Taking the numerator as H_b/T^2 and the denominator as gD/S where $S = \delta H_b$ is the suspension distance of sand particles above the bottom and $\delta \cong 1$ (Dean, 1973), this ratio is simply the steepness index squared or H_b^2/gDT^2. Values of the latter appear in Table 2 and in Figure 7 using averages of beach-face and nearshore sand samples, respectively, for the mean grain size, D. This figure shows tan β and $A^{1/m}$ plotted against H_b^2/gDT^2 with Sunamura's (1984) laboratory data curve, equation (3), superimposed in slightly modified form (constant reduction in slope of 0.03) to obtain the apparent best fit to our data. Again, the best dependent or predicted variable appears to be $A^{1/m}$ rather than tan β and the corresponding curve of best fit is

$$A^{1/m} = [0.013/(H_b^2/gDT^2)] + 0.12 \quad (5)$$

There is no overly compelling reason to select plots based on beach-face rather than nearshore sediment samples (Figures 6 - 7) as better predictors of beach slope. At first glance it is noted that nearshore sediments are generally finer (e.g., Mullet Bay, Figure 2C) and so produce slightly larger values of H_b^2/gDT^2. This is not always the case, however. Certain beaches on the windward side of Sint Maarten (Orient Bay and Dawn Beach) contained coarse sands that appear to be

rather than beach-face sand sizes would appear to rank them too low on the scale of H_b^2/gDT^2. Since beach-face sediment is affected more by the ambient wave regime and less by local variations among nearshore (and offshore) sediment sources, it is considered best to use for determinations of D in equation (5).

CONCLUSIONS

Practical applications for the above results are immediately obvious. Beach construction or restoration projects in low-energy, carbonate environments of the type described require an estimate of fill type (grain size) and amount necessary to achieve a desired two-dimensional profile over existing bottom. An equilibrium beach profile can be readily determined through use of equation (1) and the necessary estimates for the parameters A and m. Given a suitable estimate for the parameter m, the parameter A is readily calculated from the beach slope, $\tan \beta_1 = A^{1/m}$, predicted by equation (5). Pending further investigation, an appropriate average estimate for carbonate beaches on Caribbean islands appears to be m = 0.55.

It has been shown that $A^{1/m}$ provides a useful and objective estimate of beach slope on Caribbean island carbonate beaches (equation 5). The sudden inflection in the left side of the curve in Figure 7 implies that a small change in wave height or steepness in this region will be accompanied by a large change in beach slope. Beaches composed of very coarse sand will be particularly susceptible to such change. This is consistent with reports we have obtained from hotel owners on the western side of Sint Maarten that sudden and often severe subaerial erosion typically occurs in winter when Atlantic storm waves reach this side of the island.

In a definitional sense, $A^{1/m}$ is not equivalent to the mostly subaerial beach-face slope ($\tan \beta$) but is an indicator of the subaqueous nearshore slope ($\tan \beta_1$) as shown in Figure 5a. Nevertheless, on the steep, highly concave and barless beaches studied here, there appears to be ample justification to use $A^{1/m}$ in place of $\tan \beta$ in predictor equations for beach-face slope of the type presented here. Since the parameters A and m are determined by least squares fitting of the observed nearshore profile over most of its length (excluding perhaps the first 6 meters from shore), the measure of $A^{1/m}$ is not only objective but its precision can be judged by goodness-of-fit criteria.

Also, we seem to have observed near agreement between our field results and Sunamura's empirical curve for laboratory results, equation (3). Equation (3) has the same form as equation (5) but differs from it by a constant denoting a small offset in the dependent variable, $\tan \beta_1 = A^{1/m}$; With this minor adjustment, Sunamura's curve appears to fit our data reasonably well when using the squared term, H_b^2/gDT^2, as the independent variable. Given the potential use for equation 5 in field applications, it is worth noting that equation 3 is based on a very large collection of laboratory data which fit this curve quite well. Addition data from the field are needed to verify the relationship.

ACKNOWLEDGEMENTS

The authors gratefully acknowledge the assistance given by Grabowsky & Poort International N.V. of Curacao and Sint Maarten in aid of this project. We thank Lauro J. Calliari for his contribution to the field data collection effort and laboratory analyses. Contribution No. 1447 of the Virginia Institute of Marine Science and School of Marine Science, College of William and Mary, Gloucester Point, VA 23062.

REFERENCES CITED

Bailard, J.A., 1981. An energetics total load sediment transport model for a plane sloping beach. J. Geophys. Res., 86:10938-10954.

Bowen, A.J., 1980. Simple models of nearshore sedimentation; beach profiles and longshore bars. In: S.B. McCann (Ed.), The coastline of Canada, Geol. Survey of Canada, paper 80-10, pp. 1-11.

Dean, R.G., 1973. Heuristic models of sand transport in the surf zone. In: Engineering Dynamics of the Coastal Zone, First Austr. Conf. on Coastal Engineering, Sydney, Australia, pp. 208-214.

Dean, R.G., 1977. Equilibrium beach profiles: U.S. Atlantic and and Gulf Coasts. Ocean Engineering Tech. Rpt. No. 12, Dept. Civil Engineering and College of Marine Studies, Univ. of Delaware.

Dean, R.G., 1983. Principles of beach nourishment. In: P.D. Komar (Ed.), CRC Handbook of Coastal Processes and Erosion, CRC Series in Marine Science, chapter 11, pp. 217-231.

Dean, R.G., 1987. Coastal sediment processes: Toward Engineering Proceedings, Vol. I, ASCE Coastal Sediments '87, May 2-14, 1987, New Orleans, La.

Folk, R.L. and W.C. Ward, 1957. Brazos River bar: A study in the significance of grain size parameters. J. Sed. Pet., 27:3-26.

Folk, R.L. and R. Robles, 1964. Carbonate sands of Isla Perez, Alacran Reef Complex, Yucatan. J. of Geol., 72: 255-292.

Gibbs, R.J., Matthews, M.D. and Link, D.A., 1971. The relationship between sphere size and settling velocity. J. Sed. Pet., 41:7-18.

Kjerfve, B., 1981. Tides of the Caribbean Sea. J. of Geophys. Res., 86:4243-4247.

Moore, B.D., 1982. Beach profile evolution in response to changes in water level and wave height. M.S. Thesis, Univ. of Delaware, Dept. of Civil Engineering, 92p.

Netherlands Antilles Meteorological Service, 1981. Hurricanes and tropical storms of the Netherlands Antilles. Curacao, N.A.

Sunamura, T., 1984. Quantitative predictions of beach-face slopes. Geol. Soc. Am. Bull., 95:242-245.

Terwindt, J.H.J, C.H. Hulsbergen and L.H.M. Kohsiek, 1984. Structures in deposits from beach recovery, after erosion by swell waves around the southwestern coast of Aruba (Netherlands Antilles). Mar. Geol., 60:283-311.

U.S. Naval Weather Service Command, 1974. Summary of Synoptic Meteorological Observations, Caribbean and Nearby Island Coastal Marine Areas, Vol. 5, N.T.I.S., Springfield, VA.

Wilson, W.S., 1969. Field measurement of swell off the island of Aruba. Tech. Rpt. 56, Chesapeake Bay Institute, The Johns Hopkins University, 64 p.

Wilson, W.S., D.G. Wilson and J.A. Michael, 1973. Analysis of swell near the island of Aruba. J. Geophys. Res., 78:7834-7844.

Wright, L.D., A.D. Short and M.O. Green, 1985. Short-term changes in the morphodynamic states of beaches and surf zones: An empirical predictive model. Mar. Geol., 62:339-364.

Wright, L.D., 1987. Shelf-surfzone coupling: Diabathic shoreface transport. Keynote Address, Proceedings, Vol. I, ASCE Coastal Sediments '87, May 12-14, 1987, New Orleans, La.

Wright, L.D. and Short, A.D., 1984. Morphodynamic variability of surf zones and beaches. Mar. Geol., 56:93-118.

CHAPTER 121

BEACH PROFILE CHARACTERISTICS DUE TO THE INCLINED WAVES

Tai-Wen Hsu * and Shan-Hwei Ou **

ABSTRACT

The result of a theoretical approach shows that the beach profile characteristics is governed by a modified Iribarren number which includes the effects among the factors of initial beach slope, wave angle and wave steepness. A series of experiments are conducted in a three-dimensional movable bed model on the conditions of two different initial beach slopes, two incident wave angles as well as several erosive wave steepnesses. The relative importance of each factor involved in the parameter is discussed. It is shown that the modified Iribarren number is effective in the analysis of beach profile characteristics under the action of inclined waves. The empirical relationships between beach profile changes and the modified Iribarren number are proposed on the basis of experimental results.

1. INTRODUCTION

A knowledge about the characteristics of beach profiles under the action of waves is of great importance to a number of coastal engineering problems. Various factors such as the

* Graduate Student, Dept.of Hydraulics & Ocean Engineering, National Cheng Kung University, Tainan, Taiwan 70101, Republic of China.

** Professor, Ph.D., Dept.of Hydraulics & Ocean Engineering, National Cheng Kung University, Tainan, Taiwan 70101, Republic of China.

wave steepness, nearshore current, beach slope and bottom sediment are involved in the beach profile changes. Clarification of these individual factors and evaluation of their mutual interaction are quite important to the determination of the beach profile characteristics. Although much efforts have been devoted to obtain the empirical relationships between the beach profile changes and the different pertinent variables involved in the beach process, a definite solution has not yet been provided because of the complicated nature of the sediment transport in the surf zone.

Previous experimental studies were commonly performed in a two-dimensional wave flume, in which incident waves were normal to the shoreline. In the natural beach, however,the incoming wave often posseses a breaking angle to the shoreline. There are few data on the beach profile changes caused by the inclined breaking waves. Brater and Ponce-Campos(1976) performed a movable bed test in a wave basin with incoming waves oblique to the shoreline, but only the parameter of sand bluff recession was analyzed in his experiments.

The data of the beach profiles complied in this paper are all belong to the erosive type. As shown in Fig.1, the

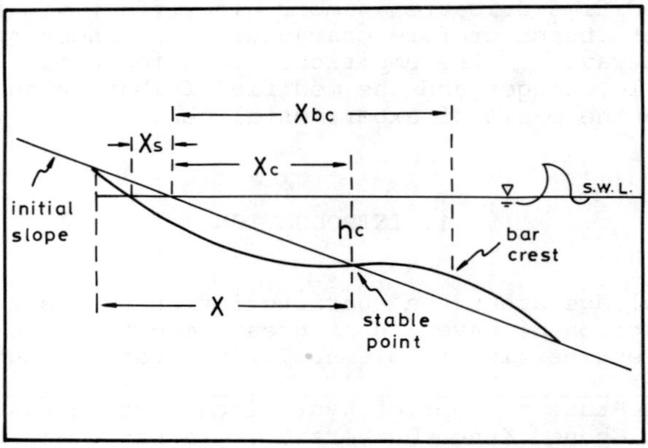

Fig.1 Definitions of beach profile characteristics

items of beach profile characteristics discussed in the present study include the erosion length X, the distance of the stable point X_c, the distance of bar crest X_{bc} and the water depth at the stable point h_c. A modified Iribarren number, $\xi_o^{1/6} \cos \alpha_b$, is proposed based on the concept of wave energy flux to correlate the beach profile changes with the inclined breaking waves, where $\xi_o = \tan\beta / (Ho/Lo)^{1/2}$ is the Iribarren number, $\tan\beta$ is the initial beach slope, Ho and Lo are the wave height and the wave length at deep water respectively, and α_b is the angle of breaking wave. The relative importance of each factor involved in this parameter is discussed. According to the experimental results, the empirical relationships between the beach profile changes and the modified Iribarren number were obtained by applying linear regression analysis.

2. THEORETICAL CONSIDERATION

A non-dimensional length L* is selected to represent the characteristics of beach profile changes. The functional relationship of L* associated with the non-dimensional sediment transport rate is expressed as

$$L^* = f_1(Q_s^*) \tag{1}$$

where Q_s^* is a dimensionless sediment transport rate.

Based on the concept of Bagnold (1963) , littoral sand transports are responsible to the wave energy flux. A linear correlation is given by

$$Q_s = k_1 I \tag{2}$$

where Q_s is the volume transport rate, I is the wave energy flux evaluated at the surf zone, and k_1 is a dimensional proportionality coefficient.

As drawn in Fig.2, the wave energy flux F between two wave crests can be written as

$$F = E_b C_{gb} \Delta\ell \cos\alpha_b \tag{3}$$

where E_b is wave energy per unit width, C_{gb} is the group velocity of waves, subscript b represents relative quantities in the breaking point, $\Delta\ell$ is the width between two wave crests. The wave energy flux to a unit shoreline length becomes

$$I = E_b C_{gb} \cos\alpha_b \tag{4}$$

Substitution of Eq.(4) into Eq.(2) gives

$$Q_s = k_1 E_b C_{gb} \cos\alpha_b \tag{5}$$

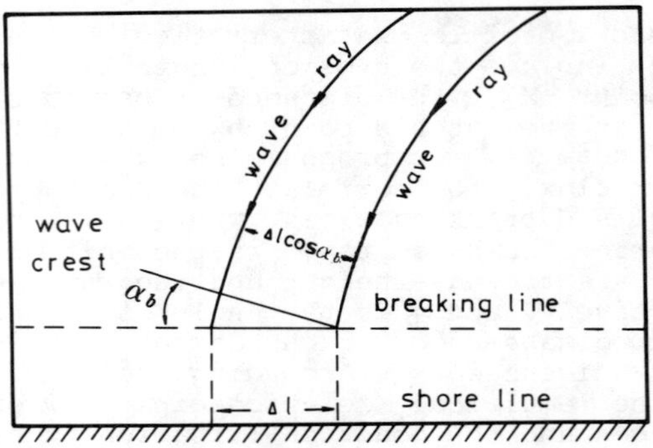

Fig.2 Sketch of the wave energy transmitted on shoreline

From the linear long wave theory, E_b and C_{gb} are evaluated by the following equations, respectively :

$$E_b = 1/8\ \rho g H_b^2 \tag{6}$$

$$C_{gb} = \sqrt{g d_b} \tag{7}$$

in which ρ is the fluid density, g is the acceleration of gravity, H_b the breaking wave height, d_b the breaking water depth. The dimensionless sediment transport rate is formulated in the form of

$$Qs^* = \frac{Qs}{\rho g u_b d_b^2} = \frac{k_1^* E_b C_{gb} \cos\alpha_b}{\rho g u_b d_b^2} \tag{8}$$

where $k_1^* = \gamma' k_1$, is a dimensionless coefficient, γ' is the submerged weight of the sand, u_b is the amplitude of the velocity at the bed in surf zone. For linear wave theory, u_b can be written as

$$u_b = \frac{\pi H_b}{T \sinh \dfrac{2\pi d_b}{L_b}} \tag{9}$$

where T is the wave period, L_b is the wave length in the surf zone. In the shallow water area, $\sinh \frac{2\pi d_b}{L_b}$ can be expressed as

$$\sinh \frac{2\pi d_b}{L_b} = \frac{2\pi d_b}{L_b} \tag{10}$$

and

$$L_b = \sqrt{g d_b}\ T \tag{11}$$

Insertion of Eqs.(10) and (11) into Eq.(9) yields

$$u_b = \frac{H_b}{2}\sqrt{\left(\frac{g}{d_b}\right)} \tag{12}$$

Combination of Eqs.(6),(7),(8) and (12) leads to

$$Qs^* = k_2 (H_b/d_b) \cos\alpha_b \tag{13}$$

where k_2 is a constant. The ratio of wave height and water depth at breaking point was found to be related to the Iribarren number(Battjes,1974). Based on available experimental data, Sunamura (1980) proposed the relationship

$$\frac{H_b}{d_b} = 1.1\,\xi_o^{1/6} \tag{14}$$

According to Eqs.(13) and (14), the transport rate can be rewritten as

$$Qs^* = k_3\,\xi_o^{1/6} \cos\alpha_b \tag{15}$$

where k_3 is a constant. By comparing Eqs.(1) and (15), we have

$$L^* = f_2(\xi_o^{1/6} \cos\alpha_b) \tag{16}$$

It is convenient to assume that the Eq.(16) has the functional form of

$$L^* = a_1 (\xi_o^{1/6} \cos\alpha_b)^{b_1} = a_1\,\xi_o^{b_2} (\cos\alpha_b)^{b_1} \tag{17}$$

where a_1, b_1 and b_2 are coefficients to be determined by experimental results. It is interesting to note that the physical parameters governing the beach profile changes include the wave steepness Ho/Lo at deep water, the initial beach slope $\tan\beta$ and the angle of breaking wave.

If the incoming waves are normal to the shoreline, Eq.(17) can be simplified as

$$L^* = a_1\,\xi_o^{b_2} \tag{18}$$

It has been recognized that the dynamics of the breaking waves in the surf zone is relevant to the Iribarren number (Battjes, 1974; Wang and yang, 1980). Eq.(18) makes it plausible that ξ_o is also of great importance for beach profile characteristics under the action of normal incident waves. The phase-difference τ/T defined by Kemp(1960) is found to be an important parameter to the beach profile changes (Sunamura and Horikawa,1974), where τ denotes the time for a wave to travel from its breaking point to the uprush limit. An empirical relationship derived by Hsu, Lee and Ou(1986) showed that the parameter τ/T is related to Iribarren number. The usefulness of Iribarren number for the analysis of beach profile changes due to normal incident waves was also confirmed in their studies.

3. EXPERIMENT AND PROCEDURE

The experiments were performed in a three dimensional wave basin of 16m long, 12m wide and 0.7m deep. The testing arrangements for both normal and inclined incident waves are shown in Fig.3 and Fig.4, respectively. The test section of movable bed model were placed with a uniform slope by using fine coal with median diameter of 0.15mm and specific gravity of 2.07.

Two different initial beach slopes (1/16 and 1/25), two wave angles(0° and 30°) as well as several erosive wave steepnesses were selected for the model tests. All experimental conditions are summarized in Table 1. The angles of breaking waves were calculated by the method proposed by Wang and Le Mehaute (1980).

Battjes(1974) clarified the types of breakers on the basis of the classification of Galvin(1968) as follows:

$$\xi_o > 3.3 : \text{surging breaker}$$

$$3.3 > \xi_o > 0.5 : \text{plunging breaker}$$

$$\xi_o < 0.5 : \text{spilling breaker}$$

The experimental data of $\tan\beta = 1/25$ falling in the domain of 0.161 ~ 0.328 belong to spilling breaker while values of $\tan\beta=1/16$ ranging from 0.256 to 0.580 correspond to transition between spilling breaker and plunging breaker.

Waves of constant characteristics were generated by a flat type wave maker. Wave characteristics were measured by capacity type wave gauges and a recorder unit. Beach profiles were measured by an electric platform at a half hour time interval along the beach. Normally, it will take three hours for the profile to reach equilibrium condition for normal incident waves. In order to examine the relations between beach profile characteristics and inclined wave behavior inside the surf zone, it is assumed inclined waves will take the same time for reaching the equilibrium conditions.

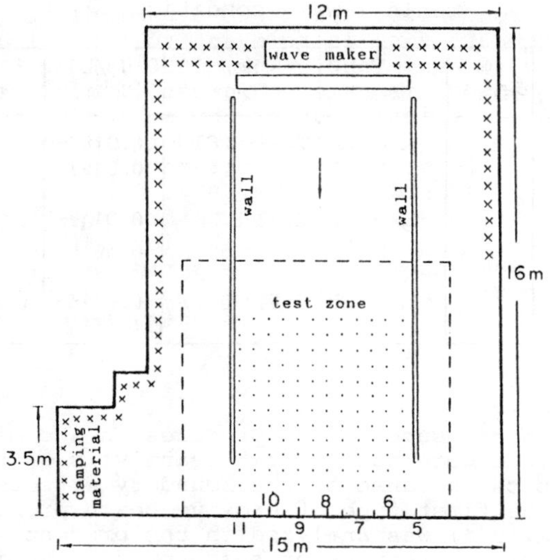

Fig.3 Arrangements of wave basin for normal incident waves

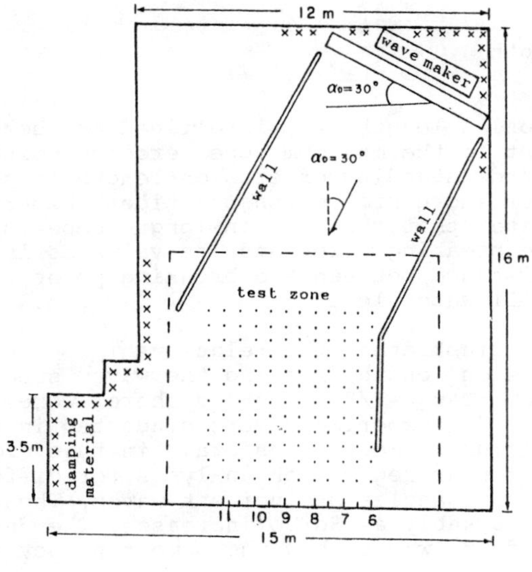

Fig.4 Arrangements of wave basin for inclined incident waves

Table 1 Wave conditions

$\tan\beta$	α_0 (deg)	T (sec)	Ho (cm)	Ho/Lo	α_b (deg)
$\frac{1}{16}$	0°	1.0~1.2	2.84~9.63	0.0182~ 0.0493	0°
$\frac{1}{25}$	0°	1.0~1.2	2.32~8.28	0.0149~ 0.0619	0°
$\frac{1}{16}$	30°	1.0~1.2	2.61~10.17	0.0116~ 0.0597	6.8~12.6

Only three successive profiles in the middle part of the movable bed were taken for the analysis of beach profile characteristics because of the boundary effects on the test zone. Only the first type of erosive beach profile (Sunamura and Horikawa,1974) was analyzed in the present study. Fiq.5 shows four typical examples of beach profile variation of four successive sections for the cases of different wave steepnesses.

4. RESULTS AND DISCUSSION

Erosion Length

The erosion length, X, is defined as the distance from stable point to the maximum upper erosion point as shown in Fig.1. The understanding of erosion length in sandy beach is essential for shore protection practices. Experiments(Chang, 1982) indicate that the erosion length depends on the location of the breaking point and its value is larger than 0.8 times the distance between the breaking point and the origin point of still water level.

Fig.6 illustrates the relationship between the dimensionless erosion length X/Lo and the wave steepness Ho/Lo. The data of $\tan_\beta = 1/30$ plotted here were compiled from Cheng(1974). His experiment were conducted in a two dimensional wave flume. Four lines drawn in this figure were obtained by a linear regression analysis for different initial beach slopes and angles of incident waves. The value of X/Lo increases remarkably as Ho/Lo increases. The data of the inclined incident waves show the same tendency for the cases

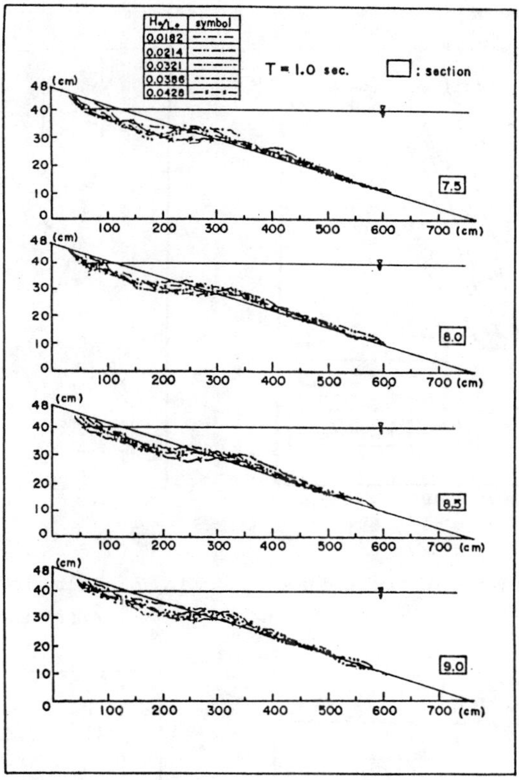

Fig.5 Typical variation of beach profiles for different wave steepnesses (tanβ =1/16)

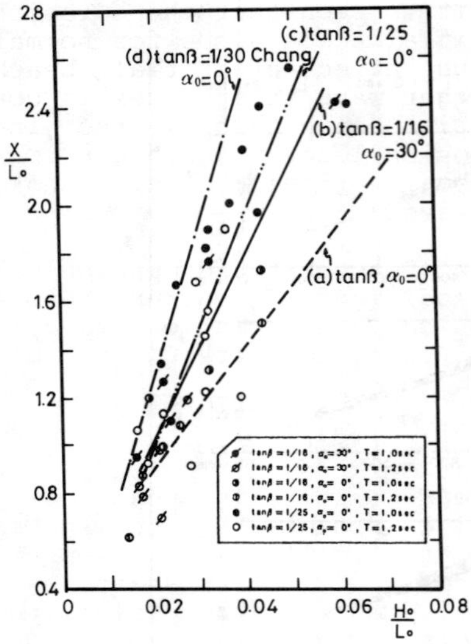

Fig.6 Relationship between X/Lo and Ho/Lo

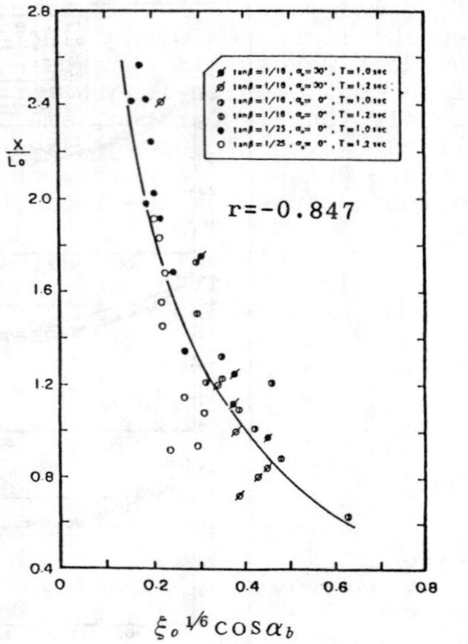

Fig.7 Relationship between X/Lo and $\xi_o{}^{1/6}\cos\alpha_b$

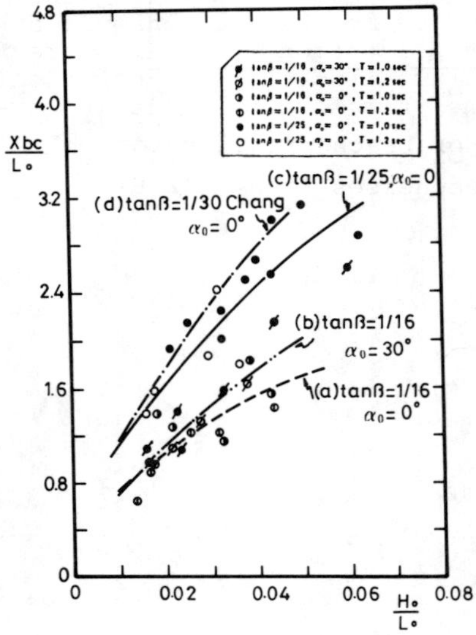

Fig.8 Relationship between X_{bc}/Lo and Ho/Lo

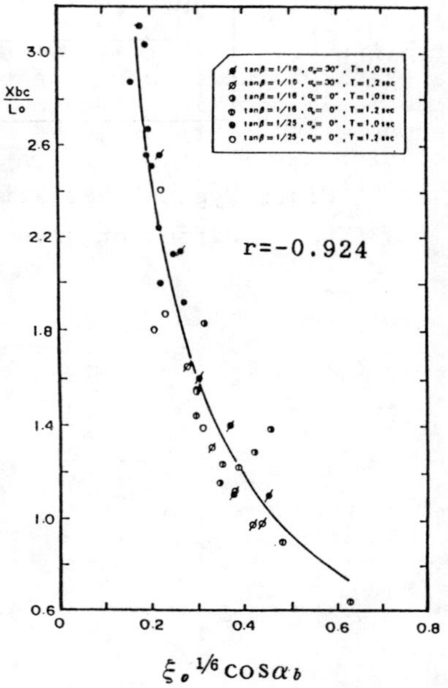

Fig.9 Relationship between X_{bc}/Lo and $\xi_o{}^{1/6}\cos\alpha_b$

of normal waves. In addition, it is seen that the erosion length increases with milder initial beach slope for normal incident waves. In this case, the effect of initial beach slope on erosion length for larger wave steepnesses is obviously found. For the same initial beach slope, the inclined waves produce greater erosion length than the normal waves. This may be due to the local variation of longshore sediment caused by the inclined waves.

The empirical relationship between the dimensionless erosion length X/Lo and the modified Iribarren number $\xi_o^{1/6} \cos\alpha_b$ as developed from theoretical consideration above is shown in Fig.7. By using linear regression method, the best fit line is given by

$$X/Lo = 0.443\xi_o^{-0.92} \cos\alpha_b^{-5.54} \quad (r = -0.847) \tag{19}$$

where r is the correlation coefficient. The points with larger values of $\xi_o^{1/6} \cos\alpha_b$ give broadly scattering in Fig.7. This is interpreted as a smaller wave steepness Ho/Lo has larger value of $\xi_o^{1/6} \cos\alpha_b$ for constant beach slope and angle of incoming waves. Under this condition, the beach profile tends to become accretion type as pointed out by Johnson(1949).

The Distance of Bar Crest

The sediment eroded from foreshore is deposited offshore as a longshore bar. The distance of the bar crest X_{bc} is defined as the distance between the bar crest and the origin point of still water level as shown in Fig.1. Laboratory studies were conducted by Horikawa, Sunamura and Kito(1973) to investigate the situation of bar crest response to various wave characteristics. An empirical relationship was proposed on the basis of available experimental data:

$$X_{bc}/Lo = B(Ho/Lo) \tag{20}$$

where B=-67.4+20.5Log(t/T) , is a dimensionless time factor, and t is the time for model test. Hughes and Chiu(1981) performed an experiment in a wave flume. A parameter $H_b/\omega T$ was proposed to analyze the location of bar crest due to severe storms, where ω is the fall velocity of a sediment particle. Experimental results by Chang(1974) indicated the location of bar crest is related to the wave breaking point. His experiments also showed a bar crest is formed outside the wave breaking point and its maximum distance is within 1.6 times the distances between the wave breaking point and the original point of still water level.

Fig.8 gives plots of X_{bc}/Lo versus Ho/Lo for the cases of different initial beach slopes and wave directions. Each curve in the figure is obtained by fitting exponential function for different initial beach slopes and angles of incident waves. The experimental data show that the ratio of

X_{bc} /Lo increases with an increase of Ho/Lo. By comparing curves (a) and (b), it is found that values of inclined waves seemed to be slightly greater than normal waves. Furthermore, it is shown that the initial beach slope has significant effect on the location of bar crest for normal incident waves.

Fig.9 depicts the relationship between X_{bc} /Lo and $\xi_o^{1/6}\cos\alpha_b$. The value X_{bc}/Lo decreases with the increase of $\xi_o^{1/6}\cos\alpha_b$. It is noted that the modified Iribarren number is shown to be highly effective in representing the data by linear regression method. The result is expressed as

$$X_{bc}/Lo = 0.458\,\xi_o^{-1.03}\quad \cos\alpha_b^{-6.20} \qquad (21)$$

Depth and Location of Stable Point

Available experimental and field data of beach profiles indicate that there exists usually a stable point on beach profiles near the breaker zone. At the stable point, sediment of a given size will be in oscillating equilibrium and without net movement. Raman and Earattupuzha(1972) found out a relationship between location of stable point and wave characteristics on the basis of experimental results. Hallermeier(1978) used a sediment entrainment parameter to calculate the water depth of the stable point. Kubo and Tomaki(1981) analyzed the water depth of the stable point from field data.

Fig.10 gives plots of the water depth of the stable point h_c against wave height Ho in deep water. The straight lines in the figure is the linear regression results. The experimental data show that the water depth of the stable point is proportional to the wave height at deep water. From lines (a) and (b), the inclined waves appear to have larger values than normal waves for the same initial beach slopes. The water depth of the stable point increases as the initial beach slope becomes smaller from 1/16 to 1/25. The line obtained by Hallermeier(1978) plotted in the figure shows the similar trend to the case of normal waves for $\tan_\beta$ =1/25.

Fig.11 is the relationship between dimensionless water depth of stable point h_c/Lo and the modified Iribarren number $\xi_o^{1/6}\cos\alpha_b$. Although the experimental data scatter broadly over a wide range of $\xi_o^{1/6}\cos\alpha_b$, it is found that a higher value $\xi_o^{1/6}\cos\alpha_b$ leads to a reduction in the value of h_c/Lo. The result of linear regression gives

$$h_c/Lo = 0.011\,\xi_o^{-1.41}\cos\alpha_b^{-8.43} \qquad (r=-0.702) \qquad (22)$$

The correlation coefficient is not good for the reason of random data base.

Fig.12 is a plot of non-dimensional location of the stable point X_c/Lo and the wave steepness Ho/Lo. The effect of the initial beach profile on the location of stable point

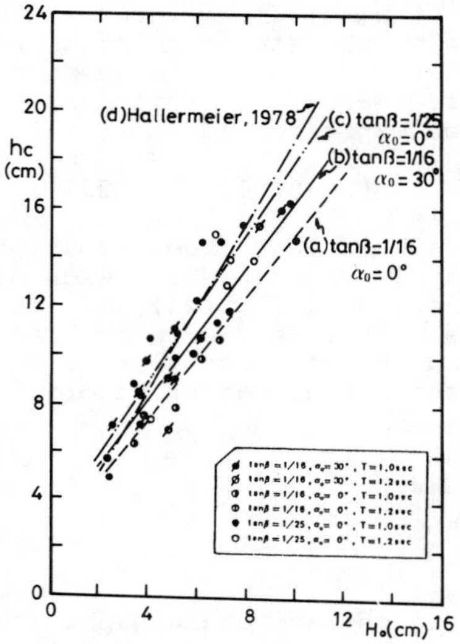

Fig.10 Relationship between h_c and Ho

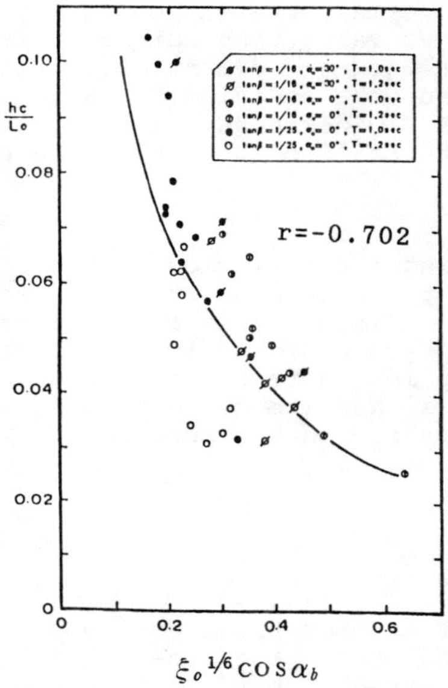

Fig.11 Relationship between h_c/Lo and $\xi_o^{1/6}\cos\alpha_b$

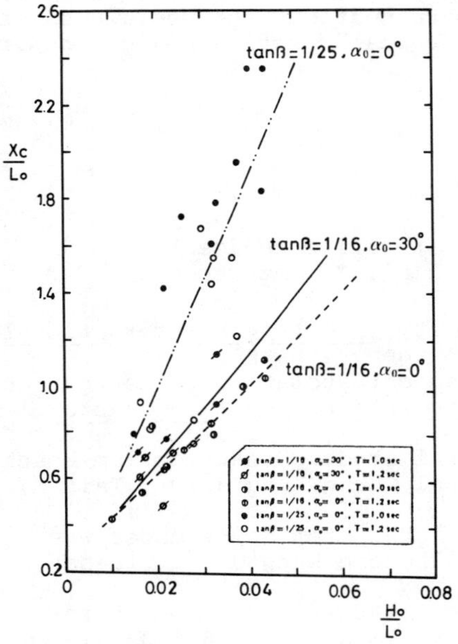

Fig.12 Relationship between Xc/Lo and Ho/Lo

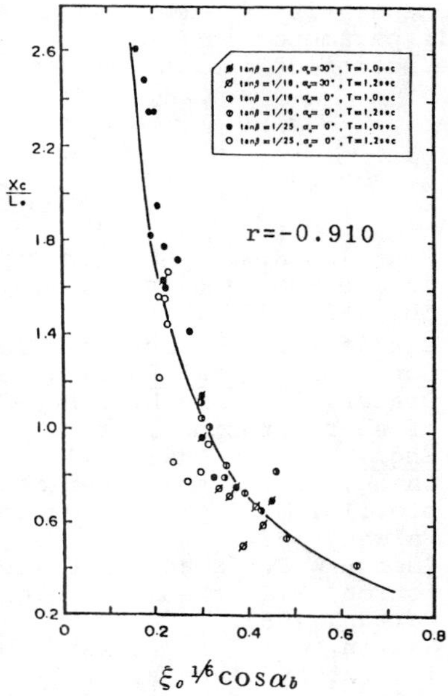

Fig13. Relationship between X_c/Lo and $\xi_o^{1/6}\cos\alpha_b$

for normal waves is obviously found. The inclined waves have larger value than normal waves for $\tan\beta = 1/16$. Fig.13 is a plot of non-dimensional location of the stable point X_c/Lo versus the modified Iribarren number $\xi_o^{1/6}\cos\alpha_b$. The value of X_c/Lo decreases as the value of $\xi_o^{1/6}\cos\alpha_b$ increases. The best fit line in this figure can be expressed as

$$X_c/Lo = 0.211\xi_o^{-1.29}\cos\alpha_b^{-7.74} \qquad (r=-0.910) \qquad (23)$$

A higher correlation coefficient reveals that the location of the stable point can be adequately described in terms of the modified Iribarren number. According Eq.(23), X_c/Lo tends to decrease as α_b increases. This is due to the fact that longshore current is generated by obliquely incident waves and its corresponding longshore sediment transport is dominated in the surf zone.

5. CONCLUSIONS

A theoretical development based on wave energy approach found that the beach profile is governed by a modified Iribarren number. This parameter covers the effects among the factors of the initial beach slope, wave angle and wave steepness. A laboratory movable bed tests were undertaken in the wave basin to find out the relationship between beach profile characteristics and the modified Iribarren number. The analysis of experimental data confirms the usefulness of this parameter in expressing the beach profile changes under the action of inclined waves.

REFERENCES

1. Bagnold, R.A. : Mechanics of marine sedimentation , in The Sea , edited by M.N. Hill, 3, Interscience Publ., 507- 528, 1963.
2. Battjes, J.A. : Surf similarity, Proc. 14th International Conf. on Coastal Eng., Copenhagen, 466-480, 1974.
3. Brater, E.F. and D. Ponce-Campos:Laboratory investigation of shore erosion , Proc. 15th Intern. Conf. on Coastal Eng., Hawaii, 1493-1511, 1976.
4. Chang, R.H. : An experimental study on equilibrium beach profile, M.S. thesis, National Cheng Kung Univ., Tainan, Taiwan, 1974.
5. Cheng, W.C.: Beach profile characteristics under wave action, M.S. thesis, National Cheng Kung Univ., Tainan, Taiwan, 1982.
6. Galvin, C.J.J.: Breaker type classification on three laboratory beaches, J. Geophys. Res., 89, 3651-3659, 1968.
7. Hallermeier, R.J. : Uses for a calculated limit depth to beach erosion, Proc. 16th Intern. Conf. on Coastal Eng., Hamburg, 1493-1512, 1978.

8. Horikawa, K., T. Sunamura and Kito H. : A study on beach transformation due to waves, Proc. 20th Conf. on Coastal Eng. in Japan, 357-363, 1973.
9. Hsu, T. W., Y. K . Lee and S.H. Ou : Beach profile characteristics under the action of normal incident waves, J. of Harbor Technology, 2, 139-154, 1986.
10. Hughes, S.A. and T.Y. Chiu : Beach and dune erosion during severe storms, Coastal and Oceanographic Engineering Dept., Univ. of Florida, 200p., 1981.
11. Johnson, J.W.: Scale effects in hydraulic models involv ing wave motion, Trans. of Amer. Geophys. Union, 30, 517-525, 1949.
12. Kemp, P.H. : The relationship between wave action and beach profile characteristics, Proc. 7th Intern. Conf. on Coastal Eng., The Hague, 262-277, 1960.
13. Kubo, M. and S. Temaki : On the stable point in sandy beach profile, Proc. 28th Conf. on Coastal Eng. in Japan., 247-250, 1981.
14. Raman H. and J.J. Earattupuzha : Equilibrium conditions in beach wave interaction, Proc. 13th Intern. Conf. on Coastal Eng., Vancouver, 1237-1257. 1972.
15. Sunamura, T.: A laboratory study of offshore transport of sediment and a model for eroding beaches, Proc. 17th Intern. Conf. on Coastal Eng., Sydney, 1051-1070, 1980.
16. Sunamura, T. and K. Horikawa : Two-dimensional beach transformation due to waves, Proc. 14th Intern. Conf. on Coastal Eng., Copenhagen, 920-938, 1974.
17. Wang, H. and W.-C. Yang : A similarity model in the surf zone, Proc. 17th Intern. Conf. on Coastal Eng., Sydney, 529-546, 1980.
18. Wang, J.D. and B. Le Mehaute : Breaking wave charac teristics on a plane beach, Coastal Eng., 4(2), 137-149, 1980.

CHAPTER 122

Sediment Suspension due to Large Scale Eddies in the Surf Zone

by

Kazuo NADAOKA, Seizo UENO and Tatsuyuki IGARASHI

Dept. Civil Eng., Tokyo Institute of Technology
O-okayama 2-12-1, Meguro-ku, Tokyo 152, JAPAN

ABSTRACT

Laboratory experiments using a fiber-optic LDV system and a small pressure transducer have been made to reveal detailed characteristics of the velocity field in the surf zone and its relationship to the sediment suspension with special reference to the three-dimensional large scale eddies referred to as "obliquely descending eddies", the existence of which was recently revealed by Nadaoka (1986). A conditional sampling technique has been used to find that the obliquely descending eddies bring highly intermittent intensive turbulence to the bottom with the large onshoreward momentum at the upper layer of the water and thus essentially characterize the turbulent flow field in the surf zone. Visual observation and concentration measurements, especially a coherence analysis of two data sets of concentration close to the bottom, have shown that the sediment suspension is mostly governed by such large scale eddies in a wide extent of the surf zone; i.e., the eddies hit the bottom and then lift up the sediment into suspension, yielding the spot-like sediment cloud in accordance with the three-dimensional eddy structure.

1. Introduction

The effect of wave breaking is one of the most significant factors for the coastal sedimentary process. In this respect, Miller (1976), Shibayama and Horikawa (1985), and others have discussed sand suspension by the action of the large scale vortices generated immediately after the wave plunging (Fig. 1). The generation of such vortices, which

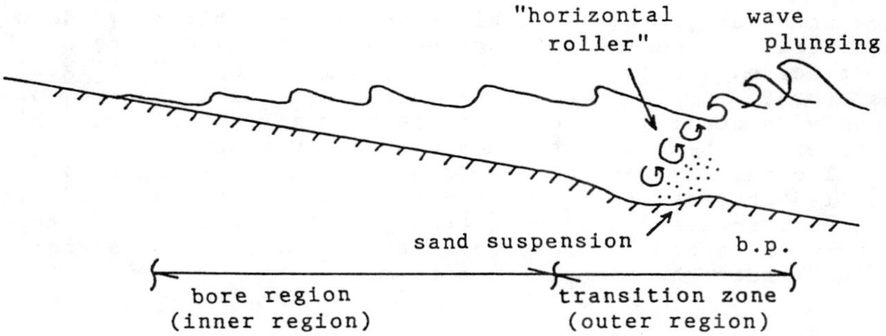

Fig. 1 "Horizontal rollers" and sediment suspension. (Miller 1976, Shibayama & Horikawa 1986, etc.)

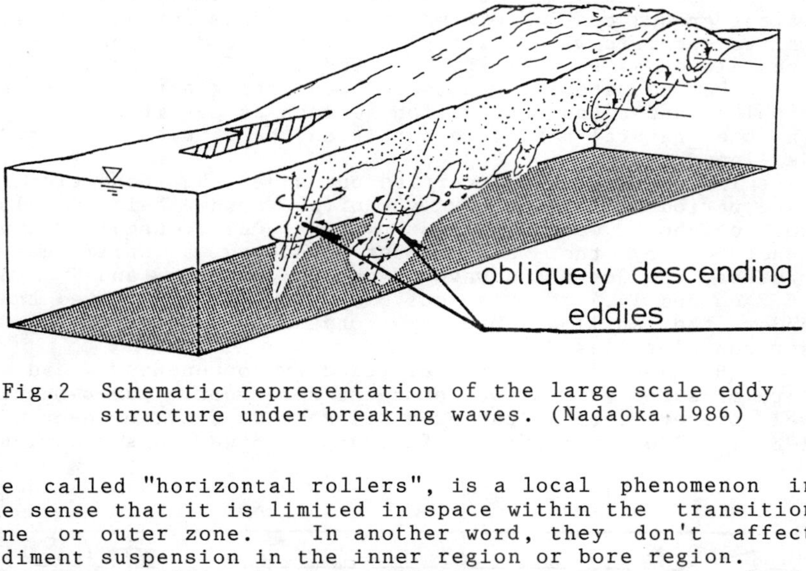

Fig.2 Schematic representation of the large scale eddy structure under breaking waves. (Nadaoka 1986)

are called "horizontal rollers", is a local phenomenon in the sense that it is limited in space within the transition zone or outer zone. In another word, they don't affect sediment suspension in the inner region or bore region.

For spilling breaker condition, on the other hand, Deigaard et al.(1986) have discussed the effect of the turbulence originated from breaking water surface on the vertical diffusion of the bottom sediment, by using the transport equation for turbulent kinetic energy. However, the real physical process of the sediment suspension under spilling breaker as well as turbulent bore has been not yet well understood, mainly because of our limited knowledge of the turbulent velocity field structure within the surf zone.

Recently, Nadaoka(1986) has found the inherent eddy structure of the breaking waves in a wide extent of the surf zone. That is, around the wave crest, dominant eddies have nearly two-dimensional flow structure the axis of which is parallel to the crest line, while behind the wave crest, the flow structure changes quickly into that of obliquely downward stretched eddies with strong three-dimensionality which is referred to as "obliquely descending eddies" (Fig.2). Therefore, it is worth to pay our attention to these obliquely descending eddies, because they may act as possible agitator of the bottom sediment in the surf zone.

The main purpose of the present study is to clarify the detailed characteristics of the turbulent velocity field in the surf zone and its relationship to the sediment suspension with special reference to the "obliquely descending eddies" under the turbulent bore.

2. Experimental Equipments and Procedure

The wave channel used for the experiments was 0.6 m deep, 0.4 m wide and 20 m long, equipped with a flap-type wave generator. The experiments were made both for a fixed bed and a movable bed. In the former case, a wooden slope of 1 on 20 was installed at one end of the channel and in the latter ground coal of 0.21mm in median diameter and of 1.45 in its specific weight was placed with the depth of 15cm on the slope, as shown in Fig.3. The use of the ground coal for the bed material was to suppress the generation of the ripples under the wave action as much as possible, because in the natural surf zone condition the ripples is hardly developed.

The experiment was carried out under the condition of wave period T=1.27s, equivalent offshore wave height H_0'=15.5 cm, offshore wave steepness H_0'/L_0=0.061. Regular waves generated by the wave maker produced a spilling-type breaker, the depth and wave height of which, h_b and H_b, were 16.7 cm and 14.7 cm, respectively. It should be noted that, under the spilling breaker condition, there exists *no* horizontal roller.

Horizontal and vertical velocity components, u and w, were measured with a two-component fiber-optic laser-Doppler velocimeter of back-scattered fringe mode (FLV, Hino et al., 1984). The water surface fluctuation η was measured with a

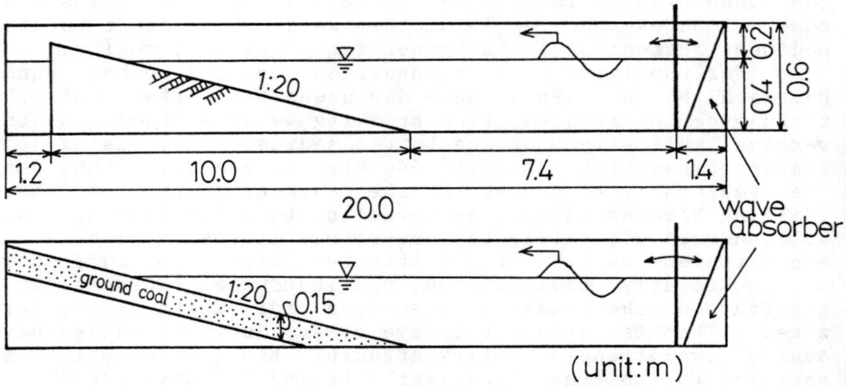

Fig.3 Experimental equipment.

capacitance-type wave gauge at right above the velocity measuring point which was located at h/h_b =0.67 (h is the still water depth). At the measuring section, breaking waves were confirmed to be well developed into so-called "turbulent bore". A small pressure transducer of 1 cm in diameter, 3 mm in thickness and 1 kHz in its frequency response was used for the measurement of bottom pressure fluctuation at the measuring section. The concentration of the suspended bottom sediment for the movable bed experiment was measured by a light-attenuation-type sensor, the position of which was manipulated to be kept at nearly the same height above the bottom through the movable bed experiment.

3. Characteristics of Velocity Field Associated with Oblique Eddies

3.1 Intermittent structure of bottom turbulence and pressure fluctuation

Figure 4 shows the phase-averaged distribution of the turbulence intensity u' and w' at the several heights z above the bottom. The origin of the horizontal axis corresponds to the phase of the zero-up crossing point of the water surface profile. The values indicated were calculated by averaging about 200 records of the high-frequency velocity component data which is obtained through a moving-average filter with a averaging time of 0.1 s. We can see in this figure that the dependence of the turbulence intensity on the phase variation as a whole is not so distinct, though the turbulence intensity at z = 6.7cm slightly increases around the phase of wave crest. Especially, the turbulence intensity close to the bottom is almost uniform except the region around the phase of the wave crest, where the intensity becomes larger probably corresponding to the increase of the orbital velocity.

Figure.5, on the other hand, is a typical example showing the raw data of the water surface fluctuation η immediately above the velocity measuring point, the bottom pressure fluctuation p, and the velocity fluctuation u and w close to the bottom (z=0.7cm). In this figure, the bottom turbulence occurs not uniformly, but appears in a quite intermittent manner with the bottom pressure disturbances at the instants indicated by the arrows. Further, we should note that such an intermittent occurrence of the bottom turbulence can be found not under the wave crest, but around the phase of the the zero-down crossing point. This means that these intermittent turbulence is not caused by the bottom shear stress due to orbital wave motion.

3.2 Conditional sampling analysis of velocity field associated with the oblique eddies

Comparison between Figs. 4 and 5 leads us to the idea that the intermittent characteristics of the bottom turbulence may be lost through the averaging process by the usual phase-averaged method because of the random occurrence of the intensive bottom turbulence, as shown in Fig.5.

Considering the occurrence of such a highly intermittent

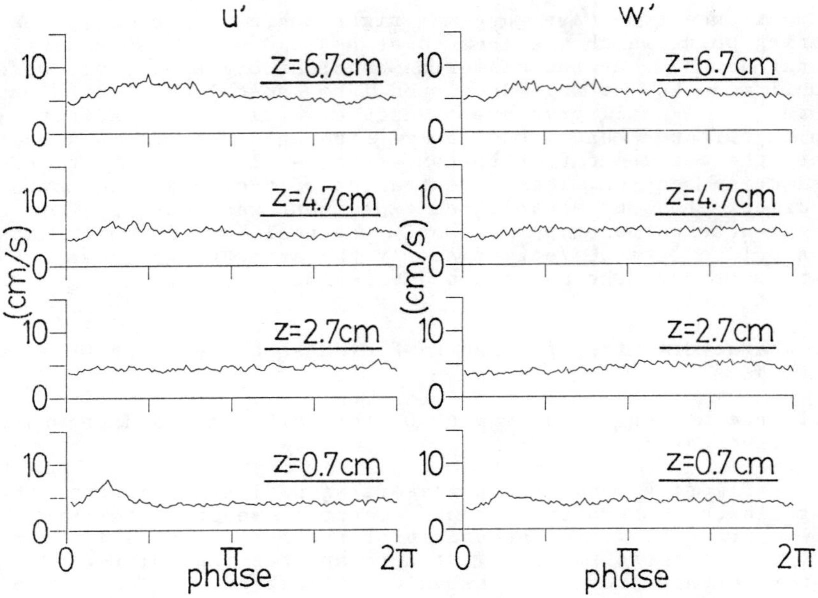

Fig.4 Distributions of phase-averaged turbulence intensity u' and w'.

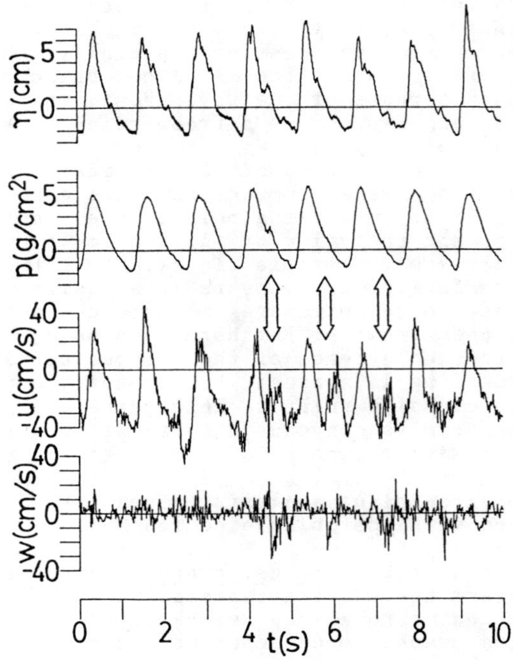

Fig.5 An example of time series of η, p, u and w.

bottom turbulence behind the wave crest, we can expect that these disturbance is caused by the action of the obliquely descending eddies mentioned first. The results of the visual observation of the eddy evolution seemed to support this expectation. To confirm this point more quantitatively, we have investigated the flow structure related to the intermittent bottom disturbances by applying a conditional sampling technique.

In Fig. 5, we can find also that the pressure disturbance occurs not so frequently as compared with the velocity disturbances. Its possible explanation is that the pressure disturbance occurs only at the instant when the pressure gage is directly hit by an obliquely descending eddy. Considering this characteristics of the pressure record, we can utilize it as a trigger signal for the conditional sampling to detect distinctively the velocity data corresponding to the intermittent bottom turbulence. Namely, by inspecting whether the magnitude of the pressure disturbance exceed the prescribed threshold level or not, we can classify each individual wave record separated by the zero-up crossing method into conditionally sampled or rejected records. The scanning of the pressure disturbance for the conditional sampling was done within the phase interval $[\pi/2, 3\pi/2]$ of each individual wave record, because the apparent pressure fluctuation calculated by the moving- average method may emerge in the phase intervals $[0, \pi/2]$ and $[3\pi/2, 2\pi]$ through the effects of the water surface fluctuation around the wave crest and of the higher harmonics due to the wave nonlinearity (Fig.6). The number of the wave records obtained through this conditional sampling is about 50, 1/10 of the total wave records.

Figure 7 represents the phase-averaged distribution of the turbulence intensity u' and w', where the solid and broken lines correspond to the conditionally sampled and rejected data, respectively. We can clearly see in this figure that the turbulence intensity of the sampled data at the most upper measuring point attains relatively large value around the wave crest and this intensive turbulence propagates obliquely downward as indicated by the arrows, while the turbulence intensity of the rejected data exhibits no appreciable dependence on the phase variation. This means that the obliquely descending eddies bring intensive turbulence to the bottom behind the wave crest.

Figure 8 shows another result of the conditional sampling analysis where the phase-averaged distributions of the mean velocity, $\langle u \rangle$ and $\langle w \rangle$, of the conditionally sampled and no-conditioned data are represented by the full and broken lines, respectively. In this figure, we can recognize that, in the large turbulence regions, both the onshoreward and the downward mean velocities are enhanced in the sampled records. This fact is directly related with the increase of the downward transport of the onshoreward momentum as described in what follows.

The left and right figures in Fig. 9 represent the phase-averaged distributions of the momentum transport for the sampled and rejected records, respectively. The solid lines indicate the phase-averaged product of the horizontal and the vertical velocity fluctuation, $\langle (u-\bar{u})\cdot(w-\bar{w}) \rangle$, where the over bar means an operator to take a time average. Comparing

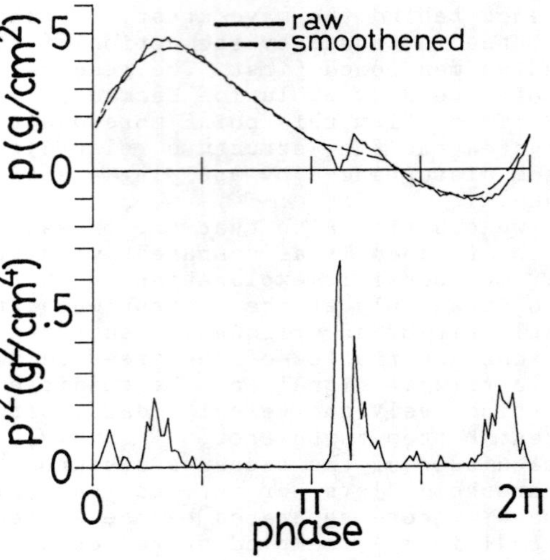

Fig.6 Raw and smoothened data of *p*. *p'* is the difference between these two data.

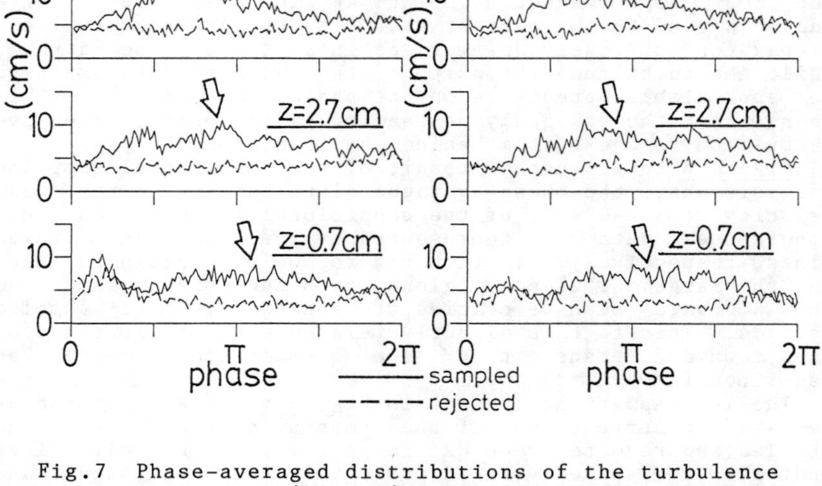

Fig.7 Phase-averaged distributions of the turbulence intensity *u'* and *w'* for conditionally sampled and rejected records.

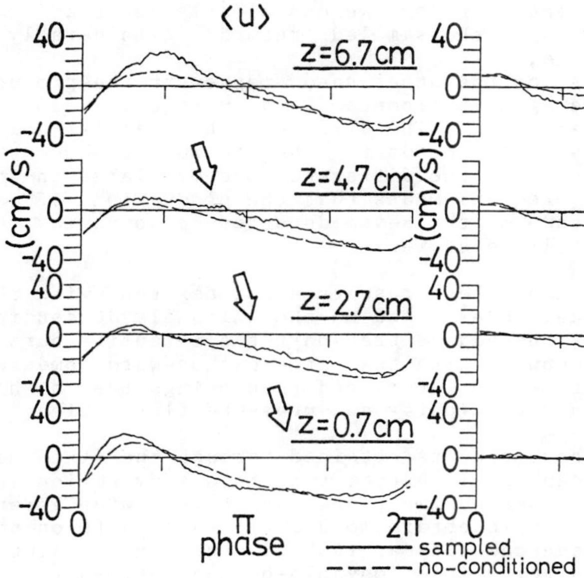

Fig.8 Distributions of phase-averaged velocity $<u>$ and $<w>$ for conditionally sampled and no-conditioned records.

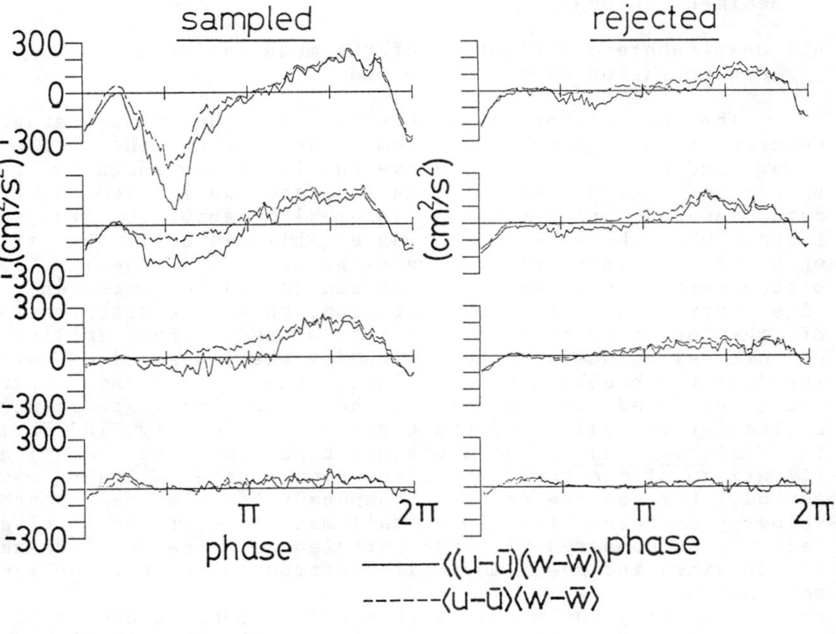

Fig.9 Distributions of phase-averaged momentum flux for conditionally sampled and rejected records.

the left and the right figures, we can clearly see that the momentum transport of the sampled records considerably increases as a whole.

The broken lines, on the other hand, represent the product of the phase-averaged horizontal and vertical velocity fluctuation, $\langle u-\bar{u}\rangle \cdot \langle w-\bar{w}\rangle$. The difference between the solid and broken lines corresponds to the turbulent momentum transport. The fact that the difference becomes large in the highly turbulent regions means that the production of an appreciable magnitude of the Reynolds stress accompanies the evolution of the oblique eddies.

In summary, all of these results are consistent with the following conceptual model based on the obliquely descending eddies. That is, the eddies transport the intensive turbulence obliquely downward with the large onshoreward momentum at the upper layer of the water, and then brings the highly intermittent turbulent velocity and pressure fluctuation to the bottom.

Considering these results obtained through the fixed-bed experiments, we can easily expect that this eddy action to the bottom plane causes sediment suspension in a wide extent of the surf zone. To obtain more definite results on the relationship between the eddy action and resultant sediment suspension, we have made movable-bed experiments, as described as follows.

4. Relation between the Obliquely Descending Eddies and the Sediment Suspension

4.1 On-offshore distribution of the mean sediment concentration close to the bed

The on-offshore distributions of the time-averaged concentration $\bar{c}$ both at z=0.5cm located within the bedload layer and at z=1.0cm just above the layer are shown in the bottom of Fig.10. The values indicated were based on the data measured during the time interval of about 70s from the instant when the wave set-up was established until the time when the sediment ripples were generated. The initial bottom surface configuration was smoothened for each run.

The upper figure, on the other hand, shows the distribution of the bottom turbulence intensity w' and of the amplitude of orbital velocity u_m normalized by the linear long wave speed at the breaking point. These values of w' and u_m are those measured at z=0.7cm in the fixed-bed experiments. Unlike for the values of the turbulence intensity indicated in Figs. 4 and 7, the averaging time for the numerical filtering of the raw data is the time of the incident wave period, because the vertical component of the near-bottom velocity contains negligibly small magnitude of the orbital velocity as compared with the turbulence. Hence w'-values so obtained include almost all contribution of the oblique eddy motion.

Comparing the shape of these distributions each other, we can recognize that the sediment concentration level $\bar{c}$ in the bed load layer correlates well to u_m, while $\bar{c}$ above the layer varies in accordance with the bottom turbulence intensity w'. This result suggests the close relationship between the sediment suspension and the action of the oblique eddies.

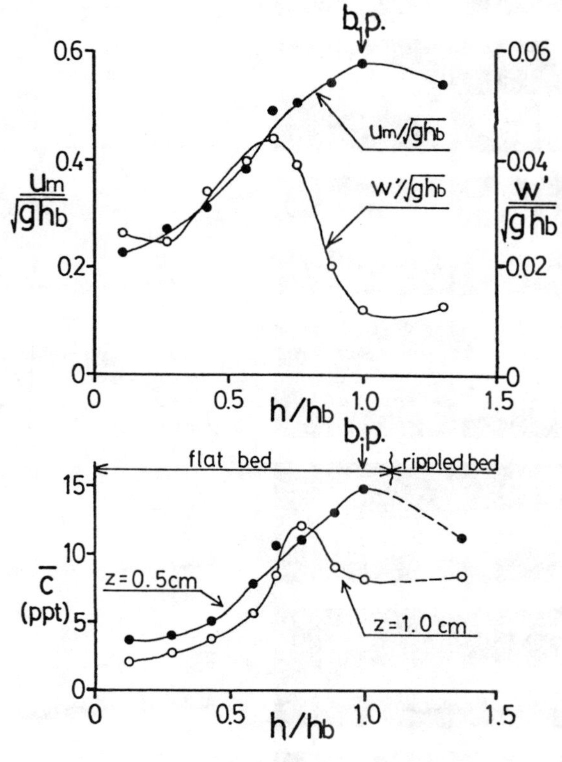

Fig. 10 Spatial distributions of mean concentration $\bar{c}$ at z=0.5 and 1.0cm and those of bottom turbulence intensity w' and amplitude of orbital velocity u_m.

4.2 Visualization of the eddy evolution and the resultant occurrence of sediment suspension

To confirm this relationship from a visual aspect, we have taken some photos from a side wall of the wave channel. Photo 1 and Figure 11 are examples showing a sequence of pictures and their tracings around the velocity measuring section taken successively by a motor-driven 35 mm camera. The vertical rod shown in the photos is the stem of the sensor to measure the sediment concentration. The instants taking the photos are indicated by the arrows denoted by the alphabets of A to H in Fig. 12, where both the water surface elevation η and the bottom sediment concentration c at z = 1.0 cm at the velocity measuring section are represented. The records shown in Fig. 12 were those obtained immediately after the starting of the wave generator; thus the first

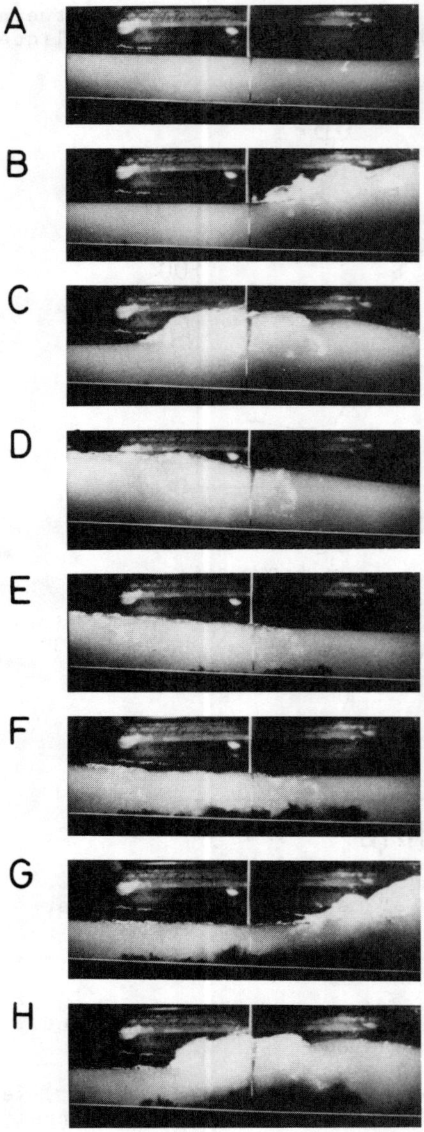

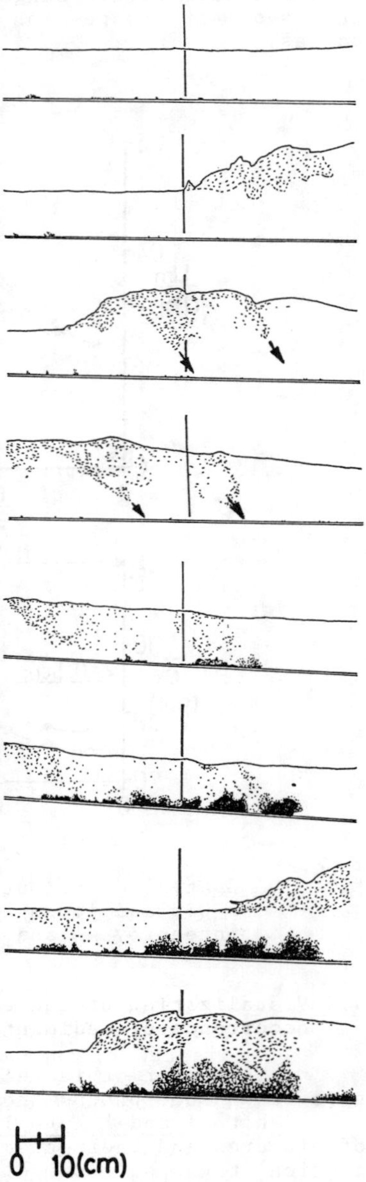

Photo 1 Development of obliquely descending eddies and sediment suspension.

Fig.11 Tracing of Photo 1.

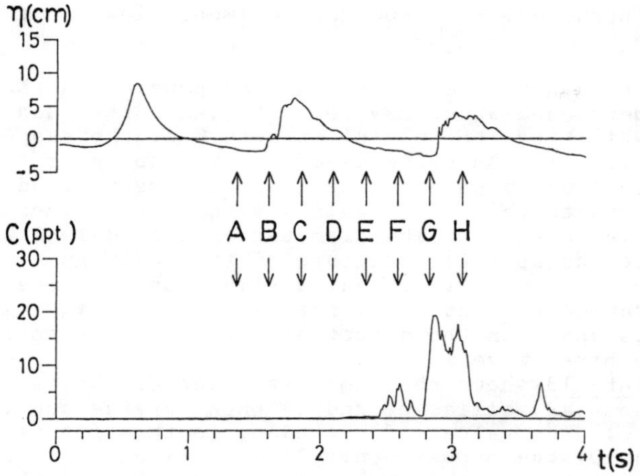

Fig.12 The instants taking the photos and the time histories of η and c.

wave indicated in the figure is a non-breaking wave. The figure indicates that the sediment concentration increases not under the first non-breaking wave, but around the phase of the trough of the second breaking wave.

The brighter regions of the photos indicate the air bubbles involved in the eddies, which are sketched in Fig.11 by the speckles. The dark regions close to the bottom in E to H, on the other hand, indicate the suspended sediment cloud. From these photos and figures, we can describe the process of the eddy development and associated sediment suspension as follows.

(A)····· No appreciable sediment suspension can be found, though there exists the sediment movement due to sheet-flow.

(B)····· The horizontal eddies grow to appear in front of the bore crest.

(C)····· The eddies develope into the obliquely descending eddies behind the wave crest. The sediment suspension, however, cannot be found yet at this stage.

(D)····· The oblique eddies stretch toward the bottom as indicated by the arrows.

(E)····· Immediately after the instant when the eddies hit the bottom surface, the sediment suspension appears in the vicinity of the bottom.

(F)····· A large amount of the sediment is lifted up into suspension by the eddy action, increasing the concentration value as shown in Fig. 12.

(G)····· The sediment clouds migrate offshore with the backwash of the bore. And then they are convected upward by the heaving orbital motion preceding to the passage of the wave crest.

(H)····· The subsequent oblique eddies evolve toward the bottom.

4.3 3-D structure of suspended sediment cloud in the surf zone

The dependence of the sand suspension on the oblique eddies mentioned above may be confirmed more quantitatively by investigating the spatial structure of the suspended sand cloud; i.e, if the oblique eddies lift up the bottom sediment into suspension, the resultant sediment cloud will show the strong three-dimensionality with the horizontal scale comparable to that of the diameter of the eddies. So to investigate the spatial structure of the sediment cloud close to the bottom, we have measured simultaneously the sediment concentration c_1 and c_2 by the two probes aligning several distances apart in the direction perpendicular to the plane of the orbital wave motion.

Figure 13 shows the time histories of the water surface elevation η as well as c_1 and c_2 both outside and inside the surf zone. The values of c_1 and c_2 in this figure are those measured by the probes manipulated to keep their hights at 1.5cm above the bottom and separated with 4cm each other in the horizontal cross-channel direction.

The values of c_1 and c_2 outside the surf zone well correlate to each other and exhibit strong dependance on the phase variation with the two marked peaks within the one

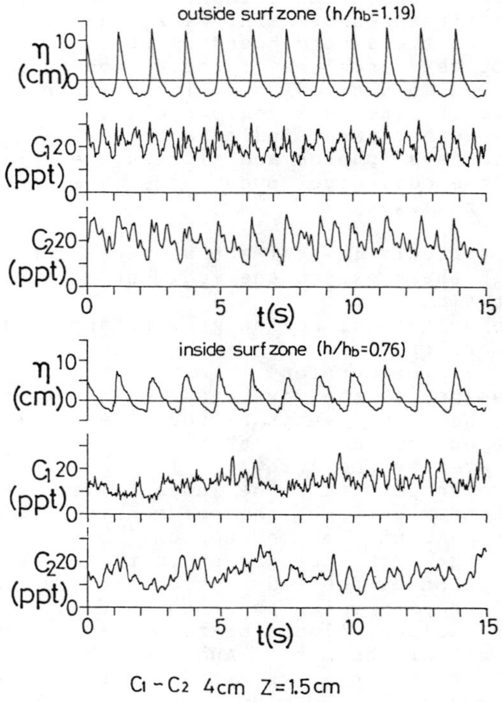

Fig.13 An example of the time histories of η as well as c_1 and c_2 both outside and inside the surf zone.

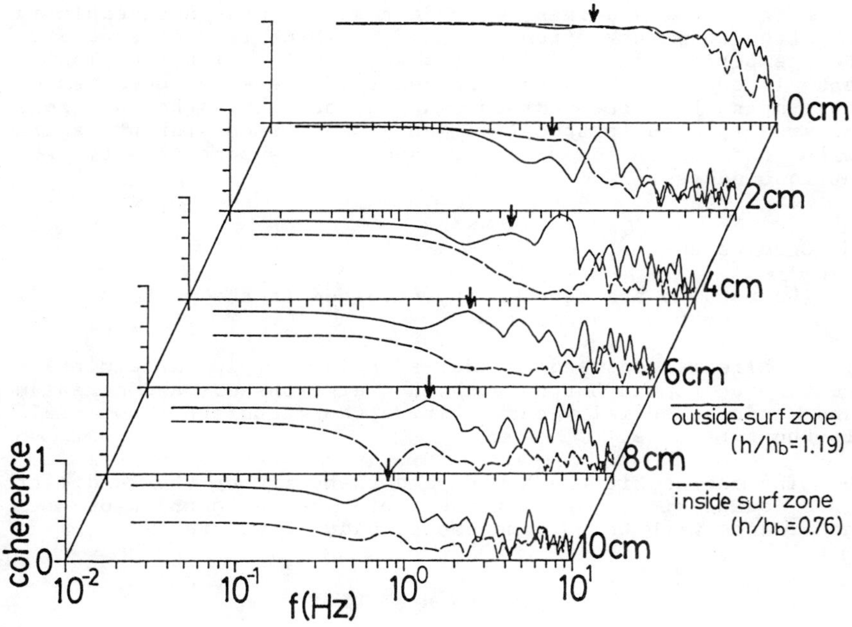

Fig.14 Coherence between bottom concentration records for several probe spacings.

wave period, though the high frequency component is slightly losed in c_2 record as compared with c_1, because the frequency response of the probe for c_2 is lower than that for c_1. The concentration variation inside the surf zone, on the contrary, shows no definite phase-dependance, and the correlation between these two records is rather poor. These facts suggest that the sediment suspension within the surf zone is mostly governed by the turbulence, especially by the obliquely descending eddies, generated with the wave breaking, though, in the outside of the surf zone, the wave fluid motion is a unique action to the sediment suspension, which is nearly two-dimensional phenomenon.

Then to investigate the correlation structure between c_1 and c_2 more quantitatively, we have calculated the coherence between c_1 and c_2. The results are shown in Fig.14, where the coherence values are indicated for several probe spacings with the horizontal axis of frequency f. The data used are those measured for the initially smoothened bed during the time interval of about 80 s from the instant when the wave set-up was established to the time when the sediment ripples were generated.

The coherence outside the surf zone denoted by full lines retain high values even for the large probe spacings, demonstrating two-dimensionality of sediment suspension outside the surf zone. On the contrary, the coherence within the surf zone represented by broken lines decreases rapidly with the probe spacings. The decrease is most noticeable in the frequency range from 0.8 to 5 Hz which is corresponding to the period of the incident waves and its higher harmonics and is comparable to the frequency of the eddy occurrence. This means that the sediment suspension inside the surf zone occurs in a quite spot-like manner and is consistent again with the proposed model of sediment suspension by the oblique eddies.

5. Conclusions

The results and conclusions obtained in the present study can be summarized as follows.

1) "Obliquely descending eddies" bring highly intermittent turbulence to the bottom with the large onshoreward momentum and thus essentially characterize the turbulent flow field in the surf zone.

2) The eddies hit the bottom and then lift up the sediment into suspension, yielding the spot-like sediment cloud in accordance with the 3-dimensional eddy structure.

Acknowledgment

This study was financially supported by a grant in aid for scientific research, No.61460165, Ministry of Education, Science and Culture, Japanese Government.

References

Deigaard, R., Fredsoe, J. and Hedegaard, I.B.(1986): Suspended sediment in the surf zone, Jour. of Waterway, Port, Coastal and Ocean Engineering, ASCE, Vol.112, No.1, pp.115-127.

Hino, M., Sawamoto, M., Yamashita, T., Hironaga, M. and Muramoto, T.(1984): Prototype 2-dimensional LDV adopting optical fibers, Laser-Doppler Velocimetry and Hot-Wire/Film Anemometry, Association for the Study of Flow Measurement.

Miller,R.L.(1976): Role of vortices in surf zone prediction: sedimentation and wave forces, Beach and Nearshore Sedimentation, Soc. Econ. Paleontol. Mineralog., Spec. Publ. No. 23, pp.92-114.

Nadaoka, K.(1986): A fundamental stdy on shoaling and velocity field structure of water waves in the nesrshore zone, Doctoral Dissertation, Tokyo Inst. of Tech. (*reproduced in Tech. Rept. of Dept. Civil Eng., Tokyo Inst. of Tech. No.36, pp. 33∿125, 1986.*)

Shibayama, T. and Horikawa S.(1985): Sediment transport due to breaking waves, Proc. of 20th Int. Conf. on Coastal Eng., ASCE, pp.1509-1522.

CHAPTER 123

Suspended Sediment Concentration in the Surf Zone

Makoto Ifuku[1) and Tadao Kakinuma[2)

Abstract

The water particle velocity,the pressure fluctuation and suspended sediment concentration near the bottom were measured in the surf zone. Data were obtained on the mean sediment concentration, phase lag between onshore velocity and suspended sediment concentration. The near-bottom velocity distributions under finite amplitude waves were calculated on the basis of the turbulent boundary layer theory by using time-independent/dependent eddy viscosities and Prandtl mixing length theory. The concentrations at reference level were estimated from the Kalkanis' theory using the velocity value at the top of a sand particle and the distributions of suspended sediment concentration were calculated on the basis of the turbulent diffusion theory by using time-independent/dependent diffusion coefficients. The computed mean values of the suspended sediment concentration agree with those observed by Noda, the authors and Nielsen and those measured by Deigaard et al.. The computed phase lag agrees well with observed by the authors.

1.Introduction

In order to investigate the sediment transport phenomena in the surf zone, it is necessary to study the sediment suspension mechanism due to waves. Many investigators have been concerned with sediment suspension. Early studies, such as those of Fairchild(1959), Homma and Horikawa(1962) and Noda(1967) were made with siphons, pumped sampler and suspended sampler. With these types of sampler it is difficult to obtain accurate records of concentration variation. For these most recent studies, such as those of Homma et al.(1965), Sleath(1982), Brennikmeyer(1974) and Wright et al.(1982) the records of concentration variation were obtained in laboratory and on site. Sleath(1982) measured the variation of sediment concentration at a fixed point above

1) Associate Professor,Department of Ocean Engineering,Ehime University, Matsuyama,Japan

2) Professor,Department of Ocean Engineering,Ehime University,Matsuyama, Japan

a ripple crest in oscillatory flow. He showed that stronger peaks in the concentration records are observed at phases where the velocity at the upper edge of the boundary layer reverses, while weaker peaks are found at phases where the velocity is approximately maximum. He expressed the characteristics of the boundary layer in oscillatory flow in terms of the vortex near the ripple and elucidated the mechanism of sediment suspension. However, in the discussion of initial motion of sediment, accurate knowledge of the near-bottom water particle velocity in the turbulent boundary layer is important in order to evaluate drag, lift, inertia and body forces acting on sediment.

The present paper deals with the details of near-bottom water particle velocity and suspended sediment concentration on the basis of field data. Meanwhile, numerical analysis has also been performed on the basis of the turbulent boundary layer theory and diffusion theory to elucidate theoretically the observed and experimental suspended sediment concentrations.

2. Field observation and data analysis

2.1 Sites and method of observations

The field observations were conducted at Baishinji coast near Matsuyama city in Ehime Prefecture, Japan, over the periods from 15 December 1982 to 19 March 1983 and from 21 December 1983 to 19 March 1984. The observation points were located 120 and 100 m off the coastal breakwater and the mean water depths at these points were 2.0 and 1.7 m (Fig. 1). The mean beach slope around both the points were approximately 0.01.

The field instrumentation system used was composed of 3 sets of two-component electromagnetic flow meters MODEL551(MARSH McBIRNEY), a pressure transduser MODEL 205-2 (Setra Systems) and a turbidimeter MODEL MA-101 RD(Hokuto Riken).

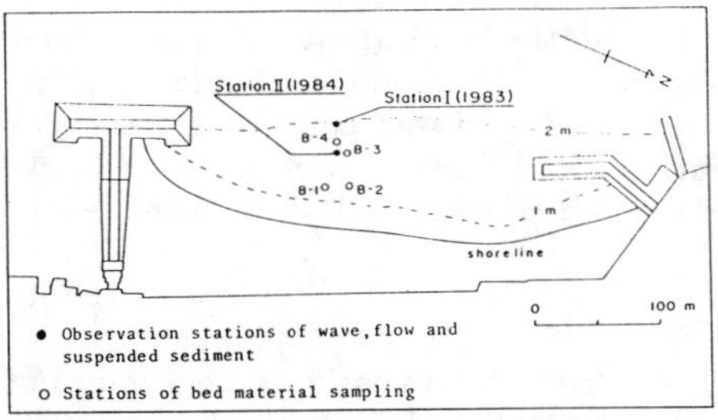

Fig.1 Observation stations of wave, flow, suspended sediment and bed material.

Station I

The instruments were mounted on a frame and fixed on the bottom using 8 concrete blocks.

The levels of electromagnetic flow meter were 0.07, 0.65 and 0.98 m above the bottom and the horizontal velocities were measured. The levels of the turbidimeter and the pressure transduser were 0.04 and 0.72 m above the bottom respectively.

Station II

An electromagnetic flow meter was set at the level of 0.3 m above the bottom and the cross-shore and vertical components of water particle velocity were measured. The other two were set at the levels of 0.45 and 0.70 m to measure the cross-shore and longshore components. The levels of the turbidimeter and the pressure transduser were 0.34 and 0.49 m above the bottom respectively.

The turbidimeter was an optical type and composed of six light sources and a receiver. Data obtained were transmitted to a shore station, where they were recorded on a data logger XR-30(TEAC).

2.2 Data analyses

Records of velocity, pressure fluctuation and turbidity (suspended sediment concentration) were digitized for a interval of 0.2 s at every 15 min. The zero-up-crossing method was employed to calculate the velocity and pressure fluctuation of the significant wave. First, spectral analyses were performed to obtain the power spectrum, cross spectrum, physical spectrum of velocity and suspended sediment concentration as well as directional spectrum.

2.3 Significant wave and bed material

During the 1983 observation the significant wave height and period were 0.41-0.53 m and 4.4-4.7 s, while in 1984 they were 0.33-0.79 m and 4.4-5.3 s. The frequency of breaker occurrence is defined as the ratio of the number of breakers to the number of waves in each data set. In both years, most of breakers visually were of the spilling type. Sampling of bed material performed in November 1983 at B-1 and 2 and in January 1984 at B-3 and 4. Bed material around the observed sites consisted entirely of sand with a median diameter of 0.18 mm.

3. Results of data analysis

3.1 Power spectrum of suspended sediment concentration

The great majority of power is in the low frequency range (0.016 Hz - 0.025 Hz), although a weaker peak appears at the peak frequency of wind waves (approximately 0.25 Hz). The power of suspended sediment concentration falls off as: $f^{-0.5}$–f^{-1}(f :frequency) and f^{-3}–$f^{-3.5}$ in the low and high frequency ranges, regardless of the frequency of breaker occurrence, while the profiles of cross-shore velocity and pressure fluctuation depend on the frequency of breaker occurrence. Fig. 2 shows the power spectrum of the suspended sediment concentration and cross-shore velocity; (a) and (b) are in the case of the low and high frequency of breaker occurrence respectively.

3.2 Suspended sediment concentration and near-bottom cross-shore velocity

Fig.3 shows the physical spectrum of cross-shore velocity observed 0.07 m above the bottom together with the time series of suspended sediment concentration observed at 0.04 m above the bottom. The uppermost figure is for the lower frequency band, 0.02-0.078 Hz; next is for 0.098-0.156 Hz; third is for the peak frequency band of wind waves, 0.176-0.234 Hz and the fourth is for 0.391-0.449 Hz. The lowest figure shows the time series of suspended sediment concentration. Energy density of the physical spectrum in the lower frequency band is approximately one-fifth as large as that in the peak frequency band. The

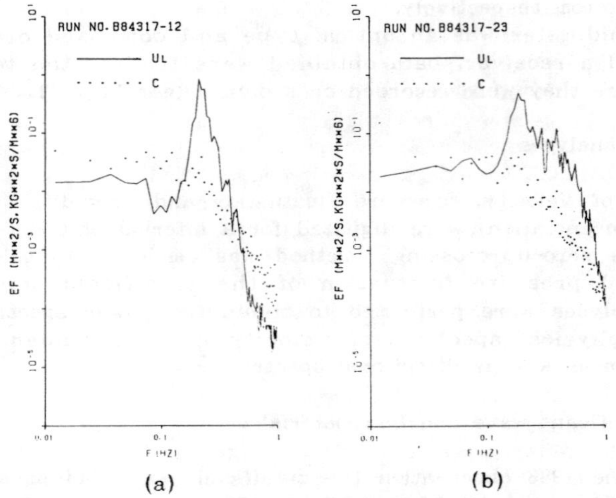

Fig.2 Power spectra of suspended sediment concentration and cross-shore velocity, where the frequencies of breaker occurrence are (a) 14 % and (b) 20 %.

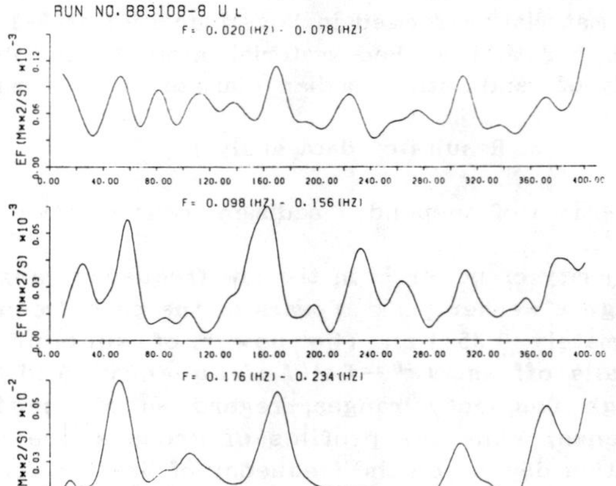

Fig.3 Physical spectra of cross-shore velocity at the level of 0.07 m and time series of suspended sediment concentration at 0.04 m above the bottom.

suspended sediment concentration is quite high when energy densities in the peak and quarter frequency bands of wind waves are high. Thus the fact that higher values of suspended sediment concentration occur at a 40-60 s time interval seems to be related to lower and peak frequency components of cross-shore velocity. The above discussion is very interesting in comparison with the results of Brennikmeyer(1974) and Wright et al.(1982), who suggested that suspended sediment concentration is related to swells or long period waves and surf beat on the basis of power spectrum analyses of wave, cross-shore velocity and suspended sediment concentration observed in the surf zone.

3.3 Phase lag between cross-shore velocity and suspended sediment concentration

Fig.4 shows the histogram of phase lag between a peak of cross-shore velocity and that of suspended sediment concentration during a wave cycle. The histograms were obtained through the analyses of 107 and 120 waves in (a) and (b) where the frequencies of breaker occurrence are 13 % and 19 % respectively. The maximum values in the histogram appear at the phases $\pi/2$, 1.3π in (a) and $\pi/2$, 1.1π in (b).

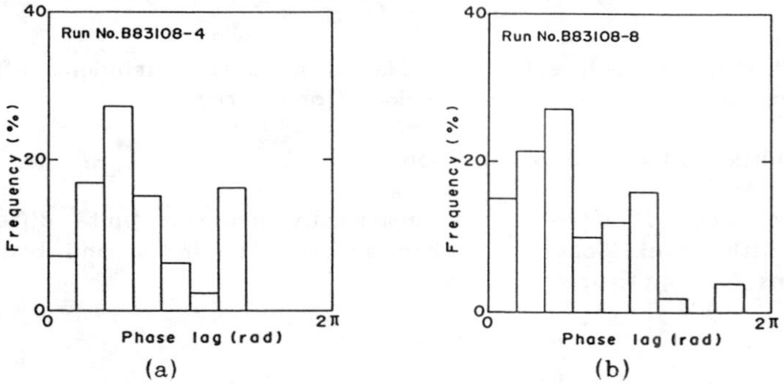

Fig.4 Histograms of phase lag between onshore velocity and suspended sediment concentration, where the frequencies of breaker occurrence are (a) 13 % and (b) 19 %.

4. Numerical analysis

4.1 Basic equations

(a) Water particle velocity in the turbulent boundary layer

The x- axis is taken along a horizontal bottom and z- axis vertically upwards. The velocity components in the x- and z- directions are denoted by u and w.

The equation of motion for a two-dimensional turbulent boundary layer is given as follows:

$$\frac{\partial u}{\partial t}+u\frac{\partial u}{\partial x}+w\frac{\partial u}{\partial z}=-\frac{1}{\rho_f}\frac{\partial p}{\partial x}+\frac{1}{\rho_f}\frac{\partial \tau}{\partial z} \quad (1)$$

in which t is the time, p is the pressure, ρ_f is the density of fluid,and τ is the shear stress. If the mean flow is parallel to the bed, Eq.(1) is reduced to

$$\frac{\partial u}{\partial t}=-\frac{1}{\rho_f}\frac{\partial p}{\partial x}+\frac{1}{\rho_f}\frac{\partial \tau}{\partial z} \tag{2}$$

The pressure gradient just outside the boundary layer is given by

$$-\frac{1}{\rho_f}\frac{\partial p}{\partial x}=\frac{\partial u_b}{\partial t} \tag{3}$$

in which u_b is the horizontal velocity at the upper edge of the boundary layer.

If the above expression also holds in the boundary layer, Eq.(2) becomes

$$\frac{\partial(u-u_b)}{\partial t}=\frac{1}{\rho_f}\frac{\partial \tau}{\partial z} \tag{4}$$

(b) Concentration

The one-dimensional turbulent diffusion equation for sediment concentration is given as follows:

$$\frac{\partial C}{\partial t}+w\frac{\partial C}{\partial z}=\frac{\partial}{\partial z}\left(K_z\frac{\partial C}{\partial z}\right)+w_0\frac{\partial C}{\partial z} \tag{5}$$

in which C is the sediment concentration, K_z is the turbulent diffusion coefficient, and w_0 is the settling velocity of sediment.

4.2 Initial and boundary conditions

Eqs.(4) and (5) were solved numerically using a finite difference method with Crank-Nicholson implicit scheme. The initial and boundary conditions for Eq.(4) are

$$\begin{aligned} &u(z,0)=0\\ &u(z_0,t)=0\\ &u(\delta_b,t)=u_b(t) \end{aligned} \tag{6}$$

in which z_0 is the roughness length and δ_b is the thickness of the turbulent boundary layer. Initial and boundary conditions for Eq.(5) are

$$\begin{aligned} &C(z,0)=0\\ &C(\delta_B,t)=C_0(t)\\ &K_z\frac{\partial C}{\partial z}+w_0C=0\,, \qquad z=h \end{aligned} \tag{7}$$

in which δ_B is the thickness of the bed-layer, $C_0(t)$ is the sediment concentration in the bed-layer, and h is the water depth.

4.3 Evaluations of parameters

The vertical distance from the bottom to the water surface is divided by 101 points. The vertical distance from the bottom to the level of three grain diameters is divided into the interval of one-half a grain diameter, and above the level the intervals are determined so as to form a geometric progression. The computational time-step is determined as T / 96 (T :the wave period).

(a) Velocity in the boundary layer

The eddy viscosities are assumed to be

$$Nz(z)=\alpha\kappa\sqrt{f}u_{bm}z \tag{8-1}$$

$$Nz(z,t)=\alpha\kappa\sqrt{f}|u_b(t)|z \tag{8-2}$$

where α is the constant of proportionality, κ is the Karman constant, f is the bottom friction factor and u_b is the maximum velocity at the upper edge of the boundary layer. Bottom friction factor is calculated from the empirical expression given by Kakinuma and Ifuku(1985):

$$f=1.95\theta^{-0.87} \quad 2\leqq\theta\leqq 40$$
$$f=0.39\theta^{-0.42} \quad 40\leqq\theta\leqq 190 \tag{9}$$

where $\theta=\rho_f u_{bm}^2/(\rho_s-\rho_f)gd_{50}$, ρ_s is the density of sediment ($\rho_s=n\rho_f+(1-n)\rho_s'$, n is the porosity, ρ_s' is the density of dry sediment), g is the acceleration gravity, and d_{50} is the median diameter of sediment.

The roughness length, according to Bakker-van Doorn(1978), is

$$z_0=\eta/33 \tag{10}$$

where η is the ripple height, which is estimated from Nielsen's empirical relationship(1981):

$$\eta=21a_m\theta^{-1.85} \tag{11}$$

where a_m is the half stroke of water particle at the bottom.

Noda(1969) evaluated the thickness of turbulent boundary layer as

$$\delta_b=25\delta \tag{12}$$

where $\delta=\sqrt{\nu T/2\pi}$ (ν : the kinematic viscosity). At each point, until the difference of normalized velocity between the times of mT(m : integer) and $(m+1)T$ becomes less than 10^{-5}, calculation is continued.

(b) Concentration

The sediment concentration in the bed-layer, according to Kalkanis(1965), is

$$C_0=2P\rho_s'd\cdot V/3\int_{z_0}^{\delta_B}udz \tag{13}$$

where d is the particle size of sediment, V is the speed of sediment propagation and P is the pick-up rate:

$$P=\frac{1}{\sqrt{2\pi}}\int_{B_*\cdot\Psi'-1/\eta_0}^{\infty}\exp(-z^2/2)dz \tag{14}$$

where $B_*=4/3\,C_L\eta_0$, C_L is the lift coefficient, $1/\eta_0$ is an empirical constant, and $\Psi'=(\rho_s-\rho_f)gd/\rho_f u_0^2$ (u_0: the local water particle velocity). It is difficult to estimate the speed of sediment propagation in Eq.(13). In the present analysis it is set to be equal to the cross-sectional mean velocity of water particles in the bed-layer.

The turbulent diffusion coefficients are assumed to be

$$K_z=\beta\alpha\kappa\sqrt{f}\,u_{bm}z \quad (15\text{-}1)$$

$$K_z=\beta\alpha\kappa\sqrt{f}\,|u_b(t)|\,z \quad (15\text{-}2)$$

$$K_z=\gamma(l_T u^2+l_L w^2)/q \quad (15\text{-}3)$$

where β, γ are the constant of proportionality, l_T and l_L are the characteristic length(l_L = $a l_T$; a: the constant of proportionality), and $q = (u^2 + w^2)^{1/2}$.

The settling velocity of sediment is calculated by

$$w_0=\sqrt{\frac{4}{3}\frac{d}{C_D}\frac{\rho_s'-\rho_f}{\rho_f}g} \quad (16)$$

where C_D is the drag coefficient. In the case of Reynolds number, R_e = $w_0 d/\nu$ being less than 5, the drag coefficient can be estimated(1968) by

$$C_D=\frac{24}{R_e}+4.5 \quad (17)$$

At each point until the difference of normalized concentration between the times of mT and $(m+1)T$ is less than 10^{-5}, calculation is continued. In the present study, the fluids are sea water for Noda's, the authors' and Nielsen's observed results and fresh water for the results obtained by Deigaard et al.. The water temperatures are assumed to be 10 °C for the authors' results and 20 °C for the others' results. The sediment is sand and density and porosity have been taken at 2.65 g/cm³ and 0.3 respectively. Sand grain sizes are 0.30, 0.18, 0.12-0.22 and 0.12 mm for Noda's, the authors', Nielsen's and Deigaard et al. results,respectively. The characteristic length, l_T, is assumed to be the distance from the bottom.

5. Numerical results

Parametric analysis is performed by using the observed data obtained by Noda, the authors and Nielsen and measured data by Deigaard et al.. The significant waves obtained through field observations by Nada, authors and Nielsen and laboratory measurement by Deigaard et al., are in the range that Stokes wave theory of second order, third order and cnoidal wave theory and Stokes wave third order are valid(1977), respectively. Einstein and El-Samni(1949) reported that the lift coefficient has a constant value 0.178 on the basis of measurement of the pressure difference. It was assumed here that the lift coefficient is 0.2 and $1/\eta_0$ is 1.5 after Kalkanis and u_0 has been taken to be the value at the top of the sand particle and the thickness of the bed-layer is twice sand particle.

5.1 Velocity in the turbulent boundary layer

The velocity profiles at various phases during a cycle of wave motion in the turbulent boundary layer are shown in Fig.5. The maximum value of the water particle velocity at phase 6 is approximately 1.03 times as large as that at the upper edge of boundary layer. In the case that the shear stress is evaluated by using time-dependent eddy viscosity or Prandtl mixing length theory the over-shooting is more significant than the case that the shear stress is evaluated by using time-independent eddy viscosity at phases 6.25, 6.375, 6.5 and 6.875 and the levels which arise the over-shooting are lower at above-mentioned phases where the water particle velocities are negative. In the cases that the shear stress is evaluated by using time-dependent eddy viscosity and Prandtl mixing length theory, both profiles show the similar trend.

Fig.6 shows the temporal variation of water particle velocity at the sand particle level. The velocity profiles are asymmetric and the phase of the maximum velocity is ahead of that at the upper edge of boundary layer by approximately $\pi/12$.

5.2 Concentration

(a) Variation of sediment concentration in the bed-layer

Fig.7 shows the variation of the sediment concentration in the bed-layer depending on the instantaneous value of $1/\psi'$. The concentration in the bed-layer is calculated on the assumption that the thickness of the bed-layer is equal to twice the grain diameter in Eq.(13). For larger value of $1/\psi'$ in the range of present analysis, the concentration in the bed-layer is asymptotic to about 0.3. This tendency is similar to the result obtained by Deigaard et al.(1986).

(b) Temporal variation of sediment concentration at several heights

Fig.8 shows the temporal variation of suspended sediment concentration; in (a)-(f) are in the cases that the relative heights are 1.67×10^{-4}, 1.44×10^{-3}, 2.44×10^{-3}, 1.22×10^{-2}, 1.93×10^{-2} and 2.41×10^{-2} respectively. With being off from the bottom,suspended sediment concentration decreases and the peak of suspended sediment concentration leads gradually. At $z/h=1.93\times10^{-2}$, is the height at which suspended sediment concentrations are observed,the peak of suspended sediment concentration appears at phase 6.25 approximately. It is the phase where the cross-shore velocity at the upper edge reverses approximately.

(c) Temporal variation of sediment concentration and phase lag between cross-shore velocity and sediment concentration

Fig.9 shows the computed temporal variations of suspended sediment concentration at 0.04 m from the bottom in comparison with the histogram of phase lag between maximum value of observed onshore velocity and maximum value of suspended sediment concentration during a wave cycle. Peaks in the histogram are found at $\pi/2$ and 1.3π in (a) and $\pi/2$ and 1.1π in (b)-corresponding to peaks in the computed temporal variation of suspended sediment concentration. The temporal variation, calculated by using time-independent turbulent diffusion coefficient, has its peaks at the above-mentioned phases; but the value of both peaks is approximately the same, and the computed temporal variation disagrees with the histogram of phase lag. The temporal variations which are computed by using time-dependent turbulent diffusion coefficient has its peaks at the above-mentioned phases and the agreement between that computed and observed is good. It is, however, worth noting that the computed temporal variation which is computed by using Eq.(15-3) could

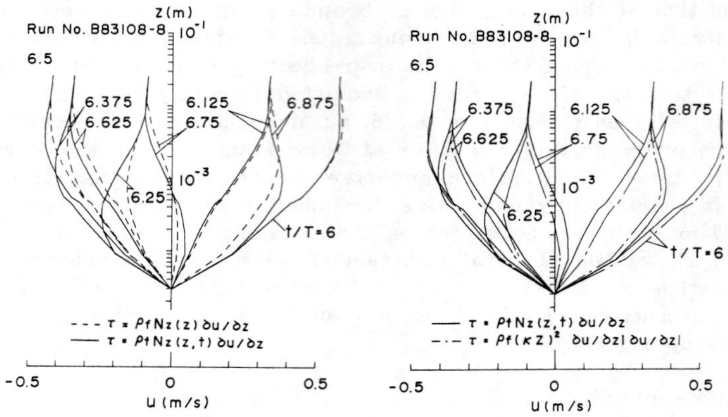

Fig.5 Velocity distributions in the turbulent boundary layer.

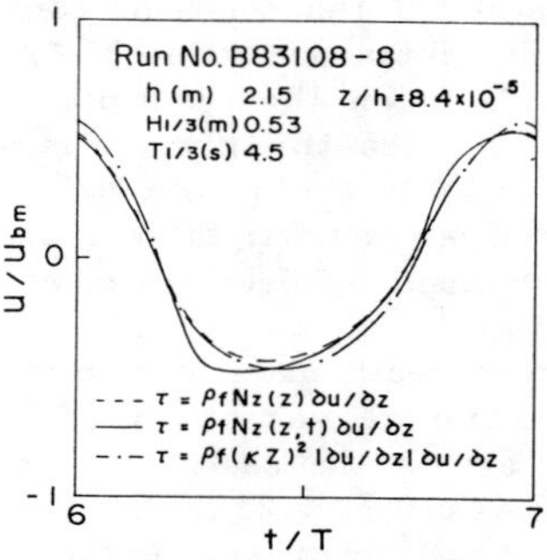

Fig.6 Temporal variations of water particle velocity at the sand particle level.

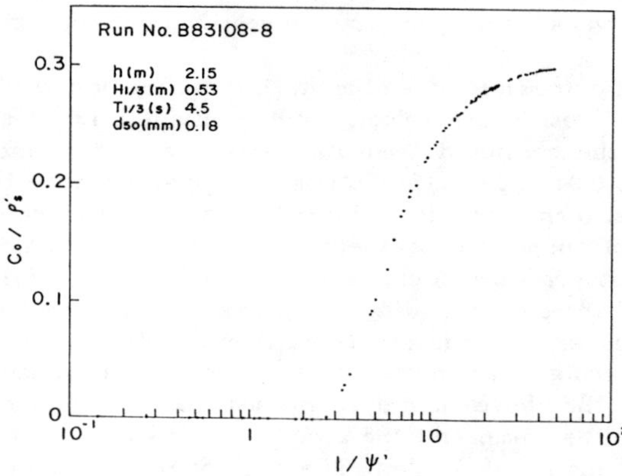

Fig.7 Relationship between the concentration in the bed-layer and $1/\psi'$.

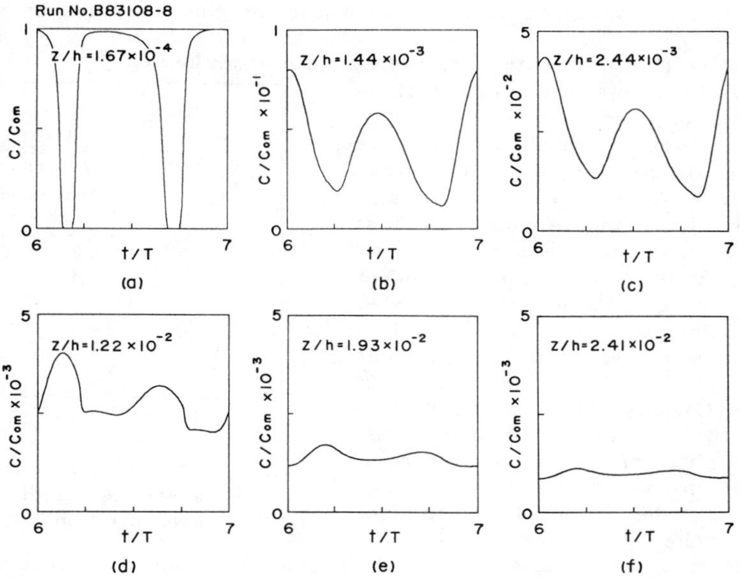

Fig.8 Temporal variations of suspended sediment concentration.

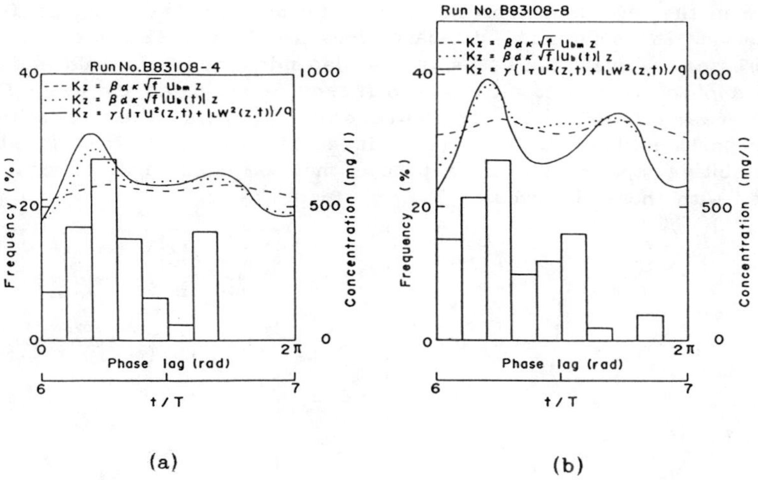

Fig.9 Computed temporal variations of suspended sediment and histogram of phase lag between maximum onshore velocity and maximum concentration observed in a wave cycle, where the frequencies of breaker occurrence are (a) 13% and (b) 19%.

elucidate the histogram of phase lag better than one computed by using Eq.(15-2).

(d) Vertical distributions of sediment concentration

Fig.10 shows the distributions of the computed suspended sediment concentration averaged over a wave cycle in comparison with those observed. The broken, solid and dotted lines are in the case that the shear stresses are evaluated by using Eq.(8-2) and the turbulent diffusion coefficients are evaluated by using Eqs.(15-1),(15-2) and (15-3) respectively. In the present analysis the constant of proportionality, α , in Eq.(8-2) is set to be equal to 0.1. The steepening of the profiles occurs at lower level above the bottom. However, the computed distributions that the turbulent diffusion coefficients are evaluated by using Eqs.(15-1) and (15-2) could not elucidate the suspended sediment concentration at higher levels above the bottom. It is,however,worth noting that the distribution of suspended sediment concentration calculated by using Eq.(15-3) could elucidate the concentration at the level farther from the bottom.

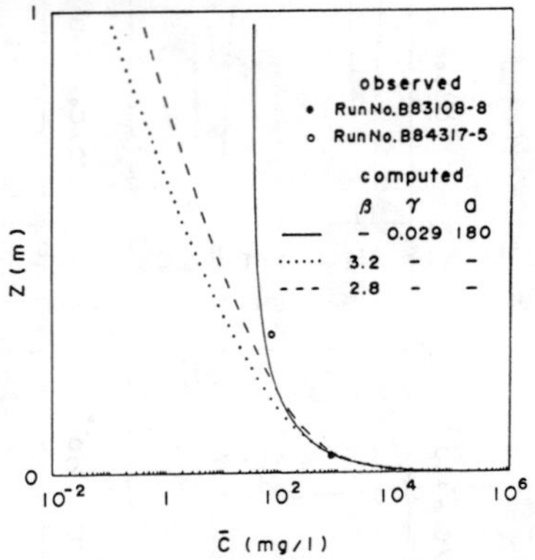

Fig.10 Distributions of mean suspended sediment concentration.

Fig.11 shows the distributions of the computed mean suspended sediment concentration in comparison with those observed; (a) and (b) are in the case of non-breaking wave by Noda and plunging breaker by Nielsen. The solid and dot-dash lines are in the case that the shear stresses are evaluated by using time-dependent eddy viscosity and Eq.(8-2) and the turbulent diffusion coefficient is evaluated by using Eq.(15-3) respectively. In the present analysis the constant of proportionality, α , in Eq.(8-2) is set to be equal to 0.1. Though exhibiting appreciable discrepancy, the computed distributions agree well with those observed.

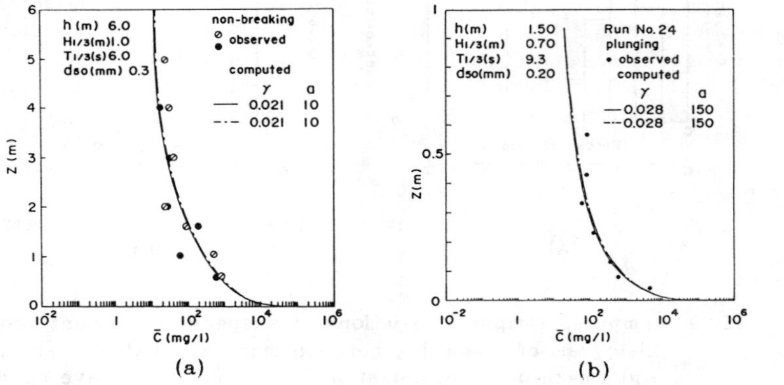

Fig.11 Distributions of mean suspended sediment concentration; (a) Noda's data, (b) Nielsen's data.

5.3 Relationship between γ , a and characteristics of wave and bed material

Fig.12 shows the relationship between the constant of proportionality γ in Eq.(15-3) and the ratio of significant wave height to the median diameter $H_{1/3}/d_{50}$. With increasing $H_{1/3}/d_{50}$, the constant of proportionality, γ ,decreases. For the same value of $H_{1/3}/d_{50}$, the constant of proportionality, γ , is largest for the plunging breaker.

Fig.13 shows the relation between the constant of proportionality, a, in Eq.(15-3) and the ratio of significant wave height to the mean water depth, the constant of proportionality, a, increases.

5.3 Turbulent diffusion coefficient

Fig.14 shows the distributions of the nondimensional turbulent diffusion coefficient which could elucidate the observed distributions of suspended sediment concentration; (a) and (b) are in the case of non-breaking wave by Noda and plunging breaker by Nielsen respectively. In this figure solid lines represent the instantaneous values of turbulent diffusion coefficient and the broken line represents the value averaged over a wave cycle. The symbols represent the value computed by Eq.(18) on the basis of the distributions of mean suspended sediment concentration. The nondimensional turbulent diffusion coefficient averaged over a wave cycle is proportional to z/h from the bottom to the level that the relative height is approximately 0.1 and proportional to $(z/h)^2$ at levels farther from the bottom.

$$\bar{K}_z = -w_0\bar{C}/(\partial\bar{C}/\partial z) \tag{18}$$

where $\bar{C}$ is the time-averaged sediment concentration.

6. Conclusions

Velocity, pressure fluctuation and suspended sediment concentration were measured near the bottom in the surf zone. The results were discussed in comparison with the calculated velocity distribution in the turbulent boundary layer and suspended sediment concentration. The following conclusions have been obtained through the present study.

(1) The histogram of phase lag between maximum concentration and maximum onshore velocity during a wave cycle exhibit peaks at phases $\pi/2$ and 1.1π or 1.3π .

(2) Good agreement between the observed results and the computed results is obtained concerning the phase lag between concentration and water particle velocity and mean value of suspended sediment concentration.

(3) The turbulent diffusion coefficient averaged over a wave cycle is proportional to z/h near the bottom and proportional to $(z/h)^2$ near the water surface.

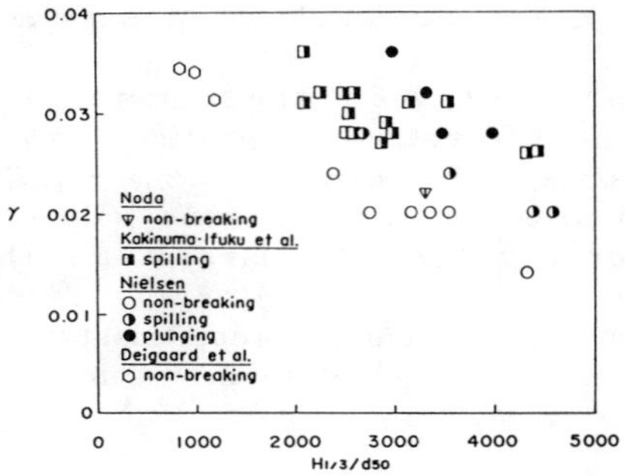

Fig.12 Proportional constant, γ as a function of $H_{1/3}/d_{50}$.

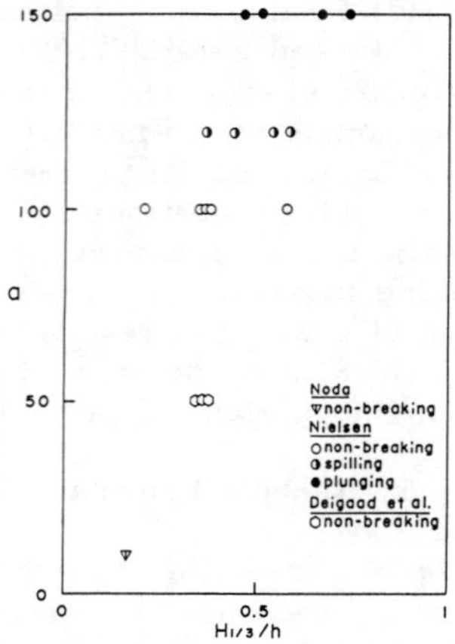

Fig.13 Proportional constant, a as a function of $H_{1/3}/h$.

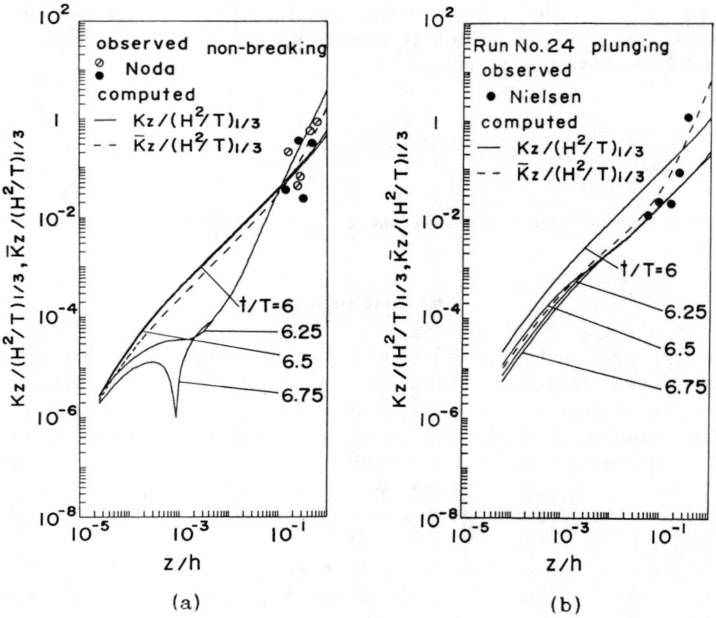

Fig.14 Nondimensional turbulent diffusion coefficient as a function of relative height.

References

1) Bakker, W.T. and Th. van Doorn(1978): Near bottom velocities in waves with a current, Proc. 16th Coastal Eng. Conf., pp.1394-1413.
2) Brennikmeyer, B.M.(1974): Mode and period of sand transport in the surf zone, Proc. 14th Coastal Eng. Conf., pp.812-827.
3) Deigaard, R., J.Fredsoe and I.B. Hedegaard(1986): Suspended sediment in the surf zone, J. Waterway, Port, Coastal and Ocean Eng., ASCE, pp.115-128.
4) Einstein, H.A. and El-Samni El. S.A.(1949): Hydrodynamic forces on a rough wall, Rev. Mod. Phys., pp.520-524.
5) Fairchild, J.C.(1959): Suspended sediment sampling in oscillatory wave action, Beach Erosion Board, Tech. Memo 115p..
6) Homma, M. and K. Horikawa(1962): Suspended sediment due to wave action, Proc. 8th Coastal Eng. Conf., pp.168-193.
7) Homma, M., K. Horikawa and R. Kajima(1965): A study on suspended sediment due to wave action, Coastal Eng. Japan, Vol.8, pp.85-103.
8) Kakinuma, T. and M. Ifuku(1985): Bottom friction factor on beaches, Proc. 32nd Japanese Conf. Coastal Eng., pp.234-237.(in Japanese)
9) Kalkanis, G.(1965): Transportation of bed material due to wave action, U.S.Army, Coastal Res.Center, Tech. Memo 2, pp.1-114.
10) Kestin, J.(1968): Boundary Layer Theory, 6th ed. McGraw Hill, p.108.
11) Nielsen, P.(1981): Dynamics and geometry of wave generated ripples, J.G.R., Vol.86, No.C7, pp.6467-6472.
12) Noda, H.(1967): Sediment suspension by waves, Proc. 14th Japanese Conf. Coastal Eng., pp.306-314.(in Japanese)
13) Noda, H.(1969): Development of turbulent boundary layer by waves, Proc. 16th Japanese Conf. Coastal Eng., pp.23-27.(in Japanese)
14) Sleath, J.F.A.(1982): The suspension of sand by waves, J. Hydraul. Res., Vol.19, pp.439-452.
15) U.S.Army Coastal Eng. Res. Center(1977): Shore Protection Manual, Vol.2, p.35.
16) Wright, L.D., R.T.Guza and A.D. Short(1982): Dynamics of high energy dissipative surf zone, Marine Geology, Vol.45, pp.41-61.

CHAPTER 124

VALIDATION OF CROSS-SHORE TRANSPORT FORMULATIONS

Richard J. Seymour, M.ASCE[1] and David Castel[2]

ABSTRACT
Seymour and King (1982) evaluated eight models for predicting cross-shore transport using beach profile data from the Torrey Pines experiment of the Nearshore Sediment Transport Study (NSTS). None of the models showed useful skill in predicting the sense, or direction, of transport. Three more data sets were acquired under NSTS and have been used in the present work to re-evaluate the original four models as well as another six not previously tested. The three new data sets include two nominally plane West Coast beaches and a barred beach on the Atlantic coast, each under a variety of wave conditions. Six of the models evaluated claimed a capability to predict the sense of the cross-shore transport, two predicted the beach slope as a result of cross-shore movement, and two gave detail predictions of changes to the beach profile position and shape.

The performance of the six models predicting direction of transport ranged from a skill factor of 0.49 (less than chance) to only 0.68. Five of the models required large changes to their calibration factor (usually based upon laboratory data) in order to have approximately the same skill in predicting erosion or accretion. One of the slope models was validated and the other gave no useful results. One of the two generalized models gave interesting results in predicting the time history of profile changes on the plane beaches for which it was developed. The other general model was not evaluated because it exhibited the lowest skill in predicting direction of transport.

INTRODUCTION
Following the first field experiment in the Nearshore Sediment Transport Study at Torrey Pines Beach, CA (see Seymour, 1983 for a review of the NSTS program) Seymour and King (1982) attempted to use the observed beach changes to evaluate the predictive capability of a number of formulations related to cross-shore transport. Of the twelve models reviewed, eight were found capable of making predictions of the transport knowing incident wave height and period - and in some cases the sand size and beach slope as well. The model performance was evaluated by determining the squared correlation coefficient between the beach face volume response and the forcing function prescribed by the model. The best of these, Dean (1973), based upon a dimensionless ratio that has come to be known as the Dean Number, and a modification by Hattori and Kawamata (1980) in which beach slope is added to the Dean Number, explained only about a third of the variance in the volume of the beach face. The other models ranged down to about 2% of the variance explained. This disappointing performance may have been due in part to the characteristics of the data set employed to test the models. A thorough analysis of this is given in Seymour (1988a), which concludes that the Torrey Pines data set was noisy, lacked significant erosionary events, and was unlikely to provide general insights into cross-shore transport mechanisms.

---

[1] Head, Ocean Engineering Research Group
[2] Senior Engineer, Ocean Engineering Research Group, Scripps Institution of Oceanography, University of California San Diego, La Jolla, CA 92093

NSTS produced three other beach profile data sets which did not suffer from the problems seen with Torrey Pines. These sets, which included data on waves and tides as well as characteristic sand sizes, were published in Seymour (1986). The three locations were Leadbetter Beach at Santa Barbara, CA (29 day series), Scripps Beach at La Jolla, CA (25 day series) and Virginia Beach, VA (23 days with a one day gap in the profiles). Santa Barbara and Scripps are typically without bars except for breaker bars which form immediately following major erosive events. Virginia Beach typically has a within-surf-zone bar that moves rapidly in response to changing wave conditions. These three data sets provided an attractive opportunity to reevaluate the cross-shore models and to include some others that had not been treated in the earlier work.

SELECTION OF MODELS

Of the eight models evaluated in Seymour and King (1982), four were in a form which would allow the prediction of a threshold point - that is, the incident wave condition above which erosion would be predicted and below which accretion would be expected, for a given beach condition. These four were Dean (1973), Hattori and Kawamata (1980) and two models contained in Short (1978). Two other threshold-type models were added for this study, Quick and Har (1985) and Sunamura and Horikawa (1974). With the exception of Quick and Har, these models were formulated to give only the direction of cross-shore transport. The formulations for the six threshold models are shown in Table I.

Two other models were added that attempted to describe the beach slope. Dalrymple and Thompson (1976), using laboratory data, established a linear relationship between the slope at mean sea level and the Dean Number. Dean (1977) established empirically an exponential shape for beach profiles, independent of wave or sediment characteristics, which takes the form

$$Y = AX^m$$

where Y = water depth and X = offshore distance

Both of these models lent themselves to testing with the present data.
Finally, two models were added that claimed to predict the time history of changes to the beach profile caused by cross-shore transport. These were Quick and Har, mentioned above, and Swart (1976). The Quick and Har model establishes an equilibrium profile based upon the Dean Number in which the major readjustment is a rotation of the profile. The model of Swart is an empirically-based, numerical formulation which allows for both rotation and translation of the beach face. A description of these models is contained in Seymour (1988b).

EVALUATING THRESHOLD MODELS

To evaluate these models, time histories were prepared of the daily sediment loss or gain for each site based upon changes in the profiles. In each case, the changes were interpreted as being caused entirely by cross-shore transport. The rationale for this interpretation and an assessment of its validity for each site is provided in Seymour (1988a). The variability is shown in Figure 1. This quantitative record could then be converted to a binary form of time series in which only the sign of the change is preserved. The six models provide just such a binary (erode/accrete) prediction and this was compared to the measured result. The skill of the predictor is represented by two numbers: the percentages of correct predictions for accretion and for erosion. It was interesting to note that the original four models used in the Torrey Pines study all exhibited an almost complete one-sidedness with these data - that is, they tended to predict either all erosion or all accretion. They had nearly perfect skill in one direction and almost zero skill in the other. This characteristic is illustrated in Figure 2, which shows the as-formulated predictions of Dean's model. The two models added for this study had approximately equal skill (as formulated) in predicting erosion or accretion - but these skill factors were less than 0.5 and therefore not as effective as random chance predictions. The thresholds for all of the models were adjusted until they had a nearly equal skill in predicting either erosion or accretion. The results are shown in Table II.

Table I

RESULTS OF THRESHOLD PREDICTION STUDIES

Dean (1973)	$(2R)\frac{H}{wT}$	<	1 ONSHORE
		>	1 OFFSHORE
Short (1979) Height model	H	<	120 cm ONSHORE
		>	120 cm OFFSHORE
Short (1979)	$\left[\frac{\rho g^2}{16\pi}\right] H^2 T$	<	30 Kw/m ONSHORE
Power model		>	30 Kw/m OFFSHORE
Hattori and Kawamata	$2\,\frac{H\beta}{wT}$	<	0.5 ONSHORE
(1980)		>	0.5 OFFSHORE
Quick and Har	$\left[\frac{H}{wT}\right]_{INITIAL}$	>	$\left[\frac{H}{wT}\right]_{FINAL}$ ONSHORE
(1985)		<	$\left[\frac{H}{wT}\right]_{FINAL}$ OFFSHORE
Sunamura	$\frac{1.845}{g^{0.33}} H \frac{\beta^{0.27}}{(Td)^{0.67}}$	<	4 ONSHORE
and Horikawa (1974)		>	8 OFFSHORE

where:

R = arbitrary constant
H = deep water significant wave height
w = sediment fall speed
T = period of spectral peak
g = gravitational constant
β = beach slope
ρ = fluid density
d = diameter of sand grains

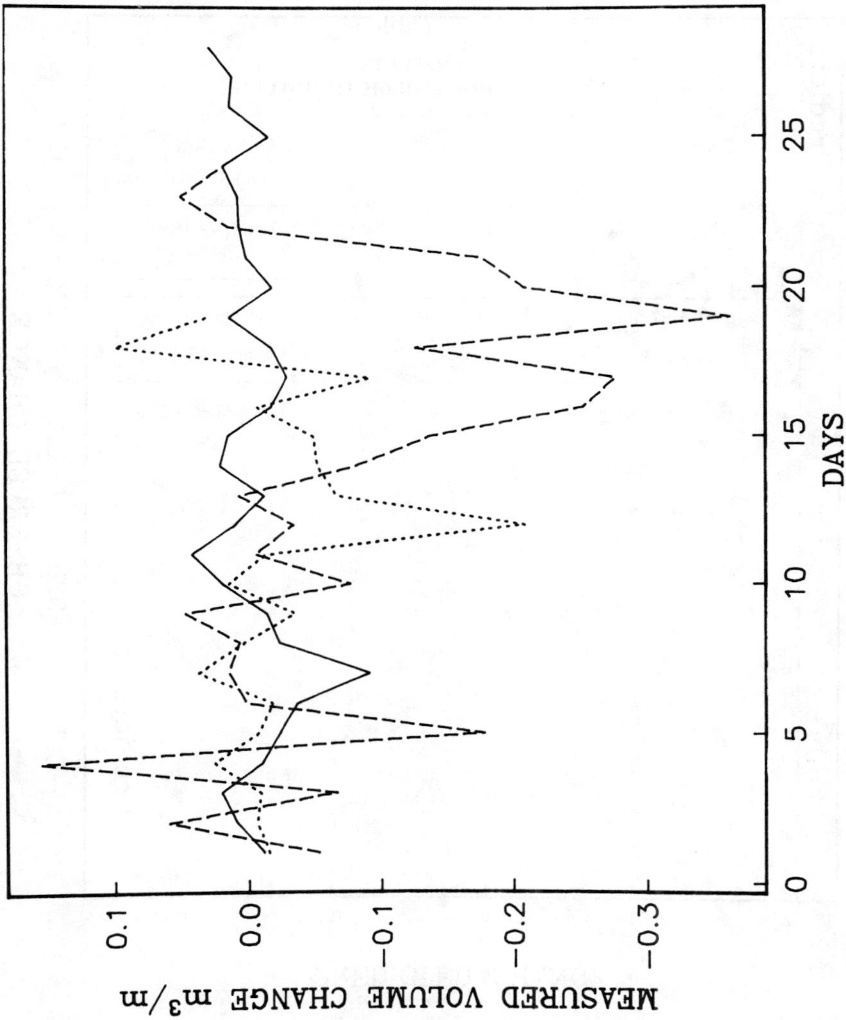

FIGURE 1

Time histories of the observed daily volume changes in the three data sets. (————) Scripps Beach, (------) Santa Barbara, (......) Virginia Beach.

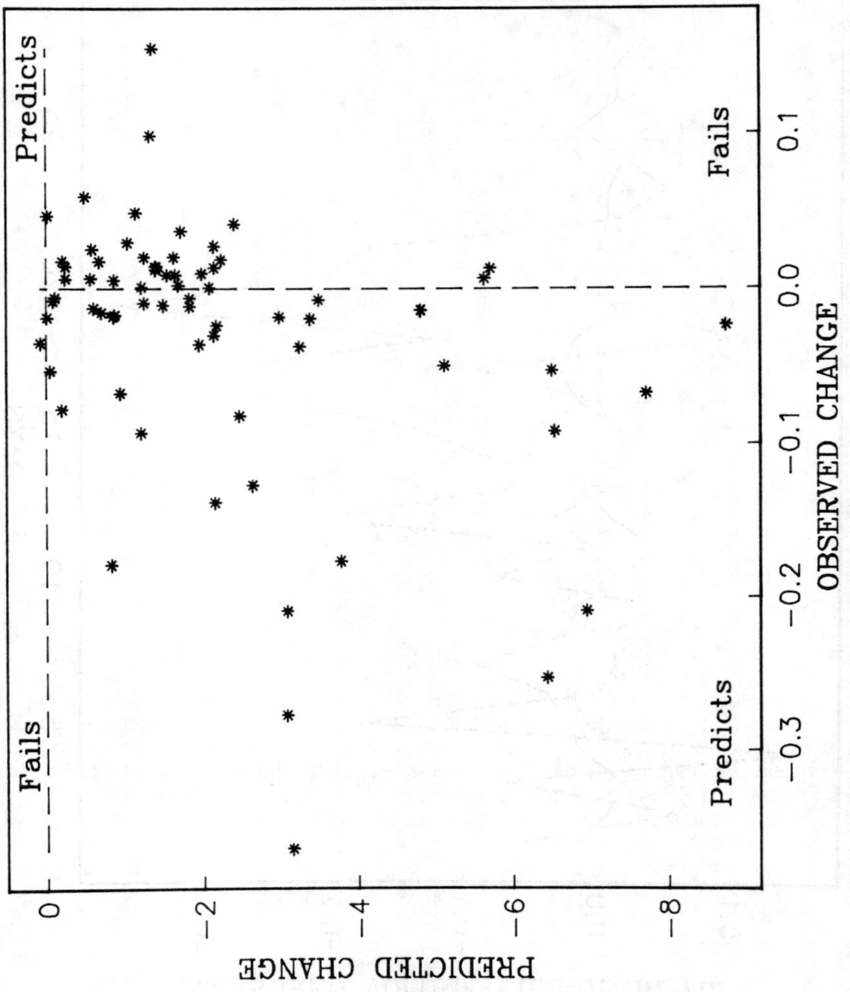

FIGURE 2

Comparisons of predictions of erosion or accretion using the model of Dean (1973) with observations. All three data sets are included.

TABLE II

RESULTS OF THRESHOLD PREDICTION STUDIES

Model	formulated threshold	adjusted threshold	skill factor
Dean	1.0	2.62	0.62
Short (height)	120	66	0.65
Short (power)	300	43	0.63
Hattori & Kawamata	0.5	0.064	0.68
Quick & Har	1.0	1.01	0.49
Sunamura & Horikawa	9-18	13.2	0.60

Inspection of Table II shows that the predictive capability of the best of these models is quite limited, with wrong estimates expected about one-third of the time. The worst of the models succeeds less often than chance alone. Therefore, they should be used with caution. Although the differences in predictive capability between models are barely significant, it is of interest to note two things. First, Short's model considering only wave height does as well as the more complex model of Dean that includes period and sediment size (indirectly, through fall speed). Second, the alteration to the Dean Number by Hattori and Kawamata in adding the beach slope causes some increase in predictive capability, adding credence to the generally-accepted idea that this parameter is a first order factor in cross-shore transport.

EVALUATING SLOPE PREDICTION MODELS

Two models that attempt to describe the general slope or shape of the beach were evaluated against this data base. The first, by Dalrymple and Thompson (1976), gives a graphical relationship between the beach slope at mean sea level and the Dean Number. The validity of this relationship was tested by calculating slopes for each profile and plotting them against the corresponding Dean Number. The results are shown in Figure 3. Inspection of this plot shows that the proportionality found under monochromatic laboratory waves is not seen in the field. There appears to be no discernible effect of Dean Number on beach slope for these data.

The second relationship tested was the exponential profile contained in Dean (1977) and described above. Best fit values for the coefficient A and the exponent m were plotted for each profile. A smoothed histogram of the distributions of these parameters are shown in Figure 4. Dean gave a typical value of 0.67 for the exponent, m, based upon Atlantic and Gulf Coast data. Figure 4 shows good agreement for these data as well. The value of the coefficient, A, is site-specific, varying from 0.065 to 0.13 and increasing with increasing sediment size, as predicted by Dean (1977). All three sites in this data set are well characterized by the $X^{0.67}$ model.

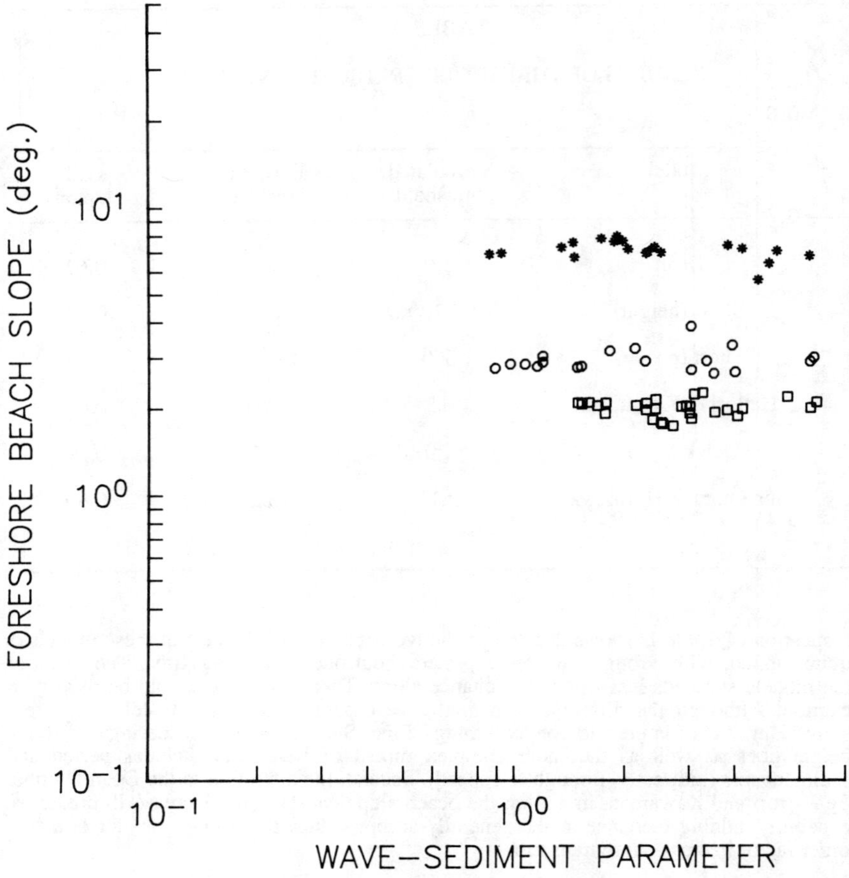

FIGURE 3

Beach slope compared to Dean Number (Wave Sediment Parameter) as described in Dalrymple and Thompson (1976). ❑ Scripps Beach, ❍ Santa Barbara, and ✳ Virginia Beach.

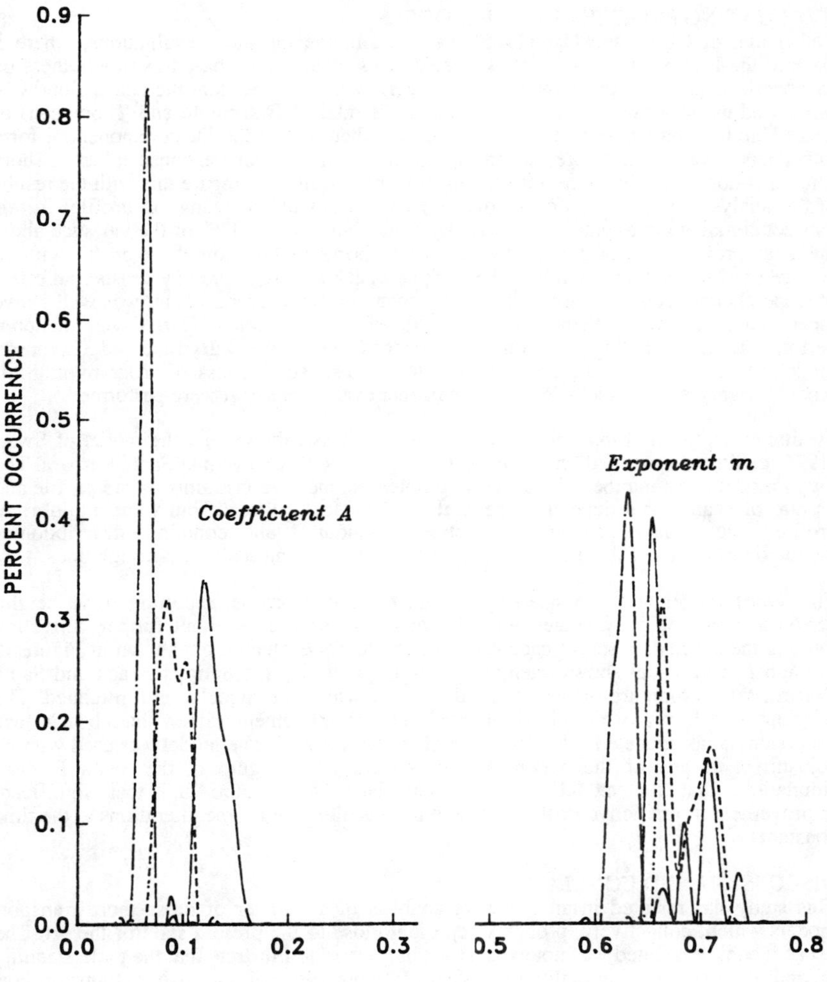

FIGURE 4

Smoothed histograms of the occurrences of the coefficients and exponents in the power law for beach contours from Dean (1977) in the form $Y = AX^m$. [Data from Santa Barbara has the shortest dash length, Scripps Beach has intermediate dash length and Virginia Beach has the largest dashes.]

EVALUATING PROFILE CHANGE MODELS

The model of Quick and Har (1985) was used in the threshold evaluations, where it showed the least skill of any of those tested. This formulation, based as most others on monochromatic laboratory moveable bed experiments, assumes that the beach slope will rapidly adjust to a new slope (based upon the model of Dalrymple and Thompson) as Dean Number changes, and that this slope will then follow the Dean exponential form into deeper water. Therefore, according to this model, beach response to cross-shore forcing is dominated by slope adjustment. It is interesting to compare this with the results of the analysis by empirical orthogonal eigenfunctions of the change in profiles in this data set contained in Seymour (1988a). This study shows that 97% of the variance about the mean profile at Santa Barbara was caused by horizontal motion of the profile without change of slope or shape. Only in the Scripps Beach set was there any substantial effect of slope change (coupled with a change in concavity) but the hinge point was well above mean sea level, rather than below as predicted by Quick and Har. Given the poor performance in predicting direction and its dependence on the Dalrymple and Thompson model, the decision was made not to undertake the substantial task of programming this model for varying tidal elevations and no further evaluations of it were performed.

Swain and Houston (1983, 1984) and Swain (1984) had shown that the model of Swart (1976) exhibited some skill in modeling the major erosionary event at Santa Barbara. The Swart model has a number of relatively complex geometric constraints on the profile and its rate of change that depend, in general, on how far it differs from some equilibrium profile (which is never achieved, in practice.) Seymour (1988b) contains a description of the major attributes of this model, which allows for changing sea level through tides.

The Swart model was evaluated by setting the initial profile, inputting wave height, period and tide height changes, and allowing it to proceed without further correction through the entire data set for each site. The results for each case are shown in Figures 5 through 7. The model shows interesting skill in predicting the Scripps Beach and Santa Barbara sets, which are of the unbarred type for which the model was formulated. The Virginia Beach set, which is dominated by bar movement rather than beach face excursion, is not modeled satisfactorily by the Swart model. The model was used without adjustment of any of the parameters. It is clear from Figure 6, the Santa Barbara simulation, that the model moves much faster than nature and that significant improvement in predictive skill could have been achieved by some alterations to the time constants.

DISCUSSION AND CONCLUSIONS

This study has resulted in an objective analysis of a number of cross-shore transport models which, collectively, probably represent close to the present state of the art. The list of models evaluated was not exhaustive, but served to illustrate that the understanding of, and the corresponding ability to make useful predictions of, cross-shore transport lags well behind that of longshore transport.

The poor performance, or outright failure, of models based upon laboratory moveable bed experiments with monochromatic waves in predicting cross-shore transport in the field has been emphatically demonstrated by these findings. Taken in its entirety, this evidence suggests that there are very substantial differences in the response of sandy beaches to monochromatic and natural waves and that empirical relationships established in the laboratory with monochromatic waves are unlikely to produce useful predictive tools for real beach profile changes.

Further, the significant differences between the Virginia Beach response (which has been verified through the findings at Duck, NC - see, for example, Mason et al., 1984) and the West Coast beaches illustrates that a simple characterization of the sediment on the beach by a single grain diameter will not be enough to allow effective predictions. It is clear that the dominant sediment transport mechanisms are quite dissimilar for the two beach types. A realistic model for cross-shore transport, involving the necessary physics, appears to be well beyond the present state of the art.

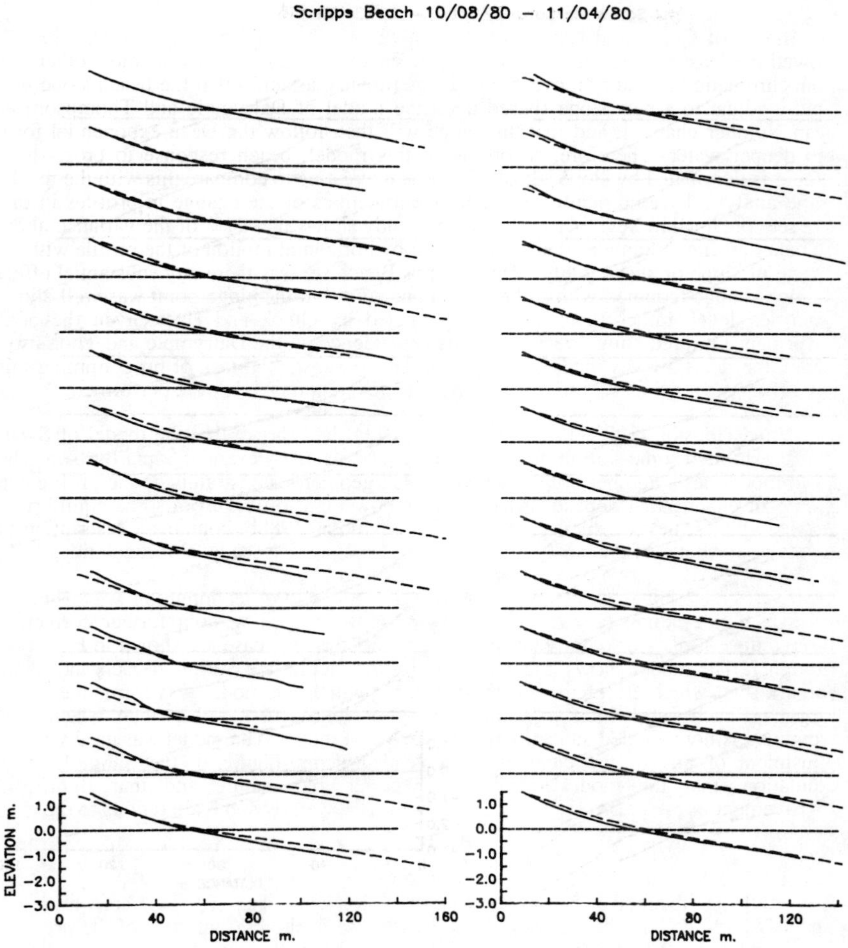

FIGURE 5

Comparison of observed beach profiles at Scripps Beach with the predictions using the model of Swart (1976). (------) indicates predictions. Sequence starts at upper left and proceeds downward in each column. Horizontal lines are mean sea level.

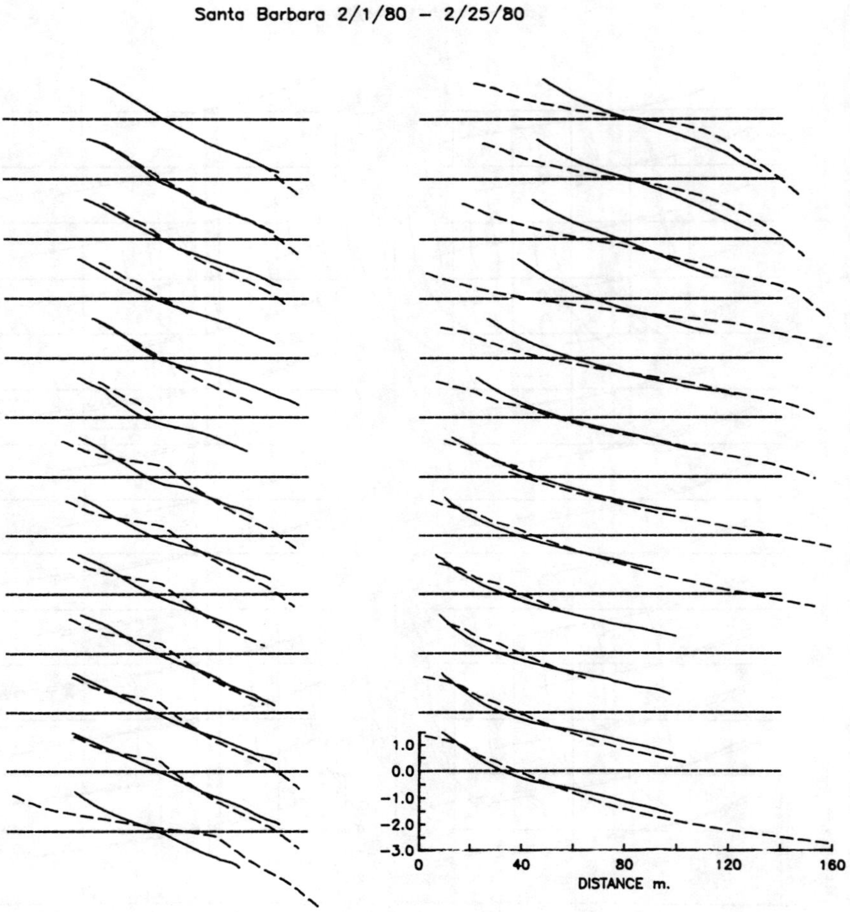

FIGURE 6

Comparisons of observed beach profiles at Santa Barbara with predictions from the model of Swart (1976).

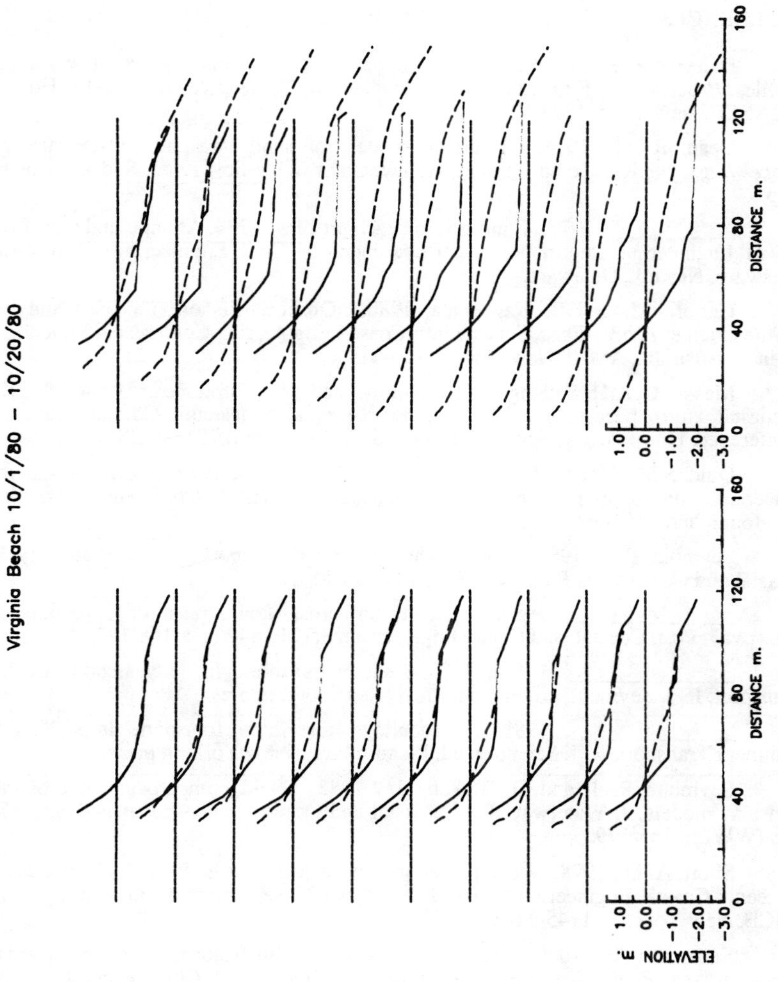

FIGURE 7

Comparisons of observed beach profiles at Virginia Beach with predictions from the model of Swart (1976).

REFERENCES

Dalrymple, R. A. and W. W. Thompson, 1976. Study of equilibrium beach profiles, *Proceedings*, Fifteenth Coastal Engineering Conference, July 11-18, Honolulu, HI, ASCE, New York: 1277-1296.

Dean, R. G., 1973. Heuristic models of sand transport in the surf zone, *Proceedings*, Conference on Engineering Dynamics in the Surf Zone, Sydney, Australia, 7 pp.

__________, 1977. Equilibrium beach profiles: U. S. Atlantic and Gulf Coasts, Ocean Engineering Report No. 12, Department of Civil Engineering, University of Delaware, Newark, Delaware.

Hattori, M. and R. Kawamata, 1980. Onshore-offshore transport and beach profile change, *Proceedings*, Seventeenth Coastal Engineering Conference, March 23-28, Sydney, Australia, ASCE, New York: 1175-1194.

Mason, C.,A.H. Sallenger, R.A. Holman and W.A. Birkemeier, 1984. DUCK82 - a coastal storm processes experiment. Proceedings, Nineteenth Coastal Engineering Conference, Houston, TX, September 3-7, ASCE, New York: 1913-1928.

Quick, M. C. and B. C. Har, 1985. Criteria for onshore-offshore sediment movement on beaches, *Proceedings*, Canadian Coastal Conference, St. Jobus, Newfoundland: 257-269.

Seymour, R. J., 1983. The nearshore sediment transport study. J. Waterway, Port, Coastal and Ocean Div., Proc. ASCE, 109 (1): 79-85.

____________, 1986. Results of cross-shore transport experiments, J. Waterway, Port, Coastal, and Ocean Engineering, ASCE, 112 (1): 168-173.

____________, 1988a. Cross-shore transport. In: Nearshore Sediment Transport, R. J. Seymour, ed. Plenum Press, New York, in press.

____________, 1988b. Modeling cross-shore transport. In: Nearshore Sediment Transport, R. J. Seymour, ed. Plenum Press, New York, in press.

Seymour, R. J. and D. B. King Jr., 1982. Field comparisons of cross-shore transport models, J. Waterway, Port, Coastal and Ocean Division, *Proceedings*, ASCE, 108 (WW2): 163-179.

Short, A. D., 1978. Wave power and beach stages: a global model, *Proceedings*, Sixteenth Coastal Engineering Conference, August 27-September 3, Hamburg, Germany, ASCE, New York, 2: 1145-1163.

Sunamura, T. and K. Horikawa, 1974. Two-dimensional beach transformation due to waves, *Proceedings*, Fourteenth Coastal Engineering Conference, June 24-28, Copenhagen, Denmark, ASCE, N.Y.: 920-938.

Swain, A., 1984. Additional results of a numerical model for beach profile development. *Proceedings* of the Annual Conference, CSCE, Halifax, Nova Scotia, Canada, May.

Swain, A. and R. J. Houston, 1983. A numerical model for beach profile development. Sixth Canadian Hydrotechnical Conference, CSCE, Ottawa, Canada, June, 2: 77.

______________, 1984. Discussion to the Proceeding Paper 17749 by Richard J. Seymour, The Nearshore Sediment Transport Study, J. Waterway, Port, Coastal and Ocean Engineering, ASCE, February, 110 (1): 130-133.

Swart, D. H., 1976. Predictive equations regarding coastal transports, *Proceedings*, Fifteenth Coastal Engineering Conference, July 11-17, Honolulu, HI, ASCE, N.Y.: 1113-1132.

CHAPTER 125

A QUASI-3D MATHEMATICAL MODEL OF COASTAL MORPHOLOGY

H.J. de Vriend*) and J.S. Ribberink*)

ABSTRACT

A semi-analytical model of 3D nearshore currents and sediment transport is presented. It describes the tidal motion, the waves, the surfzone currents and the sediment transport in complex coastal areas.
The results of a first application, to a well-documented case at the Dutch coast, indicate the potential and the shortcomings of the model. The latter are analysed and suggestions for improvement are given.

INTRODUCTION

Mathematical models are increasingly important as a tool to predict the water and sediment motion in coastal areas. Still, most of these models consider only part of the sediment transport. Roughly speaking, they describe either the "longshore" or the "cross-shore" transport. More carefully formulated, they consider either the transport along with the time- and depth-averaged current, or the transport due to waves and vertical circulations (undertow) in a more or less cross-shore profile. In complex situations, this distinction makes little sense and neither transport component can be disregarded. In these situations, the 3D time-dependent flow field and the attending sediment motion have to be described (cf. De Vriend, 1986).
In view of the wide range of time scales involved, from some seconds for the wind waves through to half a day for the tide, fully-3D time-dependent simulations are hardly feasible for practical applications. Therefore, simplified models have to be used, taking due account of the relevant phenomena.
De Vriend and Stive (1987) propose a semi-analytical model of 3D nearshore currents, based on hydrostaticity and similarity hypotheses for the velocity components and the eddy viscosity. This so-called DVS-model, in a slightly modified form, is combined here with a sediment transport model based on the Bagnold-Bailard concept (Bailard, 1981), to yield a quasi-3D coastal transport model.

After a brief description of this model, a first practical application, with both "longshore" and "cross-shore" transport mechanisms at work, is described and discussed. It concerns the rapid formation of large sand bank systems on the former ebb-tidal deltas of the estuaries closed off by the Deltaworks (Kohsiek, 1988). The results are compared with field data, and the model formulation is reconsidered.

GENERAL OUTLINE OF THE MODEL

The model consists of the following basic elements (see Figure 1).

- A regional tidal model, giving the tidal stages and the boundary conditions for the flow model in the area of interest. The velocity field

*) Senior researcher, DELFT HYDRAULICS, P.O. Box 152, 8300 AD EMMELOORD, The Netherlands

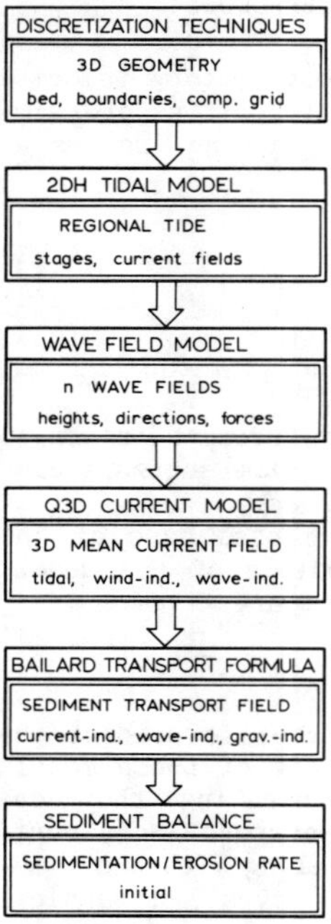

from this model can be used in the wave model, in order to have a rough indication of the effects of current refraction.

- A wave field model, describing the propagation and dissipation of wind waves in the area of interest.
- A quasi-3D current model, describing the wave orbital motion and the mean flow due to tide and wave action.
- A Bailard-type sediment transport formula, describing the near-bed transport due to waves (stirring, asymmetry transport) and currents ("convective" transport).
- The sediment balance equation, yielding the sedimentation/erosion rate under the given conditions.

These elements are put in line, as indicated in Figure 1, without any feedback mechanisms. This means that detailed current refraction, for instance, is disregarded, and that the dynamic interaction between the water and sediment motion and the bed topography changes are left out of consideration.

Figure 1 Aggregate flow chart of the quasi-3D morphological model

CURRENT MODEL

Primary and secondary flow

Key elements in the quasi-3D current model (De Vriend and Stive, 1987) are similarity approximations for the eddy viscosity, ν_t, and the wave-averaged velocity, $\vec{U}$:

$$\nu_t(x,y,z;t) = \bar{\nu}_t(x,y;t)\Phi(\zeta) \quad (1)$$

$$\vec{U}(x,y,z;t) = \bar{\vec{U}}(x,z;t)f_p(\zeta) + \sum_k \vec{U}_k(x,z;t)f_k(\zeta) \quad (2)$$

in which: x,y = horizontal co-ordinates,
z = vertical co-ordinate,
t = time,
$\zeta = \frac{z-z_b}{h}$ = boundary-fitted vertical coordinate,
z_b = bottom level,
h = water depth, and
$\overline{..}$ = depth-average.

The first product in the RHS of (2) is called the "primary flow velocity". The vertical distribution function $f_p(\zeta)$ is chosen the same as in uniform shear flow under the same conditions (turbulence, waves, bed roughness, etc.). Hence the primary flow accounts for the (as yet unknown) depth-averaged velocity as if it concerned uniform shear flow.
The remaining part of the wave-mean velocity is called "secondary flow velocity", although this name is disputable. It is supposed to consist of a number of constituents, for each of which vertical similarity is assumed. In the present version of the model the secondary flow concerns

wave-induced undertow and boundary layer streaming, but there is no reason why it should be restricted to these phenomena.
In addition to the wave-mean velocity, there is an oscillatory velocity associated with the waves. For the time being, these velocities are simply added together, although in reality they are interacting (Davies et al., 1988; Klopman and De Vriend, 1988).

Depth-averaged current model

The depth-averaged mean velocity, $\vec{\bar{U}}$, is computed with a horizontally two-dimensional shallow water model, including the effects of wind, tide and short waves. The latter exert their influence through the forcing, the bottom shear stress relationship and the mass balance (mass flux compensation). The wave influence on the horizontal eddy viscosity (Wind and Vreugdenhil, 1986; Yoo and O'Connor, 1988) is left out of consideration, as the model is meant for rather large-scale problems (model extension >> surf zone width) with irregular waves.
The system of differential equations solved by the model can be written as

$$\frac{D\vec{\bar{U}}}{Dt} = - g\, \vec{\nabla}(z_s) - \frac{\vec{\tau}_b}{\rho h} + \frac{\vec{F}_w}{\rho h} + \nu_h\, \vec{\nabla}.\vec{\nabla}(\vec{\bar{U}}) \qquad (3)$$

$$\frac{\partial z_s}{\partial t} + \vec{\nabla}.(h\vec{\bar{U}}) + \vec{\nabla}.(\frac{\vec{M}}{\rho}) = 0 \qquad (4)$$

in which: g = acceleration due to gravity,
z_s = level of the mean water surface,
ρ = mass density of the fluid,
$\vec{\tau}_b$ = bottom shear stess,
$\vec{F}_w$ = external driving force (wind, waves),
ν_h = horizontal eddy viscosity, and
$\vec{M}$ = wave-induced mass flux.

Note that this system differs from the one in the DVS-model, in that the mass flux compensation is treated as a part of the primary flow, rather than as a secondary flow constituent. Thus the mass flux compensation can take place through a horizontal circulation or a circulation in the vertical plane, depending on the geometrical situation. Besides, it is continuous in the horizontal plane, now, so that the attending bed shear stresses and sediment transports are more smoothly distributed in space.

The wave-induced forces and mass fluxes can be derived explicitly from the wave model, as they depend on wave field properties only. The following formulations are used in the DVS-model

$$\vec{F}_w = \frac{D}{\omega} \vec{k}_w \qquad (5)$$

$$\vec{M} = (1 + 7\, \tilde{Q}_b \frac{h}{\lambda}) \frac{E}{\omega} \vec{k}_w \qquad (6)$$

in which: D = energy dissipation rate per unit area,
E = energy density of the wave field,
ω = angular frequency of the waves,
$\vec{k}_w$ = wave number, as a vector in the direction of the wave energy flux,
$\tilde{Q}_b$ = fraction of the waves that is breaking ($0 \le \tilde{Q}_b \le 1$), and
λ = wave length.

The bottom shear stress consists of two parts, due to the primary and the secondary flow, respectively. In the DVS-model, the secondary shear stress is supposed to be negligible in the depth-averaged flow computation. The primary shear stress depends on the current velocity and on the waves, so it cannot be computed on the basis of wave data only. In order to close the mathematical system (3) through (6), this shear stress has to be expressed in terms of the depth-averaged velocity.
In the present version of the depth-averaged current model, Bijker's (1966) shear stress enhancement formula is used, in the approximative form

$$\frac{\vec{\tau}_b}{\rho} = C_f \, \vec{\bar{U}}|\vec{\bar{U}}| \; \left[0.75 + 0.45\left(\xi \, \frac{\hat{U}_{orb}}{|\vec{\bar{U}}|}\right)^{1.13}\right] \geq C_f \, \vec{\bar{U}}|\vec{U}| \tag{7}$$

in which: C_f = bottom friction factor for current only,
ξ = $(f_w)^{\frac{1}{2}}(2C_f)^{-\frac{1}{2}}$,
f_w = Jonsson's (1966) friction factor for waves, and
$\hat{U}_{orb}$ = amplitude of the near-bottom orbital velocity.

Following Visser's (1986) suggestion, the factor ξ is taken identically equal to 1. The orbital velocity in this expression is calculated as if it concerned regular and linear waves, with the mean amplitude and the peak period of the actual random wave field.
The shear stress relationship is closely connected with the turbulence model (vertical eddy viscosity, bottom boundary condition), and hence with the description of the primary velocity profile. In this respect, relationship (7) is not consistent with the original DVS-model (cf. Ribberink and De Vriend, 1988). Since the bed shear stress is the only aspect of the primary flow that is used in the sediment transport model, this inconsistency is not expected to have serious consequences for the model.

Primary flow profile

Consistently with the similarity assumption for the primary flow, the vertical profile function $f_p(\zeta)$ is solved from the horizontal momentum equation for uniform shear flow, rewritten to

$$\frac{\partial}{\partial \zeta}\left(\Phi \, \frac{\partial f_p}{\partial \zeta}\right) = C_\tau \qquad \text{with} \qquad C_\tau \hat{=} \frac{|\vec{\tau}_{bp}|h}{\rho \bar{\nu}_t |\vec{\bar{U}}|} \tag{8}$$

in which $\vec{\tau}_{bp}$ denotes the primary bottom shear stress.
With the parabolic/constant eddy viscosity distribution of the DVS-model and a prescribed level of zero-intersection of the (logarithmic) profile near the bottom, this leads to

$$f_p(\zeta) = \begin{cases} -\dfrac{3}{1+3\ln(2F'\zeta_o)}\{\ln\zeta - \ln(F'\zeta_o)\} & \text{for } F\zeta_o \leq \zeta \leq \frac{1}{2} \\ \dfrac{3}{1+3\ln(2F'\zeta_o)}\{-2\zeta^2 + 4\zeta - \frac{3}{2} - \ln(2F'\zeta_o)\} & \text{for } \frac{1}{2} \leq \zeta \leq 1 \end{cases} \tag{9}$$

in which: ζ_o = the level of zero-intersection for currents alone,
F' = amplification factor for the wave influence on the level of zero-intersection, and
$F\zeta_o$ = lowest level at which the logarithmic velocity profile holds good.

In the DVS-model, the amplification factor F' and F are formulated as proposed by Nielsen (1985). It is also possible, however, to choose a formulation that corresponds with (7):

$$F' = \exp\left[\kappa\, C_f^{-\frac{1}{2}}\, (1 - X_p^{-\frac{1}{2}})\right] \tag{10}$$

in which κ denotes Von Karman's constant and X_p is defined by

$$X_p = \min\left\{1,\ 0.75 + 0.45\, \left(\xi\, \frac{\hat{U}_{orb}}{|\vec{U}|}\right)\right\} \tag{11}$$

Once F' is known, F follows from the same relationship as in Nielsen's model, viz.

$$F' = F \exp\left[\frac{1}{F} - 1\right] \text{ and hence } F \approx eF' - 1 \tag{12}$$

in which e denotes the neperian.

Secondary flow

So far, the model has been elaborated for secondary flow due to wave breaking and for the residual streaming in the oscillatory boundary layer at the bottom.

Exploratory computations (De Vriend and Stive, 1987; Stive and De Vriend, 1987) indicate that the latter component is of minor importance to the short-term residual sediment transport. Therefore, only the breaking-induced secondary flow will be considered. The DVS-model uses the effective surface shear stress

$$\vec{\tau}_t = \left(\frac{1}{2} + 7\, \frac{h}{\lambda}\right) \frac{D}{\omega}\, \vec{k}_w \tag{13}$$

to drive this part of the current, assuming $\vec{\tau}_t$ to act at the wave trough level, $\zeta = \zeta_t$. The corresponding secondary velocity can be written as

$$\vec{U}_1 = \vec{U}_{1,1}\, f_{s1}(\zeta) + \vec{U}_{1,2}\, f_{s2}(\zeta) \tag{14}$$

$$\text{with: } \vec{U}_{1,1} = \frac{\vec{\tau}_t\, h}{\rho\, \bar{\nu}_t \zeta_t} \quad \text{and} \quad \vec{U}_{1,2} = \frac{\vec{\tau}_{bs} h}{\rho\, \bar{\nu}_t \zeta_t} \tag{15}$$

Here $\vec{\tau}_{bs}$ denotes the secondary part of the bottom shear stress, directed along the vertical plane through the wave rays and as yet unknown in magnitude. The vertical distribution functions are those given by De Vriend and Stive (1987), which means that $f_{s1}(\zeta)$ represents the velocity profile due to a surface shear stress and $f_{s2}(\zeta)$ closely resembles the normal shear flow profile.

The magnitude of $\vec{\tau}_{bs}$ follows from the requirement that the depth-averaged secondary flow velocity must be zero. Hence

$$|\vec{\tau}_{bs}| = -\frac{\overline{f_{s1}}\, |\vec{\tau}_t|}{\overline{f_{s2}}} = -\left(\frac{1}{2} + 7\, \frac{h}{\lambda}\right) \frac{\overline{f_{s1}}}{\overline{f_{s2}}} \frac{D}{\omega}\, |\vec{k}_w| \tag{16}$$

in which: $\overline{f_{s1}} = \frac{25}{144}$ (17)

$$\overline{f_{s2}} = -\frac{25}{144} - \frac{5}{48}\zeta_t\{-1 + 2\ln(2F'\zeta_o)\} \tag{18}$$

Note that the secondary flow is not the same as the flow due to the surface stress $\vec{\tau}_t$. The latter obviously has an f_{s1}-type profile. As $\overline{f_{s1}}$ is non-zero, however, the present framework of definitions, which forms the basis of the quasi-3D model, makes it necessary to split this flow into a primary and a secondary part, as shown in Figure 2. The secondary part is given by (14) sqq., the primary part is included in the primary flow model, with a minor inconsistency because $f_p(\zeta)$ and $f_{s2}(\zeta)$ are not exactly identical.

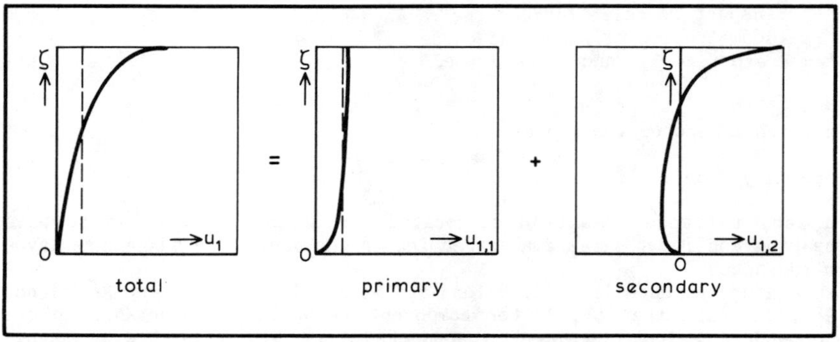

Figure 2 Splitting of the flow due to a surface shear stress

This primary flow contribution is driven by part of the wave-induced force $\vec{F}_w$ in (3). This means that there is no guarantee for this contribution to be actually present. Depending on the geometrical situation, the relevant force field can just as well generate a water level set-up, or a current in a different direction, or any combination of these. Hence the interpretation of 3D currents due to breaking waves as a superposition of a longshore current with the primary profile and a cross-shore current (undertow) with the secondary profile (Svendsen and Lorenz, 1988) is only valid for very specific situations. In more general cases, a wide range of vectorial combinations of primary and secondary flow components can be found, and the notions "longshore" and "cross-shore" make no sense for such currents.

SEDIMENT TRANSPORT MODEL

Basic concept

A sediment transport model for coastal areas should include what uses to be called "longshore" and "cross-shore" transport components. In view of the discussion in the previous section, this should rather be "convective" and "wave-asymmetry" transport components.
Most of the usual coastal transport models pertain to the convective transport, with the wave-mean current as a transporting agent and wave action as one of the stirring mechanisms.
Such models are expected to fail in the complex situations considered herein.
On the other hand, the mechanisms underlying what is summarized by the term "wave-asymmetry" transport, such as vortex shedding at rippled beds

(Nielsen, 1988) and sheet-flow over a plane bed (Bakker and Van Kesteren, 1988), are far from being fully understood, let alone that well-established models would be available.

A pragmatic way-out is to adopt a sediment transport formula that is claimed to describe the instantaneous transport rate, and formally integrate the result over the waves. This approach was followed by e.g. Madsen and Grant (1976) and Bailard (1981). The latter adopted Bagnold's (1966) energetics approach and worked it out for colinear waves and currents. Although this model concept is still subject to doubt and will not apply to every possible set of conditions (cf. De Waal, 1987; Nielsen, 1988), it has yielded quite acceptable results in large-scale laboratory tests (Roelvink and Stive, 1988) and in practical applications (Stive, 1986). Therefore, it is incorporated in the present morphological model system, at least for the time being.

The transport formulae forming the Bailard-model, generalized to any vectorial combination of waves and currents, can be written as

$$\vec{q}_{bed} = A_{bc} < |\vec{U}_t|^2 \vec{U}_t > - A_{bs} < |\vec{U}_t|^3 > \vec{\nabla}(z_b) \quad (19)$$

$$\vec{q}_{sus} = A_{sc} < |\vec{U}_t|^3 \vec{U}_t > - A_{ss} < |\vec{U}_t|^5 > \vec{\nabla}(z_b) \quad (20)$$

in which: $\vec{q}_{bed}$ = bed load transport rate,
$\vec{q}_{sus}$ = suspended-load transport rate,
$\vec{U}_t$ = equivalent instantaneous near-bed velocity,
$A_{..}$ = factors depending on the sediment properties, etc., and
$<..>$ = wave-average of the argument.

Note that this model includes both bed load and suspended-load transport. Besides, the down-slope gravitational transport component is taken into account in the bed load part. The bottom slope term in the suspended-load part represents the effect of convection by the slope-induced vertical velocity component.

Instantaneous near-bed velocity

The equivalent near-bed velocity, $\vec{U}_t$, in (19) and (20) remains to be specified. Following Bailard (1981), it consists of a mean and an oscillatory component. So, in a generalized form, this reads

$$\vec{U}_t = \vec{U}_o + \vec{U}_w \quad (21)$$

The definition of the mean flow part, $\vec{U}_o$, is not obvious, in view of the steep velocity gradients near the bottom. Bagnold's (1966) model was derived for steady uniform flow with a depth-invariant velocity and the usual quadratic friction law. Taking the bottom shear stress as determinative for the transport, the equivalent velocity can be expressed by

$$U_o = \left(\frac{\tau_b}{\rho\, C_f}\right)^{\frac{1}{2}} \quad (22)$$

or, in a generalized form for more complex steady flows,

$$\vec{U}_o = \frac{\vec{\tau}_b}{\rho\, C_f^{\frac{1}{2}}\, |\vec{U}_*|} \qquad \text{with} \qquad |\vec{U}_*| = \left(\frac{|\vec{\tau}_b|}{\rho}\right)^{\frac{1}{2}} \quad (23)$$

The oscillatory component of the equivalent near-bed velocity, $\vec{U}_w$, is filled in with the near-bed orbital velocity of waves in an inviscid fluid, as if there were no current and no bottom boundary layer. This is consistent with Bailard's (1981) approach.
Since wave-asymmetry is an important agent in wave-induced sediment transport, linear wave theory is not sufficient here. Instead, the orbital velocity is described with Rienecker and Fenton's (1981) Fourier-approximation technique for non-linear waves. Although the underlying theory is formally restricted to steady progressive waves on a horizontal bottom, the technique is applied throughout the wave field. Rienecker and Fenton showed this to be allowable for non-breaking waves, but the applicability to breaking waves is disputable. By lack of a good ready-to-use alternative, however, this point is ignored for the time being.

Elaboration of the time-mean transport

The formal algebraic elaboration of the generalized Bailard-formula, i.e. substitution of (21) into (19) and (20) and averaging over the wave period, becomes fairly complicated (cf. Guza and Thornton, 1985). Therefore, it was decided to evaluate the transport numerically in every point of the computational grid and at every time step of the computation. This turned out not to lead to prohibitive computer expenses.

VOORDELTA: A FIRST APPLICATION

Situation

The closure, as part of the Deltaworks, of several estuaries in the South-west of The Netherlands has started off a spectecular morphological evolution of the former ebb-tidal deltas (see, for instance, Kohsiek, 1988). After the in- and outgoing tide had been blocked, the dynamic equilibrium between the cross-shore actions of waves and currents was disturbed and the waves started pushing the seaward edges of the deltas onshore.
Within a decade, huge sand bank systems were formed in front of the barriers (Figure 3; also see: Van der Spek, 1987; Van den Berg, 1987), which raised questions about their future development and their impact on the coastal defence system, the local ecosystem, the possibilities of economical use, etcetera. Therefore, the Ministry of Public Works (Rijkswaterstaat) and DELFT HYDRAULICS are executing extensive studies in this area.

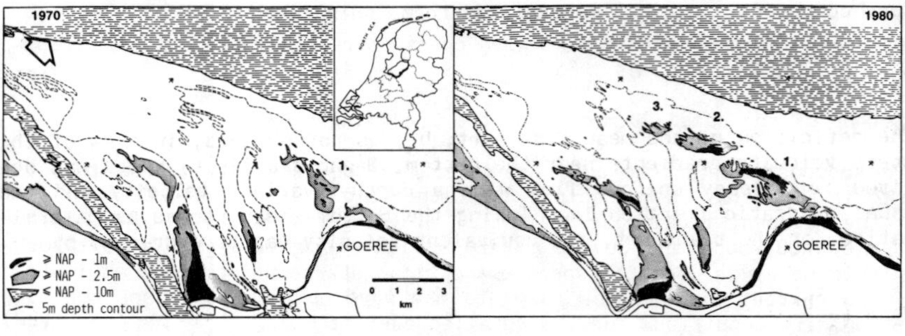

Figure 3 Deformation of the ebb-tidal delta in the mouth of the Grevelingen estuary, after closure of the Brouwersdam (from: Van der Spek, 1987)

As far as morphology is concerned, these studies are aiming at a better understanding and a hindcast of the observed phenomena, and at the prediction of future morphological developments.

As a part of these studies, a first practical application of the present model system was to hindcast and predict the morphological evolution of the ebb-tidal delta in front of the Brouwersdam, in the former mouth of the Grevelingen estuary (Figure 3).

Results of previous studies in the area

A preliminary hindcast study with a numerical model of coastal profile evolution (Stive, 1986; see Figure 4) indicated that the onshore transport due to wave asymmetry must be an important agent in the formation of the sand banks, with the offshore transport due to undertow and gravitational effects as a principal counteracting agent.
Further study of bathymetric and sedimentological data, and an extensive hindcast study with the coastal profile model (Steetzel and Stive, 1986) made clear, that the morphological processes in the later stages of development are essentially 3D, with an interaction between tidal and wave-induced currents. The following phenomena were expected to be important:

- "vertical" tide, with flooding and drying of sand banks and, correspondingly, time-varying wave penetration into the area behind the banks,
- tidal currents,
- wave-induced undertow,
- sediment stirring by waves, and
- wave-asymmetry transport.

A quasi-3D sediment transport model was set up in order to see how the combination of these phenomena works out, and as a tool in overall hindcast and forecast studies, for the Grevelingen delta as well as the adjacent Haringvliet and Eastern Scheldt deltas.

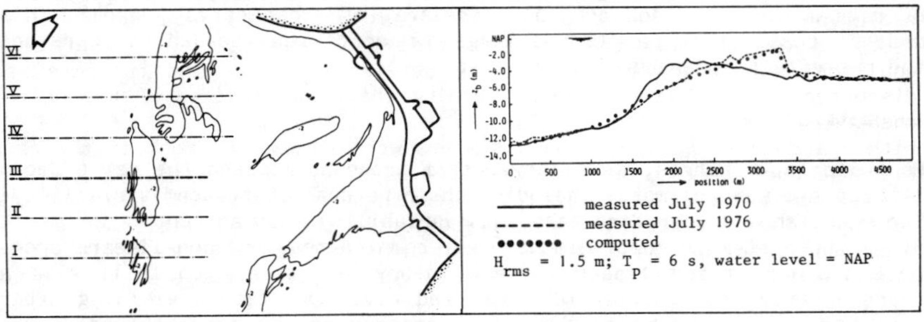

Figure 4 Results of hindcast with cross-shore profile model (from: Stive, 1986)

Set-up of the quasi-3D model for the Grevelingen delta

The composition of the quasi-3D model for the Grevelingen-area is outlined in Figure 5. The principal constituents are

- the regional tidal model RANDDELTA-II (Langerak et al., 1978), based on the WAQUA-code and developed and validated to support the Deltaworks; this model is used to generate the tidal stages and a rough estimate of the current field to be put into the wave computations, and the tidal boundary conditions for the detailed current model,
- a series of wave models, one for each incoming wave direction, based on the HISWA-code (Holthuysen and Booij, 1986; Dingemans, 1987), which takes randomness of wave height and direction into account,
- a depth-integrated detailed current model for combined tidal and wave-induced currents, based on the curvilinear version of the WAQUA-code (Willemse et al., 1985), extended with wave effects as described in the foregoing (also see: De Vriend and Van Banning, 1988),

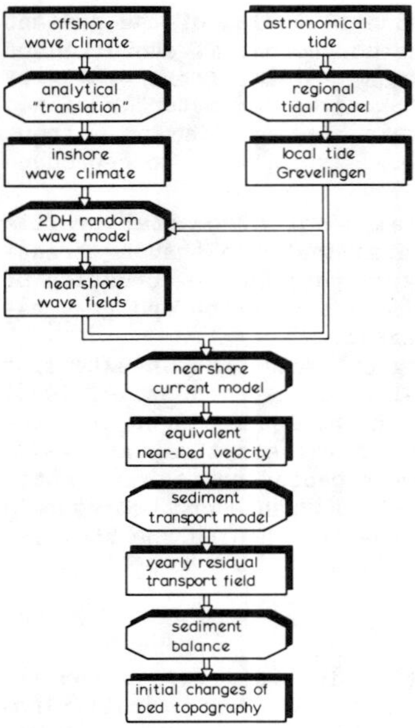

- the semi-analytical bottom shear stress model, including 3D flow effects, as described in the foregoing,
- the generalized Bailard transport formula, and
- a sediment balance module, yielding the initial rate of sedimentation and erosion for a given transport field.

Feedback mechanisms (wave-current interaction, topography-hydrodynamics interaction) are not included, for technical as well as economical reasons. This implies that the model results remain to be interpreted in terms of longer-term morphological evolutions.

The model constituents work on different types of grids and cover different areas, having the area of interest in common. This implies that a number of interfaces is needed to transform model results from one grid to another.

Figure 5 Flow chart Grevelingen model

The constituents, together with their interfaces and pre- and postprocessing facilities were brought together into an envelope system taking care of data management, job control, etcetera (cf. Boer et al., 1984). This made it possible to handle the large number of runs needed to represent the net effects of natural variability.

Schematization

Morphological changes, as a longer-term process, reflect the net effects of tide and wave climate, including their natural short-term variability. The model should therefore take this variability into account.
Since the system is multi-dimensional, complicated, and non-linear, spectral approaches and linear systems theory are likely to fail. Also a representative combination of tidal and wave conditions, yielding after e.g. one year the same morphological changes as the natural conditions, is hard to find, if it exists, at all. Therefore, a pragmatic schematization procedure was followed here (Van Banning and De Vriend, 1987; Steijn, 1988).
Key elements in this procedure are the selection of a morphologically equivalent tidal cycle and the schematization of the wave climate. The criterion for the tidal cycle selection is that the yearly transport rate, according to a given transport formula (in this case: Van Rijn's (1986) formula) should be the same for the repeated equivalent cycle and for the actual astronomical tide. In principle, this can yield a different equivalent cycle in every point of the model area. Therefore, a limited number of representative points was chosen, and the cycle with the best overall fit was selected.
The wave climate was schematized in three steps, viz.

- three representative wave directions (approximately NW, W and SW) were chosen on the basis of the available wave data,
- for each directional class, a representative wave height was chosen,
- each directional class, and the situation without waves, were attributed a certain weight, such that the weighed model results can be expected to represent the yearly transports and sedimentation/erosion rates.

Two criteria were used to determine these weights for the convective transport. The "longshore power" criterion (SPM, 1984) was used to account for the convective transport in the surf zone. The second criterion concerns the correct representation of the "stirring parameter", defined as the ratio between the primary bottom shear stress with and without waves. If this criterion is satisfied, the model is supposed to reproduce the convective transport outside the surfzone.

As the wave-asymmetry transport varies with a much higher power of the wave height than the convective transport, different weight factors were applied for this transport component. The criterion was a correct reproduction of the transport according to the Bailard formula without mean currents.

Like in the tidal schematization, full agreement can only be required in a small number of representative points.

Validation

A compound model like this is usually validated by checking the results of its principal constituents individually, and those of the model as a whole, against measured data, general observations, logical expectations, etcetera.

The available tidal data (water levels, current velocities) from the Grevelingen area were sufficient to check the detailed current model in situations without significant waves (Van der Spek and Steijn, 1988). Wave-driven current data, however, were not available in an identifiable form, so that earlier field and laboratory verifications (De Vriend and Van Banning, 1988) had to be relied upon.

Wave data from the area of interest are hardly available, but the wave model concept has been verified in the adjacent Haringvliet area (Holthuijsen et al., 1988).

Sediment transport data are lacking, too, but the available sedimentological data (Van der Spek, 1987; Van den Berg, 1987) give some insight into the net displacement of sediment. Besides, the successful hindcast with the coastal profile model (Stive, 1986) gives confidence in the Bailard-formula.

In order to check the model as a whole, the morphological evolution of the area between the closure of the dam (1971) and 1984 was hindcasted and compared with bathymetric and sedimentological data. The outcome of this comparison will be discussed in the next sections.

ANALYSIS OF HINDCAST RESULTS

Results of the 2DH model version

In order to have a reference and a back-up for the quasi-3D model, a parallel run was made with a two-dimensional horizontal (2DH) version of the model, in which no secondary bottom shear stress was incorporated in the transport model. The results, shown in Figure 6, look rather satisfactory. Although the sediment transport rate is somewhat too high all over the area, the general picture from Figure 3 is represented: the bars tend to move onshore and towards the tip of the northern island, and a second array of bars tends to form half-way the first one and the dam. Besides, though not visible in the plots, the aggradation of the area behind the outer bars is reproduced.

Results of the quasi-3D model version

The results of the full quasi-3D model, including the secondary shear stress in the transport model, are shown in Figure 7. Even though the quasi-3D model takes more of the physical phenomena (viz. undertow) into

account than the 2DH model, its results are far worse: the bars are even moving seawards now!
This unexpected result needs further analysis. If the present case allows for a 2DH approach, the additional effects introduced by the quasi-3D model must be unimportant and should therefore do no harm. The present results show otherwise, which suggests that the quasi-3D model still contains a major inconsistency.

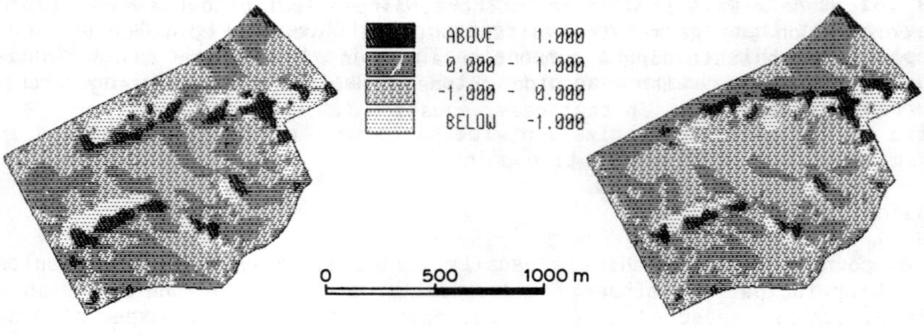

Figure 6 Sedimentation/erosion rate (2DH model version)

Figure 7 Sedimentation/erosion rate (quasi-3D model version)

Analysis and discussion

As was stated before, the hindcast studies with the cross-shore profile model indicated that the onshore wave-asymmetry transport should be the principal bank generating agent here, with the transport due to undertow and downslope gravitational effects as a counteracting agent. When introducing a net current across the bank, however, the undertow turned out to be easily dominated (Steetzel and Stive, 1986).
According to the results of the 2DH detailed current model, such across-bank currents are present, indeed.

- The (NE-going) flood current is trapped between the coastline and the outer bank array and gives rise to a strong offshore current across the banks. The (SW-going) ebb current, however, more or less follows the alignment of the bank array, which is partly dry by that time. Hence the residual tidal current on the banks is directed offshore.
- Especially for waves from NW and W, the banks act more or less as an underwater bar, attached to Goeree at the one end and ending in the open sea at the other. This situation, which is very similar to the wave tank experiment reported by Dingemans et al. (1986), gives rise to a wave-induced horizontal circulation that is directed onshore on the top of the banks and directed towards the open-sea end in the channel behind them.

The velocities due to either of these currents dominate the undertow velocity. This suggests the bank formation to be due not only to wave-asymmetry transport, but also to the wave-induced horizontal circulation, with the residual tidal current as the principal counteracting agent. Besides, if the net effect is neatly reproduced by a cross-shore profile model containing much weaker currents only, the effects of the two currents must be almost balancing. This means that a relatively weak disturbance of this balance, e.g. by introducing 3D-effects in an inconsistent manner, can have strong effects.

Since the only difference between the 2DH and quasi-3D versions of the model lies in the secondary bottom shear stress, the quasi-3D model must be inconsistent in its incorporation of this stress. Upon closer inspection, the trouble must be caused by disregarding $\vec{\tau}_{bs}$ in the depth-averaged flow equations (3). This can be illustrated by considering the strongly simplified one-dimensional momentum equation

$$0 = - g \frac{\partial z_s}{\partial x} - \frac{\tau_{bx}}{\rho h} + \frac{F_{wx}}{\rho h} \qquad (24)$$

in which the x-axis is directed onshore, perpendicular to the bank crest. Depending on the geometrical situation, the wave-induced force F_{wx} will contribute to sustaining a current or a set-up of the water level. In the Grevelingen area, with the wide channel behind the banks, only little set-up will build up. In that case, equation (24) reduces to

$$\tau_{bx} = F_{wx} \qquad (25)$$

If the secondary shear stress is disregarded, like in the present model,

$$\tau_{bpx} = F_{wx} \qquad (25)$$

In the sediment transport model, however, τ_{bs} is added to τ_{bp}, so

$$\tau_{bx} = \tau_{bpx} + \tau_{bsx} = F_{wx} + \tau_{bsx} \qquad (26)$$

Since, according to (16), τ_{bsx} is negative (directed offshore; also see Figure 2), the onshore convective transport is systematically underestimated. This explains why the quasi-3D model predicts an offshore movement of the banks. It also explains why the 2DH-model works well in this particular situation with little set-up: the correct value of τ_{bx} is F_{wx}, and hence τ_{bpx}!

Although the 2DH model suffices for the Grevelingen area, this will not be so for all practical situations: set-up and undertow are not always negligible. Therefore, the inconsistency in the quasi-3D model has to be removed. The remedy is simple and straightforward: include $\vec{\tau}_{bs}$ in the depth-averaged current model.

CONCLUSIONS

A quasi-3D model of the yearly residual sediment transport and bed topography changes in complex coastal areas was applied to a practical situations, where previous studies had indicated that 3D effects (undertow) should be important.

In this particular case, however, a depth-integrated version of the model, without 3D effects, turned out to give a good hindcast of the observed morphological evolutions, in spite of remaining doubts about the transport model.

A comparison with the much worse results of the quasi-3D model provided the possibility to trace a major inconsistency in the latter: the secondary bottom shear stress should not be disregarded in the depth-averaged current model, if it is taken into account in the sediment transport model. Once this inconsistency has been removed, the model can be a useful tool for the analysis of morphological processes in complex coastal areas.

ACKNOWLEDGEMENT

The development of a pilot version of the model, as well as its application to the Greveling area, were performed by order of the Ministry of

Transport and Public Works (Rijkswaterstaat), by a joint project-team of this organization and DELFT HYDRAULICS. The authors are members of this team.

REFERENCES

Bagnold, R.A., 1966. An approach to the sediment transport problem from general physics. USGS Prof. Paper, 422-I.

Bailard, J.A., 1981. An energetics total load sediment transport model for a plane sloping beach. J. Geoph. Res., 86, C11, p. 10938-10954.

Bakker, W.T. and Van Kesteren, W.G.M., 1986. The dynamics of oscillating sheetflow. Proc. 20th ICCE, Taipei, p. 940-954.

Bijker, E.W., 1966. The increase of bed shear in a current due to wave action. Proc. 10th ICCE, Tokyo, p. 746-765.

Boer, S., de Vriend, H.J. and Wind, H.G., 1984. A mathematical model for the simulation of morphological processes in the coastal area. Proc. 19th ICCE, Houston, p. 1437-1453.

Davies, A.G., Soulsby, R.L. and King, H.L., 1988. A numerical model of the combined wave and current bottom boundary layer. J. Geoph. Res., 93, no. C1, p. 491-508.

De Vriend, H.J., 1986. 2DH computations of transient sea bed evolutions. Proc. 20th ICCE, Taipei, p. 1698-1712.

De Vriend, H.J. and Stive, M.J.F., 1987. Quasi-3D modelling of nearshore currents. Coastal Engineering, 11, no. 5/6, p. 565-601.

De Vriend, H.J. and Van Banning, 1988. Depth-averaged tidal and wave-induced current modelling for application in the Voordelta area of the Dutch coast. DELFT HYDRAULICS, Rept. H526-3 (in preparation).

De Waal, J.C.M., 1987. Investigation on Bailard's transport formula, applied to cross-shore transport. Delft University of Technology, Dept. of Civil Engrg., M.Sc. thesis (in dutch).

Dingemans, M.W., 1987. HISWA verification in the directional wave basin. DELFT HYDRAULICS, Rept. H228-1.

Dingemans, M.W., Stive, M.J.F., Bosma, J., de Vriend, H.J., Vogel, J.A., 1986. Directional nearshore wave propagation and induced currents. Proc. 20th ICCE, Taipei, p. 1092-1106.

Guza, R.T. and Thornton, E.B., 1985. Velocity moments in the nearshore. J. Waterway, Port, Coastal and Ocean Engrg., 111, no. 2, p. 235-256.

Holthuijsen, L.B. and Booij, N., 1986. A grid model for shallow water waves. Proc. 20th ICCE, Taipei, p. 261-270.

Holthuijsen, L.B., Booij, N. and Herbers, T.H.C., A prediction model for stationary short-crested waves in shallow water with ambient currents. To be published in Coastal Engineering.

Jonsson, I.G., 1966. Wave boundary layers and friction factors. Proc. 10th ICCE, Tokyo, p. 127-148.

Klopman, G. and De Vriend, H.J., 1988. Modelling of velocity profiles due to waves and currents in the coastal zone. Proc. 21st ICCE, Malaga.

Kohsiek, L., 1987. Reworking of former ebb-tidal deltas into large long-shore bars following the artificial closure of tidal inlets in the southwest of The Netherlands. In: P.L. de Boer et al. (eds.) "Tide-influenced sedimentary environments and facies". Reidel Publ. Comp., Dordrecht.

Langerak, A., de Ras, M.A.M. and Leendertse, J.J., 1978. Adjustment and verification of the Randdelta II model. Proc. 16th ICCE, Hamburg, p. 1049-1070.

Madsen, O.S. and Grant, W.D., 1976. Sediment transport in the coastal environment. Mass. Inst. Techn., R.M. Parsons Lab., Rept. no. 209.

Nielsen, P., 1985. A short manual of coastal bottom boundary layers and sediment transport. Public Works Dept. N.S. Wales, Coastal Engrg. Branch, Rept. TM 85/1.

Nielsen, P., 1988. Three simple models of wave sediment transport. Coastal Engineering, 12, no. 1, p. 43-62.

Ribberink, J.S. and De Vriend, H.J., 1988. Sediment transport formulations for waves and currents in the Voordelta area of the Dutch coast. DELFT HYDRAULICS, H526-2 (in preparation).

Rienecker, M.M. and Fenton, J.D., 1981. A Fourier approximation method for steady waves. J. Fluid Mech., 104, p. 119-137.

Roelvink, J.A. and Stive, M.J.F., 1988. Large scale tests of cross-shore sediment transport on the upper shoreface. Proc. IAHR-Symp. Sed. Transp. Coastal Zone, Copenhagen, p. 137-147.

Shore Protection Manual, 1984. US Army Corps of Engineers, Waterways Exp. Station, Vicksburg.

Steetzel, H.J. and Stive, M.J.F., 1986. Cross-shore transport study Voordelta. Delft Hydraulics Laboratory, Report H239 (in dutch).

Steyn, R.C., 1988. Schematization of the natural conditions in multidimensional numerical models of coastal morphology. DELFT HYDRAULICS, Rept. H526-1 (in preparation).

Stive, M.J.F ., 1986. A model for cross-shore sediment transport. Proc. 20th ICCE, Taipei, p. 1551-1564.

Stive, M.J.F. and De Vriend, H.J., 1987. Quasi-3D nearshore current modelling: wave-induced secondary current. Proc.ASCE Spec. Conf. "Coastal Hydrodynamics", Delaware p. 356-370.

Svendsen, I.A. and Lorenz, R.S., 1988. Three dimensional flow profiles on littoral beaches. Proc. 21st ICCE, Malaga.

Van Banning, G.K.F.M., de Vriend, H.J. and Boer, S., 1987. Schematization and validation of 2DH mathematical models in coastal morphology. Proc. "Coastal Sediments '87 "Conf., New Orleans, p. 600-615.

Van den Berg, J.H., 1987. Note on the isallobath map Voordelta 1975-1984. Rijkswaterstaat, Directorate Zeeland, Rept. ZL 87.0020 (in dutch).

Van der Spek, A.J.F., 1987. Description of the development of the outer deltas of Haringvliet and Grevelingen. Rijkswaterstaat, Dept. Tidal Waters, Rept. GWAO-87-015 (in dutch).

Van der Spek, A.J.F. and Steijn, R.C., 1988. Confrontation of mathematical model results and measured data for the Grevelingen outer delta. Rijkswaterstaat, Dept. Tidal Waters, Rept. GWAO-88.1320 (in prep.).

Van Rijn, L.C., 1986. Three-dimensional modelling of suspended sediment transport for currents and waves. DELFT HYDRAULICS, Rept. H461/Q250/ Q422.

Visser, P.J., 1986. Wave basin experiments on bottom friction due to current and waves. Proc. 20th ICCE, Taipei, p. 807-821.

Willemse, J.B.T.M., Stelling, G.S. and Verboom, G.K., 1986. Solving the shallow water equations with an orthogonal coordinate transformation. Int. Symp. Comp. Fluid Dyn., Tokyo (also: DELFT HYDRAULICS Comm. 356).

Wind, H.G. and Vreugdenhil, C.B., 1986. Rip-current generation near structures. J. Fluid Mech., 171, p. 459-476.

Yoo, D. and O'Connor, B.A., 1988. Turbulence transport modelling of wave-induced currents. Proc. Int. Conf. Comp. Mod. Ocean Engrg., Venice, p. 151-158 (Balkema, Rotterdam, 1988).

CHAPTER 126

VERTICAL VARIABILITY OF COASTAL SEDIMENT TRANSPORT

Ryszard B. Zeidler*

ABSTRACT

The semiempirical theory of turbulent diffusion may be used as a mathematical tool for description of suspended sediment mechanics. The Author's solutions, with their exponential vertical profiles of sediment concentration, are presented in various ranges of time, space, and external factors. The inherent concept of eddy diffusivity K is shown to be ambiguous and incoherent. Therefore it is purposeless to investigate thoroughly the vertical profiles and other details of K. Accordingly, the Author makes use of his own and alien laboratory and field findings to propose such estimates of K under r e g u l a r waves and currents which are not continuous functions of the vertical coordinate but instead are locally averaged over depth layers. These estimates become more complex in real coastal zones due to the randomness of waves and currents, wave breaking, three-dimensionality and nonstationarity of nearbed turbulence, a variety of interactions, and other sources of sediment dispersion. Hence g r o s s estimates of averaged K seem even more appropriate; some formulae are put forward. The vertical profiles of sediment concentration are presented herein along with examples of sediment transport rates measured in nearshore zones.

1. SUSPENDED SEDIMENT IN TERMS OF DIFFUSION THEORIES

1.1. Analytical Solutions

The concentration of suspended sediment, C, in an unsteady two-dimensional flow with velocity u_0 may be described by the equation of turbulent diffusion

$$\frac{\partial C}{\partial t} + u_0(x,y)\frac{\partial C}{\partial x} = K_{01}(y)\frac{\partial^2 C}{\partial x^2} + \frac{\partial}{\partial y}\left(K_{02}(y)\frac{\partial C}{\partial y}\right) + w_s\frac{\partial C}{\partial y}\cdots \tag{1}$$

in which

* Professor, Polish Academy of Sciences' Institute of Hydro-Engineering (IBW PAN), 7 Kościerska, 80-953 Gdańsk

K_{01}, K_{02} = eddy diffusivities
w_s = settling velocity
t = time.

The system of coordinates is shown in Fig. 1 where the initial and boundary conditions are outlined for the two problems distinguished, A and B.

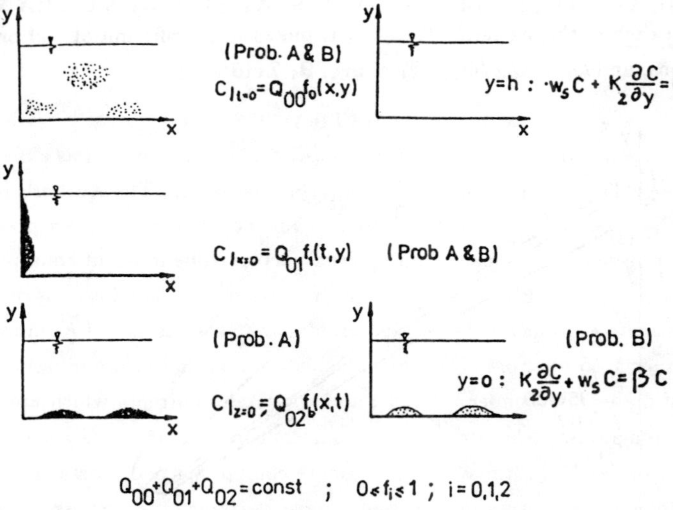

Fig. 1 Initial and Boundary Conditions

Eq. 1 in dimensionless form reads

$$\frac{\partial C^*}{\partial t^*} + 2au^*(x^*, y^*)\frac{\partial C^*}{\partial x^*} = \alpha_1^*(y^*)\frac{\partial^2 C^*}{\partial x^{*2}} + \alpha_2^*(y^*)\frac{\partial^2 C^*}{\partial y^{*2}} + (2b + \alpha_2^{*\prime}(y^*))\frac{\partial C^*}{\partial y^*} \quad (2)$$

in which

$$u_0 = Uu^*(x, y); \qquad 0 \le u^*(x, y) \le 1$$
$$K_{01} = K_1\alpha_1^*(y); \qquad 0 \le \alpha_1^*(y) \le 1$$
$$K_{02} = K_{02}\alpha_2^*(y); \qquad 0 \le \alpha_2^*(y) \le 1$$
$$a = \frac{U \cdot h}{2\sqrt{K_1 \cdot K_2}}; \qquad b = \frac{w_s h}{2K_2}; \qquad C^* = \frac{C}{Q_{02}};$$
$$x^* = \sqrt{\frac{K_2}{K_1}\frac{x}{h}}; \qquad y^* = \frac{y}{h}; \qquad t^* = \frac{K_2}{h^2}t$$

The asterisks will be neglected henceforth.

Eq. 2 has been solved for the so-called problem A_0, with constant coefficients, and subsequently envelopes have been found for other versions of the problems A and B (Nguyen and Zeidler, 1982). The solution was obtained as polynomial series, by methods of integral transforms. It assumes the form

$$C(x,y,t) = C_1(x,y) + C_2(x,y,t) + Q_0\left[C_3(y,t) + C_4(x,y,t)\right] + \\ + Q_1\left[C_5(x,y) - C_6(x,y,t)\right] \qquad (3)$$

in which each of the six terms $C_1 \cdots C_6$ is linked to specific initial and boundary conditions, and $Q_0 = Q_{00}/Q_{02}$, $Q_1 = Q_{01}/Q_{02}$.

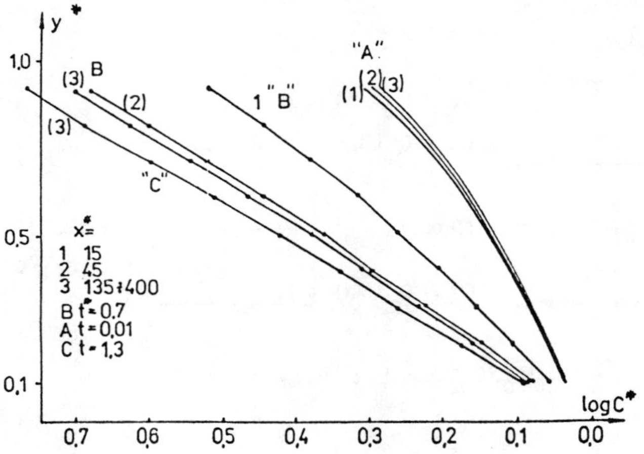

Fig. 2 Examples of Numerical Solutions to Problem A_0

The numerical results depicted in Fig. 2 make it easy to analyse Eq. 3. In the tested intervals of dimensionless $15 \leq x^* \leq 400$, $0.01 \leq t^* \leq 1.3$ one may see that the initial convex-up C-profiles, for $t^* = 0.1$, gradually transform into linear (in the semilog plot), for the longest time $t^* = 1.3$. Hence the concentration profiles in stabilized conditions (long $x \rightarrow X$ and $t \rightarrow T$) of the problem A_0 are exponential, as emphasized in Fig. 3a. Analysis of solutions to the other problems A and B (Nguyen and Zeidler, 1982) shows that they are contained within certain limits C^- and C^+, as presented in Fig. 3b.

1.2. Experimental Evidence

It is appropriate to verify how Eq. 3 agrees with laboratory and field data. The vertical profiles of C have been checked as to their character, numerical values of the exponent b, and temporal variation. Longitudinal variation, $C(x)$, has been

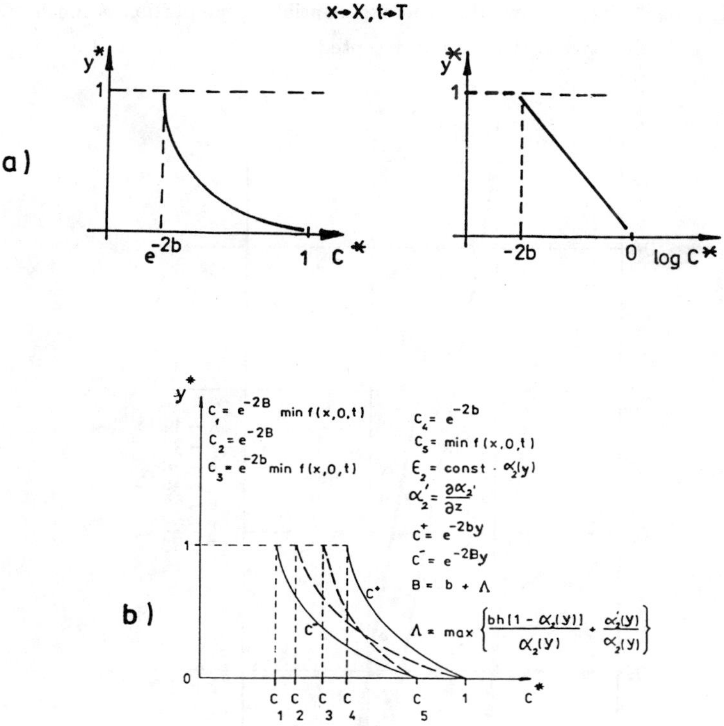

Fig. 3 Equilibrium Solutions to Problem A_0 (top) and to Problem A (bottom)

investigated, along with the effect of sand mixtures and other features. Several alien and three Polish sets of data, each embodying extensive series of C, have been employed, and many additional can be found in bibliographical sources. Some typical examples are presented below.

The numerous data utilized confirm the theoretical exponential vertical profile $C(y)$, with the following important modification. In the semilogarithmic plot, such as the one in Fig. 3a (right), the straight line is almost always replaced by two or three straight lines, in two or three layers: near-bed, central, and near-surface. The two lower layers are pronounced i.a. in the sets measured by Taggart et al (1972), who however adhered to the power-type profile, with corrections proposed by A. T. Ippen. Fig. 4 shows that the exponential representation is much more natural if only the layered structure is accepted.

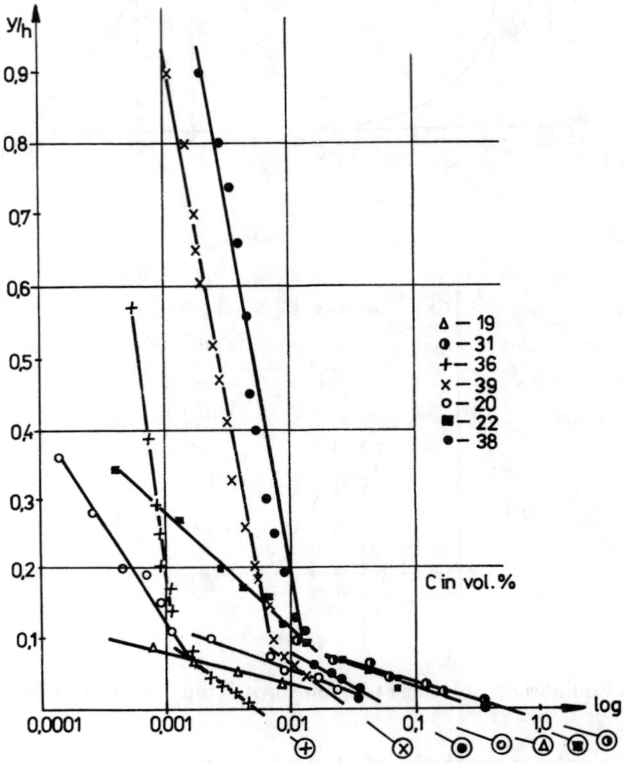

Fig. 4 Sediment Concentration Measured by Taggart et al (1972) in Runs 19...38

The theoretical exponent b agrees well with the experimental data if one assumes $b \simeq 5\frac{w_*}{u_*}$ near bed or $b \simeq 2.5\frac{w_*}{u_*}$ in the central layer; in which u_* = shear velocity. These figures are substantiated in Ch. 2.

One may postulate a few causes of the layered configuration of $C(z)$, such as sediment grading and the associated vertical variability of the von Karman constant.

However, the most important argument is due to the eddy diffusivity $K_y(y) \equiv K_{02}(y)$. It may be inferred that details of the variation of $K_y(y)$ are not essential and that a certain layer-averaged $\overline{K}_y$ is sufficient to describe the basic features of $C(y)$, cf. Fig. 5. Further comments are provided in Ch. 2.

Numerous experimental data show that the layered exponential C-profiles are characteristic not only for flows but also under waves, cf. Skafel and Krishnappan

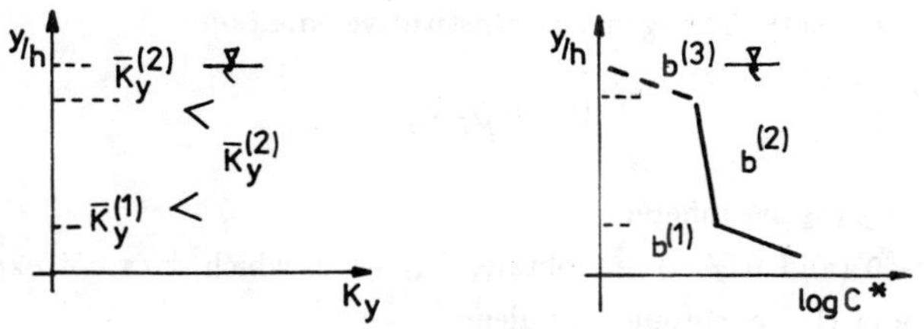

Fig. 5 Correspondence of Exponent b and Layer-Averaged $\overline{K}_y$

(1984), Antsyferov and Kosyan (1986) and many others. Some of the Author's results are presented in Ch. 3. It is also worth noting that such C-profiles are generated by different turbulence mechanisms, i.e. nearbed shear and friction, wave oscillations, wave breaking etc. Hence one may conclude that gross estimates of K, or its equivalents, provide better representation of complex coastal phenomena than detailed fields $K(y)$. These estimates are discussed in Ch. 3. The following arguments in Sec. 2.1 prove how ambiguous the detailed $K(...)$ could be.

2. EDDY DIFFUSIVITY IN THE COASTAL ZONE

2.1. Concept of Eddy Viscosity

The Reynolds equations read

$$\rho \frac{d\mathbf{u}}{dt} = \rho \mathbf{F} - \operatorname{grad} p + \mu \Delta \mathbf{u} + \operatorname{div} \Pi \tag{4}$$

in which

$\mathbf{u} = iu_x + ju_y + ku_z =$ velocity
$\mathbf{F} =$ external mass forces
$p, \rho =$ pressure and density, respectively
$\mu =$ dynamic coefficient of molecular viscosity,

while the Reynolds tensor Π is

$$\Pi = \begin{bmatrix} -\rho\overline{u_x'^2} & -\rho\overline{u_x'u_y'} & -\rho\overline{u_x'u_z'} \\ -\rho\overline{u_x'u_y'} & -\rho\overline{u_y'^2} & -\rho\overline{u_y'u_z'} \\ -\rho\overline{u_x'u_z'} & -\rho\overline{u_y'u_z'} & -\rho\overline{u_z'^2} \end{bmatrix} \tag{5}$$

One may distinguish diffusion theories, in which Π is coupled with the mean strain rate tensor S, and "differential" models, with expanded equations for Π. Let us dwell on the diffusion theory, and consider a few cases

1. Scalar eddy viscosity, homogeneous constitutive equation

$$\Pi_{ij} = \mu_T S_{ij} \tag{6}$$

The following defects are inherent

(a) For $S_{ij} \to 0$ (and $\mathbf{u} \neq 0$) one obtains $\Pi_{ij} \to 0$, which does not explain the occurrence of Π_{ij} in isotopic turbulence

(b) Each component of Π_{ij} depends on one component of S_{ij}, in disagreement with observations, e.g. spatial structure of turbulence in 1-D flows

(c) Traces of both sides in Eq. 6 disagree, i.e. $tr\Pi = -2\rho k < 0$

$$\mathrm{tr}(\mu_T S) = \begin{cases} \mu_T \mathrm{div}\mathbf{u} = 0 & \text{for } \rho = \text{const} \\ \mu_T \mathrm{div}\mathbf{u} = -\dfrac{\mu_T}{\rho}\dfrac{d\rho}{dt} \gtrless & \text{for } \rho \neq \text{const} \end{cases} \tag{7}$$

(d) By definition, Π_{ii} is not greater than zero, while it can be positive by Eq. 6.

2. Eddy viscosity is a 2-nd rank tensor; homogeneous constitutive equation

$$\Pi_{ij} = \mu_{Tik} S_{jk} \tag{8}$$

or

$$\Pi_{xx} = \mu_{Txx} S_{xx} + \mu_{Txy} S_{xy} + \mu_{Txz} S_{xz}; \; \Pi_{xy} = \mu_{Txx} S_{yx} + \mu_{Txy} S_{yy} + \mu_{Txz} S_{yz} \cdots \text{etc.}$$

The following defects surface:

(a) cf. (a) under (1)

(b) tr Π should be < 0 (see (c) under (1)) while from Eq. 8 one has tr $\Pi = \mu_{Tij} S_{ij} \gtrless 0$

(c) For 1-D flow one obtains $\Pi_{zz} = \Pi_{zy} = \Pi_{yz} = 0$, in contradiction with experiments

(d) Π components in Eq. 8 are asymmetric, in contrast to Eq. 5

(e) cf. (d) under (1)

3. Eddy viscosity is a 2-nd rank tensor; inhomogeneous & composite constitutive equation

$$\Pi_{ij} = -2\rho\alpha_{ij} k + \frac{1}{2}(\mu_{Til} S_{lj} + \mu_{Tjl} S_{li}) \tag{9}$$

in which

$$\alpha_{ij} \begin{cases} = 0 & \text{for } i \neq j \\ \neq 0 & \text{for } i = j \end{cases}$$

The following defects persist

(a) incomplete set of constitutive variables

(b) LHS & RHS traces of Eq. 9 are not necessarily equal

(c) diagonal Π_{ii} can be positive in Eq. 9

(d) cf. (b) under (1)

(e) α must satisfy $\alpha_{ij} = \alpha E_{ij}$; otherwise α_{ij} is not a tensor

(f) a new quantity, k, turbulence energy enters and must be found, so that the defects of the "differential models" are taken over.

More assumptions can be proposed, such as a 4-th rank tensor for μ_T and inhomogeneous constitutive equations etc. (Sawicki, 1988). Some corrections of the above schemes are likely but, for all possible versions, it must be concluded that the concept of eddy viscosity, and its theoretical implementation are, by and large, incoherent and vulnerable.

In addition, one must remember that eddy viscosity is not identical with eddy diffusivity

$$\mathrm{K} = \beta\mu_T\,, \qquad \text{with} \qquad \beta \neq 1 \tag{10}$$

Hence, although the concepts of eddy viscosity and eddy diffusivity are widely used, it is better to apply them with caution, and restrict oneself to gross estimates.

2.2. Formulae for K; the Coastal Zone

Our primary finding in Ch. 1 may be summarized in the following dimensional triad

$$C \sim \exp(-2b\frac{y}{h}); \quad b = \frac{w_s h}{2K_y}; \quad \text{const } K_y \text{ in } 2\cdots 3 \text{ layers} \tag{11}$$

One may make use of Prandtl's suggestion

$$K = \text{const } B(U_{\max} - U_{\min}) \tag{12}$$

in which

B = width of mixing zone

$U_{\max}$, $U_{\min}$ = maximum and minimum velocity in the mixing zone, respectively.

One may also endeavour

$$K_y = \text{const} u_* \times h \tag{13}$$

in which u_* = shear velocity.

For open-channel flow Elder gave const $\equiv 0.23$, which in our case would yield

$$2b = 4.35\frac{w_s}{u_*} \tag{14}$$

The search for a functional form of K can be more elaborate (Yaglom, 1976). Assume $\beta = 1$ and

$$K_{ij} = \begin{cases} \mu/\rho k_{ij}(\frac{\rho u_* y}{\mu}) & \text{for} \quad y_+ = \frac{\rho y u_*}{\mu} > 0 \\ \kappa_{ij} u_* y & \text{for} \quad y_+ \gg 1 \end{cases} \tag{15}$$

in which

$\kappa_{yy} \equiv \kappa_1 = \frac{\kappa}{Pr_t} \rightarrow 0.45 \cdots 0.48$

κ, Pr_t = von Karman and Prandtl coefficient, respectively

k_{ij} = dimensionless quantity.

On the other hand, the classical arguments of linear shear stress profile and logarithmic velocity profile lead to the parabolic K;

$$K_y = \kappa u_* y(1 - y/h) \tag{16}$$

Equations 15 and 16 may be compromised:

$$K_y = \begin{cases} 0.9\kappa u_* h(1 - \frac{y}{h}) \\ \kappa u_* y(1 - \frac{y}{h}) \\ \kappa,\ u_* y \end{cases} \quad \text{for} \quad y = \begin{cases} 0.85 \cdots 1.0h \\ 0.15 \cdots 0.85h \\ 0 \cdots 0.15h \end{cases} \tag{17}$$

Fig. 6 shows the above K_y together with respective values of $2b$ averaged in the three layers. The experimental data reviewed in this study, yielding $\frac{\overline{2b}}{w_s/u_*}$ about 10 near bed and 5 in the central layer (see Ch. 1) agree well with the analysis outlined in Fig. 6

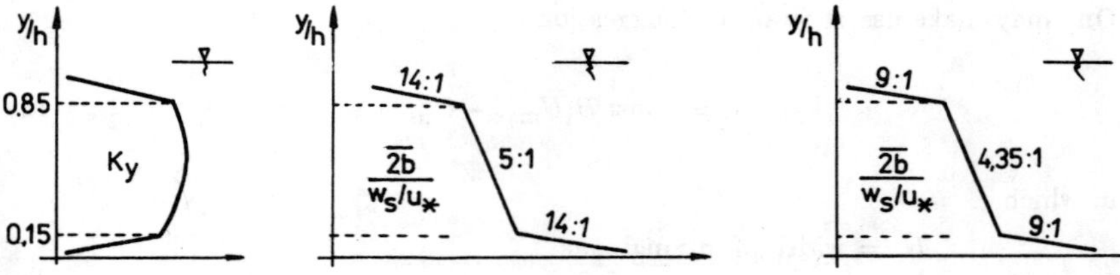

Fig. 6 Theoretically Postulated Values of b

Now examine the most common situations in the coastal zone. For waves (tilde), currents (overbar), long waves (∧) and turbulence (prime) one has

$$\mathrm{u} = \bar{\mathrm{u}} + \hat{\mathrm{u}} + \tilde{\mathrm{u}} + \mathrm{u}' \tag{18.1}$$

$$C = \bar{C} + \hat{C} + \tilde{C} + C' \tag{18.2}$$

Substituted into the diffusion equation

$$\frac{\partial C}{\partial t} + \frac{\partial u_i C}{\partial x_i} + w_s \frac{\partial C}{\partial y} + Q_c = 0 \tag{19}$$

with the source term Q_c, the components of Eq. 18 produce

$$(\frac{\partial}{\partial x_i})\ \overline{uC} = \overline{uC} + \overline{\hat{u}\hat{C}} + \overline{\tilde{u}\tilde{C}} + \overline{u'C'} \tag{20}$$

the cross-terms being uncorrelated.
One usually assumes

$$\overline{\tilde{u}\tilde{C}} = -\tilde{K}\frac{\partial \bar{C}}{\partial y}; \quad \overline{u'c'} = K\frac{\partial \bar{C}}{\partial y} \tag{21}$$

Stive and Wind (1985) neglected $\overline{u'C'} \ll \tilde{u}\tilde{C}$, which might be unjustified, as the available data shows.

Various formulae have been proposed for $\tilde{K}$ and K, e.g. Homma and Horikawa (1962) (1965), Kennedy and Locher (1972), Skafel and Krishnappan (1984), Antsyferov and Kosyan (1986). A certain insight into the possible structure of $\tilde{K}$, K... is provided by the Author's data on coastal dispersion (Sec. 2.3).

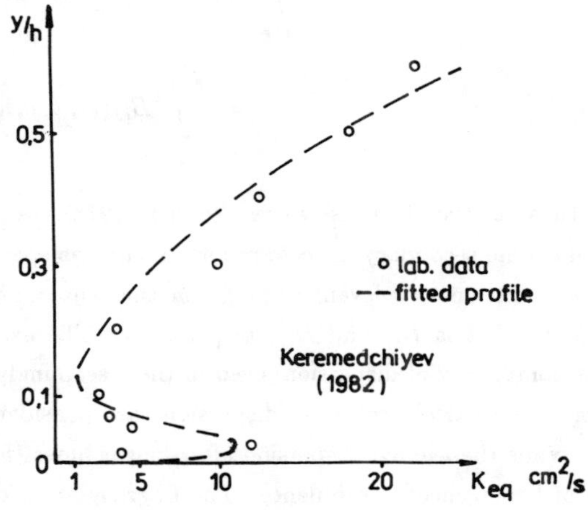

Fig. 7 Eddy Diffusivity Measured in Laboratory

It appears that the fluxes $\rho\ \overline{\tilde{u}\tilde{C}}$ and $\rho\ \overline{u'C'}$, i.e. $\tilde{K}$ and K, respectively, are diversified over depth, which is signalled in Fig. 7, with the dominance of K, due to bottom turbulence, in the nearbed layer and $\tilde{K}$, due to wave-induced mixing in the central layer.

2.3. Laboratory and Field Investigations of Eddy Diffusivity

The Author carried out two extensive studies on the dispersion of passive matter. One of them has dealt with regular waves, in a laboratory flume, and the other was conducted in the field where random waves and turbulence dominated the dispersion.

In the wave flume, naturally buoyant particles were ejected at selected segments of bed ripples, and their movements were tracked with a high-speed camera. By this means the following Lagrangian properties of the nearbed layer could be obtained in various wave phases:

$$\overline{\mathbf{Y}(\tau)} = \int_{t_0}^{t_0+\tau} \overline{\mathbf{V}(\mathbf{x},t)}dt; \tag{22.1}$$

$$\mathbf{Y}'(\tau) = \mathbf{Y}(\tau) - \overline{\mathbf{Y}(\tau)} = \int_{t_0}^{t_0+\tau} \mathbf{V}'(\mathbf{x},t)dt \tag{22.2}$$

$$D_{ij}(\tau) = \overline{\mathbf{Y}_i'(\tau)\mathbf{Y}_j'(\tau)} = \int_{t_0}^{t_0+\tau}\int_{t_0}^{t_0+\tau} \overline{\mathbf{V}_i'(\mathbf{x}_1 t_1)\mathbf{V}_j'(\mathbf{x}_1 t_2)}dt_1 dt_2 =$$

$$= \iint_{t_0}^{t_0+\tau} B_i j(t_1,\, t_2)dt_1 dt_2 \tag{22.3}$$

Since the study is described elsewhere (Zeidler, 1972), we shall not dwell on the theory and some secondary aspects reported, such as spectra, but merely confine ourselves to the most relevant findings on the tensors of displacements **Y** and velocities **V**, that is D_{ij} and B_{ij}, respectively. The examples given in Fig. 8 depict the complexity of dispersion, even in these seemingly simple regular wave conditions. The vertical profiles of dispersion vary considerably over wave period (Fig. 8a), while the average dispersion after longer times (Fig. 8b) displays certain features of homogeneous turbulence. The Lagrangian velocity tensor B_{ij} also depends on a wave phase (Fig. 8c).

The field studies embodied simultaneous measurements of waves, water velocities and sediment concentration, the latter with a radioisotopic probe, in a vertical profile with smoothly moving instrumentation. The emission noise of the radioactivity source, Am-241, was filtered out, and all signals were processed to obtain correlational and spectral characteristics. The following results can be singled out for the frequency range $0.2 \lesssim f \lesssim 2\,cps$, which may be attributed to random surface waves.

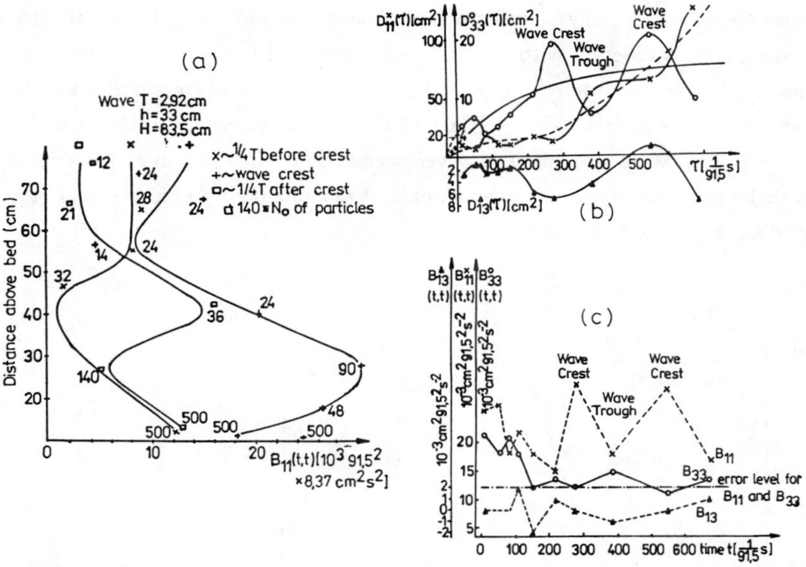

Fig. 8 Lagrangian Velocity Correlation Moments B_{ij} Across Depth (a) and in Time Domain (c) and Displacement Correlation Moments D_{ij} (b)

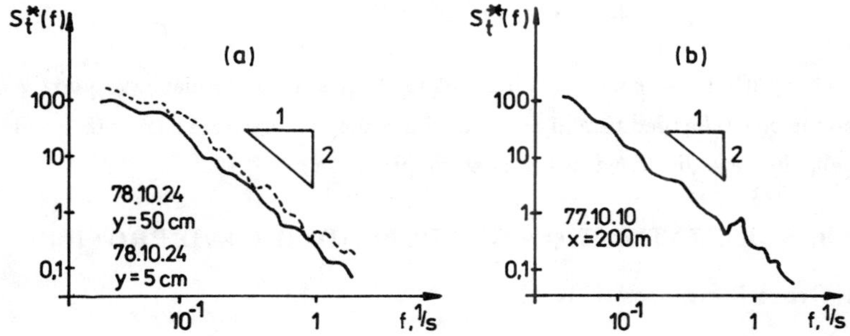

Fig. 9 Shear-Dominated Spectra of Sediment Concentration (a) and Fluorescent Dye (b) Measured in Field

Fluctuations of cross-shore water velocity, $u' \equiv u'_x$ are correlated best with surface waves, while fluctuations of the longshore velocity, $v' \equiv u'_y$, can be linked to shear flow. Fluctuations of sediment concentration measured have been found roughly analogous to concentration fluctuations of fluorescent tracers, which is illustrated in Fig. 9 showing spectra for the two kinds of passive coastal matter, i.e. suspended sediment (left) and a neutrally-buoyant tracer (right). Hence, fluorescent tracer studies may be utilized in sedimentation problems. It appears that in the situations covered, with random waves recorded in twenty-odd cycles, C' series were controlled strongly by bed shear near bottom and were dominated by waves at the free surface. The correlations $\overline{C'v'}$ are of important order of magnitude (Fig. 10). More research is however needed on both sediment concentration and turbulence. Inter alia, wave breaking still remains poorly explored as a very important source of turbulence.

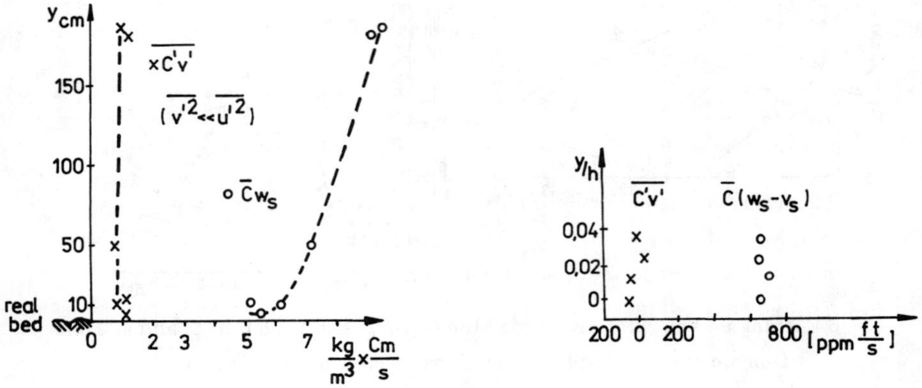

Fig. 10 Correlations $\overline{C'v'}$ for Random Waves (left) and Regular Waves (right, after Kennedy & Locher, 1972)

The Author also conducted other field studies, in particular a very extensive year-long coastal expedition in Senegal. Since they incorporate some other aspects as well, they are discussed in the next chapter.

3. SCHEMATIZATION OF SUSPENDED LOAD PROFILES

3.1. Gross Estimates of K

Chapter 1 and 2 show that vertical profiles of sediment concentration can be described in three layers, for which the exponents $2b$ in $C \sim C_0 \exp(-2by/h)$ can be postulated in various forms, depending on dispersion-controlling mechanisms.

A considerable bulk of data has been collected, which makes possible validation of the concept and formulae put forward. The data includes many alien sources, such as those mentioned above, e.g. some more recent by Skafel and Krishnappan (1984) Antsyferov and Kosyan (1986), and the Author's own data for the Baltic Sea, the Black Sea and particularly those from the coastal project off St. Louis, Senegal in 1980...81 (Zeidler, 1982 and Skaja & Zeidler, 1982). The results of the latter embody 98 vertical profiles of sediments concentration, ranging from instantaneous to averaged long-term ones, such as those exemplified in Fig. 11.

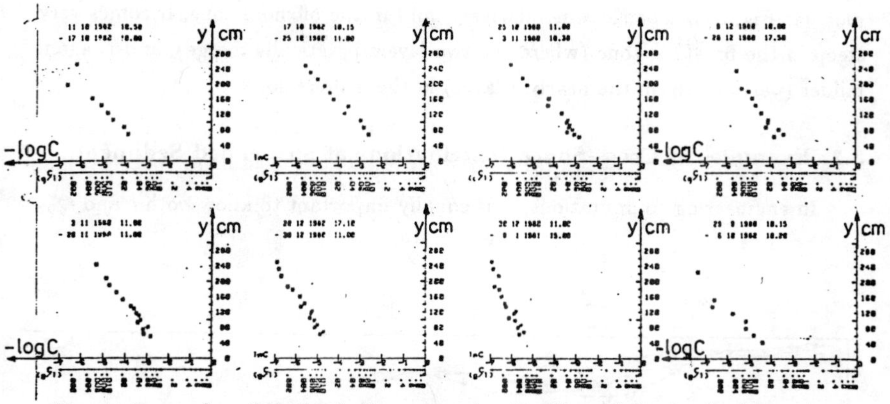

Fig. 11 Examples of Author's Concentration Profiles off St. Louis, Senegal

The available concentration profiles are exponential, so that their slopes can be identified with $2b = w_s h/2\overline{K}_y$. The following tentative conclusions may be drawn from the regression analysis attempted for $2b$.

1. In wave-dominated areas of motion, with $\widetilde{K} \gg K$ one has

$$2b \simeq 4w_s \frac{hT}{H^2}; \quad \sigma_{2b} \simeq 1 \frac{w_s Th}{H^2} \quad (R^2 \simeq 0.7) \tag{23.1}$$

2. In wind wave-induced dispersion

$$< \widetilde{K} > \simeq 2 \cdots \widetilde{K}_{\text{significant}} \qquad (\widetilde{R} \simeq 0.6) \tag{23.2}$$

3. In bed shear-controlled areas (within surf zone), with $K \gg \widetilde{K}$, one may postulate

$$2b \simeq 40 \frac{w_s}{u_*} + \quad +0(h/l) \quad (R^2 \simeq 0.6) \tag{23.3}$$

Other dispersion mechanisms can be dealt with similarly.

Extensive studies under CMEA programmes, in particular Lubiatowo'74, Lubiatowo'76, Kamchiya'77, Kamchiya'79, and other measurements in the two coastal research stations (Polish Lubiatowo and Bulgarian Kamchiya, or Shorpilovtsy) have shown that the layered structure of exponential concentration profiles persists everywhere in the coastal zone, both inshore and offshore. The profiles may be further schematized by attributing, to the nearbed and central layer, distinctly different slopes of the exponential branches (which are straight lines in the semilog plot, cf. Fig. 3a). The slope is relatively mild in the offshore zone, becomes very steep in the breaking zone (where the two layers practically merge), and is again milder (particularly in the nearbed layer) in the inshore zone.

3.2. Examples of Cross-Shore Distributions of Suspended Sediment

In engineering computations, it is equally important to know both b and Q_{02}

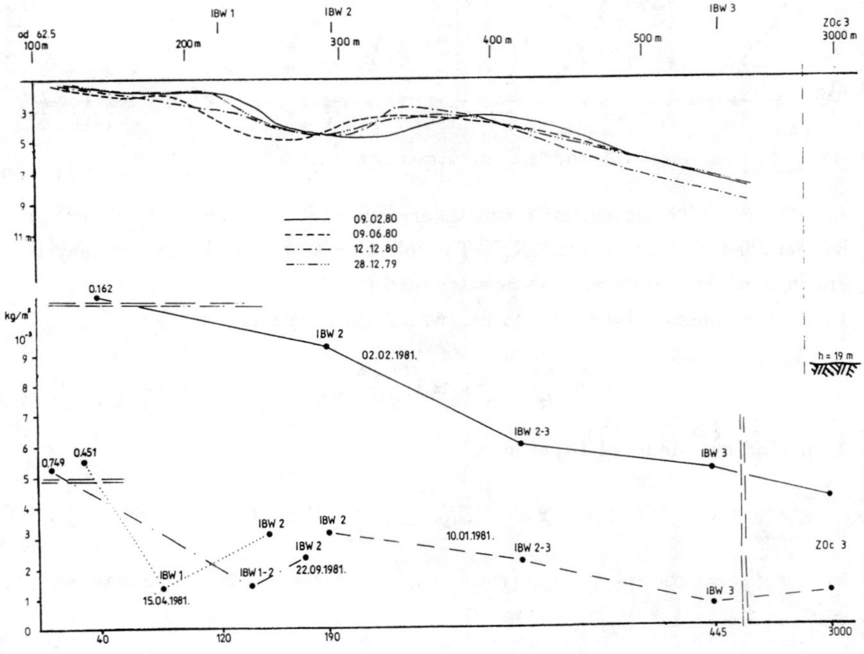

Fig. 12 Examples of Author's Cross-Shore Profiles of Suspended Mass (bottom) and Bed Topography (top) off St. Louis, Senegal

in the exponential profile $C = Q_{02} \exp(-2by/h)$. While b shows how much of the water depth is laden with sediment (or its practically important quantities), it is Q_{02} which determines the absolute magnitude of saturation with sediment. Vertical profiles of sediment concentration, such as those depicted in Fig. 11 not only provide figures for the exponent b but also yield estimates for Q_{02} and the suspended load. The data collected under the programmes mentioned in this paper gives insight into absolute magnitudes and notably the cross-shore distribution of suspended load, the latter being found through obvious integration of the concentration profile over the entire depth of water, with proper account for the layered structure of the C-profile.

Illustrated in Fig. 12 are some examples for the situations encountered off St. Louis, Senegal. It is interesting to note high figures in the zone of ultimate breaking and in the swash zone, where permanent turbidity was recorded, on most occasions, due to considerable turbulence. Although some correlations between Q_{02} and the coastal climate have already been formulated, a lot of work remains to be done before reliable estimates of b and Q_{02} will have been available for all possible coastal situations.

CLOSING REMARKS

The theoretical findings presented in this paper, supported by vast experimental evidence, point to the exponential profile as simple and well substantiated representation of the vertical distribution of suspended sediment concentration. The exponent has been correlated with external factors but more efforts are required for better quantitative description of coastal sediment fields in absolute physical terms. – A concise summary is provided in Abstract.

ACKNOWLEDGEMENTS

This study has been sponsored under the Polish Academy of Sciences' programme CPBP 02.12.2, which is gratefully acknowledged.

REFERENCES

1. Antsyferov S. M., Kosyan R. D., (1986), *Suspended Sediment in the Upper Shelf* (in Russian), Nauka, Moscow, 224p.

2. Homma M., Horikawa K., (1962), *A laboratory study on suspended sediment due to wave action*, Proceed. X Conf. IAHR, London, Vol. 1, 213-221.

3. Homma M., Horikawa K., Kajima R., (1965), *A study of suspended sediment due to wave action*, Coastal Eng. in Japan, Vol. 8, 85-103.

4. Kennedy J. F., Locher F. A., (1972), *Sediment suspension by water waves*, in *Waves on Beaches*, Acad. Press, 249-295.

5. Keremedchiyev S. D., (1982), *Morphologic analysis of equilibrium profile and wave parameters in laboratory conditions*, (in Bulgarian), Okeanologia (Sofia), No 9, 54-63.

6. Nguyen V. G. and Zeidler R., (1982), *Turbulent Diffusion of Suspended Sediment in Open-Channel Flow*, PRACE IBW PAN, Gdańsk, 160p.

7. Sawicki J. M., (1988), *Analysis of spatial models of turbulence*, Archiwum Hydrotechniki (in press).

8. Skafel M. G., Krishnappan B. G., (1984), *Suspended sediment distribution in wave field*, Journal Waterway ... Ocean Engineering, Vol. 110, No. 2, 215-230.

9. Skaja M., Zeidler R. B., (1982), *Longshore sediment transport rate*, 18th Intern. Conf. Coastal Eng. Abstracts, ASCE.

10. Stive M. J. F., Wind H. G., (1986), *Cross-shore mean flow in the surf zone*, Coastal Engineering Vol. 10, No. 4, 325-340.

11. Taggart W. C., Yermoli C. A., Montes S., Ippen A. T., (1972), *Effects of sediment size and gradation on concentration profiles in turbulent flow*, Rep. No. 152, Parsons Lab., Mass. Inst. Techn., 153p.

12. Yaglom A. M., (1976), *Semi-empirical equations of turbulent diffusion in boundary layers*, Fluid Dynamics Transactions, Vol. 7, p. I, IPPT PAN Warsaw, 99-144.

13. Zeidler R. B., (1972), *Nearbed turbulence under waves* (in Polish), Archiwum Hydrotechniki, Vol. XVIII, No. 2, 147-166.

14. Zeidler R. B., (1982), *Yearly measurements off St. Louis, Senegal*, 18th Intern. Conf. Coastal Eng. Abstracts, ASCE.

15. Zeidler R. B., (1987a), *Suspended sediment transport and beach profile evolution*, Discussion, Journal Waterway ... Ocean Eng., ASCE, Vol. 113, no. 1, 89-95.

16. Zeidler R. B., (1987b), *Concentration and transport rate of suspended sediment in marine and fluvial environments*, Euromech 215, Univ. Genoa, Ext. Abstr., 96-100.

CHAPTER 127

ABOUT THE INFLUENCE OF EROSION VOLUME ON CROSS-SHORE SEDIMENT MOVEMENT AT PROTOTYPE SCALE

by K. Uliczka[1] and H. H. Dette[2]

ABSTRACT

In order to estimate erosion of sand beaches attacked by storm surges, experiments at prototype scale are necessary, since small scale models do not reproduce physical processes in the surf zone.
The present study describes the influence of erosion volume on cross-shore sediment movement during the development of beach profile.
The discussion includes parameters which affect the erosion volume and the accuracy of assessment.

INTRODUCTION

Breaking waves of major storms cause a three-phase flow consisting of sediment, fluid and air in a complex three-dimensional pattern. In order to avoid unknown scaling effects the sediment motion under such conditions requires experimental investigations at full scale.

A set of such experiments was carried out in the "BIG WAVE CHANNEL" (GWK) - length: 324 m, depth: 7 m, width: 5 m - in Hannover. The aspects of the study described below relate to the influence of instantaneous erosion volumes on suspended sediment content from commencement of wave action to the equilibrium of beach profile.
These studies are a continuation of the series from which the analysis of sediment transport volumes was presented at 20th ICCE in Taipei.

The sediment transport calculated from suspended sediment measurements amounted to about 30 % of that given by profile surveys, Fig. 1.
In the following the influence of erosion volume on the suspended sediment at a fixed location in the surf zone is discussed.

EXPERIMENTAL CONDITIONS

Repeated test series were carried out on a beach profile with foreshore. Fig. 2 shows the initial test profile. The dune had a 1 to 4 seaward slope from 2 m above still water level (SWL) to 1 m below SWL, followed by a 1 to 20 slope down to channel floor, i. e., it represented the conditions during a storm tide.

[1] Assistant Scientist, Dipl.-Ing. [2] Chief Engineer, Dr.-Ing.
Leichtweiss-Institute, Division of Hydromechanics and Coastal Engineering, Techn. University of Braunschweig, Fed. Rep. of Germany

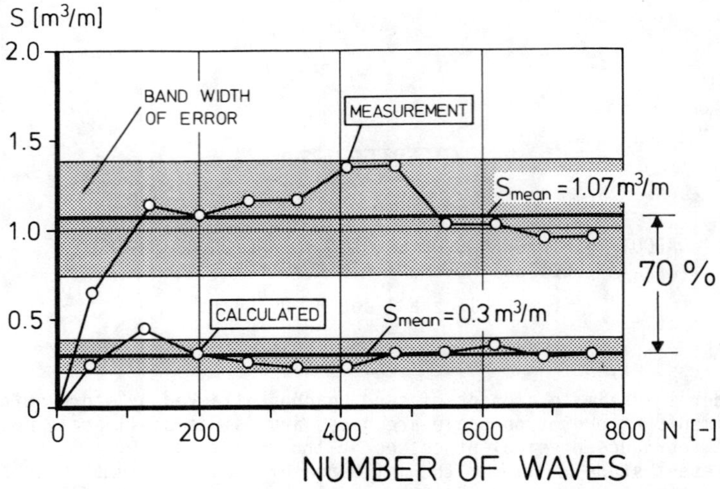

Fig. 1: Comparison of sediment transport between the calculated transport and that computed from profile changes during development phase of beach profile (Dette/Uliczka, 1986)

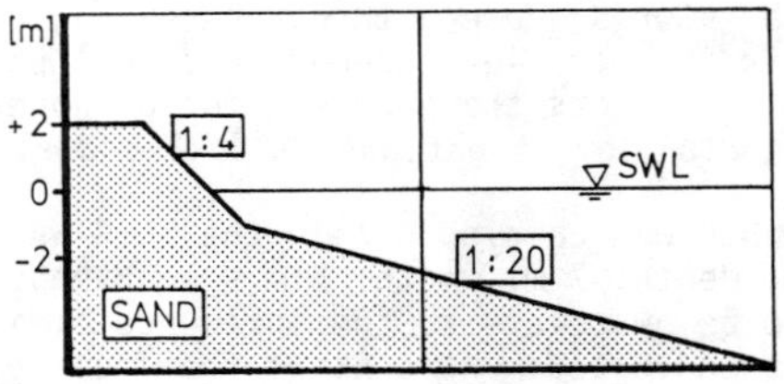

Fig. 2: Initial beach profile

The wave parameters selected were as follows:

1. Monochromatic waves: $H_{rms} = 1.5$ m, $T = 6$ s, $d = 5.0$ m
   $H_o/L_o = 0.04$, $\xi_o = 0.29$

2. Irregular waves (theoretical JONSWAP-spectrum):
   $H_{1/3} = 1.4$ m, $T_p = 6$ s, $d = 4.0$ m, $\gamma = 3.3$, $\alpha = 0.0081$,
   $\sigma_a = 0.07$, $\sigma_b = 0.09$

where ξ_o is the similarity parameter and γ, α, σ_a and σ_b are parameters of the JONSWAP-spectrum, H is the wave height, T period, L wave length, d water depth. Subscript $_o$ refers to deep water values.

The measuring station for initial wave height was at x = 150 m and x = 185 m, respectively.

The sand used for the test series had the following parameters:

- D_{50} = 330 μm, σ_g = 1.47, w = 0.0477 m/s
- D_{50} = 220 μm, σ_g = 1.48, w = 0.0284 m/s

where σ_g is the geometric standart deviation and w the mean fall velocity of sand at 18^0 C.
The measurements were carried out in test increments:

- with monochromatic waves the generator was stopped after about 80 waves when the initial wave form became significantly effected by reflection,
- with irregular waves after about 9 minutes, equal to two repetition intervals of the generated wave spectrum.

Before restarting the waves the profile was surveyed and the sediment concentration samples analysed. Restarting followed after the water surface had calmed.
The vertical distribution of suspended sediment concentration in the breaker zone was obtained by the suction method. The results of sediment concentration and wave height are time-averaged data for about 80 monochromatic waves (8 minutes) and about 9 minutes for irregular waves (about 90 waves).

The horizontal distance of the measuring stations for suspended sediment from the initial dune face was x = 26 m (D_{50} = 330 μm, monochromatic waves), and x = 22 m (D_{50} = 220 μm, location B, irregular waves), Fig. 3. The difference in x of 4 m has to be noted, when comparing data.

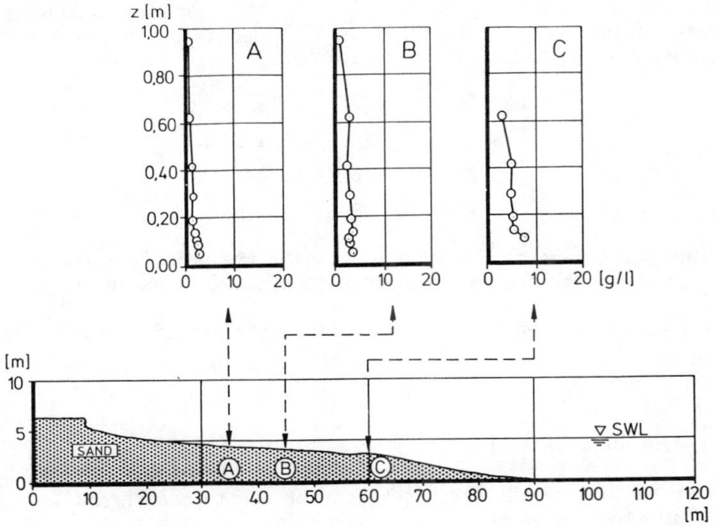

Fig. 3: Suspended sediment distribution over the equilibrium profile due to irregular waves (D_{50} = 220 μm)

At equilibrium conditions the sediment in suspension was measured at the station A, B and C, identical in Fig. 4. The measuring points for suspended sediment were vertically spaced as follows: 0.02 m, 0.02 m, 0.025 m, 0.025m, 0.05 m, 0.1 m, 0.125 m, 0.2 m, 0.325 m. On the average the lowest measuring point was 0.06 m above the bed but this varied with the movements of the bed and at times the lowest point was even burried.

Thus, the suspension was measured in the lower about 1 m thick layer.

SUSPENSION DISTRIBUTION ACROSS THE SURF ZONE AT EQUILIBRIUM CONDITIONS

The suspension distribution over the equilibrium profile due to monochromatic waves (D_{50} = 330 μm) was presented at the 20th ICCE. The sediment concentration distributions over the equilibrium profile of D_{50} = 220 μm sand after about 2,000 irregular waves are plotted in Fig. 3.

The arithmetic mean of the 33 % highest waves, evaluated for the test increment at equilibrium conditions, yielded at the ponts A, B and C:

$$A:\ H_{1/3} = 0.9\text{ m},\quad B:\ H_{1/3} = 1.1\text{ m},\quad C:\ H_{1/3} = 1.4\text{ m}$$

The water depths were:

$$d_A = 0.9\text{ m},\quad d_B = 1.3\text{ m},\quad d_C = 1.7\text{ m}$$

The water temperature was about 20° C. The breaker zone extended from about x = 65 m to nearly x = 30 m.

In order to give a description of wave conditions during the suspension measurements at equilibrium state a statistical expression for the breaker location was used. The waves were grouped into those with breakpoints seaward, landward and near (± 1 m) the measuring station. This information was obtained by analysis of video-tape and yielded the following values (% of waves at station):

	seaward	near	landward
A:	74 %	13 %	14 %
B:	65 %	16 %	19 %
C:	6 %	39 %	55 %

The high percentage of breaking waves near the point C is reflected in the high suspended sediment concentration, as well as in the wave height.

Table 1 gives the sediment concentration data associated with irregular waves.

CALCULATION OF WAVE ENERGY

In order to compare the beach development due to monochromatic and irregular waves, wave energy was calculated for each test series before the waves reached the beach profile using the socalled linear, AIRY or first order STOKE's wave theory.

DUNE WITH FORESHORE			
IRREGULAR WAVES	$H_{1/3}$ = 1.4 m	T_p = 6 s	
Height above bottom	Sediment Concentration		
	$\bar{c}$	$\bar{c}$	$\bar{c}$
m	g/l	g/l	g/l
0.05	2.8	3.7	-
0.09	2.5	2.8	-
0.11	2.2	2.7	7.6
0.14	2.0	3.6	6.2
0.19	1.8	3.2	5.3
0.29	1.7	3.1	4.8
0.42	1.4	2.5	5.1
0.62	0.9	2.4	3.1
0.94	0.8	1.8	-
Location	A	B	C

Table 1: Sediment concentration data for irregular waves (D_{50} = 220 μm)

The average wave energy per unit surface area is

$$\bar{E} = \frac{1}{8} \cdot \rho \cdot g \cdot \frac{1}{N} \sum_{i=1}^{N} H_i^2 \qquad (1)$$

The erosion and suspended sediment data for profile development are plotted against accumulated wave energy instead of time or number of waves, i. e., against $\bar{E}$ N, where N is the number of waves. For irregular waves is N = Δt/T, i. e., the duration of the test increment divided by peak wave period.
The energy flux, $\bar{E}\, c_g$, and the shape of the erosion volume and suspended sediment volume curves will be discussed in the section dealing with estimates of accuracy.

CALCULATION OF SUSPENDED SEDIMENT VOLUME

The suspended sediment mass during the development phase of beach profile was calculated from the expression

$$SV = \int_0^z \bar{c} \cdot b \cdot dz \qquad (2)$$

where b is the length of profile unit and z = 1.0 m, approx. Fig. 4 shows the suspended sediment mass in this lower about 1 m thick layer for the two grain sizes plotted against wave energy input per unit surface area seaward of breakers. Each point on the graph represents time-averaged sediment concentration volume for test increment at the energy input since commencement of test.

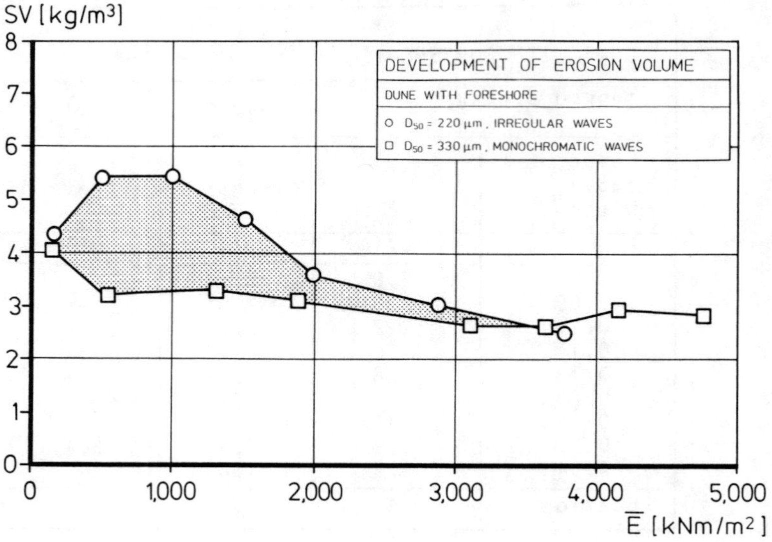

Fig. 4: Suspended sediment volume versus wave energy of monochromatic and irregular waves

The measured data on the suspended sediment volume yielded the following information:

- At the beginning of both test series the volumes coincided, but at about 1,000 kNm/m² the difference was about 40 %.
- When the equilibrium conditions of the profiles were approached suspended sediment volumes converged to about 2.5 kg/m³.
- The suspended sediment volume with monochromatic waves reached its maximum value in the first test increment whereas with irregular waves the suspension due to erosion reached its maximum after about five increments, Fig. 4.

It should be noted, that the accuracy of the sediment concentration measurements in the breaker zone was about ± 30 % related to the average (see Dette/Uliczka 1986).

Table 2 gives the data of suspended sediment volume due to monochromatic and irregular waves for the two grain sizes.

EROSION VOLUME AND SUSPENDED SEDIMENT VOLUME

The erosion volume was obtained by profile surveys after each test increment. The accuracy of profile measurements (± 0.01 m) in terms of erosion volumes is about ± 1 %. The maximum erosion volume was about 15 m³ for monochromatic waves (D_{50} = 330 μm) and about 22 m³ per metre width for irregular waves (D_{50} = 220 μm).

DUNE WITH FORESHORE					
MONOCHROMATIC WAVES			IRREGULAR WAVES		
D_{50}= 330 μm			D_{50}= 220 μm		
Test Increment -	$\bar{E}$ kNm/m^2	SV kg/m	Test Increment -	$\bar{E}$ kNm/m^2	SV kg/m
1	137	4.1	1	170	4.4
3	547	3.2	3	509	5.4
7	1313	3.3	6	1019	5.5
10	1888	3.1	9	1528	4.6
15	3119	2.6	12	1994	3.6
17	3640	2.6	18	2886	3.0
19	4159	2.9	24	3781	2.5
21	4754	2.8			

Table 2: Data on suspended sediment volumes during development of beach profile under monochromatic and irregular waves (Test increments and energy input were continually summed, whereas the suspension data were time-average for each test increment)

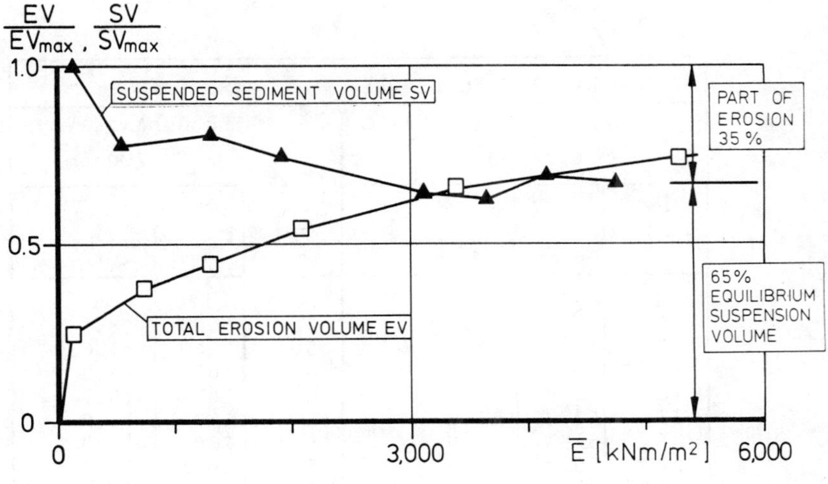

Fig. 5: Erosion volume and sediment in suspension volume as a function of wave energy input under monochromatic waves (D_{50} = 330 μm)

Erosion volume and suspended sediment volume were normalized with their maximum values. Fig. 5 and Fig. 6 show the erosion volume and suspended sediment volume as a function of wave energy under monochromatic waves (D_{50} = 330 μm) and irregular waves (D_{50} = 220 μm), respectively. The data on erosion volume are presented in Table 3.

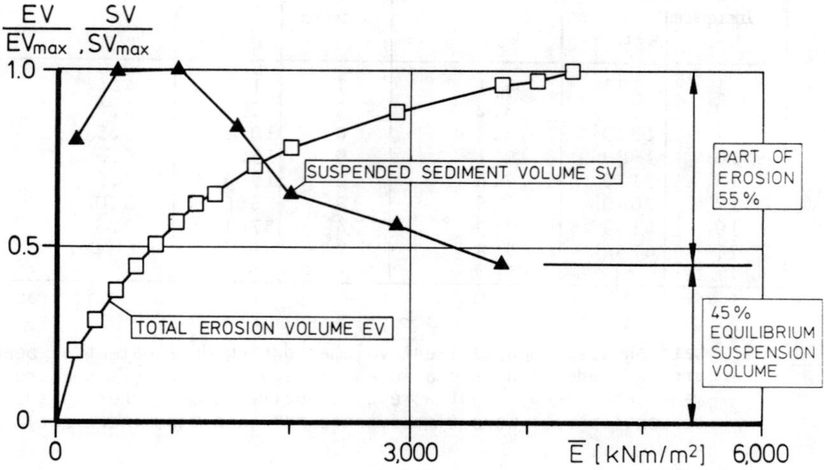

Fig. 6: Erosion volume and sediment in suspension volume as a function of wave energy input under irregular waves (D_{50} = 220 μm)

DUNE WITH FORESHORE					
MONOCHROMATIC WAVES D_{50} = 330 μm			IRREGULAR WAVES D_{50} = 220 μm		
Test Increment –	$\overline{E}$ kNm/m^2	EV m^3/m	Test Increment –	$\overline{E}$ kNm/m^2	EV m^3/m
1	137	3.8	1	170	4.4
4	739	5.7	3	509	8.1
7	1313	6.8	5	849	11.0
11	2080	8.2	6	1019	12.3
16	3379	9.9	10	1697	15.7
23	5273	11.2	12	1994	16.8
27	6437	11.9	18	2886	19.0
29	7020	12.3	25	3930	20.7
39	9889	15.1	28	4378	21.6

Table 3: Data on erosion volume during the development of beach profile under monochromatic (D_{50} = 330 μm) and irregular waves (D_{50} = 220 μm) (Test increments, energy input and erosion volume were continuously summed)

A smaller total erosion volume was generally associated with a smaller suspended sediment volume in the surf zone. This is a well known observation but some additional features immerge from the graphical presentation above:

- The erosion of the profile with sand D_{50} = 220 μm started more intensively than that of coarser sand at the same wave energy input
- After 3,000 kNm/m² about 90 % of the equilibrium erosion volume was reached for the finer sand, whereas for the coarser sand only 60 % erosion had occurred
- After 4,000 kNm/m² both suspended sand concentration profiles reached equilibrium conditions
- The effect on transport rate, described above, shows that the maximum suspension volume was reached with finer sand when about 40 % of erosion volume had been accomplished
- In terms of the equilibrium state, it is seen that 35 % and 55 %, respectively, of the maximum suspended sediment volume is transient sediment of the erosion process
- During the erosion process the suspended sediment volume at the measuring station reaches about twice the value at the equilibrium.

The 1.4 times higher suspended sediment volume of coarser sand (65 %/ 45 % = 1.4) seems to be due to the higher energy density of monochromatic waves. Under irregular waves the sediment can settle out during periods of small waves of the wave spectrum.
In order to find a direct relationship between suspended sediment volume and erosion volume the incremental data were analysed. The results are summarized in Table 4 and plotted in Fig. 7.

DUNE WITH FORESHORE			
MONOCHROMATIC WAVES D_{50}= 330 μm		IRREGULAR WAVES D_{50}= 220 μm	
EV m^3/m	SV kg/m	EV m^3/m	SV kg/m
3.8	4.1	(4.4)	(4.4)
1.4	3.2	1.8	5.4
1.6	3.3	1.3	5.4
0.7	3.1	0.5	3.6
0.6	2.6	0.4	3.0
0.7	2.9	0.3	2.5
0.7	2.8		

Table 4: Incremental data on erosion volume and suspended sediment volume under monochromatic and irregular waves

The volume in suspension and eroded for each point were determined for the same time.
There appears to be a linear relationship between erosion volume and the volume in suspension.

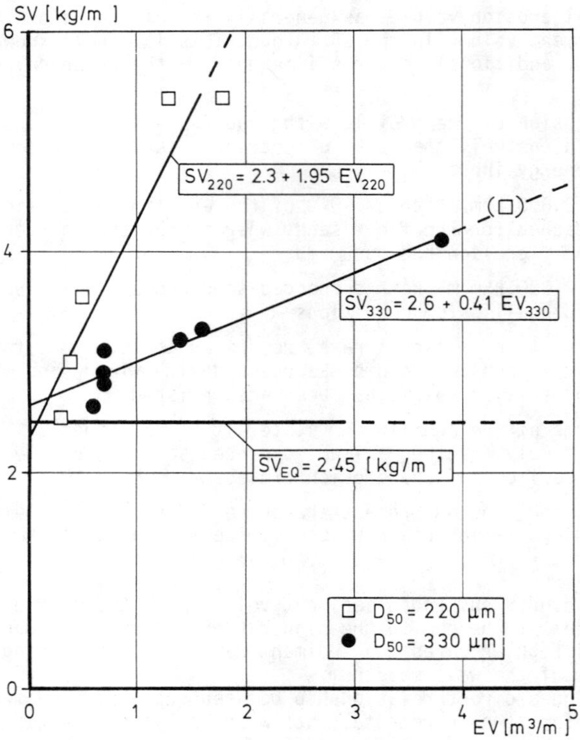

Fig. 7: Relationship between instantaneous suspended sediment volume and erosion volume for 330 μm-sand and monochromatic waves as well as 220 μm-sand and irregular waves

The suspended sediment volume at equilibrium state is about the same as presented in Fig. 4, i. e., about 2.5 kg/m after 4,000 kNm/m² energy input.
The slope of the Linear Regression Functions in Fig. 7 demonstrates the influence of grain size on suspended sediment concentration and erosion volume.

BEACH PARAMETER AND EROSION VOLUME

At first water level and wave conditions influence the erosion volume and the sediment movement as bed load and suspended sediment.
Below is a discussion of aspects of erosion as observed in full scale at the GWK.
A comparison of the growth of the erosion volumes, starting with different initial profiles, is shown in Fig. 8. The profiles were a 1 : 4 slope from channel floor to the top of dune (+ 2 m SWL) and that shown in Fig. 2, referred to as dune without and with foreshore.

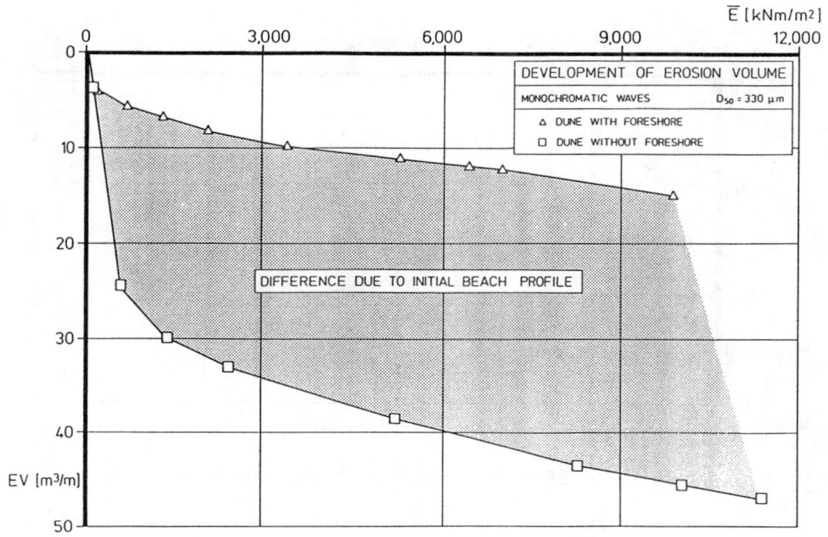

Fig. 8: Growth of erosion volume versus wave energy input for two different initial beach profiles

The erosion volume on the profile without foreshore is seen to be about three times that on the profile with foreshore. The 1 : 4 slope from channel bottom to the top of the dune lead to wave energy transformation directly in front of the dune face and therefore to large erosion volume as well as to extremely high suspended sediment concentration.

The influence of grain size on the erosion volume of a profile with dune and foreshore was determined by using two sediment sizes. A reduction of the mean diameter by about 30 % resulted in an increase of erosion volume by 100 % at energy input of 4,500 kNm/m², as shown in Fig. 9.

The comparison relates to the time, when the profile of the finer sand reached equilibrium. At the time only 70 % of the equilibrium volume of the coarser sand had been eroded.
Taking the total erosion volumes of both equilibrium profiles, after an energy input of 10,000 kNm/m², the difference was about 30 %. The energy input up to the equilibrium stage for the coarser sand was about twice that for the 220 μm-sand.

The foreshore slopes at both equilibrium profiles in the surf zone (330 μm and 220 μm sand) stabilised to about 1 : 20, whereas the slopes reported from nature for beaches with sand 220 μm are generally flatter.

DISCUSSION OF ACCURACY

The question of accuracy of the calculations and the data plots has been mentioned earlier.
If the erosion volume was plotted against the calculated energy input, both the data due to monochromatic and irregular waves should lead to

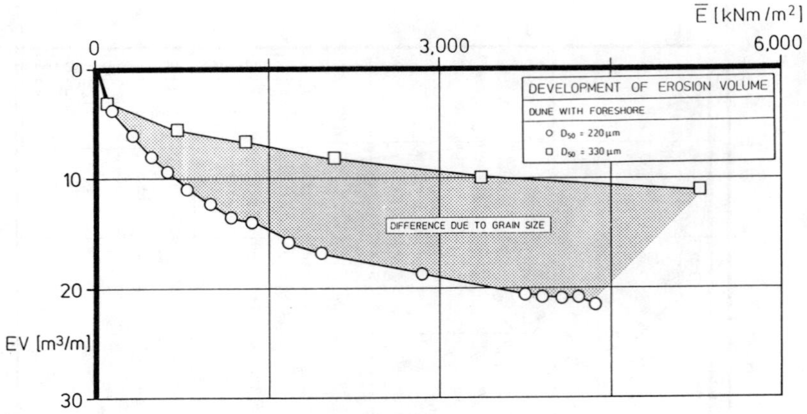

Fig. 9: Comparison of erosion volumes with different grain sizes

an identical function for equal initial profiles and the same grain size. Additional to the computation of average wave energy per unit surface area, energy flux was calculated at the measuring station in the wave channel, before breaking:

$$\overline{P} = \overline{E}\, n\, C \tag{3}$$

with

$$n = \frac{1}{2}\left[1 + \frac{4\pi d/L}{\sinh(4\pi d/L)}\right] \tag{4}$$

where d is the water depth, L the calculated mean wave length in the given depth of water, n is the relationship of wave group to wave phase velocity, and C is the mean wave propagation velocity in the given depth of water.

Fig. 10 shows the growth of the erosion volume of the profile with dune without foreshore under monochromatic and irregular waves using grain size D_{50} = 330 µm versus wave energy, and Fig. 11 the growth of erosion volume from the same tests versus energy flux.

Plotted at the same scale, Fig. 10 and 11, the differences in terms of energy flux are a maximum with 5 % at monochromatic waves, whereas at irregular waves the curves coincide. This shows that the erosion volume and suspended sediment volume could be compared both on the basis of energy input or energy flux. Nevertheless, the discrepancy of erosion volumes due to monochromatic and irregular waves is at maximum more than 50 %.
The wave conditions (monochromatic or irregular waves) have a major influence on the process of beach erosion.
The breaker form, breaker zone, rate of wave energy attenuation, velocity field etc. have a dominent influence on the sediment transport in the surf zone. As mentioned above, the higher energy density of monochro-

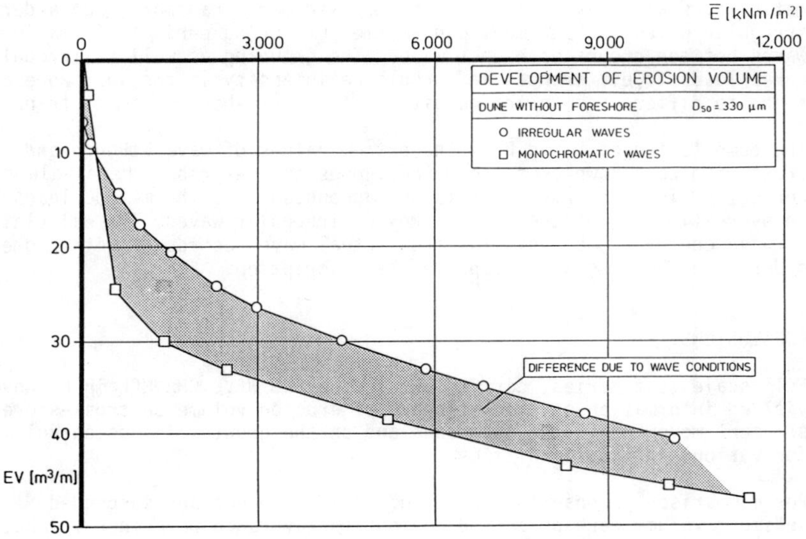

Fig. 10: Growth of erosion volume on the profile with dune without foreshore under monochromatic and irregular waves versus wave energy

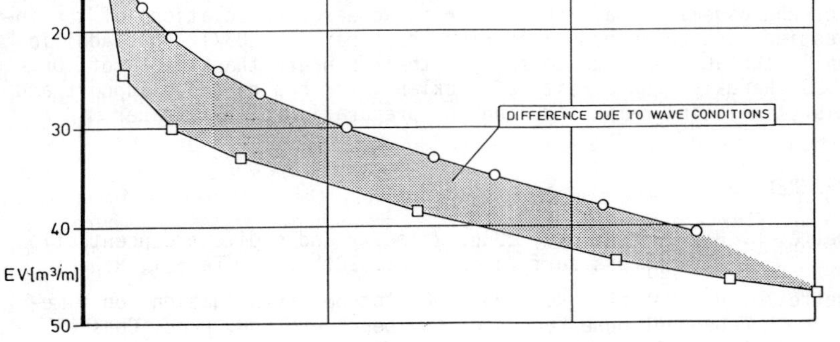

Fig. 11: Growth of erosion volume on the profile with dune without foreshore under monochromatic and irregular waves versus energy flux

matic waves at breaking leads to higher sediment transport, but under irregular waves the sediment can settle out during periods of smaller waves between groups of higher waves. The grouping as well as irregular waves leads to differences in turbulence intensity in the surf zone and hence to differences in the magnitude of the offshore sediment transport.

The computation of $\overline{E}$ and $\overline{P}$ by the extrem values of wave, trough and crest with zero down crossing method, does not describe these values correctly since the wave form is not accounted for. The method leads to an overestimating of the wave energy of irregular waves. The calculation of wave energy and energy flux from actual wave records or with higher order wave theories would improve the description.

CONCLUSIONS

Full scale test series, carried out in the BIG WAVE CHANNEL in Hannover, yielded information on the influence of erosion volume on cross-shore sediment movement in the surf zone and on the growth of erosion volumes for various initial conditions.

For comparison purpose the growth of erosion volume and suspended sediment volume were presented against the averaged wave energy input per unit surface area.

The suspended sediment volume (measured in the bottom 1 m thick layer) was related to the growth of erosion volume and was found to reach a constant value at the equilibrium conditions independent of grain size.

Initial beach profile and grain size used have a strong influence on the erosion process of a dune.

The accuracy of the results would be improved by inclusion of the detailed description of the wave form and the vortex intensity.

ACKNOWLEDGEMENT

The authors would like to acknowledge the sponsorship of the Deutsche Forschungsgemeinschaft (DFG = German Research Association) of the investigations described (Sonderforschungsbereich 205/TP A6). Additionally the authors like to express their sincere thanks to Prof. Dr.-Ing. A. J. Raudkivi, University of Auckland, for his generous support and his stimulating remarks during the preparation of this paper.

REFERENCES

Dette, H. H., Uliczka, K., 1986: Velocity and sediment concentration fields across surf zones, Proc. 20th ICCE, Taipei, R.O.C.

Dette, H. H., Uliczka, K., 1987: Prototype investigations on time-dependent dune recession and beach erosion, Proc. Coastal Sediment'87, New Orleans, U.S.A.

Raudkivi, A. J., 1976: Loose boundary hydraulics, 2nd Edition, Pergamon Press, Oxford, GB.

SPM Shore Protection Manual, Vol. I, U.S. Army, CERC, 1977

LIST OF SYMBOLS

b	m	width per unit beach profile
c	kg/m³	time-averaged sediment concentration
C	m/s	wave propagation velocity
d	m	water depth
D	µm	grain size
D_{50}	µm	mean grain size
$\bar{E}$	kNm/m²	total wave energy per unit surface area
EV	m³/m	erosion volume per unit width of beach profile
g	m/s²	gravitational acceleration
H_{rms}	m	root mean square wave height
H_o	m	deep water wave height
$H_{1/3}$	m	arithmetic mean value of 33% highest waves
L	m	wave length
L_o	m	deep water wave height
n	-	ratio of wave group and phase velocity
N	-	number of waves
$\bar{P}$	kW/m	energy flux
S	m³/m	total sediment transport
SV	kg/m	suspended sediment volume
T	s	wave period
T_p	s	peak period
w	m/s	fall velocity of D_{50}
x	m	horizontal coordinate
z	m	vertical coordinate
α	-	Phillips-parameter for JONSWAP-spectrum
γ	-	shape parameter for JONSWAP-spectrum
ξ_o	-	similarity index
ρ	kg/m³	water density
σ_a	-	shape parameter for JONSWAP-spectrum
σ_b	-	shape parameter for JONSWAP-spectrum
σ_g	-	geometric standart deviation for grain size

CHAPTER 128

Sediment Transport on Nearly-Prismatic Beaches

Marcel J.F. Stive* and J.A. Roelvink*

SUMMARY

Recent progress in the quantitative modelling of the undertow has stimulated the modelling of cross-shore sediment transport. More so than before it seems now possible to attempt the dynamic modelling of beach profile development. Also, integration of dynamic cross-shore sediment transport formulations in horizontally two-dimensional models for watermotion and sediment transport has recently been suggested. This seems to be a first step of integrating depth-averaged 2DH-modelling with 2DV-profile-modelling. Here an overview is given of these developments and the understanding gained sofar of the several current systems and the induced sediment transport and morphology that are found in the situation of random waves normally and obliquely incident on beaches which vary not or only slowly alongshore.

1. INTRODUCTION

Alongshore variations of bottom topography and/or the incident wave field (amplitude or phase) are the cause of mean nearshore currents in the form of circulation cells, nonuniform longshore currents or mixtures of these. Using the radiation stress concept many dynamics of these depth-averaged nearshore currents may be modelled (Bowen, 1969, and many others since; see Battjes, 1988, for a review).

In the case of no alongshore variations in topography and (obliquely) incident wave field the same concept yields a uniform longshore current. Here our interest is into this type of situation: viz. that of obliquely incident, horizontally uniform waves on beaches of which the alongshore variation of topography can be neglected for the horizontal current dynamics. It should be noted that in reality longshore periodicities in bottom topography and flow are generated without obvious externally imposed periodicities, particularly if the waves are of near-normal incidence (Sonu, 1972). However, we choose our longshore length scale such that these periodicities are averaged out. In this situation there are no distinct horizontal circulation cells with a depth-averaged nonzero flow, only a nearly-uniform longshore current.

In addition to this depth-averaged longshore current system we here consider the vertical circulation system resulting from the vertical imbalances of the mass and momentum fluxes, even though the depth-integrated flux balances are satisfied (cf. Stive and De Vriend, 1987, SDV furtheron). This vertical circulation system induces a near-bottom current field which in general significantly differs from the depth-averaged field.

* DELFT HYDRAULICS, P.O. Box 152, 8300 AD Emmeloord, The Netherlands

In the described case of oblique wave incidence we will find a vertically nonuniform (both in magnitude and direction) current field nearshore. In the upper half of the water column the mean velocity vector more or less follows the depth-integrated longshore current vector but close to the bottom the mean velocity vector is deviated towards the offshore due to the undertow (De Vriend and Stive, 1987, DVS furtheron and Svendsen and Lorenz, 1988). Obviously for the sediment transport the lower layers are the most relevant. Here, it is the sediment transport and eventually the resulting morphology that forms our main reason of interest in the watermotions.

In addition to these mean flow current systems the waves induce oscillatory currents in their propagation direction. Apart from their interaction with mean currents which may lead to net flows as well they may also be the cause of net sediment transport due to their asymmetrical features (cf. Bowen, 1980).

The purpose of this paper is to discuss some recent developments with respect to the understanding of the several current systems (section 2) and the induced sediment transport and morphology (section 3) that are found in the situation of random waves normally and obliquely incident on a nearly-prismatic beach.

2. CROSS-SHORE AND LONGSHORE WATERMOTIONS

2.1 Cross-shore flow due to normally incident waves

The first situation considered is that of a random wave field (with directionality and wave group effects neglected; see Section 5 Discussion) normally incident on a prismatic (i.e. alongshore homogeneous) coast of arbitrary cross-shore profile. In this situation we distinguish the following water motions which are both in cross-shore direction:
- a depth varying oscillatory flow, in phase with the random wave surface variations, and
- a depth varying mean flow, steady on the scale of many waves, which in the surfzone and below the wave trough level mainly consists of the undertow.

The oscillatory flow

For the present interest it appears furtheron that the importance of the oscillatory flow may be expressed by the following low order flow moments (cf. Bowen, 1980):
- the even moments $\langle |u|^3 \rangle$ and $\langle |u|^5 \rangle$, which are nonzero for symmetric velocities,
- the odd moments $\langle u|u|^2 \rangle$ and $\langle u|u|^3 \rangle$, which are zero for symmetric velocities.

The even moments are relatively easy to evaluate. Unlike the odd moments they do not depend on the wave asymmetry features. The theoretical estimate by Guza and Thornton (1985) for a random, linear sea (Gaussian model) gives statisfactory results. However, also a bichromatic model yields statisfactory results (Roelvink and Stive, 1988b).

The odd moments are zero for a symmetric velocity field, but can be nonzero for nonlinear waves such as occur nearshore. Specifically, in the nonbreaking cnoidal type waves the shoreward velocities are stronger and of shorter duration than the offshore flows, leading to nonzero values for the odd moments. On the other hand it appears (Stive, 1986) that the breaking skewtype waves do not contribute to the odd moments, since these waves are symmetrical around the vertical rather than the horizontal plane. An evaluation of the odd moments according to second order Stokes theory (Stive, 1986) gave a first quantitative estimate, but important improvements were obtained by introducing a more general nonlinear theory (Roelvink and Stive, 1988a,b).

The undertow

In the present case of steady (on the time-scale larger than that of the wave groups), vertically two-dimensional motion of normally incident wave groups the depth-integrated total mass and momentum equations yield a depth-mean zero flow and in essence a balance between the radiation stress and the steady wave set-up. However, locally (in the vertical) the mass and momentum fluxes need not be in balance. The potential role of this imbalance driving a seaward directed returnflow or undertow, compensating for the shoreward mass flux above trough level, was first pointed out by Dyhr-Nielsen and Sørensen (1970).

To compensate for the wave-induced non-zero mass flux above wave trough level the depth-averaged net flow below wave trough level is seaward. In the case of non-breaking waves the resulting mean flow profile appears to resemble the socalled conduction solution (Longuet-Higgins, 1953) quite well. This implies that the mean flow in the central fluid region is seaward, but near the bottom it is strongly influenced by the dynamics of the bottom boundary layer, such that a shoreward flow results. In the case of breaking waves, however, the mass flux above trough level is enhanced due to the presence of the surface roller which is one of the manifestations of the breaking process. The effects near the water surface are strong enough to dominate the flow even near the bottom such that an all seaward flow results with its maximum close to the bottom, known as "undertow".

The non-zero mass flux above wave trough level due to the waves alone is theoretically known for periodic waves of constant form relative to an irrotational wave motion below trough level. This theoretical estimate combined with the mass flux in the roller as estimated by Svendsen (1984) may be considered to give an acceptably accurate estimate of the total mass flux above trough level. If we adopt the approach of DVS (1987) for the general case of random breaking waves, characterized by -mutually not interacting- fractions of breaking and non-breaking waves, the total estimate for the total mass flux above wave trough level (m), becomes:

$$m = \left(1 + Q_b \frac{7kh}{2\pi}\right) \frac{E}{c}$$

where the second term represents the empirically derived roller contribution, in which Q_b is the breaking wave fraction.

A crude estimate of the near-bottom undertow may be derived from the above estimate of m by assuming depth uniformity of the flow below wave trough level. A more accurate description including the near-bottom details can be obtained by evaluating the accompanying horizontal momentum balance locally in the vertical, following the above mentioned suggestions of Dyhr-Nielsen and Sørensen (1970). Since 1980 several quantitative evaluations of their concept have been presented in the literature. Key papers in this context are Dally and Dean (1984), who present an internally consistent approach but neglect specific breaking wave effects, Stive and Wind (1986), who indicate the importance of the near-surface layer in breaking waves, and Svendsen et al. (1987), who present an approach for the realistic inclusion of the bottom boundary layer under breaking waves. These three contributions are integrated in a generalized approach by SDV (1987).

A recent finding is (Roelvink and Stive, 1988a,b) that in the region just after the point of initial breaking there is a spatial shift between the maximum gradient of the wave height and that of the undertow. They have found that this is due to the time needed to convert organized kinetic and potential energy into small-scale, dissipative turbulent motion. The underlying reason for this finding is that in the above described undertow model the wave breaking induced shear stress at trough level used to model the effects of the surface layer is proportional to the ratio of wave energy dissipation over wave phase speed (see SDV, 1987). This shear stress is very important in the driving of the undertow and it appears that the lag found in the predictions of the undertow gradients is due to the incorrect assumption that the dissipation source term used in the wave height prediction model is a measure of the actual dissipation. A better interpretation is that this source term is a measure of the actual production of turbulent energy. The actual dissipation may then be derived with the aid of a turbulence model. The correction to the undertow improves the prediction importantly.

2.2 Cross- and longshore flow due to obliquely incident waves

The second situation considered is that of a random wave field (with directionality and wave group effects still neglected) obliquely incident on a prismatic coast of arbitrary cross-shore profile. In this situation we distinguish the following water motions.

Watermotions which are zero when depth- and time-averaged but nonzero when considered on the timescale of the wave period or when considered locally in the vertical, viz.

- wavepath directed, depth varying oscillatory flow, in phase with the random wave surface variations, and
- depth varying mean flow, steady on the scale of many waves, compensating for the wavepath directed mass flux above wave trough level, which in the surfzone and below the wave trough level mainly consists of the undertow.

In addition to these watermotions there are nonzero, time- and depth-averaged watermotions alongshore, viz.:

- a wave-induced mean longshore current, and (potentially)
- a mean current due to an alongshore pressure gradient (e.g. due to the tide).

The wavepath directed oscillatory watermotions clearly will have components both along- and cross-shore. The evaluation of these watermotions may straightforwardly be derived taken account of the local wave angle with the parallel depth contours. Here we further concentrate on the description of the time-averaged, vertically varying current system.

The traditional depth-integrated, horizontal circulation approach obviously has a too narrow physical basis to describe the in essence three-dimensional current pattern. For this we need models which deal with the flow in three dimensions. Since the three-dimensionality is such that the variations in the horizontal directions are zero or only weak compared to that of the vertical direction, flow models using a profile technique can be useful. A first, practical formulation combining the typical cross-shore vertical flow variations with the weaker horizontal flow variations due to alongshore nonuniformities was given by DVS (1987). Their approach can be considered a natural step between 2D and fully 3D models and is therefore termed quasi-3D approach. Here we discuss this approach and some consequences for the even more simplified situation of alongshore uniformity.

In the terminology of DVS (1987) the nonzero depth-averaged timemean watermotions are termed the primary flow. The locally (in the vertical) nonzero but depth-averaged zero timemean watermotions, for instance the undertow, are called the secondary flows. As the terminology indicates this approach assumes in essence that the secondary flows are weak enough to assume that their effects on the derivation of the primary flow are negligible. Furthermore, they assume that the primary flow profile and related primary bottom shear stress are in a single vertical plane. These assumptions allow a full uncoupling between the primary and secondary flows. This approach seems a reasonable first step, but it should be mentioned that in general the uncoupling will not be permitted, especially if the secondary flows are not much smaller than the primary flows. Interactions will then result, such as for instance described by Van Kesteren and Bakker (1986), who find that a sophisticated boundary layer approach for the combination of waves and currents under an angle shows the generation of a secondary flow due to the interaction between primary flow and the oscillatory wave motion.

If the uncoupling between primary and secondary flow is accepted, following DVS (1987) the primary flow may be determined from the conventional depth- and time-averaged horizontal momentum balance equations, and the secondary flow due to e.g. the surface breaking of the wavefield may be determined from the local balances of mass and horizontal momentum in the plane with its orientation in the direction of wave energy flux propagation.

The result of the suggested modelling approach yields a vertically nonuniform (both in magnitude and direction) current field nearshore. In the upper half of the water column the mean velocity vector more or less follows the depth-integrated longshore current vector but close to the bottom the mean velocity vector is deviated towards the offshore due to the undertow. An example of the resulting (computed) vertically rotating mean velocity vector for Egmond beach is given in Figure 1. A preliminary check with nature on this phenomenon as given in DVS (1987) indicates that it resembles reality, but more definitive comparisons are awaited.

This section on the formulation of the watermotions is concluded with a finding, which has obvious implications for the modelling of both the depth- and time-mean currents and the sediment transport (also see De Vriend and Ribberink, 1988). The principle calculations of SDV (1987) indicate that the magnitude of the secondary bottom shear stress is some 10% to 20% of the magnitude of the primary shear stress. The implication of this for the resulting near-bottom shear stress orientation and closely related near-bottom flow orientation may be indicated with the following approximations.

In the considered situation of obliquely incident waves to a prismatic coast the primary shearstress direction is alongshore and of approximate magnitude (cf. Longuet-Higgins, 1970):

$$\tau_p(b) = \frac{D_{br}}{c} \sin\theta$$

where D_{br} is the breaking dissipation and c the wave phase speed (also see Figure 2 for notations). The secondary bottom shear stress is in the wave energy flux direction and of approximate magnitude (SDV, 1987):

$$\tau_s(b) \approx 0.15 \left(0.5 + 7 \frac{kh}{2\pi}\right) \frac{D_{br}}{c}$$

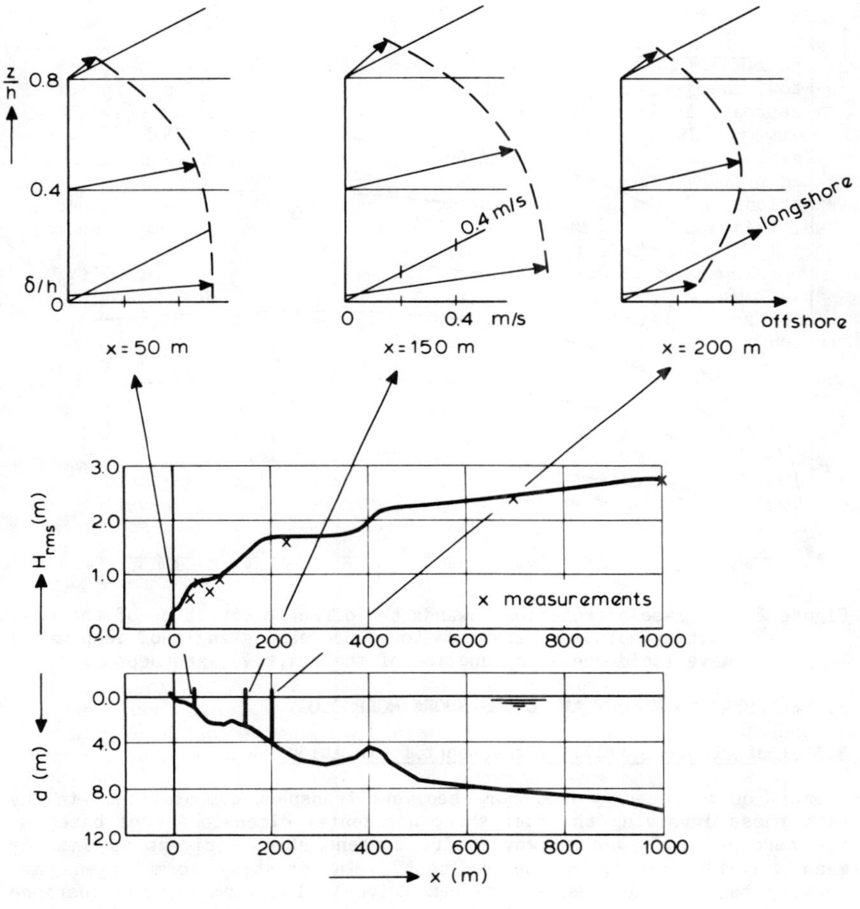

Figure 1 Model predictions of rotating mean velocity vector at Egmond beach (adapted from DVS, 1987)

The angle of the resulting velocity vector a may be derived from these aproximations straightforwardly, taking into account the refraction of the wave field over the parallel bottom contours. As a result it appears that

$$\tan\alpha = \text{function}\left(kh,\ \theta_{inc}\right)$$

which has been evaluated using Snel's law and the dispersion relationship with the result shown in Figure 2. It appears that the deviation towards the offshore direction of the resulting shearstress vector is stronger for the smaller angle of wave incidence. This is as expected since the longshore-current driving forces and resulting currents get weaker in this situation. This is in close correspondence with the field evidence gained and documented by Ingle (1966, pages 55-60), which is based on tracer experiments.

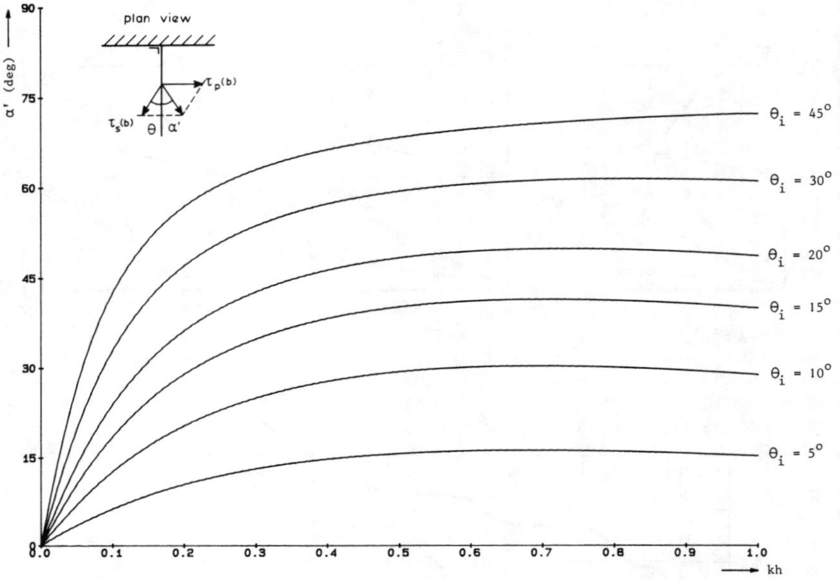

Figure 2 Increase of rotation towards the offshore direction of the resulting bottom shearstress for a decreasing angle of deep water wave incidence as a function of the relative waterdepth

3. SEDIMENT TRANSPORT AND CROSS-SHORE MORPHOLOGY

3.1 Instantaneous sediment transport formulation

Until quite recently nearshore sediment transport computations -in any rate those involving the alongshore horizontal dimension- were based on the concept of "waves or waves plus current stirr sediment up and the mean current transports the sediment". The existing formulations are usually based on an adapted uniform (river) flow formula, for instance the total bed and suspended load description after Bijker (1968) with a vertical sediment concentration description which is multiplied by the mean flow profile.

From the foregoing section it may be clear that the concept of "only the mean current transports" is considered insufficient: asymmetry-induced and wave-grouping induced effects can only be included in a formulation which is based on an instantaneous sediment transport description. The state-of-the-art in this respect is limited and with respect to accuracy not at the stage at which the above formulations are currently. It is considered essential though for the present problem. A reasonable first approximation seems to be the formulation of Bailard (1981), which is an extension of Bowen (1980) and Bailard and Inman (1981). In this formulation a distinction is made between bedload transport in a granular-fluid shear layer and suspended load transport in a layer of greater thickness, typically in the order of several centimeters. These transports are calculated as vertically integrated, but still instantaneous quantities.

Without going into the detail of the derivations and the formulations it may be schematically said that the instantaneous sediment transport is assumed to respond in a quasi-steady manner to the time-varying flow on the one hand and to the downslope gravity force on the other hand, as follows:

$$q(t) = \text{const}.u(t)\ |u(t)|^n + \text{const}\ |u(t)|^m . \partial z_b/\partial x$$

with n = 2,3, m = 3,5 and $\partial z_b/\partial x$ is the local bed slope. Note that as expected for any suitable time-varying formulation the sediment transport rate, q, contains terms proportional to some power of the near-bottom, time-varying cross-shore flow, u(t).

Field information of the last few years leads us to believe that the quasi-steady response assumption is valid for most natural beaches, but also in the laboratory it appears that under random waves the surf zone with sediments of 100 to 200 µm median grain diameter creates prevailing sheet flow conditions in the regions of significant sediment transport. Where the formulation fails is in the relatively protected areas in the trough behind breaker bars or somewhat more offshore, where one often may encounter rippled beds. For these regions a possible extension to the approach is to introduce non-steady response, for which it becomes important to include flow acceleration effects, resulting in terms with odd and even moments of the time-varying acceleration.

A further restriction to Bailards approach is the vertical integration that is performed, which assumes that the -known- vertical flow distribution is in a single plane. In the case of obliquely incident waves it was found in the foregoing that there is a vertical mean flow distribution which shows a rotation from alongshore in the upper water column towards the offshore in the lower part of the column. Clearly, a suitable formulation should also make inclusion of this effect possible.
For the moment we must approximate the rotating velocity vector effect by choosing a representative near-bottom mean velocity, which in our case of obliquely incident waves will have both a cross-shore and an alongshore component.

So, for the time being it is suggested to rely on the above approach and we note that this involves basically the evaluation of the odd and even flow moments resulting from time-averaging of the above equation. The contributions to these moments for the purely cross-shore case are identified and analyzed in several publications each extending the foregoing ones (Bowen, 1980; Bailard, 1982; Stive, 1986; Roelvink and Stive, 1988a,b).

In the case of obliquely incident waves the near-bottom mean current will be composed of both the undertow and the primary flow, which then make an angle with the oscillatory current. Bailard's approach is also then quite straightforward. In the case of a small angle of wave incidence no major principle differences are expected in the cross-shore sediment transport distribution compared to the normally incident case.
In general, the effect will be that of additional stirring for the cross-shore terms.

3.2 Cross-shore morphology of a nearly-prismatic beach

With the presented modelling approach for the near-bottom mean and oscillatory flow (section 2) and resulting sediment transport (section 3.1), which is applicable to a prismatic beach of arbitrary profile, we have available the constituents for a cross-shore sediment transport model with which profile deformations can be determined by applying the sediment balance equation.

The practical relevance of this schematized situation of a prismatic beach lies in applications dealing with the larger scale behaviour of an alongshore uniform coastal stretch which is not or only locally influenced by cross-shore structures. We assume then that on the larger -longshore- scale the local circulation systems are averaged out and that the overall behaviour of the coastal stretch may well be described by assuming that the alongshore flows are negligible for normally incident waves or that they are uniform alongshore for obliquely incident waves. Our main interest is then into the development of the cross-shore profile, since it is only in the cross-shore direction that there are sediment transport gradients present resulting in bottom topography changes.

A small extension of this approach is for situations where there is a slow variation of an alongshore feature, such as the coastline direction. This variation is then assumed to be effective on a large timescale only, such that the local coupling of the weakly dynamic alongshore processes with the strongly dynamic cross-shore processes is the only (non-dynamic) interaction that needs to be accounted for. Some consequences of this approach are that locally the primary flow follows the parallel bottom contours, while the cross-shore primary flow is neglected. Examples are weakly curved coastal stretches or weakly curved coastal regions near estuary mouths.

Some first results based on this approach lead to interesting insights into the cross-shore profile development in case there are alongshore transport gradients present. An example is given in Figure 3, where the profile developments are compared for the case of no and a weak alongshore variation in the orientation of a coastal stretch. It appears that the existence of an alongshore gradient in the sediment transport influences bar formation and the mean profile slope in the surf zone.

4 DISCUSSION

An overview is given of the developments and understanding gained sofar of the current systems and the induced sediment transport and morphology that are found in the situation of random waves normally and obliquely incident on beaches which vary not or only slowly alongshore. For a similar overview in case the horizontal variations in alongshore and in cross-shore direction are equally important reference is made to De Vriend and Ribberink (1988) in these proceedings. Some of the aspects that need extra attention from the research point of view are indicated below.

the wave field

With respect to the natural representation of the wave field it should be noted that the aspect of randomness in the wave height decay has been included but sofar the effects of wave grouping and wave directionality have been disregarded. With regard to wave grouping effects reference is made to Roelvink and Stive (1988b), who present a first modelling approach. Their results indicate that outside the surfzone wave grouping reduces wave asymmetry effects, while the opposite is true inside the surfzone. With regard to wave directionality effects reference is made to Guza and Thornton (1986), but their results are not conclusive in this respect. Lastly, it should be mentioned that the approach sofar assumes uncoupling between breaking and nonbreaking waves, using the idea of applying the theory for the separate fractions.

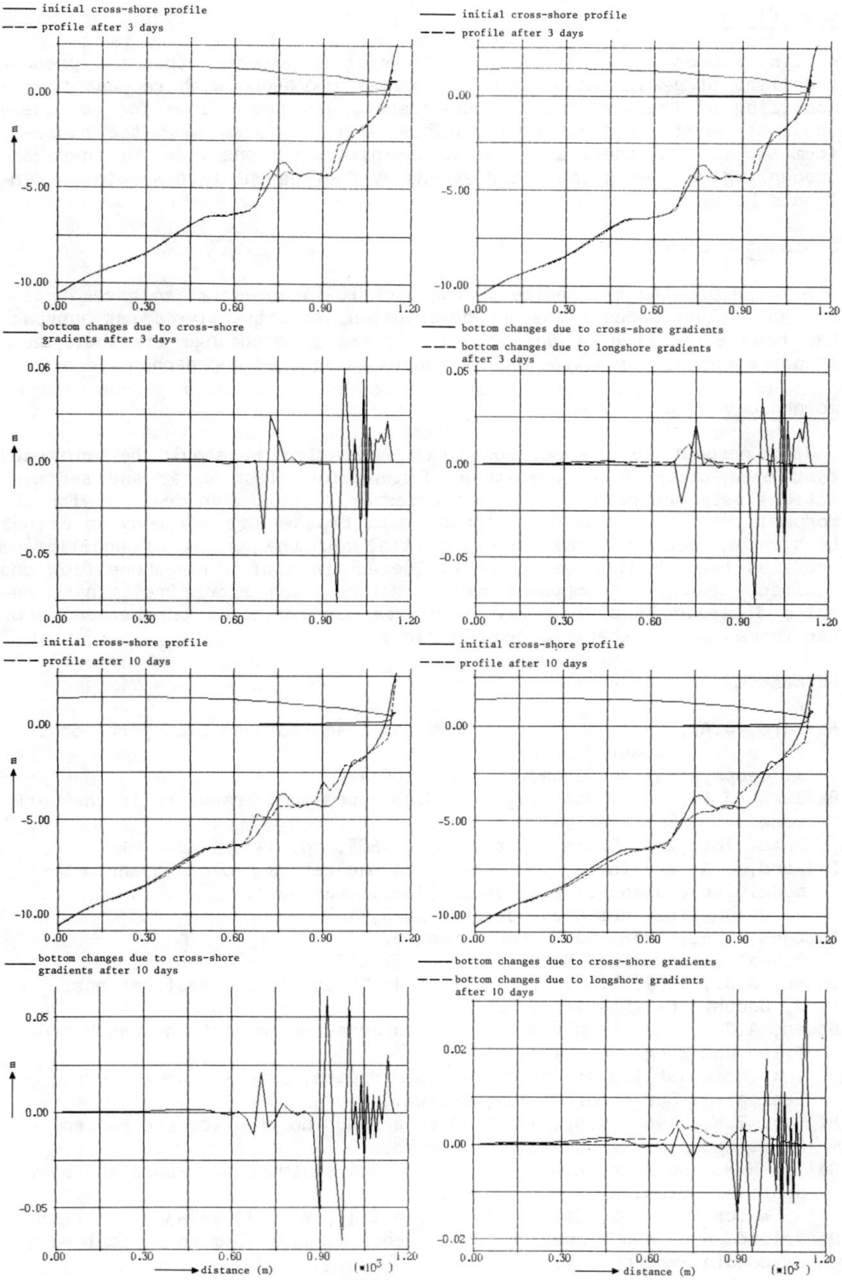

Figure 3 Cross-shore profile development and bottom changes per timestep (Δt = 0.1 days) in case of no (left) and a weak (right; radius of coastal curve of 2000 m) alongshore sediment transport variation (incident conditions H_{rms} = 2 m, T_p = 8 s, θ_{inc} = 10°)

mean flows

With respect to the cross-shore flows it is expected that the present modelling stage is satisfactory. Progress is needed with respect to the modelling of the combined longshore and cross-shore flows for obliquely incident waves. A first approach has been suggested with the quasi-3D formulation, but there are some important assumptions made. Of these the uncoupling between primary and secondary flow calculation should be mentioned here.

sediment transport

With respect to the sediment transport it is essential to have available an instantaneous transport description. An attractive first approach has been elaborated by Bailard (1981), but shortcomings are the quasi-steady response assumption and the depth-integrated approach.

morphology

With respect to the morphological evaluation it should be mentioned that even though the constituent formulations (for water and sediment motion) seem accurate this is not necessarily true when dealing with the morphological development. Progress with cross-shore morphodynamics has been made, but the predictive potential and the aspect of equilibrium profiles need further development. The addition of a longshore flow and sediment transport component needs further evaluation. Preliminary results indicate that longshore gradients can influence the mechanism of bar formation and the mean profile slope.

REFERENCES

Bailard, J.A., 1981. An energetics total load sediment transport model for a plane sloping beach.
J. Geophysical Research, 86:10938-10954.

Bailard, J.A., 1982. Modeling on-offshore sediment transport in the surf zone.
Proc. 18th Int. Conf. Coastal Eng., ASCE, pp. 1419-1438.

Bailard, J.A. and Inman, D.L., 1981. An energetics bedload transport model for a plane sloping beach; local transport.
J. Geophysical Research, 86:2035-2043.

Battjes, J.A., 1988. Surf-zone dynamics.
Annual Review Fluid Mechanics, 20:257-293.

Bowen, A.J., 1969. Rip currents, Part 1: Theoretical investigations.
J. Geophysical Research, Vol. 74.

Bowen, A.J., 1980. Simple models of nearshore sedimentation; beach profiles and longshore bars.
in: The coastline of Canada, ed. S.B.McCann, pp. 1-11.
Geological Survey of Canada, Ottawa.

Bijker, E.W., 1968. Littoral drift as a function of waves and current.
Delft Hydraulics, Publication No. 58.

Dally, W.R. and Dean, R.G., 1984. Suspended sediment transport and beach profile evolution.
J. Waterway, Port, Coastal and Ocean Eng., Vol. 110:15-33.

De Vriend, H.J. and Ribberink, J.S., 1988. A quasi-3D mathematical model of coastal morphology.
Proc. 21st Int. Conf. Coastal Eng., ASCE, to appear.

De Vriend, H.J. and Stive, M.J.F., 1987. Quasi-3D modelling of nearshore currents.
Coastal Engineering, 11:565-602.

Dyhr-Nielsen, M. and Sørensen, T., 1970. Sand transport phenomena on coasts with bars.
Proc. 12th Int. Conf. Coastal Eng., ASCE, pp. 855-866.

Guza, R.T. and Thornton, E.B., 1985. Velocity moments in nearshore.
J. Waterway, Port, Coastal and Ocean Eng., Vol. 111:235-256.

Ingle, J.C., 1966. "The movement of beach sand".
Elsevier, Amsterdam; Developments in Sedimentology, 5, 221 p.

Longuet-Higgens, M.S., 1953. Mass transport in water waves.
Philos. Transact. Roy. Soc., Ser. A, 245:535-581.

Longuet-Higgens, M.S., 1970. Longshore currents generated by obliquely incident waves.
J. Geophysical Research, 75:6778-6801.

Roelvink, J.A. and Stive, M.J.F., 1988a. Large scale tests of cross-shore sediment transport on the upper shoreface.
Symp. on Math. Mod. of Sed. Transp. in the Coastal Zone, Copenhagen, May 30-June 1.

Roelvink, J.A. and Stive, M.J.F., 1988b. Bar generating cross-shore flow mechanisms on a beach.
J. Geophysical Research, to appear.

Sonu, C.J., 1972. Field observations of nearshore circulation and meandering currents.
J. Geophysical Research, Vol. 77.

Stive, M.J.F., 1986. A model for cross-shore sediment transport.
Proc. 20th Int. Conf. Coastal Eng., ASCE, pp. 1550-1564.

Stive, M.J.F. and De Vriend, H.J., 1987. Quasi-3D nearshore current modelling: wave-induced secondary currents.
Coastal Hydrodynamics, Spec. Conf, ASCE, pp. 356-370.

Stive, M.J.F. and Wind, H.G., 1986. Cross-shore mean flow in the surf zone.
Coastal Engineering, 10:325-340.

Svendsen, I.A., 1984. Mass flux and untertow in a surf zone.
Coastal Engineering, 8:347-365.

Svendsen, I.A. and Lorenz, R.S., 1988. Velocities in combined undertow and longshore currents.
Submitted to Coastal Engineering.

Svendsen, I.A., Schaffer, H.A. and Hansen, J.B., 1987. The interaction between the undertow and the boundary layer flow on a beach.
J. Geophysical Research, 92:11845-11856.

Thornton, E.B. and Guza, R.T., 1986. Surf zone longshore currents and random waves: field data and models.
J. Physical Oceanography, Vol. 16:1165-1178.

Van Kesteren, W.G.M. and Bakker, W.T., 1984. Near-bottom velocities in waves with a current; analytical and numerical computations.
Proc. 19th Int. Conf. Coastal Eng., ASCE, pp. 1161-1177.

CHAPTER 129

Sand Ripple Geometry and Sand Transport Mechanism

Due to Irregular Oscillatory Flows

Shinji Sato*

Kiyoshi Horikawa** F. ASCE

ABSTRACT

This paper describes characteristics of sand ripples and sand transport mechanism in regular and irregular oscillatory flows on the basis of detailed laboratory measurements. A set of empirical relations were proposed to evaluate the sand ripple geometry as well as the onset of the sheet flow transport. The applicability of the proposed relationships to the irregular wave conditions with prototype scales was confirmed with existing field data.

1. INTRODUCTION

In order to understand sediment transport due to waves, it is of great significance to elucidate the geometry of bed forms as well as the mechanism of sand movement. Although a number of laboratory investigations have been performed on the sand transport due to waves, most of them have been carried out only for the condition of monochromatic oscillatory flows. Laboratory experiments for the condition of irregular oscillatory flows are necessary for better understanding of sand movement under realistic conditions encountered on natural beaches.

The sand transport mechanism and bed form geometry under irregular waves were investigated through field measurements by Inman (1957), Dingler (1974) and Nielsen (1984). Detailed laboratory measurements of the net sand transport rate under irregular waves were recently carried out by Mimura et al. (1986) and Sato and Horikawa (1986). However, the influence of wave irregularity on sand movement mechanism should be further investigated since reliable data under well-controlled irregular waves are still insufficient.

The sand transport under sheet flow condition is frequently observed on

* Research Associate, Yokohama National University, 156 Tokiwadai, Hodogaya-ku, Yokohama 240, JAPAN

** Professor, Saitama University; Professor Emeritus, University of Tokyo

natural beaches where wave-induced velocities at the bottom exceed a critical value. The evaluation of the sand transport rate under sheet flow condition is of significance since the sand transport rate becomes extremely large. Since the reproduction of the sheet flow condition is difficult in laboratory wave flumes, most of the experimental investigations on sheet flow transport have been carried out by using oscillating beds or oscillatory flow tunnels. Manohar (1955) described characteristics of bed forms as well as the disappearance of ripples on the basis of experiments by using an oscillating bed. Komar and Miller (1975) expressed the condition for the ripple disappearance in terms of the Shields parameter and the Reynolds number of oscillatory flows. Horikawa et al.(1982) evaluated sand transport rates under sheet flow condition on the basis of the measurements of the velocity and the concentration of sand particles by using an oscillatory flow tunnel. Horikawa et al.(1982) also confirmed the applicability of the criteria for the onset of the sheet flow proposed by Manohar (1957) and Komar and Miller (1975). Further investigations on the sheet flow transport were conducted by Staub et al.(1984), Bakker and van Kesteren (1986) and Sawamoto and Yamashita (1987).

The objectives of the present study are to evaluate the geometry of sand ripples in irregular oscillatory flows and to discuss the effect of wave irregularity on sand transport mechanism. Characteristics of bed forms under irregular oscillatory flows are described for a wide range of conditions corresponding to the sand ripple regime and the sheet flow regime. Empirical relationships for the sand ripple geometry as well as the critical condition for the onset of the sheet flow are proposed which are applicable to the conditions with prototype scales.

2. FACILITIES AND PROCEDURES OF EXPERIMENTS

Experiments were performed in an oscillatory flow tunnel as shown in Fig. 1. The tunnel was equipped with a hydraulically-driven piston

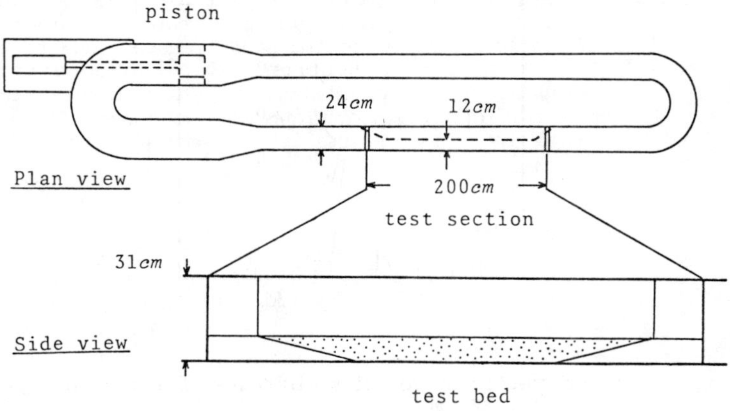

Fig. 1 Experimental facilities

controlled through an electric servo-system. The motion of the piston was controlled so as to reproduce the near-bottom water particle motion under regular and irregular waves based on the linear wave theory. Irregular oscillatory flows were simulated by assuming that the power spectrum of the irregular surface elevation was given by the Bretschneider-Mitsuyasu spectrum [Mitsuyasu (1970)]. The significant wave period was set to be in a range between 1s and 7s. Three kinds of irregular velocity histories were calculated for each wave period by assuming three different conditions for the depth to wavelength ratio h/L as illustrated in Fig. 2, in which the power spectrum density and the frequency are respectively nondimensionalized by the corresponding values at the peak of the power spectrum. Effects of the wave nonlinearity and the superimposed currents are not taken into consideration in the present study.

Two kinds of well-sorted sand were used as bed materials. The median diameters of the sands were 0.18mm and 0.56mm. The sand was filled in the test section to make initially flat test bed which was 160cm in length as shown in Fig. 3. A series of experiments (58 runs in total) were performed for the conditions of regular and irregular oscillatory flows. The width of the test section was reduced by half in some experimental runs in order to reproduce large-amplitude oscillatory

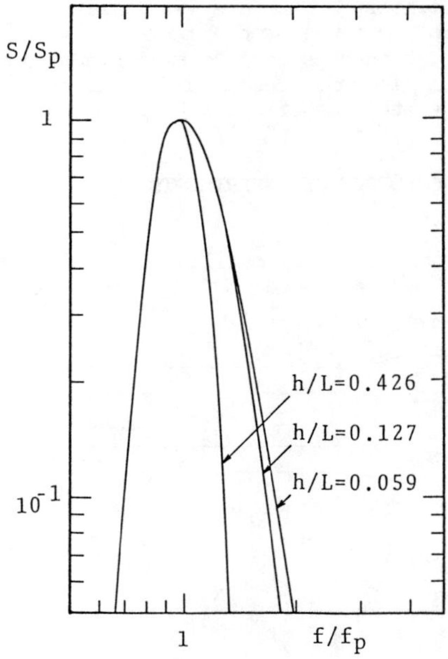

Fig. 2 Power spectra of velocity histories of irregular oscillatory flows

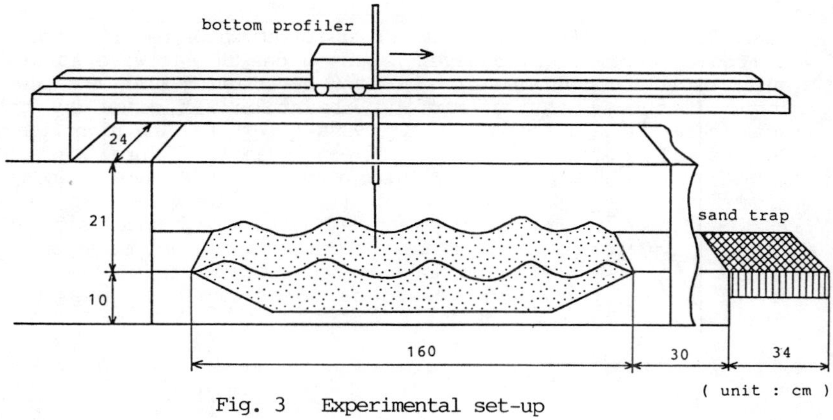

Fig. 3 Experimental set-up

flows.

Sand ripples were developed in 43 runs of the present experiments. Bottom profiles of the test bed were measured in detail by using a bottom profiler installed on the top of the tunnel. The wavelength and the wave height of the ripples were evaluated from the bottom profile measurements. In several runs of irregular oscillations, the ripple wave height were observed to vary in a range of 10% during the action of the oscillations. In such runs, ripple profiles were measured several times during the course of an experimental run. The ripple geometry was then determined as an average of these measurements. The development of bed forms and the type of sand movement for each run of the experiments were photographed through the side wall of the tunnel.

3. SAND RIPPLE GEOMETRY IN IRREGULAR OSCILLATORY FLOWS

In the analysis of several field investigations performed so far, the amplitudes of near-bottom velocity histories were calculated on the basis of the linear wave theory by using the significant wave height and period determined from the measured surface elevation. However, such evaluations sometimes induce considerable errors when the shape of the power spectrum density of near-bottom velocities is considerably different from that of the surface elevation. The great scatter of the published data in field investigations is considered to be partly attributed to the error in the evaluation of near-bottom velocities.

In order to understand the difference in the evaluation of near-bottom velocities, correlations were examined between the sand ripple geometry and hydraulic parameters evaluated in the following two methods. Figures 4 (a) and (b) show the relationship between λ/D and do/D for the condition of irregular oscillatory flows, in which λ is the ripple wavelength, do the diameter of the water particle displacement and D the sand grain diameter. The solid line indicates a regression relation for the ripple wavelength in regular oscillatory flows. The quantity do was

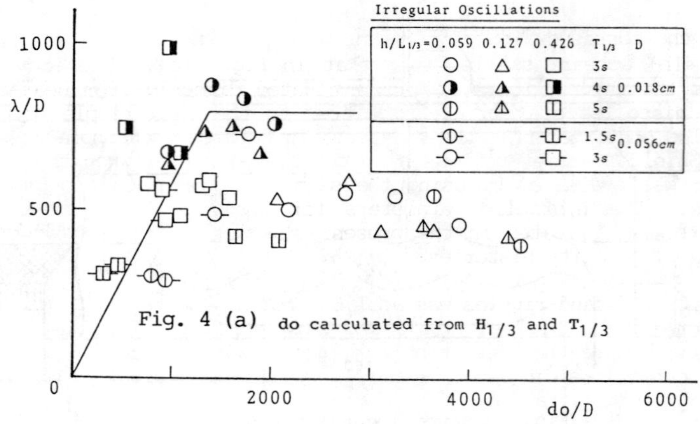

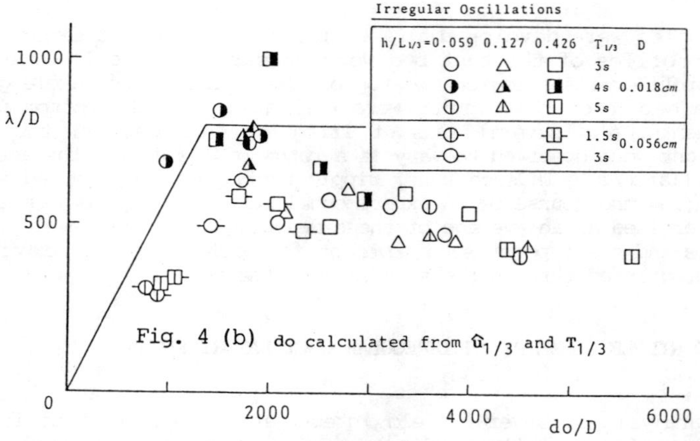

Fig. 4 Ripple wavelength in irregular oscillatory flows

evaluated in Fig. 4(a) on the basis of the linear theory by

$$do = \frac{H_{1/3}}{\sinh kh} \tag{1}$$

where *k* denotes the wave number calculated by using the significant wave period. The evaluation of *do* was based on the following equation in Fig. 4(b):

$$do = \frac{\hat{u}_{1/3} T_{1/3}}{\pi} \tag{2}$$

in which $\hat{u}_{1/3}$ was the significant amplitude of velocity histories evaluated by means of wave-by-wave analysis of the velocity history and $T_{1/3}$ was the corresponding significant wave period. The scatter of data in Fig. 4(b) appears smaller than that in Fig. 4(a). The above fact indicates that hydraulic parameters evaluated directly from near-bottom velocity histories are correlated better to the sand ripple geometry. It was also confirmed that the relationship between the ripple steepness and the Shields parameter Ψ was also correlated better when the Shields parameter was evaluated by using the significant amplitude of velocity histories. The hydraulic parameters for irregular oscillatory flows were therefore evaluated in the present study by using the significant amplitude of velocity histories.

The geometry of sand ripples was analyzed for a wide range of conditions by Brebner (1980), Nielsen (1981) and Sato et al.(1987). Sato et al. (1987) found that the sand ripple geometry was described for a wide range of conditions in terms of two hydraulic parameters, do/D and the following Shields parameter Ψ:

$$\Psi = \frac{1}{2} fw \frac{\rho \hat{u}^2}{(\rho_s - \rho)\, gD} \tag{3}$$

where fw is the friction factor, $\hat{u}$ the velocity amplitude, ρ_s and ρ densities of sand and water and g the gravity acceleration.

Figures 5 and 6 respectively show the relationship between λ/do and $do/D\Psi^{1/2}$ and the relationship between η/λ and Ψ for the ripples observed

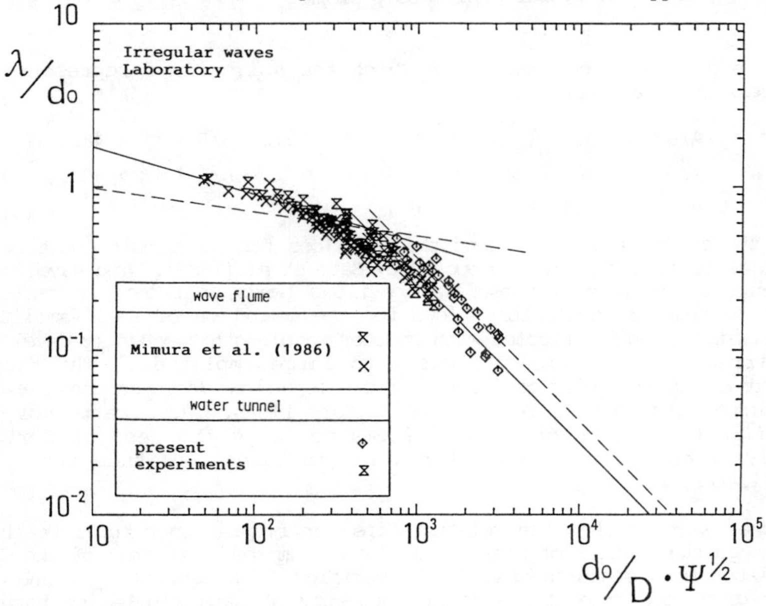

Fig. 5 Relationship between λ/do and $do/D\Psi^{1/2}$ for ripples in irregular oscillatory flows

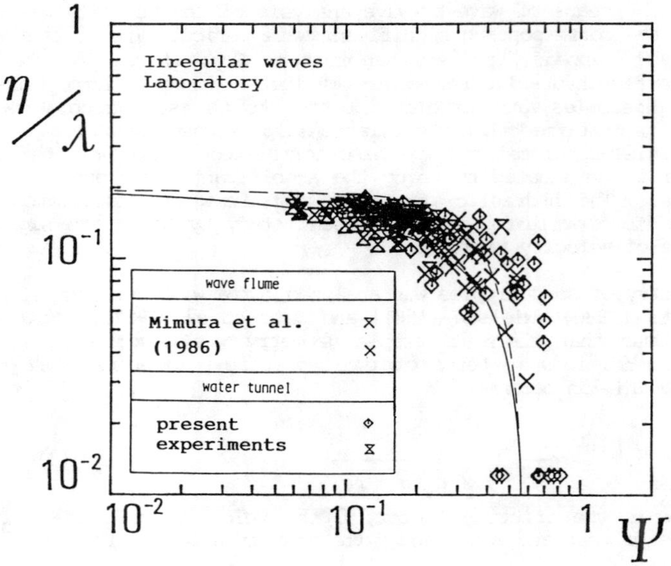

Fig. 6 Relationship between η/λ and Ψ for ripples in irregular oscillatory flows

in the present experiments, in which the solid lines represent the regression relations expressed by

$$\lambda/do = 3.55 (do/D\ \Psi^{1/2})^{-0.292} \quad \text{for } do/D\ \Psi^{1/2} < 480 \qquad (4)$$

$$\lambda/do = 280 (do/D\ \Psi^{1/2})^{-1} \quad \text{for } do/D\ \Psi^{1/2} > 480 \qquad (5)$$

$$\eta/\lambda = 0.191 (1 - \Psi/0.6) \qquad (6)$$

and the broken lines represent the relations for the ripple geometry in regular oscillatory flows proposed by Sato et al.(1987). The wavelength of ripples in irregular oscillatory flows tends to become longer than that in regular oscillatory flows for the condition of small-amplitude oscillations, while ripples with shorter wavelength frequently developed in irregular oscillatory flows with large amplitude. The ripple steepness in irregular oscillatory flows tended to decrease considerably compared with that in regular oscillatory flows. The scatter of the data in the region of large Ψ seems to imply the need of further analysis on bed form characteristics in large-amplitude irregular oscillatory flows.

Figures 7 and 8 show the relationships for ripples on natural beaches. Although the scatter of the data is large compared with that of the data in laboratory measurements, it is verified that Eqs.(4), (5) and (6) also give good predictions of the geometry of sand ripples on natural beaches.

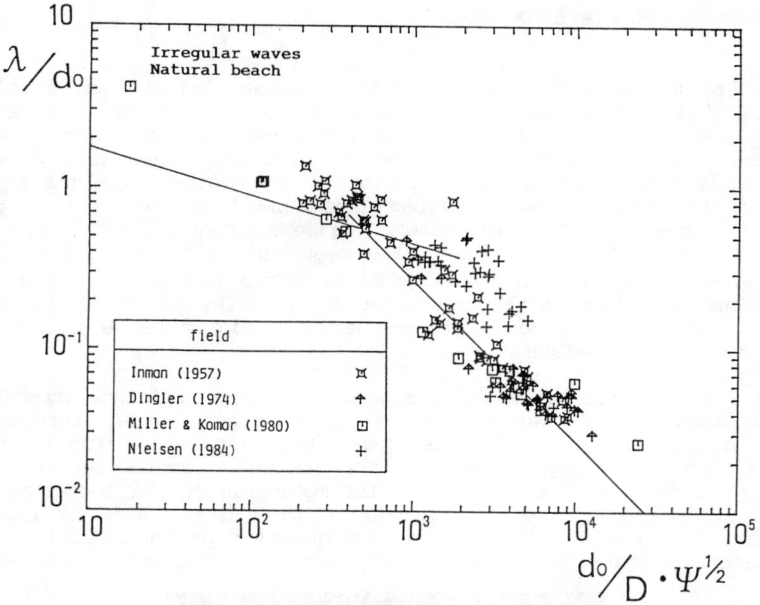

Fig. 7 Relationship between λ/do and $do/D\Psi^{1/2}$ for ripples on natural beaches

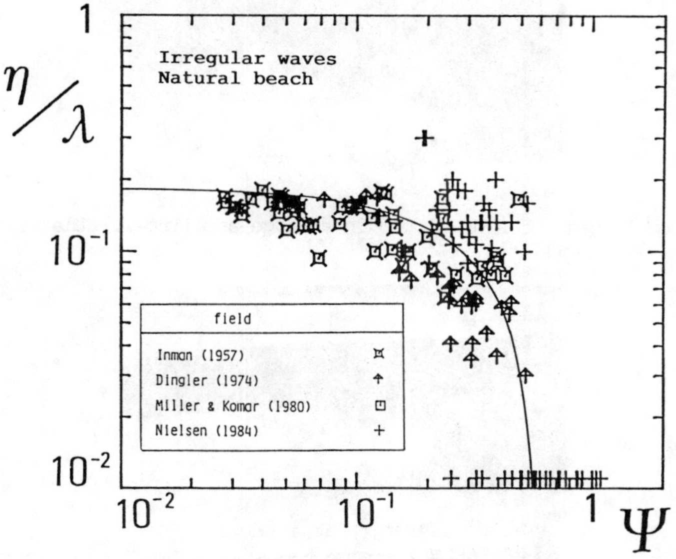

Fig. 8 Relationship between η/λ and Ψ for ripples on natural beaches

4. DISAPPEARANCE OF SAND RIPPLES

When the amplitude of oscillatory flows became large, sand ripples generally tended to disappear and sand particles began to be transported in the sheet flow. In regular oscillatory flows with large amplitudes, large-scale undulations, which were called dunes hereafter, were frequently generated as shown in Photo 1. The wavelength of the dunes was of the order of 1m. Flat bed then tended to develop with the further increase of the amplitude. The development of such dunes in regular oscillatory flows was also reported by Bosman (1981) and Sakakiyama et al. (1985). Characteristics of the dune geometry will not be further discussed in the present study since the length of the test bed in the present experiments was not considered to be enough to reproduce undisturbed dunes.

Photo 2 illustrates bed forms developed in a large-amplitude irregular oscillatory flow. Dunes were rarely generated under irregular oscillatory flows. On the other hand, small-scale two-dimensional ripples were generally observed in the transition region between the ripple and the non-ripple regime. The mechanism of the formation of these small-scale ripples was similar to the initial stage of ripple formation on a flat bed, that is, the geometry of these small two-dimensional ripples in large-amplitude irregular oscillatory flows were

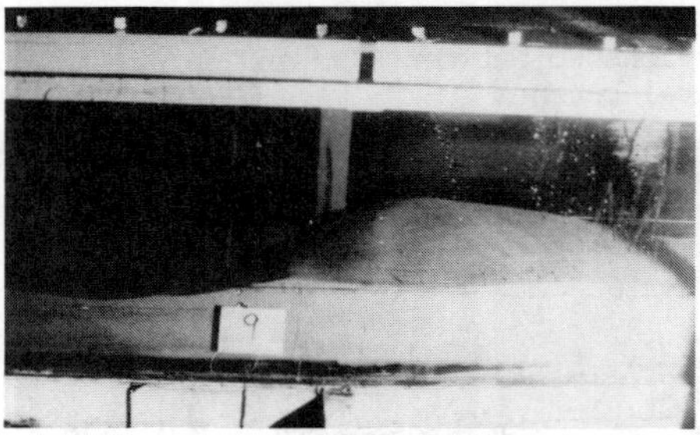

Photo 1 Bed forms developed in a large-amplitude regular oscillatory flow

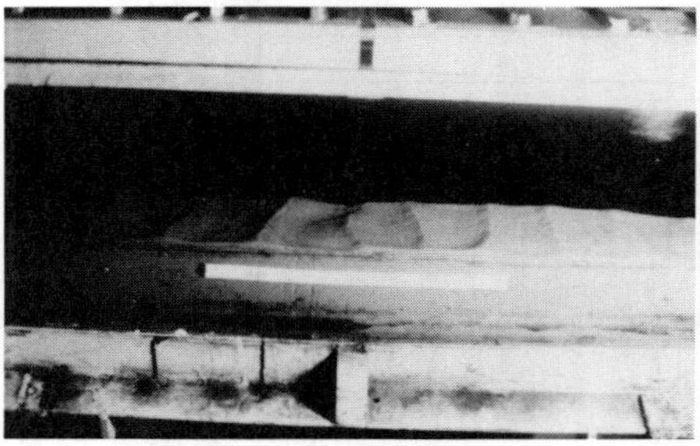

Photo 2 Bed forms developed in a large-amplitude irregular oscillatory flow

considered to be maintained by the iteration of conflicting actions of the ripple elimination due to large waves and the ripple formation due to relatively small waves.

Figure 9 shows the classification of bed forms under regular oscillatory flows. Open symbols indicate the presence of sand ripples and solid symbols indicate the disappearance of sand ripples. Data obtained by Bosman (1981) and Horikawa et al. (1982) were also plotted in the figure. Data were classified in the non-ripple regime when the ripple steepness was less than 0.05. Dunes were also classified in the non-ripple regime in the present study since sand particles on the duned bed were observed to be transported in sheet flow over the whole domain. The broken line in Fig. 9 represents the criterion for the onset of the sheet flow proposed by Komar and Miller (1975), in which ν is the kinematic viscosity of water. The relation proposed by Manohar (1955) was also examined for both regular and irregular oscillatory flows, which revealed that experimental data were not well classified by these criteria. The onset of sheet flow was therefore reanalyzed in the present study by using two parameters, do/D and Ψ, which were found to be essential parameters in describing the sand ripple geometry.

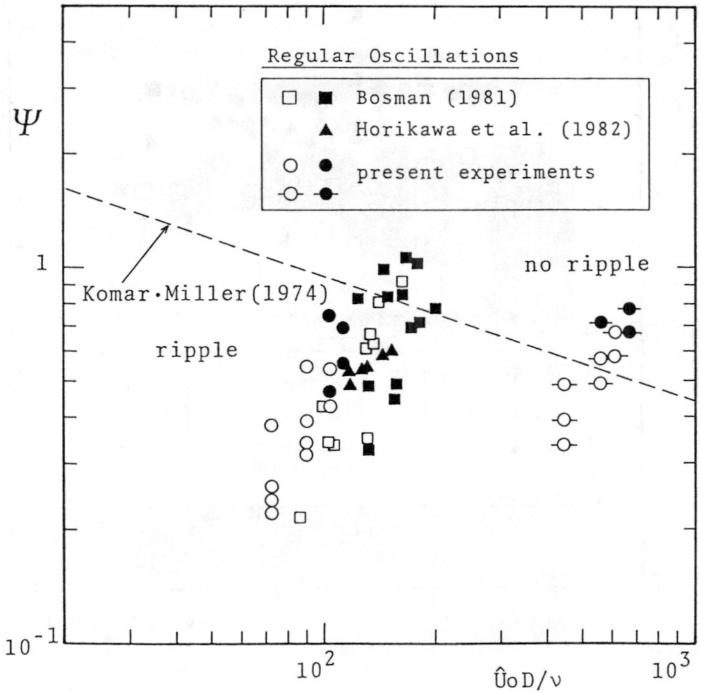

Fig. 9 Disappearance of ripples in regular oscillatory flows

Figure 10 shows the classification of bed forms under regular oscillatory flows in terms of do/D and Ψ. It is noticed that the disappearance of ripples was well described by the broken line in the figure which was expressed by

$$\Psi = 7.8 \, (do/D)^{-1/3} \qquad (7)$$

Figure 11 shows the classification of bed forms under irregular oscillatory flows. Although the number of data was not sufficient, the critical condition for the disappearance of ripples in irregular oscillatory flows appeared to be described by Eq. (7).

Figure 12 shows the classification of bed forms on natural beaches. Values of do and Ψ were calculated on the basis of the linear theory by using the significant wave height and period of measured surface elevations. Although the scatter of the data is large, Eq. (7) appears to give an appropriate boundary between the ripple and the non-ripple regimes.

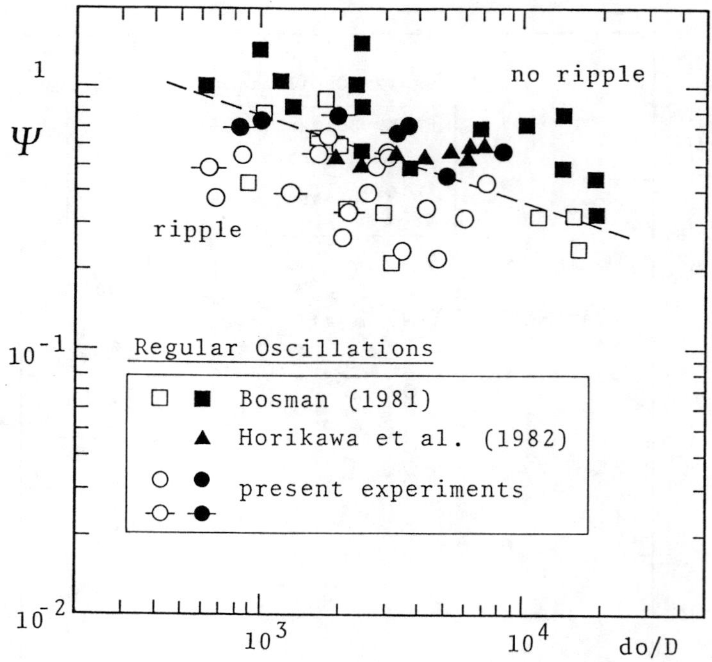

Fig. 10 Disappearance of ripples in regular oscillatory flows

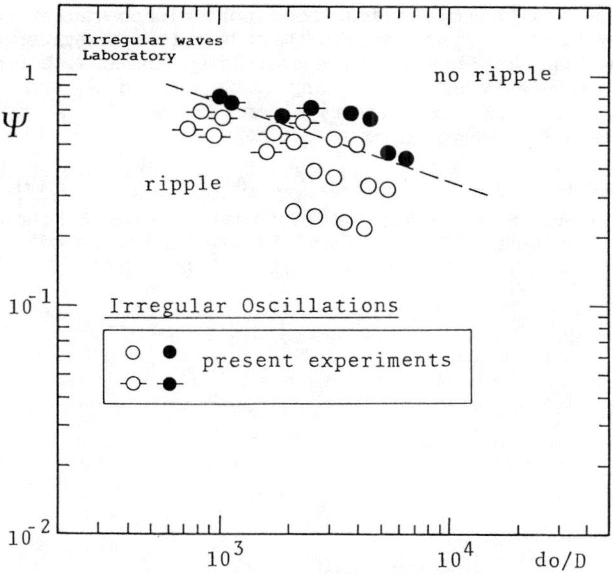

Fig. 11 Disappearance of ripples in irregular oscillatory flows

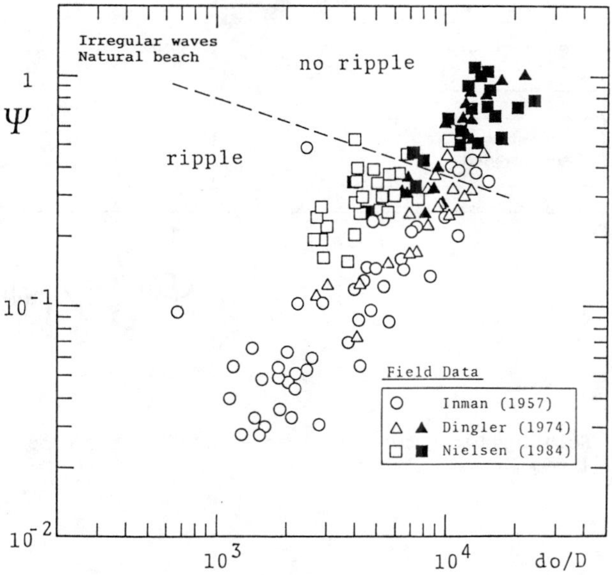

Fig. 12 Disappearance of ripples on natural beaches

Since the critical condition for the ripple disappearance was fairly described by Eq. (7), it is expected that the ripple steepness in large-amplitude oscillatory flows might be correlated better with a parameter $\Psi(do/D)^{1/3}$. Figure 13 shows the relationship between η/λ and $\Psi(do/D)^{1/3}$ for ripples in irregular oscillatory flows. The solid line represents a regression relation expressed by

$$\eta/\lambda = 0.170 \left(1 - \left(\Psi \left(do/D \right)^{1/3}/7.8 \right)^2 \right) \quad (8)$$

Figure 14 shows the steepness of ripples on natural beaches. The correlation of data in Figs. 13 and 14 in the region of large Ψ is considerably improved compared with that in Figs. 6 and 8.

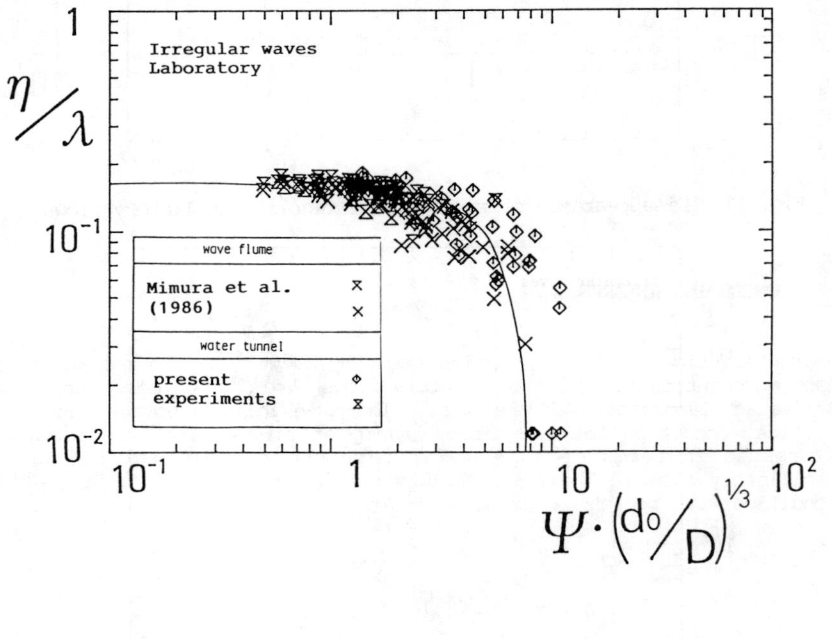

Fig. 13 Relationship between η/λ and $\Psi(do/D)^{1/3}$ for ripples in irregular oscillatory flows

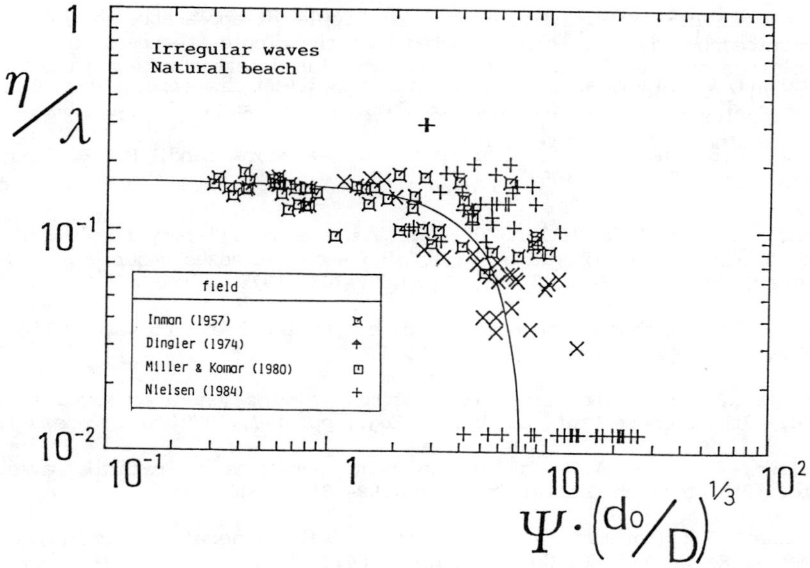

Fig. 14 Relationship between η/λ and $\Psi(do/D)^{1/3}$ for ripples on natural beaches

5. CONCLUDING REMARKS

Characteristics of the sand ripple geometry as well as the mechanism of sand movement in irregular oscillatory flows were investigated through a series of laboratory measurements. The sand ripple geometry and the critical condition for the disappearance of ripples were consistently expressed in terms of do/D and Ψ for both regular and irregular oscillatory flows. The applicability of the proposed relations to the condition with prototype scales was also confirmed with data on natural beaches.

REFERENCES

Bakker, W.T. and W.G.M. van Kesteren: The dynamics of sediment transport under oscillating sheetflow condition, Proc. 20th Conf. on Coastal Eng., pp.940-954, 1986.

Bosman, J.J.: Bed behavior and sand concentration under oscillatory water motion, Delft Hydraulics Lab., M1695-1, 49p., 1981.

Brebner, A: Sand bed-form length under oscillatory motion, Proc. 17th Conf. on Coastal Eng., pp.1340-1343, 1980.

Dingler, R.J.: Wave formed ripples in nearshore sands, Ph. D. Thesis, Univ. of California, 136p., 1974.

Horikawa, K., A. Watanabe and S. Katori: Sediment transport under sheet flow condition, Proc. 18th Conf. on Coastal Eng., pp.1335-1352, 1982.

Inman, D.L.: Wave generated ripples in nearshore sand, B.E.B. Tech. Memo., No. 100, 42p., 1957.

Komar, P.D. and M.C. Miller: The initiation of oscillatory ripple marks and the development of plane-bed at high shear stresses under waves, J. Sediment. Petrol., Vol. 45, No. 3, pp.697-703, 1975.

Manohar, M.: Mechanics of bottom sediment movement due to wave action, B.E.B. Tech. Memo., No. 75, 121p., 1955.

Mitsuyasu, H.: On the growth of spectrum of wind-generated waves (2), Proc. 17th Japanese Conf. on Coastal Eng., pp. 1-7, 1970(in Japanese).

Nielsen, A.F. and A.D. Gordon: Sediment responses to natural waves, Proc. 19th Conf. on Coastal Eng., pp.1799-1815, 1984.

Nielsen, P.: Dynamics and geometry of wave-generated ripples, J. Geophys. Res., Vol. 80, No. C7, pp.6467-6472, 1981.

Nielsen, P.: Field measurements of time-averaged suspended sediment concentrations under waves, Coastal Eng., Vol. 8, No.1, pp.51-72, 1984.

Sakakiyama, T., T. Shimizu, R. Kajima, S. Saito and K. Maruyama: Sand ripples generated by prototype waves in a large wave flume, Coastal Eng. in Japan, Vol. 28, pp. 147-160, 1985.

Sato, S., K. Mitani and A. Watanabe: Geometry of sand ripples and net sand transport rate due to regular and irregular oscillatory flows, Coastal Eng. in Japan, Vol. 30, No. 2, pp. 89-98, 1987.

Sato, S. and K. Horikawa: Laboratory study on sand transport due to asymmetric oscillatory flows, Proc. 20th Conf. on Coastal Eng., pp. 1481-1495, 1986.

Sawamoto, M. and T. Yamashita: Sediment transport in sheet flow regime, Coastal Sediments 87, pp. 415-423, 1987.

Staub. C., I.G. Jonsson and I.A. Svendsen: Variation of sediment suspension in oscillatory flow, Proc. 19th Conf. on Coastal Eng., pp. 1335-1352, 1984.

CHAPTER 130

The dynamic response of shingle beaches to random waves

K A Powell*

Abstract

An extensive laboratory investigation into the behaviour of shingle beaches has been undertaken using a large random wave flume. The study utilised a lightweight material scaled to reproduce the correct permeability of the beach, and the correct threshold and relative magnitude of the onshore/offshore movement. Results are presented describing both the wave reflection characteristics of the beach and the probabilistic distribution of wave run-up crests on the foreshore. Where possible the laboratory results are validated against field data.

1 Introduction

Although relatively scarce on a worldwide basis shingle, or gravel, beaches (D_{50} = 10 - 60mm) are a common feature around the UK coastline. However it is only in recent years that their considerable merits as coast protection structures have been fully recognised, and only over the last few years that this belated recognition has been transformed into a more widespread engineering application. This application is still restricted, however, not only by a paucity of information, regarding shingle beach processes under wave action, but also by a general lack of understanding as to how information which is currently available should be applied to a particular problem. In response to this situation an extensive series of model tests have been undertaken in the UK in recent years using both regular (Powell, 1986) and random waves. This paper summarises the procedures and results of the random wave investigation with particular emphasis on the wave reflection coefficients and wave run-up distributions.

2 Scaling Criteria for Model Beach Sediment

In order that a mobile bed physical model may accurately simulate natural beach processes it is necessary to ensure that the sediment used in the model is representative of that occurring in nature. For shingle beaches the model sediment should ideally satisfy three main criteria:

a) The permeability of the shingle beach should be correctly reproduced.

Higher Scientific Officer
* Maritime Eng. Dept., Hydraulics Research Ltd, Wallingford, OX10 8BA, UK

b) The relative magnitudes of the onshore and offshore motion should be constant.

c) The threshold of motion should be correctly scaled.

The first of these criteria basically governs the beach slope, the second determines whether the beach will erode or accrete under given wave conditions, and the third determines the wave velocity at which sediment motion will begin.

Yalin (1963) published a paper describing a method for modelling shingle beaches with the correct permeability and drag forces. For the permeability he stated that in an undistorted model the percolation slope must be identical to that of the prototype beach

i.e. $\lambda_V{}^2 \; \lambda_K \; / \; \lambda_D = 1$ (1)

where λ is the model scale
V is the flow velocity through the voids
K is permeability, a function of the voids
Reynolds number.... $V \, D_{10} \; / \; \nu$
D is sediment diameter
and ν is kinematic viscosity

Assuming that the model is operated according to Froude's Law this equation can be re-written as

$$\lambda_D = \lambda K_p \; / \; K \; (Re_p \; / \; \lambda^{\frac{1}{2}} \; \lambda_D) \tag{2}$$

where the subscript p refers to prototype values. Provided that K_p, Re_p and the form of the function K ( ) are known this equation can then be solved by successive approximation to define the particle size for the model sediment, for a given model scale.

The correct reproduction of the relative magnitudes of the onshore/offshore sediment motion requires the similitude of the dimensionless fall velocity parameter, $H_b/\omega T$, (see Shore Protection Manual, 1984).

i.e. $\lambda H_b \; / \; \lambda_\omega \; \lambda_T = 1$ (3)

where H_b is the breaking wave height
ω is the sediment fall velocity
and T is the wave period

However the settling velocity for a sphere may be approximated by

$$\omega = \left(1.33 \; gD \; (\rho_s - \rho_f) \; / \; C_D \; \rho_f\right)^{\frac{1}{2}} \tag{4}$$

where ρ_s and ρ_f are specific gravities of the sediment and fluid respectively, and C_D is the drag coefficient for the settling particles (Rouse, 1950).

For a Froudian model therefore, assuming that the beach slope is correctly reproduced, equations 3 and 4 may be combined to yield an expression for the specific gravity of the model sediment:

$$\lambda\rho_s' = \lambda\, \lambda_{C_D} / \lambda_D \qquad (5)$$

where $\rho_s' = (\rho_s - \rho_f) / \rho_f$ (6)

and λ_{C_D} is given as a non-linear function of the sediment particle Reynolds number by

$$\lambda_{C_D} = C_{D_p} / C_D\,(Rep / \lambda^{\frac{1}{2}}\, \lambda_D) \qquad (7)$$

Again, if C_{D_p} and Re_p are known, and λ_D has also been determined (i.e. equation 2), then equations 5, 6 and 7 can be solved for ρ_s, the specific gravity of the model sediment.

A second expression for the specific gravity of the model sediment is obtained through consideration of the threshold of motion of the sediment particles. For oscillating flow Komar and Miller (1973) proposed that for sediment sizes greater than 0.5mm, which is usually the case for shingle beach models, the threshold of movement would be defined by the expression,

$$U_m^2 / \rho_s' gD = 0.46\,\pi.\,(d_o/D)^{0.25} \qquad (8)$$

Where U_m is the peak value of the near bed orbital velocity at the threshold of motion and d_o is the near-bed orbital diameter. Re-working equation 8 and assuming a Froudian model yields the expression,

$$\lambda\rho_s' = \left(\frac{\lambda}{\lambda_D}\right)^{3/4} \qquad (9)$$

It should be noted, however, that equation 7 was originally developed for predicting the threshold of movement of fully submerged gravel on a horizontal sea-bed subject to non-breaking waves. This is a very different situation to an emergent, sloping shingle beach subject to breaking waves, so the accuracy of the equation cannot be assured.

Generally equations 5 and 9 give conflicting requirements for the specific gravity of the model sediment, ρ_s, and one or other of the equations usually needs to be relaxed. This complication in the modelling of beach sediments is further compounded by the fact that there is only a very limited range of specific gravities amongst the readily available materials. Frequently, therefore, the selection of the model sediment is governed as much by availability as by theoretical considerations.

For the present study the scaling requirements have been plotted in graphical format in Figure 1. These may be satisfactorily met at a scale of 1:17 by the use of crushed anthracite (coal) which has a specific gravity of 1:39, and is readily available in a range of gradings. At the selected scale the specific gravity requirements for the model sediment, based on the threshold and direction of motion criteria are respectively 1.45 and 1.35, which are acceptably close to the value for anthracite.

3 Laboratory procedures and test programme

During the course of the study four different sizes, and two different gradings, of beach material (selected to cover the range of shingle sizes commonly found around the UK coast) were used to build the model beaches. The beaches themselves were always constructed at a 1:7 slope but up to 5 different depths of beach material were tested. Each beach was subjected to up to 29 wave spectra of the JONSWAP type covering the following range of conditions:

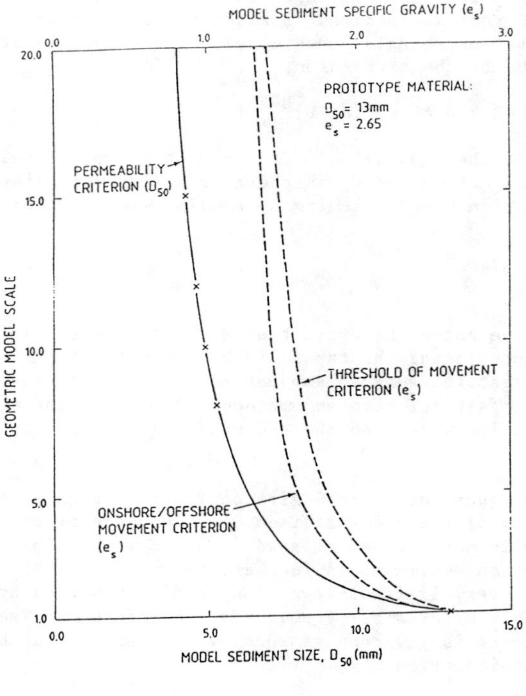

Fig 1 Sediment scaling criteria

$0.05m \leqslant H_s \leqslant 0.175m$

$0.8s \leqslant T_z \leqslant 2.7s$

$0.005 \leqslant H_s/L_z \leqslant 0.06$

The depth of water at the toe of the model was kept constant at 0.8m throughout the course of the study.

Amongst the measurements recorded during the investigation were the wave reflection coefficients for the model beaches and the wave run-up exceedance distributions on the beach face.

Wave reflections were recorded using 3 wave probes located at set positions along the centreline of the flume. The incident and reflected wave spectra were not directly measured but were calculated using an analysis program devised by Gilbert and Thompson (1978) and based on the method of Kajima (1969). The analysis method calculates values of K_r over a wide range of frequencies, but the procedure is only valid over a restricted band related to the probe spacing. For the current study, the use of three wave probes effectively provided three different probe spacings thus allowing a wide range of frequencies to be covered.

Because beaches are constantly adjusting their form in response to the incident wave conditions, it seems likely that the proportions of wave energy reflected or dissipated may also vary as the beach gradually evolves. To test this hypothesis three sets of measurements were made per experimental run, after allowing the beach an initial development period of 500 T_z.

Attempts to record wave run-up distributions in laboratory beach models usually meet with two main problems.

1) The mobility of the beach - which tends to restrict the use of instrumentation on the beach face itself.

2) The presence of edge effects along the side walls of the flume - which can affect visual recordings taken at beach level.

To overcome these problems a simple method was developed for measuring wave run-up distributions along the centre line of the flume. This involved blacking out the half of the flume furthest from the observer whilst lighting the front half from above. The image of an illuminated marker board located outside the flume, and referenced to still water level, was then reflected into the flume and projected on to the centre line boundary, between the light and dark sections of the flume. The marker board was drawn up with twelve numbered bands in such a way that its image appeared correctly orientated. With a little practise an observer was able to record the total number, and hence proportion, of wave run-ups exceeding specified levels against the image of the marker board. Initial proving of the method demonstrated a high degree of repeatability, which appeared to be independent of the observers involved.

4 Wave Run-up distributions

Generally five wave run-up recordings, of $300T_z$ duration, were taken for each test, with each recording being separated by a 200 T_z interval. Prior to any measurements being taken the profile was allowed to evolve naturally for a period of $500T_z$. This was generally long enough for the major profile features to develop and subsequent analysis of the results showed no evidence of any duration dependent trends.

On completion of a test the records were processed to provide, firstly, the cumulative number of wave run-ups exceeding a specified level per record; and then, secondly, the combined exceedance probability of wave run-up, for those levels, for all five sets of data. The resulting probability distribution was then used in all subsequent analysis.

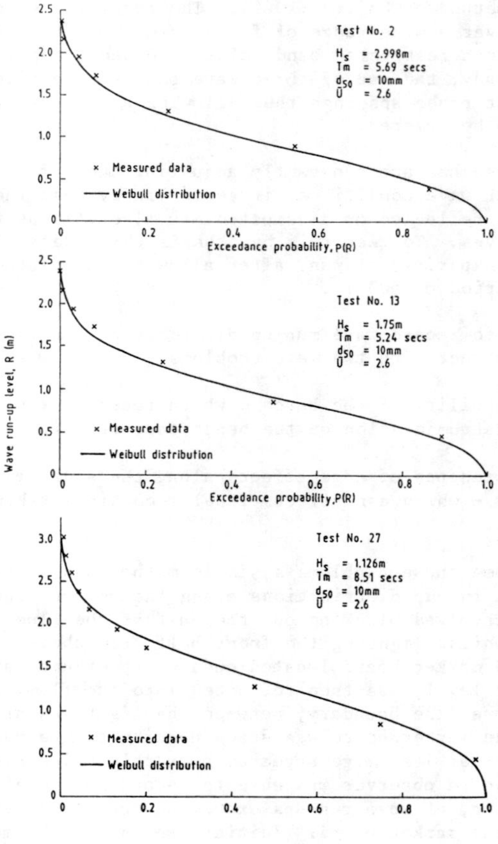

Fig 2 Typical wave run-up exceedance distributions

Typical wave run-up distributions recorded during the model tests are given in Figure 2. These measured distributions have been tested against theoretical Weibull and Rayleigh distributions of the form:

Weibull : $P(R) = \exp\,(-B\,(R-C)^{A})$ (10)

Rayleigh : $P(R) = A\,\exp\,\left(-\frac{BR^2}{2}\right)$ (11)

where A and B are curve fitting coefficients
C is a lower limiting value of R
R is a specified level relative to still water level
and P(R) is the probability of a wave run-up exceeding R

In this particular instance the lower limiting value C in the Weibull distribution has been taken as zero, and the run-up has therefore been measured relative to still water level rather than to an arbitrary mean water level which would necessarily include the component due to wave set-up.

Generally the differences between the two distributions are small when compared to the measured data. However the Weibull distribution usually returns slightly better correlation coefficients and hence provides a better fit to the data, particularly over the lower end of the range. The Weibull distribution was therefore taken as providing the best description of wave run-up on the model beaches. This is confirmed by Figure 2 where the theoretical Weibull distribution shows good agreement with the model data.

Detailed analysis of the results suggests that the Weibull coefficient B is a function of both wave height, H_s, and mean sea steepness, H_s/L_z. The precise form of the relationship is given in Figure 3 from which regression analysis yields:

$$B = 0.3\left[H_s^{1.4}\,\exp\,(-30.0\,H_s/L_z)\right]^{-1.6} \qquad (12)$$

with a correlation coefficient, r = 0.96.

As may be seen, B is therefore proportional to H_s/L_z but inversely proportional to H_s. Thus for a constant value of A, increasing B (i.e. increasing sea steepness or decreasing wave height) reduces the probability of the wave run-up exceeding a specified level.

In contrast to coefficient B, the values of coefficient A appear to be largely independent of wave climate with a mean value of 2.2 and a standard deviation, $\sigma = 0.22$. Combining this mean value with equations 10 and 12 yields an expression for determining the probable distribution of wave run-ups on a shingle beach, relative to still water level,

i.e. $P(R) = \exp\,(-BR^{2.2})$ (13)

where the value of B is given by equation 12.

Note that although equation 13 does not include any allowance for variations in beach sediment size, this was not found to be a serious handicap. Indeed throughout the test series no dependency between wave run-up and the beach material characteristics could be observed.

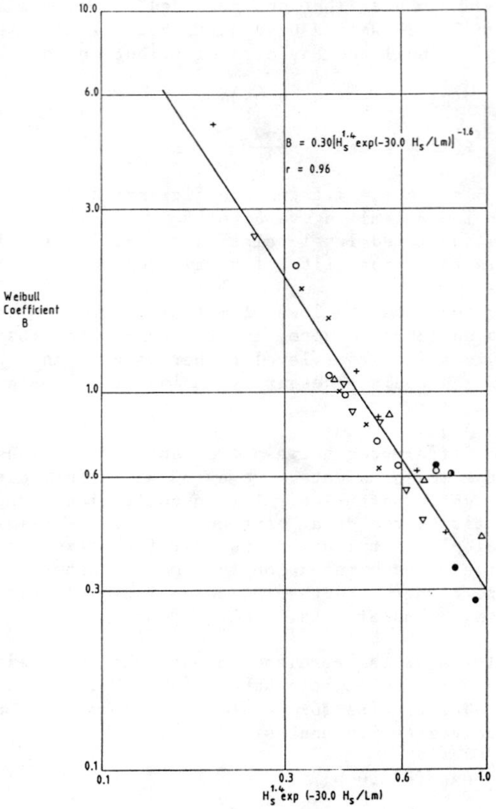

Fig 3 Weibull distribution - coefficient B

The applicability of equation 13 to the test results is confirmed by Figure 4. Here data gathered during field measurement exercises at Chesil Beach and Hurst Castle Spit (on the south coast of the UK) is compared with the predictions of equation 13 for values of P(R) = 0.5 and 0.02. As can be seen the agreement between the predicted and measured values is good.

Observations made during the course of the test programme suggested that, even when fully developed, the beach crest would be overtopped by a small percentage of the wave run-ups. Analysis of the data, based on the assumption that it fitted a Weibull distributions, allowed this percentage to be estimated for each test condition. The resulting exceedance probabilities, calculated for a crest height formed after 3000 waves, suggest that generally less than 3% of the wave run-ups overtop the beach crest, with the mean probability of overtopping $P(R \geqslant h_c) = 0.015 \pm 0.011$. No systematic variations, based on wave conditions, are apparent within the results.

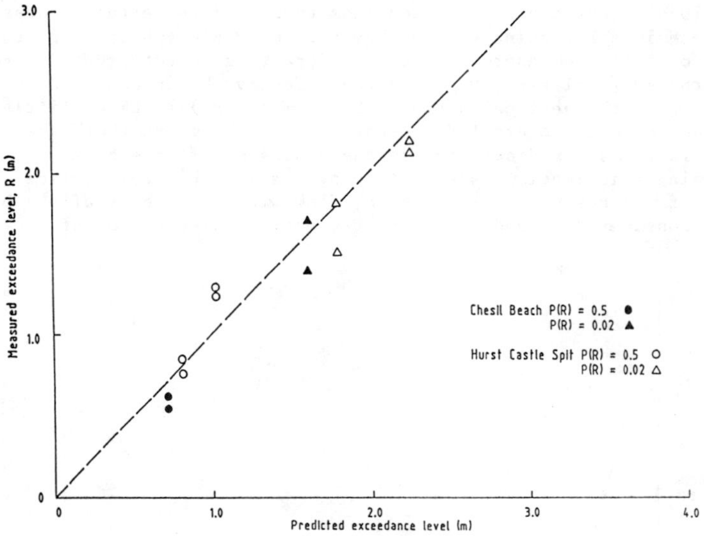

Fig 4 Comparison of measured and predicted run-up exceedance levels

5 Wave reflection coefficients

The effectiveness of a shingle beach in dissipating wave energy is an important measure of its usefulness as a coast protection structure. This is particularly true of beaches used as energy absorbing structures in enclosed waters (i.e. marina's etc) where high levels of reflected energy can have undesirable consequences for small vessels.

During the course of the present study an extensive series of measurements were made of wave reflection for a variety of wave and beach conditions. Generally three sets of reflection measurements were collected for each test, after first allowing a period of $500T_z$ for the main features of the beach to evolve. The reflection measurements were taken between 500-1000, 1500-2000 and $2500\text{-}3000T_z$ from the commencement of the test. On-line analysis of the results produced details of the incident and reflected wave spectra together with values for the reflection coefficients, both for discrete frequency bands within the spectrum and for the wave spectrum as a whole.

Values of the characteristic reflection coefficient K_r (defined as $(Sr/Si)^{\frac{1}{2}}$ where Sr and Si are respectively the reflected and incident energy densities) calculated for each wave spectrum were found to be related to the incident spectral sea steepness, H_s/L_z. This relationship is depicted in Figure 5 for all K_r values obtained. The resulting trend shows that the proportion of wave energy reflected by a shingle beach is reasonably constant, at around 10%, for all values of sea steepness greater than 0.02 (i.e. breaking wave conditions). For sea steepnesses less than 0.02 the effectiveness of the beach in dissipating wave energy reduces rapidly (i.e. higher levels of reflected energy). It is interesting to note that Figure 5 shows the material size, D_{50}, and the effective beach depth, D_B/D_{50}, (where D_B is the vertical depth of beach material) to be of little consequence to the overall K_r trend. This may suggest that wave energy is primarily dissipated in the processes of wave breaking, and overcoming frictional losses in flow over and within the surface layers of the beach. If this is so, flow within the body of the beach can be considered to add little to the overall dissipation of wave energy.

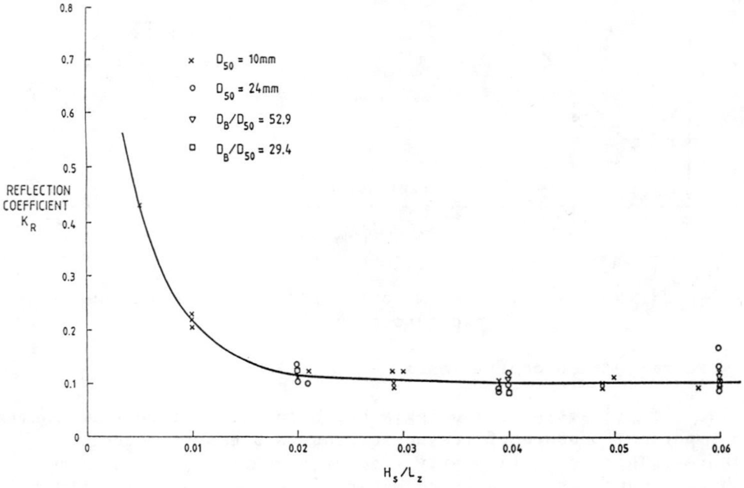

Fig 5 Wave reflection coefficients

6 Conclusions

The wave run-up and wave reflection characteristics of shingle beaches have been investigated using a scaled physical model in a random wave flume. The correct response of the model beach has achieved through the accurate reproduction of both the beach permeability, and the threshold and direction of sediment motion. So far the conclusions arising from the study may be summarised as:

1) The probabilistic distribution of wave run-up crests on a shingle beach is most closely described by a Weibull distribution.

2) The probability of the run-up exceeding any given level on a shingle beach can be adequately determined from equations 12 and 13.

3) Using these equations the proportion of wave run-ups exceeding the wave-formed beach crest is found to be generally less than 2%, regardless of beach material size or the wave conditions.

4) The proportion of normally incident wave energy reflected by shingle beaches is nearly constant, at around 10%, regardless of the size, grading or active depth of the beach material.

5) For wave steepnesses greater than 0.02 the proportion of energy dissipated is also virtually independent of the incident wave conditions.

If should be noted that these results apply only to shingle beaches subject to normally incident wave action. Further research is underway to determine the effects of oblique wave attack on the wave run-up and reflection parameters.

7 Acknowledgements

This research was funded by the Ministry of Agriculture, Fisheries and Food under contract no. CSA 1034.

8 References

1. Coastal Engineering Research Centre. 1984. Shore Protection Manual. 2 Vols, 4th edition.

2. Gilbert G and Thompson D M. 1978. Reflections in random waves - the frequency response function method. HRS Report IT 137, Wallingford.

3. Kajima R. 1969. Estimation of an incident wave spectrum under the influence of reflection. Coastal Eng. in Japan, Vol. 12.

4. Powell K A. 1986. The hydraulic behaviour of shingle beaches under regular waves of normal incidence. PhD Thesis. University of Southampton.

5. Rouse H. 1950. Fundamental principles of flow, in Engineering Hydraulics, edited by H Rouse, Wiley, New York. pp. 1-135.

6. Yalin M S. 1963. A model shingle beach with permeability and drag forces reproduced. 10th Congress Int. Ass. Hyd. Research.

CHAPTER 131

ESTIMATES OF CROSS-SHORE BEDLOAD AND BED CHANGES

Zbigniew PRUSZAK[1] and Ryszard B. ZEIDLER[2]

ABSTRACT

Four series of tracer studies employing radioisotopic sand, tracked remotely from land, were carried out at the IBW PAN's coastal research station at Lubiatowo. Incipient motion of single grains with velocity v_{cr1} has agreed with Komar and Miller's (1974) formula, and has been 5...8 times smaller than v_{cr2}, the threshold velocity of water inducing bulk motion of sediment. The ratio of average nearbed water velocity, $\bar{v}$, to the effective speed of tracer sand, in the inshore zone, $\bar{v}_s$, is estimated about 100...1,000. The cross-shore sediment transport rates measured are ten times smaller than those predicted by Pruszak's (1987) energetics formula, the discrepancy being due to the real field phenomena of mixing and overburdening in a thick bed layer. Our wave flume and wave tank studies with movable bed point to the importance of initial slope in shore evolution. The laboratory findings on bed variability are extended by our 1987 field data obtained with an ultrasonic setup. Intensive cyclic depth changes are accompanied by varying transport rates computed from continuity equation.

1. INTRODUCTION

Sediment transport in the coastal zone, particularly in the bedload mode, still requires further analytical studies and experimental data. Field facilities of our Coastal Research Station at Lubiatowo and radiotracers were used to quantify sediment motion. An ultrasonic probe was devised and used at Kamchiya (or Shkorpilovtsy on the Bulgarian Black Sea) to measure short-term variability of bed topography. The latter provided a basis for cross-shore transport computations. The figures obtained are comparable with a theoretical prediction stemming from the

Polish Academy of Sciences' Institute of Hydro-Engineering, Sen. Res. Assoc.[1] and Professor[2], 80953 Gdańsk, IBW PAN

energetics approach, which was earlier verified in a wave flume with movable bed. The three sets of the studies, together with various estimates of sediment motion measures, are discussed hereafter.

2. FIELD MEASUREMENTS OF SEDIMENT MOVEMENT

2.1. Apparatus and Environment

Four series of measurements with radioactive tracers of bed movement were conducted in 1984, 1986 and 1987 at Lubiatowo. Each series lasted several days and embodied continuous monitoring of waves, currents, bed topography and sand transport.

The instrumentation setup at Lubiatowo was installed on a steel framework , referred ta as a "Spider" , 10x10 m in plan view, with a system of three carriages and trolleys which allowed for accurate tracking,with a scintillation probe, of the movement of radioactive tracers within the area of Spider. In all experiments the latter was placed in a shallow nearshore zone on a depth of up to 1.5 m some 50-60 m offshore; Fig. 1 depicts the layout of Spider and bed topography in 1986 and 1987.

Iridium glass beads with quite uniform diameters of 0.2 mm (1984), 0.15...0.2 mm (1986), 0.088...0.15 mm (1986) and 0.2...0.25 mm (1987), in the range of the mean diameter of the Lubiatowo sand, D_{50} = 0.22 mm, were used as radiotracers. The composition of the glass was roughly 48 % SiO_2, 19 % Al_2,O_3, 17 % CaO, 6 % MgO, 5 % TiO_2, 5 % K_2O and 0.25 % IrO_2, which corresponded to the density of 2668 kg/m^{-3}. The radioisotope used, Ir-192, has the half-life period of 74.4 days. The principal radiation energy bands are 0.316 and 0.468 eV. Fifty-gram samples of radioactive sand were used in each experiment.

Waves were measured with a conductivity - type wire probe and a pressure transducer, water velocities with an electromagnetic current meter, and bed topography around Spider-manually by skin divers. The scintillation probe was moved and operated remotely from a land-based laboratory where all signals were collected and processed.

Wave parameters and water velocities measured in the

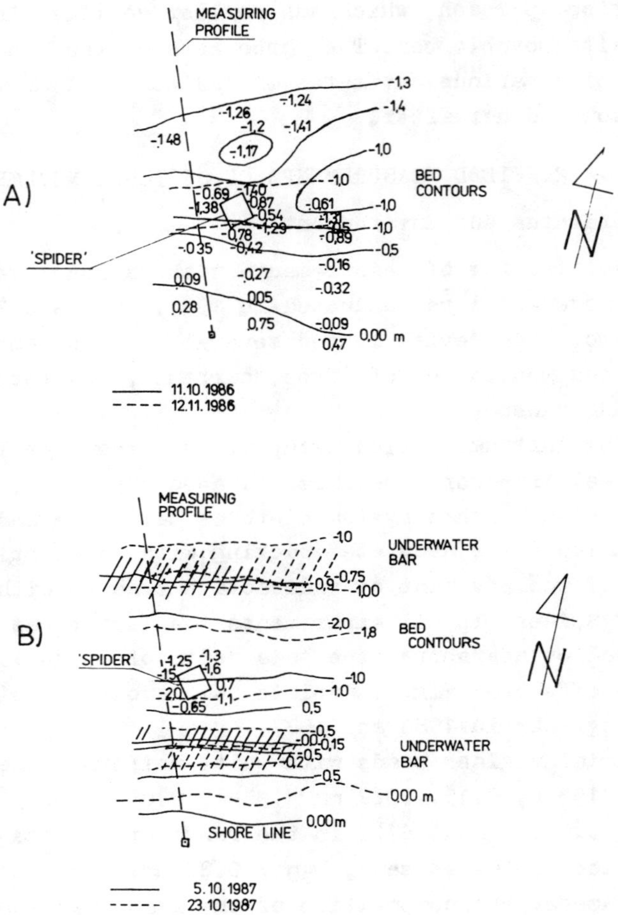

Fig. 1 Bed Topography : (A) Lubiatowo '86 ; 11 th Oct ... 12 th Nov '86
(B) Lubiatowo '87 ; 5 th Oct ... 23 d Oct '87

1986 and 1987 experiments are shown in Fig. 2 and Fig. 3, respectively. Wave crests were approximately parallel to the seaward side of Spider. In both series the prevailing longshore current was eastward.

2.2. Sediment Transport

Systematic tracking of the radiotracers within Spider has provided a background for the evaluation of various modes of sediment movement, cf. examples for 1986 and 1987 in Fig. 4, where the areas in different shades depict tracer activity, in pulses per second. Advection and dis-

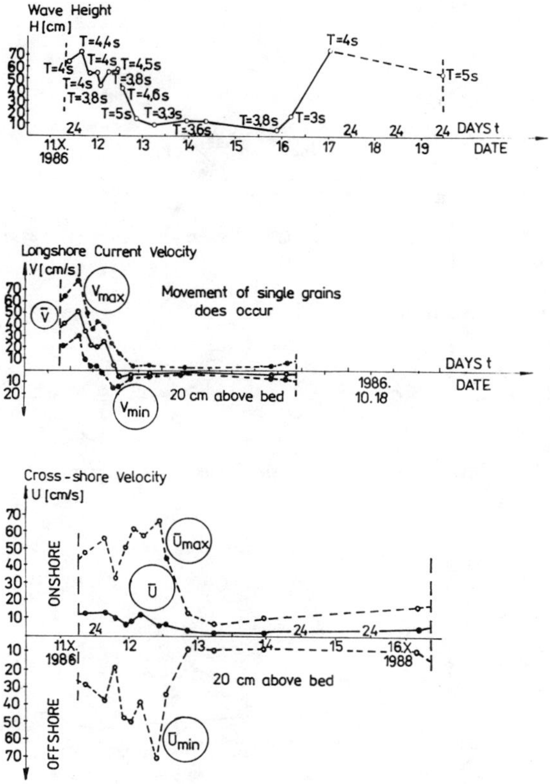

Fig. 2 Waves and Water Velocities in 1986

persion of the radiotracer provide estimates for incipient motion and bulk transport of sand.

Incipient motion has been linked to dispersion without any advection, i.e. any displacement of the centre of mass. Single grains with D_{50} = 0.2 mm begin to move under waves about H ≃ 15 cm T ≃ 3.5...4s, in water with the longshore and cross-shore velocity components of 5 and 8...10 cm s^{-1}, respectively (13th to 16th Oct '86). Similar results were obtained in 1984 and 1987. Hence the threshold velocity of water for grains with D_{50} = 0.15...0.25 mm can be given as V_{cr1} = 8...10 cm/s. These findings roughly coincide with the field data by Davis (1980) who observed no movement of 1.4-mm grains for water velocities lower than 8 cm/s, along with the incipient motion for the range of 8...22 cm/s, the number of

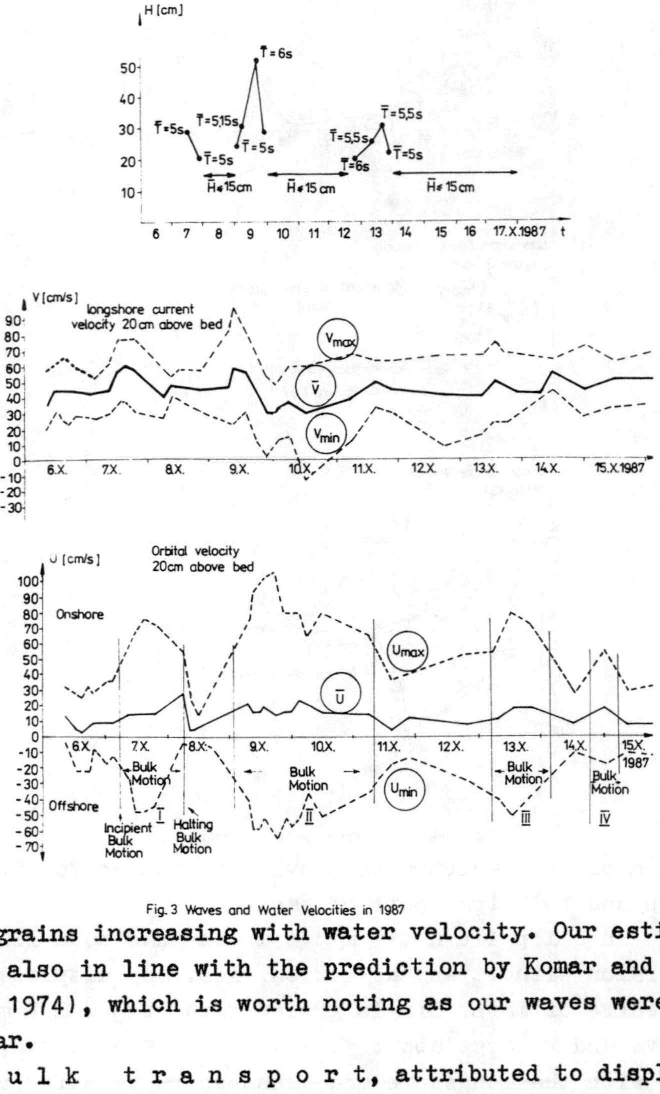

Fig. 3 Waves and Water Velocities in 1987

moving grains increasing with water velocity. Our estimates are also in line with the prediction by Komar and Miller (1974), which is worth noting as our waves were irregular.

B u l k t r a n s p o r t, attributed to displacement of the centre of mass, has been observed for water velocities about 50...60 cm/s, cf. Figures 3 and 4. All studies of 1984, 1986 and 1987 point to the same critical velocity of water V_{cr2} = 50...60 cm/s for sand grains with D_{50} = 0.15...0.25 mm.

The ratio of water velocity near bed to the effective sediment velocity i.e. the average velocity of the centre of mass is estimated about $10^3...10^4$. Some characteristic

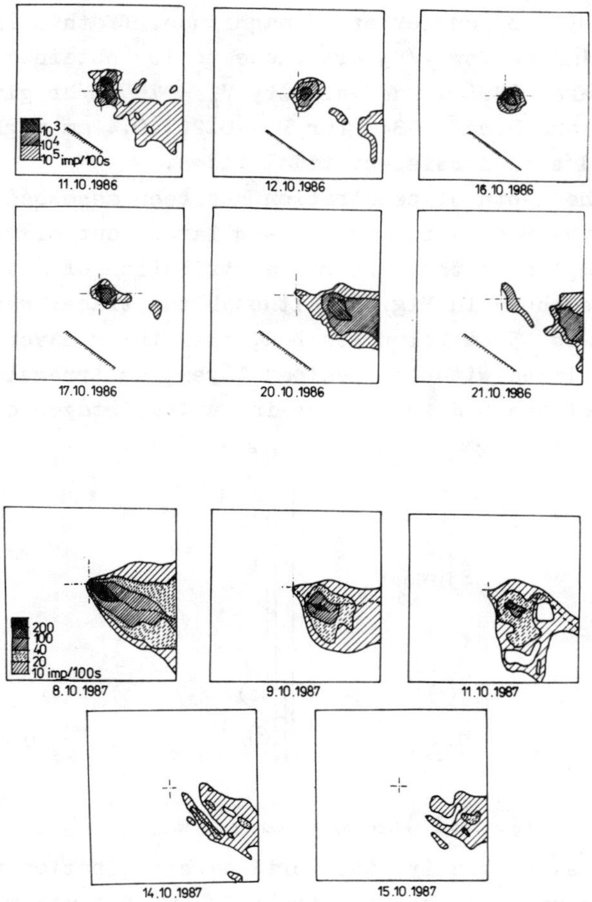

Fig.4 Plumes of Radioactive Sand Recorded within SPIDER

figures are given in Tab. 1.

Table 1.

Sand and Water Velocities

Year	D_{50} mm	V_{cr2} cm/s	V_s cm/hr	$V/\bar{V}_s$	Remarks
1984	0.2	60	12	10^4	1/curvilinear tracer trajectories
1986/1	0.15..0.25	50	12	10^3	2/bulk transport is depth-averaged
1986/2	0.088..0.15	20	8...9	10^3	
1987	0.2...0.25	55	10	$10^3..10^4$	

It should be remembered that our bulk transport is depth-averaged, so that the uppermost bed grains may move faster even by one order of magnitude. If this is so then our estimates for $V/\bar{V}_s$ are close to 10^2 obtained by Katoh and Tanaka 1986 . The velocity $\bar{V}_s$ = 75 cm/hr given by Drapeau and Long (1984) for D - 0.2...0.4 mm might be higher due to persistent tidal flows.

The depth of penetration has been assessed by measuring the radioactivity in 2-cm layers cut of core samples. Six examples of the vertical distribution of moving sediment are shown in Fig. 5. Although the tracer was detected as deep as 15 cm below sea bed, this thick layer must not be identified with the bedload layer, as irregular restructuring of sea bed took place in various stages of the ex-

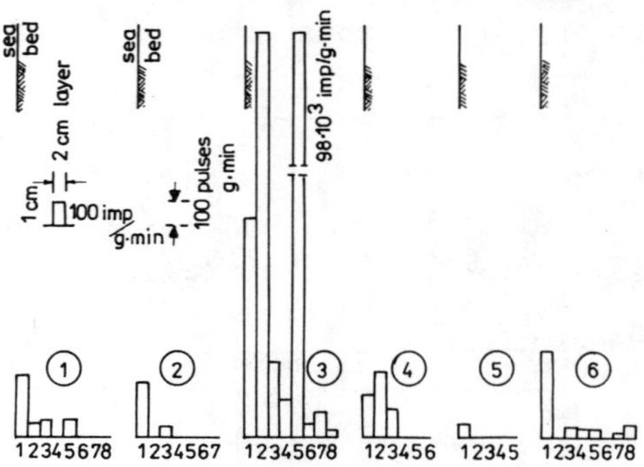

Fig. 5 Vertical Distributions of Radiotracers in 1987

periments. Generally, the vertical distribution varies from rectangular through parabolic (with a maximum at the sea bed) to Gaussian (with a maximum below bed), and seems to reflect a variety of sedimentation patterns encountered, from sheet flow through rapid overburdening to random mixing.

The absolute magnitude of the unit bedload rate at Lubiatowo was of the order of 2.5×10^{-3} kg/m·s, or 9 kg/m··hr. This value corresponds to the average effective thickness of bedload layer of 5 cm, $\bar{V}_s$ - 10...12 cm/h, and to the concentration of sand increasing sharply to about 1700 kg m^{-3} (dry weight) a few grain diameters below the mud line. If $\bar{V}_s$ is increased to 50 cm/hr due to the aforementioned reasons then the transport rate can be estimated as 40 kg/m·hr. It should be remembered that all experiments were conducted in fairly calm situations. Some other estimates cf. Drapeau and Long (1984), Lavelle et al. (1978) include higher figures, up to 60 kg/m hr, but pertain to more dynamic environments with stronger waves, currents and tides.

The critical bulk transport velocity V_{cr2} is about 5...6 times greater than the incipient motion velocity V_{cr1}. The regression analysis undertaken shows that V_{cr1} and small bedload rates are best correlated with $^H/T$ while V_{cr2} and higher bedload rates may be linked better to the mean velocity of water.

3. CROSS-SHORE SEDIMENT TRANSPORT

From among the numerous cross-shore bedload models available the one put forward by Pruszak (1987) has been taken for comparison against the field data presented above. The model is based on the assumption that the asymmetric oscillatory motion of water at seabed dominates the cross-shore bedload, and was originally tested against Pruszak's laboratory data. The formula proposed for the time - averaged cross-shore bedload rate reads

$$< q_b(x) > = \sum_{i=1}^{2} \alpha_i \frac{A_i [\rho f_w < u^3 > - < u > \tau_{cr}]}{T_i g(\rho_s - \rho)[\tan\phi + \frac{u}{|u|}\tan\beta]} \qquad 1$$

$$u_s = 0.25\left(\frac{z}{D_{50}} + 4\right) A_i < u(x) > ; \quad -4D_{50} \leq z \leq 2D_{50} \qquad 2$$

in which

$$A_i = 1 - \frac{u_{cr}}{u_{cr} + \int_{T_i} [u(x,t) - u_{cr}] dt} \qquad 3$$

α_1 = 1 for the wave crest phase i=1

α_1 = -1 for the wave trough i=2

T_i = duration of the respective phase

ϕ = internal friction angle

β = bed slope angle

u_s = speed of sand grains

Obviously, the onshore transport is assumed in the crest phase, while offshore movement for wave trough.

The laboratory findings for the bedload measured in special traps are illustrated in Fig. 6. The experiments

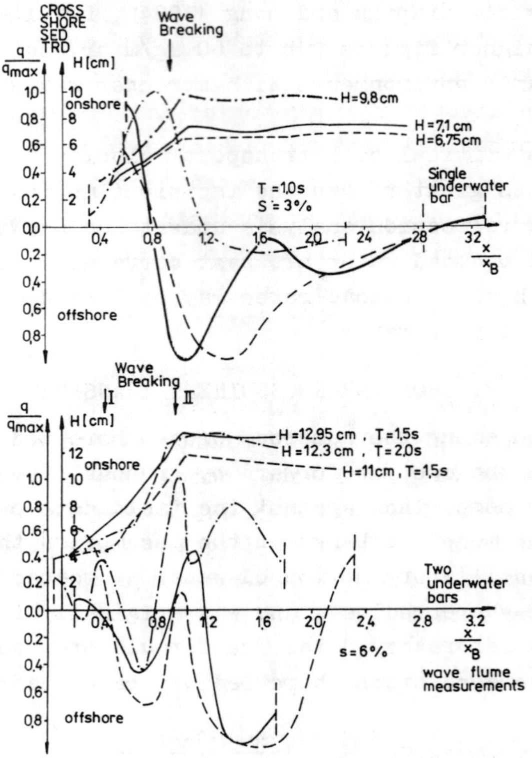

Fig. 6 Cross-Shore Transport Rate Measured in IBW PAN Wave Flume

were carried out in a 25-m long wave flume with movable bed (quartz sand with D_{50} = 0.22 mm). One or two bars were generated on initial bed slopes varying from 0 to 6 %. Offshore transport prevailed seawards of the breaking point $x=x_b$. The clearcut transition to the landward transport behind the single bar $x < x_b$ is not confirmed on the profiles with two bars.

The values computed by Eq. 1 for the reported prototype situations at Lubiatowo are of the order of 1.0...0.5 kg/m·s, hence about 10 times greater than those measured. Since Eq. 1 has been verified under laboratory conditions it must be inferred that the discrepancy is caused by different mechanics of sediment motion. Indeed, the figures given in Sec. 2 refer to bulk motion of the entire bedload layer, as judged from displacements of the radioactive plumes monitored whereas the computed quantities stem from measurements in the wave flume in which either individual grains or at most thin laminae of sand were set in motion and tracked subsequently.

The disparity of prototype and laboratory is now being compromised through an elaborate analysis of the physical factors shaping the sediment transport in both environments, followed by inclusion of all relevant quantities into the eventual version of our bedload model.

4. SHORT-TERM BED VARIABILITY

4.1. Laboratory and Field Measurements

Extensive tests in the IBW PAN wave tank with movable bed have exposed the dominance of bed slope, and not the angle of wave incidence, among the important factors of shore evolution (cf. Fig. 7). Wave-generated currents have been found to give rise to the instability of some bed features. Under oblique waves the bars migrate with meandering littoral drift.

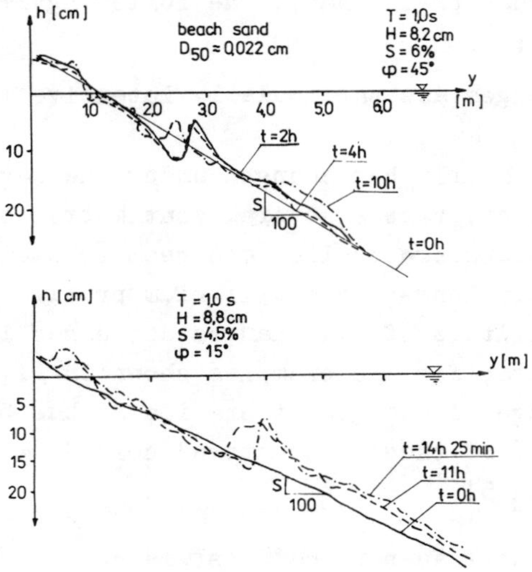

Fig.7 Bed Profiles Generated in IBW PAN Wave Tank Experiments under Oblique Waves (φ=15°; 45°)

The type of shore profile is reached rather quickly but the ultimate equilibrium requires a fairly long duration of experiments (i.e. coastal processes). In other words, the bed reacts fast to changing wave climate but adjusts slowly to the equilibrium profile. It is interesting to investigate this variability in the prototype.

Field studies on short-term variability of bed profiles were undertaken at Lubiatowo and Shkorpilovtsy. Special 210-kHz 20-W ultrasonic probes were arranged along the measuring profile to record bed changes. The absolute error is now 1 cm for 0.2...1 m elevation of the oscillators above bed. The advantages of the technique include ideal repeatibility of depth measurements, small relative error, flexibility of measurements during storms, and nondestructibility. Further improvements (40-W power, 5°=beam in lieu of 20° etc.) are under way.

The tests at Lubiatowo were primarily technological. The measurements at Shkorpilovtsy were carried out from 23d to 27th October 1987. Waves with the parameters shown in Fig. 8 (top) were normal to shore. First breaking occurred on the sand bar (between probes 17 and 19), and was usually accompanied by collapsing at the shoreline.

Although preliminary, the following findings can be singled out:

I/ bed changes are particularly intensive in the ultimate breaking zone

II/ maximum hourly bed changes under the waves experienced were 10 cm, versus several centimetres over a few hours

III/ bed changes are cyclic, and seem to oscillate about a certain long-term equilibrium profile

IV/ the amplitude of bed changes decreases in the offshore direction; for the examples shown in Fig. 8 (bottom) the range of changes at station 21 has been about 20...25% of those encountered some 20 m from shoreline (station 5).

4.2. Sediment Transport Rate Estimates

The cross-shore sediment transport rate q_x may be computed from the continuity equation

$$\frac{\partial h}{\partial t} = \frac{1}{1-p}\left(\frac{\partial q_x}{\partial x} + \frac{\partial q_y}{\partial y}\right) \qquad 4$$

in which p = porosity of bed sediment.

The bed changes have been measured, and the longshore transport q_y is neglected due to the normal wave

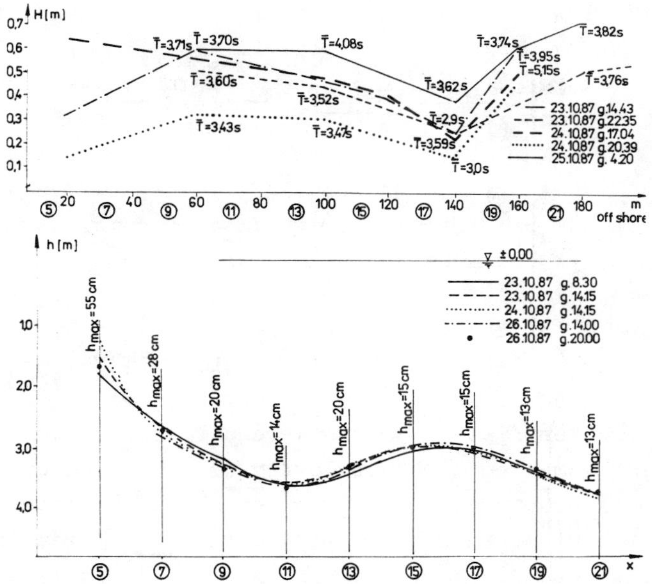

Fig. 8 Examples of Waves (top) and Bed Topography (bottom) Measured at Shkorpilovtsy

incidence. Results of the computations are illustrated in Fig. 9.

The highest transport rates have been found for the area of breaking, in agreement with our wave flume experiments. The direction of cross-shore sediment transport varies with time, and seems to depend on the actual stage of shore evolution. Further studies, including those carried on at Shkorpilovtsy in 1988, are to shed more light on this important question of cross-shore evolution.

SUMMARY and CONCLUSIONS

Field studies with radioisotopic sand tracers were carried out at IBW PAN's coastal research station at Lubiatowo in 1984, 1986 and 1987. In the conditions outlined in Figures 1...4, two critical nearbed water velocities, v_{cr1} and v_{cr2}, have been found as respective estimates for incipient motion of single grains and bulk motion of sediment. The figures measured are 8...10 cm s^{-1} and 50...60 cm s^{-1}, the first of them being in rough agreement with Komar and Miller's (1974) formula. The ratio of

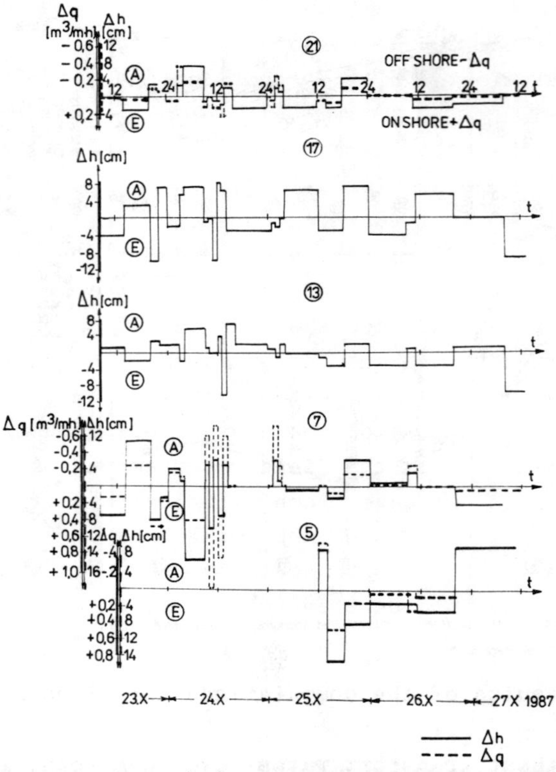

Fig. 9 Depth Increments Measured vs Sediment Transport Rates Computed

average nearbed water velocity $\bar{V}$, to the effective speed of tracer sand in the inshore zone $\bar{V}_s$, is estimated about 100...1,000, the figures in Tab.1 being divided by ten due to the noted curvilinearity of tracer trajectories. The order of magnitude of the speed $\bar{V}_s$ itself is estimated as 1 m/hr.

The cross-shore transport rates measured, 0.01... 0.015 kg/m·s for the climate encountered, are ten times smaller than the figures computed by Pruszak's (1987) formula. The latter stems from the energetics approach, and has been calibrated against wave flume data for bedload in thin seabed laminae. The discrepancy is due to the real bulk motion of the entire bedload layer in the field, subject to the mixing and overburdening effects noted.

The reported wave flume and wave tank studies with movable bed point to the importance of initial bed slope in the evolution of shore profiles. Single-bar profiles wi ness a clear-cut transition from the offshore transport seawards of the bar to the onshore transport on the opposite side, while for double-bar profiles the transitions are multiple.

The bed variability observed in the wave flume and wave tank was further explored in field studies employing special 210-kHz 20-W ultrasonic probes. The findings confirm intensive short-term bed changes in the ultimate breaker zone about 10 cm per one hour of a weak storm (cf. Fig. 8). Their amplitude decreases in the offshore direction. The short-term bed changes are cyclic, and seem to oscillate about a certain long-term equilibrium profile.

The first estimates of the cross-shore sediment transport rate obtained from the field measurements of bed changes agree with our wave flume experiments as to the occurrence of the highest rates about the breaking line. Further studies with our ultrasonic setup are expected to yield more precise data on cross-shore evolution and transport.

REFERENCES

Davis A.G. (1980). Field observations of the threshold of sand motion in a transitional wave boundary layer. Coastal Engineering 4

Drapeau G.,Long B.(1984). Measurements of bedload transport in the nearshore zone using radioisotopic sand tracers,Proc.19th Inter.Conf.Coastal Eng.,ASCE,1252-1264

Katoh K.,Tanaka N.(1986). Local movements of sand in the surf zone, Proc.20th Intern.Conf.Coastal Eng., Pap. No 72, ASCE

Komar P.D.,Miller M.G.(1974). Sediment threshold under oscillatory waves.Proc.14th Inter.Conf.Coastal Eng., ASCE

Lavelle J.W. et al.(1978). Fair weather and storm sand transport on the Long Island, New York, inner shelf. Sedimentology, Vol. 25

Pruszak Z.(1987). Some problems of mathematical modelling of total cross-shore sediment transport, Extended Abstracts, Euromech 215, S.Margherita Ligure (Genoa), September 15-19, 1987, Univ. Genoa

CHAPTER 132

Secular Change in Tidal Flat Heights

Winfried Siefert*)

Introduction

The German North Sea Coast is relatively short (370 km); its shape is extremely irregular with estuaries, extended tidal flats, partly protected by sandy islands with deep tidal gullies between them (Fig. 1).

In order to gain more detailed information and sufficient knowledge about the variety and variations in this region, the German Coastal Engineering Board (KFKI, Kuratorium für Forschung im Küsteningenieurwesen) has initiated a high-level and extensive programme, "MORAN","Morphological Analysis North-Sea-Coast". The purpose and the first ideas of the programme were presented by Siefert and Barthel (1981). Investigations are still under way.

It is of high interest whether the mean height of the tidal flats will rise, if the green house effect results in an acceleration of the mean sea level rise. In that case coastal protection in areas behind tidal flats will have to be changed completely, as then wave energy will no longer be dissipated in front of the coastline and will attack the dikes heavily.

So an analysis of the data of the tidal flat heights over more than 100 years and a comparison with the sea level rise seemed to be called for. This may give first indications to the future behaviour of the protective foreshore in times of faster rising water level.

*) Prof. Dr.-Ing.; head, Hydrol. Res. Group, Hamburg Port Authority, Lentzkai, 2190 Cuxhaven, F.R. Germany

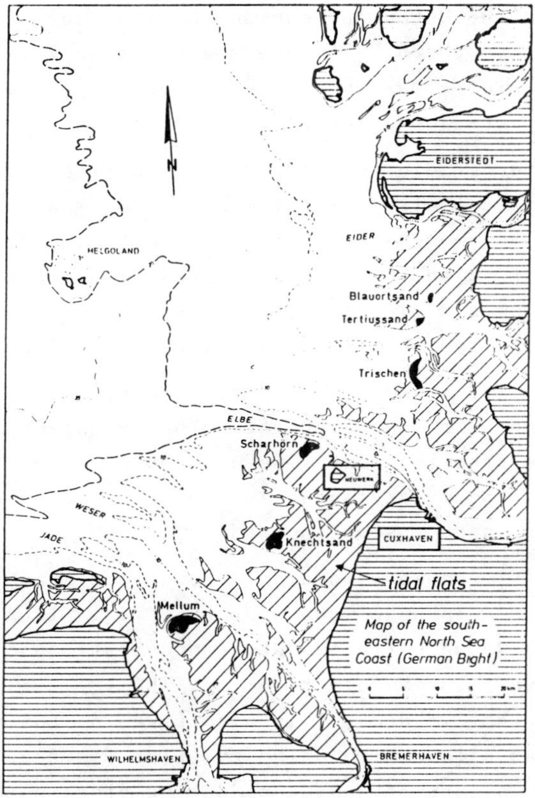

Fig. 1 Map of the German Bight with tidal flats

The stability of the wadden sea is of high interest also for other reasons: In the European tidal flats, more exactly in the channels of the tidal flats, fishing is carried out. Plaice and prawns are caught and mussels are farmed. The flats are particularly valuable as a nursery for important types of fish such as herring, plaice, sole, flounders, Kliesche and cod.

In addition the tidal flats are of great importance as a nature reserve and a health resort.

The Tidal Flats on the German North Sea Coast*)

The south and part of the east coast of the North Sea are bounded by tidal flats. They stretch for a total length of

*) Mainly based on Reineck (1978)

450 km along the Dutch, German and Danish coasts. The tidal flats of the southern North Sea measure 5 to 7 km on an average reaching a maximum of 10 to 15 km. They are intersected by rivers and numerous tidal channels.

The tides are semidiurnal. The average spring tide is approximately 2.6 m with a maximum of 4.1 m. Neap tide range is 1.8 m to 3.1 m. The tidal currents over the flats reach velocities of 30 to 50 cm/s, with a maximum (at storm surges) of 150 cm/s. In gullies the current velocities are about 100 cm/s and in large channels about 150 cm/s.

Waves are a very important factor. They come from various directions and with changing strength. The predominant wind direction is from the west. The wave movement over the tidal flats is the major factor influencing the distribution of sediments, and erosion and sedimentation.

Tidal currents and wave movement combine to transport large quantities of suspended load over the tidal flats. Measurements in the Neuwerk flat showed concentrations of suspension of 30 mg/l for a calm tide, and 300 mg/l was reached on a storm tide. If account is also taken of the shifting direction and volume with which the tide flows over the flats the transport of sediments during storm surges rises by two to three times above normal.

The water temperature of the body of water which inundates the flats fluctuates greatly and is related to fluctuations in air temperature. In the same way the salt content in the large marine bays also varies.

The present day tidal flats receive very little new sediment from outside. Most sediment changes take place within the flats in the form of redistribution, as will be shown later.

The origin of the sediments is predominantly re-worked glacial material. In addition there is a smaller proportion - especially pelitic fractions - brought down by rivers, but pelitic material can also come from older layers being eroded. The mud consists of clayey silt with a mixture of fine sand in varying proportions.

The sand on the tidal flats is fine with a mean diameter of about 0.1 mm. Subject to wave action, a fine-medium grained to medium grained sand with a mean diameter of about 0.2 mm prevails on the beaches. The commonest mineral is quartz (80%), next in order come feldspat, mica and carbonate. Some 0.2-3.5% are heavy minerals like epidot, hornblende, garnet etc.

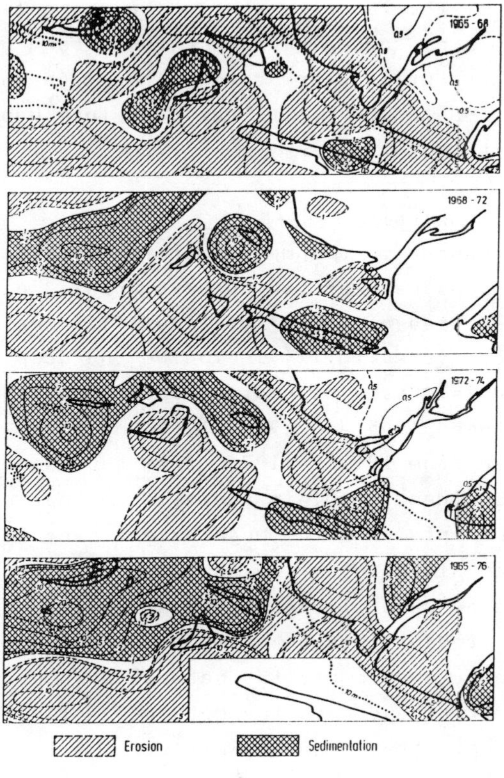

Fig. 2 Four comparisons of topographical maps

As well as the mineral component, the sediments include fecal pellets, mainly of mud. These are produced by snails, molluscs and worms. In addition the contents include foraminifera, ostracods, spines from sea urchins, peat debris and crushed shell. Gravel is mostly found in the channels, partly armouring the bed.

The lower edge of the Holocene lies at about NN-5m*) in the vicinity of the present day embankments and at the seaward end of the flats about NN-15 to 20 m*). This lower boundary fills deep channels in older layers. Some of these channels were already laid down in the Pleistocene period, and the outline of these old channels can still be traced as channels on the surface today.

The position of the larger channels changes only slowly - except when they suddenly break through sand bars or even barrier islands during storm surges. Small channels meander considerably (up to 100 m per year) and form bank cutting faces and point bars.

The Method of Analysis

Usually the changes in volume of material are evaluated by differential maps between two surveys. Fig. 2 gives an example for an area of 48 km² of tidal flat west of the island of Neuwerk (see Fig. 1) including one deep and some smaller channels. Parts with predominant erosion or sedimentation between 1965 and 1968, 1968 and 1972, 1972 and 1974, and also from 1965 to 1976 (from top to bottom on Fig. 2) can be identified. In addition to that one usually finds a graph containing the mean height over the continuous time, as given in Fig. 3. But analyses like these have to be studied carefully: Extrapolation is not reliable, the amount of redistribution of material within the area remains unknown, and so are the natural scattering and the artifical scattering due to the methods of levelling, echo sounding and data treatment as well.

That is why the MORAN project was started (Siefert, 1983 and 1987). The philosophy of the analysis is the comparison of a number of surveys (up to 16 within 35 years) with one another (giving a sum of up to 120 values) and to calculate the turnover height h_u and the net-balance height h_b as functions of wave climate and current energy during the time between the surveys. For that purpose the coastal area of 9.000 km² was divided into 9.000 units. Each of these were

*) NN equals circa mean sea level

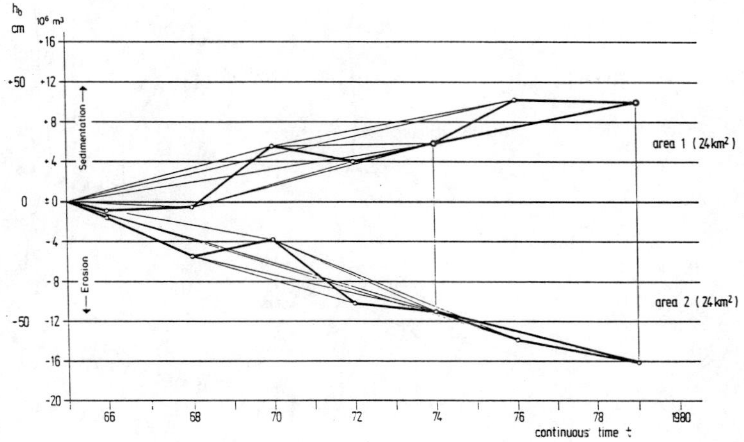

Fig. 3 Mean height changes (balance h_b) of two tidal flat areas from 1965 to 1979

subdivided into 100 small units of a size of 1 ha (Fig. 4).

Each comparison of two surveys gives a table as in Fig. 5 for the small units and mean values for the 1 km²-unit respectively, whereby

balance = mean height change h_b in dm within time interval a in years,

turnover = mean topographical height change over all small units h_u in dm, either sedimentation or erosion, within the same time interval a.

Theoretical considerations (see Siefert, 1987) lead to the representation of the turnover height h_u over the time interval a by a saturation function such as

$$h_u = \bar{h}_u \cdot (1-e^{-a/a_o})$$

as given in Fig. 6.

The morphological character of an area can thus be expressed by the two parameters $\bar{h}_u$ and a_o. The value $\bar{h}_u/a_o$ in cm/

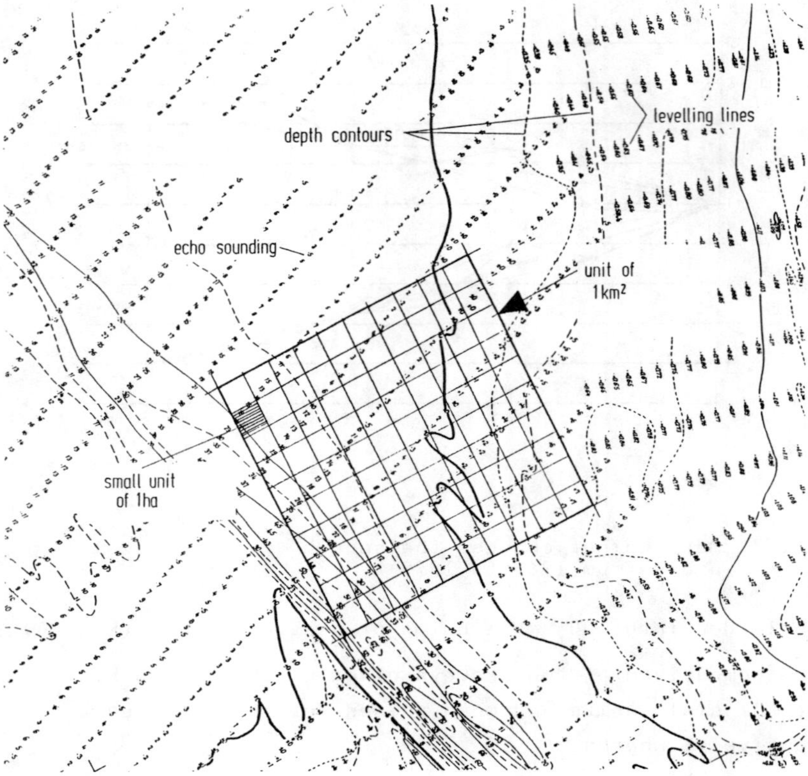

Fig. 4 Unit of 1 km² with subdivision; levelling and echo sounding data with depth contours

year represents the mean resedimentation rate; $\overline{h}_u$ is the asymptotic turnover height that turns out as the result of long-term topographical comparisons (Siefert, 1983). As

$$h_b \leqq h_u$$

by definition, the change of the mean topographical height has to be

$$h_b(a) \leqq \overline{h}_u$$

and statistically reaches values of

$$h_b \approx (0.1 \text{ to } 0.4) \cdot \overline{h}_u$$

(Siefert, 1987). Moreover, theoretical considerations lead

field 1 | field 2

	1	2	3	4	5	6	7	8	9	10
1	0	1	-1	-2	-3	-3	-3	-3	-3	-2
2	1	0	-1	-1	-2	-2	-2	-2	-2	-1
3	4	0	-1	-2	-1	-1	-1	0	0	1
4	-2	-3	0	3	1	0	0	1	0	0
5	-3	-4	-2	-2	-1	0	1	1	2	2
6	-5	-5	-4	-2	-2	-1	1	2	1	-1
7	-7	-2	0	-2	-5	-7	-6	-5	-5	-2
8	0	0	0	1	2	1	1	0	0	0
9	-1	-3	-3	-4	-4	-2	-1	-1	-1	(-1)
10	-2	-3	-4	-4	-4	-3	-2	-2	-2	-2

field 1 / field 3 (left); field 2 / field 4 (right)

$h'_b = h'_u$

top. change of small unit (1ha)

		field 1	2	3	4	1km² sum	mean
sedimentation	(dm)	10	8	3	6	27	0.27
erosion	(dm)	31	25	66	44	166	1.66
balance	(dm)	-21	-17	-63	-38	-139	$h_b = -1.39$
turnover	(dm)	41	33	69	50	193	$h_u = 1.93$
points		25	25	25	25	100	

Fig. 5 Height difference in dm for each unit between two surveys within a time interval of "a" years (example)

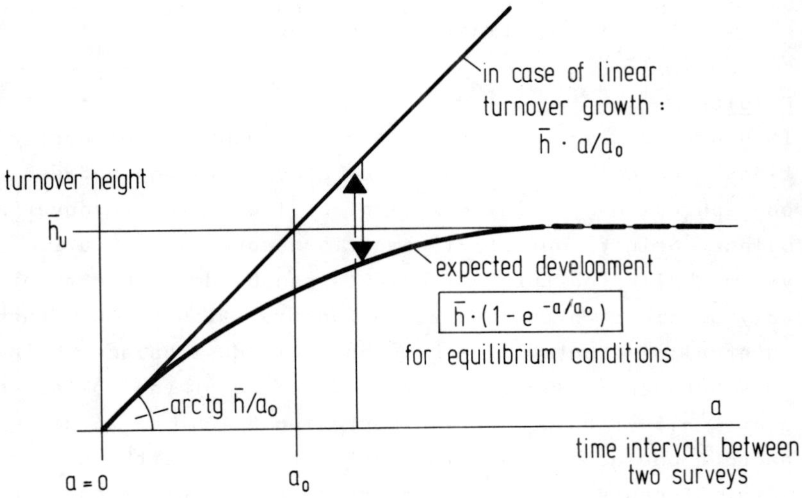

Fig. 6 Saturation function for h_u over a

to the conclusion that the mean height of such a partial area remains constant over a period if the turnover height function can be proved to be horizontally asymptotic (Fig. 7).

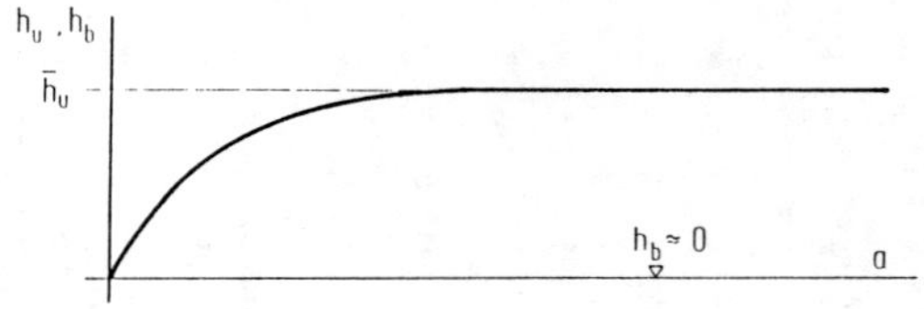

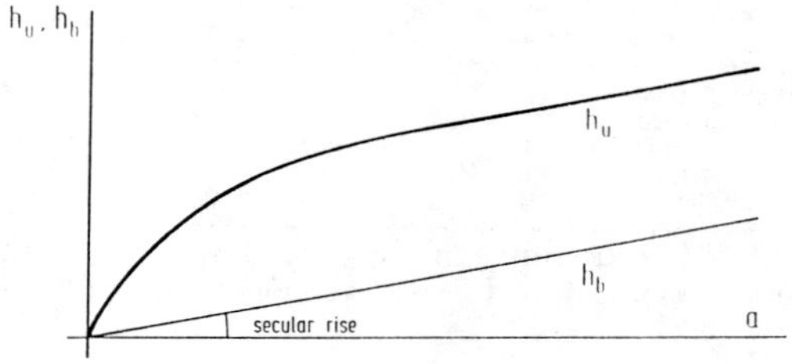

Fig. 7 Saturation function with horizontal asymptote presuming $h_b \approx 0$ for long a (top) and presuming secular rise of tidal flat height ($h_b \neq 0$ for long a) (bottom)

Results

In order to give an impression of the weakness of derivations from only a limited number of comparisons between topographical maps, Fig. 8 indicates 91 dots for turnover and balance heights out of 91 comparisons over a = 1 to 25 years. While the turnover heights can be described relatively precisely by a saturation function with a confidence interval of about $0.15 \cdot \bar{h}_u$, the individual balance heights vary between + 26 cm and - 24 cm. So one of the comparisons over a = 5 years gave a sedimentation height of 26 cm, another one gave an erosion depth of 17 cm, a third nearly zero. Of course, all of them are correct, but we have to keep in mind that each survey and therefore each comparison includes an element of chance.

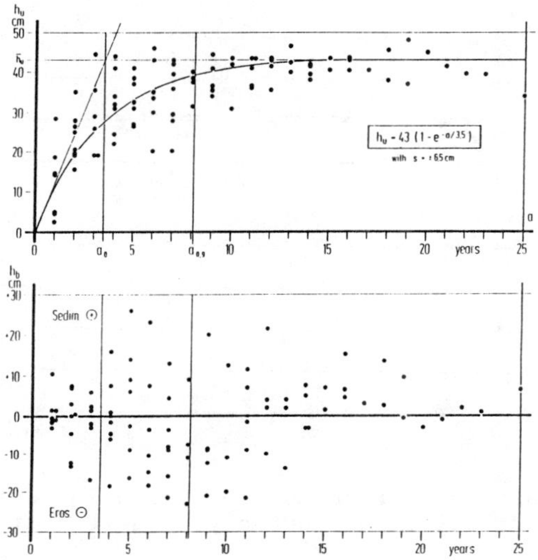

Fig. 8 Turnover and balance heights in a unit of 1 km² size for 91 topographical comparisons; time interval up to 25 years (Siefert, 1987)

Fig. 8 already indicates a horizontal asymptote over 25 years. Analyses show that especially in the higher parts of the tidal flats no secular change in height can be discerned within the last 35 years, though in the same period the mean high water level rose by about 15 cm, the mean storm surge level even more (example on Fig. 9).

Of particular interest is the evaluation of surveys of the tidal flat between Cuxhaven and Neuwerk from 1864 and 1913 (Fig. 1) and comparison with the actual situation. In this case reliable levellings across the tidal flat over a distance of about 10 km (Fig. 10) could be reconstructed by the original data set and be compared with modern data (Siefert and Lassen, 1987). The continuous development of the mean heights, including the sea level, is given in Fig. 11. It leads to the conclusion that there was a rise of tidal flat height until the 1950s, followed by an erosion period of 15 years and a slight sedimentation since. But taking into consideration that only two old surveys are being compared with 7 new levellings since 1954, it is worthwhile looking at the significance of the trends. That is possible

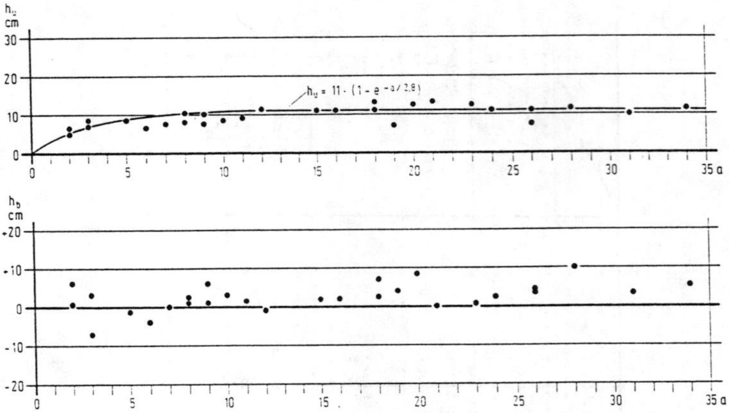

Fig. 9 Horizontal asymptote over 35 years indicates height stability without any long-term trend

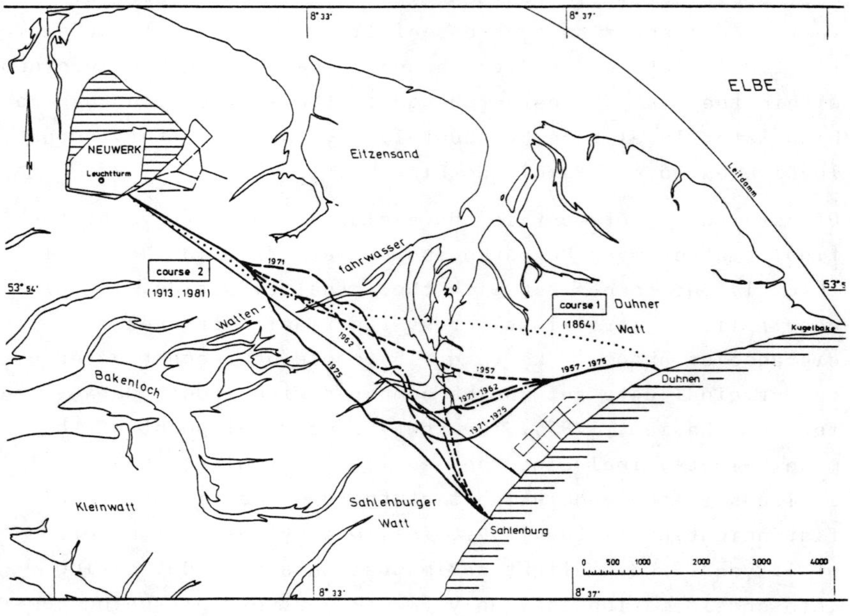

Fig. 10 Location of courses across the tidal flat from the mainland (Cuxhaven) to an island (Neuwerk)

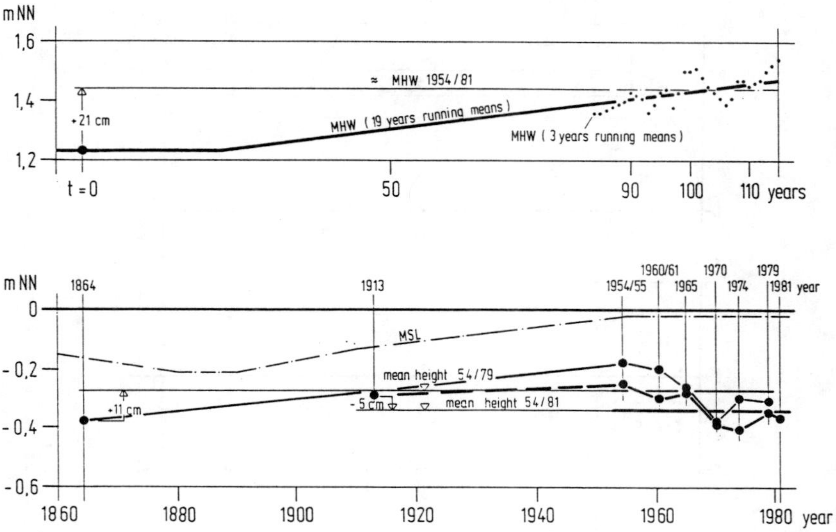

Fig. 11 Mean HW level (top), MSL and mean topographical height of the tidal flat courses of Fig. 10 from 1860 to 1980

by use of the morphological parameters on Fig. 12: The dependencies given by Siefert (1987) demanded $\bar{h}_u$ in the middle of 20 and 45 cm and a_o circa 4 years for wadden sea areas like these. The best-fit saturation function results in

$$h_u = 30 \cdot (1-e^{-a/4})$$

with a confidence interval of ± 2.0 cm. So the resedimentation is at the small rate of

$$\bar{h}_u/a_o = 7.5 \text{ cm/year}$$

and the horizontal asymptote can be proved over 120 years. So the "trends" of Fig. 11 are not at all significant; they belong to the natural scattering of yearly sedimentation-erosion-changes. It turns out that even within the last 120 years there was no rise of the mean height of the tidal flats, while the mean high water level rose by at least 20 cm and the mean storm surge level by twice that amount.

This result may be confirmed by another evaluation: The comparison of two topographical maps of an area of 4.500km² with a time interval of a = 5 years shows a mean "sedimen-

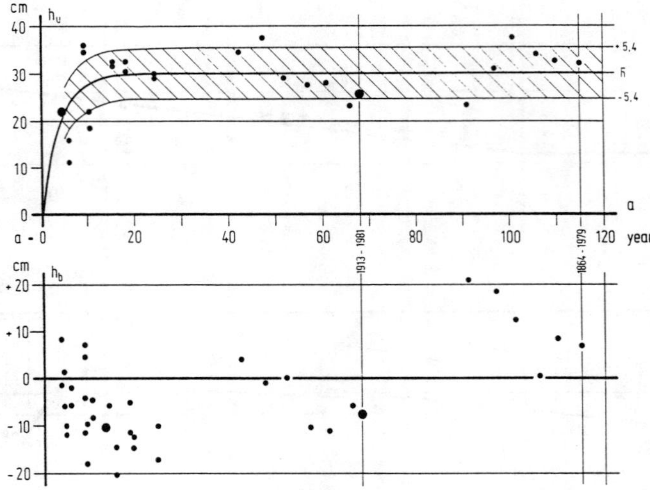

Fig. 12 Turnover (top) and balance heights for tidal flat routes of Fig. 10. The horizontal asymptote can be proved over 120 years (Siefert and Lassen, 1987)

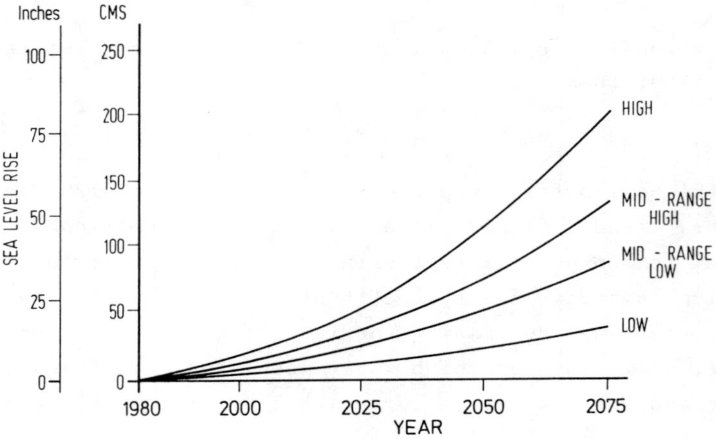

Fig. 13 Global sea level rise scenarios (Hoffman et al., 1983)

tation" of a mere 1 cm, while the turnover height is 70 cm. With a value

$$h_b / h_u = 0.014$$

this "trend" must be ignored. Any extrapolation is a matter of speculation.

Conclusions for Future Tidal Flat Development

The analysis of a large number of tidal flat units over 35 years, as well as the analysis of two courses across a tidal flat over the last 120 years, indicate that a significant height change parallel to the rising sea level has not taken place. If the future global sea level rise increases as presented in Fig. 13, the question is: Will the tidal flats rise in height with the same velocity and accelleration as the future mean sea level? If not, this poses serious problems for coastal protection work, as wave and current climates will change significantly. Moreover it is even possible that the mean tidal flat height may be reduced - due to increasing turbulence with rising water level, higher wave and current energy (see Fig. 14).

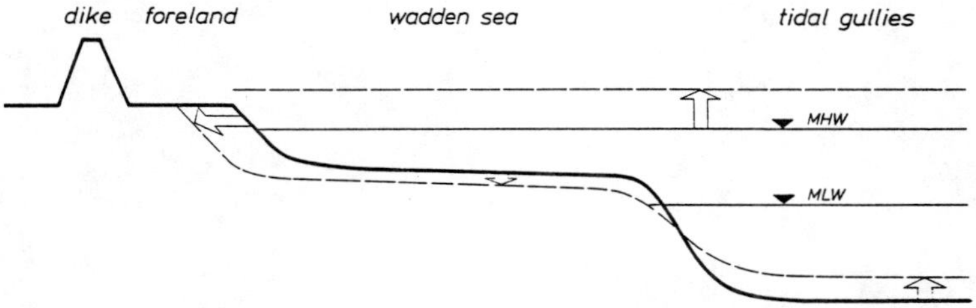

Fig. 14 Possible height development in the wadden sea due to rising sea level (Siefert and Lassen, 1987)

References

HOFFMAN, J., KAYES, D., TITUS, J.: Projecting Future Sea Level Rise. Gov. Printing Office, Washington D.C. 1983

REINECK, H.-E.: The Tidal Flats on the German North Sea Coast. Die Küste, Vol. 32, 1978

SIEFERT, W.: Morphologische Analysen für das Knechtsand-Gebiet. Die Küste, Vol. 38, 1983

SIEFERT, W.: Umsatz- und Bilanzanalysen für das Küstenvorfeld der Deutschen Bucht. Die Küste, Vol. 45, 1987

SIEFERT, W., BARTHEL, V.: The German MORAN Project. Proc., 17th ICCE, Sydney 1980, ASCE, New York 1981

SIEFERT, W., LASSEN, H.: Zum säkularen Verhalten der mittleren Watthöhen an ausgewählten Beispielen. Die Küste, Vol. 45, 1987

CHAPTER 133

BED LOAD TRANSPORT DUE TO NON-LINEAR WAVE MOTION

by

Hitoshi TANAKA*

ABSTRACT

The bed load transport rate due to wave motion is measured in a wave flume. The modified stream function theory of the author (Tanaka (1988)) is applied to the formulation of the sediment transport rate in order to include the non-linearity. The proposed formula predicts well except near the surf zone where the effect of the acceleration plays an important role.

1.INTRODUCTION

Coastal sediment movement in the onshore-offshore direction has been studied by several researchers (e.g. Horikawa (1988)). Many of empirical formulas express the sand transport rate as a function of the bottom shear stress due to wave motion, which is usually estimated through the friction coefficient proposed by Jonsson (1966). His result is, in principle, only applicable to the linear wave motion. On the other hand, the non-linearity of wave motion has a considerable effect on the sediment transport in practice, especially in regard to both the direction and the rate of the net transport. Therefore, the non-linear effect of wave motion should be included in the formulation of sediment transport rate.

Up to present, however, no practical method had been available for estimation of the bottom shear stress due to non-linear wave motion, until the present author (Tanaka (1988)) made the effect of turbulence included in the stream function theory proposed by Dean (1965) for the non-linear wave motion. This modified stream function theory which enables to predict the bottom shear stress as well as the velocity profile is applied to the formulation of the bed load transport rate due to the non-linear wave motion.

The present study is restricted to the case of the

*Lecturer, Department of Civil Engineering, Tohoku University, Aoba, Sendai 980, Japan.

wave motion alone, although the author's modified stream function theory can be extensively applied to the wave-current combined motion.

2. EXPERIMENTAL PROCEDURE

In the present study, two series of experiments were carried out: bed load transports over a flat bed (Experiment A) and over a sloping beach (Experiment B). The experimental setups are shown in Figs.1 (a) and (b).

The series of Experiment A were carried out in a wave flume, 16m long, 30cm wide and 50cm deep. A part of the bottom was lowered at the middle of the flume, on which the sand of mean diameter of 0.2 mm or 0.51 mm was filled to form a layer 5cm thick and 1.5m long. The experiments were carried out for 15 different wave and sediment conditions. The experimental condition was so carefully chosen as for the sand to be transported as bed load in the onshore direction. Then, the sand transported from the test bed onto the shoreward fixed bed was collected to obtain the net transport rate. The surface of the fixed bed was roughened with the pasted sand of the same size in order to make the mobility of the sand grain uniform along the wave flume.

In the series of Experiment B, a beach of slope 1:20 was built with the sand of diameter 0.78 mm at the end of the wave flume. The beach was exposed to the wave action for 10 minutes. The change of the beach profile was measured and used to calculate the sand transport rate. The direct measurement of the sand transport rate was also carried out by use of a sand trap originated by Horikawa, Sunamura and Shibayama (1977). In this case, the wave period and equivalent deep water wave height were fixed as 1.52 sec and 11.8 cm, respectively.

3. BOTTOM SHEAR STRESS DUE TO NON-LINEAR WAVE MOTION

Although the modified stream function theory can be referred in the preceding paper (Tanaka (1988)), a brief description of the theory and examples of the calculated result will be given below.

Combining Dean's stream function theory (Dean (1965)) and the time-invariant eddy viscosity model (Tanaka and Shuto (1981)), the stream function can be given by the following expression (Tanaka (1988)).

$$\psi(x,z) = \frac{L}{T}z - \frac{u_c^{*2}}{\kappa u_{cw}^{*}} z\{\ln(\frac{z}{z_Q}) - 1\} + \sum_{n=1}^{N} \sinh\{nk(z-z_o)\}\{X(n)\cos(nkx) + Y(n)\sin(nkx)\}$$

$$+ \mathrm{Real}\left[\sum_{n=1}^{N} \frac{ik\xi_n}{2c} \frac{H_1^{(1)}(\xi_n)}{H_o^{(1)}(\xi_{no})} \{X(n) - iY(n)\} \exp(inkx)\right] \quad (1)$$

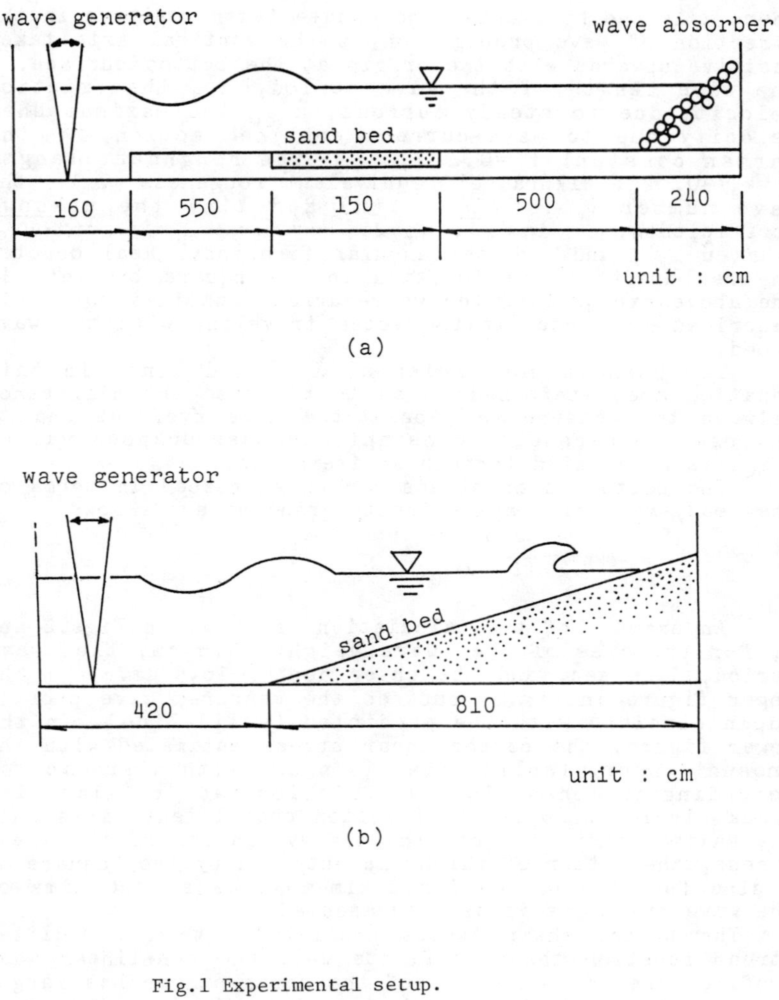

(a)

(b)

Fig.1 Experimental setup.

where x is the horizontal coordinate taken positive in the direction of wave propagation, z the vertical axis taken positive upwards with the origin at the bottom surface, L the wave length, T the wave period, u^*_c the friction velocity due to steady current, $\hat{u}^*_{cw}$ the maximum shear velocity due to wave-current combined motion, κ the Karman constant (=0.4), z_0 the roughness length ($=k_s/30$, k_s: Nikuradse's equivalent roughness), k the wave number ($=2\pi/L$), X(n) and Y(n) the unknown coefficients, $i=\sqrt{-1}$, $\xi_n=2e^{\pi i/4}\sqrt{ncz}$, $\xi_{n0}=2e^{\pi i/4}\sqrt{ncz_0}$, $c=\sigma/(\kappa\hat{u}^*_{cw})$, and σ the angular frequency. Real denotes the real part of the function in the square bracket. In the above expression, the wave-current combined motion is described in a coordinate system traveling with the wave speed.

The unknown coefficients, X(n) and Y(n), in this equation are determined so as to minimize the difference between the measured and predicted wave profiles and to uniform the Bernoulli constant. Another unknown value, $\hat{u}^*_{cw}$, is calculated through an iteration.

The bottom shear stress can be expressed in terms of the eddy viscosity and velocity gradient as follows.

$$\tau = -\rho\kappa\hat{u}^*_{cw} z \left.\frac{\partial^2\psi}{\partial z^2}\right|_{z=z_0} \tag{2}$$

An example of the calculation is shown in Figs.2 and 3, for the case of the wave height, 8.6 cm, the wave period, 1.40 sec, and the water depth, 16.5 cm. The upper figure in Fig.2 compares the measured wave profile (open circles) with the predicted (solid line). In the lower figure, the bottom shear stress estimated with the present theory (solid line) is shown with a prediction according to Jonsson's wave friction factor (line with black circles). Jonsson's friction coefficient gives only the maximum value but not the time-variation of the shear stress, the latter of which is obtained by the square of a sine function having its maximum value at the time of the wave crest, as is usually assumed.

The bottom shear stress predicted with the modified stream function theory reflects well the non-linear wave profile. The positive part of the shear stress has larger maximum value and shorter duration than the negative part. As long as the maximum value concerns, two theories give more or less similar results. However, the details are quite different, especially at the phase of the negative shear stress. This suggests that the difference between the two predictions becomes greater in the nearshore region, where waves become more sharp-crested and more asymmetric.

The vertical distribution of the estimated horizontal velocity (solid lines) is shown in Fig.3. The prediction according to Dean's stream function theory (dotted lines) is also presented. Outside the boundary layer, the

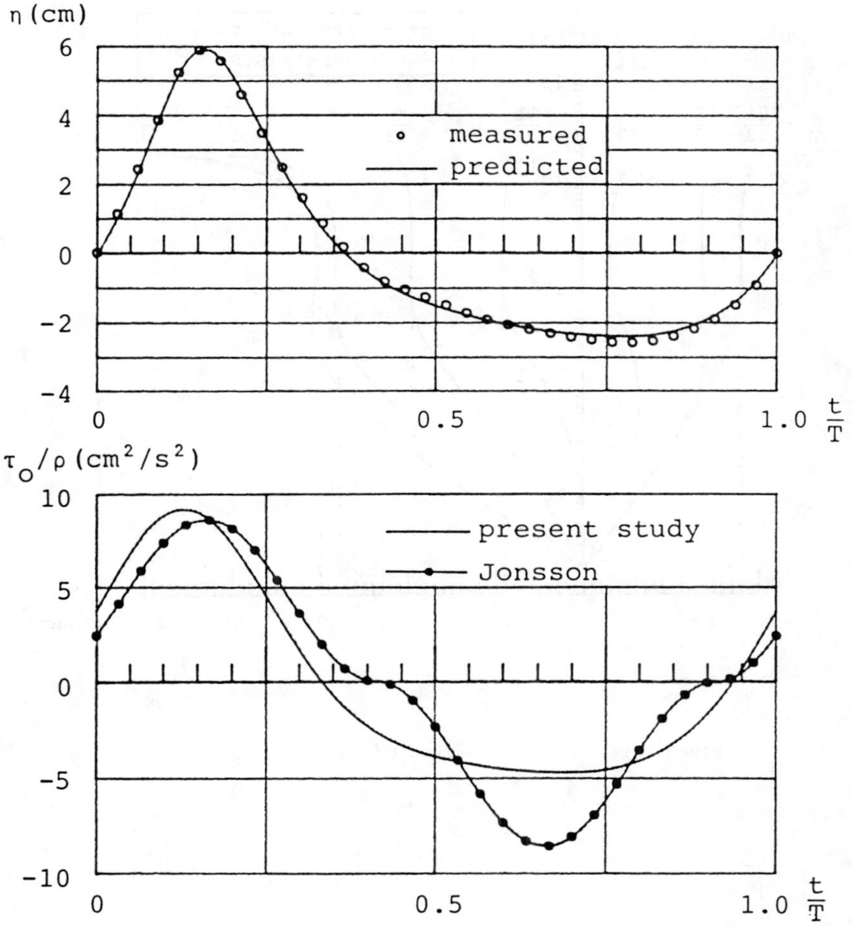

Fig.2 Time-variation of water surface level and bottom shear stress.

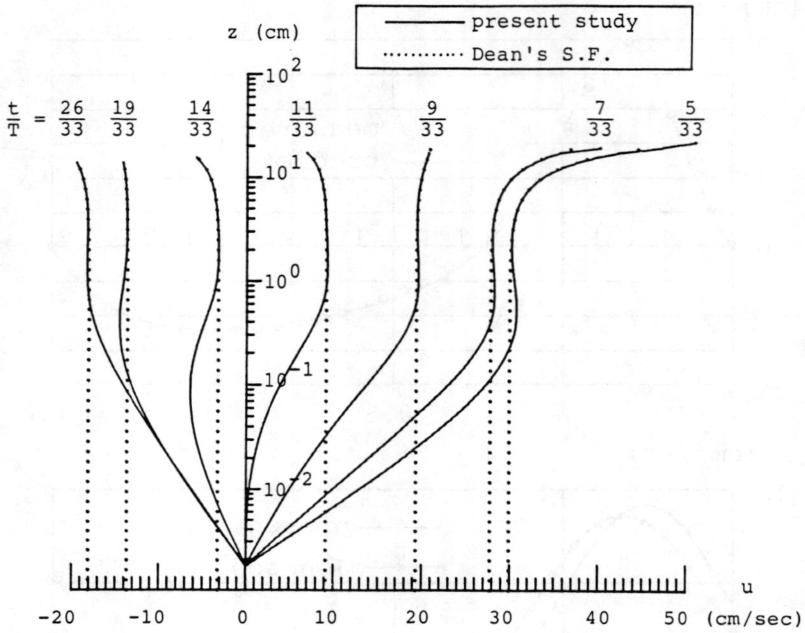

Fig.3 Vertical distribution of velocity profile.

difference between the two theories is very small. A distinct difference is found in the immediate vicinity of the bottom. This is of no wonder because Dean's method basing upon the potential theory can not satisfy the boundary condition there while the present author's theory can perfectly do.

Recent experimental measurements have revealed that Dean's stream function theory can be practically applicable to the estimation of the velocity field, except in the very vicinity of the bottom, near the surf zone over a sloping beach. Judging from this fact, it may be hopeful that the bottom shear stress obtained with the present theory can be used in the formulation of the sand transport rate not only on a flat bed but also on a sloping beach.

4.RESULTS AND DISCUSSIONS

4.1 Bedload transport over a flat bed

The instantaneous bedload transport rate is assumed to be expressed as a function of the instantaneous shear stress as follows, in an analogy to sediment movement due to steady current.

$$\frac{q_{B(t)}}{\sqrt{sgd^3}} = \alpha\, \mathrm{sign}\{\tau^*(t)\}|\tau^*(t)|^m\{|\tau^*(t)| - \tau^*_{cr}\}^n \qquad (3)$$

where q_B is the instantaneous bedload transport rate, s the immersed specific weight of sand, g the gravitational acceleration, d the diameter of sand particle, $\tau^*(t)= \tau(t)/(sgd)$, and τ^*_{cr} the critical Shields number. Values of α , m and n are to be determined from experiments. The time-variation of the bottom shear stress can be calculated with the modified stream function, Eq.(1). It should be noted again that we need the whole wave profile over one wave period for the calculation of the bottom shear stress.

On the basis of Eq.(3), the net transport rate over one wave period is given by,

$$\frac{\overline{q_B}}{\sqrt{sgd^3}} = \alpha F \qquad (4)$$

where

$$F = \frac{1}{T}\int_0^T \mathrm{sign}\{\tau^*(t)\}|\tau^*(t)|^m\{|\tau^*(t)| - \tau^*_{cr}\}^n dt \qquad (5)$$

In the above equation, the integration should be restricted to the phase when the tractive force is greater than the critical shear stress.

The exponents, m and n, and the coefficient, α , in

Eqs.(3) and (4) will be determined so as to minimize the scattering of the experimental data.

The relationship between the measured net transport rate and the function F defined by Eq.(5) is examined for a various set of the exponents m and n. It is concluded that m=0.5 and n=1.0 are the best combination of the exponents. The result is shown in Fig.4, where the net transport rate depends linearly on the function, F, as expected in Eq.4, with the coefficient α =3.5.

In conclusion, the instantaneous transport rate of bed load can be given by

$$\frac{q_B(t)}{\sqrt{sgd^3}} = 3.5\mathrm{sign}\{\tau^*(t)\}|\tau^*(t)|^{0.5}\{|\tau^*(t)|-\tau^*_{cr}\} \tag{6}$$

or by the following expression which is convenient for comparison with the existing formulas for steady current.

$$\frac{q_B(t)}{u_*(t)d} = 3.5\mathrm{sign}\{\tau^*(t)\}\{|\tau^*(t)|-\tau^*_{cr}\} \tag{7}$$

where $u_*(t)=\sqrt{|\tau(t)|/\rho}$, and ρ is the density of the fluid.

In Fig.5, Eq.(7) is compared with the bed load formula for steady current in the alluvial channel. The dashed-and-dotted line is the result of the present study, Eq.(7). The solid line show the representative bed load formulas for the alluvial channel. The present study is close to the Sato et al. formula (1958) which is often called as the PWRI (Public Works Research Institute) formula and is widely used in Japan. Therefore, the present result will be smoothly continued to the sediment transport rate in steady current, as the excursion length of water particles becomes longer.

4.2 BEDLOAD TRANSPORT OVER A SLOPING BEACH

The bed load formula introduced in the preceding section is applied to a sloping beach case.

Beach profiles measured at t=0 min and t=10 min are shown in the upper figure of Fig.6. Since the beach profiles were not always two-dimensional but variable in the transversal direction, especially near the plunging point, values measured along three parallel longitudinal lines were averaged to determine the beach profile.

A sand bar developed near the plunging point, obviously nourished by the onshore sand transport from the seaward region. The sand transport rate is calculated from the measured beach profiles and is shown by the solid line in the lower figure. The direct measurements by use of a sand trap are also given by black circles, which show a good agreement with the solid line.

In order to apply Eq.(4) to the prediction of the bed-load transport rate on the slope, wave profiles were measured at the points where sand traps were installed.

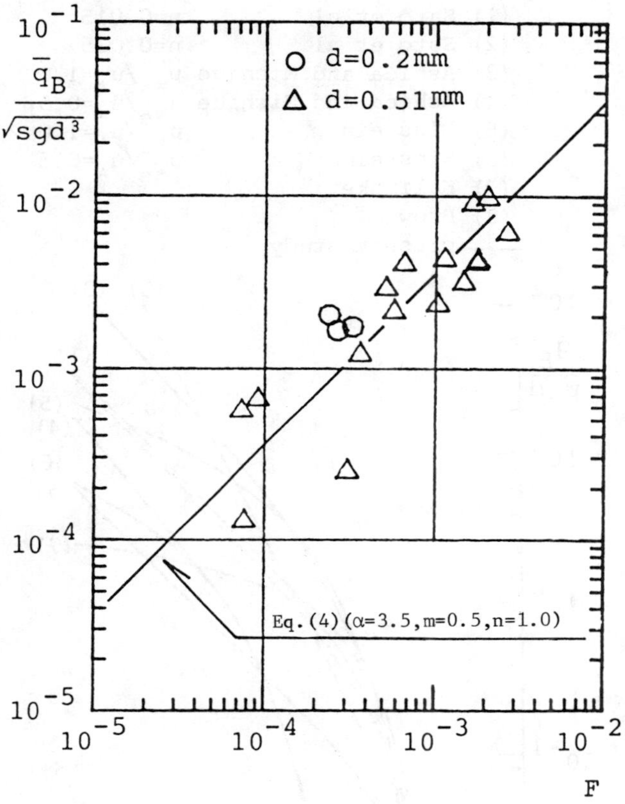

Fig.4 Bed load transport rate over a flat bed.

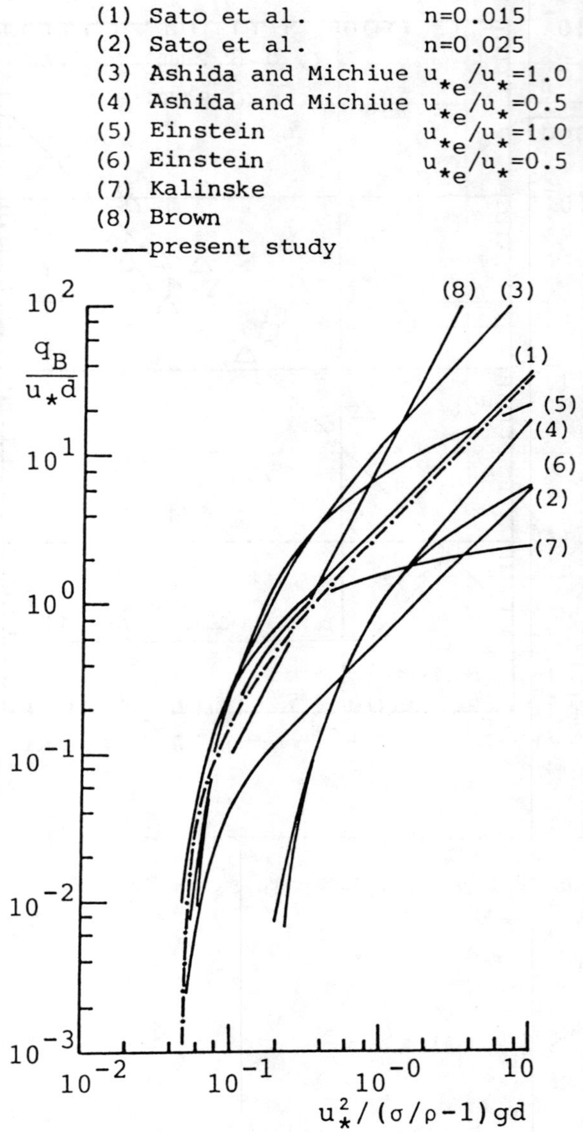

Fig.5 Comparison of bed load formulas.

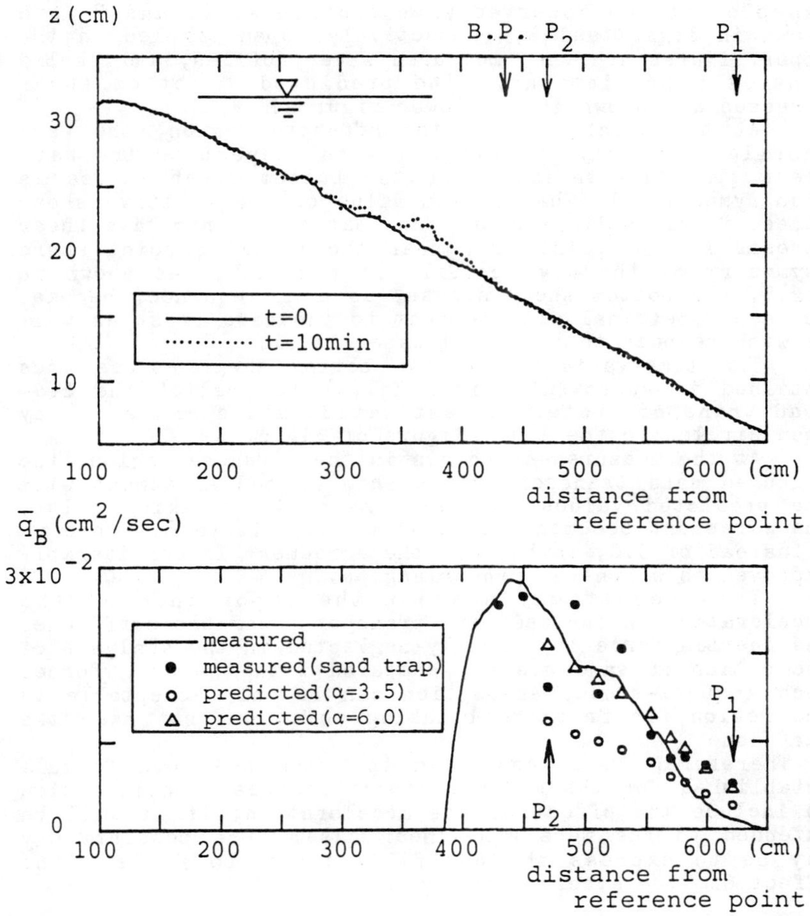

Fig.6 Change of beach profile and bed load transport rate.

Examples of the observed wave profile at P_1 and P_2 are shown in Figs.7 and 8, respectively. Open circles in the upper figures are the measured wave profiles, and solid lines the predictions. The predicted bottom shear stresses are shown in the lower figures.

At the point, P_1, in the offshore region, the wave profile is nearly symmetrical with respect to the wave crest. The time-variation of the bottom shear stress is also symmetrical. The maximum value of the positive shear stress is slightly greater than that of the negative shear stress. At the point, P_2, near the breaking point where asymmetry of the wave profile is remarkable as shown in Fig.8, the bottom shear stress, as a consequence, becomes quite asymmetrical with respect to the wave crest as well as with respect to the still water level.

The time-variation of the bottom shear stress thus obtained is substituted into Eq.(4) to predict the bed-load transport rate. The estimated values are shown by open circles in the lower figure of Fig.6.

At the measurement points in the offshore region, the measured sand transport rates show a good agreement with the predicted values. However, near the breaking point, the agreement becomes worse. If we substitute 6.0 into α, instead of 3.0, in Eq.(4), the agreement is considerably improved as shown by open triangles.

This result may suggest the importance of the acceleration in the sediment transport near the surf zone. The inertia force is usually neglected in the analysis of the motion of sand grains, compared with the drag force. Such a quasi-steady assumption can be only acceptable in the region far from the breaking point but not near the surf zone.

Therefore, it is concluded that the bed load formula established for the offshore condition needs modification to include the effect of the acceleration, if it will be extended to the more non-linear case. One possible way may be to express the coefficient α to reflect the effect of the acceleration.

5.CONCLUSION

In the previous studies on the sand movement due to wave motion, Jonsson's friction coefficient has been conveniently and widely used for the formulation of sand transport rate. This means that the non-linearity of wave motion was not adequately taken into consideration in those investigations. The present study, on the contrary, makes it possible to include the non-linear effect, by using the modified stream function theory of the author.

It is assumed that the instantaneous bed-load transport rate is expressed by Eq.(3) on the basis of the preceding studies on the sediment movement due to steady current. Exponents m and n in Eq.(3) are found to be 0.5 and 1.0, which ensure a smooth continuation from wave

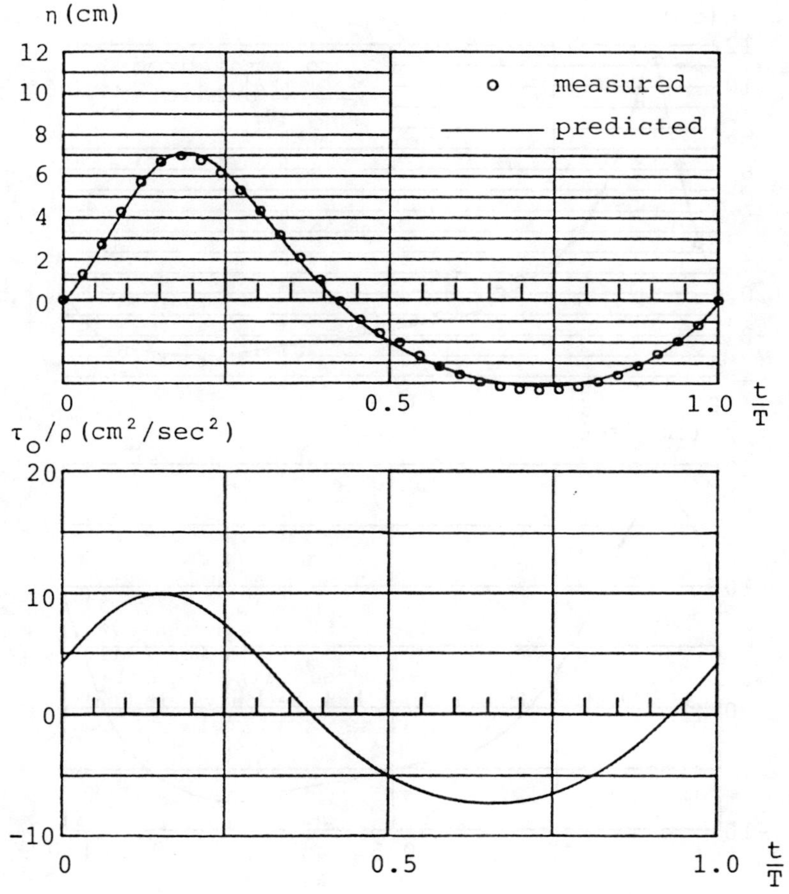

Fig.7 Time-variation of water surface level and bottom shear stress at measured point P_1.

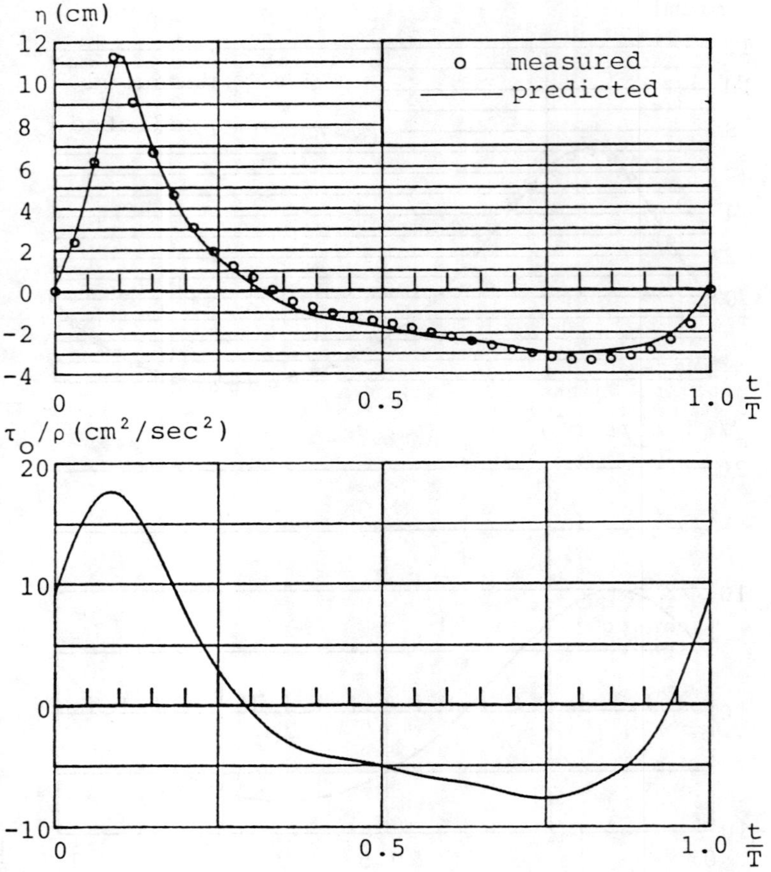

Fig.8 Time-variation of water surface level and bottom shear stress at measured point P_2.

motion to steady flow as the excursion length of water particles becomes longer. The coefficient of proportionality α is determined as 3.0 in case of a flat bottom.

The measured sediment transport rate near the surf zone is remarkably greater than the predicted. This suggests that the quasi-steady assumption can no longer be applicable in this region.

ACKNOWLEDGEMENTS

The author wishes to express his grateful thanks to Professor Shuto, Tohoku University, for his helpful advice over the course of the study. A part of this study was financially supported by the Grant-in-Aid for Encouragement of Young Scientist of the Ministry of Education, Science and Culture, Japan.

REFERENCES

Dean, R.G.(1965): Stream function representation of non-linear ocean waves, J. Geophy. Res., Vol.70, No.18, pp.4561-4572.

Horikawa, K., T. Sunamura and T. Shibayama(1977): Experimental study of two-dimensional shore transformation --- A method for measurement of sand transport rate in offshore region---, Proc. 24th Japanese Conf. on Coastal Engng., pp.170-174 (in Japanese).

Horikawa, K.(1988): Nearshore Dynamics And Coastal Process, Theory, Measurement, And Predictive Models, pp.194-217, University of Tokyo Press.

Jonsson, I.G.(1966): Wave boundary layer and friction factors, Proc. 10th Conf. on Coastal Engng., pp.127-148.

Sato, S., Kikkawa, H. and K. Ashida(1958): Study on the bed-load transportation, J. Res. Pub. Works Res. Inst., No.98, pp.1-18 (in Japanese).

Tanaka, H. and N. Shuto(1981): Friction coefficient for a wave-current coexisting system, Coastal Engineering in Japan, Vol.24, pp.105-128.

Tanaka, H.(1988): Bottom boundary layer under non-linear wave motion, Proc. ASCE, WW-Div (in press).

CHAPTER 134

Physical Modelling of Beach Erosion and Littoral Drift

Otavio J. Sayao, M.ASCE[1] and Robert B. Nairn[2]

Abstract

A new procedure for physical modelling of beach sedimentary processes is presented. It is shown that the modelling requirements proposed by Dean (1985) are necessary but not sufficient for dynamic similarity. Quantification of scale effects due to slope and relative grain size scaling conditions enables extrapolation of the physical model results to prototype situations. Selected examples of the application of the proposed model design are also given.

1. Introduction

A model is dynamically similar to its prototype if all the dimensionless variables governing the model and prototype phenomena are identical. The dimensionless functional relationship of any mechanical quantity, particularly sediment transport, is given by,

$$\Pi_A = \varphi_A (x_1, x_2, \ldots x_n) \tag{1}$$

where Π_A is the dimensionless version of A (the mechanical quantity) and x_i are the dimensionless variables. For dynamic similarity between the model and prototype, the condition to be satisfied is:

$$(\Pi_A)_p = (\Pi_A)_m \tag{2}$$

where the subscript p refers to the prototype value of Π_A and the subscript m refers to the model value of Π_A.

In a conventional physical model of beach morphology with a rectangular co-ordinate system, the general model scale n is given by the vertical scale value n_z where n is defined as the ratio of the prototype value over the model value.

---

1 F.J. Reinders and Associates Canada Limited, 201 County Court Blvd., Suite 500, Brampton, Ontario, L6W 4L2, Canada.

2 Imperial College of Science and Technology, Dept. of Civil Engineering, London, SW7 2BU, England.

Thus, Equation 2 may be written as:

$$n_{\Pi_A} = 1 \tag{3}$$

which is achieved if and only if,

$$(x_i)_p = (x_i)_m, \text{ or } n_{x_i} = 1 \tag{4}$$

is provided. If Equation 4 is satisfied then all x_i are identical in model and prototype and together with geometrical similarity, Π_A (the dimensionless version of a mechanical quantity of interest) is also identical in model and prototype.

2. Description of Littoral Processes

Sayao (1982) described sediment transport in the surf zone by the following dimensionless relationship:

$$\Pi_{Q_s} = \Phi(H_b/L_o, m_b, H_b/D, \rho_s/\rho) \tag{5}$$

where Π_{Qs} is the dimensionless littoral transport rate, H_b/L_o is the breaking wave steepness, m_b is the beach profile slope which characterises energy dissipation in the surf zone, H_b/D is the relative grain size parameter, and ρ_s/ρ is the density ratio (sediment to water) which reflects the influence of the density of the grains.

An additional dimensionless variable, especially important in situations where cross-shore transport is dominant, is the dimensionless fall time parameter, F_b or F_o,

$$F_b = \frac{H_b}{wT} \quad \text{or} \quad F_o = \frac{H_o}{wT} \tag{6}$$

where H_b is the breaker height, H_o is the deepwater wave height, T is the wave period and w is the fall velocity of the sand grains. According to Dean (1973), this parameter provides an indication of whether net onshore or offshore motion of sand will occur on a beach profile. A critical value of F_o is given in the Shore Protection Manual (US Army Corps of Engineers, 1977), for $F_o > 1$, recession of the shoreline will occur. The validity of this parameter in predicting the net cross-shore transport direction in situations where longshore transport is also appreciable can be confirmed by the laboratory data of Sayao (1982) and Readshaw (1979). Dean's fall time parameter is plotted versus deepwater wave steepness and the relative grain size parameter in Figures 1 and 2 for the results of these two three-dimensional model studies.

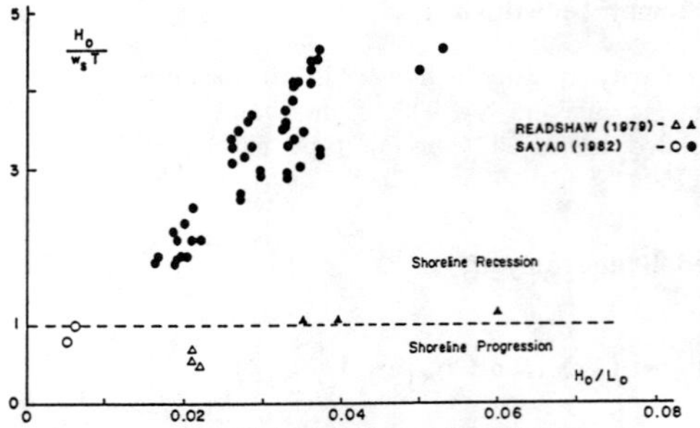

Figure 1. Dimensionless Fall Time vs. Wave Steepness

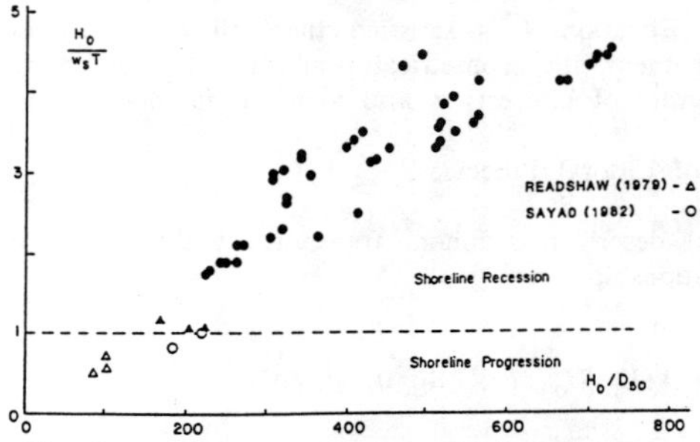

Figure 2. Dimensionless Fall Time vs. Relative Grain Size

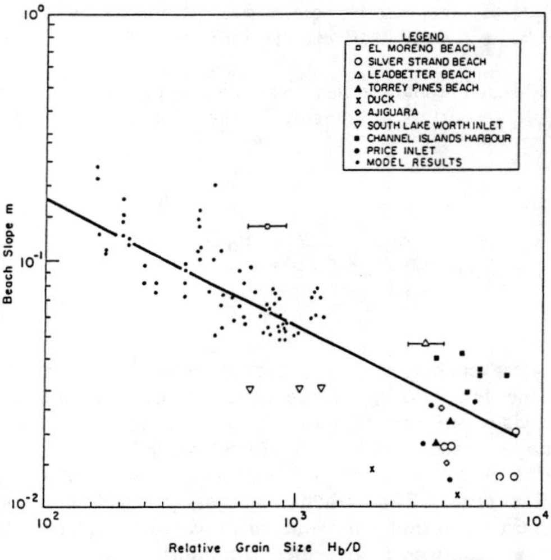

Figure 3. Beach Slope vs. Grain Size Relationship (from, Kamphuis et al., 1986)

3. Dynamic Similarity

Dynamic similarity is only achieved if all the prototype and model values of the dimensionless parameters on the right hand side of the following equation (Equation 5 extended to include the fall time parameter),

$$\Pi_{Q_s} = \varphi(H_b/L_o,\ m_b,\ H_b/D,\ \rho_s/\rho,\ F_o) \tag{7}$$

are equal, i.e. if their scales are:

$$n_{H_b/L_o} = 1;\ n_{m_b} = 1;\ n_{H_b/D} = 1,\ n_{\rho_s/\rho} = 1;\ n_{F_o} = 1.$$

Therefore, for physical model studies of surfzone processes (including cross-shore and alongshore sediment transport) in a given geometrically similar model, if the set of scaling conditions derived from Equation 7 are satisfied, then the dimensionless version of any mechanical quantity (e.g. Π_{Qs} in Equation 7) is identical in model and prototype. The implications of each of these conditions are investigated.

(i) A Wave Steepness Condition : $n_{Hb/Lo} = 1$.

The provision of this condition yields,

$$n_{H_b} = n_{L_o} \tag{8}$$

which results in a Froudian scale for the wave period as follows,

$$n_T = \sqrt{n} \tag{9}$$

This is a conventional scale for most physical models.

(ii) A Beach Slope Condition : $n_{mb} = 1$.

The representative beach slope is assumed to be,

$$m_b = d_b/\lambda_b \tag{10}$$

where the depth of breaking defines the offshore limit of the horizontal breaking distance λ_b, measured from the still water line. To satisfy this condition, the following equality must be achieved,

$$n_{d_b} = n_{\lambda_b} \tag{11}$$

This will only be possible in geometrically undistorted models with model grain size reduced geometrically as well, i.e.,

$$n = n_x = n_z = n_D \tag{12}$$

In an undistorted small scale hydraulic model all geometrical scales are the same. In a geometrically distorted model, the two horizontal scales n_x and n_y are identical but differ from the vertical scale n_z. The geometrical model distortion N may thus be defined as:

$$N = \frac{n_x}{n_z} = \frac{n_y}{n_z} \tag{13}$$

An empirical relationship for beach slope was given in Sayao (1982) as follows (see also Sayao et al, 1985),

$$m_b = 1.5(H_b/D_{50})^{-\frac{1}{2}} \tag{14}$$

valid for medium and fine sands. Figure 3 shows a plot of beach profile slope versus the relative grain size parameter for the model data of Sayao (1982) and Readshaw (1979) and field data as compiled and analyzed by Davies (1984). Equation 14 shows good agreement with the trend of the data, albeit with large scatter. Sayao et al (1985) suggested that this relationship may be improved by introducing the wave steepness. However, comparison with available data did not improve scatter, indicating a need for the analysis of further field measurements where both wave characteristics and beach profiles are recorded simultaneously. Equation 14 can also be shown to fit the data of van Hijum and Pilarczyk (1982) for gravel beaches.

The relationship described by Equation 14 can be used to develop a scaling relation for the beach slope condition as follows,

$$n_{m_b} = (n_{H_b}/n_{D_{50}})^{-\frac{1}{2}} = (n_D/n)^{\frac{1}{2}} \tag{15}$$

If the model grain size is not reduced geometrically then

$$n_{m_b} = (n_D/n)^{\frac{1}{2}} = M_1 \tag{16}$$

where $M_1 \neq 1$ is the distortion due to the beach slope condition, which leads to a scale effect.

In an attempt to compensate for this distortion, Bijker (1967) suggested that for distorted models the geometric distortion should be equal to the ratio between the equilibrium beach slopes (which according to Dean, 1977 is a function of the grain diameter).

(iii) A Relative Grain Size Condition : $n_{Hb/D} = 1$.

The imposition of this condition yields

$$n_{H_b} = n_D \tag{17}$$

which will only be possible when the sand grain size is scaled down geometrically. In practice, this is only possible for shingle beaches or beaches with coarse sand grains. Thus, an imposed grain size distortion (M_2) may be present in beach models, when $M_2 \neq 1$, and the scale effect from this non-similarity is quantified as follows,

$$n_{H_b/D} = n/n_D = M_2 \tag{18}$$

(iv) A Density Ratio Condition : $n_{\rho s/\rho} = 1$.

To satisfy this condition the following must be provided.

$$n_{\rho_s} = n_\rho = 1 \tag{19}$$

In conventional physical models using water and sand this condition is fulfilled. Departure from this condition (for example, through the use of lightweight materials) leads to significant scale effects and dissimilarity between model and prototype morphological development in the surfzone (see Kamphuis, 1975).

(v) A Fall Velocity Condition : $n_{Fo} = 1$.

The adherence to this condition leads to the Froudian scaling of fall velocity as derived from the following relationship,

$$n_{H_b} = n_w . n_T \text{ or } n_w = \sqrt{n} \tag{20}$$

Based on the work of Dean (1973), fall velocity scaling has been suggested by many researchers (e.g. Dalrymple and Thompson, 1976; Kamphuis, 1982; Hallermeier, 1984; Ito and Tsuchiya, 1984; Dean, 1985; Kriebel et al, 1986 and Yalin et al, 1986).

Several scaling laws dealing with the distortion of the fall velocity condition have also been presented in the literature (Valembois, 1961; LeMehaute, 1970; Vellinga, 1982 and 1986; Sayao and Guimaraes, 1984). For example, using the Vellinga (1982) scaling law:

$$N = n_x/n_z = (n_z/n_w^2)^{\alpha} \tag{21}$$

where by definition $N = n_x/n_z = n_y/n_z$ and the fall velocity scaling criterion, if satisfied, gives $n_x/n_z = 1$ for any value of the exponent α, i.e. no geometrical distortion. The value of the exponent α in Equation 21 varies between 0.25 (Vellinga, 1982 and 1986) and 0.50 (Valembois, 1961; LeMehaute, 1970; Sayao and Guimaraes, 1984).

4. Sediment Transport Scale and Time Scale for Littoral Processes

The dimensionless littoral transport rate (Π_{Qs}) expression depends on the units of the sediment transport rate Q_s. Adopting Q_s as a volumetric rate (m³/s), Sayao (1982) defined,

$$\Pi_{Q_s} = \frac{(1 - p)\, Q_s}{(\rho/\rho_s) H_b^2\, (H_b/T)} \tag{22}$$

where p is the beach sand porosity. Assuming $n_{(1-p)} = 1$, the scale of dimensionless sediment transport in the littoral zone is,

$$n_{\Pi_{Q_s}} = \frac{n_{\rho_s} n_{Q_s} n_T}{n_\rho\, n_H^3} = \frac{n_{Q_s}}{n^{5/2}} \tag{23}$$

for Froudian models with sand as the model material.

The time scale for littoral processes (or morphological development) may be derived from Equation 23, using the definition of Q_s (in m³/s) as,

$$n_{Q_s} = n_V/n_{t_s} \tag{24}$$

where n_V is the scale of volume of sand. Thus

$$n_{t_s} = n_V/n_{Q_s} = \frac{n_V}{n^{5/2} n_{\Pi_{Q_s}}} \tag{25}$$

Equation 25 shows that the time scale is influenced by the scale of dimensionless transport rate and by the volumetric scale. There is a possibility that the former will be different from 1 and the latter to be different than n^3 depending on the model design (i.e. whether or not there are distortions due to deviation from the conditions (i) to (v) and whether or not there is geometrical distortion).

5. Proposed Moveable Bed Model Design

The design of a moveable bed model study is not a simple task. It requires a clear understanding of the potential scale effects, as well as a thorough interpretation of model results in order that solutions for the prototype engineering problem may be proposed. Also, prototype data must be available so that prototype parameters related to the coastal phenomena at the study site may be defined.

Further, it is necessary to define the physical model scales. It is proposed here that a geometrically undistorted model is used if at all possible. Otherwise, the scaling law of Equation 21 can be used to determine an appropriate geometrical distortion and model grain size.

With reference to the model grain size it is proposed that the fall velocity criterion is used in the moveable bed model study if possible. This is preferred over other scaling criterions (such as geometrically reduced model material) because:

(a) Dean's parameter would be the same in model and prototype. This means that both model and prototype beaches will be either eroding or accreting, and dynamic similarity of erosional and accretional processes is ensured.

(b) Scale effects resulting from the non-similarity of m and H_b/D are less detrimental and more easily quantifiable than the effects resulting from the non-similarity of F_o.

If fall velocity scaling cannot be used for model design due to other constraints, it is recommended that the moveable bed model should have F_o-values in the same region as the prototype ones, i.e., that for both model and prototype F_o-values are simultaneously above or below the critical value $F_o \sim 1$.

Finally, with respect to grain size, it is strongly recommended that the model value of H_b/D exceeds 300 to ensure that suspended load is able to occur at the model scale (see Nairn, 1985).

6. Quantification of Scale Effect

The modelling design criteria are summarised in Table 1. For the quantification of scale effect, a new procedure is presented for geometrically undistorted, fall velocity scaling models. In this case, distortions are due to non-similarity of beach profile slope and relative grain size. Thus using Equations 16 and 18, the combined scale effect M due to both contributing parameters is,

$$M = M_1 . M_2 = (n_D/n)^{\frac{1}{2}} . \ (n/n_D) = (n/n_D)^{\frac{1}{2}} \qquad (26)$$

Also, $n_{\Pi Qs}$ is not equal to unity but for these types of models,

$$n_{\Pi_{Q_s}} = M = (n/n_D)^{\frac{1}{2}} \tag{27}$$

Hence, the rate of morphological development (or the littoral process time scale) may be quantified using Equations 25 and 27 as,

$$n_{t_s} = n/(n_D)^{\frac{1}{2}} \tag{28}$$

and the littoral transport rate scale becomes (using Equations 24, 27 and 28),

$$n_{Q_s} = n^2\sqrt{n_D} \tag{29}$$

Both Equations 28 and 29 are valid for geometrically undistorted, fall velocity scaling models only. If other scaling criteria are used, the derivation of these equations would be different. The expressions apply to both alongshore and cross-shore sediment transport. However, caution is advised in applying the time scale expression (Equation 28) to cross-shore problems where either the model or prototype profiles are near a stable equilibrium form.

Table 1. Geometrically Undistorted Fall Velocity Scaling Model

Scale Condition	Similarity	Comment
(i) wave steepness	$n_{H/L} = 1$	Froudian model
(ii) beach slope	$n_m = M_1 \quad (\neq 1)$	scale effect
(iii) relative grain size	$n_{H/D} = M_2 \quad (\neq 1)$	scale effect
(iv) relative density	$n_{\rho s/\rho} = 1$	Sand in model
(v) Dean's parameter	$n_{Fo} = 1$	Fall velocity scale is Froudian
time for littoral processes	$n_{ts} = n/\sqrt{n_D}$	Equation 28
littoral transport rate	$n_{QS} = n^2\sqrt{n_D}$	Equation 29

7. Applications of the Modelling Procedure

The recommended modelling procedure could be tested if a field data set were available. Unfortunately, due to the difficulties of determining sediment transport rates on beaches during storm events an ideal prototype data set is not presently available. However, the recommended procedure can be evaluated by (1) comparing pairs of flume test results at different scales (and thus regarded as model and prototype) and (2) the use of numerical models to compare profile development at different scales subject to the proposed modelling procedure.

Kriebel et al (1986) have used the fall velocity modelling procedure (originally proposed by Dean, 1985) and compared small-scale flume results (model) with the large-flume experiments (prototype) of Saville (1957). For the Kriebel et al (1986) data, the time scale for littoral processes is calculated to be, $n_{ts} \sim 6$ and Figures 4 and 5 show profile comparisons according to Equation 28. The match of model and prototype profiles in these figures is fair, but so is the match shown in Kriebel et al (1986) based on a Froude time scale. Sediment transport in this example was predominantly directed offshore.

As an example of the sediment transport rate scale, experimental data presented by Mimura et al (1986) and shown in Figure 6, is used. In this case, using their coarse sand tests (Series B) as prototype and fine sand tests as model (Series A), $n \sim 2.1$ and $n_D = 4.2$ are found for two pairs of tests, the B8 and A9 tests on a $^1/20$ initial profile, and the B2 and A3 tests on a $^1/10$ initial profile. The calculated value for the sediment transport rate scale according to Equation 29 is $n_{Qs} = 9.3$, and the experimental measurements of cross-shore sediment transport rate are in agreement with the proposed modelling procedure. (See Table 2). Also, beach profiles shown in Figure 6 are representative and show similar features. It should be noted that the tests of Mimura et al (1986) which have been investigated were characterised by onshore sediment transport.

Table 2. Cross-shore Transport Rate of Mimura et al (1986)

Initial Slope	Q_{s_p} (m^3/hr.)	Q_{s_m} (m^3/hr)	n_{Q_s}	comments
$^1/20$	0.0113	0.001	11	compares well with Equation 29
$^1/10$	0.01	0.001	10	

The recommended modelling procedure is also applicable to situations where alongshore sediment transport is predominant. Kamphuis et al (1986) have proposed an equation for alongshore transport rate as follows,

$$\Pi_{Q_s} = 0.002\ (H_{bs}/D)(m/(H_{bs}/L_o)^{\frac{1}{2}}) \quad (30)$$

Moreover, the applicability of this equation to both field results (from which its empiricity was derived) and model scale results was demonstrated by Sayao et al (1985). Equation 30 was shown to predict sediment transport rates around model circular sand islands at various scales and with different grain diameters based on the work of Nairn (1985), see Figure 7 reproduced from Sayao et al (1985). Since the variables in Equation 30 directly correspond to those of Equation 7, it follows that the time scale and sediment transport scale (Equations 28 and 29), based on the distortions of m and H_b/D respectively, will also apply to alongshore sediment transport models.

The cross-shore transport time scale was also investigated using the results of a numerical model for cross-shore sediment transport under random waves. The numerical model is based on a depth-integrated energetics approach as described by Nairn (1988). The modelling procedure is similar to that proposed by Stive (1986) based on the work of Bailard (1981). Profile change is determined from the net sediment transport rates calculated at points in a finite difference scheme

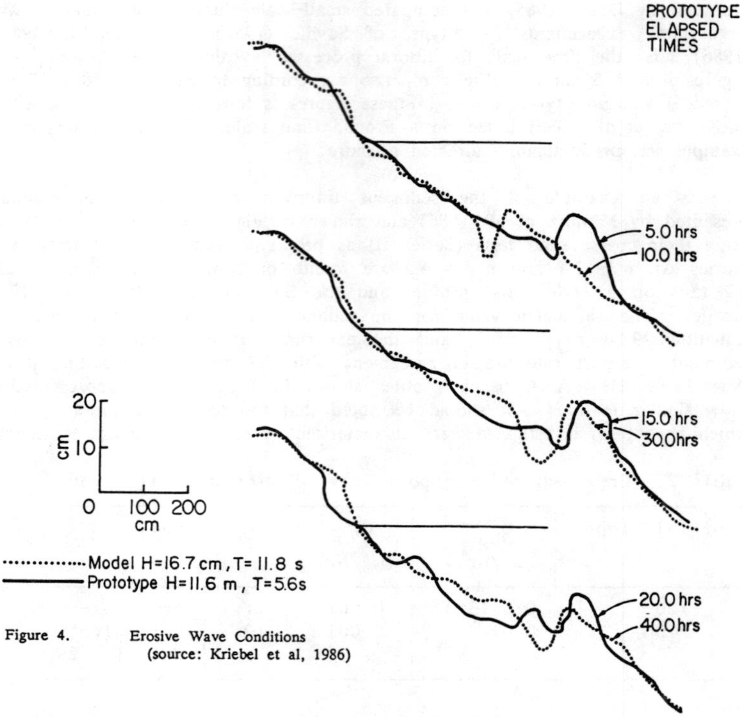

Figure 4. Erosive Wave Conditions (source: Kriebel et al, 1986)

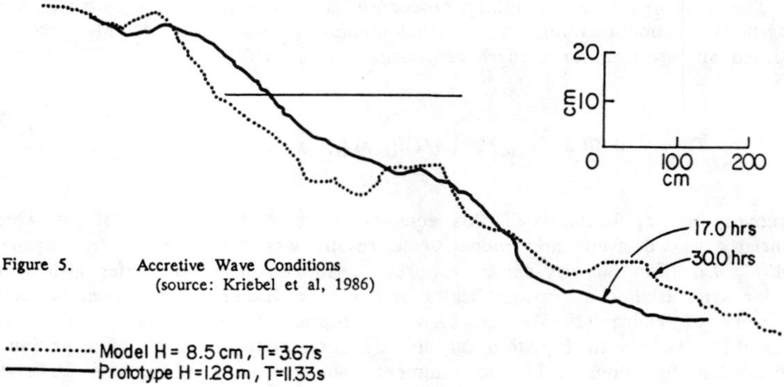

Figure 5. Accretive Wave Conditions (source: Kriebel et al, 1986)

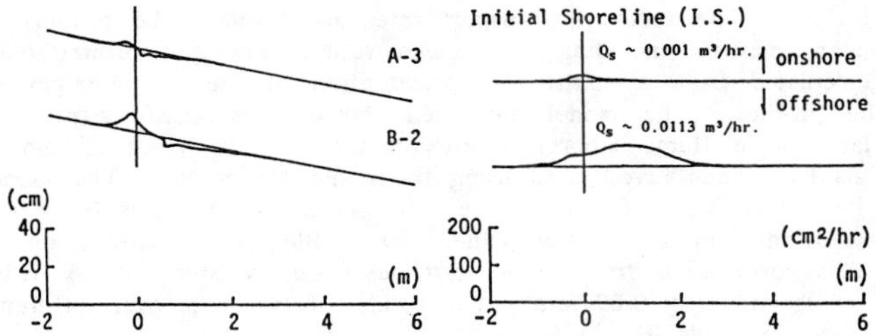

Beach profiles and cross-shore sediment transport rate
(initial slope=1/20, d_{50}=0.75mm)

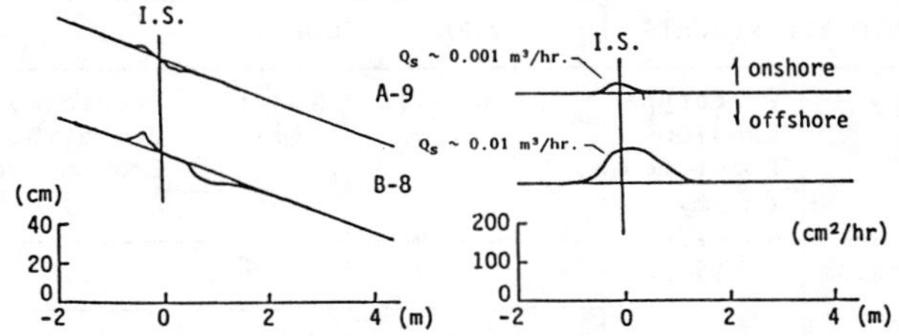

Beach profiles and cross-shore sediment transport rate
(initial slope=1/10. d_{50}=0.75mm)

Figure 6. Profiles and Cross-shore Transport
(from, Mimura et al, 1986)

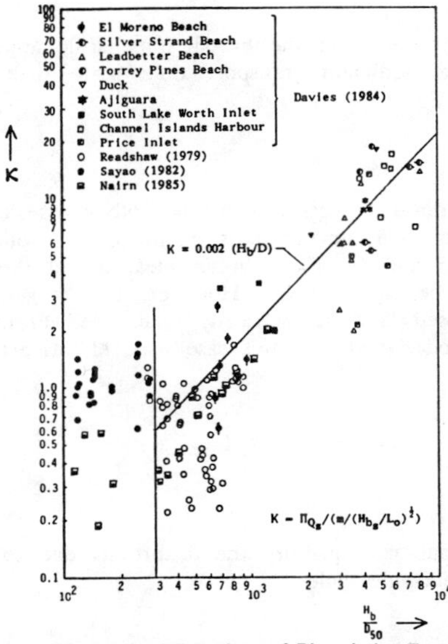

Figure 7. The Functional Dependency of Dimensionless Transport on H_b/D.
(from, Sayao et al, 1985)

across the profile. The sediment transport rates are taken to be proportional to local flow characteristics (including both mean return flow and orbital velocities) which are calculated from a statistical representation of the wave height at any point on the profile. The model has been shown to successfully predict both small and large scale flume laboratory erosion tests. Two pairs of model and prototype tests have been investigated using the numerical model. The model was assumed to be a $^{1}/10$ scale of a prototype which consisted of waves (H = 1.23 m, T = 6.33 s) incident on an initially plane $^{1}/40$ profile with a sand grain size of 0.2 mm. This corresponds to a model test described by Stive (1986) with H = 0.123 m, T = 2s and D = 0.09 mm. The results of two tests over different time periods are given in Table 3.

Table 3. Numerical Model Results.

PROTOTYPE RESULTS		MODEL RESULTS ($n = {}^{1}/10$)		
Prototype Period	Prototype Erosion Predicted (m^3/m)	Time Scales Used	Model Times	Predicted Prototype Erosion Volumes (m^3/m)
12 hrs.	15	$n_{ts} = 6.7$	1.8 hrs	7.0
		$\sqrt{n} = 3.2$	3.8 hrs	12.2
180 hrs.	86	$n_{ts} = 6.7$	27 hrs	49
		$\sqrt{n} = 3.2$	57 hrs	83

These results again demonstrate that the Froude time scale appears to be a better predictor of the offshore sediment transport time scale than that proposed by Equation 28.

8. Conclusions

The similarity of littoral processes including both cross-shore and alongshore sediment transport is due to 5 dimensionless parameters as follows (i) H_b/L_o, (ii) m_b, (iii) H_b/D, (iv) ρ_s/ρ and (v) F_o. In an ideal model the scales of each of these parameters should be equal to 1. However, this is generally not possible. The model design proposed is a geometrically undistorted Froude model with fall velocity scaling of the model sand. In this case scale effects are only due to

$$n_{m_b} \neq 1 \quad \text{and} \quad n_{H_b/D} \neq 1.$$

In the model proposed one can quantify the distortions due to the non-similarity of (ii) and (iii) by the following scales,

$$n_{t_s} = n/\sqrt{n_D} \qquad \text{(Equation 28)}$$

$$n_{Q_s} = n^2\sqrt{n_D} \qquad \text{(Equation 29)}$$

Comparisons to practical examples have successfully demonstrated the validity of these equations in situations of onshore transport in the cross-shore direction and for alongshore transport. The time scale for offshore sediment transport does however appear to be better represented by the Froude time scale than the scale proposed in Equation 28 above.

The conclusions noted above are based on comparisons to laboratory and numerical model results. Validation of the proposed model design and distortion expressions with a comprehensive field data set is still required pending the availability of an adequate data set.

9. References

Bailard, J.A. (1981). An Energetics Total Load Sediment Transport Model for a Plane Sloping Beach. J. of Geophys. Res. (86), C11, pp. 10938-10954.

Bijker, E.W. (1967). Some Considerations About Scales for Coastal Models with Moveable Bed. Delft Hydraulics Laboratory, Publication No. 50.

Dalrymple, R.A. and Thompson, W.W. (1967). Study of Equilibrium Beach Profiles. Proc. 15th Conf. on Coastal Eng. ASCE. pp. 1277-1296.

Davies, M.H. (1984). Littoral Transport Rate Prediction. M.Sc. Thesis. Queen's University, Kingston, Canada.

Dean, R.G. (1973). Heuristic Models of Sand Transport in the Surfzone. Proc. 1st Australian Conf. on Coastal Eng, Sydney, pp. 208-214.

Dean, R.G. (1977). Equilibrium Beach Profiles : U.S. Atlantic and Gulf Coasts. Ocean Eng. Rept. No. 12. U. of Delaware, Newark, 45pp.

Dean, R.G. (1985). Physical Modelling of Littoral Processes. In "Physical Modelling in Coastal Engineering". R.A. Dalrymple, Ed., A.A. Balkema, Rotterdam, pp. 119-139.

Hallermeier, R.J. (1984). Added Evidence on New Scale Law for Coastal Models. Proc. 19th Conf. on Coastal Eng. ASCE. pp. 1227-1243.

Ito, M. and Tsuchiya, Y. (1984). Scale Model Relationship of Beach Profile. Proc. 19th Conf. on Coastal Eng. ASCE. pp. 1386-1402.

Kamphuis, J.W. (1975). The Coastal Mobile Bed Model. Civil Eng. Rept. No. 75. Queen's University, Kingston, Canada.

Kamphuis, J.W. (1982). Coastal Mobile Bed Modelling from a 1982 Perspective. Civil Eng. Rept. No. 76. Queen's University, Kingston, Canada.

Kamphuis, J.W., Davies, M.H., Nairn, R.B. and Sayao, O.J. (1986). Calculation of Littoral Sand Transport Rate. Coastal Eng. 10, pp. 1-21.

Kriebel, D.L., Dally, W.R. and Dean, R.G. (1986). Undistorted Froude Model for Surfzone Sediment Transport. Proc. 20th Conf. on Coastal Eng. ASCE. pp. 1296-1310.

LeMehaute, B. (1970). Comparison of Fluvial and Coastal Similitude. Proc. 12th Conf. on Coastal Eng., ASCE, pp. 1077-1096.

Mimura, N., Otsuka, Y. and Watanabe, A. (1986). Laboratory Study on Two-Dimensional Beach Transformation Due to Irregular Waves. Proc. 20th Conf. on Coastal Eng. ASCE. pp. 1393-1408.

Nairn, R.B. (1985). Scale Series Approach to Coastal Mobile Bed Modelling. M.Sc. Thesis, Queen's U., Kingston, Canada.

Nairn, R.B. (1988). Prediction of Wave Height and Mean Return Flow in Cross Shore Sediment Transport Modelling. Proc. IAHR Symposium on Mathematical Modelling of Sediment Transport in the Coastal Zone. Copenhagen, pp. 193-202.

Readshaw, J.S. (1979). A Model Study of Alongshore Sediment Transport. M.Sc. Thesis, Queen's U., Kingston, Canada.

Saville, T. (1957). Scale Effects in Two-Dimensional Beach Studies. Proc. 17th General Meeting of IAHR. Lisbon, Vol. 1, p. A3-1 to A3-10.

Sayao, O.J. (1982). Beach Profiles and Littoral Sand Transport. Ph.D. Thesis. Queen's U., Kingston, Canada.

Sayao, O.J. and Guimaraes, J.C. (1984). Experimental Verification of Similarity Criteria for Equilibrium Beach Profiles. Proc. 19th Conf. on Coastal Eng. ASCE. pp. 1342-1357.

Sayao, O.J., Nairn, R.B. and Kamphuis, J.W. (1985). Dimensional Analysis of Littoral Transport. Proc. of the Canadian Coastal Conf., Newfoundland, pp. 241-255.

Stive, M.J.F. (1986). A Model for Cross-Shore Sediment Transport. Proc. 20th Conf. on Coastal Eng. ASCE.

U.S. Army Corps of Engineers (1977). Shore Protection Manual. Coastal Eng. Res. Centre. 3rd Ed., Washington.

Valembois, J. (1961). Estude Sur Modele du Transport Littoral, Condictions de Similitude. Proc. 7th Conf. on Coastal Eng. ASCE. pp. 277-307.

van Hijum, E. and Pilarczyk, K.W. (1982). Gravel Beaches. Delft Hydraulics Laboratory. Publication No. 274.

Vellinga, P. (1982). Beach and Dune Erosion During Storm Surges. Coastal Eng. 6., pp. 361-387.

Vellinga, P. (1986). Beach and Dune Erosion During Storm Surges. Ph.D. Thesis. Delft U. of Technology.

Yalin, M.S., Sayao, O.J. and Lai, G.Y. (1986). Physical Modelling of Suspended Solids in Open Channels. Proc. of Euromech 192. W. Betcheler, Ed., A.A. Balkema, pp. 161-169.

CHAPTER 135

Field Calculations of Wave Energy Dissipation and Related Beach Profile

By
James R. Tallent 1
Takao Yamashita 2
Yoshito Tsuchiya 3

Introduction

The process by which wave energy dissipates across the surf zone and its affect on the bed profile is, of course, a topic of immediate concern and debate. Various concepts of the wave energy dissipation process have been modeled, however, additional research is needed before confidence can be placed in a particular calculation scheme. In addition to the problems associated with proper model derivation a method of application and result interpretation of actual surf zone field data must be devised and understood. This is, of course, prerequisite to any realistic use of a wave energy dissipation model in an engineering project. The following study was therefore conducted in order to examine the applicability of surf zone field data to wave energy dissipation models and to investigate the bed profile relationship.

Two wave energy dissipation models were selected for comparison in this study, the 'Undertow Model'(UM) which is based on the conservation of wave energy flux across the surf zone (3), and the 'Turbulent Bore Model'(TBM) which is based on hydraulic jump theory (2). Individual waves were identified in the wave record by employing the zero up-crossing method, and wave energy calculations were based on small amplitude wave theory, Svendsen's non-linearity parameter **Bo** (4), and the 1/3 Significant Wave classification.

Wave elevation data which was collected before, during, and after the occurrence of a storm typical to the Ogata coast, (**H**max=5.5 m.), was used for energy dissipation calculations. Lastly, an Ogata coast average cross-shore energy dissipation estimate was used in conjunction with Dean's Equilibrium Beach Profile Equation (2), to examine the stability of Ogata Coast, Japan.

1 Visiting Researcher, D.P.R.I., Kyoto University, Japan
2 Instructor, D.P.R.I., Kyoto University, Japan
3 Professor, D.P.R.I., Kyoto University, Japan

Field Measurements and Data Processing

Measurements necessary to calculate the wave energy dissipation rate were made by instruments mounted on D.P.R.I.'s T-shaped observation pier located at Ogata coast, Japan. This pier, shown in Figure 1b, is 255.6 meters in length normal to the shoreline, and 100 meters in length at the offshore T-section. A total of 14 wave gauges were positioned along the pier, as indicated in Figure 1a. In this figure, wave gauges are referred to as '**C**' for capacitance type or as '**U**' for ultrasonic type.

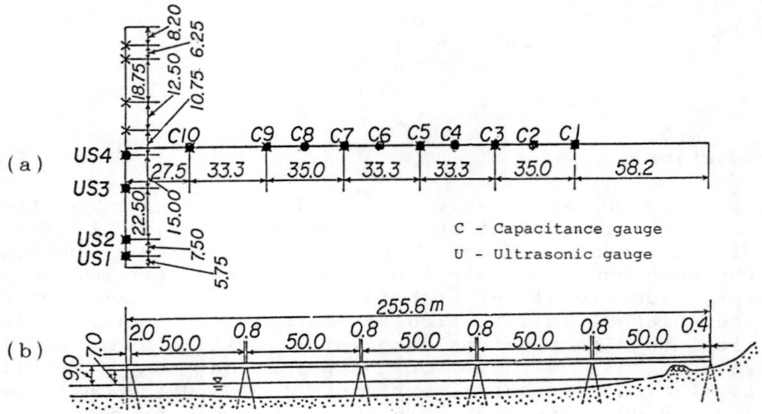

Figure 1. Ogata Coast T-shaped observation pier

Calibrations of wave gauges were continuously performed in order to determine and retain their accuracy. Wave elevation data was collected over a three day period, during which the coastal storm occurred. Wave data was stored on analogue data recorders. Digitalizing of records was accomplished by use of an A-D converter which operated at a sampling rate of 0.08 seconds. A high pass filter of 100 seconds and a low pass filter of 2.5 seconds were used to remove the long wave components and high frequency breaking wave fluctuations, respectively. Filter selection was based on record observation. The resulting wave data was then partitioned into 22 minute blocks and individual waves identified by the zero up-crossing method. Beach profile measurements were made by depth soundings conducted along D.P.R.I.'s T-shaped observation pier.

Theoretical Considerations

There are several ways by which the potential energy of irregular waves can be approximated. However, due to our data being limited to surf zone surface fluctuations, where periodic wave theories are invalid, we choose to evaluate potential energy on the basis of individual wave elevation measurements, incorporating non-linearity through Svendsen's **Bo** parameter, Equation 1. The **Bo** variation with profile shape is described in Table 1 (1). Note that if **Bo** equal 1/8 (0.125) the wave shape is of sinusoidal form and potential energy is evaluated by small amplitude wave theory.

$$Bo = \frac{1}{T} \rho g \int \frac{\eta^2}{\rho g H^2} dt = \frac{\eta^2}{H^2} \tag{1}$$

	Profiles	Bo
	Sinusoidal	0.125
	Cnoidal	$<$ 0.125
	Parabolic	0.089
	Linear	0.083

Table 1. Nonlinearity parameter **Bo** (after Basco, 1986)

Consequently, small amplitude wave theory estimates, Equation 2, and **Bo** modified potential energy estimates, Equation 3, are employed for comparison in the energy dissipation calculations.

$$E_1 = \frac{1}{16} \rho g H^2 \tag{2}$$

$$E_2 = \frac{\rho g}{2T} \int_0^T \eta^2 dt = \frac{1}{2} \rho g H^2 Bo \tag{3}$$

Where **T** is the wave period, ρ is the fluid density, **g** is the gravitational acceleration, η is the water surface elevation measured from the MWL, and **H** is the wave height.

The 'Undertow' model may be described as the conservation of wave energy flux across the surf zone. This relation may be expressed as:

$$\frac{\partial}{\partial x} (W_W + W_R) = D \tag{4}$$

where W_W and W_R represent the wave and residual (turbulent) energy fluxes, respectively. The flux of externally opposing energy sources are represented by **D**. In summing the external and residual energy fluxes an expression is derived which allows us to calculate energy dissipation across the surf zone in terms of wave elevation alone, Equation 5.

$$\frac{\partial W_W}{\partial x} = D - \frac{\partial W_R}{\partial x} = -\,DISS \qquad (5)$$

Furthermore, if we make the assumption that wave energy is traveling at group velocity **Cg**, the wave energy flux may be written as:

$$W_W = E\,C_g = E\sqrt{gh} \qquad (6)$$

where **E** is the sum of both potential and kinetic energy and **h** is the water depth. By employing the small amplitude and **Bo** modified wave energy estimates, along with Equations 2 and 3, and the 1/3 significant wave height classification, three energy flux relations are derived, Equations 7-9. These three relations will each be substituted into the dissipation relation of Equation 5, to examine computational variations.

$$W_A = \bar{E}_2\,C_g \equiv \bar{E}_2\sqrt{g\,h} \qquad (7)$$

$$W_B = (E_2)_{1/3}\,C_g = (E_2)_{1/3}\sqrt{g\,h} \qquad (8)$$

$$W_C = (E_1)_{1/3}\,C_g = (E_1)_{1/3}\sqrt{g\,h} \qquad (9)$$

The second method used to calculate the cross-shore energy dissipation rate was derived from hydraulic jump theory and is known as the 'Turbulent Bore Model' (TBM), Equation 10. This model is applicable only where the breaking wave propagates in a nearly constant form, such as that of a quasi-steady bore. Therefore, the cross-shore wave shape characteristics, described by **Bo**, must be investigated in order determine the region of model applicability.

$$DISS = \frac{\rho g h}{T}\,\frac{(d_c - d_t)^3}{4\,d_t\,d_c} \qquad (10)$$

where **h** is the mean water level, **T** is the wave period , **dc** and **dt** are the depth measurements from the crest and trough, respectively, and **DISS** is the rate of change of wave energy per unit shoreward distance expressed in the units of tons/sec/meter, Figure 2.

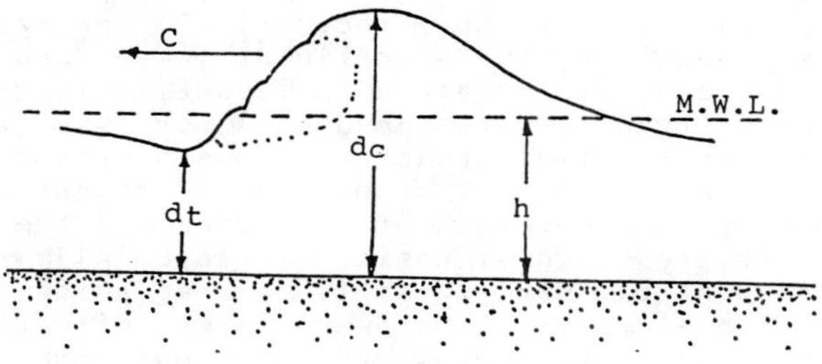

Figure 2. Turbulent Bore Model variables

The Ogata coast is best described as a high energy reflective type beach, while Dean's 'Equilibrium Profile Equation' was derived from typical low energy dissipative beaches, Equation 11. A comparison between the measured and the calculated profile will allow us to determine how unstable the profile is presently.

$$\mathbf{h} = \mathbf{A}\,\mathbf{x}^{2/3} \tag{11}$$

where **h** is the water depth, **A** is a parameter describing fluid and sediment properties, and **x** is the offshore distance. Furthermore, if: the 2/3 constant is replaced with the variable α , the spilling breaker assumption is adopted, Equation 12, and the energy dissipation relations described in Equations 5 and 6 are employed, an equation is derived which describes the cross-shore wave energy dissipation/beach profile relation, Equation 13.

$$\mathbf{H} = \gamma\,\mathbf{h} \tag{12}$$

$$\mathbf{DISS} = \frac{5}{16}\,\alpha\,\rho\,g^{3/2}\,\gamma^2\,h^{(3/2\,-\,1/\alpha)}\,A^{1/\alpha} \tag{13}$$

where γ is assumed to equal 0.8.

Results and Conclusions

All data collected before, during and after the coastal storm was divided into five time series, each representing a particular stage of the storm, and containing six hours of data. Data sets **N15**, **N18**, and **N19** correspond to sea state conditions best described as initial storm (steep short period waves), fully developed storm (well defined long crested waves), and post storm (swell), respectively. In Figure 3, cross-shore energy fluxes are shown for the relations described in Equations 7-9, and for the three storm stages described above. Several conclusions may be drawn from this figure, they include: 1) energy flux calculations based on the significant wave and small amplitude wave theory, are significantly larger and more scattered than results based on W_A or W_B , 2) the steep negative energy flux gradients seaward of wave gauge C8 indicates the region of rapid wave height decay, initial breaking, 3) shoreward of wave gauge C7 the energy flux gradients are slightly positive, especially for the fully developed storm, **N18**, possibly corresponding to wave rebuilding, and 4) from C3 shoreward large inflections in the energy flux gradient correspond to visual observations of incident and reflected wave interaction.

Because the W_B energy flux calculation, Figure 3, required the **Bo** parameter to be evaluated in terms of it's 1/3 significant value, it was necessary to investigate the **Bo**/wave-order relationship to clarify it's role. Therefore, in Figure 4, the variations in the **Bo** parameter with orders of increasing wave magnitude are presented at five surf zone locations.

The asterisk in the Figure 4 series indicates the locations of the 1/3 significant wave. To the left of the asterisk waves are of a greater order than the 1/3 significant and to the right the order is smaller. It is interesting to note that the wave shape parameter significantly fluctuates for wave orders less than the 1/3 significant, however, to the left of the asterisk the **Bo** parameter remains fairly constant. This observation indicates that the surf zone 1/3 significant **Bo** value may be used to approximate average surf zone energy characteristics.

As previously mentioned the TBM may be applied only where the breaking wave propagates in constant form, therefore, the **Bo** parameter must also be investigated in terms of it's cross-shore characteristics. In Figure 5 the cross-shore **Bo** variations are shown for the entire observation period, data blocks N15-N19, where an individual line within each data block represents a 22 minute **Bo** average.

In comparing each data block differences may be observed in the cross-shore **Bo** trends, however, between wave gauges C10 and C7 all data blocks show a similar **Bo** tendency. Therefore, within this zone of relatively constant **Bo** the TBM was applied and compared with dissipation results derived from the UM, Figure 6, where Diss-1 and Diss-2 represent TBM results based on 1/3 significant and individual wave measurements, respectively. The lower three curves are UM calculations based on the three previously defined energy flux relations.

Conclusions based on this figure include: 1) TBM calculations based on the significant wave component,

Diss-1, are approximately 2.0-2.5 times greater than calculations based on the UM, 2) TBM calculations based on individual wave measurements are of the same order of magnitude as UM calculations (approximately 0.01 tons/sec/mm) and 3) during the initial storm stages, N15-N16, maximum wave energy dissipation values are generated by the TBM, however, because maximum storm conditions were actually observed during the period corresponding to N17 we may conclude that the TBM is inapplicable for short period waves.

Finally, a comparison is made between the calculated, Equation 10, and measured beach profiles, Figure 7. Calculation assumptions include: 1) Diss = 0.01 ton/sec/m, 2) γ =0.8, and 3) h = 5.6 m. Also, the parameters **A** and were both varied in order to determine the best fit solution to the measured Ogata Coast profile.

In this study and several others, significant differences have been noted between dissipative (α = 2/3 generated) and reflective type beach profiles, suggesting that Dean's equation may be valid only under certain undefined conditions and that additional parameters are needed to describe the reflective profile.

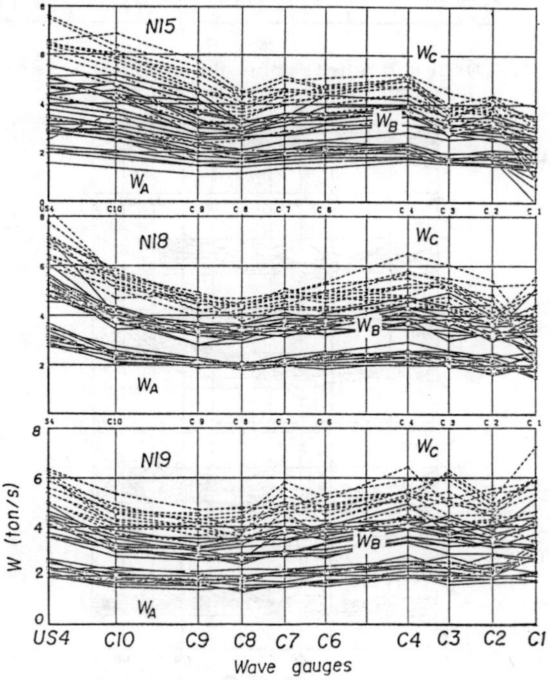

Figure 3. Cross-shore energy flux trends

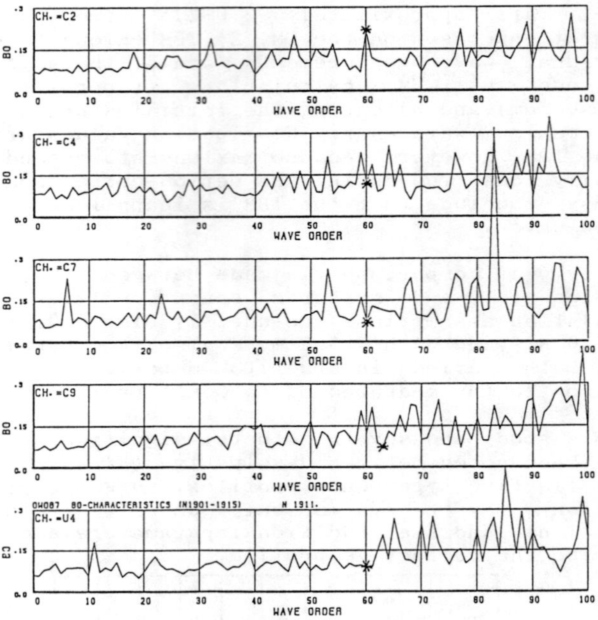

Figure 4. Ordered surf zone **Bo** characteristics

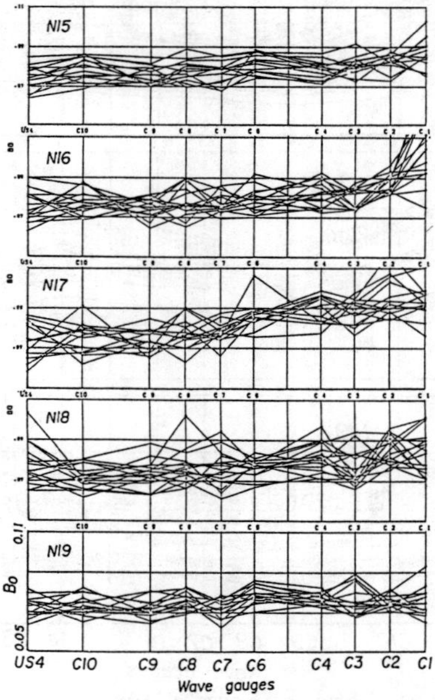

Figure 5. Cross-shore **Bo** characteristics

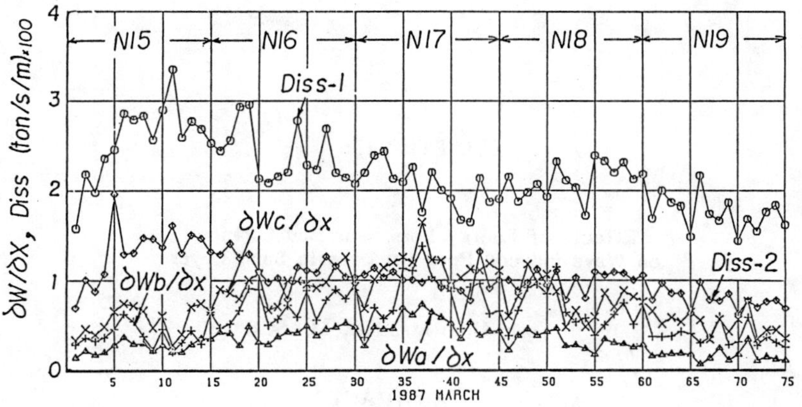

Figure 6. Wave energy dissipation comparison

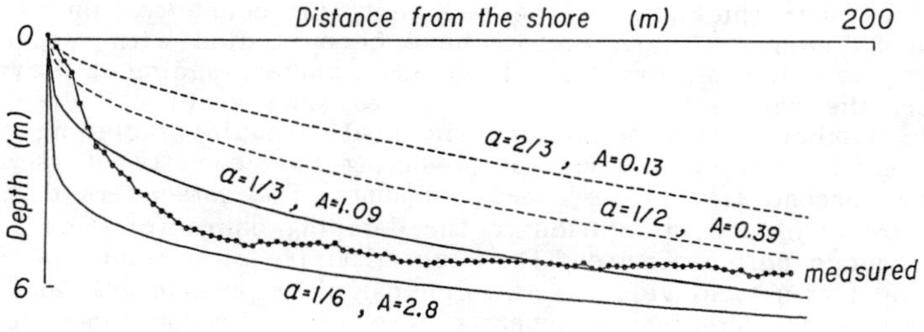

Figure 7. Comparison of measured and calculated profiles

References

1) Basco, D.R., Yamashita T.: Toward a Simple Model of the Wave Breaking Transition Region in the Surf Zones, Proc. 20th ICCE, pp. 955-970, 1986

2) Dean, R.G.: Equilibrium Beach Profiles: U.S. Atlantic and Gulf Coast, Ocean Engineering Report No. 12, Dept. Civil Engineering, University Delaware, 1977.

3) Svendsen I.A.: Mass Flux and Undertow in the Surf Zone, Coastal Engineering, Volume 8, Number 4, pp. 347-365, 1984

4) Svendsen I.A., Madsen, P.A., and Hansen, J.B.: Wave Characteristics in the Surf Zone, Proc. 16th ICCE, 1978

CHAPTER 136

Effects of Energy Loss near Bed Surface on Wave-Induced Pore Pressure in Sand Layer

Yoshi-Hiko MAENO, M.ASCE*

ABSTRACT

The effects of energy loss near the interface between the water and the sand bed on the wave-induced pore pressure are investigated by laboratory experiments. The interface between them includes the complicated factors: (1) roughness of seabed surface such as ripples; (2) boundary layer thickness in the sediment; (3) concentration of suspended sediments. These factors have been studied with respect to dimensionless parameters (e.g. Reynolds number, sediment Reynolds number, the ratio of orbital diameter to sand grain diameter, and Shields number), since many experimental formulas including these parameters have been proposed for predicting the geometry of roughness and the concentration of suspended sediments. The parameters described above are examined in relation to the damping characteristics of the wave-induced pore pressures. The transmissivity of pressure is constant until a critical value of the dimensionless parameters, and then decreases with increasing parameters. The critical values indicate the bed regime boundary of flat bed to ripple bed. Examinations of wave-induced pore pressure make it possible to predict the bed regime boundary.

INTRODUCTION

The energy losses near the interface between the water and the surface layer of the sand bed are investigated in the relation to the transmission of the wave pressures to the wave-induced pore pressures. The energy losses affect the wave-induced instability of the surface layer of the bed. Recently the wave-induced liquefaction has been studied and considered as the transient phenomena near the surface layer of the bed. On the other hand, Maeno & Hasegawa(1987c) indicated that the rapid attenuation of wave-induced pore pressures is observed in the surface layer of the bed, and that the attenuation of pore pressures in the surface layer differs from that in the sublayer. The attenuation characteristics depend on the energy losses. Since the wave-induced pore pressure relates the instability of the seabed, examinations of the energy losses are required for making clear the mechanism of wave-induced instabilities of the bed.

* Assistant Professor, Dept. of Civil Engng., Maizuru College of Technology, 234 Shiraya, Maizuru, Kyoto, 625 Japan.

The interface between the water and the bed includes the many complicated factors: e.g., (1) the roughness of seabed surface such as dunes and ripples; (2) the concentration of suspended sediments; (3) the boundary layer thickness in the sediments. Although the factors are related to each other and the other factors may exist, the selection of these factors is reasonable on the basis of the observation in laboratory experiments. The surface geometry of sand bed is not idealized flat but rippled under wave action. The roughness of bed surface causes the complicated stream regimen such as vortex. The sediments are suspended by the interaction between the severe oscillatory flow and the bed roughness.

Maeno & Hasegawa(1985a,b) had investigated the characteristics of wave-induced pore pressure in sand layer by wave tank experiments. They measured the wave-induced pore pressures, observing the bed forms of flat bed to ripple bed. The difference between the measured pore pressure and the predicted one can be explained by these factors of energy losses.

The geometry of the bed roughness and the concentration of suspended sediments have been investigated by many authors and the many empirical formulas have been proposed for the prediction of them. The intrinsic parameters also have been indicated. The empirical formulas for ripple geometry such as the wave length and the wave steepness have been proposed with reference to the Reynolds number, the Shields number and the orbital diameter. The empirical formulas for the concentration of suspended sediments also have been proposed with reference to the Shields number. This study examines the relationship between these parameters and the transmissivity of pore pressures.

INTRINSIC PARAMETERS OF ENERGY LOSSES

Tsuchiya & Banno(1987) indicated the three dimensionless parameters as the index of classification for criteria of ripple formation. Those are the Shields number, the sediment Reynolds number (or the sediment fluid number) and the ratio of the orbital diameter of water particle in the wave motion to the diameter of the sediment grain. The many formulas for the concentration of suspended sediments are related to the Shields number, and those for the geometry of bed roughness also are related to the Reynolds number, the Shields number and the ratio of orbital diameter to sediment grain diameter. Since no studies examine the bed roughness and the concentration of suspended sediments in relation to the transmissivity of the wave pressures to the wave-induced pore pressures, this study examined the relationship between the damping characteristics of wave-induced pore pressure and the following four parameters: (1) the ratio of orbital diameter of water particle; (2) the Shields number; (3) the Reynolds number; (4) the sediment Reynolds number.

First, the ratio of orbital diameter of water particle to the grain diameter of sediments is defined as follows:

$$d_0/D = H/\sinh(kh)D \quad (1)$$

where d_0 = the orbital diameter of the water particle in the wave motion; H = the wave height; k = the wave number(=$2\pi/L$); h = the water depth and D = the grain diameter of sediments and regarded as the median diameter of sediment grain, d_{50}.

Secondly, the Shields number is defined as follows:

$$\psi = \frac{u_*^2}{(\rho s/\rho w - 1)gd_{50}} \tag{2}$$

$$u_* = \sqrt{(fw/2)}u_m \tag{3}$$

$$u_m = \pi d_0/T \tag{4}$$

where u_* = the shear velocity; u_m = the maximum value of the horizontal velocity just outside the boundary layer; ρs and ρw are the densities of sand and water, respectively; g = the gravity acceleration; T = the period of the oscillatory flow; fw = the wave-current friction coefficient and Swart(1976) defined it as following form:

$$fw = 0.0025\exp[5.21(a/ks)-0.19] \quad \text{for } a/ks>1.57 \tag{5}$$

$$fw = 0.3 \quad \text{for } a/ks<1.57 \tag{6}$$

where a = the orbital amplitude of fluid just outside the boundary layer, $d_0/2$, and ks = the roughness length of the bed. Sleath(1984) suggested that the Engelund & Hansen(1967)'s formula of the roughness length of the bed is reasonable for many cases. Since Nabae sand is considerably uniform, the median grain diameter, d_{50}, substitutes for the grain size at the accumulation rate of 65 percents, d_{65}.

$$ks = 2d_{65} \tag{7}$$

Thirdly, the Reynolds number is defined as follows:

$$Re = u_m d_0/\nu = \pi d_0^2/(\nu T) \tag{8}$$

where ν = the kinematic viscosity of the fluid.

Finally, the sediment Reynolds number is defined as follows:

$$Re_D = u_* D/\nu \tag{9}$$

where D = the median grain size of sediment, d_{50}.

EXPERIMENTAL METHOD

Experiments were conducted on Nabae beach sand using the wave tank as shown in Figure 1, as well as our previous experiments (Maeno & Hasegawa; 1985a,b).

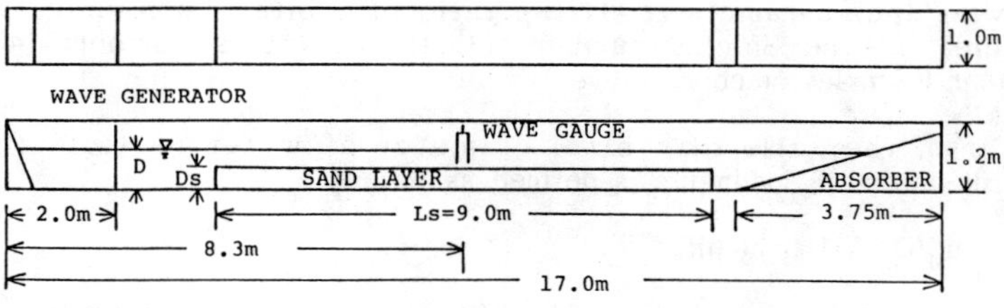

Figure 1. Experimental setup of the wave tank.

The wave tank generates the waves between the wave periods of 0.6s and 2.27s. The water depth is 0.41m over the sand bed. The length, the width and the thickness of sand bed are 9.0m, 1.0m and 0.36m, respectively. The influences of bed size effects can be disregarded in the experiments. The physical properties of Nabae sand are given in Table 1. Table 2 tabulates the experimental conditions in addition to the published cases.

Table 1. - Properties of Nabae sand.

Specific gravity	2.70
Permeability	0.023 cm/s
Uniformity coefficient	1.53
Effective grain size	0.114 mm
Average grain size	0.160 mm

Table 2. - Experimental conditions.

Case	Length of bed (cm)	Dry density (t/m^3)	Water depth (cm)
(I)	250.0	1.61	41.0
(II)	166.0	1.54	41.0
(III)	99.5	1.67	41.0
(IV)	900.0	1.55	41.0
(V)	900.0	1.57	41.0
(VI)	900.0	1.62	41.0

The wave gauge is fixed just over the center of the bed. The pore pressures were measured at various vertical positions under the mudline by four pressure transducers as shown in Figure 2, where the subscript z of P denotes the depth downward from the bed surface, and the x-axis is taken along the mud line of the sand bed.

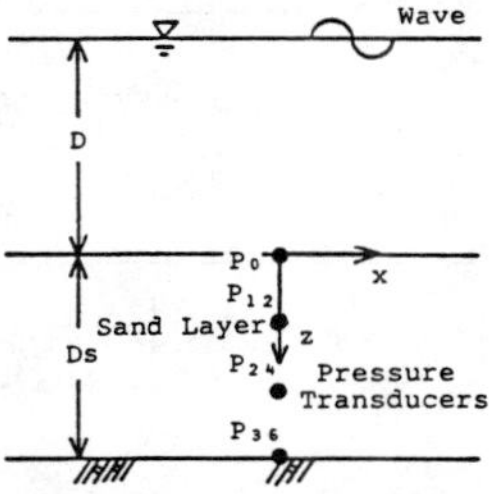

Figure 2. Definition of terms and set-up of pressure transducers.

The transmissivity of wave pressure is defined as the ratio of the measured pore pressure to the wave pressure predicted by Stokes' second order wave theory.

$$P_s=\frac{\rho g H}{2\cosh(ND)}+\frac{3\rho g N H^2 \tanh(ND)}{16\sinh^2(ND)}\left[\frac{1}{\sinh^2(ND)}-\frac{1}{3}\right]-\frac{\rho g N H^2 \tanh(ND)}{16\sinh^2(ND)} \tag{10}$$

EXPERIMENTAL RESULTS AND DISCUSSION

The relationships between the four dimensionless parameters previously described and the transmissivity of pressure in sand bed are investigated in order to study the effects of energy losses near the interface between the water and the sand bed on the damping characteristics of wave-induced pore pressures.

Relationship between Orbital Diameter and Transmissivity

Figures 3(a) and 3(b) show the relationship between the transmissivity of pressures and the ratio of the orbital diameter of water particle in the wave motion to the median grain size of sediment. The transmissivity of pressure keeps constant until a critical value, i.e., an orbital diameter of 300.0, and then decreases as the orbital diameter increases. This tendency is more clear in the wave pressure fluctuations at the bed surface and negligible in the deeper depth below mudline.

Figures 3(c) and 3(d) also show the relationship between the dimensionless orbital diameter and the transmissivities of pressure, P_0/P_s and P_{12}/P_s, for various sand bed conditions. The decreasing

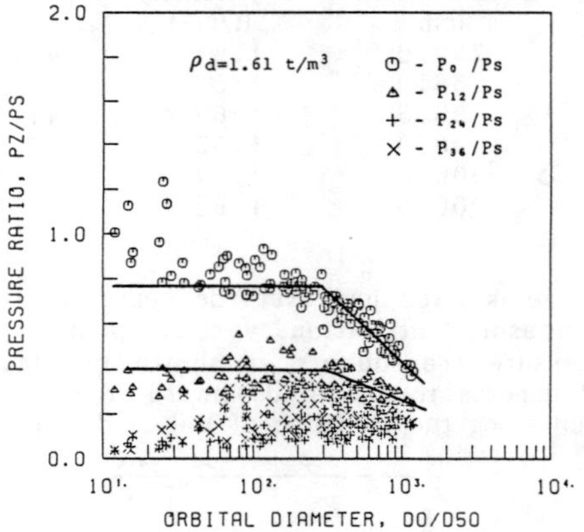

(a) ρ_d=161 t/m^3

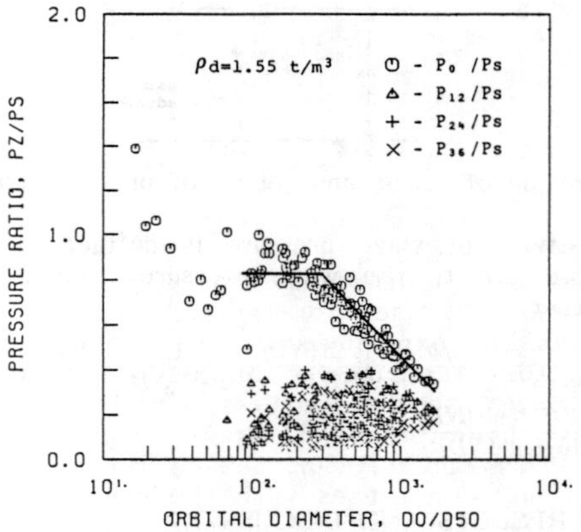

(b) ρ_d=1.55 t/m^3

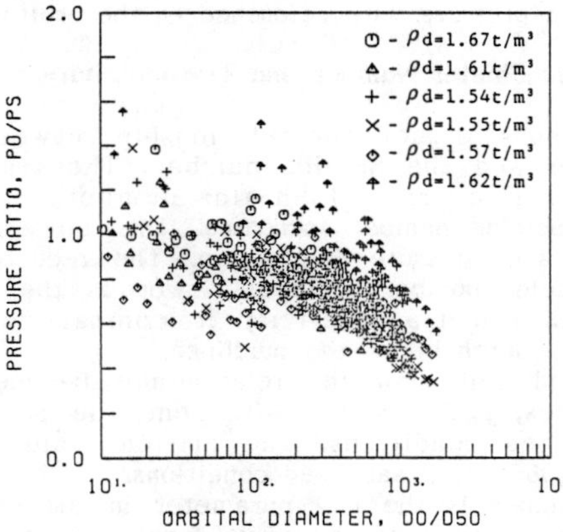

(c) wave pressure for various bed conditions

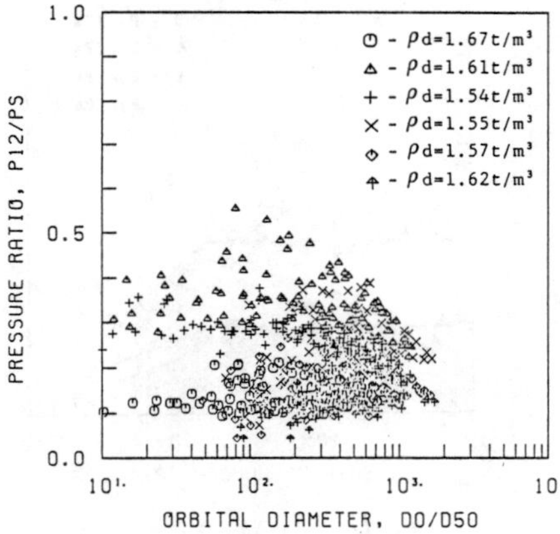

(d) pore pressure at z=12cm for various bed conditions

Figure 3. Relationship between the orbital diameter of water particle and transmissivity

tendency of the dimensionless orbital diameter is more significant in the dense sand bed than in the loose sand bed. The critical value of the dimensionless orbital diameter is constant of 300. for various bed conditions.

Tsuchiya, Ueda and Oshimo(1984) showed that the dimensionless orbital diameter should be used for the index of the ripple formation for less than 1000. The experimental results are within this range of orbital diameter. The transmissivity of pressure shows rapid reduction near $d_0/D=1000$. Using the dimensionless orbital diameter many empirical formulas have been proposed for the ripple wave length (e.g., Kaneko & Honji,1979; Kaneko,1980). Sakakiyama et al.(1986) show that the ripple wave length increases with the orbital diameter until the ratio of ripple wave length to the orbital diameter reaches a limiting value of 2/3. Based on these backgrounds it is significant to examine the transmissivity of pressure with reference to the orbital diameter.

Relationship between Shields Number and Transmissivity

Figures 4(a) and 4(b) show the relationship between the transmissivity of pressures and the Shields number. The transmissivity also keeps constant until a critical Shields number of 0.06, and then decreases as the Shields number increases. This tendency is most predominant in the wave pressure fluctuation. The reduction of transmissivity with the Shields number becomes slower as the depth below the mudline increases. The transmissivity is constant with the Shields number for the deep depth below the mudline.

Figures 4(c) and 4(d) show the relationships between the transmissivities of pressures, P_0/P_s and P_{12}/P_s, and the Shields number for the various sand bed conditions. The critical value of the Shields number is almost 0.06 for all sand bed conditions.

The Shields number is the key parameter in the empirical formulas of the ripple geometry such as the wave length and the wave steepness

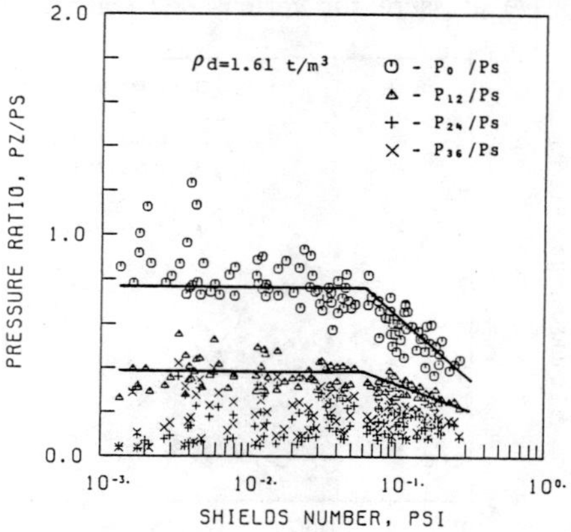

(a) ρ_d=161 t/m³

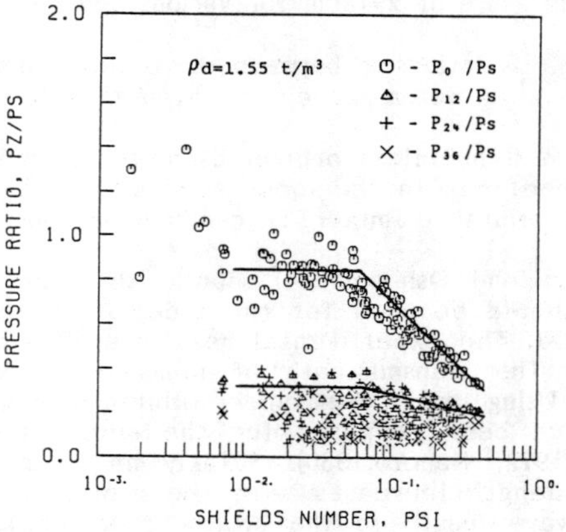

(b) ρ_d=1.55 t/m³

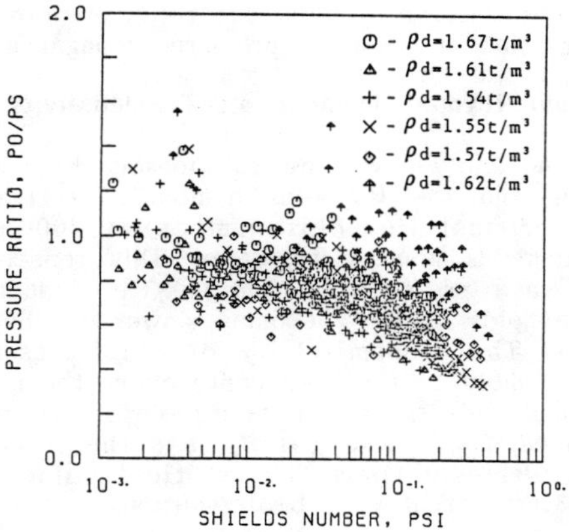

(c) wave pressure for various bed conditions

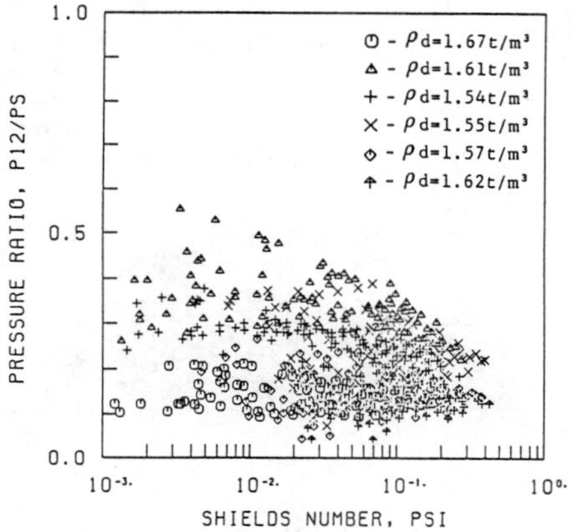

(d) pore pressure at z=12cm for various bed conditions

Figure 4. Relationship between the Shields number and transmissivity

(e.g., Sakakiyama et al.,1986; Sato, Mitani and Watanabe,1988; Sato, Sugiura and Watanabe,1987). The empirical formulas for the concentration of suspended sediments near the bed surface have proposed in relation to the Shields number (e.g., Kawamata,1982). Kawamata(1982) showed that the concentration in a vortex layer on ripples depends on cubic the Shields number.

Therefore, examinations of the relationships between the transmissivities of pressures and the Shields number enable to obtain indirectly the relationships between the transmissivities and the ripple geometry and the concentration of suspended sediments. On the basis of the previous works, the wave length of ripple becomes small and the concentration of suspended sediments near the sand bed increases as the Shield number increases. Thus, the wave-induced pressures attenuates near the sand ripple, and the generation of wave pressures and pore pressures are restricted. This inference is consistent with the experimental observation of the wave pressure propagation.

Relationship between Reynolds Number and Transmissivity

Figures 5(a) and 5(b) shows the relationship between the transmissivity of pressures and the Reynolds number. The transmissivity keeps constant until a critical Reynolds number of 3000.0, and then decreases as the Reynolds number increases. This tendency is most predominant in the wave pressure fluctuation. The reduction of transmissivity with the Reynolds number becomes slower as the depth below the mudline increases. The transmissivity of pore pressures is constant with the Reynolds number at the deep depth below the mudline.

Figures 5(c) and 5(d) show the relationships between the transmissivities of pressures, P_0/P_s and P_{12}/P_s, and the Reynolds number for the various sand bed conditions. The critical value of the Reynolds number is almost 3000. for all sand bed conditions.

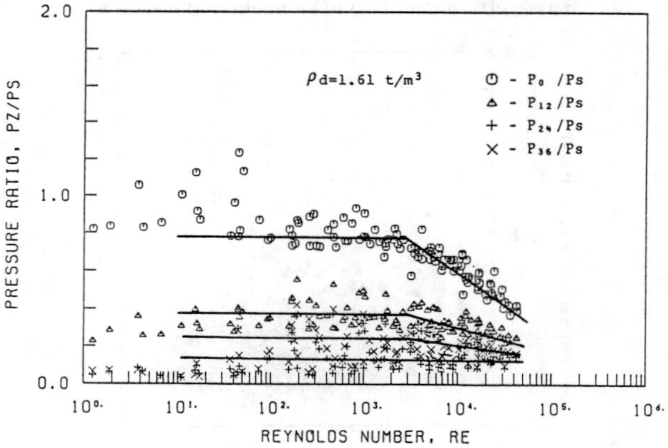

(a) ρ_d=161 t/m^3

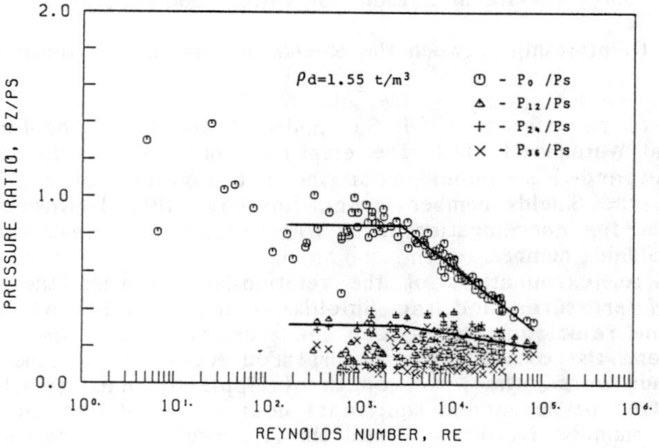

(b) ρ_d=1.55 t/m^3

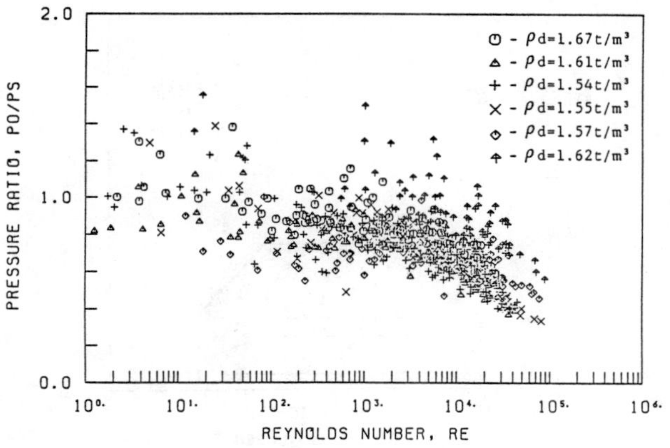

(c) wave pressure for various bed condition

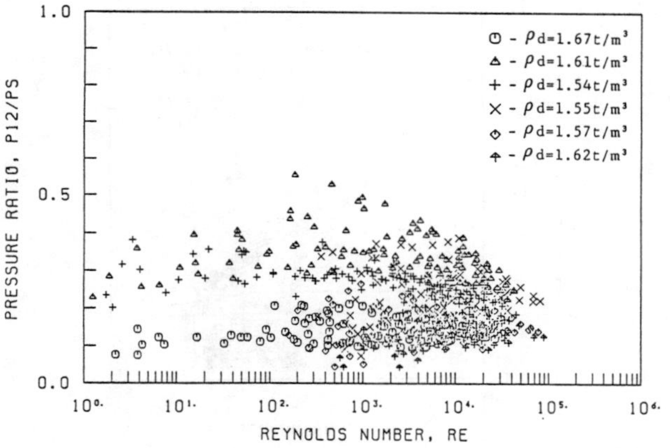

(d) pore pressure at z=12cm for various bed condition

Figure 5. Relationship between the Reynolds number and transmissivity

Relationship between Sediment Reynolds Number and Transmissivity

Figures 6(a) and 6(b) show the relationship between the transmissivity of pore pressures and the sediment Reynolds number. The transmissivity also keeps constant until a critical sediment Reynolds number of 1.5, and then decreases as the sediment Reynolds number increases. The gradient of the reduction is most steepest at the sand bed surface, and becomes gentle as the depth of sand bed increases. The transmissivity of pressures is regarded as a constant value at the deep depth of sand bed.

Figures 6(c) and 6(d) show the relationship between the transmissivities of pressure, P_0/P_s and P_{12}/P_s, and the Reynolds number for the various sand bed conditions. The transmissivities of pressure start to reduce at the critical sediment Reynolds number of almost 1.5 for all sand bed conditions.

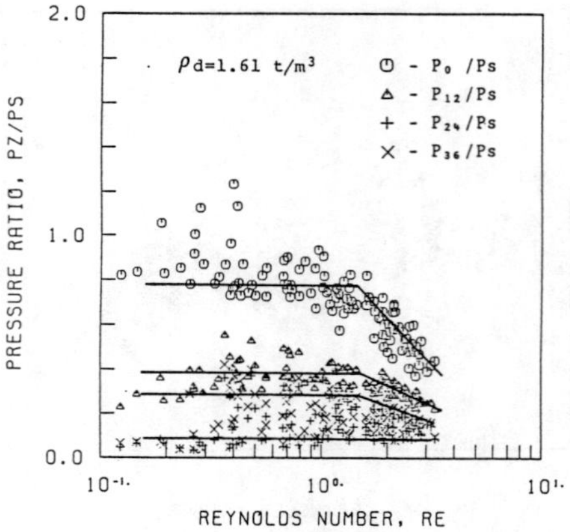

(a) ρd=161 t/m^3

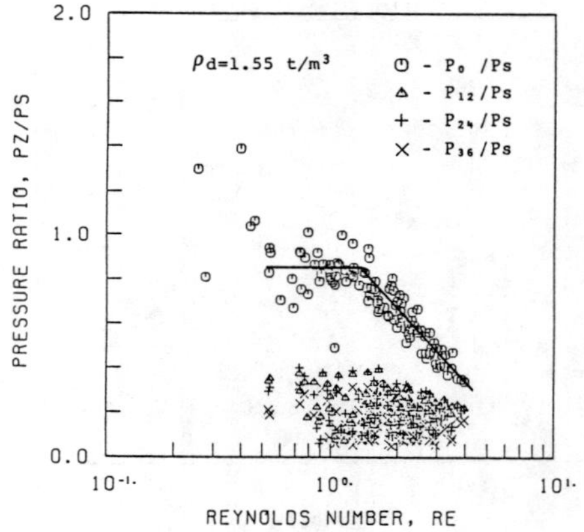

(b) ρd=1.55 t/m^3

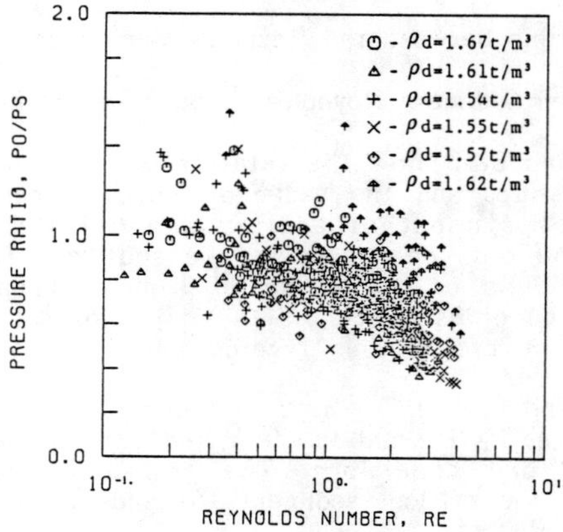

(c) wave pressure for various bed conditions

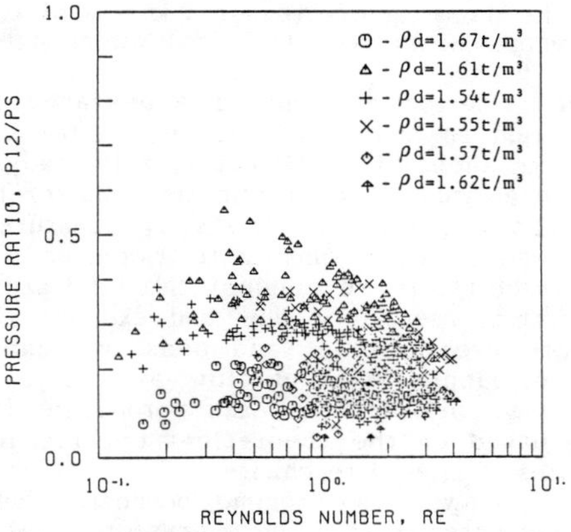

(d) pore pressure at z=12cm for various bed conditions

Figure 6. Relationship between the sediment Reynolds number and transmissivity

The relationships between the transmissivities of pressures and the sediment Reynolds number are the same as those for the Reynolds number using the index of the orbital diameter of water particle instead of the median grain diameter. Since only a single sand is used in this experiments. The sediment Reynolds number should be examine, conducting additional experiments on the other sands that show any distributions of grain particle.

Sand Ripple Geometry and Transmissivity

All four parameters described above keep the transmissivities of pressures constant until a critical value of each parameters and linearly reduce them with increasing each parameter. This tendency is predominant in the dense sand bed and at the sand bed surface and becomes weaker as the depth of sand bed increases. This fact indicates that the damping characteristics of pressures change at a critical value of each parameter. Their critical values of each parameter are 1.5 for the sediment Reynolds number, 3000. for the Reynolds number, 0.06 for the Shields number and 300. for the dimensionless orbital diameter of water particle.

Riho,Watanabe and Horikawa(1981) proposed the following bed regime boundary of flat bed to rippled bed for both fine and coarse sands:

$$d_0'/d_m > 280 \text{ and } \Psi > 0.1 \tag{11}$$

where d_0'=the orbital diameter of water particle in the finite amplitude wave action; d_m=the average grain diameter.

The critical values of both the dimensionless orbital diameter and the Shields number are in a good agreement with their bed regime boundary of flat bed to ripple bed. Since the ripple formation causes the complicated stream regimen such as vortex and increases the concentration of suspended sediments, the wave energy attenuates and decreases the rate of the wave pressure propagation into the sand bed. The damping characteristics of pressures are influenced by the energy losses near the interface between the water and the sand bed. Thus examinations of the damping of pressures is a tool to investigate the complicated phenomena near the interface between the water and sand bed surface.

Recently, the criterion of ripple disappearance in the upper regime has been focused on and usually defined as the bed regime boundary of ripple bed to sheet flow. However, this study doesn't examine the other bed regime boundaries without flat bed to ripple bed in the relation to the damping characteristics of pressures in sand beds, since waves generated by wave tank is restricted in a narrow range of oscillatory flow velocity. If we conduct the laboratory experiment in water tunnel generating oscillatory flow and examine the damping characteristics of pore pressures in sand beds, we can obtain the bed regime boundary of ripple to sheet flow as well as that of flat to ripple. Since the wave pressure can easily propagate into the sand bed without damping based on the theoretical considerations, the transmissivity of pressures is infered to change.

Therefore, the following bed regime boundaries can be determined by the damping characteristics of pore pressures: (a) no movement to flat bed; (b) flat bet to ripple bed; (c) ripple bed to sheet flow.

Boundary Layer in Sand Bed

Many types of instabilities of sand beds should be classified through their mechanisms. These mechanisms depend on their own bed conditions. For example, a bed condition is characterized by the composition and the constitution of sand beds, geotechnical properties of sediments, and so on. These bed conditions were discussed in our previous works (Maeno & Hasegawa,1987b,c). The instability near the surface layer of sand beds is influenced by the effects of energy loss near the interface between the water and the sand bed. The interface between them includes the complicated factors as previously described. The boundary layer is assumed in the sediment, defining its thickness as the depth of the surface layer which depresses the wave-induced pore pressure. It is considered that the boundary layer has any contribution to cause the instabilities of sand beds such as the score and the liquefaction.

In the boundary layer, the high erodibility may be caused, because its density is loosest; no solid skeleton in its is sufficiently composed; and it is considerably deformable. Thus, it is significant to investigate the correlation between the boundary layer thickness and the instability due to scour and erosion. The boundary layer is examined by laboratory experiments in a wave tank.

The elastic wave propagation theory predicts exponential attenuation of wave-induced pore pressures in a poro elastic media. This attenuation curve is valid for coarse sand beds (Yamamoto,1977). However, the rapid attenuation curve is observed near the surface of fine sand beds (Maeno & Hasegawa, 1985b). Figure 7 shows the relationship between the damping ratio, P_z/P_0, and the relative depth, z/Ds. This figure shows that the pore pressure is damped rapidly near the surface of a sand bed, and then is depressed slowly as the depth below the mudline increases.

The above experimental results indicate that there is a difference in mechanism of attenuation of pore pressure between near the surface and the inside of the sand bed. Since the surface layer of the sand bed is deformable; the soil skeleton in this layer is composed of the sand grains which are weakly combined with each other; wave-induced pore pressure easily squeezes out, and then attenuates greatly in the surface layer. The pore pressure, on the other hand, less attenuates

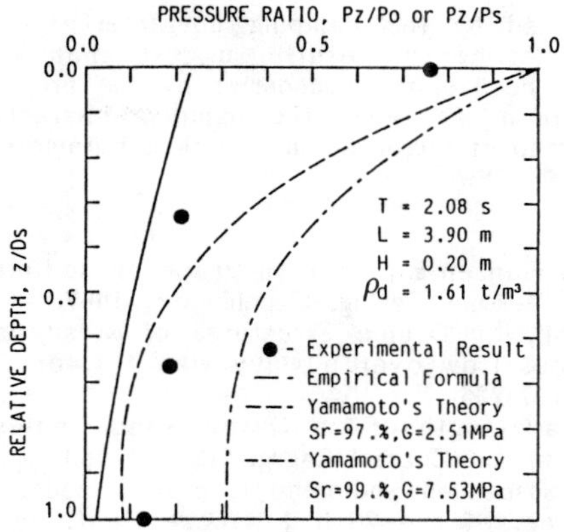

Figure 7. The relationship between the transmissivity of pressure and relative depth

in the sublayer below the surface layer, because the sublayer is relatively rigid and it is not easy for pore pressure to dissipate forward to bed surface. The above discussions suggest that the boundary layer may exists in the sand bed under wave loading. The boundary layer thickness should be predicted on the basis of laboratory experiments. Many difficulties, however, accompany these experiments. Those are the simultaneous measuring of pore pressures at various points, and the exact measuring near the surface of the bed. Since transducers disturb the flow of the pore fluid and the propagation of pressures in the beds, many miniature high quality transducers are required for these measurements. Overcoming these difficulties of laboratory experiments, the boundary layer thickness will be predicted experimentally. Moreover, the wave-induced instabilities near the surface layer will be classified based on these studies.

CONCLUSIONS

The following conclusions are obtained:

1. The energy losses near the interface between the sand bed and the water affect the wave-induced pore pressure in the surface layer of the sand bed. The effects of energy losses on the damping characteristics are examined in relation to the four dimensionless parameters: (1) the ratio of the orbital diameter of water particle to the sand grain diameter; (2) the Shields number; (3) the Reynolds number; (4) the sediment Reynolds number.

2. The transmissivity of pressure keeps constant until a critical value, and then decreases as all four dimensionless parameters increases. The rate of this reduction decreases as the depth below the mud line increases. The critical values are 300.0 for the orbital diameter, 0.06 for the Shields number, 3000.0 for the Reynolds number, and 1.5 for the sediment Reynolds number, respectively.

3. The critical values of both the dimensionless orbital diameter and the Shields number are in good agreement with Riho et al.(1981)'s bed regime boundary of flat bed to ripple bed. The flat bed to ripple bed

boundary is determined by the damping characteristics of wave-induced pore pressures. Experimental results suggest that it is possible to determine the other bed regime boundaries by the proposed method that examines the relationship between the damping characteristics of wave pressures and the group of intrinsic dimensionless parameters.

REFERENCES

Engelund,F.A., and Hansen,E., "A monograph on sediment transport in alluvial streams," Teknisk Forlag, Copenhagen, 1967.

Kaneko,A., and Honji,H., "Double Structures of steady streaming in the oscillatory viscous flow over a wavy wall," J. of Fluid Mechanics, Vol.93, 1979, pp.727-736.

Kaneko,A., "The wavelength of oscillation sand ripples," Proc. of the 27th Japanese Conf. on Coastal Engrg., JSCE, 1980, pp.207-210.

Kawamata,R., "On suspended sediment transport model under waves on rippled bed," Proc. of the 29th Japanese Conf. on Coastal Engrg., 1982, pp.218-222.

Maeno,Y., and Hasegawa,T., "Characteristics of wave-induced pore water pressure in sand layer," Proc. of International Symposium on Ocean Space Utilization, 1985a, pp.475-484, Springer-Verlag.

Maeno,Y., and Hasegawa,T., "Evaluation of wave-induced pore pressure in sand layer by wave steepness," Coastal Engrg. in Japan, JSCE, Vol.28, 1985b, pp.31-44.

Maeno,Y., and Hasegawa,T., "Evaluation of wave-induced pore pressure in saturated sand bed," Proc. of ASCE Specialty Conf. on Coastal Hydrodynamics, Coastal Hydrodynamics, 1987a, 686-700.

Maeno,Y., and Hasegawa,T., "In-situ measurements of wave-induced pore pressure for predicting properties of seabed deposits," Coastal Engrg. in Japan, JSCE, Vol.30, No.1, 1987b, pp.99-115.

Maeno,Y., and Hasegawa,T., "Effects of energy loss near interface between water and bed on propagation of wave-induced pore pressures," Proc. of Civil Engrg. in Ocean,JSCE,Vol.3,1987c,pp.101-106.

Mei,C.C., and Foda,M.A., "Wave induced stresses around a pipe laid on a poro-elastic seabed," Geotechnique, Vol.31, 1981, pp.509-517.

Riho,Y., Watanabe,A., and Horikawa,K., "Two-dimensional beach transformation and on-offshore sand transport," Proc. of the 28th Japanese Conf. on Coastal Engrg., JSCE, 1981, pp.217-221.

Sakakiyama,T., Shimizu,T., Kajima,R., Saito,S., and Maruyama,K., "Sand ripples generated by prototype waves in a large wave flume," Coastal Engrg. in Japan, JSCE, Vol.28, 1986, pp.147-160.

Sato,S., Sugiura,S., and Watanabe,A., "Sand transport mechanism and ripple disappearance in irregular oscillatory flows," Proc. of the 34th Japanese Conf. on Coastal Engrg., 1987, pp.246-250.

Sato,S., Mitani,K., and Watanabe,A., "Geometry of sand ripples and net sand transport rate due to regular and irregular oscillatory flows," Coastal Engrg. in Japan, JSCE, Vol.30,No.2, 1988, pp.89-98.

Sleath,J.F.A., "Seabed Mechanics," John Wiley & Sons, New York, 1984, xx+335 pp.

Swart,D.H., "Coastal sediment transport," Computation of Longshore Transport, Delft Hydraulics Laboratory Reports, R968, Part 1, 1976.

Tsuchiya,Y., and Banno,M.,"Seabed forms and their formation criteria," Proc. of 34th Jap. Conf. on Coastal Engrg., JSCE, 1987, pp.222-226.

Tsuchiya,Y., Ueda,Y., and Oshimo,T., "A theory of threshold of sediment movement by wave action," Proc. of the 31st Japanese Conf. on Coastal Engrg., JSCE, 1984, pp.306-310.

Yamamoto,T., "Wave induced instability in seabeds," Proc. of ASCE Specialty Conf., Coastal Sediments '77, 1977, pp.898-913.

CHAPTER 137

SIMULATION MODELING OF DUNE EROSION

M. F. Overton, Assoc. M. ASCE*

J. S. Fisher, M. ASCE**

Abstract

Using a combination of laboratory, field, and numerical techniques, we have developed a methodology for predicting the extent of dune erosion due to a given set of wave conditions and the beach and dune morphology. A basic assumption in this model is that the volume eroded from the dune is a function of the swash acting on the dune. Because the model is built on this premise, it is necessary to look at the mechanics of two ongoing processes in the storm environment. The first is the relationship between the swash characteristics and the volume of material eroded from the dune face. The second is the relationship between the time history of the swash characteristics at the dune face and the statistics of the storm event, the significant wave height and wave period.

Introduction

The process of dune erosion due to a storm event can be broken into five interacting components. Starting at the ocean side these are -- a statistical description of the storm, the individual wave, wave generated swash on the beach, the force on the dune face due to the swash and the volume eroded from the dune. Wrapped around these parameters is the fact that we are dealing with not a single event (a single wave and swash), but a series of wave and swash events generated during the storm period. However, in order to simplify the first cut description of the model and the process, we begin the discussion with our focus on an individual wave component and work from the dune face seaward.

* Associate Professor, Dept. of Civil Engineering, North Carolina State University. Raleigh, NC.
** Professor, Dept. of Civil Engineering, NCSU.

Methodology

The assumption that the volume eroded from the dune is a function of the swash acting on the dune is based on flume and field experiments. In a series of laboratory experiments, Overton, et al. (1988) showed that the volume eroded from a vertical planar dune during the action of a single swash is a function of the swash force. The swash force is defined as the product of the fluid density, the uprush velocity squared and the height of the uprush at the moment of impact. Applying this concept to the field, a set of experiments were designed to test this hypothesis for prototype scale dunes under natural swash conditions, Fisher, et al. (1987). At the Army Corps of Engineers Field Research Facility (FRF) at Duck, N. C., man-made dunes approximately one meter high and one meter wide were built on the beach and allowed to erode naturally during the rising tide. The experiments were conducted until either the dune was completely eroded (and/or overtopped) or the swash no longer hit the dune. While it is difficult to isolate the impact of a single swash on the amount of the dune eroded, it is apparent from the data that the total dune erosion during an event is linearly correlated with the summation of the swash force for each individual swash, Figure 1. This relationship seems to hold even when events are

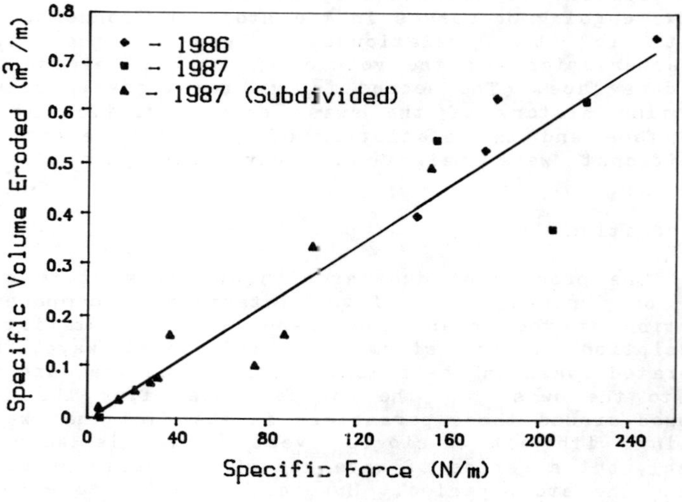

Figure 1. Specific force versus specific volume eroded from field experiments, Duck, N.C.

subdivided into smaller units of time. It is this concept, that a storm event can be represented by a single parameter (the sum of the force in each individual swash) and thus the amount of dune eroded during that event can be predicted.

Given that we can reasonably predict the volume eroded at the dune face as a function of the swash force, how can we determine the force of any arbitrary swash as it hits the dune? The laboratory experiments investigated a limited range of bore sizes. To increase the flexibility of the model, a numerical model based on the two dimensional hydrodynamics of a bore propagating over a sloping planar beach and impacting with a vertical dune was developed. Variable parameters in the model are depth of the bore at the seaward boundary, distance to the dune from the mean water level and the slope of the beach. This enabled us to determine the velocity and depth of the bore during impact with the dune for a wide range of cases. Using these two parameters, the time history of the swash force on the dune face during impact is given in Figure 2. Note that the force varies rapidly with time, rising quickly

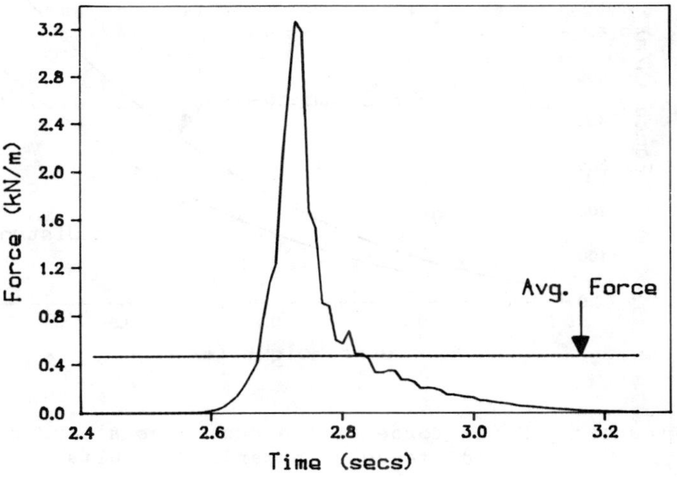

Figure 2. Swash force at the dune versus time - numerical results.

to its maximum and then decaying. In order to be compatible with the laboratory results for the prediction of the volume eroded, the average force during impact is used to represent the effect of a single swash, as discussed in Overton et al. (1987).

In order to reduce computation time in the simulation model, it is desirable to find a direct relationship between the input parameters in the hydrodynamic model and the output parameter, swash force. Therefore, the relationship between the height of the bore, the distance to the dune, the slope of the beach, and the swash force was investigated. A limited range of parametric studies have been conducted to determine this relationship. First, the height of the bore was varied holding slope of the beach and distance to the dune constant. This relationship is nonlinear, as shown in Figure 3. Repeating this test for the same slope but for a greater distance (7 m compared to 3 m) to the dune yields similar results but with smaller magnitudes. The cut off at the low end of the curve (small bore heights) indicates that bores of this magnitude do not reach the dune with measurable force. Second, the impact of slope was investigated. Bore height versus swash force is plotted in Figure 4. For a

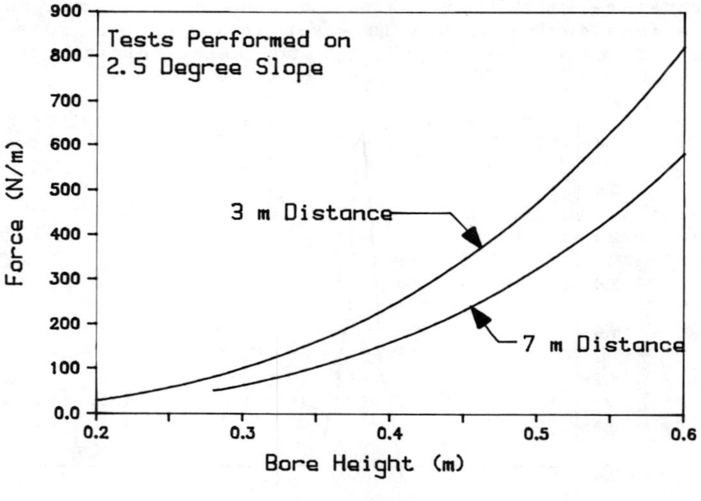

Figure 3. Swash force at the dune versus bore height, two distances - numerical results.

given distance to the dune, the steeper beach reflects more of the swash energy and the force at the dune is less for the same bore height when compared to that on a more shallow slope. Once a full set of these parametric curves has been generated for the range of applicable values, these relationships can be easily incorporated into the simulation model. This avoids having to run the hydrodynamic model each time a bore is generated on the beach.

How then are the bore characteristics (height and distance from the dune) generated from the wave information? Wave transformation from deep water to a point of breaking is well described using any number of

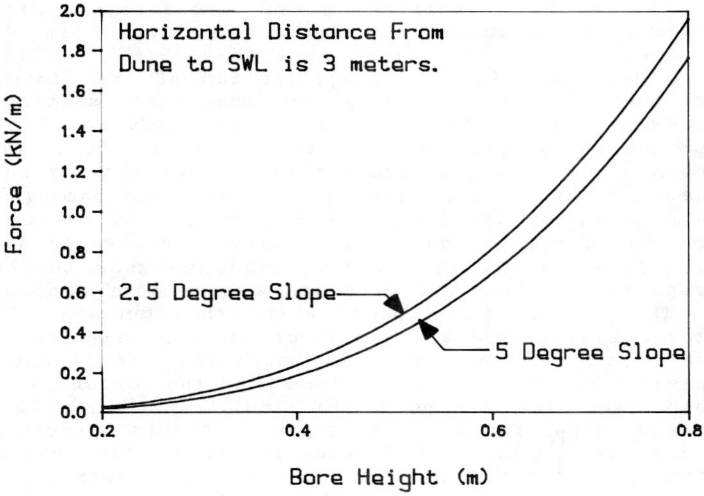

Figure 4. Swash force at the dune versus bore height, two slopes - numerical results.

accepted wave theories. It is the transformation of the breaking wave from a wave of propagation to a wave of translation in the surf zone which is not well understood but which is currently being studied by many investigators. Of particular interest for our application, is a recent set of laboratory experiments conducted by Papanicoloau and Raichlen (1987). Their primary interest was to focus on the breaking and subsequent landward propagation of solitary waves. One piece of information extracted from this set of experiments indicates that after breaking, the wave takes on a borelike form at a certain point on the beach. From this work, it is determined that the wave decays to about one half of its breaking height at the point in which it becomes borelike. The distance traveled from the breaking point is about 20 times the depth at breaking. Similar relationships were also shown to exist in an earlier paper by Svendsen et al., 1978. Therefore, given the height and depth at breaking, the height of the bore and the position along the beach can be determined.

What then are the deep water wave characteristics which represent the given storm event? A storm event can be specified in terms of its significant wave height, period and duration. At this point in the model development, we can no longer ignore the fact that a storm consists of a number of wave events which interact and alter the characteristics of a single swash. Going back through the model, how then do we account for the fact that there is a series of interacting wave components which characterize the storm environment and not a set of isolated waves propagating without interference to impact with the dune.

Starting with the specification of the storm, we see that it is necessary to take the statistical representation of the storm and break this down into a representative set of individual waves, that is, a record of wave height versus time. Theoretically we can take each individual wave up to breaking, through the swash zone, up the beach as a bore and calculate the force of impact at the dune. However, because of swash-swash interactions in the surf zone, we know that each offshore wave does not generate a bore that propagates up the beach to impact with the dune. These interactions may be a function of the distance from mean water level to the dune, the dominant period and the distribution of wave periods in the offshore wave field, and the steepness of the beach as well as, perhaps, other factors. At this point this interaction, or the prediction of the swash period at the dune as a function of the offshore parameters and geometry, is not well understood. In order to better understand this phenomena for application to this model, we analyzed the 1986 and 1987 Duck data with respect to swash period. Data from eight individual field experiments were available for analysis. The offshore period, which was obtained from the offshore data which the FRF routinely collects, for these experiments ranged from about 6 seconds to 14 seconds. Defining the swash period as the

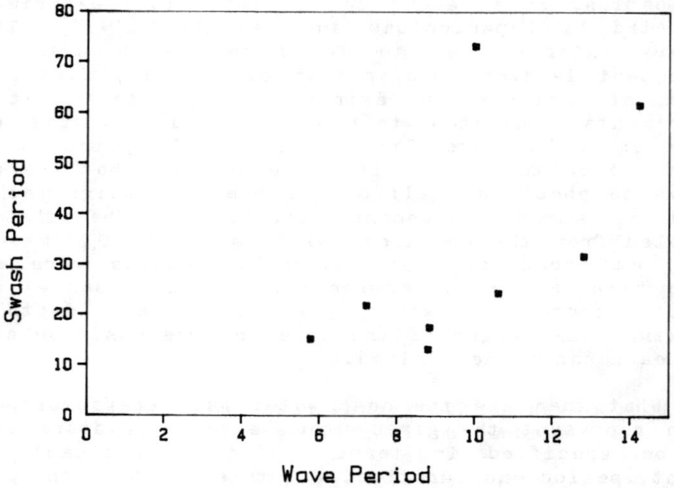

Figure 5. Offshore wave period versus swash period for field experiments at Duck, N.C.

duration of the experiment divided by the number of swash hits on the dune face during this period, the wave period can be plotted against the swash period for these eight experiments, Figure 5. There is no definitive correlation between these two periods as evidenced by the wide range of swash periods for a given wave period. A 10 second significant wave period produced swash periods from 10 to 70 seconds during different experiments. Attempting to determine whether distance from the dune was a dominating factor, the swash period versus horizontal distance from the dune is plotted in Figure 6. The different symbols correlate to different experiments. Four experiments were subdivided into smaller time increments and plotted as swash period versus distance for that subinterval. Distance from the dune changes during the experiment as the tide rises yielding a corresponding change in swash period. However, the offshore wave period is assumed to be constant for the entire experiment. Again, there is no definitive correlation between parameters, however, the trend of the data is intuitively correct. The swash period increases for greater distances to the dune and for the same offshore period.

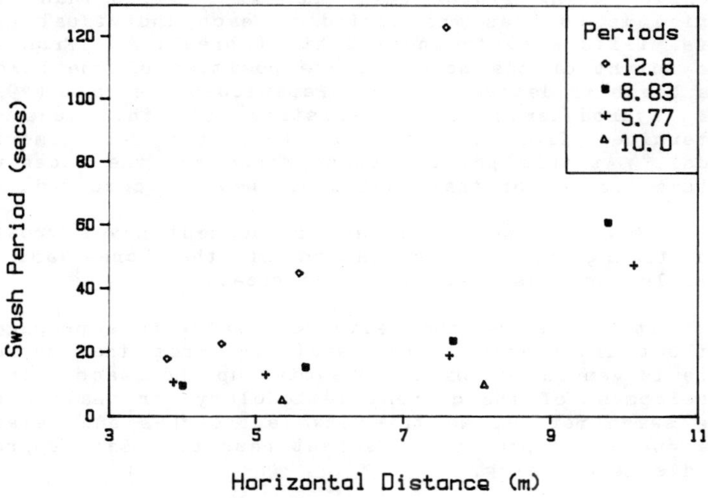

Figure 6. Swash period versus horizontal distance from the dune for four different offshore periods.

The question remains as to how to put the model together for multiple swashes. Starting at the oceanside, we will discuss how each component has been treated in this current version of the model.

Model Development

The assumed input information to the simulation model at the ocean boundary is the significant wave height, the wave period and the duration of the storm event. Using a Rayleigh distribution for the wave heights, we can generate a set of wave heights corresponding to a given significant wave height. The draw back to this approach is that it does not generate a wave time series for which a wave period for each individual wave can be determined. An alternative approach to adjust for this problem is to first use the significant wave height and period and an assumed spectral shape to form a wave spectra. The wave spectra can be represented by the summation of N number of individual wave components specified by a wave height and a wave period. While this technique may not produce a truly random data set, it does identify an individual wave height with a wave period. However, at this point in the model development, we have not tried this alternative approach.

Given a Rayleigh distribution of wave heights, each wave is propagated from deep water to the point of breaking using linear wave theory. The dominant wave period is used as the period of each individual wave. This yields a depth and height at breaking. From that the height of the bore and the position of the bore on the beach is determined from Papanicoloau et al. (1987). The hydrodynamic characteristics at the dune are determined from the results from the 2-D numerical model. At this point a swash force and the subsequent volume eroded for that individual wave is computed.

How are the second and subsequent waves treated? The timing of the generation of the bores and the correlating swash period are critical.

It is assumed that each deep water wave propagates without interference to breaking. From that point a bore is generated which propagates up the beach. In the development of the current methodology for dealing with the swash period, we tried two approaches and selected the one which produced the best results. Each approach is discussed below.

The first approach uses field results to calibrate the model in the following manner. One, a wave height fitting a Rayleigh distribution is generated from a random number generator and the specified significant wave height. Two, using results from the hydrodynamic model, check whether the bore hits the dune. If not, generate a new wave height and test again. If so, calculate swash force and advance in time an increment equal to the given swash period. For this approach, the swash period for the storm event must be known.

The second approach is let the algorithm determine the swash period. For example, for any given bore depth generated the swash may or may not reach the dune given a particular beach slope and the distance from the dune. Therefore, one alternative is to simply generate the set of bore depths and let the numerical model determine whether the bore reaches the dune. If not, advance in time a time step equal to the wave period without adding to the summation of the swash force term. If so, calculate the swash force, add to the summation series and advance in time. Obviously, using this approach there is no guarantee that the swash period of the model equals that of the field results.

Results

Four simulation tests were conducted using the 1987 Duck field data. Field data used as input in the simulation model were the offshore significant wave height and wave period, the slope of the beach, the distance from the dune to the mean water level, the change in mean water level due to the rising tide, the duration of the erosion event, and the swash period. Data from three experiments were used. Two experiments were simulated in their entirety while the third was divided into two sub-time periods and simulated as different events. Each set of test data was processed using both methods for dealing with the swash period. The results, erosion predicted versus erosion measured, are shown in Figure 7 using method one, in which swash

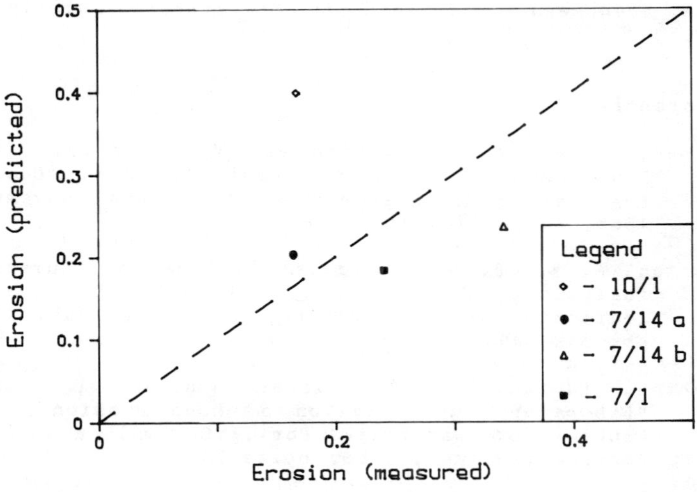

Figure 7. Predicted versus measured erosion.

period is taken as a known. As shown, three of the points are predicted fairly well, though with a tendency to underestimate the erosion, while one point is over predicted by a factor of two. When comparing these results to those using method two, in which the swash period is determined by the algorithm, there is very good correlation between the predicted and measured for two of the points, 7/14a and 10/1. These are also the experiments which are best predicted in method one. Examining the output of this approach it is seen that the swash period as determined by the simulation algorithm closely approximates the measured swash from the field data. While it is not clear at this point why this correlation is occurring, it does point out that swash period is a significant factor in the simulation model.

Conclusions

The two key components to the simulation model are one, the relationship between the swash characteristics and the volume of dune eroded and two, the relationship between the statistics of the storm event and the time history of the swash characteristics at the dune face. While the simulation model presented herein has met with limited success and therefore has potential as a predictive tool for dune erosion, several components of the model must be refined. The most obvious of these is the swash period. While we have built a basis for predicting the magnitude of the swash force, we know little about the relationship between the timing of offshore waves and the timing of those waves which impact with the dune. Advances in the understanding of this phenomena will lead to an improved simulation model.

References

Fisher, J. S., M. F. Overton and T, Chisholm, "Field Measurements of Dune Erosion", 20th Coastal Engineering Conference, Taipei, Taiwan, November, 1986, pp. 1107-1115.

Overton, M. F., J. S. Fisher and T. Fenaish, "Numerical Analysis of Swash Forces on Dunes", ASCE Coastal Sediments '87, New Orleans, Louisiana, May, 1987, pp. 632-641.

Overton, M. F., J. S. Fisher and M. A. Young, "Laboratory Investigation of Dune Erosion", ASCE Journal of Waterway, Port, Coastal and Ocean Engineering, May, 1988, pp. 367-373.

Papanicoloau, P. and F. Raichlen, "Wave Characteristics in the Surf Zone", ASCE, Conference on Coastal Hydrodynamics, Newark, Delaware, June 1987, pp. 765-780.

Svendsen, I. A., P. A. Madsen and J. B. Hansen, "Wave Characteristics in the Surf Zone", ASCE, 16th Coastal Engineering Conference, Germany, August, 1978, pp. 520-539.

CHAPTER 138

The Criterion of Ripple Formation by Wave Action

Yoshito Tsuchiya * and Masato Banno **

Abstract

In this paper, we emphasize that there must exist a common mechanism governing ripple formation in the case of both waves where the acceleration effect exists, and unidirectional flow, where the acceleration effect is usually ignored. Using a light sediment with an immersed specific gravity of about 0.8, which is nearly one-half that of natural sand, a series of ripple formation experiments were performed in order to determine the formation criterion. Using the experimental data, along with data collected by other researchers the ripple-formation criterion was investigated in terms of the ratio of water particle orbital diameter to sediment grain diameter as an acceleration parameter, on the so-called Shields diagram. Emphasis is placed on the importance of this ratio in the criterion of ripple formation in relation to that by unidirectional flow.

1. INTRODUCTION

Sand transport due to wave action or by combined wave & current action is one of the transport phenomena which takes place as prescribed by the laws governing sediment and fluid interation. Many investigations on such a phenomenon have been actively carried out. One of the most salient futures pertaining to sediment transport by waves and by flowing water is the threshold of sediment movement. All the experimental data for the threshold of sediment movement can be plotted on a Shields diagram, but there exists an acceleration effect in wave motion. In relation to sediment transport, the relevant phenomena by waves and by flowing water must be identified and qualified in the mechanics of sediment transport.

The seabed sediment ripple is of great interest not only in understanding the formation phenomena, but also in establishing the mechanics of sediment transport by wave action. Many physical model experiments of beach change have been performed, but none have introduced seabed forms in their comparisons between model and prototype phenomena, as well as in their similitude. Physical model experiments should be carried by placing major emphasis on bed form similitude, which in turn results in a similitude of the governing physical phenomenon.

---

* Professor, Disaster Prevention Research Institute, Kyoto University, Uji, Kyoto 611. JAPAN.
** Civil Engineer, Division of Civil Engineering Design, Penta Oceans Construction Ltd., Tokyo 112. JAPAN.

When no bed form considerations are made severe scale effects may result in the comparison between model and prototype. It is therefore noted that, in physical model experiments seabed forms and their formation criteria are required in order to determine what seabed forms occur in the prototype.

Thus far, many attempts have been made to classify seabed forms, by Bagnold(1946), Manohar(1955), Komar & Miller(1975), Kaneko(1980) and Horikawa & Shibayama(1982), in terms of Shields number and other parameters such as sediment Reynolds number and the ratio of water particle orbital diameter to sediment grain diameter. As previously mentioned, the classification of bed forms by waves should be made in relation to that by unidirectional flow. It is however further mentioned that a common formation mechanism must exist between the sea and river beds, even though additional parameters may exist which specify a specific phenomenon. In 1975 Komar and Miller examined the seabed forms and their formation criteria in relation to those by unidirectional flow. In the criterion, experimental data of the sea bed forms were plotted on a Shields diagram where the sediment Reynolds number is defined using maximum water particle velocity. An empirical formula of the criterion from ripple to sheet flow condition was then proposed, however, no considerations were made on the acceleration effect due to wave motion on the criterion of ripple formation. Kaneko proposed a criterion of seabed forms in terms of the sediment mobility parameter and the ratio of boundary layer thickness to sediment grain diameter. In the criterion, seabed forms are classified into five such as no ripple, two-dimensional ripple, brick pattern, irregular ripple and sheet flow. Sunamura introduced the asymmetry of wave motion into the criterion of seabed forms and proposed a dimensionless sediment-wave parameter which was given from the threshold condition by Komar and Miller. Horikawa and Shibayama, on the other hand, proposed a criterion of sediment transport types using the Shields number and the ratio of sediment fall velocity to maximum water particle velocity, but not shear velocity. This criterion is not for seabed classification and its formation criterion, but the dimensionless parameters used are worthy of being considered in relation to the parameters employed by the other researchers.

In this paper, using a light sediment with an immersed specific gravity of about 0.8 which is nearly one-half that of natural sand, a series of ripple formation experiments were carried out in order to determine the formation criterion in the case of small sediment Reynolds number. Using the experimental data, along with data collected by other researchers the ripple formation criterion is investigated based on the ratio of water particle orbital diameter to sediment grain diameter on the so-called Shields diagram. This ratio expresses the effect of wave acceleration on sediment transport. Emphasis is placed on the importance of the ratio in the criteria of ripple formation in relation to that by unidirectional flow.

2. SAND TRANSPORT TYPES AND BED FORMS

In general, there are usually two states of sand transport which are only for convenience in their investigation. Recently, Shibayama and Horikawa(1982) proposed a classification of sand transport types in relation to the ripple formation. Four states are used to classify sediment transport, they are bed load, transition from bed load to suspended load, suspended load and sheet flow. Their classification is shown in Figure 1 where τ^* is the Shields number and u_0/w_0 the ratio of maximum water particle

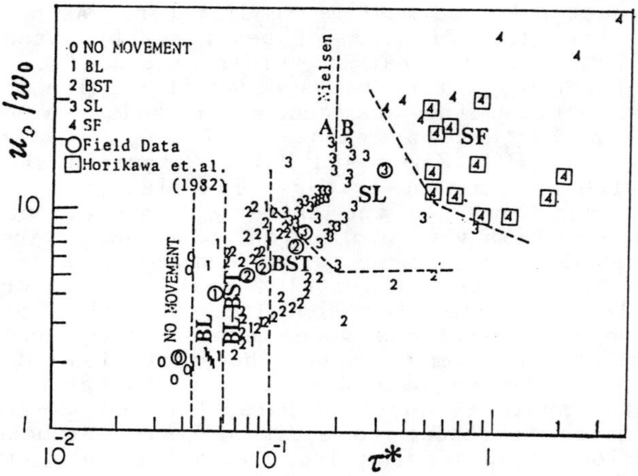

Figure 1. Classification of sand transport types by Shibayama & Horikawa

velocity at the sea bottom to fall velocity of the sand. These transport states are identified according to the resulting bed forms. In the case of no movement, sand does not apparently move by the wave action, but at the threshold condition a critical state is reached such that sediment movement is initiated. The transport state identified as flat bed describes a condition where sediment movement and transport exist yet the bed remains flat. In the state, sand moves mainly as bed load. At a certain state when the shearing stress acting on the sand bed becomes large, ripples appear. In this state, sand moves in suspension around the ripples, and this corresponds to suspended load in the classification of sand transport types by Shibayama and Horikawa. As the shearing stress becomes larger, a flate bed again results, and it is usually identified as sheet flow, first named by Bagnold(1946). The flat bed states occurring under conditions just after the movement threshold is reached and after ripple formation correspond to the lower and upper regimes of flow in the case of river bed mechanics.

3. THE CRITERIA FOR SEABED FORMS

The most important factors governing seabed formation are first considered by referring to both the previous investigations of seabed forms, their formation criteria and the threshold of sand movement. Secondly, using the main factors selected the criteria of seabed forms are proposed.

3.1 Classifications of Seabed Forms and Their Main Factors

As already mentioned, Komar & Miller(1975) classified seabed forms into three groups, no movement, ripple and sheet flow as shown in Figure 2 where u_0D/ν is the sand grain Reynolds number relative to the maximum water particle velocity at the bottom, not to the shear velolcity. The sand transsnport phenomenon is well governed by the shearing stresses acting on the seabed, it is questionable as to why the sand grain Reynolds number u_0D/ν is used instead of u^*D/ν where u^* is the shear velocity.

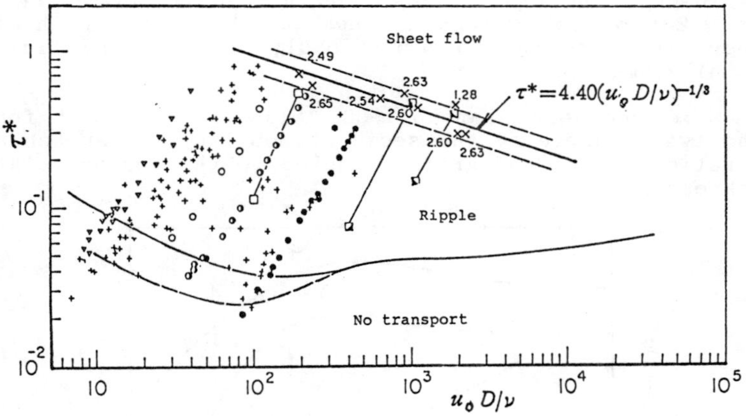

Figure 2. Classification of ripple formation by Komar & Miller(1975)

In Figure 2, it is clearly shown that the criterion for separation between no-movemnt and ripple is not well established, however, in the upper regime where sheet flow initiates, the criterion operates well, and results in an empirical formula in the form

$$\tau^* = 4.40(u_0D/\nu)^{-1/3} \tag{1}$$

In addition to the ripple to sheet flow classification scheme employed by Komar & Miller, Manohar(1955) considered the sand grain size effect in addition to the Shields number. Carsterns, Neilson & Altinbilek(1969) introduced the ratio of the water particle orbital diameter to the sand grain diameter into the criterion, however no graphical presentation was presented. In 1976, Dingler and Inman proposed an emprical criterion in the form

$$u_0^2/(\sigma/\rho-1)gD = 240 \quad (2)$$

In 1979, Nielsen proposed an empirical criterion in the form

$$\tau^* = 0.83 \quad (3)$$

And more recently Shibayama and Horikawa(1982) proposed the critial Shields number as 0.5 to 0.6., and 0.4 in the experiment by Shibayama(1984). The criteria proposed by Nielsen and Shibayama & Horikawa are independent of the sand grain Reynolds number. It is generally difficult to qualify the resistance law in the sheet flow. As stated by Komar and Miller(1975), the constitutive equation for two-phase flow by Bagnold(1956) is given by

$$\tau^* = C \tan\varphi \quad (4)$$

where C is the static volume concentration being about 0.6 to 0.7 for natural sands and $\tan\varphi$ is the coefficient of solid friction which varied from 0.375 to 0.75 in Bagnold's experiment. The critical Shields number therefore varies from 0.23 to 0.53. These values of the critical Shields number are close to those previously mentioned by Nielsen and Shibayama & Horikawa.

More recently, Kaneko(1980) classified seabed forms using two parameters, the sediment mobility paprameter and the ratio of the sand grain diameter to the boundary layer thickness in Figure 3.

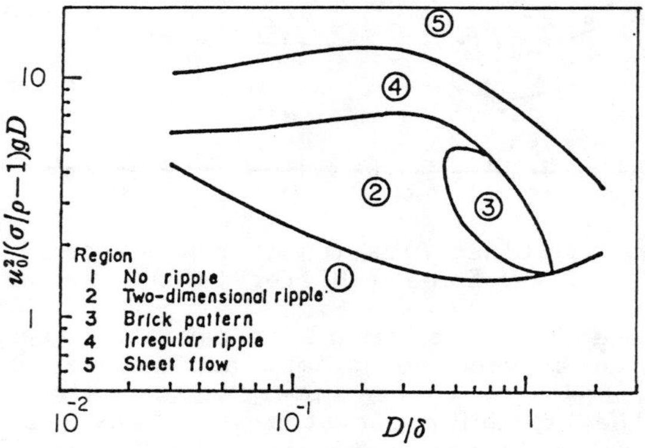

Figure 3. Classification of seabed forms by Kaneko(1980)

In this figure, the ratio is rewritten as

$$D/\delta = \sqrt{\pi D^2/\nu T} \quad (5)$$

where T is the wave period and ν the kinematic viscosity of water. The ratio, which is in the form of a Reynolds number, considers the effect of boundary shearing stress acting on the seabed. He classified seabed froms into the following five stages, 1) no ripple, 2) two-dimensional ripple, 3) brick pattern, 4) irregular ripple and 5) sheet flow. The classification shows that a brick pattern forms under limited conditions defined by the two parameters, furthermore, the region over which the brick pattern forms is small relative to the regions occupied by the other four bed form classifications. In this classification, however, no consideration is made of the threshold of sand movement and the flat bed in the lower regime.

Furthermore, Sunamura(1981) introduced the asymmetry effect of the wave field on the bed forms using the Ursell parameter $U_r = HL^2/h^3$, where H is the wave height, L the wave length and h the water depth and a new parameter F, where

$$F = \frac{u_0^2}{(\sigma/\rho - 1)\, g D^{1/2} d_0^{1/2}} \tag{6}$$

where d_0 is the water particle orbital diameter. This parameter was introduced by Komar and Miller(1975) as a relation describing the threshold of sand movement. His classification is expressed as an empirical formula in the form

$$F = K U_r^{1/4} \tag{7}$$

where the coefficient K specifies seabed forms in the following five stages: 1) no movement, 2) flat bed, 3) ripple bed, 4) flattened-ripple bed and 5) sheet flow. The asymmetry of the wave field was shown to influence seabed formation, however, this effect will not be considered any further.

From the above considerations of seabed forms and their formation criteria, the expressions for the criteria from flat bed to ripple bed and ripple bed to sheet flow are not so clear, however, the dimensionless parameters for specifying the formation criteria are useful. River bed forms and their formation criteria are considered, and in this criteria the ratio of water depth to sand grain diameter is assumed to be very large, while ignored in the criterion for seabed forms. In the theory of the threshold of sand movement by Tsuchiya, Ueda and Oshimo(1984), as shown in Figure 4, the theoretical curves of the sand movement threshold are expressed by three parameters, the Shields number, sand grain Reynolds number or sediment-fluid number and the ratio of water particle orbital diameter to sand grain diameter. And in this paper, specified are four stages; they are no sand movement, flat bed, ripple and sheet flow, flat bed in the upper regime. Based on these considerations, the formationcriterion of seabed can be expressed by

$$F_1(\tau^*,\ u^*D/\nu \text{ or } D_\nu^*,\ d_0/D) = 0 \tag{8}$$

where the ratio does not remarkably influence the threshold

of sand movement as shown in Figure 4, but as already mentioned by Tsuchiya (1987), the ratio effect is remarkable in the geometry of ripples when the ratio becomes greater than about 1000. Therefore an influence of the ratio on the criteria from flat bed to ripple and ripple bed to sheet

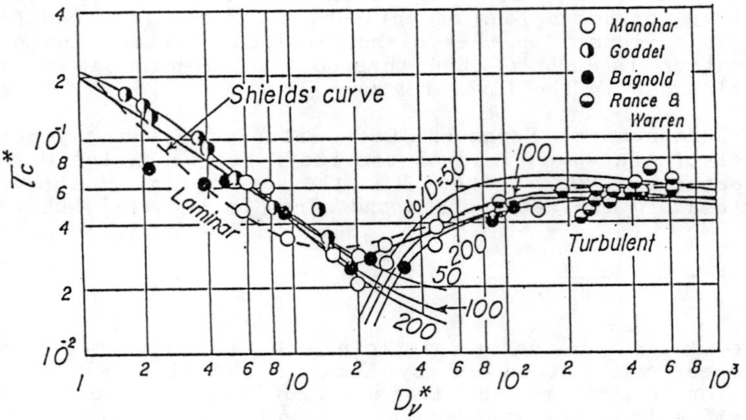

Figure 4. Criterion for the threshold of sand movement

flow may be expected to exist. It is well-known that the ratio expresses the acceleration effect in the wave motion. Physically, the seabed formation when the ratio becomes infinity may correspnds to that by unidirectional flow having an infinite value of the ratio of water depth to sand grain diameter. From this fact the flat beds in the lower and upper regimes may be defined as lower-flat bed and upper-flat bed, respectively.

The upper-flat bed corresponds to the so-called sheet flow. In these regimes, however, the essential difference between sea and river bed formation is the free surface effect which is described by the Froude number, and exists only in the river bed formation.

Additionally, dimensional analysis is used to describe the seabed forms and their formation criteria by Yalin(1964), Mogridge & Kamphuis(1973), Dingler & Inman(1976), and more recently Kaneko 1980) in the form

$$F_2\{u_0^2/(\sigma/\rho-1)gD,\ D^2/\nu T,\ d_0/D\} = 0 \quad (9)$$

Using expression (9) Kaneko derived Figure 3. Both expressions (8) and (9) are physically related to the shear velocity.

3.2 Experimental Data and Their Arrangement

Using a double-deck wave tank 38m long, 0.7m wide and 0.7m deep and a light sediment with an immersed specific gravity of about 0.8 and a diameter of 0.085cm, several experiments were carried out to determine seabed forms and their formation criteria in the region of small sediment Reynolds number. Experimental data from this and other experiments are shown in Table 1 for the immersed specific gravities and diameters of the sediment used.

Table 1. The experimental data used

Researchers (Year published)	Immersed specific gravity $(\sigma/\rho-1)$	Grain diameter D in cm
Manohar(1955)	1.60	0.101
	1.65	0.028
Lofquist(1968)	1.65	0.055
	1.65	0.021
		0.018
Carstens, Neilson & Altinbilek(1969)	1.47	0.030
	1.62	0.059
	1.66	0.019
Horikawa & Watanabe (1967)	1.65	0.020
Sunamura(1981)	1.65	0.020
	1.65	0.070
	1.65	0.156
Tsuchiya & Banno (1988)	0.80	0.085

In the expression of the formation criterion by (8), the shear velocity, which results from the interaction between waves and sediment, should be carefully estimated. In this paper, formulas for the friction factor by Jonsson(1966) and Madsen & Grant(1980) were used. Practically, the friction factor can be expressed by Nielsen(1981) as

$$f_w = \exp\{5.12(5D/d_0)^{0.194}-5.98\} \tag{10}$$

Using the experimental conditions of waves and the sediment properties shown in Table 1, the friction factor was calculated by (10) and Madsen & Grant' formula and then the shear veloctiy was obtained in relation to the maximum water particle velocity at the sea bed. Finally, all the dimensionless parameters in (8) were calculated.

Figure 5 shows an example of the changing stages of bed forms in relation to the Shields number and sediment grain Reynolds number or sediment-fluid number. In the figures, based on the bed forms specified by the researchers bed forms are shown with symbols which are initial movement, flat, initiation of ripple, ripple and sheet flow. All the experimental data were arranged in the sediment conditions as shown in the figure. From these figures, criteria for the four stages in seabed forms can be obtained. The criterion from no-movement to flat-bed in the lower regime is greatly due to the findings of observers, and as stated by Sleath (1984), the experimental data are very scattered.

However, the experimental data for the other criteria are not so scattered because observational findings of the criteria are generally obvious.

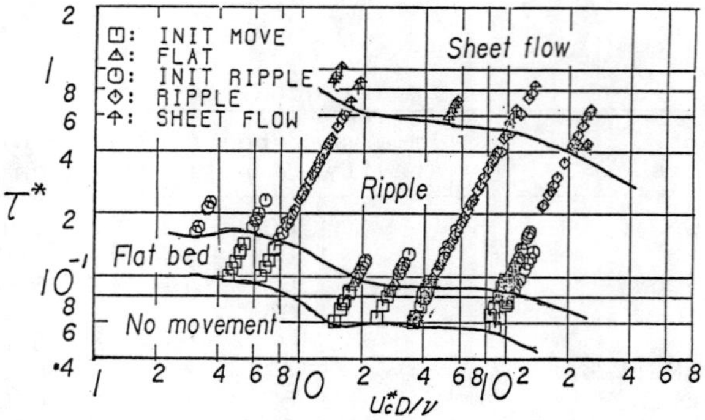

(a) In the case of sediment grain Reynolds number

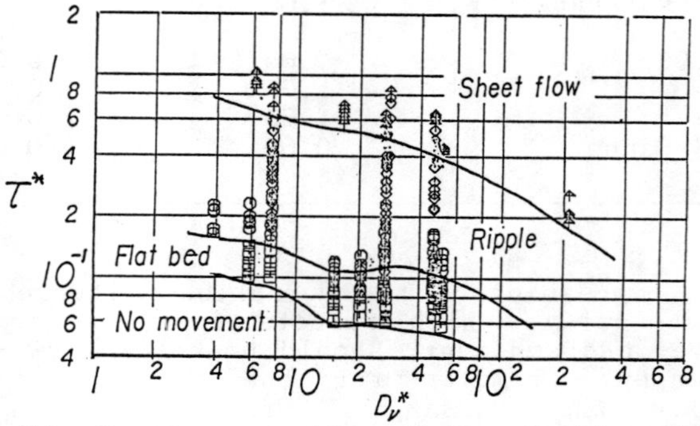

(b) In the case of sediment-fluid number

Figure 5. Determination of the criteria from experimental data (1)

3.3 The Formation Criterion of Seabed Forms (1)

The formation criterion for seabed forms by (8) is shown in Figure 6 where only the Shields curve is shown for the threshold of sand movement. In the figure, the criteria from lower-flat bed to ripple bed and ripple bed to upper-flat bed, sheet flow are shown with the ratio of water particle orbital diameter to sand grain diameter. Using (10) Komar & Miller' empirical formula (1) can be transformed to

$$\tau^* = 3.22 f_w^{1/7} D_\nu^{*-3/7} \tag{11}$$

where in the estimation of the friction factor the value of d_0/D in (10) was evaluated as 3,400 by Carstens, Neilson & Altinbilek(1969) and 16,000 by Kennedy & Falcon(1965) for the criterion from ripple to sheet flow.

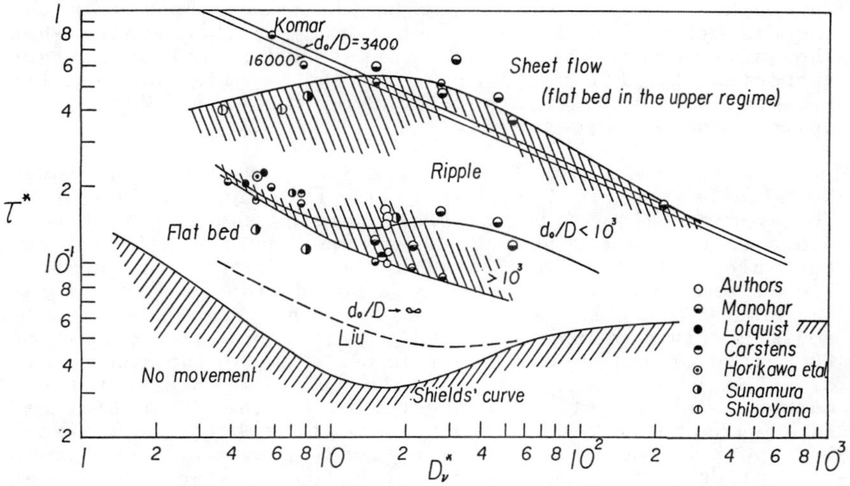

Figure 6. The formation criterion of seabed forms by (8)

In the formation criteria shown in Figure 6, no experimental data for the threshold of sand movemnt are shown due to the high degree of scatter. In Figure 1, the general trend of the experimental data are shown to be similar to the curves by Tsuchiya & Banno. It is noted that the criteria from lower-flat bed to ripple and from ripple to upper-flat bed can be described by the Shields number, the sediment grain Reynolds number or sediment-fluid number and the ratio of water particle orbital diameter. The experimental data is not enough to definitely specify the criteria, so that further experiments are needed especially for the regions where the sedimnt-fluid number are very small or greater than about 100, and for the case when light sediment is used in the wide range of d_0/D. However, it is noted from the figure that in the criterion from lower-flat bed to ripple the effect of the ratio of water particle orbital diameter to sand grain diameter clearly exists, that is, the values of the critical Shields number in the case of d_0/D values larger than 1,000 are plotted lower than those in the case of values smaller than 1,000. This tendency increases with the increase in the sediment-fluid number. And, the criterion changes with the sediment-fluid number. In the figure Liu's curve for the criterion from lower-flat bed to ripple by unidirectional flow is also shown. In this case, the value of d_0/D may be considered as infinity in relation to the criterion of ripple formation by waves. These tendencies show that the the critical Shields number for larger sediment is smaller than that smaller sediment, that is, ripples formed in course sediment are generated by a smaller Shields number than that in fine sediment. Comparing Liu's curve for unidirectional flow with the curves for the criterion from lower-flat bed to ripple, the curve for the d_0/D values larger than 1,000 which are not so comparable to infinity may tend to approach Liu's curve.The range of lower-flat bed becomes narrow with the increase of sediment-fluid number and may then vanish when the value becomes about 100, this tendency may be further supported by Liu's curve. In the region of samller sediment-fluid number, the effect of the ratio d_0/D is not obvious and may disappear.

In the criterion from ripple to upper-flat bed, Komar & Miller's empirical formula (11) is shown in Figure 6. The average value of the critical Shields number may be very close to the values used by Shibayama & Horikawa(1982), but the Shields number changes with the sediment-fluid number. The curve shown in the figure by a solid line is very similar to Komar & Miller's, except for the region of small sediment-fluid number. In this criterion, the effect of the ratio of water particle orbital diameter to sand grain diameter disappears in the whole range of sediment-fluid number. The formation of sheet flow in the upper flat-bed is of course caused by high shear flow both in sediment transport by waves and unidirectional flow, however, it is recognized that in the upper-flat bed no interaction between the free surface and seabed forms by wave action exists, but pronounced interaction between the free water surface and river bed forms exists where the ratio of water depth to sediment grain diameter is most predominant. Therefore, the formation criterion of sheet flow by wave action can not be directly compared with that by unidirectional free surface flow.

3.4 The Formation Criterion of Seabed Forms (2)

By use of (9) the formation criterion of seabed forms is considered. As already described, Kaneko(1980) proposed the criteria of seabed forms classifying them into his own five stages, as shown in Figure 3. Using the experimental data shown in Table 1 with the same dimensionless parameters as those by Kaneko, but classifying only four stages of seabed forms as used in the previous criterion, the formation criterion was examined.

Using Manohar's data shown in Table 1, all the dimensionless parameters in (9) were calculated and the results are shown in Figure 7 where (a) and (b) represent values of d_0/D smaller than 1,000 and larger than 1,000, respectively. It is obvious that the formation criterion of seabed forms can be made, but the effect of the ratio of water particle orbital diameter to sediment grain diameter is not so remarkable. In the region of large Reynolds number $D^2/\nu T$ in (a) and around the transition from lower-flat bed to ripple, larger values of d_0/D may result in larger values of the sediment mobility parameter. In the criterion from ripple to sheet flow, however, the influence of the ratio is hardly found. This may be supported by the result obtained by Figure 6.

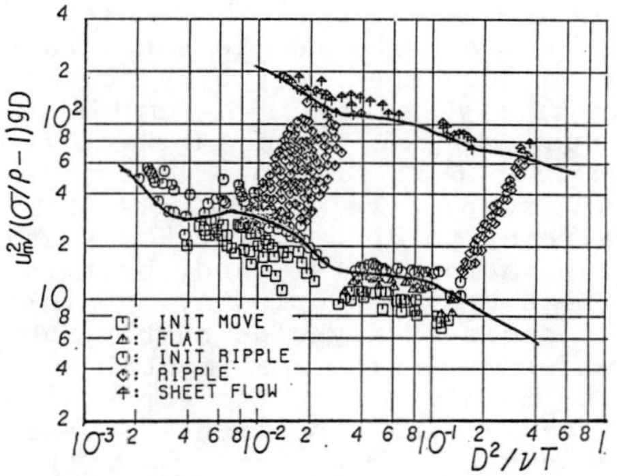

(a) In the case where d_0/D<1,000

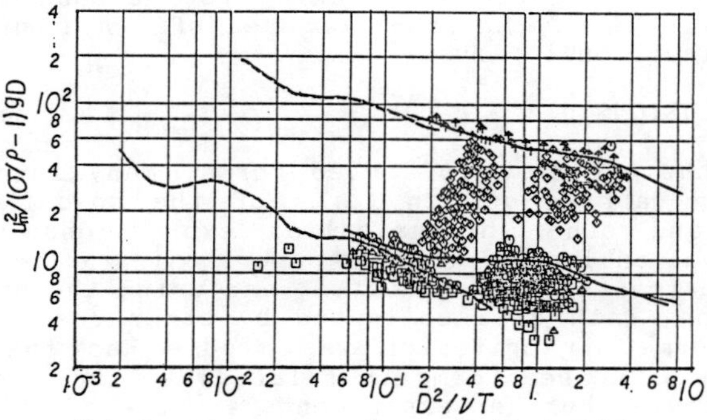

(b) In the case where d_0/D>1,000

Figure 7. Determination of the criteria from experimental data (2)

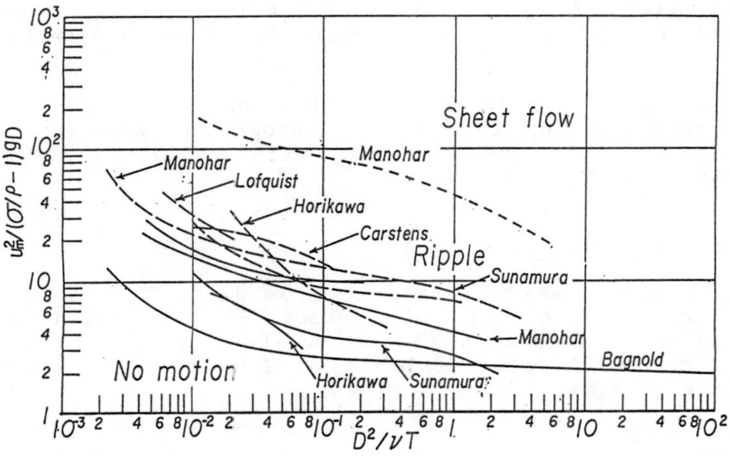

Figure 8. The formation criterion of seabed forms by (9)

In the figure, neglecting the effect of the ratio of water particle orbital diameter to sand grain diameter, the criteria are obtained separately by researchers in terms of the sediment mobility parameter $u_0^2/(\sigma/\rho-1)gD$ and the sediment Reynolds number $D^2/\nu T$. Figure 8 shows the criteria from no movement to flat bed, flat bed to ripple and ripple to upper-flat bed, respectively. On the threshold of sand movement, as already pointed out by Komar and Miller(1975), an effect of the ratio exists. And, by this expression the criteria from flat bed to ripple have a similar tendency in relation to the sediment Reynolds number, however, there is a large variation due to experimental conditions. This variation may be mainly due to the expression of the formation criteria, as well as experimental conditions.

It is recongnized from this fact that the previous expression of the formation criterion of seabed forms is better than the latter one taking into account the mutual relation in the formation prosseses of bed forms by waves and unidirectional flows.

4. CONCLUSIONS

In the criterion of sebed forms, many investigations have been carried out in terms of the so-called Shields diagram and other dimensionless expressions derived by dimensional analysis. In this paper, using a light sediment some experiments were conducted, and using the experimental data, along with data collected by other researchers, the ripple formation criterion was studied as the formation criterion of seabed forms in relation to that by unidirectional flows. Further experiments are needed, but from the formation criteria of seabed forms the main conclusions are:

1) As shown in Figure 6, the formation criterion of seabed forms can be made on the Shields diagram by introducing the ratio of water particle orbital diamter to sand grain diameter which acts as an acceleration effect in the wave motion. When the ratio becomes infinity the critical Shields number for the criterion tends to the unidirectional one.

2) The criterion of sheet flow, upper-flat bed can be expressed both by the Shields number and sediment-fluid number. In this criterion, the ratio has little influence on the formation criterion.

Acknowledgement

We appreciate the help of Mr. James R. Tallent in the preparation of this manuscript. Part of this investigation was founded by a Grant-in-Aid for Scientific Research (No. 62420040) from the Ministry of Education, Science & Culture of Japan.

References

Bagnold, R.A. (1946): Motion of waves in shallow water,

interaction between waves and sand bottoms, Proc. Roy. Soc. London, Series A, Vol. 187, pp. 1-15. Tech. Memo. No. 28, 16pp.

Grant, W.D. & O.S. Madsen (1982): Movable bed roughness in unsteady oscillatory flow, Jour. Geophy. Res., Vol. 87, No. C1, pp. 469-481.

Horikawa, K. & A. Watanabe (1967): A study on sand movement due to wave action, Coastal Engg. in Japan, Vol. 10, pp. 39-57.

Inman, D.L. (1957): Wave-generated ripples in nearshore sands, CERC, US Army Corps of Enggr., Tech. Memo., No. 100, 14pp.

Jonsson, I.G. (1966): Wave boundary layer and friction factor, Proc. 10th ICCE, ASCE, pp. 127-148.

Kaneko, A. (1980): Wave length of ripples formed in oscillatory flow, Proc. 27th Japanese Conf. Coastal Engg., JSCE, pp. 207-210 (in Japanese).

Kennedy, J.F. & M. Falcon (1965): Wave generated sediment ripples, MIT Hydrodynamics Lab., Rept. No. 86, 86pp.

Komar, P.D. & M.C. Miller (1973): The threshold of sediment movement under oscillatory water waves, Jour. Sed. Pet., Vol. 43, No. 4, pp. 1101-1110.

Komar, P.D. & M.C. Miller (1975): The initiation of oscillatory ripple marks and the development of plane-bed at high shear stresses under waves, Jour. Sed. Pet., Vol. 45, No. 3, pp.697-703.

Liu, H.K. (1957): Mechanics of sediment ripple formation, Proc. ASCE, Vol. 83, No. HY2, pp. 1-23.

Lofquist, K.E.B. (1978): Sand ripple growth in an oscillatory flow water tunnel, CERC, US Army Corps of Enggr., TP78-5.

Manohar, M. (1965): Mechanics of bottom sediment movement due to waves action, CERC, US Army Corps of Enggr., Techn. Memo. No. 75, 121pp.

Mogridge, G.R. & J.W. Kamphuis (1973): Experiments on bed form generated by wave action, Proc. 13th ICCE, ASCE, pp. 1123-1142.

Nielsen, P. (1981): Dynamics and geometry of wave-generated ripples, Jour. Geophy. Res., Vol. 86, No. C7, pp. 6467-6472.

Sleath, J.F.A. (1978): Measurement of load in oscillatory flow, Jour. ASCE, No. WW4, pp. 291-307.

Sunamura, T. (1981): Bedforms generated in a laboratory wave tank, Sci. Rept. Inst. Geoscience, Univ. Tsukuba, Sect. A, Vol. 2, pp. 31-43.

Swart, D.H. (1976): Predictive equations regarding coastal transport, Proc. 15th ICCE, ASCE, pp. 1131-1139.

Tsuchiya, Y., Y. Ueda & T. Oshimo (1984): A theory of the the threshold of sediment movement by wave action, Proc. 31st Japanese Conf. Coastal Engg., JSCE, pp. 272-276(in Japanese).

Yalin, M.S. & R.C.H. Russell (1962): Similarity in sediment transport due to waves, Proc. 8th ICCE, ASCE, pp. 151-167.

CHAPTER 139

Extreme Erosion Event on an Artificial Beach

J P Möller* and D H Swart**

Abstract

At Oranjemund just north of the mouth of the Orange on the South-West African/ Namibian coastline the Consolidated Diamond Mines (Pty) Limited (CDM) is mining for diamonds in the inshore area. They use an artificially-built seawall of sand to keep the sea out of the paddocks which are being stripped and mined at bedrock level, which is well below sea level. The seawall runs parallel to the original shoreline at a distance of up to 350 m offshore. The beach profile is correspondingly very steep and under most conditions offshore sediment losses occur, which are compensated for by artificial nourishment. During February 1987 offshore losses of about 120 m^3 per running metre of seawall were recorded from above the waterline. In this paper a data set is presented to serve as a basis for the calibration of on-offshore sediment models and some simulations are reported on.

1. Background

Diamonds have been mined on the west coast of South-West Africa/Namibia for more than 50 years (see Figure 1). As mining techniques have improved more and more diamonds have been taken out of what used to be the surf-zone area. CDM has perfected the technique of inshore mining by using seawalls or bunds constructed of sand (Figure 2). Details of the coastal engineering aspects of the mining technique have been given by Möller **et al** (1986). In essence the construction of a seawall of sand in the nearshore zone and its maintenance against wave attack amounts to a large-scale sand suppletion or beach management project. Figure 3 shows how a typical nearshore beach profile which has a nearly horizontal platform up to 400 m wide at 6 m below MSL has been modified by the construction of the seawall to a beachface with a nearly uniform slope of 1 in 25. At the top of this very steep beach profile sits the seawall proper with its foot at about 4 m above MSL and its crest - at about 10 m above MSL. The crest width of the wall

----

*Coastal Processes and Management Advice, EMA, CSIR, Box 320, Stellenbosch, Republic of South Africa.
**Programme Manager, Coastal Processes and Management Advice, EMA, CSIR, Box 320, Stellenbosch, Republic of South Africa.

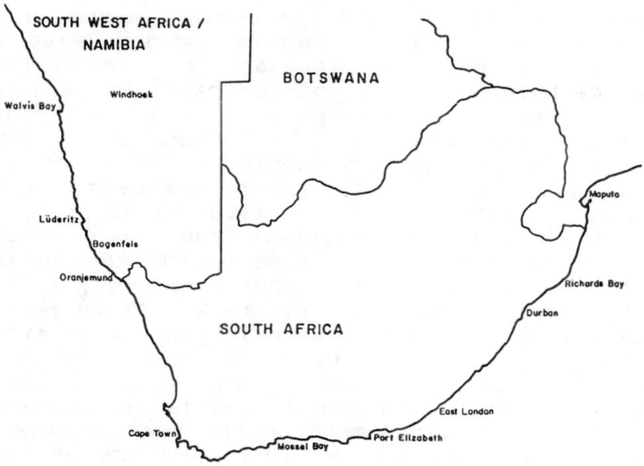

FIG. 1 LOCALITY MAP

FIG. 2 AERIAL VIEW OF MINING AREA BEHIND SEAWALL

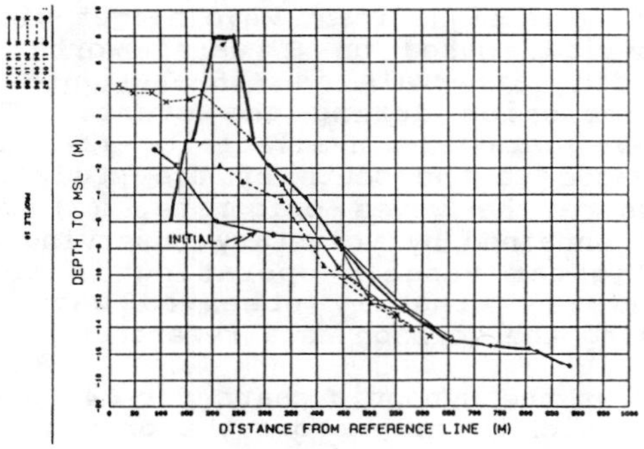

FIG. 3 BEACH AND NEARSHORE PROFILE DEVELOPMENT DUE TO SEAWALL CONSTRUCTION

varies between 17 and 25 m and the seaward slope is at the angle of repose or steeper. During extreme wave events classical dune erosion takes place and the seawall ("dune") has to be maintained on a round-the-clock basis by large earthmoving equipment. During periods when the earthmoving equipment can concentrate solely on beachwall building and maintenance it is possible to supply 16 000 m^3 of sand to the seawall per day. The seawall is being extended northwards in the direction of the littoral drift by active dumping at a rate of approximately 800 m per year. To minimise maintenance costs the area behind the seawall is divided into paddocks which are, once mined out, left to be inundated by the sea again (see Figure 2). This means that only a localised area needs to be maintained. Nevertheless the length to be maintained is of the order of 800 m at times but could also be as short as 200 m if localised erosion problems are encountered. This means that the volume of artificial nourishment to the seawall can amount to between 20 and 80 m^3/day per running metre of seawall. Under "normal" storm conditions this should be sufficient to counteract wave-induced erosion.

During a routine visit to Oranjemund on 24 and 25 February 1987 extreme wave conditions were observed. Four cross-sections were taken in front of the seawall during a 23-hour period to record the wave-induced erosion of the beach in front of the seawall and the seaward face of the seawall. This paper discusses the results obtained and relates them to available predicted techniques in literature.

2. Data

The data relevant to the particular storm is summarised here to serve as basis for use by researchers/engineers who want to verify onshore-offshore sediment models. The data can be subdivided into four types, namely, wave data, profile data, sand suppletion rates and grain size data.

2.1 Wave data

An offshore Waverider station at the site is situated in 106 m of water. The significant wave height with a 1 in 20 year return period, based on 8 years' worth of data at 6 hourly recording intervals is 4.5 m. Unfortunately it was out of commission during the event. Three other sources of wave data are available; (1) data from a Waverider situated to the south of the study area in the region from where the waves originate, (2) data on the format of that obtained by voluntary observing ships (VOS) gathered by lighter vessels operating in the area off Oranjemund and (3) visually observed wave heights at breaking by using the horizon as a reference.

Figure 4 shows the synoptic chart for 14h00 on the 22nd February 1987, which clearly shows the origin of the storm event.

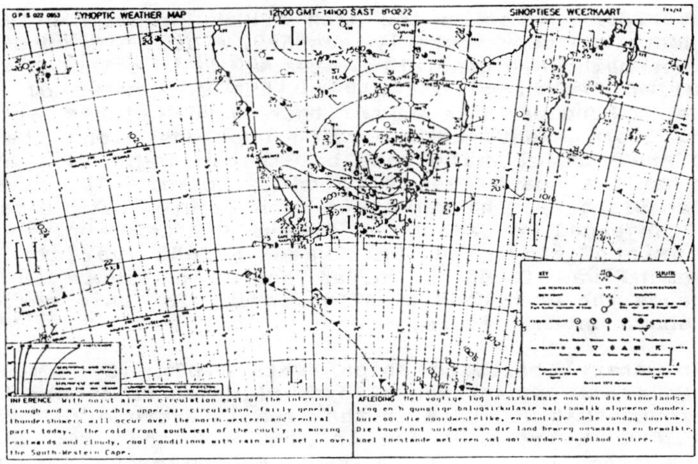

FIG. 4 SYNOPTIC CHART 14H00 : 22.02.1987

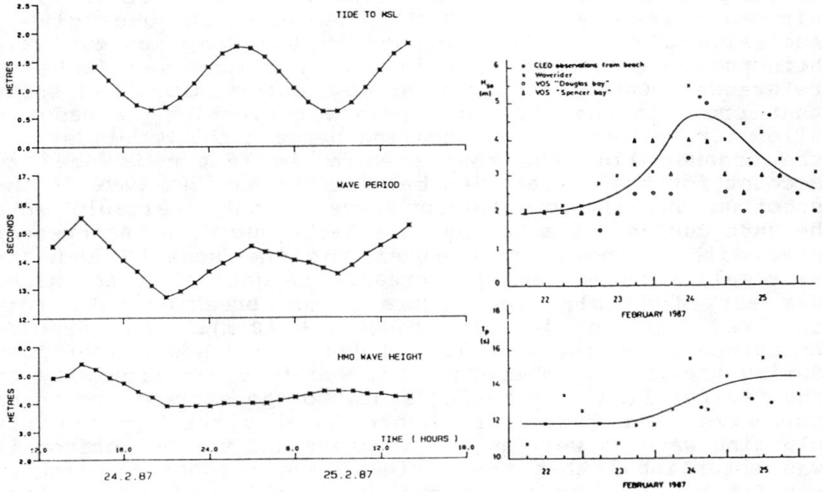

FIG. 5 WAVE DATA FROM VARIOUS SOURCES FOR FEBRUARY 1987 STORM AT ORANJEMUND

Waverider Waves in the area arrive predominantly from the south to southwesterly sectors. This is consistent with the main wave generating area which is situated in the so-called "Roaring Forties" to the southwest of South Africa. Studies have shown that wave heights attained during storm events at recording stations along the west coast are virtually the same but that a lag occurs as one moves up the coast, as is to be expected. The mean lag between Slangkop near Cape Town (see Figure 1 for location) and Oranjemund is about 24 hours. After applying the appropriate time lag the wave height variation at Oranjemund was found to be as shown in Figure 5. For reference the predicted tidal variation for this period is also shown on the figure.

Voluntary Observing Ships' data (VOS) By comparison with Waverider measurements for earlier storm events it has been found that VOS data in the area is extremely reliable. This is most probably due to the fact that two lighter vessels owned by to the De Beers Marine group operate in the area on a regular basis. The mariners have a long experience of working in this particular area. The VOS data obtained from these lighters for the February storm are also shown in Figure 5.

Visually observed maximum wave breaker heights (CLEO) As part of a low-level monitoring programme (CLEO) along similar lines as the COPE programme in Queensland, Australia and the LEO programme in the USA, maximum wave heights are estimated by using the horizon as a frame of reference. Comparison to Waverider observations has shown that provided that the appropriate conversion is made to allow for refraction and shoaling between the Waverider and the breaker line and that a suitable factor is used to account for the correlation between the maximum wave in the spectrum and the significant wave height, reliable wave heights can be estimated by this technique. In the present case with a known crest level for the seawall and the extremely high maximum pre-breaker heights this technique was easy to apply (see Figure 6 for breaking wave fury observed from an elevation of MSL + 10 m). The exposed beachface below the seawall did not exceed a width of 20 m during the event. Therefore one was in close proximity to the breaker line. By taking shots to the extreme trough of the wave just before it broke as a virtually totally plunging wave as well as to the crest and to the horizon it was established that the maximum wave height at breaking was 8.5 m. Obviously the maximum height varied with time but during the late afternoon on 24 February, the maximum height varied between 7.5 m and 8.5 m. It would appear that these heights are far in excess of those values quoted earlier although the wave periods are roughly in line. The wave heights observed in this manner represents a maximum height over a 2-4 minute period. It is possible to use the relationship developed by Longuett-Higgins (1952) to transfer the observed wave heights to significant wave heights at breaking for 13 sec waves. On the basis of the synoptic chart on Figure 4 it is possible to calculate a combined shoaling/refraction coefficient of 1.1 between

FIG. 6 BREAKING WAVE FURY AS OBSERVED FROM SEAWALL (MSL + 10M)

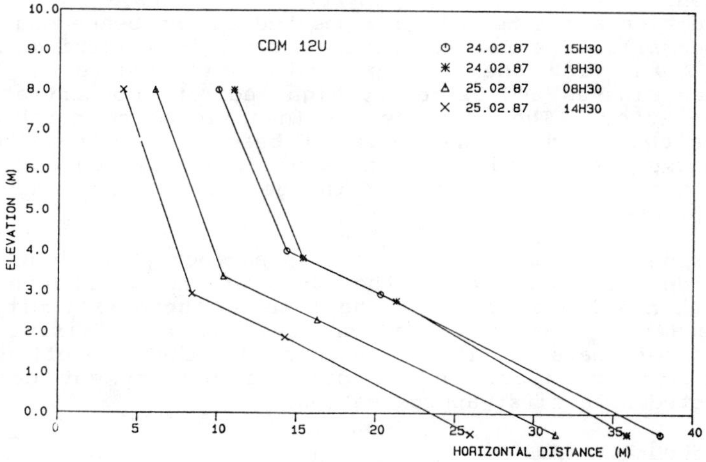

FIG. 7 BEACH FACE PROFILES MEASURED ON 24/25 FEBRUARY 1987

deep-sea and a water depth of 10 m. Therefore the wave height variation observed at the breaker line can be transferred to deep-sea significant wave heights. The results obtained in this manner are shown in Figure 5.

Summary It is clear that the data from the three sources show a similar trend, although there is some scatter, as is to be expected. The solid lines through the data represent the best estimate of the mean trend depicted in the data.

2.2 Profile data

As can be seen in Figure 3 the beach profiles are steep and the exposed beachface is at best fairly restricted. At 4 occasions during the 23-hour period a cross-section of the seawall and beachface was taken. This was done at extreme risk to the observers, because the beach face was only exposed for about two minutes at a time during which the staff-bearer had to go down a 6 m unconsolidated sand-slope onto the beach, taking off only metres ahead of the next onrushing wave. The four profiles are given in Figure 7. In the assessment of the profile changes observed due cognizance should be taken of the general morphological behaviour of this area.

The dominant wave direction in the area is approximately south-westerly which leads to a net northerly drift of sediment of about 1.4 x 10^6 m^3/yr. The seawall acts as an obstruction and material is deposited on the beach south of the seawall. The fact that the seawall is maintained more than 200 m from the original waterline results in high sediment losses and extremely high sediment concentrations in the water. The suspended sediment is moved northwards and is deposited on the downdrift beach. This results in the unusual situation that the shoreline grows on both the updrift and downdrift ends of the seawall in the long term (see Figure 8).

During the event the seawall was perfectly straight. In addition, the area where the observations were done was more than a kilometre from the then northern extremity of the seawall. Waves broke parallel along a fairly long section of seawall. It can be concluded that the effect of longshore transport on the profile development can be neglected as a first approximation.

2.3 Sand nourishment

For the reasons outlined above, the nourishment rates along the seawall are fairly high, as can be seen on Figure 9, which contains monthly nourishment rates over a three year period. The volume of material given in this figure is the total volume needed for (1) so-called seawall building to advance the seawall to the north and (2) for maintenance purposes.

It was obvious that the seawall was steadily being eroded. Therefore although the general policy has been to

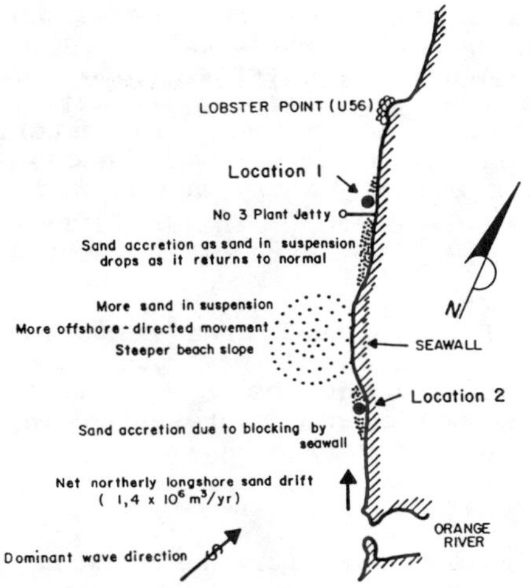

FIG. 8 SCHEMATIC REPRESENTATION OF NEARSHORE PROCESSES ADJACENT TO SEAWALL

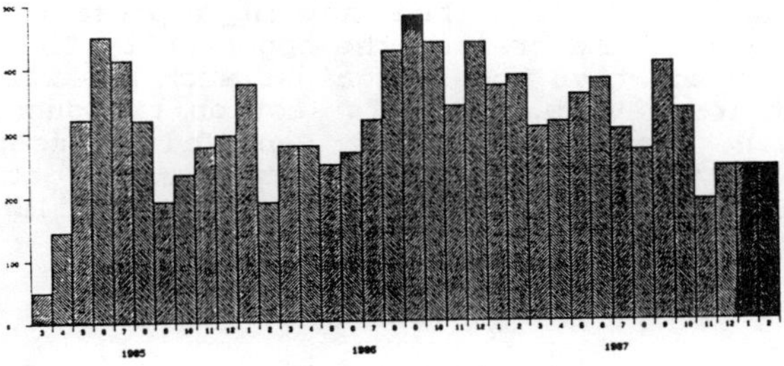

FIG. 9 INSHORE MINING PROJECT : MONTHLTY FEEDING RATES

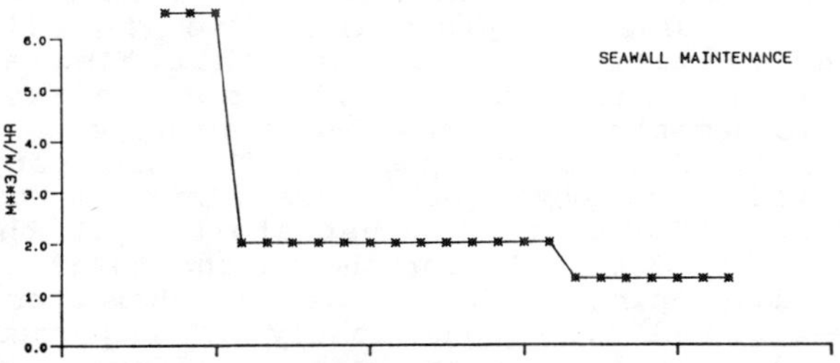

FIG. 10 SEAWALL MAINTENANCE DURING FEBRUARY 1987 STORM EVENT

work on a 24-hour a day basis with the effort being split between beachwall building and beachwall maintenance, all earth-moving equipment was diverted onto maintaining the paddock being stripped and mined at the time. The seaward face of this paddock is 500 m long. The material was being actively dumped on 400 m of sea wall along this paddock during the 24-hour period to try and curtail wave-induced erosion. A total of 1021 15 m^3 truckloads of sand was dumped onto the sea wall and dozed into the water by bulldozer.

On the basis of the actual area renourished over time during the 24 hour period it was possible to construct Figure 10, which is the best estimate of seawall maintenance during the event, represented in cubic metres maintenance per metre of wall.

2.4 Grain size data

Available grain size data for the seawall area indicate that the median grain size is about 200 microns.

3. Comparison with predictive models

The data set described herein represents an extreme erosion event. It creates the opportunity to test onshore-offshore predictive models against such an extreme offshore sediment loss event. In this section two such comparisons are given, namely, with the D-profile concept of Swart (1974a, 1974b, 1976, 1986) and with the energetics approach of Nairn (1988). The full data set is available from the authors.

3.1 Profile concept of Swart

The empirical approach of Swart has been well documented. It uses of the concept that an equilibrium profile will eventually be formed if a given wave condition persists long enough. At any time during the transition from the initial conditions to the final (equilibrium) condition the quantity of onshore or offshore sediment transport taking place is proportional to the difference between the schematized profile at that time and the schematized equilibrium profile. In order to facilitate the profile schematization, the beach profile is broken into essentially three parts, namely, (1) that portion of the profile which is above (or landward of) the point of maximum uprush, the so-called **backshore**, (2) the area between the point of maximum uprush and the lower limit of the actively developing profile, which is characterized by very active sediment transport (mostly as suspended load) and by well-developed bars and other bedforms, the so-called developing profile or D-profile, and (3) the area below (or seawards of) the lower limit of the D-profile, where a transition basically takes place between the newly developing profile and the original profile. This last area is called the transition area and transport here usually takes place as bed load. Swart (1974a) compared his theoretically based framework with extensive model and field data on profile development and presented empirical relationships for all parameters required to apply the technique in a predictive mode.

Over the past 14 years the technique has been extensively applied to a variety of coastal applications, with a large degree of success. Seymour (1988) concluded in a paper presented at this (ICCE) conference that the technique has the highest success rate of the onshore-offshore predictive models tested in his study.

Using the data described above a set of model input was prepared. Application of the model yielded satisfactory profile remoulding (Figure 11) and loss rates. Figures 12 and 13 which are highly compatible with the observed values, as deduced from the data presented earlier. Volumetric loss from above the water line is exceptionally close to the observed value, namely 130 m^3/m as opposed to 123 m^3/m over the 23 hour period. However, variations in the 0 m and +5 m contours are not as accurately predicted, specifically over the last 5 to 10 hours of the period. Nevertheless, the overall comparison is good and it can be concluded that this fairly simplistic method can give reliable results. A strong disadvantage of this technique is the fact that large-scale bedforms such as profile perturbations and breaker bars cannot be predicted.

3.2 Energetics Approach of Nairn

Bailard (1981) developed a model for the prediction of cross-shore sediment transport which utilised an energetics approach. Bailard assumed that instantaneous sediment transport is proportional to some power of the instantaneous orbital velocity at the bed. Stive (1987) used the same technique for random waves, and used the Stokes II wave theory to compute odd velocity moments needed as input for the model. Both bed load and suspended load are catered for, in three modes of transport, namely, asymmetric (onshore) transport, mean flow (offshore) transport and downslope (offshore) transport due to gravity. Nairn (1988) used the same techniques as Stive. As far as waves are concerned, the dissipation model for random breaking waves of Battjes and Janssen (1978) was used, whilst transfer of individual regular wave components in a probability density function was done according to the method of Dally (1987). Bottom changes were calculated using a Lax-Wrendroff second-order scheme based on the flux of sediment across a finite difference section of the model (Nairn 1988). Rob Nairn graciously ran his model on the Oranjemund data set. The results obtained when applying the Nairn model are given in Figure 14. It shows that a very realistic profile readjustment took place, although reference to Figure 15 shows that the volumetric loss rates predicted by the model are too low by a factor of approximately two. This is not considered an insurmountable problem as the loss rates would be appreciably improved if the model could cater adequately for sediment removal from above the water line, which is obviously not a function of the mathematical technique used

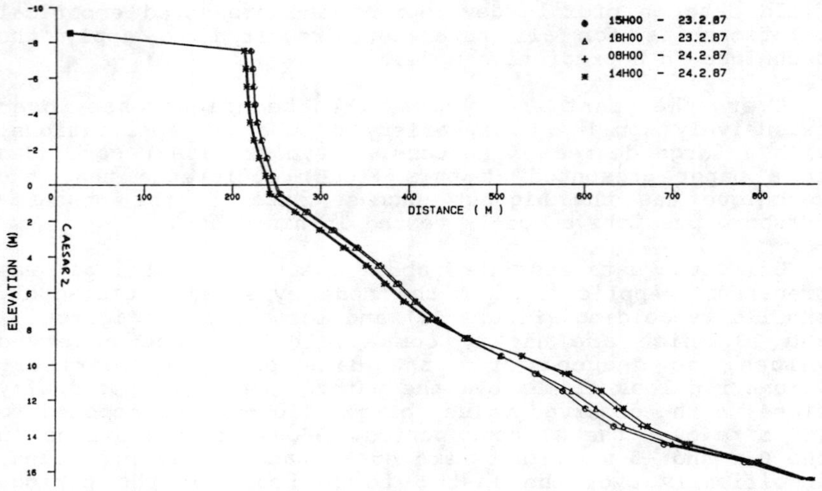

FIG. 11 MODEL PROFILE REMOULDING

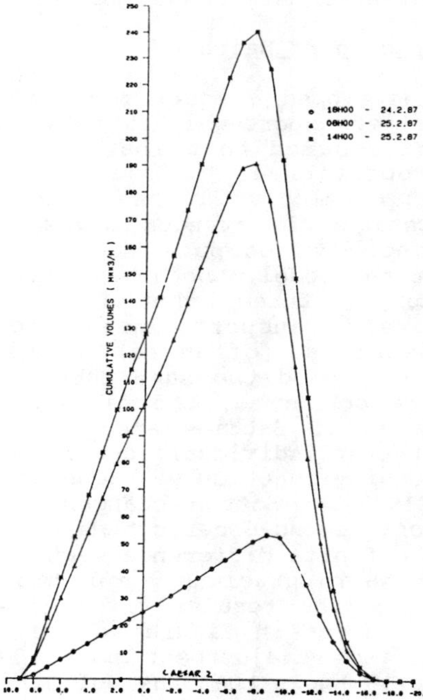

FIG. 12 MODEL EROSION RATES

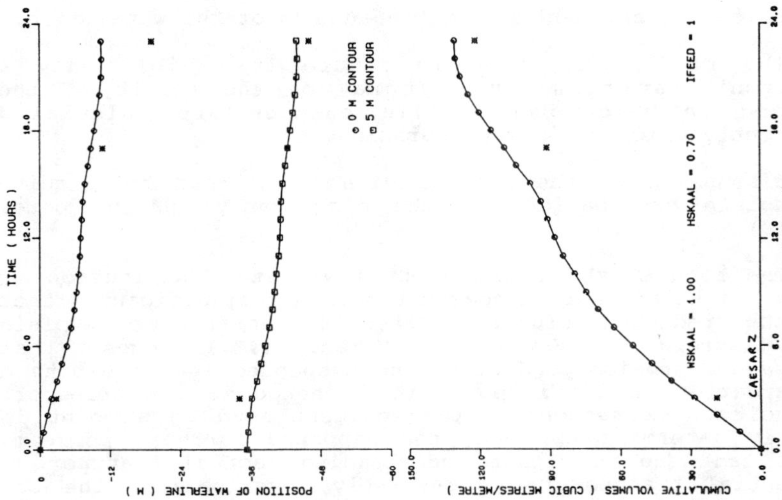

FIG. 13 WATERLINE VARIATION AND VOLUMETRIC LOSS

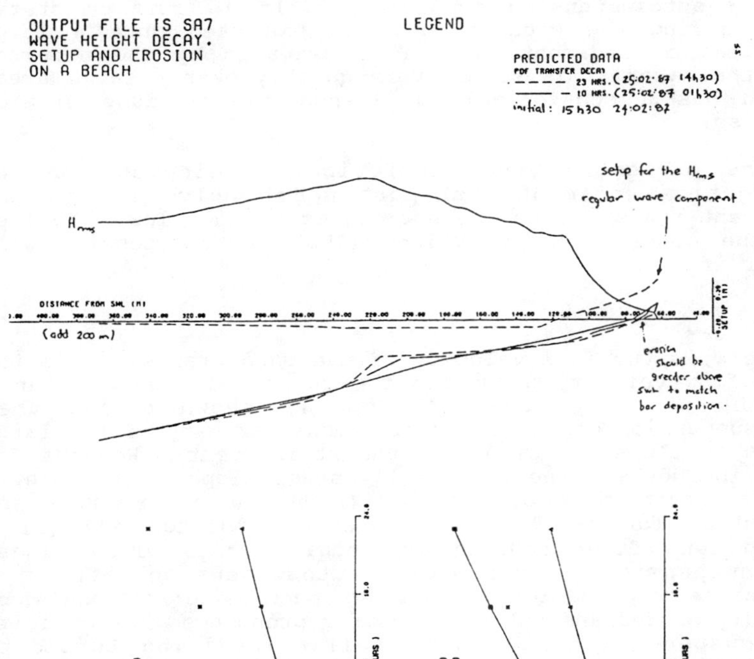

FIG. 14 MODEL PROFILE REMOULDING (NAIRN)

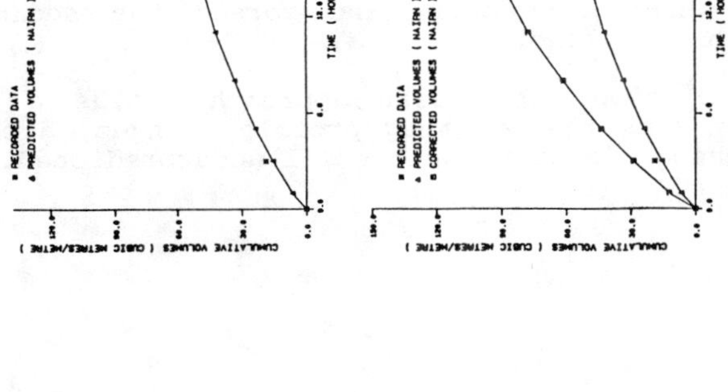

FIG. 15 RECORDED AND PREDICTED EROSION RATES

to calculate sediment movement seawards of the water line.

The results depicted in Figure 14 exhibit only a monotonic bar formation. It may be the result of the method in which bed profile changes are calculated presently which needs some pursuing.

Discussion of the results with Nairn lead to a quite plausible explanation for the discrepancy (Nairn, pers. comm.).

The band width of the input bivariate distribution of wave height and period does not have any significant effect on the predicted volumetric rates and can thus not explain the discrepancy. However, de Vriend (1987) suggests that in severe erosion conditions the suspended load exhibits a retardation in its response to a change in the transport capacity. Consequently, the sediment load at a point is partly determined by what has happened further upstream. This can lead to either underloading (actual transport < potential transport) or conversely, overloading (the so-called autosuspension condition). It is further stated that during the erosion of an unprotected dune a large overload of sediment is carried seawards by the undertow and the material is spread very quickly over a large area, which has a considerable influence on the dune erosion process.

The presently used theoretical development of the energetics sediment transport model only relates the sediment transport to the energetics of the flow. In lieu of the comments by de Vriend (1987) autosuspension will occur if

$$w < \bar{u}_s \tan \beta$$

where w is the fall velocity of the sand grains, $\bar{u}_s$ is the mean flow velocity advecting the suspended load and tan ß is the profile slope. It can be shown easily that autosuspension conditions could occur for many of the large waves in the spectrum during the storm event. Reasons for this include (1) the unnaturally steep slope of 1:20, even for a grain size of 0.5 mm (2) the wave exposure and potential for rapid return flows or undertow and (3) a steep dune face inducing avalanching which will trigger autosuspension. Moreover, autosuspension offers a plausible explanation of how so much material could be rapidly eroded and not show up as a pronounced bar deposit. Autosuspension can lead to self-maintaining turbidity currents which will carry and spread the sediment well outside the surfzone.

In conclusion, the Nairn approach yields a promising comparison with the observed profile changes, although the rate of change is at present still underpredicted.

4. Conclusions

* A data set was presented which represents an extreme wave-induced erosion event, with a volumetric loss rate from above the water line of approximately 130 m^3/day. The data are available on request.

* Computations with the semi-empirical D-profile approach of Swart (1986) and with the energetics model of Nairn (1988) show that it is possible to predict such extreme events and open some interesting avenues for more detailed research on the mechanism of extreme wave-induced sediment transport.

5. Acknowledgments

The logistic support of the Consolidated Diamond Mines (Pty) Ltd in compiling the data set is gratefully acknowledged. Thanks are also due to Rob Nairn of the Imperial College of Science and Technology who spent a substantial amount of time and effort to model this event with his technique.

6. References

BAILARD, J.A. (1981). An energetics bedload model for a plane sloping beach: Local transport. Journal of Geophysical Research. Vol. 86, No C3

BATTJES, J.A. and JANSSEN, J. (1978). Energy loss and setup due to breaking of random waves. Proc 17th International Conference on Coastal Engineering

DALLY, W.R. (1987). Wave transformation in the surf zone. Ph.D. Thesis, University of Florida

de VRIEND, H.J. (1987). Two and three-dimensional modeling of coastal morphology. Recent developments at the Delft Hydraulics Laboratory. Report No 377

LONGUETT-HIGGINS, M.S. (1952). On the statistical distribution of the heights of sea waves. Journal of Marine Research, Vol XI, 1952

MÖLLER, J.P., OWEN, K.C. and SWART, D.H. (1986). Coastal engineering studies for inshore mining of diamonds at Oranjemund. Proc. 20th International Conference on Coastal Engineering. Taiwan

NAIRN, R.B. (1988). Prediction of wave height and mean return flow in cross shore sediment transport modeling. IAHR Symposium on Mathematical Modeling of Sediment Transport in the Coastal Zone. Copenhagen

SEYMOUR, R.J., CASTEL, D. (1988). Validation of cross-shore transport formulations. Proc. 21st International Conference on Coastal Engineering, Malaga

STIVE, M.J.F. (1987). A model for cross-shore sediment transport. Proc. 20th International Conference on Coastal Engineering, Taipei

SWART, D.H. (1974a). Offshore sediment transport and equilibrium beach profiles. Doctoral thesis, Technische Hogeschool, Delft

SWART, D.H. (1974b). A schematization of on-offshore transport. Proc. 14th International Conference on Coastal Engineering

SWART, D.H. (1976). Predictive equations regarding coastal transports. Proc. 15th International Conference on Coastal Engineering. Honolulu, Hawaii

SWART, D.H. (1986). Prediction of beach changes and equilibrium beach profiles. Lecture notes for short course on : Dynamics of Sand Beaches. Taipei, Taiwan

CHAPTER 140

A MODEL FOR BREACH GROWTH IN A DIKE-BURST

Paul J. Visser*

ABSTRACT

A mathematical model for sand-dike breach erosion is presented. The heart of the model is a modified Bagnold (1963) energetics-based sand transport conception combined with a simplified Galappatti and Vreugdenhil (1985) pick up mechanism for the suspended load. The model has been tested to three laboratory experiments. The agreement between model predictions and experimental results is surprisingly good. Prototype calculations are presented for the 73 m high sand-dike of a proposed pumped-storage plant in the Netherlands.

1. INTRODUCTION

One of the possibilities for storage of energy is a water reservoir with a relatively high water level. In a flat country as The Netherlands a pumped-storage plant can be built by constructing an artificial basin surrounded by dikes. A proper location for the plant is along the coast against the Brouwersdam in the south-western part of The Netherlands. Recent feasibility studies conclude that the optimum dimensions of the reservoir are an area of about 15 km^2 and a water depth of about 70 m (resulting in a storage capacity of 20000 MWh and an electric power of 2000 MW). The ring-dike of the basin should be constructed with local sand ($D_{50} \approx 0.20$ mm).

Failure of the basin dike can result in a major disaster for the surroundings of the plant with losses of both human life and property. The possible consequences of an eventual dike-burst are important aspects for the determination of the exceptable failure probability of the basin. These consequences will depend on the behaviour of the flood wave, which in its turn will depend on the reservoir outflow hydrograph, particularly the maximum rate of outflow. The outflow hydrograph is governed largely by the growth of the breach with time.

The problem is comparable with that of earth-dam failures, for which the last 20 years a number of mathematical models has been developed, see Singh and Scarlatos (1988). These models, as also the model of Fujita and Tamura (1987) for the breaching of bursts in river

---

* Scientific Officer, Department of Civil Engineering, Delft University of Technology, P.O. Box 5048, 2600 GA Delft, The Netherlands.

dikes, assume bed load as the prevailing transport mechanism. But in case of failure of a high sand-dike suspended load transport will dominate. Since this results in a completely different breach erosion, these models cannot be applied here.

The aim of the investigation is a mathematical model which describes the growth of the breach and the discharge with time. The study was done for the NOVEM (Netherlands Company for Energy and Environment).

Starting point is a basin with an area of 15 km² and a water depth H_w = 70 m. It is assumed that the 73 m high ring-dike is

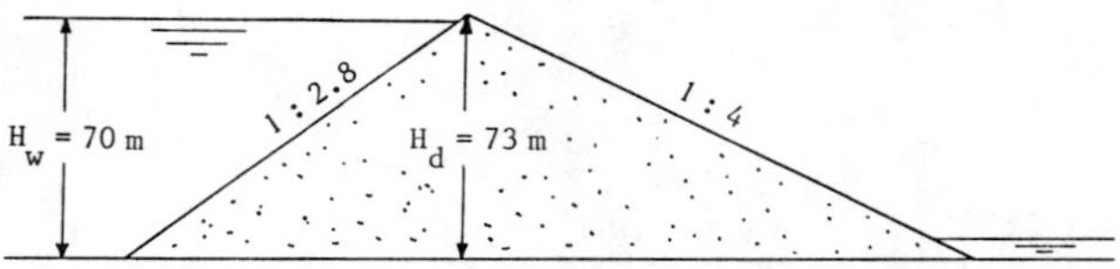

Fig. 1 Schematization of the cross-section of the high sand-dike

completely built up with sand (D_{50} = 0.20 mm), see Fig. 1. The inner and outer slopes are 1 : 2.8 and 1 : 4, respectively. All these quantities can also be varried. It is further assumed that the dike-burst starts with a small initial breach at the crest of the dike, for instance due to piping.

2. DESCRIPTION OF SAND TRANSPORT

A proper description of the pick up and transport of sediment is essential for the simulation of breaching processes. In the case of failure of the sand dike of a 70 m high reservoir, the outflow velocities will become very large (order of magnitude 20 m/s). None of the existing sediment transport formulae has been verified for these velocities. Hence, an energetics-based approach is applied in which the sand transport is expressed directly in the available power for it from the flow. Thus, a huge overestimation of the sand transport (as by some other formulae) is prevented.

The present approach is based on the energetics-based sediment transport conception of Bagnold (1963, 1966). The incorporation of the slope effect differs ,however, from Bagnold's (1963) approach.

Bagnold expressed the sediment transport in the available energy ω per unit time and unit area of the bed surface:

$$\omega = \tau_b = C_f \, \rho \, u^3 \qquad (1)$$

where τ_b is the bed shear stress, u is the depth-averaged current velocity, C_f is the drag coefficient for the bed ($\approx$ 0.01) and ρ is the water density.

The Bagnold conception includes both bed load transport (capacity s_b) and suspended load transport (capacity s_s). Here the order of magnitude of the ratio s_b/s_s will be:

$$\frac{s_b}{s_s} \approx \frac{13}{tg\phi}\frac{W}{u} \approx \frac{13}{0.6}\frac{0.02}{20} \approx 0.02 << 1 \qquad (2)$$

where ϕ is the internal angle of friction of the sediment in the bed load ($\phi \approx 32^0$ for sand) and W is the fall velocity of the sediment. So the bed load transport can be neglected with respect to the suspended load transport.

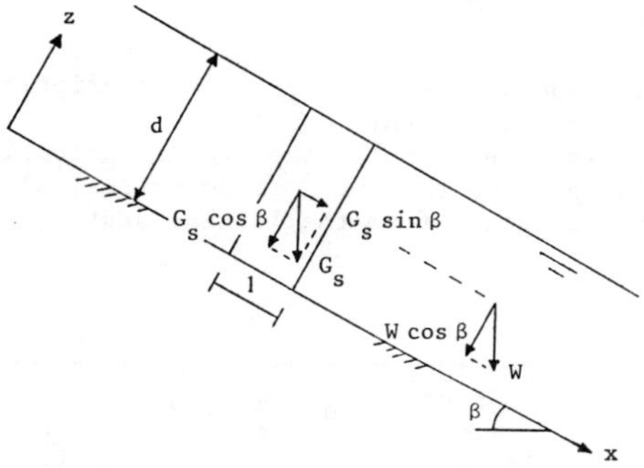

Fig. 2 Components G_s cos β and W cos β normal to the slope of G_s and W, respectively

Consider a column of water in the flow on a slope (with angle β) with an unit area of 1 m^2 (along the slope) and a height d equal to the water depth, see Fig. 2. If G_s is the immersed weight of the suspended material within this column and u_s is the average transport velocity of these solids, then the suspended load transport s_s, can be written as

$$s_s = G_s \, u_s \qquad [N/ms] \qquad (3)$$

The lifting work P (per unit time and unit width) by the turbulence which is required to keep the solids in suspension is

$$P = (G_s \cos \beta)\,(W \cos \beta) \qquad (4)$$

This work is delivered by ω; however, only a (small) part of ω is used for this:

$$P = e\,\omega \qquad (5)$$

where e is the efficiency factor for the suspended load transport (e = 0.01 according to Bagnold, 1966).

Combining (1), (3), (4) and (5) yields

$$s_s = \frac{0.01\, C_f\, \rho\, u_s\, u^3}{W \cos^2 \beta} \qquad [N/ms] \qquad (6)$$

The effect of the weight component G_s sin β is that the suspended sediment will push the fluid along instead of vice versa. So $u_s > u$, but the difference will be small ($u_s \approx u$). Hence it is assumed that the contribution of G_s sin β to the lifting work P is negligible. This approach differs from that of Bagnold (1963) and Bailard (1981) who assumed contributions to P of $G_s \sin \beta\, u_s$ and $e\, G_s \sin \beta\, u_s$, respectively.

The weight of 1 m³ sand under water is: $(1 - p)(\rho_s - \rho)\, g$ [N], in which p is the porosity, ρ_s is the sediment density and g is the acceleration of gravity. Now (6) can be written as:

$$s_s = \frac{0.01\, C_f\, u^4}{(1 - p)\, \Delta\, g\, W \cos^2 \beta} \qquad [m^2/s] \qquad (7)$$

in which $\Delta = (\rho_s - \rho)/\rho$. Equation (7) applies to the capacity of the suspended load transport. There remains the description of the process of pick up of sediment up to this capacity.

Fig. 3 shows schematically the two-dimensional situation of flow and sand transport on the outer or downstream slope in a stage which differs only slightly from the initial state (slope 1 : 4).

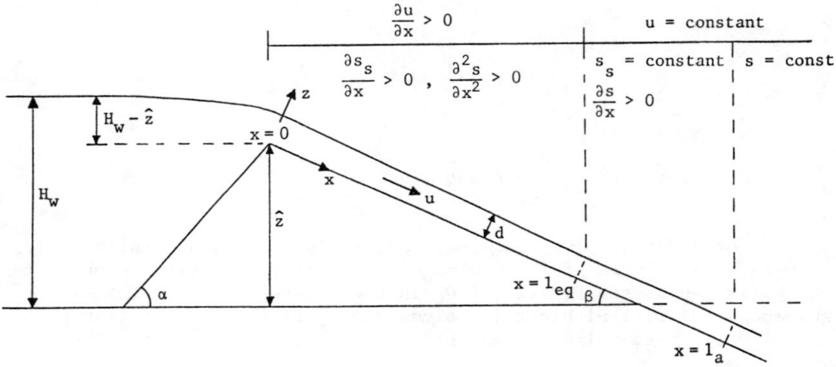

Fig. 3 Schematization of flow and sand transport on the downstream slope

The crest acts as a weir, the flow over it is critical. On the downstream slope, where the flow is super-critical, the current accelerates up to a distance $x = \ell_{eq}$ from the crest (x is the coordinate along the slope, x = 0 at the crest); there the current velocity reaches the equilibrium value for uniform flows

$$u_{eq} = \left(\frac{g\, d \sin \beta}{C_f} \right)^{1/2} \qquad (8)$$

Upstream from the crest the sand transport is very small (in the experiments) or nothing at all (in practice, due to the revetment of the slope). The entrainment of sediment in suspension starts at the crest. For $x = \ell_a$ (= adaption length) the suspended load transport reaches its capacity; naturally ℓ_a can be larger than L (= length of downstream slope), and always $\ell_a \gtrsim \ell_{eq}$ due to the adaption process of the suspended load.

A description of the entrainment process for sediment in suspension is given by Galappatti and Vreugdenhil (1985). For small values of W/u_* as in the present situation, i.e. $W/u_* = W/(\sqrt{C_f}\, u) \approx 0.01$ (see (2), u_* is the bed shear velocity), their theory can simply be approximated as

$$s(x) = \frac{x}{\ell_a}\, s_s \qquad \text{for} \qquad 0 \le x \le \ell_a \qquad (9)$$

where (with q is the outflow rate per unit width):

$$\ell_a = \frac{u\ d}{W \cos \beta} = \frac{q}{W \cos \beta} \tag{10}$$

3. MATHEMATICAL MODEL

3.1 Outflow rate

The maximum outflow rate Q_{max} will occur for large values of the water depth H_w in the reservoir, see Fig. 11. Then the flow in the breach is still critical, so the equation for the rate of outflow is:

$$Q = q\ b = m_o \frac{2}{3} (\frac{2}{3} g)^{\frac{1}{2}}\ b\ H^{3/2} \tag{11}$$

where m_o is the discharge coefficient (≈ 1.0), b is the average breach width and $H = H_w - \hat{z}$ ($\hat{z}$ is the height of the breach top above the reservoir bottom, see Fig. 3).

3.2 Growth of breach in depth

The equation describing the erosion of the downstream slope is

$$\frac{\partial z_b}{\partial t} + \frac{\partial s}{\partial x} = 0 \tag{12}$$

where t is the time and $z_b = z_b(x, t)$ is the position of the sloping bottom in z-direction (z is the coordinate normal to the slope).

The suspended load capacity s_s increases for $0 < x < \ell_{eq}$ since u accelerates here. This means with (9) and (12)

$$\frac{\partial}{\partial x}\left|\frac{\partial z_b}{\partial x}\right| = \frac{\partial^2 s}{\partial x^2} > 0 \quad \text{for} \quad 0 < x < \ell_{eq} \tag{13}$$

i.e. the rate of erosion increases along the slope. Hence the downstream slope becomes steeper with increasing x and in time, see Fig. 4.

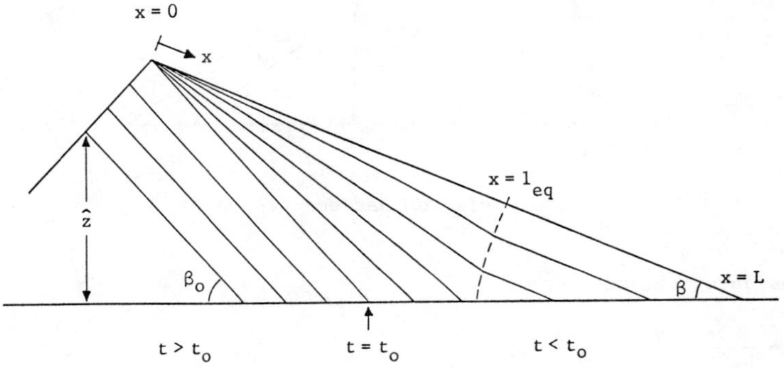

Fig. 4 Erosion of the downstream slope

If it is assumed that C_f and β are constant, then s_s is constant for $\ell_{eq} < x < \ell_a$ since u is constant here, see (7). This means that

$$\frac{\partial}{\partial x}\left|\frac{\partial z_b}{\partial t}\right| = 0 \qquad \text{for} \qquad \ell_{eq} < x < \ell_a \tag{14}$$

i.e. the rate of erosion is constant along this part of the slope, see Fig. 4.

Fig. 4 shows also the result of the above described development: the downstream slope becomes steeper and steeper. The steepness of the slope will not exceed, however, a certain limit tg β_o. If this limit has been achieved on the entire slope (on $t = t_o$), then the erosion rate becomes also constant for $0 < x < \ell_{eq}$, see the straight lines in Fig. 4 for $t > t_o$. Thus, for $t \geq t_o$ the erosion of the downstream slope is completely determined by the erosion at $x = \ell_{eq}$, i.e. by the erosion at the toe of the slope (x = L):

$$- L \frac{dz_b}{dt} = s_L \qquad \text{for} \qquad t \geq t_o \tag{15}$$

where with (9):

$$s_L = \frac{L}{\ell_a} (s_s)_L \tag{16}$$

Substitution of d = q/u into (8) gives for the current velocity at x = L:

$$u_L = \left(\frac{g\ q \sin \beta}{C_f}\right)^{1/3} \tag{17}$$

The relation between the fall dz of the breach top and the erosion dz_b of the downstream slope follows from a simple geometrical

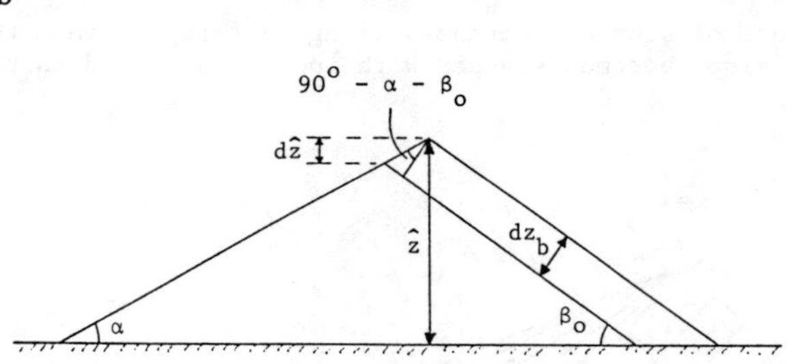

Fig. 5 Relation between $d\hat{z}$ and dz_b

consideration, see Fig. 5:

$$\frac{d\hat{z}}{dt} = \frac{\sin \alpha}{\sin (\alpha + \beta_o)} \frac{dz_b}{dt} \tag{18}$$

Substitution of (15), (16), (7), (10), (17), $\beta = \beta_o$ and (11), successively, into (18) gives:

$$\frac{d\hat{z}}{dt} = -k\,H^{\frac{1}{2}} = -k\,(H_w - z)^{\frac{1}{2}} \quad \text{for } t \geq t_o \tag{19}$$

$$\text{where} \quad k = 0.0082\,\frac{(g)^{\frac{1}{2}}}{\Delta\,(1-p)}\,\left(\frac{m_o}{C_f}\right)^{1/3}\,\frac{\sin\alpha\,(\sin\beta_o)^{4/3}}{\sin(\alpha+\beta_o)\,\cos\beta_o} \tag{20}$$

3.3 Growth of breach width

Consider the small initial breach $A_oB_oC_oD_o$ of Fig. 6 with an average width b_o, a bottom width a_o and a height h_o. It is assumed that the angle of the side slopes is somewhat smaller than the angle of repose ϕ. The direction of the flow through the breach is perpendicular to the paper.

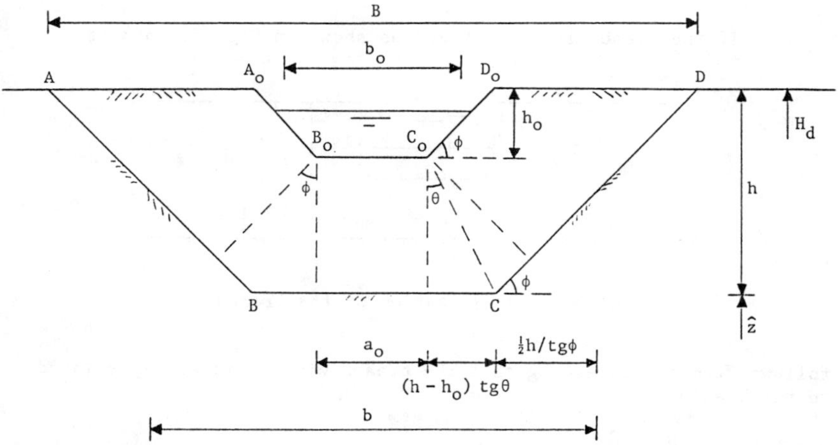

Fig. 6 Assumed growth of breach ABCD from initial $A_oB_oC_oD_o$

The rate of erosion down below the side slopes is larger (due to larger current velocities) than higher on the slopes. This means that the slopes become steeper and steeper, until the situation gets unstable and the sides will slide into the channel. This is in accordance with the observations (which are described in chapter 4). Thus, it is assumed that the angle of the side slopes is constant and equal to ϕ. Hence, if the sand dike is homogeneous then the initial breach $A_oB_oC_oD_o$ will develop into the breach ABCD with an average width b:

$$b = a_o + 2\,(h - h_o)\,tg\,\theta + \frac{h}{tg\,\phi} \tag{21}$$

and width B at the crest of the dike:

$$B = a_o + 2\,(h - h_o)\,tg\,\theta + \frac{2h}{tg\,\phi} \tag{22}$$

in which h is the breach depth and $0 < \theta < \phi$.

If $b_o << b$ and $h_o << h$, then (21) and (22) can be written as:

$$\frac{b}{h} = 2 \text{ tg } \theta + \frac{1}{\text{tg } \phi} \tag{23}$$

$$\frac{B}{h} = 2 \text{ tg } \theta + \frac{2}{\text{tg } \phi} \tag{24}$$

For sand: $\phi \approx 32^o$ Thus, with $0 < \theta < \phi$:

$$1.6 < \frac{b}{h} < 2.9 \tag{25}$$

$$3.2 < \frac{B}{h} < 4.4 \tag{26}$$

If the rate of erosion just left from the point C_o equals to that just right from this point then $\theta = \phi/2$ and

$$\frac{b}{h} = 2.2 \text{ and } \frac{B}{h} = 3.8 \tag{27}$$

If the breach is schematized as shown in Fig. 7, then it

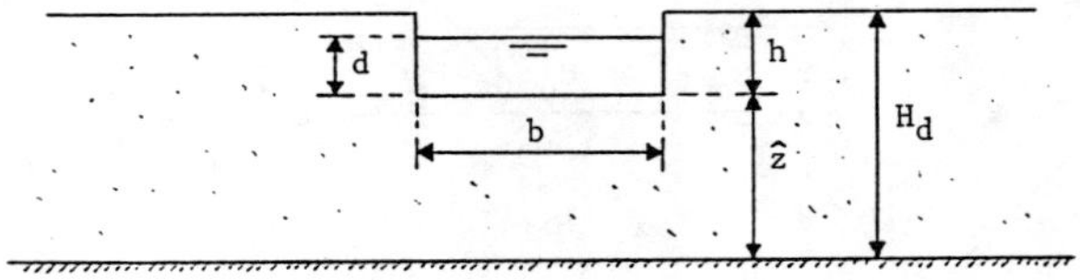

Fig. 7 Schematization of the breach

follows from the foregoing that the breach width b is simply related to the breach depth

$$b = r\,h = r\,(H_d - \hat{z}) \tag{28}$$

where r is a constant with a theoretical value of about 2.2 and H_d is the height of the dike.

The continuity equation for the total sediment transport S through the breach channel reads

$$\frac{\partial S}{\partial x} - \frac{\partial bh}{\partial t} = 0 \tag{29}$$

where (see Fig. 7):

$$S = s(b + 2d) \tag{30}$$

Substitution of (28) and (30) into (29) gives with $\partial h/\partial t = - \partial z_b/\partial t$ and the assumption that b and d are nearly constant along the slope:

$$\frac{b + 2d}{2b}\frac{\partial s}{\partial x} + \frac{\partial z_b}{\partial t} = 0 \tag{31}$$

i.e. (19) should be corrected to

$$\frac{d\hat{z}}{dt} = - \frac{b + 2d}{2b} k H^{\frac{1}{2}} = - f k H^{\frac{1}{2}} \tag{32}$$

Thus, $d\hat{z}/dt$ slows down with a factor f compared with the 2-dimensional situation. The factor f can be calculated for any $\hat{z}$ from $d = q/u_L$, (11), (17) and (28).

3.4 Decrease of reservoir water depth

The decrease of the water depth in the reservoir follows simply from the equation describing continuity of mass:

$$A \frac{dH_w}{dt} = - Q \tag{33}$$

or with $H_w = H + \hat{z}$:

$$\frac{dH}{dt} + \frac{d\hat{z}}{dt} = - \frac{Q}{A} \tag{34}$$

The equations (32) and (34) form a system of coupled differential equations for H and $\hat{z}$.

4. EXPERIMENTS

Three experiments were performed for the verification of the model. Fig. 8 (lines for t = 0) shows the cross-section of the dike at

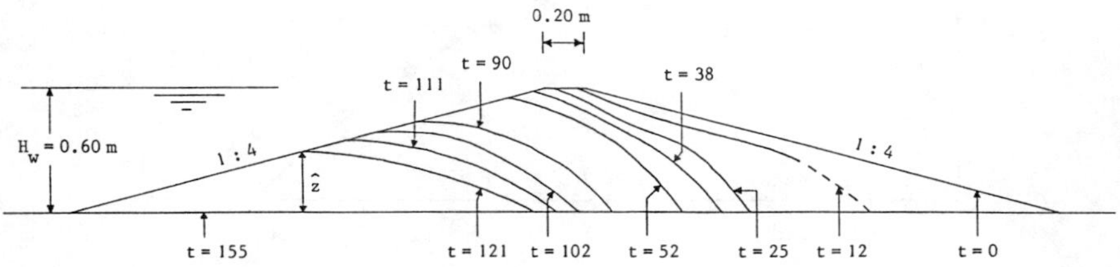

Fig. 8 Cross-section of the dike in the breach on different times t (in seconds) in Exp. 3

the start of each experiment, completely built up with sand: D_{50} = 0.10 mm in Exp.'s 1 and 3, D_{50} = 0.20 mm in Exp. 2. The upstream water depth was kept constant at 0.60 m, the downstream water depth at the start of the experiments was nil. The breach growth was video taped because of the very rapid process of breaching.

Exp. 1 was done in a large (wave) basin in which a 6 m long dike was constructed. The dike-burst started at t = 0 with the flow through a small initial channel in the crest of the dike. Fig. 9 shows the top width B(t) of the breach as observed from the video tape.

Exp.'s 2 and 3 were performed in a flume with a width of 0.50 m. Exp. 3 is a two-dimensional experiment: the erosion of the dike took place over the total width of the flume. The observed development of the downstream slope (see Fig. 8) is in agreement with the reasoned one (see Fig. 4). The downstream slope in Exp. 3 became steeper until

$t = t_o = 25$ s without any appreciable drop of the crest. Fig. 10 shows the observed top height $\hat{z}(t)$ in Exp. 3.

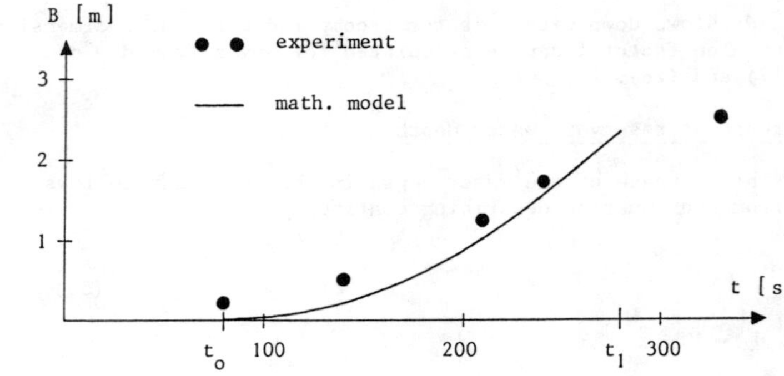

Fig. 9 Observed (Exp. 1) and theoretical top width B(t) of the breach

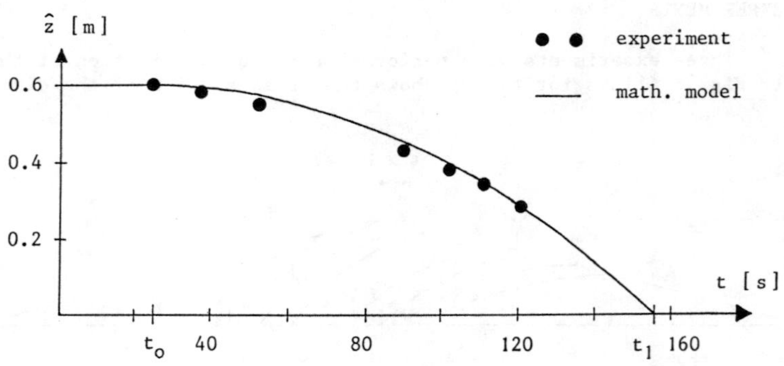

Fig. 10 Observed (Exp. 3) and theoretical top height $\hat{z}(t)$

5. VERIFICATION AND RESULTS

5.1 Comparison with experiments

The present breach erosion model has been verified to the data of the three experiments. In this paper only the testing for Exp.'s 1 and 3 is briefly described. As stated in chapter 3, the model can be applied after the steepening of the downstream slope, i.e. for $t > t_o$.

Fig. 10 shows the comparison of the computed and observed crest height $\hat{z}(t)$ of the dike in the breach. The calculation was done for H_w = constant = 0.60 m , t_o = 25 s , $\alpha = 14^o$, $\beta_o = \phi = 32^o$, $m = 1.2$, $\Delta = 1.65$, $p = 0.4$, $f = 1$ (2-dimensional) and $C_f = 0.06$. The rather large value of 0.06 for C_f was computed with (17) and $u = q/d$ using the observed water depths on the crest of the weir (for q) and on the downstream slope in an initial stage of Exp. 3. On $t = t_1 = 155$ s the dike in the flume was completely washed away.

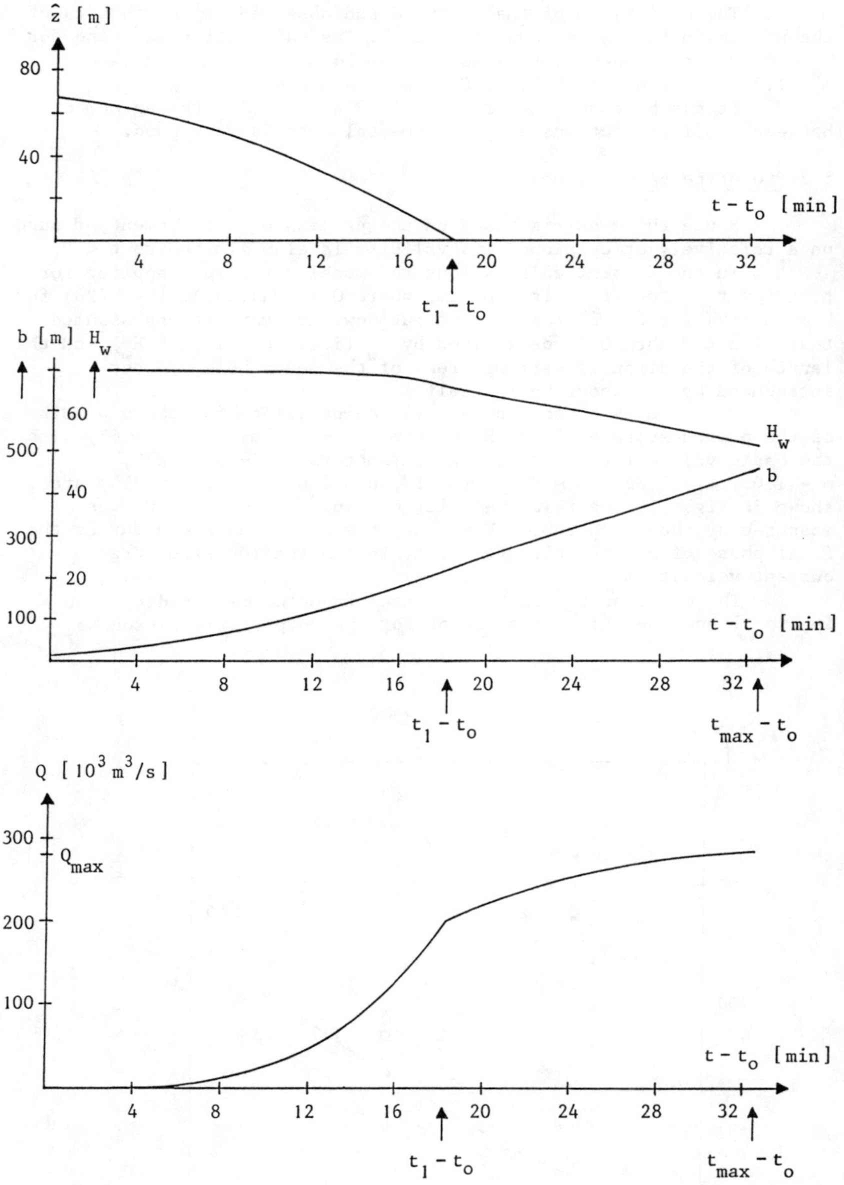

Fig. 11 Top height $\hat{z}(t)$, average breach width $b(t)$, reservoir water depth $H_w(t)$ and outflow rate $Q(t)$ as computed for the high sand-dike of Fig. 1.

The comparison of the computed and observed top width B(t) of the breach in Exp. 1 is shown in Fig. 9. The calculation was done for H_w = constant = 0.60 m , t_o = 80 s , $\alpha = 14^o$, $\beta_o = \phi = 32^o$, $m = 1.2$, $\Delta = 1.65$, $p = 0.4$, $C_f = 0.06$ and $B/h_o = 3.8$.

It can be concluded from Fig.'s 9 and 10 that the agreement between model predictions and experimental data is very good.

5.2 Prototype calculations

Since the model is based on the process of entrainment of sand on a relatively steep slope, its validity is also limited to $t < t_1$.

In the present calculations the model was simply applied for $t_1 \leq t \leq t_{max}$ too (t_{max} is the time where Q is maximal), i.e. (28) for b with (19) for $\hat{z}$. For the rate of outflow, however, it was assumed that if $\hat{z} < 0$ then Q is determined by H_w (i.e. in (11) $h = H_w$) and the length of the circular weir upstream of the scour hole and not influenced by the scour hole itself.

The results of the (numerical) computations for the sand-dike of the pumped-storage plant (H_w = 70 m , A = 15 km² , $\alpha = 19.6^o$) with the basic values for the different parameters ($\beta_o = \phi = 32^o$, $m = 1.0$, $\Delta = 1.65$, $p = 0.4$, $r = 3$, $e = 0.01$ and $C_f = 0.015$) are shown in Fig. 11. The relatively large value of 0.015 for C_f was measured by the Dutch Public Works Department in the field during the final phase of a "sand-closure", i.e. in a situation with large current velocities.

The duration t_o can be estimated from the measured t_o = 80 s in Exp. 1 and the time-scale factor for the morphological process.

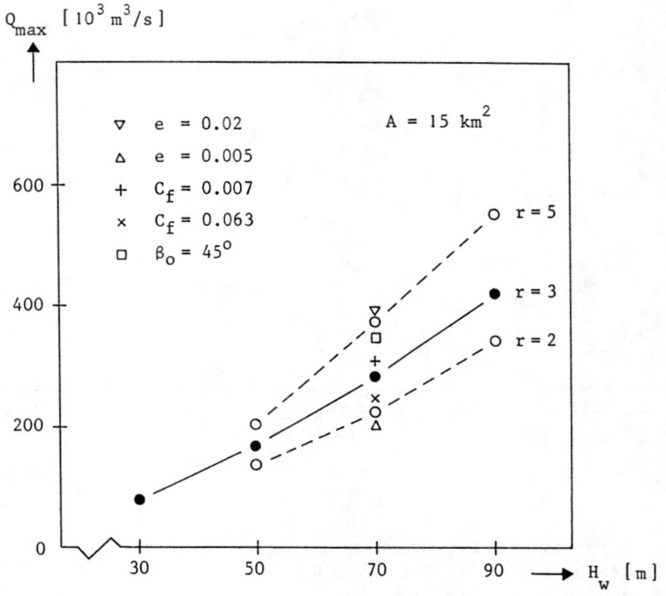

Fig. 12 Sensitivity of the computed Q_{max} for the different parameters

This gives t_o = 560 s = 9.3 min., see Visser, Ribberink and Kalkwijk (1986).

Thus, the model predicts for the outflow rate a maximum value:

$$Q_{max} = 2.8 * 10^5 \ m^3/s \quad \text{for} \quad t = t_{max} = 2.5 * 10^3 \ s = 42 \ min.$$

Fig. 12 shows the sensitivity of the computed Q_{max} for the different parameters, i.e. for the width/depth-ratio r of the breach, the efficiency factor e for the suspended load transport, the drag coefficient C_f for the bed, the critical slope angle β_o and the reservoir water depth H_w. The model is especially sensitive for e and r.

6. DISCUSSION

The output of the study is a relatively simple model for the simulation of sand-dike breach erosion. The investigation was done for the feasibility study of a pumped-storage plant in the Netherlands. The model predicts for the 73 m high sand-dike of the water reservoir the following failure scenario in case a small initial gap springs up at the crest on t = 0:

- o $0 < t < t_o$: the slope angle β of the breach channel in the downstream slope will become larger,
- o $t = t_o$ = 9 min.: β will reach a critical value β_o , the flow rate will be still small and the erosion of the top will start (the duration t_o is very sensitive to the initial conditions),
- o $t_o < t < t_1$: erosion of the dike in the breach,
- o $t = t_1$ = 28 min.: the dike in the breach will be completely washed away, $Q_1 \approx 2.0 * 10^5 \ m^3/s$,
- o $t_1 < t$: the breaching process will continue in both vertical and horizontal direction; for the outflow rate Q the increase of the breach width is of importance,
- o $t = t_{max}$ = 42 min.: $Q_{max} \approx 2.8 * 10^5 \ m^3/s$.

The model is based on the erosion process of a relatively steep slope ($t_o < t < t_1$) and was also tested as such to the results of three experiments. The agreement between model and experiments is surprisingly good.

For much lower dikes, as river or sea dikes, Q_{max} will generally be much larger than Q_1. It is clear that the modelling of the breach growth for $t > t_1$ should be improved before the model can be released for these dikes.

A sensivity analysis indicates that the efficiency factor e for the suspended load transport and the width/depth-ratio r of the breach are the most sensitive parameters in the model; e is the most uncertain parameter. Measurements on the pick up and transport of sediment at high velocities ($\approx$ 10 m/s) should be done to reduce this uncertainty.

REFERENCES

Bagnold, R.A., 1963. Mechanics of marine sedimentation. In The Sea: Ideas and Observations, vol. 3, Interscience, New York, pp. 507–522.

Bagnold, R.A., 1966. An approach to the sediment transport problem from general physics. Geological Survey Professional Paper 422–I , U.S. Government Printing Office.

Bailard, J.A., 1981. An energetics total load sediment transport model for a plane sloping beach. Journ. Geophysical Res., vol. 86, pp. 10938–10954.

Fujita, Y. and Tamura, T., 1987. Enlargement of breaches in flood levees on alluvial plains. Natural Disaster Science, vol. 9, pp. 37–60.

Galappatti, R. and Vreugdenhil, C.B., 1985. A depth–integrated model for suspended sediment transport. Journ. Hydraulic Res., vol. 23, pp. 359–377.

Singh, V.P. and Scarlatos, P.D., 1988. Analysis of gradual earth–dam failure. Journ. Hydraulic Eng., vol. 114, pp. 21–42.

Visser, P.J., Ribberink, J.S. and Kalkwijk, J.P.Th., 1986. The development of the breach and rate of outflow in case of dike failures (in Dutch). Rep. 8–86, Hydraulic Eng. Group, Fac. of Civil Eng., Delft Univ. of Techn.

ACKNOWLEDGEMENT

The work upon which this paper is based was supported in part by the NOVEM B.V. (Dutch Company for Energy and Environment).

CHAPTER 141

SHORELINE AT JETTY DUE TO CYCLIC AND RANDOM WAVES

by

Todd L. Walton, Jr.[1], M. ASCE
Philip L-F. Liu[2], M. ASCE
Edward B. Hands[3], M. ASCE

ABSTRACT

This paper examines the effects of random and deterministic cycling of wave direction on the updrift beach planform adjacent to a jetty. Results provided using a simplified numerical model cast in dimensionless form indicate the importance of the time series of wave direction in determining design jetty length for a given net sediment transport. Continuous cycling of wave direction leads to the expected analytical solution. Simplications in the numerical model used restrict the applications to small wave angles, no diffraction, no reflection of waves off structure, no refraction, and no sand bypassing at jetty. The concept can be extended to more sophisticated numerical models.

INTRODUCTION

Coastal navigation structures such as jetties serve a number of practical purposes, two of which are: keeping the navigation channel open by prevention of sand transport into the channel, and providing a sand storage reservoir for bypassing to downdrift beaches (to prevent erosion due to the interruption of the natural longshore sand transport).

An important jetty design criteria is the determination of jetty length which is equivalent to

[1] Research Hydraulic Engineer, Coastal Engineering Research Center, U.S. Army Engineer Waterways Experiment Station, Vicksburg, MS 39180-0631.
[2] Professor of Civil and Environmental Engineering and Associate Dean of Engineering College, Cornell University, Ithaca, NY 14853.
[3] Research Physical Scientist, Coastal Engineering Research Center, U.S. Army Engineer Waterways Experiment Station, Vicksburg, MS 39180-0631.

determination of the updrift fillet sand storage area necessary between dredging intervals. Under constant wave action (i.e. constant wave height, wave period, and wave breaker angle) simple design formula such as provided by the analytical solution of Pelnard-Considere (1956) can be used to provide an answer to this design question. Under more general conditions of changing wave climate, an analytical solution is not possible.

This paper examines the effects of changing wave climate (in particular wave direction deterministic cycling and random effects) on the updrift beach planform prior to the onset of natural bypassing around the jetty. Results of the type in this paper show the importance of the time history of wave climate on design jetty length for a given sand storage requirement. Simplifications in the numerical model used restrict the applications to: (1) small wave angles ($A_b \leq 15$ degrees), (2) no reversing wave climate, i.e. no diffraction effects, (3) no reflection of waves off structure, (4) no refraction, (5) no sand bypassing at jetty, and (6) no change in profile shape during shoreline advance. Each of these simplifications can be overcome by a more sophisticated numerical model than used in the present analysis.

MODEL

Two equations can be used to describe shoreline shape (e.g. Hanson and Kraus, 1980; Walton and Chiu, 1979). The first equation describes the continuity of sediment transport in the longshore direction:

$$(1/D)\ (\partial Q/\partial x) + \partial y/\partial t = 0 \tag{1}$$

where D=closure depth, Q=volumetric longshore sediment transport rate, and the coordinate system is shown in Fig. 1. The second governing equation relates the longshore sediment transport rate to wave and current parameters. The equation used in this note is that of the CERC type formulation (see CERC, 1984):

$$Q = K_c\ H_b^{5/2}\ \sin\ (2A_b) \tag{2}$$

where

$$K_c = [K\ g^{1/2}]/[16(s_s/s_f - 1)\ (1-a)] \tag{3}$$

with

$$A_b = A - \partial y/\partial x \tag{4}$$

where A = deep water wave angle; A_b = breaking wave angle; H_b = breaking wave height; K = a dimensionless constant (= 0.39 in CERC (1984)) relating the immersed weight sediment transport rate to the "longshore energy flux"; g = acceleration of gravity; s_s = specific gravity of sediment; s_f = specific gravity of fluid; and a = porosity of sediment (assumed = 0.6).

By assuming small wave angles and constant wave

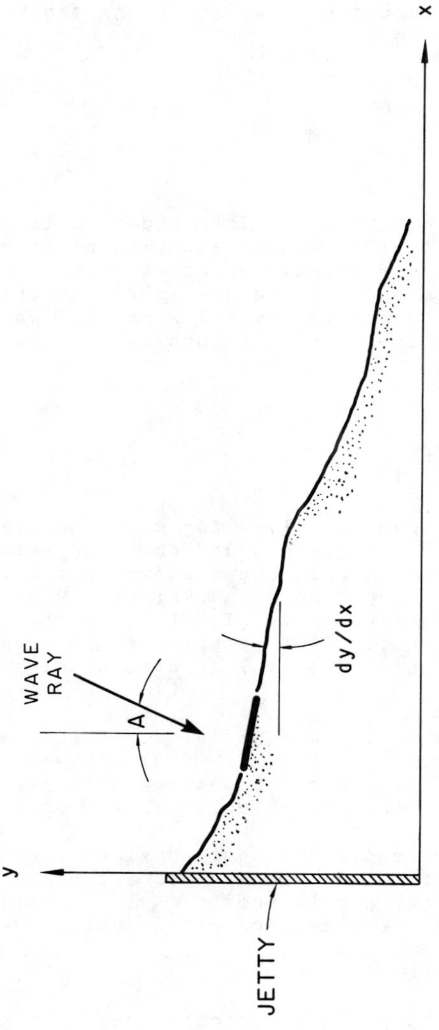

Fig. 1. - Shoreline Coordinate System

height along the beach the two equations can be reduced to the following form:

$$\partial y/\partial t = \varepsilon\ (\partial^2 y/\partial x^2) \qquad (5)$$

where

$$\varepsilon = (2\ K_c/D)\ H_b^{5/2} \qquad (6)$$

with units of length^2/time. This equation is a parabolic partial differential equation often referred to as the heat (or mass) diffusion equation.

This equation can be recast in dimensionless form by dividing the quantities x,y by $2(\varepsilon t_s)^{1/2}$ where t_s is the desired solution time to obtain:

$$\frac{\partial \hat{y}}{\partial \hat{t}} = \frac{\partial^2 \hat{y}}{\partial \hat{x}^2} \qquad (7)$$

where

$$\hat{t} = t/4t_s \qquad (8a)$$
$$\hat{y} = ty/2(\varepsilon t_s)^{1/2} \qquad (8b)$$
$$\hat{x} = x/2(\varepsilon t_s)^{1/2} \qquad (8c)$$

Equation 7 can be solved for given initial conditions and boundary conditions and changing wave climate via an implicit numerical scheme known as the Crank-Nicholson method (see Burden and Faires, 1985) which ensures numerical stability. Initial conditions and boundary conditions for the jetty shoreline system remain the same in dimensional or dimensionless form and are given as:

$\hat{y} = 0.0$ at $\hat{t} = 0.0$, $\hat{x} \geq 0$ (initial conditions) (9a)
$\hat{y} = 0.0$ at $\hat{x} = \infty$, $\hat{t} \geq 0$ (b.c. far from jetty) (9b)
$\partial\hat{y}/\partial\hat{x} = A$ at $\hat{x} = 0.0$ (b.c. at semiinfinite jetty) (9c)
(or equivalently $A_b = 0.0$ at jetty)

The Crank-Nicholson scheme provides for stability under all conditions of time step Δt and distance step Δx. Numerical accuracy is provided via a sufficiently large number for the dimensionless quantity r where

$$r = \varepsilon \Delta t/\Delta x^2 \qquad (10)$$

In the present study the numerical model used r = 1.0 to provide acceptable accuracy (error 0.5 percent). In the case of constant wave conditions the model was verified by comparing the numerical model solution to the non-dimensional form of the analytical solution provided by Walton and Chiu (1979) as follows:

$$\tilde{y} = \hat{y}\ \sqrt{\pi}/A = \exp\ (-\hat{x}^2) - \hat{x}\ \sqrt{\pi}\,\mathrm{erfc}(\hat{x}) \qquad (11)$$

RESULTS FOR DETERMINISTIC WAVE CYCLING

The following cases of deterministic wave cycling (where only wave direction is changing) were run on the model:

case 1a - 1 cycle, 10 - 0 square waveform
case 2a - 2 cycles, 10 - 0 square waveform
case 3a - 4 cycles, 10 - 0 square waveform
case 4a - 8 cycles, 10 - 0 square waveform
case 5a - 16 cycles, 10 - 0 square waveform
case 6a - 80 cycles, 10 - 0 square waveform

A pictorial representation of the temporal wave direction change of cases 1a through 5a is shown in Fig. 2. Cases 1b-6b are similar only with reversed sequencing of wave direction (i.e. starting with A = 0 degrees at time t = 0.0). Time T represents a characteristic time period (the solution time), for example, a year if case 1a is used and there are two predominant wave directions lasting for 6 months each.

For each case the mean wave angle was $\bar{A}$ = 5 degrees while the actual wave angle fluctuated about the mean from 0 to 10 degrees. Cases 1a,b and 3a,b are plotted on Fig. 3 in the non-dimensional format of Eq. 8 where the mean wave angle $\bar{A}$ was used for the wave angle of Eq. 9. Note that the integrated shoreline change via this approach is the same in all cases, only the alongshore distribution of these changes differs between the different cases. Figure 3 also provides the analytical solution (Eq. 11) in the case of constant wave direction. Cases 6a and 6b have not been shown in Fig. 3 as they are essentially that of the analytical solution for the same average wave angle of $\bar{A}$ = 5 degrees. Due to plotting resolution cases 2a,b; 4a,b; and 5a,b are not provided but can be interpolated between adjacent cases in Fig. 3. As would be expected for rapidly fluctuating direction around a mean wave direction, the numerical solution approaches the analytical solution with the same mean wave angle.

As an example of how such information might be used, consider a situation where dredging to initial conditions might occur each year (i.e. dimensional time period T = 1 year). If the wave climate is two seasonal (1 cycle) in the year with the wave angles shown as in case 1a (or 1b), an increase of jetty length on the order of 40 percent above that for constant wave conditions is necessary to prevent sand from spilling into the channel. Case 1b differs from case 1a only in phase (6 months for the example under consideration). The intent here is to show the importance of wave sequencing on the design considerations which often is overlooked.

Although the solution curves are non-dimensionalized, it must be recognized that different mean climate levels (i.e. different $\bar{A}$) will provide different sets of curves. This is shown in Fig. 4 where cases 1a and 1b (1 cycle) are shown along with cases 7a and 7b (1 cycle with wave direction shifting between 15 and 5 degrees, mean wave angle $\bar{A}$ = 10 degrees). In case 7a it is noted that the increase in jetty length necessitated by the cyclic wave direction is only 20 percent above the non-dimensionalized analytical solution where $\bar{A}$ = 10 degrees, but recall that the mean wave angle has doubled (10 versus 5 degrees) requiring the <u>dimensional</u> jetty length of the analytical solution to double.

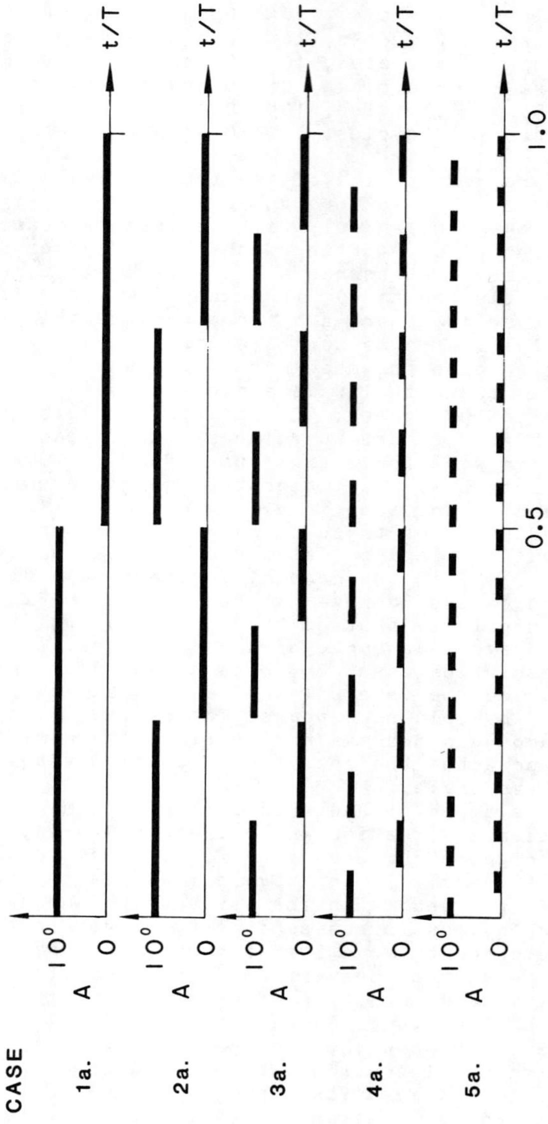

Fig. 2. - Temporal Representation of Cyclic Wave Direction

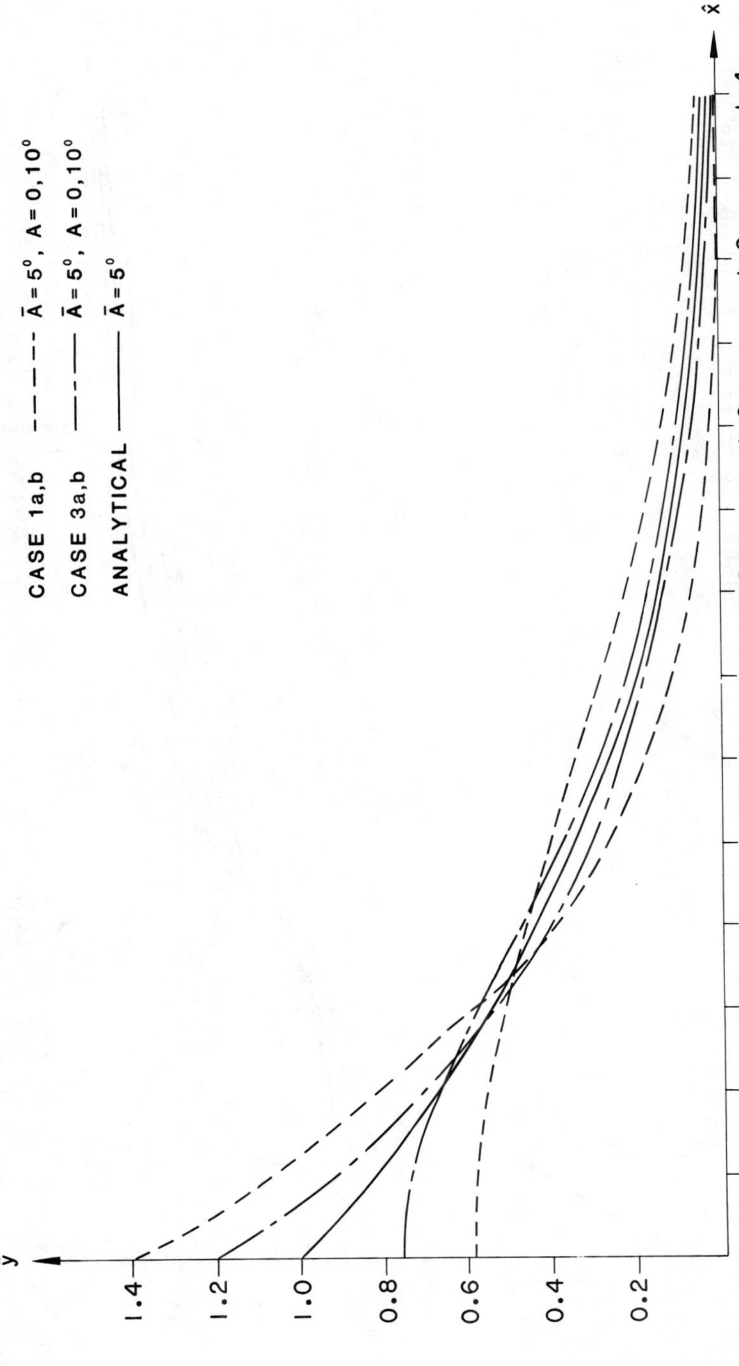

Fig. 3. - Results of Numerical Model for Cases 1a,b through 6a,b

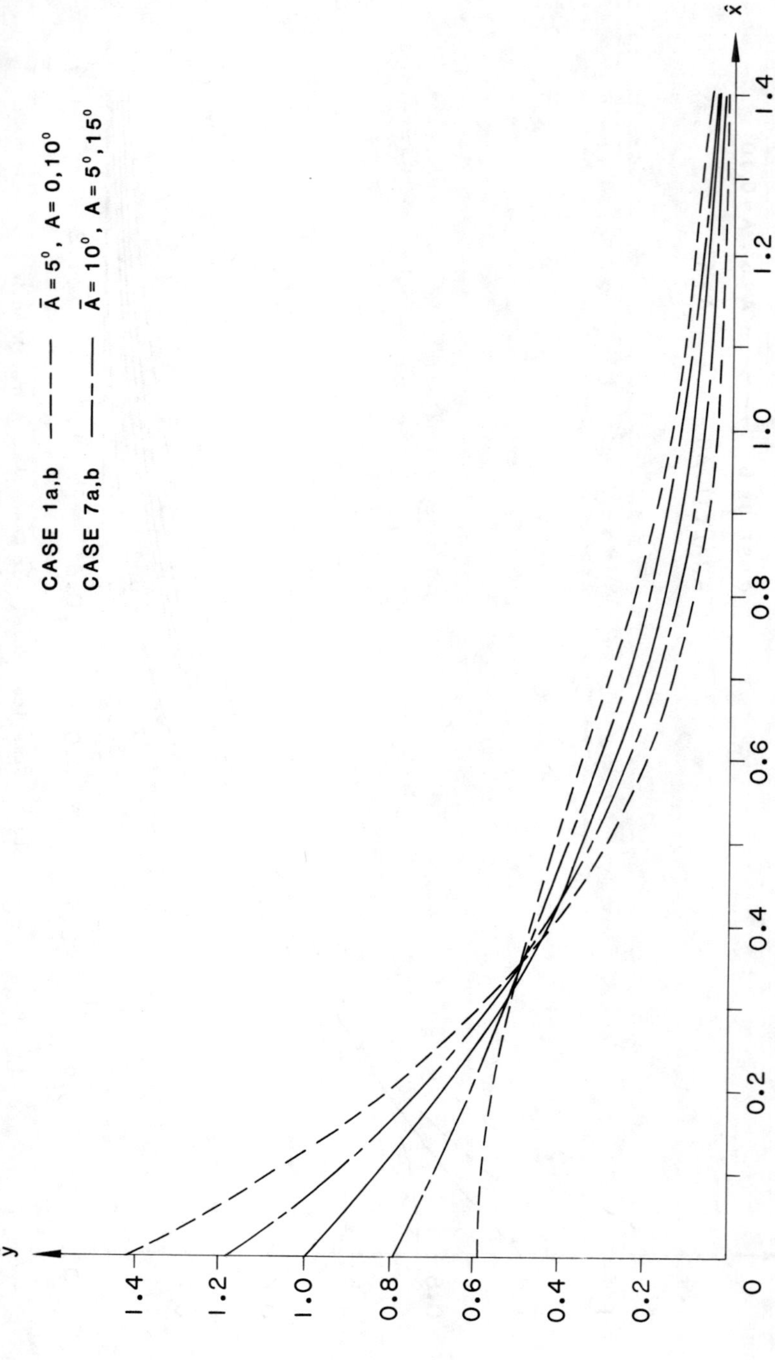

Fig. 4. - Results of Numerical Model for Cases 1a,b versus 7a,b

RESULTS UNDER DIRECTIONAL UNCERTAINTY

LeMehaute and Wang (1983) have investigated the problem of shoreline response under random wave action via simulation of wave climate with known probability distribution. In the semi-infinite jetty and straight shoreline with constant wave condition case where the only random variable is wave direction, an expected shoreline shape as well as confidence intervals on shoreline shape can be obtained very simply. Such solutions are of value in evaluation of design parameters (such as jetty length) when wave climate uncertainty exists.

The expected shoreline shape in dimensionless form can be reexpressed as follows:

$$E[\hat{Y}] = (E[A]/\sqrt{\pi})\ (\exp(-\hat{x})-\hat{x}\ \sqrt{\pi}\ \mathrm{erfc}(\hat{x})) \tag{12}$$

where E[ ] represents the expected value and $\hat{y}$ is now replaced by the random variable $\hat{Y}$. The previous analytical solution, eq. 11, as provided in Fig. 3 can be used to find the expected shoreline by replacing A with E[ ] where A is also a random variable.

In the general case where wave direction is unknown to within $R = A_{max} - A_{min}$ degrees, the directional probability distribution function (PDF) can best be expressed as the uniform PDF where:

$$F(A) = (A - A_{min})/R \tag{13}$$

with F(A) = probability that wave angle < A. For this particular PDF, the $A = A_u$ corresponding to an upper confidence limit with $1.-F(A_u)$ as the probability of exceedance (i.e. $\hat{Y} \leq \hat{y}$) is $A_u = R\ F(A_u) + A_{min}$. The $A = A_l$ corresponding to a lower confidence limit with probability $F(A_l)$ is $A_l = R\ F(A_l) + A_{min}$. The lower and upper confidence interval shorelines can then be expressed by Eq. 11 and the Fig. 3 analytical shoreline using $A = A_u$ and $A = A_l$. Confidence limits for shorelines based on other PDF's can be found in a similar manner. The important point to note from this simplified analysis is that uncertainty in wave direction can lead to shoreline realizations having wide discrepancies from those projected based on expected wave conditions. In the case of a uniform PDF for wave direction as postulated, a 90 percent confidence interval shoreline can lead to a necessary increase of jetty length 80 percent above that required for the expected wave angle (for the given case of $A_{min} = 0.0$).

CONCLUSIONS

A brief examination of the shoreline shape due to jetty construction under deterministic cyclic wave action and under uncertainity in wave direction has been provided for some simple hypothetical wave direction scenerios. Shorelines due to cyclic effects in wave direction show considerable deviation from solutions with the same mean wave direction. Time sequencing of wave direction is thus seen to be of considerable importance in design studies. Uncertainty in wave direction also plays an important role in shoreline fluctuation and should be factored into design considerations. In realistic cases of design a more detailed model (i.e. Hansen and Kraus (1980)) might be used to optimize jetty length.

ACKNOWLEDGMENTS

The work presented herein was conducted under the Evaluation of Shore Protection Structures work unit of the Shore Protection and Restoration Program, Coastal Engineering area of Civil Works Research and Development being executed by the Coastal Engineering Research Center, U.S. Army Engineer Waterways Experiment Station, Vicksburg, MS. Permission was granted by the Chief of Engineers to publish this information.

APPENDIX 1. - REFERENCES

Burden, R. L., and Faires, J. D., Numerical Analysis, PWS Publishing Company, Boston, MA, 1985.

CERC, Shore Protection Manual, Coastal Engineering Research Center, U.S. Army Engineer Waterways Experiment Station, U.S. Government Printing Office, 1984.

Hanson, H. and Kraus, N., "Numerical Model for Studying Shoreline Change in the Vicinity of Coastal Structures," Report No. 3040, Dept. of Water Resources Engineering, University of Lund, Lund, Sweden, 1980.

LeMehaute B., Wang, J. D., and Liu, C. C., "Wave Data Discretization for Shoreline Processes," Journal of Waterway, Port, Coastal and Ocean Engineering, ASCE, Vol. 109, No. 1, 1983, pp. 63-78.

Pelnard-Considere, R., "Essai de Theorie de l'Evolution des Formes de Rivage en Plages de Sable et de Galets," 4th Journees de l'Hydraulique, Les Energies de la Mer, Question III, Rapport No. 1, 1956.

Walton, T. L., Jr., and Chiu, T. Y., "A Review of Analytical Techniques to Solve the Sand Transport Equation and Some Simplified Solutions," Proc. of Coastal Structures 79 Conference, May 14-16, 1979, Alexandria, VA, American Society of Civil Engineers, New York, N.Y., 1979.

APPENDIX II. - NOTATION

The following symbols are used in the paper:

a = porosity of sediment/sand, (dimensionless);
g = acceleration of gravity;
s_f = specific gravity of seawater, (dimensionless);
s_s = specific gravity of sediment/sand, (dimensionless);
t = time;
t_s = solution time;
x = distance alongshore direction, ($x = 0.0$ at jetty);
$\hat{x}$ = dimensionless distance alongshore
y = distance to shoreline from baseline in offshore direction;
$\hat{y}$ = dimensionless distance to shoreline from baseline in an offshore direction;
A = offshore reference wave angle;
A_b = breaking wave angle;
Amax = maximum wave angle where uncertainty exists;
Amin = minimum wave angle where uncertainty exists;
A_l = wave angle corresponding to lower confidence limit;
A_u = wave angle corresponding to upper confidence limit;
D = water depth at offshore limit of significant longshore sediment transport;
E[ ] = expectation operator;
F( ) = probability of random variable ( ) being less than or equal to value given;
K = dimensionless constant;
K_c = dimensional constant, (length/time);
PDF = abbreviation for probability distribution function;
Q = volumetric sand/sediment transport rate, (length/time);
R = range of uncertainty in wave direction;
T = characteristic time of wave direction cycling scenerio;
$\hat{Y}$ = random variable equivalent of $\hat{y}$;
ε = dimensionless "diffusion" coefficient.

CHAPTER 142

CHARACTERISTICS OF SHINGLE BEACHES WITH REFERENCE TO CHRISTCHURCH BAY, S. ENGLAND

by Robert Nicholls* and Norman Webber+

ABSTRACT

The rapid recession of the shingle bank of Hurst Beach (up to 3.5m/yr) makes it an excellent natural laboratory for the study of the factors which influence the stability of shingle beaches. Studies have included: the significance of long period, high energy, swell waves - the classification and quantification of overwash processes - run-up and seepage characteristics - the effect of settlement of the underlying strata - and the implications for practices in shingle nourishment. The studies have revealed the distinctive character of shingle beaches as compared with the more fully researched sand beaches. More detailed research on shingle beaches is justified particularly in relation to (i) the run-up characteristics including its interaction with swash cusps and (ii) the influence of the subsidiary sand fraction on the beach characteristics.

INTRODUCTION

Pebble and cobble (henceforth called shingle) beaches, or at least beaches where the upper foreshore is predominantly shingle, are to be found in many parts of the world, but are nowhere of more coastal engineering significance than in Britain (Carr, 1983a). Whilst for recreational activities such beaches have less appeal than their sand counterparts, they are nonetheless one of the most efficient forms of coast protection.

In Southern and Eastern Britain, maintenance usually entails regular profile monitoring and, when needed, the construction of groynes and the recycling of shingle transported by littoral drift (e.g. Foxley & Shave, 1983). Projects involving shingle nourishment, usually from marine sources, are becoming increasingly favoured (e.g. Hayling Island, Hampshire and Seaford, Sussex).. This has raised important questions concerning the field characteristics of shingle beaches. The extensive literature on sand beaches is often misleading when applied to shingle beaches because of fac-

*Institute of Marine Studies, Plymouth Polytechnic, Drake Circus, Plymouth PL4 8AA, U.K.
+Department of Civil Engineering, The University, Southampton, SO9 5NH, U.K.

tors such as permeability (e.g. Muir-Wood, 1970).

The authors have over a number of years examined the processes operating on Hurst Beach, the shingle bank at the eastern end of Christchurch Bay between Saltgrass Lane and Hurst Castle (Fig. 1). Hurst Beach is being starved of littoral drift due to updrift coast protection works, comprising seawall and groynes at Milford-on-Sea and recently has experienced rapid recession of up to 3.5m/yr. (Nicholls & Webber, 1987a; 1987b). Thus, Hurst Beach makes an excellent natural laboratory where the processes which influence the stability of shingle beaches can be examined and quantified. Much of the work has been described in detail by Nicholls (1985). Some of the conclusions of pertinence to the stability of shingle beaches are summarised in this paper.

THE STUDY AREA

The beaches in the eastern half of Christchurch Bay (Fig. 1) are composed of shingle with a mean size in the range -2.5 to -5.5 phi (6 to 45mm) with subsidiary fine to coarse sand (125 to 1000μm) which mainly occurs on the foreshore. The form of the beach profiles on Hurst Beach are fairly typical of many of the shingle banks in southern Britain. The foreshore (average slope about 8°) is backed by a supratidal beach face (average slope about 11 to 13°) and a beach crest (Fig. 2). Landward of the crest the beach drops with a variable slope (typically 5 to 13°) to a saltmarsh. The crest height varies both temporally and spatially with a maximum elevation in the period 1980 to 1982 of 4.6m. O.D., 3.6m. above the highest predicted tide (O.D. approximates mean sea level). The net littoral drift is eastwards, but

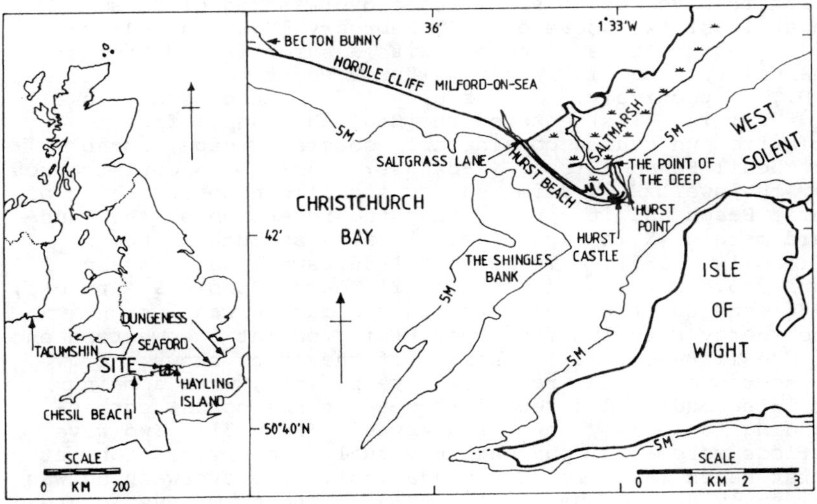

Fig. 1. Location Plan. Crown Copyright. Reproduced from Admiralty Chart 2219 with the permission of the Controller of Her Majesty's Stationery Office.

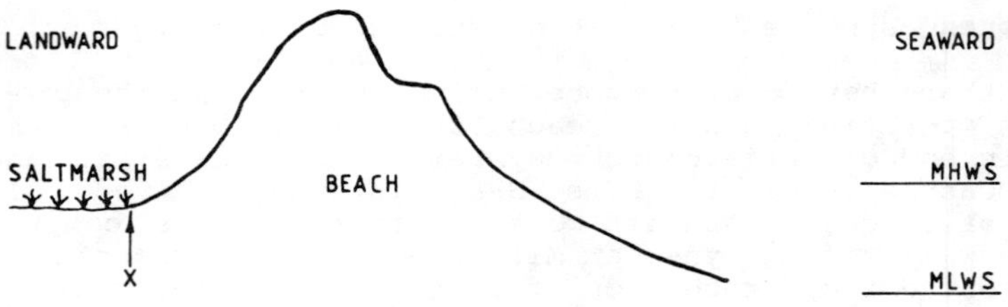

Fig. 2. A typical beach profile across Hurst Beach (vertical exaggeration X5)

with a sub-cell boundary (i.e. a minimum in the rate of net littoral drift) at Hordle Cliff, (Nicholls & Webber, 1987a).

The study area is exposed to waves generated in the Atlantic and the western English Channel. However, the Isle of Wight and the shallow shoal of the Shingles Bank, have a major effect on the energy and direction of waves impinging on the shoreline. In contrast to the British Isles generally, the tidal range is quite small, being 2.2m at springs. Storm surges can add up to 1m to water levels in this part of the English Channel. This means that, even during a neap tide, the highest predicted tide (HAT: 1.05m O.D.) may be exceeded. There are fast tidal currents off Hurst Castle in the narrow entrance to the West Solent, attaining a maximum of 2.3m/s. on mean spring tides.

LONG PERIOD WAVES

Long period, high energy swell waves originating in the Atlantic have caused significant overwashing of Hurst Beach on at least two occasions, 13 February 1979 and 2 January 1986. Both events coincided with a surge in the English Channel raising high water at Hurst Point to 1.4 and 1.0m. O.D., respectively. In February 1979, a high energy swell (H_s = 7m, T_z = 18s) entered the English Channel from the Atlantic and caused considerable coastal damage, notably at Chesil Beach (Draper & Bownass, 1983). In Christchurch Bay the swell ($H_s \lesssim$ 2.5m) caused significant overwashing of Hurst Beach and it appears that the recession of the landward margin (X in Fig. 2) was locally as much as 60m (Nicholls, 1985). In January 1986, swell waves again caused overwashing of both Chesil Beach (Dobbie & Partners, 1986) and Hurst Beach, although the damage was less severe. The energy spectrum from West Bexington (at the western end of Chesil Beach) at the height of the event shows energy present over a wide range of wave periods, with a main peak at 16 seconds and a secondary peak at 9 seconds, corresponding to locally generated waves. (Fig. 3). Two wave periods were also apparent in visual wave observations at Hurst Beach (H_b = 2m, T = 9 and 25s). The overwashing was evidently due to the longer waves. These two events are considered in more detail in subsequent sections.

Clearly, swell waves are an important factor when considering the stability of Hurst Beach. Indeed, when combined with a surge, they may, like Chesil Beach (Carr, 1983b) produce the greatest run-ups and this situation may apply generally wherever beaches are exposed to long fetches.

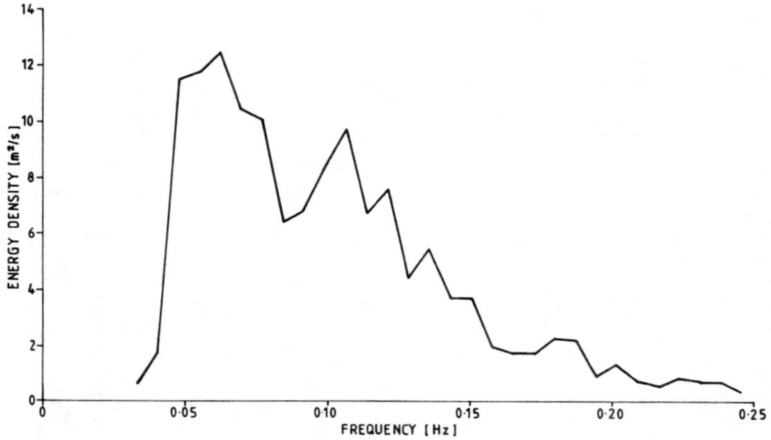

Fig. 3. The energy spectrum at West Bexington at 1030 GMT on 2 January 1986. (Measured by the Institute of Oceanographic Sciences).

OVERWASH PROCESSES

Overwash is defined as any swash that passes over the highest point of a barrier beach, which in the case of a shingle bank is the crest. Thus, overwashing causes sediment transport from the seaward to the landward side of the barrier. Overwash processes have received most attention on the sandy barrier islands of the East and Gulf Coasts of the USA (e.g. Leatherman, 1979; Oertel & Leatherman, 1985), but are also of importance on mixed sand and shingle barriers (Orford & Carter, 1982) and shingle banks. For instance the landward recession of Hurst Beach is caused by overwashing (Nicholls & Webber, 1987b) and a number of washover fans (the product of overwashing) can be seen on its landward margin. In Britain, such processes have often been described as 'overtopping' but it is best if this term is only used to describe processes which increase the height of the crest (cf Orford & Carter, 1982). Thus, overwashing may be seen as the 'failure' of a shingle bank.

On Hurst Beach two distinct forms of overwashing can be distinguished:

(i) Crest-maintaining overwashing. The overwashing occurs without the height of the crest being reduced, e.g. Fig. 4 (small changes in the height of the crest of about 0.1m may occur);

(ii) Throat-confined overwashing. The overwashing locally reduces the height of the crest to form what is usually termed a 'throat' or occupies a pre-existing topographic low. A washover fan is deposited on the landward side of the throat.

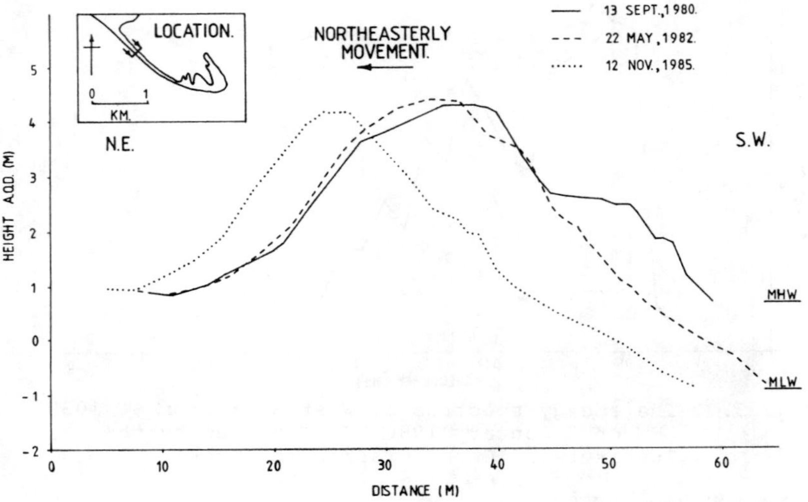

Fig. 4. Recession of a sample cross-section of Hurst Beach due to a series of crest-maintaining overwashing events (after Nicholls & Webber, 1987b).

Only the latter process has been described on sand barriers (e.g. Leatherman, 1979) or the mixed sand and shingle barriers of SE Ireland (Orford & Carter, 1982). The occurrence of crest-maintaining overwashing on shingle beaches is attributed largely to their high surface permeability. These two forms of overwashing may occur simultaneously at different locations (Fig. 5) and are clearly related: as the volume of swash passing over the crest increases so does the likelihood of the crest being reduced in height, causing a transition from crest-maintaining to throat-confined overwashing. The surface permeability must also be of significance in this transition. An extension of throat-confined overwashing to more widespread failures of the crest occurred on 23 November 1984, but individual throats were still present within the low sections of beach (see below).

On Hurst Beach, overwashing is most frequent at Saltgrass Lane, where the beach is most depleted. Throat-confined overwashing occurred on about 6 occasions per year between 1980 and 1982. More recently, shingle nourishment has reduced its occurrence. Elsewhere, overwashing occurred about twice a year between 1981 and 1987, although not all the beach was affected in any one event. Overwashing usually occurs in response to storm waves combined with a surge, such that the tidal elevation approaches or exceeds the highest astronomical tide (1.05m O.D.). Two important exceptions are the overwashing of Hurst Beach by large swell waves, on 13 February 1979 and 2 January 1986, (see previously). Several large washover fans thought to have been deposited by throat-confined overwashing during the former event were still present on the back of Hurst Beach in May 1988.

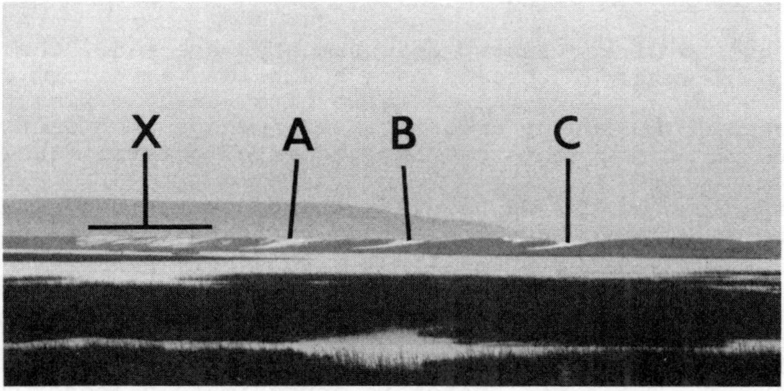

Fig. 5. Well-developed active regularly-spaced washover throats (A, B and C) on Hurst Beach on 2 January 1986. Crest-maintaining overwashing is occurring on the left of the picture (X).

Crest-maintaining overwashing is by far the more common process. Between 1980 and 1984, with the exception of three throats, all throat-confined overwashing occurred at Saltgrass Lane. However, on 23 November 1984, there was significant storm wave activity accompanied by a surge raising tidal levels to an estimated height of 1.6m O.D. The crest was reduced in height over a total distance of about 350m, comprising two lows containing 7 and 9 throats, respectively (Fig. 6).

Every throat formed in the period being considered was filled artificially in an attempt to reduce recession, usually using the washover deposits. Therefore, the long term evolution of the throats was not observed.

Crest-maintaining overwashing is a less efficient mechanism for net onshore sediment transport than throat-confined overwashing: the maximum measured volume of onshore transport over the crest during a single storm was $6m^3/m$ as compared to $150m^3/m$ due to throat-confined overwashing. An overwash budget for Hurst Beach is not available, but it is clear that infrequent throat-confined overwashing events could equal or exceed the net onshore transport caused by more numerous crest-maintaining overwash events.

Therefore it is worth considering if the prevention of overwashing should always be a principal aim in shingle beach management. Throat-confined overwashing is clearly unacceptable. In contrast, the shingle bank remains an effective barrier during crest-maintaining overwashing. Very few, if any, swashes reach the landward margin. The resulting landward movement of the bank is easily monitored, although as Fig. 4 illustrates, it can be quite rapid (3m/yr). The main benefit is the smaller volume of beach fill required. This approach may be useful as a short or medium term solution to problems at sites where some recession is permissible.

RUN-UP CHARACTERISTICS

Although no direct run-up measurements were made, the formation of washover throats on a beach with a uniform longshore section allows the position of a run-up maxima to be inferred. The run-up associated with these high energy events is of most interest in studies of coastal stability.

The spacings of the washover fans and overtop deposits formed on 23 November 1984 are shown in Fig. 6. In the case of the overtop deposits, the longshore peaks of run-up were indicated by a topographic maximum where vertical accretion had been most pronounced, rather than a topographic minimum as indicated by the throats (Fig. 6B). The sample comprises 27 measurements with a range between 10.3 and 45.7m. The distribution is approximately lognormal. The washover throats show two distinct modes at about 13 and 23m with a few larger spacings. The overtop deposit data is more scattered with the mode occurring at about 19m. The scatter is to be expected as the overtop deposits have a volume a tenth or less than the washover throats and as such will not 'record' the run-ups so clearly. Taking the dataset as a whole, there is a significant (sig. > 95%) decline in the spacings along Hurst Beach (Fig. 6C). The interpretation of this result is not clear. Not surprisingly, taking the scatter and this factor into account, autocorrelation showed no significant evidence of memory. However, when the results were compared with a Poisson distribution for longshore spacings of 10, 15 and 20m, the observed distribution was non-random in all cases (sig > 95%). Thus, the washover throats and overtop deposits show a preferred longshore spacing, but cannot be considered to be periodic.

On 2 January 1986 the throat-confined overwashing of Hurst Beach by the long period swell produced four small throats with a regular longshore spacing (mean = 37m, standard deviation = 2.69m) (Fig. 5). The adjacent crest-maintaining overwashing showed similar longshore spacing, although it was not possible to take direct measurements. This observation indicates that the run-up was exhibiting regularly-spaced maxima along the shore.

Thus two distinct overwash events of Hurst Beach have produced morphological evidence indicating non-randomly spaced longshore maxima of run-up. Orford & Carter (1984) have described over 60 regularly-spaced washover throats on the mixed sand and shingle barrier of Tacumshin, S.E. Ireland. The modal spacing was about 40m and appeared to be contolled by high elevation beach cusps.

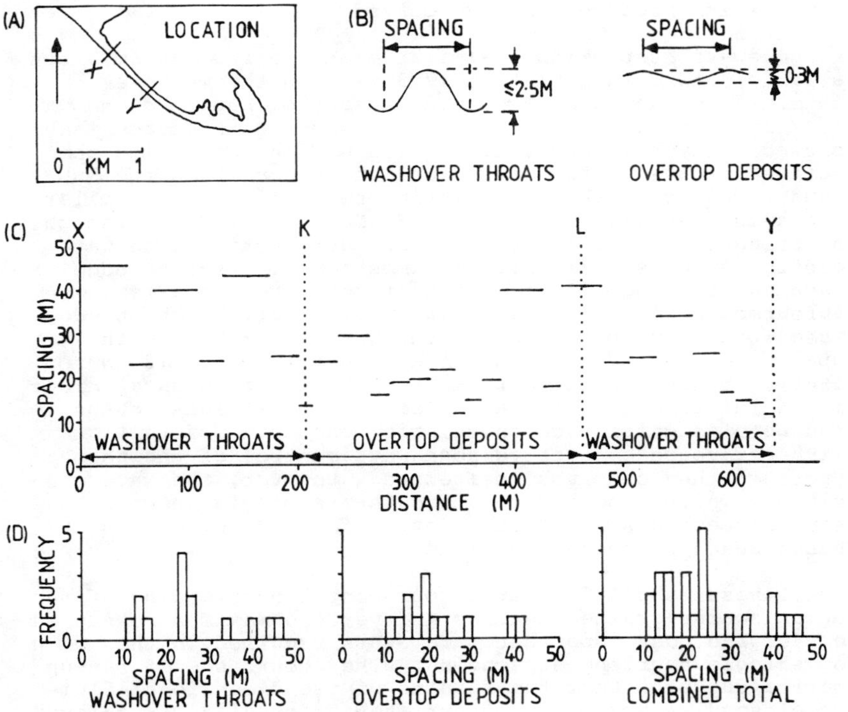

Fig. 6. The longshore spacing of the washover throats and overtop deposits formed on 23 November 1984. (A) Location Plan. (B) Definition sketches of spacing as measured on washover throats and overtop deposits. (C) The location of each longshore spacing between X and Y together with its magnitude. Spacings K and L are transitional between the two morphologies. (D) The numerical frequency distribution of the washover throats, the overtop deposits and their combined total.

Two distinct hypotheses may explain regularly-spaced run-up maxima:

(i) longshore variation in the wave field e.g. phase-coupled edge waves;

(ii) interaction of the run-up with pre-existing rhythmic three-dimensional beach morphology e.g. run-up amplification at swash cusps.

To some extent these two hypotheses are linked as it is generally considered that many three-dimensional beach features are a product of edge waves e.g. swash cusps (Inman & Guza, 1982). However, it is important to distinguish between active longshore variation in the wave field and the influence of three-dimensional morphology which may modify the run-up a considerable period of time after the waves which produced it have ceased.

Large swash cusps with a similar mean spacing (36.6m), but a larger standard deviation (7.38m) than the washover throats are known to have been present on Hurst Beach prior to 2 January 1986. During the overwashing the beach face showed a rhythmic undulation which was in phase with the active throats. These are presumed to have been the subdued remnants of the pre-existing beach cusps. A similar mechanism may have occurred on 23 November 1984, although no evidence is available. By contrast, active edge wave control appears unlikely. The most easily excited edge wave and the one with the largest amplitude is a zero mode subharmonic edge wave (Inman & Guza, 1982). Such an edge wave would have produced longshore run-up maxima with a spacing of about 44 (23 November 1984) and 112m (2 January 1986), showing little agreement with the field data. Thus it is inferred that run-up amplification at swash cusps can be of significance on shingle beaches. Orford & Carter (1984) favoured an active edge wave control of the throat spacings they observed in Ireland. However, the wave conditions which produced the throats are poorly known and a similar mechanism to that inferred for the two events at Hurst Beach is quite plausible.

These results all illustrate our poor understanding of run-up on coarse-grained beaches and yet the maximum run-up is one of the most important parameters required in the design of shingle nourishment schemes. The occurrence of run-up maxima suggests that the results of two-dimensional flume model studies may be inappropriate for studies of maximum run-up. Even in the absence of the evidence presented, the formation of a single washover throat on a previously uniform length of beach leads to the same conclusion. Clearly, wave basin model studies combined with accurate field measurement of run-up characteristics on shingle beaches are required, (see Holman & Guza, 1984). These should include an analysis of the longshore structure of run-up and its interaction with three-dimensional topography. From the evidence presented, it is apparent that even with a uniform incident run-up, amplification of that run-up at swash cusps may occur and this process requires quantification. The common occurrence of swash cusps on shingle beaches may make this a significant factor. The run-up characteristics of long period high energy swell waves must also be considered.

SEEPAGE CHARACTERISTICS

Shingle beaches are highly permeable when compared to their sand equivalents. The piezometric and seepage characteristics of Hurst Beach were investigated along one cross-section using a line of standpipes (Fig. 7).

The constriction of the tidal flow between Hurst Point and the Isle of Wight causes a differential water level of up to 0.7m across Hurst Beach on the rising stage of equinoxial tides, i.e. higher water levels in Christchurch Bay than in the West Solent. This might be expected to produce landward seepage of seawater through the beach. However, this is not the case, and the water table shows its maximum amplitude at the landward margin (see Day 2 - Fig. 7). This apparently anomalous behaviour can be mainly attributed to the large permeability contrast, estimated to be of the order of 10^3, between the relatively impermeable veneer of sand over the foreshore and the more permeable much less sandy washover deposits on the landward side. This example illustrates that while shingle beaches are highly permeable, the sedimentology, particularly the amount and location of sand, has a profound influence on the seepage characteristics of such beaches.

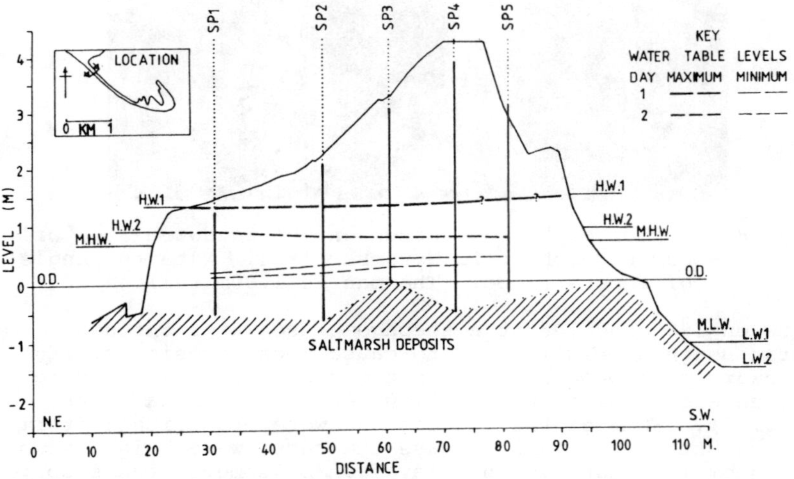

Fig. 7. Maximum and minimum water levels at five standpipes (SP1 to SP5) in Hurst Beach in response to an equinoxial tide and storm waves. (Day 1 - 8 March 1981) and an equinoxial tide and calm conditions (Day 2 - 5 April 1981). The elevation of high water for Day 1 in Christchurch Bay (on the south-west side) includes an estimate of wave setup. The elevations of low water are estimated and only shown for guidance.

Landward seepage of seawater through the beach does occur, but only in association with wave activity. For instance on 8 March 1981, with breaking waves of up to 1.8m, the water table dipped landward throughout the tidal cycle (Fig. 7) with flows estimated to be in the range 0.3 to $1.7 m^3/h/m$. The differing behaviour may be attributed to:

(a) Offshore sediment movement of the sand veneer on the foreshore exposing more permeable sediments;

(b) Increased differential water levels across the beach due to wave set-up, plus a contribution to the water table due to swash infiltration.

These effects are reinforced by high water levels produced by storm surges, because it appears that, in bulk terms, there is a vertical increase in the permeability of the beach.

Fig. 8. A seepage hollow formed on the landward side of Hurst Beach on 23 November 1984. Note the angle of repose slope. The man is 1.80m tall.

Between 1980 and 1986 the seepage of seawater through Hurst Beach was sufficient to cause local erosion of its landward margin on at least six occasions. This produces paired erosional seepage hollows and depositional fans (Fig. 8). Some of these occasions were significant storm surges but this was not always the case, wave height also being of importance. The elevation of some of the seepage hollows (up to 1.7m O.D.) demonstrates the high water levels which may be developed within the beach. These features display a wide range of morphology, some being similar, but on a smaller scale, to the so-called 'cans' of Chesil Beach (Carr & Blackley, 1974). The maximum volume of shingle eroded from a seepage hollow during a single storm was $10m^3$ (or about $5m^3/m$). Once formed a seepage hollow may be active again producing an increase in size. However, the rapid recession of Hurst Beach leads to the regular infilling of the seepage hollows with washover deposits.

A feature of all the seepage hollows observed was an angle of repose slope (e.g. Fig. 8). Shingle is removed from the base of the hollow resulting in a slope failure migrating into the beach. Clearly if this process were to continue until the seepage hollow intersected the crest it would result in failure of the beach, creating a site for throat-confined overwashing. The size distribution of the seepage hollows suggests that this did not happen on Hurst

Beach during the period of observation. However, at Dungeness, Kent (Eddison, 1983) and in S.E. Ireland (Carter, Johnston & Orford, 1984) seepage erosion appears to aid overwashing. On a nourished shingle bank which will usually be designed to be static, the attritional losses caused by seepage erosion could ultimately have similar consequences. Thus, any beach profiling/monitoring should include any landward slope in addition to the seaward portion of the beach.

SETTLEMENT BENEATH BEACHES

As Hurst Beach moves northeastwards, due to overwashing, so it is moving across the Spartina saltmarsh which occurs in its lee. Saltmarsh deposits can often be seen on the foreshore but are rapidly eroded. The best such exposure known to the authors occurred between 1979 and 1980 when several hundred square metres of former saltmarsh surface was clearly exposed, including the dead and crushed Spartina stems and leaves (Fig. 9). The surface was horizontal as in the living saltmarsh, but levelling demonstrated that it was about 1.0m lower. After careful examination of a number of hypotheses it was concluded that the difference in elevation was due to the settlement of the saltmarsh deposits beneath the weight of the beach. The Spartina surface of the saltmarsh provides a useful datum for the direct measurement of such changes.

Cartographic evidence demonstrates that the settlement must have occurred in less than 10 years. The thickness of the deposits behind the beach exceeds 5m and includes an unknown thickness of highly compressible peat. The thickness of the beach sediments shows both temporal and spatial variation with a probable maximum of 4m at this site. A preliminary geotechnical analysis demonstrates that the settlement will not have reached completion in 10 years, although there is some uncertainty about long-term values.

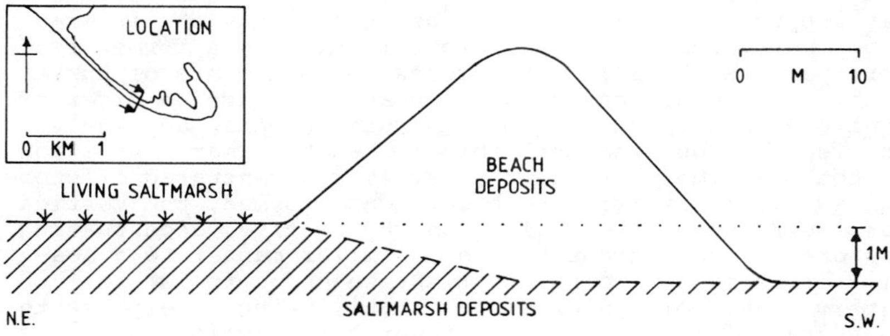

Fig. 9. A diagrammatic cross-section illustrating the settlement of the salt-marsh deposits which was visible across Hurst Beach between 1979 and 1980.

It is concluded that the geotechnical properties of the deposits beneath beaches may have significance when considering beach stability. This factor is independent of grain size and also applies to sand beaches (e.g. Dean, 1987). Rapid settlement will reduce the elevation of the overlying beach and thus on an eroding coastline will accentuate shoreline recession. In the static situation of nourishment, settlement must be considered when calculating volumes of beach fill. In effect, settlement provides a third potential loss of sediment in addition to offshore and longshore losses.

SHINGLE NOURISHMENT

In recent years, marine-dredged beach fill has become the dominant source for shingle nourishment in Britain on a tonnage basis. However, it normally contains considerable quantities of sand due to: (i) source and (ii) the method of delivery to the beach. Split bottom barges drop the dredged gravel as high on the foreshore as possible during the rising tide. On the falling tide the gravel is recovered and placed where required. This method inevitably mixes the sandy foreshore sediments with the beach fill which itself may contain finer material to assist pumping. Such mixtures of shingle and sand are rarely present on the supratidal portion of 'natural' shingle beaches, largely because they are eroded by waves. Thus, marine-dredged beach fill is usually unstable on the supratidal portion of a shingle beach. For instance, at Hurst Beach and Hayling Island marine-dredged beach fill was imported as described. During storms, offshore sediment movement occurred and vertical supratidal beach scarps, locally up to 2m high, developed. (Fig. 10). While subsequent onshore transport deposited shingle berms, the scarps persisted for months at both sites, migrating onshore during storms, before being removed artificially. At Hayling Island they reformed suggesting that they may persist for years without human interference.

Beach scarps up to 3m high are common features on some sandy beaches (Sherman & Nordstrom, 1985), but shingle will only support a vertical slope when sand fills the interstices, e.g. beach scarps up to 0.1m high have formed for short periods (hours) in the foreshore sediments of Hurst Beach. Thus, the scarps in the beach fill are an order of magnitude larger and much more persistent than any equivalent features on a natural shingle beach. Scarp formation was not observed, but clearly involves accentuated offshore sediment movement from the base. Once formed, the vertical scarp must cause wave reflection when active, accentuating offshore sediment transport in a similar manner to a seawall. This is not to say that the beach scarp is the primary cause of the erosion, it only being a response to the hydraulically unstable sediment distribution.

Therefore, it is best to use land-derived beach fill which usually contains little or no sand for shingle nourishment. However, the large scale of many recent and proposed nourishment schemes probably makes such sources inappropriate.

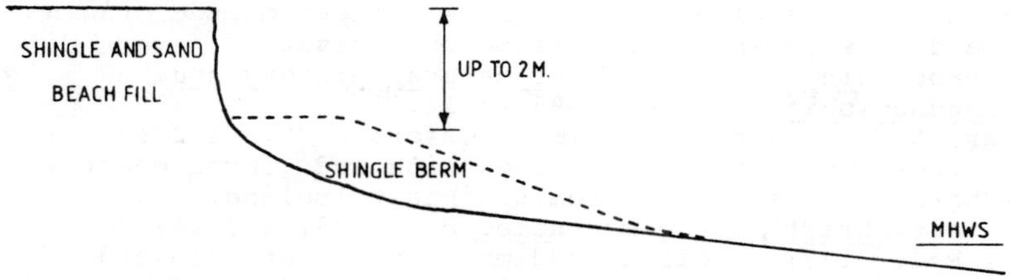

Fig. 10. A diagrammatic cross-section through a beach scarp on a nourished shingle beach.

Some improvements in marine delivery methods may be possible. If not, suitable allowance for beach scarping must be made during design. Assuming that the volume of beach fill is appropriate and longshore losses are small, in the long-term, scarping is not a problem; rather it should be seen as sorting of the beach fill. The shingle moved offshore is returned to the beach while the sand will remain on the foreshore and ultimately the beach will display a surface sorting similar to that seen on a 'natural' shingle beach.

CONCLUSIONS

Shingle beaches exhibit characteristics which are somewhat distinct from sand beaches. These include a far superior performance under wave action for equivalent volumes of beach material, so it is not surprising that shingle is becoming increasingly favoured as a nourishment material. However, field studies indicate a number of problems which require more research. Of particular importance is the observation that run-up shows longshore maxima. There is inadequate understanding of run-up on shingle beaches and its interaction with the swash cusps which are so often present. The role of the subsidiary sand component of the beach sediments is often ignored, but is of importance with regard to factors such as permeability and beach scarping. It is also important that existing shingle nourishment projects are monitored and studied.

ACKNOWLEDGEMENTS

The authors thank Dave Coats, Alan Carr, and Julian Orford for useful discussions, Laurie Draper, IOS for providing wave data, and Dobbie and Partners and their clients, Weymouth and Portland Borough Council and the Wessex Water Authority for permission to use unpublished work at Chesil Beach.

REFERENCES

Carr, A.P. 1983a. Shingle beaches: aspects of their structure and stability. p 97-104. In Shoreline Protection. Thomas Telford Ltd., London.

Carr, A.P. 1983b. Chesil Beach: Environmental, economic and sociological pressures. Geographical Journal. 149, 53-62.

Carr, A.P. & Blackley, M.W.L. 1974. Ideas on the origin and development of Chesil Beach, Dorset. Proceedings of the Dorset Natural History and Archeological Society, 95, 9-17.
Carter, R.W.G., Johnston, T.W. & Orford, J.D. 1984. Stream outlets through mixed sand and gravel coastal barriers: examples from southeast Ireland. Zeitschrift fur Geomorphologie, NF 28, 427-442.
Dean, R.G. 1987. Coastal sediment processes: Toward engineering solutions. p 1-24. In: Proceedings of Coastal Sediments 87. ASCE, New York.
Dobbie, C.H. & Partners. 1986. Chesil sea defence scheme: Report on flood event 2 January 1986. Unpublished.
Draper, L. & Bowmass, T.M. 1983. Unusual waves on European coasts, February 1979. p 270-281. In: Proceedings of 18th Coastal Engineering Conference, ASCE, New York.
Eddison, J. 1983. Flandrian barrier beaches off the coast of Sussex and southeast Kent. Quaternary Newsletter, 39, 25-29.
Foxley, J.C. & Shave, K. 1983. Beach monitoring and shingle recharge. p 163-170. In: Shoreline Protection. Thomas Telford Ltd., London.
Holman, R.A. & Guza, R.T. 1984. Measuring run-up on a natural beach. Coastal Engineering, 8, 129-140.
Inman, D.L. & Guza, R.T. 1982. The origin of swash cusps on beaches. Marine Geology, 49, 133-148.
Leatherman, S.P. (ed). 1979. Barrier Islands. Academic Press, New York.
Muir-Wood, A.M. 1970. Characteristics of shingle beaches: the solution to some practical problems. p 1059-1076. In: Proceedings 12th Coastal Engineering Conference, ASCE, New York.
Nicholls, R.J. 1985. The stability of the shingle beaches in the eastern half of Christchurch Bay. Ph.D. Thesis, Dept. of Civil Engineering, University of Southampton, 468 pp.
Nicholls, R.J. & Webber, N.B. 1987a. Coastal erosion in the eastern half of Christchurch Bay. p 549-554. In: Culshaw, M.G. et al (eds) Planning and Engineering Geology. Geological Society, Engineering Geology Special Publication No. 4, London.
Nicholls, R.J. & Webber, N.B. 1987b. The Past, Present and Future Evolution of Hurst Castle Spit, Hampshire. Progress in Oceanography, 18, 119-137.
Oertel, G.F. & Leatherman, S.P. (eds). 1985. Barrier Islands. Special Issue. Marine Geology, 63.
Orford, J.D. & Carter, R.W.G. 1982. Crestal overtop and washover sedimentation on a fringing sandy gravel barrier coast, Carnstone Point, S.E. Ireland. Journal of Sedimentary Petrology. 52, 265-278.
Orford, J.D. & Carter, R.W.G. 1984. Mechanisms to account for the longshore spacing of overwash throats on a coarse clastic barrier in southeast Ireland. Marine Geology, 56, 207-226.
Sherman, D.J. & Nordstrom, K.F. 1985. Beach scarps. Zeitschrift fur Geomorphologie. NF 29, 139-152.

CHAPTER 143

PREDICTION OF 3-D BEACH CHANGES ON THE FUJI COAST

OHISHI,H*., T.UDA**, Y.MURAKAMI*** AND

M.WATANABE***

ABSTRACT

Beach changes around Tagonoura Port on the Fuji Coast were investigated by using the topographic survey data collected in the 15 years between 1968 and 1983. Temporal and spatial changes of the offshore distance from a reference point to certain contours were analyzed to elucidate the actual situation of topographic changes on the Fuji Coast, and the littoral transport rate on the downdrift coast of Tagonoura Port was found to be about $1.17 \times 10^5 m^3/yr$ on the average between 1973 and 1983. It is concluded that a large amount of sediment eroded from the downdrift beach of Tagonoura Port is carried away toward a zone deeper than 20m. A numerical model predicting three-dimensional beach changes was also developed, taking account of both the vertical distribution of littoral transport and the effect of the wave-dissipating breakwaters.

I. INTRODUCTION

In recent years beach erosion has become severe along Japan's coasts. The Fuji Coast, surrounding Suruga Bay and formed by the fluvial sediment supply from the Fuji River, is one of these eroded coasts. Since the early 1960s sediment supply to this coast has decreased due to the influence of extensive excavation of the river bed, eroding the coast around the river mouth. Since 1959 the breakwater at Tagonoura Port has been constructed in the middle of the coast, and then the downdrift beach of the harbor was further eroded. As a countermeasure against beach erosion, wave-dissipating breakwaters, composed of concrete blocks parallel to the shoreline, have been constructed since 1973 on the downdrift coast of the harbor. The beach erosion on the Fuji Coast seems to be a typical example of beach erosion occurring along Japanese coasts from a caused point of view. For this reason, it is of primary importance to investigate the actual situation of the beach erosion and the effectiveness of the countermeasures along this coast in order to understand the beach erosion problems in Japan.

Along this coast, topographic surveys have been conducted once a year since 1968. This study aims to investigate beach changes around the harbor in the 15 years between 1968 and 1983 by using the topographic survey

* Head, Numazu Work Office, Chubu Regional Construction Bureau, Ministry of Construction, Shizuoka 410 Japan.

** Dr.Eng., Head, Coastal Eng. Div., Public Works Research Institute, Ministry of Construction, Ibaraki 305 Japan.

*** Numazu Work Office.

data. In order to analyze the field data, temporal and spatial changes of the offshore distances from a reference point to certain contours are studied. The validity of this method has been proved when applied to the analysis of beach profile changes on the Suruga Coast (Uda and Takeuchi, 1985) and Kochi Coast(Uda et al.,1986).

The temporal and spatial beach changes on the Fuji Coast are elucidated through this study, and the littoral transport rate on the downdrift coast of Tagonoura Port is evaluated. In particular, it is concluded that a large amount of sediment eroded from the downdrift beach of Tagonoura Port is carried away toward a zone deeper than 20m. The second aim of this study is to develop a numerical model which can predict three-dimensional beach changes. For this purpose, a model in which the vertical distribution of littoral transport is taken into account and the effect of the wave-dissipating breakwaters on the beach changes can be evaluated is developed.

II. METHOD OF INVESTIGATION

The Fuji Coast surrounds Suruga Bay facing the Pacific Ocean to the south (Fig.1). The water depth of Suruga Bay is so deep that the Fuji Coast has a steep slope as shown in Fig.1. At a location about 5.5km east of the mouth of the Fuji River lies Tagonoura Port, which was constructed in 1959. Topographic surveys have been conducted once a year since 1968 in the region between survey line No.0, adjacent to Numazu Port, and No.85, next to the mouth of the Fuji River. The survey line interval for the bottom sounding is 250m. In this study the bottom sounding data obtained over the 15 years between 1968 and 1983 are analyzed in order to investigate the beach changes on the Fuji Coast. For this purpose, temporal and spatial changes of the offshore distance from the reference point to certain contours between 2m above and 30m below the MSL are investigated.

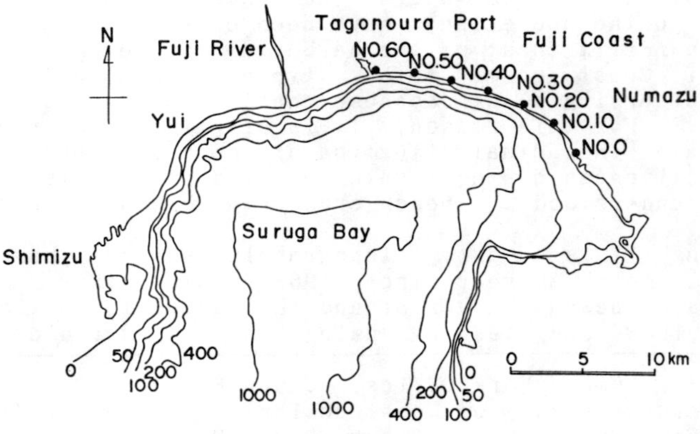

Fig.1 Location of the Fuji Coast

III. BEACH EROSION ON THE FUJI COAST

3.1 Spatial and temporal changes of shoreline configuration

First, the actual situation of the beach erosion on this coast will be studied focusing on shoreline changes before advancing the detailed analysis with respect to the movement of bottom contours in the shallow water zone. The changes in the shoreline configuration on the Fuji Coast are shown in Fig.2. The abscissa is taken for the number of the survey lines set along the coastline, in which No.0 and No.85 are located next to Numazu Port and the Fuji River, respectively. The locations of various facilities are denoted by the serial number in the figure. Since the interval of the survey lines represents 250m, the entire distance in Fig.2 includes 20.25km of shoreline. The ordinate expresses the change of shoreline positions measured in each year since 1968 until 1983 relative to the shoreline configuration in 1968.

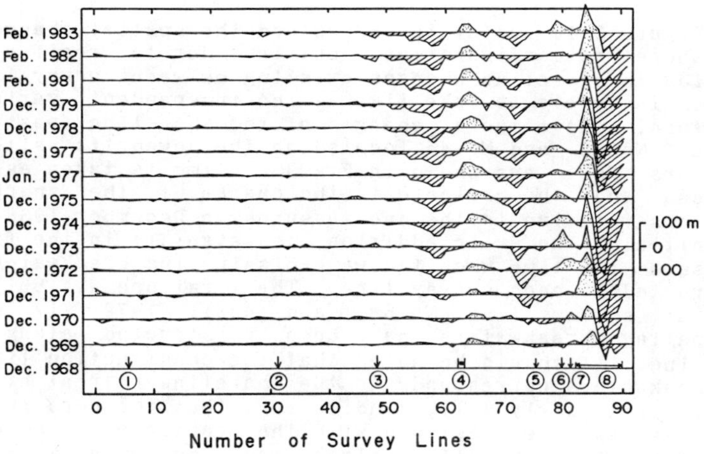

Fig.2 Temporal and spatial changes of the shoreline position on the Fuji Coast (1: Shinnakagawa Flood Way, 2: Showa No.2 Flood Way, 3: Showa Flood Way, 4: Tagonoura Port, 5: Nyudo Drainage channel, 6: Motofugi No.2 Drainage Channel, 7: Motofuji Drainage Channel, 8: Fuji River).

After the construction of the harbor breakwater in 1959, the east coast of the harbor has been gradually eroded. The eroded region extended from No.62 up to No.49, or 3.5km, until 1983. These features of the shoreline retreat in a region next to the harbor breakwater obstructing littoral transport show that the dominant direction of littoral drift on this coast is eastward. As the beach was eroded, wave-dissipating breakwaters, composed of the concrete blocks, have been constructed to counteract beach erosion, and at present, the whole coastline between Tagonoura Port and Showa Flood Way is covered by the concrete blocks. However, many concrete blocks have been scattered in this region due to the scouring at the foot of the breakwaters and due to the wave action on the blocks(Kohno et al., 1986).

As is clearly understood in Fig.2, beach erosion was observed between Tagonoura Port and Showa Flood Way. Nevertheless, the shoreline configuration between Showa Flood Way and Numazu Port at No.0 has been of stable form during the same period. This implies that the eroded materials from the updrift coast were carried away in the offshore direction. This offshore movement of sand is important when studying the littoral drift on the Fuji Coast, so that this point will be discussed in detail in the later section. Furthermore, it is found from Fig.2 that beach erosion is also severe next to the mouth of the Fuji River. Just west of Tagonoura Port, where the shoreline has advanced in recent years in contrast with the shoreline retreat on the east coast, littoral sand has been lost partly by the offshore movement in front of the tip of the breakwater, and partly by the excavation of sand for use in the construction industry.

Figure 2 is useful to understand the spatial changes of the shoreline configuration, but in order to realize the time lag in shoreline changes existing between survey lines, it is favorable to take time as the independent variable. Therefore, the temporal changes of the shoreline positions between No.62 and No.45 located on the downdrift site of Tagonoura Port are shown in Fig.3. Time is taken as the abscissa and the ordinate is the change of the shoreline positions relative to the one measured in December,1968 for the initial value. In addition, the sign(□) in the figure expresses the time when the wave-dissipating breakwater was constructed at each survey line. The first one was built in 1973 at survey line No.58, and until 1978, 17 wave-dissipating breakwaters have been constructed along the shoreline. It should be noted that the construction time of the breakwaters corresponds to the shoreline retreat at each survey line fairly well, indicating that the construction site was extended eastward with the continuous shoreline retreat. Maximum shoreline retreat was observed at No.58 and it reaching about 50m in two years between 1971 and 1973. Thereafter the rate of the shoreline retreat became small by virtue of the wave-dissipating breakwater constructed in 1974. It is found that there exists a time lag in the initiation time of the shoreline retreat with the distance from No.62 next to Tagonoura Port. It was not until 1979 that the shoreline retreat began at No.48. In addition, it is understood that the rate of the shoreline change at the initial stage tends to decrease with the distance from the port, when compared with the shoreline changes observed at two survey lines located far from the port such as No.48 and in the vicinity of the port such as No.58.

3.2 Numerical simulation of change of shoreline configuration

Some features of the changes of the shoreline configuration observed on the coast just east of Tagonoura Port can be predicted by using the one-line model of shoreline evolution as described in the following, if simple assumptions were made. First, it is assumed that the beach changes are caused only by the imbalance of littoral transport and do not depend on cross-shore sand transport. The definition sketch and the initial conditions are shown in Fig.4. Secondly, it is assumed that there exists a vertical wall totally obstructing littoral transport, like a jetty, at the righthand side of the region, and that the distance from the initial shoreline to the wave-dissipating breakwater of continuous type is equal to ℓ . Furthermore, the wave incidence is assumed to be from clockwise direction as shown in Fig.4. In the calculation, the diffraction effect of the jetty is neglected.

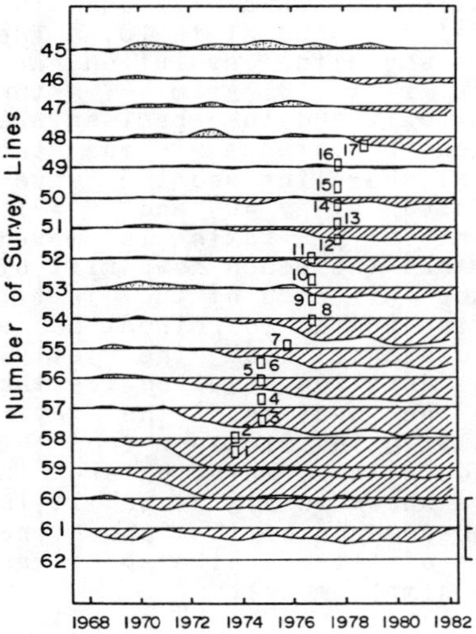

Fig.3 Temporal changes of the shoreline positions in the region east of Tagonoura Port.

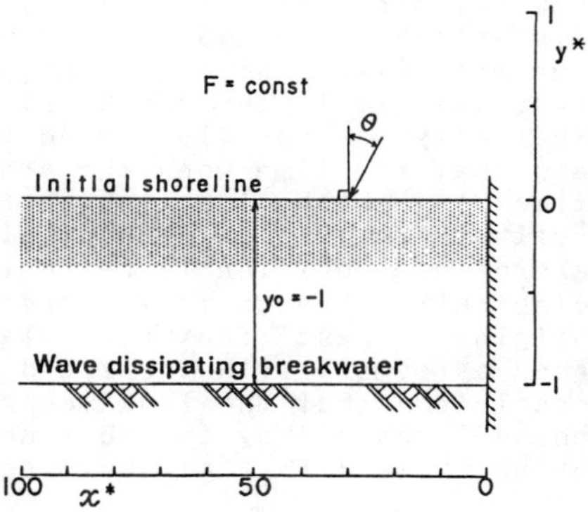

Fig.4 Definition sketch of shoreline simulation model.

Fundamental equations of the one-line model of the shoreline evolution are expressed by Eqs. (1) and (2).

$$\frac{\partial q}{\partial x} + h\frac{\partial y}{\partial t} = 0 \qquad (1)$$

$$q = F\left(-\frac{\partial y}{\partial x} + \tan\theta\right) \qquad (2)$$

Where q: littoral transport rate, h: characteristic height of beach changes caused by littoral transport, x: longshore distance, t: time, F: coefficient depending upon the wave characteristics. In this study coefficient F and h are assumed to be constant in order to understand the essence of the phenomenon as easily as possible. Next, in order to simplify the analysis some dimensionless variables are introduced as follows:

$$y=\ell y^*,\ x=\ell x^*,\ t=\frac{\ell^2 h}{F}t^*,\ q=Fq^* \qquad (3)$$

where ℓ is a characteristic length scale, and variables with the asterisk denote the dimensionless forms. Substituting these forms into Eqs. (1) and (2), then dimensionless forms of the equations are obtained.

$$\frac{\partial q^*}{\partial x^*}+\frac{\partial y^*}{\partial t^*}=0 \qquad (4)$$

$$q^*=-\frac{\partial y^*}{\partial x^*}+\tan\theta \qquad (5)$$

In this calculation it is assumed that the shoreline does not retreat if the shoreline position coincides with the location of the wave-dissipating breakwater, and the breaker angle normal to the shoreline is equal to 10°. The results of the simulation of the shoreline evolution are expressed in Fig.5 showing the x^*-t^* diagram of the shoreline changes. The vertical axis and the abscissa are taken for the dimensionless longshore distance x^* and the dimensionless time t^*. The initial shoreline and the wave-dissipating breakwater are located at $y^*=0$ and $y^*=-1$, respectively, and the base line of y^* axis is moved upwards. It is found from Fig.5 that the beach downdrift of the jetty was rapidly eroded by the influence of the jetty, and that the time when the shoreline begins to retreat has a time lag with the longshore distance from the jetty. Furthermore, it is found that the duration required for the shoreline position to retreat to the location of the wave-dissipating breakwaters increases with the distance from the origin. These features obtained in the simulation are in good agreement with the field data shown in Fig.3, and it is concluded that the fundamental phenomenon of the shoreline changes caused by the obstruction of littoral transport can be understood through the present simple model.

3.3 Temporal and spatial changes of offshore distances

In the previous section, only the temporal and spatial changes of the shoreline configuration were discussed. In this section, temporal and spatial changes of the offshore distances from the reference point to some contours are studied in detail. First, the longshore distribution of the offshore distances from the reference points, which were determined along the foot of the coastal dike, to some contours selected at 5m intervals are shown in Fig.6. It can be clearly seen from Fig.6 that the bottom slope is steep between No.5 and No.25 because of the small intervals of the offshore distance to each contour, and that there exists a comparatively mild slope between No.25 and No.45. Beyond No.46, the contours have irregular shapes, expressing the irregular bottom topography.

Next, the changes of the offshore distance at some typical survey lines are investigated. For the typical survey lines, No.54 and No.58, located east of Tagonoura Port where the beach was eroded severely in recent years, are selected. Figure 7 shows the temporal change of the offshore distance from the reference point to some contours

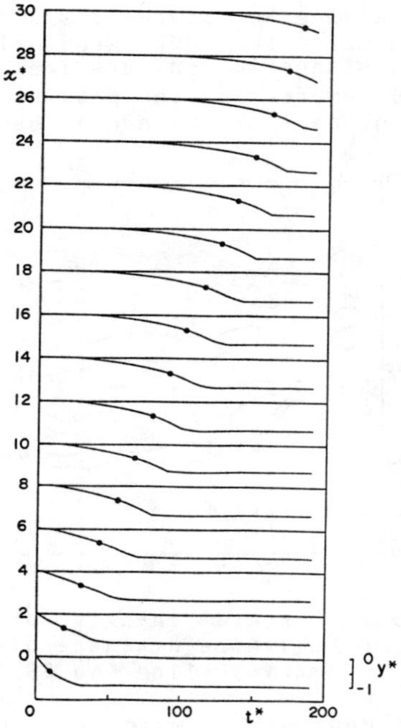

Fig.5 x^*-t^* diagram of change of shoreline position.

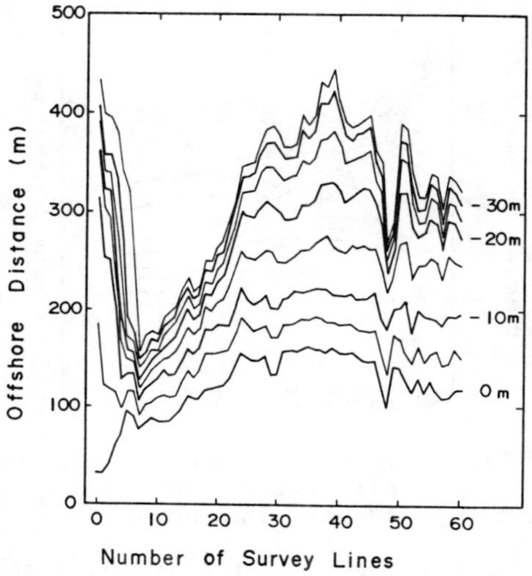

Fig.6 Longshore distribution of offshore distance to some contours.

at No.54. The abscissa is the elapsed time, and the vertical axis is the offshore distance. The shoreline position began to retreat rapidly from Dec., 1975 and the contour lines between 2m above and 10m below the MSL retreated simultaneously with the shoreline change. This means that the beach profile in this region has changed its shape as a whole. By contrast, the change in offshore distance to contours deeper than 16m is small, except for the tendency of a gradual advance in the region deeper than 20m. This fact implies that part of sand eroded from the nearshore zone was carried away in the offshore direction, and therefore net loss of sand was caused on this beach, although the mechanism of the movement of sand is not known at present.

Similarly, the temporal changes of the offshore distance at survey line No.58, located 1km west of No.54 and 1km east of Tagonoura Port, are shown in Fig.8. It is found that the shoreline retreat at No.58 began faster than that at No.54 when Figs.7 and 8 are compared. Rapid retreat of the shoreline position began in 1971, 4 years earlier than that at No.54. The reason that the beach changes at No.58 were observed faster than the changes at No.54 is because of the proximity of survey line No.58 to the harbor structure obstructing littoral transport. The existence of the time lag in shoreline change shows that the beach erosion on this coast mainly depends upon the littoral transport. Furthermore, it is found from Fig.8 that the offshore distance defined between 2m above and 10m below the MSL gradually decreased, whereas the offshore distance to contours deeper than about 16m gradually increased, expressing the possibility of the offshore sand movement of part of the eroded sand.

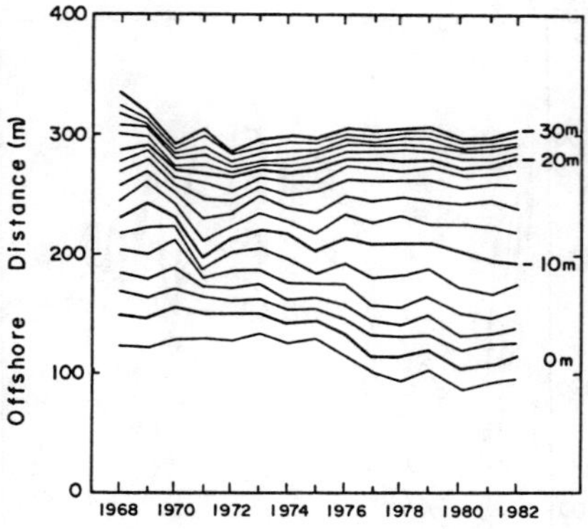

Fig.7 Temporal changes of offshore distance at survey line No.54.

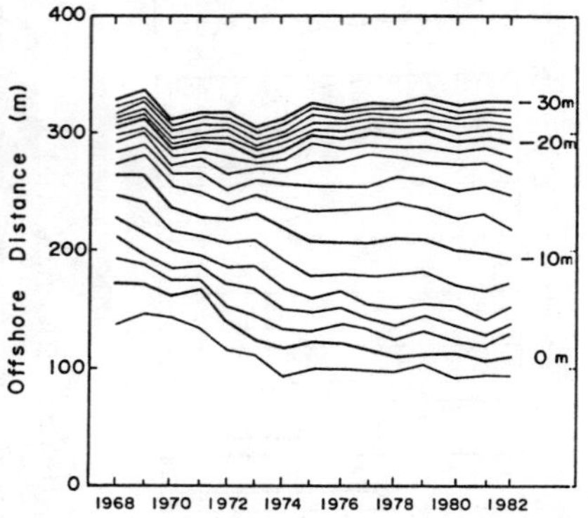

Fig.8 Temporal changes of offshore distance at survey line No.58.

Knowledge of the offshore movement of bed materials is of vital importance in order to formulate a plan to counteract beach erosion on this coast, because this movement results in the net loss of sand in the nearshore zone, and the beach will be eroded to a large extent in the long term. For the purpose, the cumulative value of the area surrounded by the contours and the abscissa in Fig.6 is calculated first, and then the change in the cumulative value (ΔS) with reference to the one measured in 1974 is obtained (Fig.9). For the calculation, 1983 bottom sounding data is selected, and the origin is set at No.60.

Cumulative area corresponding to the shoreline change shows the net loss in area in the longshore direction, and ΔS decreased about $5x10^4 m^2$ in 9 years up to No.0, adjacent to the breakwater of Numazu Port. Between No.60 and No.45, the rate of the decrease of ΔS is large, but thereafter it becomes small. In Fig.9, the changes in the area surrounded by the 5-m-interval contours are shown as well. It is found that ΔS corresponding to 5 and 10m deep contours undergoes a change similar to that corresponding to the shoreline. Namely, the absolute values of ΔS increase abruptly between No.60 and No.45, and they show a net loss in area at No.0. The losses of the areas attain about $7x10^4 m^2$.

On the contrary, the changes of the area surrounded by the 20,25 and 30m deep contours contrast sharply with those measured in shallower zone. In the offshore zone ΔS increases with the distance from No.60, whereas ΔS in the shallow zone decreases. However, it should be noted that the eroded area in the shallow zone increased rapidly between No.60 and No.45 , while the accreted area in the offshore

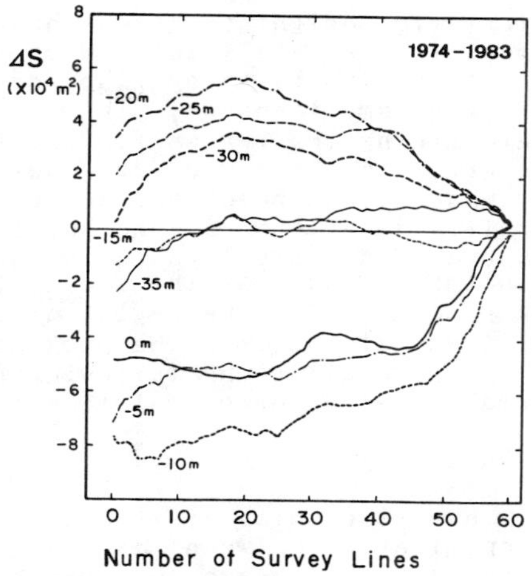

Fig.9 Longshore distribution of change of cumulative value of the area surrounded by offshore distance to some contours as shown in Fig.6.

zone tends to increase gradually from No.60 up to No.18. The increase rate of ΔS is especially large in the region where mild bottom slope was observed in the offshore zone as shown in Fig.6. These features indicate that a large amount of the sediment eroded on the downdrift beach of Tagonoura Port will be finally carried away toward the zone whose depth is between 20 through 30m. This kind of offshore sand movement along this coast was also confirmed through field investigations, using a radioactive tracer(Shuto et al., 1977). The results of the present study agree well with the findings of this field test.

3.4 Characteristic height of beach changes and littoral transport rate

On the east coast of Tagonoura Port, where beach erosion has been mainly due to the imbalance of littoral drift in recent years, temporal changes in the beach sections have been measured once a year so that the relationship between the changes in the sectional area and those of the shoreline position can be investigated. The calculation is carried out between No.61 and No.45 , where beach changes caused by littoral transport are more dominant than those caused by cross-shore transport, judging by the movement of contours shallower than 10m which are very similar to each other, and the amount of sand carried in the offshore direction being comparatively small in this region, as shown in Fig.9. The characteristic height of beach changes caused by littoral transport can be determined empirically from the relationship between the change in the sectional area and that of the shoreline position. If the change in the sectional area (ΔA) can be expressed in terms of the change in shoreline position (Δy) for a linear relationship, then the regression coefficient becomes equal to the characteristic height of beach changes caused by littoral transport (Uda et al.,1986). When the regression coefficient is calculated by using the data set, whose absolute values are small enough, the accuracy of the prediction in determining the regression coefficient will be lowered. Therefore, the profile data between No.52 and No.59 are used, where large beach changes are observed. In addition, the profile data at No.60 and No.61 are neglected, because these survey lines are located in the vicinity of the harbor structures, so that the local influences are considered to be predominant. The region of the calculation of the sectional area extends from the backshore to the 20m-deep location. As a result, the following relation stands between both variables with the correlation factor of 0.80.

$$\Delta A = 13.3 \cdot \Delta y + 150 \qquad (6)$$

Where ΔA and Δy have the units of m^2 and m, respectively. The regression coefficient between ΔA and Δy is equal to the characteristic height of beach changes (h). Due to the relation in Eq.(6), h becomes 13.3m.

Next, littoral transport rate at a typical location east of Tagonoura Port will be evaluated based upon the field data of beach changes. The littoral transport rate can be evaluated on the basis of the continuity relation of sand volume by analyzing the temporal change of the total sand volume of the eroded beach located on the downdrift side of the coastal structure obstructing the continuous longshore movement of sand (Uda et al., 1986). First, the contour map on the downdrift coast of Tagonoura Port is shown in Fig.10.

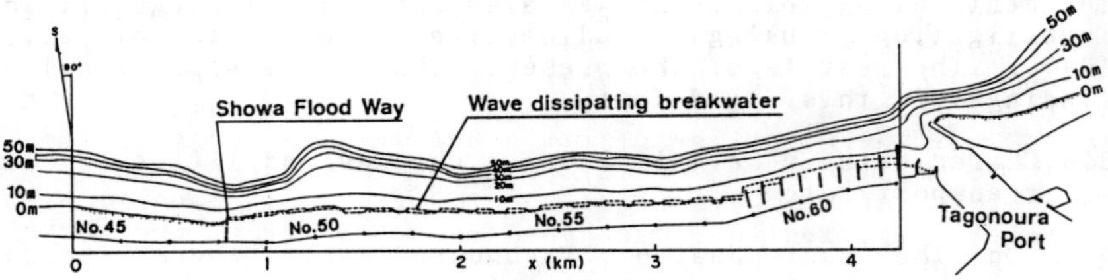

Fig.10 Contour map on the downdrift coast of Tagonoura Port and coordinate system for shoreline simulation.

There are 11 groins and 17 wave-dissipating breakwaters along the shoreline. In this map the region between No.45 and No.62 is selected for the calculation zone. It should be noted that at survey line No.45, far from the harbor, the littoral transport rate has kept constant to the dominant wave incidence on this coast, because at this site the shoreline configuration has not changed since 1968, as shown Fig.3. On the other hand, the littoral transport rate rounding the tip of the breakwater of Tagonoura Port is assumed to be negligibly small, because the water depth at the entrance channel to the harbor is sufficiently deep like as about 10m as shown in Fig.10. The total amount of littoral transport passing through No.45 is approximately equal to the volume change of sand in the test area between

No.45 and No.62. Strictly speaking, there may be a small amount of sand movement toward the offshore zone . If this is correct, such loss of sand should be added to the volume change to give the exact transport rate. However, in this study, it was impossible to measure this value reliably because of the relative low accuracy of the bottom sounding data in the far offshore zone, so that in the present study this is assumed to be negligible.

The change in the total sand volume in the eroded area can be approximately obtained from the change of the plane area in the eroded zone multiplied by the characteristic height of beach changes(h=13.3m). Because the characteristic height of beach changes is the vertical height of the rectangle whose sectional area and horizontal scale are equivalent to the measured sectional area and the change of the shoreline position, respectively. Therefore the multiplication of this vertical height by the change in the plane area in the eroded region results in the volume change. The result of this calculation is shown in Fig.11. The total volume of sand eroded has increased with time, although there is some scatter. The time derivative of the total sand volume approximately becomes equal to littoral transport rate at the downdrift end of the area examined as shown in Fig.10. The time derivative of the sand volume change becomes $Q=1.17x10^5 m^3/yr$ between 1973 and 1983. This value is equal to the average littoral transport rate at No.45 in a period between 1973 and 1983.

IV. PREDICTIVE MODEL OF THREE DIMENSIONAL BEACH CHANGES

In this section a new one-line model of shoreline evolution is developed in order to predict three-dimensional beach changes with some improvements on the former model(Uda and Saito, 1987). In this model the vertical distribution of littoral transport is taken into account, and not only the change of the shoreline position but also the change of the offshore distance to certain contours are predictable by using this vertical distribution of littoral transport. The vertical distribution can be determined from the beach profile changes measured with time at a site, assuming that the beach slope is sufficiently small and the vertical distribution of beach changes during a short period is similar to that of littoral transport. For example, Figure 12 shows the results. To determine the distribution , the reference years selected are 1970 at No.58 and 1973 at No.54. Then, the changes in the offshore distance are divided by the change in the shoreline position to obtain the normalized distribution. There is scatter to some extent, but the vertical distribution of horizontal changes in the beach topography has a triangular shape, taking a maximum value at the MSL and decreasing monotonously above and below the MSL as shown in Fig.12. This distribution of littoral transport is approximately equal to the vertical distribution of littoral transport under the assumption that the vertical distribution of littoral transport does not vary so much, even if the beach profile changes are caused by the littoral transport over a short period.

In the calculation, the equation given by Savage was used for the littoral transport formula. The Savage coefficient is assumed to be 0.2. The littoral transport rate was vertically divided at each depth relative to the average of the normalized vertical distribution as shown in Fig.12 except above the MSL, where the maximum value was used to obtain the best fit with the measured beach changes.

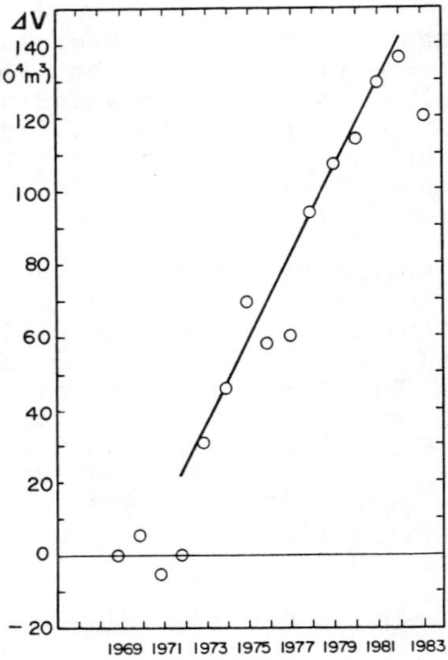

Fig.11 Temporal change of total eroded volume of sand on the downdrift coast of Tagonoura Port.

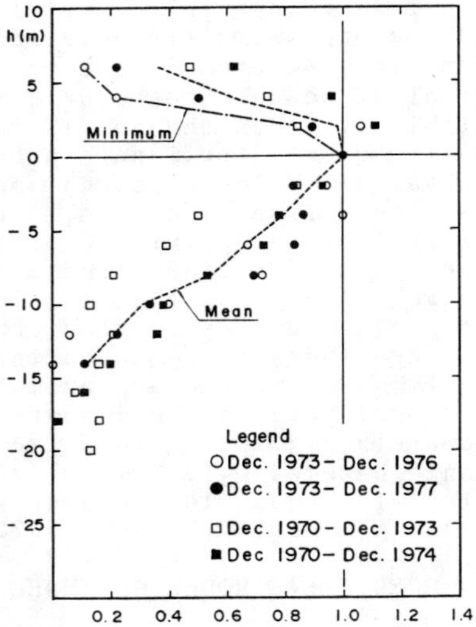

Fig.12 Vertical distribution of normalized horizontal distance in each depth.

On the Fuji Coast there are wave-dissipating breakwaters placed along the shoreline discontinuously as shown in Fig.10, and therefore a model of beach changes around the wave-dissipating breakwater is required as follows. First, in the simulation, the wave-dissipating breakwater is assumed to be of continuous type because the longshore distance in the opening is narrow compared with the longshore length of the wave-dissipating breakwater. Then the depth at the foot of the wave-dissipating breakwater is determined from the given beach profiles, because the location of the breakwater is already known. The wave-dissipating breakwaters, as shown in Fig.3, have been built since 1973, and therefore the total length of the wave-dissipating breakwater from No.62 was chosen so as to satisfy the actual situation of the construction. As is mentioned in Fig.10, between No.59 and No.62 there are groins but the coast between them is covered with the concrete blocks. Therefore the physical condition at this site can be assumed to be the same as at the wave-dissipating breakwater.

If a wave-dissipating breakwater is constructed along the coast, the beach behind the wave-dissipating breakwater is usually protected against beach erosion, and therefore littoral transport rate at this site will be reduced depending upon the location of the wave-dissipating breakwater. Here, the reduction of the littoral transport rate, named by the cutting ratio of littoral transport, is taken into acount. The cutting ratio of littoral transport can be obtained, assuming that the littoral transport rate decreases with the ratio of the area above the depth of the tip of the breakwater with respect to the total area in the vertical distribution of littoral transport. The greater this point depth is, the more the littoral transport rate decreases. The beach changes are assumed to be observed only in the zone deeper than the point depth of the wave-dissipating breakwater, and the horizontal change of the beach topography is calculated from the vertical distribution of littoral transport.

For the initial condition, the bottom topography measured in 1970 was used, and the beach changes until 1983 are predicted. It is assumed that the incident wave height and period at the far offshore zone are equal to $H=0.85$m and $T=9$sec, and the waves are incident from $\alpha=S13^\circ W$ based upon the field observation conducted on the coast. Secondly, the longshore distributions of wave height and wave direction at a location of 10m deep along the coast are evaluated by wave refraction calculation using this incident wave. Furthermore, wave height and wave direction at the breaking point are calculated, assuming that the bottom contours in the shallow water zone are parallel at each location. Wave refraction calculation was performed repeatedly in a shallow water zone under the assumption that the incident wave at a location of 10m deep did not vary as much, even if beach changes occurred in the zone.

As the beach erosion proceeds, the bottom profile gradually tends to reach a stable profile, which is mainly determined by wave characteristics, bottom slope and grain sizes of the bed materials. In this study this stable profile is empirically determined. The beach profile at No.58 measured in 1982 may be selected for such a profile approximately, because the eroded beach section next to Tagonoura Port should approach the stable profile more rapidly. If the beach profile during the initiation time of the construction of the wave-dissipating breakwater has retreated compared with the stable beach profile, and if a part of the beach profile differed from the stable beach profile due to the difference of the bottom profile, their part of the retreated profile was used instead of the stable beach profile. After the beach profile reached to the stable form, it was assumed that the littoral transport rate is equal to the input data at the updrift end, and the same amount of sand can pass through the site with the stable profile.

The predicted changes of the offshore distance to some bottom contours after 4 and 8 years from the beginning are compared with the measured values in Figs.13 and 14. The

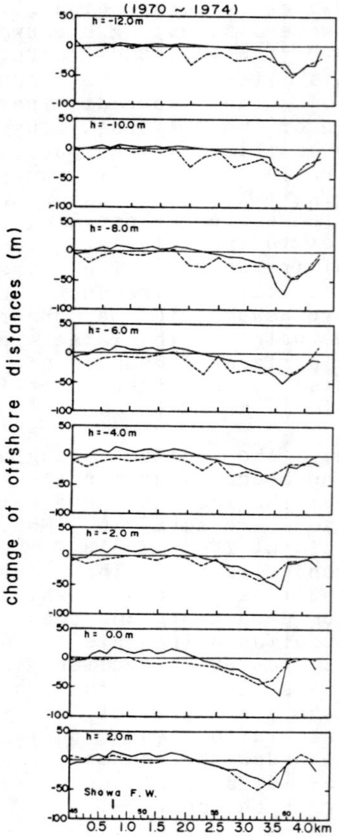

Fig.13 Comparison of measured and predicted changes of offshore distance after 4 years from the beginning.

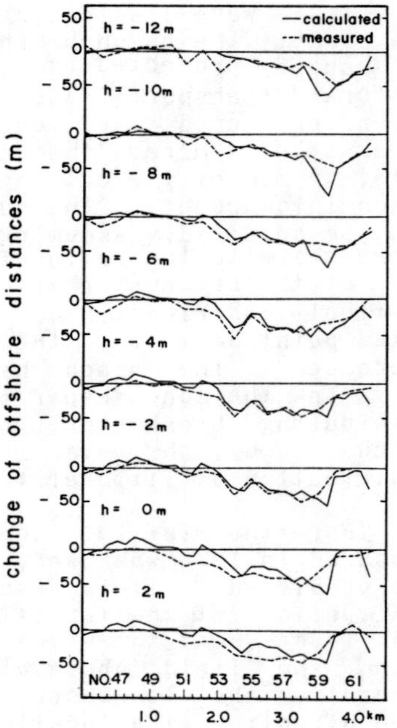

Fig.14 Comparison of measured and predicted changes of offshore distance after 8 years from the beginning.

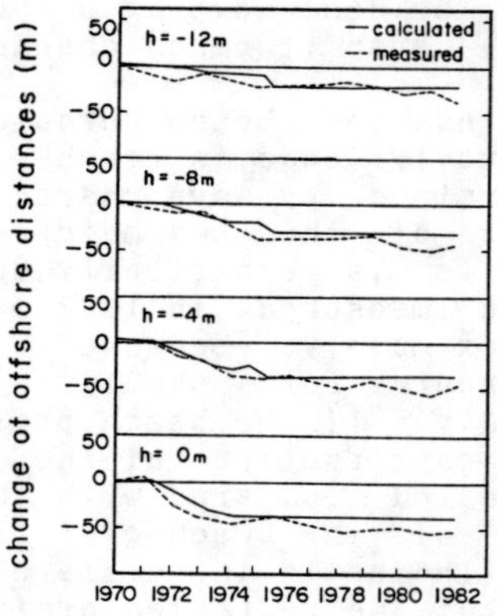

Fig.15 Comparison of temporal changes of measured and predicted offshore distances.

longshore distance along the coast is taken for the abscissa, and the changes in the offshore distances are compared. It is found that the predicted and measured changes of the offshore distances to some contours agree well. It is difficult to understand the process of the temporal beach changes from Figs.13 and 14 alone, although they are useful in understanding the spatial beach changes. Therefore, temporal changes of the offshore distances to some contours are expressed in Fig.15, where survey line No.58 is selected as an example. It is concluded that the process of beach changes is well predicted from the point of view of temporal beach changes.

V. CONCLUSIONS

1) The littoral transport heading eastward was obstructed due to the construction of the breakwater of Tagonoura Port located in the middle of the Fuji Coast, and the downdrift coast of the harbor was severely eroded. The maximum retreat of the shoreline between 1968 and 1983 was about 60m at survey line No.58.

2) Beach erosion was severe between No.60 and No.45, and recently, retreat of the shoreline is tending to decrease. This is due to the decrease of the littoral transport itself on the coast and due to the effect of the construction of wave-dissipating breakwaters.

3) The littoral transport rate passing through No.45 located east of Tagonoura Port was found to be about $1.17 \times 10^5 m^3/yr$ on an average between 1973 and 1983.

4) It is found that large part of the sediment eroded on the downdrift beach of Tagonoura Port is carried away toward the offshore zone, whose depth is between 20 through 30m.

5) A predictive model of three-dimensional beach changes was developed, in which the vertical distribution of littoral transport is taken into account and the effect of wave-dissipating breakwaters on beach changes can be evaluated. The predicted beach changes were in good agreement with observed ones.

REFERENCES

Kohno,H., T.Uda and Y.Yabusaki (1986) : On the scattering of concrete armour units of detached breakwaters due to waves, Proc. 20th Coastal Eng. Conf., ASCE, pp.1972-1986.

Shuto,N., J.Taguchi and T.Endo (1977) : On the movement of gravels on the Fuji Coast, Proc. 23th Japanese Conf. on Coastal Eng., pp.221-225.(in Japanese)

Uda,T. and T.Takeuchi (1985) : An analysis of beach erosion on Suruga Coast, Coastal Eng. in Japan, Vol.28,pp.161-172.

Uda,T., C.Agemori and N.Chujo (1986) : Beach changes caused by offshore dredging, Coastal Eng. in Japan, Vol. 29, pp. 215- 226.

Uda,T. and H.Saito (1987) : Beach erosion on the Ogawarako Coast and prediction of shoreline evolution, Proc. Coastal Sediments '87, ASCE, pp.484-499.

CHAPTER 144

TOWARDS MODELLING COASTAL SEDIMENT TRANSPORT

By Peter Nielsen*

Abstract

Sediment transport data from the field and laboratory tests are used to gain insight into two fundamental questions. Firstly: What is the relative importance of coexisting waves and currents for the resulting sediment transport? Secondly: Is the influence of grain size as strong as traditional models predict, or is it as weak as the empirical CERC-formula indicates?

Wave tank data reveal that the oscillatory velocity will in most cases determine the direction as well as the magnitude of the shore normal sediment transport, and wave flume data on shore normal transport as well as field data on littoral drift show weaker grain size dependance than traditional sediment transport models predict. It is suggested that wave dominance as well as weak grain size dependence are manifestations of the fact that the dominant transport mechanisms are often more organised than the diffusion process on which many traditional models are based.

1. Introduction

The relative importance of waves, or more specifically the oscillatory velocity component $\tilde{u}(z,t)$, and steady currents $\bar{u}(z)$ has been the subject of considerable research and discussion. The conclusion of most of the theoretical work has been that the currents should be the most important with respect to the actual transport while u may well be the dominant factor in the entrainment process. Recent experimental work has shown however that the transport component resulting from the product of the oscillatory components of concentrations and velocities ($\overline{\tilde{u}\tilde{c}}$) is often more important than $\bar{u}\,\bar{c}$. Examples of this evidence will be discussed below.

In this context it is of course important to acknowledge the fact that different types of steady currents will have different potential for transporting sediments depending on their strength close to the bed where the sediment concentrations are largest. For example, the steady boundary layer currents which are generated by the waves will generally be much stronger at the bed than currents which are driven by a uniform pressure gradient such as tidal currents or undertow, but they may not be apparent from measurements taken above the wave boundary layer.

*: Coast and Rivers Branch, Public Works Department, 140 Phillip Street, Sydney 2000, Australia.

Some insight into the problem of grain size dependance was gained by the study of Nielsen (1988) which tested three different models against wave flume data on shore normal transport. The first of these models is a diffusion model developed along the lines suggested by Nielsen et al (1978) and by Nielsen (1979). The second model is inspired by the "heuristic model" of Dean (1973) and the third is based on a simple "grab and dump" concept. The third model is much simpler than the two others, but it is also more reliable. More specifically, it shows a more realistic grain size dependence of the resulting transport than the two more traditional models. The traditional models tend to predict by far too strong decline of the sediment transport rate with increasing grain size.

Although the logical connection is not very direct it is interesting to note that a similar situation can be observed with respect to the total longshore transport or littoral drift. That is, existing traditional models like that of Deigaard et al (1987), which should be valid for a fairly common type of surf zones, predict a strong grain size dependence while the data, as presented in the Shore Protection Manual show no grain size dependence at all (see Figure 1).

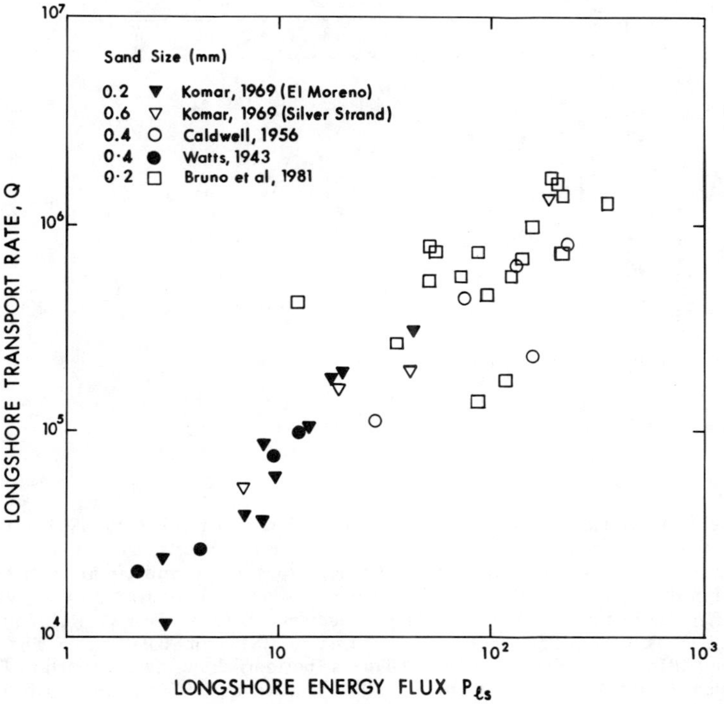

Figure 1: Littoral drift data as function of the longshore wave energy flux (from CERC 1984). The data shows no significant grain size dependance for the measured littoral drift.

To this which we may call "The C.E.R.C. formula paradox" one of the following explanations may apply:

1: The data is unreliable.

2: The processes modelled by Deigaard et al (1987) were not dominant in the surf zones where the measurements were made.

2. Relative Importance of uc and uc

Figure 2 is an illustration of the relative importance of waves ($\tilde{u}(z,t)$) and a superimposed steady current ($\bar{u}(z)$) for the sediment transport along a wave flume.

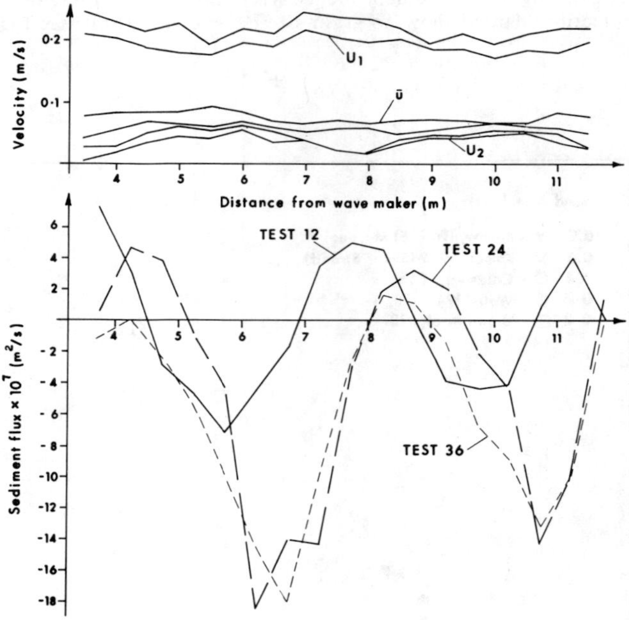

Figure 2: Wave flume sediment transport experiments by Schepers (1978). The hydrodynamic conditions (top) were similar in the three tests: T=1.5 s, and D=0.30 m. The envelopes of the velocity magnitues u, U_1, and U_2 are plotted as function of distance from the wave maker and were all measured 0.1m above the bed. The sediments had the following characteristics (d,w): Test 12: (0.125mm, 0.010m/s), Test 24: (0.250mm, 0.028 m/s), Test 36: (0.465mm, 0.050 m/s). The corresponding measured net sediment fluxes (bottom) show three important facts: (1) Even though the measured current u(0.1) is always positive (shoreward), the sediment transport is predominantly negative. (2) The measured transport rates are of very similar magnitude despite the large difference in grain sizes. (3) The transport pattern for the coarser sand (d/A>0.004) is shifted relative to that of the finest, probably because acceleration effects become more important relative to velocity magnitude with increasing grain size.

In the top part of the figure are shown the measured velocity components $\overline{u}$, U_1 and U_2, where U_1 is the amplitude of the fundamental mode of the wave motion and U_2 is the amplitude of the second harmonic. The velocities were measured ten centimetres above the bed. The bottom part of the figure shows the corresponding, measured sediment transport rates. The experiments were carried out at the Delft University of Technology and have been described in more detail by van de Graaff and Tilmans (1980).

The first thing to note about the measured sediment transport rates is that although the steady current ten centimetres above the bed was at all points shoreward, the measured transport rates were predominantly offshore. Thus, the current velocity ten centimetres above the bed is no indicator for the direction of the resulting sediment transport.

For three dimensional situations there is also evidence that the waves are more dominant than one might have expected with respect to transport capacity. For example Davison (1987) has reported on tracer experiments where the centroid velocity as well as the larger axes of the "tracer ellipses" always tended to be in the wave direction rather than in the direction of the tidal currents. It is not clear from these experiments however if the transport is actually carried by u or by the steady wave induced boundary layer drift.

3. Sediment Transport in Pure Oscillatory Flow

In order to be able to model coastal sediment transport in general, it is important to understand how sand is transported in a purely oscillatory flow, i.e. a flow where $\overline{u} \equiv 0$; but it can be difficult to generate such a flow in wave tanks because of wave induced mass transport. Most important in this context is the boundary layer drift under progressive waves which was described by Longuet-Higgins (1956).

However, in oscillating water tunnels, where the water motion is horizontally uniform this drift should not occur. Therefore, the oscillatory water tunnel (OWT) provides the best experimental facility for studying sediment transport in purely oscillatory flow.

Sato (1986) performed sediment transport experiments in an OWT which are of interest in this context. He found that with water velocities of the "Stokes-wave-type" i.e. with

$$u(t) = U_1 \cos \omega t + U_2 \cos 2\omega t \qquad (1)$$

the resulting sediment transport was always "seaward" that is in the direction opposite to the largest instantaneous velocity.

This result is to some extent in conflict with the field observations of Doering and Bowen (1988) who found that the correlation between instantaneous velocities and concentrations was positive which means that the net sediment flux at the measuring point tended to be in the direction of the largest instantaneous velocities.

When the two authors found opposite results, it is most likely due to the fact that the bed forms were different in their experiments.

In Sato's tests the bed was covered by regular, sharp crested ripples while Doering and Bowen's field tests would have involved bed forms that were flatter and more rounded.

The trend found by Sato for sediment transport over sharp crested ripples is also apparent in the numerous wave flume tests of Vellinga (1975), Schepers (1978) and Tilmans (1979) which were analysed and modelled by Nielsen (1988). That is, those test show a clear tendency for the resulting transport to go in the direction opposite to that of the largest velocities. The reason for this is that the transport over sharp crested ripples is a two step process where the sand is not carried very far by the velocities which activate it. It is trapped behind the ripple crest and then later released to be transported in the opposite direction after the flow reversal.

While the results from Sato (1986) and those analysed by Nielsen (1988) show that velocity-moment-models like that of Bailard and Inman (1981) predict transport in the wrong direction over sharp crested ripples, such models might apply over more rounded bed forms.

4. Grain Size Dependence of Shore Normal Transport

Nielsen (1988) used three different approaches for modelling the Dutch flume experiments. The first was the traditional diffusion approach, the second was a heuristic entrainment and settling approach similar to that of Dean (1973) and the third was an ultra simplistic "grab and dump" model.

All of the models were quantified in terms of parameters from the steady concentration profiles only; that is, the reference concentration Co and the vertical length scale L_s defined by:

$$\bar{c}(z) = C_o \exp[-z/L_s] \qquad (2)$$

All of the three modelling approaches were reasonably successful and this shows the comforting fact that the immense complexity of the variation of c(z,t) needs not be considered explicitly in order to obtain reasonable results. Only the time averaged concentration profile $\bar{c}(z)$ is needed, and fortunately experimental data on $\bar{c}(z)$ is abundantly available.

The model results can be summarised as follows:

Diffusion model: $$Q = C_o L_s U_1 S\left(\frac{Umax}{Umin}\right) F_d\left(\frac{Ls}{wT}\right)$$

Heuristic entrainment model: $$Q = C_o L_s U_1 S\left(\frac{Umax}{Umin}\right) F_e\left(\frac{Ls}{wT}\right)$$

Grab & Dump model: $$Q = C_o wAS\left(\frac{Umax}{Umin}\right)$$

Where U_1 is the velocity amplitude of the fundamental mode of the wave motion, U_{max} and U_{min} are the extreme shoreward and seaward velocities respectively, w is the sediment settling velocity, A is the water particle semi excursion and T is the wave period.

The skewness function S(Umax/Umin) was the same in all models and the two functions F_d and F_e of the time scale ratio L_s/wT are quantitatively similar.

From these formulae we can see how the resulting sediment flux Q depends on grain size with all other things being equal. Because C_o and L_s are both decreasing functions of the grain size we see that the two first models predict a fairly strong decrease in Q for increasing grain size, while the "grab and dump" model predicts a more moderate grain size dependance. It turns out that the data basically confirm the weak grain size dependance predicted by the grab and dump model, see Figure 2 and Nielsen (1988) for more detail.

In essence the "grab and dump model" differs from the two more traditional models by assuming a more organised mode of transport. It basically assumes that the sand is moved in parcels. Inspection of the transport process for fairly coarse sand ($U_1/w<10$) over rippled beds confirms that it does in fact have this kind of nature. For finer sand, the organised nature is not obvious, but the "grab and dump model" will still work (see Nielsen 1988).

The trapping mechanisms by which a vortex dominated flow can keep sand in suspension and transport it very effectively were described by Nielsen (1984). However, for fine sand ($U_1/w>10$) the sand will on the average stay in suspension much longer than the life time of the vortices, and the success of the "grab and dump model" may thus be somewhat fortuitious for fine sand.

5. CONCLUSIONS

From the presently available data the following conclusions can be drawn about the relative importance of steady and oscillatory flow components with respect to sediment transport. For non breaking waves over horizontal beds the influence of, even fairly strong, currents is insignificant, see Figure 2.

Davison (1987) found that in an environment with tidal currents of similar though slightly smaller magnitude running perpendicular to the waves, the sediment transport inferred from radioactive tracers was always in the wave direction.

The role of the wave induced boundary layer drift over natural sand beds is however virtually unresolved. This is basically because these current themselves are not well known. The model of Longuet-Higgins (1956) is not known to be valid for turbulent boundary layers, and more recent "turbulent" models like that of Jacobs (1984) are not known to be valid for the relative roughness range which is typical for natural sand beds.

The transport due to purely (or almost purely) oscillatory flows over sharp crested bed forms goes in the direction apposite to that of the largest instantaneous velocities. This tendency may be reversed for rounded bed forms and flat beds so that velocity-movement-models like that of Bailard and Inman (1981) could apply to such situations.

Laboratory data on shore normal sediment transport (Figure 2) show a weaker grain size dependance than classical transport models would predict, and interestingly, the same is true for the litoral drift data shown in Figure 1. Judging from the results of Nielsen (1988) it might be inferred that the weak grain size dependence is due to the efficiency of highly organised transport modes which dominate the transport of coarser sediments.

It is not obvious that the implications from Nielsen (1988)'s study can be applied directly to the "CERC formula paradox" for littoral drift but it is possible that more organised (than diffusion) modes of transport tend to dominate for coarser sand and that such dominance leads to weak grain size dependance for littoral drift as well as for shore normal transport.

REFERENCES

Bailard, J.A. and D.L. Inman (1981): An energetics bedload model for a plane, sloping beach : local transport. Journal of Geophysical Research, Vol 86, pp 2035-2043.

C.E.R.C. (1984): Shore Protection Manual. Fourth Edition, Vol.1.

Davison, A. (1987): A radioactive tracer study of sand-seagrass dynamics, in the nearshore region of Holdfast Bay, South Australia. Australian nuclear science and technology organisation. Lucas Heights Laboratories. ANSTO/C67.

Dean, R.G., (1973): Heuristic model of sand transport in the surf zone. Proc. First Australasian Conf. Coastal Eng., Sydney, pp. 208-214.

Deigaard, R., Fredsoe, J. and Hedegaard I.B., (1986): Mathematical model for littoral drift. Journal of waterway, port, coastal and ocean engineering. (A.S.C.E.), Vol. 112, No. 3, pp. 351-369.

Doering, J.C. and A.J. Bowen (1988): Wave induced flow and nearshore suspended sediment. Proc. 21st Int. Conf. on Coastal Engineering, Torremolios.

Graaff, J. van de and Tilmans, W.M.K., (1980): Sand transport by waves. Proc. 17th Int. Conf. Coastal Eng., Sydney, pp. 1140-1157.

Jacobs, S.J., (1984): Mass transport in a turbulent boundary layer under a progressive water wave. Journal of Fluid Mechanics, vol 146, pp 303-312.

Longuet-Higgins, M.S., (1956): The mechanics of the boundary layer near the bottom in a progressive wave. Proc. 6th Inc. Conf. on Coastal Engineering, Florida, pp 184-193.

Nielsen, P., (1979): Some basic concepts of wave sediment transport. Series Paper 20, Institute of Hydrodynamics and Hydraulic Engineering (ISVA), Technical University of Denmark.

Nielsen, P., (1984): On the motion of suspended sand particles. Journal of Geophysical Research, Vol 89, No C1, pp 616-626. Erata: Vol 90, No C2, pp 3255-3256.

Nielsen, P., (1988): Three simple models of wave sediment transport. Coastal Engineering Vol. 12, pp.43-62.

Nielsen, P., Svendsen, I.A. and Staub, C., (1978): Onshore-offshore sediment transport on a beach. Proc. 16th Int.Conf. Coastal Eng., Hamburg, pp. 1475-1492.

Sato S. (1986): Oscillatory boundary layer flow and sand movement over ripples. Ph. D. Thesis, Dept. of Civil Engineering, University of Tokyo.

Schepers, J.D., (1978): Zandtransport onder invloed van golven en een eenparige stroom bij varierende korreldiameter. Masters Thesis, Delft University of Technology.

Tilmans, W., (1979): Zandtransport in de golfrichting in relatief ondiep water. Masters Thesis, Delft University of Technology.

Vellinga, P., (1975): Zandtransport door golven en de invloed van stoorgolven.

CHAPTER 145

LONGSHORE SEDIMENT TRANSPORT RATE vs. CROSS - SHORE DISTRIBUTION OF SEDIMENT GRAIN SIZES

C.I.Moutzouris (*)

ABSTRACT

Existing models for longshore sediment transport rate computations assume the sediment grain size and grain size-related parameters to be uniform in both the cross-shore and longshore directions. Field results from tideless beaches, which are briefly described in the paper, show that the latter change in both directions due to changing wave energy-levels. The sensitivity analysis described in the paper shows that both longshore current and transport rate computations are sensitive to the cross-shore changes in grain size. Finally, a modified linearity coefficient for the wave power equation is proposed based upon the cross-shore distributions of grain size as found in nature.

1. INTRODUCTION

It is now widely believed that sediment grain size is an important parameter in wave-induced sediment transport processes in the surf zone. As a result, attempts are increasingly being made to incorporate the grain size in models of longshore sediment transport rate computations. Some examples are as follows. Dean et al.(1983) plotted results from various previous studies and found that the linearity coefficient K of the empirical wave power equation increases with decreasing sediment size D. Theoretical predictions by Deigaard et al.(1986) indicate a similar decrease in K with increasing D. Detailed predictor models, such as the Bijker (1971) formula, have already incorporated the grain size and a number of other grain-related parameters.

In all existing models, however, grain size and grain size-related parameters are assumed to be uniform in both the cross- and longshore directions in the surf zone. Recent results from field measurements on tideless beaches along the Greek coast conducted by the team of the Civil Engineering Department, National Technical University of Athens, show clearly that grain size and grain size-related parameters

(*)National Technical University of Athens, Civil Engineering Department, Patission 42, 10682 Athens, Greece.

change in both directions, due to changing wave energy-levels. Changes are more predominant in the cross-shore direction. Maximum grain size is observed in zones of maximum energy, such as the breaker zone and zones of wave convergence. Changes are in most cases so drastic that selecting the values to be introduced into the models is a difficult task.

It therefore becomes clear that grain size and grain size-related parameters can no longer be considered to be uniform across the surf zone. Realistic models should not only introduce the influence of grain size but also the cross-shore changes.

In the present paper, a synopsis of the above-mentioned field results is presented and a sensitivity analysis of longshore sediment transport computations on the cross-shore changes of grain size is made.Finally, a modified linearity coefficient for the wave power equation is proposed, which is based on the cross-shore distribution of grain sizes, as found in nature.

2. CROSS-SHORE DISTRIBUTION OF GRAIN-RELATED PARAMETERS

Field measurements have been/are being made along various Greek beaches:Kokkino Limanaki,Marathonas,Linoperamata,Platanias,Plakias and Rio. Each beach has its own specific characteristics:Kokkino Limanaki is a small-scale pocket beach with a strongly arcuate coastline 310m long. It is exposed to medium wave energy from two dominant directions. Marathonas has a straight coastline and is exposed to medium wave energy from one direction only. Linoperamata is also a straight beach but receives high wave energy. Longshore transport occurs to the left and to the right. Platanias has a slightly arcuate coastline with medium to high wave energy. The equilibrium of the latter beach was recently disturbed by the construction of a small fishing harbour,which caused local erosion and accretion. Plakias is a large-scale pocket beach exposed to high wave energy. Finally, Rio is a shingle beach along a protruding headland which receives medium wave energy from both directions. Characteristics which are common include microtidal environments (maximum range:25 to 30 cm) and locally generated sea states. Therefore, sediment distributions on the sea bed are almost exclusively wave-induced, which is considered to be highly advantageous.

Cross-shore distributions of grain size and grain size-related parameters are obtained from spot sediment samples collected with hand-operated grab samplers. Samples are limited to the upper 5cm of bed material so as to collect the sediment deposited during only the previous sea state.Samples are analysed in the laboratory. Grain sizes are computed by sieving according to the ASTM standards.

In the following paragraphs, a synoptical overview of the main results is presented.They concern wave-induced cross-shore distributions of grain size-related parameters, which reflect the hydrodynamic conditions prevailing across-shore.

A typical grain size distribution is shown in Fig.1, as found on the Marathonas beach.Noticeable differences in the

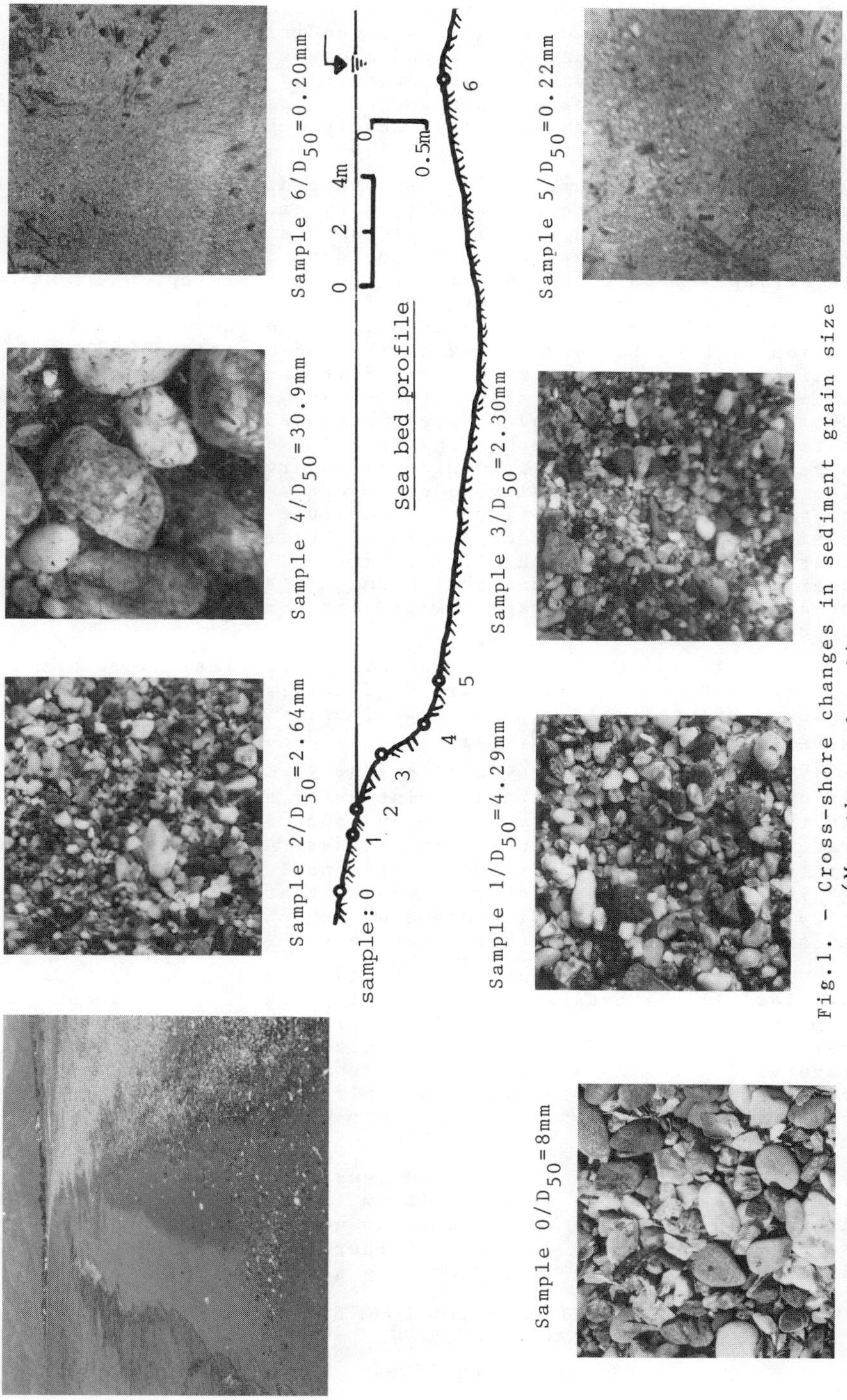

Fig.1. - Cross-shore changes in sediment grain size (Marathonas beach)

cross-shore sediment size distribution are observed throughout the beaches and under all wave conditions. A regular sediment sorting is detected in the cross-shore direction due to differing hydrodynamic loading. Grains are found to be coarser and worse sorted in zones of increased wave energy than in zones with lower energy levels, where grains are finer and better sorted.

Fig.2 shows a number of cross-shore distributions of statistical mean diameters for various types of beaches. D_{50} is the sediment grain diameter exceeded in size by 50% by weight in the sample population. M_z is the graphic mean diameter, as defined by Folk and Ward,1957. The plunge step, where wave breaking occurs, is always found to be composed of the coarsest and worst sorted material, due to the increased action of breaking waves (Moutzouris and Kypraios,1987). Grains are coarser and worse sorted at the toe of the step than at the crest. Therefore, the toe of the plunge step and the zone immediately seawards are composed of the coarsest material across the beach. From the plunge step shorewards, a regular decrease in grain size is observed. Sediment on the beach face is normally found to decrease in size and improve in sorting. Exceptions to this rule are due to the occasional presence of beach cusps, the terrigenous supply of coarser material, and the berm formation. In these cases, coarser material is found than normally. The offshore zone is covered with fine and rather well sorted material due to the decreased wave action.

Our field data indicate that all statistical diameters D_n (sediment grain diameter exceeded in size by 1-n% by weight) have a qualitatively similar distribution in the cross-shore direction as D_{50} (see Fig.3).

The settling velocity w of grains in water depends mainly upon the physical characteristics of the particles (such as the diameter, density, shape and roundness). Water temperature has a minor influence in beaches. As density is more or less constant across a beach, the most characteristic is grain size. It is, therefore, no wonder that the settling velocity of grains with median diameter D_{50} shows a qualitatively similar distribution across-shore as D_{50} (see Fig.3).

The porosity p is found to depend mainly upon the mode of flow. In the breaker zone, wave impacts on the bed cause compaction of the sediment and porosity decreases. At the beach face, water flow is mostly parallel to the sediment layer and porosity decreases. As a result, porosity is found to change across-shore. The maximum values observed were of the order of 0.40, which is the most common value of p in models (see Fig.3).

The distribution of the mean density ρ_s of the bed material across-shore is almost uniform (see Fig.3), which means that grain size and mean density do not appear to be inversely proportional to one another. Therefore, ρ_s is not influenced by the wave-energy changes across-shore.

3. SENSITIVITY OF LONGSHORE TRANSPORT RATES ON CROSS-SHORE CHANGES OF GRAIN-RELATED PARAMETERS

In the third part of the paper, the sensitivity of two

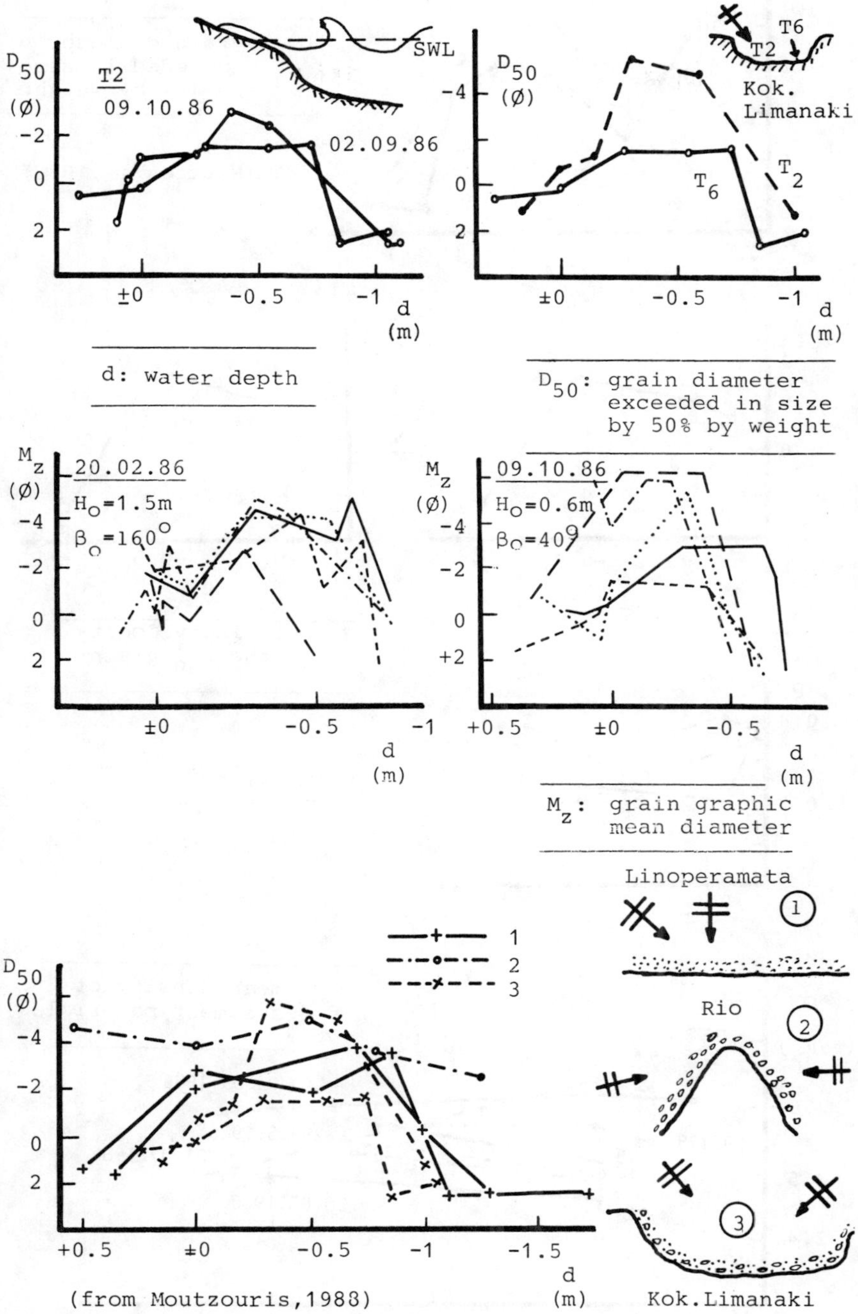

Fig.2.- Cross-shore distrbution of characteristic statistical mean grain diameters

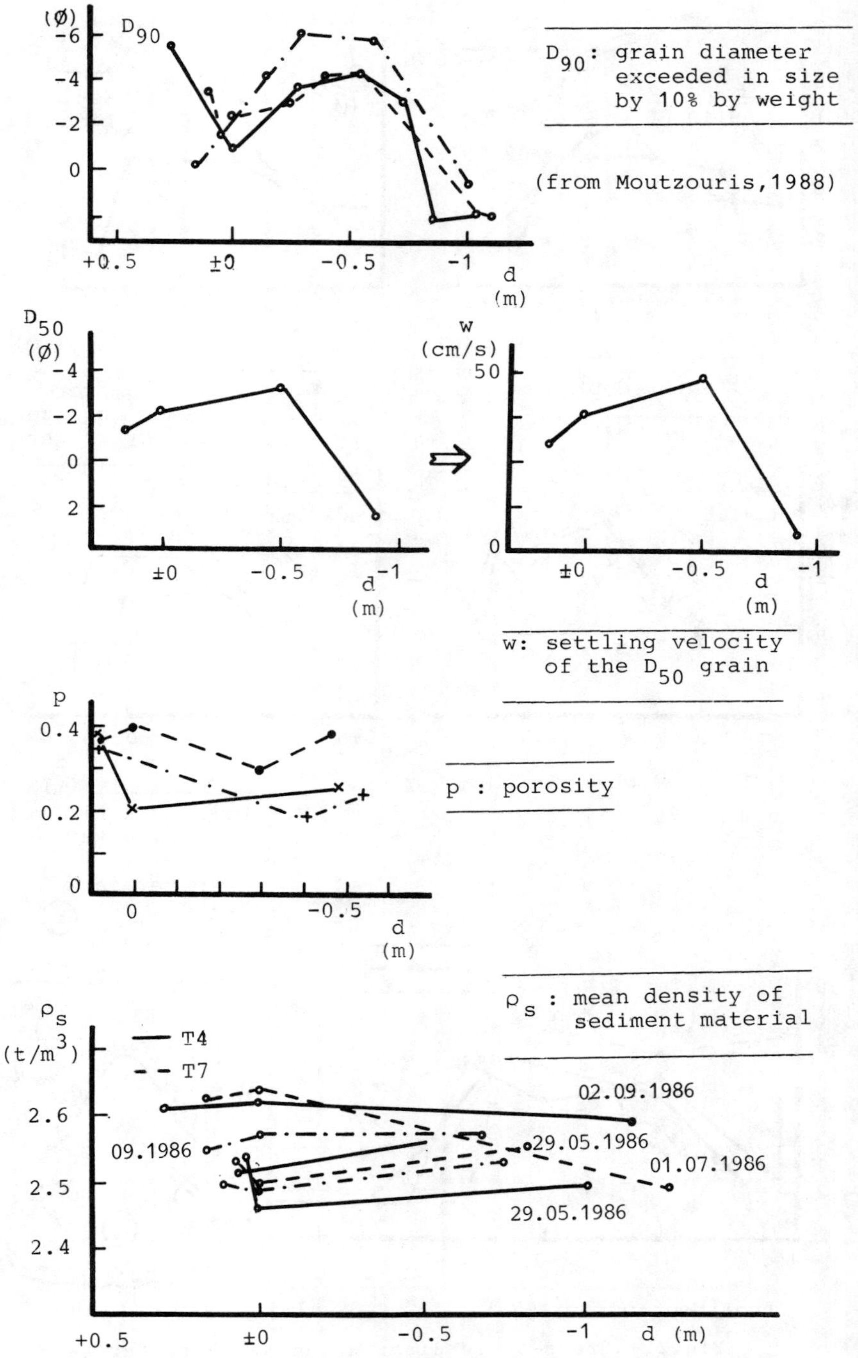

Fig.3.-Cross-shore distribution of various grain size-related parameters (Kok.Limanaki beach)

commonly used longshore sediment transport predictors on the cross-shore distribution of grain size and grain-related parameters is analysed. The predictors examined are the wave power equation and the Bijker formula.

The empirical wave power equation for the total longshore sediment transport rate S_t gives the spatially integrated immersed weight rate as a linear function of the total wave power available for transporting sediments.The best known and most widely used value of the non-dimensional linearity coefficient K is reported by Komar and Inman (1970) and is independent of the sediment characteristics. Their value was derived for sand size within the range of 0.18 and 0.6mm.

It has been increasingly argued recently that K could not be a constant but should depend upon some other parameters and most probably upon the sediment size. Dean et al (1983) plotted the values of K from various studies (Watts, Galdwell,Bruno and Gable, Komar and Inman,Johnson and Galvin, Moore and Cole, and Dean et al.) and found that K increases with decreasing sediment size D. Theoretical predictions by Deigaard et al.(1986) also indicate a decrease of K with increasing D. Komar (1988) does not seem to agree with such a dependency.

Fig.4 shows the cross-shore distribution of D_{50},as found on one of our tideless beaches and the corresponding values of K, as computed from Dean et al (1983). X denotes the distance offshore, measured from the still water line. It can be seen that the longshore rate decreases by a factor of almost 3 if the largest diameter (0.75mm) is taken into account instead of the smallest one (0.37cm). The smallest diameter on the beach (0.37mm) gives almost 3 times higher transport rates than the largest diameter (0.75mm). It is therefore concluded that longshore transport rates computed with the wave power equation are very sensitive to the cross-shore changes in sediment grain size, if the linearity coefficient K is taken to be grain size-dependent. The Komar and Inman (1970) model would give a constant transport rate independent of the grain size distribution across-shore.

The Bijker formula computes the total longshore sediment rate as the sum of the bed and the suspended loads. Numerous grain size-related parameters are introduced in the formula.

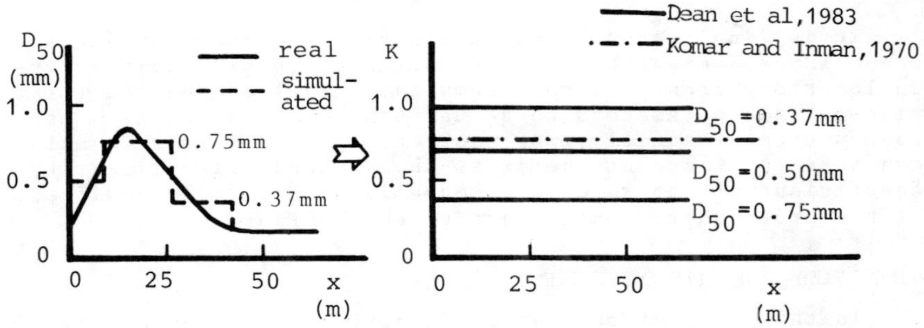

Fig.4. - Cross-shore distribution of the linearity coefficient

Among others are the D_{50} and D_{90} diameters, the Chezy roughness and wave friction coefficients (or factors), the grain settling velocity, the longshore current velocity, etc.

The sensitivity of the Bijker formula on the cross-shore distribution of grain size-related parameters is shown in Fig.5, where S_t is the volumetric transport rate. These rates were computed with (a) a uniform grain size across the surf zone, as is common practice in transport rate computations, and (b) changing distributions, as found in the beaches of our field measurements. The results show that the Bijker formula is very sensitive to the cross-shore changes in size-related parameters. Of major influence upon the results is the sensitivity of the longshore current model introduced in the Bijker formula on the cross-shore changes in the grain size-related parameters. Fig.6 shows the longshore current velocity according to the Longuet-Higgins (1970) and Komar (1975) models, as computed with grain size distributions observed in the field.

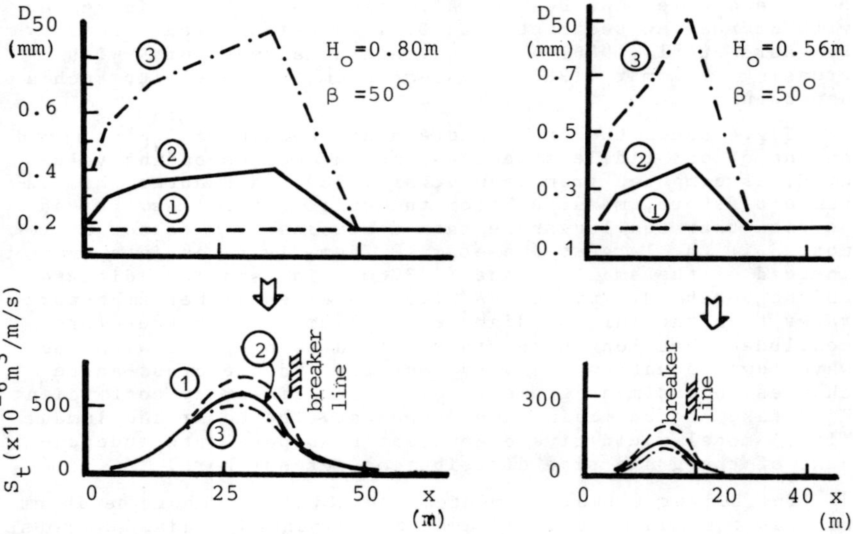

Fig.5. - Sensitivity of the Bijker formula on the cross-shore changes in grain size-related parameters

It is finally concluded that neglecting the sediment size on a beach might be a first source of considerable errors in longshore transport rate computations. Assuming the grain size-related parameters to be uniform across-shore is a second source of equally important errors. The Bijker formula can take into account the cross-shore changes. The linearity coefficient of the wave power equation needs to be modified if the cross-shore changes are to be incorporated.

4. MODIFIED LINEARITY COEFFICIENT

In this part of the paper, a modified linearity coefficient K_m is proposed for use in the computations with the wave power equation. The modification seems necessary in view of

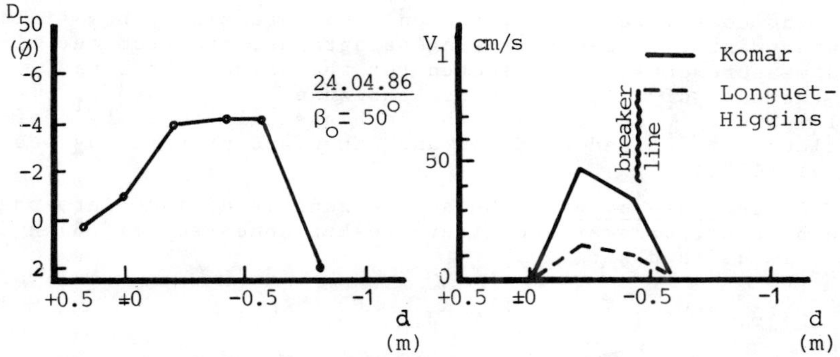

Fig.6.- Sensitivity of longshore current on the cross-shore changes in grain size

the results presented above.

The active zone of a beach is divided into a number of n successive sub-zones, with almost uniform grain size in each one. K_m is defined as the sum of the ponderated coefficients K, as proposed by Dean et al.(1983). In the general case of n sub-zones, the resulting $K_{m,n}$ is defined as:

$$K_{m,n} = \sum_{i=1}^{n} C_{n,i} \; K_{n,i}$$

with: $\sum_{i=1}^{n} C_{n,i} = 1$ and $C_{n,i} = \frac{l_{n,i}}{l_a}$

$l_{n,i}$ is the width of sub-zone i and l_a is the length of the active zone (see Fig.7).

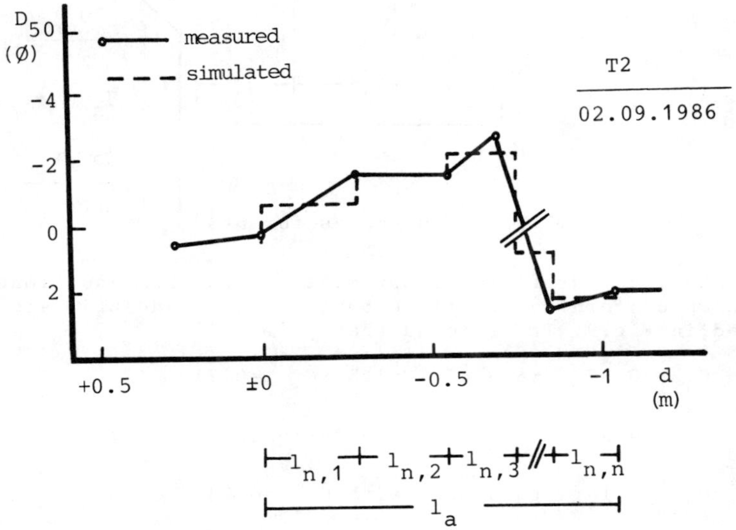

Fig.7.- Definition sketch

The coefficients $C_{n,i}$ can only be determined from field measurements. In the following paragraphs, data from our field measurements are presented for the cases of n=3 and n=2. In the case of n=1, the unique sub-zone has a width $l_{1,1}$ ($=l_a$) and $K_{m,1}$ coincides with the classical linearity coefficient, as defined by Komar and Inman and proposed by Dean et al. (1983).

In the case of n=3, the active zone is divided into the pre-breaker, breaker and after-breaker zones (see Fig.8) and $K_{m,3}$ is defined as:

$$K_{m,3} = C_{3,1}\,K_{3,1} + C_{3,2}\,K_{3,2} + C_{3,3}\,K_{3,3}$$

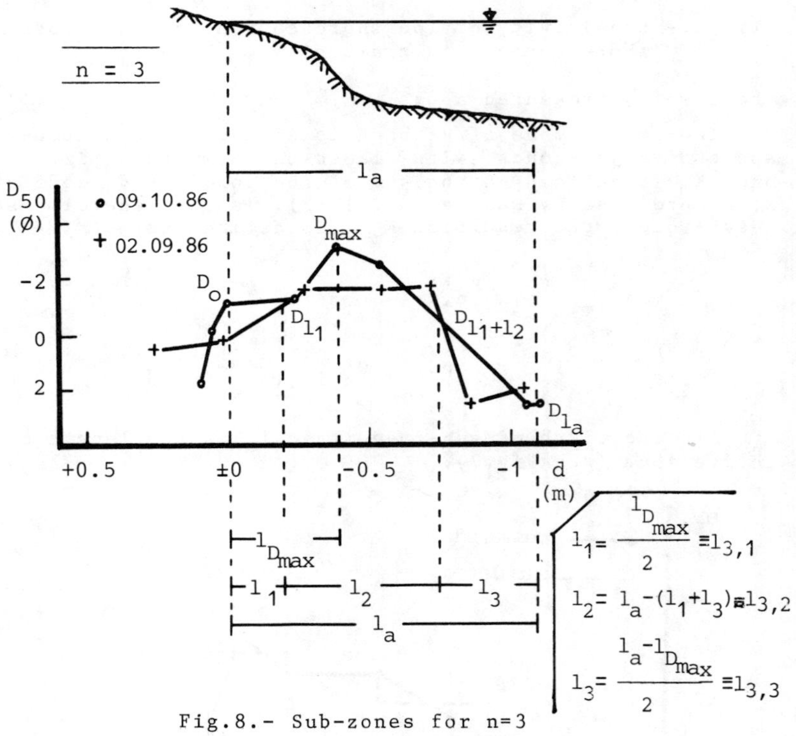

Fig.8.- Sub-zones for n=3

Data from our field measurements on the Kokkino Limanaki beach concerning coefficients $C_{3,1}$, $C_{3,2}$ and $C_{3,3}$ are plotted in Fig.9. It is found that:

$C_{3,1}=0.07$ to 0.47 with $\bar{C}_{3,1}=0.21$

$C_{3,2}=0.50$ (by definition)

$C_{3,3}=0.16$ to 0.38 with $\bar{C}_{3,3}=0.28$

The following modified linearity coefficient could be proposed for the cases presented in Fig.9:

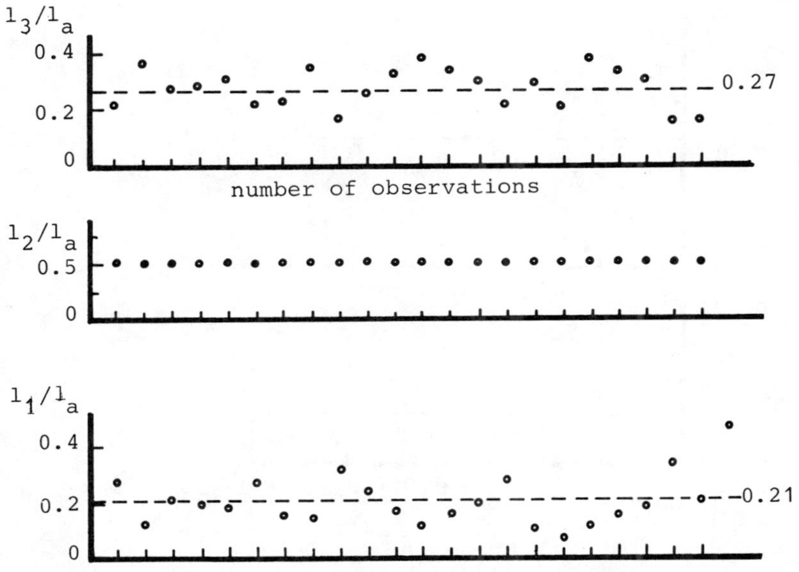

Fig.9. - Lengths of the sub-zones (n=3)
(Kokkino Limanaki beach)

$$K_{m,3}=0.2K_{3,1}+0.5K_{3,2}+0.3K_{3,3}$$

$K_{3,1}$, $K_{3,2}$ and $K_{3,3}$ are linearity coefficients in sub-zones with almost uniform grain size. By adopting the results of Dean et al (1983), $K_{3,1}$, $K_{3,2}$ and $K_{3,3}$ are obtained as functions of the uniform grain size.

An estimation of $K_{m,3}$ is now made for the same beach considered above. The field measurements show the following range of values (see Fig.10):

$$\frac{D_{3,1}}{D_{3,2}}=0.11 \quad \text{to} \quad 0.33 \quad \text{with} \quad \overline{\left(\frac{D_{3,1}}{D_{3,2}}\right)}=0.23$$

$$\frac{D_{3,3}}{D_{3,2}}=0.04 \quad \text{to} \quad 0.20 \quad \text{with} \quad \overline{\left(\frac{D_{3,3}}{D_{3,2}}\right)}=0.10$$

It is repeated that these values reflect conditions, which prevailed along the beach during the period of field measurements. Therefore, it is necessary to examine the sensitivity of $K_{m,3}$ on the changes in $D_{3,i}$. Fig.11 shows the changes in $K_{m,3}$. It is found that $K_{m,3}$ decreases with increasing grain diameters and/or cross-shore size sorting. The value proposed by Komar and Inman,1970,for the linearity coefficient is found to be rather higher than $K_{m,3}$. The values proposed by Dean et al,1983, coincide with $K_{m,3}$ in the special case of $D_{3,1}=D_{3,2}=D_{3,3}$.

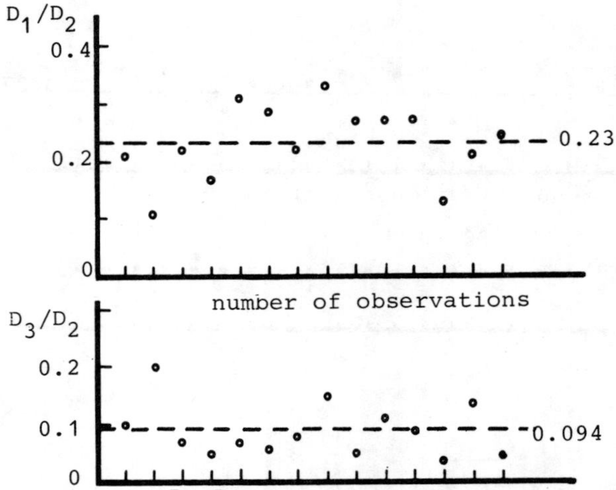

$$D_1 = \frac{D_o + D_{l_1}}{2} \qquad D_2 = \frac{1}{2}\left(\frac{D_{l_1} + D_{max}}{2} + \frac{D_{l_1+l_2} + D_{max}}{2}\right)$$

$$D_3 = \frac{D_{la} + D_{l_1+l_2}}{2}$$

Fig.10. - Grain diameters in the sub-zones (n=3)

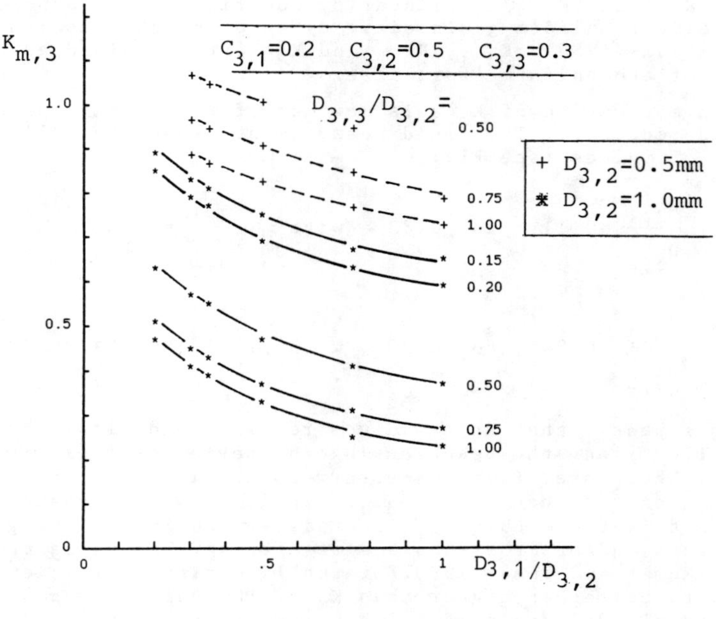

Fig.11. - Modified linearity coefficient (n=3)

In the case of n=2, it is proposed to divide the active zone into a pre-breaker zone and an after-breaker zone. The limit between the two zones is situated in the transition zone at the toe of the plunge step (see Fig.12). $K_{m,2}$ is now defined as:

$$K_{m,2}=C_{2,1}\ K_{2,1}+C_{2,2}\ K_{2,2}$$

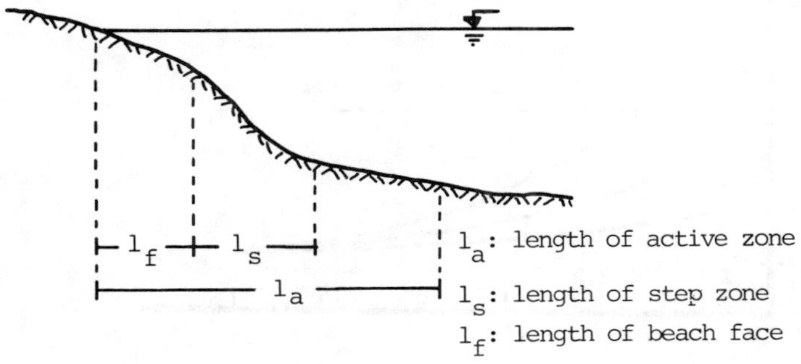

Fig.12. - Sub-zones for n=2

The after-breaker zone is comprised of the plunge step and the beach face with widths l_s and l_f, respectively. Data from the Kokkino Limanaki beach show that (see Fig.13):

$$\frac{l_s}{l_a} = 0.02 \quad \text{to} \quad 0.14 \quad , \qquad \frac{l_f}{l_a} = 0.04 \quad \text{to} \quad 0.41$$

and

$$\frac{l_s+l_f}{l_a} = 0.08 \ \text{to} \ 0.48$$

Therefore:

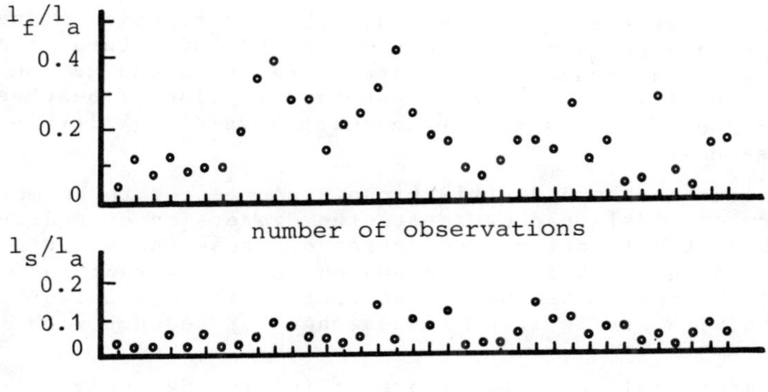

Fig.13. - Lengths of sub-zones (n=2)

$C_{2,1} = 0.1$ to 0.5 and $C_{2,2} = 0.9$ to 0.5

The sensitivity of $K_{m,2}$ on $D_{2,1}/D_{2,2}C_{2,1}$ and $C_{2,2}$ is examined in Fig.14.

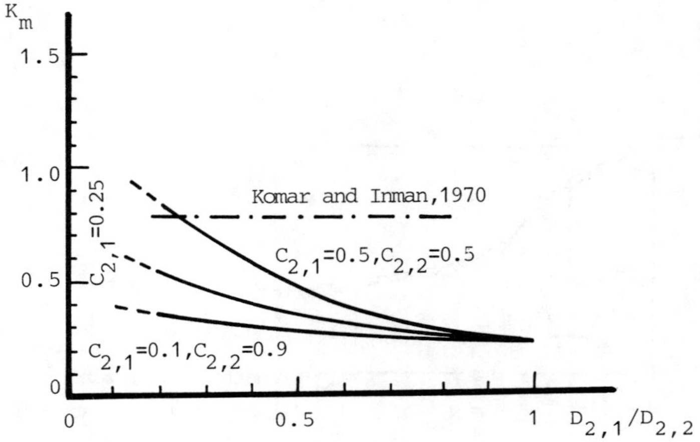

Fig.14. - Modified linearity coefficient (n=2)

5.DISCUSSION

The field results briefly presented in the paper are from tideless beaches with mixed sedimentary environments. Although beaches in most parts of the world are of this type, the research effort devoted to them has been too small compared to the extensive work carried out for beaches covered with fine sand. Special interest seems to have been given recently from researchers around the world (f.i.Komar,1988, Moutzouris,1988,Nicholls and Webber,1988, van der Meer,1988, etc) to beaches which are not exclusively sandy.

The results clearly show that the distribution of sediment grain size is non-uniform on the bed of natural beaches. Grain size and grain size-related parameters change considerably, especially in the cross-shore direction of beaches.The main reason for this is the changing wave-energy levels across-shore.

The cross-shore distributions are expected to be more uniform on tidal beaches due to the dispersion of sediment caused by tidal action. The degree of cross-shore grain uniformity on a beach is also dependent upon the range of sizes met on the beach. Sediment distribution is more uniform for beaches with narrow size ranges than for beaches with broad ranges.

Hydrodynamic conditions prevailing in the active zone of a beach is the main driving mechanism for longshore sediment transport. Transport rates are also believed to depend upon the beach slope. Sediment size distribution found on a seabed immediately after a sea state reflects the hydrodynamic conditions during the sea state. Introducing the cross-shore distribution into a model for sediment transport computations is in a way equivalent to introducing the cross-shore hydrodynamic conditions and the beach slope because slopes are sediment size-dependent.

The sensitivity of longshore sediment transport models described in this paper is a consequence of the above statements. It is believed that no model can give realistic results for beaches with mixed sedimentary environments unless the cross-shore changes in grain size and grain size-related parameters are taken into account. Further research is necessary in relation to the latter considerations, as has been attempted in the present paper.

REFERENCES

Bijker,E.W.(1971).Longshore transport computations.Jour.Wat. Harb.Coast. Engin.Div.,ASCE,Vol.97,WW4, pp.687-701

Dean,R.G.,Berek,E.P.,Gable,C.G.,and Seymour,R.J.(1983).Longshore transport determined by an efficient trap.Proc. 18th ICCE, Cape Town,Vol.II, pp.954-968

Deigaard,R.,Fradsøe,J. and Hedegaard,I.B.(1986).Mathematical model for littoral drift.Jour.Wat.Port Coast and Ocean Engin. ,Vol.112, No.3, pp.351-369

Folk,R.L.and Ward,W.C.(1957).Brazos River Bar:A study of the significance of grain-size parameters.Jour. Sedim.Petrol., Vol.44, pp.40-53

Komar,P.D. and Inman,D.L.(1970).Longshore sand transport on beaches. Jour.Geoph.Res.,Vol.75, No.30, pp.5914-5927

Komar,P.D.(1975).Longshore currents and sand transport on beaches.Proc. Civ.Engin.in the Oceans Conf.,ASCE,pp.333-354

Komar,P.D.(1988).Environment controls on littoral sand transport.Abstracts 21th ICCE, Torremolinos, pp.17-18

Longuet-Higgins,M.S.(1970).Longshore currents generated by obliquealy incident sea waves-Part I.Jour.Geoph.Res.,Vol.75, pp.6778-6789

Moutzouris,C.I.(1988).Wave-induced distribution of sediment size-related parameters on the nearshore sea bed.Proc. Seminar on Coastal Data, Grenoble, pp.88-99

Moutzouris,C.I.and Kypraios,N.(1987).Temporal and spatial grain size distribution of sediment in a tideless pocket beach.Proc.Coastal Sediments'87,ASCE,N.Orleans,Vol.II,pp. 1909-1924

Nicholls,R. and Webber,N.(1988).Characteristics of shingle beaches with reference to Christchurch Bay,S.England.Abstracts 21 th ICCE, pp.498-499

Van der Meer,J.W.(1988).Rock slopes and gravel beaches under wave attack. Ph.D.thesis,Delft University of Technology, Delft

CHAPTER 146

The theorical temporal structure of the longshore currents

JOSE AGUILAR[1]
JOSE JAVIER DIEZ[2]

1) INTRODUCTION

For a better Knowledge of the causes about the littoral solid transport is essential to experience advances in the littoral hydrodynamic. In this field the longshore currents are may be who play the most important role due to the intensity, the distribution and the relative duration of them, causing the movement of the solid particles due to the suspension of them and the friction in bottom.

The first theorical formulation about the longshore currents were established by Bowen (1969), Longet-Higgins (1970) and Thornton (1970). Previously another works trated an aproximation about the problem, obtaining a little precise results basing his works in simplest formulations, it shows in Sonu et alt. (1966).

2) AN OBSERVATION ABOUT THE EXISTING SOLUTIONS.

The basic equations of motion are:

- The Conservation of mass equation
- The Conservation of energy equation
- The Conservation of momentum equation.

The conservation of energy equation is not used because two reasons:

1. It is a complex formulation
2. It utilize disipation mechanism that they are not simples and precises for his formulation.

With the proposit of obtain a simple expression about the phenomenon there are realized series of simplifications, where the most important are:

a) Vertical and temporal averages in all the equations.
b) Elimination of the "Y" variable considering the case like a quasi-orthogonal problem; in this way X is the independent variable.

---

(1) - Prof. Titular de Puertos
Universidad Politécnica de Valencia.
(2) - Catedrático de Oceanografía e Ingeniería de Costas
Universidad Politécnica de Madrid.

c) Utilization of Monochromatic waves.
This simplification together with the others represent waves like a curve H(x).

d) Utilization of ideal beach profiles, terms linealization and others.

Not all the models use the same simplifications, but the final results is always a distribution V(x) (in some cases V(x,y)), that doesn't have in count the temporal nature of the phenomenon, eliminated by one of the simplifications before mentioned.

Battjes (1974) tryed to improve this lack trough relations of the parallel velocity V with formulations of the radiation stress in truncated function of the Rayleigh distribution of the waves in a sea state.

3) A NEW ECUATION DEDUCTION.

The objetive is to establish a new ecuation were we will find the local longshore current velocity value V(x,t), like a consecuence of waves gived by the displacement of the sea surface, like a time serie.

In this way, the longshore current will be a variable in time, which structure is possible to study.

For it we have to revert to the classic basic equations of Navier-Stokes, and make a parallel deduction to them (Phillips 1966, Mei 1983) but without the temporal average.

In conclusion the most important Hypothesis are:

1) Admiting the "Shallow Water wave" simplification we have

$$U = \frac{C}{\delta}\zeta \qquad (U_x, U_y) = U\,(-\cos\alpha, \sin\alpha)$$

$$P_z = \rho g\,(-z+h+\zeta)$$

$$C = \sqrt{g\delta} \qquad \delta = h + d \qquad \text{(Figure 1)}$$

2) Accepting the Snell's Law :

$$\frac{C}{\sin\alpha} = cte$$

$$\frac{\partial\alpha}{\partial x} \simeq \frac{\sin 2\alpha}{2\delta}\,\frac{\partial\delta}{\partial x} \qquad \frac{\partial\alpha}{\partial y} = 0 \qquad \frac{\partial\alpha}{\partial t} = 0$$

2) Admiting the hypothesis of "long crested waves" we have:

$$\frac{\partial \zeta}{\partial y} = -tg\alpha \frac{\partial \zeta}{\partial x} \simeq - \sin x \frac{\partial \zeta}{\partial x}$$

3) For the internal stresses representation we can introduce it by the temporal adverage values:

$$\tau_{xy} \cong \rho \frac{gTH^2}{8\pi^2 \delta} \frac{\partial V}{\partial x}$$

Although the others hypothesis will guide us to a temporal equation, the inexistence of this equations take us to the mentioned expression. In this way we can interpret beginning with other previous models results, that the turbulent tensions have only a distribution effect on to the X axis; it has make that some works leaving aside this term presented models with anyone inclusion of this.

4) The Set- up will be the proposed by Bowen (1968)
5) The section could be in any form d= d (x) (Figure 1).
6) The wave representation in breaking zone is in the following form (Figure 1):

$$\zeta_{Xi,Yi,t} = \zeta_{Xo,Yo,t} \; f_{fi} \; f_{ri}$$

... were

$\zeta_{Xi,Yi,t}$ is the sea's displacement in (Xi,Yi) point at instant t.

$\zeta_{Xo,Yo,t-\Delta t}$ is the sea's displacement in a reference point (Xo, Yo).

Δt is the wave's time employed to go through the points (Xo,Yo) and (Xi,Yi)

$$t \cong - \int_{X_o}^{X_i} \frac{dx}{C}$$ ... in the hypothesis of "quasi-orthogonal" incidence

$$f_{fi} = \frac{H_i}{H_o} = \sqrt{\frac{\sin 2\alpha_0}{\sin 2\alpha_i}}$$ is the refraction factor at the point i

$$f_{ri} = \begin{cases} 1 & \text{if} \quad Ho\, f_{fi} <= \gamma\delta \\ \dfrac{}{Hof_{fi}} & \text{if} \quad Ho\, f_{fi} > \gamma\delta \end{cases}$$ is a breaking form factor at the (Xi,Yi) point

With it we are looking for a temporal series representation "Sea's displacement at (Xi,Yi) point" in relation with the correspondent at (Xo,Yo)

With all is possible formulate an ecuation of the type:

$$\frac{dV}{dt} \simeq f\left(V, \frac{\partial V}{\partial X}, \frac{\partial^2 V}{\partial x^2}, x, t, \zeta\right)$$

This equation is :

$$\frac{\partial V}{\partial t} + \frac{\partial V}{\partial x}\left[-c\frac{\zeta}{\delta} - \left(\frac{gTH^2}{8\pi^2\delta D}\left(\frac{\partial\zeta}{\partial x} + \frac{\partial\delta}{\partial x}\right) + \frac{\partial}{\partial x}\left(\frac{gTH^2}{8\pi^2\delta}\right)\right)\right] +$$

$$+ \frac{\partial V}{\partial y}\left[2\frac{c\zeta}{\delta}\sin\alpha + 2V\right] + \frac{\partial^2 V}{\partial x^2}\left[-\frac{gTH^2}{8\pi^2\delta}\right] +$$

$$+ V\left[\frac{1}{D}\frac{\partial\zeta}{\partial t} + \frac{\partial\delta}{\partial x}\left(\frac{c\zeta^2}{D\delta} - \frac{c\zeta}{2\delta^2}\right) + \frac{\partial\zeta}{\partial x}\left(-\frac{c(D+\zeta)}{D\delta}\right)\right]$$

$$+ V^2\left[-\frac{\partial\zeta}{\partial x}\frac{\sin\alpha}{D}\right] + \frac{1}{\rho D}\tau^B =$$

$$= \frac{\partial\zeta}{\partial x}\left[\frac{c^2\zeta}{D\delta^2}(2D+\zeta)\sin\alpha + g\sin\alpha\right] + \frac{\partial\zeta}{\partial t}\left[-\frac{c(D+\zeta)}{D\delta}\sin\alpha\right] +$$

$$+ \frac{\partial\delta}{\partial x}\sin\alpha\left[-\frac{c^2\delta^2(D+\zeta)}{D\delta^3} + \frac{g\zeta^2}{\delta^2} + \frac{c^2\zeta^2}{2\delta^3}\right] + \frac{1}{\rho D}\tau^S$$

$$D = \delta + \zeta$$

$$\tau^B = \rho\, C_F \sqrt{\left(\frac{c\zeta}{\delta}\right)^2 + \left(\frac{c\zeta\sin\alpha}{\delta} + V\right)^2}\left(V + \frac{c\zeta}{\delta}\sin\alpha\right)$$ ... is the stress in the bed

$$\tau^S = \rho\, k\, W^2 \cos\Theta$$ ... or similar, is the stress due to the wind action

$$k \begin{cases} 1.1\ 10^{-6} & W \le 7.2\ m/s \\ 1.1\ 10^{-6} + 2.5\ 10^{-6}\left(1 - \frac{7.2}{W}\right) & W > 7.2\ m/s \end{cases}$$

For the explicit form resolution we most utilize a small Δt. If we admit that the time discretization error is the same order of magnitude then the spatial discretization, beginning with the wave phases term we can obtain:

$$\frac{2\pi x}{L} + \frac{2\pi t}{T} \qquad\qquad \Delta x = C \Delta t$$

CONCLUSIONS

As a conclusion in accord with all before mentioned, we can conclude that we have basis for the study of the theorical temporal structure of the longshore currents. This is based in anew deduction of the movement equation where we don't have to make a temporal averages in it development and let the inoident wave comes from a unic time serie $\zeta(t)$ of the superficial sea displacement at (Xo,Yo) point of reference in the contour, with discretization in Δt intervals.

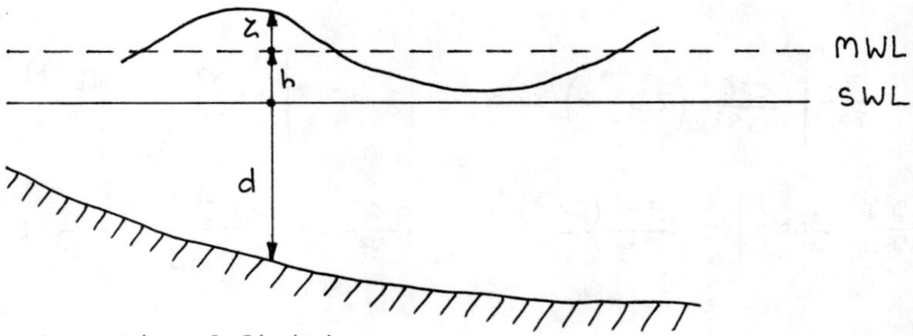

Figure 1: A section definition.

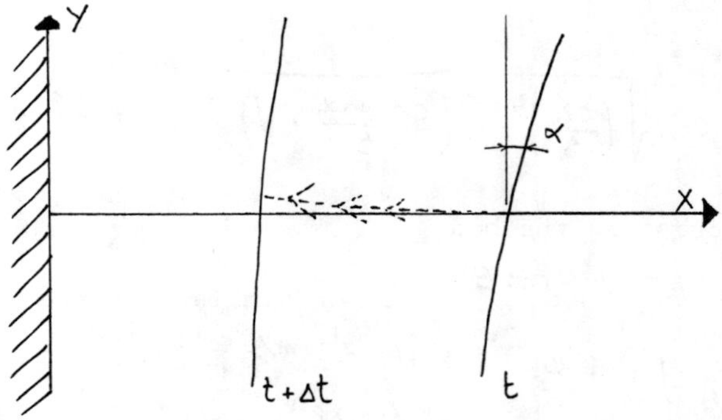

Figure 2: A Plant Definition.

REFERENCIAS

BATTJES, J.A. (1974). "Computation of Set-up, Longshore Currents, Run-up and Overtopping due to Wind-generated waves". Delft University

BOWEN, A. (1969). "The generation of Longshore Currents on a plane beach". J. of Marine Research Vol. 27, 2, pp. 206-215

LONGHUET-HIGGINS, M.S. (1970). "Longshore Current Generated by Obliquely incident Sea Waves. 2". J.of Geophysical Research. Vol. 75, nº 33, pp. 6790-6801.

MEI C. (1983). "The Applied Dynamics of Ocean Surface Waves". John Wiley & Sons. N.Y.

PHILLIPS, O.M. (1966). "The Dynamics of the Upper Ocean". Cambridge University Press. London

SONU, C.; Mc. CLOY, J.M. y Mc. ARTHUR, D.S. (1966) "Longshore Currents and Nearshore Topographies" I.C.C.E., Chp. 32, pp. 325-349.

THORNTON, E.B. (1970). "Variation of Longshore Currents Across the Surf Zone". I.C.C.E., Chp. 18, pp. 291-308.

SUBJECT INDEX

Page number refers to first page of paper.

Volume 1 – 1-998 Volume 2 – 999-1980 Volume 3 – 1981-2998

AUTHOR INDEX

Page number refers to first page of paper.